AA001005

2006 25th International Conference on Microelectronics

Serbia and Montenegro
14-17 May 2006

Volume 1 of 2

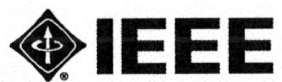

IEEE Catalog Number: 06TH8868
ISBN: 1-4244-0116-X

Copyright © 2006 by The Institute of Electrical and Electronics Engineers, Inc.
All Rights Reserved

Copyright and Reprint Permissions: Abstracting is permitted with credit to the source. Libraries are permitted to photocopy beyond the limit of U.S. copyright law for private use of patrons those articles in this volume that carry a code at the bottom of the first page, provided the per-copy fee indicated in the code is paid through Copyright Clearance Center, 222 Rosewood Drive, Danvers, MA 01923.

For other copying, reprint or republications permission, write to IEEE Copyrights Manager, IEEE Operations Center, 445 Hoes Lane, Piscataway, New Jersey USA 08854. All rights reserved.

IEEE Catalog Number:	06TH8868
ISBN:	1-4244-0116-X
LOC:	2005938573

Additional Copies of This Publication Are Available from:

IEEE Service Center
445 Hoes Lane
Piscataway, NJ 08854
IEEE Service Center
445 Hoes Lane
Piscataway, NJ 08854

Phone:	(800) 678-IEEE
	(732) 981-1393
Fax:	(732) 981-9667
E-mail:	customer-service@ieee.org

MIEL 2004 Best Paper Award Winners

Left to right: F. De Paola, J. Millan, N. Stojadinović (Conference Chairman),
A. Paskaleva, H. Hein

MIEL 2004 Best Oral Paper Award

X. Perpina, X. Jorda, P. Godignon, J. Millan, H. Von Kiedrowski,
J. Vobecky, N. Mestres

for

Direct Measurement of Self-Heating Effects at the Drift Region of 600V PT-IGBTs

J. Millan accepts the award from N. Stojadinović

MIEL 2004 Best Poster Paper Award

P. Baureis, H. Hein, M. Peter, F. Oehler

for

A Fully Integrated 0.35 μm CMOS MMIC Amplifier for Short Range 443 MHz ISM Band Transceiver Applications

H. Hein accepts the award from N. Stojadinović

MIEL 2004 Best Student Paper Award

F. De Paola, V. Aliberti, B. Rajaei, N. Rinaldi, J. Burghartz

for

A Scalable Physical Model for Coplanar Waveguide Transition in Flip-Chip Applications

F. De Paola accepts the award from N. Stojadinović

MIEL 2004 Best Paper Award
on behalf of
Microelectronics Reliability

A. Paskaleva, M. Lemberger, S. Zurcher, A.J. Bauer

for

Electrical Properties and Conduction Mechanisms in $Hf_xTi_ySi_zO$ Films Obtained from Novel MOCVD Precursors

A. Paskaleva accepts the award from N. Stojadinović

STEERING COMMITTEE

Asia & Pacific

C. Y. Chang, National Chiao-Tung University, Taiwan
Y.-I. Choi, Ajou University, Korea
S. Dimitrijev, Griffith University, Australia
H. Iwai, Tokyo Institute of Technology, Japan
C. Jagadish, Australian National University, Australia
B. Z. Li, Fudan University, China
T. Ohmi, Tohoku University, Japan
M. K. Radhakrishnan, Philips Electronics, Singapore
H. Wong, City University of Hong Kong, Hong Kong

Europe

H. Detter, Technical University of Vienna, Austria
G. Ghibaudo, ENSERG, France
G. Golan, ATCT Ltd., Israel
V. Liberali, University of Milan, Italy
M. Ostling, Royal Institute of Technology, Sweden
V. Pershenkov, Moscow Physics Engineering Institute, Russia
R. Popović, EPFL, Switzerland
S. Selberherr, Technical University of Vienna, Austria
N. Stojadinović, University of Niš, Serbia and Montenegro

America

V. Arora, Wilkes University, PA, USA
J.-P. Colinge, University of California at Davis, CA, USA
M. Estrada, CINVESTAV, Mexico
J. Liou, University of Central Florida, FL, USA
A. Nathan, University of Waterloo, Canada
V. Oklobdžija, University of California, CA, USA
A. Ortiz-Conde, University Simon Bolivar, Venezuela
K. Shenai, University of Illinois at Chicago, IL, USA
R. Singh, Clemson University, SC, USA

PROGRAMME COMMITTEE

Chairman:
N. Stojadinović, University of Niš, Serbia and Montenegro

Vice-Chairmen:
S. Dimitrijev, Griffith University, Australia
H. Iwai, Tokyo Institute of Technology, Japan
S. Selberherr, Technical University of Vienna, Austria
J. Liou, University of Central Florida, FL, USA

Scientific Secretary:
I. Manić, University of Niš, Serbia and Montenegro

Organization Secretary:
T. Pešić, University of Niš, Serbia and Montenegro

Corresponding Members:

I. Adesida, University of Illinois at Urbana-Champaign, USA
K. Arshak, University of Limerick, Ireland
E. Atanassova, Bulgarian Academy of Sciences, Bulgaria
V. Benda, Technical University of Prague, Czech Republic
M. Blyzniuk, Ivan Franko National University of L'viv, Ukraine
T. Brožek, PDF Solutions, Inc., USA
J. Burghartz, IMS Chips, Germany
I. De Wolf, IMEC, Belgium
A. Dziedzic, Technical University of Wroclaw, Poland
Z. Đurić, IHTM-CMTM, Serbia and Montenegro
D. Fleetwood, Vanderbilt University, USA
D. Flores, National Centre of Microelectronics, Spain
G. Ghibaudo, ENSERG, France
P. Hagouel, Aristoteles University of Thessaloniki, Greece
S. Haque, Philips, USA
S. Horita, Japan Advanced Institute of Science and Technology, Japan
K. Itoh, Hitachi, Ltd., Japan
H. Ishiwara, Tokyo Institute of Technology, Japan
N. Janković, University of Niš, Serbia and Montenegro
Z. Jakšić, IHTM-CMTM, Serbia and Montenegro
M. Jevtić, Institute of Physics, Serbia and Montenegro
E.D. Kim, Korea Electrotechnology Research Institute, Korea
K. Kim, Samsung, Korea
S. Koshevaya, National Institute of Astrophysics, Mexico
D. Kouvatsos, NCSR Demokritos, Greece
J. Kuo, National Taiwan University, Taiwan
K. Lee, KAIST, Korea
V. Litovski, University of Niš, Serbia and Montenegro
J. Lutz, Technical University of Chemnitz, Germany
P. Mawby, University of Warwick, United Kingdom
S. Mijalković, Delft University of Technology, The Netherlands
E. Miranda, University of Buenos Aires, Argentina
T. Mouthaan, University of Twente, The Netherlands
A. Napieralski, Technical University of Lodz, Poland
J. Nicolics, Vienna University of Technology, Austria
P. Nikolić, University of Belgrade, Serbia and Montenegro
N. Novkovski, Faculty of Natural Science and Mathematics, Macedonia
M.K. Radhakrishnan, Philips Electronics, Singapore
R. Ramović, University of Belgrade, Serbia and Montenegro
G. Reeves, Royal Melbourne Institute of Technology, Australia
N. Rinaldi, University of Naples, Italy
A. Rusu, Technical University of Bucharest, Romania
N. Shammas, Staffordshire University, United Kingdom
E.M. Shankar, De Montfort University, United Kingdom
D. Tjapkin, University of Belgrade, Serbia and Montenegro
V. Tvarožek, Slovak Technical University, Slovak Republic
R. Ubar, Tallin Technical University, Estonia
V. Vashchenko, National Semiconductor Corporation, USA
J. Vobecky, Czech Technical University in Prague, Czech Republic
G. Wachutka, Technical University of Munchen, Germany
P. Yu, University of California, San Diego, USA
M. Zwolinski, University of Southampton, United Kingdom
Lj. Živanov, University of Novi Sad, Serbia and Montenegro

CONTENTS

VOLUME 1

Workshop
Nanotechnologies

Invited Keynote Papers

Semiconductor Manufacturing in the Nanotechnology World of the 21st Century
R. Singh, A. Venkateshan, K. Poole, D. Mohan, P. Chatterjee 3

Nanoscience and Nanotechnologies in Serbia
Z. Djurić 11

Failure of Ohm's Law: Its Implications on the Design of Nanoelectronic Devices and Circuits
V.K. Arora 17

Device Integration Issues Towards 10 nm MOSFETs
M. Ostling, B. Gunnar Malm, M. Von Haartman, J. Hallstedt, Z. Zhang, P.-E. Hellstrom, S. Zhang 25

Nanoscale Materials, Devices, and Systems for Sensing, Detection, and Environmental Pollution Monitoring and Mitigation
A. Vaseashta 31

The Role of Non-Equilibrium Plasmas and Micro-Discharges in Top Down Nanotechnologies and Selforganized Assembly of Nanostructures
Z. Petrović, G. Malović, M. Radmilović-Radjenović, N. Pauč, D. Marić, P. Maguire, C. Mahony 39

Challenges of Ta$_2$O$_5$ as high-k Dielectric for Nanoscale DRAMs
E. Atanassova, A. Paskaleva 47

Electrically Active Defects at the Interface between (100)Si and Hafnium Dioxide Thin Films
P. Hurley, K. Cherkaoui 55

Capacitance Behavior of Nanometer FD SOI CMOS Devices with HfO$_2$ High-K Gate Dielectric Considering Gate Tunneling Leakage Current
J.B. Kuo, C.H. Lin 61

Plenary Sessions

Invited Keynote Papers

Defect Engineering and Stress Control in Advanced Devices on High-Mobility Substrates
C. Claeys, G. Eneman, E. Simoen 67

Reviews and Prospects of Low-Voltage Nano-Scale Embedded RAMs
K. Itoh 77

Gate Oxide Reliability for Nano-Scale CMOS
J. Stathis...83

Effects of Device Aging on Microelectronics Radiation Response and Reliability
D. Fleetwood, M.P. Rodgers, L. Tsetseris, X.J. Zhou, I. Batyrev, S. Wang, R.D. Schrimpf,
S.T. Pantelides...89

Session
Nanotechnologies

Nonlinear Model of Microtubule Dynamics and Its Impact on Kinesin Motion
M.V. Satarić...99

Adsorbed Mass Fluctuations of a Micro/Nanoresonator Surrounded by an Arbitrary Gas Mixture
Z. Djurić, O. Jakšić, I. Jokić, M. Frantlović...103

Scanning Probe-Shaped Nanohole Arrays with Extraordinary Optical Transmission as Platform for Enhanced Surface Plasmon-Based Biosensing
Z. Jakšić, M. Maksimović, D. Vasiljević-Radović, D. Tanasković, M. Sarajlić...107

Synthesis and Optical Properties of CdSe Nanocrystals Obtained from $CdCl_2$ and Na_2SeSO_3 Aqueous Solutions in the Presence of Gelatine
A.E. Raevskaya, A.L. Stroyuk, S.Ya. Kuchmii, Yu.M. Azhniuk, V.M. Dzhagan,
V.O. Yukhymchuk, M.Ya. Valakh...111

Growth and ac-Properties of Lanthanum-Manganese Oxide Films on Si Substrates
A.A. Dakhel...115

Low - k Polyimide/Silica Nanocomposites for Microelectronics Applications
E. Logakis, D. Fragiadakis, P. Pissis...119

Investigation of Electrode Patterns Suitable for Nano-Litre Drop Coated Conducting Polymer Composite Sensors
K. Arshak, C. Cunniffe, E. Moore, L. Cavanagh...123

A Novel Vertical Impact Ionisation MOSFET (I-MOS) Concept
U. Abelein, M. Born, K.K. Bhuwalka, M. Schindler, M. Schmidt, T. Sulima, I. Eisele...127

Tunnel FET: A CMOS Device for High Temperature Applications
M. Born, K. Bhuwalka, M. Schindler, U. Abelein, M. Schmidt, T. Sulima, I. Eisele...131

Poster Session
Nanotechnologies

Analysis of Nonlinear Effects in Nanometer-Scale Silicon-on-Insulator Rib Waveguides
V.M.N. Passaro, F. De Leonardis, G.Z. Mashanovich...137

Analysis of Pulsed Excitation in Small Silicon-On-Insulator Microring Resonators
F. De Leonardis, V.M.N. Passaro, G.Z. Mashanovich...141

xi

Silicon Surface Exfoliation Under Compression Plasma Flow Action
I.P. Dojčinović, M.M. Kuraica, D. Randjelović, M. Matić, J. Purić 145

Magnetic Field Influence on Silicon Surface Periodic Structures Obtained by Plasma Flow Action
I.P. Dojčinović, M.M. Kuraica, M. Mitrović, D. Randjelović, M. Matić, J. Purić 149

A Consideration of Transparent Metal Structures for Subwavelength Diffraction Management
Z. Jakšić, M. Sarajlić, M. Maksimović, D. Vasiljević-Radović, D. Jovanović 153

The Fabrication of Single Electron Transistor by Polysilicon Thin Film and Point-Contact Lithography
K.-D. Huang, J.-T. Lin, S.-F. Hu, C.-L. Sung 157

Impact of Rapid Photothermal Processing on Properties of ZnO Nanostructures for Solar Cell Applications
S. Shishiyanu, R. Singh, T. Shishiyanu, O. Lupan 161

Session
Power Devices and ICs

Invited Keynote Paper

Evolution of Silicon Power Devices and Challenges to Material Limit
A. Nakagawa 167

Invited Keynote Paper

High Speed Electro-Thermal Models for Inverter Simulations
P.A. Mawby, A.T. Bryant, P.R. Palmer, E. Santi, J.L. Huidgins 175

Influence of Small Emitter and Contact Non-Uniformities on the Current Filamentation in 3.3-kV p^+-n^--n^+ Silicon Diodes
H.P. Felsl, E. Falck, F.-J. Niedernostheide, J. Lutz 183

Schottky Contacts on Single-Crystal CVD Diamond
D. Doneddu, O.J. Guy, D. Twitchen, A. Tajani, M. Schwitters, P. Igić 189

Dynamic Thermal Simulation of Power Devices Operating With PWM Signals
Z. Zhou, M.S. Khanniche, P. Igić, N. Janković, S.G. Batcup, P.A. Mawby 193

Dynamic Characteristics of Novel Super-Junction LDMOS Switches under Charge Imbalance Conditions
K. Permthammasin, G. Wachutka, M. Schmitt, H. Kapels 197

Self-Heating Effects in SOI NLDEMOS Power Devices
F. Dieudonne, S. Haendler, S. Chouteau, J. Rosa, P. Waltz, A. Perrotin, L. Boissonnet, B. Rauber, C. Schaffnit, C. Raynaud 201

Poster Session
Power Devices and ICs

An Alternative Process Architecture for CMOS Based High Side RESURF LDMOS Transistors
P.M. Holland, P. Igić 207

Comparison of Measured and Simulated Characteristics of Boron Implanted 4H-SiC DiMOSFET
S.C. Kim, W. Bahng, K.H. Kim, I.H. Kang, S. J. Kim, N.K. Kim 211

IGBT Behavioral PSPICE Model
K. Asparuhova, T. Grigorova 215

Numerical Analysis of a Trench Gate FLIMOSFET with No Quasi-Saturation, Improved Specific On-Resistance and Better Synchronous Rectifying Characteristics
R. Vaid, N. Padha 219

Session
Microsystem Technologies

Invited Keynote Paper

Wireless Ultra-Low Power Smart Data Acquisition System for Pressure Sensing in Medical Application
K. Arshak, E. Jafer 225

In-Line Concentration Measurement of Nanoliter Liquid Sample using Low-Coherence Spectral Interferometry
M. Tomić, Z. Djinović, A. Vujanić 233

Using Design of Experiment to Investigate the Effects of Conducting Polymer Composite Sensor Composition on the Response to an Homologous Series of Alcohols
K. Arshak, E. Moore, C. Cunniffe, L. Cavanagh 237

IR Bimaterial Detectors Performance
Z. Djurić, D. Randjelović, J. Matović, J. Lamovec 241

Dynamic Elastic Bending in Optically Driven Microcantilever: Surface Strain Effects
D.M. Todorović, T. Grozdić 245

Real-Time Tracking of a Moving Object Inside a Tube
K. Arshak, F. Adepoju 249

xiii

Poster Session
Microsystem Technologies

A Simplified Method of Analysis of MEMS Bimaterial Cantilever Element
J. Matović ... 255

Design Parameters Optimization Using Process Variations of the Pull-In Voltage for MEMS
R. Voicu, C. Tibeica, M. Bazu ... 259

Development of a Portable Gamma Radiation Monitoring System
K. Arshak, D. Buckley, O. Korostynska ... 263

Thermal Characterization of the Micro Bonding Process Using a Hot Air Stream
D. Andrijašević, I. Giouroudi, W. Smetana, W. Brenner, D. Esinenco ... 267

Designing and Modeling MEMS Resonator in VHF Range
M. Al Khusheiny, B.Y. Majlis ... 271

Session
Opto and Microwave Devices and ICs

Invited Keynote Paper

Silicon Integrated Photonics for Microelectronics Evolution
H. Wong, V. Filip, C.K. Wong, P.S. Chung ... 277

Diagnostics of Homogeneity of Individual Layers of Large-Area Silicon Solar Cells Using Local Irradiation
V. Benda, Z. Machacek, J. Salinger ... 285

Improved Dual Grating-Assisted Directional Coupler for Silicon Nanophotonics
G. Mashanovich, V. Passaro, G. Ensell, F. Gardes, G. Reed ... 289

High Speed Modulator in Q-Band Range on 4H-SiC p-i-n Diodes
R. Kakanakov, L. Kolaklieva, L. Romanov, A. Kirillov, A. Lebedev, M. Boltovets, K. Zekentes ... 293

Low-Frequency-Noise Characteristic of Quasi-Enhancement-Mode HEMT Using a Selectively Hydrogen-Pretreatment
I.H. Kang, S.C. Kim, W. Bahng, N.K. Kim ... 297

Hyper Sound Amplification
S. Koshevaya, V. Grimalsky, M. Tecpoyotl-Torres, J. Escobedo-Alatorre, M. Diaz-Ayala, A. Garcia-B ... 301

Poster Session
Opto and Microwave Devices and ICs

Noise as a Diagnostic Tool for Quality of GaSb Laser Diodes
Z. Chobola, J. Vanek, E. Hulicius, T. Šimeček 307

Nonlinear Surface Ultrasonic Monopulses in Solid Film-Substrate System
V. Grimalsky, S. Koshevaya, E. Gutierrez-D., A. Garcia-B. 309

Accurate Noise Modeling of HEMT for Low-Noise Applications
I.H. Kang, S.C. Kim, W. Bahng, N.K. Kim 313

Modeling and Constituent Design of SPR Based and/or Waveguide Type Sensors
N. Dmitruk, O. Mayeva, O. Korovin, M. Sosnova, O. Lytvyn, V. Min'ko 317

Effect of Interface Microrelief on Optical, Electrical and Photoelectric Characteristics of Heteroepitaxial $Al_xGa_{1-x}As$/GaAs Structures
N. Dmitruk, O. Borkovskaya, R. Konakova, A. Korovin, I. Mamontova, O. Kondratenko 321

Session
Physics and Modeling

Invited Keynote Paper

a-$Si_{1-x}C_x$:H TFTs: Fabrication and Modeling
M. Estrada, A. Cerdeira, R. Garcia, B. Iniguez 327

Boron Redistribution During SOI Wafers Thermal Oxidation
Z. Djurić, M.M. Smiljanić, K. Radulović, Ž. Lazić 333

Transport of Non-Equilibrium Charge Carriers in Bipolar Semiconductors Materials
Y. Gurevich, J.E. Velazquez-Perez 337

A High-Frequency Extension of a Surface-Potential-Based Substrate Model for Noise Coupling Analysis
N. Simić, F. Ingvarson, S. Kristiansson, M. Zgrda, K. Jeppson 341

Analytical Modeling of the Triggering Drain Voltage at the Onset of the Kink Effect for PD SOI NMOS
M. Sarajlić, R. Ramović 345

Author Index 349

VOLUME 2

Session
Processes and Technologies

P-Type Conduction in Sputtered ZnO Thin Films Doped by Nitrogen
K. Shtereva, V. Tvarozek, I. Novotny, J. Kovac, P. Sutta, A. Vincze 357

Investigation of Metal-Polycrystalline Silicon Carbide Bonding While Metallization
G. Golan, V. Manevych, I. Lapsker, B. Gorenstein, A. Axelevitch 361

PVD Aluminium Nitride as Heat Spreader in Silicon-on-Glass Technology
L. La Spina, H. Schellevis, N. Nenadović, L. Nanver 365

Reliability Issues Related to Laser-Annealed Implanted Back-Wafer Contacts in Bipolar Silicon-on-Glass Processes
G. Lorito, V. Gonda, S. Lui, T.L. Scholtes, H. Schellevis, L. Nanver 369

A Novel Bottom Gate Polysilicon Thin-Film Transistor with Smart Body Tie
J.-T. Lin, K.-D. Huang, S.-T. Lin 373

Poster Session
Processes and Technologies

Electrical Properties of HfO$_2$ Films Formed by Ion Assisted Deposition
K. Cherkaoui, A. Negara, S. McDonnell, G. Hughes, M. Modreanu, P. Hurley 379

PECVD Growth of Thick Silicon Oxynitride for On-Chip Optical Interconnects Applications
C.K. Wong, H. Wong, C.W. Kok, M. Chan 383

The Characterization and Optimization of the Thermal Oxidation Process Equipment Using Experimental Design and Data Transformation
H. You, X. Jia, S. Wang 387

PTC Effect and Structure of Polymer Composites Based on Polypropylene/Co-Polyamide Blend Filled with Dispersed Iron
A. Kanapitsas, C. Tsonos, E. Logakis, C. Pandis, P. Pissis, E. Kontou, Y.P. Mamunya, E.V. Lebedev, C.G. Delides 391

A New Method of Evaluation of Liquidus Temperatures of Ternary Alloys
A. Prijić, Z. Prijić, B. Pešić 395

Session
Circuit Design

Invited Keynote Paper

System-on-Chip Design: Engineering of Art
Z. Stamenković 401

Invited Keynote Paper

A New Circuit Model for Designig Fully Integrated Class-A Power Amplifier
X. Guan, H. Feng, A. Wang, L. Yang 409

Gate Layout Improvement Aimed at Testability
M. Blyzniuk 413

An Improved Write Driver for Miniaturized Hard Disk Drives
F. Tamigi, I. Marano, A. Venca, N. Rinaldi 417

Optical Receiver with Voltage-Controlled Transimpedance in BiCMOS Technology
N. Tadić, H. Zimmermann 421

Improved Linearity Active Resistor Using Equivalent FGMOS Devices
C. Popa 425

A Stochastic Approach to Crosstalk Analysis in Mixed-Signal ICs
G. Boselli, G. Trucco, V. Liberali 429

The 4-Phase Frame Partitioning Circuit
S. Poriazis 433

Fault Diagnosis in Digital Part of Mixed-Mode Circuit
M. Andrejević, V. Litovski, M. Zwolinski 437

Poster Session
Circuit Design

Simulation of Crosstalk between Several Interconnection Lines in CMOS Integrated Circuits
M. Petković 443

Reordering in Topology Decision Diagram Method for Symbolic Circuit Analysis
S. Djordjević, P.M. Petković 447

An Improved Performance FGMOS Voltage Comparator for Data Acquisition Systems
C. Popa 451

On Silicon Timing Validation of Digital Logic Gate "A Study of Two Generic Methods"
A.P. Singh, N.S. Panwar 455

High Speed Low Power CMOS Comparator for Pipeline ADCs
M.B. Guermaz, L. Bouzerara, A. Slimane, M.T. Belaroussi, B. Lehouidj, R. Zirmi 459

xvii

Session
Modeling and Simulation

Invited Keynote Paper

Optimization Issue in Interconnect Analysis
S. Holzer, S. Selberherr 465

Invited Keynote Paper

Advanced Circuit and Device Modeling with Verilog-A
S. Mijalković 471

Process and Device Simulation With a Generic Scientific Simulation Environment
M. Spevak, R. Heinzl, P. Schwaha, T. Grasser 475

Design and Simulation of Thick Film Thermistors Using Commercial Simulation Tools
V. Marić, O. Aleksić, Lj. Živanov 479

Fully Automated Electrothermal Simulation using Standard CAD Tools
F. De Paola, J. Nowakowski, V. d'Alessandro, N. Rinaldi 483

A Novel 3D Embedded Gate Field Effect Transistor: Device Concept and Modelling
K. Fobelets, P.W. Ding, J.E. Velazquez-Perez 487

A Modified EKV PDSOI MOSFETs Model
J. Alvarado, A. Cerdeira, V. Kilshytska, D. Flandre 491

Poster Session
Modeling and Simulation

Monte Carlo Analysis of Voltage-Current Characteristic Nonlinearity and Harmonic Generation in Submicron Semiconductor Structures
D. Persano Adorno, M. Capizzo, M. Zarcone 497

Diode I-U Curve Fitting with Lambert W Function
P. Hruska, Z. Chobola, L. Grmela 501

A New Treshold Voltage Analytical Model of Strained Si/SiGe MOSFET
P. Lukić, R. Ramović, R. Šašić 505

Graded-Channel SOI nMOSFET Model Valid for Harmonic Distortion Evaluation
M. de Souza, M. Pavanello, A. Cerdeira, D. Flandre 509

A Simple Polysilicon Thin-Film Transistor SPICE Model
I. Pappas, A.T. Hatzopoulos, D.H. Tassis, N. Arpatzanis, S. Siskos, A.A. Hatzopoulos,
C.A. Dimitriadis, G. Kamarinos 513

Modeling and Simulation of Submicron MOSFETs with Alternative Gate Dielectrics for DRAM Cells
N. Konofaos, G. Alexiou 517

Investigation of the Novel Attributes of a Vertical MOSFET with Internal Block Layer (bVMOS): 2-D Simulation Study
J.-T. Lin, K.-C. Lin, T.-Y. Lee, Y.-C. Eng ... 521

Filling of Learning Process in Electronics/Microelectronics Studies by Teaching for Understanding Properties Using Open Training CAD Tools
M. Blyzniuk ... 525

Distributed Framework for Teaching Fundamentals of Heat Transfer in Electronic Circuits
M. Paszkowski-Rogacz, P. Januszkiewicz, P. Gocek, G. Jablonski, M. Janicki,
G. De Mey, A. Napieralski ... 529

Charge Carriers Distributions in Rectangular Quantum Rod
J. Šetrajčić, B. Tošić, V. Sajfert, D. Ilić, S. Jaćimovski, S. Vučenović ... 533

Numerical Investigation of O_2 Adsorption Effect in Si-CF_4/O_2 System
Y. Grigoryev, A. Gorobchuk ... 537

Dispersion Relation for Two-Valley Quasi-Hydrodynamic Models in SCWs Propagation in n-GaAs Thin Films
A. Garcia-B., V. Grimalsky, E. Gutierrez-D., S. Koshevaya ... 541

Diffusion-Drift Model of Fully-Depleted SOI MOSFET
G. Zebrev, M. Gorbunov ... 545

Session
Reliability Physics

Invited Keynote Paper

Low Frequency Noise and Fluctuations in Sub 0.1 μm Bulk and SOI CMOS Technologies
G. Ghibaudo, J. Jomaah ... 551

Analysis of Surface Generation Mechanisms in MOS Capacitors
F. Kong, P. Tanner, L. Hold, J. Han, S. Dimitrijev ... 557

Effects of NO Annealing on the Characterizations of GaN MIS Capacitor
L. Lin, P.T. Lai, K.M. Lau ... 561

Long Range Statistical Lifetime Prediction of Ultra-Thin SiO_2 Oxides: Influence of Accelerated Ageing Methods and Extrapolation Models
D. Pic, D. Goguenheim, J.-L. Ogier ... 565

Low Frequency Noise of GaAs MESFET Degraded in ESD Test
M. Jevtić, J. Hadži-Vuković ... 569

Dependence of DGMOSFET 1/f Noise on Transistor Geometry and Technology Parameters
M. Videnović-Mišić, M.M. Jevtić ... 573

xix

Use of RADFETs for Quality Assurance of Radiation Cancer Treatments
A. Jakšić, K. Rodgers, C. Gallagher, P.J. Hughes 577

Effect of the Metal Electrode on the Characteristics of Ta_2O_5 Capacitors for DRAM Applications
E. Atanassova, D. Spassov, A. Paskaleva 581

Reliability Properties of Ta_2O_5 Films Grown on N_2O Plasma Nitrided Silicon
N. Novkovski, E. Atanassova 585

Stress Induced Leakage Currents and Charge Trapping in Thin Zr- and Hf-Silicate Layers
A. Paskaleva, M. Lemberger, A.J. Bauer 589

Reliability Investigation of NLDEMOS in 0.13μm SOI CMOS Technology
D. Lachenal, Y. Rey-Tauriac, L. Boissonnet, A. Bravaix 593

Physics and Electrical Characterization of Excimer Laser Crystallized Polysilicon TFTs
L. Michalas, M. Exarchos, G.J. Papaioannou, D. Kouvatsos, A. Voutsas 597

Reliability Assessment of a RF PA Assembly With Embedded Coin Construction
J. Zhou, H. Lu, M. Zhou, B. Inkman, D. Anderson, B. Johnson 601

The Effect of Number of Zincation in Electroless Nickel Immersion Gold (ENIG) Under Bump Metallurgy (UBM) on Reliability in Microelectronics Packaging
S.H. Abdullah, I. Ahmad, A. Jalar 605

Poster Session
Reliability Physics

Investigation of the Ion Defect States by Photoacoustic Spectroscopy
D.M. Todorović, V. Jović, M. Smiljanić, T. Grozdić 611

Relationship between Intrinsic Breakdown Field and Bandgap of Materials
L-M. Wang 615

Electrode Effect on NTC Planar Thermistor Volume Resistivity
O. Aleksić, B. Radojčić, R. Ramović 619

Performances of Conventional Thick-Film Resistors After Multiple High-Voltage Pulse Stressing
I. Stanimirović, M.M. Jevtić, Z. Stanimirović 623

Influence of Simultaneous Mechanical and Electrical Straining on Conventional Thick-Film Resistors
Z. Stanimirović, M.M. Jevtić, I. Stanimirović 627

Influence of Electrode Materials and the Manner of Electrode Surface Processing on Gas-Filled Surge Arresters Relevant Characteristics
B. Lončar, P. Osmokrović, A. Vasić, R. Šašić 631

New Insights into Tunneling Current through Oxynitride/Oxide Stack for MOSFETs
L.F. Mao, Z.O. Wang 635

Spontaneous Recovery in DC Gate Bias Stressed Power VDMOSFETs
I. Manić, S. Djorić-Veljković, V. Davidović, D. Danković, S. Golubović, N. Stojadinović 639

Lifetime Estimation in NBT Stressed P-Channel Power VDMOSFETs
D. Danković, I. Manić, S. Djorić-Veljković, V. Davidović, S. Golubović, N. Stojadinović 645

Evaluation of Reflow Ovens for Lead-Free Soldering
J. Lempinen, A. Tuominen 649

Session
System Design

Word-Length Oriented Multiobjective Optimization of Area and Power Consumption in DSP Algorithm Implementation
A. Ahmadi, M. Zwolinski 655

A Mission Level Design Language Based on AleC++
B. Andjelković, V. Litovski, V. Zerbe 659

Clocking Challenges in High Speed Source Synchronous Interfaces
V. Zlatković 663

An Adaptive Pulse-Width Control Loop
G. Jovanović, D. Mitić, M. Stojčev 667

Low Power Computing for Secure and Reliable Sensor Networks
R. Agarwal, E.M. Popovici, C. O'Keeffe, B. O'Flynn, S.J. Bellis 671

Design and Implementation of an FPGA Based High-Speed Data Buffer for Optical Interconnects
S.-H. Voss, M. Talmi, J. Saniter 675

DefSim: Measurement Environment for CMOS Defects
T. Borejko, A. Jutman, W.A. Pleskacz, R. Ubar 679

Poster Session
System Design

NNARX Model of Speech Signal Generating System: Test Error Subject to Modeling Mode Selection
D. Protić, M. Milosavljević 685

The Electronic and Program Control System of X-Ray Ion Mobility Spectrometer
V.S. Pershenkov, A.U. Razvalyaev, A.D. Tremasov 689

The Innovative Method for Determining Characteristics of Over-Voltage Protection Elements
P. Osmokrović, B. Lončar, S. Stanković, A. Vasić 693

Approach to Partially Self-Checking Finite State Machine Design
G. Djordjević, T. Stanković, M. Stojčev 697

Laboratory ADC Tester Based on NI-6251 Acquisition Card
M. Nikolić, M. Sokolović, P. Petković 701

Late Submission Paper

Hybrid Empirical-Neural Model of the Loaded Microwave Cavity Applicators
Z. Stanković, B. Milovanović, M. Milijić 705

Author Index 709

Workshop
Nanotechnologies

2006 25th International Conference on Microelectronics

Semiconductor Manufacturing in the Nanotechnology World of the 21st Century

R. Singh, A. Venkateshan, K. F. Poole, D. Mohan and P. Chatterjee

Abstract - The term nanotechnology has different meanings and expectations to different people. In this paper we have defined and examined what really is nanotechnology. Based on the use of fundamental knowledge of science and system level engineering and business knowledge that we know as of today, the role of nanotechnology in the semiconductor manufacturing in the 21st century is described. As of today, there is no other technology that can replace silicon CMOS based integrated circuits as the feedstock of electronics industry. Key trends in semiconductor manufacturing are predicted that will drive the growth of global electronics industry.

I. INTRODUCTION

Global electronics industry with annual sales of about $1.5 trillion stands next to agriculture industry with annual sales of about $2 trillion. Semiconductor products with annual sale of about $235 billion, serve as feedstock to the electronics industry. With continued innovations in semiconductor and electronics industries (via cost and size reduction and creation of new applications with improved functionality), and almost flat or very little growth in agriculture industry, it may be no surprise if in the next one or two decades, the electronics industry surpasses the agriculture industry.

The prefix "nano" and the word "technology" are very well defined; however the combination of the two leading to the term nanotechnology has no clear definition and has different meanings and expectations to different people in academia, business, industry and government. It is common practice to find articles like," Next Big Thing Is Very, Very Small" [1]. Nanotechnology has been claimed to usher us into the next industrial revolution and replace our entire manufacturing base with a new, radically precise, less expensive, and more flexible way of making products [2]. There are also individuals like Eric Drexler, who treats nanotechnology like something so different that for him and others it has become like a new religion. According to Drexler, nanotechnology will enable us to reconstruct reality from the atomic scale on up in ways limited only by our imaginations [3].

R. Singh, A. Venkateshan, K. F. Poole and D. Mohan are with the Center for Silicon Nanoelectronics and Holcombe Department of Electrical and Computer Engineering Clemson University Clemson, SC 29634, Phone: 864-656-0919, Fax: 864-656-5910, E-mail: srajend@clemson.edu
P. Chatterjee is with i2 Technology, Inc. Dallas, TX 75234

Given the importance of semiconductor industry in the global economy, it is quite natural to examine the future of semiconductor manufacturing in the nanotechnology world of the 21st century. We have used fundamental knowledge of science, system level engineering and business knowledge to predict the future of semiconductor manufacturing in the context of the nanotechnology world of the 21st century.

II. WHAT IS NANOTECHNOLOGY?

The reason that nanotechnology has different meanings to different people is due to the simple fact that there is no real invention associated with nanotechnology that can be attributed as the starting point of nanotechnology. The advocates of nanotechnology give credit to the famous physicist Feynman for his lecture in 1959 popularly known as **"There's plenty of room at the bottom."** According to Feynman, *"The principles of physics, as far as I can see, do not speak against the possibility of maneuvering things atom by atom. It is not an attempt to violate any laws; it is something, in principle that can be done; but in practice, it has not been done because we are too big."* If really Feynman was the driver of nanotechnology, one has to ask the simple question that why nanotechnology did not become a common word of researchers just after his lecture in 1959. A careful examination of the advancements in the silicon IC industry and related semiconductor products (e.g. detectors) as well advancements in the signal and image processing has led to the development of electronic instruments with improved signal to noise ratio. In 1982 Binning and Rohrer at IBM invented scanning tunneling microscope (STM) to manipulate conducting materials at atomic scale. In 1985 they invented atomic force microscope to study non-conducting materials and for the first time they were able to write IBM by placing one atom at a time. For the invention of STM, Binning and Rohrer won Noble Prize in Physics in 1986. Immediately, after the discovery both STM and AFM became very important material synthesis and diagnostic tool. In 1985, Curl, Smalley and Kroto invented a new modification of pure carbon known as fullerenes (also called the buckyballs) and won Nobel Prize in Chemistry in 1996. A new form of carbon-the nanotube was discovered by Sumio Ijima in 1991. Unlike the discovery of transistor and integrated circuits, these discoveries are not directly responsible for creating

1-4244-0116-X/06/$20.00 ©2006 IEEE

nanotechnology as an entity. Due to the increase in US government R & D funding ($116 million in 1997 to a $961 million in 2004), it is only in the last couple of years that nanotechnology is getting the attention of academia, government industry and business. The US patents awarded annually for nanotechnology inventions have tripled since 1996 [4].

According to National Technology Initiative (NTI) [5], the research and technology development at the atomic, molecular or macromolecular levels, in the length scale of approximately 1 - 100 nm range, to provide a fundamental understanding of phenomena and materials at the nanoscale and to create and use structures, devices and systems that have novel properties and functions because of their small and/or intermediate size is called nanotechnogy. This definition is indeed slippery. As stated above, nanotechnology, in most cases, isn't technology. Rather it involves basic research on structures having at least one dimension of about one to several hundred nanometers. It is worth mentioning that some nanotechnology has been around for a long time: nano-size carbon black particles have been used into tires for 100 years as a reinforcing additive, long before the prefix "nano" ever created a stir. A vaccine, which often consists of one or more proteins with nanoscale dimensions, might also qualify as nanotechnology. Due to these fundamental problems associated with the use of the nanotechnology as stated in NTI, we will use nanotechnology in the context of semiconductor products where at least one side of the device used in the fabrication of integrated circuits is less than 100 nm. We will not differentiate between the approaches used in the fabrication of these systems. The issue of "bottom up" (atom by atom approach to build the required < 100 nm dimension) or "top down" (current lithography techniques used to reduce material dimension < 100 nm) approach will be dictated by throughput and cost related manufacturing issues and will be discussed in later sections.

III. FUNDAMENTAL SCIENCE AND TECHNOLOGY

The Planck distance (4.05×10^{-35} m) is the shortest length scale in physics, below it quantum mechanics and gravity merge and the structure of the space breaks down [6]. The Planck time (1.35×10^{-43} s) is when quantum correction ceases to be important to general relativity following the Big Bang [6]. As pointed out by Lloyd [7], these Planck scale numbers set the ultimate limits on the performance of a computer. From practical point of view, the smallest device that can be used in the future generation of computing system can be based on a single atom [8]. The typical size and bond length of an atom are of the order of 0.1 nm. Thus the minimum dimension of future device can be of the order of 0.2 nm. Future nanotechnology based computing systems can not have devices below the dimension of 0.2 nm. The switching speed of any man-made computing system will be in the range of 10^{-18} s and 10^{-3} s. The faster limit represents the time scale of all the known phenomena that can be used to manufacture a computing system. Temporal probing of a number of fundamental dynamical processes requires intense pulses at femtosecond or even attosecond (1 as $=10^{-18}$ s) [9]. A number of biological materials including DNA are being considered as future nanotechnology materials. Switching speed in biological system is of the order of 10^{-3} s. Thus from the fundamental point of view, the use of nanotechnology can provide computing devices of the size 0.2 nm -100 nm and speed in the range of 1 ms to 1 as. As we will see in the following section, the laws of thermodynamics and quantum mechanics sets further limit on these length and time scales.

IV. COMPUTER PERFORMANCE LIMITS SET BY THERMODYNAMICS AND QUANTUM MECHANICS

The rules of thermodynamics are applicable to all the dimensions of nanotechnology. The computing device (irrespective of the device dimension) generates heat. The energy dissipated and the heat generation is governed by the Heisenberg uncertainty principle. In two previous publications [8, 10] we have discussed at length that heat dissipation is the first fundamental limit that provides an insurmountable barrier in the use of high speed switching devices for next generation of computing systems. The energy dissipated and the heat generation in a logic device is governed by the Heisenberg uncertainty principle. In addition to the heat generated in the switching operation, additional heat will be generated in the interconnect system. The faster computers generate more heat than the slow one. As shown in Fig.1 [11], the anisotropy of the material poses a practical limit in heat removal. Based on the best estimates of today, we are removing 300 to 400 watts per square centimeter with our current prototype, all locally, on the chip [12]. The material limitations are posing fundamental limits in the use of computing devices that will require heat removal of the order of kilo watts per square centimeter and higher power densities. In order to reduce size, various devices tout the concept of a single entity to define a state. For example a "single electron" device, these devices will never provide a reliable system. A property of a reliable system is the ability to assign a state with certainty. Based on quantum mechanical considerations, reliable systems operating at low voltages of about 1 V and below will require nodes which consist of >40 particles, charges or any other entity [8].

V. REVERSIBLE COMPUTATION AND NANOMOTORS

The literature on the topic of reversible computers and nanotechnology has a fundamental flaw. A careful examination of the published work shows that this flaw originated in the paper published by Bennett [13] in 1973. Based on the notion of zero energy dissipation, Bennett [13] mathematically demonstrated that general purpose computing machines might be made logically reversible at every step. In support of his theoretical work, Bennett provided the biological system as a physical example of reversible computation.

The point missed by Bennett is the fact that biological systems capable of processing context-related events have [13] to be open, nonlinear and operate at non-equilibrium or preferably far from equilibrium conditions at slow speeds of the order of milli seconds [14]. The stability of the system is constantly affected by external disturbances, dissipative exchanges with the environment and structural changes within the system [14]. All of these requirements are not met in non-biological closed systems and reversible computation with no energy loss is fundamentally impossible. Silicon IC designers have known for long time that it is impractical to maintain both high performance and low levels of heat generation and energy uses with reversible computing. At best the concept of reversible logic or adiabatic switching applied to integrated circuits saves power at the expense of speed. This is already in practice in a number of applications.

Similar to reversible computers, there are many claims that one can operate a nano-motor based on the exploitation of random motion, with no use of external source of energy. The molecular motors on which life depends are open systems and are driven by Brownian motion. These motors work at physiological temperatures and thus experience thermal motion, yet their function, which requires correct alignment of parts and steering of reactions, is executed with precision [15]. For closed systems with no external source of energy, one can never exploit "random motion" into useful work. Only by supplying external energy source, the nano-machines can operate in an open system mode. The question of interfacing these man-made non-biological motors at nano-scale with the micro-scale and macro scale world is a practical engineering problem and does not pose any fundamental issue.

VI. PROCESSING OF NANOMATERIALS

It has been known for long time that the properties of ultra-thin films differ from the bulk values. As shown in Figure 2, research in the nanodimension materials in the last decade has established that the properties of nanodimension material differ from the corresponding thin film or bulk materials. This is because at nanoscales, the properties of material depend on the quantum-confined effects. The one to three dimensional nanomaterials (~0.1-4 nm) provides a unique opportunity to optimize a number of physical, chemical, biological, mechanical, electrical,

optical and thermal properties of interest for future computing systems. A number of current materials (e.g. gate dielectric stack of CMOS, barrier layer materials of the interconnect system etc) used in silicon ICs are exploiting the unique properties offered by nano-materials.

The one-dimensional material of about 0.1-4 nm poses a number of fundamental challenges for practical use in the manufacturing of future computing systems. As an example, carbon nano tubes (CNTs) with diameter as small as 0.3 nm are being investigated as an electronic material. The practical use of these one dimensional materials is like finding a needle in a bundle of hay [16]. Conventional "top down" approaches are being used in the generation of CNTs. The "bottom-up" approach is used in placing one CNT at one time to make a functional device. Thus the issue of throughput is a fundamental issue in the use of these "bottom-up" approaches for any practical computing system. The two CNTs will always be separated by a certain distance. Quite often, the literature ignores this fact and the properties measured for a single CNT are over estimated. Being a one dimensional material, such materials do not have isotropic (same in all 3 dimensions) properties. This is a fundamental limit and relates to defects that control performance, reliability and yield of a device. Recent work of Snow and co-workers [17] shows that the device properties of a random network of carbon nanotubes are far inferior to those devices where CNTs are placed one by one by the "bottom up" approach. The anisotropic nature of the properties of carbon nanotubes is undesirable for optimum performance and reliability of future computing systems. It is worth mentioning here that the lack of isotropic properties was one of the reasons that high-temperature superconductors investigated vigorously in mid to late 80s were never used in the manufacturing of high performance and low-cost computers.

Nano-diamond (~1-4 nm) is a more promising material for future smaller devices because of its excellent properties such as wide band gap, high carrier transport speed, high breakdown field and negative electron affinity, coupled with high thermal conductivity, high optical transparency, high hardness and chemical inertness. Moreover diamond has μ_n= 2200 cm^2/V-s and μ_p= 1600 cm^2/V-s which shows that it is very much in line with the legacy of silicon CMOS. Nano-diamond offers the potential of fabricating ultra high-speed and smallest size computing systems using silicon substrates and CMOS technology. There is no fundamental barrier in the realization of nano-diamond based CMOS technology. However, a number of material and processing issues need to be resolved before the realization of the new technology.

In terms of processing technique, it is worth mentioning here that the "top down" approach is already being used in the manufacturing of 90 nm feature size computing systems. The most advanced photo-mask manufacturing today is already exploiting the ability to manipulate material at atomic level and generate virtually defect free photo-masks. However, there is no bottom up technique invented as of today that can be used directly in the manufacturing of future nanosystems.

The advocates of "bottom up" approach are using "Self-Assembly" to understand and control the intra-molecular quantum behavior of specifically designed and synthesized molecules using a surface to localize and stabilize them. Their goal is to interconnect, assemble and test nano devices and nano-machines starting from atomic or molecular parts. Due to the limitations of throughput and control of defects the so called "Self-Assembly" will not be making any contribution to the future computing systems. In our opinion, the meaning of "Self Assembly" also has been taken wrongly. All the work reported to date exploits selective chemistry. True self assembly process involves programmed cell death or apoptosis [18].

VII. System Level Engineering and Business Issues

For practical reasons, the potential role of nanotechnology in computing systems has to be examined from system level engineering and business point of view. There are applications that will require highest possible speed of computing systems. A typical example is simulation of nuclear explosions, and more accurate weather prediction. Similarly, applications like detection of gravitational waves require highest possible signal to noise ratio. For these applications expensive cooling systems leading to low temperature electronics is required. We will focus on applications that require ambient temperature operation and mass scale global applications. The semiconductor industry is moving in the direction where one asks the question, "What can be accomplished in one logical operation?" It is obvious that pushing nanoelectronics to ultimate highest possible speed is not going to provide high return of investment.

Any industry has three major periods, i.e. emerging period, growth period and mature period. Depending upon the success of each industry we can add another period which comes right after mature period that is decline period. After the rapid growth period the industry most likely exhausts its period of rapid growth in revenues and earnings then it eventually moves into maturity which closely resembles the overall rate of growth of the economy. Though the earnings and cash flow are still likely positive for these companies, their products and services have become less distinguishable from those of their competitors. Price competition becomes more vicious, taking profit margins along with it, and companies begin to explore other areas for products or services with potentially higher margins. Though mature industries mark time in mature stage, still investments in these companies stock can remain very attractive for many years. Share prices within mature industries tend to grow at a relatively stable rate that can often be predicted with some degree of accuracy based on sustainable growth prospects from historical trends. Perhaps even more importantly, companies in mature industries are able to withstand economic downturns and recessions better than growth companies because of their strong financial resources.

On one hand, the semiconductor industry is a matured industry, it is also the most innovative industry we have on earth today. Due to the profitability issue, the hardware industry will be forced to find new ways to cut the cost of manufacturing and increase the profitability. The results presented in this paper show that future progress in semiconductor manufacturing will continue to be delivered by "top down" approach and not by the "bottom up" approach of the nanotechnology.

VIII. Nanotechnology and Disruptive Nanoelectronics

In light of the fundamental knowledge of science and system level engineering and our business knowledge, we can examine the potential of nanotechnology in the practical realization of disruptive Nanoelectronics. Other than the conventional silicon based nanoelectronics, current research is being carried out primarily in the development of following new disruptive nanotechnologies: (i) quantum computing, (ii) molecular computing, and (iii) DNA computing.

Quantum computers are based mostly on using the properties of coherence leading to entangled states. Thus it is important to consider the time scale of the survival of entangled states. Phenomena such as decoherence, dephasing, and relaxations operate on the time scale of picoseconds and are largely responsible for deentanglement[7] and quantum computers have to be faster than picoseconds [8]. Since error correction is part of quantum computing, more heat will be generated in a quantum computer than a Von-Neumann architecture based computer operating at the same speed. Assuming that researchers can lengthen the decoherence interval to 10 microseconds; quantum computer chips will still consume more than 100 megawatts [19]. The facts presented here indicate that quantum computers if ever realized will involve massive size and extra ordinary cooling techniques (e.g. cooling of a nuclear reactor) needs to be implemented.

The literature is full of conceptual errors about the potential of molecular computing systems. First of all a two terminal logical molecular device will never find any practical applications in realizing a practical system. The second claim by the advocates of molecular computers is that components as large 10^{24} can be accommodated in the design of molecular nanoelectronics. The power density values [8] for such a molecular computer (molecular device area of the order of $0.025nm^2$) operating at switching speed of 1 ps will be of the order of $10^{29} W/cm^2$. Thus due to heat dissipation problem, it is fundamentally impossible to envision general purpose molecular computers.

The switching speed in biological systems is of the order of 10^{-3} s. The neuron diameter is 2 nm. Our brain with an aerial neural density of about $10^6/cm^2$ and a slow switching speed of the order of 10^{-3} s handles a power density of 2.5 nW/cm^2 very well. It is possible that in future such man made DNA computing systems can be interfaced to human beings to help the physically challenged human beings or those involved in accidents or brain related diseases. However, the claim that DNA based computing systems can be used as general purpose computing systems is fundamentally impossible.

For logical devices, the other approaches involving optical and spin based computing systems also have same problems of heat dissipations as we have discussed for other devices. The passive devices do not pose any fundamental issue of heat generation. However, the integration of passive device fabricated by "bottom up " approach with logical devices resulting into practical system faces the same manufacturing problems as are faced by conventional silicon nanoelectronics.

IX. SILICON CMOS BASED NANOELECTRONICS AS THE DRIVER OF GROWTH IN 21ST CENTURY

The following key trends are predicted for the 21st century:

(1) As of today, there is no other technology that can replace silicon CMOS (including SOI, strained silicon and various alternate transistor structures) ICs as the feedstock of electronics industry. Nano-diamond based CMOS structure grown on silicon substrate has the potential to provide future generations of semiconductor products. However, a number of materials and processing issues must be resolved before the mass scale manufacturing of electronics based nano-diamond.

(2) Barring any unpredicted global economic catastrophic disaster, the historical "boom and bust cycles" of the industry may become history and a lower but steady growth is expected. There are two primary reasons for this prediction. The first reason is that semiconductor industry executives are cautious about capital investment. The second reason is that the supply chain management has improved significantly and except for a few cases the inventory build-up is minimum. The RFID tags are creating a revolutionary opportunity to increase efficiency of supply chain management. For global applications, there are many challenges still to be solved for RFID technology. The availability of low-cost RF ID will provide growth in semiconductor industry and supply chains will have visibility to where inventory is, when orders are arriving. The improved supply-chain management due to RFID technology will pave way to "just in time" delivery of semiconductor products.

(3) The 300 mm wafer manufacturing is forcing more and more semiconductor companies to become Fabless. Except for a few major semiconductor manufacturers, top leaders are using foundries to meet manufacturing needs. The foundry model will continue to flourish and logic type 300 mm Foundries at lowest technology nodes will continue to be in great demand.

(4) In designing 90nm/65-nm semiconductor products, process variation is playing a more important role than in the past. For technology nodes at 90 nm and below, the physics of manufacturing silicon ICs plays a critical role on circuit design. As a result, design for manufacturing has become a necessity for cost-effective manufacturing of advanced semiconductor products. The process and tool developer can ease the burden in reducing the design cost of next generations of semiconductor products by reducing the process variation and incorporation in circuit design. For each technology node, the process window is shrinking. In the selection of processing tools, process variation must be given a higher priority. With 300 mm manufacturing and 90 nm technology nodes, design for manufacturing (DFM) has become a necessity for all nodes below 90 nm. In other words, the designer, process engineer and mask maker must have an open communication. This provides a unique opportunity for countries with lot of design expertise to have a successful 90 nm and lower technology nodes Fabs. Fig. 3 shows the key principles of DFM.

(5) System level innovation will be growth driver. Silicon is increasingly becoming commodity, and the software is becoming a differentiator between solutions. Countries with strength in software will provide higher growth opportunity of designing and manufacturing semiconductor and electronic products.

(6) Due to innovation, the industry will continue to deliver products with lower cost, smaller sizes and improved functionality. However, in future due to heat dissipation issues, marginal improvement in speed of the silicon circuits will be observed.

(7) The integration of micromechanical systems (MEMS) and Nanomechanical systems (NEMS) based active and passive devices with silicon CMOS either through system on chip or system in package will continue to provide growth opportunity. However, reliability and packaging

will remain the most challenging aspect of the integrated systems.

(8) Most of the 300 mm fabs employ about 80-90% of the wafer manufacturing steps by single wafer processing and other 10-20% by batch processing. The cycle time for such fabs is about 60-75 days. On the other hand for single wafer processing fabs the cycle time is about 30 days. Packaging has become more important than ever. For advanced semiconductor products, wafer level packaging is gaining ground. System on Chip is a long term universal solution, and system in package is a near term profitable solution. The use of single wafer manufacturing in both wafer processing and wafer level packaging at the same location has the potential to meet the goal of close to perfect supply chain management. These issues are discussed at length our recent publications. [20-22].

(9) Semiconductor industry is already a global industry. With India and China having sustained higher growth for the last few years than the rest of the world, these two countries will play a significant role in the design and manufacturing of semiconductor products. China is discounted as a major manufacturing force in the foreseeable future. China is now the world's largest IC market at $40.8 billion, representing 21% of the world's $192.4 billion appetite for ICs. However, exploding demand in China won't spur similar expansion in the domestic IC manufacturing base. China's IC production was just 6% of its consumption in 2005 ($2.6 billion), and despite 36% CAGR it'll meet just 10% of domestic IC demand in 2010 ($12.1 billion), which would be only 4% of the total forecasted $319 billion worldwide IC production (IC Insights Inc, Jan 16, 2006). India is already playing a significant role as the semiconductor design hub and in future it may play a very important role in advanced semiconductor manufacturing. Other countries from other parts of the globe (not having any significant part at the moment) will have the opportunity to create semiconductor growth in their countries.

(10) The big players of the industry will try to bring the 450 mm wafers into manufacturing. The equipment suppliers as well as most of the industry will try to oppose the introduction of 450 mm wafers. The economic health of the industry will dictate whether the introduction of 450 mm wafers will take place or not. However, if 450 mm wafer manufacturing becomes a reality, the semiconductor industry will be run by few giants like the current global petrochemical industry.

X. CONCLUSION

In this paper, we have described nanotechnology in the context of semiconductor manufacturing in the 21st century. For ultra high speed computing systems, the heat dissipation problem is a fundamental issue and nanotechnology does not provide any breakthrough that can solve the packaging problem of these advanced computing systems with heat dissipation capability of the order of kilo watts per square centimeter and higher power densities. The "bottom up" approach of nanotechnology has fundamental limits of throughput and defects. As a result, "top down" approach will continue to provide next generations of semiconductor products unless hit by economic constraints even before the issue of fundamental limits will arise.

REFERENCES:

[1] E. Vonderheid, *The Institute,* Vol. 27, No. 3. 1, 2003.

[2] "The business of nanotech," Business Review, Cover Story, Feb. 14, 2005. http://www.businessweek.com/magazine/content/05_07/b392000 1_mz001.htm

[3] K. E. Drexler, *Engines of Creation: The Coming Era of Nanotechnology,* Anchor Press, New York, NY, 1986.

[4] A. Regalado, "Nanotechnology Patents Surge As Companies Vie to Stake Claim," The Wall Street Journal, Vol. CCXLIII. No. 119, A1, 2004.

[5] National Nanotechnology Imitative, http://www.nano.gov/

[6] N. Gershenfeld, *"The Physics of Information Technology,"* Cambridge University Press, 2000.

[7] S. Lloyd, *Nature,* 406, 1047, 2000.

[8] R. Singh, J. O. Poole, K. F. Poole and S. D. Vaidya, *J. Nanosci. Nanotech,* Vol. 2, No. ¾, pp. 363-368, 2002.

[9] P. Zallas, D. Charalambldis, N. A. Papadoglannis, K. Witte and G. D. Taklrls, *Nature,* 426, 267 (2002).

[10] R. Singh et al, MIEL 2002

[11] I. Sauciuc, "Investigating Electronics Cooling Technologies," Advanced Packaging, http://ap.pennnet.com/Articles/Article_Display.cfm?Section=AR CHI&ARTICLE_ID=243253&VERSION_NUM=2&p=36

[12] P. Ross, "Beat the Heat," *IEEE Spectrum,* May 2004, pp 39-43.

[13] C. H. Bennett, *IBM J. Res. Develop.* 17,525 (1973)

[14] S. N. Rao and R. Singh, *Physics Education,* 12 (3), 206 (1995).

[15] M. Schliwa and G. Woehlke, *Nature,* 422, 769 (2003).

[16] Prof. J. Narayan, NC State University, Raleigh, NC, personal communication, June 27, 2004.

[17] E. S. Snow, J. P. Novak, P. M. Campbell, and D. Park, *Appl. Phys. Lett.,* 82, 2145, (2003).

[18] M. T. Heemels, *Nature,* 407, 769, 2000.

[19] R. C. Johnson "Error Correction May Stall Quantum Computing," http://www.eetimes.com/at/news/OEG20021202S0077

[20] R. Singh, M. Fakhruddin and K. F.Poole, *IEEE Transactions on Semiconductor Manufacturing,* 16, 96, 2003.

[21] R. Singh, A.Venkateshan, M. Fakhruddin, K. F.Poole, N. Balakrishnan & L. D. Fredendall, *Semiconductor Fabtech, 19th Edition,* 85, 2003.

[22] R. Singh and R.Thakur, IEEE Spectrum, February 2005.

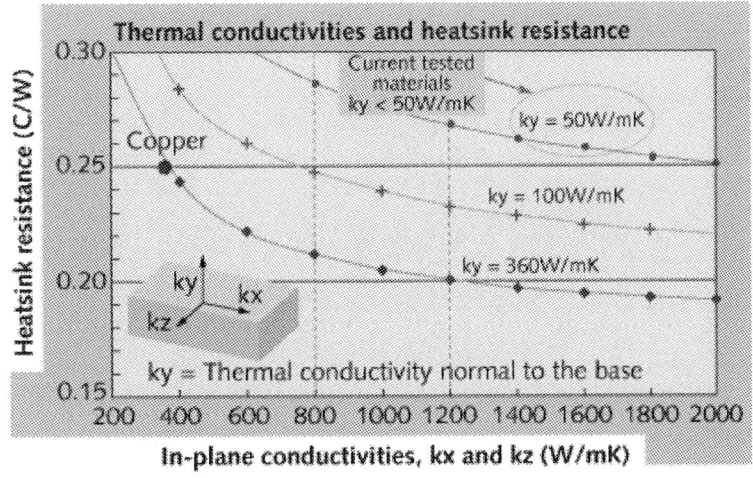

Fig. 1. Effect of anisotropy of thermal conductivity on the heat sink resistance [from Ref. 11]

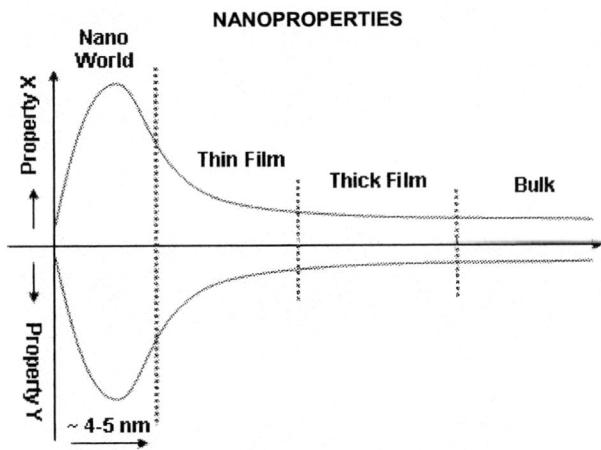

Fig. 2. Properties of Nanofilm as a function of thickness.

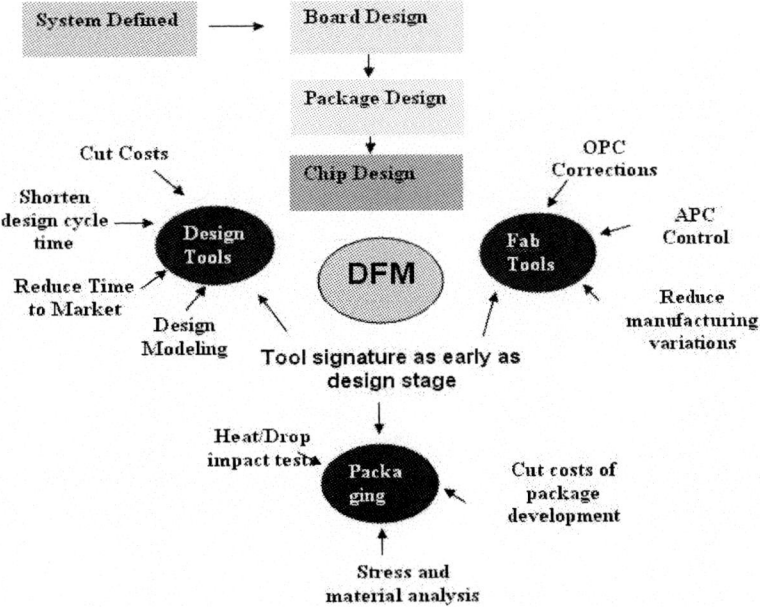

Fig. 3. Key principles of design for manufacturing (DFM)

10

2006 25th International Conference on Microelectronics

Nanoscience and Nanotechnologies in Serbia

Zoran Djurić

Abstract – The paper shows that the large interest researchers throughout the world for the multidisciplinary fields of nanoscience and nanotechnologies reflected very fruitfully in the science in Serbia. The paper presents some topics of nanoscience and nanotechnology researched in Serbia, outlines the organization of research in this field of global importance and mentions some of the problems encountered along the way. A personal view to some of the current matters is also given. In spite of financial difficulties, Serbian science has shown that it is able to take part in one of the most propulsive and attractive fields of science.

I. INTRODUCTION

During the last decade, as is well known, a large interest arose among the research workers for the insight into the so-called nanoworld – the world whose dimensions are measured in nanometers. Bearing in mind that fundamental processes reflect in this dimension range both in living and non-living matter, an interest for the field of nanoscience and nanotechnologies arose in physics, chemistry, biology, medicine, etc.

On the other hand, great expectations arose as well. The supporters of this direction in science and technology expect from nanotechnologies a help in solving the most important problems of the humanity, for instance the problems of energy crisis, elimination of incurable diseases, improvement of environmental protection and prevention of its further degradation. Shortly, the supporters expect from nanotechnologies to revolutionize life on earth [1-2], also [3] and references cited therein.

Developed countries already perceived potential benefits of nanotechnologies and their investments into this field are steadily increasing. Practically all countries with organized science today have their national strategies in the field of nanotechnologies. These programs define the priority research topics, and in accordance to these, national and international research networks are being planned and established, as well as specialized centers for specific fields of nanotechnologies.

The same as any other technology expected to bring large profits, nanotechnologies may be abused, and one of the possible ways is to use them to increase still more the economic gap between the rich and the poor.

A special danger is hidden in claiming those branches of knowledge which are in the foundations of creation of life. Also, there are problems connected with the fact that these technologies may influence the environment and the human health. For instance, simple and harmless compounds may become dangerous if their dimensions are reduced to the nanometer level. Nanoparticles readily permeate body cells and are able to cross biological barriers, such as the blood-brain barrier.

On ethics and health issues in nanotechnologies see refs. [2], [4-5].

Connected with the above, one of the problems is a lack of experts in the field, since conventional education in Europe and in the U.S.A. is not convenient for nanotechnologies. Bearing in mind its cross-disciplinary nature, i.e. that one may be required at the same time to master knowledge and expertise in physics, chemistry, medicine and biotechnologies, microelectronics and microsystem technologies, some countries start education for nanotechnologies already at the high school level. A special attention is given to the education of wider population, in order to avoid misinterpretation of nanotechnologies and a negative attitude of the society to them.

II. NANOSCIENCE & NANOTECHNOLOGY DEFINITIONS AND POTENTIAL IMPACT

Although it is impossible to give an overall definition of nanotechnologies, we will assume the following [2]

Nanoscience is the study of phenomena and manipulation of materials at atomic, molecular and macromolecular scales, where properties differ significantly from those at large scale.

Nanotechnologies are design, characterization, production and application of structures, devices and systems by controlling shape and size at nanometer scale.

The prefix "nano" is derived from the Greek word for dwarf. One nanometer (nm) is equal to one-billionth of meter, 10^{-9} m. A human hair is diameter is approximately 80,000 nm, and that of a red blood cell is approx. 7000 nm. Fig. 1 shows proportions of the micro and nanoworld.

Nanotechnologies do not represent a completely new discipline of science and technology. What makes them specific is their work on discovering, understanding and applying different phenomena appearing on the nanolevel and the possibility to control these phenomena, regardless of the fact if their origin is physical, chemical or biological and regardless of that if they belong to the living or non-living world [3].

Similar to other contemporary technologies (for instance information technologies), nanotechnologies did not appear by accident. They appeared when conditions were secured to "see" nanostructure and to manipulate them in a controllable manner (the appearance of the AFM, nanolithography, molecular beam epitaxy, nanotransistors, etc).

Zoran Djurić is with the IHTM – Institute of Microelectronic Technologies and Single Crystals, Njegoševa 12, 11000 Belgrade, Serbia and Montenegro

Fax +381 11 182 995, e-mail zdjuric@nanosys.ihtm.bg.ac.yu

1-4244-0116-X/06/$20.00 ©2006 IEEE

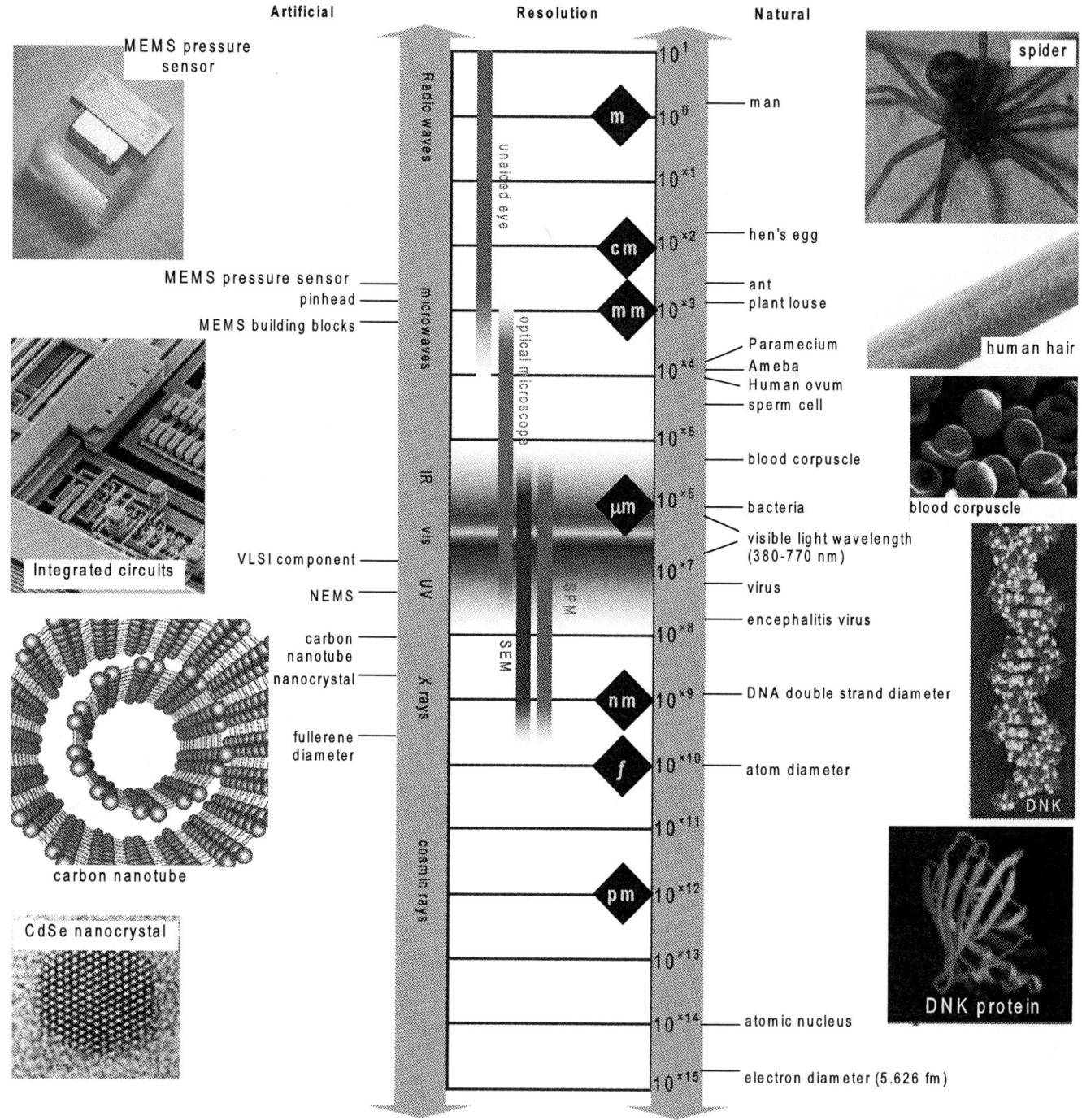

Fig. 1. Proportions of the micro- and nanoworld (figure partly based on [3])

It is generally held today that nanotechnologies are among the key breakthrough of the 21st century [2]. It is expected that they will enable the development of a number of different fields which could improve the competitiveness of industry and thus improve the quality of living of all citizens.

Nanotechnology is real – the questions are generally when, not if. Building at the nanoscale enables new interactions in materials, semiconductors and biological agents. The new scale allows manipulations on the cellular level, which should enable new discoveries in pharmaceuticals, biodefense and health care. The U.S.A. National Science Foundation (NSF) sees a potential market totaling US$ 1 trillion in the next 10-12 years, as outlined in Table 1.

TABLE 1. POTENTIAL IMPACT OF NANOTECHNOLOGY

Sector	Size (US$ B)	Share (%)
Materials	340	31
Electronics	300	28
Pharmaceuticals	180	17
Chemical manufacture	100	9
Aerospace	70	6
Tools	20	2
Improved healthcare	30	3
Sustainability	45	4
Total	1085	100

III. FROM THE HISTORY OF NANOTECHNOLOGY

Our material world consists of atoms. This claim was spoken by the Greek philosopher Democritus.

It is remarkable that three papers published in 1905 by Einstein and Sutherland laid such a foundation for science that they lie at the base of our understanding of nanostructures, are useful to biologists and are essential to professional investigators [6-9].

However, many authors believe that everything started in 1959, when the Nobel prize winner Richard Feynman held his famous speech "There is Plenty of Room at the Bottom". At this occasion he said [10]

The principles of physics, as far as I can see, do not speak against the possibility of maneuvering things atom by atom. It is not an attempt to violate any laws; it is something, in principle, that can be done, but in practice, it has not been done because we are too big... At the atomic level, we have new kinds of forces and new kinds of possibilities, new kinds of effects... But it is interesting that it would be in principle possible (I think) for a physicist to synthesize any chemical substance that chemist writes down..."

TABLE 2. CHRONOLOGY OF THE DEVELOPMENT OF NANOTECHNOLOGIES (PARTLY BASED ON [3])

1959	Nobel prize winner, physicist Richard Feynman held his famous speech "There's plenty of room at the bottom", where he described the potentials of atom engineering in the future.
1974	Norio Taniguchi from *Tokyo Science* University utilizes for the first time the word "nanotechnologies"
1982	Gerd K. Binning and Heinrich Rohrer from the IBM Zurich Research Laboratory invented the scanning tunneling microscope) which enabled researches to see atoms for the first time and to manipulate with them. Some time later, the invention of the atomic force microscope (AFM) enabled the advent of a number of different kinds of scanning probe microscope which became the eyes to the nanoworld. In 1986 these researchers were awarded a Nobel prize for physics..
1985	R. F. Curl Jr, H. W. Kroto and R. E. Smalley

	discovered fullerenes with the dimensions ~1 nm
1987	Eli Yablonovitch from the UCLA discovered photonic bandgap materials (photonic crystals)
1989	Physicists from the IBM draw the logo of their company by precision positioning of 35 xenon atoms
1991	Sumio Ijima, a physicist from the NEC research Labs in Japan, discovered multilayered carbon nanotubes
1993	Rice University established the first nanotechnology lab in the USA
1998	Researchers from the Delft University of Technology (Netherlands) fabricated the first transistor based on carbon nanotubes
2000	Si technology reaches nanodimensions, while still obeying the Moore's Law
2000	John Pendry from the Imperial College, UK proposed the "perfect lenses: based on optical metamaterial with negative refractive index
2000	The president of the U.S Bill Clinton announced the National nanotechnology Initiative (NNI). After this 40 other countries included nanotechnologies among their priority programs
2001	Mitsui & Co, Japan announce their plans for mass production of nanotubes
2002	Nanotechnologists from the IBM demonstrate the possibility to write data on a square inch corresponding to a 100 GB hard disk – it is possible to write 25 millions of pages on the area of a postage stamp.
2002	Mean cost of a transistor reaches a value of 100 n$, which is 7 orders of magnitude less than in 1968
2002	EU starts its Sixth Framework Program (FP6) with total funds of 17.5 billion €, while for the thematic priority 3 – nanotechnologies and nanoscience, knowledge-based multifunctional materials and new production processes and equipment the funds amount 1.3 billion €.
2005	EU proposes the Seventh Framework Program (FP7). Its thematic priority "Nanoscience, nanotechnologies and novel production techniques" is planned to receive triple funds, 4.8 billion €.

It appears to the author that the decisive impulse, which brought to the overall interest for nanotechnologies, arrived from the integrated circuits technology. As a consequence of a constant need to decrease prices, the dimensions of the elementary components of the integrated circuits decreased. This required the introduction of the technologies of nanolithography, ion etching, etc. which not only enable the fabrication of nanotransistors (per a price of 100 n$/piece) but almost the whole top-down approach in nanotechnology fabrication.

The importance of the semiconductor industry is huge.

- The semiconductor industry, with its annual sales of about 250 billion €, represents the foundation of high-tech economy
- Semiconductors, on the other hand, enable electronics to reach the consumer market – further 1100 billion €

- Electronics further enables various services (telecommunication, Internet, radio, TV) to reach 5000 billion € market.

During the past period of more than 4 decades, the semiconductor industry was driven by a "smaller-cheaper-better" synergy model into nanotechnology. Throughout this period it followed pretty closely the law of G. E. Moore [11] (this trend of the development of integrated circuits anticipates the decrease of dimensions and price of components and the increase of the number of components per chip) – see Fig. 2.

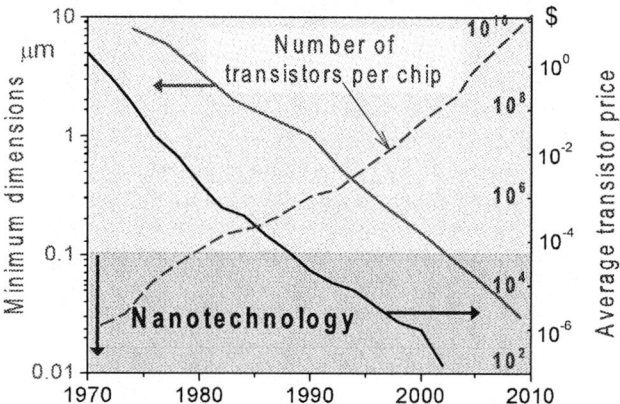

Fig. 2. Trends set by Moore's law and the advent of nanotechnologies

IV. CURRENT STATUS OF SERBIAN SCIENCE

Research in Serbian science is divided into two large groups, fundamental research and research in technological development.

Fundamental science is further divided into biology (607 researchers), physics (464), chemistry (849), mathematics and mechanics (489), geosciences and astronomy (244), humanities (973), history, archeology and ethnology (377) and language and literature (344). Bracketed values denote the number of registered researchers in each field. Thus the total number of researchers in Serbia in fundamental science is about 5070.

The first 6 fields have more or less (through the projects proposed and granted funding in 2006-2010 period) expressed an interest for work in nanotechnologies. These 6 fields engage a total of 3505 researchers. The number of full time researchers is about 2320 (the term "full time researcher" is used here to denote a researcher engaged 12 months on a funded science project). The difference between these numbers is a consequence that not all researchers are fully engaged during the whole year.

The author processed in more detail only the data on nanoprojects in the fields of physics and chemistry, since their results and fields of research are within the scope of this text and this conference.

The Serbian Ministry of Science and Environmental Protection plans to fund 48 projects in physics in the period 2006-2010. Of these 48 projects, 21 gravitate to nanotechnologies, while 12 out of these 21 are practically fully dedicated to nanotechnologies. A total of 120 full time researchers is engaged in these 12 projects, or about 1/3 of all researchers in physics.

A total of 77 projects will be funded in chemistry. 27 of these gravitates to nanotechnologies, while 22 projects are fully dedicated to nanotechnologies. About 160 full time researchers are engaged in these 22 projects, i.e. about 28% of all engaged researchers in chemistry.

Research in the technological development part is divided into the following fields: Electronics and Telecommunication (457), Industrial Software (220), Mechanical Engineering (279), Traffic Engineering (115), Urbanism and Civil Engineering (126), Power Engineering and mining (232), Materials and Chemical Engineering (310) and Biotechnologies (993). The total number of registered researchers is 2732. Full time researchers amount about 1400.

In a certain manner, the topics and the content of nanotechnology research reflects world research in this field. Basically, there are two cases: some projects worked in the previous period in the fields close to nanotechnologies, while some projects introduced the prefix nano as a fashion detail only. The reasons for this are rather obvious.

The most important themes (projects) in physics are

- Carbon and inorganic nanotubes
- Nanostructures and nanodevices in physical electronics
- Physics of low dimensional and nano-sized structures and materials
- Preparation and characterization of nanostructured material surfaces
- Physics, modeling and characterization of phenomena in nanolayers of advanced MOS devices
- Modification, synthesis and analysis of nanostructured materials by ion beams, gamma irradiation and vacuum deposition
- Physical basis for applying nonequilibrium plasmas in nanotechnologies and treatment of materials
- nano-sized and bulk rare earths and 3D based oxides – synthesis, structural and magnetic properties
- Investigation of electronic structure, spin and orbital magnetism and optical properties of crystals and systems with reduced dimensionality
- Amorphous and nanostructured chalcogenides
- Theoretical study of the properties of strongly correlated systems with complex structures
- Advancement of the physical characteristics of nanostructured materials

The most important research topics in chemistry are

- Chemistry of ions in the gas phase: fullerenes and atomic clusters
- Nanostructured solid solutions for application in electronics and alternative energy sources
- Synthesis, characterization and application of colloidal dispersions of inorganic nanomaterials
- Synthesis of functional materials with controlled structure on molecular and nanolevel

- Synthesis, characterization and activity of organic and coordination compounds and their use in (bio)nanotechnology
- Nanostructured non-oxide ceramics and carbon-based materials and their composites
- The synthesis and characterization of polymers and polymer (nano)composites of defined molecular and supermolecular structure
- Deposition of ultra-fine powders of metals and allows and nanostructured surfaces by electrochemical techniques
- Design of nanocrystalline (NdPr)-Fe-B magnetic materials and components based on "smart" magnetic materials
- Finely dispersed systems: micro-, nano- and ato engineering
- Synthesis, characterization and use of new fullerene derivates
- Synthesis, characterization and application of nanostructured catalysts at different catalyst supports in a fuel cell, water electrolysis and electro-organic synthesis
- Structural modification and reactions of micro porous and mesoporous materials
- Advanced materials for proton exchange membrane fuel cell applications
- Synthesis of nanopowders and processing of ceramics and nanocomposites for application in novel technologies
- Synthesis and characterization of nanoparticles and nanocomposites
- Synthesis, structure, properties and application of nanostructured functional ceramics and bioceramics
- Synthesis of bioactive fullerene molecules and nanomedical research
- Electrochemical characterization of oxide polymer coatings on modified metal surfaces.

Regretfully, only two projects in Technological Development belong to nanotechnologies, with a total of 28 researchers – 20 full time researchers. These projects handle the following topics
- Micro and nanosystem technologies
- Nanostructured surfaces in pulsed plasma – technological and ecological advancement.

Bearing in mind that there are nanotechnology projects in medicine and biology as well, dedicated to the possibilities of cancer treatment, to biosensors, etc., a conclusion is that the listed project themes show that Serbian science really follows the state of the art in world research in this field.

In 2005 the Ministry of Science and Environmental Protection of Serbia established a Council for Nanoscience and Nanotechnology as an official body to deal with the strategic tasks posed by the advent of nanotechnologies.

It should be mentioned here that during 2005 several national and international topical scientific meetings fully dedicated to nanoscience were held in Serbia and Montenegro, including the International Workshop on Nanoscience and Nanotechnology, Belgrade [12], also nanoETRAN workshop within the framework of the 49th ETRAN conference in Budva [13], nanoworkshop in YUCOMAT [14].

V. FUNDING OF SCIENCE IN SERBIA

The funding of public services in Serbia decreased steadily in the period 1991-2000, parallel with the drop of industrial production. In average, the rate of drop of gross product was 7.2%, while the funding of public activities decreased 8.5% a year. In the period 2000-2005 the budget of the Republic Serbia increased almost 15 times, while the funds allotted for science, research and development increased 6.8 times.

Large changes occurred in the structure of public expenses in 2003 as compared to 1990. The share of budget funding for education increased (from 27% to 32.2%), also that of culture, information and arts (11% to 12.4%), of social programs (8.6% to 9.5%), while the funding of research decreased from 8.9% to 4.6%, i.e. almost 50%.

A real level of total funding in research in 2003 was 29.1% of that in 1990, meaning that funding of science in Serbia dropped at a yearly rate of 10% (total public services dropped at a rate of 4.9%). This trend continued in 2004 and 2005, and regretfully in 2006 as well.

As another indicator we quote here that the share of budget funding for science in Serbia decreased from 2.45% in 2000 to 1.1% in 2005.

All the relevant data on funding of other public services in Serbia point to the fact that funding of research and development in Serbia significantly lags behind all other public services (i.e. education, culture, sport, health, etc.)

Observed in absolute amounts, funding of science in this moment shows that Serbia is a really very poor country. As an example, we quote that in 2001 Sweden funds for science amounted about 10 billions of US$, which is only slightly less than the total gross national product of Serbia. This means that Sweden's funds for science are more than 150 times larger than those of Serbia. At the same time, the total population count in both countries is comparable.

A price of a single published paper in an international science journal in Sweden is more than 10 times larger than that in Serbia.

The most frustrating detail in the whole picture is the restrictive national policy of funding science. The Serbian government in average decreased its funding of science to about 50% compared to other public services. Fig. 3 shows – without any special comments – the official data (as published online at the Ministry of Science and Environmental Protection web portal) on planned and realized funding of Serbian science.

It is clear that the current state of the affair strongly intensifies the brain drain in Serbian science and enlarges the already large gap between the country and the developed economies. On the other hand, it is well known that there is an explicit dependence between the funding of research and the overall advancement of economy. This is also readily seen on the example of literally all developed countries. Thus the long term consequences of the current negligence to the advancement of science in Serbia could prove disastrous to the whole country's economy.

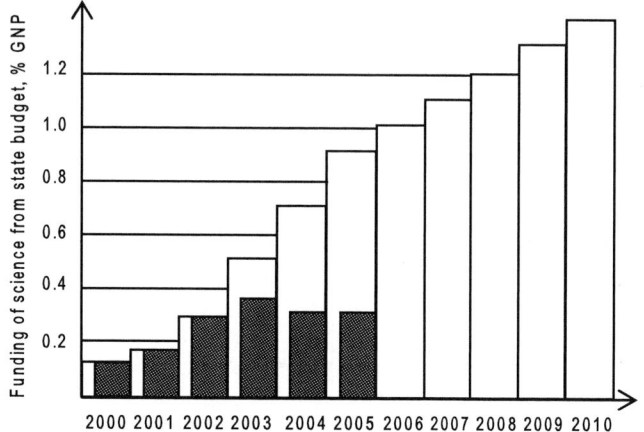

Fig. 3. Funding of Serbian science from the state budget, % of gross national product (investment to infrastructure included). Planned: empty rectangles, realized: shaded

Here we give two quotations of people responsible for research in the U.S.A. because in their opinion their government did not allot sufficient funds for science.

"Not only our economy and quality of life depend critically on a vibrant R&D enterprise, but so do our national and homeland security."

(Hart-Radman Commission on National Security, 2001)

"It's a creeping crisis, and it's not something the American psyche responds too well. It's not a Sputnik shot, it's not a tsunami..."

(Craig Barrett, Chairman, Intel Corporation)

It may be also interesting to analyze the relation of the European Union to Serbian science. It appears to the author that the help of the EU to our science is significantly smaller than its help to other public services.

A large difference between the number of full-time funded researchers and the total number of people engaged in science leads one to believe that there is a large and rather unused potential even among those already engaged in science.

VI. CONCLUSION

The author has the pleasure to note that a substantial number of nanotechnology projects in Serbia will be funded in the period 2006-2010 through the Ministry of Science and Environmental Protection. In spite of the effort of the said Ministry, the funding of these projects, the same as the overall investments to science and development in Serbia, remain on an extremely low level compared both with the surrounding countries and the Europe as a whole. In this moment the investments into research infrastructure and lab equipment are practically nonexistent. It is the opinion of the author that the developed world, including the EU, could make more use of our research results than

our country itself, since our industry is in a rather poor state. In spite of this, it appears to the author that their help offered to our science is less than adequate, although, if nothing else, it could work strongly toward the prevention of any "knowledge apartheid". Besides an increased funding, it is necessary to find a convenient organizational form for national nanoscience and science generally to enable a larger coordination between projects and ensure more efficient and productive research. In spite of the present difficulties and less than desirable condition, Serbian research teams continue to generate non-negligible output, giving place to the hope to a brighter future.

VII. ACKNOWLEDGEMENTS

The author is indebted to dr. Stana Petrović, who kindly enabled the access to various data quoted in Section V. Also, the author is thankful to Dr. Zoran Jakšić and Dr. Dana Vasiljević-Radović for their help and discussions, without which this work would not be possible.

This work has been partially supported by the Serbian Ministry of Science and Environmental Protection within the framework of the project IT.6151.B.

REFERENCES

[1] M. C. Roco, *"National Nanotechnology Initiative"*, Overview, ASME Workshop, September 22, 2004. (http://nano.asme.org/nationalnanoinitiative.pdf)

[2] Nanoscience and Nanotechnologies: Opportunities and Uncertainties, Report 29. July 2004. (http://www.nanotec.org.uk/finalReport.htm)

[3] Z. Djurić, "Nanotechnologies as a global task of researchers in 21st century (plenary lecture)", (in Serbian language) Proc. 49th Conf. ETRAN, Budva, June 5-10. 2005,vol. 1,2,3,4, pp. 13-27

[4] W. L. Robison, "Nano-Ethics", in Discovering Nanoscale, eds. A. Mordman, J. Schummer, IOS Press, Amsterdam, 2004.

[5] Z. Popović, "Nanoethics", Invited Paper (in Serbian language), Proc. 49th Conf. ETRAN, Budva, June 5-10. 2005,vol. 4, pp. 271-274.

[6] W. Sutherland, "A Dynamic theory of diffusion for non-electrolytes and the molecular mass of albumin," Phylosophical Magazine, S.6.9, pp. 781-785, 1905

[7] A. Einstein, "Brownian Motion,", Analen deo Physik, 17, 891, 1905

[8] A. Einstein, "Molecular Size," Einstein thesis, Annalen der Physik, 19, 289, 1906

[9] B. H. J. McKeller, "How the mass movement of trillions of atoms changed the world – Einstein, Sutherland and Brownian Motion"

[10] R. Feynman, "There's plenty of room at the bottom: an invitation to enter new field of physics", Engineering and Science, 23, 5, pp. 22-36, 1960, http://www.zyvex.com/nanotech/feynman.

[11] G. E. Moore-a. "Cramming more component onto integrated circuits," Electronics, vol.38, No 8, april 19., 1965.

[12] http://www.nanosys.ihtm.bg.ac.yu/nanosci/cosent2005.htm

[13] http://etran.etf.bg.ac.yu

[14] http://www.yu-mrs.org.yu/conf05.htm

2006 25th International Conference on Microelectronics

Failure of Ohm's Law:
Its Implications on the Design of Nanoelectronic Devices and Circuits

Vijay K. Arora

Abstract - Impact of Ohm's law failure on the design of twenty-first-century nano-circuits is evaluated and discussed. The direct and differential resistance is shown to rise dramatically in the nonohmic regime, where applied voltage V >> V_c, the critical voltage for the onset of nonohmic behavior. The velocity-field characteristics that affect transport in quantum nanostructures are reviewed. The impact on the familiar current and voltage division laws and CMOS design is indicated. The results presented are useful in characterizing and evaluating performance of nano-devices and related circuits.

I. INTRODUCTION

The analysis of electronic devices and circuits relies on the validity of Ohm's law. This paradigm is based on the velocity response to an applied electric field (E) that is linear:

$$v = \mu_o E \qquad (1.1)$$

where μ_o is the ohmic mobility that describes the ease with which itinerant carriers (electrons or holes) are able to roam through a given device by undergoing the accelerating process of an electric field and decelerating process of the collisions with the lattice ions.

In a typical experimental setup (Fig. 1), a resistor of length L and area of cross-section A $(A=W\ d)$ is stimulated by a voltage V applied across its length. The response is the current I through the resistor.

For a resistor with electron concentration n and drift velocity v, the current is $I = n\ q\ v\ A$, where q is the electronic charge. The linear I-V characteristics, with the application of Eq. (1.1), are given by:

$$I = \frac{V}{R_o} \qquad (1.2a)$$

with

Vijay K. Arora is with the Division of Engineering and Physics, Wilkes University, Wilkes-Barre, PA 18766, U. S. A., E-Mail: varora@wilkes.edu

$$R_o = \frac{1}{nq\mu_o}\frac{L}{A} = \rho\frac{L}{A} = \rho_s\frac{L}{W} \qquad (1.2b)$$

where $\rho = \dfrac{1}{nq\mu_o}$ is the resistivity (Ω–m) of the resistor and

$\rho_s = \dfrac{\rho}{d}$ ($\Omega/\square$) is its sheet resistivity. Ohm's law, given by Eq. (1.2a), has been the basis on which electronic circuits are designed, characterized, and their performance evaluated. It worked very well for macro resistors (L >100 μm). Ohm's law does not hold for micro/nano-resistors and circuits when L is few nanometers and electric field is high in scaled-down dimensions.

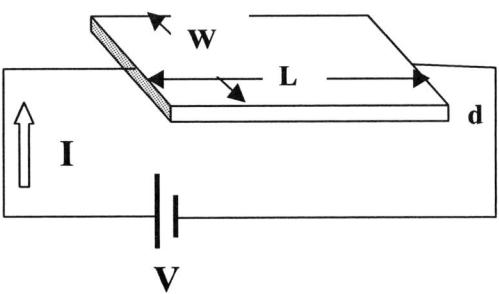

Fig. 1. The current flow in a conducting sheet due to the presence of itinerant carriers.

II. THE ULTIMATE VELOCITY

Equation (1.1) is not valid for electric fields exceeding the critical value $E_c = \dfrac{V_t}{\ell_o} = \dfrac{v_{th}}{\mu_o}$, which is 2.59 kV/cm for room-temperature thermal voltage V_t of 25.9 mV and a typical Ohmic mean free path ℓ_o of 0.1 μm [1]. The ultimate saturation velocity v_{sat} is limited by the thermal velocity

$v_{th} = \sqrt{\dfrac{2k_B T}{m^*}}$ for electrons with energy $\varepsilon = k_B T$ in a

nondegenerately-doped regime. The published literature is inconclusive in predicting the dependence of saturation velocity on low-field mobility. Higher mobility will make velocity approach towards saturation faster.

In the high electric field regime ($E \gg E_c$), Eq. (1.1) modifies to an empirical relation [2]:

$$v = \frac{\mu_o E}{\left[1 + \left(\dfrac{E}{E_c}\right)^{\gamma}\right]^{1/\gamma}}$$

$$= v_{sat} \frac{\dfrac{E}{E_c}}{\left[1 + \left(\dfrac{E}{E_c}\right)^{\gamma}\right]^{1/\gamma}} \tag{2.1a}$$

with

$$v_{sat} = \mu_o E_c \tag{2.1b}$$

$\gamma = 2$ for electrons and $\gamma = 1$ for holes are found to fit the experimental data well [2]-[3].

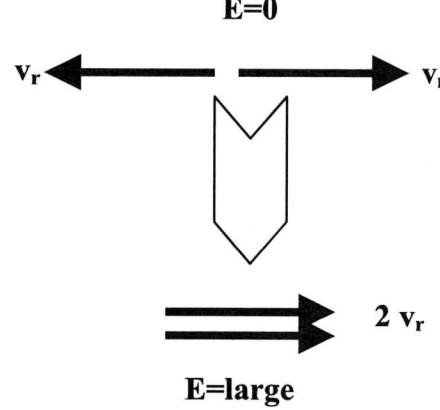

Fig. 2. Random motion transforms to a streamlined one as extremely high electric fields are encountered.

Considering the velocity limitations above, the current in a resistor (channel) is limited by its saturation value $I_{sat} = n\, q\, v_{sat}\, A$, which in turn depends on the doping concentration n and saturation velocity v_{sat}. It is, therefore, essential to assess the magnitude of this ultimate velocity that results in the current saturation. In equilibrium, the band diagram is flat and randomly oriented velocity vectors cancel each other, as shown in Fig. 2 for one-dimensional motion. As the applied electric field tilts the band diagram, an electron traveling in the direction of electric field find it difficult to surmount the barrier, while that traveling in the opposite direction accelerates in a mean free path and collide, randomizing its velocity and restarting its journey for

another mean free path. The net result is that random vectors $\mathbf{v_r}$ streamline in a very high electric field (See Fig. 2).

In the presence of a very high electric field, all electrons are streamlined opposite to the direction of the applied electric field. The ultimate unidirectional drift velocity is the average of its absolute value $|v| = \sqrt{\dfrac{2\varepsilon}{m*}}$. When this averaging is taken, the ultimate velocity for a semiconductor of a given dimensionality, 3 (bulk), 2 (quasi-2-dimensional Q2D) or 1(quasi-1-dimensional Q1D), is given by

$$v_{uj} = v_{thj} \frac{\mathscr{F}_{\frac{(j-1)}{2}}\left(\eta_j\right)}{\mathscr{F}_{\frac{(j-2)}{2}}\left(\eta_j\right)} \tag{2.2}$$

$$j = 3(bulk), 2(Q2D), 1(Q1D)$$

with

$$v_{thj} = v_{th} \frac{\Gamma\left(\dfrac{j+1}{2}\right)}{\Gamma\left(\dfrac{j}{2}\right)} \tag{2.3a}$$

$$v_{th} = \sqrt{\frac{2 k_B T}{m*}} \tag{2.3b}$$

$$\mathscr{F}_j(\eta) = \frac{1}{\Gamma(j+1)} \int_0^\infty \frac{x^j\, dx}{1 + e^{x-\eta}} \tag{2.4}$$

$$\eta_j = \frac{E_F - E_{cj}}{k_B T} \tag{2.5}$$

v_{thj} is the average thermal velocity for j-dimensionality, $\Gamma(j)$ is the gamma function, $\mathscr{F}_j(\eta)$ is the Fermi integral of order j, and η_j is the normalized Fermi Energy E_F with respect to band edge E_{cj} (including the zero point energy) for a given configuration. $\Gamma(n+1) = n!$ for an integer n. $\Gamma(3/2) = \sqrt{\pi}/2$, and $\Gamma(1/2) = \sqrt{\pi}$. The Fermi energy η_j is obtained from

$$n_j = N_{cj}\mathscr{F}_{\frac{(j-2)}{2}}\left(\eta_j\right) \tag{2.6}$$

where n_3 is the volume concentration of electrons in bulk materials, n_2 is the surface density in Q2D materials, and n_1 is the linear density in Q1D materials. N_{cj} for j =3, 2, and 1 are given by

$$N_{c3} = 2\left(\frac{m^* k_B T}{2\pi\hbar^2}\right)^{\frac{3}{2}} \tag{2.7a}$$

$$N_{c2} = \frac{m^* k_B T}{\pi\hbar^2} \tag{2.7b}$$

$$N_{c1} = \sqrt{\frac{2m^* k_B T}{\pi\hbar^2}} \tag{2.7c}$$

Fermi integral $\mathscr{F}_{\pm 1/2}(\eta)$ cannot be analytically obtained. $\mathscr{F}_0(\eta) = ln(1+e^\eta)$. For nondegenerately-doped semiconductors, the Fermi integral is always e^η, independent of order j. Hence the ultimate velocity is always the thermal velocity $v_{uj} = v_{thj}$ as appropriate for a given dimension. However, if there is a possibility of quantum emission, this ultimate velocity will be lower. As an example, in a bulk (3D) semiconductor, the saturation velocity is given by [4]

$$v_{sat3} = v_{th3}\mathscr{L}\left(\delta_Q\right) \tag{2.8a}$$

$$v_{th3} = \frac{2}{\sqrt{\pi}} v_{th}, \quad \delta_Q = \frac{\hbar\omega_o}{k_B T} \tag{2.8b}$$

$$\mathscr{L}(x) = coth(x) - \frac{1}{x} \tag{2.8c}$$

Here $\mathscr{L}(x)$ is the Langevin function and $\hbar\omega_o$ is the energy of quantum emitted in a high electric field. The exact nature of this quantum could be spacing between the quantum levels, spacing between lower and the next higher valley, the energy of the Brillouin zone as an electron traverses the zone, or the energy of a phonon.

Figure 3 shows the saturation velocity as a function of temperature with and without quantum emission for the Γ-valley of GaAs (m*=0.067 m_o) under nondegenerate conditions. Quantum is taken to be the energy of an optical phonon ($\hbar\omega_o = 36\,meV$). As expected the ultimate velocity increases as $\sqrt{T}$ with temperature. However, the onset of quantum emission may lower that ultimate velocity. If somehow the quantum emission is inhibited, higher velocities are possible.

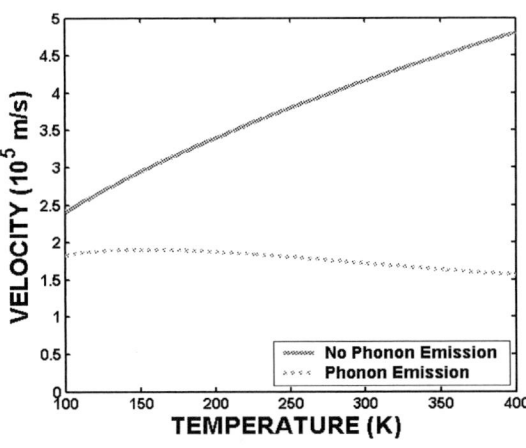

Fig. 3. Saturation velocity versus temperature, with and without phonon emission, for nondegenerate bulk (3D) GaAs (Γ-valley). $\hbar\omega_o = 36$ meV.

Similar results exist for Q2D and Q1D nanostructures. The ultimate velocity is lower for Q2D as compared to the bulk (3D) structure by a factor $v_{th2}/v_{th3} = \pi/4 = 0.79$ and the quantum-emission-limiting factor is $I_1(\hbar\omega_o/k_B T)/I_0(\hbar\omega_o/k_B T)$, where I_j (x) is the modified Bessel function of order j. Similarly, for Q1D nanostructures, the ultimate velocity factor is $v_{th1}/v_{th3} = 1/2 = 0.5$ and the quantum-emission-limiting factor is $tanh(\hbar\omega_o/k_B T)$.

In the degenerate regime, the ultimate velocity approaches the Fermi velocity that itself is a function of carrier concentration. Taking the case of bulk (3D) ultimate velocity for a single valley is given by

$$v_{u3} = v_{th3} \frac{\mathscr{F}_1(\eta_3)}{(n/N_{c3})} \tag{2.9}$$

where the normalized Fermi energy, $\eta_3 = \left(E_F - E_{co}\right)/k_B T$, is evaluated from

$$n = N_{c3}\,\mathscr{F}_{\frac{1}{2}}(\eta_3) \tag{2.10}$$

$$\eta_3 \simeq \frac{ln\,u}{1-u^2} + $$
$$\frac{(3\sqrt{\pi}u/4)^{2/3}}{1+\left[0.24+1.08*(3\sqrt{\pi}u/4)^{2/3}\right]^{-2}} \tag{2.11a}$$

$$u = \frac{n}{N_{c3}} \tag{2.11b}$$

Equation (2.11a) is an approximation for finding the Fermi energy [5].

The dependence of the ultimate velocity given by Eq. (2.9), as a function of temperature, for n = 10^{17}, 10^{18}, and 10^{19} cm^{-3} is shown (not including the effect of quantum emission) in Fig. 4. The graph for the nondegenerate statistics where n $\ll$ N$_c$, is also shown. In the nondegenerate regime, the ultimate velocity, v_{th3}, is given by

$$v_{th3} \simeq \frac{\Gamma(2)}{\Gamma\left(\frac{3}{2}\right)} v_{th} = \frac{2}{\sqrt{\pi}} \sqrt{\frac{2k_B T}{m^*}} \quad (2.12)$$

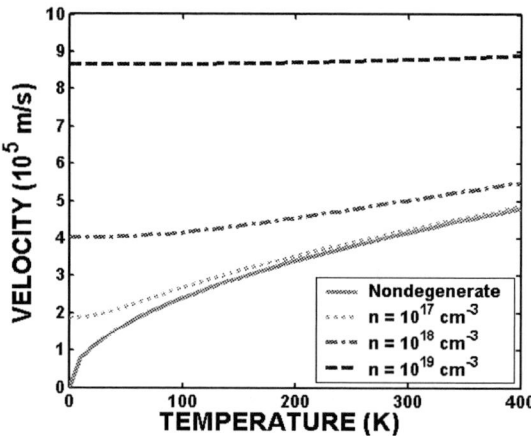

Fig. 4. The ultimate velocity versus temperature for the Fermi-Dirac statistics for a bulk (3D) GaAs (Γ-valley).

In the degenerate regime, n > N$_c$, the ultimate velocity is given by

$$v_{u3} \simeq \frac{3}{4} \sqrt{\frac{2E_F}{m^*}} = \frac{3}{4} v_{F3}$$
$$= \frac{3}{4} \frac{h}{m^*} \left(\frac{3n}{8\pi}\right)^{1/3} \quad (2.13)$$

The ultimate velocity is now independent of the lattice temperature. This is the case when doping concentration is high and/or temperature low.

III. VELOCITY-FIELD CHARACTERISTICS

A sound basis for the limiting velocity was developed earlier [1] by modifying the Fermi-Dirac function, where the Fermi energy (chemical potential) E_F is replaced by the electrochemical potential $E_F \pm q\vec{E}\cdot\vec{\ell}$ (+ for electrons and − for holes or vice versa depending on the direction of the applied electric field) [6]. The velocity field profiles, in the nondegenerate limit, are re-written in a slightly modified form to include the quantum emission in the saturation regime [6]-[8].

$$v_3(E) = v_{sat3}\, \mathcal{L}\left(\frac{E}{E_{c3}}\right) \quad (3.1)$$

$$v_2(E) = v_{sat2}\, \frac{I_1\left(\dfrac{E}{E_{c2}}\right)}{I_o\left(\dfrac{E}{E_{c2}}\right)} \quad (3.2)$$

$$v_1(E) = v_{sat1}\, tanh\left(\frac{E}{E_{c1}}\right) \quad (3.3)$$

The saturation velocity for each of the three dimensions is given by

$$v_{sat3} = v_{th3}\mathcal{L}(\delta_Q) \quad (3.4a)$$

$$v_{sat2} = v_{th2}I_1(\delta_Q)/I_o(\delta_Q) \quad (3.4b)$$

$$v_{sat1} = v_{th1}\, tanh(\delta_Q) \quad (3.4c)$$

The critical electric field E_{cj} (j=3, 2, 1) for the onset of nonlinear behavior is given by

$$E_{c3} = E_{co}\mathcal{L}(\delta_Q)/3 \quad (3.5a)$$

$$E_{c2} = E_{co}I_1(\delta_Q)/2I_o(\delta_Q) \quad (3.5b)$$

$$E_{c1} = E_{co}\, tanh(\delta_Q) \quad (3.5c)$$

with

$$E_{co} = \frac{V_t}{\ell_o} \quad (3.6)$$

Figure 5 shows the normalized velocity-field profiles for the bulk, Q1D, and Q2D nanostructures. In either dimensionality, quantum emission does not affect the ohmic mobility. It only affects the saturation velocity by lowering it, consistent with the results presented in Fig. 3. Figure 5 gives a comparison of the modulating factors with E_{cj} so defined that the slope in the low-field limit is the same as the empirical Eq. (2.1). The empirical factor of Eq. (2.1) for $\gamma = 2.8$ is also shown [9]. In each case, the mobility is $\mu_o j = vsat/E_{cj}$. The empirical factor of Eq. (2.1) with $\gamma = 2.8$ gives the best fit for Q1D profiles. For other dimensions, except for the knee portion, the empirical equation is suitable with appropriate value of γ, depending on the dimensionality.

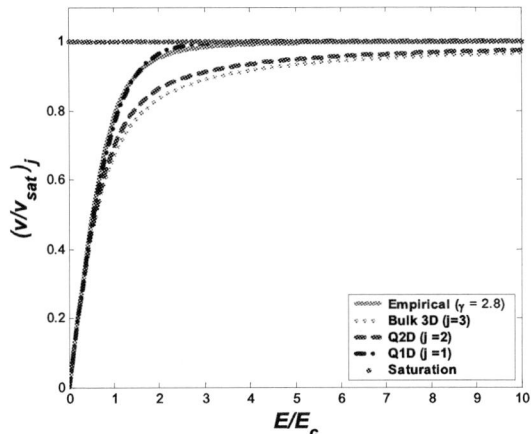

Fig. 5. Normalized velocity-field profiles for the bulk, Q2D, and Q1D samples. Also shown is graph of the empirical relation of Eq. (2.1) with $\gamma = 2.8$.

IV. I-V CHARACTERISTICS

Using Eq. (2.1), the nonlinear current-voltage are obtained as

$$I = \frac{V}{R_o}\frac{1}{\left[1+\left(\frac{V}{V_c}\right)^{\gamma}\right]^{\frac{1}{\gamma}}} = I_{sat}\frac{\frac{V}{V_c}}{\left[1+\left(\frac{V}{V_c}\right)^{\gamma}\right]^{\frac{1}{\gamma}}} \quad (4.1)$$

with

$$I_{sat} = nqv_{sat}A = n_s qv_{sat}W = \frac{V_c}{R_o} \quad (4.2)$$

I_{sat} is the saturation current for the resistor that depends on the volume concentration n or surface concentration n_s of the electrons. $V_c = E_c L$ is the critical voltage at the onset of nonohmic behavior.

Greenberg and De Alamo [9] measured I-V characteristics of InGaAs HFET structure. Their results indicate a direct experimental verification of Eq. (4.1). In their measurements, good fit to Eq. (4.1) is obtained for E_c= 3.8 kV/cm, I_{sat} = 565 mA/mm, and γ =2.8. With

this value of E_c, $V_c = 0.38\ V$ for a 1-μm resistor. For a macro-resistor of L = 1 cm =10,000 μm, this value becomes $V_c = E_c L$ =3.8 kV. Therefore, Ohm's law is valid up to 3.8 kV, an unapproachable high voltage. However, with $V_c = 0.38\ V$ for a 1-μm resistor, the current is close to the saturation value for a reasonable voltage applied. Figure 6 is a plot of Eq. (4.1) for resistors of length L = 5 μm, 20 μm, and 80 μm with all other dimensions and material properties the same. Solid curve is the replication of the experimental data [9] for a 5-μm resistor. When plotted on a scale extending to 10 V, 5-μm resistors clearly shows deviation from Ohm's law for relatively low voltages above V_c = 1.9 V. For 20-μm resistor, V_c =7.6 V and for L = 80 μm resistor V_c = 30.4 V. On a scale of 10 V, 80-μm resistor appears to follow Ohm's law. For 80-μm resistor, nonlinear behavior will become apparent if voltage is extended beyond 30.4 V. $V = V_c$ marks a transition from ohmic to nonohmic regime. Figure 6 and Eq. (4.1) indicate the validity of Ohm's Law in the regime where V<< V_c. In the other extreme where $V \gg V_c$, the current saturation occurs. Current saturation value (I_{sat}/ W) depends on the doping density and the saturation velocity, but not on the length of the resistor or scattering-limited mobility, as many earlier works have conjectured. All three curves approach the same saturation value. 5-μm resistor appears to approach that saturation at relatively low voltage compared to 80-μm resistor.

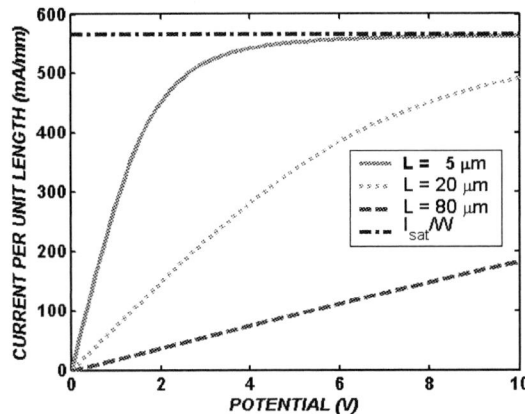

Fig. 6. I-V characteristics of micro-resistor of length L = 5 μm, 20 μm, and 80 μm, approaching the same saturation current.

The direct resistance, following the traditional model, R=V/I, can now be obtained from Eq. 4.1 as follows

$$R = \frac{V}{I} = R_o\left[1+\left(\frac{V}{V_c}\right)^{\gamma}\right]^{\frac{1}{\gamma}} = R_o\frac{1}{\left[1-\left(\frac{I}{I_{sat}}\right)^{\gamma}\right]^{\frac{1}{\gamma}}}$$

$$(4.3)$$

The incremental, differential, or small-signal resistance is given by r =dV/dI. For signals to propagate through the microresistor, the dc bias point determines its actual resistance. For an empirical relation of Eq. (4.1), the incremental resistance is given by

$$r = \frac{dV}{dI} = R_o \left[1 + \left(\frac{V}{V_c} \right)^{\gamma} \right]^{1+\frac{1}{\gamma}}$$

$$= R_o \frac{1}{\left[1 - \left(\frac{I}{I_{sat}} \right)^{\gamma} \right]^{1+\frac{1}{\gamma}}} = \frac{R^{\gamma+1}}{R_o^{\gamma}} \qquad (4.4)$$

Figures (7) and (8) show the resistance blow-up indicated by Eqs. (2.4) and (2.5) for $\gamma = 2.8$.

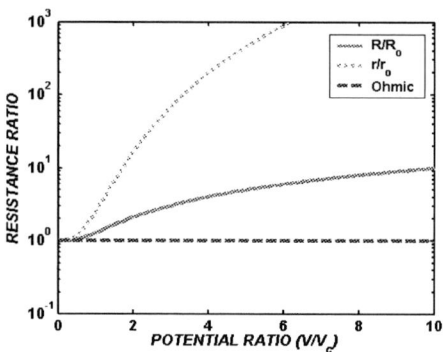

Fig. 7. Direct (R/R_o) and incremental (r/R_o) resistance ratio as a function of voltage ratio V/V_c. $\gamma = 2.8$.

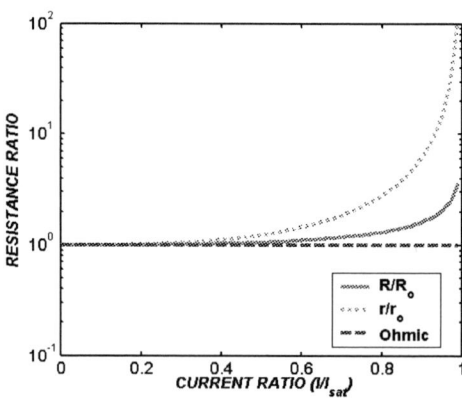

Fig. 8. Direct (R/R_o) and incremental (r/R_o) resistance ratio as a function of current ratio I/I_{sat}. $\gamma = 2.8$.

As expected, when $V/V_c < 1$, the resistance is ohmic. However, it dramatically rises in the regime $V/V_c > 1$. Incremental resistance rises much faster than the direct resistance. Similarly, when the current in the channel is more than roughly 50 % of its ultimate saturation value, the resistance blows up. This resistance blow-up of parasitic regions in a field-effect-resistor, for example, may degrade the channel behavior. Also, the parasitic resistances that are extracted under ohmic conditions may have substantially enhanced value if the current is closer to the saturation value of the parasitic resistor.

V. NONOHMIC CIRCUIT BEHAVIOR

In a one-dimensional model [8], the nonlinear current-voltage characteristics are given by

$$I = \frac{V_c}{R_o} tanh\left(\frac{V}{V_c} \right) = I_{sat} tanh\left(\frac{V}{V_c} \right) \qquad (5.1)$$

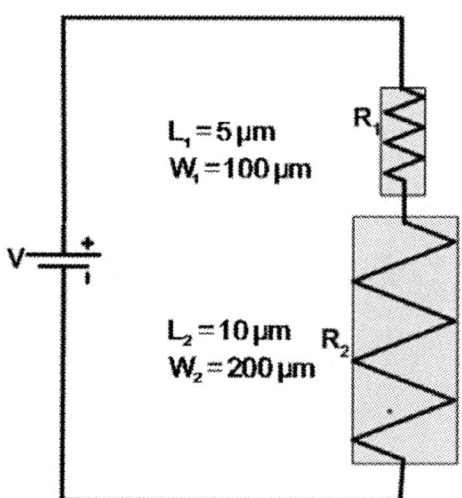

Fig. 9. Voltage divider circuit with two micro-resistors of the same ohmic value ($R_{o1} = R_{o2}$).

As discussed earlier, Eq. (5.1) is a compact one-dimensional model for analysis. The length of a resistor plays predominant role in transforming *I-V* and resistive behavior. Equation (1.2) shows that the ohmic resistance is a function of W/L provided the thickness d is the same. For illustration purpose, a resistor R_1 with $L_1 = 5$ μm and $W_1 = 100$ μm and another R_2 with $L_2 = 10$ μm and $W_2 = 200$ μm is taken. Both resistors have the same Ohmic values ($R_{o1} = R_{o2}$) as W/L is the same. When connected in series, as in Fig. 9, an applied voltage V will be equally divided when Ohm's law is applicable. However, when Eq. (5.1) is used in place of Ohm's law (I=V/R_o), the voltage V_1 across R_1 is obtained from

$$V_{c1} tanh\left(\frac{V_1}{V_{c1}} \right) = V_{c2} tanh\left(\frac{V - V_1}{V_{c2}} \right) \qquad (5.2)$$

V_{c1} = 1.9 V for the 5-μm resistor and V_{c2} = 3.8 V for the 10-μm resistor. The output V_1 across the 5-μm resistor is given in Fig. 10 when the divider circuit is excited by a voltage source of 0-10 V. Also, shown are the results expected from Ohm's law. The voltage across the shorter resistor is larger than its Ohmic value while that for the longer resistor is less than the Ohmic value.

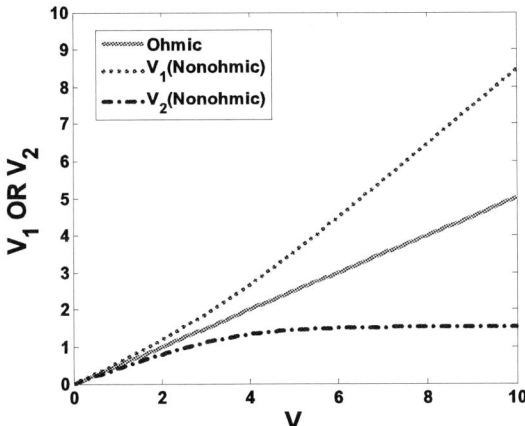

Fig. 10. Voltage division in two micro-resistors with the same ohmic value (R_{o1}=R_{o2}).

Figure 11 shows a current-divider circuit where two resistors with the same resistance R_{o1}= R_{o2} = 33.6 Ω are considered.

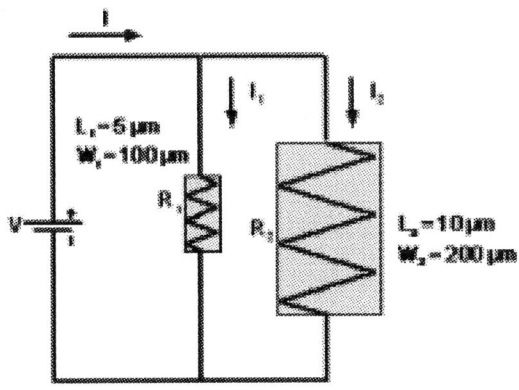

Fig. 11. Current divider circuit with two micro-resistors (R_{o1}=R_{o2}).

With the current per unit length of 565 mA/mm, the saturation current for the resistor 1 is I_{sat1} = 56.5 mA and that for resistor 2 is I_{sat2} = 113 mA. As before, the critical voltages V_{c1}= 1.9 V and V_{c2} = 3.8 V. Figure 12 shows that the resulting current in each resistor is substantially below its ohmic value. Only when V < V_c, validity of Ohm's law is visible. As V increases beyond

V_c, the maximum current that can be drawn from the voltage source is 170 mA as compared to 595 mA predicted by Ohm's law at V = 10 V. When two parallel channels are conducting, the more current will channel through the higher length resistor even if both resistors have the same ohmic value. The example presented demonstrates vividly the need for close scrutiny of the current and voltage division laws in integrated circuits where micro-resistors are necessarily present. This is particularly important when the parameters extracted under Ohmic conditions are used in predicting the circuit behavior operating near the saturation regime.

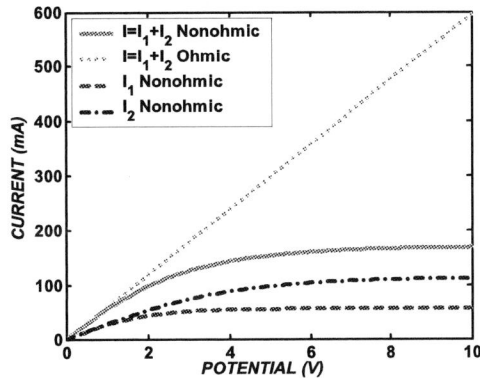

Fig. 12. Ohmic and nonohmic currents in a current divider circuit of Fig. 11.

It is now natural to re-evaluate the power consumption principle $P = V I$ when nonlinear I-V characteristics are considered. With Eq. (5.1) applied, the power law becomes

$$P = VI = \frac{VV_c}{R_o} tanh\left(\frac{V}{V_c}\right) = VI_{sat}\, tanh\left(\frac{V}{V_c}\right) \quad (5.3)$$

Power consumption is $P = \dfrac{V^2}{R_o}$ in the ohmic regime ($V<V_c$),

as expected. In $V>>V_c$ regime, the power is $P = \dfrac{VV_c}{R_o}$. The

power consumption is substantially lower than predicted by the application of Ohm's law. The power consumption will not only be lower, but also will have a linear behavior, in direct contrast to the quadratic behavior predicted by Ohm's law.

VI. CMOS CIRCUIT DESIGN

The basic element of modern digital circuits is the Complementary Metal-Oxide-Semiconductor (CMOS) inverter. The circuit inverts the logical value of the input voltage. In a CMOS inverter, the gate of n- and p-channel MOSFETs are tied together. A positive gate bias, which turns the NMOS on, turns off the PMOS, and vice versa. In the presence of high input, the output is connected to ground (that is when input is high, output is low). When input is low, the output is connected to the high end. Both stable states correspond to a very low current consumed from the power supply. In a CMOS circuit, switching speed is enhanced when the current $I_{Dn(p)}$ in NMOS and PMOS transistors are the same [3]:

$$I_{Dn(p)} = \frac{W_{n(p)} C_{ox} \mu_{on(p)}}{L_{n(p)}} \cdot \frac{\left(V_{GTn(p)} - \frac{1}{2} V_{DSn(p)} \right) V_{DSn(p)}}{1 + \frac{V_{DSn(p)}}{V_{cn(p)}}} \qquad (6.1)$$

with

$$C_{ox} = \frac{\varepsilon_{ox}}{t_{ox}} \qquad (6.2a)$$

$$V_{GTn(p)} = V_{GSn(p)} - V_{Tn(p)} \qquad (6.2b)$$

$$V_{cn} = \frac{v_{satn(p)}}{\mu_{on(p)}} L_{n(p)} \qquad (6.2c)$$

In the fabrication process, the oxide thickness t_{ox} is the same for both transistors and the threshold voltage is adjusted so that $V_{Tn} = V_{Tp}$. In a traditional design based on Ohm's law ($V_{DS} \gg V_c$), the length for both the NMOS and PMOS is equal ($L_n = L_p$). The width is then scaled with the mobility so $(W\mu_o)_{n(p)}$ product is the same for both transistors. Hence to account for the lower mobility of PMOS, the width of p-channel is scaled so that $W_p = W_n \frac{\mu_{on}}{\mu_{op}} \approx 2$ for a typical case of $\mu_{on} = 2 \mu_{op}$. However, in a submicron length channel as shown above, we need to match $V_{cn(p)}$ for both channels. Assuming that the saturation velocity v_{sat} is same for both the channels, an alternate design for the current to be the same in both channels is

$$W_p = W_n \qquad (6.3a)$$

$$L_n = \frac{\mu_{on}}{\mu_{op}} L_p \approx 2L_p \qquad (6.3b)$$

VII. CONCLUSIONS

A new paradigm that includes the impact of the failure of Ohm's law is presented. The appropriate average thermal velocity for a given dimension limits the drift velocity and hence the drift current for nondegenerate samples. The appropriate Fermi velocity is the limiting drift velocity and hence limits the drift current for the degenerate samples. The saturation velocity may be lower than the ultimate velocity due to the onset of a quantum emission that depends on a particular experimental setup or design. Based on the velocity-field profiles, the deviation from Ohm's law is evaluated. The saturation velocity limits the ultimate current. When applied to the voltage and current division circuits, the lower-length resistors are found to have higher resistance compared to higher-length resistor even if their ohmic values are the same. When applied to CMOS circuit design, new scaling law that keeps width the same and scales the length of the transistors is found to give balanced currents for the two complementary transistors. The results presented can have profound effect on characterization and performance evaluation of nanostructures being considered.

ACKNOWLEDGEMENTS

The assistance of Ms. Avanthi Mantrala and Mr. Ankur Sharma in preparing and reviewing the manuscript is greatly appreciated.

REFERENCES

[1] Vijay K. Arora, *Applied Physics Letters,* **80**, 3763(2002).
[2] Ankur Sharma and Vijay K. Arora, *Journal of Applied Physics,* **97**, 093704 (2005).
[3] V. K. Arora and M. B. Das, *Electronics Letters,* **25**, 820(1989).
[4] Vijay K. Arora, David S. L. Mui and Hadis Morkoc, *IEEE Transactions on Electron Devices,* **ED-34**, 1231 (1987).
[5] J. S. Blakemore, Solid State Electronics, **25**, 1067(1982).
[6] V. K. Arora, *Japanese Journal of Applied Physics,* **24**, 537(1985).
[7] Vijay K. Arora, David S. L. Mui and Hadis Morkoc, *Applied Physics Letters,* **50**, 1080(1987).
[8] V. K. Arora, *Microelectronic Journal,* **31**, 853(2000).
[9] David R. Greenberg and Jesus A. del Alamo, *IEEE Transactions on Electron Devices,* **41**, 1334 (1994).

2006 25th International Conference on Microelectronics
Device Integration Issues Towards 10 nm MOSFETs

Mikael Östling, B. Gunnar Malm, Martin von Haartman, Julius Hållstedt,
Zhen Zhang, Per-Erik Hellström and Shili Zhang

Abstract – An overview of critical integration issues for future generation MOSFETs towards 10 nm gate length is presented. Novel materials and innovative structures are discussed. Implementation of high κ gate dielectrics is presented and device performance is demonstrated for TiN metal gate surface channel SiGe MOSFETs with a gate stack based on ALD-formed HfO_2 / Al_2O_3. Low frequency noise properties for those devices are also analyzed. A selective SiGe epitaxy process for low resistivity source/drain contacts has been developed and implemented in pMOSFETs. A spacer pattering technology using optical lithography to fabricate sub 50 nm high-frequency MOSFETs and nanowires is demonstrated, Finally ultra thin body SOI devices with high mobility SiGe channels are demonstrated.

I. INTRODUCTION

The ITRS roadmap [1] identifies a number of challenges for continued successful scaling of MOSFET technology. It is clear that new materials and process modules will be needed to meet the ITRS roadmap requirements and enhance performance for a given technology node. The roadmap also indicates that the requirements for low standby power, active power and high-performance applications are different. For example low-power applications such as battery operated handheld devices require a reduced gate leakage current. To reduce the gate leakage, standard oxynitride gate insulators will be replaced by high-κ dielectrics. Among the promising candidates for the 45 nm technology node [2] are hafnium oxides (HfO_2) and hafnium silicates HfSiON with a high κ-value in the range 10-15, which should be compared to 3.9 for SiO_2 and 6-7 for the oxynitrides. This leads to significantly reduced gate leakage for the same equivalent oxide thickness (EOT). The main issues related to these types of dielectric materials, which still have to be addressed by researchers, are the high number of fixed/trapped charges and interface states. Both threshold voltage stability and low-field mobility are negatively impacted by the high amount of charge present in the high-κ oxides [3]. While the reduced mobility can be partially offset by strain enhancement techniques, the poor threshold voltage control and possible reliability problems can not be accepted. An additional complication is the poor thermal stability of high-κ materials. The dielectrics should be stable during high temperature processing steps (mainly source/drain activation anneals), since e.g. re-crystallization can increase the gate leakage current. For the ultimate scaling of CMOS, below 10 nm gate length,

The authors are with the School of Information Technology, KTH – Royal Institute of Technology, P.O. Box 229, 164 40 KISTA, Sweden, E-mail: ostling@imit.kth.se

other high-κ materials such as La_2O_3, with a larger κ-value might be of interest [4, 5].

High-κ materials are often used in combination with different metal gate electrodes e.g. TiN, TaN [6, 7]. Metal gates are important for several reasons, including the ability to control threshold voltage by tuning the work function of the gate electrode. For nitrided metal gates the tuning can be done either during the reactive sputter deposition or by subsequent nitrogen ion implantation[8, 9]. This allows reduced channel doping and hence higher mobility in both bulk and thin body SOI devices. Furthermore, metal gates do not suffer from depletion, which in turn increases the EOT, compared to the case with a highly doped polysilicon gate electrode. For successful metal gate integration, selective etching processes, with high anisotropy, need to be developed for patterning of 10 nm gate lengths. The use of fully nickel silicided (FUSI) polysilicon gates offers a more straightforward approach in this respect, since the patterning of polysilicon gates is more mature. In this case, the work function control can be achieved by dopant pile up at the metal gate/oxide interface. The Combination of FUSI and high-κ has generated a lot of attention recently [10, 11].

For high-performance applications the challenge is mainly to maintain sufficiently high drive current for short-channel devices which suffer from short channel effects (SCE) and high parasitic resistances. For higher drive current and increased switching speed the focus is on different methods of mobility enhancement, using strain. For CMOS applications both higher hole and electron mobility are desired. For PMOS the first attempts at increased channel mobility were done by selective SiGe epitaxy in the channel region [12]. However, from the 90 nm technology node, selective SiGe growth in the source/drain has emerged as the preferred method to create compressive strain in the PMOS channel [13]. A significantly increased electron mobility has also been demonstrated in NMOS devices, where a dielectric capping layer, commonly silicon nitride, introduces a tensile strain in the channel region. The strain in both PMOS and NMOS channels is uniaxial, which is beneficial compared to biaxaial strain. Alternatively ultra-thin body SOI MOSFETs with lowly doped channels also offer higher mobility as well as better control of SCE and junction leakage. SOI substrates are also used for novel multi-gate structures e.g. the FinFET [14]. An innovative device structure, featuring an inverted-T channel, was recently demonstrated [15]. This device combines the thin body SOI and the FinFET structures to achieve better on-current to area ratio. For the ultimate CMOS, silicon nanowires are promising, thanks to the optimized SCE control, using a gate-all-around structure [16]. The ITRS roadmap clearly indicates that the main obstacle for the ultimate scaling towards 10 nm is the source/drain and contact resistance, which cannot be reduced enough in relation to the increasing current densities in sub 50 nm multi-gate devices. A possible solution to this, is the use of complementary Schottky contacts to PMOS and NMOS respectively [17].

Lithography of 10 nm gate lines will be a serious challenge, due to effects such as line edge roughness (LER). Another interesting development, which will be discussed in this paper, is nano-structures defined by a combination of conventional lithography and spacer etching techniques. The spacer concept has some advantages over the traditional e-beam definition including faster processing and reduced LER [18].

II. HIGH-K AND METAL GATE INTEGRATION

In this section we discuss the performance of high-κ gate dielectric materials for use in surface channel SiGe MOSFETs. Traditionally, the gate oxide thickness has been scaled down with gate length, to maintain good short channel control. However, for low standby power operation it is predicted that SiO_2 and oxynitrides will reach a limitation at an EOT of about 1.5 nm, which is already used in volume production for the 90 and 65 nm technology nodes. Therefore, materials with a higher dielectric constant will be used to achieve low gate leakage at comparable EOT. By considering the increased fringing field, due to the higher κ-value, the influence on short channel effects and switching speed can be analyzed to find an optimum κ-value close to 30 [19]. High-κ materials are often used in combination with different metal gate electrodes [6], e.g. TiN and TaN. In this work TiN and poly-SiGe have been used as mid-band gap gate materials. The dielectrics discussed here are different combinations of Al_2O_3 with κ of 9 and HfO_2 with κ of 25 deposited by atomic-layer-deposition (ALD). The ALD technique allows arbitrary combinations of films to be deposited using a range of different precursors [20]. To obtain good interface properties a sandwiched structure of $Al_2O_3/HfO_2/Al_2O_3$ was investigated and compared to Al_2O_3 only and $HfAlO_x$. The effective dielectric constant is reduced to about 10 but the properties for integration into a standard CMOS process are much more promising. In this approach the aluminum oxide forms the interface to the channel region (either Si or strained SiGe) as well as to the poly-silicon gate electrode. Another important aspect of the ALD technique is the (in-situ) surface treatment before dielectric deposition [21]. The presence of a surface oxide will influence the final film quality and especially for SiGe channels a high number of interface states might be observed. While it is possible to remove the native oxide using HCl vapour an amorphous oxide might be beneficial for the formation of the Al_2O_3

interface layer. In Fig 1. a TEM cross-section of a transistor with a surface SiGe channel and a high-κ gate stack is shown. Note that a SiO_2 interface layer can be observed, especially close to the gate edge. The EOT (including the interface layer) was determined from CV measurements and was found to be 1.9 nm. The TEM also shows Ni(SiGe) on both the source/drain and gate-poly regions.

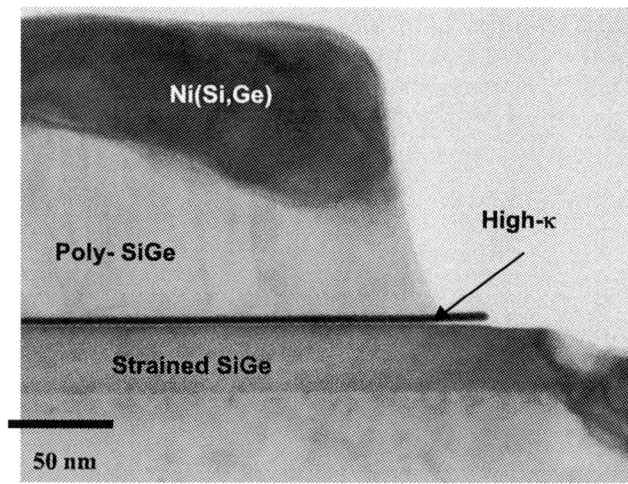

Fig. 1. TEM of a surface channel SiGe MOSFET with high-κ $HfO_2/Al_2O_3/HfO_2$ gate stack and nickel salicided source/drain region and gate electrode.

The interface state density (D_{it}) was extracted, to examine to the quality of the high-κ/strained SiGe channel interface. A relatively high value of D_{it} 7×10^{12} $cm^{-2}eV^{-1}$ was obtained. This suggests that high-temperature process steps after the high-κ deposition might have degraded the film properties. For the TiN metal gate devices with a reduced thermal budget excellent Dit values of 1.6×10^{12} $cm^{-2}eV^{-1}$ was obtained for a SiGe channel device and 3.3×10^{12} $cm^{-2}eV^{-1}$ for a Si-channel device with identical gate stack.

The purpose of introducing a surface SiGe channel is to obtain high hole mobility and carrier confinement. In Fig. 2 mobility values are compared for Si and SiGe channel devices. The SiGe devices show significantly better mobility than the Si control, which is very close to the theoretical curve. It is interesting to observe that the TiN metal gate devices compare favorably to the device with poly-SiGe gate. This can be related to a reduced phonon scattering due to screening by the metal gate.

Another issue with the increased D_{it} and the number of fixed charges (N_t) in the high-κ dielectrics is the possible influence on the low-frequency noise [22]. Compared to the case for buried channel SiGe devices, where the carriers are physically separated from the (oxide) interface, a much stronger influence on carrier mobility due to coulomb scattering and trapping/de-trapping in slow states will be observed for surface channel devices. In Fig. 3. low-frequency noise spectra for devices with different gate stacks on SiGe channels are compared to a SiO_2 reference.

The highest noise (top curve) is observed for the case with a single layer Al_2O_3 gate. Gate stacks with either HfO_2 or $HfAlO_x$ sandwiched between Al_2O_3 layers show better noise performance. The SiO_2 reference shows the best noise performance, indicating that further optimization of the high-κ gate stack is needed for low noise applications.

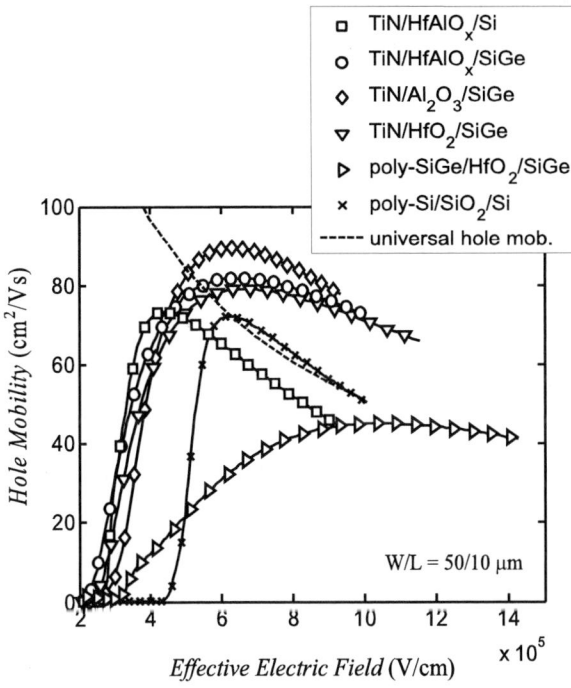

Fig. 2. Hole mobility for Si and SiGe surface channel devices with different high-κ gate stacks and SiGe/TiN gate electrodes.

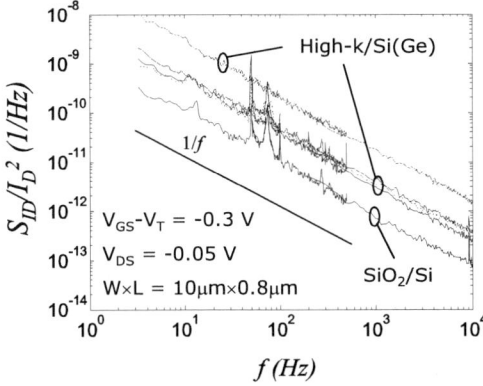

Fig. 3. Drain current noise power spectral density for high-κ surface channel SiGe devices and a SiO_2/Si channel control device.

III. SELECTIVE SiGe EPITAXY FOR S/D REGIONS

Control of the extension doping and junction depth is a key requirement to control short channels effects in nano-scale MOSFETs. We have investigated the use of selectively deposited SiGe layers in the source/drain regions of pMOSFETs. This solution was originally proposed by Özturk et al.[23]. We have demonstrated a

novel process concept where a Si recess and highly boron-doped SiGe regrowth are obtained in a non-interupted process sequence by using in-situ HCl etching [24, 25]. This type of process has also been discussed by Loo et al. [26]. Among the key issues is good epitaxial quality, combined with a high amount of substitutionally placed, electrically active boron. To achieve this there is a trade off against loss of selectivity and amorphous deposition instead of growing single crystalline layers. It also important to maintain the integrity of the gate oxide during the Si recess etch. We have successfully demonstrated pMOSFETs. After gate-poly patterning and oxide spacer-definition the wafer is loaded into a ASM epsilon CVD reactor where a HCl etching step at elevated temperature is performed to form the recessed region. After this step careful treatment of the exposed silicon surface is needed to for successful epitaxial growth and to avoid loss of selectivity. The solution employed here is a surface treatment using NF_3 at low power. The SiGe deposition was optimized to achieve a very high boron concentration, preferably above the solid solubility limit. The maximum doping level that could be obtained in high quality selective epitaxial growth (taking the loading effect into account) was 6×10^{20} cm^{-3}. Figure 4 shows the TEM cross-section of the full transistor structure. Epitaxial growth of SiGe is seen in the source/drain regions. Some nucleation of poly-SiGe is also seen on the gate spacer.

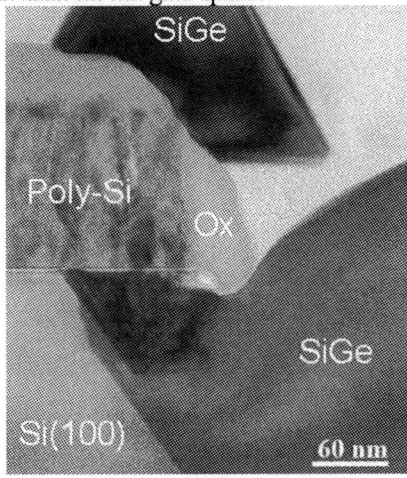

Fig. 4. TEM cross-section of high-quality, boron doped SiGe epitaxial layer in the source/drain region of a pMOSFET.

IV. SPACER PATTERNING TECHNOLOGY FOR SUB 50 NM CMOS

The spacer patterning technology flow uses two standard optical lithography steps to produce nano-sized features and CMOS type devices [27, 28]. In our approach I-line lithography was used to produce sub-50 nm MOSFETs with 1, 2 or multiple gate fingers. By using more advanced DUV lithography the pitch of polysilicon gates fingers can be further reduced. The spacer process flow utilizes a combination of deposited films, which enables etching/removal with high selectivity. A SiO_2 hard-mask and a SiGe poly layer are deposited on top of the gate poly silicon. This structure is shown in Fig. 5. A window is then defined in the SiGe film by standard lithography and dry etching. Next follow a silicon-nitride (Si_3N_4) deposition and dry-etching, selective to the oxide hard mask, this step forms the Si_3N_4 spacer, which defines the gate length. Hence, the dimension of the final nano-structure is controlled by the Si_3N_4 film thickness, which can be scaled down to less than 10 nm in our current process. A wet etch removes the SiGe film from the oxide surface. Fig 6 shows the wafer at this step. A second lithography step is then used to define contact pads in the oxide hard mask layer. A final dry etching step then transfers the pattern to the poly silicon gate layer.

Fig. 5. Deposited Layer stack used for definition of spacer gates.

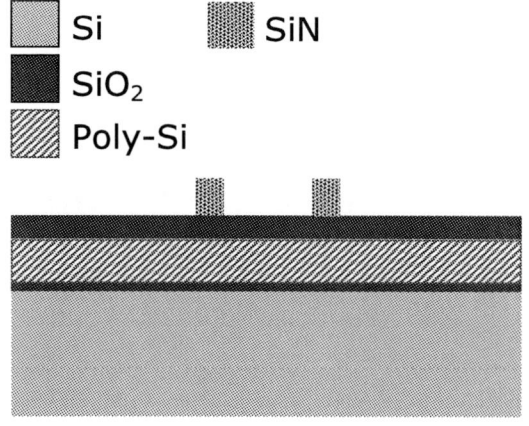

Fig. 6 Device structure after definition of SiN spacers, this pattern will be transferred to the poly silicon gate film.

An example of a 50 nm transistor fabricated using the spacer process flow is shown in Fig 7. NiSi was used for the gate and source/drain contacts. In this process, Ni-silicided nano-wires with excellent electrical properties, i.e. low resistivity, have also been demonstrated for line widths of less than 25 nm, as shown in Fig. 8.

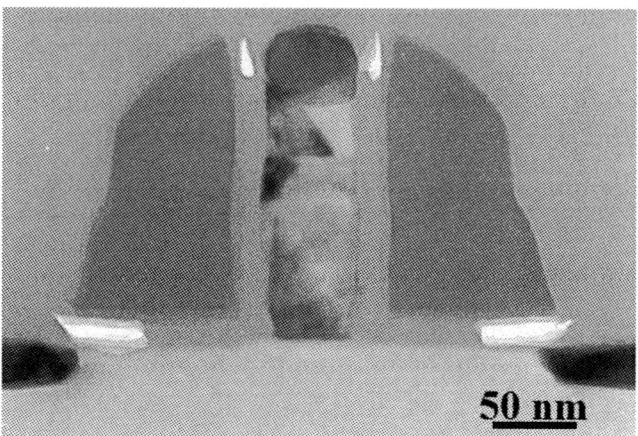

Fig. 7. TEM of a 50 nm NMOS transistor fabricated using the spacer gate process flow.

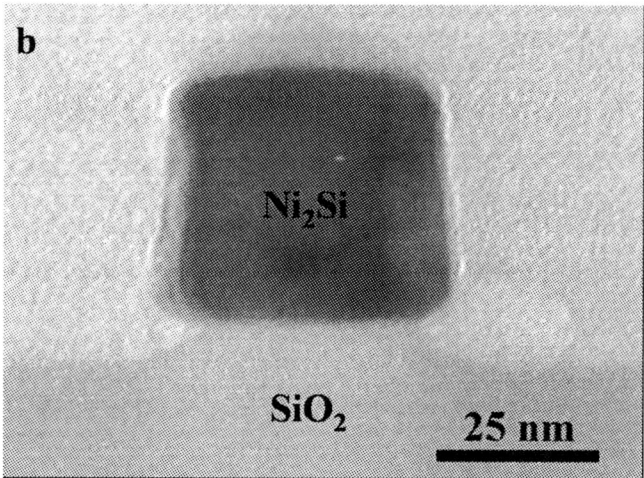

Fig. 8. TEM of a 25 nm silicided nanowire fabricated using the spacer gate process flow.

V. ULTRA THIN BODY SOI DEVICES

In this section we discuss ultra thin body (UTB) SOI devices. The UTB devices offer significantly improved control of the short channel effects, compared to bulk devices, with the same gate length. No intentional substrate doping is needed in fully depleted devices and hence the threshold voltage is controlled only by the silicon body thickness and the gate work function. According to the scaling rules, the thickness of the thin silicon body layer should be less than one third of the gate length. Therefore, typical silicon layer thicknesses are in the order of 10 - 15 nm for a 50 nm gate length device. Such thin layers can be achieved from a starting material (SOI wafer) with several 100 nm silicon thickness by a combination of sacrificial oxidation [29] and silicon etching in HCl chemistry. Compared to a bulk device with similar SCE control the UTB SOI devices have higher channel mobility thanks to the low doping. To further enhance mobility, strained channels can be incorporated on SOI [29], either by wafer bonding or epitaxial techniques. We have successfully implemented compressively strained SiGe and SiGeC layers in UTB SOI pMOSFETs. An example of an 80 nm gate device is shown in Fig. 9. A high quality SiGe 8 nm layer has been grown by RPCVD on top of the thinned down silicon. The fabricated devices show good performance, in terms of IV-characteristics. A significantly increased effective hole mobility, extracted from long channel split-CV measurements is demonstrated in Fig. 10. Compared to the Si-control the effective hole mobility is increased by approximately 60 %.

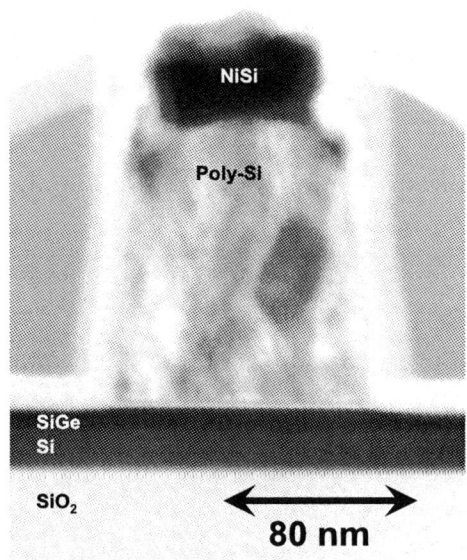

Fig. 9. Strained SiGe channel on ultra thin body SOI substrate.

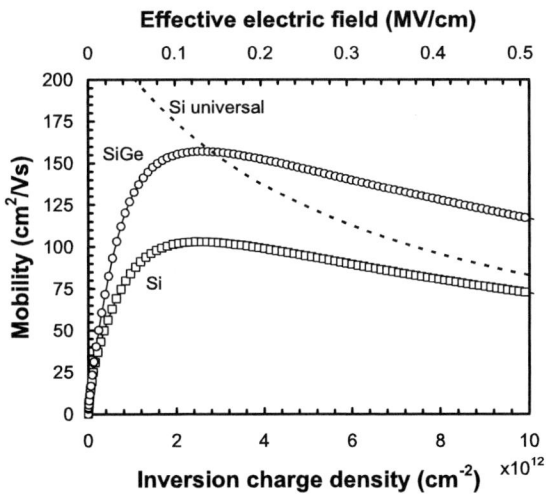

Fig 10. P-channel mobility in Si and strained SiGe transistors on ultra thin body SOI.

VI. CONCLUSIONS

New materials and innovative device structures suitable for the ultimate scaling of CMOS to 10 nm gate lengths have been discussed. A combination of strained channels and hafnium based oxides/silicates will fulfill the requirements for the 45 nm node. For future scaling an appropriate structure based on multiple gates will probably be needed. Among the challenges for the research community are low-resistive contact metallization and gate dielectrics with higher κ-values. Patterning of 10 nm gates will also be an important topic, especially if metal gates are adopted instead of the proven poly-silicon technology

.

ACKNOWLEDGEMENT

The author wishes to thank the researchers and PhD students at ICT/EKT for their valuable contributions to this paper. In particular we would like to mention Drs H. H. Radamson, Y.-B. Wang, J. Seger, D.-P. Wu, C. Isheden. We wish to thank G. Sjöblom, J. Westlinder and Dr J. Olsson, Uppsala University, for their contribution to the high-κ work. The nano-scale CMOS development was also supported in the EU Network of Excellence SiNano..
SSF and Vinnova are acknowledged for financial support in the High-Frequency Silicon and the High-speed/frequency and Optoelectronics program respectively.

REFERENCES

[1] "International Technology Roadmap for Semiconductors (ITRS) 2005 update http://www.itrs.net/Common/2005ITRS/Home2005.htm ," 2005.

[2] M. A. Quevedo-Lopez, S. A. Krishnan, P. D. Kirsch, H. J. Li, J. H. Sim, C. Huffman, J. J. Peterson, B .H. Lee, G. Pant, B. E. Gnade, M. J. Kim, R. M. Wallace, D. Guo, H. Bu, and T. P. Ma, "High Performance Gate First HfSiON Dielectric Satisfying 45nm Node Requirements," Tech. Dig. IEDM, pp. 237-240, 2005.

[3] M. von Haartman, J. Westlinder, D. Wu, B. G. Malm, P. E. Hellstrom, J. Olsson, and M. Ostling, "Low-frequency noise and Coulomb scattering in Si/sub 0.8/Ge/sub 0.2/ surface channel pMOSFETs with ALD Al$_2$O$_3$ gate dielectrics," Solid-State Electronics, vol. 49, pp. 907-14, 2005.

[4] H. Iwai, S. Ohmi, S. Akama, C. Ohshima, A. Kikuchi, I. Kashiwagi, J. Taguchi, H. Yamamoto, J. Tonotani, Y. Kim, I. Ueda, A. Kuriyama, and Y. Yoshihara, "Advanced gate dielectric materials for sub-100 nm CMOS," San Francisco, CA, USA, 2002.

[5] A. C. Jones, H. C. Aspinall, P. R. Chalker, R. J. Potter, K. Kukli, A. Rahtu, M. Ritala, and M. Leskela, "Recent developments in the MOCVD and ALD of rare earth oxides and silicates," Materials Science & Engineering B (Solid-State Materials for Advanced Technology), vol. 118, pp. 97-104, 2005.

[6] D. Wu, A.-C. Lindgren, G. Persson, G. Sjöblom, M. von Haartman, J. Seger, P.-E. Hellström, J. Olsson, H.-O. Blom, S.-L. Zhang, M. Östling, E. Vainonen-Ahlgren, W.-M. Li, E. Tois, and M. Tuominen, "A Novel Strained Si$_{0.7}$Ge$_{0.3}$ Surface-Channel pMOSFET With an ALD TiN/Al$_2$O$_3$//HfAlO$_x$/Al$_2$O$_3$ Gate Stack," IEEE Electron Device Lett., vol. 24, pp. 171-173, 2003.

[7] L-Å Ragnarsson, S. Severi, L. Trojman, D. P. Brunco, K. D. Johnson, A. Delabie, T. Schram, W. Tsai, G.

Groeseneken, K. de Meyer, S. de Gendt, and M. Heyns, "High performing 8 Å EOT HfO2/TaN low thermalbudget n-channel FETs with solid-phase epitaxially regrown (SPER) junctions," *Tech. Dig. VLSI Symposium*, pp. p234, 2005.

[8] H.-C. Wen, K. Choi, P. Majhi, H. Alshareef, C. Huffman, and B. H. Lee, "A systematic study of the influence of nitrogen in tuning the effective work function of nitrided metal gates," *IEEE International Symposium on VLSI Technology*, pp. 105 - 106 2005.

[9] H. Wakabayashi, Y. Saito, K. Takeuchi, T. Mogami, and T. Kunio, "A dual-metal gate CMOS technology using nitrogen-concentration-controlled TiNx film," *IEEE Trans Electron Devices*, vol. 48, pp. 2363 - 2369, 2001.

[10] Y. H. Kim, C. Cabral. Jr., E. P. Gusev, R. Carruthers, L. Gignac, M. Gribelyuk, E. Cartier, S. Zafar, M. Copel, V. Narayanan, J. Newbury, B. Price, J. Acevedo, P. Jamison, B. Linder, W. Natzle, J. Cai, R. Jammy, and M. Ieong, "Systematic Study of Workfunction Engineering and Scavenging Effect Using NiSi Alloy FUSI Metal Gates with Advanced Gate Stacks," *Tech. Dig. IEDM*, pp. 657-660, 2005.

[11] A. Lauwers, A. Veloso, T. Hoffmann, M. J. H. van Dal, C. Vrancken, S. Brus, S. Locorotondo, J.-F. de Marneffe, B. Sijmus, S. Kubicek, T. Chiarella, M. A. Pawlak, K. Opsomer, M. Niwa, R. Mitsuhashi, K. G. Anil, H.Y. Yu, C. Demeurisse, R. Verbeeck, M. de Potter, P. Absil, K. Maex, M. Jurczak, S. Biesemans, and J. A. Kittl, "CMOS Integration of Dual Work Function Phase Controlled Ni FUSI with Simultaneous Silicidation of NMOS (NiSi) and PMOS (Ni-rich silicide) Gates on HfSiON," *Tech. Dig. IEDM*, 2005.

[12] M. von Haartman, A.-C. Lindgren, P.-E. Hellström, M. Östling, T. Ernst, L. Brévard, and S. Deleonibus, "Influence of gate width on 50 nm gate length $Si_{0.7}Ge_{0.3}$ channel PMOSFETs," *Proceedings of the 33rd European Solid-State Device Research - ESSDERC '03*, pp. 529-532, 2003.

[13] S. E. Thompson, M. Armstrong, C. Auth, S. Cea, R. Chau, G. Glass, T. Hoffman, J. Klaus, Z. Ma, B. Mcintyre, A. Murthy, B. Obradovic, L. Shifren, S. Sivakumar, S. Tyagi, T. Ghani, K. Mistry, M. Bohr, and Y. El-Mansy, "A logic nanotechnology featuring strained-silicon," *IEEE Electron Device Lett.*, vol. 25, pp. 191-193, 2004.

[14] D. Hisamoto, W.-C. Lee, J. Kedzierski, H. Takeuchi, K. Asano, C. Kuo, E. Anderson, T.-J. King, J. Bokor, and C. Hu, " FinFET-a self-aligned double-gate MOSFET scalable to 20 nm," *IEEE Trans Electron Devices*, vol. 47, pp. 2320 - 2325, 2000.

[15] L. Mathew, M. Sadd, S. Kalpat, M. Zavala, T. Stephens, R. Mora, S. Bagchi, C. Parker, J. Vasek, D. Sing, R. Shimer, L. Prabhu, G.O. Workman, G. Ablen, Z.Shi, J.Saenz , B. Min, D. Burnett, B.-Y. Nguyen, J. Mogab, M.M. Chowdhury, W. Zhang, and J. G. Fossum, "Inverted T channel FET (ITFET) - Fabrication and Characteristics of Vertical-Horizontal, Thin Body , Multi-Gate , Multi-Orientation Devices, ITFET SRAM Bit-cell operation. A Novel Technology for 45nm and Beyond CMOS.," *Tech. Dig. IEDM*, 2005.

[16] Sung Dae Suk, Sung-Young Lee, Sung-Min Kim, Eun-Jung Yoon, Min-Sang Kim, Ming Li, Chang Woo Oh, Kyoung Hwan Yeo, Sung Hwan Kim, Dong-Suk Shin,

Kwan-Heum Lee, Heung Sik Park, Jeong Nam Han, C.J.Park, and J.-B. Park, "High Performance 5nm radius Twin Silicon Nanowire MOSFET(TSNWFET): Fabrication on Bulk Si Wafer, Characteristics, and Reliability," *Tech Dig. IEDM*, pp. 735-738, 2005.

[17] J. Kedzierski, P. Xuan, E. H. Anderson, J. Bokor, T.-J. King, and C. Hu, " Complementary silicide source/drain thin-body MOSFETs for the 20 nm gate length regime," *Tech Dig. IEDM*, pp. pp. 57 - 60, 2000.

[18] Y.-K. Choi, Lindert, N., P. Xuan, S. Tang, D. Ha, E. Anderson, T.-J. King, J. Bokor, and C. Hu, "Sub-20 nm CMOS FinFET technologies," *Tech. Dig. IEDM*, pp. 421-414, 2001.

[19] N. R. Mohapatra, M. P. Desai, S. G. Narendra, and V. R. Rao, "The effect of high-K gate dielectrics on deep submicrometer CMOS device and circuit performance," *IEEE Trans Electron Devices*, vol. 49, pp. 826-831, 2002.

[20] M. Leskelä and M. Ritala, "Atomic layer deposition (ALD): from precursors to thin film structures," *Thin Solid Films*, vol. 409, pp. 138-146, 2002.

[21] D. Wu, H. Radamson, P.-E. Hellström, S.-L. Zhang, M. Östling, E. Vainonen-Ahlgren, E. Tois, and M. Tuominen, "Influence of Surface Treatment Prior to ALD High-K Dielectrics on the Performance of SiGe Surface-Channel pMOSFETs," *IEEE Electron Device Lett.*, vol. 25, pp. 289-291, 2004.

[22] M. von Haartman, D. Wu, B. G. Malm, P. E. Hellstrom, S. L. Zhang, and M. Ostling, "Low-frequency noise in Si/sub 0.7/Ge/sub 0.3/ surface channel pMOSFETs with ALD HfO/sub 2//Al/sub 2/O/sub 3/ gate dielectrics," *Solid-State Electronics*, vol. 48, pp. 2271-5, 2004.

[23] S. Gannavaram, N. Pesovic, and C. Ozturk, "Low temperature (800°C) recessed junction selective silicon-germanium source/drain technology for sub-70 nm CMOS," *Tech. Dig. IEDM*, pp. 437 - 440, 2000.

[24] C. Isheden, P.-E. Hellström, H. H. Radamson, S.-L. Zhang, and M. Östling, "MOSFETs with recessed SiGe source/drain junctions formed by selective etching and growth," *Electrochemical and Solid State Lett.*, vol. 7, pp. G53-G55, 2004.

[25] C. Isheden, H. H. Radamson, E. Suvar, P.-E. Hellstrom, and M. Östling, "Formation of shallow junctions by HCl-based Si etch followed by selective epitaxy of B-doped Si1-xGex in RPCVD," *J. Electrochem. Soc.*, vol. 151, pp. C365-C368, 2004.

[26] R. Loo, M. Caymax, P. Meunier-Beillard, I. Peytier, F. Holsteyns, S. Kubicek, P. Verheyen, R. Lindsay, and O. Richard, "A new technique to fabricate ultra-shallow-junctions, combining in situ vapour HCl etching and in situ doped epitaxial SiGe re-growth," *Applied Sujrface Science*, vol. 224, pp. 63-67, 2004.

[27] Y.-K. Choi, T.-J. King, and C. Hu, "A spacer patterning technology for nanoscale CMOS," *IEEE Trans. Electron Devices*, vol. vol. 49, pp., pp. 436-441, 2002.

[28] Y.-K. Choi, T.-J. King, and C. Hu, "Spacer FinFET: nanoscale double-gate CMOS technology for the terabit era," *Solid State Electronics*, vol. 46, pp. 1595-1601, 2002.

[29] J. Seger, P. E. Hellstrom, J. Lu, B. G. Malm, M. von Haartman, M. Ostling, and S. L. Zhang, "Lateral encroachment of Ni-silicides in the source/drain regions on ultrathin silicon-on-insulator," *Applied Physics Letters*, vol. 86, pp. 253507-1, 2005.

2006 25th International Conference on Microelectronics

Nanoscale Materials, Devices, and Systems for Sensing, Detection, and Environmental Pollution Monitoring and Mitigation

A. Vaseashta

Abstract – The objective of this investigation is to utilize nanoscale materials, devices, and systems as in-situ sensors and detectors of chemical and biological agents coupled with remote satellite image processed data for environmental air pollution monitoring. The efficacy of nanoparticles based environmental sensing using commercially available nanoscale metal-oxide based gas detectors is augmented by image processed satellite data to monitor local and regional air pollution dispersion. To extend our research to other forms of pollution, we present preliminary investigations of nanotechnology based sensors towards mitigating air and water based contaminants. Similar extensions of preliminary research results towards the detection of food and air-borne pathogens using nanoparticles based diagnostic tools are also presented.

I. INTRODUCTION

Pollution crisis poses a major challenge all around the world, adversely affecting the lives of millions of people by contributing to debilitating and deadly health disorders. Pollution, in general, is contamination by a chemical or other agent that renders part of the environment unfit for intended or desired use. Natural processes release toxic chemicals into the environment, as a result of ongoing industrialization and urbanization. Some of the major causes of the pollution crisis are – deforestation, polluted rivers, environmental contamination, and soil pollution. Other sources of pollution include iron and steel mills; zinc, lead, and copper smelters; municipal incinerators; oil refineries; cement plants; and nitric and sulphuric acid producing plants. Of the group of pollutants that contaminate urban air, fine suspended particulate matter, sulphur dioxide (SO_2), and ozone pose the most widespread and acute risks. Recent studies on the effects of chronic exposure to air pollution have singled out particulate matter as the pollutant most responsible for the life-shortening effect of unhealthy air, although other pollutants may also play a vital role. These pollutants cause respiratory and other health disorders. Environmental contaminants in soil enter into the watershed to further exacerbate wide spread pollution. Consequently, rapid detection of contaminants in the environment by emerging technologies is of paramount

A. Vaseashta - Senior Member-IEEE, is with the Nanomaterials Processing & Characterization Laboratories, Department of Physics and Graduate Program in Physical Sciences, Marshall University, One John Marshall Drive, Huntington, WV 25755-2570, USA.
E-mail: prof.vaseashta@marshall.edu

significance. The pollution in the environment in some developing countries have reached to an alarming level and hence real-time pollution monitoring sensors, sensor networks, and real-time monitoring stations need to be employed to gain a thorough understanding. A tool providing interactive qualitative and quantitative information about the pollution is essential for policy makers to protect massive populations, especially in developing countries.

The physical and chemical properties of materials in reduced dimensions show size dependence and may exhibit properties different from the bulk. There are several examples of rather remarkable properties attributed to the reduced dimensions [1]. The study of nanoscale systems has become one of the most promising disciplines in science and technology, as it refers to the fundamental understanding and resulting technological advances arising from the exploration of new physical, chemical, and biological properties of systems that are intermediate in size, between the isolated atoms and molecules, and bulk material, where the transitional characteristics between the two can be understood, controlled, or manipulated. Nanoscale materials, devices, and systems are at the convergence of the smallest of structures with dimensions comparable to most molecules of the living systems. Recently, various nanoscale materials, devices, and systems with remarkable properties have been developed, with numerous unique applications in chemical and biological sensors, nanophotonics, nanobiotechnology, and in-vivo analysis of cellular processes at nanoscale levels.

Although different, yet on a scientifically related note, the potential and risk for inadvertent or deliberate contamination of the environment, food and agricultural products has recently increased, rendering decentralized sensing as an important issue for several federal agencies. In clinical medicine, the current trend is to decentralize laboratory facilities and conduct clinical trials employing direct reading, portable, and lab-on-chip (LOC) systems. A nanotechnology based sensor platform will enable the direct electrical detection of biological and chemical agents in a label-free, highly multiplexed format over a broad dynamic range. This platform utilizes functionalized nanotubes, nanowires, or nanoparticles to detect molecular binding with high sensitivity and selectivity. The platform is capable of detecting broad range of molecules, viz., DNA, RNA, proteins, ions, small molecules, cells and even the pH values. Detection is possible in both liquid and gas

1-4244-0116-X/06/$20.00 ©2006 IEEE

phase and is highly multiplexable, allowing for the parallel detection of multiple agents. Recent progress in nanostructured materials and its possible applications in chemical and biological sensors could have a significant impact on efficient and accurate data collection, processing, and recognition. Furthermore, nucleic acid layers combined with nanomaterials based electrochemical or optical transducers produce a new kind of affinity biosensors as "DNA Biosensor" or "Genosensor" for small molecular weight molecules. Genosensors are attractive devices for converting the hybridization event into an analytical signal for acquiring sequence-specific information in connection with clinical, environmental or forensic investigations. The development of novel and sensitive assays for DNA hybridization and detection has become an increasingly important research field. Continued development through combined efforts in microelectronics, surface/interface chemistry, molecular biology, and analytical chemistry will further the establishment of genosensor technology as a major component of analytical biochemistry. The design and fabrication of DNA-modified surfaces and materials which are reproducible, stable and selective to complementary DNA sequences are crucial in the development of emerging analytical tools such as DNA chips or simple diagnostic devices for detecting few DNA sequences such as electrochemical genosensors [2]. These devices have been extensively used for the fast, cost-effective and simple diagnosis of inherited or infectious diseases, but also for the early determination of infectious agents in various environments.

Substantial segments of the scientific community are confident that nanoscience and nanotechnology will revolutionize research on and applications in the areas of biology, medicine, human health, and provide unprecedented means to forewarn and/or provide protection against the potential and risk for inadvertent or deliberate contamination of the environment, food and agricultural products. Many technologists have developed new tools, yet they often have limited understanding of the needs of the biomedical communities or the restrictions that biology places on the proper design of nanotools and nanosystems. Additionally, the future success of bio-defense in the world will demand the development of a novel and complex set of materials and devices. The nanoscale materials, devices and systems will have an influence on virtually all industries from civilian to defense. More specifically, the physico-chemical kinetics in nanostructured systems explains the interaction of gases, liquids and biological media, leading to its application for a wide variety of environmental and biomedical functions. Hence, these new materials promise a cornucopia of new products with superior performance characteristics that will dramatically transform the markets in a number of key industries. However, further advances are necessary, to entirely understand, characterize, develop, and optimize the properties of these recent nanoscale materials, devices and systems.

II. NANOPARTICLES BASED DIAGNOSTIC PLATFORMS

A. Environmental Pollution Detection & Monitoring

Ground-based measurements are usually used to monitor urban pollution. Satellite data, however have traditionally been untapped by environmental pollution scientists. With the advancement of remote sensing technologies with high spatial and spectral resolution, it is now possible to model urban pollution using satellite images. Availability of newly launched nanotechnology based NO_x sensors can also give accurate ground pollution concentrations. One of the objectives of this investigation is to develop a unique capability to acquire, display and assimilate these valuable sources of data to accurately assess urban pollution by real-time monitoring using nanotechnology based sensors and satellite imagery. This invaluable and integrated tool will be beneficial towards prediction processes to support public awareness and establish (policy) priorities for air pollution in polluted areas. The data, as observed from satellite is processed employing the Multivariate techniques, which are based on groupings in a multivariate data set. ER Mapper™ has the ability through a series of Cluster algorithms. Figure 1 shows various sources of pollution, monitoring devices linked to a remote server, and authentication of the ground based data by satellite data. Urban pollution concentration of NO, NO_2, CO_2, SO_2, CH_4, CO, and aerosols are modeled using NASA TERRA mission ASTER satellite with 14 bands and high spatial resolutions, i.e. 15-90 m. To distinguish between pollution concentration and atmospheric scattering by large sized dust particle created by urban sprawl, change detection analysis (CDA) on temporal data from Landsat™ 1-7 is used in the investigation. GIS Software ArcGIS (v. 9.x) and ArcInfo Workstation from Environmental System Research Institute (ESRI) are employed to process the satellite data into useable format. Remote sensing software ER-Mapper™ (v.7.0) is further employed to accurately model the data through feature extraction processes for pattern recognition. Several digital imaging processes such as geometric registration, radiometric normalization, and unsupervised classification and accuracy assessment of results are conduced. Multidate correlations and regression analysis methods are used for individual gases with satellite-recorded reflectance of bands.

The task of any remote sensing system is to detect radiation signals, determine their spectral character, derive appropriate signatures, and inter-relate the spatial positions of the classes they represent. The remote sensing of atmospheric pollution over land, however, is a challenging task. Multispectral satellite data over land have different responses at different wavelengths, as gases absorb electromagnetic energy in very specific regions of the spectrum. We have employed ASTER, an imaging instrument flying on the Terra satellite as part of NASA's

Earth Observing System (EOS). It is the only high-spatial resolution instrument on the satellite that has 14 bands from the visible to the thermal infrared region. In the visible green and near-infrared (V-NIR) between 0.52 μm – 0.86 μm, there are three bands with 15 m resolution (see the compilation of bands in Table 1). In the short-wave infrared (SW-IR) between 1.6 μm – 2.43 μm, there are six bands with 30 m resolution. In the thermal infrared (Th-IR) between 8.125 μm-11.65 μm, there are five bands with a resolution of 90 m. ASTER acquires data over a 60 km swath whose center is pointable.

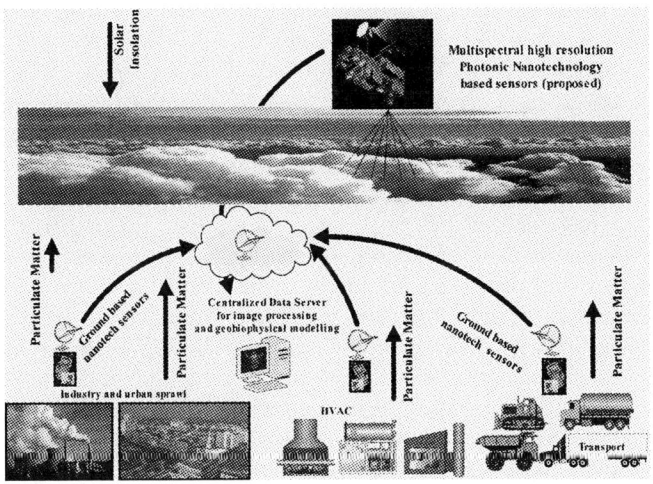

Fig. 1: Conceptual schematic displaying the experimental framework. [3].

TABLE 1: ASTER SENSOR SYSTEMS: BASELINE PERFORMANCE REQUIREMENTS

Sub Sys-tem	Band No.	Spectral Range (μm)	Radiometric Resolution	Abso-lute Accuracy (σ)	Spatial Reso-lution	Signal Quanti--zation Levels
VNIR	1	0.52 - 0.60	NE Δp≤ 0.5 %	≤± 4 %	15 m	8 bits
	2	0.63 - 0.69				
	3N	0.78 - 0.86				
	3B	0.78 - 0.86				
SWIR	4	1.600 - 1.700	NE Δp≤ 0.5 %	≤± 4 %	30 m	8 bits
	5	2.145 - 2.185	NE Δp≤ 1.3 %			
	6	2.185 - 2.225	NE Δp≤ 1.3 %			
	7	2.235 - 2.285	NE Δp≤ 1.3 %			
	8	2.295 - 2.365	NE Δp≤ 1.0 %			
	9	2.360 - 2.430	NE Δp≤ 1.3 %			
TIR	10	8.125 - 8.475	NE Δp≤ 0.3 %	≤ 3K (200 ·	90 m	12 bits
	11	8.475 - 8.825		≤ 2K (240 ·		
	12	8.925 - 9.275		≤ 1K (270 ·		
	13	10.25 - 10.95		≤ 2K (340 ·		
	14	10.95 - 11.65				

Figure 2 shows the satellite images for three cities (Los Angeles, USA, San Francisco, USA; and Calcutta, India) at 60 km swath. The images illustrate haze layers or smog related to anthropogenic activities. Reflectance, scattering, and absorption of particulates and aerosols can range in size from approximately 0.2 μm onwards, based on the

wavelength ranges of the satellite sensors. The TERRA platform has one sensor system with a range of wavelength values from 0.52 μm and instantaneous field of view (IFOV) of 15 m in band one to 11.65 μm and IFOV of 90m in band 14. Fifteen bands are the result of bands 3N and 3B in the NIR, which tends to penetrate light cloud cover but can be more reflective if it contains particulate matter. The areas in each of the images trend to show more reflective 'smog' with an increase in industrialization and or urbanization.

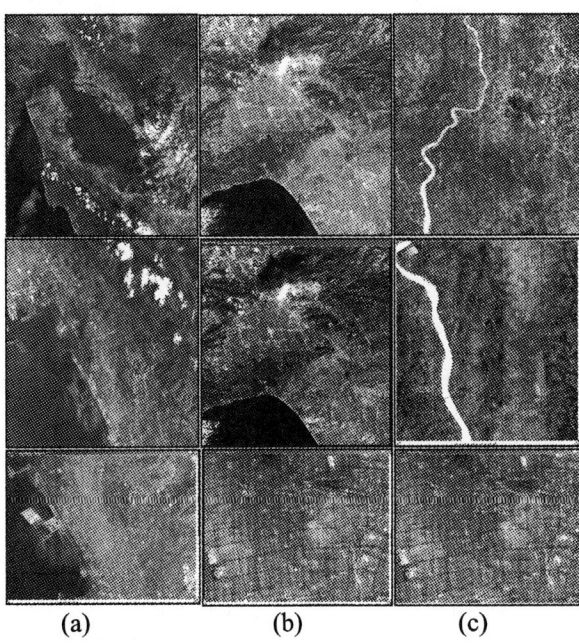

(a) (b) (c)

Fig. 2: Terra ASTER Satellite image of three cities at 60 km swath ((a) Los Angles, US; (b) San Francisco, US; and (c) Calcutta, India).The bottoms frames are images processed using ER Mapper™ (v.7.0) showing smog and pollution related to anthropogenic activities.

Commercially available nanotechnology based NOx sensors provide accurate ground pollution concentration. Ground based sensors and processed satellite imagery provide qualitative as well as quantitative environmental pollution information. Conventional, nanotechnology-based tungsten oxide sensors are used for *in-situ* urban pollution measurements. The WO_3 sensor is a low-cost, commercial device that contains a proprietary blend of dopants and catalysts to optimize the sensitivity (<0.5-10 ppm) and selectivity for NO_X gases. The sensor was calibrated specifically for NO_X to conduct single gas studies, albeit measurements for other pollutants, e.g. volatile organic compounds (VOC) were also conducted. The data acquired from monitoring sites are uploaded via personal digital assistant (PDA), linked with global positioning system (GPS) capability. The information on the resulting air quality levels can operate as a monitoring system and be displayed in the form of geographical information systems (GIS) databases. The WO_3 sensors were adjusted for zero to maximum scale using a feedback circuit. Conventional sensors were used to measure the VOC and certain other common pollution particles. The WO_3 sensor shows an increasing voltage output with increasing NO_x concentrations, as shown in Figure 3. The response and recovery time of the sensor is dependent on the NO_x concentrations. The circuitry in the sensor refreshes itself periodically. An internet GIS framework is employed to disseminate real-time data. Design considerations that optimize performance, reliability, and security of Internet GIS systems for environmental monitoring and data dissemination are under consideration.

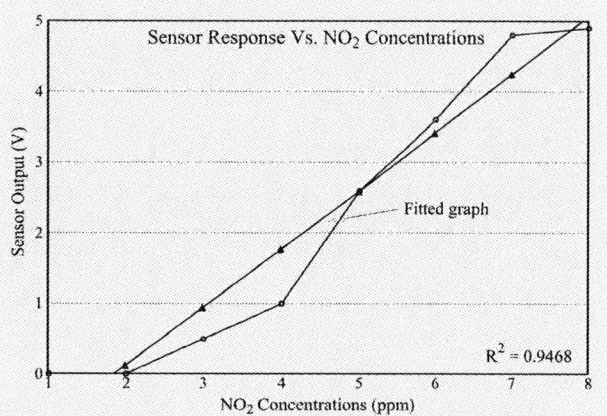

Fig. 3: Typical response of a gas sensor to NO2 concentrations.

Nanotubes can be used as conductor between two metal electrodes and the conductance between the electrodes can be measured as a function of gate bias voltage. Since the electrical characteristics are strong function of its atomic structure, mechanical deformation, and chemical doping can induce changes in conductance, thus making such devices as small and sensitive sensors to their chemical and mechanical environment. Chemical sensors based on an individual or ensembles of SWNTs detected 200 ppm of NO_2 and $< 2\%$ of NH_3 in a few seconds [4]. Hence, sensors made from SWNT have high sensitivity and fast response time even at room temperature. The first principle calculations using density functional theory (DFT) on several molecules, such as CO, NH_3, NO_2, O_2, and H_2O, show the direction of the charge transfer and hence the doping of the semiconductor tube [5], which results in change in conductivity. For H_2O, a simulated molecular configuration shows repulsive interaction and no charge transfer is observed in the presence of water molecule [6], which offers an important option of using SWNTs in water as biochemical sensors. Gas sensors employing ionization work by detecting the ionization characteristics of distinct gases; however they are limited by the size and high voltage operation and large power consumption. The CNTs exhibit excellent field emission characteristics due to the existence of a very large field at the tips even at very low voltages, which could produce compact, battery powered gas ionization sensors. The field emission based ionization gas sensors are expected to show good sensitivity and selectivity, and are

unaffected by the factors such as temperature, humidity, and gas flow. The CNTs have demonstrated electric field induced change in band gap, which can lead to detection of ionic species without altering charge or doping of the SWNTs.

B. Chemical and Biological Sensors

The upsurge in terrorist activity has generated tremendous demands for innovative tools capable of detecting major explosives and chemical warfare agents (CWAs) in a faster, simpler and more reliable manner at the site of terrorism. Rapid identification of the appropriate explosive(s) and CWA(s) will allow first responders and emergency personnel to make important decisions concerning barricading, evacuating, or efficient decontamination of a particular site and will prevent responders from becoming victims themselves. The ability to monitor major explosives or nerve agents at the originating source should offer significant advantages in terms of providing a timely warning, an efficient and inexpensive response, and an ability to process a smaller sample size than current technologies allow. The chem.-bio sensors developed using nanoscale materials offer unprecedented ability to detect the chemical signature in a label-free, highly multiplexed format over a broad dynamic range. In an attempt to probe chemical and biological molecules, techniques such as electrochemical biosensors, surface plasmon resonance (SPR), Atomic force microscopy (AFM) are employed.

Electrochemical sensors: Traditional methods for DNA sequencing, based on the coupling of electrophoretic separations and radioisotopic (^{32}P) detection, are labor intensive and time consuming, and are thus not well suited for routine and rapid medical analysis, particularly for point-of-care tasks. Recent advances in biosensors based on nucleic acid have led to the development of genosensor technology for gene sequence analysis and for nucleic-acid ligand binding studies [7]-[12]. Electrochemical DNA biosensors (genosensors) offer promise for obtaining knowledge necessary for development various areas such as biomedical and environmental research. In particular, electrochemical monitoring of DNA hybridization has recently been an attractive research area [13]-[23]. Direct electrochemical DNA analysis based on guanine signal has been reported in the literature [13]-[14]. Immediate applications include directly quantifying DNA samples for use in sequencing or polymerase chain reactions (PCR), or pharmaceutical testing and quality control. Eventually, they could be applied for clinical applications such as detection of pathogenic bacteria, tumors, and genetic disease, or for forensics. Our approach to investigating biosensors include the application of many different electrochemical transducers based on nanomaterials to choose the best surface material for the immobilization of genetic material, DNA. The electrochemical characterization of these working electrodes is conducted by using different

34

electrochemical techniques such as differential pulse voltammetry (DPV). A schematic of an electrochemical DNA sensor with hybridization sensing strategies is shown in Figure 4a. In these sensors, a probe sequence is immobilized within the recognition layer, where a base-pairing interaction recruits a target molecule to the sensor. It is vital to develop sensing strategies to maintain critical dynamics of target capture, to generate a sufficient recognition signal. These devices can be used for monitoring sequence-specific hybridization events directly, based on the oxidation signal of guanine, DNA intercalators (metal coordination complexes, antibiotics etc.) or simply using some metal tags viz. gold or silver nanoparticles or magnetic particles, for detecting the oxidation signal of gold or silver, or electro-active DNA bases in the presence of DNA hybridization.

Surface Plasmon Resonance: Use of SPR to detect *E-coli* O157:H7 bacteria by immobilizing antibodies is accomplished by a coupling matrix on the surface of a thin film of precious metal such as gold, deposited on the reflecting surface of an optically transparent wave-guide. When visible or near-infrared radiation is made to totally internally reflect at the interface of the metal and the reflecting surface by a prism, SPR occurs. When the antigens interact with antibodies, the refractive index of the medium surrounding the sensor changes, which in turn causes a shift in the angle of resonance. The change in resonance angle is proportional to the change in the concentration of antigens bound to the surface. In order to investigate the capability of self-assembled monolayers (SAM) based SPR biosensor for pathogen detection, an initial baseline was established and various experiments were conducted. Immobilization using *Escherichia coli* O157:H7 polyclonal antibodies suspended in NaOAc (pH 5.5) produced a pixel change of 12, while *Escherichia coli* O157:H7 (pH 7.4) polyclonal antibodies suspended in PBST produced a pixel change of 7. Increasing the concentration of antigen in direct assay resulted in an increased response. Upon passing the antigen sample at a concentration of 4×10^8 CFU/ml for 10 min, a change in the pixel value of 0.1 was noted, while a concentration of 7×10^9 CFU/ml produced an average change in the pixel value of 0.1667. One pixel change is equivalent to a 0.006° change in angle, as shown in figure 4(b). For the same concentrations of antigens and different concentrations of secondary antibodies in the sandwich assay, the response obtained was almost 30 times higher than for the direct assay. Preliminary studies clearly showed that the SAM based LEICA SR 7000 SPR biosensor can be used to monitor biomolecular interactions, and to rapidly detect pathogens. With improvements in the detection protocol and sensor chips, the sensitivity and specificity can be enhanced considerably. As an extension to the detection, and to increase the sensitivity and specificity of the binding assays, particularly in the case of antibody binding, surface plasmon fluorescence spectroscopy (SPFS) will be applied simultaneously with SPS. As the reflectivity monitored at different incident angles reaches a minimum corresponding to the excitation of the surface plasmons, the surface field intensity is maximized, yielding a mirror image of the plasmon reflectivity curve. This intensity enhancement can be employed to increase the fluorescence emission of surface bound dye molecules excited by the amplified electromagnetic field. Figure 4(b) shows a schematic of SPR and a typical response curve.

Atomic Force Microscopy: Figure 4(c) shows an AFM image of a rod shaped *Escherichia coli*, bound to its corresponding antibody on a SAM based gold chip. Similar projects for other pathogens, proteins, and biological molecules are in progress. Similar to SPR, improvements in the detection protocol, sensitivity and specificity can be enhanced considerably by improving upon functionalization strategies.

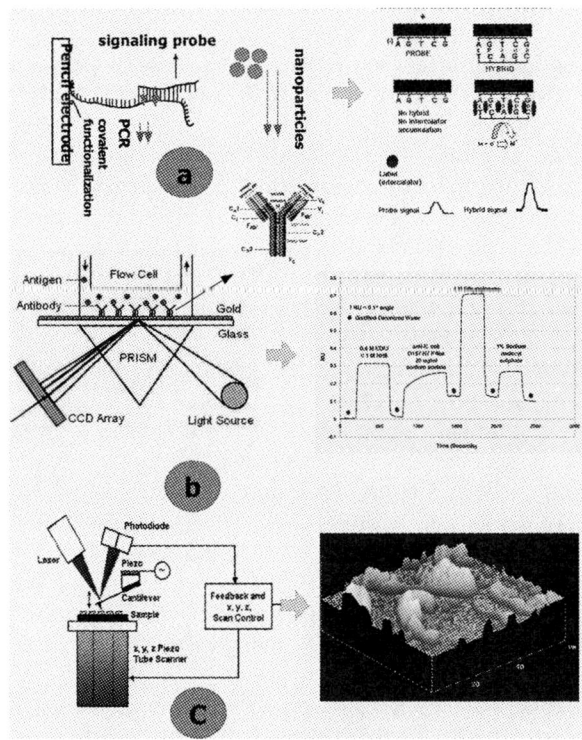

Fig. 4: Detection of chemical/biological agents using (a): Electrochemical sensors, and hybridization sensing strategies, (b): Surface Plasmon Resonance, and a typical response of a flow cell and, (c): Atomic Force Microscopy on *Escherichia coli*, bound to its corresponding antibody on a SAM based gold chip

C. Water and Soil Contamination Remediation

Pure water is a very limited natural resource; in many cases there is not enough water supply of appropriate quality for industrial and home use. Many pollutants in water streams have been identified to be toxic and harmful for human health. Among them arsenic is considered as a high priority one. Arsenic is an element that occurs naturally in rocks and soil, water, air, plants, and animals. Volcanic activity, the erosion of rocks and minerals, and forest fires are natural sources that can release arsenic into the environment. Human activities and industrial products are also responsible for arsenic release into the environment. Wood preservatives, paints, drugs, dyes, soaps, metals and semi-conductors contain arsenic. Agricultural applications, mining, and smelting also contribute to arsenic releases. Arsenic is found at higher levels in underground sources of drinking water than in surface waters, such as lakes, reservoirs, and rivers. Most of the arsenic in drinking waters is in inorganic form either as As (III) or As (V). Studies have linked long-term exposure to arsenic contamination with cancer and cardiovascular, pulmonary, immunological, neurological and endocrine effects. The US-EPA has proposed a new MCL for arsenic in drinking water of 0.010 mg/L effective from Jan. 23, 2006. It is believed that with the new standard around 5000 community water systems of the 74000 in total would need to take corrective actions to lower arsenic levels in drinking water. Ninety-four percent of these water systems serve fewer than 10,000 people each [24]. A variety of treatment processes have been studied for arsenic removal from water. The major technologies include precipitation–coagulation followed by filtration, lime softening, reverse osmosis (RO), electrodialysis (ED), nanofiltration (NF) and ion exchange/sorption [25]-[26]. These have been evaluated by EPA for possible application to small community systems. Reverse osmosis and electrodialysis have been found more effective, but they are costly and water recovery is not optimised. Also, both RO and ED are more effective in removing As (V) than As (III) and the oxidising agent which should be added to improve efficiency is harmful to membranes. Sorption on iron (III) oxides, such as amorphous hydrous ferric oxide (FeOOH), poorly crystalline hydrous ferric oxide (ferrihydride) and goethite (α-FeOOH) have been found to be effective in removing both As (V) and As (III) from aqueous solutions [27]-[30]. These sorbents can also be easily synthesized from cheap raw materials and possess fairly uniform chemical and physical properties [31]. As sorption in columns is an established process, the preparation of an effective and relatively cheep sorbent based on iron oxides would make sorption attractive for use in small community systems. Preliminary results with synthetic akaganeite (β-FeOOH) have shown that the maximum capacity of the sorbent is around 75 mg of As per gm of akaganeite, while the results reported in the literature so far have an average value of less than 20 mg As/g of sorbent. This shows that akaganeite is an effective material for the removal of arsenic from water streams. Also, quite good results were obtained with synthetic magnetite/maghemite, with a maximum capacity of ca. 40 mg of As per gm of sorbent. The main problem at the moment is that these are nanomaterials with small particle size (around 20-50 nm) and are not suitable for sorption columns. However, with magnetite/maghemite, which is a magnetic material solid/liquid, separation can be easily done in high gradient magnetic field. One of the methods under investigation is

to incorporate the nanomaterials into a well organized host. Nanoporous and microporous crystals, such as molecular sieves (zeolites) are ideal hosts for accommodation of organic and inorganic molecules, polymer chains, etc. due to their uniform pore size and their ability to sorb molecular species, as shown in Figure 5. The incorporation of iron oxides/oxyhydroxides based nanostructures to zeolites would produce a novel sorbent, which will have both the high capacity for arsenic removal and suitable size for use in sorption columns. Our preliminary results with magnetically modified natural zeolite have shown that our new composite material has significant sorption capacity of arsenic ca. 50 mg of As per gm of sorbent.

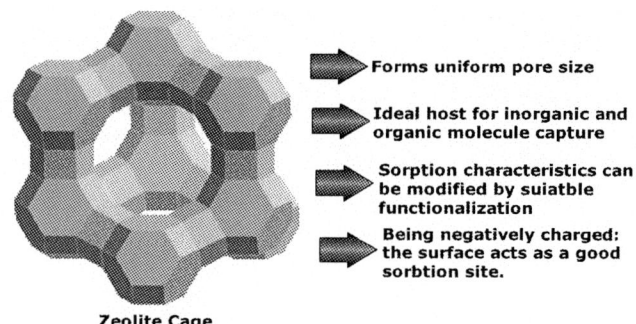

Fig. 5: Zeolite as an ideal hosts for accommodation of organic and inorganic molecules, polymer chains, etc.

III. CONCLUSION

Nanoscience and nanotechnology addresses some of the greatest challenges of the 21st century by providing routes to synthesize materials by design and establishing connection between structure and function of the biomolecules to human physiology. Since, nanomaterials are ideal building blocks for the fabrication of structures, nano-devices, and functionalized surfaces; one of the key challenges for such applications is to precisely control over the growth of these materials at desired sites with a desired structure and orientation. The present investigation presents a convincing argument for an imminent and ongoing need to develop nanoscale materials, devices, and system for sensing, detection, and environmental pollution monitoring and mitigation. Not only from scientific standpoint, the investigation also presents suitable tools for policy makers to introduce appropriate guidelines to reduce pollution to improve general health across the globe. The sensors suggested in this investigation can be integrated into the next-generation `gene-chips', especially where detection of less than an attomolecule, such as amino acids or DNA, is critical. Recent discovery of quantum confined particles or quantum dots (QDs) having unique optical and electronic properties, such as size and composition-tunable fluorescence emission from visible to infrared wavelength, large absorption coefficient across a wide spectral range and very high level of brightness and photo-stability, will lead to the development of multifunctional nanoparticles probes for cancer targeting and real-time in-vivo imaging in living cells. The broad excitation profiles and narrow, symmetric emission spectra in high quality QDs are well suited to optical multiplexing, in which multiple colors and intensities are combined to encode genes, proteins, and small molecule libraries. QDs encapsulated with ABC triblock copolymer and linking this polymer to tumor-targeting ligands and drug-delivery functionalities, in-vivo targeting studies of human prostate cancer growing in nude mice suggests that the QD probes accumulate at tumors sites [32]. These results suggest new possibilities for ultra sensitive and multiplexed imaging of molecular targets. Other nano-technology projects include a nanoscale barcode for genome-wide screening such as disease susceptibility and therapeutic responses, blood fingerprinting, and development of a technology capable of directing nerve growth through scar tissue for spinal chord regeneration, to name a few. Furthermore, advances in nanophotonics will enhance the resolution limit of satellites, thus providing details and enhanced feature extraction capability.

ACKNOWLEDGEMENT

The author would like to acknowledge the contribution of Ms. O. Pummakaranchana of Asian Institute of Technology, Bangkok, Thailand; Dr. A. Erdem from Ege University, Izmir, Turkey; and Mr. P. Roy, Mr. K. Casto, Mr. J. Barrios, and Prof. J. Brumfield of Geobiophysical Modeling Group at Marshall University, WV, USA.

REFERENCES

[1] A. Vaseashta, D. Dimova-Malinovska, and J. M. Marshall, *Nanostructured and Advanced Materials*, Dordrecht, The Netherlands: Springer, 2005.

[2] T. Drummond, M. G. Hill, and J. K. Barton "Electrochemical DNA Sensors", *Nature Biotechnology*, 2003, vol. 21(10), pp. 1192-1199.

[3] A. Vaseashta, "Functionalized Nanoparticles for Sensing, Detecting, and Monitoring", in *Proc. 1ˢᵗ International Workshop on Semiconductor Nanocrystals, SEMINANO 2005*, Budapest, 2005, pp. 139-143.

[4] Y. M. Wong, W. P. Kang, J. L. Davidson, A. Wistsora-at, and K. L. Soh, "A novel microelectronic gas sensor utilizing carbon nanotubes for hydrogen gas detection", *Sensors and Actuators*, 2003, vol. B 93, pg. 327.

[5] J. Zhao, A. Buldum, J. Han, and J. P. Lu, "Gas molecule adsorption in carbon nanotubes and nanotube bundles", *Nanotechnology*. 2002, vol. 13, pg. 195.

[6] G. C. Philip, K. Bradley, M. Ishigami, and A. Zettl, "Extreme oxygen sensitivity of electronic properties of carbon nanotubes", *Science*, 2000, vol. 287, pg. 1801.

[7] M. Yang, M. E. McGovern and M. Thompson, "Genosensor technology and the detection of interfacial nucleic acid chemistry", *Analytica Chimica Acta*, 1997, vol. 346, pg. 259.

[8] J. Wang, "From DNA Biosensor to Gene Chips," *Nucleic Acid Research*, 2000, vol. 28, pg. 3011.

[9] E. Palecek and M. Fojta, "Detecting DNA hybridization and damage", *Analytical Chemistry*, 2001, vol. 73. pp. 75A-83A.

[10] A. Erdem, M. Ozsoz, "A review: Electrochemical DNA biosensors based on DNA-Drug Interactions' *Electroanalysis*, 2002, vol. 14, pg. 965-974.

[11] J. Wang, "Towards Genoelectronics: Electrochemical Biosensing of DNA Hybridization", *Chemistry European J.*, 1999, vol. 5, pg. 1681.

[12] J. Wang, G. Rivas, J. R. Fernandes, J. L. L. Paz, M. Jiang, R. Waymire, "Indicator-free electrochemical DNA hybridization biosensor", *Anal.Chim Acta*, 1998, vol. 375, pg. 197.

[13] M. I. Pividori, A. Merkoçi, S. Alegret, "Electrochemical genosensor design: immobilization of oligonucleotides onto transducer surfaces and detection methods", *Biosensors and Bioelectronics*, 2000, vol. 15, pp. 291-303.

[14] J. Wang, A. N. Kawde, A. Erdem, M. Salazar, "Magnetic bead-based label-free electrochemical detection of DNA hybridization" *Analyst*, 2001, vol. 126 (11), pg. 2020-2024.

[15] Erdem, A., Pividori, M., del Valle, M., and Alegret, S., "Rigid carbon composites: a new transducing material for label-free electrochemical genosensing", *J. Electroanal. Chem.*, 2004, vol. 567, pp. 29-37.

[16] A. Erdem, K. Kerman, B. Meric, U.S. Akarca, M. Ozsoz, "DNA Electrochemical Biosensor For The Detection of Short DNA Sequences Related To The Hepatitis B Virus", *Electroanal.*, 1999, vol. 11(8), pg. 586.

[17] A. Erdem, K. Kerman, B. Meric, U.S. Akarca, M. Ozsoz, "A Novel Hybridization Indicator Methylene Blue for the electrochemical Detection of Short DNA Sequences Related to the Hepatitis B Virus", *Anal. Chim. Acta*, 2000, vol. 422 (2), pp. 139-149.

[18] S.O. Kelley, E.M. Boon, J.K. Barton, N.M. Jackson, M.G. Hill, "Single-base mismatch detection based on charge transduction through DNA", *Nucleic Acids Res.* 1999, vol. 27, pg. 4830.

[19] E. M. Boon, D. M. Ceres, T. G. Drummond, M. G. Hill, J. K. Barton, "Mutation Detection by Electrocatalysis at DNA-Modified Electrodes", *Nature Biotechnology*, 2000, vol. 18, pp. 1096-1100.

[20] M. Ozsoz, A. Erdem, K. Kerman, D. Ozkan, B. Tugrul, N. Topcuoglu, H. Erken, M. Taylan, "Electrochemical Genosensor Based on Colloidal Gold Nanoparticules for the Detection of Factor V Leiden Mutation Using Disposable Pencil Graphite Electrodes", *Analytical Chemistry*, 2003, vol. 75, pp. 2181-2187.

[21] G. Marrazza, I. Chianella, M. Macsini, "Disposable DNA electrochemical biosensors for environmental monitoring", *Anal. Chim. Acta*, 1999, vol. 387, pg. 297-307.

[22] J. Wang, G.-U. Flechsig, A. Erdem, O. Korbut, P. Gründler, "Label-free DNA Hybridization based on Coupling of a Heated Carbon Paste Electrode with Magnetic Separations", *Electroanalysis*, 2004, vol. 16, pg. 928.

[23] J. Wang, D. Xu, R. Polsky, "Magnetically-Induced Solid-State Electrochemical Detection of DNA Hybridization", *J. Am. Chem. Soc.*, 2002, vol. 124, pg. 4208.

[24] http://www.epa.gov/safewater/arsenic.html

[25] M. R. Jekel "*Removal of Arsenic in Drinking Water Tretament*". Ed. O Nriagu Jerome, "*Arsenic in the Environment, Part I: Cycling and Characterization*". New York, Wiley, 1994.

[26] E. O. Kartinen, C. J. Martin. "An overview of arsenic removal process", *Desalination*, 1995, Vol. 103, pp. 78–88.

[27] L. A. Zeng, "A method for preparing silica-containing iron (III) oxide adsorbents for arsenic removal", *Water Research*, 2003, vol. 37, pp. 4351–4358.

[28] J. A. Wilkie, and J. G. Hering. "Adsorption of arsenic onto hydrous ferric oxide: effect of adsorbates/adsorbent ratio and co-occurring solutes", *Colloid Surf A: Physico-chem. Eng. Aspects*, 1996, vol. 107, pp. 97-110.

[29] K. P. Raven, A. Jain, and R. H. Loeppert, "Arsenite and arsenate adsorption on ferrihydrite: kinetics, equilibrium, and adsorption envelopes", *Environment Science Technology*, 1998, vol. 32, pp. 344–349.

[30] X. Sun and H. E. Doner, "Adsorption and oxidation of arsenite on goethite", *Soil Science*, 1998, vol. 163(4), pp. 278–287.

[31] U. Schwertmann and R. M. Cornell. *Iron Oxides in the Laboratory: Preparation and Characterization,* 2nd completely revised and extended edition, Wiley–VCH, Weinheim, 2000.

[32] X. H. Gao, Y. Cui, R. M. Levenson, L. W. K. Chung, and S. M. Nie, "In vivo cancer targeting and imaging with semiconductor quantum dots", *Nature Biotechnology*, 2004, Vol. 22, pg. 9669.

2006 25th International Conference on Microelectronics

The Role of Non-Equilibrium Plasmas and Micro-Discharges in Top Down Nanotechnologies and Selforganized Assembly of Nanostructures

Z.Lj. Petrović, G. Malović, M. Radmilović-Radjenović, N. Puač, D. Marić, P. Maguire and C. Mahony

Abstract - We have reviewed the role of plasma technologies in future development of nanoelectronics and in the development of other nanotechnologies. First we address the problems in application of the standard plasma etching procedure. We proceed to discuss the development of damage free plasma processes, in particular the etching by fast neutral beams. Fast neutral beams have a potential for implementation beyond the standard application in production of Integrated Circuits (Ics). In the second half of the paper we discuss some of the applications of micro discharges. The dc Volte Ampere characteristics should be analyzed in order to reveal the predominant physical processes by studying E/N, pd and jd^2 scaling. We will also give some results related to rf microdischarge operatting at atmospheric pressure also known as plasma needle. It gives rise to new possibilities for medical treatments (including surgery) with minimum tissue damage. Finally we briefly describe other emerging applications of plasmas in nanotechnologies.

I. INTRODUCTION: PLASMAS AND NANOTECHNOLOGIES

Nanotechnologies appear to be the current fashion of the science funding agencies and editors of scientific journals. The field is still open to new ideas and approaches but two general strategies are well defined. Bottom-up approach is to use the best of our abilities and facilities to produce some structures with extremely small dimensions, even at the level of single atoms. While that is a major achievement in terms of human abilities an even more challenging step will be to transfer this ability into industrially applicable and economically viable technology. For example if we were able to develop a transistor that has a miniscule size equal to one molecule and we are able to put it on a surface in a precise position and prove its functionality we are still far from a useful technology. We need to put one billion transistors like that, in only few minutes, and to connect them in such a way that the whole system functions as an integrated circuit (IC). In addition, all that should cost only few hundred dollars per IC. It is

Z.Lj.Petrović, G.Malović, M.Radmilović-Radjenović, N. Puač, D. Marić are with the Institute of Physics University of Belgrade, POB 68, 11080 Zemun, Serbia and Montenegro, E-mail: zoran@phy.bg.ac.yu.
P. Maguire and C.Mahony are with Nanotechnology research Institute, University of Ulster, Newtownabbey, BT37 0QB United Kingdom, E-mail: pd.maguire@ulster.ac.uk

thus obvious that the critical research in this approach will not be in achieving a result but in providing the technology for its practical implementation.

The second approach is to follow the path set out by Moore's law (more than 40 years ago) and by the roadmaps outlined by the microelectronic (now already nanoelectronic would be justified) industry. That is to follow the existing and well proven technologies and to allow them to achieve further miniaturization through research requiring a large degree of ingenuity.

The former approach is labeled bottom up and the latter top down technologies. While the former is in the domain from 1 nm to 10 nm it is not in industrial applications and at the same time the latter is at the moment at 90 nm in mass production while 65 nm technology has been finalized in laboratories. One is left with the research for future incarnations labeled 45 nm, 32 nm and 22 nm technologies which will no doubt need to change the basic principles as well as predominantly used materials in present day electronics industry.

While having in mind the holy grail of the modern applied research in physics which is to maintain Moore's law, we can safely predict that most of the research that is carried under the banner of bottom up nanotechnology will actually yield applications in completely different, still unforeseen, areas. On the other hand the ICs of the near and not so near future will still be made by using the same principles that are implemented today. Two steps that are critical for further miniaturization are photolithography and plasma etching.

Whatever the approach, plasmas, in particular the low temperature plasmas, have a large role to play in bringing the technologies from the scientific laboratories to mass production. While that may be expected in the top-down approach the bottom up achievements may also need plasma processing, in particular plasma etching for achieving self organized and functional assembly of nanostructures in industrial environment.

II. PLASMA ETCHING

The physical basis for application of plasmas in IC manufacturing arises from the reactive ion etching (RIE) which was observed empirically by Hosokawa and coworkers in 1974 [1] and was explained through direct experimentation by Coburn and Winters [2] in 1979. The synergism between physical etching (as seen in Figure 1 by the effect of Ar^+ ions) and chemical etching (as shown by the effect of XF_2) leads to at least ten times faster etching in the combination.

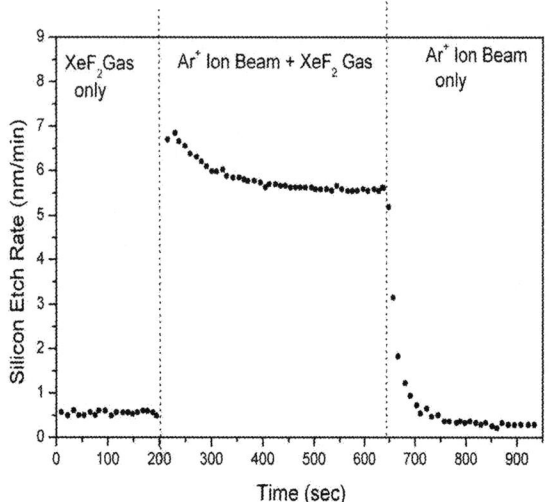

Fig. 1. Synergism between reactive and physical etching by combined effects of radicals and high energy ions [2].

At the same time application of the nonequilibrium plasma means that the mean energy of ions in the bulk of the plasma is low (close to thermal) and that ions gain most of their energy in the sheath. Thus their trajectory is perpendicular to the surface of the wafer, allowing a high degree of anisotropy. At the same time, by selecting proper plasma chemistry, selectivity of etching is achieved whereby a great difference in the etching rate exists between the mask (photoresist) and the substrate (poly Si or SiO_2).

In nonequilibrium plasmas one uses the great mass ratio between electrons on one side and positive ions and background gas molecules on the other to decouple transfer of energy to electrons by electric field from the subsequent transfer of kinetic energy to ions and molecular particles. Thus electrons may achieve high energies and start dissociation and ionization while at the same time ions in the bulk of the plasma have low mean energies.

Etching is, of course, a much more complex process involving redeposition, polymerization and, also, the ions usually deposit most of their energy deep inside the substrate which may lead to the damage of the substrate. Damage may also occur due to high energy photons and also, as often discussed, due to charging of the substrate especially in case of etching of the dielectrics.

The property of the non-equilibrium plasmas that the ions are at the room temperature while the electrons are at a much higher temperature is the basis for anisotropic etching that has enabled emergence of submicron technologies and has fueled the growth of electronic

industry for the past 30 years as predicted by the Moore's law [3].

Another property of the non-equilibrium plasmas is that gas molecules and ions have low energies which gives an opportunity to treat thermally unstable materials including polymers, organic materials, even the living matter. Plasmas are also versatile and lead to numerous applications in addition to etching,, such as: cleaning, polymerization, thin film deposition, plasma chemical reactors of different types, even plasma surgical equipment [4,5].

Plasmas used for etching invariably operate at radio frequencies between 13.56MHz and 200 MHz and in recent years two modes of operation have become dominant: capacitively coupled plasmas (CCP) and inductively coupled plasmas (ICP). Plasma sources may also be enhanced by application of an external magnetic fields.

A. Capacitively Coupled Plasmas: modeling and 2 frequency operation

The CCPs are mainly maintained by the input of energy to electrons from the moving sheath boundaries and by the effect of collisions on electron motion in rf field [6,7]. The electrons formed at the instantaneous cathode which dominate the maintenance of DC and low frequency plasmas may also participate though it is possible to sustain a plasma without their contributions. In case when negative ions are present double layer formation may lead a dominant source of ionization [8], while in case of rare gas buffer gases metastables may play a significant role in plasma maintenance [9]. In order to be able to claim that one understands the basic kinetics of such complex systems as rf plasmas, one needs to achieve a very good qualitative and quantitative agreement between measurements and predictions of the models and such models may prove to be the basis for designing plasma reactor tools [8,10], both CCPs and ICPs [11-12]. Development of comprehensive tools for plasma modelling [13-14] is one of the goals which need to be completed in order to improve reactors for nano-electronic applications.

Open issues in basic development of the comprehensive models are the following:
- interpretation and accurate representation of fluid equations [15];
- interpretation of the transport data [15];
  - understanding of kinetic phenomena in charge particle transport [16] such as
  - negative differential conductivity [17],
  - anomalous diffusion [18],
  - absolute negative mobility [19] and
  - the role of magnetic field in dc transport of electrons [20];
- lacking of data for numerous important processes;
- data for ion transport and collisions [21]
- data for fast neutral collisions [22]
- shortage of data for negative ions [23];

- appropriate data for time resolved transport are required [24,25];
- understanding of non-hydrodynamic transport in proximity of boundaries or for temporal transients should become part of the models after some detailed benchmark calculations to verify the procedures [26];
- shortage of data for electron radical collisions; and most importantly
- the need for data for reactions of plasma particles with surfaces and for processes occurring on the surfaces.

In principle we need the cross sections for electrons, ions and fast particles for all gases used in IC manufacturing, critical shortage exists for the electron transport data and cross sections for gases such as HBr, BF_3 and for almost all the radicals[27]. In addition to all this new research is required into optimizing performance of plasmas and producing the necessary data bases for plasma etching in environmentally friendly gases or to develop optimal processes for treatment of new materials such as low-k dielectrics [28].

It seems that CCP reactors are the most interesting type of plasma reactors for the industry at the moment. Their operation under 2 frequencies and even 3 frequencies allows a good and functionally separated control of plasma maintenance on one hand and the biasing on the other [29]. Such performance of plasma reactors allows also a better control of the ion energy distribution function and flux by varying only the biasing voltage. Nevertheless, for further miniaturization it will be necessary to revisit the same old issues that have been resolved on a level of larger dimensions. Those would include: etch stop, charging and other types of damage, non-uniform etching (especially close to the boundaries of reactor), increasing of the size of wafers while maintaining uniformity and many more. The new issues will be on the other hand dictated by the choice of materials. If all those issues are resolved the CCP reactors may continue to be the most useful tool even for dimensions of the order of 45 nm and less.

B. Inductively Coupled Plasmas

ICPs appear to have been used very much in recent times. Their main advantage is a high density of charged particles due to reduced losses [30,31]. As a result radicals and excited states [32] are present at higher densities as compared to CCPs. This changes the properties of the plasma facilitating ionization and thereby reducing the mean electron energy [32]. ICPs allow special design of the coils that make it possible to achieve uniformity over large areas. On the other hand complex geometry makes it more difficult to model ICPs unless simplifying assumptions are made [33].

Typical issues that require further notice in the vase of ICPs are the E-H transition and its understanding, the role of excited species and radicals, 2 frequency operation and

achievement of functional separation [34]. Nevertheless the ICP reactors seem to play a secondary role in the current plans of the semiconductor industry and IC manufactures. While suffering of the same limitations as CCPs, ICPs may also be useful even for significantly smaller dimensions.

C. Pulsed operation of discharges and charging free processing

Present day ICs consist of a large number of interconnect layers (typically 7-9). Contacts between different levels and across the surface of the IC provide functionality to the IC. As a result of the need to produce very complex ICs and at the same time to achieve minimal dimensions, the connections have to be very narrow and deep, or in other words of a very high aspect ratio. Etching of the connections is the most difficult process determining the limits of the resolution and defining the technology. Numerous „aspect ratio" related problems may occur during the production of ICs..

Necessarily, due to the varying solid angle of the plasma from within the nanostructure, the etch rate will be aspect ratio dependent and that is undesirable effect. In addition, there are other associated issues, such as microloading, etchstop, notching and many more. Most of these problems have been strongly associated with charging of the bottom of the nanostructures [35]. In principle we want to make a nanostructure of a very high aspect ratio in a dielectric. As the number of positive ions deposited at the bottom of the nano structure increases, the potential at the bottom also increases and eventually it will become equal to the potential that corresponds to the energy of ions. At that point no ions will reach the surface and etching will stop. This effect is illustrated in Fig. 2 where we show the potentials at the bottom of a nanostructure [35] as a function of the aspect ratio. It was found that for the aspect ratios greater than 7 the potential at the bottom of the structure becomes equal to the potential corresponding to the ion energy. At the same time the potential at the side walls of the structure closer to the top is small as it is defined by the energy of electrons reaching the side-walls. The predictions of calculations shown in Fig. 2 were made for the ion beam energy of 300 eV and the resulting critical aspect ratio of 7 is in excellent agreement with experimental observations. In addition to etch stop other forms of the damage exist. For example high energy ions and photons may damage the substrate, and in particular may affect the operation of small transistors. Charging of the bottoms of the wells leads to an increase of the potential on the dielectrics of the transistors and due to very small dimensions even a small amount of charge may induce too high voltage and cause the breakdown in the dielectric of the MOSFET transistor.

Understanding of the causes of IC damage during the process of plasma etching, which necessarily requires both models and experiments, may provide us with a strategy to avoid charging dependent and other problems in plasma etching of ICs. For example an obvious solution would be to have pulsed operation of plasma so that in the afterglow we may discharge the bottoms of nanostructures. After the collapse of the sheath electrons may be able to arrive at the surface and eventually reach the bottom of the structure. For that purpose we may observe temporal development of the charging potential [35] as shown in Fig. 3.

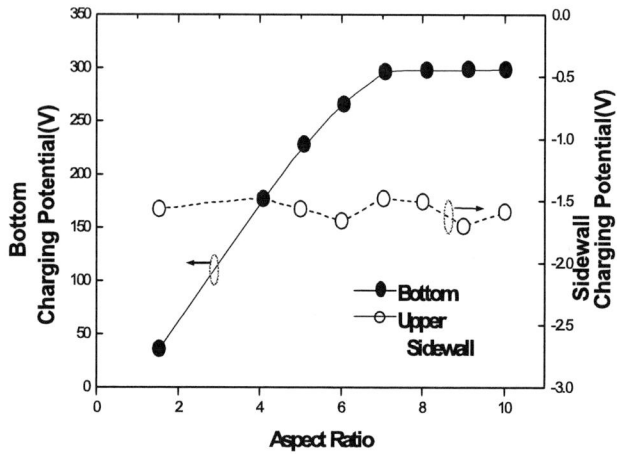

Fig. 2. Charging potential (at the bottom of a trench and at the top of the sidewalls) as a function of the aspect ratio for structures in dielectric etched by ions with 300 eV energy [35].

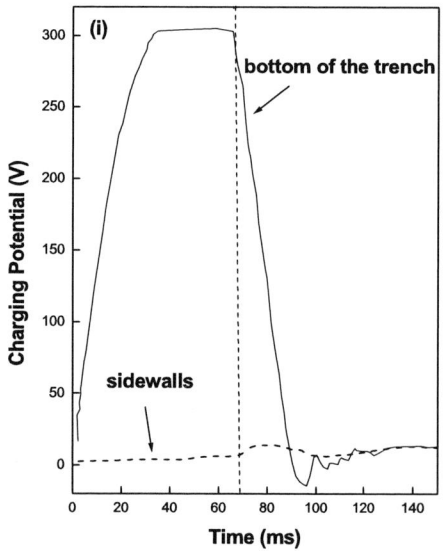

Fig. 3. Temporal development of charging potentials at the bottom of the trench and at the sidewalls [35].

Experiments with pulsed plasmas gave excellent improvement regarding the defects that may be associated with charging and therefore it became necessary to study the operation pulsed plasmas[36] where plasma sustaining source is pulsed while biasing voltage remains continuous. Makabe and the coworkers have been able to show that in

the afterglow during a short period a double layer forms that may push negative ions into the nanotrench [37] and will improve neutralization of positive charges.

As the damage induced by charging both through etch stop and through the breakdown of the very thin dielectrics are the principal cause of defects of ICs at the moment techniques for avoiding these problems should be high one of the highest priorities for further research.. Especially it will be critical to reduce the charging damage if one wants to proceed to even smaller dimensions to true nanotechnologies. In addition it has become necessity to use multiple levels of transistors and realization of their interconnects may become of the critical issues in future.

D. Fast neutral etching and sources of fast neutrals

As sizes of the opening decrease and as aspect ratio increases in order to have more interconnect layers, it may prove that the pulsing strategy is not sufficient. A general solution to the problem of charging would be to use fast neutrals instead of ions. Those fast neutrals may be formed in the gas phase by charge transfer collisions or may be produced by surface neutralization (of both positive and negative ions). Contribution of hyperthermal ions to the etching has been well recognized [38] but it has been proposed in 1991 that very high energy fast neutrals may be produced by the charge transfer in sheaths and may contribute significantly to plasma etching [39]. When charging was identified as a major source of IC damage it was proposed that fast neutral etching could solve these problems [22]. The idea of this proposal was to take advantage of charge transfer collisions to form a beam of fast neutrals and to apply a grid to stop the ions. A similar proposal which relied more heavily on additional neutralization at surfaces of narrow tubes was given more recently and was shown to give excellent results without the damage due to charging [40,41]. These techniques have been tested in production of ICs and were shown to reduce the damage. However pulsing and other techniques were also sufficiently efficient for the same purpose at the present level of technologies (90 nm-65 nm). Thus fast neutral etching has not been employed on the industrial level at the moment. It is however believed that it will have to be implemented for smaller dimensions which will be much more susceptible to the charging damage.

It has been shown recently that fast neutral sources reduce damage both due to charging and due to UV photons and thus provide essential tool for etching of the structures on nanometer scales. It was shown that 8 nm columns could be made by using fast neutral beam etching [42]. This is a good indication that fast neutral beams may become essential in manufacturing of nanostructures smaller than 45 nm.

We have studied the efficiency of neutralization in a realistic system similar to that proposed by Samukawa and coworkers [40] by considering collisions in the gas phase along with the ion neutralization at the surface. Surface neutralization of ions and collisions of ions in the gas were analyzed first separately and then their joint effect was determined. Calculations [43] indicated that for grazing incidence conditions it is safe to assume that the neutralization efficiency is 100%. However it was found that the effect of gas phase charge transfer and surface collisions in converting ions into fast neutrals are approximately equal for teh conditions of the system used in [40]. These calculations were coupled with a plasma model based on Particle In cell (PIC) method [44]. In Fig. 4 we show our calculations for the energy distribution of fast neutrals in the gas phase and by surface collisions for the geometry of the in [40] and for two different pressures. The point of our work was to implement knowledge on atomic gas phase collisions and on surface collisions with aplasma model for realistic conditions in order to verify the efficiency of the technique and then to study its possible practical implementations.

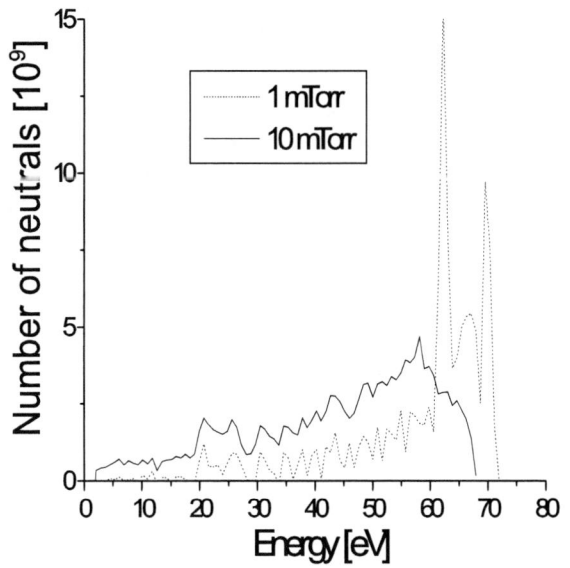

Fig. 4. Energy dependence of fast neutrals originating from a set of parallel tubes as described by Samukawa and coauthors [40]. The realistic ion distribution functions were calculated for an rf discharges at two pressures and ions were allowed to have collisions with neutrals in the gas phase and to have grazing incidence collisions with the walls of the tubes [44]

III. MICRODISCHARGES AND THEIR APPLICATIONS

Non-equilibrium plasmas produced easily only at low pressures. Atmospheric pressures for gaps of the order of 1 cm operate far from the minimum in the so called Paschen curve. At atmospheric pressure growth of ionization is very rapid and highly ionized plasma is formed very easily. This results in the tendency to produce thermal plasma due to relatively strong coupling between ions and electrons.

Techniques to to produce non-equilibrium plasmas consist of interrupting the discharge either temporally, or by inhomogeneous electric field or by a dielectric barrier. Another way to achieve atmospheric pressure discharges which are still nonequilibrium is to operate at very small gaps which have to be of the order of few microns or less in order to achieve minimum breakdown voltage and stable operation under low temperature conditions. Technology of plasma etching has enabled us to produce discharges with dimensions of the order of few microns. However the term micro discharges covers all sub millimeter discharges and this field has grown into the currently most interesting activity of the physics of collisional nonequilibrium plasmas.

The initial motivation for these studies came from the need to optimize plasma screens but new applications were developed very rapidly. Regarding the nanotechnologies and naoelectronics the application of micro discharges are in localized diagnostics of ICs during their manufacture, in choosing appropriate conditions for electro mechanical micro systems which may eventually lead to nanomachining, in localized treatment of materials and assembly of nanostructures and in micro and nano-biological processing and diagnostics.

A variety of microplasma sources with potential use in different portable devices have been reported:

- DC plasma sources operating in helium at atmospheric pressure have been used for optical emission detection [45] and localized micromachining of silicon
- Microhollow cathodes that operate at ~300V DC, have been fabricated to create plasma at atmospheric pressure [46] and have been used as sources of UV light [47];
- Dielectric barrier discharges have also been miniaturized and used for gas analysis [48].
- RF discharges, capacitively coupled mode [49] and inductively coupled plasma sources [50] have been studied with numerous applications in mind, including micro analytical applications;
- Microwave plasma sources have been also miniaturized [51].

A. Scaling of the properties of micro discharges

It is very hard to establish very good understanding of the operating conditions in micro discharges. The best way is perhaps to start from the low pressure discharges and employ standard scaling laws. Discharges should scale according to:

- E/N-electric field to gas number density ratio-proportional to the energy gain from the field between two collisions;
- Pd- pressure (or Nd by using gas number density) times the characteristic distance between two electrodes-proportional to the number of collisions;

- Jd^2-current density normalized by the geometric dimension to the square- describing the space charge effects and
- ω/N-frequency normalized by gas number density for rf discharges.

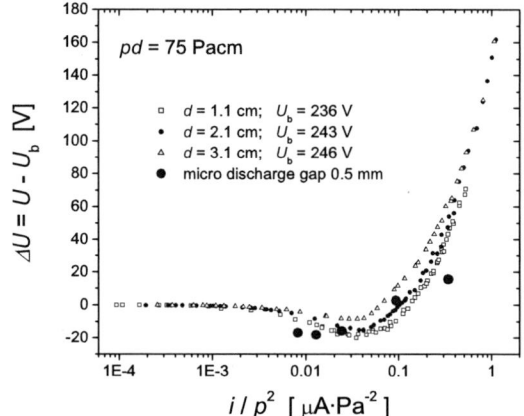

Fig. 5. Volt-Ampere characteristics of a dc discharge for normal conditions [54] and for the microdischarge conditions. All results show reasonable agreement with jd^2 (j/p^2) scaling, but also that for the microdischarge low current conditions have not been achieved.

In studies of the breakdown characteristics of gas discharges it is customary to measure Paschen curves [52,53] (function of the breakdown voltage on pd). However the procedure is valid only if one operates in the low current limit (or in other words in Townsend's (diffuse) dark discharge). The attention is usually no paid to the scaling so measurements of the Voltage Ampere characteristics in such systems are seldom made and it is not certain from which conditions such breakdown curves are derived. As one example we show a VI characteristics (as shown in Fig. 5) measured under low pressure conditions [54] compared to the results obtained at 0.5 mm [55]. Since the gap is 20 times smaller than that in the low pressure conditions the low current limit should be achieved for 400 times smaller currents and to stay in the Townsend regime one needs to cut the current down to several nA.

B. Plasma needle

Medicine and biology are among the most promising, but the same time most challenging, fields of application of nanotechnologies, plasmas and in particular micro discharges. Four important areas of application of MST in medicine and biology may be specified: 1) diagnostics, 2) drug delivery, 3) neutral prosthetics and tissue engineering and 4) minimally invasive surgery. Micro discharges may play and important role especially in tissue engineering and in minimally invasive surgery.

One example of a versatile micro discharge that has been developed recently for such applications is the so called plasma needle [56]. The dimensions of the discharge are of the order of 1 mm and the plasma is sustained by an RF field. The plasma provides a facility to attack living matter by radicals formed in the plasma while gas heating does not exceed the acceptable limit for the living organisms. The system has been used to induce localized sterilization, apoptosis, and even separation of the cells from the tissue that may be assembled back into the tissue [56]. At the same time the plasma may be used for localized treatment of the materials and more powerful versions are being used for minimally invasive surgery. In Fig. 6 we show a plasma needle developed at the Institute of Physics in Belgrade.

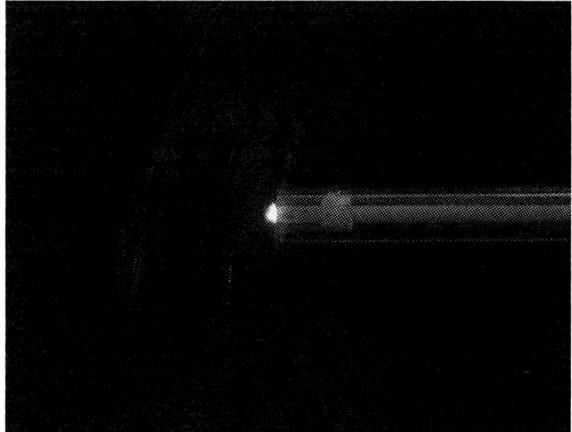

Fig. 6. Plasma needle (Institute of Physics, Belgrade) [57].

The benefits from the studies of microdischarges may also be indirect. The understanding and control of electrical breakdown across the gap is critically important for the microelectronics industry. With each new generation of devices, the gap spacing is often reduced. The integrated circuit, MEMS, hard disk drive magnetic recording industry and flat panel display industries may benefit from understanding of the breakdown mechanisms in micro gaps under atmospheric pressure and some of these processes may limit the deposition of energy and some applications of nano and micro structures.

IV. Conclusion

Applications of plasmas in nanotechnologies are not limited only to the specific cases of plasma etching micro discharges and applications that were mentioned above. For example production of some typical nanostructures such as nanotubes proceeds through application of plasmas and in particular some applications such as that of the secondary electron sources involve gas discharges. In addition the best way to achieve functionalization of nanotubes is by employing plasmas. Plasmas are also used for thin film deposition for analysis of surfaces and for treatment and changing of the properties of materials and surfaces. Nanostructured materials such as hyper-hydrophobic surfaces rely on plasma etching of nano structures and possible plasma polymerization. Applications in biology were shown to be numerous and to open new possibilities. Most importantly plasmas are well integrated in the present day top down technologies that have been tested in industrial environment.

While applications in nano-electronics have been the driving force behind the development of nano technologies, one may actually expect first applications in intelligent nanostructuring and in biomedical applications. The electronics will most probably continue along its top-down route and will be open to new possibilities.

The applicability of nanotechnologies will depend on the ability to achieve massively parallel production within the acceptable economical limits and that is where application of non-equilibrium plasmas will show its potential. Of course plasmas are not panacea for nanotechnologies but are fully integrated in their development.

ACKNOWLEDGEMENT

This work was supported by MNZZS project 141025. We thank all our colleagues who contributed to the work on topics covered here.

REFERENCES

[1] N. Hosokawa, R. Matsuzaki, and T. Asamaki, "RF Sputter-Etching by Fluoro-Chloro-Hydrocarbon Gases ", *Jap. J. Appl. Phys.*, Suppl. 2, Pt. 1, pp. 435, 1974

[2] J. W. Coburn and Harold F. Winters "Ion- and electron-assisted gas-surface chemistry: An important effect in plasma etching", *J. Appl. Phys.*, vol. 50, pp. 3189-3196, 1979.

[3] G.E. Moore, "Cramming more components onto integrated circuits", *Electronics*, Vol. 38, pp. 1-4, 1965.

[4] R.Hippler, S.Pfau, M.Schmidt and K.H. Shoenbach (Eds.), *Low Temperature Plasma Physics*, Wiley-VCH, Berlin 2001.

[5] T. Makabe (Ed.) *Advances in Low Temperature RF Plasmas: Basis for Process Design* North Holland, Amsterdam 2002.

[6] Z. Lj. Petrović, F. Tochikubo, S. Kakuta, and T. Makabe, "Spatiotemporal optical emission spectroscopy of RF discharges in SF6," *J. Appl. Phys.*, vol. 73, pp. 2163–2172, 1993.

[7] A. V. Phelps and Z. Lj. Petrović, "Cold cathode discharges and breakdown in argon: Surface and gas phase production of secondary electrons," *Plasma Sources Sci. Technol.*, vol. 8, pp. R21–R44, 1999.

[8] N. Nakano, N. Shimura, Z. Lj. Petrović, and T. Makabe, "Simulations of RF glow discharges in SF6 by the relaxation continuum model: Physical model and function of the narrow gap reactive-ion plasma etcher," *Phys.Rev. E*, vol. 49, no. 5b, pp. 4455–4465, 1994.

[9] Z. Lj. Petrović, S. Bzenić, J. Jovanović, and S. Djurović, "On spatial distribution of optical emission in radio frequency discharges," *J. Phys.D: Appl. Phys.*, vol. 28, pp. 2287–2293, 1995.

[10] N.Nakano, Z.Lj.Petrović and T. Makabe, "The radical transport in the narrow- gap- reactive- ion etcher in SF6 by the relaxation continuum model", *Jpn. J. Appl. Phys.*, vol. 33, pp. 2223-2228, 1994.

[11] P. L. G. Ventzek, R. J. Hoekstra, and M. J. Kushner, "Two-dimensional modeling of high plasma density inductively coupled sources for material processing," *J. Vac. Sci. Technol. B,* vol. 12, pp. 461–477, 1994.

[12] W. Z. Collison and M. J. Kushner, "Conceptional design of advanced inductively coupled plasma etching tool using computer modeling," *IEEE Trans. Plasma Sci.*, vol. 24, pp. 135–136, 1996.

[13] K.Maeshige, M.Hasebe, Y.Yamaguchi and T.Makabe, "Predictive study of plasma structure and function in reactive ion etcher driven by very high frequency", *J. Appl. Phys.*, vol. 88, pp. 4518-4524, 2000.

[14] T.Makabe and K.Maeshige, „vertically integrated computer-aided design for device processing" *App. Surf. Sci.* Vol. 192 pp. 176-200 2002.

[15] R.E. Robson, R.D. White and Z.Lj.Petrović, „Physically based modelling of collisionally dominated low temperature plasmas", *Rev. Modern. Phys.* Vol. 77 pp. 1303-1320 2005.

[16] Z.Lj. Petrović, Z.M. Raspopović, S. Dujko and T. Makabe Kinetic Phenomena in Electron Transport in Radio Frequency Fields *Appl.Surf. Sci,.* vol. 192, pp.1-25 2002.

[17] S. B. Vrhovac and Z. Lj. Petrović, "Momentum transfer theory of nonconservativeparticle transport in mixtures of gases: General equations and negative differential conductivity," *Phys. Rev. E,* vol. 53, pp. 4012–4025, 1996.

[18] K. Maeda, T. Makabe, N. Nakano, S. Bzenić, and Z. Lj. Petrović, "Diffusion tensor in electron transport in gases in a radio frequency fields," *Phys. Rev. E.,* vol. 55, pp. 5901–5908, 1997.

[19] M. Šuvakov, Z. Ristivojević, Z.Lj. Petrović, S. Dujko, Z.M. Raspopović, N.A. Dyatko, A.P. Napartovich, "Spatial Profiles of Electron Swarm Properties and Explanation of Negative Mobility of Electrons", *IEEE Trans. Plasma Sci.*, vol. 33 pp. 532-533, 2005.

[20] Z M Raspopović, S Dujko, T Makabe and Z Lj Petrović, "Transport coefficients for electrons in argon in crossed electric and magnetic rf fields", *Plasma Sources Sci. Technol.*, vol. 14 pp. 293-300, 2005.

[21] A I Strinić, G N Malović, Z Lj Petrović and N Sadeghi, "Electron excitation coefficients and cross sections for excited levels of argon and xenon ions", *Plasma Sources Sci. Technol.*, vol. 13 pp 333-342 2004.

[22] Z.Lj.Petrović and V.D.Stojanović, "The role of heavy particles in kinetics of low current discharges in argon at high electric field to gas number density ratio" *J.Vac.Sci. Technol.* A, vol. 16, pp.329-336, 1998.

[23] Z. Lj. Petrović, Z.M.Raspopović, V.Stojanović, J.Jovanović, T.Makabe and J.de Urqijo, „Data and Modeling of Negative Ion Transport in Gases of Interest for Production of Integrated Circuits and Nanotechnologies" *Appl.Surf. Sci,.* To be published 2006.

[24] S. Bzenić, Z. Lj. Petrović, Z. M. Raspopović, and T. Makabe, "Drift velocities of electrons in time varying electric fields," *Jpn. J. Appl. Phys.*, vol. 38, pp. 6077–6083, 1999.

[25] Z.Raspopović, S.Sakadžić, Z.Lj.Petrović and T.Makabe, "Diffusion of electrons in time-dependent E(t)xB(t) fields", J. Phys. D vol.33 pp.1298-1302, 2000.

[26] S.Bzenić, Z.M.Raspopović, S.Sakadžić and Z.Lj.Petrović, "Relaxation of Electron Swarm Energy Distribution Functions in Time-Varying Fields", *IEEE Trans. Plasma Sci.*, vol. 27 pp. 78-79 1999.

[27] I Rozum, N J Mason and J. Tennyson, „Electron collisions with the CF3 radical using theR-matrix method, *New J. Phys*, vol. 5, Pp. 155.1-155.12 2003.

[28] M.Miyauchi, Z. Lj. Petrović and T.Makabe, Optical Emission Diagnostics of Etching of low-k Dielectrics in a Two Frequency Inductively Coupled Plasma, submitted 2005.

[29] T.Kitajima, Y.Takeo, Z.Lj.Petrović and T.Makabe, "Functional separation of biasing and sustaining voltages in two frequency capacitively coupled plasma", *Appl.Phys.Lett.*, vol. 77, pp. 489-491, 2000.

[30] J. Hopwood, "Review of inductively coupled plasma for plasma processing," *Plasma Sources Sci. Technol.*, vol. 1, pp. 109–116, 1992.

[31] A. Okigawa, M. Tadokoro, A. Itoh, N. Nakano, Z. Lj. Petrović, and T.Makabe, "Three-dimensional optical emission tomography of an inductivelycoupled plasma," *Jpn. J. Appl. Phys.*, vol. 36, pp. 4605–4616, 1997.

[32] T. Makabe and Z. Lj. Petrović Development of optical computerized tomography in capacitively coupled plasmas and inductively coupled plasmas for plasma etching, *Appl.Surf. Sci.*, vol. 192, pp. 88-114, 2002;

[33] U. Kortshagen and L. D. Tsendin, "Fast two-dimensional self-consistent kinetic modeling of low-pressure inductively coupled RF discharges," *Appl. Phys. Lett.*, vol. 65, pp. 1355–1357, 1994.

[34] T. Denda, Y.Miyoshi, Y.Komukai, T.Goto, Z.Lj.Petrović and T.Makabe, "Functional separation in two frequency operation of an inductively coupled plasma", *J. Appl. Phys.*, vol. 95, pp.870-876, 2004.

[35] J. Matsui, N.Nakano, Z.Lj.Petrović and T. Makabe, "The effect of topographical local charging on the etching of deepsubmicron structures in SiO2 as a function of aspect ratio," *Appl. Phys. Lett.*, vol.78, pp. 883-885, 2001.

[36] K.Hioki, N.Itazu, Z.Lj.Petrović and T. Makabe, "Optical Emission Spectroscopy of Pulsed Inductively Coupled Plasma in Ar" *Jpn. J. Appl. Phys.*, vol. 40, pp. L1183-L1186, 2001.

[37] T.Ohmori, T.Goto, T.Kitajima and T.Makabe, „Negative charge injection to a positively charged SiO_2 hole exposed to plasma etching in a pulsed two-frequency capacitively coupled plasma in CF_4/Ar", *Appl.Phys. Lett.*, vol. 88, pp. 4637-4639, 2003.

[38] K. P. Giapis, T.A. Moore and T.K. Minton, „Hyperthermal neutral beam etching", J. Vac. Sci. Technol. A vol. 13, pp. 959-965, 1995.

[39] Z.Lj. Petrović and A.V.Phelps, Heavy Particle "Excitation and Ionization in Low Pressure Discharges", Proceedings of the International Seminar on Reactive Plasmas, 1991, Ed. T.Goto, Nagoya, pp. 351-360.

[40] S. Samukawa, K. Sakamoto, K., Ichiki, "Generating high-efficiency neutral beams by using negative ions in an inductively coupled plasma source", *J. Vacuum Sci. and Technol. A,* vol.20, pp.1566-1573, 2002.

[41] S. Panda and D.J. Economou, "Anisotropic etching of polymer films by high energy (100s of eV) oxygen atom neutral beams", *J. Vac. Sci. Technol.* A, vol. 19, pp. 398-404, 2001.

[42] S.Samukawa, Proc. 4th International workshop on basic aspects of Nonequilibrium Plasmas Interacting with Surfaces, Fujikyu, Ed. T. Makabe, 2006.

[43] A.Stojković, M.Radmilović-Rađenović and Z.Lj. Petrović, "Neutralization of Ion Beams for Reduction of Charging Damage in Plasma Etching", *Materials Science Forum,* vol. 494, pp. 297-302, 2005.

[44] M. Radmilović-Rađenović, A. Stojković, A. Strinić, V. Stojanović, Ž.Nikitović, G.N. Malović and Z.Lj. Petrović, "Modeling of a plasma etcher for charging free processing of nanoscale structures ", *Materials Science Forum, accepted,* 2006.

[45] J.C.T. Eijkel, H. Stoeri and A. Manz, "A DC microplasma on a chip employed as an optical emission detector for gas chromatography", *Anal. Chem.,* vol. 72, pp. 2547-2552, 2000.

[46] R.H. Stark and K.H. Schoenbach, "Direct current glow discharge in atmospheric air", *Appl. Phys. Lett.,* vol. 74, pp.3770-3772, 1999.

[47] P. Kurunczi, J. Lopez, H. Shah and K. Becker, "Excimer formation in high-pressure microhollow cathode discharge plasmas in helium initiated by low energy electrons collisions", *Int. J. Mass Spectrom.,* vol.205, pp. 277-283, 2001.

[48] T. Shirafuji, T. Kitagawa, T. Wakai and K. Tachibana, "Observation on self-organized filaments in a dielectric barrier discharge of Ar gas", *Appl. Phys. Lett.,* vol.83, pp. 2309-2311, 2003.

[49] M. Radmilovic-Radjenovic, J. K. Lee, F. Iza and G.Y. Park, "Particle-in-cell simulation of gas breakdown in microgaps", *J. Phys. D: Appl. Phys,.* vol.38, pp. 950-954, 2005.

[50] O.B. Minayeva and J. A. Hopwood, "Microfabricated inductively coupled plasma-on-a-chip for molecular SO2 detection: a comparison between global model and optical emission spectrometry", *J. Anal. At. Spectrom.,* vol. 18, pp. 856-863, 2003.

[51] A. M. Bilgic, U. Engel, E. Voges, M. Kuckelheim and A. C. Broekaert, "A new low-power microwave plasma source using microstrip technology for atomic emission spectrometry", *Plasma Sources Sci. Technol.,* vol. 9, pp. 1-4, 2000.

[52] D Mariotti, J A McLaughlin and P Maguire,"Experimental study of breakdown voltage and effective secondary electron emission coefficient for a micro-plasma device" *Plasma Sources Sci. Technol.* Vol. 13 pp.207-212, 2004.

[53] G.Malović, A Strinić, S Živanov, D Marić and Z Lj Petrović, "Measurements and analysis of excitation coefficients and secondary electron yields in Townsend dark discharges"*Plasma Sources Sci. Technol.* Vol. 12 pp. S1-S7, 2003.

[54] D Marić, P Hartmann, G Malović, Z Donkó and Z Lj Petrović, "Measurements and modelling of axial emission profiles in abnormal glow discharges in argon: heavy-particle processes"*J. Phys. D: Appl. Phys.* vol.36 pp.2639-2648, 2003.

[55] G.Malović, A Strinić, S Živanov, D Marić and Z Lj Petrović, P.Maguire and C.Mahony, unpublished.

[56] E.Stoffels, I.E. Kieft, R.E.J. Sladek, "Superficial treatment of mammalian cells using plasma needle", *J.Phys. D: Appl. Phys,.* vol. 36, pp. 2908-2913, 2003.

[57] N.Puac, G.Malović, and Z Lj Petrović, unpublished.

2006 25th International Conference on Microelectronics

Challenges of Ta_2O_5 as High-k Dielectric for Nanoscale DRAMs

E. Atanassova, A. Paskaleva

"Yesterday is history
Tomorrow is mystery
Today is a gift….."
Eleonor Roosevelt

Abstract – The present status, successes, challenges and future of Ta_2O_5, and mixed Ta_2O_5-based high-k layers as active component in storage capacitors of nanoscale DRAMs are discussed. The engineering of new Ta_2O_5-based dielectrics (doped Ta_2O_5 and multicomponent high-k dielectrics) as well as of metal/ high-k interface in MIM capacitor configuration are identified as critical factors for further reduction of EOT value below 1 nm.

I. INTRODUCTION

The continuous scaling of Si devices has guided the microelectronics into the era of nanoelectronics when the components are with nanoscale dimensions, (the so-called top-down approach of nanotechnology which is a characteristic of microelectronics). While manufacture aspects dominated the problems of scaling in the past, one now faces the first fundamental physical limitations as structures approach atomic dimensions. The thickness of SiO_2 used now is about and below 1.4 nm, (the theoretical limit enough to reach the band gap of SiO_2 and to show characteristics close to these of bulk oxide is about two monolayers). The most effective approach to overcome the fundamental limits for SiO_2-device scaling is the use of new, alternative dielectrics with high dielectric constant k^*. With high-k dielectrics the desired equivalent oxide thickness (EOT)** can be achieved concurrently with a reduced current by increasing the physical thickness. There are urgent questions facing the Si industry today: when high-k? what k and what dielectric? The answer to this complex question depends on both the technology node and the device type, DRAM or MOSFET. Here we will focus only on the case of DRAMs. The current densities corresponding to various thicknesses of SiO_2 as a function of applied voltage are nearly the same for DRAMs and FETs. For ultra thin dielectric the leakage mechanism is quantum mechanical tunneling which ultimately cannot be mitigated by improvement of dielectric quality. This results in a real limit to the minimum useful dielectric thickness for SiO_2 independently of the device type. For memory

reduction of capacitor area in DRAMs has made it difficult to maintain the storage charge needed for stable operation of memory cell to prevent soft errors caused by α-particles and cosmic rays radiation. The capacitance should not be less than ~25 fF/cell so that stable cell operation can be guaranteed. For nanoscale memories there is not much room for achieving high capacitance by scaling down the dielectric thickness or expanding the capacitor area. The concept of using high-k dielectrics to obtain the same EOT was presented more a decade ago, and the research efforts have continued ever since. The alternative insulators must meet a set of criteria. It is essential to distinguish the requirements for DRAMs and FETs. Memory capacitors require low leakage current and high capacitance density for charge storage, while the interface quality is not that critical. Memory capacitors require control of the interface primary to limit interfacial reactions to keep the total capacitance high. Furthermore, no electric field penetration is required below the bottom electrode, so that it may be metal or poly-Si. Therefore, *the key parameters* to be optimized for dielectric in DRAMs are *the leakage current and the dielectric constant*, and the purpose is to obtain films with high-k and extremely low leakage current. Naturally the dielectric selection for storage capacitors began by focusing on candidates with the highest ε. Consequently in the course of this process it became clear that the relative impact of the other films' related parameters has to be considered. Although the understanding of the high-k dielectrics is increasing at a very rapid pace, a lot of unresolved problems on material stability and electrical performance are still laying ahead. The high-k films enable reduction of currents to tolerable levels but the preparation of such dielectric with a good uniformity and composition control at an atomic level is proving to be a very challenging problem which is directly related to the major challenge with high-k layers – achieving EOT of at most 1 nm. Finding a material to replace SiO_2 is a formidable challenge because SiO_2 is a nearly perfect dielectric. The only notable drawback of

E. Atanassova and A. Paskaleva are with the Inst. of Solid State Phys., Bulg. Acad. Sci., 72 Tzarigradsko Chaussee, 1784 Sofia, Bulgaria, E-mail: elenada@issp.bas.bg

* the relative permittivity is often given by ε; since always $k\sim\varepsilon$, k and ε can be used interchangeable without loss of meaning.

** thickness of any dielectric scaled by ratio of its ε to that of SiO_2.

1-4244-0116-X/06/$20.00 ©2006 IEEE

issue giving rise to the end of the long successive SiO$_2$ story at the entrance of microelectronics in the nanoscale. *At present no material can fully replace SiO$_2$ for DRAM applications.* Among the metal oxides Ta$_2$O$_5$ is the strongest candidate and Ta$_2$O$_5$-based memories are already in production. Ta$_2$O$_5$ with high ε, high breakdown fields, low leakage current tolerable for Gigabit DRAMs and excellent step coverage characteristics can be obtained by a number of methods compatible with Si technology [1,2]. The essential parameter, however, which favors Ta$_2$O$_5$ in terms of memory applications is its value of stored charge, usually several times higher than other candidates. (Maximum stored charge can be represented by the permittivity and breakdown field, $Q_{max} = \varepsilon E_{bd}$). The purpose of this paper is to illustrate the existing complex issues for the use of Ta$_2$O$_5$ as high-k oxide in storage capacitors; to point out its present status and potency for future applications, as well as the extent to which Ta$_2$O$_5$ is different to SiO$_2$ and to other high-k insulators.

II. FACTORS AFFECTING THE KEY PARAMETERS OF TA$_2$O$_5$ AS A DIELECTRIC IN STORAGE CAPACITOR

The major focus of Ta$_2$O$_5$ research is to improve ε and leakage currents by appropriate technological steps. The later one has stronger impact on the memory effect because not only a layer with high k is needed but the layer also has to keep the current within reasonable limits. ε and current depend on films related parameters which reflect on one hand fundamental materials properties (band structure, barrier height at Ta$_2$O$_5$/Si interface, density and nature of bulk traps and slow states, thermal stability in contact with Si) and on the other hand device performance issues such as the quality of two interfaces (at gate and at Si) and long term reliability. Since leakage is the primary reason for switching to high-k dielectric it is necessary to understand the current transport mechanisms in a given dielectric. The conduction process identified in Ta$_2$O$_5$ is generally attributed to Poole-Frenkel (PF) (including modified one) and Schottky emission. The current is rather bulk- than electrode-limited and the component corresponding to hopping of electrons from one to another state is essential in many cases. Space charge limited current is more probable for thick films. Since the specific intrinsic properties of Ta$_2$O$_5$ are completely different from those of SiO$_2$, the nature of leakage current should be studied in close relation to the preparation conditions considering information on bulk traps and slow states. The detailed understanding of the conductivity is complicated by the presence of interfacial layer which is typically SiO$_2$-like, lower-k layer, (k values are between those of SiO$_2$ and Si). A lack of reliable data for barrier heights for electrons and holes causes additional difficulties in data analysis. Which are the key parameters that could determine the success in improving dielectric constant and the leakage current? Several fundamental ones are considered:

Film microstructure and stoichiometry: One of the advantages of Ta$_2$O$_5$ is its high ε. There are efficient technologies giving films with a dielectric constant of bulk oxide from 20 to 40 and even higher, up to 50-55 for specially doped Ta$_2$O$_5$. Although it is possible to obtain films with high ε the presence of interfacial layer compromises the benefits of Ta$_2$O$_5$ as a high-k material. By relevant annealing [3,4] it is possible to improve the film stoichiometry, to make more abrupt interfaces at two electrodes, to reduce structural non-perfections, and by this way to control and maximize ε even for films obtained by conventional methods (bulk dielectric constant ε_t of sputtered Ta$_2$O$_5$ can be increased from ~25 to ~37 after O$_2$ annealing at 850° C, [5]).

Band gap and band offsets: The higher ε comes at the expense of a smaller band gap and lower conduction and valence band offsets between Si and Ta$_2$O$_5$. The small ΔE_c value correlates with high leakage current, i.e. the trade-off between ε and band offsets tend to limit the advantage of *pure* Ta$_2$O$_5$ as high-k oxide. However, the values of barrier heights are not a critical factor for dynamic memories, in contrary to FET where the transistor performance depends fundamentally on the quality of the interface that determines carrier mobility and device stability. Since Ta$_2$O$_5$ does not have reported reliable values of ΔE_c, the closest most readily alternative indicator for the band offsets is the band gap E_g. The value of E_g~4.5 eV ($\pm$ 0.3 eV) has been obtained, i.e. relatively low. At the same time Ta$_2$O$_5$ capacitors typically exhibit low leakage currents, up to several orders of magnitude lower currents than electrically equivalent SiO$_2$, indicating that Ta$_2$O$_5$ has fewer problems than some other materials. It is not possible to realize simultaneously high k and wide E_g because k varies roughly inversely with the band gap [2]. The relatively small values of E_g and ΔE_c of high-k insulators originate from their electronic structure which is qualitatively different from that of SiO$_2$ and Si-oxynitride, (the lowest conduction band offset energy is determined by the d-state energy of metal atom). When discussing tolerable levels of leakage currents it should be noted that leakage limits are application dependent and the aim has to

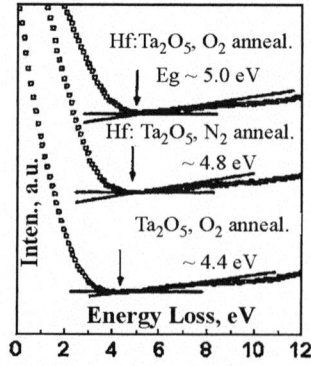

Fig. 1 O1s energy loss spectra for 10 nm layer, [6]

be by balancing ε and E_g to achieve desirable low current and stable stored charge. E_g could be increased by doping of pure Ta$_2$O$_5$ with Hf and Zr (Fig.1). Since many factors affect on both k and E_g the detailed relationship between permitivitty and E_g is not trivial. The band offsets to Si can be additionally influenced by the interfacial layer. The local bonding at the interface could add an extra interfacial dipole, which modifies barrier heights [2,7]. Our results revealed that by preliminary nitridation of Si surface the conduction band offset in the $Ta_2O_5/SiON/Si$ stack can be increased to 1.3 eV, which is significantly higher than the theoretically predicted value of ~0.4 eV for Ta_2O_5/Si interface.

Ta_2O_5/Si interface: SiO_2–like layer inevitably presents between Si and high-k film due to the thermodynamic instability of high-k materials in direct contact with Si, (thermodynamic considerations show that Si reduces Ta_2O_5 forming an interfacial layer [8]). The growth of this layer is favored by the active oxidizing ambient during film deposition. This lower-k layer decreases the global dielectric constant, controls the leakage current and may have a large impact on the carrier transport through the capacitor. Its general obstacle is compromising the minimal EOT according to the relation, EOT=d_s + ($\varepsilon_s/\varepsilon_{high-k}$)$d_{high-k}$, i.e. EOT will never be less than the thickness d_s of the interfacial layer which means that for non-optimized films the interface can even nullify the benefits of Ta_2O_5 as a high-k film, particularly if its thickness is comparable to the active Ta_2O_5 thickness. The microstructure and composition of interfacial layer depend strongly on the fabrication conditions. Very often the layer is a mixed one of SiO_2 and suboxides of both Ta and Si; the later are located nearer the Si surface [4]. At the beginning of high-k film investigations a common view was that this layer has to be removed. **To combat the nature, however, is awfully hopeless!** Actually, *it is not necessary to eliminate this layer completely*. The efforts to eliminate this layer did not result in better system parameters. The more successful approach is whether the stack can exhibit acceptable leakage and desirable ε. The solution here is to suppress the negative effect of this layer and take advantage of its presence. Thus the interfacial region could be exploited to obtain desirable properties; the actual interface will be controlled by the better Si/SiO_2–like one instead of that of Ta_2O_5/Si. The challenge lies in finding relevant annealing steps without other non-desirable subtle effects. Another option to achieve this is to introduce nitrogen at the Si surface. The formation of SiON interface layer enhances not only the stacked capacitance but also reduces the strain at the interface, minimizes charge traps, increases immunity to hot electron degradation and improves reliability. SiON layers grown by thermal methods are lightly "doped" with N and thus offer only a moderate increase of k of the interface region. A higher doping is possible with ion implantation. The surface nitridation, however, strongly affects the electrical behavior of the capacitor; sometimes additional problems appear. Fig. 2 shows representative data of layers on N^+ implanted Si. The data suggest the existence of a high density of negatively charged traps in the nitrided layer which act as slow states. This charge modifies the cathode field and hinders electron injection from Si. The temperature stimulated detrapping of electrons results in a strong increase of leakage current. The trapping/detrapping

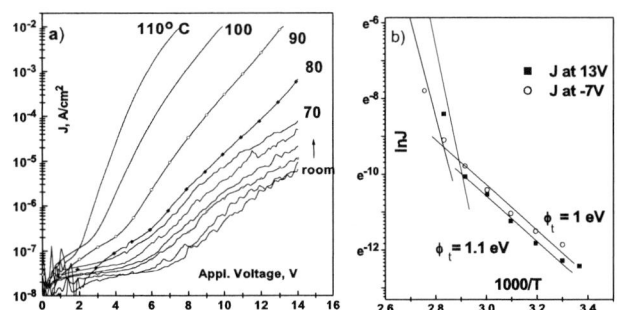

Fig. 2 a) Temperature dependent I-V curves at accumulation; b) Arrhenius plot of the current for the two gate polarities [9].

processes in the interface layer are not reversible and cause much instability of electrical performance. The conduction process is dominated by PF emission from traps at ~1 eV below the conduction band of Ta_2O_5. Fig. 3 illustrates the beneficial effect of N_2 post-fabrication annealing at 850 °C on 15 nm Ta_2O_5. The annealing improves the stoichiometry and microstructure of both the bulk oxide and the interfacial region which manifests as a reduced amount of suboxides. This is not accompanied either by any oxidation of Si or nitridation processes in the stack. A real reduction of the interfacial region width is detected, making the interface more abrupt.

Bulk defects and slow states: Owing to the low temperature of deposition high leakage current is very often observed in as-fabricated films. The current originates from bulk traps and slow states. The structural nature of electrically active centers is not clarified in details because their parameters are strongly dependent on the preparation conditions. The obvious source of electrically active centers is the intrinsic defects (oxygen vacancies and interstitials, suboxides of Ta and Si, strained and broken Ta-O and Si-O bonds) because high-k oxides contain much more defects than SiO_2 and exactly they are responsible for the domination of the bulk-limited conduction mechanism(s) in these oxides. The cause of oxide charge, for example, could be the intrinsic defects as well as extrinsic ones such as hydrogen involved during post-deposition steps. The interfacial region is a preferential location of charge trapping. The source of the charge close to the interface with Si (usually positive but not always) originates from the specific bonding of Ta atoms in this region. Due to the high coordination of Ta, Ta_2O_5 has high number of bonds, forms an overconstrained interface with Si, and therefore degradation in current is expected. This is why a presence of SiO_2-like interfacial layer is not only desirable to gain high-quality interface, but it is even obligatory. This is why it is not necessary to remove this layer – contrary, the challenge here must be addressed to its control so that the interface to be close as possible to this of SiO_2/Si. The covalent bonding with a low coordination makes SiO_2 an excellent glass former, so that SiO_2 is amorphous. The bonding can relax locally to minimize the defect concentrations. The greater ionic character of the

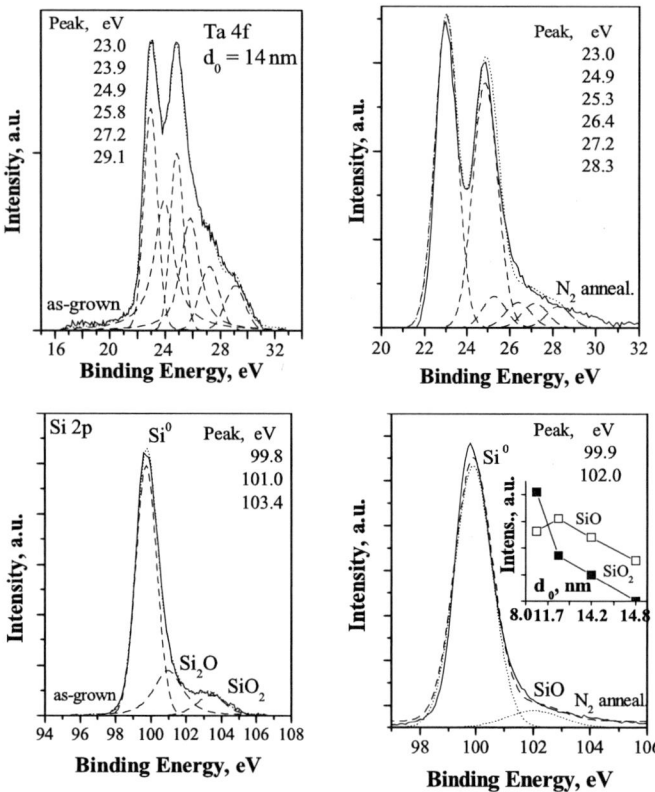

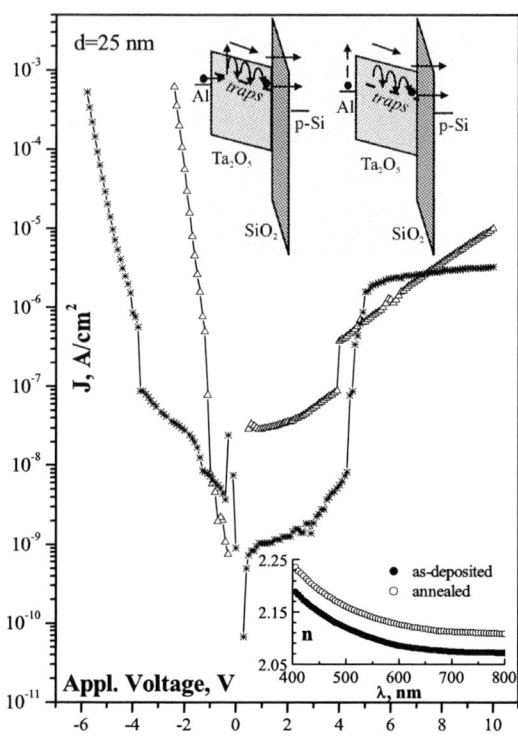

Fig. 3 Decomposition of Ta 4f and of Si 2p spectra at a distance 14 nm from the surface [10].

Fig. 5 I-V curves of rf sputtered Ta_2O_5 before (8) and after (*) O_2 anneal at 850 °C. The inset, refractive index vs. λ, [5].

bonding and the higher atomic coordination numbers mean that the high-*k* oxides are poor glass formers (it is difficult to maintain these oxides amorphous during high temperature processing). The effect of poor glass forming ability and high coordination is that the oxides have higher defect concentrations. The non-equilibrium concentration of defects is high because the oxide network is not so able to relax and rebond to remove defects. C-V hysteresis of high-*k* oxides is one of the potential problems limiting the application. The type of slow states and their parameters can be estimated similarly as in devices with SiO_2. Dedicated investigations concerning slow states in Ta_2O_5 stack capacitors practically are missing. As structural imperfections act as electrically active centers they must be effectively annealed. Figs. 4,5 illustrate the effect of O_2 treatment at 850 °C. The process is beneficial for the dielectric constant of bulk oxide and the reduction of oxide charge to ~10^{10} cm^{-2}. The annealing is not accompanied by additional oxidation of Si; it could effectively reduce oxygen vacancies, leading to ~3-4 orders of magnitude lower current; it stimulates the

increase of optical thickness and of refractive index due to the improved density of the films. A transition from PF mechanism to Schottky emission limited current after annealing is established, i.e. by appropriate technological removal of bulk traps the transition to the desirable electrode limited mechanism is easily achieved. It suggests that there is a potential to control additionally the current by using relevant top electrodes. The annealed films have higher breakdown fields and the immunity to radiation damage (γ irrad. 10^6–10^8 rad) is also improved (Fig.6). Since the Ta_2O_5 could be used in radiation rich conditions, it is critical to have knowledge of radiation effects in Ta_2O_5 devices. At present these effects are very poorly studied. First experiments show that the irradiation will be a

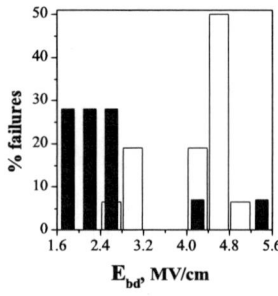

Fig. 4 Breakdown histograms of Ta_2O_5 before (!) and after (□) O_2 anneal at 850°C, [11].

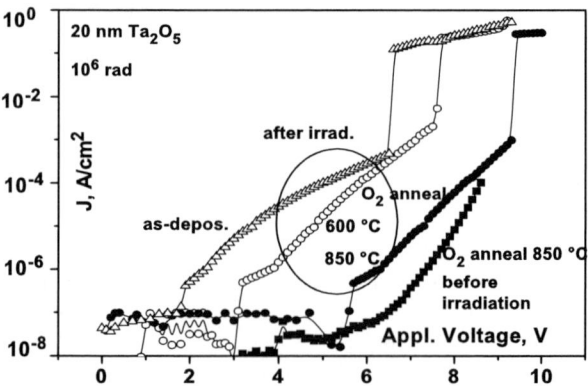

Fig.6 The effect of γ irradiation on I-V curves, [12]

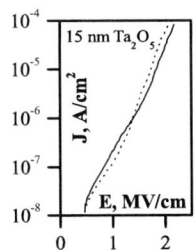

Fig. 7 J vs. E before (-) and after (..) PMA, [11]

problem in terms of leakage current – the extent of the current increase depends on the layer thickness, and the effect is stronger for thinner films and higher doses.

Ta_2O_5 is often heated in H_2 in order to check the effect of this commonly used annealing ambient. At present, clear answer of the question what is the effect of H_2 on high-k dielectrics is missing. There is a limited amount of data for the energy level of interstitial atomic hydrogen in some high-k insulators, with respect to the oxide band edges indicating that H acts as a shallow or deep donor level [7]. Systematical results for the impact of hydrogen on the system Ta_2O_5/Si also are missing. All the same, how does hydrogen behave in Ta_2O_5? Our results [11] imply that post-metal annealing (PMA) in H_2 reduces the oxide charge in thermal Ta_2O_5 (up to ~10^{10} cm^{-2}) but definitely gives rise to C-V hysteresis indicating a generation of slow states with a density of ~$5x10^{11}$ cm^{-2}; ε_t increases to 35 and the breakdown field slightly improves. The leakage current characteristics are not really affected (Fig. 7) suggesting that even if net annealing of oxide charge occurs the current is not sensitive to this process. It emerges that the annealing has a beneficial effect of reducing both oxide and interface traps; it influences mainly C-V curves while interface states have no effect on the I-V curves. Hydrogen undoubtedly involves slow states, i.e. worsens the interfacial region, (most likely the part closer to Ta_2O_5 interface). PMA has rather deleterious than beneficial effect on sputtered Ta_2O_5 (it makes the oxide charge more positive). It seems the effect of hydrogen strongly depends on the initial parameters of the stack – in one case PMA generally improves whereas in another case degrades the parameters. Evidently, it is difficult to make an explicit inference and to identify the effect of hydrogen on the behavior of the capacitors without a detailed consideration of the processing conditions and of the charge state of hydrogen taking part in that process. Every interpretation of data must also consider the key role of SiO_2-like interfacial layer and its strong sensitivity to hydrogen. Usually hydrogen tends to reduce high-k oxides. Thus the final effect of PMA can be a result of two competing processes (reduction of bulk Ta_2O_5 and annealing of interface region) and the domination of one of them. Obviously a lot of unresolved "hydrogen" problems are still laying ahead.

Crystallization effects: So far we have discussed how to improve the electrical characteristics of stack by relevant annealing steps. As a typical high-k material, however, Ta_2O_5 crystallizes if subjected to relatively high temperatures (600-700 °C). Although crystalline Ta_2O_5 has usually a greater ε, its current can be higher because the grain boundaries serve as an additional leakage path. On the other hand, it is generally recognized that the annealing repairs oxygen vacancies and can in some cases effectively reduce leakage current, (stimulating annealing as it was mentioned above and established by many authors). Despite of the intensive investigations the exact relationship between the leakage current and crystalline structure of Ta_2O_5 has not been clarified and seems that *it is not possible.* This circumstance is related to the fact that the crystallization effects depend by a complex manner on film parameters and annealing conditions, and each combination film parameters/thermal processing should be studied individually. The exact temperature of crystallization depends on both the thickness and the stoichiometry of the layer, (the thinner layer maintains amorphous phase even at ~800°C [2]). Figs.8 illustrate the crystallization effects in sputt. 20 nm Ta_2O_5, [13]. As deposited and O_2 annealed at ~800 °C films are amorphous whereas the heated ones at 850°C crystallize in orthorhombic β phase. The crystallization process starts inhomogeneously at the interface SiO_2/Ta_2O_5 and is stimulated by the presence of compositional and structural defects. It implies that the crystallization may not be limiting factor for achieving tolerable level of leakage current and after appropriate technological steps stimulating net annealing and suppressing crystallization, it is possible to reach low enough current. The 850 °C anneal favors a larger ε at a smaller leakage current, but the crystal phase manifests electrically as slow states. Therefore, the results could be discussed in the terms of relative impact of two concurrent mechanisms during treatment – an appearance of crystal phase and real annealing.

Generally the crystallization is undesirable feature since it introduces non-uniformity in the range of grain size. Poly-crystalline dielectrics have non-uniform leakage distribution and can give rise to large statistical variation for nano-meter devices across the chip. It is essential to

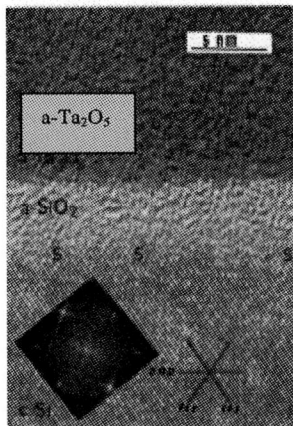
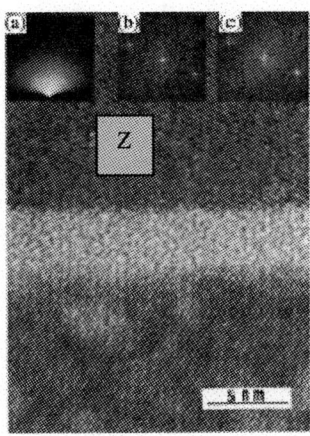

Fig. 8. HRTEM micrograph in (110) projection: a) as-depos. films; the steps at Si/SiO$_2$ interface are labeled with S; inset – FFT from Si matrix; b) after anneal at 850 °C; Z indicates fringes at Si/SiO$_2$ interface, 'a' RHEED pattern after etching; 'b' FFT of Ta_2O_5 image, two spots correspond to (001) interplanar distances in orthorhombic Ta_2O_5; 'c' FFT from the region Z, (001) and (1100) spots of orthorhombic Ta_2O_5, [12].

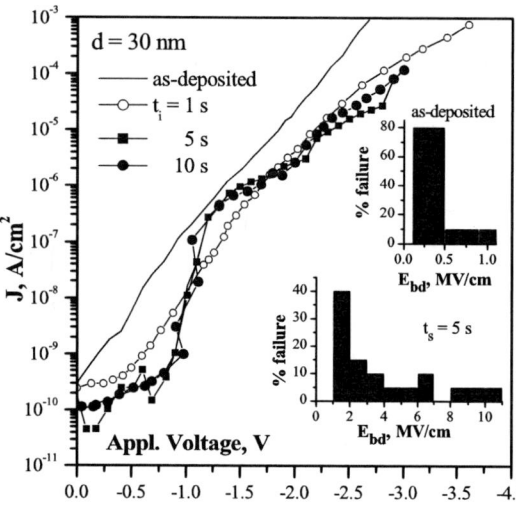

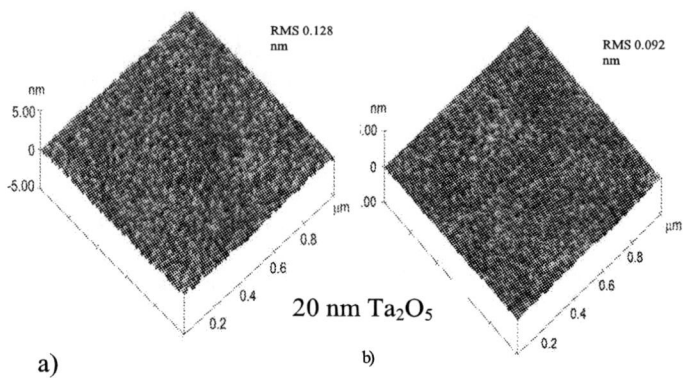

Fig. 10 AFM 3D topography before a) after b) 5 s μw irrad., [15].

Fig.9 I-V curves before and after μw irrad.; inset - breakdown hystograms, [15].

introduce alternative dielectric remaining amorphous after device integration. Strategies have been proposed to prevent crystallization and deleterious effects of electrical transport along grain boundaries. Doped Ta_2O_5 and/or bi-layers containing Ta_2O_5 are likely to meet these challenges. (10 nm Ta_2O_5/~5 nm Nb_2O_5 bi-layer has another temperature of crystallization as compared with pure Ta_2O_5 [14]). The concept of doping high-k oxides is that dopants act as network modifiers, can reduce current and interface state density, can also stabilize the amorphous phase. Acceptable capacitor characteristics are expected and more important they intact after standard device integration, i.e. these dielectrics will not require replacement technology, a circumstance of a first meaning for Si industry. Another solution is to find alternative annealing method(s) to replace high temperature heating and consequently to avoid crystallization. We have used microwave radiation at room temperature as an annealing process. The μw radiation can provide low leakage current by improving the film microstructure at strongly reduced thermal budget (room temperature and extremely short times). An improvement of a number of parameters (stoichiometry and microstructure of the layer; leakage current and breakdown fields; surface morphology) (Figs. 9,10) is established after ~5 s μw irradiation. The capacitance increases with ~30 % and oxide charge reduces below 10^{11} cm^{-2}. The irradiation is not accompanied either by crystallization effects or additional oxidation of Si both circumstances indicating that *short time μw radiation* has a potential to replace high-temperature annealing processes at least for Ta_2O_5.

Metal gates, (integration challenges): The top electrode comprises the next part of challenges in capacitor stack. Due to the incompatibility of Si with high-k dielectrics, metal electrodes are proposed to change poly-Si gate and even to develop capacitor in MIM instead MOS configuration. Adding to this challenge is the problem of integrating a metal-gate process in the conventional process flow, (the incorporation of metal gates increases the integration complexity). The introduction of metal electrodes introduces its own set of manufacturing and reliability challenges and requires the development of specific process modules and adequate integration schemes. The usefulness of each of metal(s) candidates or memory design depends on their ability to achieve the desired work function and respectively leakage level. One of the key issues here is the control of the electrode work function throughout processing. For example, the effect of various electrodes (Al, W, Au, TiN, TiN/W) on the characteristic of capacitors with Ta_2O_5 [3] shows that some parameters such as interface state densities, breakdown fields and charge trapping are defined by the properties of Ta_2O_5 itself. The dielectric constant, oxide charge, leakage current at high fields, charge to breakdown and stress-induced leakage currents are remarkably affected by the electrode. The nature and spatial distribution of gate-deposition-induced defects are sensitive to the technological process - the effect is so strong that it tends to outweigh the effect of the intrinsic properties (work function) of the gate material. All capacitors demonstrated low leakage current (~10^{-7} A/cm^2 up to 1 MV/cm) suggesting that these gates have a potential as upper electrodes of Ta_2O_5 capacitors. At higher fields the current becomes dependent on the electrode material regardless that PF mechanism appears to be dominant in the most of the cases. The sputtered TiN is responsible for the higher leakage current observed for TiN and TiN/W gates, i.e. the deposition condition is critical in this case. The low leakage current detected for W and Au gate capacitors (Fig.11) at higher fields is a natural consequence of the known inertness of these two materials, i.e. in this case the intrinsic properties of the electrode are essential. W deposition is not accompanied by an introduction of a detectable damage. Different defect creation mechanisms during sputtering of W and TiN (two component material) are most likely responsible for the observed behavior of the two type capacitors. The results show that the sputtering technique is a beneficial technology for W deposition ensuring stable gate contact to Ta_2O_5 and tolerable value of the current, while radiation induced damage during

52

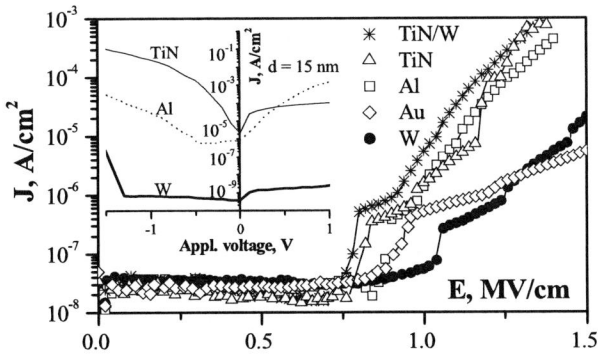

Fig. 11 I-V curves for capacitors with various gates, [3,16]

deposition of TiN make this method less favorable in the case of TiN gate. Therefore, the choice not only of the gate material but its deposition technique remains a concern for Ta_2O_5. In fact, many of the metals considered for high-*k* oxides tend to be quite reactive with the dielectric. As a result, the electrical thickness, leakage current and the work function of the electrode itself may be uncontrollably modified during processing. Metal nitrides (TiN_x, TiAlN) are interesting candidates since they tend to be more stable in contact with high-*k* dielectrics than the transition metals.

Long-term reliability: Long-term reliability is directly related to the leakage current and to all problems concerning its reduction. Many questions on the physics of degradation mechanisms in Ta_2O_5, which are typical of all high-*k* materials, remain unanswered. Although the reliability research on high-*k* dielectrics is steadily growing, due to the growing importance of these materials, their detailed reliability characterisation and charge trapping issues are still poorly understood at this stage. There are predictions regarding the intrinsic high-*k* degradation in terms of extrinsic factors (breakdown is provoked by process-induced defects). The analyses and models are made using mainly the knowledge of SiO_2. Many of them, however, are too speculative and are not based on any deeper understanding of the differences between the degradation of SiO_2 and high-*k* materials. For example, the specific structural properties of high-*k* materials raise the question whether the existing methodology for reliability assessments developed for SiO_2, can be applied. In the case of high-*k* materials the effect of high and low electrical field stress may be very different. Some of the defect-related issues may be addressed by applying a post-deposition anneal to high-*k* films. Calculations indicate that N_2 and NH_3, their ions and derivatives can bond to oxygen vacancies in various charge states or initiate an oxygen substitution process [17] which may passivate the oxide charges and electron-trapping centers. For Ta_2O_5 (as for many high-*k* oxides) a strong polarity dependence of the electrical properties was found and explained by the specific band structure of the stack. Electron and hole trapping lead to a distortion in the energy band of the high-*k* dielectrics, which enhance the dielectric internal electric field and further induces preferential

breakdown either in the interfacial layer or in the bulk oxide. This model is able to explain the polarity dependent breakdown observed in high-*k* stacks [18]. The reliability characteristics of Ta_2O_5 capacitors are also sensitive to the metal gate [16]: the capacitors with sputtered W; TiN or TiN/W gates are very vulnerable to a CCS degradation; Al capacitors are less sensitive to a stress-induced traps generation; CCS does not create measurable SILC in Au-capacitors. The highest charge to breakdown is observed for Al-gate samples. It emerges that the traps responsible for the bulk trapping and those leading to breakdown are not one and the same. An important part of DRAM development is the ability to predict device lifetimes and failure rates. Ta_2O_5 failure mechanisms are entirely different from those of SiO_2 and the reliability models cannot be simply transferred. Regardless of the progress made in this field many crucial questions remain unanswered, (for example, whether or not the oxygen species have a direct role in breakdown). There still is a lack of consensus concerning the mechanism that causes the breakdown in high-*k* insulators. In summary, the understanding of physical differences between reliability phenomenon of Ta_2O_5 as high-*k* oxide and SiO_2 is still a future task.

III. Conclusions and Future Predictions

The same electron feature of Ta_2O_5 as a high-*k* oxide, that *facilitates its desirable k values – the d electron origin of Ta-O bonds – is also responsible for its intrinsic limitations.* Besides intrinsic and integration issues the performance of Ta_2O_5-based capacitors depends strongly on the fabrication process. In contrast to high quality SiO_2 which behaves very much alike regardless of the growth method, *the properties of Ta_2O_5 are very sensitive to the deposition process.* The challenge is to prepare films with uniformity and compositional control on the atomic level. However, there are *no principal problems limiting the use of Ta_2O_5 as a dielectric in storage capacitors of nanoscale DRAMs.* To gain success in achieving tolerable level of leakage current and high enough capacitance a number of film-related parameters must be carefully and simultaneously optimized. The overcoming of these issues requires persistent efforts in both the fundamental understanding of stack properties and the capacitor engineering. Ta_2O_5 studies give *hope that pure Ta_2O_5 can be further extended* for applications in memory devices of sub-70 nm generations. A variety of options exist in this field: chemical modification of Ta_2O_5 matrix by adding suitable metal agents, development of novel dielectrics using a layered combination of Ta_2O_5 and other high-*k* dielectrics. Continuous optimization of the key parameters of these mixed oxides may give an answer of the question how far can we go with Ta_2O_5. Multicomponent dielectrics based on Ta_2O_5 emerge as promising contenders to extend the potential of pure Ta_2O_5. Lower leakage current, higher ε, lower EOT and stable amorphous phase can be achieved by controlling the composition and bond structure, (more or less oxidized or silicate rich layers can be obtained). On the way of development of these new dielectrics a number of challenges in changing research environment have to be overcome. There is a debate about the various solutions. One point of agreement, however, emerges – *a philosophy of continual improvement of the storage capacitor properties*. The requirements of rapid introduction of the modified high-k dielectrics into effective production technologies additionally multiply the challenges. Needless to remember the known enormous task of development of all high-*k* dielectrics – in a very short time frame to replace the nearly perfect SiO_2/Si system with a high-*k* dielectric based one for the sole reason higher ε. A strategy of the investigations in the next stage will be characterized with *harmonization* of variety of advanced *fabrication methods* and improved *characterization techniques* (electrical and microstructural, considering the specificity of multicomponent high-*k* layers on Si or metal layer). New technological schemes and deeper understanding of fundamental properties of these materials are necessary to develop real solution for stack capacitor that complies with future roadmap requirements. The techniques currently available for compositional and electrical analysis need to be improved; the complexity of high-*k* materials requires characterisation with high depth resolution, (non-destructive excellent quantitative depth profiling). A reliable methodology for the physical characterisation of very thin high-*k* layers by combination of techniques with potential for in-line use and classical off-line techniques is necessary. Thus by optimization of the analysis conditions a strong improvement of the data interpretation is expected. Contrary to FETs where high-*k*/Si interface remains the critical stopper for integration of high-*k* materials at circuit level, the introduction of MIM configuration of the capacitors in DRAMs will completely eliminate issues with thermodynamical stability of Ta_2O_5–based high-*k* layer on Si. Engineering of metal electrode/high-*k* interface, the choice of electrodes and how to tune their work functions remain challenging but essential tasks to final solution for successful capacitor integration. The electrode deposition induced damage and respectively poor control of capacitor formation resulting in reduced yield compound the problem with metal gates. In exploring the concept of metal gate and mixed Ta_2O_5–based dielectrics first results give encouraging indications to believe that the roadmap requirements for future technology nodes will be reached. Although the interface Si/high-*k* will be replaced by Me/high-*k* one, to achieve sub-1 nm EOT scaling, the growth of the high quality very thin high-*k* films and respectively appropriate interface control are required. In terms of trapped charges and reliability the performance of high-*k* devices is still rather poor compared to those with SiO_2. Insufficient reliability data and a lack of microscopic understanding of the mechanism(s) of degradation in high-*k* films will motivate future intensive work in this field. It is of paramount importance, however, to first establish an accurate methodology for reliability at circuit level of high-*k* materials, i.e. the reliability concern of Ta_2O_5–based layers is a part of the reliability topic of high-*k* materials.

ACKNOWLEDGEMENT

The work was supported by Bulgarian National Science Foundation, contract F1508.

REFERENCES

[1] *Intern. Techn. Roadmap for Semicond.* http://public.itrs.net.

[2] R.M. Wallace, and G. Wilk, Eds., *Alternative gate dielectrics for microelectronics,* MRS Bulletin, 2002, vol.27.

[3] E. Atanassova, and A.Paskaleva, "Effect of metal electrode on characteristics of Ta_2O_5 for DRAMs", in *New processes and mater. for the next microel. generations,* S. Duenas, and H. Castan, Eds., Trans World Res. (in press).

[4] E.Atanassova, and T.Dimitrova, "Ta_2O_5 as alternative to SiO_2 for DRAMs", in *Handbook of Surf. and Interf. of Mater.*, H. Nalwa, Ed., Vol 4, pp 439-79, San Diego: Acad. Press, 2001.

[5] E. Atanassova, N.Novkovski, A.Paskaleva, et al., "Conduction mechanisms in Ta_2O_5 on Si", *Sol. St. Electr.,* 2002, vol. 46, pp.1887-98.

[6] J. Lu, and Y. Kuo, "Hf-doped tantalum oxide with sub-2 nm EOT", *Appl. Phys. Lett.,* 2005, vol. 87, pp. 232906 - 1-3

[7] J. Robertson, and P.W. Peacock, "Atomic structure, band offsets, growth and defects at high-K oxide:Si interfaces", *Microel. Eng.*, 2004, vol. 72, pp. 112-20.

[8] K.J. Hubbard, and D.G. Schlom, "Thermodynamic stability of binary oxides in contact with silicon", *J. Mater. Res.*, 1996, vol. 11, pp. 2757-76

[9] A. Paskaleva, E. Atanassova, and M. Georgieva, "Charge trapping in Ta_2O_5 on nitrided Si", *J.Phys.D:Appl.Phys.*, 2005, vol. 38, pp. 4210-16.

[10] E.Atanassova, G.Tyuliev, A. Paskaleva, et al., "XPS study of N_2 anneal effect on Ta_2O_5", *Appl. Surf. Sci.,* 2004, vol. 225, pp.86-99.

[11] E. Atanassova, and D. Spassov, "H_2 annealing effects on thermal Ta_2O_5 on Si", *Microel. J.*, 1999, vol. 30, pp.265-74.

[12] E. Atanassova, A.Paskaleva, R. Konakova, et al., "Influence of γ-radiation on Ta_2O_5", *Microel. J.,* 2001, vol. 32, pp. 553-62.

[13] E.Atanassova, M.Kalitzova, G.Zollo, A.Paskaleva, et al, "High temperature-induced crystallization in Ta_2O_5", *Thin Solid Films*, 2003, vol. 426, pp.191-99.

[14] K.Cho, J. Lee, J.-S., Lim, H. Lim, J. Lee, S. Park, C.-Y. Yoo, et al., "Low temperature crystallized Ta_2O_5/Nb_2O_5 bi-layers integrated into RIR capacitor for 60 nm generation and beyond", *Microel. Eng.*, 2005, vol. 80, pp. 317-20.

[15] E.Atanassova, R.V.Konakova, V.F.Mitin, et al., "Microwave radiation on Ta_2O_5/Si", *Microel. Reliab.,* 2005, vol. 45, pp.123-35.

[16] E.Atanassova, A.Paskaleva, N. Novkovski, et al., "Conduction and reliability behavior of Ta_2O_5-Si structures", *J. Appl. Phys.*, 2005, vol. 97, pp. 094104 - 1-11.

[17] J. Greer, A.Korkin, and J. Labanovski, Eds., *Nano and Giga Challenges in Microelectronics*, Elsevier, 2003.

[18] W.Y. Loh, B.C. Cho, M.S. Joo, et al., "Charge Trapping and Breakdown Mechanism in High-K Dielectrics with Metal Gate", *Techn. Dig.–Intern. El. Dev. Meet.,* 2003, pp. 927-30.

2006 25th International Conference on Microelectronics

Electrically Active Defects at the Interface between (100)Si and Hafnium Dioxide Thin Films

P.K. Hurley and K. Cherkaoui

Abstract – The density and energy distribution of electrically active interface defects in the (100)Si/SiO$_x$/HfO$_2$ system are presented. Experimental results are analysed for HfO$_2$ thin films deposited by atomic layer deposition and metal-organic chemical vapour deposition on (100)Si substrates. The paper discusses the origin of the interface states, and their passivation in hydrogen over the temperature range 350-550°C.

I. INTRODUCTION

The Si/SiO$_2$ system has been at the heart of silicon based integrated circuit technology since the first developments of metal-oxide-semiconductor (MOS) processes over 40 years ago. Scaling of the minimum dimensions of MOS based field effect transistors MOSFET's has been tracked by a corresponding reduction of the thickness of the SiO$_2$ or SiON gate dielectric. Currently the gate oxide thickness has a physical thickness around 1.1nm, and further reduction is restricted by gate oxide leakage current densities from direct quantum mechanical tunneling and by the number of silicon and oxygen atoms constituting the oxide layer. This has prompted the requirement for materials of a higher dielectric constant than SiO$_2$ or silicon oxynitride to be used as the gate dielectric [1].

The use of high dielectric constant (high-*k*) materials in the gate stack of MOSFET's introduces many new technological issues, such as: mobility reduction in *n* channel MOSFET's [2], transient instability [3], fixed oxide charges in the high-*k* layer [4], interface layer formation and control [5], gate work function tuning [6] and the understanding and passivation of electrically active interface states [7]-[10]. This paper provides an overview of recent work on the measurement and analysis of the interface state density and energy profile of the interface states in the (100)Si/SiO$_x$/HfO$_2$ system. The results presented are for high-*k* metal-insulator-semiconductor (MIS) samples with relatively high interface state densities (> 1x10^{11}cm^{-2}), where the interface states result in measurable distortion in the capacitance-voltage (CV) response of metal gate (100)Si/SiO$_x$/HfO$_2$ structures. Lower interface state densities can be achieved in high-*k* MIS structures [11], but are not the main focus this paper.

Authors are with Tyndall National Institute, University College Cork, Cork. Ireland, E-mail: phurley@tyndall.ie

A range of possible explanations can be put forward for the physical origin of the high interface state densities detected in high-*k* MIS systems. Taking the case of HfO$_2$ as the high-*k* layer, possibilities include: Hf-Si bonds at the silicon dielectric interface, isolated Hf atoms in the SiO$_x$ interlayer, Hf bonding to the substrate impurity atoms (P or B), contaminants such as C or Cl arising from the MOCVD or ALD processes, or dangling bond defect sites (P$_{b0}$/P$_{b1}$) as detected in the (100)Si/SiO$_2$ system following hydrogen detachment. In this paper experimental results and analysis are presented for HfO$_2$ MIS systems formed by ALD, and MOCVD, with the view to providing further insight into the origin of the interface states in (100)Si/SiO$_x$/HfO$_2$ systems. Results are also presented for the response of the interface state density to annealing in N$_2$/H$_2$ over the temperature range 350-550°C.

II. EXPERIMENTAL OBSERVATIONS

A range of publications have reported non-ideal CV behaviour for (100)Si/high-*k* structures, where a frequency dependent distortion is observed in the region between accumulation and strong inversion for high-*k* samples on *n* and *p* type silicon substrates. These features have been reported for ZrO$_2$, HfO$_2$, La$_2$O$_3$ and other high-*k* MOS structures [12]-[17].

An example of such a CV response is shown in Figure 1, where the CV and conductance-voltage (GV) behaviour are shown for a (100)Si/SiO$_x$/HfO$_2$/TiN gate stacks over *p* type silicon. The HfO$_2$ film was formed by atomic layer deposition from HfCl$_4$ and H$_2$O at 300°C, and the wafer received no high temperature post deposition annealing or forming gas annealing after TiN deposition. The interface SiO$_2$ layer is ~ 1nm formed during a standard chemical cleaning process. Full processing details can be found in [9]. The frequency dependent distortion in the CV response around 0 volts, and the associated peak in the conductance, is characteristic of defects at the (100)Si/insulator interface which exhibit a peak density value at a specific energy level in the band gap at the (100)Si/insulator interface [18], [19]. Corrections for series resistance and parasitic capacitance, which results in a frequency dispersion of the accumulation capacitance, were performed on the measured capacitance and conductance data using the method outlined in [20].

1-4244-0116-X/06/$20.00 ©2006 IEEE

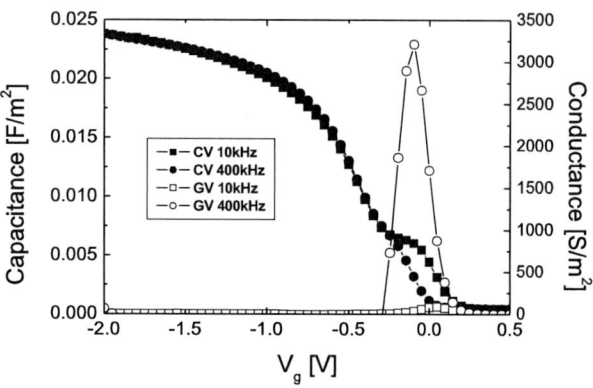

Fig. 1. Corrected CV and GV characteristics at 10kHz and 400kHz for a (100)Si/SiO₂/HfO₂/TiN gate stacks over p type silicon. Equivalent oxide thicknesses (E_{ot}) in the range 1.4 to 1.6 nm. Frequency dependent capacitance-voltage (CV) and conductance–voltage (GV) analysis using a HP4284A LCR meter following open calibration and bond pad correction. Measurements at 22°C.

The frequency dependent CV distortion (referred to subsequently as a peak), is also observed in HfO₂ thin films deposited on (100)Si substrates by metal-organic chemical vapour deposition (MOCVD). An example CV response as a function of ac signal frequency for a (100)Si/SiOₓ/HfO₂/HfSiₓOᵧ/TiN gate stack is shown in Figure 2. For these samples the 3 nm HfO₂ films were deposited by MOCVD using a liquid [(C₂H₅)₂N]₄Hf precursor and O₂ at 485°C, followed by a 1 nm hafnium silicate cap. TiN electrodes were deposited by ALD. The final process step was an 800°C spike anneal in N₂. The CV characteristic exhibits a peak between accumulation and inversion, which increases with reducing ac signal frequency, as expected for an interface defect response.

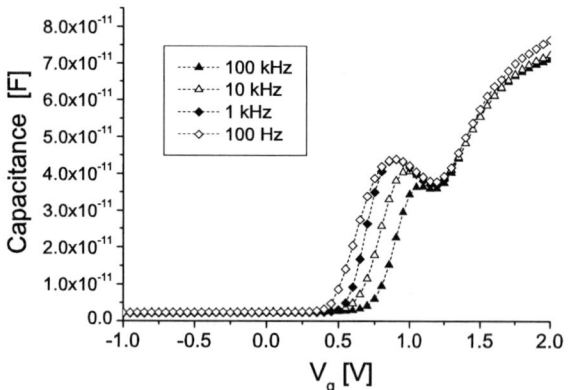

Fig. 2. CV characteristics at 100 Hz, 1kHz, 10kHz and 100kHz for a (100)Si/SiOₓ/HfO₂/HfSiₓOᵧ/TiN gate stacks over n type silicon. HfO₂ films by MOCVD. Capacitor area is 65μm x 65μm LOCOS isolated structures, with fully overlapped TiN. Measurements at 22°C.

As the HfO₂ thin films are formed by ALD and MOCVD for the results presented in Figure 1 and Figure 2 respectively, it is possible that the interface states are a consequence of contamination from Cl in the case of the ALD or defect energy levels due to C or N in the case of the MOCVD process. However, the CV and GV peak response is also observed (not shown here) for (100)Si/SiOₓ/HfO₂/Ni structures, where the HfO₂ layer (2.5nm) has been deposited on n type (100)silicon (HF last) surfaces at 150°C from HfO₂ monoclinic pellets. For this deposition approach from a solid HfO₂ source, C, Cl and N contaminations levels should be negligible. The observation of the peak feature for ALD, MOCVD and e-beam evaporated HfO₂ thin films is not consistent with contamination induced interface defect levels, and points to a common interface defect intrinsic to the high-k MIS structure.

III. INTERFACE STATES: ORIGIN, ENERGY PROFILE AND PASSIVATION

The interface state density profile across the energy gap $D_{it}(E_g)$ can provide further evidence relating to the origin of the interface defects in the (100)Si/SiOₓ/HfO₂ system. The $D_{it}(E_g)$ profiles have been extracted from the corrected CV characteristic using the Berglund integral method. Taking as an example the (100)Si/SiOₓ/HfO₂/TiN gate stacks samples presented in Figure 1, the $D_{it}(E_g)$ profiles across the full energy gap region can be obtained by concatenation of the energy regions of E_v to E_i and E_i to E_c determined from $D_{it}(E_g)$ obtained from p and n substrate samples respectively. The full energy gap distribution obtained using this approach for the (100)Si/SiOₓ/HfO₂/TiN gate stacks prior to the FGA are shown in Figure 3. The Figure also shows the $D_{it}(E_g)$ profile (upper energy gap region only) determined for the (100)Si/SiOₓ/HfO₂/HfSiₓOᵧ/TiN gate stacks shown in Figure 2. These results are based on $D_{it}(E_g)$ extractions from the CV responses at 10kHz. The ALD and MOCVD films (upper gap) demonstrate the same general distribution, with the higher peak D_{it} values for the MOCVD HfO₂ films.

Peak $D_{it}(E_g)$ values prior to FGA passivation, are in the range $(4 - 12) \times 10^{12}$ cm⁻² eV⁻¹. The form of the $D_{it}(E_g)$ profile, and the peak densities, are consistent with measurements of (100)Si/SiO₂/HfO₂/Au formed by MOCVD [17] and with the electrical signature of silicon dangling bonds at the interface in the thermally grown (100)Si/SiO₂ system following hydrogen detachment by vacuum annealing at 700°C for 1 hour [21]. It is noted that the peak value in the interface state density in the upper portion of the energy gap (0.86 to 0.92 eV above E_v), is higher than the peak value in the lower portion of the energy gap (0.24 to 0.28eV above E_v). This asymmetry is also observed for the thermally grown (100)Si/SiO₂ system following hydrogen detachment [18,21].

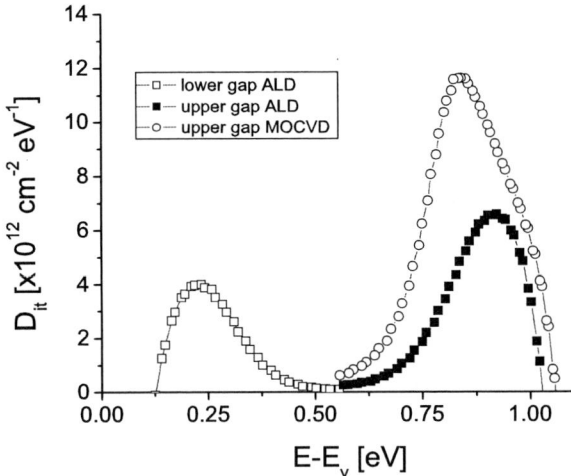

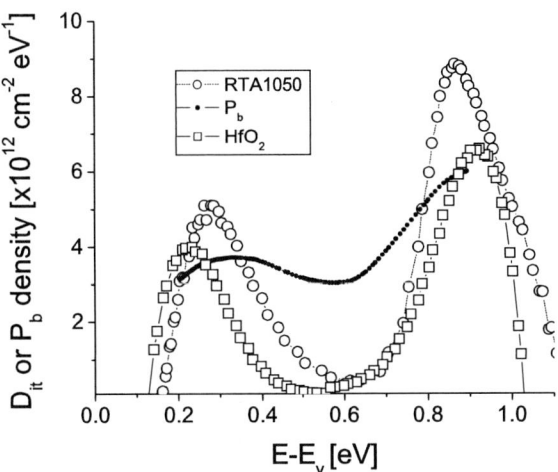

Fig. 3. D_{it} (E_g) profile for (100)Si/SiO$_x$/HfO$_2$/TiN gate stack MIS samples obtained from p (open squares) and n (filled square) samples (as in Figure 1). Samples without post deposition annealing or FGA. The Figure also shows the $D_{it}(E_g)$ profile for the (100)Si/SiO$_x$/HfO$_2$/HfSi$_x$O$_y$/TiN gate stack in the upper energy gap (open circles). All $D_{it}(E_g)$ are from the 10 kHz characteristics. The energy (E) is with respect to the highest energy in the valence band, E_v.

The similarity of the $D_{it}(E_g)$ profile and the electrical signature of silicon dangling bonds at the interface in the thermally grown (100)Si/SiO$_2$ system following hydrogen detachment by vacuum annealing, is also consistent with the measurements of (100)Si/SiO$_x$/HfO$_2$ structures using electron spin resonance (ESR).

The ESR analysis reveals an essentially (100)Si/SiO$_2$-like interface for HfO$_2$, ZrO$_2$ and Al$_2$O$_3$ MIS structures, indicating the overlying high-k layer does not impact on the dominant interface defects in high-k MIS systems. These measurements were performed on both as-deposited high-k films on (100)/Si [22], and Al$_2$O$_3$, ZrO$_2$ and HfO$_2$ on (100)/Si following hydrogen detachment by VUV (8.48 eV) irradiation [23, 24].

Figure 4 plots the $D_{it}(E_g)$ profile for the (100)Si/SiO$_x$/HfO$_2$/TiN gate stacks from Figure 3, together with the $D_{it}(E_g)$ profile obtained for the (100)Si/SiO$_2$ interface following hydrogen detachment by a 1050°C rapid thermal anneal (RTA) in N$_2$ [18]. For the (100)Si/SiO$_2$ system RTA in N$_2$ is an alternative method to vacuum annealing or rapid pull from a furnace oxidation for the creation of samples with un-passivated P_{bo} and P_{b1} centres. The Figure also illustrates the energy distribution of the combined density of P_{bo} and P_{b1} centres as determined by Gerardi [21]. The similarity of the energy distribution for the (100)Si/SiO$_x$/HfO$_2$/TiN and (100)Si/SiO$_2$ structures suggests P_{bo} and P_{b1} centres as the dominant defects responsible for the high interface state densities often observed in (100)Si/SiO$_x$/HfO$_2$ structures.

Fig. 4. D_{it} (E_g) profile for (100)Si/SiO$_2$/HfO$_2$/TiN gate stacks MIS samples as in Figure 1 (open squares), the (100)Si/SiO$_2$ interface D_{it} (E_g) profile following hydrogen detachment by a 1050°C RTA in N$_2$ (open circles) and the density of P_b defects determined in [21] (filled circles).

To evaluate the interface state density [cm^{-2}] from the $D_{it}(E_g)$ profiles [cm^{-2}eV^{-1}] shown in Figures 3 and 4, integration limits of 0.10 eV from the band edges (E_v and E_c) to the midgap energy (E_i) are used. Using this approach the influence of HfO$_2$ deposition temperature and inert (N$_2$) post deposition RTA on the density of states in the lower (E_v +0.1 eV < E < E_i)) and upper energy gap regions (E_i < E < E_c-0.1 eV) can be investigated. Figure 5 presents the density values in the lower (hashed) and upper (open) energy gap regions for a range of processes.

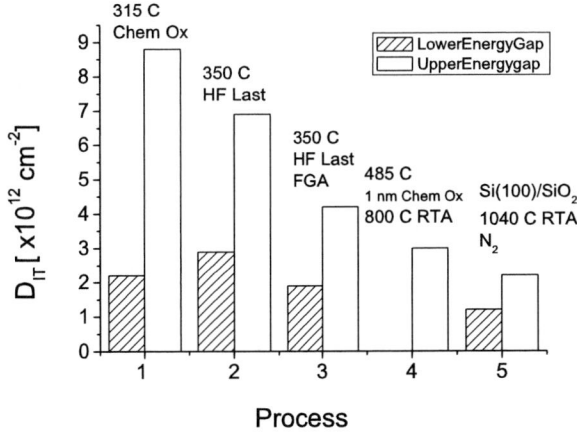

Fig 5. The density of interface states D_{it} [cm^{-2}], for a range of metal gate (100)Si/SiO$_x$/HfO$_2$ gate stacks. The Figure presents the integrated density values in the lower (hashed) and upper (open) energy gap regions. Details for the various processes are included in Table 1.

TABLE I
DETAILS OF THE DEPOSITION CONDITIONS, PRECURSOR AND
GATE TYPE FOR THE SAMPLES PRESENTED IN FIGURE 5.

Process	Deposition/Growth Process/Precursor	Gate Type
1	MOCVD $Hf[O\text{-}C\text{-}(C\text{-}H_3)_3]_4$ 315°C	Au thermal evaporation
2	MOCVD $Hf[O\text{-}C\text{-}(C\text{-}H_3)_3]_4$ 350°C	Au thermal evaporation
3	MOCVD $Hf[O\text{-}C\text{-}(C\text{-}H_3)_3]_4$ 350°C + 450°C for 30 minutes 10%H_2/90%N_2	Au thermal evaporation
4	MOCVD $Hf[(C_2H_5)_2N]_4$ 485°C 800°C spike RTA in N_2	TiN electrodes (ALD)
5	Thermal oxidation 850°C RTA (1040°C, 20s,N_2),	Hg probe

From Figure 5 (processes 1 and 2) the defect densities obtained at the (100)Si/SiO$_x$/HfO$_2$ interface from the CV/GV analysis are in the range (2-8)x10^{12} cm^{-2} for HfO$_2$ films deposited at temperatures ≤ 350°C. This is a high density compared to the inherent density which is typically obtained (~1x10^{12}cm^{-2}) for the thermally grown (100)Si/SiO$_2$ system for oxidation temperatures >800°C (following hydrogen detachment by vacuum annealing). However, the samples (process 1 and 2) in Figure 5 experienced no thermal budget following the HfO$_2$ deposition. Forming gas annealing at 450°C for 30 minutes in a 10%H_2/90%N_2 ambient (Process 3) results in a reduction of (30-35)% in the defect densities in the lower and upper energy gap regions. The incomplete passivation of the interface defects with a standard FGA has been reported in other works [25]. The incomplete passivation resulting from the forming gas anneal could result from the higher initial defect density at the (100)Si/SiO$_x$/HfO$_2$ interface compared to the (100)Si/SiO$_2$ interface, and/or to issues relating to hydrogen diffusion through the HfO$_2$ film and interface strain. Performing a post deposition rapid thermal anneal in nitrogen or oxygen at temperatures of ~800°C, (process 4) results in a reduction of the electrically active defect densities to values which become comparable to those observed at the standard (100)Si/SiO$_2$ interface, following hydrogen detachment.

Hydrogen passivation of the interface defects was examined in more detail by forming gas annealing experiments (5%H_2/95%N_2) over the temperature range 350°C-550°C for the (100)Si/SiO$_2$/HfO$_2$/TiN gate stacks (GV/CV results for no forming gas annealing shown in Figure 1). The FGA (5%H_2/95%N_2) was performed at 350, 400, 450, 500 and 550°C for 30 minutes. The cool down

ambient was N_2 or 5%H_2/95%N_2 and the samples were unloaded from the furnace at 350°C.

The effect of FGA with cooling in N_2 on the peak D$_{it}$ values in the lower and upper energy gap regions is shown in Figure 6. The peak density in the lower energy gap is located between 0.24 to 0.28eV above the valence band, and the upper energy gap peak is located 0.86 to 0.92 eV above the valence band edge. Interface defect passivation is evident in the range 350-400/450°C, with lower energy gap values < 1x10^{12} cm^{-2}eV^{-1} achieved at 400 and 450°C FGA (no RTA in N_2). This corresponds to integrated densities of ~ 2x10^{11} cm^{-2}. At FGA temperatures ≥ 450°C there is an increase in D$_{it}$ values, and the effect is more prominent in the upper energy gap, with the 550°C FGA peak density exceeding the as-deposited case.

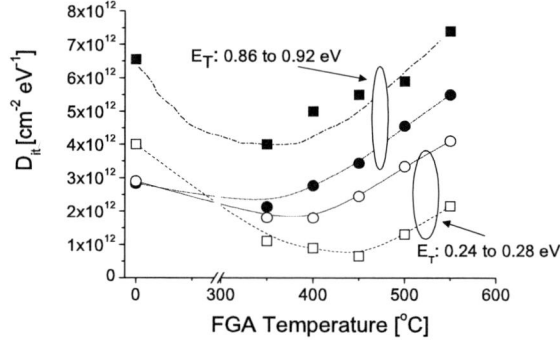

Fig. 6. Peak D$_{it}$ values [cm^{-2} eV^{-1}] versus FGA temperature for cooling in N_2. Filled symbols (upper energy gap), open symbols (lower energy gap). Samples with (circles) and without (squares) 500°C N_2 RTA prior to TiN gate deposition shown. E$_T$ is the energy level of the peak density with reference to E$_v$. (Lines are guide to the eye).

The corresponding results for FGA's where the cooling ambient is H_2/N_2 are shown in Figure 7.

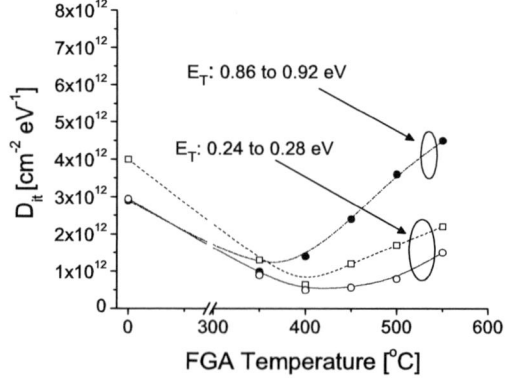

Fig. 7. Peak D$_{it}$ values [cm^{-2} eV^{-1}] versus FGA temperature for cooling in H_2/N_2. Symbols as in Figure 6

From Figures 6 and 7, there is no significant effect of the cooling ambient on the peak densities of interface states in the lower and upper energy gap regions. In addition, the increase in the interface state density for FGA's $\geq 450^{\circ}C$ is also observed for cooling in H_2/N_2 or N_2 ambients.

Further experiments were performed with varying H_2/N_2 anneal times at the optimum temperature of $400^{\circ}C$, and the results are presented in Figure 8. It is important to note here that the maximum thermal budget experienced by these samples prior to the FGA process in a $500^{\circ}C$ RTA step in N_2. Further details can be found in [9].

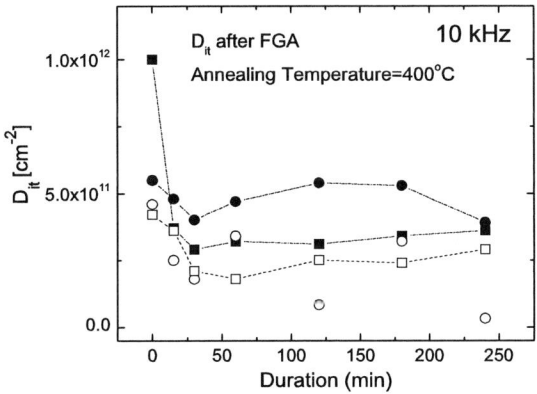

Fig. 8. Evolution of the integrated interface state density D_{it} [cm^{-2}] with time (15 to 240 minutes) for FGA at $400^{\circ}C$ for $(100)Si/SiO_x/HfO_2/TiN$ gate stacks. Integration limits of 0.10 eV from the band edges (E_v and E_c) to the midgap energy (E_i) are used. Symbols as in Figure 6.

It is interesting to note that the optimum FGA conditions ~$400^{\circ}C$ reported here, do not correspond to polysilicon gate $(100)Si/SiO_x/HfO_2$ stacks [7], or to $(100)Si/SiO_2/HfO_2/TiN$ stacks [8], where optimum values in the range $520-530^{\circ}C$ are reported. However, the results reported in [7] and [8] are for $(100)Si/SiO_x/HfO_2$ structures with post deposition anneals in the range $900^{\circ}C$ after HfO_2 deposition, which could explain the difference in the optimum FGA temperature.

IV. CONCLUSION

Electrically active interface defects have been examined for $(100)Si/SiO_x/HfO_2$ stacks formed by ALD and MOCVD processes. The results presented have focused on high-k metal-insulator-semiconductor (MIS) samples with relatively high interface state densities ($> 1 \times 10^{11}$cm^{-2}), where the interface states result in measurable distortion in the capacitance-voltage (CV) response of metal gate $(100)Si/SiO_x/HfO_2$ structure. The interface state density over the energy gap extracted from the CV and GV

response has the electrical signature characteristic of the P_{b0} centre for the $(100)Si/SiO_2$ interface. Performing a post deposition rapid thermal anneal in nitrogen or oxygen at temperatures of ~$800^{\circ}C$, results in a reduction of the electrically active defect densities to values which become comparable to those observed at the standard $(100)Si/SiO_2$ interface (following hydrogen detachment). These results indicate P_{b0} centres as the dominant interfacial defect responsible for the high interface states and CV distortion often reported in HfO_2 MIS structures. A quantitative agreement between the electrically evaluated interface state density from CV analysis and density of P_{b0} centres in the HfO_2 MIS structures has also been obtained [10]. This conclusion does not rule out the possibility that effects such as Hf-Si bonds at the silicon dielectric interface, isolated Hf atoms in the interface SiO_x layer [26] or contaminants (C, Cl), contribute to lower levels ($< 1 \times 10^{11}$cm^{-2}) of interface states.

Examination of the response of the peak interface state densities to FGA yielded optimum conditions around $400^{\circ}C$, with an increase in the peak interface state densities for FGA at temperatures $\geq 450^{\circ}C$. The cooling ambient (N_2 or H_2/N_2) from the forming gas anneal was found to have no significant effect on the peak density of states. Isothermal anneals at $400^{\circ}C$ indicate that D_{it} values saturate after 30 minutes. The optimum anneal temperature of ~$400^{\circ}C$ is different to other works for $(100)Si/SiO_x/HfO_2$ stacks, and this is an area which requires further study.

ACKNOWLEDGEMENT

The authors would like to acknowledge the Science Foundation Ireland (05/IN/1751) and INTEL Ireland/Enterprise Ireland (IP/2003/0178) for financial support of the work.

REFERENCES

[1] G. D. Wilk, R. M. Wallace and J. M. Anthony, "High-k gate dielectrics: Current status and materials properties considerations", *J. Appl.. Phys.*, vol. 89 (10), pp. 5243-5275, 2001
[2] K. Torii, Y. Shimamoto, S. Saito, O. Tonomura, M. Hiratani, Y. Manabe, M. Caymax, and J. W. Maes, "The mechanism of mobility degradation in MISFETs with Al_2O_3 gate dielectric", in *Symp. VLSI Tech. Dig.*, pp. 188–189, 2002
[3] A. Kerber, E. Cartier, L. Pantisano, R. Degraeve, T. Kauerauf, Y. Kim, A. Hou, G. Groeseneken, H. E. Maes, U. Schwalke, "Origin of the threshold voltage instability in SiO_2/HfO_2 dual layer gate dielectrics", *IEEE Electron Device Lett.*, vol. 24, pp. 87-89, 2003
[4] J. Buckley, B. De Salvo, D. Deleruyelle, M. Gely, G. Nicotra, S. Lombardo, J. F. Damlencourt, P. Hollinger, F. Martin and S. Deleonibus, "Reduction of fixed charges in atomic layer deposited Al_2O_3 dielectrics", *Microelectronic Engineering*, vol. 80 pp. 210–213, 2005

[5] W. Tsai, R. J. Carter, H. Nohira, M. Caymax, T. Conard, W. Cosnier, S. DeGendt, M. Heyns, J. Petry, O. Richard, W. Vandervorst, E. Young, C. Zhao, J. Maes, M. Tuominen, W. H. Schulte, E. Garfunkel, T. Gustafsson, "Surface preparation and interfacial stability of high-k dielectrics deposited by atomic layer chemical vapor deposition", *Microelectronic Engineering*, vol. 65 (3) pp. 259–272, 2003

[6] G. Pourtois, A. Lauwers, J. Kittl, L. Pantisano, B. Sorée, S. De Gendt, W. Magnus, M. Heyns and K. Maex, "First-principle calculations on gate/dielectric interfaces: on the origin of work function shifts", *Microelectronic Engineering,* vol. 80 pp. 272–279, 2005

[7] R. J. Carter, E. Cartier, A. Kerber, L. Pantisano, T. Schram, S. De Gendt and M. Heyns, "Passivation and interface state density of SiO₂/HfO₂-based polycrystalline-Si gate stacks", *Appl. Phys Lett.,* vol. 83(3), pp. 533-535, 2003

[8] X. Garros, G. Reimbold, D. Duret, C. Leroux, B. Guillaumot, O. Louveau, C. Hobbs, F. Martin, "Interface states in HfO₂ stacks with metal gate: Nature, Passivation, Generation", *Proceedings 43rd Annual IEEE International Reliability Physics Symposium,* , pp 55 – 60, 2005

[9] M. Schmidt, M. C. Lemme, H. Kurz, T. Witters, T. Schram, K. Cherkaoui, A. Negara and P. K. Hurley, "Impact of H₂/N₂ annealing on interface defect densities in Si(100)/SiO₂/HfO₂/TiN gate stacks", *Microelectronics Engineering,* vol. 80, pp. 70-73, 2005

[10] P. K. Hurley, B. J. O'Sullivan, V. V. Afanas'ev and A. Stesmans, "Interface States and P$_b$ Defects at the Si(100)/HfO₂ interface", *Electrochemical and Solid-State Letters*, vol. 8 (2) G44-G46, 2005

[11] J. H. Sim, S. C. Song, P. D. Kirsch, C. D. Young, R. Choi, D. L. Kwong, B. H. Lee and G. Bersuker, "Effects of ALD HfO₂ thickness on charge trapping and mobility", *Microelectronic Engineering*, vol. 80, pp.218–221, 2005

[12] S. Mudanai, F. Li, S. B. Samavedam, P. J. Tobin, C. S. Kang, R. Nieh, J. C. Lee, L. F. Register, and S. K. Banerjee, "Interfacial Defect States in HfO₂ and ZrO₂ nMOS Capacitors", *IEEE Electron Device Letters*, vol. 23 (12) pp. 728-730, 2002

[13] W. Zhu, T. P. Ma, T. Tamagawa and Y. Di, J. Kim, R. Carruthers, M. Gibson and T. Furukawa, "HfO2 and HfAlO for CMOS: Thermal Stability and Current Transport", in *Proceedings of the International Electron Device Meeting* (IEDM), Washington, DC pp. 463-466, 2001

[14] P. Masson, J. L. Autran, M. Houssa, X. Garros and C. Leroux, "Frequency characterization and modeling of interface traps in HfSi$_x$O$_y$/HfO₂ gate dielectric stack from a capacitance point-of-view", *Appl. Phys. Lett.*, 81 (18) , pp. 3392-3394, 2002

[15] A. Callegari, E. Cartier, M. Gribelyuk, H. F. Okorn-Schmidt, and T. Zabel, "Physical and electrical characterization of Hafnium oxide and Hafnium silicate sputtered films", *J. Appl. Phys.*, vol. 90 (12), pp. 6466-6475 2001

[16] J. B. Cheng, A. D. Li, Q. Y. Shao, H. Q. Ling, D. Wu, Y. Wang, Y. J. Bao, M. Wang, Z. G Liu and N. B. Ming "Growth and characteristics of La₂O₃ gate dielectric prepared by low pressure metalorganic chemical vapor deposition", *Applied Surface Science* vol. 233 (1-4) pp. 91–98, 2004

[17] B. J. O'Sullivan, P. K. Hurley, E. O'Connor, M. Modreanu, H. Roussel, C. Jimenez, C. Dubourdieu, M. Audier and J. P. Sénateur, "Electrical Evaluation of Defects at the Si(100)/HfO₂ Interface", *Journal of the Electrochemical Sociey,* vol. 151 (8), G493-G496, 2004

[18] B. J. O'Sullivan, P. K. Hurley, C. Leveugle and J. H. Das, "Si(100)–SiO₂ interface properties following rapid thermal processing" *J. Appl. Phys.*, vol. 89 (7), pp. 3811-3820, 2001

[19] J. H. Stathis, D. A. Buchanan, D. L. Quinlan, A. H. Parsons and D. E. Kotecki, "Interface defects of ultrathin rapid-thermal oxide on silicon", *Appl. Phys. Lett.* 62 (21), pp. 2682-2684, 1993

[20] K. S. K. Kwa, S. Chattopadhyay, N. D. Jankovic, S. H. Olsen, L. S. Driscoll, A. G. O'Neill, "A model for capacitance reconstruction from measured lossy MOS capacitance-voltage characteristics" *Semiconductor Science and Technology*, vol. 18 (2), pp. 82-87 2003,

[21] G. J. Gerardi, E. H. Poindexter, P. J. Caplan, N. M. Johnson, "Interface traps and P$_b$ centres in oxidised (100) silicon wafers ", *Appl. Phys. Lett.*, vol. 49 (6), pp. 348-350, 1986.

[22] B. J. Jones and R. C. Barklie, "Analysis of defects at the interface between high-k thin films and (100) silicon", *Microelectronic Engineering*, vol. 80, pp. 74-77 2005

[23] A. Stesmans and V. V. Afanas'ev, " Si dangling-bond-type defects at the interface of (100)Si with ultrathin layers of SiO$_x$, Al₂O₃, and ZrO₂", *Appl. Phys. Lett.*, vol. 80 (11) pp.1957-1959, 2002

[24] A. Stesmans and V. V. Afanas'ev, "Si dangling-bond-type defects at the interface of (100)Si with ultrathin HfO₂", *Appl. Phys. Lett.*, vol. 82 (23), pp. 4074-4076, 2003

[25] J. L. Cantin and H. J. von Bardeleben, "An electron paramagnetic resonance study of the Si(100)/Al₂O₃ interface defects", *Journal of Non-Crystalline Solids* vol. 303 (1) pp. 175–178, 2002

[26] S. N. Rashkeev, K. van Benthem, S. T. Pantelides, S. J. Pennycook, "Single Hf atoms inside the ultrathin SiO₂ interlayer between a HfO₂ dielectric film and the Si substrate: How do they modify the interface?", *Microelectronic Engineering.*, vol. 80, pp. 416-419, 2005

2006 25th International Conference on Microelectronics

Capacitance Behavior of Nanometer FD SOI CMOS Devices with HfO$_2$ High-K Gate Dielectric Considering Gate Tunneling Leakage Current

J. B. Kuo, Fellow IEEE, C. H. Lin

Abstract - This paper reports the C_{SG}/C_{DG} capacitance behavior of 100 nm fully-depleted (FD) SOI CMOS devices with HfO$_2$ high-k gate dielectric considering gate tunneling leakage current. According to the 2D simulation results, a unique two-step C_{SG}/C_{DG} versus V_G curve exists for the device with the 1.5nm HfO$_2$ gate dielectric due to the vertical displacement effect. Gate tunneling leakage current has a more impact on C_{DG} as compared to C_{SG}.

I. INTRODUCTION

SOI technology has been a mainstream technology for CMOS VLSI [1]. HfO$_2$ high-k dielectric material has been used to implement nanometer SOI CMOS devices with a thin gate dielectric [2]. For nanometer CMOS devices with a thin gate oxide, the gate leakage due to tunneling current is important [3][4]. Until now the capacitance behavior of the nanometer FD SOI CMOS devices using high-k gate dielectric has not been reported. In this paper, the C_{SG}/C_{DG} capacitance behavior of 100nm fully-depleted (FD) SOI CMOS devices with HfO$_2$ high-k gate dielectric considering vertical displacement effect and gate tunneling leakage current is described. It will be shown that, a unique two-step C_{SG}/C_{DG} versus V_G curve exists for the device with the 1.5nm HfO$_2$ gate dielectric due to the vertical displacement effect. Gate tunneling leakage current has a more impact on C_{DG} as compared to C_{SG}. In the following sections, the capacitance behavior is described first, followed by discussion and conclusion.

II. CAPACITANCE BEHAVIOR

Fig. 1 shows the FD SOI NMOS device under study. It has an n$^+$ poly gate, a thin-film of 20nm doped with a p-type density of 2×10^{18}cm^{-3}, a 50nm LDD region doped with an n-type density of 8×10^{18}cm-3 under a sidewall spacer, and a buried oxide of 310nm. In addition to 7.7nm HfO$_2$, 1.5nm oxide and 1.5nm HfO$_2$ are used as the gate dielectric; oxide, nitride, and air are used as the sidewall spacer.

Authors are with Department of Electrical Engineering, BL-528, National Taiwan University, Roosevelt Rd. Taipei, Taiwan 106-17, Email: jbkuo@cc.ee.ntu.edu.tw

2D simulation with the tunneling parameters as described in Ref. [4] has been used to carry out the study. Fig. 1 shows the electric field contours in the 100nm FD SOI NMOS device with 1.5nm (a)HfO$_2$/(b) oxide gate dielectric, biased at V_D=1V and V_G =0.8V. As shown in the figure, the vertical electric field and the fringing electric field near the sidewall spacer may cause a dramatic impact on the capacitance behavior.

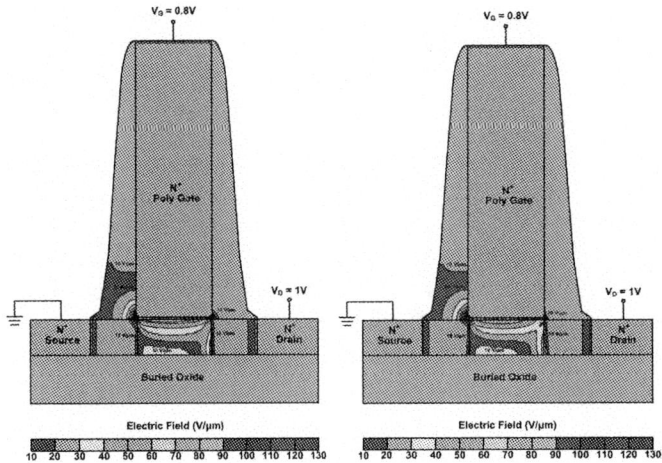

Fig. 1. Electric field contours in the 100nm FD SOI NMOS device with 1.5nm (a)HfO$_2$/ (b) oxide gate dielectric, biased at V_D=1V and V_G =0.8V considering gate tunneling leakage current.

Fig. 2 shows the C_{SG}/C_{DG} versus V_G-V_{TH} of this 100nm FD SOI NMOS device with gate dielectric of 1.5nm oxide and 7.7nm/1.5nm HfO$_2$ and sidewall spacer of nitride, oxide and air, biased at V_D=1V, considering the gate tunneling leakage current and without. As shown in the figure, when a higher dielectric material such as nitride is used for the sidewall spacer, C_{SG}/C_{DG} becomes slightly larger. In addition, there is a unique two-step curve phenomenon with the C_{SG}/C_{DG} curves for the case with 1.5nm HfO$_2$ (red) as the gate dielectric. In contrast, for the 1.5nm oxide case (green) and its effective thickness counterpart-7.7nm HfO$_2$ case (blue), their C_{SG}/C_{DG} curves

1-4244-0116-X/06/$20.00 ©2006 IEEE

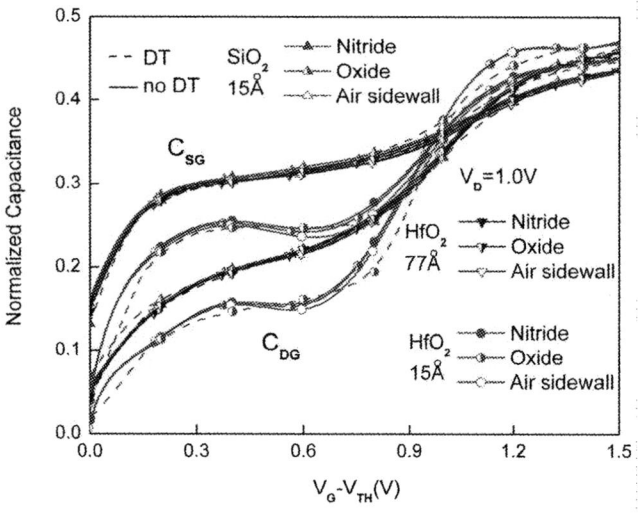

oxide gate dielectric. As shown in the figure, at the silicon side of the gate dielectric/silicon interface the vertical

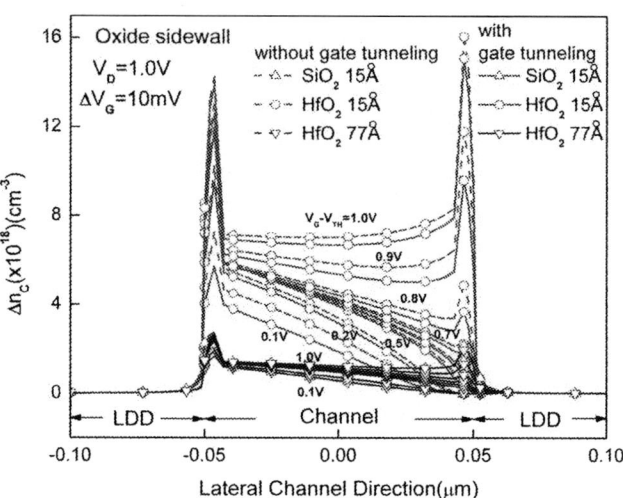

Fig. 2. C_{SG}/C_{GD} versus V_G-V_{TH} of the 100nm FD SOI NMOS device with gate dielectric of oxide and HfO$_2$ and sidewall spacer of nitride, oxide and air, biased at V_D=1V, considering the gate tunneling leakage current and without.

Fig. 3. Increment of electron density (Δn_c) profile in the lateral channel of the FD SOI NMOS device with the 1.5nm/7.7nm HfO$_2$ gate dielectric, biased at V_D=1V and ΔV_G of 10mV.

are similar without the two-step curve behavior. As shown in the figure, considering the gate leakage current (dashed lines), the C_{DG} curves perturb much more as compared to the CS_G ones.

The two-step curve behavior of the 1.5nm HfO$_2$ case could be understood by studying the increment of the electron density (Δn_c) profile in the lateral channel of the FD SOI NMOS device with the 1.5nm/7.7nm HfO$_2$ gate dielectric, biased at V_D=1V and ΔV_G of 10mV, as shown in Fig. 3. When V_G-V_{TH} increases from 0V to 0.5V, the device is in saturation, the increment of the electron density (Δn_c) profile more quickly spreads from the source side to the drain side for the 1.5nm HfO$_2$ case (red) as compared to the 7.7nm case (blue). When V_G-V_{TH} is equal to 0.5V, the Δn_c profile reaches the drain side. After 0.5V, a further increase in V_G-V_{TH} does not cause a rapid increase in the Δn_c profile, which is strongly correlated to the first-stage saturation of the C_{SG}/C_{DG} curves as shown in Fig. 2. The first-stage saturation continues until V_G-V_{TH} reaches 0.8V. As shown in Fig. 3, when V_G-V_{TH} >0.8V, a further rapid increase in the Δn_c profile could be seen, which is also related to the rapid rise in the C_{SG}/C_{DG} curves in the linear region. As shown in the figure, the gate tunneling leakage current affects the cases with the 1.5nm gate oxide the most.

Fig. 4 show the electric field profile in the lateral channel at the dielectric side and the silicon side of the FD SOI NMOS device with 1.5nm/7.7nm HfO$_2$ and 1.5nm

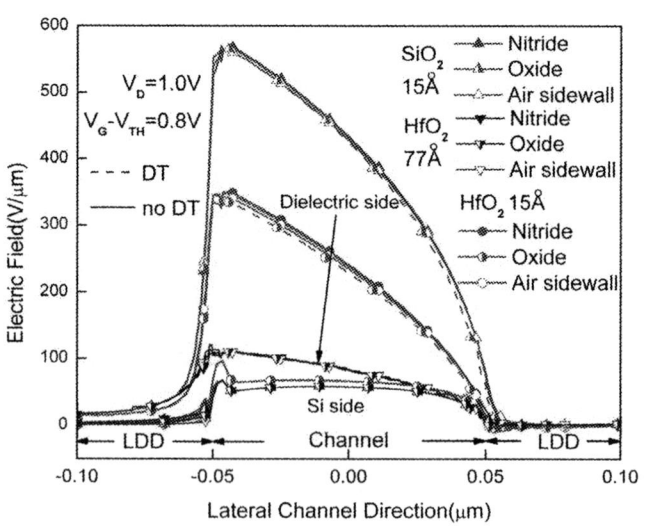

Fig. 4. Electric field profile (a) along the U-shape edges of the poly gate at the dielectric side and (b) in the lateral channel at the dielectric side and the silicon side of the FD SOI NMOS device with 1.5nm/7.7nm HfO$_2$ and 1.5nm oxide gate dielectric.

electric field is small for all three cases. In addition, the 1.5 nm HfO$_2$ case (red) has a smaller vertical electric field in the lateral center at the gate dielectric side as compared to the 1.5nm oxide case (green). From Fig. 4, the two-step C_{SG}/C_{DG} curve behavior is not just due to the vertical electric field alone. It is due to combination of the gate dielectric and the vertical electric field - the displacement effect. Although the 1.5nm HfO_2 case has a smaller vertical electric field, its displacement, which is the product of the dielectric constant the electric field, is much larger as compared to the 1.5nm oxide case. Therefore, the two-step C_{SG}/C_{DG} curve behavior is caused by the vertical displacement effects. In addition, considering the gate tunneling leakage current (dashed lines), the vertical electric field becomes smaller.

III. DISCUSSION

More insight into the device behavior could be obtained by studying the peculiar two-step curve of the C_{SG}/C_{DG} curves for the 1.5nm HfO_2 case. Fig. 5 shows C_{SG} and C_{DG} versus V_G-V_{TH} of the FD SOI NMOS device with HfO_2 gate dielectric for various thicknesses, biased at V_D=1V. As shown in the figure, when the thickness of the HfO_2 gate dielectric increases from 1.5nm, the peculiar two-step C_{GS} curve becomes less noticeable. When the thickness reaches 4.6nm, it disappears. Therefore, the peculiar two-step capacitance curve is due to the combined effects of the vertical electric field and the dielectric constant of the gate dielectric-the vertical displacement effect. As shown in the figure, the gate leakage current has a more impact on the C_{DG} curves as compared to the C_{SG} cases.

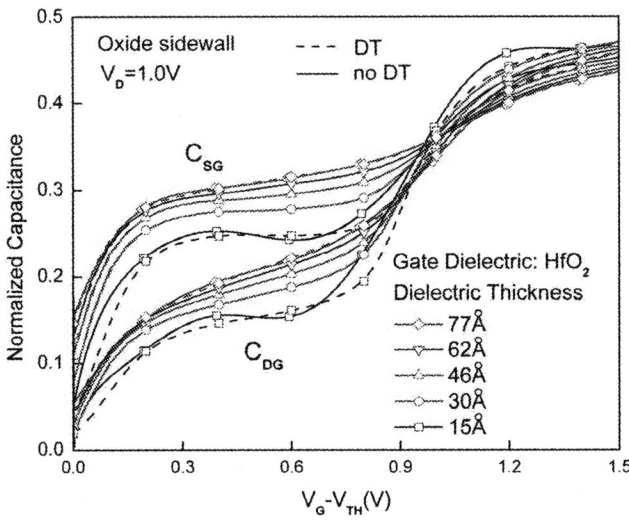

Fig. 5. C_{SG} and C_{DG} versus V_G-V_{TH} of the FD SOI NMOS device with HfO_2 gate dielectric for various thicknesses, biased at V_D=1V.

IV. CONCLUSION

In this paper, the C_{SG}/C_{DG} capacitance phenomenon of 100nm fully-depleted (FD) SOI CMOS devices with HfO_2 high-k gate dielectric considering the vertical displacement effect and the gate tunneling leakage current has been reported. According to the 2D simulation results, a unique two-step C_{SG}/C_{DG} versus V_G curve exists for the device with the 1.5nm HfO_2 gate dielectric due to the vertical displacement effect. Gate tunneling leakage current has a more impact on C_{DG} as compared to C_{DG}.

ACKNOWLEDGEMENTS

This project is supported under research grants from NSC and TSMC. The authors wish to thank Mr. Y. S. Lin of NTUEE for his help during the initial phase of the project.

REFERENCES

[1] J. B. Kuo, "Low-Voltage SOI CMOS VLSI Devices and Circuits," Wiley, New York, ISBN 0471417777, 2001.

[2] A. Vandooren, S. Egley, M. Zavala, T. Stephens, L. Mathew, M. Rossow, A. Thean, A. Barr, Z. Shi, T. White, D. Pham, J. Conner, L. Prabhu, D. Triyoso, J. Schaeffer, D. Roan, B. Y. Nguyen, M. Orlowski, J. Mogab, "50-nm FD SOI CMOS Technology with HfO2 Gate Dielectric," IEEE Trans. Nano-technology,, pp. 324-328, Vol. 2, No. 4, Dec. 2003

[3] W. C. Lee and C. Hu, "Modeling CMOS Tunneling Currents Through Ultrathin Gate Oxide Deu to Conduction- and Valence-Band Electron and Hole Tunneling," IEEE Trans. Electron Devices, Vol. 48, No. 7, pp. 1366-1373, July 2001

[4] Y. C. Yeo, T. J. King, and C. Hu, "MOSFET Gate Leakage Modeling and Selection Guide for Alternative Gate Dielectrics Based on Leakage Considerations," IEEE Trans. Electron Devices, Vol. 50, No. 4, pp. 1027-1033, April 2003

Plenary Sessions

66

2006 25th International Conference on Microelectronics

Defect Engineering and Stress Control in Advanced Devices on High-Mobility Substrates

C. Claeys, G. Eneman and E. Simoen

Abstract - The downscaling of CMOS below 45 nm has triggered the use of high-mobility substrates in order to compensate the mobility degradation related to the implementation of high-*k* dielectrics. Strain engineering has become a very popular technique to boost up the mobility and drive current.

This paper discusses the electrical performance of junctions and transistors processed in strained Si on thin (250-350 nm) strain relaxed SiGe buffer (SRB) layers. The impact of the substrate (misfit and threading dislocation density, use of a C-rich layer in the SiGe buffer, global or locally epitaxial layer) and processing parameters (anneal conditions and doping type) on a variety of performance parameters such as transconductance, mobility, diode leakage current, minority carrier lifetime and low frequency noise will be investigated. Some physical models are proposed to explain the experimental observations. Finally, the potential and issues with alternative high-mobility substrates will be briefly highlighted.

I. INTRODUCTION

For a long period, shrinking device dimensions formed the basis for improving the electrical performance in line with the ITRS roadmap requirements. As the practical limits of device scaling becomes hampered by physical and economical limits, extensive research is focusing on novel solutions. This includes the implementation of a variety of new materials for gate stack (high-*k* dielectrics, fully silicided (FUSI) and metal electrodes like NiSi, PtSi), interconnects (low-*k* dielectrics based on organic or inorganic films, Cu, barrier layers) and in some cases also alternative substrates (SOI, Ge, GeOI).

The main challenge to enhance device performance is to achieve an improvement of the drive current I_{on} without sacrificing the off-state leakage I_{off}. Viable process solutions should also not impact the cost and the device reliability. A key control parameter that received much attention during the last decade is the carrier mobility, whereby the used process technology has a great influence. For scaled down technologies, the I_{off} can be reduced by using high-*k* gate dielectrics [1]. However, it is well known that high-*k* dielectrics degrade the channel mobility due to enhanced phonon and remote Coulomb scattering [2]. An improvement in carrier mobility can be achieved by the use

Authors are with IMEC, Kapeldreef 75, B-3001 Leuven, Belgium E-mail: c.claeys@ieee.org

C. Claeys and G. Eneman are also with Electrical Eng. Dept., KU Leuven, Kasteelpark Arenberg 10, B-3001 Leuven, Belgium

G. Eneman is also a Research assistant of The Fund for Scientific Research – Flanders (Belgium)

of fully silicided (FUSI) or metal gates, as clearly illustrated in Fig. 1.

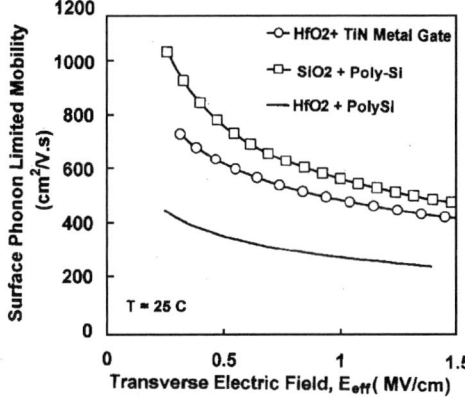

Fig. 1. Surface phonon limited mobility component at room temperature for devices using concencional poly-Si gate, high-*k* gate dielectric with poly-Si gate and high-*k* gate dielectric with midgap TiN metal gate, respectively [3].

The most common approach to boost up the low-field carrier mobility, which is already used successfully for a 65 nm CMOS platform technology [4], is to rely on strain engineering, whereby the changes in the lattice structure cause a splitting in the valence and conduction bands. The suppression of the intervalley and interband scattering associated with the band splitting and/or the reduction of the in-plane effective mass may lead to an enhanced mobility [5-6]. The carrier mobility enhancement is different for holes than for electrons and depends on the value and the direction of the uniaxial or biaxial stress and on the channel orientation. Biaxial and uniaxial strains have a similar impact on the conduction band, but uniaxial strain significantly warps the valence band thereby resulting in a larger hole mobility enhancement compared to biaxial strain [7]. An example of biaxial strain in a device is a channel region with strained silicon on a relaxed buffer layer, whereas uniaxial strain can e.g. be obtained by SiGe recessed source/drain regions.

There exist two main methods for strain engineering, i.e., the global and the local strain approach. The first can be achieved by using epitaxial SiGe layers, whereby the stress in the film will increase the electron and hole mobility in the devices. As an alternative one can use a strained Si on a relaxed SiGe buffer layer or even directly on a buried oxide. Whether or not a strain-relaxed buffer layer is used underneath will have an impact on the density

1-4244-0116-X/06/$20.00 ©2006 IEEE

of both misfit (MD) and threading dislocations (TDs). As will be discussed further, TDs can enhance the diode leakage current, degrade the carrier lifetime and increase the low frequency noise.

Local strain approaches are based on dedicated processing steps and/or modules such as e.g. shallow trench isolation, the use of liners and capping layers, silicidation and/or metal gate electrodes, dry etch processes, contact etch stop layers (CESL), and source/drain engineering. In the latter case the source/drain regions may be formed in strained Si or SiGe recessed areas that are created by selective epitaxial growth. Recently, some reviews on different strain engineering approaches have been published [8-9].

The epitaxial growth of SiGe over a strain relaxed buffer (SRB) of SiGe to enhance the MOSFET performance has received much attention. Si-rich layers lead to a biaxial tension in the film, while Ge-rich layers are under biaxial compression. Originally, strained Si was fabricated starting with a strain-relaxed step-graded buffer, i.e. in the buffer the Ge concentration increases from 0 up to 20 or 30% [9]. This approach leads to rather thick SRBs (> 1 μm), which have the drawback of self-heating during device operation due to the low thermal conductivity of SiGe. Thin layers can overcome this problem and give additional economical benefit due to the higher growth rate and the lower material consumption. Recently, a thin SRB (200-400 nm) approach has been developed whereby the strain-relation in the buffer is enhanced by the implementation of a thin SiGeC layer [10].

This paper will mainly review the defect formation and identification in strained Si on virtual SiGe substrates. Much attention will be given to the impact of defects on the electrical performance parameters such as diode leakage current, carrier lifetime and low frequency noise. The importance of some processing induced local strain approaches will also be addressed. Finally, the potential of other high-mobility substrates will be briefly discussed.

II. EXPERIMENTAL

The thin SRB layers used in this study have been fabricated by CVD epitaxy in a standard ASM Epsilon reactor on 200 mm wafers. A schematic view of the used structure, together with an indication of the thickness of the different layers is given in Fig. 2. The process starts with an undoped 215 nm thick SiGe layer with 22% Ge, in which a 5 nm thick 0.75 at % C-doped layer is incorporated. Before deposition of the top $Si_{0.8}Ge_{0.2}$ layer, a temperature spike anneal is done to optimize the relaxation. The last step is the epitaxy of a strained Si layer (typical 8-12 nm). Typical TD densities are in the range of a few 10^6 cm^{-2}. In addition, pile-up and misfit dislocations can be observed after defect etching under a microscope.

These substrates are then used for the fabrication of n^+p or p^+n junction diodes in a p- or n-well, respectively. In case of transistors, standard processing steps such as 2 nm

gate oxides, highly doped drain and halo implants, spike activation anneals and silicidation were used. More details on the used processing is given elsewhere [11]. Standard Czochralski Si and wafers with a graded thick buffer layers have been used as a reference.

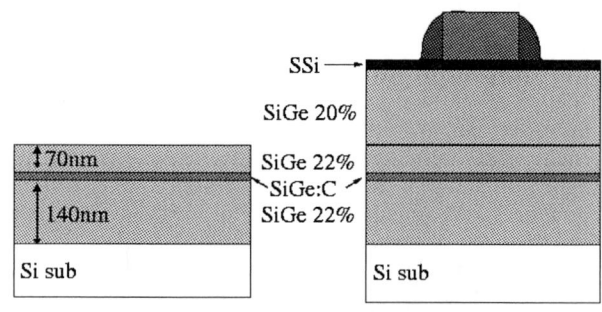

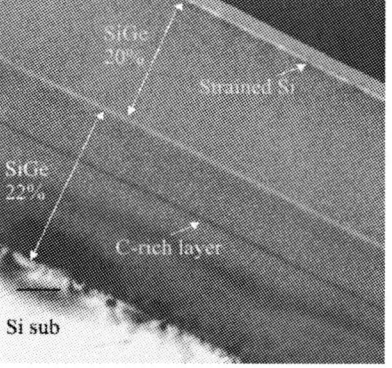

Fig. 2. View of the strained Si and SiGe virtual substrates (top). XTEM micrograph showing the different regions (bottom).

III. DEFECT CHARACTERISATION

In these hetero-epitaxial structures the observed defects are misfits, related to the plastic relaxation of lattice mismatch stress at the interface between layers, and threading dislocations, crossing the film. The MD formation depends on the critical film thickness h_c and is obeying the van der Merwe law [12]

$$h_c = \frac{(1-2\nu)a}{4\pi(1-\nu)^2 f_0}\left[\ln\frac{1-\nu}{2\pi f_0}+1\right] \quad (1)$$

where ν is the dislocation velocity, f_0 is the mismatch of the atom spacing between the two layers, measured at the interface and expressed as the fraction $\Delta a/a$ with a the lattice spacing of the substrate. For strained Si film thicknesses below the critical value for layer relaxation, thermal treatments can lead to the formation of MDs at the strained-Si/SRB interface [13]. This is illustrated in Fig. 3, showing the MD density as function of the anneal temperature for structures with different degrees of relaxation and different film thickness. For a thick graded buffer no MDs are seen, even not after annealing at 1000°C. For 80% relaxation and a 12 nm film, the structure is stable up to 800°C while for higher temperatures MDs are created. The higher the degree of relaxation, the lower the driving force for TD movement.

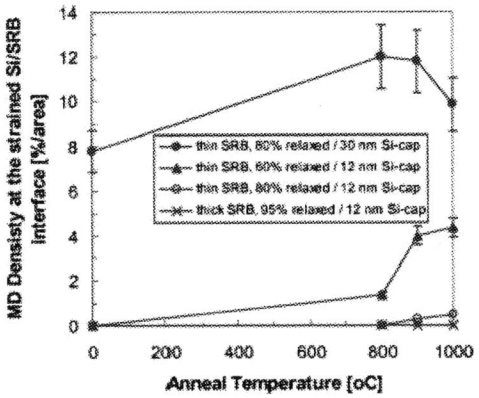

Fig. 3. Evolution of MD at the SRB/strained Si interface for different Si film thickness as function of the 3 min anneal temperature [13].

The dislocations can be revealed by a variety of techniques such as defect etching and optical microscopy, transmission electron microscopy, electron beam induced current (EBIC) measurements [14] and emission microscopy (EMMI) on Schottky diodes [15]. The latter technique allowed pointing out that whereas for a reference diode junction breakdown occurs near the edges, for devices on SSi the breakdown pattern can be associated with MDs [16-17]. There is still a debate on whether or not a dislocation has to be contaminated before becoming electrically active [18].

Another important parameter is the surface roughness. AFM measurements indicate that for thin SRB layers below the critical thickness the roughness is about 1 nm, independent of the layer thickness, while a higher surface roughness is observed for thicker SRBs [19]. For commercial thick SRBs, CMP techniques are used to strongly improve the surface roughness.

IV. ELECTRICAL PERFORMANCE OF DEVICES IN SSi ON SRB SUBSTRATES

Both n^+p and p^+n junction diodes and transistors have been characterized in detail in order to study the impact of the processing parameters, and the associated defect formation, on electrical performance parameters such as diode leakage current, minority carrier lifetime and low frequency noise. The main conclusions of this systematic study are summarized in the following subsections.

A. Diode Leakage Current

The reverse current density of junction diodes depends on a variety of parameters such as thickness of the SRB layer and Si film, location of the C-rich defect layer in

relation to the junction depth, distribution and density of dislocations, thermal budget of the junction anneal, doping type of the well, the mutual interaction between different types of defects, etc. A systematic investigation has been performed, resulting in the following conclusions

1) The junction leakage current density is several orders of magnitude higher for SSi than for reference Si, pointing out the role played by the defects [16].

2) The leakage current is correlated with the dislocation density as shown in Fig. 4 [17]. For n^+p junctions a current of 10 pA/TD at a reverse voltage $V_r=-1V$ and 25°C is found. The different behavior for p^+n junction is still unclear.

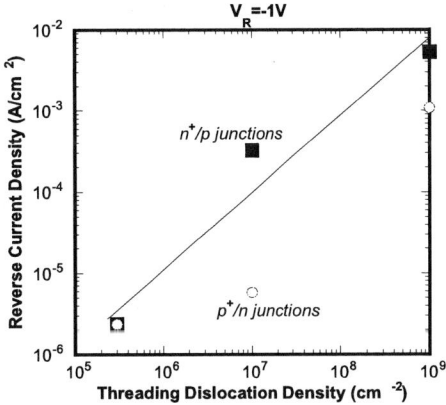

Fig. 4. Reverse current density in function of TDD for n^+/p and p^+/n junctions. T=300 K, $V_R=-1$ V. The C-rich layer is 270 nm away from the wafer surface.

3) The activation energy of the leakage current is different for a thin (250 nm) or a thick (350 nm) SRB structure, as can be seen in Fig. 5. For a thin SRB the activation energy is about half the bandgap, pointing towards a defect controlled mechanism. However, for a thick structure the process becomes diffusion controlled.

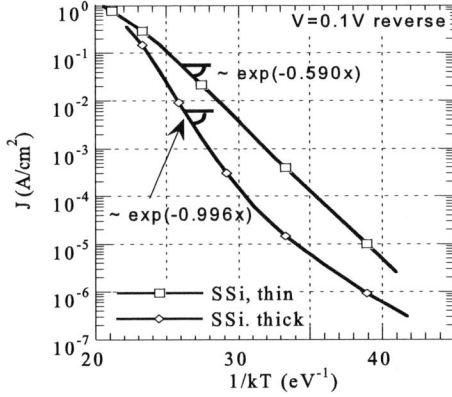

Fig. 5. Activation energy of the leakage current density for n^+p diodes processed in a thin (250 nm) and a thick (350 nm) SRB structure.

4) The type of defects has a strong impact on the leakage current [20]. One has to differentiate between the dislocations associated with the epitaxial structure, the defects coming from the carbon-rich layer, and the residual implantation damage.

- Threading dislocations increase the trap-assisted tunneling current at room temperature, while above 100°C the diffusion current overwhelms the TD generation current.
- The defects associated with the C-layer induce a high relaxation of the SiGe virtual substrate. Bringing the carbon-induced defects inside the electrically active part of the junction results in a higher leakage, caused by an increase in generation current, visible over the studied temperature and voltage range of the experiments
- The residual implantation damage has a pronounced effect making the junction leakage more sensitive towards the anneal temperature. For the junctions inside SiGe buffers, however, the leakage induced by implantation damage is negligible as compared to the leakage caused by the other defect types.

B. Carrier Lifetime

The electrical activity of the threading and misfit dislocations control the lifetime value in the SRBs. The impact of a TD on the generation lifetime τ_g is given by the simple expression [21]

$$\tau_g = 1/(\sigma_s v_{th} N_{TD} n_D) \qquad (2)$$

with σ_n the capture cross section for electrons, v_{th} the thermal velocity, N_{TD} the threading dislocation density and n_D the number of traps per length of dislocation. This correlation is also found experimentally as shown in Fig. 6 [17, 22].

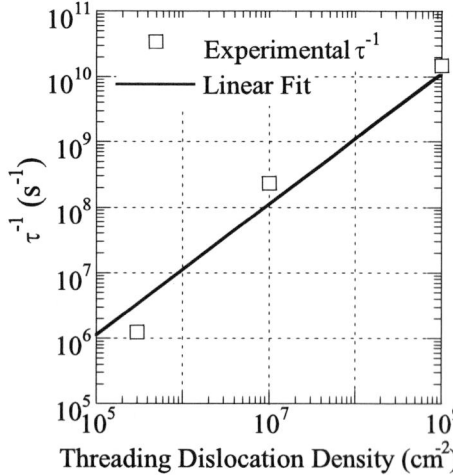

Fig. 6. Inverse generation lifetime versus TD density. The data points are calculated from the reverse current measurements and the solid line is a fit according to Eq. (2).

The lifetime was not only calculated from the leakage current but has also been determined by Microwave Absorption (MWA), using different excitation wavelengths [23]. Both techniques are giving similar trends.

For both thin (250 nm) and thick (350 nm) SRB wafers the impact of a high temperature spike anneal on the generation lifetime has been studied [24]. As can be seen in Fig. 7, for the reference wafer the lifetime increases for increasing anneal temperature due to the annealing out of the residual implantation damage associated with the well implant. The dotted line corresponds with a defect density of 10^7 cm^{-2} calculated with Eq. (2). For the thick SRB the dominant generation centers are related to TDs, while for the thin SRB wafer the influence of the defects associated with the C-rich layer reduces the lifetime with about a decade.

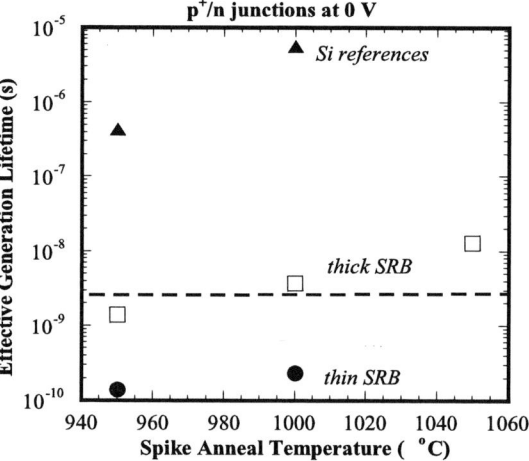

Fig. 7. Effective generation lifetime at 0 V versus spike anneal temperature for p$^+$n junctions made on a thick (350 nm) SRB, a thin (250 nm) SRB and a reference Si wafer.

C. Low Frequency Noise

Low frequency noise has evolved into a powerful diagnostic tool to obtain information on the quality of semiconductor materials and the impact of advanced processing modules [25]. Although the origin of the low frequency noise can be associated with fluctuations in either the number of carriers, which is controlled by trapping/de-trapping phenomena, or in the mobility, one can also analyze the data using the unified or correlated mobility model obeying the expression [26]

$$S_{VG} = S_{VGF} [1 \pm \alpha_{sc} \mu_0 C_{EOT} (V_{GS} - V_T)^2] \qquad (3)$$

with S_{VG} the input-referred voltage noise spectral density, V_{GS} the gate voltage, V_T the threshold voltage, C_{EOT} the equivalent gate oxide capacitance density, and α_{sc} the scattering coefficient related to the effect of Coulombic scattering at the charged oxide traps on the low-field mobility μ_0. The sign in Eq. (3) depends on the nature of the responsible traps (+ repulsive, − attractive). Dependent on the dominant fluctuators only part of the formula has to be used.

In the literature, it is for SiGe SRB devices reported that the noise might increase due to i) the presence of threading dislocations, and/or ii) the indiffusion of Ge into the gate dielectric [27]. An illustration of the influence of a TD on the normalized current noise spectral density S_{id}/I_D^2 is shown in Fig. 8. For good quality SRB material the dislocation density is in the order of 10^5-10^6 cm^{-2}, so that for small geometry device it is possible to have a defect-free transistor. The latter can be checked by the techniques outlined in section III.

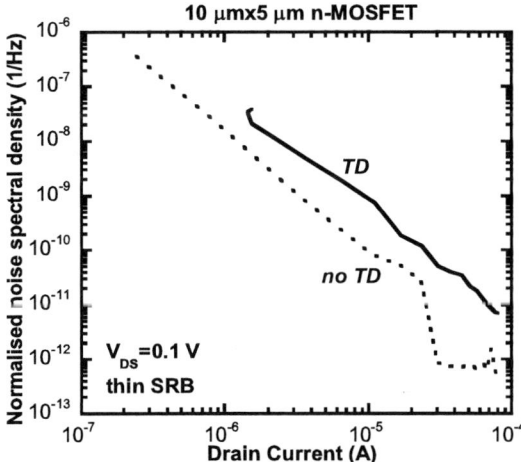

Fig. 8. Normalized current noise spectral density versus drain current for a 10 μm x 5 μm n-MOSFET with and without a threading dislocations.

For SSi devices without TDs in the active region and in case of a control of the Ge outdiffusion, it has been observed that the strain has a beneficial impact on the noise performance [28]. The origin of the noise improvement can be related to i) a better quality, from a viewpoint of surface roughness and/or grown-in crystal defects, of the epitaxially grown Si layer compared to the Czochralski substrate [29], or ii) a reduction of the bulk oxide defects due to the strain. It has been shown by EPR studies that SiO$_2$ oxides grown under tensile stress have a lower density of dangling-bond related P_b-centers [30].

An important observation is that there is not only a reduction of the LF noise for SSi n-channel transistors but there exists also a correlation between the noise and the carrier mobility, as shown in Fig. 9 [31]. Compared to the Si reference, SSi devices have a lower input-referred voltage noise and a higher low-field mobility.

The authors have recently reviewed the LF noise in n-MOSFETs on SSi [32]. It was demonstrated that for thick SRB devices the LF noise spectra may degrade due to the presence of a generation-recombination (GR) component, caused by processing-induced local strain relaxation resulting in defect-assisted noise generation.

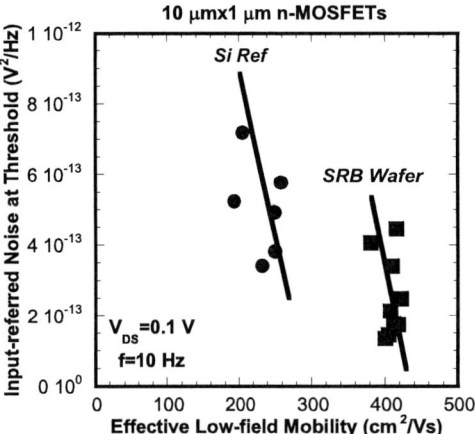

Fig. 9. Empirical correlation between the input-referred voltage noise spectral density at V_T and f=10 Hz and the low-field mobility for a set of 10 μm x 1 μm n-MOSFETs fabricated on standard silicon (Si Ref) or a SSi wafer on a thin SRB.

V. PROCESSING-INDUCED STRESSORS

As already mentioned in the introduction, there is a variety of stressors that can be used to achieve local strain. The choice will not only depend on the process simplicity, the economical aspects, manufacturing issues and the reliability, but also whether a tensile or a compressive stress is required. For stress parallel to the current flow, the first enhances the electron mobility while the second one is needed for boosting up the hole mobility [33]. A different behavior has been observed for stress applied perpendicular to the current flow [34]. There are also differences between uniaxial and biaxial stress effects [35] and the stress behavior is different in long- and short channel devices [36].

The interest in using local strain engineering is very high and the number of publications is increasing rapidly. This is illustrated by some recently reported approaches including the use of tensile SiC source/drain regions instead of SiGe, giving the additional benefit that there is no Ge outdiffusion [37], epitaxial techniques so that n-channels are in the <110> direction and p-channels in the <100> one for optimal drive current [38], local Ge condensation to avoid a recess etch for source/drain regions [39], a buried strained SiGe layer in the channel region in combination with shallow trench isolation [40]. An interesting feature to mention is that there also exists a stress memorization effect, whereby one can continue to rely on the stress after its source has been removed such as e.g. in the case of disposable stress liners [41].

The contact etch stop layers (CESL) used in advanced processes also give an interesting approach for strain engineering. As illustrated in Fig. 10 for triple gate

FinFET technology, the use of a tensile SiN CESL resulted in a 20% improvement for n-MOS while for p-MOS the increase in I_{on} was 10% for a compressively strained SiN CESL [42]. The observed improvement in mobility is depending on the gate length and will for shorter lengths saturate or even decrease. For stress simulations of these devices one has to take into account that FinFETs are 3D structures whereby both the top and the sidewalls contribute to the current. As mentioned before, the mobility is orientation dependent.

the channel) strongly depends on the etch depth as shown in Fig. 12 [44].

It is of course possible to implement more than one stressor in a process flow. The overall stress, impacting the electrical performance will then be the sum of the different components. Recent investigations were aimed at studying the combined impact of different stressors on the low frequency noise behavior [45]. As illustrated in Fig. 13, the noise is higher for devices with a SiGe stressor than with Si_3N_4 liner stress. In case that both stressors are used only the impact of the SiGe stressor is observed.

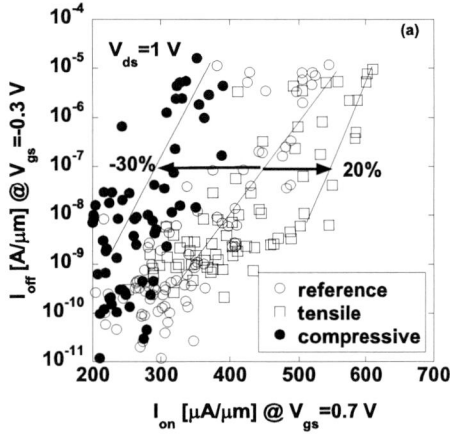

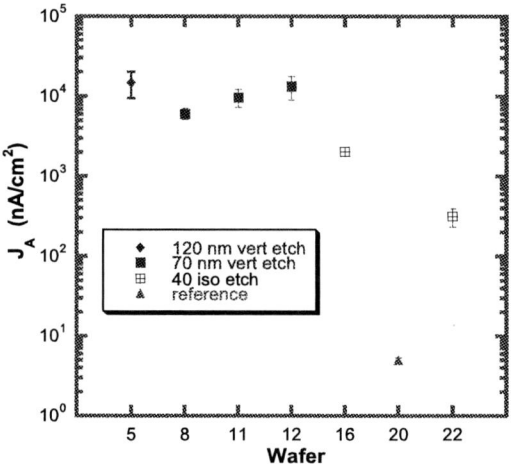

Fig. 11. Area leakage current density at -1 V for SiGe junctions with different etch characteristic (depth and profile).

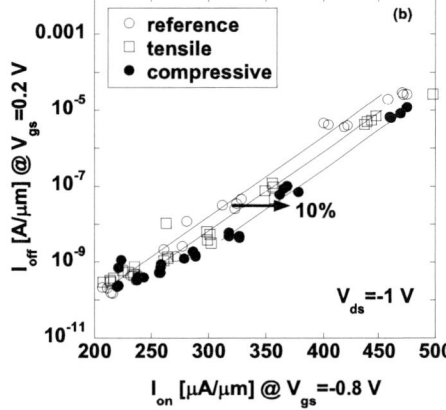

Fig. 10. I_{on} –I_{off} behavior of (a) nMOS devices and (b) pMOS devices; W = 35 nm; the strained layers obtained by SiN CESL have an intrinsic stress of 800 Mpa [42].

It is obvious that for each of the possible stressors a detailed evaluation of its impact on the electrical performance has to be done in combination with an in-depth defect characterization. A systematic study of the leakage current in recessed SiGe junctions pointed out that both the area and the perimeter leakage current density are strongly depending on the etching characteristics (depth and profile) as shown in Fig. 11 for the area component [43]. The results can be explained based on the dislocations present at the SiGe interface. Simulations point out that the parallel stress component (i.e. parallel to the current flow in

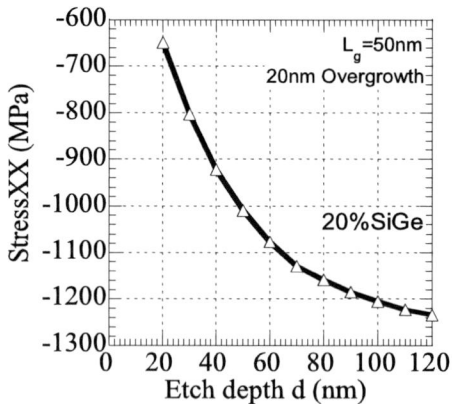

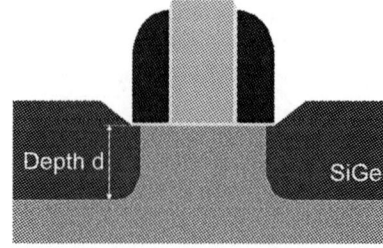

Fig. 12. Simulation of the parallel stress versus the etch depth

72

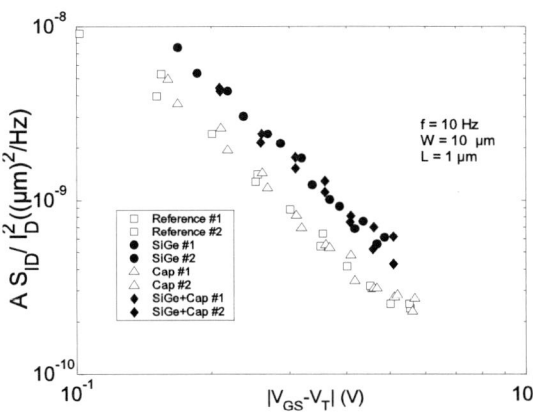

Fig. 13. Normalized drain current spectral density versus gate voltage overdrive for p-MOSFETs with different stressor studied at f=10 Hz and V_{DS}=-0.05 V: unstressed reference (□);15% SiGe S/D (●); cap layer (△) and 25% SiGe S/D+cap layer (♦). For each wafer, data for two 10 μmx1 μm devices are displayed.

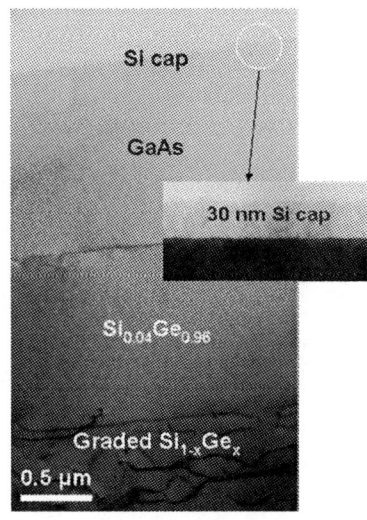

Fig. 14. TEM image of a GaAs on Si structure with a graded SiGe buffer to reduced the threading dislocations and a top Si cap [46].

VI. ALTERNATIVE SUBSTRATES

For sub 32 nm technology nodes there is a strong interest in looking at alternative substrates, such as Ge, Ge-on-insulator (GeOI), strained Ge-on-insulator (SGeOI), III-V hetero-epitaxy on local Ge on Si, and other relaxed III-V materials using SiGe and InGaAs graded buffers. Figure 14 illustrates a GaAs on Si structure, in which a SiGe graded layer was used to lower the TD density. On top a Si cap is deposited [46]. These types of strain engineering approaches allow the on-chip integration of advanced microelectronic and photonic components and circuits [47]

Defect engineering is based on a good fundamental insight into defect and dislocation theory. The strain energy can be reduced by the dislocation formation energy, which will only occur above a critical film thickness and at higher temperatures. Structures can be designed to have dislocations only nucleated in pre-defined areas. However, it should be taken into account that dislocations also act as regions of enhanced diffusion or so-called diffusion pipes which locally enhance the doping concentration. The dislocation engineering is also influenced by the wafer orientation [46].

Recently, there has been a renewed interest in using bulk Ge and GeOI substrates. For Ge crystals, both the grown-in and the processing-induced defects have to be taken into account [48]. During crystal pulling 30, 60 and 90° dislocations may be introduced, which can act as sinks for vacancies and the formation of V_2-H complexes. The latter will degrade the electrical device performance. The role of intrinsic point defects, void formation and the behavior of oxygen in Ge has recently been reviewed [49]. Notwithstanding strong similarities with Si, more effort is needed for a good understanding and effective defect engineering.

From a processing viewpoint, the main obstacles to overcome are the deposition of high-k dielectrics, the control of implantation processes (residual implantation damage, dopant activation), surface passivation issues, and the reduction of the contact resistance by using Ge-silicides.

An important issue is the formation of shallow junctions, whereby especially the diffusion control of the n-type dopants (P, As) is problematic [50-52]. As illustrated in Fig. 15, the lowest sheet resistance (Rs) values can be obtained for a $5 \cdot 10^{15}$ at/cm² P implantation, activated by a 600°C 1s Rapid Thermal Annealing [51]. The corresponding junction depth is around 100 nm. Extrapolating this value to the ITRS specification for the 22 nm extensions, this is still one order of magnitude too high. The use of even higher implantation doses is prohibited by the occurrence of precipitate formation, as illustrated by Fig. 16 [53]. This means that a defect engineering approach is required to control the diffusion and precipitation of ion-implanted P.

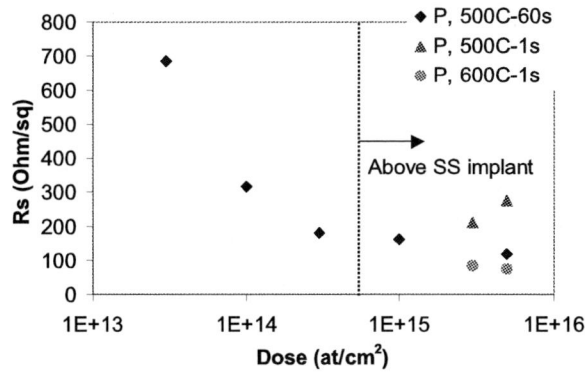

Fig. 15. Sheet resistance versus implantation dose for 15 keV P implanted p-type Ge. No SiO₂ cap was used [51].

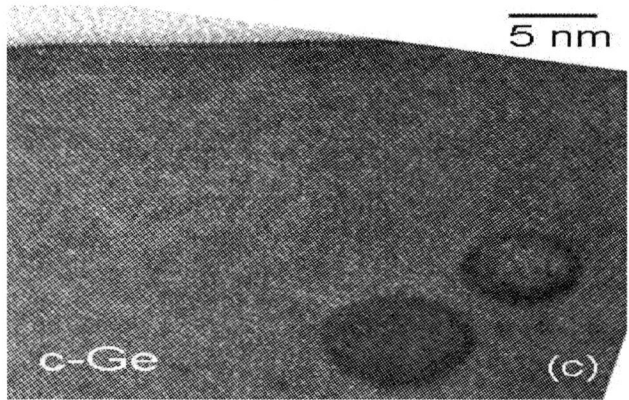

Fig. 16. TEM cross-section micrographs of Ge crystal as-implanted with 15 keV of $5 \cdot 10^{15}$ atoms/cm^2 P, showing P precipitates located within the first 30 nm from the Ge surface, after annealing at 50°C for 60 s.

To overcome the leakage problem associated with the lower bandgap, there is a good potential for GeOI substrates, which can be fabricated by a variety of techniques such as smart-cut, epitaxial growth, Ge condensation approaches, etc. In case of the smart-cut process, which is based on hydrogen implantation, the hydrogen can form extended defects and nanobubbles [54]. A high-resolution TEM micrograph of a {311} defect in Ge is shown in Fig. 17. The larger atomic mass and the better contrast make it easier to reveal these defects in Ge compared to Si.

Although good device performances of fully silicided NiSi/Hf-LaAlO$_3$ devices on GeOI smart-cut substrates have been reported [55], much research is still needed before Ge-based technologies are introduced in the market. Another promising approach could be the use of a Si top layer on a strained Ge/relaxed Si$_{0.3}$Ge$_{0.7}$ in order to fabricate buried channel strained Ge MOSFETs [56].

Figure 18 points out that for TiN-TaN/HfO$_2$/GeOI n-channels the low frequency noise is higher than for their Si counterparts [57]. The origin of the higher noise is related to the poorer quality of the high-k dielectric on Ge than on Si.

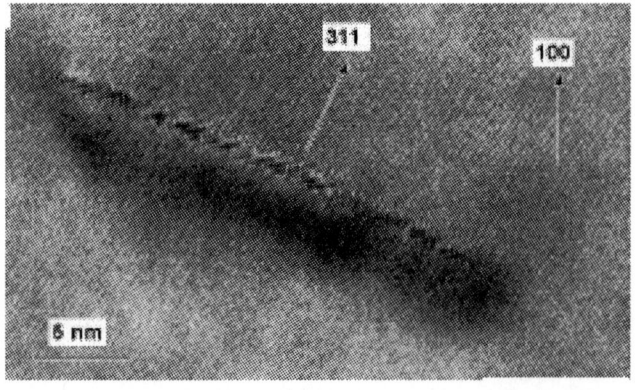

Fig. 17. High resolution TEM image of a {311} defect in Ge [54].

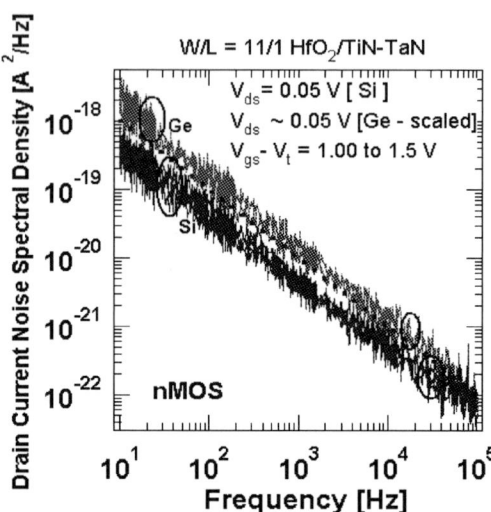

Fig. 18. Comparison of LF noise spectra at $|V_{ds}|$=0.05 V and $|V_{gs}-V_t|$ = 1.0 to 1.5 V for HfO$_2$/TiN-TaN n-MOSFETs on GeOI and Si substrates.

III. CONCLUSION

The importance of defect engineering and stress control has been reviewed with emphasis on the electrical activity of the defects and their impact on the device performance. Attention was given to the leakage current, the carrier lifetime and the low frequency noise. Although the main focus was on strained Si on SRB layers, other approaches were also briefly reviewed, including the potential use of Ge and GeOI. Much efforts are still needed in order to come to high-performance high-yield commercial technologies for future technology nodes.

ACKNOWLEDGEMENT

The authors also want to thank M. Bargallo, F. Crupi, M. Caymax, R. Delhougne, G. Giusi, R. Loo, R. Rooyackers, A. Satta, P. Srinivasan, J. Vanhellemont, P. Verheyen, for stimulating discussions and the use of co-authored results

REFERENCES

[1] G.D. Wilk, R.M. Wallace and J.M. Anthony, "High-k gate dielectrics: Current status and materials properties considerations", *J. Appl. Phys.*, vol. 89, pp. 5243-5275, 2001.

[2] M. Fischetti, D. A. Neumayer and E.A. Cartier, "Effective electron mobility in Si inversion layers in metal-oxide-semiconductor systems with a high-k insulator: the role of remote phonon scattering", *J. Appl. Phys.*, vol. 90, pp. 4587-4608, 2001.

[3] R. Chau, S. Datta, M. Doczy, B.J. Doyle, K. Kavalieros and M. Metz, "High-κ/metal-gate stack and its MOSFET characteristics", in *IEEE Electron Dev. Lett.*, vol. 25, pp. 408-410, 2004.

[4] C.-H. Han *et al.*, "A 65 nm ultra low power logic platform technology using uni-axial strained silicon strained transistors", in *IEDM Techn. Dig.*, pp. 65-69, 2005.

[5] B.M. Haugerud, L.A. Bosworth and R.E. Belford, "Elevated-temperature electrical characteristics of mechanically strained Si devices", *J. Appl. Phys.*, vol. 95, pp. 2792-2796, 2004.

[6] M.V. Fischetti, Z. Ren, P.M. Solomon, M. Yang and K. Rim, "Six-band k.p calculation of the hole mobility in silicon inversion layers: Dependence on surface orientation, strain, and silicon thickness", *J. Appl. Phys.*, vol. 94, pp. 1079-1091., 2003.

[7] T. Ghani, M. Armstrong, C. Auth, M.D. Giles, K. Mistry, A. Murthy, L. Shifren, S. Thompson and M. Bohr, 'Uniaxial strained silicon CMOS devices for high performance logic nanotechnology", in *Proc. SiGe: Materials, Processing and Devices*, D. Harame *et al.*, Eds., The Electrochem. Soc. PV 2004-07, Pennington, NJ, pp. 681-692, 2004.

[8] D.K. Sadana, S.W. Bedell, A. Reznicek, J.P. de Souza, K.E. Fogel and H.J. Hovel, "Strain engineering for silicon CMOS technology", in *Proc. ULSI Process Integration IV*, C. Claeys, F. Gonzalez, S. Zaima, D.A. Buchanan and J.O. Borland, Eds., The Electrochem. Soc. PV 2005-06, Pennington, NJ, pp. 360-382, 2005.

[9] M.L. Lee, E.A. Fitzgerald, M.T. Bulsara, M.T. Currie and A. Lochtefeld, "Strained Si, SiGe, and Ge channels for high-mobility metal-oxide-semiconductor field-effect transistors", *J. Appl. Phys.*, vol. 97, pp. 011101-1/27, 2005.

[10] R. Delhougne, G. Eneman, M. Caymax, R. Loo. P. Meunier-Beillard, P. Verheyen, W. Vandervorst and K. De Meyer, "Selective epitaxial deposition of strained silicon: a simple and effective method for fabricating high performance MOSFET devices", *Solid-State Electron.*, vol. 48, pp. 1307-1316, 2004.

[11] G. Eneman, P. Verheyen, R. Rooyackers, R. Delhougne, R. Loo, M. Caymax, P. Menier-Beillard, K. De Meyer and W. Vandervorst, "Fabrication of strained Si nMOSFET transistors on thin buffer layers with selective and non-selective epitaxial growth technique", *Mat. Sci. Semicond. Proc.*, vol. 8, pp. 337-341, 2005.

[12] J.H. Van der Merwe, "Crystal Interfaces. Part I. Semi-infinite crystals", *J. Appl. Phys.*, vol. 34, pp. 117-122, 1963.

[13] R. Loo, R. Delhougne, M. Caymax and M. Ries, "Formation of misfit dislocations at the thin strained Si/strain relaxed buffer interface", *App. Phys. Lett.*, vol. 87, pp. 182108-1/3, 2005

[14] X.L. Yuan, T. Sekiguchi, S.G. Ri and S. Ito, "Detection of misfit dislocations at the interface of strained $Si/Si_{08}Ge_{0.2}$ by electron-beam induced current technique", *Appl. Phys. Lett.*, vol. 84, pp. 3316-3318, 2004.

[15] M.S. Rasras, I. De Wolf, G. Groeseneken and H. Maes, "Spectroscopic identification of light emitted from defects in silicon devices", *J. Appl. Phys.*, vol. 89, pp. 249-258, 2001.

[16] G. Eneman, E. Simoen, R. Delhougne, E. Gaubas, V. Simons, P. Roussel, P. Verheyen, A. Lauwers, R. Loo, W. Vandervorst, K. De Meyer and C. Claeys, "Defect analysis of strained silicon on thin strain-relaxed buffer layers for high mobility transistors", *J. Phys. Condens. Matter*, vol. 17, pp. S2197-S2210, 2005.

[17] E. Simoen, G. Eneman, C. Claeys, M. Scholz, R. Loo, P. Verheyen and K. De Meyer, "Defect engineering considerations for strained silicon substrates", in *Proc. Semiconductor Silicon 2006*, H. Huff, H. Iwai and H. Richter, Eds., The Electrochem. Soc. Trans., in press, 2006.

[18] W. Seifert, M. Kittler, J. Vanhellemont, E. Simoen, C. Claeys and F.G. Kirscht, "Recombination activity of oxygen precipitation related defects in Si", *Inst. Phys. Conf. Ser.*, vol. 149, pp. 319-324, 1996.

[19] E. Escobedo-Cousin, S. H. Olsen, S.J. Bull, A.G. O'Neill, H. Coulson, C. Claeys, R. Loo, R. Delhougne and M. Caymax, "Study of surface roughness and dislocation generation in strained Si layers grown on thin strain-relaxed buffers for high performance MOSFETs", to be presented at ISTDM, Princeton, USA, May 2006.

[20] G. Eneman, E. Simoen, R. Delhougne, P. Verheyen, R. Loo, M. Caymax, C. Claeys, W. Vandervorst and K. De Meyer, "Analysis of the leakage current origin in thin stain relaxed buffer substrates", *J. Electrochem. Soc.*, vol. 153, in press, 2006.

[21] L.M. Giovani, H.S. Luna, A.M. Agarwal and L.C. Kimerling, "Correlation between leakage current density and threading dislocation in SiGe p-i-n diodes grown on relaxed graded buffer layers", *Appl. Phys. Lett.*, vol. 78, pp. 541-543, 2001.

[22] G. Eneman, E. Simoen, R. Delhougne, P. Verheyen, R. Loo and K. de Meyer, "Influence of dislocations in strained Si/relaxed SiGe layers on n+/p junctions in a metal-semiconductor field –effect transistor technology", *Appl. Phys. Lett..*, vol. 87, pp. 192112-1/3, 2005.

[23] E. Gaubas, R. Tomasiunas, G. Eneman, R. Delhougne and E. Simoen, "Study of recombination and transport characteristics in strain-relaxed Si-SiGe layers", *Semiconduct. Sci. and Techn.*, vol. 20, pp.1052-1063, 2005.

[24] E. Simoen, G. Eneman, S. Shamuilia, V. Simons, E. Gaubas, R. Delhougne, R. Loo, K. De Meyer and C. Claeys, "On the electrical activity of misfit and threading dislocations in p-n junctions fabricated in thin strain-relaxed buffer layers", *Solid State Phenomena*, vol. 118-119, pp. 285-290, 2005.

[25] C. Claeys, E. Simoen and A. Mercha, "Low-frequency noise assessment for deep submicrometer CMOS technology nodes", *J. Electrochem. Soc.*, vol. 1571, pp. G307-G318, 2004.

[26] K.K. Hung, P.K. Ko, C. Hu and Y.C. Cheng, "A unified model for the flicker noise in metal-oxide-semiconductor field-effect transistors", *IEEE Trans. Electron Dev.*, vol. 37, pp. 654-665, 1990.

[27] W.C. Hua, M.H. Lee, P.S. Chen, S. Maikap, C.W. Liu and K.M. Chen, "Ge outdiffusion effect on flicker noise in strained Si nMOSFETs", *IEEE Electron Dev. Lett.*, vol. 25, pp. 693-695, 2004.

[28] E. Simoen, G. Eneman, P. Verheyen, R. Delhougne, R. Loo, K. De Meyer and C. Claeys, "On the beneficial impact of tensile-strained silicon substrates on the low-frequency noise of n-channel metal-oxide-semiconductor transistors", *Appl. Phys. Lett.*, vol. 86, pp. 223509-1/3, 2005.

[29] H.S. Momose, "Si channel surface dependence of electrical characteristics in ultra-thin gate oxide CMOS", in *Proc. ULSI Process Integration III*, C. Claeys, F. Gonzalez, J. Murota, P. Fazan and P.R. Singh, Eds., The Electrochem. Soc. Ser., PV 2003-06, pp. 361-374, 2003.

[30] A. Stesmans, D. Pierreux, R.J. Jaccodine, M.-T. Lin and T.J. Delph, "Influence of in situ applied stress during thermal oxidation of (111)Si on P_b interface defects", *Appl. Phys. Lett.*, vol. 82, pp. 3038-3040, 2003.

[31] E. Simoen, G. Eneman, P. Verheyen, R. Delhougne, R. Rooyackers, R. Loo, W. Vandervorst, K. De Meyer and C. Claeys, "The low-frequency noise of strained silicon n-

MOSFETs", in *Proc. 18th ICNF*, T. Gonzalez, J. Mateos and D. Pardo, Eds., Am. Inst. Phys., pp. 187-190, 2005.

[32] E. Simoen, G. Eneman, P. Verheyen, R. Loo, K. De Meyer and C. Claeys, "Processing aspects in the low-frequency noise of n-MOSFETs on strained silicon substrates", *IEEE Trans. Electron Dev.*, vol. 53, in press, May 2006.

[33] S.E. Thompson *et al.*, "A logic nanotechnology featuring strained silicon", *IEEE Electron Dev. Lett.*, vol. 25, pp. 191-193, 2004.

[34] W. Zhan, A. Sebaugh, V. Adams, D. Jovanovic and B. Winstead, "Opposing Dependence of the electron and hole gate current in SOI MOSFETs under uniaxial strain", *IEEE, Electron Dev. Lett.*, vol. 26, pp. 410-412, 2005.

[35] S.E. Thompson, G. Sun, K. Wu, J. Lim and T. Nishida, "Key difference for process-induced uniaxial versus substrate-induced biaxial stressed Si and Ge channel MOSFETs", in *IEDM Tech. Dig.*, pp. 221-224, 2004.

[36] G. Eneman *et al.*, "Layout impact on the performance of a locally strained p-MOSFET", in *Symp. VLSI Tech. Dig.*, pp. 22-23, 2005.

[37] K.-W. Ang *et al.*, "Thin body silicon-on-insulator n-MOSFET with silicon-carbon source/drain regions for performance enhancement", in *IEDM Tech. Dig.*, pp. 503-506, 2005.

[38] T. Sanuki *et al.*, "Uniaxial–biaxial stress hybridization for super-critical strained-Si directly on insulator (SC-SOI) PMOS with different channel orientations", in *IEDM Tech. Dig.*, pp. 515-518, 2005.

[39] K.-J. Chui *et al.*, "Source/drain germanium condensation for p-channel strained ultra-thin body transistors", in *IEDM Tech. Dig.*, pp. 499-504, 2005.

[40] D. Chanemougame *et al.*, "Performance boost of scaled Si pMOS through novel SiGe stressor", in *Symp. VLSI Tech. Dig.*, pp. 180-181, 2005.

[41] C.-H. Chen *et al.*, "Stress memorization technique (SMT) by selectively strained-nitride capping for sub 65 nm high performance strained-Si device applications", in *Symp. VLSI Tech. Dig.*, pp. 56-59, 2004.

[42] N. Collaert *et al.*, "Performance improvement of tall triple gate devices with strained SiN layers", *IEEE Electron Dev. Lett.*, vol. 26, pp. 820-022, 2005.

[43] M.B. Gonzalez, G. Eneman, P. Verheyen, C. Claeys, H. Bender, A. Benedetti, K. De Meyer, E. Simoen, R. Schreutelkamp, F. Nouri and L. Washington, "Impact of etching depth on the leakage current of recessed SiGe junctions", to be presented at ISTDM, Princeton, USA, May 2006.

[44] G. Eneman, unpublished.

[45] G. Giusi, E. Simoen, G. Eneman, P. Verheyen, F. Crupi, K. De Meyer, C. Claeys and C. Ciofi, "The low-frequency (1/f) noise behavior of locally stressed HfO$_2$/TiN gate stack p-MOSFETs", submitted to *IEEE Electron Dev. Lett.*

[46] E.A. Fitzgerald, M.L. Lee, B. Yu, K.E. Lee, C.L. Dohrman, D. Isaacson, T.A. Langdo and D.A. Antoniadis, "Dislocation engineering in strained MOS materials", in *IEDM Tech. Dig.*, pp. 519-522, 2005.

[47] A. Chin, H.L. Kao, D.S. Yu, C.C. Liu, C. Zhu, M.-F. Li, S. Zhu and D.-L. Kwong, "High performance metal-gate/high

–k MOSFETs and GaAs compatible RF passive devices on Ge-on-insulator technology", in *Proc. 7th Intern. Conf. On Solid-State and Integrated Circuits Techn. – ICSICT 2004*, R. Huang, M. Yu, J. Liou, T. Hiramoto and C. Claeys, Eds., pp. 302-305, 2004.

[48] J. Vanhellemont, O. De Gryse, S. Hens, P. Vanmeerbeek, D. Poelman, P. Clauws, E. Simoen, C. Claeys, I. Romandic, A. Theuwis, G. Raskin, H. Vercammen and P. Mijlemans, "Grown-in defects and diffusion in Czochralski-grown germanium" in *Defects and Diffusion in Semiconductors-An annual retrospective VII*, D.J. Fisher, Ed., Defect an Diffusion For., Trans. Tech. Publ. Inc., vol. 130-132, pp. 149-176, 2004.

[49] J. Vanhellemont, S. Hens, J. Lauwaert, O. De Gryse, P. Vanmeerbeek, D. Poelman, P. Spiewak, I. Romandic, A. Theuwis and P. Clauws, "Recent progress in understanding of lattice defects in Czochralski grown germanium: catching-up with silicon", *Solid-State Phenomena*, vol. 108-109, pp. 683-690, 2005.

[50] C. Claeys and E. Simoen, Eds., *"Germanium Based Technologies: From Materials to Devices"*, Ch. 3, to be published by Elsevier, Sept. 2006.

[51] E. Simoen, A. Satta, M. Meuris, T. Janssens, T. Clarysse, C. Demeurisse, B. Brijs, I. Hoflijk, W. Vandervorst and C. Claeys, "Defect removal, dopant diffusion and activation issues in ion-implanted shallow junctions in crystalline germanium substrates", *Solid State Phenomena*, vol. 108-109, pp. 691-696, 2005.

[52] A. Satta, E. Simoen, T. Janssens, T. Clarysse, B. De Jaeger, A. Benedetti, I. Hoflijk, B. Brijs, M. Meuris and W. Vandervorst, "Shallow junction ion implantation in Ge and associated defect control", *J. Electrochem. Soc.*, vol. 153, in press, 2006.

[53] A. Satta, E. Simoen, T. Janssens, T. Clarysse, R. Duffy, W. Vandervorst, "Diffusion and activation behavior of high dose P implants in Ge", submitted for publication in *Appl. Phys. Lett.*

[54] T. Akatsu, K.K. Bourdelle, C. Richtarch, B. Faure and F. Letertre, "Study of extended defect formation in Ge and Si after H ion implantation", *Appl. Phys. Lett.*, vol. 86, pp. 181910-1/3, 2005.

[55] D.S. Yu, K.C. Chiang, C.F. Cheng, A. Chin, Z. Zhu, M.F. Li and D.K. Kwong, "Fully silicided NiSi:Hf-LaAlO$_3$/SG-GOI n-MOSFET with high electron mobility", *IEEE Electron Dev. Lett.*, vol. 25, pp. 559-561, 2004.

[56] H. Shang, E. Gousev, M. Gribelyuk, J.O. Chu, P.M. Mooney, X. Wang, K.W. Guarini and M. Ieong, "Fabrication, device design and mobility enhancement of germanium channel MOSFETs", in *Proc. 7th Intern. Conf. On Solid-State and Integrated Circuits Techn. – ICSICT 2004*, R. Huang, M. Yu, J. Liou, T. Hiramoto and C. Claeys, Eds., pp. 306-309, 2004.

[57] P. Srinivasan, E. Simoen, D. Misra and C. Claeys, to be presented at the European MRS meeting, Symposium on Ge-based semiconductors… from materials to devices, Nice, France, June 2006.

2006 25th International Conference on Microelectronics

Reviews and Prospects of Low-Voltage Nano-Scale Embedded RAMs

Kiyoo Itoh

Abstract–Low-voltage nano-scale embedded RAMs are described, focusing on RAM cells and peripheral circuits. First, challenges and trends of low-voltage RAM cells are discussed in terms of signal charge, signal voltage, and noise. ECC to cope with the ever-increasing soft-error rate, power-supply controls to widen the voltage margin of cells, and a fully-depleted SOI to reduce V_T-variation are also investigated. Then peripheral circuits are explained in terms of leakage reduction and compensation for speed variations. Based on this, it is concluded that low-voltage RAMs cannot be achieved without reducing speed variations caused by variations in V_T, thus resulting in a further need for compensation circuits and new devices with reduced V_T variation.

I. INTRODUCTION

Low-voltage nano-scale embedded (e-) RAMs are becoming increasingly important because they play critical roles in reducing power dissipation and chip size of MPUs/MCUs/SoCs. Thus, sub-1-V RAMs have been actively researched and developed [1-4], including a 0.6-V 16-Mb e-DRAM [5], a 1.2- to 1-V 16-Mb e-DRAM [6, 7], and a 24-MB SRAM cache in a 0.8- to 1-V MPU [8]. To create such e-RAMs, however, many challenges [1-4] remain with RAM cells and peripheral circuits. In addition to being the smallest cells possible, they are high signal-to-noise-ratio (S/N) designs for stable and reliable RAM cells, and reductions in the ever increasing leakage and speed variation of RAM cells and/or peripheral circuits as devices and V_{DD} are scaled down.

This paper describes circuit and device designs for low-voltage e-RAMs using the one-transistor one-capacitor (1-T) DRAM cell and the six-transistor (6-T) SRAM cell. First, a high S/N design for RAM cells is discussed in terms of signal charge, signal voltage, and noise. Then peripheral circuits are investigated in terms of leakage and speed variation. Finally, future prospects are given. In this paper, leakage currents denote subthreshold currents.

II. RAM CELLS

A. Signal Charge

The ever decreasing signal charge (Q_S) of non-selected cells (Fig. 1) [1-4] restricts low-voltage operations with increased soft-error rate (SER), because Q_S is almost equal to the soft-error critical charge. The Q_S of DRAM cells is

Kiyoo Itoh is with Central Research Laboratory, Hitachi, Ltd., Kokubunji, Tokyo 185-8601, Japan,
E-mail: k-itoh@crl.hitachi.co.jp

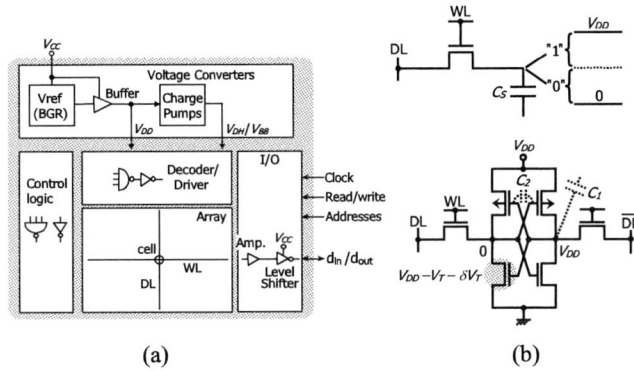

(a)   (b)

Fig. 1. RAM chip (a) and RAM cells (b) with DRAM cell for the upper and SRAM cell for the lower. δV_T: V_T-mismatch between paired MOSTs.

given by $C_S V_{DD}/2$, while that of SRAM cells is approximately given by $(C_1 + 2C_2) V_{DD}$ [9]. Here, C_S is the cell-node capacitance of DRAM cells, and C_1 and C_2 are the parasitic cell-node capacitances of SRAM cells. Due to ($C_1 + 2C_2$) « $C_S/2$, the SER of SRAM cells is always much larger than that of DRAM cells, and it rapidly increases with device scaling [2] because of the rapid decrease in C_1 and C_2, although spatial scaling reduces the collected charge. In contrast, the SER of 1-T DRAM cells gradually decreases because C_S needs to be gradually decreased to maintain a large signal voltage. One remedy for SRAM cells is to add a large capacitance to ($C_1 + 2C_2$) [10], even though this requires more complicated processes and a longer write cycle. A triple-well structure shielding the cell array, as a soft-error barrier, is also effective in reducing the SER of RAM cells. The most effective way is to use ECC, as will be explained later.

B. Signal Voltage of DRAM Cells

The small signal voltage and the ever slower sense-amplifier operation of a half-V_{DD} sensing prevent a low-voltage operation. The floating signal voltage v_S developed on the data line (DL) with capacitance C_D is given by $v_S \cong (V_{DD}/2) C_S/C_D$ for C_D » C_S. It is successfully sensed if $v_S > \delta V_T + v_N$, where δV_T is the offset voltage (i.e., V_T-mismatch between paired MOSTs) of sense amps, and v_N is the noise at sensing. To lower V_{DD} with a C_S small enough to accept a planar capacitor, which is a key to e-DRAMs, reducing C_D, δV_T, and v_N is essential. The C_D is reduced adequately enough by having a short DL, as exemplified by a 32-cell-

1-4244-0116-X/06/$20.00 ©2006 IEEE   77

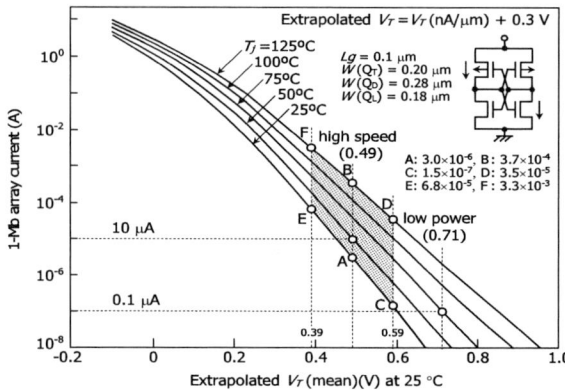

Fig. 2. 1-Mb array current vs. V_T of cross-coupled MOSTs [3].

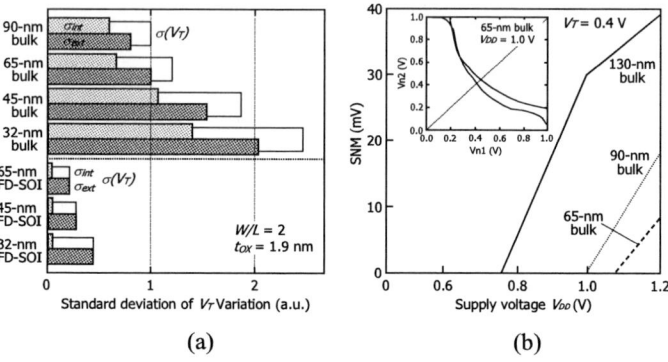

(a)   (b)

Fig. 3. Standard deviations of V_T variation, $\sigma(V_T)$, and intrinsic (σ_{int}) and extrinsic (σ_{ext}) V_T variations (a) [3, 11, 20, 21], and SNM of the 6-T bulk-CMOS SRAM cell taking 6σ of V_T variation into consideration (b) [11].

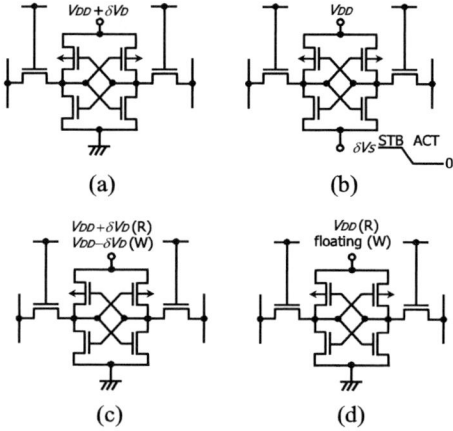

Fig. 4. Power controls of SRAM cells [4].

connected DL and 5-fF C_S for a 1.2-V 322-MHz 16-Mb e-DRAM [6]. The δV_T can be reduced considerably if the largest MOST possible is used for sense amps despite an area penalty. Reducing v_N is extremely important not only for successful sensing, but also for fast sensing despite a half-V_{DD} data-line precharge (i.e., mid-point sensing). The mid-point sensing always involves the slowest speed in a chip despite distinctive features of a quiet array and halved data-line charging power [1]. This is because the gate-over-drive of turned-on MOST in cross-coupled MOSTs in sense amps is the lowest in a chip, which is given by $V_{DD}/2 - (V_T + v_N)$. Thus, to lower V_{DD}, the value of $(V_T + v_N)$ must be minimized for fast sensing, which calls for reducing v_N and finding the lowest V_T possible while reducing the subthreshold currents involved. The noise v_N consists of many components [1]: an inaccurate $V_{DD}/2$-level setting caused by a DC level fluctuation of the $V_{DD}/2$ generator and a capacitive coupling to the DL from the precharge circuit and equalizer; a capacitive imbalance between a pair of DLs; the word line to DL coupling; and the adjacent DL coupling. Recently, an accurate $V_{DD}/2$-level setting with fuse trimming, a differential driving of the precharger and equalizer, and DL transpositions without area penalty have been reported [6, 7]. The V_T of the sense-amp MOST was reportedly lowered to 0.2 V, coupled with power switches [5], enabling 0.6-V V_{DD} sensing.

C. Signal Voltage of SRAM Cells

A large necessary V_T and a large V_T variation of cross-coupled MOSTs in the 6-T SRAM cell are major obstacles to low-voltage operation. The V_T of cross-coupled MOSTs must be quite high to reduce the leakage that rapidly increases as V_T decreases, as shown in Fig. 2 [3]. Here, the V_T is the average of the cross-coupled MOSTs in a chip, because it is the average V_T that determines chip leakage. For example, for a low-power 1-Mb e-SRAM that allows a leakage of 0.1 μA at T_{jmax} = 75°C, the V_T at 25°C might be higher than 0.71 V. For a high-speed 1-Mb e-SRAM that can tolerate a leakage of 10 μA at T_{jmax} = 50°C, the V_T can be as low as 0.49 V. In addition to such high V_Ts, the intra-die or inter-die V_T variation (ΔV_T) that increases as devices

get smaller (Fig. 3(a)) [3, 4, 11] reduces the signal voltage on the data line because of a reduction in the drive current of on-MOST in cross-coupled MOSTs, which is proportional to the gate-over-drive $(V_{DD} - V_T - \delta V_T)$. Here, the V_T is also the average V_T in a chip, and $\delta V_T (= 2\sqrt{\Delta V_T})$ is again the V_T-mismatch of cross-coupled n-MOSTs. Hence, even for a fixed $V_{DD} - V_T$, each cell can have a different drive current, depending on its δV_T. For $V_T = 0.71$ V and $\delta V_T = 0.1$ V, the minimum V_{DD} (V_{DDmin}) for a successful operation is as high as 0.81 V. In practice, the V_{DDmin} must be higher than this value to suppress the variation of access time of cells that is prominent at around $V_{DD} \cong V_T + \delta V_T$ [12]. The continually increasing V_T variation also degrades the static noise margin (SNM): The V_{DDmin}, defined as the V_{DD} for an SNM of 0, becomes higher with device scaling, as shown in Fig. 3(b) [11].

To solve the problems many power-supply controls for the cells (Fig. 4[4]) have been proposed. Although they are effective only for high V_{DD} over 1 V, or they prevent MOSTs from being scaled down, they nevertheless reduce leakage, widen the voltage margin, or compensate for the V_T variation. Type (a) features the raised supply ($V_{DD} = V_{DD} + \delta V_D$) and dual-$V_T$ scheme [3, 13, 14]. The raised supply

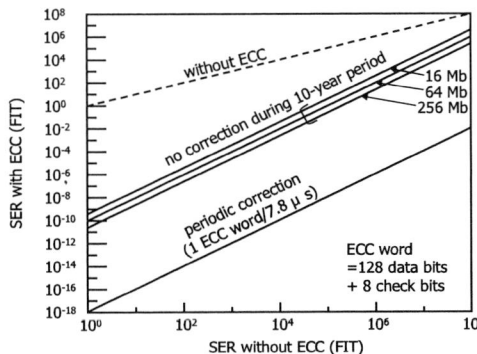

Fig. 5. SER reduction through on-chip ECC.

offsets the high V_T and the δV_T of cross-coupled MOSTs, though MOSTs are unscalable due to their need for a high stress-voltage. Moreover, the well known negative word line scheme [1-4] applied to low-V_T transfer MOSTs cuts leakage during non-selected periods, while increasing the cell read current. Type (b) features the source offset driving [15]. In the active to standby mode transition, it lowers the data-line voltage from 1.5 V to 1 V to relax the electric field of all MOSTs, and raises the ground line to 0.5 V to increase the V_T of off-MOST. However, reducing the supply voltage by δV_S in the standby mode restricts the low-voltage operation with increased SER. Types (c) and (d) feature the dynamic control of cell-power supply. Type (c) switches the power supply [16] to a lower level in the write mode and a higher level in the read mode to widen the write and read margins. The voltage difference between read and write has been reported to be 200 mV at $V_{DD} = 1.1$ V. Type (d) leaves the supply line at a floating level during the write mode [17]. Raising both the power supply and word line in the read mode has also been proposed for a TFT load cell [18].

D. On-chip ECC

On-chip error checking and correcting (ECC) is the key to RAMs in the future, especially to SRAMs with their inherent small signal charge, as explained previously. The SER reduction of RAM by using a single-error correcting code is expressed by the equation $E = (N^2 T / 2N_0^2 W) E_0^2$, [23], where E is SER with ECC, E_0 is SER without ECC, N is the number of bits of an ECC word including check bits, N_0 is the number of data bits of an ECC word, T is the correction period, and W is the number of ECC words in a RAM (i. e., $W = M/N_0$, where M is the memory capacity). Figure 5 shows the SER reduction using a code of $N = 136$ and $N_0 = 128$. Even if SER without ECC is as high as 10^6 FIT (one upset per 1,000 hours) and the errors are not corrected at all during a ten-year period ($T = 10$ y), the SER is improved by four to five orders of magnitude through ECC. If periodic error correction (one ECC word every 7.8 μs) is performed like a DRAM refresh operation ($T = 7.8$ μs × W), the resulting SER becomes as low as 10^{-6} FIT. ECC is also effective for hard errors, especially for random single-cell faults, such as the V_T mismatch described above.

In addition, combining with redundancy produces a synergistic effect [24], which results in a drastic increase in fault tolerance.

III. PERIPHERAL LOGIC CIRCUITS

Obstacles to low-voltage operation of peripheral circuits are the ever-increasing subthreshold leakage and speed variations. The subthreshold leakage can be sufficiently reduced as far as RAMs are concerned. The speed-variation issue that is common to all nano-meter LSIs necessitates compensation circuits for, and new devices against the variation.

A. Leakage

Reducing leakage in the active mode is especially important, although this is more difficult than in the standby mode because leakage needs to be controlled much faster. Fortunately, leakage currents can be quickly, simply, and drastically reduced by utilizing RAM's features [1-4]. The basic reduction concept is to use a high-V_T MOST that is achieved with a high actual V_T or effectively with a low actual V_T MOST. Of many proposals, the gate-source offset driving, the gate-source self-back-biasing, power-switches with a level holder, and multi-static V_T [1-4] are practical for RAMs. In fact, they sufficiently reduced the standby and/or active leakage of a 0.6-V 16-Mb DRAM [5], 256-Mb DRAMs [1], a hypothetical 1-V 16-Gb DRAM [1], and a 1.2-V SRAM [19].

B. Speed Variation

Inter-die and intra-die V_T variations increase not only variations in leakage, but also variations in speed. For example, for the high-speed SRAM design with $V_T = 0.49$ V, the leakage varies as much as four orders of magnitude for a V_T variation of ± 0.1 V and a temperature variation of 100°C, as shown in Fig. 2 [3]. Such is the case for peripheral circuits. For any ΔV_T, the degree of speed variation, $\Delta V_T / (V_{DD} - V_T)$, increases with lower V_{DD}. It is enhanced by device scaling involving the ever larger ΔV_T (Fig. 3). Figure 6(a) shows delay versus feature size, F, for a low-power design with V_{DD}-scaling based on ITRS 2003[22]. Delay times for ± $3\sigma(V_T)$ are normalized by that for the average V_T (i.e., $V_{T0} = 0.3$ V) for each generation. For the bulk CMOS, the speed spread is from 1.19 to 0.86 in the 90-nm generation. However, it rapidly increases with device scaling, reaching as large as 3.76 to 0.53 in the 32-nm generation. This is an unacceptable increase. A 2.5-time increase in V_T variation and a decrease in V_{DD} from 0.9 V to 0.6 V are responsible for the increase. If V_{DD} is scaled down as for F, the speed spread increases to an unacceptable level, as in Fig. 6(b), although such V_{DD} scaling is ideal in terms of low power and ease of device development. For an excessive intra-die variation, there may be no solution without new devices with less V_T

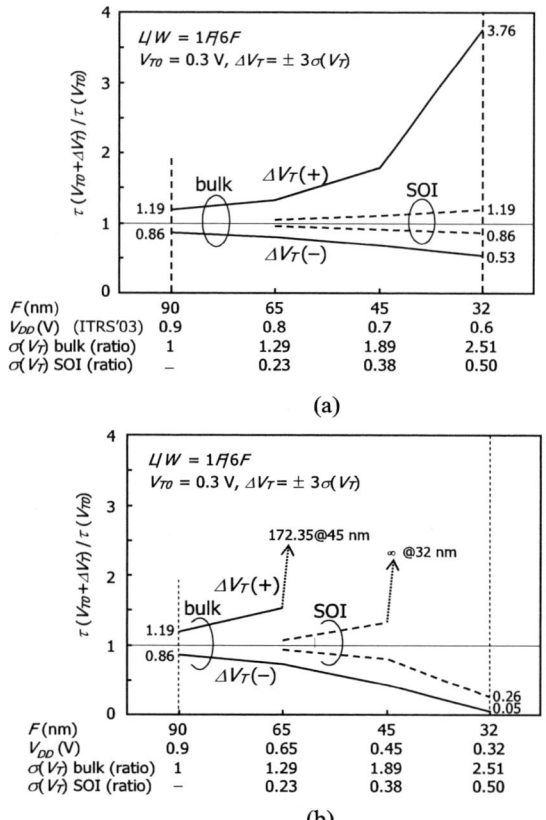

(a)

(b)

Fig. 6. Speed variations of an inverter for the V_{DD} projected by ITRS 2003 [22](a), and the V_{DD} scaled down as for F(b). The delay time is assumed to be proportional to $V_{DD}/(V_{DD}-V_T)^{1.25}$ [25].

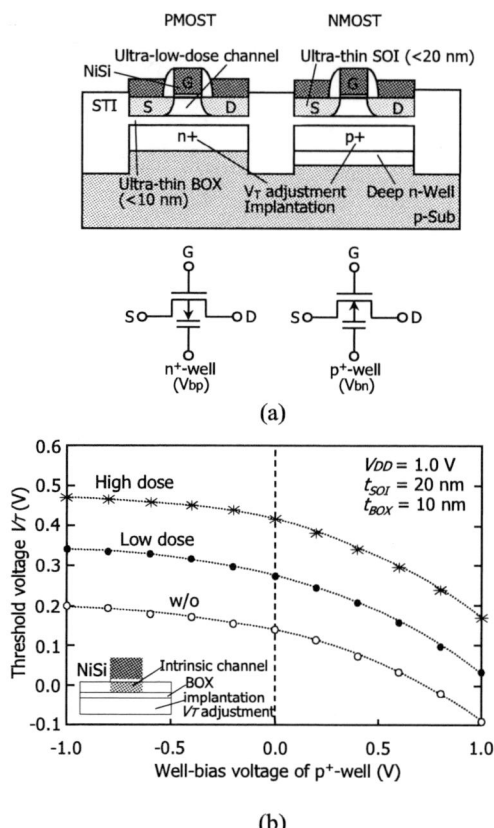

(a)

(b)

Fig. 7. Structure of double-gate FD-SOI MOSTs (a) and V_T characteristics of NMOS (b) [11].

variation, as explained later. For the inter-die variation, compensation by controlling the substrate voltage is unavoidable. It has been reported that positive body bias improved the speed by 63 % in slow process conditions, and negative body bias reduced leakage by 75 % in fast process conditions [5].

IV. FUTURE PROSPECTS

For RAMs, low-voltage operations are eventually restricted by the retention characteristics and signal-to-noise-ratio of RAM cells, even if coupled with ECC, a fast and reliable sensing, and the leakage current and speed variation of peripheral circuits, as described previously. For existing RAM cells, a high and unscalable V_T, which requires a high V_{DD}, is necessary to ensure a small retention current and a long enough refresh time [1]. Thus, new RAM cells, which do not rely on the charge and are thus insensitive to leakage, such as nonvolatile RAMs, will be strongly needed for lower-voltage operation. Even if a dual V_{DD} scheme [4] with a high V_{DD} for the cell array and a low V_{DD} for the peripheral circuits is used, ever increasing speed variations with device scaling will eventually limit low-voltage operations of the whole chip. If V_T variations continue to get larger and larger, as in existing bulk

CMOSTs, ultra-low voltage RAMs could not be achieved without reducing the resultant speed variations. Note that even if the inter-die variations can be compensated for by the improving the existing circuits, intra-die speed variations cannot be compensated for. In this case, the V_{DD} of peripheral circuits must get higher and higher to offset the speed variation. Alternatively, new MOSTs with small V_T variations, such as a fully depleted SOI [11] despite expensive wafers, will be necessary.

A double-gate fully depleted (FD) SOI [11, 20, 21], as shown in Fig. 7(a), is promising in the nano-meter era if the expected features are fully verified. An ultra-thin BOX (buried oxide) layer allows the V_T to be widely controlled by positive and negative back-bias controls (Fig. 7(b)), making it possible to create new low-voltage circuits such as a dynamic-V_T MOS circuit and new SRAM cells [20, 21]. Figure 8 depicts an example applied to an SRAM cell [26].

The cell widens the voltage margin for read and write: During read operation the well-biases of the accessed column, Vbn [n] and Vbp [n], are controlled to 0 V, so the V_T of PMOSTs, M_5 and M_6, is lowered with a forward bias. For example, when a cell with an N_1 voltage of 0 V and N_2 voltage of V_{DD} is read, the N_1 voltage raised as a result of a ratio of M_1 and M_3 tends to lower the N_2 voltage. However, the lowered V_T of M_6 prevents the lowering, enabling to

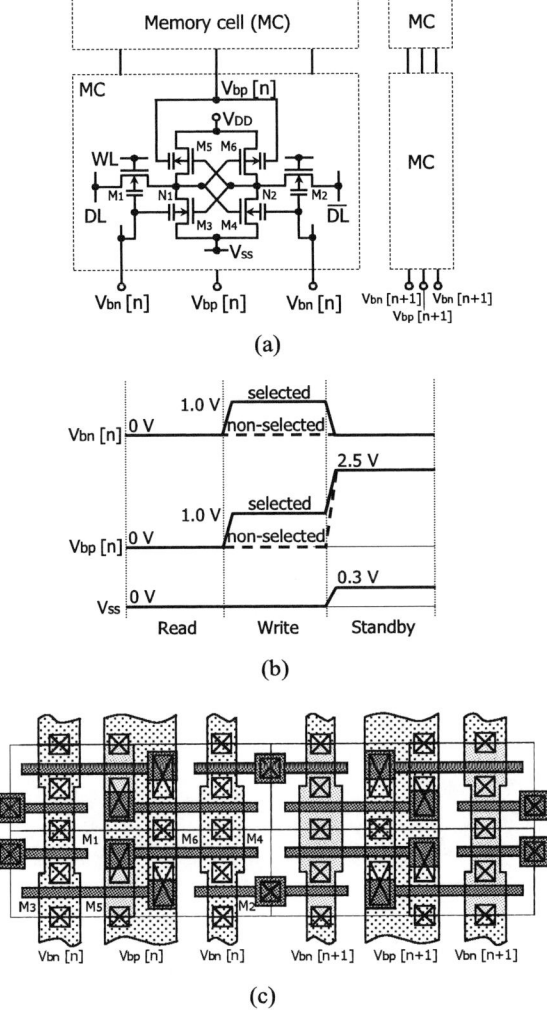

(a)

(b)

(c)

Fig. 8. Double-gate FD-SOI SRAM cell [26].
Memory cells (a), operating timing (b), and cell layout (c).

manage the speed variation, calling for new techniques. Here, reducing the gate tunneling current is also vital: it is not discussed in this paper, though, because this kind of reduction is the intended responsibility of the process and device designers [3]. In any event, two approaches in the nanometer era can be envisioned based on the above discussion: One is high-V_{DD} bulk-CMOS e-RAMs for low-cost applications, and the other is low-V_{DD} FD-SOI e-RAMs for high-speed and low-power applications.

V. CONCLUSION

Ultra-low voltage nano-scale embedded RAMs were described, focusing on RAM cells and peripheral circuits. First, challenges and trends of low-voltage RAM cells were discussed in terms of the signal charge, signal voltage, and noise, clarifying the importance of ECC in coping with the ever increasing soft-error rate and a fully depleted SOI to reduce variations in V_T. Then, peripheral circuits were discussed in terms of leakage reduction and compensation for speed variations, and it was concluded that ultra-low voltage RAMs would not be achieved without reducing speed variations caused by the V_T variation, thus prompting a need for compensation circuits and new devices with reduced V_T variation.

ACKNOWLEDGEMENT

The author is grateful to Dr. Osada, Mr. Yamaoka, and Dr. Tsuchiya for their discussion and support with valuable data.

REFERENCES

[1] K. Itoh, *VLSI Memory Chip Design*, Springer-Verlag, 2001.
[2] Y. Nakagome, IBM J. R & D, 47, no.5/6, pp. 525-552, 2003.
[3] K. Itoh, CICC Dig., pp. 339-344, 2004.
[4] K. Itoh, ICICDT Dig., pp. 235-242, 2005.
[5] K. Hardee, ISSCC Dig., pp. 494-495, 2004.
[6] M. Iida, ISSCC Dig., pp. 460-461, 2005.
[7] M. Shirahama, ISSCC Dig., pp. 462-463, 2005.
[8] S. Naffziger, ISSCC Dig., pp. 182-183, 2005.
[9] P. Carter, IEEE J. SSC, 22, No.3, pp. 430-436, 1987.
[10] S-M Jung, IEDM Dig., pp. 289-292, 2003.
[11] R. Tsuchiya, IEDM Dig., pp. 631-634, 2004.
[12] J. Wuu, ISSCC Dig., pp. 488-489, 2005.
[13] K. Itoh, Symp. VLSI Circuits Dig., pp. 132–133, 1996.
[14] K. Itoh, PATMOS Dig., pp. 3-15, 2004.
[15] K. Osada, ISSCC Dig., pp. 302–303, 2003.
[16] K. Zang, ISSCC Dig., pp. 474-475, 2005.
[17] M. Yamaoka, ISSCC Dig., pp. 480-481, 2005.
[18] H. An, Symp. VLSI Circuits Dig., pp. 282-283, 2004.
[19] M. Yamaoka, ISSCC Dig., pp. 494-495, 2004.
[20] M. Yamaoka, Symp. VLSI Circuits Dig., pp. 288-291, 2004.
[21] M. Yamaoka, Int'l SOI Conf. Dig., pp. 109-111, 2004.
[22] ITRS 2003 EXECUTIVE SUMMARY Table 6a, p. 57.
[23] M. Horiguchi, IEEE J. SSC, 23, p. 27, Feb. 1988.
[24] H. L. Kalter, IEEE J. SSC, 25, p. 1118, Oct. 1990.
[25] K. Chen, Trans. on Electron Devices, pp. 1951-1957, 1997.
[26] M. Yamaoka, A-SSCC Dig., pp. 109-112, 2005.

improve the SNM. During write operation, the well-biases are controlled to a high voltage, so the V_T of NMOSTs, M_1-M_4, is lowered with a forward-bias, while the V_T of M_5 and M_6 is raised with no bias. Thus, the above voltage combination for N_1 and N_2 is more easily changed to the opposite one when 0 V and V_{DD} are applied to /DL and DL, respectively, because N_2 is more easily discharged with increased M_2 current and reduced M_6 current. During standby, the NMOS source line and the Vbp [n] are raised to 0.3 V and 2.5 V (the I/O supply), respectively, so subthreshold currents of PMOSTs and NMOSTs are reduced with increased V_T. Moreover, the ultra-thin and lightly-doped channel of the SOI structure suppresses the V_T variation (Fig. 3). As a result, even in the 32-nm generation the speed spread remains in the same range as that for the 90-nm bulk CMOS, as seen in Fig. 6(a). This implies that the SOI would extend the low-voltage limitation of bulk CMOS by at least three generations. This is true for a relatively high V_{DD}. If V_{DD} is scaled down, as in Fig. 6(b), however, not even the SOI will be able to

82

Gate Oxide Reliability for Nano-Scale CMOS

J. H. Stathis

Abstract - The reliability of the gate oxide in microelectronics, i.e., the ability of a thin film of this material to retain its excellent dielectric properties while subjected to high electric fields, has been a perennial concern over the last 40-45 years. Two dominant gate oxide failure mechanisms, dielectric breakdown and the negative bias instability, have continued to cause concern as MOSFET devices have scaled to nanometer dimensions.

I. OXIDE BREAKDOWN

The effect of gate oxide reliability on nano-scale CMOS circuits may be expressed as the maximum allowable voltage that can be applied to the total gate area on a chip, such that no more than a specified failure rate will result. Fig. 1 shows a compendium of various predictions concerning oxide breakdown from different research groups [1-5] for the maximum operation voltage, V_{max}, as a function of gate oxide thickness (t_{ox}). (For a detailed discussion of this figure see [4] and [6].)

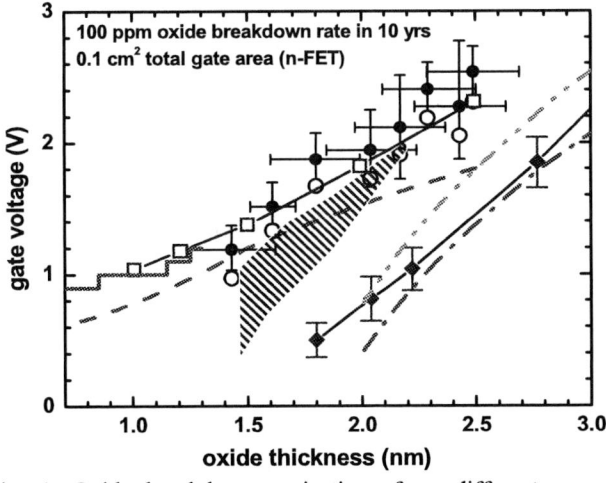

Fig. 1. Oxide breakdown projections from different research groups during the period 1998-2002. This figure shows the maximum allowable voltage that can be applied to the total gate area on a chip, such that no more than a specified failure rate will result. The failure rate in this case is defined as the fraction of chips that will experience one or more oxide breakdown events. Also shown are industry roadmaps for gate oxide thickness and operation voltage from 1999 (dashed) and 2002 (stepped). From [6].

As shown in Fig. 1, the predicted reliability limits due to oxide breakdown have moved toward thinner oxides and higher operation voltage as more data have been collected and new analysis approaches have been applied. Indeed,

J. H. Stathis is with the IBM Research Division, Yorktown Heights, NY, 10598 USA, e-mail: stathis@us.ibm.com

the most optimistic projection, based on an empirical power-law voltage dependence [7], supports the use of oxynitride gate dielectrics down to ~1nm at 1V. However, more aggressive scaling, or operation at higher voltage for improved speed, makes it increasingly likely that one or several oxide breakdown (BD) events may expected over the life of a chip. The earlier oxide reliability projections were based on the assumption that a single breakdown (soft or hard) on a chip would cause circuit failure, which is no longer believed to be correct. For accurate reliability projections it is necessary to better understand the nature of the BD event [8] and the effect of BD on circuits. Therefore it has become necessary to look in more detail at the nature of the breakdown event and the behavior of devices and circuits after oxide failure [6,9,10].

A. Progressive Breakdown

Several groups [11-13] have pointed out that "hard" BD is not a sudden, catastrophic process, as previously thought. BD occurs gradually over a measurable time scale. For poly-gate CMOS at present-day oxide thickness, the growth of the gate leakage through the BD spot can be very slow at low stress voltage. This phenomenon is called "progressive" breakdown (PBD) [11]. PBD is a *gradual* hard BD, and is distinct from "soft" BD, which is a stable, low current that is typically not observed in small devices. An example of a current-vs.-time time trace is shown in Fig. 2 [13].

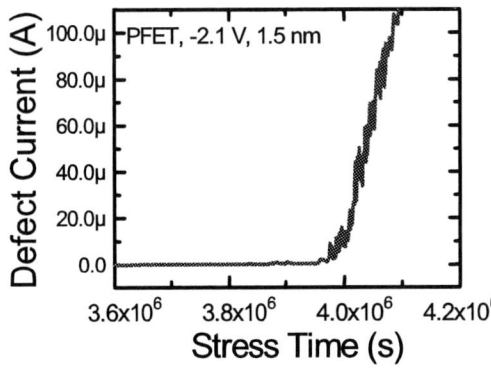

Fig. 2. Breakdown transient for 1.5nm oxide stressed at -2.1V. After [13].

The post-BD leakage growth rate can be quantified in various ways [6,10-14]. Fig. 3 shows the voltage dependence of the progressive breakdown rate, R_D, for $t_{ox}=1.5$nm [13]. This is similar to the voltage dependence of the trap generation prior to breakdown, suggesting that the same defect generation process that controls the initial breakdown time also drives the growth of the BD spot.

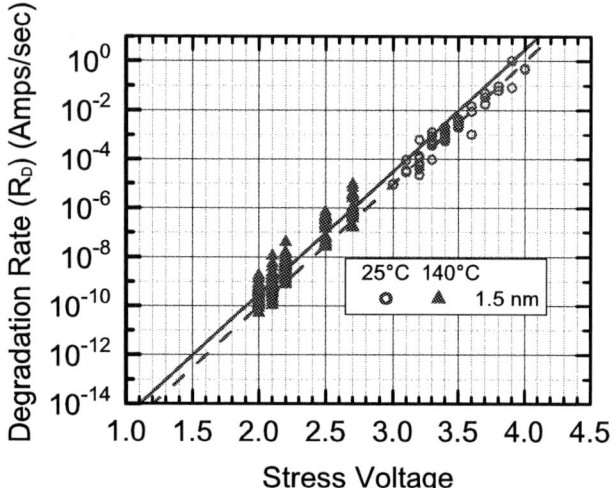

Fig. 3. The rate of increase of stress current for a 1.5 nm oxide after the beginning of breakdown, showing an exponential dependence for 10 orders of magnitude over a wide range of voltages. The degradation rate R_D is defined as the average rate of increase from $10\mu A$ to $100\mu A$. After ref. [13].

B. Effect of Progressive Breakdown on Circuit Operation

Because the breakdown process is gradual and continuous, the chip or circuit failure will not coincide with the onset of breakdown, but instead will occur at a later time when a critical breakdown current is reached. Rather than first (soft) BD, the appropriate failure criterion is the leakage current that disrupts circuit operation. This criterion can increase lifetime estimates by several orders of magnitude over traditional projections [15]. The new oxide failure criterion has two key elements: Understanding and characterizing the post-breakdown defect growth, and understanding and characterizing the circuit sensitivity to leakage currents in gates that have experienced BD. Circuit simulations can be used to estimate circuit sensitivity to BD, by adding a voltage-dependent current source between the gate and one diffusion of a transistor as illustrated in Fig. 4 [9,16,17].

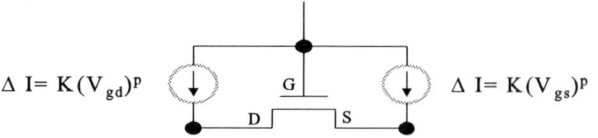

Fig. 4. Circuit model of gate-to-diffusion breakdowns. After ref. [17].

The effect of progressive BD has been studied experimentally using inverters in a $0.13\mu m$ technology (t_{ox} = 1.5nm) [17,18]. Constant voltage stress at 2.6–3.9 V of

either polarity was applied from input to output, with V_{dd} and ground terminals floating. In this way a BD was made to occur at the drain side of either the n-FET or p-FET. Progressive BD was stopped at various stages by a current compliance [19].

The transfer characteristics of the broken inverters (Fig. 5) exhibit a combination of V_t shifts due to the voltage stress and reduced output swing due to post-BD leakage. The characteristics of the BD spot are different depending on stress polarity and whether the inverter output voltage is higher or lower than the input. In this figure the transfer curves show additional shift in switching point due to threshold voltage shifts in the n-FET and p-FET. These shifts occur prior to the BD event.

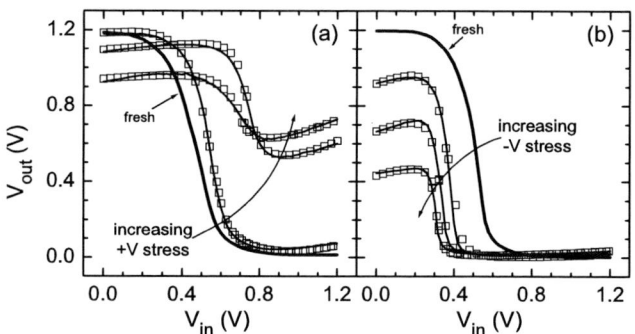

Fig. 5. Transfer curves of inverters after BD to various levels. (a) Positive stress on inverter input. (b) Negative stress on inverter input. Lines are experiment, symbols are model. For positive/negative stress, the leakage is highest when the input is higher/lower than the output. After ref. [18].

Calculated transfer characteristics using the same gate-to-drain leakage current model but without the V_t shift (Fig. 6) illustrate the influence of the oxide BD leakage current alone in the inverter transfer curve, to more accurately represent the effect of early BD under circuit operation conditions. The inverter transfer curves shown in Fig. 6 are the expected characteristics for chips in the field, where the earliest oxide breakdown may occur prior to significant V_t shift.

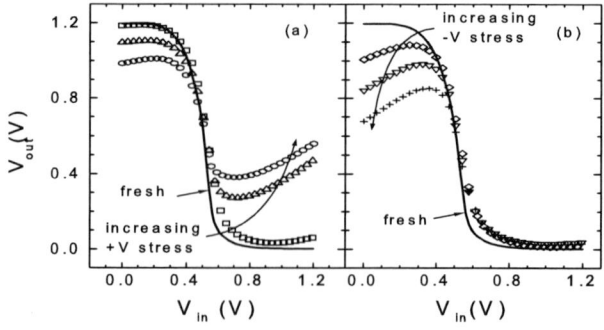

Fig. 6. Simulated inverter transfer curves with oxide leakage currents between the inverter input and output using same leakage as in Fig. 3, but with no V_t shifts (a) positive stress on the inverter input, (b) negative stress. After ref. [18].

84

Fig. 7 shows calculated transfer curves for two inverters in series with a drain (input-output) breakdown in the second inverter. The output of the first inverter is degraded, even though there is no breakdown in this stage. This is because the BD leakage in the second stage loads the first stage. Subsequent logic stages will restore the correct logical "1" and "0" states as long as the output of the broken stage is on the correct side of the crossover voltage V_{co}.

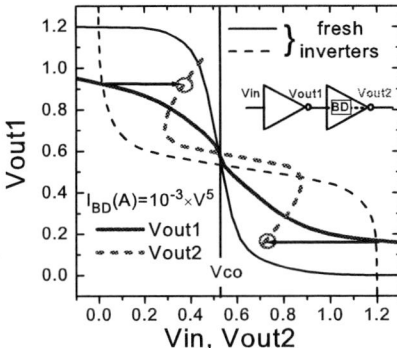

Fig. 7. Transfer curves for two inverters in series, with a drain (input-output) breakdown in the second stage. Small circles indicate the output states of the second inverter. Inverter chains transmit the correct logic state as long as output of broken stage is on the correct side of the crossover voltage. Thin (solid and dash) black lines represent Vout1 and Vout2 respectively without BD. After ref. [18].

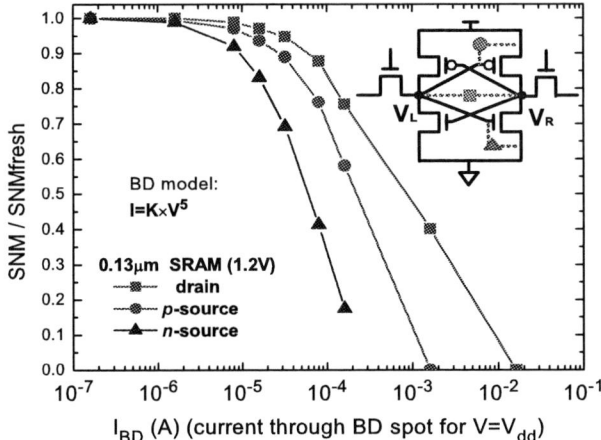

Fig. 8. Normalized SNM from circuit simulations as a function of BD leakage at V_{dd}, for 6-T SRAM cells with various BD locations. Symbols indicate BD locations. After ref. [16].

In an SRAM cell (Fig. 8, inset) oxide BD in either inverter of the cell loads the other inverter. Gate-to-source BD does not affect the transfer curve of an inverter, to first order. However, it does perturb the voltage at the output of the opposite inverter. A p-(n-) source BD raises (lowers) the voltage at the output of the opposite inverter, which must then supply current through the channel resistance of the on-state n-(p-)FET of the intact inverter. In order to quantify the cell stability we extract the worst-case static noise margin (SNM). This is the minimum DC noise voltage necessary to flip the state of the cell during a "read" operation, where the word line is pulled high while the bitlines are pre-charged high.

For fixed leakage, BD at p-source has less effect than n-source, because the opposing n-FET is stronger (relative to the p-FET). Fig. 8 shows the SNM, normalized to the SNM of the fresh cell, as a function of I_{BD} in a 0.13μm technology [16]. These results were obtained from circuit simulations. For the cells considered in this work, a 50% degradation in SNM results from oxide BD when the current through the BD spot reaches ~20–50 μA for the worst-case n-source breakdown [20]. Pass-gate or p-source breakdown may tolerate higher leakage, up to ~500 μA. These values are comparable with the on-currents of the fresh p-FET and n-FET respectively used in this SRAM cell, and may decrease with device widths, e.g., for smaller SRAM cells.

C. Proposal for an Improved Breakdown Terminology

Oxide breakdown events are typically classified as "soft" or "hard" depending on the magnitude of the post-breakdown conduction. There is some confusion in the literature over the characterization of breakdown modes because of the lack of a precise definition of the terms and because for some experimental conditions the detection of one or the other breakdown mode may be difficult. For example, when testing a large area structure or a very thin oxide where the initial current is larger than the breakdown current, a "soft" event could be missed, or a "hard" event could be interpreted as soft. Although some of these problems can be overcome with careful experimental design, the recent understanding of PBD has made these earlier terminologies less satisfactory.

Various schemes have been devised to characterize the BD "hardness," e.g., the post-breakdown resistance (V_{dd}/I_{BD}) or conductance (dI_{BD}/dV), [21-24] however this designation is often ambiguous because there is no universally accepted criterion. The result is that one author may refer to a given post-BD current level as SBD while another might characterize the same event as HBD.

Furthermore, the realization that the initial breakdown event ("first BD") may not disrupt circuit functionality has led to another usage of the terms SBD and HBD depending on the intended operation conditions of the MOSFET [25]. This operational definition obscures the physical nature of the BD. Here we describe a new view of BD characterization with a simpler, more physically meaningful terminology [8].

The steep voltage dependence of the post-breakdown degradation rate leads to an important implication for the BD characterization. As earlier pointed out by Monsieur, [11] if the oxide is stressed at a high voltage where the post-breakdown degradation rate is fast compared to the experimental sampling time (typically longer than ~tens of milliseconds) then the breakdown will appear as "hard" according to the typical usage of this term. Likewise, if the oxide is stressed at a low voltage where the degradation rate is slow compared to the experimental sampling time, then the breakdown will appear as "soft".

This implies that there is no distinct physical characteristic which we can use to classify HBD vs. SBD. Rather, it is the exponential voltage and thickness dependence of the PBD growth time τ_D which causes a BD to appear as HBD for thick oxides and/or high voltage and as SBD for thin oxides and/or low voltage. This is illustrated schematically in Fig. 9, where the dashed line corresponds to a constant value of τ_D on the order of the experimental sampling time. Below this line the BD appears soft, while above the line the BD appears hard in a typical experiment. The hatched region corresponds to the domain which is accessible to experiment, i.e. within this band the time to first BD is of order seconds to hours. As oxide thickness is reduced the time to BD decreases rapidly because of the rapid increase in tunneling current [1], which requires the use of lower V_{stress} to keep the time to first BD within measurable range.

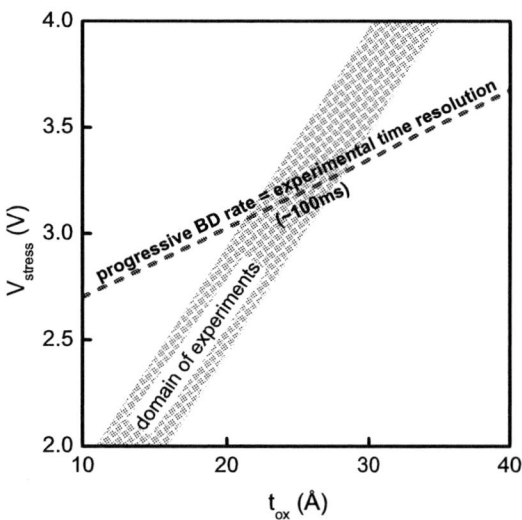

Fig. 9. The dashed line corresponds to a constant value of τ_D on the order of the experimental time resolution, for oxides of thickness t_{ox} stressed at voltage V_{stress}. Below this line the BD appears soft, while above the line the BD appears hard in a typical experiment. The hatched region corresponds to the domain which is accessible to experiment. After a similar figure by Monsieur [11].

This figure explains the observed trend that HBD is more prevalent in thicker oxides, i.e. this is really an effect of the changing V_{stress}, more so than t_{ox}. This also provides an explanation for the so-called HBD prevalence ratio which shows a rapid transition from ~0% to ~100% over a narrow voltage range, moving to higher voltage with increasing thickness [10]. The transition from SBD-like to HBD-like is not completely abrupt because of the existence of a statistical distribution in τ_D.

According to the new viewpoint, HBD and SBD are really just different manifestations of the same PBD mode, and the distinction between HBD and SBD depends mostly on measurement conditions. The PBD growth rate depends exponentially on voltage and thickness. This causes a BD to appear as HBD for thick oxides and/or high voltage and as SBD for thin oxides and/or low voltage. Physically, all BD should be described as PBD and characterized by its post-breakdown degradation rate.

We prefer to avoid the terms "soft" and "hard" because of their vague meaning. A more accurate term to describe cases of low or moderate post-breakdown conduction, such as results from removing the stress during the BD transient (either intentionally, e.g. by a compliance limit, or unintentionally, e.g. by series resistance) is arrested BD ("a-BD").

To describe the impact of breakdown on device and circuit functionality, different terminology should be used to clearly distinguish from the physical description. A BD which disrupts device or circuit functionality can be called disruptive or destructive. Of course, this is an application-specific description [9]. For example, a BD with ~ 50μA leakage at operation condition may be destructive in an SRAM application [16] but not in logic [6,9,26]. It is important to realize also that a less severe BD (i.e. non-destructive) cannot be assumed to be completely innocuous, since the initial BD spot may grow progressively into a destructive one. This terminology is summarized in Table 1 [27].

TABLE I
OXIDE BREAKDOWN TERMINOLOGY

Old terminology:	Soft (SBD)	Hard (HBD)
Improved physical description:	Arrested BD	
	Progressive Breakdown, described by post-breakdown rate of current growth	
Improved operational definitions (circuit dependent):	Non-destructive	Destructive

II. NBTI

The threshold voltage (V_t) shift in p-FETs caused by the negative-bias-temperature instability (NBTI) has emerged as one of the more interesting and potentially serious reliability limiters for state-of-the-art CMOS technology [28]. Although recognized for many years [29], NBTI has increasing significance for newer technologies that operate with lower supply voltage, because of the inability to fully scale the device threshold voltage, leading to reduced headroom. In addition, the introduction of nitrogen in the gate oxide, for control of dopant penetration and gate current, causes an increase in NBTI for the same physical oxide thickness and voltage condition [30,31]. The origin of this nitrogen-enhancement effect is still under investigation. Fig. 10 shows an example of this effect [32].

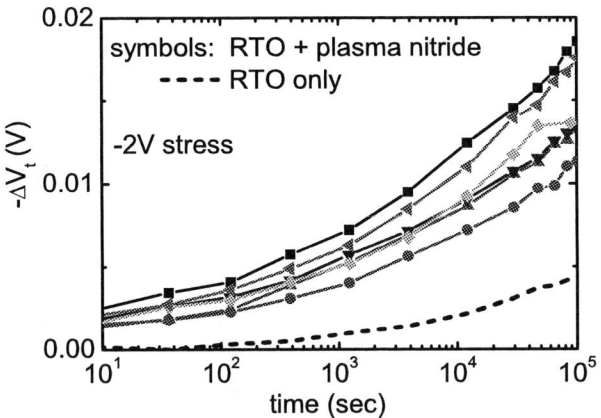

Fig. 10. Threshold voltage shift vs. time for 1.4nm oxide with various nitrogen concentrations. From [32].

Among the many interesting physical phenomena involved in NBTI, one aspect is the relative contribution of interface states and bulk traps to the net V_t shift [33]. The interface states near the middle portion of the Si band gap can be measured by capacitance-voltage, charge pumping, gated-diode, etc. These techniques fail close to the band edge, but in ultra-thin oxides the leakage current near flat-band condition (so-called low-voltage stress-induced leakage current, LV-SILC) is sensitive to the interface states at the conduction band edge [34-37]. Figs. 11 and 12 compare the contributions of mid-gap states (measured with gated diode) and conduction band-edge states (from LV-SILC) in pure SiO_2 vs. plasma-nitrided oxides [38].

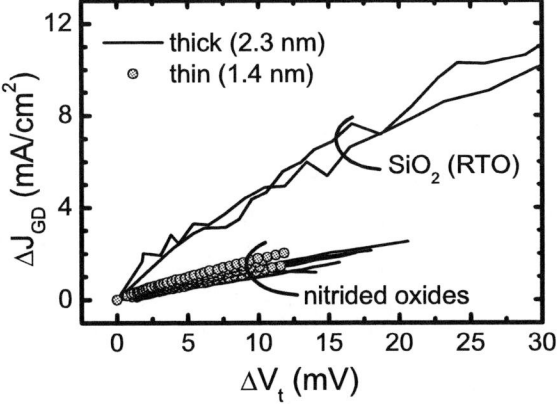

Fig. 11. Mid-gap recombination center density (gated diode current) vs. V_t shift for SiO_2 and oxynitride pFETs under negative bias stress (~10MV/cm). The gated diode current could not be measured in the 1.4nm SiO_2 sample because of high direct tunneling current. From [38].

Fig. 11 shows that the defect density at mid-gap is about four times greater for SiO_2 compared to oxynitride, for the same ΔV_t. Fig. 12 shows that pure oxide has negligible generated interface states at the conduction band

edge, in contrast to oxynitride which shows a significant density of stress-induced interface states at this position. Together these two figures demonstrate that the defects associated with the nitrogen-enhanced NBTI [30] are different from those in pure oxide, and in particular that these nitrogen-associated defects have electrical levels in the upper portion of the Si band gap [32,38].

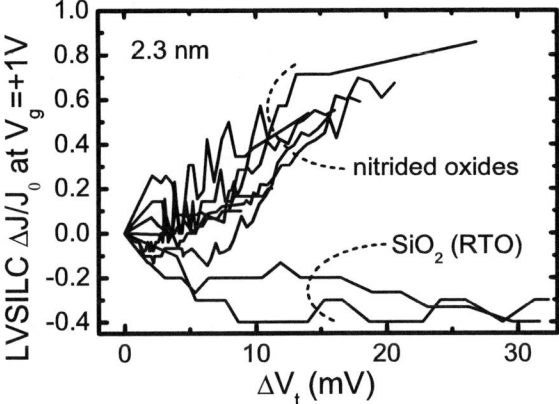

Fig. 12. Conduction band-edge interface state density (LV-SILC) vs. V_t shift for 2.3nm SiO_2 and oxynitride pFETs stressed at -10MV/cm. From [38].

III. CONCLUSION

Oxide breakdown and NBTI are two physical failure mechanisms in ultra-thin gate oxide which continue to generate interest. Ongoing work in these two subjects is needed to ensure the reliability of nano-scale CMOS circuits.

ACKNOWLEDGEMENTS

This paper is based on significant contributions from B.P. Linder (IBM) and R. Rodríguez (U. Autónoma de Barcelona, Spain). We also acknowledge helpful discussions with S. Lombardo (CNR-IMM, Catania, Italy), S. Zafar (IBM) and E.Y. Wu (IBM).

REFERENCES

[1] J. H. Stathis and D. J. DiMaria, "Reliability projection for ultra-thin oxides at low voltage," *Digest of the 1998 International Electron Devices Meeting*, pp. 167-170, 1998.

[2] R. Degraeve, B. Kaczer, and G. Groeseneken, "Reliability: a possible showstopper for oxide thickness scaling?," *Semicond. Sci. Technol.*, vol. 15, pp. 436-444, 2000.

[3] M. A. Alam *et al.*, "Physics and prospects of sub-2nm oxides," in *The Physics and Chemistry of SiO₂ and the Si-SiO₂ Interface - 4*, vol. 2000-2, H. Z. Massoud, I. J. R. Baumvol, M. Hirose, and E. H. Poindexter, Eds. Pennington, NJ: The Electrochemical Society, 2000, pp. 365-376.

[4] J. H. Stathis, "Reliability limits for the gate insulator in CMOS technology," *IBM J. Res. Develop.*, vol. 46, pp. 265-286, 2002.

[5] E. Y. Wu, E. J. Nowak, A. Vayshenker, W. L. Lai, and D. Harmon, "CMOS scaling beyond the 100-nm node with sili-

con-dioxide-based gate dielectrics," *IBM J. Res. Develop.*, vol. 46, pp. 287-298, 2002.

[6] J. H. Stathis, B. P. Linder, R. Rodríguez, and S. Lombardo, "Reliability of ultra-thin oxides in CMOS circuits," *Microelectron. Reliab.*, vol. 43, p. 1353, 2003.

[7] E. Y. Wu et al., "Voltge-dependent voltage-acceleration of oxide breakdown for ultra-thin oxides," *Digest of the 2000 International Electron Devices Meeting*, pp. 541-544, 2000.

[8] S. Lombardo et al., "Dielectric breakdown mechanisms in gate oxides," *J. Appl. Phys.*, vol. 98, pp. 121301-35, 2005.

[9] J. H. Stathis, R. Rodríguez, and B. P. Linder, "Circuit implications of gate oxide breakdown," *Microelectron. Reliab.*, vol. 43, pp. 1193-1197, 2003.

[10] E. Wu, J. Suñé, B. P. Linder, J. H. Stathis, and W. L. Lai, "Critical assessment of soft breakdown stability time and the implementation of new post-breakdown methodology for ultra-thin gate oxides," *Digest of the 2003 International Electron Devices Meeting*, pp. 319-322, 2003.

[11] F. Monsieur et al., "A thorough investigation of progressive breakdown in ultra-thin oxides. Physical understanding and application for industrial reliability assessment.," *2002 International Reliability Physics Symposium Proceedings*, pp. 45-54, 2002.

[12] T. Hosoi, P. Lo Re, Y. Kamakura, and K. Taniguchi, "A new model of time evolution of gate oxide leakage current after soft breakdown in ultra-thin gate oxides," *Digest of the 2002 International Electron Devices Meeting*, pp. 155-158, 2002.

[13] B. P. Linder, S. Lombardo, J. H. Stathis, A. Vayshenker, and D. J. Frank, "Voltage dependence of hard breakdown growth and the reliability implication in thin dielectrics," *IEEE Electron Device Lett.*, vol. 23, pp. 661-663, 2002.

[14] J. S. Suehle, B. Zhu, Y. Chen, and J. B. Bernstein, "Detailed study and projection of hard breakdown evolution in ultra-thin gate oxides," *Microelectron. Reliab.*, vol. 45, pp. 419-426, 2005.

[15] B. Kaczer, R. Degraeve, R. O'Connor, P. Roussel, and G. Groeseneken, "Implications of progressive wear-out for lifetime extrapolation of ultra-thin (EOT~1nm) SiON films," *Digest of the 2004 International Electron Devices Meeting*, pp. 713-716, 2004.

[16] R. Rodríguez et al., "The impact of gate oxide breakdown on SRAM stability," *IEEE Electron Device Lett.*, vol. 23, pp. 559-561, 2002.

[17] R. Rodríguez, J. H. Stathis, and B. P. Linder, "A model for gate oxide breakdown in CMOS inverters," *IEEE Electron Device Lett.*, vol. 24, pp. 114-116, 2003.

[18] R. Rodríguez, J. H. Stathis, and B. P. Linder, "Modeling and experimental verification of the effect of gate oxide breakdown on CMOS inverters," *2003 International Reliability Physics Symposium Proceedings*, pp. 11-16, 2003.

[19] B. P. Linder et al., "Gate oxide breakdown under current limited constant voltage stress," *Digest of the 2000 Symposium on VLSI Technology*, pp. 214-215, 2000.

[20] K. Mueller, S. S. Gupta, S. Pae, M. Agostinelli, and P. Aminzadeh, "6-T cell circuit dependent GOX SBD model for accurate prediction of observed Vccmin test voltage dependency," *2004 International Reliability Physics Symposium Proceedings*, pp. 426-429, 2004.

[21] H. Satake and A. Toriumi, "Dielectric breakdown mechanism of thin-SiO_2 studied by the post-breakdown resistance statistics," *IEEE Trans. Electron Devices*, vol. 47, pp. 741-745, 2000.

[22] K. Okada, "The gate oxide lifetime limited by 'B-mode' stress induced leakage current and the scaling limit of silicon dioxides in the direct tunneling regime," *Semicond. Sci. Technol.*, vol. 15, pp. 478-484, 2000.

[23] R. Degraeve, B. Kaczer, A. De Keersgeiter, and G. Groeseneken, "Relation between breakdown mode and breakdown location in short channel nMOSFETs and its impact on reliability specifications," *2001 International Reliability Physics Symposium Proceedings*, pp. 360-366, 2001.

[24] B. Weir, M. A. Alam, P. J. Silverman, and Y. Ma, "Low voltage gate dielectric reliability," in *Semiconductor Silicon/2002*, vol. 2002-2, H. R. Huff, L. Fabry, and S. Kishino, Eds. Pennington, New Jersey: The Electrochemical Society, Inc, 2002, pp. 365-374.

[25] J. Suñé, E. Wu, and W. Lai, "Limits of the successive breakdown statistics to assess chip reliability," *Microelectronic Engineering*, vol. 72, pp. 39–44, 2004.

[26] B. Kaczer et al., "Impact of MOSFET oxide breakdown on digital circuit operation and reliability," *Digest of the 2000 International Electron Devices Meeting*, pp. 553-556, 2000.

[27] J. H. Stathis, "Gate oxide reliability for nano-scale CMOS," Proc. International Symposium on the Physical and Failure Analysis of Integrated Circuits, pp. 127-130, 2005.

[28] J. H. Stathis and S. Zafar, "The negative bias temperature instability in MOS devices: A review," *Microelectron. Reliab.*, 2006.

[29] Y. Miura and Y. Matukura, "Investigation of silicon-silicon dioxide interface using MOS structure," *Jpn. J. Appl. Phys.*, vol. 5, p. 180, 1966.

[30] N. Kimizuka et al., "NBTI enhancement by nitrogen incorporation into ultrathin gate oxide for 0.10μm gate CMOS generation," *Digest of the 2000 Symposium on VLSI Technology*, pp. 92-93, 2000.

[31] Y. Mitani, M. Nagamine, H. Satake, and A. Toriumi, "NBTI mechanism in ultra-thin gate dielectric: Nitrogen-originated mechanism in SiON," *Digest of the 2002 International Electron Devices Meeting*, pp. 509-512, 2002.

[32] J. H. Stathis, G. LaRosa, and A. Chou, "Broad energy distribution of NBTI-induced interface states in p-MOSFETs with ultra-thin nitrided oxide," *2004 International Reliability Physics Symposium Proceedings*, pp. 1-7, 2004.

[33] V. Huard et al., "A thorough investigation of MOSFETs NBTI degradation," *Microelectron. Reliab.*, vol. 45, pp. 83-98, 2005.

[34] P. E. Nicollian et al., "Low voltage stress-induced-leakage-current in ultrathin gate oxides," *1999 International Reliability Physics Symposium Proceedings*, pp. 400-404, 1999.

[35] A. Ghetti, E. Sangiorgi, J. Bude, T. Sorsch, and G. Weber, "Low voltage tunneling in ultra-thin oxides: a monitor for interface states and degradation," *Digest of the 1999 International Electron Devices Meeting*, pp. 731-734, 1999.

[36] N. Kimizuka et al., "The impact of bias temperature instability for direct-tunneling ultra-thin gate oxide on MOSFET scaling," *Digest of the 1999 Symposium on VLSI Technology*, pp. 73-74, 1999.

[37] F. Crupi et al., "On the role of interface states in low-voltage leakage currents of metal-oxide-semiconductor structures," *J. Appl. Phys.*, vol. 80, pp. 4597-4599, 2002.

[38] J. H. Stathis et al., "Interface state generation in pFETs with ultra-thin oxide and oxynitride on (100) and (110) Si substrates," *Microelectronic Engineering*, vol. 80, pp. 126-129, 2005.

2006 25th International Conference on Microelectronics

Effects of Device Aging on Microelectronics Radiation Response and Reliability

D. M. Fleetwood, M. P. Rodgers, L. Tsetseris, X. J. Zhou, I. Batyrev, S. Wang, R. D. Schrimpf, and S. T. Pantelides

Abstract – Recent work is reviewed that shows that MOS and bipolar device radiation response can change significantly with aging time after device fabrication and/or packaging. Effects include changes in radiation response due to burn-in, pre-irradiation elevated temperature stress, and/or long-term storage. These changes are attributed experimentally and theoretically to the motion and reactions of water and other hydrogen-related species. Similar hydrogen-related reactions can also affect the long-term reliability of MOS devices and integrated circuits, as illustrated in detail here for negative-bias temperature instability.

I. INTRODUCTION

It is usually assumed implicity that the radiation response and long-term reliability of microelectronic devices and integrated circuits (ICs) do not change significantly during the time between when devices are manufactured and/or packaged and when they are tested. This assumption is based upon the belief that the defects and impurities that affect radiation response and reliability are affected much more by the higher temperatures (400 to 1000 °C) of device processing than the lower temperatures for packaging and storage (0 to 300 °C). After packaging, the highest temperatures to which devices typically are exposed is ~ 125 to 150 °C; the higher temperatures in this range typically are encountered if devices are burned-in before being incorporated into systems of interest.

In 1994, Shaneyfelt et al. showed that burn-in can significantly affect the radiation response of MOS devices [1]. Changes in radiation response were observed in gate and field oxides of devices manufactured and packaged at Sandia National Laboratories, with a reduction in interface (but not oxide) traps observed in burned-in devices (Fig. 1). No significant changes were observed in preirradiation characteristics. These results were extended by Clark and co-workers using devices manufactured by National Semiconductor [2]. Changes in post-irradiation leakage were observed that were higher for burned-in devices than non-burned-in devices, especially for plastic packages. Effects of burn-in on radiation response also have been reported for power MOSFETs [3]. Thus, manufacturing and packaging processes can play significant roles in determining the burn-in sensitivity of MOS devices.

D. M. Fleetwood, M. P. Rodgers, L. Tsetseris, X. J. Zhou, I. Batyrev, S. Wang, R. D. Schrimpf, and S. T. Pantelides are with Vanderbilt University, P. O. Box 92, Station B, Nashville, TN 37235 USA. E-mail: dan.fleetwood@vanderbilt.edu.

The effects of electric field (Fig. 2) and temperature (Fig. 3) during elevated-temperature reliability screens were assessed by Shaneyfelt et al. for devices built in a 2-µm technology at Sandia National Laboratories [4]. Based on a limited data set, an effective activation energy for the observed increase in post-irradiation field-oxide leakage of ~ 0.38 eV was estimated for this technology. It was noted that this activation energy is similar to that of the diffusion of molecular hydrogen [4]. Consistent with the possibility that hydrogen diffusion might be associated with the changes in device radiation response, Shaneyfelt et al. also found that the changes in MOS radiation response observed in the field oxides studied were independent of bias, at least for the range of biases (-5 V to 20 V) examined [4].

These results were extended to full static random access memories (SRAMs) by Shaneyfelt et al. in [5], which evaluated the effects of pre-irradiation elevated temperature stress (PETS) on devices manufactured by Paradigm and by Cypress Semiconductor. Burn-in or other PETS treatments were observed to affect both the leakage currents and the doses to functional failure for these technologies. Failure doses dropped by ~ 40% (Fig. 4) for Paradigm SRAMs that were burned-in for 802 hours, as compared to failure doses of devices that received no burn-in treatment. Shaneyfelt et al. hypothesized that the

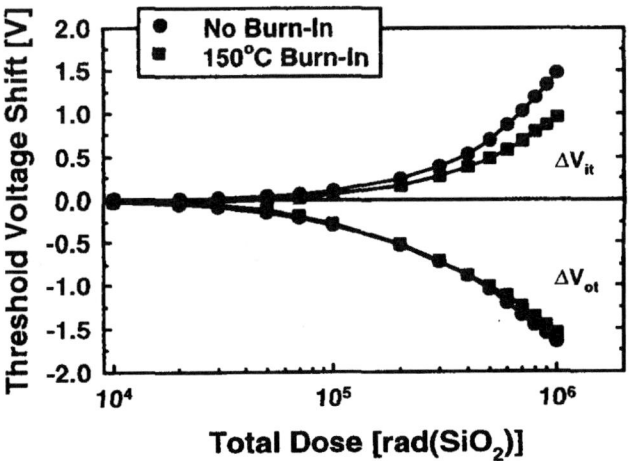

Fig. 1. Threshold-voltage shifts due to interface traps and oxide trap charge for 2 x 16 µm *n*MOS transistors with 32 nm oxides fabricated at Sandia National Laboratories. The devices were irradiated with 10-keV X rays at a dose rate of 167 rad(SiO_2)/s and a gate bias of 5 V. (After [1], © IEEE, 1994).

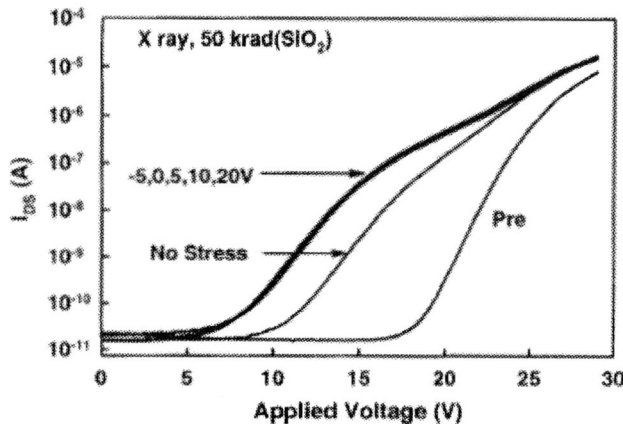

Fig. 2. Current as a function of gate voltage and pre-irradiation elevated-temperature stress bias (applied for one week at 150 °C) for field oxide transistors with 800 nm oxides fabricated in a 2 μm technology at Sandia National Laboratories. The devices were irradiated with 10-keV X rays at a dose rate of 1670 rad(SiO$_2$)/s and 5 V bias. (After [4], © IEEE, 1996).

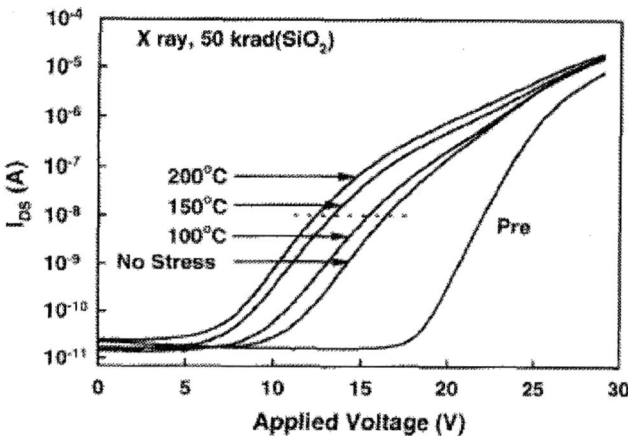

Fig. 3. Current as a function of gate voltage and pre-irradiation elevated temperature stress (one week at 5 V) for field oxide transistors with 800 nm oxides fabricated in a 2 μm technology at Sandia National Laboratories. The devices were irradiated with 10-keV X rays at a dose rate of 1670 rad(SiO$_2$)/s and a 5 V gate bias. (After [4], © IEEE, 1996).

mechanisms responsible for changes in radiation response with burn-in were similar to effects that might occur during long-term use or storage, and proposed a test methodology (Fig. 5) that attempts to use high-temperature bakes to identify parts that may show significant aging effects [5]. This assumption is most directly applicable for Arrhenius processes with single, effective activation energies. However, no aging data were presented in [5], and later work (below) shows that aging effects in microelectronics often are more complex than this test sequence presumes.

Bipolar linear microcircuits were also identified as sensitive to PETS by Pease et al. for LT 1014 and LM 111 devices [6]. It was suggested that a reduction in interface-trap density and an increase in oxide-trap charge density

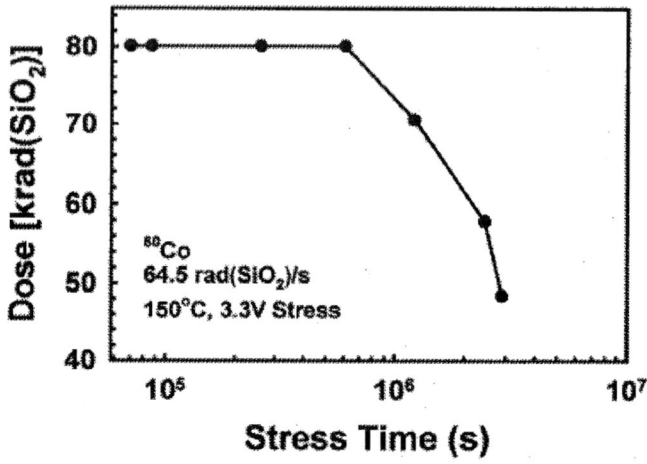

Fig. 4. Dose to first functional failure versus preirradiation elevated-temperature stress time for 3.3 V Paradigm Static Random Access Memories. (After [5], © IEEE, 1997).

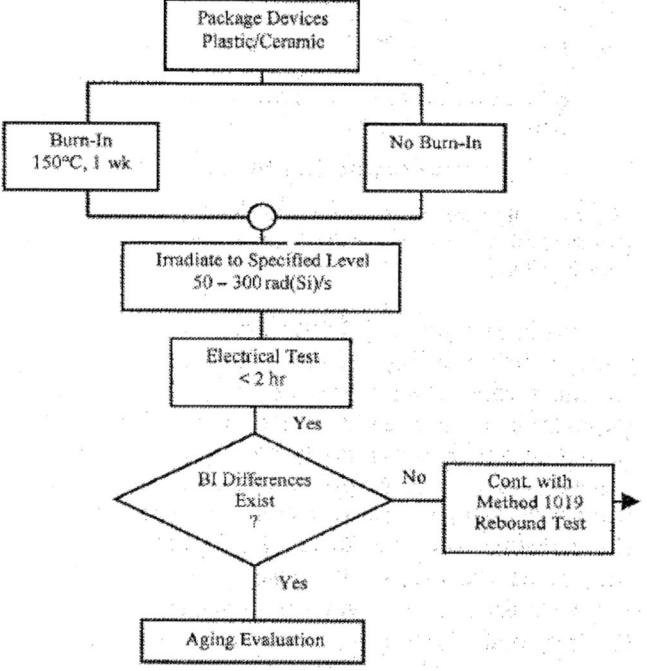

Fig. 5. Proposed test method to identify devices potentially sensitive to aging effects (after [5], © IEEE, 1997).

were responsible for the effects observed in these devices. An extreme sensitivity to PETS effects was observed by Shaneyfelt et al. in linear bipolar ICs that also exhibited enhanced low-dose-rate sensitivity (ELDRS) associated with enhanced interface-trap formation during low-dose-rate irradiation, relative to higher-rate irradiation [7]. Remarkably, baking devices at modest temperatures (100 to 200 °C) was found to almost entirely remove ELDRS for LM111s manufactured by National Semiconductor (Fig. 6). It was suggested that the inter-relation between PETS and ELDRS might be attributable to a mutual dependency of the two phenomena on hydrogen in the thick, low-quality oxides that passivate base-emitter junctions [7]. Baking devices at 450 °C for 200 s was found, in remarkable contrast, to *increase* the radiation sensitivity of the same LM 111s (Fig. 7) at higher dose rates. These results suggest that a complex interaction likely occurs between hydrogen species and precursor defects to radiation damage, e.g., oxygen vacancies in SiO_2 and the dangling bonds at the Si/SiO_2 interface that typically are passivated by hydrogen prior to radiation exposure or high-field stress [8].

The first study of the effects of long-term storage on MOS radiation response was performed by Karmarkar, et al. in 2001 [9]. In this work the measured radiation response of MOS capacitors in 2001 was compared to the radiation response of capacitors from the same wafer that

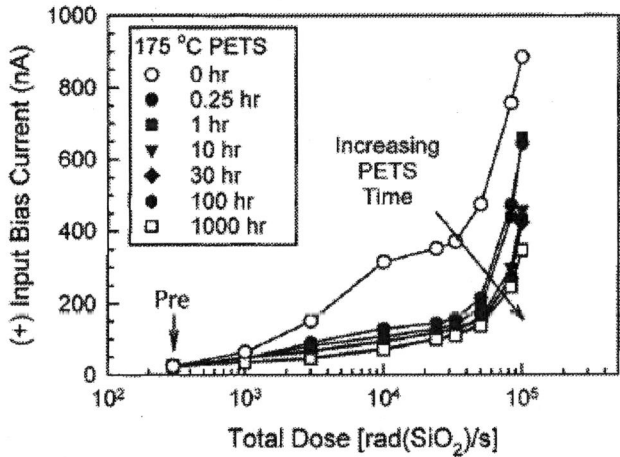

Fig. 6. Input bias current as a function of dose and pre-irradiation elevated temperature stress (PETS) time for LM111 devices from National Semiconductor irradiated at 0 V at a dose rate of 0.1 rad(SiO_2)/s. Increasing the duration of the PETS treatment reduces the enhancement of damage at low dose rate in these devices. (After [7], © IEEE, 1997.)

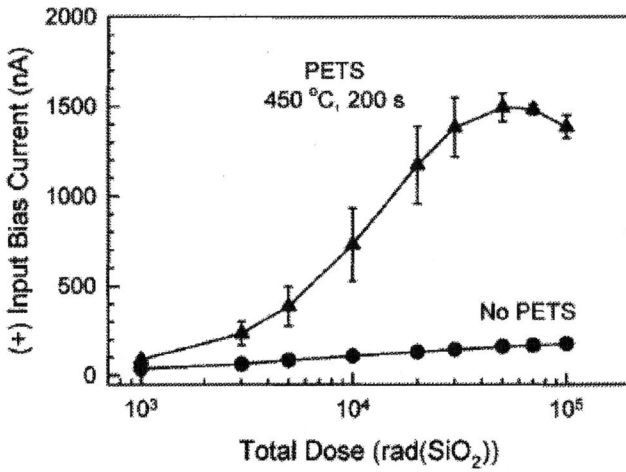

Fig. 7. Input bias current as a function of dose and pre-irradiation stress for LM111 devices from National Semiconductor irradiated at 0 V at a dose rate of 50 rad(SiO_2)/s. (After [7], © IEEE, 1997.)

been irradiated in 1986. These devices had no passivating overlayers, and were stored in room ambient conditions. For Al-gate devices, a significant decrease in the rate of buildup of radiation-induced oxide-trap charge was observed after aging (Fig. 8), with much less change seen for devices with poly-crystalline Si gates in [9]. Baking the devices of Fig. 8 at 205 °C for ~ 18 h restored the original radiation response (Fig. 9). These effects were attributed to moisture absorption in the SiO_2 during storage, leading to passivation of oxide-trap precursors; this process is reversed with a pre-irradiation bake [9]. Effects of aging on interface traps were not assessed in [9], owing to limitations of the availability of data from 1986.

The aging results of Figs. 8 and 9 were obtained on simple capacitors that did not experience full CMOS processing. In the next section, we discuss aging experiments that were performed on fully processed MOS devices with P-glass (3% P-doped SiO_2) passivation layers [11]. We summarize their responses as functions of irradiation and post-irradiation biased annealing, and present evidence that the absorption and interactions of water can play a critical role in these results as well.

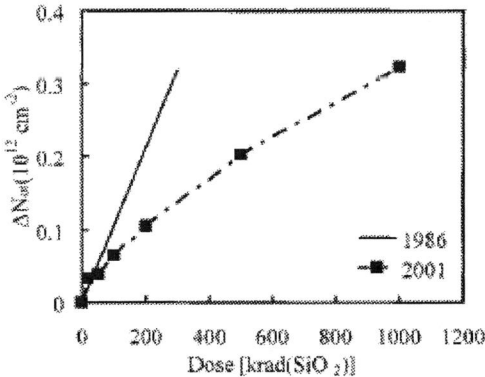

Fig. 8. Oxide-trap charge densities for Al-gate capacitors with 33 nm non-radiation-hardened oxides irradiated in 1986 [10] and 2001. Devices were irradiated with 10-keV X rays at a bias of 5 V. (After [9], © IEEE, 2001.)

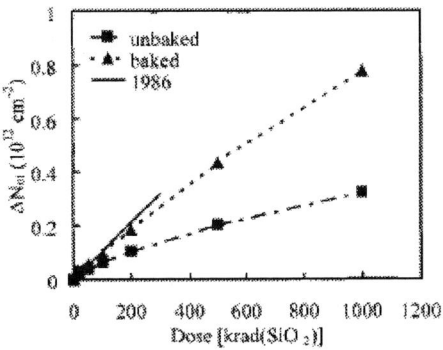

Fig. 9. Oxide-trap charge densities for Al-gate capacitors with 33 nm non-radiation-hardened oxides irradiated in 1986 [10] and 2001. The baked devices were heated to 205 °C for 18 hours at 0 V bias. Devices were irradiated with 10-keV X rays at a bias of 5 V. (After [9], © IEEE, 2001.)

II. AGING EFFECTS IN MOS TRANSISTORS

Devices tested for aging effects in [11] include nMOS transistors processed at Sandia National Laboratories in 1984, with poly-Si gates, oxide thicknesses of 32 nm or 60 nm, and P-Glass chip passivation. All irradiations were performed with a 10-keV X-ray irradiator at room temperature at 6 V bias. Some devices were baked before irradiation. Devices were annealed at room temperature and then 100 °C at 6 V bias. Fig. 10(a) compares threshold voltage shifts for nMOS transistors for 32 nm oxide parts irradiated in 1988 [12] and in 2005 [11]; the latter exposures were performed with and without exposure to PETS. The threshold-voltage rebound for parts irradiated in 2005 is much larger than for parts irradiated in 1988 (Fig. 10(a)). Fig. 10(b) shows the estimated threshold-voltage shifts due to interface-trap charge ΔV_{it} for the 32 nm parts. By the end of the post-irradiation room-temperature anneal, the values of ΔV_{it} for parts not exposed to PETS are ~ 67% greater than the maximum ΔV_{it} experienced by these parts in 1988. However, when parts are exposed to PETS prior to irradiation, these shifts in magnitude decrease substantially, but are still greater than the 1988 values [11],[12]. Fig. 10(c) shows the shifts in threshold voltage due to oxide-trap charge ΔV_{ot} for the 32 nm gate oxide transistors. Much less change is observed with aging or baking.

Also tested in [11] were 60 nm gate oxide transistors stored in hermetically sealed packaged since 1987. These parts were irradiated to 100 krad(SiO$_2$); results are plotted in Fig. 11. Fig. 11(a) plots ΔV_{th} for nMOS transistors. The parts not exposed to PETS show an increase in magnitude of ΔV_{th} of ~0.2 V during the post-irradiation biased anneal, as compared to the 1988 results [11],[12]. The magnitude of this increased shift during annealing is reduced by approximately 50% when parts are baked prior to irradiation. Fig. 11(b) shows ΔV_{it} for the 60 nm gate oxide transistors. Less change is observed with aging than for the non-hermetically sealed devices of Fig. 10(b). Fig. 11(c) shows the values of ΔV_{ot}. To within experimental uncertainty, there is little difference in oxide-trap charge among the devices irradiated in 2005, with or without PETS, compared to the 1988 values [11],[12].

The combinations of aging and PETS effects illustrated in Figs. 10 and 11 also may lead to significant complications in evaluations of dose-rate and temperature effects in MOS and linear bipolar devices. As one potential past example, Winokur et al. reported a larger increase in the enhancement of interface-trap buildup at low dose rates in [13] on MOS transistors from different processing and packaging lots than in follow-on work reported in [12]. The enhancement in interface-trap density was especially large at zero bias (Fig. 12). While factors related to dosimetry were identified in [12] that may account for some of the differences in results between [12] and [13], the work reported in [11] suggests that differences in aging responses of the two different types of devices in [12] and

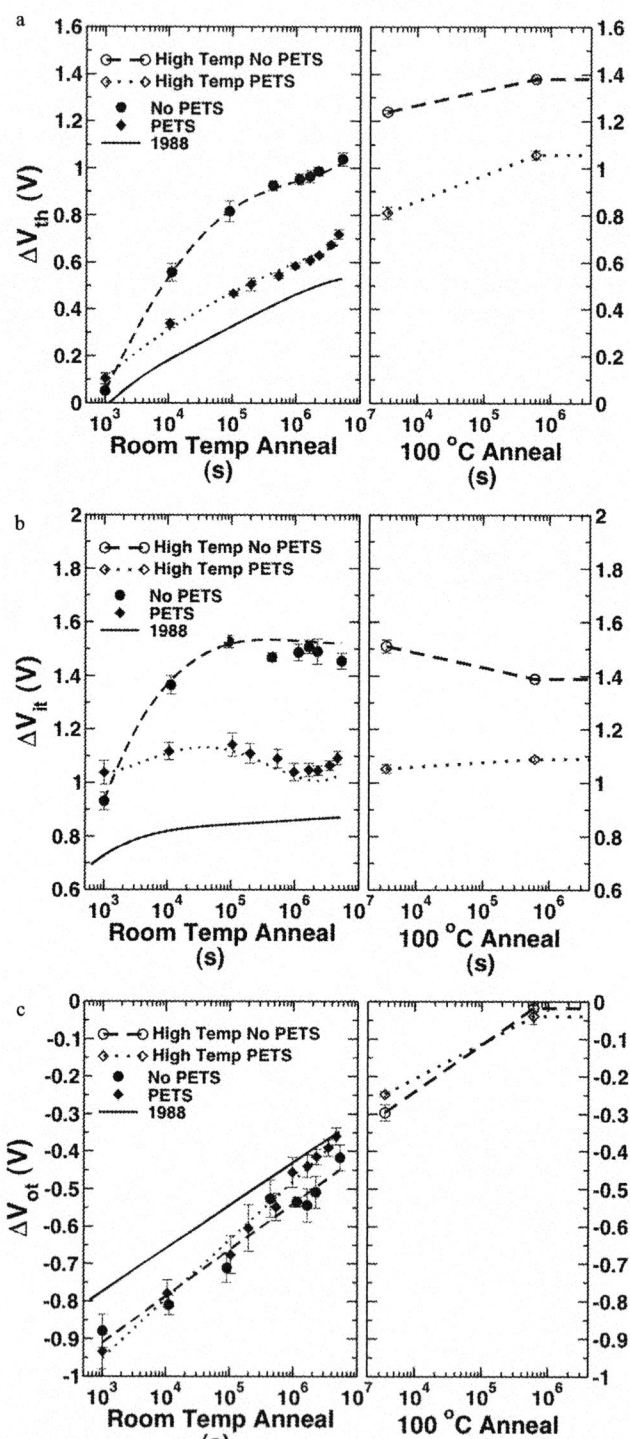

Fig. 10. ΔV_{th} (a), ΔV_{it} (b), and ΔV_{ot} (c) for 32 nm gate oxide nMOS transistors stored since 1987 in a non-hermetic environment vs. postirradiation anneal time for exposures to a dose of 500 krad(SiO$_2$). The irradiation and anneal bias was 6 V. The 1988 data are from [12]. (After [11], © IEEE, 2005.)

[13] may also have contributed to the observed differences in apparent dose-rate dependence of interface-trap buildup. Thus, it can be difficult to distinguish true or apparent dose rate effects from changes in device radiation response due to aging effects during a low-dose-rate radiation exposure.

Moreover, the complex interactions between aging and PETS effects observed in Figs. 6-7 and 10-11 may account at least in part for difficulties in applying elevated temperature irradiation as a hardness assurance test for enhanced low-dose-rate sensitivity in linear bipolar devices and ICs [14], as emphasized by Shaneyfelt et al. [8]. In radiation hardness assurance testing, all of these effects

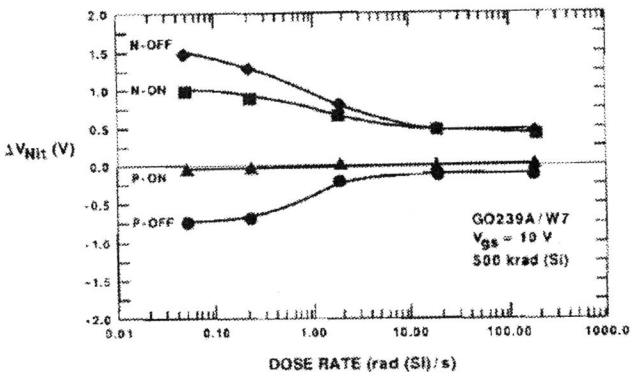

Fig. 12. Threshold voltage shifts due to interface traps as a function of radiation dose rate for 3 μm x 16 μm transistors with radiation-hardened 45 nm oxides that were fabricated, packaged, and tested at Sandia National Laboratories in 1987. Interface-trap buildup with 0 V gate bias is observed to be strongly enhanced at low dose rates, consistent with enhanced low-dose-rate sensitivity and/or potential aging effects in these devices. (After [13], © IEEE, 1987, used with permission.)

must be accounted for, preferably by testing as close to system use conditions as possible [1],[7],[14],[15].

III. THEORY OF H_2O IN SiO_2

Especially for parts stored in a non-hermetic environment, it is likely that moisture is associated with the observed effects of aging on radiation response [9]. Calculations reported in [11] show it is energetically favorable for water molecules to be absorbed in SiO_2 as interstitials. The energy gain is 0.3-0.5 eV per molecule, depending on topology. Bakos et al. [16] showed an interstitial water molecule in amorphous SiO_2 can break into two bonded SiOH (silanol) groups (Fig. 13 (a)), but the resulting complex is 0.3-0.7 eV higher in energy; the range is due to different local topologies in the amorphous network. Recently [11], a lower-energy configuration has been found, as shown in Fig. 13(b). The key difference between the structures is that Fig. 13(a) entails a broken ring [16], whereas in Fig. 13(b) there are no broken rings [11]. The energy is 0.3 eV lower than for free interstitial H_2O in SiO_2 (some variation with topology is expected). This complex can release hydrogen during subsequent irradiation or high-field stress.

Other hydrogen-related reactions also are important. Our preliminary results suggest O vacancies in SiO_2 can efficiently crack H_2O. This leads to the elimination of the O vacancy and the formation of excess hydrogen, which contributes to enhanced interface-trap buildup. These and other [8],[9] results suggest strongly that hydrogen-related species in SiO_2 cause many of the observed changes in radiation response with aging. Because MOS long-term reliability is also known to be quite sensitive to hydrogen [17],[18], it is quite likely that many long-term reliability issues in MOS devices and ICs are affected similarly – for example, bias-temperature instabilities, as we now discuss.

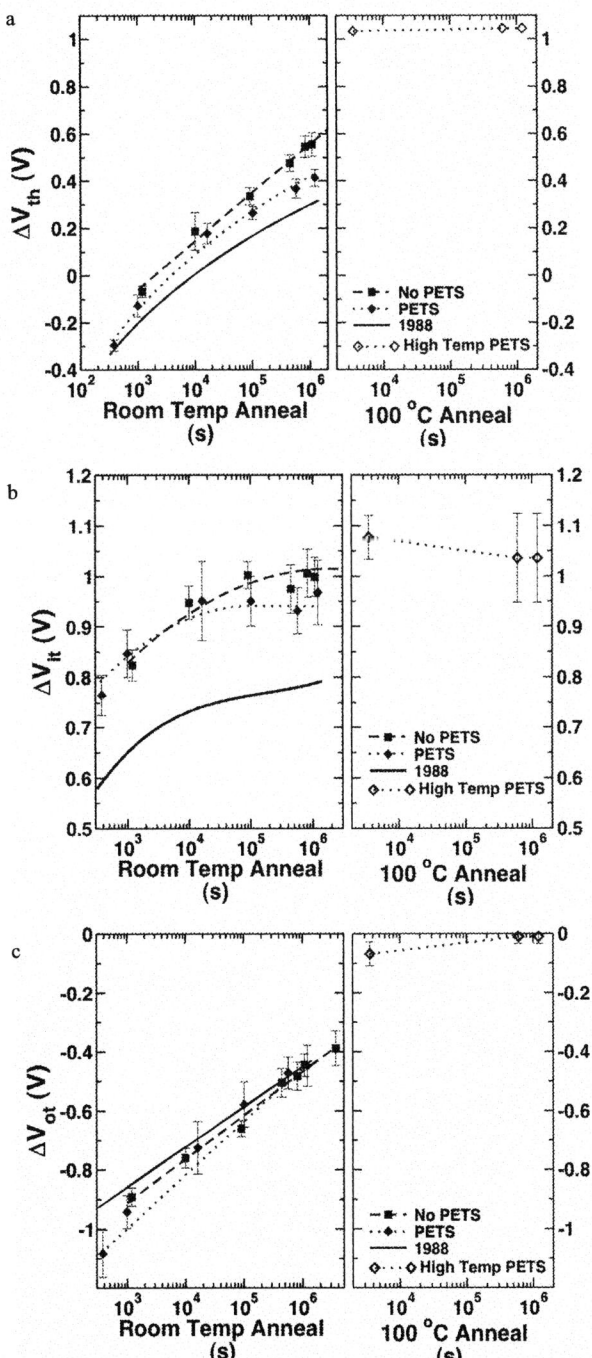

Fig. 11. ΔV_{th} (a), ΔV_{it} (b), and ΔV_{ot} (c) for 60 nm gate oxide nMOS transistors stored hermetically since 1987 vs. postirradiation anneal time for exposures to 100 krad(SiO$_2$). The irradiation and anneal bias was 6 V. The 1988 data are from [12]. (After [11], © IEEE, 2005.)

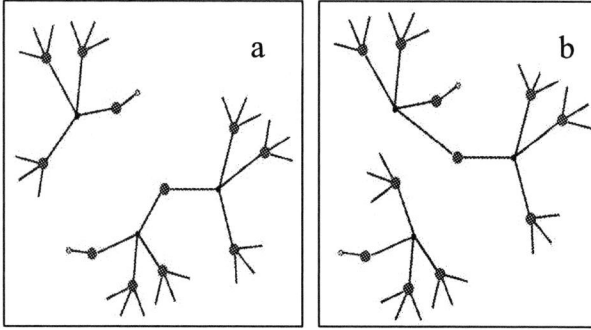

Fig. 13. Schematic diagram of (a) a water complex consisting of two SiOH (silanol) groups and a broken ring [energy = +(0.3-0.7) eV]; (b) a water complex consisting of two SiOH groups and no broken ring [energy = -0.3 eV]. The large dots denote O atoms; the small dots are Si atoms; and the gray circles are H atoms in SiO_2. (After [11], © IEEE, 2005.)

IV. BIAS-TEMPERATURE INSTABILITIES

Recently NBTI has been compared in SiO_2 and HfO_2 gate dielectrics; in the work reported in [19] the HfO_2 capacitors also have a thin oxynitride layer at the Si interface. The physical thickness of the high-κ layer was 6.8 nm; the interfacial oxynitride layer was 1.0 nm. Oxide and interface-trap charge densities, ΔN_{ot} and ΔN_{it}, are shown as functions of NBTI stress temperature in Fig. 14. For SiO_2, the effective activation energy in Fig. 14(a) for ΔN_{ot} is 0.27 ± 0.03 eV, and for ΔN_{it} is 0.31 ± 0.04 eV. Fig. 14(b) shows that, for SiO_xN_y/HfO_2 devices, the effective activation energy for ΔN_{ot} is 0.35 ± 0.04 eV, and for ΔN_{it} is 0.22 ± 0.03 eV [19]. These results are comparable to the range of effective activation energies that have been reported in the literature for thermal SiO_2 [18], and are also consistent with recent work on HfO_2 by Houssa et al. [20].

The direct dissociation of the Si-H bond at the Si/SiO_2 interface has been suggested as a mechanism for generation of traps during NBTI, for example, in the presence of holes at the interface [21]. Density functional theory (DFT) calculations have been performed to evaluate this and other potential processes that may lead to negative-bias temperature instabilities (NBTI) [22]. DFT calculations show that the removal of the H atom from the Si-H bond to a remote Si-Si bond raises the energy of the system by about 1.9 eV. When this value is added to the associated migration barrier, ~ 0.5 eV, the result leads to a dissociation activation energy E_a of ~ 2.4 eV, in close agreement with experimental measurements (2.6 eV) [23]. When holes are present, as is the case for NBTI in pMOS transistors, the energy required to remove the H from a Si-H bond decreases to 1.6 eV, and E_a decreases to 2.1 eV [22]. These values are well above those expected to affect NBTI, at least in the absence of very high electric fields and/or significant hot carrier effects.

Given that direct depassivation of Si-H bonds is not the cause for NBTI for typical experiments at low electric

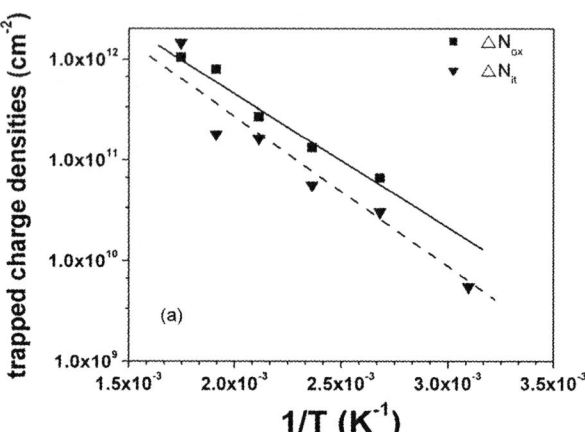

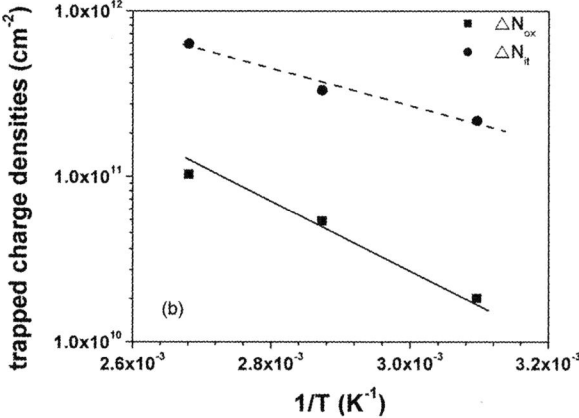

Fig. 14. Oxide- and interface-trap charge densities vs. stress temperature for (a: upper) SiO_2 capacitors stressed for 20 min at an electric field of -1.54 MV/cm; and (b: lower) SiO_xN_y/HfO_2 capacitors stressed for 20 min at -1.15 MV/cm. (After [19], © The American Institute of Physics, 2004.)

fields, a plausible alternative is the *depassivation* reaction [22],[24],[25]:

$$Si\text{-}H + H^+ \rightarrow Si^+ + H_2 . \qquad (1)$$

First-principles DFT calculations find the reaction energy and barrier for reaction (1) to be 0.5 eV and 0.95 eV, for n-type Si in inversion. Hence, this process is energetically preferred to direct depassivation of Si-H bonds.

Excess hydrogen above and beyond levels required to passivate fully the dangling bonds at the Si/SiO_2 interface is a likely origin of this reliability problem. For example, in n-type Si (i.e., pMOS transistors), H primarily binds to P dopant atoms [22]. The energy to move H from a P-H bonding site to a remote position in Si is 0.6 eV, in strong contrast to the value of 1.9 eV for Si-H bonds. The P-H binding energy, when combined with the migration barrier of 0.7 eV for H$^-$ in Si [26], gives an activation energy of 1.3 eV for dissociation of the P-H complex, in agreement with the experimental value of ~ 1.2 eV [27]. So, this process also is significantly preferred energetically to direct Si-H bond breaking.

The stability of the P-H complex changes dramatically when the charge state of hydrogen changes to neutral (H^0). In this case, the value of the binding energy Δ_{PH} drops from 0.6 eV to 0.2 eV. Combined with the small diffusion barrier of H^0 in Si (0.1-0.2 eV [28]), the activation energy (E_a^0) to release H^0 from a P-H complex is only 0.3-0.4 eV. Because of the very small activation energy, all of the hydrogen trapped in P-H complexes in the depletion region can be released, and can migrate rapidly with a small barrier for H^0 (0.1-0.2 eV). At temperatures of 100-200 °C, some hydrogen can be released from complexes deeper in the Si substrate [22]. Hydrogen that arrives or is released in the inversion layer becomes positively charged (H^+) by trapping a hole under negative bias during NBTI; this H^+ is swept to the interface by the negative bias. The inversion layer and the depassivation reaction act as sinks for hydrogen in Si. Because of the relatively high density of dopant atoms in the substrate, only a small percentage of P atoms must bind a hydrogen atom prior to NBTI to produce the densities of interface traps typically measured in NBTI experiments, emphasizing the consistency with data of this explanation. Other mechanisms may be significant for NBTI and/or PBTI in *n*MOS devices [22], which typically is not as significant a problem as is NBTI in *p*MOS transistors [29].

Protons that reach the interface from the Si side encounter a barrier of 1 eV to enter the SiO_2 [30]. In contrast, the barrier against lateral migration is only ~ 0.3 eV [31]. Thus, protons can migrate rapidly along the interface. Because both transporting protons and interface traps are positively charged at negative bias, reactions of H^+ with Si^+ are not allowed due to simple electrostatics, so these protons cannot passivate pre-existing interface traps. In addition, the energy for H^+ at the center of a Si-Si bond next to a Si-H entity is lower by 0.2 eV with respect to other sites for H^+ in Si. Therefore, hydrogen can easily find Si-H bonds and activate reaction (1), leading to interface-trap formation. At 100-200°C, this reaction reaches quasi-equilibrium quickly.

The dynamic balance of interface-trap formation and passivation is controlled by the diffusion of the product H_2. In the diffusion-limited regime, E_a is given by [32],[33]:

$$E_a = \tfrac{1}{2}\Delta E + \tfrac{1}{4}\Delta_D . \qquad (2)$$

In Eq. (2), ΔE is the reaction energy of H_2 with a pre-existing interface trap, and Δ_D is the diffusion barrier of the migrating species. Using the value for $\Delta E \approx 0.5$ eV, and $\Delta_D = 0.45$ eV [34] for H_2 diffusion in SiO_2, it is found that $E_a \approx 0.36$ eV [22], in very good agreement with the low effective activation energies reported in many NBTI experiments [18]-[20].

Protons that arrive at the interface, preferentially in the vicinity of a Si-H bond, can also eventually migrate into the oxide where they contribute to the buildup of oxide trapped charge. Significantly, oxide-trap charge in the form of a proton is much more difficult to neutralize via electron

tunneling than trapped holes, leading to the relative stability of trapped H^+ in SiO_2 under NBTI stress conditions [19],[24]. DFT calculations suggest $E_a = 0.2\text{-}0.3$ eV for the increase of oxide trapped charge [22], in agreement with measured values [18]-[20]. As for interface traps, the small E_a observed in experimental studies is only an *apparent* activation energy, observed in the large stress time limit, after the migration of H^+ from Si to SiO_2 reaches quasi-equilibrium. Combined irradiation and bias-temperature effects are even more complex, and involve not only hydrogen, but also electron and hole trapping and de-trapping in the gate dielectric layer [35].

CONCLUSION

Oxide and interface traps in MOS devices are affected strongly by hydrogen-related species. These species can migrate and react during device storage or use, especially when hermeticity is not maintained. As a result, the radiation response and long-term reliability of MOS and bipolar devices can change significantly with time. As a result, measurements performed early in a device lifetime may not be able to easily predict the significant consequences of these processes, which naturally occur over a wide range of effective activation energies. Hence, the radiation response of a MOS device can be affected significantly by aging, and the inferred device reliability also likely varies significantly with aging time. For devices that are sensitive to these kinds of aging effects, additional margins against failure are required.

ACKNOWLEDGMENT

This work was supported in part by the US Navy and by the Air Force Office of Scientific Research through a Multidisciplinary University Research Initiative.

REFERENCES

[1] M. R. Shaneyfelt, D. M. Fleetwood, J. R. Schwank, T. L. Meisenheimer, and P. S. Winokur, "Effects of Burn-In on Radiation Hardness," *IEEE Trans. Nucl. Sci.,* vol. 41, pp. 2550-2559, 1994.

[2] S. D. Clark, J. P. Biggs, M. K. Williams, D. R. Alexander, M. C. Maher, and R. L. Pease, "Plastic Packaging and Burn-In Effects on Ionizing Dose Response in CMOS Microcircuits," *IEEE Trans. Nucl. Sci.,* vol. 42, pp. 1607-1614, 1995.

[3] S. Djoric-Veljkovic, I. Manic, V. Davidovic, S. Golubovic, and N. Stojadinovic, "Effects of Burn-In Stressing on Post-Irradiation Annealing Response of Power VDMOSFETs," *Microelectron. Reliab.,* vol. 43, pp. 1455-1460, 2003.

4] M. R. Shaneyfelt, P. S. Winokur, D. M. Fleetwood, J. R. Schwank, and R. A. Reber, Jr., "Effects of Reliability Screens on MOS Charge Trapping," *IEEE Trans. Nucl. Sci.,* vol. 43, pp. 865-872, 1996.

[5] M. R. Shaneyfelt, P. S. Winokur, D. M. Fleetwood, G. L. Hash, J. R. Schwank, and F. W. Sexton, "Impact of Aging on Radiation Hardness," *IEEE Trans. Nucl. Sci.,* vol. 44, pp. 2040-2047, 1997.

[6] R. L. Pease, M. R. Shaneyfelt, P. S. Winokur, D. M. Fleetwood, J. Gorelick, S. McClure, S. Clark, L. Cohn, and D. Alexander, "Mechanisms for Total Dose Sensitivity to Preirradiation Thermal Stress in Bipolar Linear Microcircuits," *IEEE Trans. Nucl. Sci.,* vol. 45, pp. 1425-1430, 1998.

[7] M. R. Shaneyfelt, J. R. Schwank, S. C. Witczak, D. M. Fleetwood, R. L. Pease, P. S. Winokur, L. C. Riewe, and G. L. Hash, "Thermal Stress Effects and Enhanced Low-Dose-Rate Sensitivity in Linear Bipolar ICs," *IEEE Trans. Nucl. Sci.,* vol. 47, pp. 2539-2545, 2000.

[8] P. M. Lenahan and P. V. Dressendorfer, "Hole Traps and Trivalent Silicon Centers in MOS Devices," *J. Appl. Phys.,* vol. 55, pp. 3495-3499, 1984.

[9] A. P. Karmarkar, B. K. Choi, R. D. Schrimpf, and D. M. Fleetwood, "Aging and Baking Effects on the Radiation Hardness of MOS Capacitors," *IEEE Trans. Nucl. Sci.,* vol. 48, pp. 2158-2163, 2001.

[10] D. M. Fleetwood, P. S. Winokur, L. J. Lorence, Jr., W. Beezhold, P. V. Dressendorfer, and J. R. Schwank, "The Response of MOS Devices to Dose-Enhanced Low-Energy Radiation," *IEEE Trans. Nucl. Sci.,* vol. 33, pp. 1245-1251, 1986.

[11] M. P. Rodgers, D. M. Fleetwood, R. D. Schrimpf, I. G. Batyrev, S. Wang, and S. T. Pantelides, "The Effects of Aging on MOS Irradiation and Annealing Response," *IEEE Trans. Nucl. Sci.,* vol. 52, No. 6, Dec. 2005.

[12] D. M. Fleetwood, P. S. Winokur, and J. R. Schwank, "Using Laboratory X-ray and Cobalt-60 Irradiations to Predict CMOS Device Response in Strategic and Space Environments," *IEEE Trans. Nucl. Sci.,* vol. 35, pp. 1497-1505, 1988.

[13] P. S. Winokur, F. W. Sexton, G. L. Hash, and D. C. Turpin, "Total-Dose Failure Mechanisms of ICs in Laboratory and Space Environments," *IEEE Trans. Nucl. Sci.,* vol. 34, pp. 1448-1454, 1987.

[14] R. L. Pease, L. M. Cohn, D. M. Fleetwood, M. A. Gehlhausen, T. L. Turflinger, D. B. Brown, and A. H. Johnston, "A Proposed Hardness Assurance Test Methodology for Bipolar Linear Circuits and Devices in a Space Ionizing Radiation Environment," *IEEE Trans. Nucl. Sci.,* vol. 44, pp. 1981-1988, 1997.

[15] D. M. Fleetwood and H. A. Eisen, "Total-Dose Radiation Hardness Assurance," *IEEE Trans. Nucl. Sci.,* vol. 50, pp. 552-564, 2003.

[16] T. Bakos, S. N. Rashkeev, and S. T. Pantelides, "H_2O and O_2 Molecules in Amorphous SiO_2: Defect Formation and Annihilation Mechanisms," *Phys. Rev. B,* vol. 69, Article No. 195206, 2004.

[17] D. M. Fleetwood, "Effects of Hydrogen Transport and Reactions on Microelectronics Radiation Response and Reliability," *Microelectron. Reliab.,* vol. 42, pp. 523-541, 2002.

[18] D. K. Schroder and J. A. Babcock, "Negative Bias Temperature Instability: Road to Cross in Deep Submicron Silicon Semiconductor Manufacturing," *J. Appl. Phys.,* vol. 94, pp.1513-1530, 2003.

[19] X. J. Zhou, L. Tsetseris, S. N. Rashkeev, D. M. Fleetwood, R. D. Schrimpf, S. T. Pantelides, J. A. Felix, E. P. Gusev, and C. D'Emic, "Negative Bias-Temperature Instabilities in MOS Devices with SiO_2 and SiO_xN_y/HfO_2 Gate Dielectrics," *Appl. Phys. Lett.,* vol. 84, pp. 4394-4396, 2004.

[20] M. Houssa, S. D. Gendt, J. L. Autran, G. Groeseneken, and M. M. Heyns, "Role of Hydrogen on Negative Bias Temperature Instability in HfO_2-based Hole Channel Field-Effect Transistors," *Appl. Phys. Lett.,* vol. 85, pp. 2101-2103, 2004.

[21] M. A. Alam and S. Mahapatra, "A Comprehensive Model of pMOS NBTI Degradation," *Microelectron. Reliab.,* vol. 45, pp. 71-81, 2005.

[22] L. Tsetseris, X. J. Zhou, D. M. Fleetwood, R. D. Schrimpf, and S. T. Pantelides, "Physical Mechanisms of Negative-Bias Temperature Instability," *Appl. Phys. Lett.,* vol. 86, Article No. 142103, 2005.

[23] K. L. Brower, "Kinetics of H_2 Passivation of P_b Centers at the (111) Si-SiO_2 Interface," *Phys. Rev. B,* vol. 38, pp. 9657-9666, 1988.

[24] S. N. Rashkeev, D. M. Fleetwood, R. D. Schrimpf, and S. T. Pantelides, "Proton-Induced Defect Generation at the Si-SiO_2 Interface," *IEEE Trans. Nucl. Sci.,* vol. 48, pp. 2086-2092, 2001.

[25] L. Tsetseris and S. T. Pantelides, "Migration, Incorporation and Passivation Reactions of Molecular Hydrogen at the Si-SiO_2 Interface," *Phys. Rev. B,* vol. 70, Article No. 245320 (2004).

[26] N. M. Johnson and C. Herring, "Diffusion of Negatively Charged Hydrogen in Silicon," *Phys. Rev. B,* vol. 46, pp. 15554-15557, 1992.

[27] J. Zhu, N. M. Johnson, and C. Herring, "Negative Charge State of Hydrogen in Silicon," *Phys. Rev. B,* vol. 41, pp. 12354-12375, 1990.

[28] C. G. van de Walle, P. J. H. Denteneer, Y. Baryam, and S. T. Pantelides, "Theory of Hydrogen Diffusion and Reactions in Crystalline Silicon," *Phys. Rev. B,* vol. 39, pp. 10791-10808, 1989.

[29] D. M. Fleetwood, X. J. Zhou, L. Tsetseris, S. T. Pantelides, and R. D. Schrimpf, "Hydrogen Model for Negative-Bias Temperature Instabilities in MOS Gate Insulators," PV 2005-01: ISBN 1-56677-459-4: *Silicon Nitride and Silicon Dioxide Thin Insulating Films and Other Emerging Dielectrics VIII,* edited by R. E. Sah, M. J. Deen, J. Zhang, J. Yota, and Y. Kamakura, pp. 267-278, 2005.

[30] S. T. Pantelides, S. N. Rashkeev, R. Buczko, D. M. Fleetwood, and R. D. Schrimpf, "Reactions of Hydrogen with Si-SiO_2 Interfaces," *IEEE Trans. Nucl. Sci.,* vol. 47, pp. 2262-2268, 2000.

[31] S. N. Rashkeev, D. M. Fleetwood, R. D. Schrimpf, and S. T. Pantelides, "Dual Behavior of H^+ at Si-SiO_2 Interfaces: Mobility versus Trapping," *Appl. Phys. Lett.,* vol. 81, pp. 1839-1841, 2002.

[32] K. O. Jeppson and C. M. Svensson, "Negative Bias Stress of MOS Devices at High Electric Fields and Degradation of MNOS Devices," *J. Appl. Phys.,* vol. 48, pp. 2004-2014, 1977.

[33] S. Ogawa and N. Shiono, "Generalized Diffusion-Reaction Model for the Low-Field Charge Buildup Instability at the Si-SiO_2 Interface," *Phys. Rev. B,* vol. 51, pp. 4218-4230, 1995.

[34] M. L. Reed and J. Plummer, "Si-SiO_2 Interface Trap Production by Low-Temperature Thermal Processing," *Appl. Phys. Lett.,* vol. 51, pp. 514-516, 1987.

[35] X. J. Zhou, D. M. Fleetwood, J. A. Felix, E. P. Gusev, and C. D'Emic, "NBTI and Radiation Effects in High-K Alternative Dielectrics," *IEEE Trans. Nucl. Sci.,* vol. 52, No. 6, Dec. 2005.

Session
Nanotechnologies

Nonlinear Model of Microtubule Dynamics and Its Impact on Kinesin Motion

Miljko V. Satarić

Abstract – In this paper we elaborate the nonlinear model based on ferro-electric properties of microtubules (MTs) and triggered by hydrolysis of GTP (guanosine triphosphate) followed by conversion of chemical energy into large conformational rotation of corresponding tubulin dimer. We attempted to elucidate some functional properties of micro-tubules pertaining kinesin motor protein on the basis of this elegant model.

I. INTRODUCTION

Microtubules are actually nanotubes with diameter of 25 nm, (Fig. 1). They are found in nearly all eukariotic cells and are polymers of filaments consisting of tubulin globular protein.

MTs serve as tracks on which motor proteins may carry materials about the cell and serve as scaffolding to maintain the cell shape since they are among the most rigid structures within a typical cell.

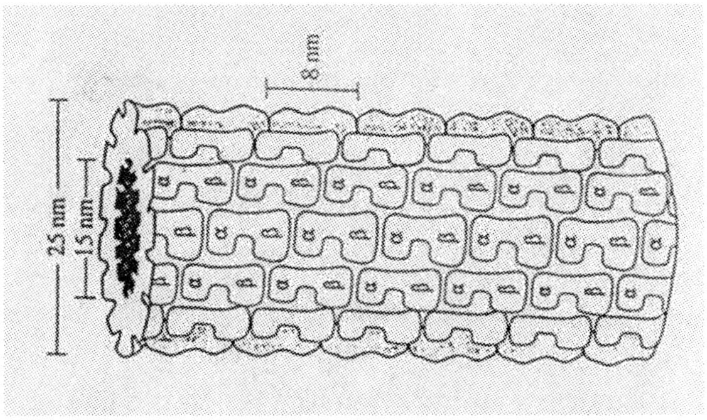

Fig. 1. A cartoon showing the construction of a MT from individual tubulin dimers.

The building block of a MT is a tubulin dimer consisting of two slightly different globular proteins named α and β tubulin, respectively. Tubulin protein contains

---

Miljko V. Satarić is with the Faculty of Technical Sciences, University of Novi Sad, Trg D. Obradovića 6, 21000 Novi Sad, Serbia and Montenegro, E-mail: bomisat@nspoint.net

approximately 900 amino acids comprising some 14.000 atoms with combined mass of 110 kDa [1].

Tubulin's secondary structure is of crutial significance to the model discussed in this paper. The core of tubulin globula consists of β-sheet surrounded by several α-helices terminated with loops protruding partially through the globular structure of the protein, (Fig. 2.a).

These loops are important either for the formation of weak bonds between neighbouring proteins bringing about the polymerization of tubulin into microtubule; see one filament on (Fig. 2.b). Additionally, they provide a binding site for the energy giving molecule GTP.

These flexible loops are idealy suited for adjusting the positions of the many amino acid residues that participate in binding between two neighbouring dimer in accordance with so called "hand and glove" strategy.

II. FERROELECTRIC PROPERTIES OF MTs AND KINK EXCITATIONS DUE TO GTP HYDROLYSIS

The basic motivation for our model is the secondary structure of tubulin protein. Virtually every peptide group in an α-helix possesses a considerable dipole moment on the order of $p_0 = 1:2 \ 10^{-29}$C m = 3; 5D. All these dipoles are almost parallel to the helical axis, giving rise to an overall dipole moment of this particular helix. It is generally accepted that this large dipole moment of an _-helix has an important biological role. The idea that MTs are ferroelectrics was proposed some 30 years ago and farther quantitatively elaborated by Satarić et al [2,3,4,5], on the basis of their piezoelectric properties. It is apparent that due to the strong curvature of an MT cylinder (see Fig. 3.a) the inner parts of the globular tubulin structure are compressed while the outer ones are stretched by a substantial amount of tension.

This leads to additional redistribution of excess negative charge enhancing the transversal component of the net dipole moment of every dimer comprising a MT. This was corroborated by a detailed map of the electric charge distribution for the tubulin dimer [6], (Fig. 3b).

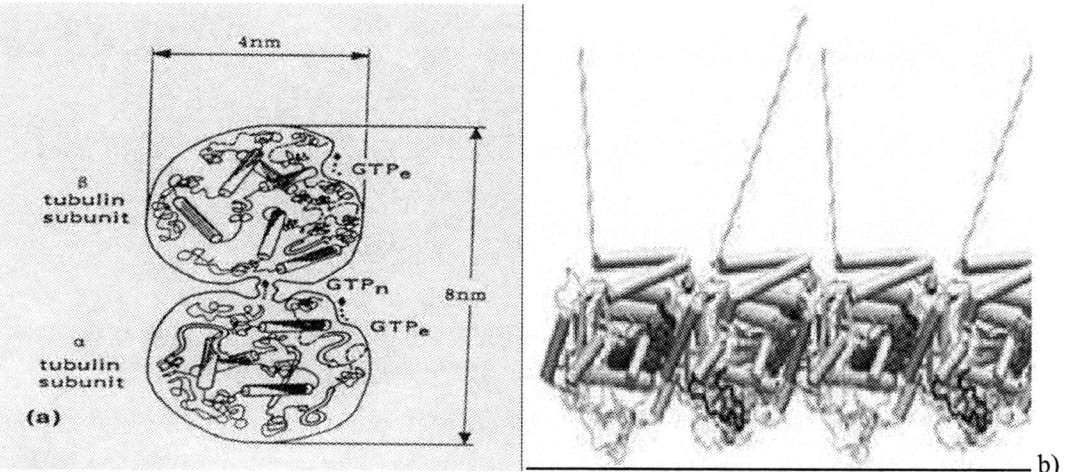

Fig. 2. a) The secondary structure of tubulin protein; cylinders represent α-helices. b) The sketch of one MT filament consisting of tubulin dimers joined by protruding loops.

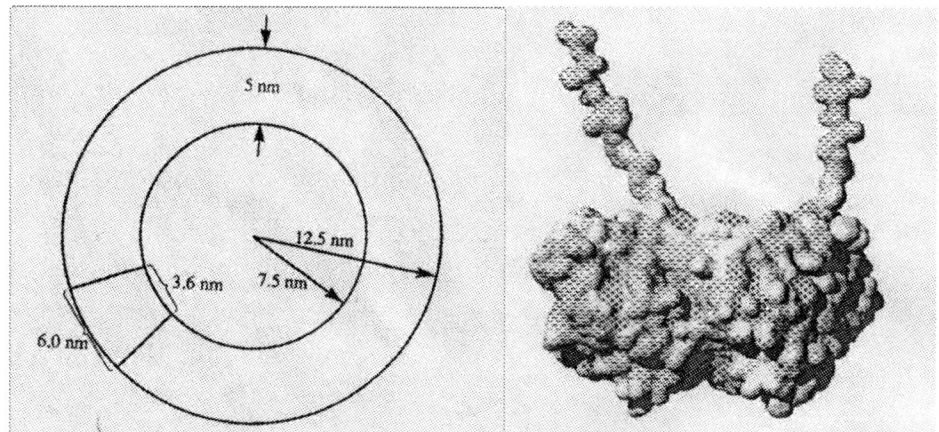

Fig. 3. a) The cylindrical geometry of a MT; b) The detailed map of the electric charge distribution in a dimer.

Taking into account the above arguments it is highly plausible to envisage the MT structure as a ferroelectric system with dipole moments of tubulins oriented as shown in (Fig. 4).

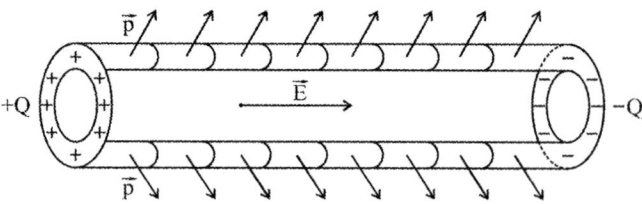

Fig. 4. The MT ferroelectric scheme with overal electricaly charged tips.

The tips of a MT therefore possess overall net charges +Q and –Q respectively. We therefore conclude that MT cylinder supports an endogenous electric field $\vec{E}$ parallel with axis of MT.

III. THE ROLE OF GTP HYDROLYSIS IN NONLINEAR MT DYNAMICS

In the case under consideration of tubulin and MT a β-tubulin in dimeric structure has an exchangeable GTP site that is an active domain. The chemical potential energy of GTP hydrolysis is converted into mechanical motion as follows. The dissociation of the inorganic phosphate group (P_γ) in the reaction (GTP → GDP + P_γ) causes a shift of a few angstroms in the GTP binding site that contains a polypeptide loop called a sensor or switch loop. This displacement causes a further conformational change that propagates along a tightly associated α-helix that is called realy-helix. The relay-helix serves as a "piston" that transmits sensor stimulus to adhere to a specific site on the oposite side of the same β-tubulin latching it in a "closed" conformation.

100

This state of the relay-helix pertains to a prestressed spring, to use a mechanical analogy. The next stage of the conformational scenary is triggered by the release of GDP and causes the relay-helix to unlatch alowing tubulin to tilt rotating by angle θ in (x; r) plane in the filament direction as shown in Fig. 5.

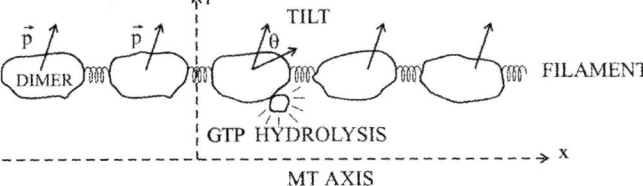

Fig. 5. The tilt deformation of MT dimers caused by GTP hydrolysis.

This transformation is evolutionary consistent with the conformational strategies exibited by motor proteins due to ATP hydrolysis [7].

The force estimated elsewhere [8] results in the tilt of the tubulin dimer by an angle

$$\theta_0 \simeq 0,5 rad. \tag{1}$$

Consequently we calculated [8] the work done by the pistonlike movement of relay-helix to be $2 \cdot 10^{-20}$J = 0; 13eV . Since the energy released in one GTP hydrolysis event is approximately 0; 25eV it follows that the most part of it is utilized in above tilt of tubulin dimer.

Therefore we treat a tubulin dimer with just hydrolysed GTP and released GDP as rotational pendulum with respect to tilt angle θ and with corresponding rotational inertia per unit volume J, possesing kinetic energy

$$w_{kin} = \frac{1}{2} J \left(\frac{\partial \theta}{\partial t} \right)^2 . \tag{2}$$

The splay energy density has the form

$$w_{sp} = \frac{1}{2} k \left(\frac{\partial \theta}{\partial x} \right)^2 \tag{3}$$

where k denotes splay elastic modulus. The crucial part of total energy is the energy of dimer-dimer coupling within a protofilament which we expand in terms of tilt angle θ as even function up to fourth order as follows

$$w_{el} = \frac{1}{r^2} (-A\theta^2 + B\theta^4) \tag{4}$$

where A and B are coeficients that are expressed in terms of cylindrical components of the elasticity tensor.

Why besides quadratic term we keep fourth order? Since the triggering tilt angle ranges to $\theta_0 \sim 0,5$ rad, we have $\theta^4 \sim 1/16$ which is not negligible in comparisson with $\theta^2 \sim 1/4$.

The last but not the least important part of total energy of pendulum is the density of polarization energy

$$w_{pol} = \frac{1}{2} \left(\frac{p_l^2}{\chi_l} + \frac{p_t^2}{\chi_t} \right) - p_l E - \mu_p p_t \theta \tag{5}$$

where p_l, p_t are longitudinal and transversal projections of tubulin (pendulum) polarization, χ_l, χ_t the anizotropic dielectric susceptibilities of MT, while μ_p is phenomenological constant of model and E is the constant intrinsic electric field within a MT.

We also accounted the presence of viscosity of the medium surrounding the MT (cytosol). This effect was modeled by including a friction term in the equation of motion by corresponding torque.

$$\tau_{vis} = -\Gamma \frac{\partial \theta}{\partial t} \tag{6}$$

that is subsequently involved in the equation of motion using Euler-Lagrange procedure. Γ depends both on the viscosity coefficient and the structural details of the MT through the Stokes-Einstein formula.

Combining all terms in Eqs (2-6) in the context of Euler-Lagrange equation, then using scaled variables and traveling wave form, one obtains the following nonlinear ordinary differential equation

$$\frac{d^2\eta}{d\xi^2} + \beta \frac{d\eta}{d\xi} - \eta^3 + \eta + \varepsilon = 0 \tag{7}$$

with abreviations

$$\eta = \frac{\theta}{\theta_0}; \quad \xi = \alpha(x - vt)$$

θ_0 is the amplitude of tilt; v is the velocity of kink excitation and α is the kink's wave number. β and ε involve elastical and ferroelectrical parameters and intrinsic electric field E.

The most important consequences of Eq (7) are kink and antikink excitations resembling to domino-effect in the well-known tangens hyperbolic shape, (Fig 6).

$$\theta(\xi) = \pm\theta_0 tgh(\sqrt{2}\xi) \tag{8}$$

which propagate in oposite directions with constant terminal velocity v_t obeying Ohm's law

$$v_t = \mu E. \tag{9}$$

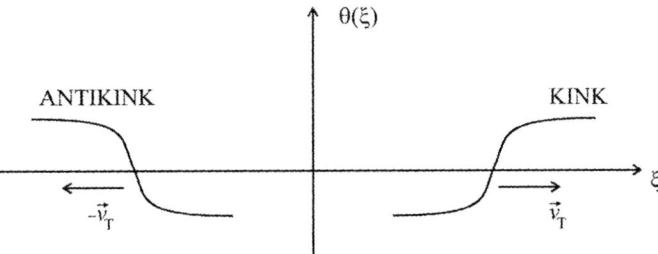

Fig. 6. The kink and antikink excitations propagating in oposite directions along MT filaments.

The kink's mobility μ depends on the model parameters including the ferroelectric character of MT and of its elastic properties as well as of viscous disipations on the very natural way. This linear response holds even for very strong fields of the order of 10^5 V/m.

IV. BIOLOGICAL IMPLICATIONS OF MODEL

In this paper we analyzed an interesting mechanism of utilization of GTP hydrolysis energy in MTs for creation of highly localized kink excitation which propagates along a MT filament combining mechano-elastic and dipolar energy and being controled by an intrinsic electric field provided by ferroelectric character of MTs. The intensity of this field depends on the length of the MT as well as on the physiological conditions of the solution (pH, ion concentration and temperature), due to screening effects of counter-ions. The shorter the MT's the stronger the field and greater the kink's terminal velocity, (Eq. 9).

The interesting case are MTs in long neuron cells. The axon potential propagating along neuron carries electric field of the order of 10^5 V/m. This field even partially screened along MTs causes high terminal velocities of kinks involved reaching several meters per second. If such fast (hot) kink encounters a motor protein (kinesin) bound to the same filament it may cause a fug-of-war [8] effect on the motor disrupting its two directional random walk. Since much faster than loaded kinesin, hot kink goes farther pushing many motors in the same direction.

Let us now estimate the kink motion in MTs in non neuronal cells. For a stable MT with a typical length L = 4 μm the intrinsic electric field in the central region of MT can be calculated to be

$$ E = \frac{Q}{\pi \varepsilon_0 \varepsilon_r L^2}. \tag{10} $$

Taking Q = 13e (e = $1{,}6 \cdot 10^{-19}$C), i.e., one excess charge per filament tip, (Fig. 4) and letting the relative dielectric constant to lie in the range 10 < ε_r < 80 as being the function of solution's composition, one finds the values of intrinsic electric field in the range (0; 5 < E < 5) V/m resulting in the window of kink terminal velocities given by

$$ (0,2 < v_t < 2)\mu m/s. \tag{11} $$

Experimental evidencies show that depending on the ATP and salt concentrations and the load placed on the kinesin molecule it propagates along the MT with velocities ranging in the window (0 < v_t < 1)μm/s, i.e. there is a good overlap in the velocities of these two types of biological motions that may interact with each other.

Using an easy-to-grasp analogy, the kinesin motor can be viewed as a surfer and the moving kink as a water wave. Thus, when properly matched, the motor may be able to "catch the wave" and travel on its cress for a while, (Fig. 7).

Our new findings [9] about kink's dynamics show that even in the case where a MT is subjected under the harmonic force (sinusoidal electric field) kinks and antikinks are pulled to travel in the opposite directions. The kink's velocity depends on amplitude and frequency of applied field.

At some critical values of the frequency of the driving force the directions of kink (and antikink) motions change to the opposite ones.

This is very promising fact which means that intrinsic electric fields, possibly generated by cell membrane or other parts of cell, could play the decisive role as controlling mechanism for motor protein activities.

It is also likely that applied external electromagnetic fields interferes cell's processes through stimulating or impeding activities of motor proteins.

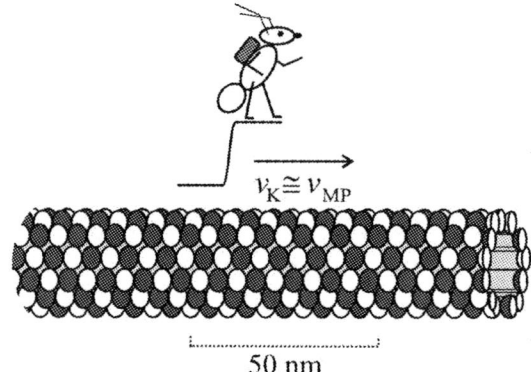

Fig. 7.The kink is catching kinesin with load

REFERENCES

[1] L.A. Amos, The Microtubule Lattice - 20 years on Trends Cell. Biol. 5, 48-51 (1995)

[2] M.V. Satarić, J.A. Tuszyński and R.B. Žakula, "Kinklike Excitations as an Energy-Transfer Mechanism in Microtubules", Phys. Rev. E48, 589-597(1993)

[3] M.V. Satarić, S. Zeković, J.A. Tuszyński and J. Pokorni, Mossbauer Effect as a Possible Tool in Detecting Nonlinear Excitations in Microtubules, Phys. Rev. E58, No 5, 6333-6339 (1998)

[4] M.V. Satarić and J.A. Tuszynski, Impact of Regulartoy Proteins on the Nonlinear Dynamics of DNA, Phys. Rev. E65, 051901-10 (2002)

[5] S. Portet, J.A. Tuszynski, J.M. Dixon and M.V. Satarić, Models of Spacial and Orientational Self-Organisation of Microtubules under the Influence of Gravitational Fields, Phys. Rev. E68, 021903-9 (2003)

[6] J.A. Tuszynski, E. Craford, E.J. Carpenter, M.L.A. Nip, J.M. Dixon and M.V. Satarić, Molecular Dynamics Stimulations of Tubulin Structure and Calculations of Electrostatic Properties of Microtubules, Math. Comput. Modelling 41, No 10, 1055-1070 (2005)

[7] R.D. Vale and R.A. Milligan, The Way Things move; Looking Under the Hood of Molecular Motor Proteins, Science, 288, 88-95 (2000)

[8] M.V. Satarić, B.M. Satarić and J.A. Tuszynski, Nonlinear Model of Microtubule Dynamics, Electromagnetic Biology and Medicine,24,255-264(2005)

[9] M.V. Satarić, Lj. Budinski-Petković, I. Lončarević, Electric Fields as Control Mechanism for Dynamics of Motor Proteins along Microtubules, Manuscript in preparatio

2006 25th International Conference on Microelectronics

Adsorbed Mass Fluctuations of a Micro/Nanoresonator Surrounded by an Arbitrary Gas Mixture

Z. Djurić, O. Jakšić, I. Jokić and M. Frantlović

Abstract - Micro/nanoresonator mass and frequency fluctuations caused by sorption processes on a resonator surface are investigated. Arbitrary gas mixture is considered, assuming that particle arrivals at the surface are all poissonian in nature, independent from each other, and that sorption dynamics follows the Langmuir isotherm. Power spectral density for mass fluctuations are derived using the analytical Langevin approach. The results of simulations are given for a silicon micro/nanocantilever in the atmosphere of four gases.

The presented analysis is useful for calculation of the ultimate performance of MEMS/NEMS sensors and oscillators, as well as for investigation of the possibilities of their parameters optimization by choosing the gas mixture in the cantilever's atmosphere. Another objective is to consider the possibility of identification of gases in the mixture, based on the power spectral density of the adsorbed mass fluctuation.

I. INTRODUCTION

The microcantilevers used in AFM have proven themselves as highly sensitive transducers of not only the inter-atomic forces. They can also convert the temperature change, adsorbed mass change or other external influences into either measurable deflection or oscillation parameter change. Thus the idea arose to develop a new group of MEMS/NEMS sensors with a micro/nanocantilever as a fundamental building block [1].

The micro/nanocantilevers are used as highly sensitive and selective detectors of presence and concentration of both chemical and biological substances, as well as for measurement of numerous physical parameters: force, adsorbed mass, temperature, infrared radiation, etc. [1,2].

Fast response, high sensitivity, miniature dimensions, low production cost, possibility of in situ monitoring in real time, as well as the capability of operation in the air, vacuum, gas mixtures and liquids make them suitable for many applications in the industry, environmental protection and research in both natural and technical sciences [2].

The principle of operation of a large group of micro/nanocantilever sensors is based on the adsorption of particles of a specimen on the cantilever surface. The particles can be, for example, gas atoms or molecules, biomolecules or microorganisms. In order to provide a specific response to certain substances (i.e. to achieve

selectivity), the cantilever surface is covered by a thin layer that has the affinity only for the certain kind of molecules.

Adsorption process affects mechanical properties of the cantilever: stiffness constant and mass. This results in change of the cantilever's deflection (static mode) or resonant frequency (dynamic mode), thus enabling both detection and determination of the number of adsorbed particles (i.e. concentration) in the analyzed sample [3,4].

If we observe, for example, the adsorption of gas particles, then besides the adsorption of particles which are to be detected, a spontaneous adsorption occurs of the particles of other gases from the surrounding atmosphere.

If we consider the physical adsorption only, it is known that after a certain time interval, a spontaneous desorption of the adsorbed particle occurs. As both the adsorption and desorption are random processes, the number of the adsorbed particles fluctuates. The fluctuations of the adsorbed mass cause the frequency fluctuations of the system consisting of the cantilever together with the adsorbed particles on its surface. The frequency fluctuations are the source of the adsorption-desorption (AD) noise (one of the total frequency noise components), which becomes increasingly prominent as the cantilever's dimensions decrease [5].

The analysis of both the AD and total noise is important for determination of the minimal detectable signal and the sensitivity of the sensor. The atmosphere around the sensor is a mixture in a majority of cases. The objective of our research is to enable the calculation of the ultimate sensor performance under the given conditions, but also to consider the possibility of the performance improvement (i.e. lowering the noise level) by adding a certain amount of a gas to the mixture. However, the attempts have already been made to identify the gases in the mixture based on the AD noise power spectral density in the case when this noise component dominates. Therefore the analysis presented here has the similar additional objective.

Besides being the sensitive elements of sensors, the micro/nanocantilevers are used as the frequency determining components of the MEMS/NEMS oscillators [6]. Due to the possibility of integration with the other components of the system within the same semiconductor chip, MEMS/NEMS oscillators enable miniaturization of the portable equipment. By decreasing the cantilever's dimensions, which can be of the order of 100 nm or lower, these oscillators can generate the frequencies higher than 100 MHz. In that case the analysis of the AD and the total

Z. Djurić, O. Jakšić, I. Jokić and M. Frantlović are with the Department of Microelectronic Technologies and Single Crystals, Institute of Chemistry, Technology and Metallurgy, Njegoševa 12, 11000 Belgrade, Serbia & Montenegro, E-mail: zdjuric@nanosys.bg.ac.yu

1-4244-0116-X/06/$20.00 ©2006 IEEE

frequency noise is also needed, since the frequency noise affects the performance of the oscillator.

In this paper we analyze the AD process of a gas mixture in the space around a cantilever. It is assumed that the AD process in the thermodynamic equilibrium can be described by the Langmuir isotherm.

II. THEORETIC ANALYSIS

An isolated system, consisting of a micro/nano-cantilever and surrounding mixture of n gases, is analyzed under following conditions:

- Single layer adsorption.
- All adsorption sites are equivalent.
- Both the adsorption and desorption rate are independent of the population of neighboring sites.

The analysis follows the same procedure as the one used while deriving the Langmuir's adsorption isotherm for the case of a single gas atmosphere [7].

For each gas in the mixture, the change of the number of adsorbed gas particles is given by the expression [8]

$$dN_i / dt = \alpha_{Si} C_{1i} p_i \theta_{free} A_{eff} - C_{2i} N_{mi} \theta_i A_{eff} , \quad i = 1, \dots, n \quad (1)$$

where N_i is the number of adsorbed gas particles, α_{Si} is the sticking probability that indicates adsorbent affinity towards the specified gas specimen, p_i is the partial pressure and θ_{free} is the fraction of the surface without the adsorbed particles. The surface coverage, θ_i, is

$$\theta_i = N_i / (N_{mi} A_{eff}) , \quad (2)$$

where N_{mi} is the total number of possible sites per unit area for the specified gas specimen, and A_{eff} is the surface exposed to the adsorption processes.

C_{1i} and C_{2i} are gas dependent constants

$$C_{1i} = \frac{1}{\sqrt{2\pi k_B M_{ai} T}} = \frac{2.635 \cdot 10^{24}}{\sqrt{M_i T}} \left[\frac{Pa}{m^2 s} \right], \quad (3)$$

$$C_{2i} = 1/\overline{\tau}_i \; [1/s], \quad \overline{\tau}_i = \tau_{0i} e^{E_{di}/(RT)} . \quad (4)$$

In previous expressions, k_B is the Boltzmann constant, R is the universal gas constant, T is the absolute temperature, E_{di} is the desorption energy, and τ_{0i} is the thermal vibration period of the adsorbed particles of the respective gas. M_i is the molar mass of the gas, $Ma_i[kg]=M_i[g]\cdot 1.66 \cdot 10^{-27}$ is the mass of a single particle and $\overline{\tau}_i$ is mean sojourn time for the adsorbed particle.

The equation (1) represents a system of n first order mutually dependent ordinary differential equations. Their dependence is determined by the relationship between surface coverages:

$$\sum_{i=1}^{n} \theta_i + \theta_{free} = 1 . \quad (5)$$

A. Equilibrium

The equilibrium is reached when the resonator's overall mass reaches its steady value. That means that the overall adsorbed mass remains constant. Since the molecule masses differ for the different gases, each poissonian process must be in equilibrium itself: the adsorption and desorption rates for each gas must be equal.

Using (5) and (1), the system of n equations for the equilibrium written in the matrix form becomes

$$\begin{bmatrix} \theta_{1e} \\ \theta_{2e} \\ \vdots \\ \theta_{ne} \end{bmatrix}^{\mathbf{T}} \cdot \left(\mathbf{I} + \begin{bmatrix} \dfrac{C_{21}N_{m1}}{\alpha_{S1}C_{11}p_1} & 1 & \cdots & 1 \\ 1 & & & 1 \\ \vdots & & & \vdots \\ 1 & 1 & \cdots & \dfrac{C_{2n}N_{mn}}{\alpha_{Sn}C_{1n}p_n} \end{bmatrix} \right) = \begin{bmatrix} 1 \\ 1 \\ \vdots \\ 1 \end{bmatrix}^{\mathbf{T}} . \quad (6)$$

In (6) the operator $\mathbf{T}$ means transposition, and the index e refers to the equilibrium. Now, partial surface coverages in equilibrium can be easily derived for each gas and then, the number of adsorbed particles in equilibrium, N_{ie}, can be calculated. These values will be needed later.

B. Fluctuations

We focused on deviations from the equilibrium values N_{ie} and θ_{ie}, in the same manner as in our previous work [8]

$$N_i(t) = N_{ie} + \Delta N_i(t), \qquad \theta_i(t) = \theta_{ie} + \Delta \theta_i(t) . \quad (7)$$

The equations (1) can be transformed using (5)-(7) and rewritten in a Langevin form. In the matrix form these equations are given as

$$d\Delta \mathbf{N} / dt = -\mathbf{M} \, \Delta \mathbf{N}(t) + \xi(t) , \quad (8)$$

where $\Delta \mathbf{N}(t)$ is a column matrix with the elements $\Delta N_i(t)$, $\mathbf{M}$ is a square matrix of coefficients, and $\xi(t)$ is a column matrix of stochastic source functions, each of them having a δ–function as a correlation function.

It is possible, by performing the Langevin procedure, to link the spectrum of the Markovian random variables N_i to that of purely random variables $\xi(t)$.

We are interested in the spectra of $\langle \Delta N_i \cdot \Delta N_j \rangle$,

$i,j=1,2,\dots n$, where ΔN_i and ΔN_j are correlated since different particles are approaching the same surface.

In the complex domain, (8) can be transformed to:

$$\Delta \mathbf{N}(j\omega) = \left(\mathbf{M} + j\omega\mathbf{I}\right)^{-1}\boldsymbol{\xi}(j\omega). \qquad (9)$$

The unknown spectra $\left\langle \Delta N_i(j\omega)\cdot\Delta N_j(-j\omega)\right\rangle$ are now possible to determine as the elements of a nxn matrix

$$\Delta \mathbf{N}(j\omega)\cdot\Delta \mathbf{N}^\mathbf{T}(-j\omega) = \left(\mathbf{M}+j\omega\mathbf{I}\right)^{-1}\mathbf{G}\left(\left(\mathbf{M}-j\omega\mathbf{I}\right)^{-1}\right)^\mathbf{T}, (10)$$

where $\mathbf{G}$ is the matrix whose elements are the spectra $\left\langle \xi_i(j\omega)\xi_j(-j\omega)\right\rangle$, $i,j=1,2,\dots n$.

Quite generally, it can be shown that the spectra $\left\langle \xi_i(j\omega)\xi_j(-j\omega)\right\rangle$ are related to the equilibrium [9]. In our case, stochastic sources are independent and hence all the cross-spectra equal zero. For $i=j$, all spectra are white

$$\left\langle \xi_i\,\xi_i\right\rangle = 4\,A_{eff}\,N_{mi}\,\theta_{ie}\,C_{2i}, \qquad (11)$$

and the matrix $\mathbf{G}$ is diagonal.

For the purpose of estimating the power spectral density of mass fluctuations, we must take into account the mass of the particles, M_{ai}. The total adsorbed mass is

$$\Delta m_{ads}(t) = M_{a1}\Delta N_1(t)+\dots+M_{an}\Delta N_n(t) = \mathbf{M}_\mathbf{a}^\mathbf{T}\Delta\mathbf{N}, \quad (12)$$

where $\mathbf{M}_\mathbf{a}$ is a column matrix.

Again, since we are dealing with random functions, we must apply Wiener-Khinchine theorem and do the averaging in a complex domain.

The spectral density of the total mass fluctuations is obtained using (10) and (12) in the form of a 1x1 matrix

$$\mathbf{S}_{\Delta m}(j\omega) = \mathbf{M}_\mathbf{a}^\mathbf{T}\left(\mathbf{M}+j\omega\mathbf{I}\right)^{-1}\mathbf{G}\left(\left(\mathbf{M}-j\omega\mathbf{I}\right)^{-1}\right)^\mathbf{T}\mathbf{M}_\mathbf{a}. \quad (13)$$

III. RESULTS

The results of calculations will be given for a 7 MHz silicon cantilever sized 2μm x 0.5μm x 20nm.

For every gas in the Table I, $\alpha_S=1$ and $\tau_0=10^{-13}$ s at $T=300$ K. The pressures are varied. In the literature [10], the desorption energies for clusters of particles are given in the form of distribution of desorption energy instead of a single value for each adsorbate/adsorbent combination. However, we presumed a non porous, homogeneous surface with one value of desorption energy for each gas. The desorption energies of gases in the Table I are chosen with the only purpose to investigate adsorption processes on silicon micro/nanoresonators. At the moment, the accurate values of desorption energies of different gases on micro/nanostructures are generally not available. Their importance will certainly invoke further experimental investigations. The higher the gas desorption energy, the stronger its influence, as we will see.

TABLE I
GAS PARAMETERS [5]

No.	gas	M	N_m [particles/m^2]	E_d [kcal/mol]
1	H_2	2	$2\cdot10^{19}$	5
2	N_2	28	$8\cdot10^{18}$	15
3	O_2	32	$8\cdot10^{18}$	21
4	He	4	$2\cdot10^{19}$	10

Let's assume that the cantilever is surrounded by a mixture of gases from the Table I, with partial pressures of 10^{-6} Pa. In the equilibrium, oxygen occupies about 40% of the cantilever surface, and nitrogen, hydrogen and helium together occupy less than 1%, leaving approximately 60% of the surface free. Under the partial pressures of 10^{-4} Pa, the adsorbed oxygen covers 99% of the surface, and at 10^{-2} Pa almost entire cantilever surface is covered by oxygen. Although the partial pressures of the gases in the mixture are equal, oxygen is the most adsorbed gas due to its highest desorption energy.

The spectral densities of adsorbed mass fluctuations are shown in Fig. 1. The following cases are illustrated: when only one of the gases surrounds the cantilever and when the atmosphere is the mixture of gases. The pressure of each gas in a single gas atmosphere equals the partial pressure of the same gas in the gas mixture. In the Fig. 1a, 1b and 1c the pressure of each gas is 10^{-6} Pa, 10^{-4} Pa and 10^{-2} Pa, respectively.

It can be seen from the diagrams (Fig. 1) that the adsorbed mass spectral density in the case of a single gas is of the lorenzian type (one "knee" exists). However, in the case of the mixture, the total adsorbed mass spectral density may have one or more "knees" (less or equal to the number of gases), depending on the type of gases, their amount in the mixture and dominantly of their desorption energies. Hence, number and position of "knees" in a noise spectrum could be missleading when gas identification is concerned.

Also, these diagrams show that at both the same temperature and pressure, in the case of a single gas atmosphere, the highest adsorbed mass fluctuations belong to the gas with the highest desorption energy.

When the cantilever is surrounded with the mixture of gases, and the partial pressure of each gas is 10^{-6} Pa, the total adsorbed mass fluctuation spectral density is dominantly determined by the adsorbed mass fluctuation of the gas with the highest desorption energy. It can also be seen that at higher partial pressure values, the resulting adsorbed mass fluctuations in the case of the mixture can be lower than in the case of the single gas at the same pressure. This means that it is possible to achieve lower AD noise, and therefore lower total noise in MEMS/NEMS devices by adding a certain gas mixture to the cantilever's atmosphere.

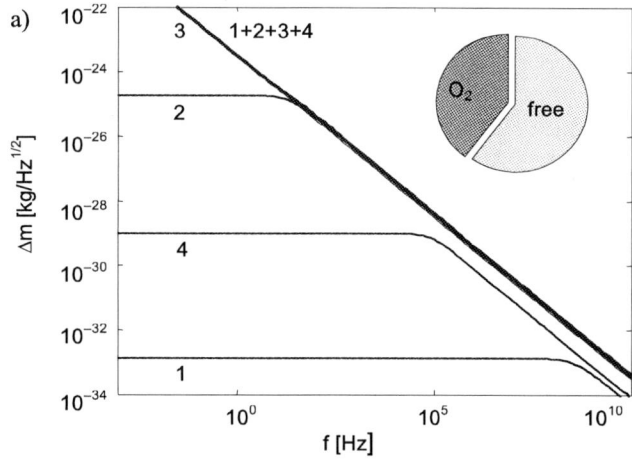

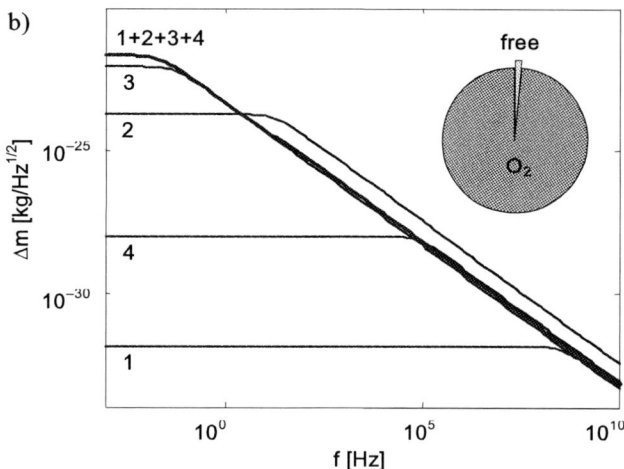

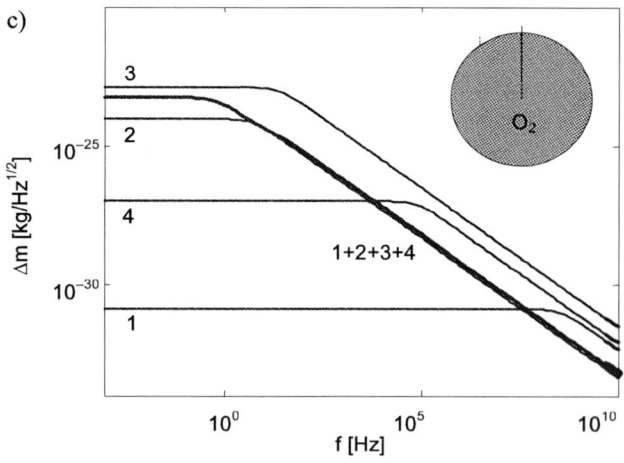

Fig. 1. Spectral densities of adsorbed mass fluctuations at room temperature on the cantilever surface (2μm x 0.5μm x 20nm). Line 1: H_2 only, line 2: N_2 only, line 3: O_2 only, line 4: He only, and line 1+2+3+4: the mixture of equal amounts of H_2, N_2, O_2 and He. The pressure of each gas is: a) 10^{-6} Pa, b) 10^{-4} Pa, c) 10^{-2} Pa. The pie-charts represent coverages of the cantilever surface in the equilibrium.

IV. CONCLUSION

In this paper a theoretic model is presented for determination of the spectral density of the adsorbed mass fluctuation in the case of arbitrary gas mixture that surrounds the micro/nanoresonator. Besides enabling investigation of the AD processes in the micro/nano-structures in general, this model is useful for both qualitative and quantitative analysis of the AD noise and also for estimation of its contribution to the total noise in MEMS/NEMS sensors and oscillators. Based on this model it is possible to determine both the minimal detectable signal and sensitivity of the micro/nanocantilever sensors.

The possibility of developing the method for identification of the gases in mixtures based on the AD noise spectral density is analyzed. Minimization of the AD noise (and therefore the total noise) in MEMS/NEMS sensors and oscillators by choosing the mixture of the surrounding gasses is also considered.

In our further research we intend to experimentally verify the presented theoretic model.

ACKNOWLEDGEMENT

This work has been partially supported by the Serbian Ministry of Science and Environmental Protection within the framework of the project TP-6151.

REFERENCES

[1] P.G. Datskos et al., "Micro and nanocantilever Sensors", in *Encyclopedia of Nanoscience and Technology*, Ed. H.S. Nalwa, American Publishers, 2004.

[2] N.V. Lavrik et al., "Cantilever transducers as a platform for chemical and biological sensors", *Rev. Sci. Instr.*, 2004, vol. 75, pp. 1-25.

[3] M.J. Sepaniak et al., "Microcantilever transducers: a new approach in sensor technology", *An.Chem.*, 2002, vol.74, 568.

[4] B. Ilic et al., "Enumeration of DNA molecules bound to a nanomechanical oscillator", *Nano Letters*, 2005, vol. 5, pp. 925-929.

[5] Z. Djurić et al., "Adsorption–desorption noise in micro-mechanical resonant structures", *Sens. and Actuators* A, 2002, vol. 96, pp. 244-251.

[6] C.T.-C. Nguyen, "Micromechanical resonators for oscillators and filters", Proc. 1995 IEEE International Ultrasonic Symposium, Seattle, WA, 1995, pp. 489-499.

[7] I. Langmuir, "Vapor pressures, evaporation, condensation and adsorption", *J.Amer.Chem.Soc., 1932,* vol. 54, pp. 2798-2832.

[8] Z. Djurić et al., "Adsorbed mass and resonant frequency fluctuations of a microcantilever caused by adsorption and desorption of particles of two gases", in *Proc.* 24th MIEL, Niš, 2004, vol. 1, pp. 197-200.

[9] K.M. van Vliet, J.R. Fasset, "Fluctuations due to electronic transitions and transport in solids", in *Fluctuation Phenomena in Solids*, ed. by R.E.Burgess, Academic Press, N.Y. , 1965.

[10] M. Jaroniek, M. Kruk, "Standard nitrogen adsorption data for characterization of nanoporous silicas," *Langmuir*, 1997, vol. 15, pp. 5410-5413.

2006 25th International Conference on Microelectronics

Scanning Probe-Shaped Nanohole Arrays with Extraordinary Optical Transmission as Platform for Enhanced Surface Plasmon-Based Biosensing

Zoran Jakšić, Milan Maksimović, Dana Vasiljević-Radović,
Dragan Tanasković, Milija Sarajlić

Abstract – We analyzed the use of subwavelength ordered nanohole patterns for detection of miniscule amounts of biological analytes. Owing to their surface plasmon resonance (SPR) operation, such structures show an extraordinarily high optical transmission in visible spectrum, at the same time concentrating the illumination to the very small area of the nanoholes and being much more sensitive to refractive index changes than the conventional SPR sensors. We applied an approximate analytical approach to design our V-shaped, nanometer-sized hole arrays. We utilized scanning-probe nanolithography to fabricate the designed nanoholes in silver substrate. We investigated the topography of the obtained patterns by atomic force microscopy. Compared to conventional SPR optical sensors, the nanohole array-based structures allow for the detection of smaller quantities of analytes, enhance nonlinear effects and ensure improved sensitivities to different analytes.

I. INTRODUCTION

Sensors based upon surface plasmon resonance (SPR) are among the most important devices for highly selective all-optical sensors (e.g. [1]-[3]. They offer simple design, extremely high sensitivity to refractive index changes (below 10^{-5}), response to minute amounts of analytes (typically layers with a thickness below 200 nm, down to several tens of nm) and real time operation. Sensor surface functionalization enables the fabrication of selective chemical sensors and biosensors.

Among the SPR-based structures are subwavelength hole arrays with extraordinary optical transmission. These were first described in the seminal paper by Ebbesen et al [1] and subsequently analyzed in many papers (e.g. [4]-[6]). If an optically thick (opaque) film perforated with an ordered or disordered array of holes is illuminated by optical radiation with a wavelength much larger than the holes, then according to the Bethe-Bowcamp theory [7], [8] no radiation should be transmitted at all. However,

---

Zoran Jakšić, Dana Vasiljević-Radović, Dragan Tanasković, and Milija Sarajlić are with the IHTM – Institute of Microelectronic Technologies and Single Crystals, Njegoševa 12, 11000 Belgrade, Serbia and Montenegro
Fax +381 11 182 995, E-mail jaksa@nanosys.ihtm.bg.ac.yu
Milan Maksimović is with the IHTM – Institute of Microelectronic Technologies and Single Crystals, Njegoševa 12, 11000 Belgrade, Serbia and Montenegro, now with the Department of applied mathematics, University of Twente, The Netherlands; E-Mail: m.maksimovic@math.utwente.nl

Ebessen et al found that extraordinarily large portion of the incident radiation is transmitted through subwavelength nanoholes if the screen is built of metal with good conductance.

The use of holes with extraordinary optical transmission (EOT) as sensors was described in [9], [10]. Nanocavity arrays were also analyzed in [11]. The main property of these structures is that the analyte is located within a small volume near the subwavelength hole (where illumination is confined), i.e. its dimensions are constrained along all three axes, while in the case of conventional SPR sensors the analyte dimensions are constrained only along the z-axis. This means that much smaller amounts of analyte should be detectable in the case of EOT sensor, which is an important advantage in the case of chemical sensors and biosensors. Another property of the EOT structures is a high field confinement within the nanoholes which results in enhancement of nonlinear effects.

Various parameters of the EOT structures influence their electromagnetic behavior, including nanohole shape, dimensions and layout, refractive index of the metal and the nanohole interior, etc. Typically higher field localizations are encountered if angular nanoholes are used instead of rounded shapes [12], [13].

In this paper we investigate the use of arbitrary shaped nanoholes in silver films as biosensors. We designed V-shaped nanoholes arranged in a square pattern and utilized scanning probe-based nanolithography to fabricate them. We assessed their use for recognition of miniscule amounts of biological analytes.

II. THEORY AND PRINCIPLE OF OPERATION

There are several complementary theories published which describe the origins of the EOT, including those utilizing surface plasmon resonance [14] and dynamical diffraction resonances [15]. Probably the most widely accepted among these is that the surface plasmon states at the front and at the back surface of the metal film become coupled for a certain set of geometry and material parameters, while the enhanced transmission of light itself proceeds through tunneling of the evanescent waves along the walls of the nanoholes. The interference of the plasmon states at the opposite surface results in re-creation of propagating modes.

1-4244-0116-X/06/$20.00 ©2006 IEEE

A metal sample with an array of arbitrarily shaped subwavelength holes is shown in Fig. 1.

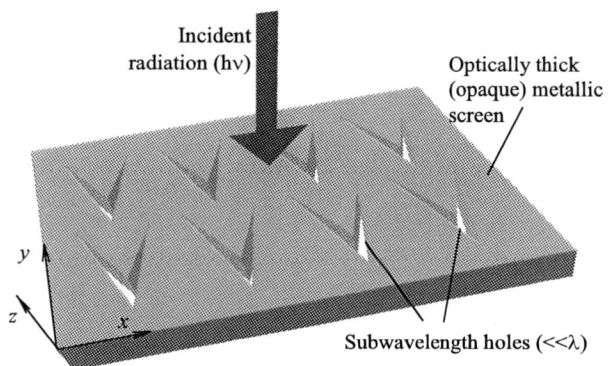

Fig. 1: Ordered subwavelength array with V-shaped holes

The in-plane wave vector of surface plasmons k_x must be continuous at the interface between metal and dielectric

$$k_x = \frac{\omega}{c}\sqrt{\frac{\varepsilon_1\varepsilon_2}{\varepsilon_1+\varepsilon_2}} \qquad (1)$$

where ε_2 is the real permittivity of the dielectric which can be approximated as constant while ε_1 is metal permittivity given by the Drude model

$$\varepsilon_1 = 1 - \frac{\omega_p^2}{\omega^2} \qquad (2)$$

where ω_p is the plasma frequency for the given metal.

If the structure with holes is dipped into a solution or a gaseous mixture containing an analyte or the analyte layer is deposited on its surface, the dielectric permittivity above the surface is changed from ε_2 to a value of ε_3. Thus the wavelength of the peak transmission through the structure is shifted to a different value.

The peak wavelength can be quantitatively determined in the following straightforward manner. To this purpose the nanoholes are considered to be periodic corrugations enabling coupling of propagating photons through the diffractive grating effect. A diffractive grating is described by

$$k_x = \frac{\omega}{c}\sin\theta \pm 2m\pi/a \qquad (3)$$

where a is the lattice parameter (the distance between neighboring holes), θ is incident angle which has to be 0 for in-plane coupling, and m is an integer. Since the transmission maximum peak wavelength is $\lambda_{peak} = 2\pi/k_x$, it is calculated as

$$\lambda_{peak} = \frac{a}{m}\sqrt{\frac{\varepsilon_1(\lambda_{peak})\varepsilon_2}{\varepsilon_1(\lambda_{peak})+\varepsilon_2}} \qquad (4)$$

and thus λ_{peak} is straightforwardly determined as a solution of the biquadrate equation obtained from (4)

Another point of interest is the response to an analyte film with a given thickness and uniform composition. If the analyte layer permittivity is ε_2 and the permittivity of the surrounding medium ε_3, the effective permittivity of the layer which takes into account the thickness of the film can be determined by properly weighting the average of ε_2, ε_3. The evanescent electromagnetic field decays away from the metal surface into the medium with a characteristic decay length, l_d. It may be assumed that this length is about 30%-40% of the operating wavelength. Since the intensity of light decays with z as $[\exp(-z/l_d)]^2$, the weight factor is $\exp(-z/l_d)$ and the effective permittivity is calculated as

$$\varepsilon_{eff} = \frac{1}{l_d}\int_0^\infty \varepsilon(z)\exp(-z/l_d)dz \qquad (5)$$

III. SENSING APPLICATION ISSUES

Fig. 2. shows the extraordinary transmission peak calculated according to the analytical approach published in [16]. The calculation was done for the case of lattice parameter $a=1$ μm, the ratios $d/l=0.08$, $h/l=0.6$ (d is the metal layer thickness, h is the distance from the surface, l is the hole characteristic dimension (equal to diameter for round holes), p is the ratio of hole surface to the whole sample surface. It can be seen that the transmission peak is narrower for smaller number of nanoholes on the surface.

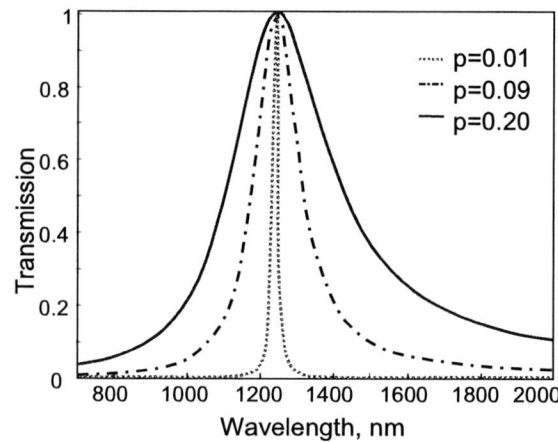

Fig. 2. Spectral transmission of a nanohole array with square matrix layout of circular holes, $d/l=0.08$, $h/l=0.6$

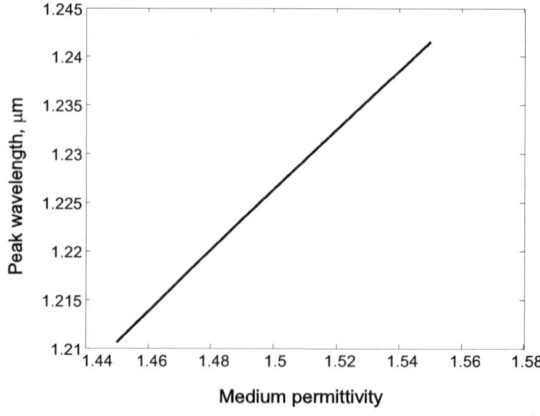

Fig. 3. Reflection dip wavelength versus permittivity of analyte

Fig. 3 shows the dependence of the peak wavelength on the permittivity of analyte.

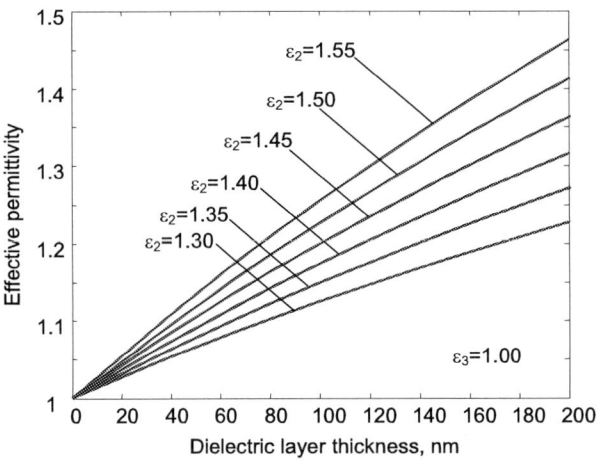

Fig. 4. Effective permittivity versus analyte layer thickness

Fig. 4 shows the calculated effective permittivity on the thickness of the analyte on the metal surface. This is useful for the case when the analyte has the same refractive index value as the sensitizing layer on the surface, and its presence is detected through the measurement of the layer thickness changes.

IV. EXPERIMENTAL

The substrate for our samples were double polished single crystalline silicon wafers. As shown in Fig 5, the first technological step was to rinse the wafer and to spin-coat it by positive photoresist 400 nm thick. The resist layer was dried without baking. A 20 nm thick silver layer was RF sputtered over the photoresist. The flatness of the obtained surface was better than 2 nm.

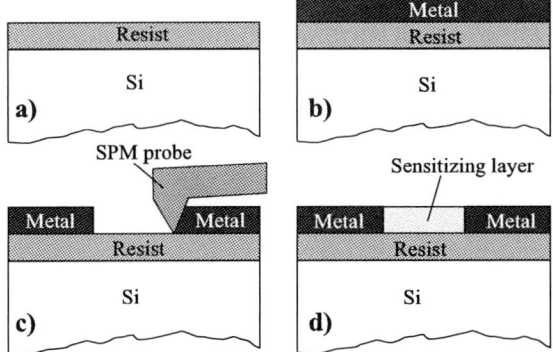

Fig. 5. Technological steps for fabrication of EOT nanohole array-based sensor. a) spin-coating of Si wafer by resist; b) sputtering of metal layer; c) SPM nanolithography for nanohole formation; b) functionalization by depositing a sensitizing layer

The nanolithography experiments were done under normal atmospheric pressure and at room temperature. No humidity control was done. Vibration- and shock-free conditions were ensured by an anti-vibration table unit with active oscillation dumping.

To fabricate the nanohole arrays we utilized scanning probe nanolithography on our Veeco Autoprobe CP-Research atomic force microscope (AFM). We utilized the z-scanner movement ('scratching' mode). A silicon nitride microcantilever was used for the operation. The scanner base position was adjusted to 0.9 μm below the zero position, i.e. the scanning probe tip was pressed against the sample surface with a force of several nN. This force was sufficient to 'dig' a groove in the silver layer, while the resist sublayer served as an elastic limiter to the probe tip.

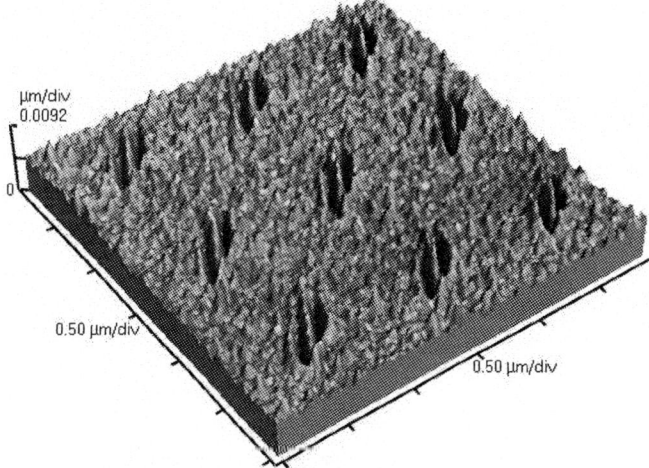

Fig. 6. AFM image of our square array of V-shaped nanoholes fabricated by SPM in silver

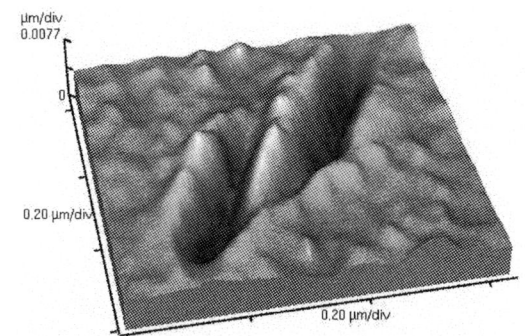

Fig. 7. An AFM magnified image of a single nanohole

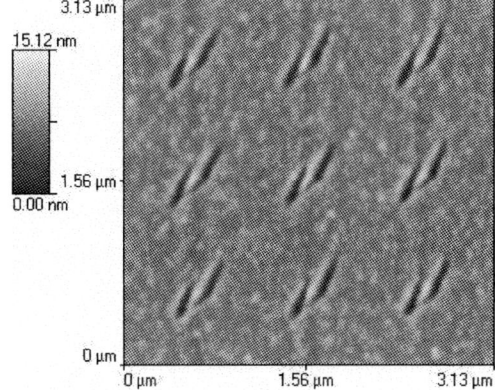

Fig. 8. 2D scan of V-shaped nanoholes fabricated in sputtered Ag. Different shades denote various elevations

The structures we fabricated were characterized by atomic force microscope. Fig. 6 shows s 3D presentation of a nanohole array we fabricated in sputter-deposited silver layer. Each hole is roughly V-shaped, as shown in the magnified detail in Fig. 7. This ensures the possibility to achieve larger field concentrations within the holes.

V. Conclusions

We investigated the applicability of arbitrarily-shaped nanoholes with extraordinary optical transmission for sensing of chemical and biological analytes. In our experiments we used scanning-probe nanolithography to produce V-shaped nanoholes in 20 nm thick silver substrate with a lattice parameter of 1 μm and characterized them by AFM. The groove depth in different samples ranged from 4 nm to 80 nm, while the line width was 80-120 nm. A problem with the SPM method is a relatively poor aspect ratio of the obtained structures, but the obtained depths show that the fabricated structures can be applied in sensing. We believe the SPM approach to nanofabrication could be useful as a simple and low-cost tool for the assessment of various nanohole geometries before larger arrays are fabricated using more sophisticated and complex modern methods, e.g. by ion beam lithography.

The advantages of the use of nanohole arrays in sensing over the conventional SPR-based structures are that much smaller amounts of analytes are necessary (very important both in chemical and biological sensing), sensitivities are improved, field localizations are enhanced and thus nonlinear effects are more pronounced and can be more easily utilized. Finally, the small dimensions of the structures allow for easier implementation of parallel sensing and multiple signal readout.

Acknowledgements

This work has been partially supported by the Serbian Ministry of Science, Technologies and Development within the framework of the project IT.6151.B.

References

[1] J. Homola, S. S. Yee, G. Gauglitz, "Surface plasmon resonance sensors: review," *Sensors and Actuators B* 54 pp. 3–15, 1999

[2] J. Homola, H. Vaisocherová, J. Dostálek, M. Piliarik, "Multi-analyte surface plasmon resonance biosensing," *Methods* 37 pp. 26–36, 2005

[3] L. S. Jung, C. T. Campbell, T. M. C., M. N. Mar, S. S. Yee, "Quantitative Interpretation of the Response of Surface Plasmon Resonance Sensors to Adsorbed Films," *Langmuir*, 14 (19), pp. 5636 -5648, 1998.

[4] T. W. Ebbesen, H. J. Lezec, H. F. Ghaemi, T. Thio, P. A. Wolff, "Extraordinary optical transmission through subwavelength hole arrays," *Nature* 391, 667–669, 1998.

[5] A. Degiron, T.W. Ebbesen, "The role of localized surface plasmon modes in the enhanced transmission of periodic subwavelength apertures," *J. Opt. A: Pure Appl. Opt.* 7 S90–S96, (2005)

[6] F. J. Garcia-Vidal, L.Martín-Moreno, J. B. Pendry, "Surfaces with holes in them: new plasmonic metamaterials," *J. Opt. A: Pure Appl. Opt.* 7 S97–S101, 2005.

[7] H. A. Bethe, "Theory of Diffraction by Small Holes," *Phys. Rev.* 66, 163-182 (1944).

[8] C. J. Bouwkamp, "On the diffraction of electromagnetic waves by small circular disks and holes," *Philips Res. Rep.* 5, pp. 401–22, 1950

[9] A.G. Brolo, R. Gordon, B. Leathem, K. L. Kavanagh, "Surface Plasmon Sensor Based on the Enhanced Light Transmission through Arrays of Nanoholes in Gold Films," *Langmuir*, 20 pp. 4813–4815, 2004

[10] P. R. H. Stark, A. E. Halleck, D. N. Larson, "Short order nanohole arrays in metals for highly sensitive probing of local indices of refraction as the basis for a highly multiplexed biosensor technology," *Methods* 37 pp. 37–47, 2005

[11] Y. Liu, J. Bishop, L. Williams, S. Blair, J. Herron, "Biosensing based upon molecular confinement in metallic nanocavity arrays," *Nanotechnol.* 15 pp. 1368–1374, 2004

[12] J. A. Matteo, D. P. Fromm, Y. Yuen, P. J. Schuck, W. E. Moerner, L. Hesselink, "Spectral analysis of strongly enhanced visible light transmission through single C-shaped nanoapertures," *Appl. Phys. Lett.* 85, 4, pp. 648-650, 2004

[13] K. J. Klein Koerkamp, S. Enoch, F. B. Segerink, N. F. van Hulst, L. Kuipers,"Strong Influence of Hole Shape on Extraordinary Transmission through Periodic Arrays of Subwavelength Holes," *Phys. Rev. Lett.* 92, 18, pp. 183901-1-4, 2004

[14] L. Martín-Moreno, F. J. García-Vidal, H. J. Lezec, K. M. Pellerin, T. Thio, J. B. Pendry, and T. W. Ebbesen, "Theory of extraordinary optical transmission through subwavelength hole arrays," *Phys. Rev. Lett.* 86, 1114–1117, 2001.

[15] K. L. van der Molen, K. J. Klein Koerkamp, S. Enoch, F. B. Segerink, N. F. van Hulst, L. Kuipers, "Role of shape and localized resonances in extraordinary transmission through periodic arrays of subwavelength holes: Experiment and theory," *Phys. Rev. B* 72, pp. 045421-1/9, 2005

[16] A. K. Sarychev, V. A. Podolskiy, A. M. Dykhne, V. M. Shalaev, "Resonance Transmittance Through a Metal Film with Subwavelength Holes", *IEEE J. Quant. Electr.* 38, 7, pp. 956-963, 2002.

2006 25th International Conference on Microelectronics

Synthesis and Optical Properties of CdSe Nanocrystals Obtained from CdCl₂ and Na₂SeSO₃ Aqueous Solutions in the Presence of Gelatine

A. E. Raevskaya, A. L. Stroyuk, S. Ya. Kuchmii, Yu. M. Azhniuk, V. M. Dzhagan, V. O. Yukhymchuk, and M. Ya. Valakh

Abstract - CdSe nanocrystals were obtained in relatively mild conditions from Na_2SeSO_3 and $CdCl_2$ in aqueous gelatine solutions. Crystal parameters of gelatine-embedded CdSe nanoparticles, obtained from optical absorption, luminescence, and Raman spectra, are correlated to synthesis conditions: gelatine concentration, $CdCl_2$ excess and temperature of ageing.

I. INTRODUCTION

CdSe nanocrystals, possessing a number of unique size-dependent physical properties, have become an object of extensive research in view of new fundamental physical effects observed as well as targeted applications as active media for optical and optoelectronic devices, solar cells, fluorescent labels in biophysical experiments etc (See e.g. [1] and references therein). Size-dependent physical characteristics of CdSe nanocrystals, especially luminescence, whose spectral position can be effectively varied practically across the whole visible spectral range, can be tailored using a variety of techniques elaborated to obtain both diluted and close-packed nanocrystals.

Among numerous synthetic approaches to metal selenide nanoparticles and nanocomposites the most widely used methods are arrested precipitation [2, 3], electrodeposition from solutions [4], photochemical [5], microwave-induced deposition [6], solvothermal synthesis [3] etc. The disadvantages of these methods consist in utilization of toxic selenium and cadmium sources, in particular, organometallic cadmium (II) and selenium (II) compounds as well as application of elevated pressures, temperatures or high-energy irradiation.

Here we report on synthesis of CdSe nanoparticles

A. E. Raevskaya, A. L. Stroyuk, and S. Ya. Kuchmii are with the Department of Photochemistry, Institute of Physical Chemistry, Ukrainian National Academy of Sciences, Prospect Nauky, 31, 03028 Kyiv, Ukraine, E-mail: stoyuk@inphyschem-nas.kiev.ua

Yu. M. Azhniuk is with the Department of Crystal Physics, Institute of Electron Physics, Ukrainian National Academy of Sciences, Universytetska Str., 21, 88017 Uzhhorod, Ukraine,, E-mail: azhn@ukrpost.net

V. M. Dzhagan, V. O. Yukhymchuk, and M. Ya. Valakh are with the Department of Optics, Institute of Semiconductor Physics, Ukrainian National Academy of Sciences, Prospect Nauky, 41, 03028 Kyiv, Ukraine, E-mail: yukhym@isp.kiev.ua

with good optical characteristics at relatively mild conditions from Na_2SeSO_3 and $CdCl_2$ in aqueous gelatine solutions. Kinetics of formation and growth of CdSe nanocrystals and the effect of the reacting mixture parameters on the nanocrystal size were studied. Raman, optical absorption, and luminescence spectra of CdSe nanocrystals, encapsulated in gelatine films, were analyzed.

II. EXPERIMENTAL

Stock solutions of $CdCl_2$, Na_2SO_3, and Na_2SeSO_3 were prepared from photographic gelatine and high-purity reagents. Sodium selenosulfate solutions were obtained by dissolution of elemental Se in hot (60–80°C) 1.0 M solution of Na_2SO_3. $[Na_2SeSO_3]/[Na_2SO_3]$ concentration ratio was adjusted to 1:4 in the stock solutions of sodium selenosulfate. The reacting mixtures for the synthesis of CdSe nanocrystals were prepared from 0.1 M stock $CdCl_2$ and 0.1 M stock Na_2SeSO_3 solutions in 1–15 mass % gelatine solution in distilled water. CdSe formation was studied in quartz parallel-sided 1.0–10.0-mm thick optical cuvettes in the dark in the presence of air oxygen.

Films with incorporated CdSe nanocrystals were prepared from parental 10–15% gelatine solutions. Following a typical procedure, viscous gelatine (10–15 %) colloidal CdSe solution was deposited onto a glass plate preliminarily treated successively with concentrated hydrogen peroxide and sulfuric acid. Plates with as-deposited humid gelatine films were dried in a drying box at 15–20°C in the dark at natural ventillation for 3–5 days. The film thickness within 0.18 to 0.27 mm was measured using a micrometer with the uncertainty of ±0.01 mm.

Optical absorption spectra of CdSe colloids were recorded on a Specord M40 spectrophotometer (the duration of one measurement was 20–30 s), while those of the gelatine films with CdSe nanocrystals were obtained using a LOMO MDR-23 monochromator, the instrumental width not exceeding 0.5 nm. Raman and photoluminescenc spectra were measured on a LOMO DFS-24 double grating monochromator with a cooled FEU-136 phototube, 476.5-nm and 488.0-nm lines of an Ar^+ laser being used for excitation. Spectral resolution of Raman measurements was not worse than 3 cm^{-1}. All measurements were carried out at room temperature.

1-4244-0116-X/06/$20.00 ©2006 IEEE

III. RESULTS AND DISCUSSION

Figure 1 illustrates the evolution of the optical absorption spectra of aqueous gelatine solutions of $CdCl_2$ and Na_2SeSO_3 after mixing the reagents. The reaction between Cd (II) salts and sodium selenosulfate in basic aqueous solutions is known to result in CdSe formation [2, 3]. Polymer gelatine molecules act as a stabilizer of CdSe crystals being formed, interrupting their growth at the stage of nanoparticle formation and ensuring long-term stability of colloidal CdSe solution towards aggregation.

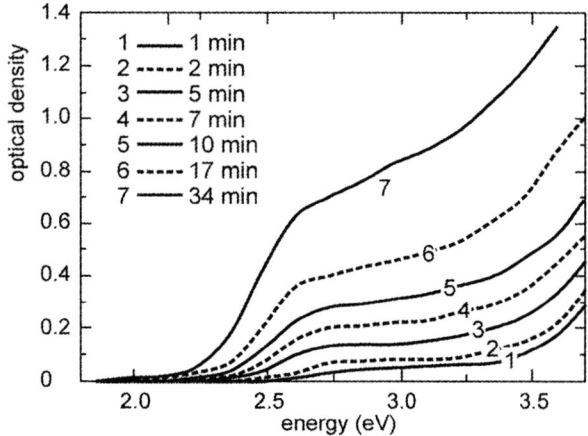

Fig. 1. Evolution of the absorption spectrum of 1% gelatine solution, containing $[CdCl_2] = 3 \cdot 10^{-3}$ M, $[Na_2SeSO_3] = 3 \cdot 10^{-3}$ M and $[Na_2SO_3] = 1.2 \cdot 10^{-2}$ M. Optical path $l = 1.0$ cm. The ageing duration is indicated in the figure.

With the growth duration the optical density of the solution increases while the absorption edge energy position of the growing CdSe nanoparticles decreases from about 2.6 to 2.2–2.3 eV, in all cases noticeably exceeding the value of bulk CdSe band gap ($E_g^{bulk} = 1.84$ eV [7]). Moreover, for the longer exposition values a maximum near the absorption edge is clearly observed, indicating the formation of CdSe nanocrystals with the size corresponding to spatial exciton confinement. The average radius of CdSe nanoparticles can be estimated in the effective mass approximation [7]:

$$\Delta E_g = \frac{\hbar^2 \pi^2}{2R^2} \left(\frac{1}{m_e^*} + \frac{1}{m_h^*} \right) \qquad (1)$$

where ΔE_g is the difference between the bulk CdSe E_g value and the energy position of the long-wave edge of the absorption band of CdSe nanocrystals, $m_e^* = (0.11 \div 0.13) m_0$ [8, 9] and $m_h^* = (0.44 \div 0.63) m_0$ [9, 10], m_0 being electron rest mass. In our case the uncertainty invoked by the difference in the effective mass values [9, 10] was much less than that due to the error of determination of the absorption maximum position. It is seen from Fig. 2 that the average radius of CdSe nanoparticles, calculated from the absorp-

tion spectra, increases with the duration of the reaction between $CdCl_2$ and Na_2SeSO_3.

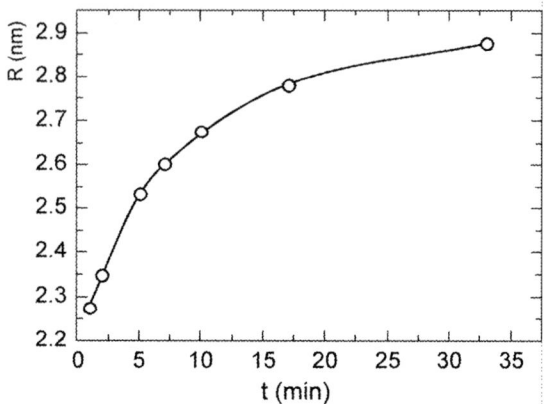

Fig. 2. Dependence of the average radius of CdSe nanocrystals on the duration of the reaction between Cd Cl_2 and Na_2SeSO_3 in 1% gelatine solution. $[CdCl_2] = 3 \cdot 10^{-3}$ M, $[Na_2SeSO_3] = 3 \cdot 10^{-3}$ M, $[Na_2SO_3] = 1.2 \cdot 10^{-2}$ M.

As seen from Fig. 3, the average size of CdSe nanocrystals decreases with the increase of gelatine concentration.

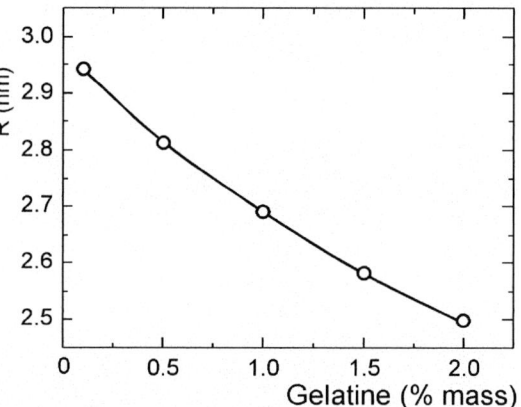

Fig. 3. Correlation between the average radius of CdSe nanocrystals grown at 18 °C (duration of the reaction 30 min, ripening 45 h) and gelatine concentration. $[CdCl_2] = 3 \cdot 10^{-3}$ M, $[Na_2SeSO_3] = 3 \cdot 10^{-3}$ M, $[Na_2SO_3] = 1.2 \cdot 10^{-2}$ M.

Optical absorption and luminescence spectra of gelatine films with CdSe nanocrystals, deposited from the aqueous solutions, are shown in Fig. 4. The parental solutions, used for the film preparation, were synthesized in 15% gelatine at 18 °C at different ratios of the initial concentrations of $CdCl_2$ and Na_2SeSO_3. It is seen that with the increase of $[CdCl_2]$ a typical confinement-related near-edge absorption maximum shifts towards higher energies and smears. Simultaneously the luminescence maximum also exhibits a blue shift. Such behaviour is the evidence for the nanocrystal size decrease. It also correlates with the similar behaviour of the absorption spectra of the colloidal solutions with different $[CdCl_2]/[Na_2SeSO_3]$ ratio.

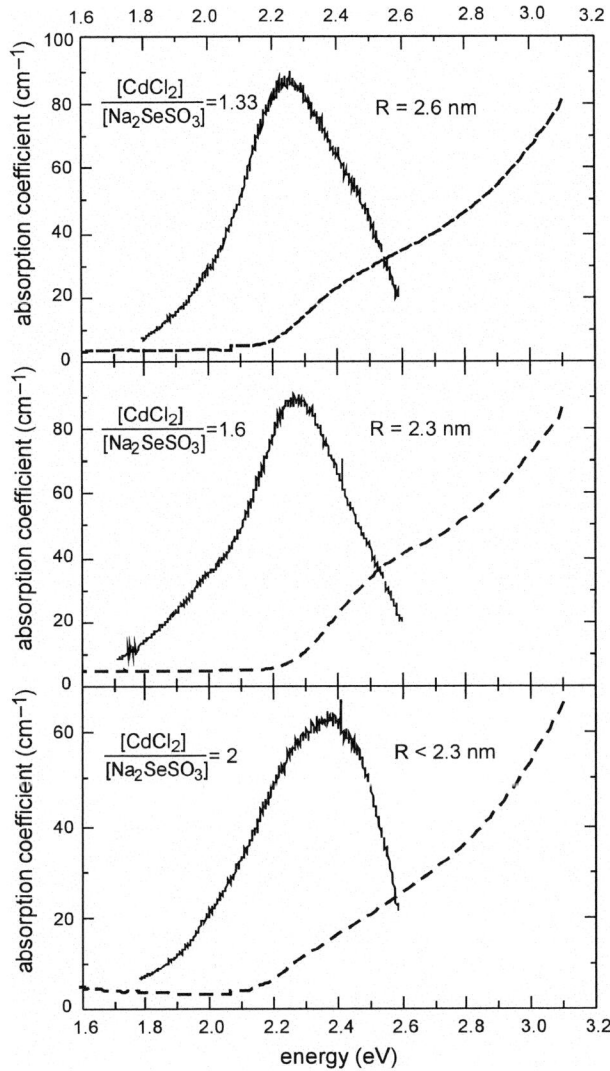

Fig. 4. Photoluminescence (solid curves) and optical absorption (dashed curves) spectra of gelatine-encapsulated CdSe nanocrystals of different size obtained at different excess of $CdCl_2$ over Na_2SeSO_3 which was taken in the concentration of $3 \cdot 10^{-3}$ M.

The smaller size of CdSe nanoparticles in the presence of excessive Cd(II) may be associated with the growth of nucleation rate at virtually constant rate of nuclei growth, the latter being determined predominantly by the rate of selenosulfate hydrolysis. This effect also determines, in our opinion, the decrease of the size of CdSe nanocrystals with increasing gelatine concentration (Fig. 3).

It is seen from Fig. 5 that the parental solution ageing at elevated temperatures results in a considerable growth of the average nanocrystal size, observed as a red shift of the absorption edge and the confinement-related maximum which also becomes more pronounced. Simultaneously the luminescence band essentially narrows, its halfwidth decreasing from 0.41 to 0.20 eV as a result of the 2-h ageing. This fact can be the evidence for the more uniform size distribution of the nanocrystals achieved in the course of the ageing.

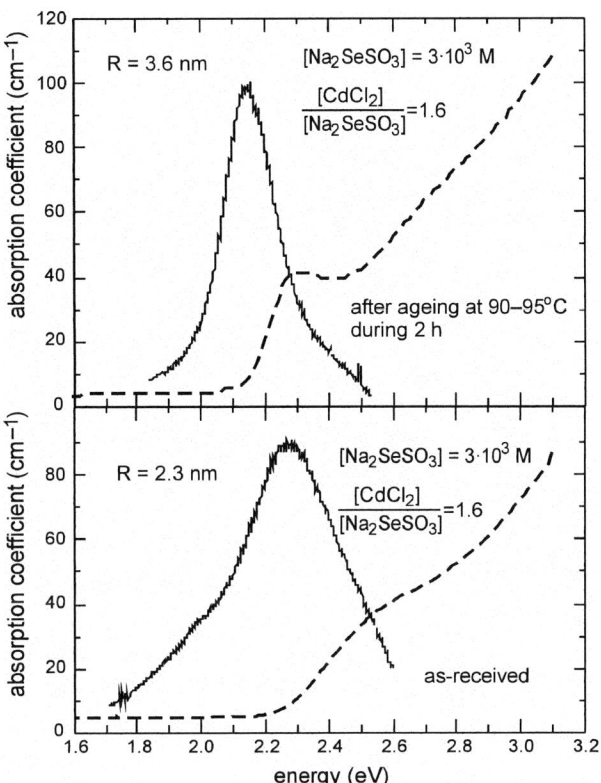

Fig. 5. Effect of ageing at 90–95 °C on the photoluminescence (solid curves) and optical absorption (dashed curves) spectra of gelatine-encapsulated CdSe nanocrystals.

Hence, control of gelatine concentration, $CdCl_2$ concentration excess over Na_2SeSO_3, and higher-temperature ageing duration can be used for tailoring the CdSe nanocrystal size in order to obtain the desired optical characteristics, especially the luminescence peak position which in our case is, similarly to [11, 12], observed by about 0.1–0.2 eV below the lowest-energy absorption peak. Besides, the above ageing control can result in the luminescence maximum narowing what can be significant for possible applications.

Raman spectra of the gelatine films with CdSe nanocrystals were observed at the background of the much broader luminescence band (Fig. 6). A distinct CdSe LO phonon band at 209 cm^{-1} and weaker second-order 2LO phonon maximum near 420 cm^{-1} clearly indicate the presence of crystalline CdSe. No dependence of the LO phonon band frequency position on the nanocrystal size was observed what is in agreement with the analysis of the main factors, affecting Raman frequency and lineshape in CdSe-type nanocrystals of similar size range [13].

The halfwidth of the CdSe LO phonon peak is practically the same (17–19 cm^{-1}) for all the measured samples, independently of the average size what enables one to assume practically the same average size dispersion for the whole sample series. It should be noted that the LO phonon halfwidth value practically coincides with that in borosilicate glass-embedded CdSe nanocrystals with average radii of 3.2 nm (20 cm^{-1}) [14] for which the size dispersion of about 20 % is reported. Hence, the size dispersion of the same order can be assumed for our case.

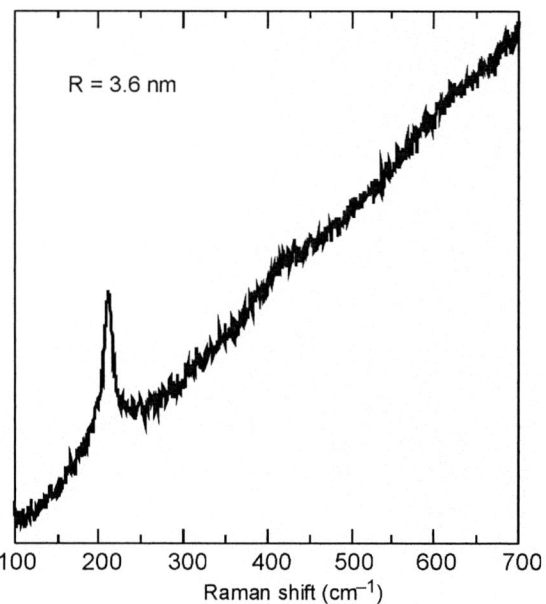

Fig. 6. A typical Raman spectrum of gelatine-encapsulated CdSe nanocrystals at the background of an intense photoluminescence band.

IV. CONCLUSION

Formation of CdSe nanocrystals at the interaction between CdCl$_2$ and Na$_2$SeSO$_3$ in aqueous gelatine solutions was studied. It was found that the size of the CdSe nanocrystals being formed can be tailored via variation of the conditions of this reaction, i.e. initial concentrations of the reagents and the ratio of these concentrations, gelatine concentration, temperature and duration of heat treatment for growing CdSe nanocrystals. Absorption and Raman scattering spectroscopy of thin gelatine films, containing CdSe nanocrystals of various size, enabled us to confirm the presence of CdSe, characterize the average nanocrystal size and roughly estimate the size dispersion. It was shown that not only luminescence peak position, but also its halfwidth can be effectively controlled by thermal treatment conditions.

REFERENCES

[1] C. B. Murray, C. R. Kagan, and M. G. Bawendi, "Synthesis and characterization of monodisperse nanocrystals and close-packed nanocrystal assemblies", *Annual Review on Materials Science*, 2000, vol. 30, p. 545-610.

[2] L. Xu, L. Wang, X. Huang, J. Zhu, H. Chen, and K. Chen, "Surface passivation and enhanced quantum-size effect and photo stability of coated CdSe/CdS nanocrystals", *Physica E*, 2000, vol. 8, p. 129-133.

[3] R.B. Kale and C.D. Lokhande, "Band gap shift, structural characterization and phase transformation of CdSe thin films from nanocrystalline cubic to nanorod hexagonal on air annealing", *Semiconductor Science and Technology*, 2005, vol. 20, p. 1-9.

[4] K. Rajeshwar, N.R. de Tacconi, and C.R. Chenthamarakshan, "Semiconductor-Based Composite Materials: Preparation, Properties, and Performance", *Chemistry of Materials*, 2001, vol. 13, p. 2765-2782.

[5] A.E. Raevskaya, A.L. Stroyuk, and S.Y. Kuchmii, "Photocatalytic Synthesis of Composite CdSe/CdS Nanoparticles", *Theoretical and Experimental Chemistry*, 2005, vol. 41, p. 181-186.

[6] O. Palchik, R. Kerner, A. Gedanken, A. M. Weiss, M. A. Slifkin, and V. Palchik, "Microwave-assisted Polyol Method for the Preparation of CdSe "Nanoballs", *J. Mateials Chemistry*, 2001, vol. 11, p. 874-878.

[7] S.V. Gaponenko, *Optical Properties of Semiconductor Nanocrystals*, Cambridge: Cambridge University Press, 1998.

[8] A.I. Ekimov, F. Hache, M.C. Schanne-Klein, D. Ricard, C. Flytzanis, I.A. Kudryavtsev, T.V. Yazeva, A.V. Rodina, and A.L.Efros, "Absorption and intensity-dependent photoluminescence measurements on CdSe quantum dots: assignment of the first electronic transitions, *J. Optical Society of America B*, 1993, vol. 10, p. 100-107.

[9] R.Cohen and M.D. Sturge, "Fluorescence line narrowing, localized exciton states, and spectral diffusion in the mixed semiconductor CdS$_x$Se$_{1-x}$", *Phys. Rev. B*, 1982, vol. 25, p. 3828-3840.

[10] C. Trallero-Giner, A. Debernardi, M. Cardona, E. Menendez-Proupin, and A.I. Ekimov, "Optical vibrons in CdSe dots and dispersion relation of the bulk material", *Phys. Rev. B*, 1998, vol. 57, p. 4664-4669.

[11] S.A. Empedocles, R. Neuhauser, K. Shimizu, and M.G. Bawendi, "Photoluminescence from Single Semiconductor Nanocrystals", *Advanced Mateials*, 1999, vol. 11, p. 1243-1256.

[12] D. Nesheva, C. Raptis, Z. Levi, Z. Popovic, and I. Hinic, "Photoluminescence of CdSe nanocrystals embedded in a SiO$_x$ thin film matrix", *J. Luminescence*, 1999, vol. 82, p. 233-240.

[13] A.V. Gomonnai, Yu.M. Azhniuk, V.O. Yukhymchuk, M. Kranjčec, and V.V. Lopushansky, "Confinement-, surface- and disorder-related effects in the resonant Raman spectra of nanometric CdS$_{1-x}$Se$_x$ crystals", *Physica Status Solidi (b)*, 2003, vol. 239, p. 490-499.

[14] Yu.M. Azhniuk, A.G. Milekhin, A.V. Gomonnai, V.V. Lopushansky, V.O. Yukhymchuk, S. Schulze, E.I. Zenkevich, and D.R.T. Zahn, "Resonant Raman studies of compositional and size dispersion of CdS$_{1-x}$Se$_x$ nanocrystals in a glass matrix", *J. Physics: Condensed Matter*, 2004, vol. 16, p. 9069-9082.

2006 25th International Conference on Microelectronics

Growth and ac-Properties of Lanthanum-Manganese Oxide Films on Si Substrates

A. A. Dakhel

Abstract: The prepared (La-Mn) oxide thin films on glass and p-Si substrates have been characterised by UV-VIS absorption spectroscopy, energy dispersion X-ray fluorescence (EDXRF) and X-ray diffraction (XRD). The XRF spectrum was used to determine the weight fraction ratio of Mn to La in the prepared samples. The optical studies determine the optical bandgaps of the prepared samples and its variation with crystallisation. XRD shows that La oxide and Mn oxide do not prevent each other to crystallise alone and do not form a solid solution. However, grains of $LaMnO_3$ compound were observed to be formed through a solid-state reaction for T > 800 °C. The ac-conductance and capacitance were studied, as a function of frequency and gate voltage. It was found that the "correlated barrier hopping" CBH model controls the frequency dependence of the conductivity, while the Kramers-Kronig (KK) relations explain the frequency dependence of the relative permittivity. The parameters of CBH model were determined showing that the ac-conduction is realised by a single-polaron hopping mechanism.

I. INTRODUCTION

Due to their electrical properties, rare-earth oxides (REOs) are attractive materials for the production of electronic elements. They have high quality of dielectric properties and stable chemical and thermal properties [1], so that they can accomplish with the necessary requirements [2] to be alternative to replace SiO_2 [3]-[5]. Two major issues must be taken into the consideration while dealing with REOs. First issue is their reactivity to the atmospheric substances like H_2O, CO_2, H, etc. The hygroscopic nature creates a structural-incorporated hydroxyl groups in the oxide [6]-[9]. The second issue is the possibility of formation of silicates as interlayer for REO films grown on Si substrate. It was observed that the formation of RE-O-Si bond increases with increasing of the post-annealing temperature depending on the RE ionic radius, so that La can easier form silicate interlayer [10,11]. Because of its highest relative permittivity (RP), La_2O_3 is a very attractive material but, on the other hand, it is the most unstable REO against the formation of silicates and the interaction with the ambient atmosphere contents (humidity, CO_2, etc) [12,13].

The electrical and structural study of pure Lanthanum oxide as an insulator thin film prepared by different methods has been studied by several investigations [14]-[18]. The plan of the present study is to investigate the effect of structural variations because of the annealing at

A. A. Dakhel is with the Dept. of Physics, College of Science, University of Bahrain, P.O. Box 32038, Kingdom of Bahrain. Email adakhil@sci.uob.bh

different temperatures, on the optical, structural and ac-electrical properties of (La–Mn) compound-oxide thin films grown on Si and glass substrates. One of the possible compounds, which will appear by solid-state reaction as consequence of annealing is lanthanum manganate $LaMnO_3$ compound, whose powder crystalline structure was studied in ref.[19].

II. EXPERIMENTAL DETAILS

The starting materials are a pure Lanthanum element (from Fluka A. G.) and a MnO_2 (from BDH chemical Ltd). For the preparation of (La-Mn) oxide films, the alternating thermal deposition method was used to deposit the starting materials on cleaned silicon and glass substrates held at room temperature in an oxygen atmosphere of about 1.3×10^{-2} Pa. All the as-grown films were totally oxidized by means of annealing at 400 °C in pure dry oxygen for 20 min. Some of these samples were investigated under name AM400. Other samples annealed at 600 °C for 20 min. in air atmosphere were named CR600, at 800 °C – CR800. Thickness was measured by Gaertner L117 ellipsometer to be around 108 nm. Finally, aluminum film electrodes of about 150 nm were deposited to form a gate and back for each Al/oxide/p-Si structure.

The spectral transmittance $T(\lambda)$ was measured by UV-VIS-Shimadzu double beam spectrophotometer. The composition of the annealed (La-Mn) oxide film was studied by XRF method using an Amptek XR-100CR detector. The crystal structures were investigated by Philips PW 1710 X-ray diffractometer. The electrical measurements were carried out using a Keithley 3330 LCZ instrument and a Keithley 614 electrometer.

III. CHARACTERISATION OF THE SAMPLES

Fig.1 shows the XRF spectrum of thin (La-Mn) oxide film on Si substrate. The observed signals are due to Si (K_α of 1.74 keV), Mn (K_α of 5.89 keV and K_β of 6.49 keV), and La (L_α of 4.65 keV and $L_{\beta 1}$ of 5.04 keV). The weight fraction ratio (x) of Mn to La in (La-Mn) oxide film was determined by the known method of micro radiographic analysis [20,21] to be 96.3 ±0.3 %.

XRD patterns of (La-Mn) oxide film samples are depicted in Fig.2 with the patterns of α-Mn_2O_3 and La_2O_3 powders. The XRD of AM400 shows its amorphous structure. Then, each La oxide and Mn oxide has crystallised alone at the annealing conditions of CR600. Mn oxide in CR600 sample crystallised forming a cubic

1-4244-0116-X/06/$20.00 ©2006 IEEE

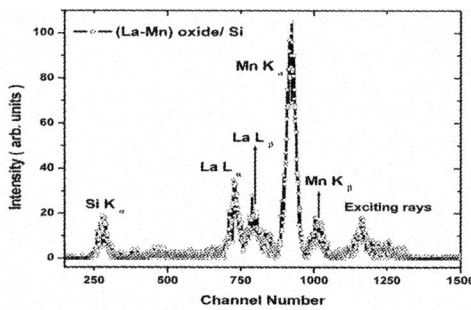

Fig. 1. XRF spectrum of (La-Mn) double oxide thin film grown on Si substrate. The exciting line was Ni-filtered Cu K_α -line.

α- Mn_2O_3 of lattice parameter 0.937 nm, which is almost identical with the literature value [22]. The peak at 29.7° was identified as (101) peak of the A-type La_2O_3 of parameters a= 0.39373 nm and c=0.613 nm [23]. The average grain size (gs) was estimated from (222) peak of α-Mn_2O_3 to be about 69 nm and from the (101) peak of La_2O_3 to be about 25.7 nm. This means that Mn oxide and La oxide in (La-Mn) compound oxide do not prevent each other to crystallize alone at 600 °C.

The XRD pattern of CR800 shows the disappearance of La oxide reflections and the appearance of a strong peak at 32.73 °, which identified to be the (002) reflection from $LaMnO_3$ of orthorhombic system with lattice parameters of a=0.57046 nm, b= 0.77029 nm, and c= 0.55353 nm [19].

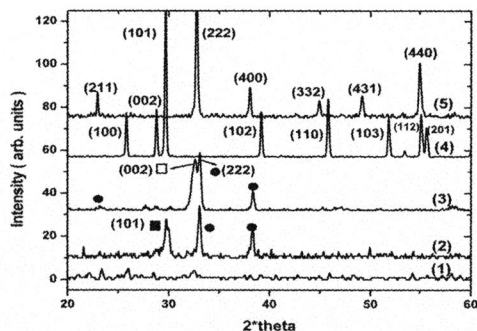

Fig. 2. X-ray diffraction pattern from powder La_2O_3 (4), powder α-Mn_2O_3 (5), and (La-Mn) oxide films prepared under different annealing temperatures: AM400 (1), CR600 (2), and CR800 (3). The scan speed was 0.01 °/s. The filled circles are for α-Mn_2O_3, open square are for $LaMnO_3$, and filled squares are for La_2O_3.

The gs of $LaMnO_3$ was calculated from the (002) peak to be about 27.2 nm. The access of Mn oxide in the CR800 sample has crystallized in a cubic α-Mn_2O_3 with gs of about 69 nm.

The normal spectral transmittances of (La-Mn) oxide films grown on glass substrates in the transparent and absorption region (200 –1100 nm) are shown as inset of Fig.3. The investigated samples have high transparency T > 0.85 in the transparent region. The spectral T(λ) shows

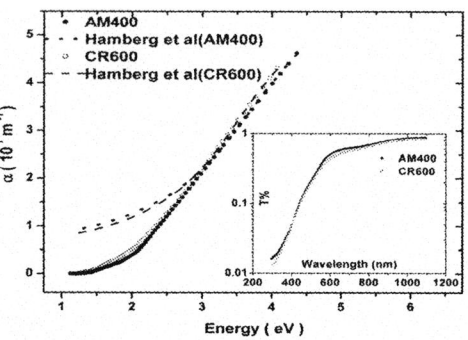

Fig. 3. The calculated spectral optical absorption coefficient (α) as a function of energy for (La-Mn) oxide films on glass substrates. The fitting lines to the high-energy part of the spectral α(E) were made according to Hamberg et al model [25]. The inset shows the experimental spectral normal transmittance T(λ).

no any special feature in the studied energy range. The non-sharp absorption edge is attributed due to either the mixture of phases. The absorption coefficient α was calculated by a special computer program [24]. Then, the energy gap was calculated by using Hamberg et al equation [25], to be about 3.70 eV for AM400 and 3.53 eV for CR 600. These values are near each other to within 0.2 eV (about 5%), and being between the bandgap values of pure La_2O_3 and Mn_2O_3 (the bandgap for La_2O_3 is 5.5 eV [17] or 6.4 eV [26] and for α-Mn_2O_3 is 2.6 eV [our data]). This result is expected since it was observed, in general, that the bandgap value of the double-oxides film sample lies in between the two bandgap values of the constitute oxides [27,28].

IV. CAPACITANCE-GATE VOLTAGE CHARACTERISATION

The parameters concluded from the gate-voltage $C(V_g)$ dependence of the capacitance measured at 100 kHz are given in table 1. The flat-band voltage shift (ΔV_{FB}) is reduced by the post-preparation annealing compared to that of the as-prepared (AM400) oxide film. The relative permittivity RP (ε_{ox}) is lower than that of pure La_2O_3 film, which was (10-17) [29]. However, we have observed that RP of AM400 is near the value (5.1) measured by us for a crystalline Mn_2O_3 film. This means that the Mn-oxide content in the AM400 sample mainly controls the RP at this stage of (La-Mn) oxide film growth. But, when the film crystallised, the value of RP increased depending on the formed crystalline structure. From the results one can conclude that the value of RP of the double-oxide film is being in between the values of RP of the constituent oxides or the addition of La oxide to Mn oxide increases the resultant RP relative to that of pure Mn oxide. This experimental fact might be useful to use in the field of Si-oxide substitution technique.

The density per unit area of the charges (Q_{ox}) [30] in the oxide sample was estimated and the results are given in table 1 with the interface trap density D_{it}, which was estimated by Hill's method [31]. The values of Q_{ox} and D_{it} vary following the structural and interface-state variations due to the annealing. However, the values of Q_{ox} and D_{it} were found to be within the known device-grade of $10^{10} - 10^{11}$ cm^{-2}. The reduction of D_{it} in CR600 relative to that of AM400 can be explained due to the formation of ultra-thin Si oxide interlayer, which makes good interface characteristics; such notice was also mentioned in [14]. Then the increase of D_{it} for CR800 relative to that of CR600 is attributed to the structural and phase change in the interface layer. Following the experimental conclusions given in ref.[32], the lanthanum-silica interaction creates amorphous La-Si-O silicate, embedded in the silica surface and modifying it.

TABLE 1

THE PARAMETERS CONCLUDED FROM $C(V_g)$ MEASUREMENTS AT 100 kHz AND 293 K FOR (La-Mn) OXIDE SAMPLES PREPARED AT DIFFERENT ANNEALING CONDITIONS.

Sample	$\Delta V_{FB}(V)$	$Q_{ox}(10^{15}$ Charges/m^2)	$D_{it}(10^{15}$ eV^{-1}/m^2)	ε_{ox}
AM400:	0.92	-2.23	10.1	4.74
CR600 :	0.77	-3.74	2.7	6.96
CR800:	0.13	-0.75	13.5	11.53

V. AC-CONDUCTION MEASUREMENTS

The measured ac conductivity (σ_{ac}) of insulators is expressed as; $\sigma_{ac} = \sigma_{dc} + \sigma_{ac}(\omega)$ [33], where σ_{dc} is the dc-conductivity of the sample and $\sigma_{ac}(\omega)$ is the frequency-dependent part of the conductivity. For conduction by hopping, $\sigma_{ac}(\omega)$ is expressed by a power law: $\sigma_{ac}(\omega)=A_\sigma \omega^s$ [34]. According to the correlated barrier hopping (CBH) model, A_σ is given by [35,36]: $A_\sigma = k\,(\alpha N_{LS})^2 / \tau_0^\beta \varepsilon_{ox}^s W_M^6$, where $k = e^6/24\pi^3\varepsilon_0^5 = 0.415 \times 10^{-60}$ C.V^5.m^5, W_M is the maximum barrier height for hopping in eV, N_{LS} is the trap concentration in m^{-3}, $\alpha = n_{el}^{7/2}$, n_{el} is the number of simultaneously hopped electrons between centres, τ_0 is the effective relaxation time (about 10^{-13} s [36]), and $\beta=1$-s. The exponent s is given at temperature T for randomly distributed hopping centres by : $s = 1 - 6k_B T\,[W_M + k_B T \ln(\omega\tau_0)]^{-1}$. According to the CBH model, the energy W_M is equal to the bandgap (E_g) of the insulating material for bipolaron hopping and $E_g/4$ for a single-polaron transport. Experimentally, W_M has any value less than E_g depending on the film's microstructure.

Fig. 4 shows the room-temperature frequency dependence of ac-conductivity $\sigma_{ac}(\omega)$ at a gate voltage of -1 V in a frequency range of 1 - 100 kHz. The analysis show that $\sigma_{ac}(\omega)$ data of all samples follow the CBH model especially for f >10 kHz with different parameters (table 2) depending on the microstructure. The calculated value of σ_{dc} was almost equal to that value measured directly for each sample by the usual dc-technique of order of magnitude 10^{-8} S/cm at -1 V.

TABLE 2

THE DC-CONDUCTIVITY σ_{dc} AND THE CALCULATED PARAMETERS (s, s', W_M, αN_{LS}, and R_{min}) OF ELLIOTT CBH MODEL AT 293 K FOR (La- Mn) OXIDE SAMPLES PREPARED UNDER DIFFERENT ANNEALING CONDITIONS.

Sample :	σ_{dc}(S/m)	s	s`	W_M (eV)	N_{LS} (m^{-3})	R_{min} (nm)
AM400:	3.34×10^{-6}	0.50	0.50	0.72	1.31×10^{24}	1.68
CR600 :	8.63×10^{-6}	0.73	0.25	0.98	7.9×10^{25}	0.83
CR800 :	3.20×10^{-6}	0.65	0.65	0.86	1.5×10^{27}	0.58

For the ac-conduction part, the calculated W_M was not far from the value $0.25E_g$ to within 20% difference, which means that the conductivity is realised by a single polaron hopping mechanism with a minimum (cut-off) jump (R_{min}) [37] calculated and given in table 2. The fitting of the experimental results to the CBH model gives the values of

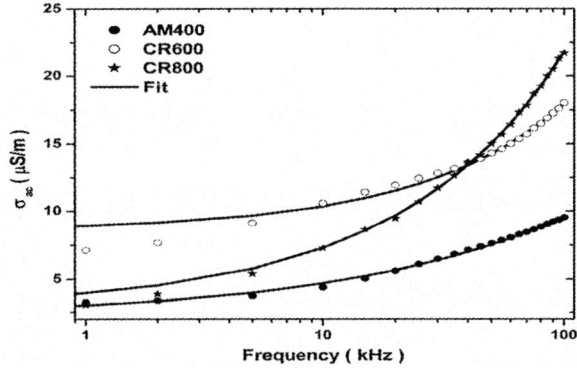

Fig. 4. The dependence of ac-conductivity σ_{ac} on signal frequency at room temperature in accumulation state. The lines represent the theoretical simulation according to Elliott CBH model.

N_{LS}. The large increase of N_{LS} for CR800 to 10^{27} m^{-3} is attributed mainly to the relatively high D_{it} (as back conduction doors) rather than to the hopping centres inside the sample. The value of N_{LS} and hence the concentration of the hopping centres is proportional to the conductivity.

117

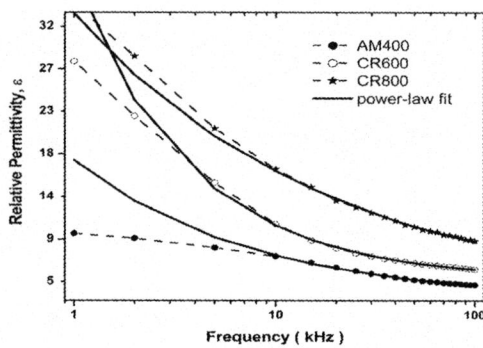

Fig. 5. The dependence of relative permittivity on signal frequency at room temperature at accumulation state. The lines represent the theoretical simulation according KK relations.

The capacitance decreases with increasing signal frequency due to the effect of charge redistribution by carrier hopping on centres [38]. However, it was proved by Kramers-Kronig (KK) relations that the power-law following of the ac conductivity of an insulator causes its RP to follow another power law of a form [33]: $\varepsilon(\omega) \propto \omega^{s`-1}$. Fig. 5 shows that this relation is adequate especially for f >10 kHz with the exponent "s`" value in most cases is equal to "s".

VI. CONCLUSIONS

Double (La-Mn) oxide films of fraction ratio Mn to La of 96.3% were grown on glass and p-Si substrates. The samples were annealed at different conditions in order to prepare different structures and agitate a solid-state reaction. The prepared compound oxide films have almost identical optical bandgaps. The XRD shows that La oxide and Mn oxide do not prevent each other to crystallise alone and not forming a solid solution. However, grains of perovskite $LaMnO_3$ compound was formed through a solid-state reaction for T > 800 °C. The values of Q_{ox} and D_{it} were within the device-grade of $10^{10} - 10^{11}$ cm^{-2}. The ac-conduction studies as a function of frequency suggest that the conduction mechanism in the double-oxide is due to hopping of current carriers following the correlated barrier-hopping (CBH) model. Hence, the parameters W_M, s and N_{LS} were determined. The effects of different annealing conditions on the relative permittivity, density of interface charges, and density of fixed charges in the oxide were demonstrated.

REFERENCES

[1] V. A. Rozhkov, A. Yu. Trusova, I. G. Berezhnoy, Thin Solid Films 325 (1998) 151.
[2] K. Cho, Computational Mater. Sci. 23 (2002) 43.
[3] M. Laskela, M. Ritala, J. Solid State Chem.171 (2003) 170.
[4] Y. Yeo, T. King, C. Hu, Appl. Phys. Lett. 81 (2002) 2091.

[5] E. Miranda, J. Molina, Y. Kim, H. Iwai, Microelectron. Reliability 45 (2005) 1365.
[6] A. A. Dakhel, J. Alloys and Compounds 388 (2005) 177.
[7] J. Kwo, M. Hong, B. Busch, D. A. Muller, Y. J. Chabal, A. R. Kortan, J. P. Mannaerts, B. Yang, P. Ye, H. Grossmann, A. M. Sergent, K. K. Ng, J. Bude, W. H. Schulte, E. Garfunkel, T. Gustafsson, J. Cryst. Growth 251(2003) 645.
[8] H. B. Lal, J. Phys. C.: Solid State Physics 13 (1980) 3969.
[9] S. Bernal, F. J. Botana, R. Garcia, J. M. Rodriguez-Izquierdo, Reactivity of solids 4 (1987) 23.
[10] K. J. Hubbard, D. G. Schlom, J. Mater. Res. 11 (1996) 2757.
[11] H. Ono, Appl. Phys. Lett. 78 (2001) 1832.
[12] M. Laskela, M. Ritala, J. Solid State Chem.171 (2003) 170.
[13] M. Suzuki, M. Kagawa, Y. Syono, T. Hirai, J. Cryst. Growth 112 (1991) 621.
[14] M. Laskela ,M. Ritala, J. Solid State Chem.171 (2003) 170.
[15] Y. Kim, S. –I. Ohmi, K. Tsutsui, H. Iwai, Solid-State Electron. 49 (2005) 825.
[16] Y. Kim, K. Miyauchi, S. Ohmi, K. Tsutsi, H. Iwai, Microelectron. J. 36 (2005) 41.
[17] S. Ohmi, C. Kobayashi, I. Kashiwagi, C. Ohshima, H. Ishiwara, H. Iwai, J. Electrochem. Soc. 150 (2003) F134.
[18] Y. Wu, M. Yang, A. Chin, W. Chen, C. Kwei, IEEE EDL 21 (2000) 341.
[19] P. Norby, I. G. Krogh Anderson, E. Krogh Anderson, J. Solid State Chem. 119 (1995) 191.
[20] M. J. Anjos, R. T. Lopes, E. O. F. de Jesus, R. Cesario and C. A. A. Barradas, Spectrochemica Acta B55 (2000) 1189.
[21] S. Hayakawa, Jia Xiao-Peng, M. Wakatsuki, Y. Gohshi and T. Hirokawa, J. Crystal Growth 210 (2000) 388.
[22] K. J. Kim, Y. R. Park, J. Cryst. Growth 270 (2004) 162.
[23] K. A. Gschneidner, LeRoy Eyring (eds.) "Handbook on physics and Chemistry of rare earths", North Holland pub. Company, Amsterdam, 1982, p.573.
[24] E. G. Birgin, I. Chambouleyron, J. M. Martinez, J. Comput. Phys. 151 (1999) 862.
[25] I. Hamberg, C. G. Granqvist, K. -F. Berggren, B. E. Sernelius, L. Engstrom, Phys.Rev. B 30 (1984) 3240.
[26] H. Nohira, T. Shiraishi, K. Takahashi, T. Hattori, I. Kashiwagi, C. Ohshima, S. Ohmi, H. Iwai, S.Joumori, K. Nakajima, M. Suzuki, K. Kimura, Appl. Surf. Sci. 234 (2004) 493.
[27] A. A. Dakhel, Appl. Phys. A 77 (2003) 677.
[28] N. Barreau, S. Marsillac, D. Albertini, and J. C. Bernede, Thin Solid Films 403 (2002) 331.
[29] P. Pisecny, K. Husekova, K. Frolich, L. Harmatha, J. Soltys, D. Machajdik, J. P. Espinos, M. Jergel, J. Jakabovic, Materials Science in Semiconductor Processing 7 (2004) 231.
[30] D. A. Neamen, Semiconductor Physics and Devices- Basic properties, 2nd edition, Irwin/McGraw- Hill Inc,1997. P. 434.
[31] W. H. Hill, Solid State Electron. 23 (1980) 987.
[32] H. Vidal, S. Bernal, R. T. Baker, D. Finol, J. A. P. Omil, J.M. Pintado, J. M. Rodriguez-Izquierdo, J. Catalysis 183 (1999) 53.
[33] A. Ghosh, Phys. Rev. B 41 (1990) 1479.
[34] R. M. Hill, A. K. Jonscher, J. Non-Cryst. Solids 32 (1979) 53.
[35] S. R. Elliott, Adv. Phys. 36 (1987) 135.
[36] S. R. Elliott, Phil. Mag. 36 (1977) 1291.
[37] R Salam, Phys. Stat. Sol. (a) 117 (1990) 535.
[38] A. Vasudevan, S. Carin, M. A. Melloch, S. Harmon, Appl. Phys. Lett. 73 (1998) 671.

2006 25th International Conference on Microelectronics

Low – k Polyimide/Silica Nanocomposites for Microelectronics Applications

E. Logakis, D. Fragiadakis, P. Pissis

Abstract – The temperature and frequency dependence of dielectric permittivity of hybrid polyimide/silica nanocomposites, prepared by the in situ generation of crosslinked organosilicon nanophase by sol–gel techniques, was investigated by broadband dielectric relaxation spectroscopy. Dielectric permittivity decreases and its frequency and temperature dependence becomes weaker in the hybrids as compared to pure polyimide. The results are explained in terms of loose inner structure of the spatial aggregates of the organosilicon nanophase and reduced polymer dynamics in the hybrids. Water sorption measurements from the vapor phase show reduced water contents of the nanocomposites with respect to pure polyimide. These results suggest that the hybrids under investigation may have a reasonably good potential as low-k materials for microelectronics applications.

I. INTRODUCTION

There is an increasing demand in microelectronics for materials with low values of the real part of dielectric permittivity ε' (better known as low–k materials) to be used as inter-metal and inter-layer dielectrics. Candidate materials should combine low ε' values (below 3.0, possibly even below 2.5) with several other good properties, including good thermal stability, high thermal conductivity, chemical resistance, low water absorption and good processability [1,2]. Polymers and polymer-based materials, in particular polyimides (PIs), have attracted much interest in recent years for such applications [2]. However, ε' of the starting polymer (typically in the range 3.0 to 3.5 for polyimides) should be further reduced. Introduction of porosity into the polymer (ε' of air being practically 1.0) [2, 3] and reinforcement by inorganic nanoparticles, resulting in overall decrease of molecular mobility [4 - 6], are perspective routes for ε' reduction. The second route seems more effective, as other properties are at the same time also improved. It is essential for that improvement and the reduction of ε' that the filler is in the form of nanoparticles distributed in the matrix giving rise to high values of the surface-to-volume ratio. The development of chemical bonds between the filler and the matrix has a beneficial effect in that respect [4-6].

We reported recently the preparation of PI–silica nanocomposites by the in situ generation of cross–linked

E. Logakis, D. Fragiadakis and P. Pissis are with the Department of Physics, National Technical University of Athens, Zografou Campus, GR 157 73 Athens, Greece, E-mail: ppissis@central.ntua.gr

organosilicon nanophase (ON) through the sol–gel process (PI–ON hybrids) and their characterization by various techniques [7, 8]. Measurements by dielectric relaxation spectroscopy (DRS) showed a non-additive decrease of ε' in the nanocomposites, which was treated in terms of effective medium theories (EMT) and attributed to a loose inner structure of the spatial aggregates of ON, in agreement with the results of density measurements [7]. DRS measurements were performed at room temperature (20 ^{0}C) and EMT calculations were based on the ε' data at 1 KHz. However, knowledge of the frequency and temperature dependence of ε' over wide ranges, including those of potential applications in microelectronics devices, is essential for assessing the suitability of the nanocomposites prepared and for designing new materials with improved performance characteristics. To that aim DRS over wide frequency and temperature ranges was employed in the present work. In addition, water absorption from the vapour phase was employed to measure water contents of the nanocomposites.

II. EXPERIMENTAL

The PI-ON hybrids were prepared from polyamic acid of molar mass 5.000 or 10.000 or 15.000 (series 5, 10, 15, respectively) with ethoxysilane end groups (PAAS) and methyl triethoxysilane (MTS). The PAAS/MTS mass ratio was systematically varied from 100/0 to 100/120, corresponding to PI/ON mass ration varying from 100/0 to 64.4/35.6. Details of preparation have been given elsewhere [7-9]. Samples are coded by the PASS/MTS ration followed by the series number 5, 10 or 15, e.g. 100/70-15.

A Novocontrol Alpha Analyser in combination with the Novocontrol Quatro Cryosystem were used for broadband DRS measurements in the frequency range 10^{-1} – 10^6 Hz and the temperature range from -150 to 220 ^{0}C.

Equilibrium water sorption measurements from the vapor phase at ambient relative conditions and at a relative humidity of 98% (desiccator with a saturated K_2SO_4 aqueous solution) were performed at 20 ^{0}C. The water content h, defined as mass of absorbed water divided by the mass of dry sample, was determined by weighing (Mettler Toledo AX 105, accuracy 10 µg). Dry masses have been determined by drying the samples in vacuum (5 x 10^{-3} Torr) at 120 ^{0}C for 48 h.

1-4244-0116-X/06/$20.00 ©2006 IEEE   119

III. RESULTS AND DISCUSSION

Figure 1 shows results for the frequency dependence of ε' of the hybrids of series 10 at 20 °C. We observe an overall decrease of ε' with increasing ON content. Similar results have been obtained for the hybrids of the other two series. At first glance the results look surprising in terms of EMT, as ε' of compact silica is in the rage of 3.8-4.0. Two effects may contribute to the reduction of ε' in the nanocomposites: porosity of the ON [7] and decrease of ε' of the PI matrix as a result of reduced molecular mobility due to formation of chemical bonds with the ON. Measurements in another PI/silica system, prepared also by sol-gel techniques and characterized by the presence of chemical bonds between the two components, have indicated considerable suppression of molecular mobility and of cooperativity in the temperature region of the glass transition [10]. In that respect it is interesting to note that the drop of $\varepsilon'(f)$ at higher frequencies in Fig. 1 is due to the γ relaxation of the PI matrix. Preliminary results have indicated that the magnitude (dielectric strength) of the γ relaxation increases on hybridization in series 5, whereas it decreases in series 10 and 15 [8]. These results can be discussed in terms of two opposite effects: decrease of molecular mobility due to constraints imposed by the presence of the ON particles on the one hand and, on the other hand, increase of free volume resulting from loosened packing of the PI chains due to tethering on the ON particles. The results suggest that the second effect dominates over the first one for short PI chains, whereas the opposite is true for longer chains. It would be interesting to further follow this point in future, as well as to extend EMT calculations on the basis of effective ε' values at subzero temperatures where the γ relaxation does not make any contribution to the measure ε' values.

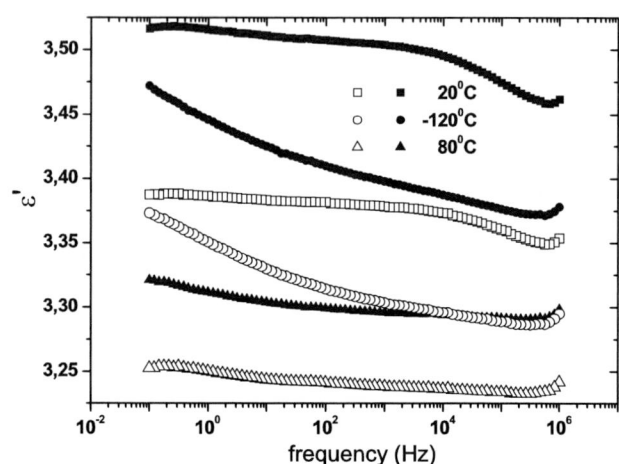

Fig.2. Real part of dielectric permittivity ε' against frequency f at various temperatures for the samples PI (filled symbols) and 100/70-15 (open symbols).

Results for $\varepsilon'(f)$ for PI and for a hybrid at -120 , 20 and 80 °C in Fig.2 show, next to the reduction of ε' in the hybrid at each temperature, lower ε' values for both samples at -120 °C as compared to 20 °C. Also the drop in ε' related with the γ relaxation is shifted to lower frequencies, with respect to measurements at 20 °C, as expected for a thermally activated relaxation process, whereas at 80 °C the drop is shifted to higher frequencies out of the frequency range of measurements. The overall decrease of $\varepsilon'(f)$ with decreasing temperature from 20 to -120 °C in Fig.2 is reasonable for a glassy polymer [11], whereas the further decrease at 80 °C may be related with loss of water, this point deserving further investigation in future work.

Figures 3 and 4 show, for selected samples, results for the temperature dependence of ε' and of loss tangent (tanδ = $\varepsilon''/\varepsilon'$), respectively, at a fixed high frequency of 560 KHz (to be completed in future with more samples, more frequencies and higher temperatures). The dispersion of ε' and the corresponding maximum of tanδ at temperatures around 0 °C is due to the γ relaxation of the PI matrix. The further decrease of ε' at higher temperatures in Fig.3 may be related with water loss. The dispersion and the loss peak shift to higher/lower temperatures with increasing/decreasing frequency. With respect to pure PI matrix, the dispersion increases at low silica contents and decreases significantly at higher silica contents. These effects can not be predicted merely on the basis of measurements at room temperature and a fixed frequency (typically 1 MHz), often used for a quick characterization of low-*k* materials for microelectronics applications.

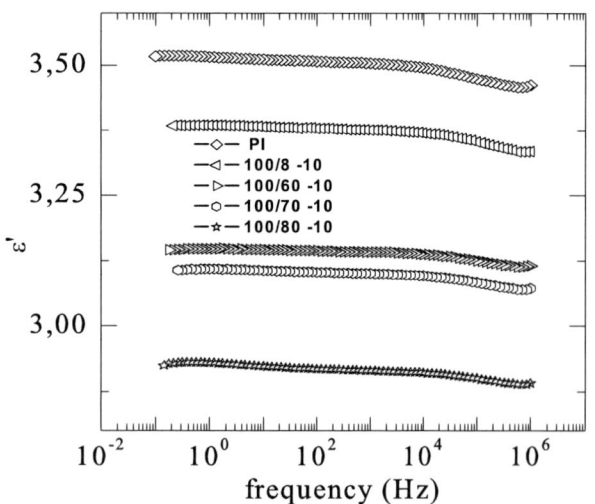

Fig.1. Real part of dielectric permittivity ε' against frequency f for the samples indicated on the plot at 20 °C.

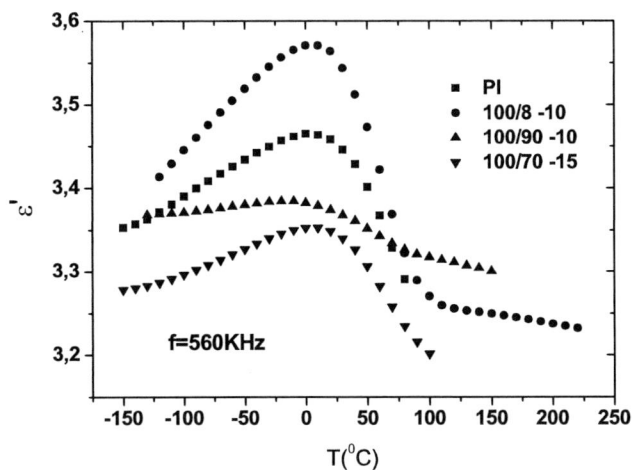

Fig.3. Real part of dielectric permittivity ε′ against temperature T for the samples indicated on the plot at 560 KHz.

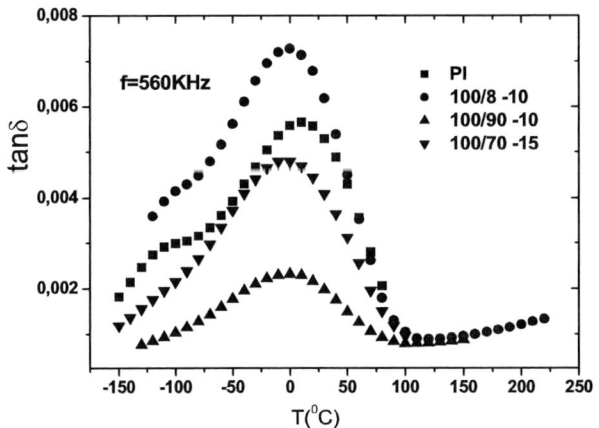

Fig.4. Loss tangent tanδ against temperature T for the samples indicated on the plot at 560 KHz.

It is interesting to note the presence of a faster and weaker, with respect to the γ, relaxation at about -100 °C in Fig.4, more pronounced in PI and the sample with the lowest silica content. This dispersion, more clearly observed in the temperature than in the frequency scans, may be related with the presence of water in the samples, a point to be discussed later. It would be interesting to further follow in future work this relaxation by measuring samples at various water contents.

Table 1 lists results for water sorption of selected samples from the vapor phase: water content h at ambient conditions and also at 98% relative humidity. At ambient conditions water sorption is clearly reduced in the hybrids, the reduction being stronger at higher ON contents. The results are less clear at the higher value of relative humidity of 98%, where minor changes of water content between the samples are observed. This different behavior may be related with the expected different state of water: molecularly distributed water at ambient conditions against, mostly, clustered water at 98% relative humidity.

The relatively high values of water content at 98% relative humidity are consisted with this interpretation. It should be mentioned here that water content measurements in various PIs have indicated molecular distribution of water even at high values of relative humidity and negligible clustering [12,13]. Water clustering in the hybrids under investigation here may be related with the porous structure of silica and/or the increase of free volume due to loosened packing of the chains. Bearing in mind the hydrophilicity of silica, these preliminary results of reduced water sorption in the hybrids (Table I) may be interpreted in terms of reduced mobility of PI chains and reduced permeability of the hybrids. It would be interesting to further follow this point by water sorption measurements at various relative humidity values, both dynamic (to determine diffusion coefficients) and in equilibrium.

TABLE I

Sample	Water content at ambient conditions	Water content at relative humidity 98%
PI	0,0072	0,0262
100/8 – 10	0,0080	0,0314
100/60 – 10	0,0066	0,0271
100/90 – 10	0,0050	0,0228
100/8 – 15	0,0060	0,0254
100/70 – 15	0,0039	0,0250

IV. CONCLUSION

The results suggest that the hybrid polyimide/silica nanocomposites studied in this work are interesting materials for microelectronics applications. Not only the values of dielectric permittivity ε′, but also the temperature and frequency dependence of ε′ are reduced in the nanocomposites, as compared to pure polyimide. Preliminary results of water sorption measurements indicate that also water contents are reduced in the nanocomposites. Further work is needed to understand better the dependence of these effects on molar mass of polyimide and their relation to porosity of the organosilicon nanophase and to reduction of polymer dynamics.

REFERENCES

[1] Semiconductor Industry Association, International Technology Roadmap for Semiconductors (ITRS) (http://www.itrs.net/ntrs/publntrs.nsf).

[2] G. Maier, ''Low dielectric constant polymers for microelectronics'', Prog. Polym. Sci., 26:3–65, 2001.

[3] M.S. Silverstein, M. Shach-Caplan, B.J. Bauer, R.C. Hedden, H.J. Lee, B.G. Landes, ''Nanopore formation in a polyphenylene low-k dielectric'', macromolecules, 2005, Vol.38, No.10, pp.4301-4310.

[4] Chyi-Ming Leu, Yao-Te Chang and Kung-Hwa Wei, ''Polyimide-Side-Chain Tethered Polyhedal Oligomeric

Silsequioxane Nanocomposites for Low-Dielectric Film Applications'', Chem. Mater., 2003, 15, 3721-3727.

[5] M. Vasilopoulou, S. Tsevas, A.M. Douvas, P. Agritis, D. Davazoglou and D. Kouvatsos, ''Characterization of various low-k dielectrics for possible use in applications at temperatures below 160 ^{0}C'', *Second Conference on Microelectronics, Microsystems and Nanotechnology*, Journal of Physics: Conference Series 10 (2005) 218–221.

[6] Y.-H. Zhang , S.-G. Lu , Y.-Q. Li , Z.-M. Dang , J. H. Xin , S.-Y. Fu , G.-T. Li , R.-R. Guo , L.-F. Li ,''Novel Silica Tube/Polyimide Composite Films with Variable Low Dielectric Constant'', Adv. Mater., 2005, 17, No. 8, pp. 1056 – 1058.

[7] V.Y. Kramarenko, T.A. Shantalil, I.L. Karpova, K.S. Dragan, E.G. Privalko, V.P. Privalko, D. Fragiadakis, and P. Pissis, ''Polyimides reinforced with the sol-gel derived organosilicon nanophase as low dielectric permittivity materials'', *Polym. Adv. Technol.*, 15:144–148, 2004.

[8] D. Fragiadakis, E. Logakis, P. Pissis, V. Yu. Kramarenko, T. A. Shantalii, I. L. Karpova, K. S. Dragan, E. G. Privalko, A. A.Usenko and V. P. Privalko, ''Polyimide/silica nanocomposites with low values of dielectric permittivity'', *Second Conference on Microelectronics, Microsystems and Nanotechnology*, Journal of Physics: Conference Series 10 (2005) 139–142.

[9] T.A. Shantalii, I.L. Karpova, K.S. Dragan, E.G. Privalko, and V.P. Privalko, ''Synthesis and thermomechanical characterization of polyimides reinforced with the sol-gel derived nanoparticles'', *Sci. Technol. Adv. Mater.*, 4:115–119, 2003.

[10] V.A. Bershtein, L.M. Egorova, P.N. Yakushev, P. Pissis, P.Sysel, and L. Brozova, ''Molecular dynamics in nanostructured polyimide-silica hybrid materials and their thermal stability'', J. Polym. Sci. Pt. B-Polym. Phys., 40:1056–1069, 2002.

[11] M. G. McCrum, B. E. Read, and G. Williams, Anelastic and Dielectric Effects in Polymeric Solids ~Wiley, London, 1967 and Dover, New York, 1991.

[12] K. I. Okamoto, N. Tanihara, H. Watanabe, K. Tanaka, H. Kita, A. Nakamura, Y. Kusuki, K. Nakagawa, ''Sorption and diffusion of water vapor in Polyimide films'', J Polym Sci Polym Phys 1992, 30, 1223.

[13] G. Dlubek, R. Buchhold, Ch. Hübner, A. Nakladal, ''Water in local free volumes of polyimides: A positron lifetime study'', *Macromolecules* **32,** 2348 (1999).

2006 25th International Conference on Microelectronics

Investigation of Electrode Patterns Suitable for Nano-Litre Drop Coated Conducting Polymer Composite Sensors

K.I. Arshak, C. Cunniffe, E.G. Moore and L.M. Cavanagh

Abstract— **This study presents an analysis of electrode patterns suitable for use with drop coated conducting polymer gas sensors. A thin-film technique was used to efficiently fabricate the copper electrode patterns [1]. Conducting Polymer Composite (CPC) materials were deposited using a 500 nano-litre syringe onto the electrode patterns to produce an array of sensors for organic solvent vapour detection. The sensors were exposed to propanol vapour in steps of 3000 ppm from a minimum concentration of 5000 ppm up to a maximum concentration of 20,000 ppm. Empirical results showed that a non-parallel electrode configuration produces a marginally larger responce and is also less noisy than the interdigitated or parallel electrode configurations. Results show that increasing the baseline resistance of the sensing material gives a larger responce.**

I. INTRODUCTION

Much research has been carried out in the area of electrode geometry for use in gas sensors in the past [2], [3]. Previous works investigated the effect of geometry and position of electrodes for semi-conductor gas sensors [3]. It was discussed in the conclusion that placing electrodes beneath the sensing layer is not the optimal site, but if they are placed as such a wider electrode gap increases sensitivity [3]. It was also observed in a recent work [4] that noise levels decrese as electrode gap distances increase with electrode gaps ranging from 20μm -140μm where it was stated that the underlying physics causing this was still being investigated. Previous work on electrode patterns for use with ploymer carbon-black composites was carried out where spray coating was used to deposit the sensing material resulting in a homogenous sensing layer [2]. The electrode patterns investigated included 42 circular configurations. It was shown in that work that the electrode geometry did not have an effect on sensor responce magnitude but noise properties are strongly effected by electrode configuration [2]. It is the aim of this work to investigate the optimal electrode pattern for use beneath a drop coated conducting polymer composite sensing material. The material was deposited using a drop coating technique. Upon deposition of the

K.I. Arshak, C. Cunniffe, E.G. Moore and L.M. Cavanagh are with the Microelectronic and Semiconductor Research Group, Department of Electronic and Computer Engineering, University of Limerick, National Technological Park, Limerick, Ireland. Email: khalil.arshak@ul.ie

material drop the conducting polymer composite dispersed to from a ring structure. A number of different patterns (Fig. 2) were investigated to show which patterns yeild noisy responces. Different gap widths in parallel electrode configuration were tested to find a relationship between gap width and baseline resistance. A range of sensor baseline resistances were also used to find correlations between baseline resistance and sensor responce. The sensor arrays were exposed to solvent concentrations of 5000ppm to 20000ppm in increments of 3000ppm.

II. EXPERIMENTAL

The sensors were produced on an alumina substrate, which was coated with a layer of copper using an Edwards Thermal Evaporation unit. The resulting substrate was then coated with photo resist using a spin coater. Patterns were designed using Eagle PCB software and printed on acetate. The pattern was UV exposed onto the substrate, and then subsequently dipped in developer and etched. The remaining photoresist was then striped from the pattern on the substrate. A more detailed description of this process is described in [1]. The sensing material composed of carbon black, polyethylene adipate, and surfactant as described earlier [5] and drop coated using a 500 nano litre syringe set to deposit 100 nano litres. The drop coating apparatus is shown in Fig. 1

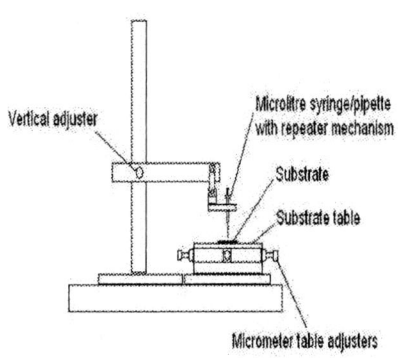

Fig. 1.   Setup Used For Drop Coating Sensor Materials.

The 25.4mm x 12.7mm alumina substrate may be inserted into the testing equipment to enable the patterns to the left of the substrate (Fig. 2) to be tested and then inserted to allow the patterns on the right of the substrate (Fig. 2) to be tested. The sensor arrays were placed in a dynamic flow gas test chamber, which permitted the arrays to be exposed to specific vapour concentrations in a controlled manner. The chamber used a Bronkhorst EL-Flow mass flow meter/controller to control the carrier gas and a μ-Flow liquid mass flow meter with a Controlled Evaporator Mixer (CEM). An EZ-7000 controller unit was used to manually operate the liquid and gas flow controllers. The system was serially connected to a PC for automatic operation. The sensors were exposed to propanol vapour in steps of 3000 ppm from a minimum concentration of 5000 ppm up to a maximum concentration of 20,000 ppm. The exposure cycle consisted of a 30 seconds flush period followed by 60 seconds exposure to the solvent vapour and another 30 seconds flush. The array responses were recorded using a National Instruments data acquisition card (Model No: PCI-MIO-16E-4) and LabVIEW software.

III. RESULTS AND DISCUSSION

Electrode patterns (Fig. 2) were drafted using Cadsoft Eagle PCB design software such that two different configurations may be tested using one drop of sensing material to eliminate sensor-to-sensor reproducibility issues. This allowed for direct comparisons between electrode configurations for each drop of sensing material deposited across each electrode. The electrode configurations consisted of varying gaps between the points and different angles of attack of electrode into the drop of sensing material as well as an interdigitated configuration (Fig. 2). Table I details the electrode gaps at each stage of the manufacturing process and Table II details the baseline resistance of the sensors.

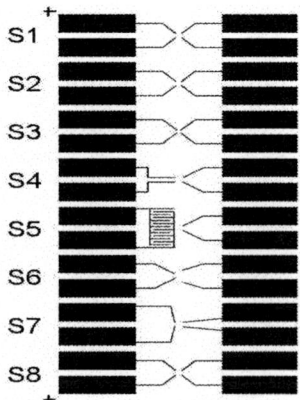

Fig. 2. Electrode pattern allowing for the testing of two electrode patterns.

TABLE I
ELECTRODE GAPS SPECIFICATION (ALL FIGURES IN μM)

Sensor	Process Stage	Left	Right
	CAD Output	200	200
S1	Mask	97.9	116.6
S1	Etched Pattern	102.6	166.6
	CAD Output	150	150
S2	Mask	0	0
S2	Etched Pattern	0	0
	CAD Output	100	100
S3	Mask	0	0
S3	Etched Pattern	0	0
	CAD Output	200	200
S4	Mask	121.2	116.6
S4	Etched Pattern	107.2	135.2
	CAD Output	200	200
S5	Mask	~107	111.9
S5	Etched Pattern	~121	121.2
	CAD Output	200	700
S6	Mask	135.3	587.4
S6	Etched Pattern	139.9	596.7
	CAD Output	200	200
S7	Mask	172.6	103.4
S7	Etched Pattern	158.5	111.9
	CAD Output	200	200
S8	Mask	103	107.3
S8	Etched Pattern	93.24	95

These configurations presented a method of determining the important features of the dropped sensing material. A stereomicroscope image of the acetate mask focused on S4 is presented in Fig. 3, and the etched pattern of S4 is displayed in Fig. 4.

A volume of 100 nano-litres of Polyethylene adipate\carbon black composite material was deposited onto the electrodes using a nano-litre drop coating technique. Fig. 5 shows a stereomicroscope image of the resulting sensor with the deposited material focusing on S4. The image shows the electrodes beneath the material in both a parallel and non parallel configuration. The ring structure which is formed on deposition and is the most significant constituent element of the material is also visable in Fig. 5.

By varying the structure of the electrode patterns the resultant data showed that the most important structure of the sensing material is the ring structure around the edge of the dropped material. The electrode gap in the middle of the sensing material had little effect whereas the distance between the electrodes where the ring crossed over the electrode defined the baseline resistance of the sensor. Sensors with a higher baseline resistance boasted a larger and less noisy response while the sensors with electrodes, which lay parallel to each other,

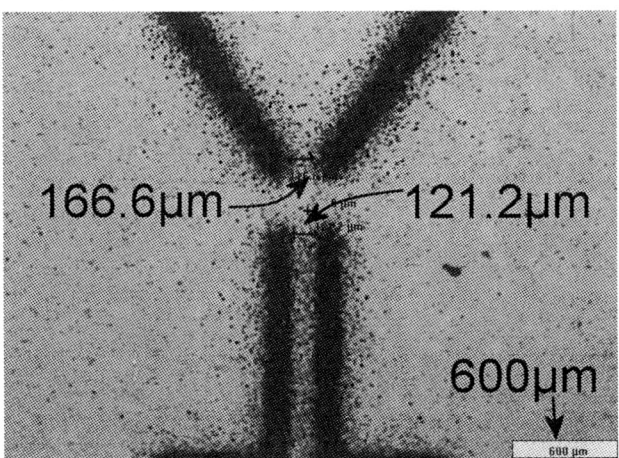

Fig. 3. Stereomicroscope image focusing on S4 of mask printed on acetate.

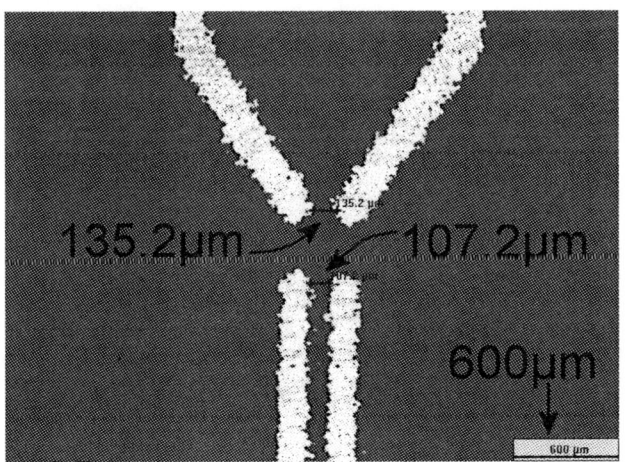

Fig. 4. Stereomicroscope image focusing on S4 of etched pattern.

TABLE II
BASELINE RESISTANCES OF SENSORS (ALL FIGURES IN kΩ)

Sensor	Baseline Resistance	
	Left	Right
S1	145.5	186.5
S2	0	0
S3	0	0
S4	29	34.75
S5	20	138
S6	48.2	64
S7	75.5	31.56
S8	69.45	59.3

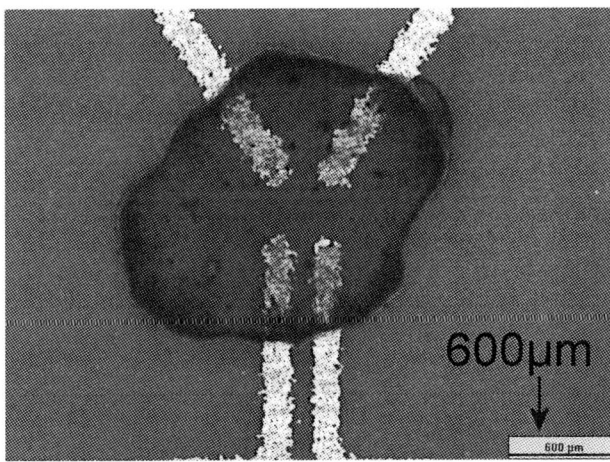

Fig. 5. Stereomicroscope image focusing on S4 of etched pattern with deposited material showing ring structure.

produced noisy response such as the patterns S4 and S5 in Fig. 2. The sensor responses were pre-processed using fractional baseline manipulation, shown in Equation. (1) which produces a normalised response and can enhance contrast and reduce drift effects [6].

$$V = \frac{V_{gas} - V_{air}}{V_{air}} \qquad (1)$$

Where V_{gas} is the voltage drop across the sensor in responce to the vapourised solvent and V_{air} is the voltage drop across the sensor in responce to the flush gas.

A typical response for sensor pattern S4 is displayed in Fig. 6 showing that the parallel electrodes produces a noisy response and the alternative pattern on the right hand side of S4 yielded a cleaner response and also exhibited a marginally larger response. Fig. 7 shows a typical responce from an interdigitated electrode pattern.

To test the effect of electrode gap width in a parallel configuration an array was designed with increasing gap sizes from 200μm to 800 μm. The maximum $\triangle V/V\%$

was extracted from the raw data. Fig. 8 shows the resulting graph of baseline resistance versus percentage change of voltage the seven parallel electrode configurations with varying gaps. The $\triangle V/V\%$ vs R_0 graph shows a trend in the data illustrating that the percentage voltage change increases as the baseline resistance increases, however the electrode Gap vs R_0 graph also shows there was no correlation between the baseline resistance of the sensor and the electrode gap width. Employing interdigitated electrodes for this applications didn't enhance the sensor responce. Due to the parallel nature of the interdigitated fingers the sensor exhibited a similar but more exaggerated noisy responce illustrated in Fig. 7 to that of the parallel electrodes. The interdigitated configuration contribute to lowering the baseline resistance but as shown in Fig. 8 a higher baseline resistance yields a better percentage voltage change.

IV. CONCLUSION

Results show that incresing the baseline resistance of the sensing material gives a larger responce and there is

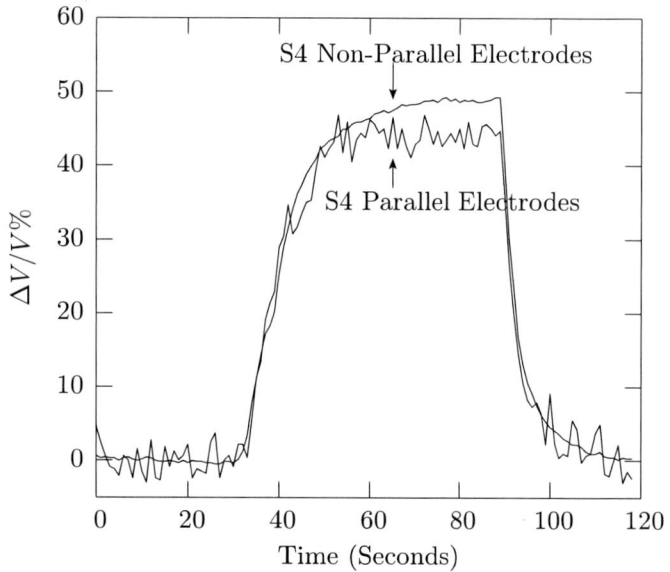

Fig. 6. Graph showing the typical response of PEA sensors response to 20000ppm of Propanol using electrode pattern S4.

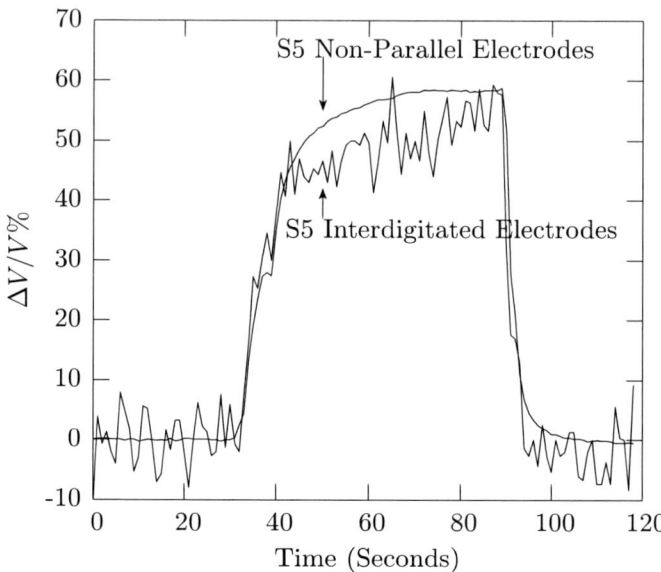

Fig. 7. Graph showing the typical response of PEA sensors response to 20000ppm of Propanol using electrode pattern S5.

no correlation between the baseline resistance and the electrode gap in a parallel configuration. Empirical results showed that a non-parallel electrode configuration produces a marginally larger responce and is also less noisy than the interdigitated or parallel electrode configurations. The electrode gaps are less important due to the material deposition method as the principle components of the material disperse to the edges to form a ring structure which governs the base line resistance. This work shows that manipulating the electrode pattern can improve the sensitivity and stability of these drop coated conducting polymer composite sensors for use in electronic nose applications.

Acknowledgements

This work was conducted as part of a collaborative project between AMT Ireland, University of Limerick and University College Cork, and is funded by Enterprise Ireland under project ref. no. ATRP/2002/427 (Intelli-SceNT).

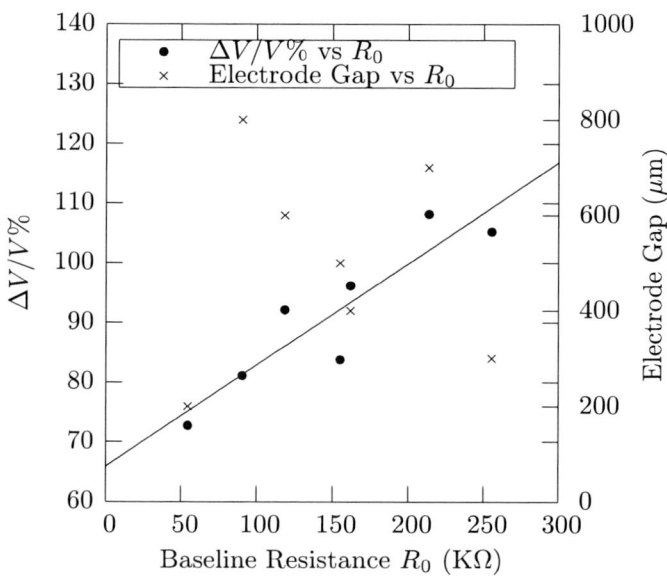

Fig. 8. Plot of Baseline Resistance vs Change in Voltage and Electrode Gap for Seven Parallel Plate Electrodes

References

[1] K. Arshak, C. Cunniffe, E. Moore, L. Cavanagh, and J. Harris, "A novel approach to electronic nose-head design, using a copper thin film electrode patterning technique," in *28th International Spring Seminar on Electronics Technology*, Austria, 2005, pp. 185–190.

[2] B. Matthews, J. Li, S. Sunshine, L. Lerner, and J. Judy, "Effects of electrode configuration on polymer carbon-black composite chemical vapor sensor performance," *Sensors Journal, IEEE*, vol. 2, no. 3, pp. 160–168, 2002.

[3] X. Vilanova, E. Llobet, J. Brezmes, J. Calderer, and X. Correig, "Numerical simulation of the electrode geometry and position effects on semiconductor gas sensor response," *Sensors and Actuators B: Chemical*, vol. 48, no. 1-3, pp. 425–431, May 1998.

[4] V. T. Wong, A. Huang, and C.-M. Ho, "Towards high density silicone polymeric chemical vapor sensor arrays," in *Proceedings of 11th International Symposium on Olfaction and Electronic Nose (ISOEN'05)*, vol. -, no. -, Barcelona, Spain, Apr. 2005, pp. 398 – 401.

[5] K. Arshak, E. Moore, L. Cavanagh, J. Harris, B. McConigly, C. Cunniffe, G. Lyons, and S. Clifford, "Determination of the electrical behaviour of surfactant treated polymer/carbon black composite gas sensors," *Composites Part A: Applied Science and Manufacturing*, vol. 36, no. 4, pp. 487–491, 2005.

[6] R. Gutierrez-Osuna, "Pattern analysis for machine olfaction: a review," *Sensors Journal, IEEE*, vol. 2, no. 3, pp. 189–202, 2002.

2006 25th International Conference on Microelectronics

A Novel Vertical Impact Ionisation MOSFET (I-MOS) Concept

U. Abelein, M. Born, K. K. Bhuwalka, M. Schindler, M. Schmidt, T. Sulima, I. Eisele

Abstract – This paper presents experimental results of a novel vertical impact ionisation MOSFET (I-MOS). The device consists of a vertical gated triangular barrier diode (TBD), also know as planar doped barrier MOSFET (PDBFET). At low drain-source voltages the behaves like a conventional MOSFET. Drain-source voltages of more than 1.5 V activate gate controlled impact ionization in the sub-50 nm n-channel device, resulting in a subthreshold swing of 20 mV/decade at room temperature. The device shows an excellent I_{ON}/I_{OFF} ratio of 2.5×10^8 in this mode.

I. INTRODUCTION

The well known lateral impact ionisation MOSFET (I-MOS) concepts are based on a gated p-i-n diode [1-3]. We present experimental results of a novel vertical I-MOS. The device structure is a vertical gated triangular barrier diode (TBD), which is also known as planar doped barrier MOSFET (PDBFET). The advantage of this concept is the arbitrary choice of doping profiles between source and drain. High $\delta p+$ doping in the channel region is chosen to achieve impact ionisation (II) in the subthreshold region. Depending on the drain voltage the device can be operated in the conventional MOSFET as well as in the II mode. In the II mode at room temperature the device shows a subthreshold swing of 20 mV/dec. Due to the delta-doping profile the OFF currents are very low which results in an I_{ON}/I_{OFF} ratio $> 10^8$.

The presented device is stable and compared to the lateral I-MOS [1-3] it is less sensitive to hot carrier degradation effects. Due to its planar delta doping the electric field strength in the PDBFET is only half of that of a conventional MOSFET [4]. The delta acts as a carrier injector because it determines the threshold voltage. The resulting voltage drop occurs between the delta layer and drain (see Fig. 6). This allows higher supply voltages and increases reliability.

II. DEVICE FABRICATION

The devices as shown in Fig. 1 were fabricated by growing a sequence of doped silicon layers using molecular beam epitaxy (MBE) on an antimony doped n^+ <100> silicon substrate ($\rho < 20$ mΩcm) which is used as source

The authors are with the Institute of Physics, Universitaet der Bundeswehr Munich, 85577 Neubiberg, Germany, Phone: +49-89-60043971, Fax: +49-89-60043877, Email: ulrich.abelein@unibw-muenchen.de

electrode. The stack consists of a 40 nm intrinsic silicon layer (unintentional n-doping $< 10^{16}$ cm^{-3}) followed by a highly boron doped 3 nm $\delta p+$ layer ($>10^{19}$ cm^{-3}) and again a 40 nm intrinsic silicon layer. We will refer to this thickness of the intrinsic layers as $L_i = 40$ nm in section III of this paper, as this region forms the channel of the device. On top a 300 nm phosphorus doped n^+ layer (4×10^{18} cm^{-3}) has been grown, which acts as drain electrode. This layer stack was patterned by anisotropic low damage reactive ion etching. A 4.5 nm gate oxide was thermally grown by RTP (wet oxide, 800 °C, 300 s). Afterwards 250 nm n+ polysilicon were deposited and patterned to form the gate electrode. The devices were then passivated by a LPCVD silicon nitride. Finally the devices were finished by contact trench opening and metallization. Fig. 2 shows a secondary ion mass spectrometry (SIMS) profile of the silicon mesa stack after complete processing.

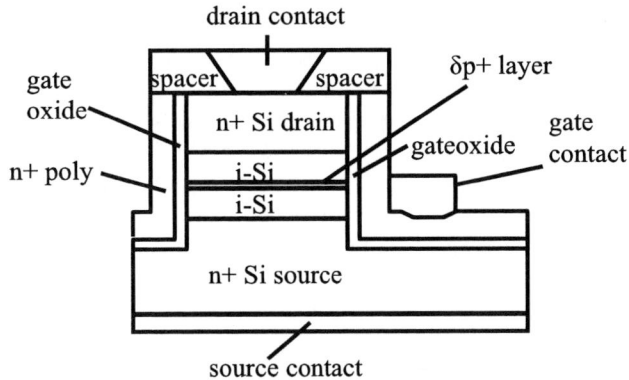

Fig. 1: Schematic drawing of the vertical I-MOS.

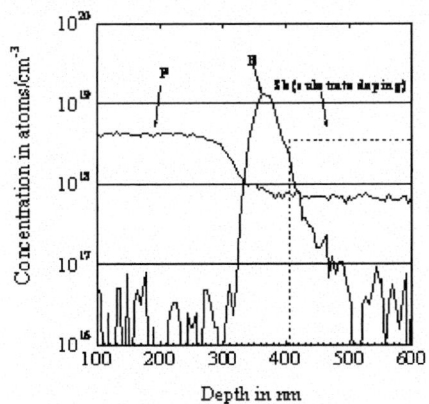

Fig. 2: SIMS profile of the vertical I-MOS layer stack.

1-4244-0116-X/06/$20.00 ©2006 IEEE

III. RESULTS AND DISCUSSION

The device can be operated as a conventional MOSFET as well as an I-MOS depending on the applied drain-source voltage.

For $V_{DS} < 1.5$ V the device acts as a conventional short channel MOSFET. The subthreshold slope S observed in this region is 130 mV/decade and the threshold voltage V_T is 2.2 V due to the oxide thickness and the high delta doping, respectively [6]. Fig. 3 shows a typical experimental input characteristic of the device in the conventional MOSFET mode. I_{OFF} is below the noise level of 10^{-7} μA/μm, I_{ON} is 200 μA/μm, which leads to an I_{ON}/I_{OFF} ratio in the conventional MOSFET mode of 2×10^{10}.

For drain-source voltages above $V_{DS,II} = 1.5$ V the device operates in impact ionisation mode. The subthreshold slope S lowers with rising V_{DS} down to 20 mV/decade. The on current I_{ON} reaches 500 μA/μm and the device shows an excellent I_{ON}/I_{OFF} ratio of 2.5×10^{8}, which is the highest ever observed for an I-MOS. Fig. 4 presents a typical experimental input characteristic of the vertical I-MOS in the impact ionisation region. Note that for all measurements the gate voltage V_G was varied from 0 V to 4.5 V in both directions. Compared to [5] the device shows no appreciable hysteresis.

Unlike the lateral I-MOS [7] the vertical I-MOS does neither show a shift of V_T nor a change in the subthreshold slope due to hot carrier damage with repeating measurements. To test the reproducibility of the characteristics, the input characteristic for a fixed drain voltage in the impact ionisation region was measured 120 times by driving the gate voltage from 0 V to 4.5 V and back. Neither the off current nor the subthreshold swing changed during all measurements. Also no shift of the threshold voltage or any other degradation by hot carriers was observed.

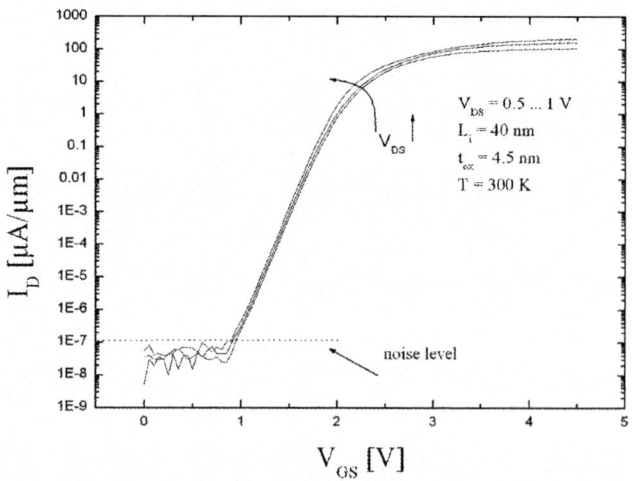

Fig. 3: Typical I_D-V_G input characteristics as a function of drain voltage V_D of the vertical I-MOS in conventional MOSFET mode at room temperature (2 μm gate width)

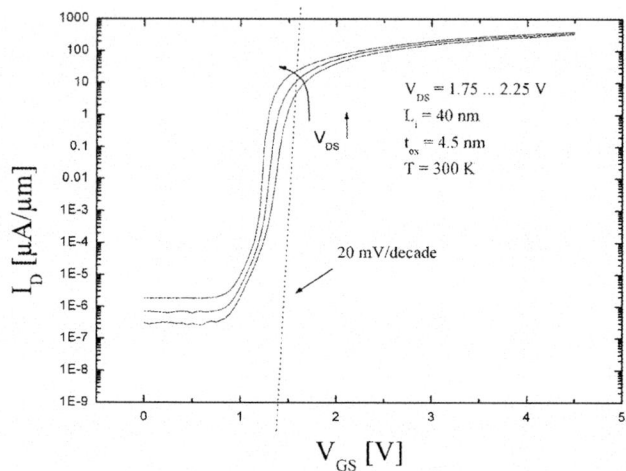

Fig. 4: Typical I_D-V_G input characteristics as a function of drain voltage V_D of the vertical I-MOS in II mode at room temperature (2 μm gate width)

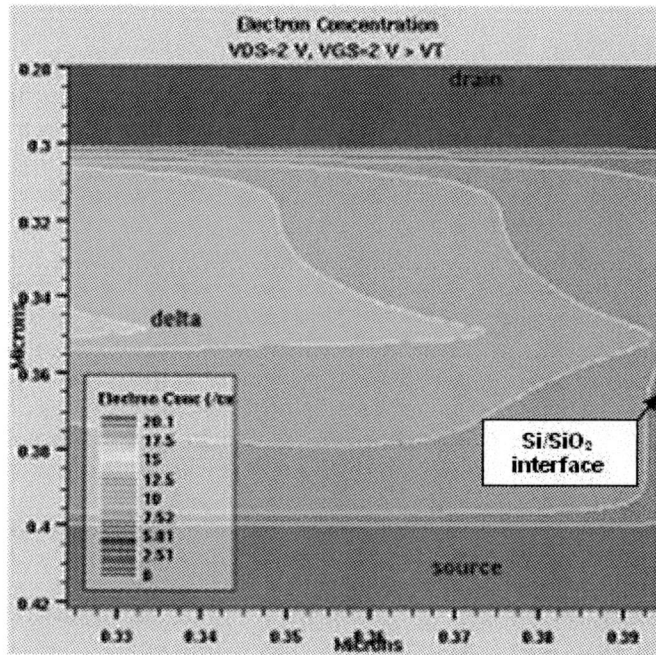

Fig. 5: 2D ATLAS simulation of the electron density in a vertical n-channel I-MOS.

The reason for the stability of V_T and subthreshold slope is the carrier transport, which is not confined to the oxide-silicon interface like in the lateral I-MOS where hot carriers are injected into the oxide and cause traps that lead to the shift of V_T. In the vertical I-MOS the channel spreads out into the bulk [6] where the impact ionisation takes place. Fig. 5 shows a 2D ATLAS [8] simulation of the electron density in a vertical n-channel I-MOS. Only in the region of the delta layer high carrier densities are observed near the oxide. The main carrier transport between the delta layer and drain takes place in the bulk region. This is obviously the main reason for the stability of the device.

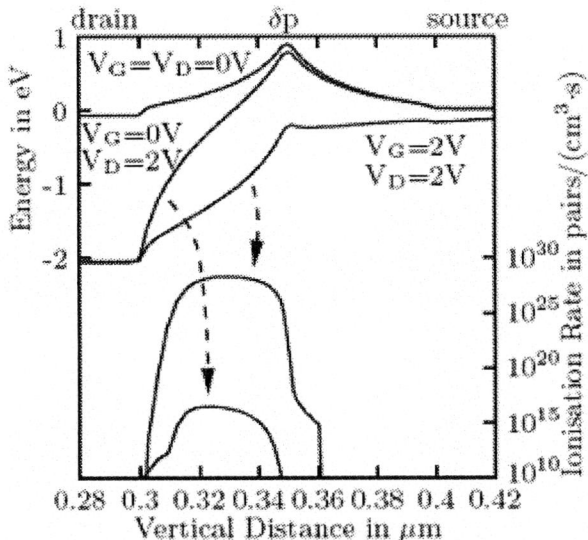

Fig. 6: ATLAS simulation of the conduction bands of the vertical n-channel I-MOS (Deltadoping $N_{\delta p}= 2 \times 10$ cm^{-3}) for different bias conditions. II is confined to the drain side intrinsic region and the II rate is controlled by gate voltage.

Fig. 6 shows ATLAS simulations of the dependency of the II rate from V_G and V_D together with their effect on the conduction band in an vertical n-channel I-MOS. It can be clearly seen that applying a positive gate as well as drain voltage at the same time lowers the barrier at the delta layer and increases the II rate by orders of magnitude compared to applying only a positive drain voltage. This proofs the gate control of the II. Fig. 6 shows, that the II is confined to the intrinsic region at the drain side of the vertical I-MOS. The length L_i of this region, which is 40 nm in the device presented here, determines the channel length L_{ch} of this device.

TABLE I

COMPARISON OF DIFFERENT I-MOS CONCEPTS

	Vertical I-MOS	[3]	[7]
Type	n	n/p	p
L_{ch} (nm)	40	100	300
$V_{DS|II}$ (V)	1.5	5.5	15
I_{ON}/I_{OFF}	2.5×10^8	28.9/23	10^5
I_{ON} (μA/μm)	500	81.1/78.2	~100
S (mV/dec)	20	11.8	10

Table 1 compares the main results achieved with the vertical device with those of the lateral concepts [3] and [7]. The drain-source voltage necessary to achieve impact ionisation is reduced up to a factor of 10 and at the same time the on current I_{ON} observed is significantly higher in the vertical device compared to the lateral concepts.

Finally the vertical device's low OFF currents lead to the highest I_{ON}/I_{OFF} ratio ever observed for an I-MOS device, although the channel length of 40 nm is the shortest realized so far.

IV. CONCLUSION

We demonstrated gate controlled II in a sub-50 nm n-channel PDBFET. The present device was significantly improved by choosing a higher doping in the delta layer.

In contrast to a p-i-n structure the device is less sensitive to degradation caused by hot carriers. The highly doped delta layer results in an excellent I_{ON}/I_{OFF} ratio, the highest ever observed so far. The leakage current is very low in contrast to [3]. The II rate can be controlled by the delta doping and the channel length.

An even higher δ-doping could be used to suppress the normal MOSFET operation mode and use the device only in impact ionisation mode.

REFERENCES

[1] Gopalakrishnan, K., Griffin, P. B., and Plummer, J. D., "I-MOS: a novel semiconductor device with a subthreshold slope lower than kT/q", *IEEE Int. Electron Devices (IEDM) Tech. Dig.*, pp. 289-292, 2002.

[2] Gopalakrishnan, K., Griffin, P. B., and Plummer, J. D., "Impact Ionization MOS (I-MOS) – Part I: Device and Circuit Simulations", *IEEE Trans. On Electron Devices*, vol. 52(1): pp. 69-76, 2005.

[3] Choi, W. Y., Song, J. Y., Lee, J. D., Park, Y. J. and Park, B.-G., "100 nm n-/p-Channel I-MOS Using a Novel Self-Aligend Structure", *IEEE Electron Devices Letters,* vol. 26(4): pp. 261-263, 2005

[4] Hansch, W., Ramgopal Rao, V., Fink, C., Kaesen, F., and Eisele, I., "Electric Field Tailoring in MBE-grown vertical sub-100 nm MOSFETs", *Thin Solid Films*, vol. 321: pp. 206-214, 1998

[5] Rao, V. R., Wittmann, F. Gossner, H. and Eisele I., "Hysteresis Behavior in 85-nm Channel Length Vertical n-MOSFETs Grown by MBE", *IEEE Trans. on Electron Devices*, vol. 43(6): pp. 973-976, 1996

[6] Born, M., Abelein, U., Bhuwalka, K.K., Schindler, M., Schmidt, M., Ludsteck, A., Schulze, J. and Eisele, I., "Sub-50 nm High Performance PDBFET with Impact Ionization", 4th Int. Conf. on Silicon Epitaxy and Heterostructures, p. 308-309, Hyogo, Japan, 2005.

[7] Gopalakrishnan, K., Woo, R., Jungemann, C., and Plummer, J. D., "Impact Ionization MOS (I-MOS) – Part II: Experimental Results", *IEEE Trans. On Electron Devices*, vol. 52(1): pp. 77-84, 2005.

[8] *ATLAS User Manual, 2-D Device Simulation Software*, SILVACO International, Santa Clara, CA.

2006 25th International Conference on Microelectronics

Tunnel FET: A CMOS Device for High Temperature Applications

Mathias Born, Krishna Kumar Bhuwalka, Markus Schindler, Ulrich Abelein, Matthias Schmidt, Torsten Sulima, Ignaz Eisele

Abstract— **This paper presents experimental data on the temperature dependence of silicon tunnel field effect transistors (FETs) and corresponding simulations. It shows that the characteristics of tunnel transistors depend only weakly on temperature and that the "subthreshold" swing is temperature independent. The behavior is compared to conventional MOSFETs.**

I. INTRODUCTION

Gated p-i-n diodes operating as surface tunnel transistors have been proposed as solution of the many problems encountered in scaling down the conventional MOSFET, thereby turning the parasitic effect of tunneling into an operating principle [1–6]. Outstanding features of this device are a very low off current, determined by the p-i-n leakage current, and a swing independent of kT/q which can be optimized to be less than 60 mV/dec [7]. Very high I_{ON}/I_{OFF} ratios can be achieved, independent of geometrical scaling [8]. In this work we present experimental data, which show that in addition the I-V characteristics of tunnel FETs are only weakly temperature dependent.

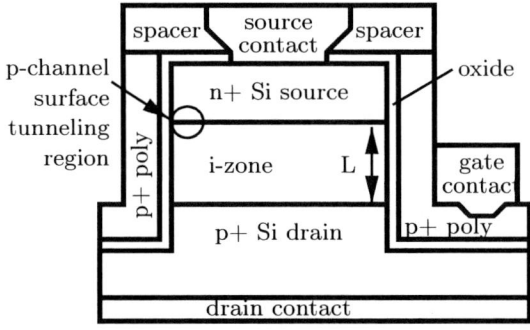

Fig. 1. Schematic view of a vertical tunnel FET. The channel length is determined by the thickness of the i-zone. The assignment of drain and source reflects the p-channel operating mode.

II. THE TUNNEL FIELD EFFECT TRANSISTOR

Figure 1 shows a schematic view of the vertical tunnel FET. The substrate is used as p+ region. The 70 nm intrinsic region and the 300 nm n+ region are grown in a

The authors are with the Institute of Physics, Universität der Bundeswehr München, 85577 Neubiberg, Germany, contact e-mail: Mathias.Born@unibw.de

commercially available LPCVD tool. The device mesas are defined by reactive ion etching (RIE). A 4.5 nm thermal wet oxide, grown at 800 °C, is used as gate oxide with p+-polysilicon as gate electrode [9]. Every device is contacted through the top of its mesa (source in Figure 1) and the substrate backside (drain in Figure 1).

Application of a negative gate voltage induces a hole channel in the i-zone and leads to a tunnel junction at the source end. This is the p-channel operating mode. In the n-channel operating mode the n+ region acts as drain while the p+ region acts as source. The electron channel induced by the positive gate bias creates a tunnel junction at the source. Figure 2 shows experimental transfer characteristics in both operating modes. The asymmetry results from the work function of the p+-polysilicon gate material. The characteristics would be symmetric with a midgap gate material. The drain current changes exponentially as a function of gate voltage. A high on-off-ratio of more than six orders of magnitude is observed. Corresponding output characteristics can be seen in Figure 3. Since the sharpness of the doping profile of the devices presented here is not sufficient and since the gate oxide thickness is not low enough [9], only quasi saturation can be observed. This can be improved if abrupt dopant changes and a thinner gate oxide are used according to the simulation in Figure 4.

Figure 5 shows experimental transfer characteristics in p-channel operating mode at different temperatures. Raising the temperature from 33K to 373K increases the drain current by a factor of 4.4, which is mainly due to the variation of band gap with temperature. Note that the temperature dependence is weak *even though* tunneling in silicon is phonon assisted, because only phonon emission is required, which can take place nearly independent of temperature. As can be clearly seen, the "subthreshold" swing S is independent of temperature and thus independent of the kT/q thermal limit. Note that the very high swing values indicate a poor gate control which can be greatly improved by decreasing the gate oxide thickness and, more important, by increasing the doping profile sharpness which is the most critical parameter for tunnel FETs. Simulations predict that in this case values below 60 mV/dec can be achieved independent of temperature [10].

The characteristics of a tunnel FET can be described

1-4244-0116-X/06/$20.00 ©2006 IEEE

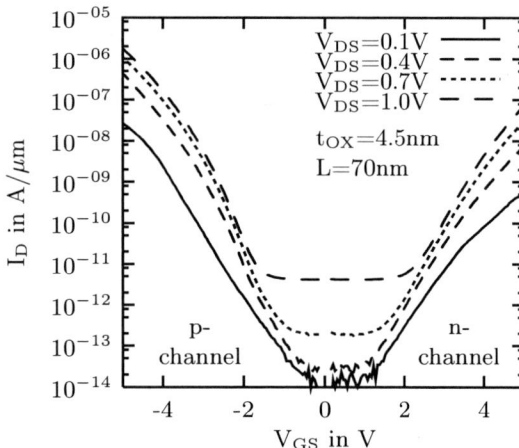

Fig. 2. Experimental transfer characteristics (drain current I_D versus gate voltage V_{GS}) in p- and n-channel operating mode at different drain voltages. Note that $10^{-14} A/\mu m$ represent the noise level of the measurement equipment.

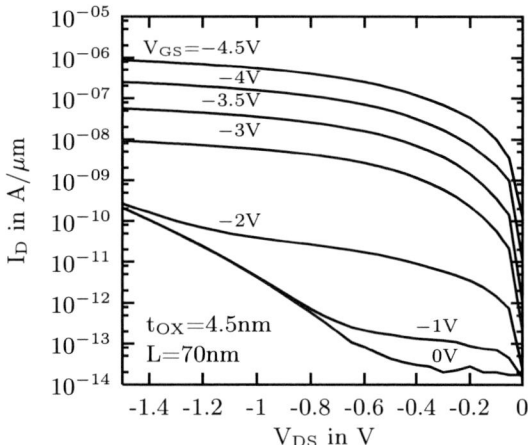

Fig. 3. Experimental output characteristics (drain current I_D versus drain voltage V_{DS}) in p-channel operating mode at different gate voltages.

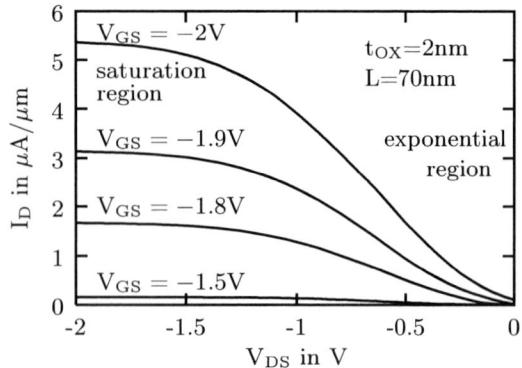

Fig. 4. Simulated output characteristics (drain current I_D versus drain voltage V_{DS}) in p-channel operating mode at different gate voltages.

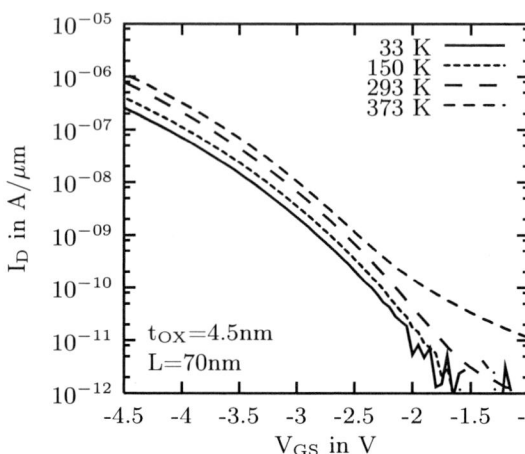

Fig. 5. Experimental transfer characteristics (drain current I_D versus gate voltage V_{GS}) in p-channel operating mode at different temperatures. Note: the data for 33 K - 293 K and for 373 K were taken from two different devices.

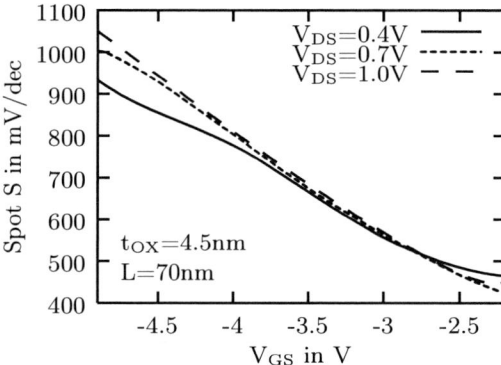

Fig. 6. Experimental spot swing S in the "subthreshold" region as a function of gate bias in p-channel operating mode at different drain voltages. A strong dependence can be observed even in non-saturation.

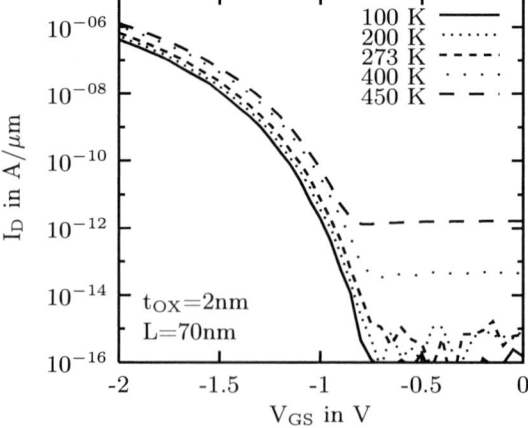

Fig. 7. Simulated transfer characteristics (drain current I_D versus gate voltage V_{GS}) in p-channel operating mode at different temperatures.

132

by a single equation [10]

$$I_{DS} = A_{KANE}D^2W_g^{-1/2}V_{GS}^2 \times$$
$$\exp\left(-\frac{B_{KANE}W_g^{3/2}}{V_{GS}D}\right) \qquad (1)$$

where D, A_{KANE} and B_{KANE} are constants and W_g is the bandgap energy. This predicts a strong dependence of swing on V_{GS} in the quasi-saturation region as shown in Figure 6

$$S_{TUNNEL} = \frac{V_{GS}^2 \cdot \ln 10}{2V_{GS} + B_{KANE}W_g^{3/2}/D} \qquad (2)$$

in contrast to the conventional MOSFET, where the subthreshold swing is

$$S_{MOSFET} = \ln 10 \cdot \frac{nKT}{q} \qquad (3)$$

Here, n is a device geometry parameter which is set to one for further discussion. The main temperature effect in the tunnel FET is the temperature dependence of the band gap, which is weak [11]:

$$W_g(T) = W_g(0) - \frac{\alpha T^2}{T + \beta} \qquad (4)$$

α and β are fitting constants. Its influence on the drain current results in a parallel shift of the I-V-characteristics as can be seen in Figure 5. Figure 7 shows a simulation of the transfer characteristics at different temperatures. It agrees qualitatively well with the experimental data in Figure 5. Although the bulk leakage current of the p-i-n diode is raised significantly by temperature, the tunnel FET can still be used at temperatures of 450K and above.

Note that the conventional "threshold" voltage concept of a MOSFET cannot be applied. It describes the transition between the diffusion limited subthreshold behavior and the drift limited on-behavior. According to (1) this is not the case for the tunnel FET because the gate controlled current is determined by only one mechanism. However, one can arbitrarily *define* a threshold voltage using the constant current method [10]. Since the I-V-characteristics are slightly shifted by temperature, the threshold voltage defined by this method shows a temperature dependence in the range of 1-2 mV/K which is similar to the conventional MOSFET. However, it can be clearly seen from Figures 5 and 7 that this temperature coefficient depends on the constant current used for defining the threshold voltage. The slope is not constant but depends on the gate voltage. It also depends on the slope itself which in turn is determined by the doping profile.

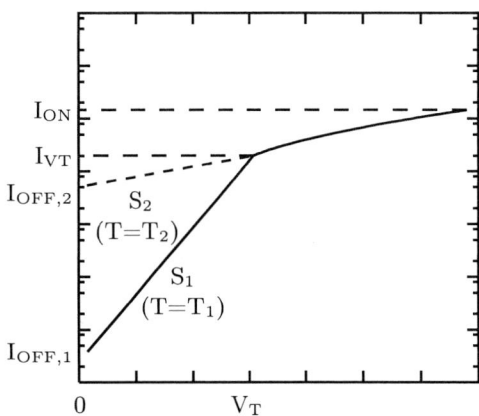

Fig. 8. Schematic transfer characteristics (logarithmic scale) of conventional MOSFETs. Two different subthreshold slopes $S_1 < S_2$ for two different temperatures $T_1 < T_2$ are shown. The off-current I_{OFF} degrades quickly with increasing temperature.

III. Tunnel FET versus conventional MOSFET

For a comparison of the tunnel FET with the conventional MOSFET it is worth reviewing the latter one briefly. Temperature mainly influences three parameters in conventional MOSFETs [11–13]. The threshold voltage shows a positive temperature coefficient in the range of 1-5 mV/K which is due to a shift of the Fermi level with temperature. The surface carrier mobility decreases monotonically with increasing temperature which can be approximated by $\mu \propto T^{1-2}$. These opposite temperature coefficients result in an isothermal point which the tunnel transistor does not have. Differentiating (1) with respect to W_g yields

$$\frac{\partial I_{DS}}{\partial W_g} = -\frac{1}{2}A_{KANE}DV_{GS} \times$$
$$\exp\left(-\frac{B_{KANE}W_g^{3/2}}{V_{GS}D}\right) \times$$
$$(DW_g^{-3/2}V_{GS} + 3B_{KANE}) < 0 \qquad (5)$$

Thus, the drain current increases monotonically with decreasing bandgap energy, that is with increasing temperature.

The third and most important aspect is the subthreshold slope of the conventional MOSFET. It is the main limiting parameter at high temperatures as can be seen in Figure 8. Here, constant current I_{VT}, and constant threshold voltage V_T are assumed for simplicity. The on-off-ratio is dominated by V_T and by the subthreshold slope S:

$$\frac{I_{VT}}{I_{OFF}} = 10^{V_T/S} \qquad (6)$$

The ratio between the off-currents at different temperatures is then

$$\frac{I_{OFF,2}}{I_{OFF,1}} = 10^{V_T/S_1 - V_T/S_2} \qquad (7)$$

For example, with $V_T = 0.5V$, $T_1 = 300K$, and $T_2 = 450K$ this ratio is $0.5V/60mV - 0.5V/90mV \approx 2.8$ orders of magnitude. In addition the off-current of the tunnel transistor is a diffusion current and thus depends strongly on temperature. However, it is several orders of magnitude lower at room temperature. According to Figure 7 this as well as the temperature independent slope enable the tunnel transistor to function at very high temperatures, which render a conventional MOSFET inoperable.

IV. Conclusion

Tunnel FETs can be designed to work over a wide temperature range without significant variation in performance. Band-to-band tunneling in the tunneling junction determines the device performance, unlike the conventional MOSFET, where the properties of the channel, e.g. the carrier mobility μ, are dominant. This makes it possible to scale down the channel length of the tunnel FET to 10 nm without decreasing its performance.

Future work will have to address two important aspects: The on-currents of all tunnel transistors still do not fullfill the ITRS requirements [14]. One reason is the smear out of the doping profile due to dopant diffusion during the device processing. But even if this problem is solved by lowering the processing thermal budget, the on-current of silicon tunnel FETs will not be sufficient because it is limited by the band gap of silicon [10]. It can be improved by using SiGe to decrease the band gap, as proposed in [7]. Pure n- and p-channel operating mode tunnel FETs are needed for CMOS-like circuits. The n- operating mode in the p-FET and the p- operating mode in the n-FET can be suppressed by workfunction engineering and by changing the doping profile appropriately [15].

V. Summary

Experimental data on the temperature dependence of silicon tunnel field effect transistors have been presented. It has been found that the characteristics of tunnel transistors depend only weakly on temperature and that the subthreshold swing is temperature independent. This has been compared to the conventional MOSFET. Simulation results raise the expectation that the tunnel transistor can be operated at 450 K and beyond.

References

[1] Baba, T. Proposal for surface tunnel transistors. *Jpn. J. Appl. Phys.*, 31:L455–L457, 1992.

[2] Reddick, W. M. and Amaratunga, G. A. J. Silicon surface tunnel transistor. *Appl. Phys. Lett.*, 67(4):494–497, 1995.

[3] Koga, J. and Toriumi, A. Three-terminal silicon surface junction tunneling device for room temperature operation. *Electron Device Letters*, 20(10):529–531, 1999.

[4] Hansch, W., Fink, C., Schulze, J., and Eisele, I. A vertical MOS-gated Esaki tunneling transistor in silicon. *Thin Solid Films*, 369:387–389, 2000.

[5] Bhuwalka, K. K., Sedlmaier, S., Ludsteck, A., Tolksdorf, C., Schulze, J., and Eisele, I. Vertical tunnel field-effect transistor. *IEEE Trans. Electron Devices*, 51(2):279–282, 2004.

[6] Aydin, C., Zaslavsky, A., Luryi, S., Cristoloveanu,S., Mariolle, D., Fraboulet, D., and Deleonibus, S. Lateral interband tunneling transistor in silicon-on-insulator. *Appl. Phys. Letters*, 84:1780–1782, 2004.

[7] Bhuwalka, K. K., Schulze, J., and Eisele, I. Performance enhancement of vertical tunnel field-effect transistor with SiGe in the δp^+ layer. *Jpn. J. Appl. Phys.*, 43(7A):4073–4078, 2004.

[8] Bhuwalka, K. K., Schulze, J., and Eisele, I. Scaling the vertical tunnel FET with tunnel bandgap modulation and gate workfunction engineering. *IEEE Trans. Electron Devices*, 52(5):909–917, 2005.

[9] Bhuwalka, K. K., Born, M., Schindler, M., Schmidt, M., Sulima, T., and Eisele, I. P-channel Vertical Tunnel Field-Effect Transistor Down to Sub-50 nm Channel Lenght. In *Int. Conf. Solid State Devices and Materials*, pages 288–289, Kobe, Japan, 2005.

[10] Bhuwalka, K. K., Schulze, J., and Eisele, I. A simulation approach to optimize the electrical parameters of a vertical tunnel field-effect transistor. *IEEE Trans. Electron Devices*, 52(7):1541–1547, 2005.

[11] Sze, S. M. *Physics of Semiconductor Devices*. Wiley, New York, 1981.

[12] Nishida, M. and Ohyabu, H. Temperature Dependence of MOSFET Characteristics in Weak Inversion. *IEEE Trans. Electron Devices*, ED-24(10):1245–1248, 1977.

[13] Tewksbury, S. K. N-Channel Enhancement-Mode MOSFET Characteristics from 10 to 300 K. *IEEE Trans. Electron Devices*, ED-28(12):1519–1529, 1981.

[14] ITRS Road-Map, International technology road-map for semiconductor, 2005. Available from: http://public.itrs.net.

[15] Wang, P. F., Hilsenbeck, K., Nirschl, T., Oswald, M., Stepper, C., Weis, M., Schmitt-Landsiedel, D., and Hansch, W. Complementary tunneling transistor for low power applications. *Solid-State Electronics*, 48(12):2281–2286, 2004.

Poster Session
Nanotechnologies

2006 25th International Conference on Microelectronics

Analysis of Nonlinear Effects in Nanometer-Scale Silicon-on-Insulator Rib Waveguides

V. M. N. Passaro, *Senior Member, IEEE*, F. De Leonardis and G. Z. Mashanovich

Abstract- Raman gain, two photon absorption, free carrier absorption, self and cross phase modulation induced by Kerr effect, plasma dispersion, walk-off effect and nonlinear polarization coupling are investigated theoretically in nanometer-scale silicon-on-insulator waveguides. The influence of the rib waveguide dimensions on the time-space evolution of both pump and Stokes pulses is presented.

I. INTRODUCTION

The need for low cost photonic devices has stimulated a significant amount of research in silicon photonics [1]. Silicon can be considered as an ideal platform for integrated optics and optoelectronics due to the constant improvement of the quality-cost ratio of commercial silicon wafers driven by the electronic IC industry. While a wide variety of passive devices were developed in the 1990's, recent activities have focused on achieving active functionality, mostly light amplification and generation, in silicon on insulator (SOI) waveguides. One approach that has been experimentally investigated for light amplification and generation is the Raman effect [2-3]. This approach relies on the fact that the gain coefficient for Stimulated Raman Scattering (SRS) is approximately 10^4 times higher in silicon than in silica. Moreover, (SOI) waveguides can confine the optical field to an area that is approximately 100 times smaller than modal area in a standard single mode optical fibre. The combination of these properties makes SRS observable over the millimetre-scale interaction length, usually encountered in integrated optical devices. In addition to light generation and amplification, the Raman effect can also perform wavelength conversion [4] that is of paramount importance in optical networks.

In our work, we propose, for the first time, at the best of our knowledge, a complete model for the design of all-optical AND gates based on Raman effect in nanophotonic SOI waveguides. Our model is general, and it considers SRS, Two Photon Absorption (TPA), Free Carrier Absorption (FCA), free-carrier dispersion, polarization

V. M. N. Passaro is with the Elettrotecnica ed Elettronica Dept., Politecnico di Bari, via Edoardo Orabona n. 4, 70125 Bari, Italy, E-mail: passaro@deemail.poliba.it

F. De Leonardis is with the Ingegneria dell'Ambiente e per lo Sviluppo Sostenibile Dept., Politecnico di Bari, viale del Turismo n. 8, 70100 Taranto, Italy, E-mail: f.deleonardis@poliba.it

G. Z. Mashanovich is with the Advanced Technology Institute, School of Electronics and Physical Sciences, University of Surrey, Guildford, United Kingdom, E-mail: G.Masanovic@surrey.ac.uk.

coupling, and self-phase-modulation (SPM) and cross-phase-modulation (XPM) effect induced by the Kerr effect. The model includes four partial differential nonlinear equations. Three equations take into account the coupling between the pump pulse and Stokes waves (quasi-TE or quasi-TM polarization), while the fourth equation takes into account the time evolution of the TPA-induced free carrier density. This general approach enables analysis of several effects often neglected in other models, such as the walk-off effect, the time evolution, the plasma dispersion, and the Kerr effect.

Using this model, the optimization of nanophotonic SOI waveguides has been carried out to increase the SRS effect. We particularly focus our theoretical analysis on the influence of the waveguide dimensions on the time-space evolution of fast pulses in all-optical AND gates.

II. THEORY

The study of the most nonlinear effects in an SOI waveguide involves the use of short pulses with widths ranging from 10 ns to 10 fs. When such optical pulses propagate inside a waveguide, both dispersive and nonlinear effects significantly influence their shape and spectrum. In this section, we derive a basic system of equations that governs the propagation of optical pulses in the presence of these effects.

In a single-mode SOI rib waveguide we consider the two propagating modes as a quasi-TE (x-component of electric field dominant) and a quasi-TM (y component dominant). We assume the pump as aligned with TE polarization, and the fundamental Stokes wave as the linear combination of the lowest-order TE and TM modes. With these assumptions, which fit a large number of experimental conditions, the electric field of interacting pump (p) and Stokes (s) waves can be represented as:

$$\mathbf{E}(x,y,z,t) = \hat{\mathbf{x}}\left[\begin{array}{c} C_1 A_1(z,t)F_1(x,y)e^{j(\beta_{01}z-\omega_p t)} \\ +C_2 A_2(z,t)F_2(x,y)e^{j(\beta_{02}z-\omega_s t)} \end{array}\right] \\ + \hat{\mathbf{y}}\left[C_3 A_3(z,t)F_3(x,y)e^{j(\beta_{03}z-\omega_s t)}\right] \tag{1}$$

where subscripts 1, 2 and 3 are related to the pump, TE and TM-polarized Stokes waves, respectively. A_k, $F_k(x,y)$ and β_{0k} (k=1,2,3) are the slowly-varying amplitudes, transverse field distributions of the waveguide modes, and propagation constants in the z-direction, respectively, where (x,y) denotes the cross section plane. The coefficients C_k are $C_k = \left(\int\int\limits_{-\infty}^{+\infty} \left|F_k(x,y)\right|^2 dxdy \right)^{-1/2}$. Finally, ω_p and ω_s are the angular pulsations of the pump and fundamental Stokes waves, respectively.

Similarly to [5], the partial differential equations for the slowly varying amplitudes can be written as:

$$
\frac{\partial A_1}{\partial z} + \beta_{11}\frac{\partial A_1}{\partial t} + j\frac{1}{2}\beta_{21}\frac{\partial^2 A_1}{\partial t^2} = -\frac{\left(\alpha_1^{(prop)} + \alpha_1^{(FCA)}\right)}{2}A_1 - 0.5\beta^{(TPA)}f_{11}\left|A_1\right|^2 A_1 \tag{2}
$$
$$
+ j\gamma_{11}\left|A_1\right|^2 A_1 + j2\gamma_{1k}\left|A_k\right|^2 A_1 + j\frac{2\pi}{\lambda_p}\Delta n A_1 - \frac{1}{2}g_R f_{1k}\frac{\omega_p}{\omega_s}\left|A_k\right|^2 A_1
$$

$$
\frac{\partial A_k}{\partial z} + \beta_{1k}\frac{\partial A_k}{\partial t} + j\frac{1}{2}\beta_{2k}\frac{\partial^2 A_k}{\partial t^2} = -\frac{\left(\alpha_k^{(prop)} + \alpha_k^{(FCA)}\right)}{2}A_k - \beta^{(TPA)}f_{k1}\left|A_1\right|^2 A_k
$$
$$
+ j2\gamma_{k1}\left|A_1\right|^2 A_k + j\gamma_{kk}\left|A_k\right|^2 A_k + j2\gamma_{kp}\left|A_p\right|^2 A_k + j\frac{2\pi}{\lambda_s}\Delta n A_k + \frac{1}{2}g_R f_{k1}\left|A_1\right|^2 A_k
$$
$$
+ jk_{kkpp}A_p A_k^* A_p e^{j(2\beta_{0p}-2\beta_{0k})z} \tag{3}
$$

where $k = 2,3$ and $p = 3,2$.

This system includes all nonlinear effects and is valid if the pulse width involved in the process is ≥ 1 ps, i.e. quite longer than the response time of the Raman effect (100 fs). The coefficients present in the system (2)-(3) are defined as:

$$
\gamma_{ii} = \frac{\omega_i}{2cn_{eff,i}}f_{ii}\left(\chi_{xxxx}^{NR} + \chi^R(0)\right)
$$

$$
\gamma_{ij} = \begin{cases} \dfrac{\omega_i}{2cn_{eff,i}}f_{ij}\left(\chi_{xxyy}^{NR} + \chi^R(0)\right); \\[2mm] \dfrac{\omega_i}{2cn_{eff,i}}f_{ij}\left(\chi_{xxyy}^{NR}\right) \end{cases} \qquad k_{iijj} = \frac{\omega_i}{2cn_{eff,i}}f_{iijj}\chi_{xxyy}^{NR}
$$

where $i,j = 1,2,3$. The silicon non-resonant susceptibility assumes value of $\chi_{xyyx}^{NR} = 0.25 \times 10^{-18}$ (m²V⁻²) [4] ($\chi_{xyyx}^{NR} = \chi_{xyxy}^{NR} = \chi_{xxyy}^{NR} \cong 0.5\chi_{xxxx}^{NR}$), and the subscripts x,y indicate the two possible polarisations (quasi-TE or quasi-TM). The Raman–resonant susceptibility can be written as:

$$
\chi^R\left(\omega_s; \omega_p - \omega_s, \omega_s\right) = \frac{2\Omega_R\Gamma_R\xi_R}{2j\Gamma_R\Delta\omega + \Omega_R^2 - \Delta\omega^2}
$$

where $\xi_R = 11.2 \times 10^{-18}$ (m²V⁻²) [4] is the value of the Raman susceptibility under resonance condition ($\Delta\omega = \Omega_R$), $\Gamma_R = 2\pi \times 53$ GHz is the resonance half-width, $\Omega_R = 15.6$ THz is the frequency shift between the pump and the Stokes waves and $\Delta\omega = \omega_p - \omega_s$. The overlap integrals f_{ij} and f_{ijkl} are given by:

$$
f_{ij} = \frac{\int\int\left|F_i(x,y)\right|^2 \left|F_j(x,y)\right|^2 dxdy}{\int\int\left|F_i(x,y)\right|^2 dxdy \int\int\left|F_j(x,y)\right|^2 dxdy} \qquad i,j = 1,2,3
$$

$$
f_{iijj} = \frac{\int\int\limits_{-\infty}^{+\infty}F_i^*F_i^*F_jF_j\,dxdy}{\sqrt{\left(\int\int\limits_{-\infty}^{+\infty}\left|F_i\right|^2 dxdy\right)\left(\int\int\limits_{-\infty}^{+\infty}\left|F_i\right|^2 dxdy\right)\left(\int\int\limits_{-\infty}^{+\infty}\left|F_j\right|^2 dxdy\right)\left(\int\int\limits_{-\infty}^{+\infty}\left|F_j\right|^2 dxdy\right)}} \qquad i,j = 2,3
$$

In particular, f_{ii}^{-1} represents the effective core area of the optical modes and is a very important parameter because it determines the efficiency of non-linear devices. In the previous equations, c is the light velocity, and $n_{eff,i}$ are the effective indices of the waveguide modes for the pump and Stokes waves. Term $g_R = 4\omega_s\chi^R(\Omega_R)/(cn_{eff,s})$ is the Raman gain. It is clear that in the system (2)-(3), terms including $+g_R$ determine the SRS effect, while terms with $-g_R$ represent, together with the TPA effect (see term in $\beta^{(TPA)}$), the contribution for the pump depletion. Coefficients γ_{ii} take into account the SPM induced by the Kerr effect while coefficients $\gamma_{i,j}$ (with $i \neq j$) represent the relevant terms for the XPM effect. The terms in k_{iijj} in the system of equations (2)-(3) represent the coherent coupling between the two polarization components for the Stokes waves. It is important to note the time derivatives in Eq. (2)-(3), take into account the walk-off effect induced by the different group velocity (see terms in β_{11} and β_{1k}) and the group velocity dispersion (see terms in β_{21} and β_{2k}). The total optical loss coefficient can be written as summation of two contributions $\alpha_i^{(T)} = \alpha_i^{(prop)} + \alpha_i^{(FCA)}$, where $\alpha_i^{(prop)}$ is the propagation loss coefficient in the rib waveguide, which depends on the fabrication process. Moreover, $\alpha_i^{(FCA)}$ is the contribution to the total loss due to the free carrier absorption (FCA), which is induced by the change in the free carriers generated mainly by the TPA of the pump pulse. According to [2], we evaluate $\alpha_{p,s}^{(FCA)}$ as:

$$
\alpha_i^{(FCA)} = \sigma_i \cdot N_c = \sigma_0 \cdot \left(\frac{\lambda_i}{1.55}\right)^2 N_c
$$

where N_c is the density of electron-hole pairs generated by the TPA process. The coefficient $\sigma_0 = 1.45 \times 10^{-17}$ cm⁻² is the FCA cross section measured at $\lambda = 1.55$ μm, and λ_i is the wavelength of the relevant wave (pump or Stokes).

In Eqs. (2)-(3) $\Delta n = -1.66 \cdot \delta_i \cdot N_c$ is the change of the effective refractive index due to the plasma dispersion effect induced by the free carriers, where $\delta_i = 8.8 \cdot 10^{-22} \cdot \lambda_i/(1.55)$. To obtain a consistent mathematical model, Eqs. (2)-(3) need to be coupled to the rate equation governing the free carrier dynamics in the waveguide core, given by [3]:

$$
\frac{dN_c}{dt} = -\frac{N_c}{\tau_{eff}} + \frac{\beta^{(TPA)}}{2\hbar\omega_p}\left(\left|A_1(z,t)\right|^2 f_{11}\right)^2 \tag{4}
$$

where τ_{eff} is the carrier effective recombination lifetime.

For numerical purposes, it is useful to introduce the normalized variables $U_i = A_i / \sqrt{P_0}$ and $\tau = t/T_0$, where P_0 and T_0 are the input Gaussian pulse peak power and width, respectively ($T_0 = T_{FWHM} / 1.665$).

To solve numerically the coupled system of partial differential equations, we used the collocation method [6] by a set of Hermite-Gauss basis functions, i.e. $\phi_m(\tau) = H_{m-1}(\tau) \exp(-0.5\tau^2)$ and up to $M = 35$ collocation points. It has been already demonstrated [6] that this method can improve the accuracy by about three orders of magnitude compared to the well-known split-step Fourier method [5] for the same computation time. Applying the collocation approach, the system of partial differential equations (2)-(4) becomes a matrix system of ordinary differential equations which has been solved using a fourth-order Runge-Kutta procedure.

III. NUMERICAL RESULTS

A. Time Evolution

In this sub-section the main features of the time evolution of Stokes pulses propagating in a nanometer-scale SOI rib waveguide are given. Fig. 1 shows the waveguide geometry involved in the analysis.

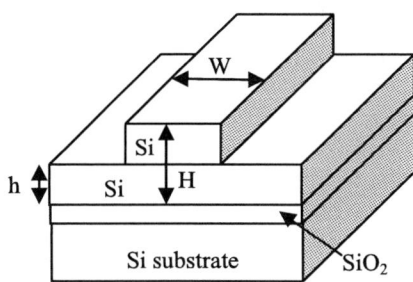

Fig. 1. Schematic diagram of the SOI waveguide.

High index contrast between waveguide cladding and core gives high light confinement in submicron structures, however, it makes the control of waveguide birefringence challenging. In addition, high net Raman gain in silicon can be difficult to achieve due to the losses induced by free carriers because of the TPA process in silicon. One way for diminishing these losses is to reduce the free carrier lifetime τ_{eff} through lateral scaling of the waveguide modal area [2]. Unlike the work in [2] where the effect of the cross-section scaling has been demonstrated for waveguide widths around 1 μm, in this paper we investigate the influence of both the waveguide nano cross section and very short pulse width on the SRS effect. Therefore, it is an important requirement to design the waveguide as polarization insensitive. We have assumed the Stokes wavelength of $\lambda_s = 1.5487$ μm for pulses travelling along a nanometric waveguide having the total rib height of $H = 500$ nm ($\sim \lambda_s/3$). Our simulations, performed by a full-vectorial finite element method,

demonstrate that the polarization insensitivity can be satisfied for $r \leq 0.3$ where $r = h/H$. Furthermore, the calculations show that the optimum value of the rib width W increases by decreasing r. Therefore, the best trade-off between polarization insensitivity and small carrier recombination lifetime is obtained for $r = 0.3$ and $W = 395.6 \pm 5$nm. Fig. 2 shows the space-time evolution of both the pump and Stokes pulses assuming $P_0 = 1.5$ W, $T_{FWHM} = 1$ ps and $g_R = 10.5$ cm/GW. The power of the probe has been kept 30 dB below the pump power.

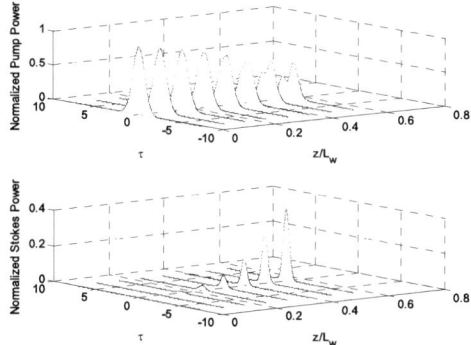

Fig. 2. Time-space evolution of pump and Stokes pulse.

Several features of Fig. 2 are noteworthy. The Raman pulses build up after $z / L_w = 0.2$, where L_w is the walk-off length. The energy transfer from the pump to the Stokes pulse induces formation of two peaks in the pump pulse shape, which is result of the pump depletion. The relevant dip corresponds exactly to the location of the Stokes wave. The time locations of the pump and Stokes pulses are separated due to the walk-off effect (evaluated considering the material dispersion of the silicon). The Raman pulse is also narrower than the input pulse. This is a consequence of the spectral broadening induced simultaneously by SPM, XPM, and plasma dispersion. It can be noted from Eqs. (2)-(3), that SPM, XPM and plasma dispersion coefficients make the complex amplitude of the electric field time-dependent. Fig. 3 shows better the influence of the pump pulse power and FWHM width on the time-space evolution of the Stokes wave. Again, the power of the probe is kept 30 dB below the pump power and the waveguide length is $L = 10$ mm. For $P_0 = 1.5$ W and $T_{FWHM} = 1$ ps it is not possible to neglect both the pump depletion effect induced by the SRS effect and, above all, the walk-off effect shown by the time shift between the pump and Stokes pulse. This means that in such small structures the time derivative in the model cannot be neglected, which usually occurs in the literature. The second subplot shows the time evolution of pump and Stokes waves for $P_0 = 0.5$ W and $T_{FWHM} = 100$ ps. In this case the walk-off effect can be neglected as well as the pump depletion effect induced by the Raman scattering. However, the depletion induced meanly by TPA effect that becomes dominant when reducing the pulse FWHM, is weak.

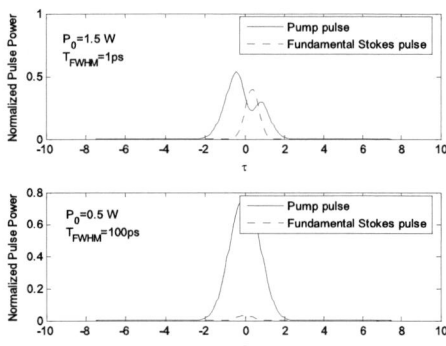

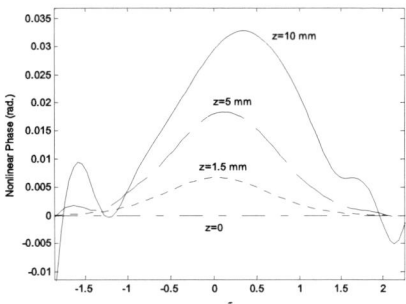

Fig. 3. Time evolution of pump and Stokes pulse evaluated at L=10 mm and for different values of P_0 and T_{FWHM}.

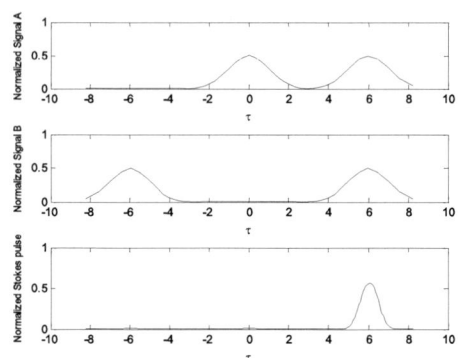

Fig. 4. Time evolution of Stokes pulse nonlinear phase at different z positions.

Fig. 4 shows the time-space evolution of the nonlinear phase induced by the Kerr and plasma dispersion effects at different z positions, assuming $P_0 = 1.5\,\text{W}$ and $T_{FWHM} = 1\,\text{ps}$. It is possible to observe that the nonlinear phase has an asymmetrical profile propagating along the waveguide, mainly due to the walk-off effect.

B. All-Optical AND Gate

In this sub-section we show the model applied to the investigation of an all-optical Raman AND gate. The device is constituted by an SOI Y-junction. The two logic signals are inputs for the Y junction arms (A, B) that recombine at the output. The waveguide length after the Y junction is selected to increase the net Raman gain and the sidelobe suppression (SLS) ratio. However, the maximum waveguide length cannot be larger than that required for exciting higher order Stokes waves when both input signals are at "1" logic. The waveguide nano cross-section results in lifetime of $\tau_{eff} \approx 1\,\text{ns}$, the TE-polarized input signals are pulsed at 1.4332 μm and the output Stokes signal is centred at 1.5487 μm (Raman shift). Unlike for the other approaches in GaInAsP/InP AND gates based on Kerr effect [7], it is demonstrated that the peak power of each input signal induce a net Raman gain weakly above the threshold, relatively to the fundamental Stokes wave. In this situation, it is possible to significantly enhance the output signal. Fig. 5 shows the temporal profiles of input and transmitted pulses for $T_{FWHM} = 10\,\text{ps}$.

Fig. 5. All-optical AND gate operation with L=10 mm and $T_{FWHM} = 10\,\text{ps}$.

It is clearly shown that all-optical AND gate operation is obtained, with SLS ratio larger than 18 dB. The performance of this AND gate can be improved by increasing the waveguide length and reducing T_{FWHM} down to 10 ps.

III. CONCLUSION

The general model presented in this paper allows accurate prediction of the space and time evolution of fast pulses in SOI nano-scale rib waveguides, by taking into account all relevant linear and non-linear physical effects and without any a-priori assumption. Our model particularly shows the walk-off effect, which is dominant at the picosecond scale and generally neglected in other models. In addition, in this work we propose the guidelines for the design of polarization insensitive SOI nanophotonic waveguides with high SRS efficiency.

REFERENCES

[1] G. T. Reed and A. P. Knights, *Silicon Photonics: An Introduction*, John Wiley, Chichester, 2004.
[2] R. Claps, V. Raghunathan, D. Dimitropoulos, and B. Jalali, "Influence of nonlinear absorption on Raman amplification in silicon waveguides," *Opt. Express*, vol. 12, pp. 2774-2780, 2004.
[3] A. Liu, H. Rong, M. Paniccia, O. Cohen, and D. Hak, "Net optical gain in a low loss silicon-on-insulator waveguide by stimulated Raman scattering," *Opt. Express*, vol. 12, pp. 4261-4268, 2004.
[4] D. Dimitropoulos, V. Raghunathan, R. Claps, and B. Jalali, "Phase-matching and nonlinear optical processes in silicon waveguides", *Opt. Express*, vol. 12, pp. 149-160, 2004.
[5] G. P. Agrawal, *Nonlinear fiber optics*, Academic Press, London, 1989, pp. 26-36.
[6] S. Deb and A. Sharma, "Nonlinear pulse propagation through optical fibers: an efficient numerical method," *Opt. Eng.*, vol. 32, pp. 695-699, 1993.
[7] S. H. Jeong, T. Mizumoto, M. Takenaka, Y. Nakano, "All-optical Wavelength Conversion in a GaInAsP/InP Optical Gate Loaded with a Bragg Reflector," *Applied Optics*, vol. 42, pp. 6672-6677, 2003.

2006 25th International Conference on Microelectronics

Analysis of Pulsed Excitation in Small Silicon-on-Insulator Microring Resonators

F. De Leonardis, V. M. N. Passaro, *Senior Member, IEEE*, and G. Z. Mashanovich

Abstract – In this paper, we investigate optical properties of silicon-on-insulator microring resonators with submicron cross sections, at the wavelength around 1550 nm and under pulsed excitation.

I. INTRODUCTION

Channel dropping filters that access one channel of a wavelength division multiplexed (WDM) signal and do not disturb the other channels are useful elements for WDM communication systems. Ring or disk resonators are attractive candidates for this application because they can potentially realize the narrowest linewidth for a given device size. In addition, a cascade of coupled resonators can further modify the simple Lorentzian response of a single resonator, giving desirable higher order filter characteristics [1]. Potentials of microring resonators are related to the fact that they support traveling wave resonant modes. Unlike the standing wave resonators, any ring or disk may completely extract particular wavelength from the input bus waveguide, offering superior performance. In applications such as WDM, the ring resonators have to be small enough so that the spacing of resonant frequencies accommodates the set of WDM channels within a telecommunications window. The communication window supported by erbium amplifiers is 30 nm. Rings with free spectral range FSR > 30 nm require radii R ≤ 5 µm, which is possible by state-of-the-art fabrication technologies. Hence, these microring structures require high index contrast to make the bending losses negligible. In that case, microring performance is limited by scattering due to sidewall roughness. Such scattering leads to two detrimental effects: energy loss due to the radiation continuum and scattering into the counter-propagating mode. Contra-directional coupling induced by the surface roughness can be more problematic than the radiation loss, because the periodic nature of the ring leads to a natural

F. De Leonardis is with the Dipartimento di Ingegneria dell'Ambiente e per lo Sviluppo Sostenibile, Politecnico di Bari, viale del Turismo n. 8, 70100 Taranto, Italy, E-mail: f.deleonardis@poliba.it

V. M. N. Passaro is with the Dipartimento di Elettrotecnica ed Elettronica, Politecnico di Bari, via Edoardo Orabona n. 4, 70125 Bari, Italy, E-mail: passaro@deemail.poliba.it

G. Z. Mashanovich is with the Advanced Technology Institute, School of Electronics and Physical Sciences, University of Surrey, Guildford, United Kingdom, E-mail: G.Masanovic@surrey.ac.uk.

phase matching between forward and backward modes due to any possible perturbation [2].

Recent research activities have focused on achieving better filter functionality based on microrings [3]-[5]. Microring resonators can be used in all-optical applications [6]-[7]. Majority of models and experiments proposed in the literature have investigated III/V compound semiconductors. In this paper, we propose a theoretical analysis of a microring filter based on Silicon-on-Insulator (SOI) technology, where high index contrast between waveguide cladding and core facilitates light guiding in structures with nanometric dimensions, thus making the bending loss in the SOI microring negligible. However, when the ring cross section becomes very small, some effects induced by the Two Photon Absorption (TPA) cannot be considered negligible.

II. THEORY

In Figs. 1(a) and 1(b) the single ring resonator, evanescently coupled to two external waveguides, and the ring cross section are depicted, respectively.

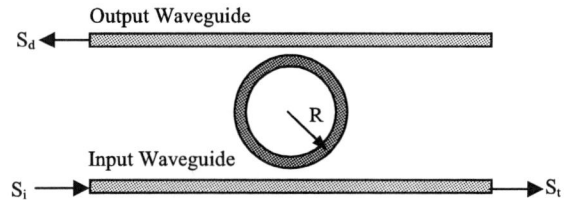

Fig. 1(a). Evanescently coupled microring resonator.

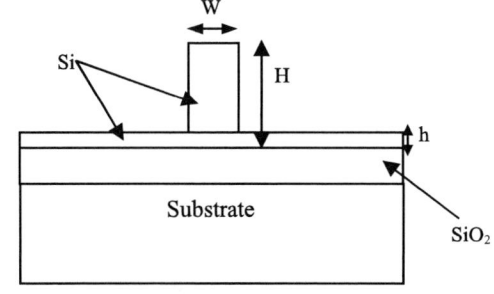

Fig. 1(b). Cross section of the ring waveguide.

We assume that the ring supports a traveling wave of amplitude $A(t)$, such that $|A(t)|^2$ represents the total power flowing through each ring cross section at time instant t. With this assumption the time evolution of the field envelope amplitude is governed by means of the following differential equation [1]:

$$\frac{dA(t)}{dt} = j(\omega_0 - \omega)A(t) - \frac{1}{\tau}A(t) - v_g \alpha^{(FCA)}A(t) - jv_g\frac{2\pi}{\lambda}\Delta nA(t) +$$
$$- v_g\frac{\beta^{(TPA)}}{A_{eff}}|A(t)|^2 A(t) - j\sqrt{\frac{2}{\tau_e}}\sqrt{\frac{v_g}{2\pi R}}S_i$$

(1)

where ω_0 is the angular frequency at the resonance, ω is the angular frequency of the input beam, v_g is the group velocity and S_i denotes the input wave amplitude. In Eq. (1) τ is the ring cavity photon lifetime defined as:

$$1/\tau = 1/\tau_e + 1/\tau_d + 1/\tau_l$$

(2)

where $1/\tau_e$ and $1/\tau_d$ are related to the power leaving the ring due to the external coupling with the input and output waveguides, respectively, and $1/\tau_l$ represents the power loss due to other intrinsic effects such as propagation, bending and scattering losses. The last term is given by: $1/\tau_l = \alpha_{loss}v_g$, where α_{loss} is the linear loss coefficient. Eq. (1) has been obtained from the wave equation under a slowly-varying amplitude approximation. In particular, the terms with $\alpha^{(FCA)}$ (free carrier absorption), Δn (effective index change due to plasma dispersion) and $\beta^{(TPA)}$ (two photon absorption) have been derived by introducing linear and third order polarization vectors in the wave equation. We evaluate $\alpha^{(FCA)}$ and Δn by the following expressions:

$$\alpha^{(FCA)} = \sigma_0 \cdot \left(\frac{\lambda}{1.55}\right)^2 N_c$$

(3)

$$\Delta n = \left(-8.8 \times 10^{-22} N_c - 8.5 \times 10^{-18} N_c^{0.8}\right)$$

(4)

where N_c is the density of electron-hole pairs generated by TPA. $\sigma_0 = 1.45 \times 10^{-17}$ cm^{-2} is the FCA cross section measured at $\lambda = 1.55$ μm, and λ is the input signal wavelength.

To obtain the complete mathematical model, Eq. (1) has been coupled to the rate equation governing the free carrier dynamics into the waveguide core, given by [8]:

$$\frac{dN_c}{dt} = -\frac{N_c}{\tau_{eff}} + \frac{\beta^{(TPA)}}{2\hbar\omega}\left(|A(z,t)|^2 / A_{eff}\right)^2$$

(5)

where τ_{eff} is the effective recombination lifetime of free carriers and A_{eff} is the effective area of the optical mode inside the ring.

III. NUMERICAL RESULTS

In this section we show the influence of the TPA on the microring resonator time evolution. It is worth noting that TPA could be an undesirable effect in the design of a ring resonator filter or, on the other side, an advantage in the design of some nonlinear all-optical devices. Hence, sometimes TPA has to be avoided, while in other cases it has to be controlled and excited. We focus our analysis on the limits induced by the TPA on the performance of filters based on small ring resonators. It is evident from Eq. (5) that the TPA effect can be reduced by decreasing the optical power and the effective recombination lifetime of free carriers. It is not trivial to find the maximum input power in order to neglect the TPA, because it depends on the ring enhancement factor. This factor depends also on a number of different parameters such as ring radius, cavity photon lifetime and FWHM of the input pulse. On the contrary, the reduction of τ_{eff} can be carried out by an appropriate design of the ring cross section. To obtain an estimation for τ_{eff}, free carrier diffusion needs to be considered in addition to the recombination lifetime. If diffusion carriers move out from the modal area, this results in an effective lifetime that can be shorter than the recombination lifetime in SOI structures, τ_r. If τ_t is the transit time, then $1/\tau_{eff} = 1/\tau_r + 1/\tau_t$. It has been demonstrated [9] that τ_r depends on the surface-recombination velocity ($S = 10^3 cm/s$) at the interface between the top silicon and buried oxide, with a typical value of about 100 ns. Transit time τ_t also depends on the optical mode size $\tau_t = 0.5w\sqrt{H/(SD)}$, where w is the optical mode width, H is the total height of the SOI rib waveguide and D is the ambipolar diffusion coefficient [9]. Thus, the previous relationship shows that a small cross section is necessary to obtain a low value for τ_{eff}. To enhance the effect of scaling, we consider a ring with cross section of few hundreds of nanometres. We have fixed the rib total height to $H = 500$ nm ($\sim \lambda/3$), $r = 0.1$ and $W = 300$ nm in order to achieve very high confinement in the rib and negligible bending losses. For these values, τ_{eff} has an estimation of about 1 ns (large lifetime). The parameter used in the simulations are $R = 5$ μm, $\alpha_{loss} = 2$ dB/cm, $\kappa^2 = 2\%$ (power fraction coupled from the input waveguide to the ring), The effective area, calculated by a full-vectorial finite element method, is $A_{eff} = 0.16$ μm^2, and $\beta^{(TPA)} = 0.7$ cm/GW [9].

A. CW operation.

Fig. 2 shows the spectrum of the normalized power at the end of the output waveguide (Reflectivity) and of the input waveguide (Transmittivity) versus the wavelength detuning for different values of CW input power $|S_i|^2$. The plot has been obtained by solving the system (1)-(5) for each wavelength. The wavelength detuning is referred to the value of $\lambda_0 = 1.5450$ μm (TPA-free resonance condition). For comparison reason, solution without the TPA is also plotted by solid line. The structure has been designed such to obtain optimum traveling behaviour without the TPA. By setting the matching condition to:

$$\frac{2}{\tau_e} = \kappa^2 v_g \left(2\pi R\right)^{-1} \qquad (6)$$

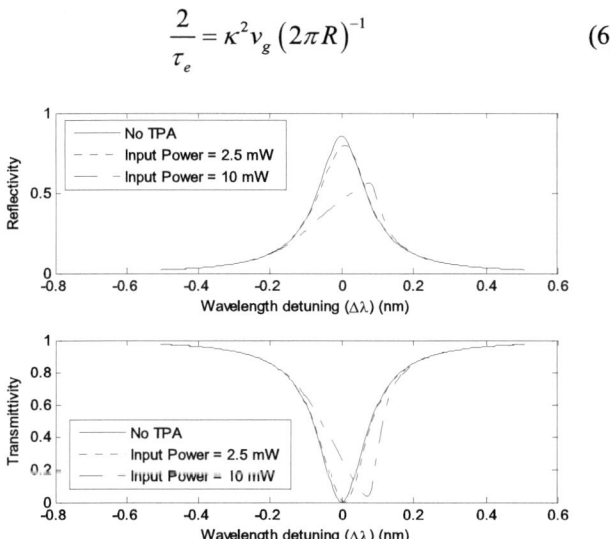

Fig. 2. Reflectivity and Transmittivity versus the wavelength detuning for different values of input power.

the total signal power could be extracted from the resonator at the resonance even in the presence of losses in the ring (see the dotted curve in the second subplot) [1]. Thus Eqs. (2) and (6) require that the input and output waveguides are no longer symmetrically coupled to the ring ($\tau_d \neq \tau_e$). Therefore, Fig. 2 shows that the power at the end of output waveguide decreases by increasing the input power (dash-dotted line). This is mainly due to the FCA induced by the nonlinear TPA effect, leading to a decrease in the photon lifetime inside the cavity. In this condition, Eq. (6) does not hold and the microring is unmatched with the waveguide, i.e. the transmittivity at the resonance is different from zero. Moreover, it can be seen that for these relatively high levels of input power both broadening of the transmittivity and reflectivity curves, and shift of the peak position towards a larger wavelength occur. These effects depend mainly on the plasma dispersion effect induced by the free carrier excess and on other nonlinear terms in Eq. (1). Our calculations show that the microring considered could work as a TPA-free CW optical filter for input powers lower than 2.5 mW (15.6 mW/μm²).

B. Pulsed operation

In this section we focus our analysis on pulsed excitation, condition never considered for SOI resonant microrings. We first simulate the nonlinear propagation in the microring assuming a Gaussian input pulse with FWHM much larger than the ring cavity photon lifetime, i.e. $T_{FWHM} = 200$ ps $\gg \tau_e = 31.41$ ps (evaluated without the TPA). The input pulse has been applied with a delay of 400 ps. Fig. 3 has three subplots: normalized power inside the ring, and those at the output and at the input waveguides versus time, respectively. Each normalization is referred to the peak input power ($|S_i|^2_{peak} = 250$ mW).

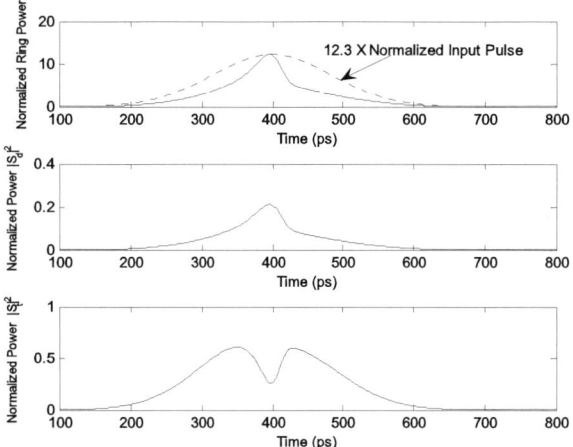

Fig. 3. Normalized power versus time with input pulse $T_{FWHM} = 200$ ps.

The wavelength of the input pulse was set at the value detuned by 0.2253 nm with respect to the cavity resonance mode (this is more clear later). Several features of Fig. 3 can be noticed. The power cavity enhancement factor is 12.3 (see the peak of the first subplot). The power at the input waveguide end initially rises with the ring power, but then drops to a minimum as the microring resonance becomes closer to the input pulse wavelength. We note that this dip exactly corresponds to the power peak in the ring and at the end of output waveguide. This dip occurs at t = 400 ps where the carrier concentration of 6.17×10^{17} cm^{-3}, which is generated by the TPA, causes a net decrease of the refractive index of 4.3834×10^{-4}, and a relevant shift of the resonance mode wavelength by precisely 0.2253 nm (i.e. the same amount of initial detuning). Thus, the shifted wavelength is locked to the resonance condition and the power can exit from ring and be detected at the output waveguide. In addition, at the falling edge of the input pulse, the power in the ring begins to decrease, and the power at the input waveguide end rises again as the cavity resonance mode returns to its initial position.

In Fig. 4, time evolution of the normalized power is shown, assuming Gaussian input pulse with $T_{FWHM} = 1$ ps and peak input power of 250 mW. In this case, the power enhancement factor falls to 0.37, and the time evolution is quite different than in the previous case. As the input pulse is narrower than the time constant of the evanescent coupling process ($\tau_e = 31.41$ ps), the field in the ring cannot be charged up. This is evident by observing the power level and the pulse shape at the end of the input waveguide. Thus a stringent condition for pulsed excitation of nano-section microrings is that they can work as filters with a high enhancement factor only if the FWHM of the input pulse is significantly larger than τ_e.

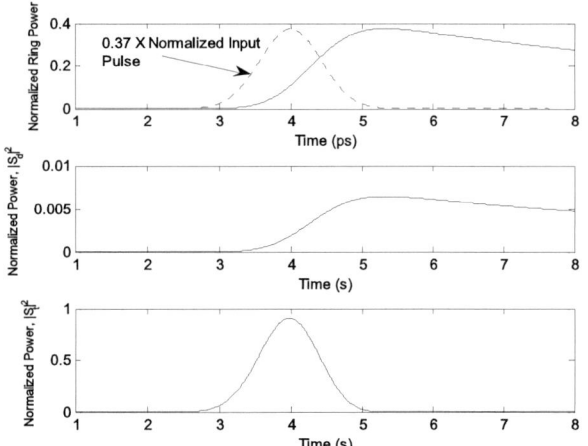

Fig. 4. Normalized power versus time with input pulse $T_{FWHM} = 1$ ps.

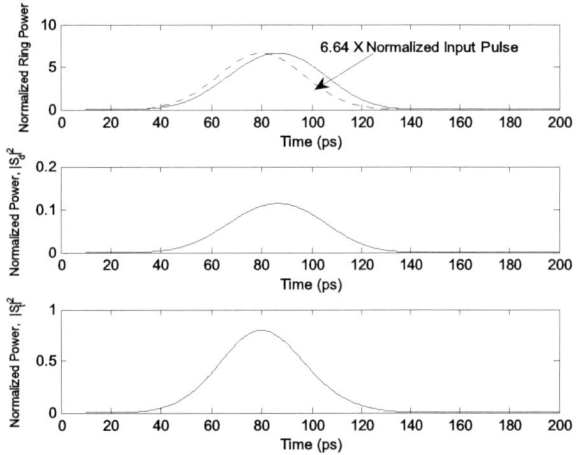

Fig. 5. Normalized power versus time with input pulse $T_{FWHM} = 40$ ps.

Thus, it is necessary to reduce the coupling factor $\kappa^2 \leq 1\%$ for the operation in the fast pulse regime. For comparison the time evolution in the microring for $T_{FWHM} = 40$ ps (closer to $\tau_e = 31.41$ ps) and input peak power $|S_i|^2_{peak} = 250$ mW is shown in Fig. 5. In this case the power enhancement factor is 6.64, larger than in the previous case but almost half of the power enhancement factor calculated for $T_{FWHM} = 200$ ps. In conclusion, our calculations indicate that for pulsed excitation maximum TPA-free input power is about 6.4 mW larger than in CW operation (55.6 mW/μm^2, 256% more), due to smaller free carrier density generated by the TPA in this regime. Thus,

assuming input peak power of $|S_i|^2_{peak} = 6.4$ mW and $T_{FWHM} = 200$ ps, undistorted Gaussian pulse power can be achieved at the output waveguide edge as 20 dB larger than the power at the input waveguide end. For $T_{FWHM} = 40$ ps and same input power, the power improvement at the output with respect to the input waveguide is only about 8 dB, although the generated free carrier density is lower than in the previous case. This depends mainly on the input pulse with FWHM closer to τ_e, leading the microring resonator to a strongly unmatched condition.

IV. CONCLUSION

The microring resonant filters are very interesting devices that can contribute to increase the integration level in photonic systems. In this paper we present, for the first time to the best of our knowledge, a theoretical investigation of the influence of input power levels and FWHM widths in pulsed excitation of SOI microring resonators with nanometric-scale cross section. Maximum peak powers are obtained in order to maximize the ring enhancement factor without exciting nonlinear effects due to two photon absorption.

REFERENCES

[1] B. E. Little, S. T. Chu, H. A. Haus, J. Foresi, and J. P. Laine, "Microring resonator channel dropping filters", *J. Lightwave Technol.,* vol. 15, pp.998-1005, 1997.

[2] M. N. Armenise, V. M. N. Passaro, F. De Leonardis, and M. Armenise, "Modeling and Design of a Novel Miniaturized Integrated Optical Sensor for Gyroscope Applications", *J. Lightwave Technol.,* vol. 19, pp. 1476-1494, 2001.

[3] V. Van, P. P. Absil, J. V. Hryniewicz, and P. T. Ho, "Propagation loss in single mode GaAs-AlGaAs microring resonators: Measurement and model," *J. Lightwave Technol.,* vol. 19, pp.1734-1739, 2001.

[4] B. Liu, A. Shakouri, and J. E. Bowers, "Wide tunable double rong resonator coupled lasers," *IEEE Photon. Technol. Lett.,* vol.14, pp. 600-602, 2002.

[5] Y. Yanagase, S. Suzuki, Y. Kokubun, and S. T. Chu, "Box-like filter response and espansion of FSR by a vertically triple coupled microring resonator filter," *J. Lightwave Technol.,* vol. 20, pp.1525-1529, 2002.

[6] P. P. Absil, J. V. Hryniewicz, B. E. Little, P. S. Cho, R. A. Wilson, L. G. Joneckies, and P. T. Ho, "Wavelength conversion in GaAs micro-ring resonators," *Opt. Lett.,* vol. 25, pp. 554-556, 2000.

[7] V. Van, T.A. Ibrahim, P. P. Absil, F. G. Johnson, R. Grover, and P. T. Ho, "Optical signal processing using nonlinear semiconductor microring resonators," *IEEE J. of Selected Topics in Quantum Electron.,* vol. 8, pp.705-713, 2002.

[8] A. Liu, H. Rong, M. Paniccia, O. Cohen, and D. Hak, "Net optical gain in a low loss silicon-on-insulator waveguide by stimulated Raman scattering," *Opt. Express,* vol. 12, pp. 4261-4268, 2004.

[9] R. Claps, V. Raghunathan, D. Dimitropoulos, and B. Jalali, "Influence of nonlinear absorption on Raman amplification in Silicon waveguides," *Opt. Express,* vol. 12, pp. 2774-2780, 2004.

2006 25th International Conference on Microelectronics

Silicon Surface Exfoliation under Compression Plasma Flow Action

I.P. Dojčinović, M.M. Kuraica, D. Randjelović, M. Matić and J. Purić

Abstract - Silicon single crystal surfaces have been modified by supersonic compression plasma flows (CPF) action. Triangular and rhombic regular fracture features are obtained on the Si (111), while rectangular ones are produced on Si (100) surface. Some of these regular structures can become free from the underlying bulk, formed as blocks ejected from the surface. Surface cleavage and exfoliation phenomena as the results of specific conditions during CPF interaction on silicon surface are, also, observed. Such conditions are the results of rapid heating and melting of surface layer, long existence of molten layer (~40 μs) and fast cooling and recrystalisation taking place under the high dynamic pressure and thermodynamic parameters gradients.

I. INTRODUCTION

Material surface can be modified by pulsed energy beams, such as pulsed laser, electron beams, ion beams, plasma sample treatment etc. Surface and interface properties are very important for semiconductor devices and their engineering applications. During surface treatment by high-power energy beams the material may be removed from surface in the form of vapor, liquid droplets, or solid flakes due to evaporation, sputtering, ablation, explosive boiling, and exfoliation [1,2].

High-power pulsed energy streams interaction with material surfaces results in rapid melting and resolidification of surface layer. High temperatures and consequent thermal stresses, as well as mechanically strained surface during treatment, result in significant deformation and fracture of the layer, induced defects, cracking and exfoliation of the coating. During high intensity pulse plasma beam (HIPPB) treatment of single crystal Al_2O_3 and B_4C surfaces, irregular cracks [3] and flakes [4] on the coated target surfaces are obtained, respectively. During Ti:sapphire picosecond and femtosecond laser interaction with CaF_2 (111) surface, strong fracturing of the crystal is obtained. Irregular cracks are developed on treated surfaces and the upper layer breaks off in places [5]. Furthermore, Ti:sapphire and excimer laser interactions with MgO single crystal caused regular fracture features [6,7]. Plastic deformation in MgO

I.P. Dojčinović, M.M. Kuraica, and J. Purić are with the Faculty of Physics, University of Belgrade, Studentski trg 12, 11000 Belgrade, Serbia & Montenegro, E-mail: ivbi@ff.bg.ac.yu

D. Randjelović and M. Matić are with the Institute of Chemistry, Technology and Metallurgy – Department of Microelectronic Technologies and Single Crystals, Njegoševa 12, 11000 Belgrade, Serbia & Montenegro, E-mail: danijela@nanosys.ihtm.bg.ac.yu

typically involves the growth of a relatively few dislocations that form extremely long, convoluted structures known as dislocation "multiplication" [7]. In case of MgO (100) irradiated area is covered with a pattern of rectangular structures, with typical dimension of the order of 10×20 μm^2. Many of these rectangles become free from the underlying bulk as formed blocks eventually ejected from the surface. During laser interaction with MgO (111) surface triangular features are typical fracture patterns. On the laser-irradiated silicon surface, after 20-50 laser pulses, regular fractures are developed [8]. On Si (001) wafers two sets of fracture lines intersecting at 90° form a grid that divides the surface into rectangular blocks with sides 10-40 μm long. In the case of Si (111) surface fracture lines definitely intersect at 60° and 120° angles.

In this experiment silicon single crystal surfaces have been modified by supersonic compression plasma flows action. Triangular and rhombic regular fracture features are obtained on the Si (111), while rectangular ones are produced on Si (100) surface. It was found that some of these regular structures can become free from the underlying bulk, formed as blocks ejected from the surface. Surface cleavage and exfoliation phenomena as the results of specific conditions during CPF interaction on silicon surface are, also, observed and studied.

II. EXPERIMENTAL SETUP

Si (100) and Si (111) surface of single crystal was treated with quasistationary compression plasma flow (CPF) made by magnetoplasma compressor (MPC). The MPC used in this experiment was described elsewhere [9-11], therefore only a few details are given here for the sake of completeness.

The electrode system of MPC consists of the conically shaped copper cathode and cylindrical anode (outer electrode), made of 8 symmetrically positioned copper rods (Fig. 1). Conically shaped cathode of MPC defines the profile of acceleration channel. The discharge device of MPC is situated in a vacuum chamber filled with nitrogen at 400 Pa pressure. Using 800 μF, 4 kV capacitor banks, the obtained current maximum was up to 100 kA and time duration up to 150 μs with current half period ~70 μs. In the MPC interelectrode region the plasma is accelerated due to the Ampere force. The plasma flow is compressed due to interaction between longitudinal component of current swept-away, and intrinsic azimuth magnetic field. The stable CPF is formed 20 μs after the

1-4244-0116-X/06/$20.00 ©2006 IEEE

beginning of the discharge. During a quasistationary phase the plasma flow parameters are slowly changing in time within certain volume [9]. This is a consequence of an ion-drift acceleration of magnetized plasma realised using specially shaped accelerating channel [12].

Fig. 1. Magnetoplasma compressor (MPC). A typical discharge image is given in the corner.

The advantages of MPC, as compared to other types of plasma accelerators, are high stability of generated CPF, controllability of their composition (different gasses and their mixtures), size (CPF up to 6 cm in length and 1 cm in diameter), and high plasma parameters (electron density up to $4 \cdot 10^{17}$ cm^{-3} and temperature up to 2-3 eV), as well as the CPF time duration (quasistationary stable phase is 40-50 μs) and large flow velocity (~40 km/s) sufficient for material surface modification. Beside that, the operation in the ion current transfer mode [12] with the minimization of the electrodes erosion represents the main advantage of the quasistationary plasma accelerators in comparison with the classical ones. Magnetic flux conservation is a particular characteristic of CPF. During the action of CPF on a sample surface, due to CPF deceleration current loops (vortices) are formed [13], and magnetic field 10^{-3}-10^{-2} T is induced at the surface [17].

For the studies of CPF interaction with silicon surfaces, commercial one-side polished n-type silicon wafers (100 and 111 orientation) 300 μm thick and 10 mm in diameter were used. The sample is mounted on the cylindrical brass holder of the same diameter, and placed in front of the MPC cathode at the distance of 6 cm. To investigate the morphology of treated silicon surface, optical microscopy (OM) and atomic force microscopy (AFM) were used.

III. RESULTS

OM micrographs of treated Si (111) and Si (100) surfaces are given in Fig. 2 and Fig. 3, respectively. Triangular and rhombic regular fracture features are obtained in the case of Si (111) (Fig. 2), as expected for threefold symmetry [14,15]. Some of the blocks are ejected from the surface, and large holes at the surface emerged. On Si (100) surface treated by CPF, rectangular structures are obtained (Fig. 3). Some of the blocks, as in the case of Si (111), are ejected from the surface. Thickness of the

ejected block is 6-10 μm. At the bottom of arisen holes ripple structures are observed (Fig. 3).

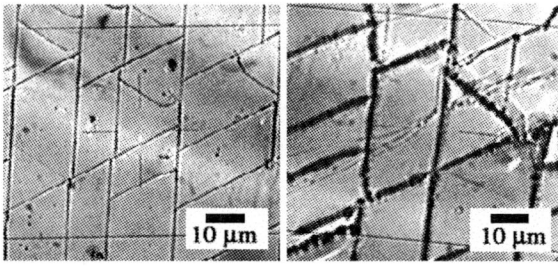

Fig. 2. Si (111) surface after CPF treatment by six plasma pulses.

It was found that regular fracture features are not obtained on Si (111) surface, treated by one pulse at distance of 6 cm, as was obtained for Si (100) (Fig. 3). However, during the set of six pulses under the same other conditions, on Si (111) surface regular fracture features were obtained.

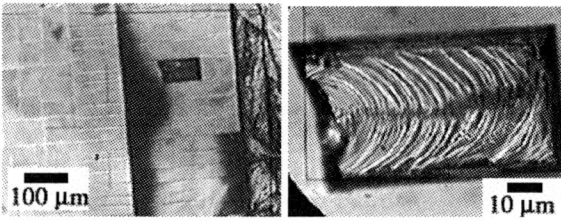

Fig. 3. Si (100) surface after CPF treatment by single plasma pulse. Right figure is an enlarged part of the left.

Histograms of the regular fracture features (blocks) area obtained by CPF treatments of the Si (111) and Si (100) sample surface are shown in Fig. 4. In the case of Si (111), surface area of 300 μm^2 corresponds to equilateral triangles, with ~25 μm side length (Fig. 2). It may be noted that rhombic regular fracture features are a multiple of two or four areas of characteristic triangles. On Si (100) surface treated by CPF, the area of 5000 μm^2 corresponds to rectangular structures with typical dimensions of the order of 120×40 μm^2 and squares with ~70 μm side length (Fig. 3). Observed length of cleavage along crystal planes on Si (100) surface is up to 1 mm.

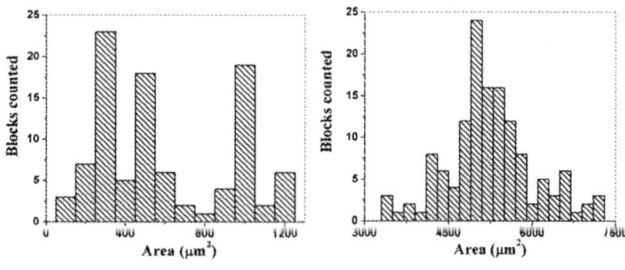

Fig. 4. Histograms of regular fracture features (blocks) area obtained by six (left) and single (right) plasma pulse treatments of the Si (111) and Si (100) sample surface, respectively.

AFM micrographs of treated Si (100) surfaces are given in Fig. 5. Surface profile indicates cleavage along crystal plane (Fig. 5b). Cleavage height is (1.3 ± 0.1) μm. Estimated angle between Si (100) surfaces and cleavage plane is (54 ± 3) °. This indicates that cleavage plane is Si (111).

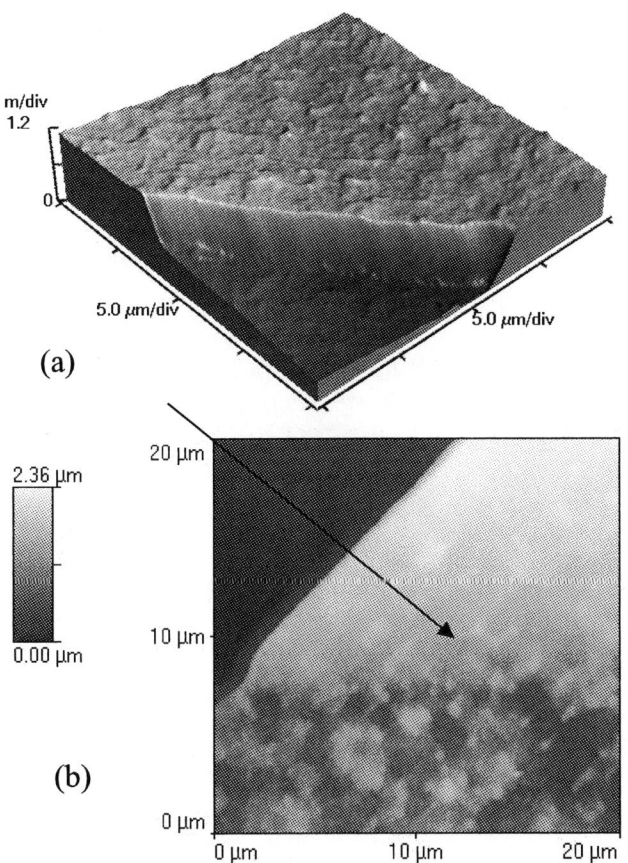

(a)

(b)

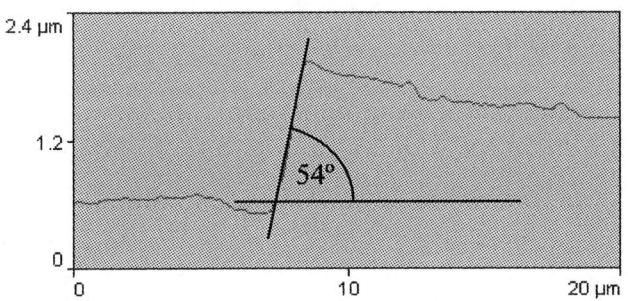

Fig. 5. Si (100) surface after CPF treatment: a) Three-dimensional view of the surface morphologies; b) Two-dimensional view of the treated Si surface with the surface profile along the arrow.

The formation of observed surface features may be explained by energetic action of CPF on the surface (absorbed energy 10-15 J per pulse, flow power density $\sim 1 \cdot 10^5$ W/cm²) [17]. Interaction of CPF with silicon sample surface causes a formation of a shock-compressed plasma layer (plasma plume) [16], which results in the shielding of a processed surface. CPF kinetic energy thermalization is

causing the heating of shock-compressed plasma layer. Target surface is heated by convective and radiative heating and melt-solid zone penetrates into sample depth. Between the formed molten layer and underlying cold bulk a zone is present that suffers a thermal influence from the molten layer. Subsequent fast cooling and recrystalisation is performed in conditions of high dynamic pressure of CPF (of the order of several atmospheres) [17], with high gradient of thermodynamic parameters and induced magnetic field.

Using the high speed camera time of interaction was estimated to be ~40 μs. It may be taken that molten layer exists on the target surface during the interaction. This period is much longer when compared to HIPPB sample treatment, in which duration of interaction is ~1 μs, and molten layer exists for a period of several microseconds [3,4]. Ti:sapphire and excimer laser energy densities are ~10 J/cm², i.e. of the same order as energy density of CPF, but pulse duration is very short, up to picosecond range [5-7].

By measuring the thickness of ejected silicon blocks typical depth of melted near-surface layer is estimated to be 6-10 μm, while the HIPPB experiment has a depth of melted layer of ~1 μm.

IV. DISCUSSION

Process of crystallization begins at a maximal depth of the molten zone penetration. The crystallization of molten silicon features is occurring under the fast cooling. Cooling rate is of the order of 10^{10} K/s (quenching effect). Rapid solidification can produce serious cracking problems due to shrinkage and differential stresses. During the cooling process thermoelastic stresses appeared in the surface layer. Gradient of the thermoelastic stresses caused bending of silicon wafer, due to which the dislocations and changing in silicon lattice occurred and this led to the plastic deformation. Thermal stresses can result in significant fracture, cracking and exfoliation of the near-surface layer [6]. During plasma treatment surface layers were mechanically strained and lattice defects where produced. Clustering of these defects can also induce cracking and exfoliation of the layer.

Similarly, in the case of MgO (100) and (111) laser irradiation [7], lattice defects production at high fluencies is attributed to mechanical processes involving deformation and fracture with accompanying dislocation motion. Formation of extremely long, convoluted structures is explained by "multiplication" of these dislocations [7].

During crystallization, silicon layer is grown on silicon substrate. Also a strong anisotropy of the absorbed energy in the layer is present. Formation of regular fracture features can be explained by high temperature annealing leading to coalescence and ordering of vacancies, and also by quenching effects like quenched growth processes. Effect of ordering of vacancies on Si (001) occurred during annealing at sufficiently high temperature because vacancy clusters line up into vacancy line defects (VLD) [18]. Due to motion of mobile vacancies, at elevated temperatures, they coalesce into elongated vacancy islands. Although VLD is not deeper than one atomic layer, this ordering of defects at high temperature in the zone of thermal influence can initiate a formation of regular fracture along the crystal planes.

Low adhesion between blocks and silicon bulk, and eventual ejection of blocks from the CPF treated surface, can be explained by development of subsurface fracture, parallel to the surface. Cracking between the block and the bulk is growing due to local energy absorption.

Accumulation or "incubation" of defects [19] is an effect that occurs when single shot fluence is below the damage threshold, i.e. defects are increasing by increasing number of laser beam shots. In case of Si (111) surface treatment by plasma flow, which is also below single shot damage threshold, mechanical defects are accumulated during several shots (Fig. 2).

V. Conclusion

Regular fracture features and exfoliation occurred on silicon single crystal surface treated by CPF. Surface modification is performed by fast heating and melting, long existence of melted layer (~40 µs) and annealing of underlying zone of thermal influence. Fast cooling (quenching) and recrystalisation is, also, present. These processes are occurring in the presence of high pressure, temperature gradient and induced magnetic field from CPF. As the results of all of these processes triangular and rhombic regular fracture features are obtained on the Si (111). Equilateral triangles have ~25 µm side length, and rhombic regular fracture features are a multiple of two or four areas of characteristic triangles. Rectangular fracture features are produced on Si (100) surface with typical dimensions of the order of 120×40 µm^2 and squares with ~70 µm side length. Some of these regular structures can become free from the underlying bulk, formed as blocks ejected from the surface.

References

[1] R. Kelly and A. Miotello, "On the mechanisms of target modification by ion beams and laser pulses", *Nucl. Instrum. Methods B*, 1997, vol. 122, pp. 374-400.

[2] J. H. Yoo, S. H. Jeong, X. L. Mao, R. Greif, and R. E. Russo, "Evidence for phase-explosion and generation of large particles during high power nanosecond laser ablation of silicon", *Appl. Phys. Lett.*, 2000, vol. 76, pp. 783-785.

[3] J. Piekoszewski, E. Wieser, R. Grotzschel, H. Reuther, Z. Werner, and J. Langner, "Pulsed plasma beam mixing of Ti and Mo into Al$_2$O$_3$ substrates", *Nucl. Instrum. Methods B*, 1999, vol. 148, pp. 32-36.

[4] J. Langner, J. Piekoszewski, M. Sadowski, G. P. Glazunov, E. D. Volkov, O. S. Pavlichenko, O. Motojima, V. I. Tereshin, and J. Stanislawski, "Erosion behaviour of boron carbide

under high-power pulsed fluxes of hydrogen plasma", *Fusion Eng. Des.*, 1998, vol. 39-40, pp. 433-437.

[5] D. Ashkenasi, H. Varel, A. Rosenfeld, F. Noack, and E. E. B. Campbell, "Pulse-width influence on the laser-induced structuring of CaF$_2$ (111)", *Appl. Phys. A*, 1996, vol. 63, pp. 103-107.

[6] A. Rosenfeld, D. Ashkenasi, H. Varel, M. Wahmer, and E. E. B. Campbell, "Time resolved detection of particle removal from dielectrics on femtosecond laser ablation", *Appl. Surf. Sci.*, 1998, vol. 127-129, 76-80.

[7] R. L. Webb, L. C. Jensen, S. C. Langford, and J. T. Dickinson, "Interactions of wide band-gap single crystals with 248 nm excimer laser radiation. I. MgO", *J. Appl. Phys.*, 1993, vol. 74, pp. 2323-2337.

[8] D. H. Lowndes, J. D. Fowlkes, and A. J. Pedraza, "Early stages of pulsed-laser growth of silicon microcolumns and microcones in air and SF$_6$", *Appl. Surf. Sci.*, 2000, vol. 154-155, pp. 647-658.

[9] J. Purić, I. P. Dojčinović, V. M. Astashynski, M. M. Kuraica, and B. M. Obradović, "Electric and Thermodynamic Properties of Plasma Flows Created by Magnetoplasma Compressor", *Plasma Sources Sci. Technol.*, 2004, vol. 13, pp. 74-84.

[10] I. P. Dojcinovic, M. R. Gemisic, B. M. Obradovic, M. M. Kuraica, V. M. Astashinskii, and J. Puric, "Investigation of Plasma Parameters in a Magnetoplasma Compressor", *J. Appl. Spectrosc.*, 2001, vol. 68, pp. 824-830.

[11] S. I. Ananin, V. M. Astashinskii, G. I. Bakanovich, E. A. Kostyukevich, A. M. Kuzmitski, A. A. Man'kovskii, L. Ya. Min'ko, and A. I. Morozov, "Study of the Formation of Plasma Streams in a Quasistationary High-Current Plasma Accelerator", *Sov. J. Plasma Phys.*, 1990, vol. 16, pp. 102-107.

[12] A. I. Morozov, "Principles of coaxial (quasi-) steady-state plasma accelerators", *Sov. J. Plasma Phys.*, 1990, vol. 16, pp. 69-78.

[13] S. I. Ananin, V. M. Astashinskii, E. A. Kostyukevich, A. A. Man'kovskii, and L. Ya. Min'ko, "Interferometric Studies of the Processes Occurring in a Quasi-steady High-Current Plasma Accelerator", *Plasma Phys. Rep.*, 1998, vol. 24, pp. 936-942.

[14] H. Rauscher, "The interaction of silanes with silicon single crystal surfaces: microscopic processes and structures", *Surf. Sci. Rep.*, 2001, vol. 42, pp. 207-328.

[15] D. D. D. Ma, C. S. Lee, F. C. K. Au, S. Y. Tong, and S. T. Lee, "Small-Diameter Silicon Nanowire Surfaces", *Science*, 2003, vol. 299, pp. 1874-1877.

[16] J. Purić, V. M. Astashynski, I. P. Dojčinović, and M. M. Kuraica, "Creation of Silicon Submicron Structures by Compression Plasma Flow Action", *Vacuum*, 2004, vol. 73, pp. 561-566.

[17] V. V. Uglov, V. M. Anishchik, V. V. Astashynski, V. M. Astashynski, S. I. Ananin, V. V. Askerko, E. A. Kostyukevich, A. M. Kuz'mitski, N. T. Kvasov, and A. L. Danilyuk, "The effect of dense compression plasma flow on silicon surface morphology", *Surf. Coat. Technol.*, 2002, vol. 158-159, pp. 273-276.

[18] H. J. W. Zandvliet, "Ordering of vacancies on Si(001)", *Surf. Sci.*, 1997, vol. 377 379, pp. 1 6.

[19] M. Henyk, R. Mitzner, D. Wolfframm, and J. Reif, "Laser-induced ion emission from dielectrics", *Appl. Surf. Sci.*, 2000, vol. 154-155, pp. 249-255.

2006 25th International Conference on Microelectronics

Magnetic Field Influence on Silicon Surface Periodic Structures Obtained by Plasma Flow Action

I.P. Dojčinović, M.M. Kuraica, M. Mitrović, D. Randjelović, M. Matić and J. Purić

Abstract – External magnetic field influence on silicon submicron surface periodic structures obtained by the action of nitrogen supersonic quasistationary compression plasma flow (CPF) is studied. CPF is generated by magnetoplasma compressor. It was found that, without external magnetic field, highly-oriented silicon periodic cylindrical shape structures are produced during single pulse surface treatment. Silicon periodic structures were modified by external constant magnetic field. Hexagonal structures of the side length about 500 nm and height of the order of 10 nm, are obtained. Morphology investigation was made by SEM and AFM microscopy.

I. INTRODUCTION

Surface and interface properties are very important for semiconductor devices and their surface engineering applications. Many properties of the materials of interest depend on size, shape and regularity of the surface structures [1]. Silicon surface modification by pulsed energy beams such as pulsed laser, ion beams and plasma flows are of great importance, especially for formation of grating-like patterns, i.e. wave-like periodic structures.

Laser-induced periodic surface structures (LIPSS) induced on the silicon surface due to the incident laser light, were previously described [2]. LIPSS effect corresponds to a inhomogeneous energy deposition associated with the interference of the incident beam with a surface scattered field [3]. At normally incident laser beam, periodical surface structures are perpendicular to the laser beam polarization and have a period which corresponds to laser wavelength. In some cases, structures have a period which is much higher than laser wavelength. This structures are described as laser induced capillary waves [3].

Ripple structures on silicon surface, created by ion bombardment [4], are based on the interplay between the sputtering and surface diffusion smoothing processes. The interplay between these two effects is responsible for the

creation of cones, dots and holes on surfaces at normal ion beam incidence, and especially for ripple-like and wave-like surface morphologies, when the direction of the ion beam is tilted to the surface normal (off-normal incidence) [4,5]. In the case of grazing-incidence sputtering geometries orientation of the surface structures is forced to be parallel to the ion beam orientation. For angles smaller then the critical incident angle structures become perpendicular to the ion beam orientation. In non-metal substrates, including silicon, at normal incidence ripples are not observed [5]. Ripples can be created by bombardment at normal ion beam incidence in the case of some metals, due to anisotropic surface diffusion [5].

In the case of surface treatment by pulsed plasma streams, periodical formation was not observed so far. In several type of plasma sources (coaxial plasma gun [6], rod plasma injector [7], thermal plasma jet [8]), rapid melting and resolidification have occurred, but wave-like structures have not been observed. Quasistationary plasma flow action with silicon solid surface was studied [9-10], and submicron cylindrical periodic structures were obtained and studied. Here, the silicon periodic structures are modified by the simultaneous action of quasistationary plasma flow action and the constant external magnetic field. Morphology obtained under such conditions was investigated by SEM and AFM microscopy.

II. EXPERIMENTAL SETUP

In this experiment, Si (100) surface of single crystal has been treated with quasistationary compression plasma flow (CPF) produced by magnetoplasma compressor (MPC). The MPC was described elsewhere [11,12], therefore only a few details are given here for the sake of completeness. MPC is a quasistationary plasma accelerator (plasma gun). A capacitor bank of 800 μF, 4 kV was used as an energy source. The MPC vacuum chamber was filled with nitrogen at 400 Pa pressure. With nitrogen as a working gas, and the maximum discharge current of 70 kA, about 40 μs long steady state CPF is generated.

The plasma acceleration by the Ampere force in MPC interelectrode gap is accompanied by formation of CPF at the outlet of the discharge device. The plasma flow is compressed due to interaction of longitudinal current component with intrinsic azimuthal magnetic field (pinch effect). During a quasistationary phase the plasma flow parameters are slowly changing. Namely, the continual

I.P. Dojčinović, M.M. Kuraica, M. Mitrović and J. Purić are with the Faculty of Physics, University of Belgrade, Studentski trg 12, 11000 Belgrade, Serbia & Montenegro, E-mail: ivbi@ff.bg.ac.yu

D. Randjelović and M. Matić are with the Institute of Chemistry, Technology and Metallurgy – Department of Microelectronic Technologies and Single Crystals, Njegoševa 12, 11000 Belgrade, Serbia & Montenegro, E-mail: danijela@nanosys.ihtm.bg.ac.yu

1-4244-0116-X/06/$20.00 ©2006 IEEE

ionization processes are taking part in working gas introduced in interelectrode region. The plasma is steadily accelerated and permanently compressed.

The advantages of MPC, as compared to other types of plasma accelerators are: high stability of generated CPF, controllability of their composition (different gasses and their mixtures), size (up to 6 cm in length and 1 cm in diameter), and high plasma parameters (electron density up to $4 \cdot 10^{17} \text{cm}^{-3}$ and temperature up to 2-3 eV), as well as the CPF time duration and large flow velocity (~40 km/s) sufficient for material surface modification [11]. Magnetic flux conservation is a particular characteristic of CPF. During the action of CPF on a sample surface current loops (vortices) can arise.

For the interaction studies of the CPF with silicon surfaces, commercial one-side polished n-type silicon wafer (100 orientation) 1 mm thick and 10 mm in diameter were used. The sample is glued to the cylindrical brass holder of the same diameter with conductive carbon paste, and mounted perpendicularly in front of the MPC cathode at the distance of 5 cm. A permanent magnet was inserted between the brass holder and the Si sample during the investigation of magnetic field influence on the Si surface structure characteristics. It gives 0.46 T and 0.48 T magnetic field perpendicular to the Si sample surface.

To investigate the morphology of treated silicon surface, scanning electron microscopy (SEM) (JEOL 840A) and atomic force microscopy (AFM) (AutoProbe CP-Research SPM, TM Microscopes - Veeco) were used.

III. RESULTS

As example of the obtained silicon surface cylindrical structures (SCS) [9-10] without external magnetic field, SEM micrographs of treated Si (100) surfaces are given in Fig. 1. It is worth emphasizing that these highly oriented periodic structures are obtained by single plasma pulse treatment of the silicon sample surface. SCS are obtained in the periphery part of the target surface.

Typical wavelengths (hill-to-hill distance) of SCS formed on the treated silicon surfaces are about 2-3 µm, amplitude (half hill-to-valley distance) 0.2 µm and length up to 200 µm. AFM micrographs were used for SCS surface investigation. The SCS surfaces are found to be smooth, homogenous and sinusoidally shaped.

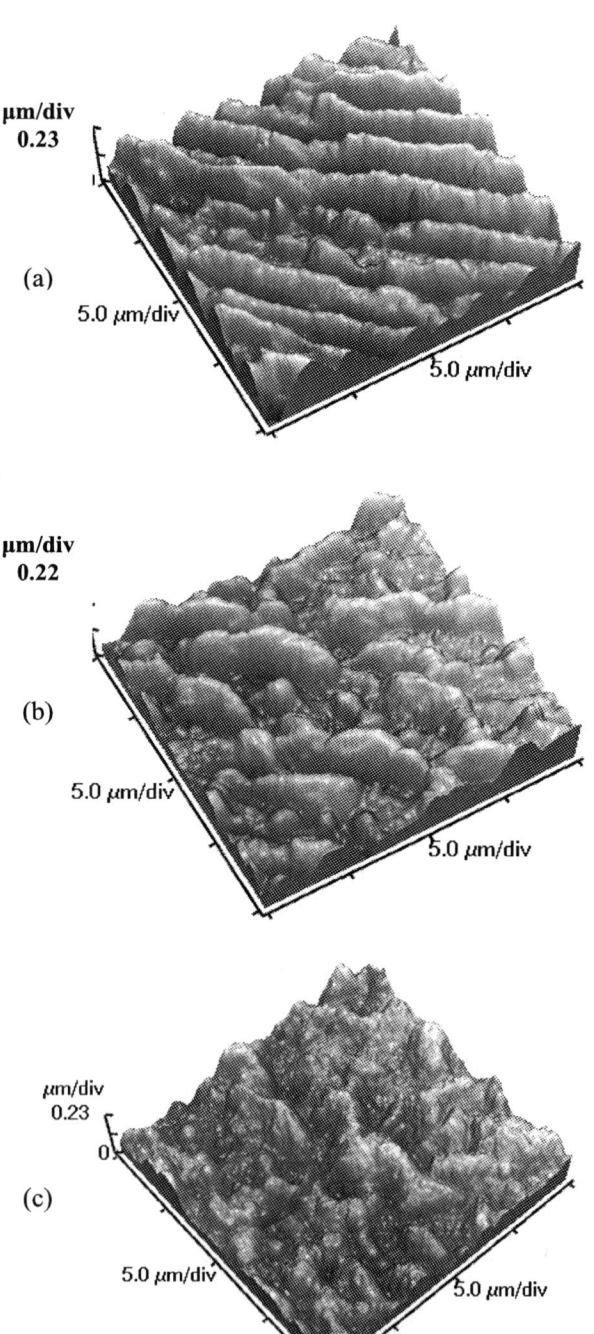

Fig. 1. SEM micrograph of surface cylindrical structures obtained in single pulse treatment of silicon sample surface by CPF without external magnetic field.

Fig. 2. AFM micrographs of silicon cylindrical structures with applied external magnetic field B = 0.46 T. Plasma energy density increase from (a) to (c).

Compression plasma flow action on Si sample in the presence of the steady external magnetic field results in a increase in plasma energy density and destruction of cylindrical structures. Typical representatives of the obtained AFM micrographs, used for SCS surface investigation, were given in Fig. 2.

Also, hexagonal structures were obtained on SCS surface with applied external magnetic field. Typical AFM micrographs of these structures were given in Fig. 3.

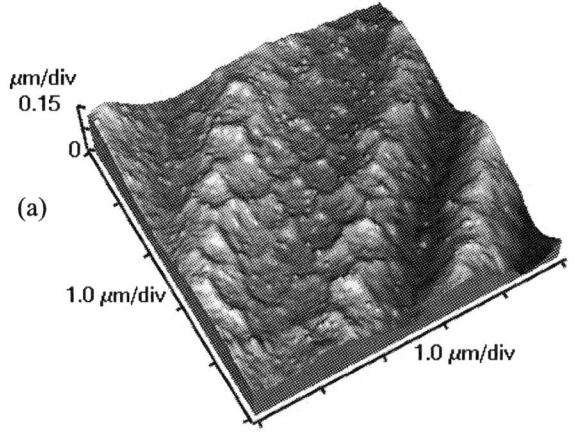

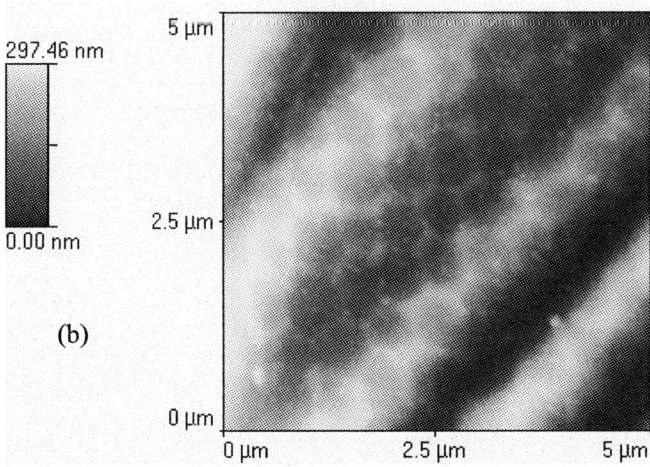

Fig. 3. AFM micrographs of hexagonal structures obtained on SCS with applied external magnetic field B = 0.48 T: 3D (a) and 2D (b) topography.

Typical side length of hexagonal structures obtained on SCS surface is 500 nm. Their height are of the order of 10 nm.

With applied external magnetic field, on silicon SCS surface small and large holes are obtained. Typical SEM micrographs of these holes were given in Fig. 4. Typical dimensions of these holes are 1 to 10 µm, at periphery and central zone of silicon target, respectively.

IV. DISCUSSION

Interaction of CPF with silicon sample surface causes a formation of a shock-compressed plasma layer

(SCPL), plasma plume [9] which results in the shielding of a processed surface from a direct action of a CPF. Plasma plume has been observed so far in plasma and laser surface interaction [13]. Thickness of SCPL is about 1 cm. CPF kinetic energy thermalization caused the SCPL heat up. Deceleration of the CPF results in the formation of current loops (vortices), due to freezing of magnetic field into plasma, and magnetic field 10^{-3}-10^{-2} T is induced at the surface [10].

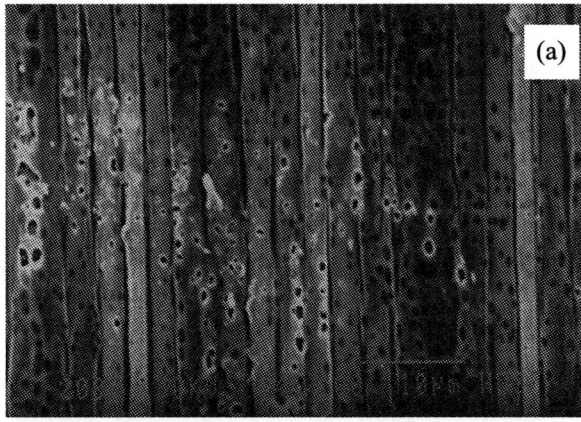

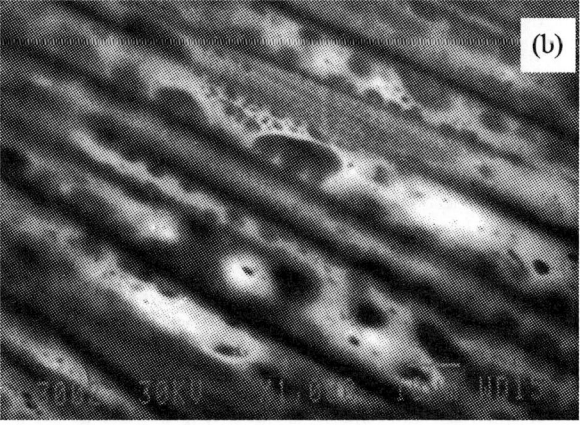

Fig. 4. SEM micrographs of holes obtained on SCS surface with applied external magnetic field B = 0.46 T. Plasma energy density increase from (a) to (b).

Energetic action of CPF on the surface (absorbed energy 10-15 J per pulse, flow power density ~$1 \cdot 10^5$ W/cm^2) [10], is causing the fast heating and melting of the surface layer. Target surface is heated by convective and radiative heating. Using the high speed camera, time of interaction was estimated to be ~40 µs. It may be taken that molten layer exists on the target surface during the interaction. By analyzing the cross section of treated silicon sample, thickness of near-surface molten layer is estimated to be 6-10 µm.

Melted layer at the silicon surface is subjected to surface or volume forces, like surface tension, gradients of thermodynamic parameters and Lorentz force. It is known that these forces, together with pressure of the plasma flow on the melted layer, cause the development of various instabilities [10]. At the completely liquid surface, in the presence of high dynamic pressure of CPF of the order of several atmospheres [10], perturbation is occurring. The restoring force is provided by surface tension, and wave motion is induced and established at fluid surface. These are plasma flow induced capillary waves [14]. This process may be compared with laser induced capillary waves [3,13]. Capillary effects as well as disturbance of the molten layer, due to the recoil pressure, may generate ripple structure [13]. Laser induced capillary waves are smooth and sinusoidally shaped structures [3], as obtained SCS by CPF silicon sample treatment without external magnetic field.

Obtained periodic cylindrical structures on silicon surface are plasma flow induced capillary waves frozen during fast cooling and recrystalisation phase (quenching effect). Melted layer is exhibited to rapid cooling because of the heat sink into the bulk of the target. Recrystalisation occurred in condition of high dynamic pressure of CPF and high thermodynamic parameters gradients.

External magnetic field was impressed to plasma dynamics and thermal convection in the molten silicon layer. With magnetic field increasing, plasma flow density and surface absorbed energy are, also, increasing. High local absorbed energy leads to destruction of cylindrical structures (Fig. 2). Moreover, on silicon SCS surface large holes are obtained (Fig. 4). This holes along with many droplets obtained in sample periphery part, suggest mechanism like explosive boiling. In this case rapid emission of vapor and liquid droplets are occurred [15]. Hot region near the surface relaxes explosively, i.e. breaks down in a very short time into vapor plus equilibrium liquid droplets. Vapor bubble grows to a critical radius, and then rapid expansion leads to the violent ejection of mass.

Heat and mass transfer includes buoyancy convection within molten silicon, thermocapillary (Marangoni) convection at the free surface and diffusion. Thermocapillary flow, due to the temperature dependence of surface tension, forces the surface melt to flow. Magnetic field was impressed on the heat and mass transfer in the melt during the crystal-growth processes. Magnetic fields are suppress convective flow in electrically conductive melt. Silicon being highly conductive in the molten state and flow can be damped by the use of static magnetic fields [16].

V. CONCLUSION

On silicon single crystal surface treated by CPF without external magnetic field, highly-oriented periodic cylindrical structures are obtained and analyzed. These structures are capillary waves induced by the plasma flow and then quenched from the molten state during fast cooling and recrystalisation. Silicon periodic structures were modified by external constant magnetic field. On silicon SCS surface hexagonal structures and large holes are obtained. It is found that these hexagonal structures of

the following dimensions: side length about 500 nm and height of the order of 10 nm.

REFERENCES

[1] R. Gago, L. Vazquez, R. Cuerno, M. Varela, C. Ballesteros and J. M. Albella, "Production of ordered silicon nanocrystals by low-energy ion sputtering", *Appl. Phys. Lett.*, 2001, vol. 78, pp. 3316-3318.

[2] M. S. Trtica and B. M. Gakovic, "Pulsed CO_2 Laser Surface Modifications of Silicon", *Appl. Surf. Sci.*, 2003, vol. 205, pp. 336-342.

[3] J. F. Young, J. E. Sipe and H. M. van Driel, "Laser-induced periodic surface structure. III. Fluence regimes, the role of feedback, and details of the induced topography in germanium", *Phys. Rev. B*, 1984, vol. 30, pp. 2001-2015.

[4] S. Habenicht, K. P. Lieb, J. Koch and A. D. Wieck, "Ripple propagation and velocity dispersion on ion-beam-eroded silicon surfaces", *Phys. Rev. B*, 2002, vol. 65, pp. 115327, 1-6.

[5] U. Valbusa, C. Boragno and F. B. de Mongeot, "Nanostructuring surfaces by ion sputtering", *J. Phys.: Condens. Matter*, 2002, vol. 14, pp. 8153-8175.

[6] B. Liu, C. Liu, D. Cheng, R. He and S. Z. Yang, "Pulsed high energy density plasma processing silicon surface", *Thin Solid Films*, 2001, vol. 390, pp. 149-153.

[7] J. Piekoszewski, Z. Werner, J. Langner and M. Janik-Czachor, "Irradiation of silicon with a pulsed plasma beam containing Mo ions", *Surf. Coat. Technol.*, 1997, vol. 93, pp. 258-260.

[8] H. Kaku, S. Higashi, H. Taniguchi, H. Murakami and S. Miyazaki, "A new crystallization technique of Si films on glass substrate using thermal plasma jet", *Appl. Surf. Sci.*, 2005, vol. 244, pp. 8-11.

[9] J. Purić, V. M. Astashynski, I. P. Dojčinović, and M. M. Kuraica, "Creation of Silicon Submicron Structures by Compression Plasma Flow Action", *Vacuum*, 2004, vol. 73, pp. 561-566.

[10] V. V. Uglov, V. M. Anishchik, V. V. Astashynski, V. M. Astashynski, S. I. Ananin, V. V. Askerko, E. A. Kostyukevich, A. M. Kuz'mitski, N. T. Kvasov, and A. L. Danilyuk, "The effect of dense compression plasma flow on silicon surface morphology", *Surf. Coat. Technol.*, 2002, vol. 158-159, pp. 273-276.

[11] J. Purić, I. P. Dojčinović, V. M. Astashynski, M. M. Kuraica, and B. M. Obradović, "Electric and Thermodynamic Properties of Plasma Flows Created by Magnetoplasma Compressor", *Plasma Sources Sci. Technol.*, 2004, vol. 13, pp. 74-84.

[12] I. P. Dojcinovic, M. R. Gemisic, B. M. Obradovic, M. M. Kuraica, V. M. Astashinskii, and J. Puric, "Investigation of Plasma Parameters in a Magnetoplasma Compressor", *J. Appl. Spectrosc.*, 2001, vol. 68, pp. 824-830.

[13] T. Y. Choi and C. P. Grigoropoulos, "Plasma and ablation dynamics in ultrafast laser processing of crystalline silicon", *J. Appl. Phys.*, 2002, vol. 92, 4918-4925.

[14] L. D. Landau, E. M. Lifshitz, *Gidrodinamika* (in Russian), Moskva: Nauka, 1988.

[15] R. Kelly and A. Miotello, "On the mechanisms of target modification by ion beams and laser pulses", *Nucl. Instrum. Methods B*, 1997, vol. 122, pp. 374-400.

[16] A. Croll and K. W. Benz, "Static magnetic fields in semiconductor floating-zone growth", *Prog. Crystal Growth Charact. Materials*, 1999, vol. 38, pp. 133-159.

2006 25th International Conference on Microelectronics

A Consideration of Transparent Metal Structures for Subwavelength Diffraction Management

Zoran Jakšić, Milija Sarajlić, Milan Maksimović, Dana Vasiljević-Radović, Dušan Jovanović

Abstract – We considered theoretically and experimentally one-dimensional multilayered metallodielectric nanofilms with nanometric thickness for imaging below the diffraction limit. We investigated their behavior in the ultravioled and visible spectrum from the point of view of near field optics, but also considered some of their properties in the far field. We designed our structures using the transfer matrix method and utilized RF sputtering to fabricate them. We consider some possible approaches to extract optical information from such multilayers.

I. INTRODUCTION

The use of metamaterials with negative refractive index (NRM) [1] or, alternatively, with negative dielectric permittivity (single-negative – SNG material) in imaging below the diffraction limit was first proposed by Pendry in his seminal paper [2]. In this paper he described the use of a NRM plane-parallel slab as an optically complementary medium to that of the free space. He concluded that it could be used to extract complete image data from the object, including evanescent waves connected with the object features with subwavelength dimensions. A consequence would be the reconstruction of its perfect image. In this manner such structures overcome the well-known diffraction limit and thus offers a plethora of novel possibilites in photolithography, microscopy, sensorics, etc. Currently a large number of teams performs research of the so-called perfect lenses and their version which includes absorptive losses and dispersion, the superlenses – for reviews see e.g. [3], [4]. The physical origin of the amplification of the evanescent waves inside the NRM slabs is related to the coupled surface plasmons (SP), excited by the evanescent waves at the two surfaces of the NRM slab.

If the observed geometrical distances are much smaller than the wavelength, then the quasi-static (extreme near-field) approximation is valid. In that case the electric field is dominated by p-polarized radiation, and magnetic fields are s-polarized. It is then only important that the lens material has $\varepsilon < 0$, while μ may remain larger than zero. Good metals (e.g. copper, silver, gold) have negative dielectric permittivity in a range of frequencies in the UV and visible part of the spectrum. Thus it is possible to use a planar slab of metal as a superlens for the optical range. The experimental use of silver superlenses was described in [5]-[8] and it has been showed that resolution enhancements of at least four times are possible by using simple silver SNG nanofilms.

The monolithic NRM and SNG superlenses pose the problems of large localizations of field within their structures. Another problem is the unavoidable existence of large absorption losses, which is the reason why metal layers in such structures must be very thin. A solution to overcome these obstacles is to slice the original solid metal slab into thin sublayers divided by dielectric strata and thus form a metallodielectric multilayer [9]. Historically, metallodielectrics were first used for optical filtering [10], [11]. Their use as "transparent metals" with adjustable transmission band was proposed by [12]-[13] and applied in the UV in [14]. The most recent considerations of metallodielectrics include an analysis of their application in superlensing [15]-[17]. Considerations dedicated to generalization of the super-resolution lenses to far field operation have been published in [18], [19].

In this paper we describe the design and fabrication of silver-silicon dioxide nanofilm multilayers intended for the use in sub-diffraction limited imaging. We consider some possibilities for their coupling to the far field modes. We use transfer matrix technique to design our structures and radiofrequent sputtering to fabricate our samples.

Zoran Jakšić, Milija Sarajlić, Milan Maksimović and Dana Vasiljević-Radović are with the IHTM – Institute of Microelectronic Technologies and Single Crystals, Njegoševa 12, 11000 Belgrade, Serbia and Montenegro

Fax +381 11 182 995, E-mail: jaksa@nanosys.ihtm.bg.ac.yu

Dušan Jovanović is with the IHTM – Institute of Catalysis and Chemical Engineering, Njegoševa 12, 11000 Belgrade, Serbia and Montenegro, E-mail: dusanmj@nanosys.ihtm.bg.ac.yu

Milan Maksimović is now with the Department of Applied Mathematics, University of Twente, The Netherlands; E-Mail: m.maksimovic@math.utwente.nl

II. THEORY

We consider a metallodielectric multilayer as shown in Fig. 1. Metal nanofilms maintaining plasmon modes (in our case silver) and those of dielectric (silicon dioxide) are alternating on a transparent substrate (fused silica). This represents the standard configuration for multilayer superlenses [3], [9]. The object with dimensions smaller than the operating wavelength is positioned near the top surface.

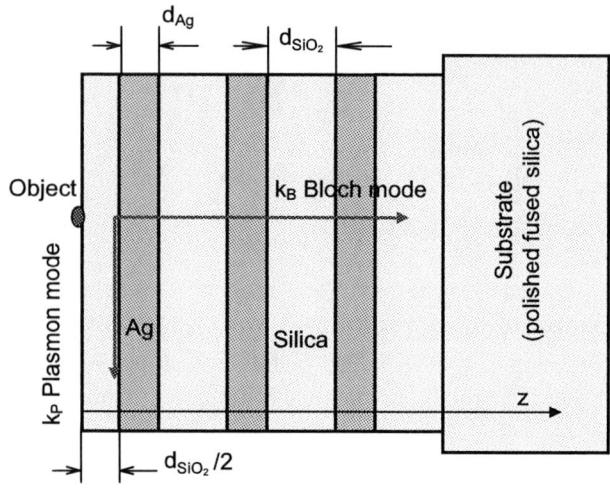

Fig. 1. Silver-silica metallodielectric multilayer.

The object wavevector can be represented by an in-plane component and an evanescent (exponentially decaying) component in the z-direction (perpendicular to the multilayer plane). The in-plane component couples with the surface electromagnetic waves propagating along the metal-dielectric boundary. These waves resonantly couple with the waves on the further metal layers and the result is the Bloch propagation of the evanescent modes [15].

We assume that magnetic permeability is constant and equal to unity throughout the multilayer. We used the following expressions for the silver extinction coefficient [14], valid in the range 200 nm to 1 μm:

$$
k_{Ag} = \begin{cases} \sqrt{\dfrac{a_1 + c_1\lambda^2 + e_1\lambda^4}{1 + b_1\lambda^2 + d_1\lambda^4 + f_1\lambda^6}}, & \lambda < 318\,\text{nm} \\[2ex] a_2 + \dfrac{b_2}{\lambda} + \dfrac{c_2}{\lambda^{3/2}} + \dfrac{d_2}{\lambda^2}, & \lambda > 318\,\text{nm} \end{cases} \quad (1)
$$

where $a_1 = 1.1716163$, $b_1 = -26.831136$, $c_1 = -23.250175$, $d_1 = 244.03336$, $e_1 = 115.58375$, $f_1 = -747.72811$. For wavelengths above plasma frequency, $a_2 = 23.880509$, $b_2 = -46.806634$, $c_2 = 39.418581$, $d_2 = -9.708619$.

The real part of the silver refractive index is

$$
n_{Ag} = \begin{cases} a_3 + b_3\lambda^{3/2} + c_3\lambda^2 + \dfrac{d_3}{\lambda}, & \lambda < 305\,\text{nm} \\[2ex] \exp\left(a_4 + \dfrac{b_4}{\lambda^{3/2}} + \dfrac{c_4}{\lambda^2} \right), & \lambda > 305\,\text{nm} \end{cases} \quad (2)
$$

where $a_3 = 8.4544209$, $b_3 = -90.780355$, $c_3 = 118.66256$, $d_3 = -0.80076022$, and $a_4 = 0.48332493$, $b_4 = -4.2648334$, $c_4 = 2.3417131$.

We consider two-dimensional waves propagating along the z direction. The dispersion relation for the Bloch waves is obtained in the form

$$
\cos(k_B d) = \cosh(\alpha_{SiO_2} d_{SiO_2}) \cosh(\alpha_{Ag} d_{Ag}) + \\ + \frac{\alpha_{SiO_2}^2 \varepsilon_{Ag}^2 + \alpha_{Ag}^2 \varepsilon_{SiO_2}^2}{2\alpha_{SiO_2}\alpha_{Ag}\varepsilon_{SiO_2}\varepsilon_{Ag}} \times \\ \times \sinh(\alpha_{SiO_2} d_{SiO_2}) \sinh(\alpha_{Ag} d_{Ag}) \quad (3)
$$

where

$$
\begin{aligned} \alpha_{SiO_2}^2 &= k_x^2 - \left(\frac{\omega}{c}\right)^2 \varepsilon_{SiO_2}\mu_{SiO_2} \\ \alpha_{Ag}^2 &= k_x^2 - \left(\frac{\omega}{c}\right)^2 \varepsilon_{Ag}\mu_{Ag} \end{aligned} \quad (4)
$$

The condition for the existence of propagating Bloch modes is given as

$$
\left|\cos(k_B d)\right| \le 1 \quad (5)
$$

The Bloch modes are propagating if $\alpha_i^2 < 0$ and evanescent if $\alpha_i^2 > 0$.

We designed our multilayer superlens in a manner similar to that described in [17]. We calculated the transmission properties of the metallodielectric stack using the transfer matrix technique for lossy/absorptive case, which includes both propagating and evanescent modes. For our structures we chose a thickness of 26 nm for silver layers and 40 nm for silicon dioxide.

III. COUPLING TO FAR-FIELD MODES

It has already been said that one of the problems with the use of both bulk and multilayer superlens structures, and indeed with the plasmonics generally, is that their propagating modes are evanescent, which is the reason why these structures are applicable only in the near field. While this is sufficient for many applications (including contact lithography), others (e.g. sensing, microscopy) require a degree of freedom more since their image details need to be transferred to distances much larger than the wavelength without losing the subwavelength precision.

Here we propose a generalization to the multilayer case of the recently proposed method [18] to couple evanescent electromagnetic field components to far field modes. To this purpose corrugations are introduced to one or more plasmonic surfaces within the multilayer.

The metal surface corrugations may be either ordered nanopatterns (two possibilities are illustrated in Fig. 2 and include a matrix of nanospherical pits and an array of tetrahedra) or a random roughness.

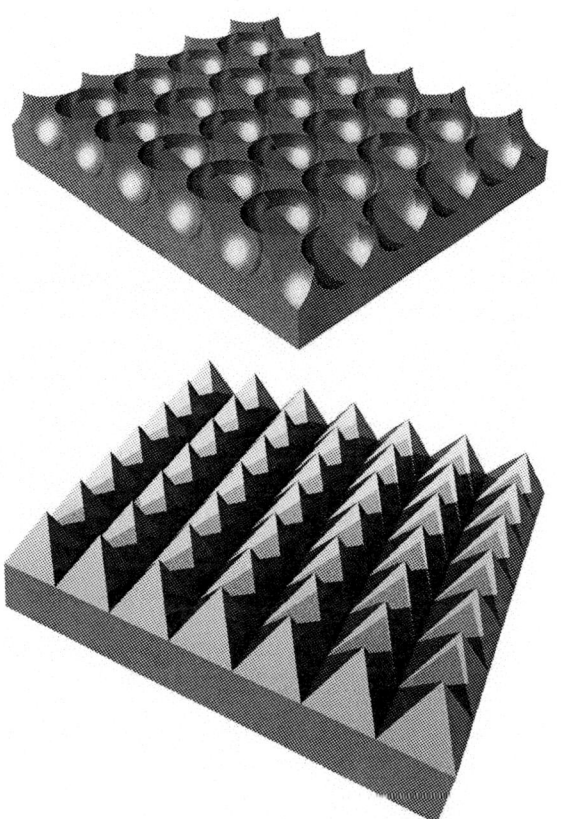

Fig. 2. Periodic subwavelength corrugations of the top stratum in multilayer configuration. Top: nanospherical pit array; bottom: array of tetrahedra.

A further improvement of this approach would be to use an in-plane diffractive optical element (DOE) on the surface of the metallodielectric multilayer to tailor redistribution of electromagnetic energy and directly obtain a far field image of nanometer-sized details. This would serve at the same time both to couple plasmon modes to Bloch waves and to form the far field image utilizing diffractive beam shaping.

The case of random surface corrugations is shown in Fig. 3 for the case when several strata are involved.

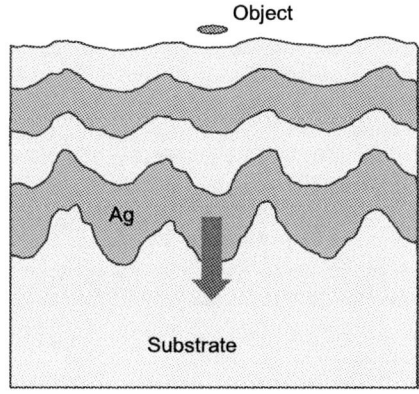

Fig. 3. Random surface corrugation; the process of nanofilm deposition smoothens top layers to a certain degree.

Fig. 3 shows a realistic case of rough (or roughened) substrate surface to which alternating metal and dielectric nanofilms are deposited, e.g. by sputtering technique. The roughness becomes less pronounced farther from the substrate surface, which is a consequence of "smoothing" of layers in the deposition of subsequent layers.

IV. EXPERIMENTAL

For our experiments we used unpolished fused silica plates as substrates. Thus we had a surface with random corrugations to begin with. After rinsing and drying the substrates we used radiofrequent sputtering to deposit alternating layers of silver (26 nm thickness) and silicon dioxide (40 nm thickness) on our Perkin Elmer sputtering equipment. We maintained the Ar pressure at $2 \cdot 10^{-5}$ bar and used the operating mode without substrate rotation. Silicon dioxide was deposited for 4 min. at a cathode voltage of 1200 V and at 1 kW rf power. Silver deposited for 25 seconds at a cathode voltage of 1100 V and at 0.5 kW rf power. The substrate has been kept cold during deposition. We sputtered samples with two, three and four layer pairs, and the final nanofilm was always 20 nm thick silicon dioxide (a half of the thickness used in the pairs). Fig. 4 shows surface morphology of the substrate prior deposition (top) and after that; the smoothing of top layers is clearly visible.

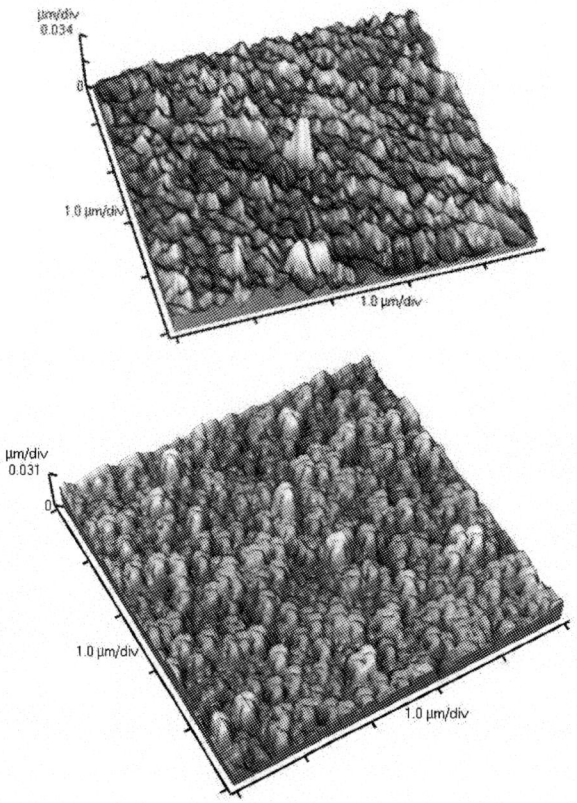

Fig. 4. Atomic force microscope images of the surface of the used fused silica substrate surface (top) and of the metallodielectric 4½ layer

We characterized transmission of our samples on a UV-vis spectrophotometer Thermo-Nicolet. The measured transmission spectra are shown in Fig. 5. As expected, a large transmission peak is visible near the silver plasma frequency (320 nm). The suppression of longer wavelengths is significant even for a small number of pairs and amounts 4 to 5 orders of magnitude in the red part of the spectrum.

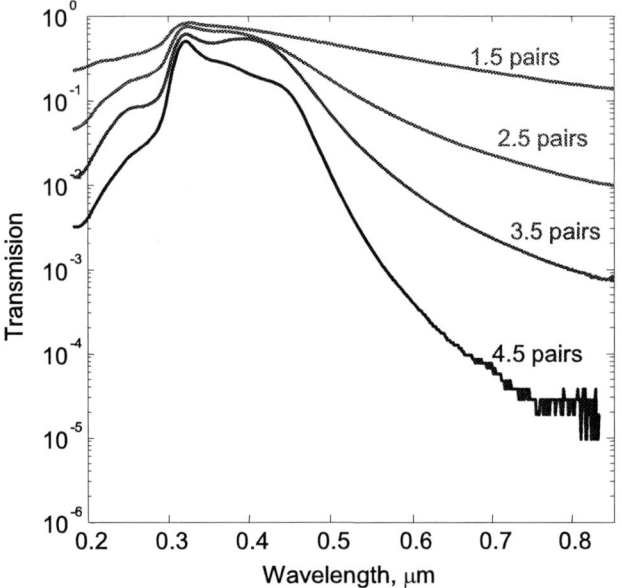

Fig. 5. Measured transmission of experimental silver-silica multilayers for different numbers of layer pairs, d_{Ag}=26 nm, d_{SiO2}=40 nm

In the presented experimental structures the wavelengths above 320 nm are usable for superlensing. Typical wavelength for photolithography is 365 nm and it is near the optimum value. The built-in layer corrugations enable coupling (albeit unoptimized) to the far field modes.

V. CONCLUSIONS

We used the concept of multilayer superlenses and diffractive optical elements to define superresolution structures with their plasmonic modes coupled to propagating Bloch waves. As practical materials we used silver and silicon dioxide, sputtered to a fused silica substrate. We utilized the surface roughness of the silica plate as the natural coupling corrugation of the structure. We believe the described structures could find application in sensing and microscopy.

VI. ACKNOWLEDGEMENTS

This work has been partially supported by the Serbian Ministry of Science, Technologies and Development within the framework of the project IT.6151.B.

REFERENCES

[1] V. G. Veselago, "The electrodynamics of substances with simultaneous negative values of epsilon and mu," Sov. Phys. Uspekhi, 10, pp. 509-514, 1968.

[2] J. B. Pendry, "Negative refraction makes a perfect lens," Phys. Rev. Lett. 85, 3966–3969, pp. 2000

[3] S. A. Ramakrishna, "Physics of negative refractive index materials," Rep. Prog. Phys. 68, pp. 449–521, 2005.

[4] Z. Jakšić, N. Dalarsson, M. Maksimović, "Electromagnetic Structures Containing Negative Refractive Index Metamaterials", Proc. 7th Conf. TELSIKS, Sep. 28-30, 2005, Niš, Vol. 1, pp. 145-154

[5] F. Fang and X. Zhang, "Imaging properties of a metamaterial superlens," Appl. Phys. Lett., 82, pp. 161–163, Jan. 2003.

[6] N. Fang, H. Lee, C. Sun, X. Zhang, "Sub-diffraction-limited optical imaging with a silver superlens," Science, 308 (5721), pp. 534-537, Apr 2005

[7] D. O. S. Melville, R. J. Blaikie, C. R. Wolf, "Submicron imaging with a planar silver lens," Appl. Phys. Lett. 84, 22, pp. 4403-4405, 2004.

[8] D. O. S. Melville, R. J. Blaikie, "Super-resolution imaging through a planar silver layer, " Opt. Expr. 13, 6, pp. 2127-2134, March 2005

[9] E. Shamonina, V.A. Kalinin, K.H. Ringhofer, L. Solymar, "Imaging, compression and Poynting vector streamlines with negative permittivity materials," Electron. Lett. 37, pp. 1243-4, 2001

[10] H. A. Macleod, "Thin-Film Optical Filters", Institute of Physics Publishing, London, 2001

[11] H. A. Macleod, "A New Approach to the Design of Metal-Dielectric Thin Film Optical Coatings", Opt. Acta 25 pp. 93-106, 1978

[12] M. Bloemer, M. Scalora "Transmissive properties of Ag/MgF2 photonic band gaps", Appl. Phys. Lett. 72, pp. 1676-78, 1998

[13] M. Scalora, M. Bloemer, A. C. Pethel, J. P. Dowling, C. M. Bowden, A. S. Manka, "Transparent, metallo-dielectric, one-dimensional, photonic band-gap structures", J. Appl. Phys. 83 pp. 2377-2383, 1998

[14] Z. Jakšić, M. Maksimović, M. Sarajlić, Silver-silica transparent metal structures as bandpass filters for the ultraviolet range," J. Opt. A, 7, 1, pp. 51-55, 2005

[15] S. Feng, J. M. Elson, P. L. Overfelt, "Optical properties of multilayer metaldielectric nanofilms with all-evanescent modes," Opt. Express 13, p. 4113-4124, 2005.

[16] S. Feng, J. M. Elson, P. L. Overfelt, "Transparent photonic band in metallodielectric nanostructures," Phys. Rev. B 72, 085117, 2005.

[17] S. Feng, J. Merle Elson, "Diffraction-suppressed high-resolution imaging through metallodielectric nanofilms", Opt. Expr. 14, 1, pp. 216-221, Jan 2006

[18] I.I. Smolyaninov, J. Elliot, A.V. Zayats, C.C. Davis, "Far-field optical microscopy with a nanometer-scale resolution based on the in-plane image magnification by surface plasmon polaritons", Phys. Rev. Lett. 94, 057401, 2005.

[19] I. I. Smolyaninov, C. C. Davis, J. Elliott, G. A. Wurtz, A. V. Zayats, "Superresolution optical microscopy based on photonic crystal materials", Phys. Rev. B 72, 085442 2005.

2006 25th International Conference on Microelectronics

The Fabrication of Single Electron Transistor by Polysilicon Thin Film and Point-Contact Lithography

Kuo-Dong Huang, Jyi-Tsong Lin, Shu-Fen Hu and Chin-Lung Sung

Abstract - In this paper, point-contact lithography depended on the proximity effect is employed to fabricate the single electron transistor (SET) by using polysilicon thin film which is deposited upon an insulation layer (POI or TFT). The electrical characteristics of the SET fabricated, such as Coulomb black and Coulomb oscillation, are observed and discussed appropriately. It can be operated beyond 180 °K and the SET characteristics can be still observed.. In addition, the channel width of the SET below 20 nm has been also fabricated.

I. INTRODUCTION

Recently, single electron transistor (SET) has been attended to many applications in future ultra-low power and ultra-high density integrated circuits [1]. The dramatic developments in SETs have been seen based on the controllable transfer of single electrons between the conducting "island" and source/drain, and several important scientific experiments of it have been already enabled. Recent research in SET field has presented some exciting concepts, which exploits the quantum effect and the feature of ultra-low power consumption making the SET be feasible for microelectronic and nanoelectronic circuits in the nearest future [2]-[5]. However, the main limitation of SET is how to obtain a very small and high quality island for operating in room temperature. The observations of Coulomb blockade oscillations in silicon SETs at room temperature have been reported by many groups [6]-[9]. In this case, several reports utilized geometric patterns and advanced processes had proved that the good quantum dot structure could be achieved, such as the process by PADOX (pattern dependant on oxidation) [10], the method by the controllable electrical field [11], the vertical cavity [12], and the lithography by point contact [13].

Employing a simple method based on electron-beam overlapping and overexposure techniques on the single crystalline SOI was developed to fabricate sub-10 nm electrode gaps with good electrical properties [14]. However, no SET research is found based on the poly-

Kuo-Dong Huang and Jyi-Tsong Lin are with the Department of Electrical Engineering, National Sun Yat-Sen Univ., 70 Lien-hai Rd, Kaohsiung 804, Taiwan ROC, E-mail: d933010002@student.nsysu.edu.tw, jtlin@ee.nsysu.edu.tw

Shu-Fen Hu and Chin-Lung Sung are with the National Nano Device Laboratories, 1001-1, Ta Hsueh Road, Hsinchu, Taiwan, 300, R.O.C., E-mail: sfhu@mail.ndl.org.tw, clsung@mail.ndl.org.tw

crystalline film by the same techniques. In this paper, the method is used for the polysilicon film upon the insulation layer (POI or TFT) for the first time. The cost advantage of the polysilicon film over the corresponding single crystalline is one of considerations. Besides, the room-temperature operation of SET is called for a sub-10 nm islands which is surrounding with tunneling barriers so that the electron transferring into and out of the island can be prohibited by the Coulomb charging energy. So it can be evaluated that the tunneling barrier can be automatically formed by the grain boundary of polysilicon film instead of the single crystalline film. Therefore, the pattern of the point-contact SET shown in Figure 1 is prepared for the over exposure by electron-beam lithography. The electron-beam dosage of lithography is of importance the same as the pattern, and to adjust these two factors is a challenge for the performance optimization. The point-contact width can hardly be controlled after the development. The etching process except the non-uniform thickness of the poly film and the unsteadiness of the electron-beam can be overcome suitably. In this paper, the electrical characteristics resulted from the point-contact quantum dot of the SET produced are observed beyond 180 °K. The low temperature characteristic is, however, presented with complicated curves undesired. This can be attributed to the multiple dots system and traps formed after the lithography of the polysilicon. The phenomenon will be discussed in the following section.

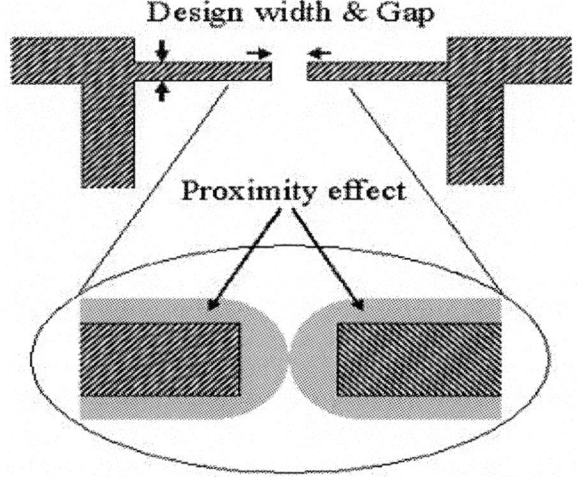

Fig. 1. The point-contact SET fabricated by utilizing the principle of proximity effect. The designed pattern of the width and gap is adjusted suitably for the optimization.

1-4244-0116-X/06/$20.00 ©2006 IEEE

II. FABRICATION

At first, the p-type (100) silicon substrate was prepared. The buried layer of 200 nm was fabricated with the wet oxidation on the substrate at 850 °C. Then, the polysilicon film of 25 nm was deposited by LPCVD at 620 °C for 30 min. In this way, the phosphorus was in-situ diffused into the polysilicon film that the donor concentration is about 1×10^{16} cm^{-3}. For activating the dopant, the rapid thermal annealing (RTA) was used at 925 °C for 20 min with N$_2$ gas. Then the active region was patterned by the electron-beam lithography based on the proximity effect with the electron-beam energy of 40 KeV, and dosage of 13 μC/cm^2. The transformer-coupled etching (TCP) with the gas of Cl$_2$ was used for the etching of the polysilicon film. By this way, the self-aligned point-contact islands were adjusted so that the point-contact channel was about 20 nm. Then, the gate oxide layer of 5 nm was oxidized just on the 25 nm polysilicon film, and the polysilicon thickness was reduced to 20 nm spontaneously. After that, the 200 nm passivation layer was deposited by LPCVD at 700 °C using the macromolecular compounds (TEOS). Then, the reactive ion etching (RIE) with the gas of CHF$_3$ and CF$_4$ was used to shape TEOS and form the contact holes. Next, the metal layer (Al-Si-Cu alloy) was deposited by the sputtering deposition (PVD), and the metal etching was used by RIE with the mixed gases of BCl$_3$ and Cl$_2$.

III. RESULTS AND DISCUSSION

Coulomb blockade and Coulomb oscillation characteristics of SET are indicative phenomena for the electron quantum transportation. In Figure 2, the Coulomb blockade voltage (V_b) versus temperature (T) is shown, the drain current (Id) is symmetric versus the drain (Vd) biases as shown in the inlet of Figure 2. The blockade voltage is the drain threshold voltage at which the SET can be turned on as shown in the figure. While the temperature is decreased, the blockade voltage is raised. According to the illustration, we can estimate the total charging energy of the quantum dot by following the equation of $E_C = e/2V_b$ [15], where e is the electronic charge unit, E_C is the charging energy in the quantum dot.

In Figure 3, the I_d-V_g characteristic of SET at 180 °K is presented. The Coulomb effect is proved by the step curve, which is different from the linear ohmic effect. The transconductance property is shown as the gray curve (right scale), and the negative differential conductance (NDC) is observed. Even though the multiple dots structure is fabricated from the polysilicon film in this work, fortunately, the complicated curves resulted from the different grain sizes and traps are disappeared at the high temperature. It is considered that the temperature gives the electron enough thermal energy to overcome the lower barrier height of the potential well, and the multiple

junctions resulted from the grain boundary can be also summarized for the total junction barrier from source to drain.

The complicated curve resulted from the different grain size and traps can be also observed at low temperature. The complex period of the Coulomb oscillation shown in Figure 4 is definitely observed at 40 °K. The multiple dots comprise more than one size of grains, the different charging confinement results in complicate oscillation events. According to the equation of $\Delta V_g = e / C_g$ [15] and of the spherical capacitance, it is possible to estimate the grain size, and is reasonable for the fabrication. Based on the ΔV_g of 0.1 V, 0.5 V and 0.75 V, the grain size can be roughly calculated as 20 nm, 4 nm and 2.6 nm respectively. The maximum size of 20 nm is consistent with the fabrication in which the polysilicon film is 20 nm and the point-contact width is also about 20 nm. The period of Coulomb oscillation will be determined by the quantum dot size. Therefore, it is reasonable that the Coulomb oscillation property is shown for the resonance point at the certain gate voltage on the multiple dots of SET with polysilicon The grain resonance phenomenon shows that the periodic waves of the oscillation resulted from different grains are resonated at the gate voltage about 1.0 V and 1.3 V. It is believed that this phenomenon is observed for the first time on the polysilicon SET.

Figure 5 shows the 3D profile of the I_d-V_g-V_d characteristic. Concerning the I_d-V_d aspect, the blockade voltage is tuned from the gate voltage. Whereas, as the I_d-V_g aspect is concerned, the amplitude of oscillation is also tuned from the drain bias. The Coulomb confinement area which is called Coulomb diamond indicates that no electron flow is transported into and out of the quantum dot.

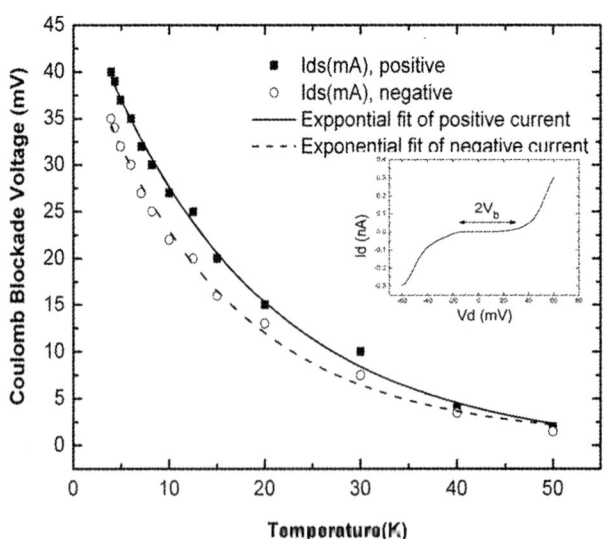

Fig. 2. The Coulomb blockade voltage versus temperature. The drain current is symmetric as the drain bias is between negative and positive voltages.

158

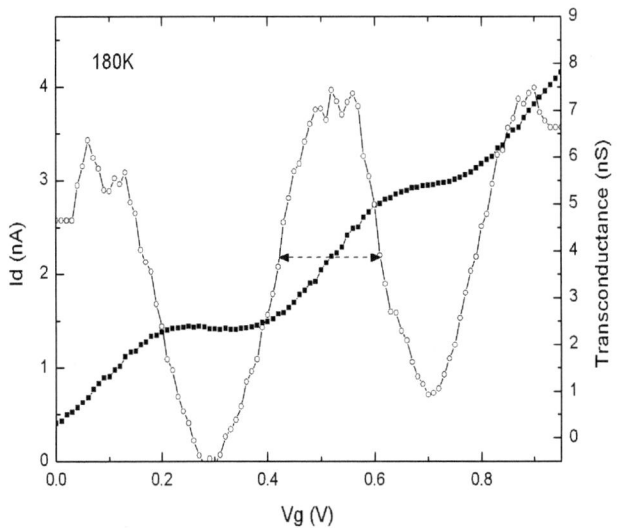

Fig. 3. The Coulomb effect and the negative differential conductance (NDC) of the *Id-Vg* curve of SET at 180 °K. The double arrow indicates the full width half maximum (FWHM).

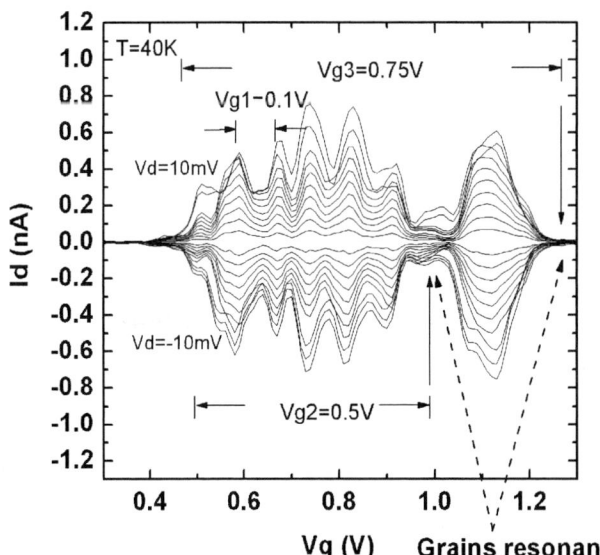

Fig. 4. The complex periods of Coulomb oscillation. Tree kinds of period can be found in this figure. The "grain resonance" can be observed at the gate voltage about 1.0 V and 1.3 V on the polysilicon SET.

IV. CONCLUSION

In summary, the single electron transistor fabricated with the polysilicon thin film, had been investigated. Utilizing the point-contact lithography based on the proximity effect, the channel width could be shrunk below 20 nm. The *I-V* characteristics, such as Coulomb blockade and Coulomb oscillation, were definitely observed. The high temperature operation of SET had been performed at 180 °K. The complex oscillation at low temperature of

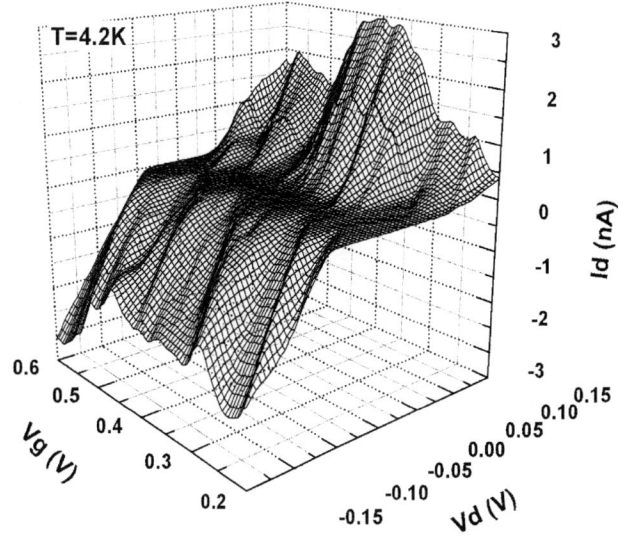

Fig. 5. The 3D profile of the I_d-V_g-V_d characteristic. The flat area near the value of $I_d = 0$ A is the Coulomb confinement area where no electron flow is transported into and out of the island.

40°K had been also explained for the multiple dots system and traps, which was observed on the polysilicon SET. Furthermore, the process and material were fully compatible with the present CMOS technology, which is worthwhile for further research.

ACKNOWLEDGEMENT

This work was performed under the contract number NSC93-2112-M-492-002 supported by the National Science Council, Republic of China. Technical supports from the members at National Nano Device Laboratories are acknowledged.

REFERENCES

[1] K. K. Likharev, "Single-electron devices and applications," *Proc. IEEE*, vol. 87, p606–632, Apr. 1999.

[2] C. P. Heij, P. Hadley, and J. E. Mooij, "Single-electron inverter", *Appl. Phys. Lett.*, Vol. 78, p1140, Feb. 2001.

[3] Y. Takahashi, A. Fujiwara, K. Yamazaki, H. Namatsu, K. Kurihara, and Katsumi Murase, "Multigate single-electron transistors and their application to an exclusive-OR gate", *Appl. Phys. Lett.*, Vol. 76, p637, Jan. 2000.

[4] A. Dutta, S. P. Lee, S. Hatatani, and S. Oda, "Silicon-based single electron memory using a multiple-tunnel junction fabricated by electron-beam direct writing", *Appl. Phys. Lett.*, Vol. 75, p1422, Sep. 1999.

[5] K. Yano, T. Ishii, T. Sano, T. Mine, F. Murai, T. Hashimoto, T. Kobayashi, T. Kure, and K. Seki, "Single-Electron Memory for Giga-to-Tera Bit Storage," *Proc. IEEE*, vol. 87, p633, Apr. 1999.

[6] Y. T. Tan, T. Kamiya, Z. A. K. Durrani, H. Ahmed, "Room tem-perature nanocrystalline silicon single-electron transistors", *J. of Appl. Phys.*, vol.94, p633, July 2003.

[7] M. Saitoh, T. Murakami, and T. Hiramoto, "Effects of Oxidation Process on the Tunneling Barrier Structures in RoomTemperature Operating Silicon Single-Electron Transistors", *IEEE Trans. On Nanotechnology*, vol. 1, no. 4, Dec.2002.

[8] L. Zhuang, L. Guo, S. Y. Choub, "Silicon single-electron transistor switch operating at room temperature", *Appl. Phys. Lett.*, vol. 72, p1205, March 1998.

[9] B. H. Choi, S. W. Hwang, I. G. Kim, H. C. Shin, Y. Kim, and E. K. Kim, "Fabrication and room-temperature characterization of a silicon self-assembled quantum-dot transistor," *Appl. Phys. Lett.*, vol. 73, p3129–3131, Nov. 1998.

[10] Y. Ono, Y. Takahashi, K. Yamazaki, M. Nagase, H. Namatsu, K. Kurihara, and K. Murase, "Fabrication Method for IC-Oriented Si Single-Electron Transistors", *IEEE Trans. On Electron Device*, vol. 47, p147, Jan. 2000.

[11] D. H. Kim, S. K. Sung, K. R. Kim, J. D. Lee, and B. G. Parkb, "Fabrication of single-electron tunneling transistors with an electrically formed Coulomb island in a silicon-on-insulator nanowire" , *J. Vac. Sci. Technol.*, B 20(4), Jul/Aug 2002.

[12] H. T. Lin , Y. M. Wan, S. F. Hu and C. L. Sung, "Fabrication and Development of Silicon nitride/Poly Silicon/Silicon nitride Single Electron Transistors", *SNDT2004*, Taiwan, p383, 2004.

[13] T. H. Wang, "Si single-electron transistors with in-plane point-contact metal gates", *Appl. Phys. Lett.*, vol. 78, p2160, April 2001.

[14] K. Liu, Ph. Avouris, J. Bucchignano, R. Martel, and S. Sun, "Simple fabrication scheme for sub-10 nm electrode gaps using electron beam lithography", *Appl. Phys. Lett.*, vol. 80, p865, Feb 2002.

[15] D. K., "Transport in nanostructures", *Cambridge university*, 1997.

2006 25th International Conference on Microelectronics

Impact of Rapid Photothermal Processing on Properties of ZnO Nanostructures for Solar Cell Applications

S. Shishiyanu, R. Singh, T. Shishiyanu, and O. Lupan

Abstract - The Nanotechnology with chemical deposition (CD) and Rapid Photothermal Processing (RPP) of nanostructured ZnO thin films for solar cell applications was elaborated. The influence of growth processes and the impact of RPP on surface morphology, particles size and resistivity values are presented and discussed. The ZnO thin films were deposited on silicon substrates by chemical deposition method at room temperature and normal pressure. The obtained thin films were rapid photothermal processed in vacuum and N_2 ambient. Nanostructures of the deposited films were optimized by adjusting various growth parameters: concentration of zinc complex solution, temperature of aqueous solution of anions and RPP regimes. Structural and electrical properties were investigated by Energy Dispersive X-ray (EDX) spectroscopy, scanning electron microscopy (SEM), electrical resistivity measurements. Electrical resistivity measurements showed that the room temperature resistivity of 10^5 $\Omega\cdot$cm for as-deposited ZnO, decreased to 10^3 $\Omega\cdot$cm after rapid photothermal processing. The impact of RPP temperatures was found to have an important role in the formation of ZnO nanostructures properties for solar cells applications and photoluminescence enhancement. The highest intensity of photoluminescence was obtained at 650°C RPP temperature. The experimental results shown that by RPP is possible to control the surface morphology, electrical properties and photoluminescence of nanostructured zinc oxide thin films as active component and antireflection coating of the solar cells.

I. INTRODUCTION

Extensive research has been done on the growth and characterization of zinc oxide (ZnO) semiconducting oxide thin films due to their versatile applications in electronic and optoelectronic devices [1-3]. Zinc oxide nanostructures are generally used in solar cells, as a buffer layers ZnO is a technologically important material who's electrical and optical properties are size and shape –dependent [4]. A large number of methods for the deposition of the high-quality ZnO films have also been published, which are based either on wet chemical [1-3] or vapor phase deposition [4].

S. Shishiyanu, T. Shishiyanu, and O. Lupan are with the Department of Microelectronics and Semiconductor Devices, Faculty of Computers, Informatics and Microelectronics, Technical University of Moldova, Blvd. Stefan cel Mare 168, MD-2004 Chisinau, Moldova, E-mail: sergeteo@mail.utm.md.
R. Singh is with the Center for Silicon Nanoelectronics, Holcombe Department of Electrical and Computer Engineering, Clemson University, Clemson, SC 29634-0915, USA.
E-mail: srajend@ces.clemson.edu

The chemical bath deposition (CBD) of is an aqueous-chemical method which involves the hydrolysis of metal ions in solutions, followed by controlled heterogeneous precipitation on a substrate. The CBD is attractive due to its low cost, simplicity and ease in adaptation to large-area deposition.

Thin film and nanostructures can be deposited at the relatively low temperatures on any substrate material (insoluble) and surface profile. Deposited zinc oxide may be easy controlled by adjusting the composition of chemical solutions and the durations of maintenance of the substrate in bath.

The properties of the zinc oxide films, especially the structural and electrical ones, can be formed and improved by proper post-deposition rapid photothermal processing under suitable conditions - temperature and ambient. Rapid photothermal processing (RPP) is based on rapid radiative heating and cooling of substrates in vacuum or in the presence of inert atmosphere. The RPP systems halogen lamps provide both heating and radiations effects due to the wider spectrum from 0.4 µm to 1 µm. The principal difference between conventional furnace annealing (CFA) and RPP is that the RPP photo-spectrum and CFA are different [5].

In this research, the experimental results of the electrical resistivity, crystalline structure and photoluminescence of zinc oxide nanostructures after post-deposition RPP process have been investigated, and we expect to find out the optimum RPP regimes to obtain the high quality ZnO for solar cell applications.

II. EXPERIMENTAL PROCEDURE

A. Materials

Reagent grade zinc sulphate, natrium hydroxide and ethanolediamine (EN) were used without further purification. Corning glass was used as a substrate. For comparison, some experiments were also performed using silicon substrates. Substrates were first cleaned in dilute HCl (1:5 by volume) for 15 min and then rinsed in deionised water (DI). Second, were rinsed in ethanol acetone (1:1) mixture, DI and dried in an inert gas flux.

B. Deposition Process

Fresh aqueous zinc-complex solution (0.12M) of the 0.5 M zinc sulphate ($ZnSO_4$), natrium hydroxide and

1-4244-0116-X/06/$20.00 ©2006 IEEE   161

ethanolediamine were prepared immediately before deposition process. The addition of ethanolediamine enhances the wetability of the substrates and provides a slow controlled deposition. The freshly prepared solutions were mixed thoroughly until complete dissolution. The starting chemical bath prepared in the described manner initially appeared turbid before the addition of sodium hydroxide.

The cleaned substrate was suspended in 200 ml of the precursor solution (0.12 M zinc complex) and the temperature was kept at room temperature or at 60 °C during the deposition. The thin films were grown for 25 min at 70 °C. and irradiated with ultraviolet (UV) light during the growth process. Irradiation powers was 125-150W employed for 2-5 s time intervals during the growth cycle. After the chemical bath deposition, substrates were recovered from the solution and rinsed with DI water for 30 sec in order to detach unreacted species and reaction byproduct. Finally, the substrates were dried in air at 150 °C for 5 min prior to characterization. The mechanism of ZnO formation by chemical deposition is in our further attentions.

C. Rapid Photothermal Processing

Annealing is the most important process for defect removing and improvement of the zinc oxide properties. Rapid photothermal processing system has been introduced as an alternative thermal annealing equipment solution. The halogen lamp-based RPP system provides short cycle time, reduced exposure and flexibility compared to conventional annealing furnaces. Very strong demand in thermal budget and cycle time reduction make RPP a very popular processing method in recent years [5-7]. Schematic diagram of a RPP system used in these investigations have been presented previously in [8].

The chemical deposited zinc oxide was processed in a RPP IFO-6 system for 15 s at temperatures of 300-700 °C under 1 atm dry N_2 ambient. The comparison of resulting characteristics after RPP shows that the optimal RPP temperature is 650 °C and durations is 15-20 s. In [8] was presented the typical substrate temperature profile. Thus RPP annealed ZnO and the effect of annealing on structural, photoluminescence and electrical properties were studied.

D. Methods of investigations

The surface morphologies of the obtained ZnO nanocomposites were characterized by scanning electron microscopy (SEM) VEGA TS 5130 MM, 20keV equipped with an Energy Dispersive X-ray (EDX) system for chemical composition microanalysis. The ratio of Zn and O in the ZnO was investigated by EDX analysis made in plane detection mode.

To investigate the electrical resistivity two-point probe method where used where ohmic contacts were ensured by the Al film vacuum evaporated.

Photoluminescence (PL) have been excited by the 351.1 nm line of an Ar^+ Spectra Physics laser and analyzed in quasibackscattering geometry through a double spectrometer with 1200 grooves/mm gratings assuring a linear dispersion of 0.8 nm/mm. The spectrometer was equipped with a photomultiplier with SbKNaCs photocathode working in a photon counting mode. The spectral resolution was higher than 0.5 meV. The samples were mounted on the optical cryogenic systems cold station.

III. RESULTS AND DISCUSSIONS

Fig. 1 shows the impact of RPP treatment on the surface morphology of a ZnO chemically deposited under UV irradiation from zinc complex solution 0.12 M, temperature of aqueous solution of anions 98 °C. A continuous coverage of the substrate surface is observed in Fig 1. The as-grown ZnO microcrystallites mean size is ~300nm. RPP leads to the increase of the grain sizes by a factor of 1.5.

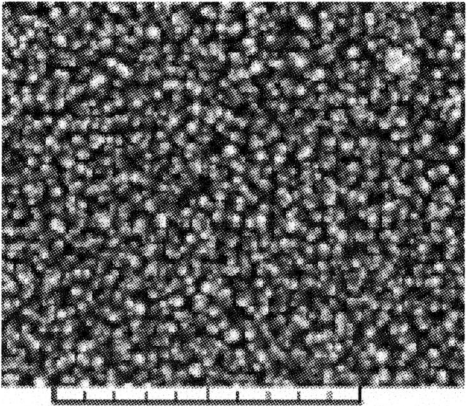

Fig. 1. SEM micrograph of ZnO nanocrystals deposited on Corning glass substrate precipitated from an aqueous zinc complex solution 0.12 M irradiated UV with 150W power for deposition. Samples were subjected to RPP at 650C, 20 sec.

Fig. 2 shows the EDX spectrum of zinc oxide. EDX analysis was performed of the RPP at 650C, 15 s. It is apparent that Zn/O atomic ratios were approximatively 1:1 for scanned within a selected area (50μm×50μm) on the material.

The EDX measurements showed that there was no contamination or impurities in the ZnO and suggest that our technological route is useful for pure ZnO nanocomposite at relatively low temperatures.

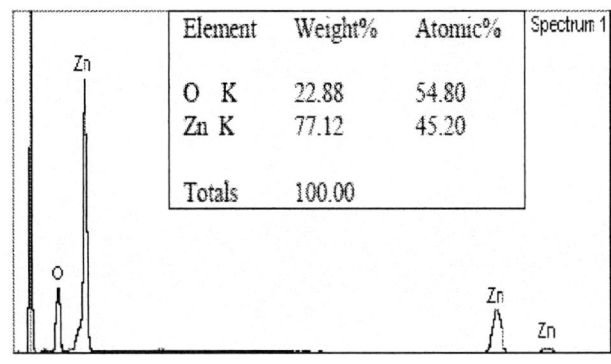

Fig. 2. EDX spectrum of a ZnO on glass obtained by chemical deposition and RPP. The ration of Zn:O is 45:55 (at.%). Only three characteristic X-ray peaks originating from Zn, O are observed; each peak is indexed in the figure.

Fig. 3 shows the electrical resistivities variation with different RPP atmosphere of ZnO films investigated. The experimental results shown that the lowest values of resistivity were for ZnO annealed in mixture of Air + N_2.

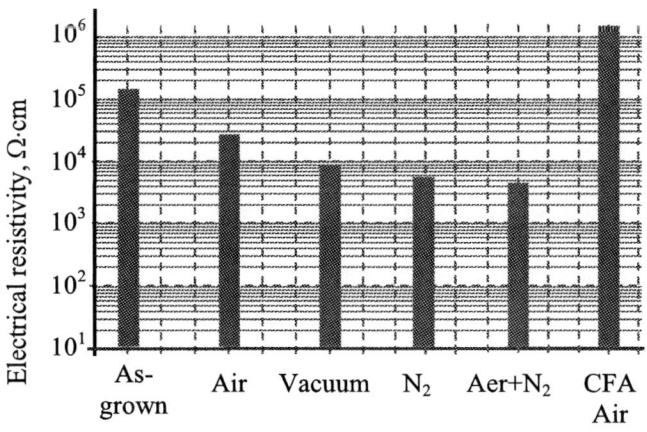

Fig. 3. Variation of the electrical resistivity of ZnO with the RPP treatment atmosphere.

The increase of resistivity value is due to desorption of the O_2 on the grain boundaries, leading to annihilation of oxygen acceptor states at these places, which act as traps for electrons. This desorption mechanism is more evident when samples are rapid photothermal processed in the Air+N_2.

In order to investigate the impact of the RPP on the photoluminescence properties at ZnO we annealed at different RPP temperatures for 15 s (Figs. 4,5): sample 1 is the as grown film; 2—treated by RTA at 400 °C; 3—treated by RTA at 500 °C; 4—treated by RTA at 650 °C (the morphology is presented in Fig. 1)

Fig. 4 shows UV-PL emission spectra of different ZnO nanocomposite fabricated by chemical deposition and rapid photothermal processing measured at 10 K. The emission from the as grown sample is dominated by resonant Raman scattering (RRS) labeled as 2LO – 5LO (longitudinal - optical) peaks, i.e. the near band gap luminescence is weaker that the RRS. Differences between the position of the $n \cdot LO_{RRS}$ peaks and the excitation energy (3.53 eV) is n times the energy of the A_1 (LO) phonon in wurtzite-type ZnO (72 eV). It indicates that the emission is from the n-order Raman scattering. RRS from solids is observed if the energy of the incoming or scattered photons matches real electronic states in the material. Resonant Raman scattering from solids can be observed if the energy of the incoming or scattered photons matches real electronic states in the material. One refers to incoming and outgoing resonance (see, e.g., [9]).

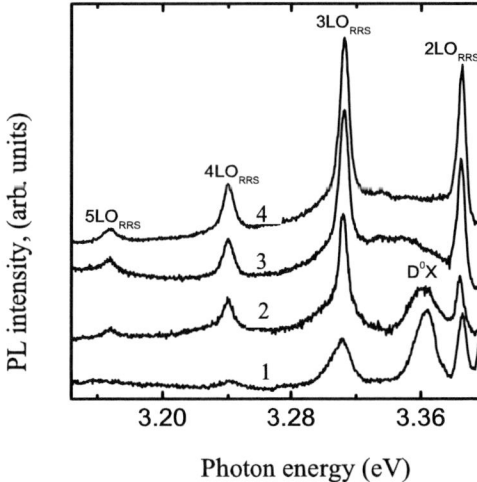

Fig, 4. UV emission spectra of different ZnO nanocomposite fabricated by chemical deposition and rapid photothermal processing measured at 10 K. Numbers in the figure indicate the sample. ZnO films as labeled: sample 1 is the as grown film; 2-treated by RTA at 400 °C; 3-treated by RTA at 500 °C; 4-treated by RTA at 650 °C (the morphology is presented in Fig. 1). The spectra are shifted on the Y axis for the sake of clarity.

Fig. 5 shows the visible PL emission spectra of as grown ZnO and subjected to RPP at different temperatures. The PL spectra of the as-grown ZnO show a typical green-red deep-defect level emission at 1.8 eV without near-band edge emission. One can see that the emission from the as grown film represents a combination of luminescence dominated by a PL band associated with the recombination of excitons bound to a neutral donor (D^0X) and resonant Raman scattering (RRS). Deep-defect-level emission is essentially reduced and newly emerging near-band edge emission is improved in the ZnO RPP annealed at the 400, 500 °C, 15 s (lines 2 and 3). This PL characteristic is signified enhanced after annealing at a RPP temperature of 650 °C (line 4).

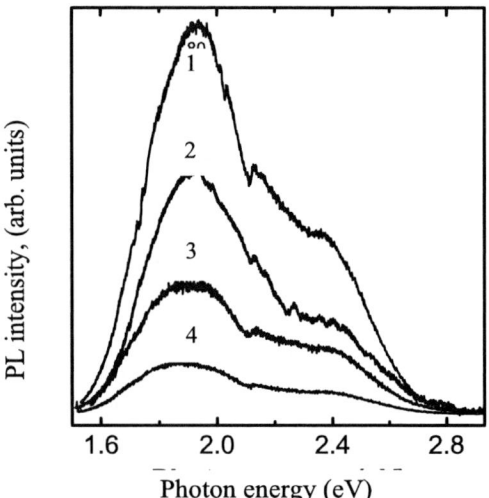

Fig, 5. Visible PL spectra of different ZnO films as labeled: sample 1 is the as grown film; 2—treated by RTA at 400 °C; 3—treated by RTA at 500 °C; 4—treated by RTA at 650 °C (the morphology is presented in Fig. 1) measured at T = 10 K.

The ZnO annealed at 650 °C shows a near-band edge emission at 3.38 eV and much lower emission peak related to deep-defect-level. The reduction of deep-defect-level emission in the rapid photothermal processed ZnO indicate that the concentration of defects responsible for the deep-defect-level emission is reduced by RPP.

The energy position of the luminescence peak in the as grown film (3.363 eV) is close to the previously reported I_4 line (see, e.g., [10]) associated with excitons bound to a H-related neutral donor. It could be that our layers are unintentionally doped with hydrogen. Another possibility is that the D^0X PL band in our films is related to an exciton bound to some structural defects. In such a case, the suppression of the D^0X PL band with increasing the RPP temperature is explained by the annealing of defects which localize the excitons. Apart from this UV luminescence two PL bands centered at 2.4 and 1.8 eV are present in the visible spectral range as illustrated in Fig. 5. The visible luminescence is also strongly suppressed by increasing the RTA temperature and is in accordance with our previously results [11,12].

IV. CONCLUSIONS

In summary, ZnO nanocomposites were successfully synthesized by chemical deposition, UV radiation and rapid photothermal processing for solar cell applications.

EDX and PL measurements showed that there were no contaminated or impurities in the ZnO, but there are deep-level-defect. By annealing with RPP at 650 °C the sample possessed improved electrical and luminescent properties. These results suggest that ZnO can be used as buffer layer in solar cell applications.

ACKNOWLEDGEMENT

The authors wish to acknowledge the financial support of MRDA-CRDF project MOE2-3052-CS-03 and Prof. Dr. hab. I. Tigineanu, Dipl. Eng. E. Monaico for their helpful assistance with SEM and EDX analyses and Dr. hab. V. Ursaki for PL measurements and fruitful discussions.

REFERENCES

[1] B. Rech, O. Kluth, T. Repmann, T. Roschek, J. Springer, J. Müller, F. Finger, H. Stiebig, and H. Wagner, "New materials and deposition techniques for highly efficient silicon thin film solar cells", *Solar Energy Materials and Solar Cells,* 2002, vol. 74, pp. 439-447.

[2] D. Hariskos, S. Spiering, and M. Powalla, "Buffer layers in Cu(In,Ga)Se$_2$ solar cells and modules", *Thin Solid Films,* 2005, vol. 480-481, pp. 99-109.

[3] A. Ennaoui, M. Weber, R. Scheer, and H. J. Lewerenz, "Chemical-bath ZnO buffer layer for CuInS$_2$ thin-film solar cells", *Solar Energy Materials and Solar Cells,* 1998, vol. 54, pp. 277-286.

[4] K.Govender, D.S.Boyle, P.O'Brien, D.Binks, D.West, and D.Coleman, Adv.Mater. 14 (2002) 1221.

[5] R. Singh, M. Fakhruddin, and K. F. Poole, "Rapid photothermal processing as a semiconductor manufacturing technology for the 21st century", *Applied Surface Science*, 2000, vol. 168, pp. 198-203.

[6] S. T. Sisianu, *Nonconventional Technologies in Microelectronics with Rapid Photon Annealing and Stimulated Diffusion,* Tehnica, Chisinau, p. 221 (in Romanian), 1998.

[7] R. Singh, V. Parihar, S. Venkataraman, K. F. Poole, R. P. S. Thakur, and A. Rohatgi, "Changing from rapid thermal processing to rapid photothermal processing: what does it buy for a particular technology?", *Materials Science in Semiconductor Processing,* 1998, vol. 1, pp. 219-230.

[8] S. T. Shishiyanu, O. I. Lupan, T.S. Shishiyanu, V. P. Sontea, and S.K. Railean, "Properties of SiO$_2$ thin films prepared by anodic oxidation under UV illumination and rapid photothermal processing", *Electrochimica Acta*, 2004, vol. 49, pp. 4433–4438.

[9] P. Y. Yu, M. Cardona, *Fundamentals of Semiconductors,* Berlin/Heidelberg/New York (Springer-Verlag) 1996.

[10] B. K. Meyer, H. Alves, D. M. Hofmann, W. Kriegseis, D. Forster, F. Bertram, J. Christen, A. Hoffmann, M. Strassburg, M. Dwworzak, U. Haboeck, and A.V. Rodina, Phys. Status Solidi B 241 (2004) 231.

[11] S. T. Shishiyanu, O. I. Lupan, E. Monaico, V. V. Ursaki, T. S. Shishiyanu, I. M. Tiginyanu. Photoluminescence of chemical bath deposited ZnO:Al films treated by rapid thermal annealing. *Thin Solid Films*, 2005, vol. 488, pp. 15-19.

[12] S. T. Shishiyanu, T. S. Shishiyanu, O. I. Lupan. Nanotechnology for nanostructured and nanocomposite materials fabrication. *MD Patent № 4489, Issued 07/2005.

Session

Power Devices and ICs

2006 25th International Conference on Microelectronics

Evolution of Silicon Power Devices and Challenges to Material Limit

Akio Nakagawa

Abstract The author first briefly reviews recent success of MOS gate power devices. The main objective is to predict, for the first time, the silicon limit characteristics of IGBTs for its on-resistance and SOA. The author also proposes ideal gate drive in order to realize the ultimate limit of high speed switching of MOS gate power devices. The results lead to new FOM, characterizing the high speed switching capability of various power devices.

Silicon devices have still great potential, competing with emerging new material devices.

I. EVOLUTION OF POWER DEVICES

Power devices have evolved so rapidly that 3.3kV IGBTs have even replaced 4.5kV GTOs, which was developed in the late 80's for traction control of bullet trains. Figure 1 and 2 show application fields of power devices in 1997 and 2005, respectively. The distinguished difference of the two figures is that most of the applications of GTO and BTr have been occupied by IGBT and its module.

Figure 3 shows the evolution of high voltage large current power devices in Toshiba. The lifetime of GTOs was as short as only 12 years. Nowadays, MOS gate devices are predominantly used in almost all of the application fields, including LDMOS in power ICs, MOSFETs for low voltage and medium voltage applications and IGBTs for high power applications.

New material SiC and GaN devices are being developed in order to break through the silicon limit. In the mean time, super junction devices were proposed and developed in 1998[1]. The super junction MOSFETs already broke through the so called silicon unipolar device limit in the voltage range from 200V to 700V and significantly enhanced the potential of silicon devices.

Another recent remarkable advancements are high speed trench power MOSFET, intelligent power module(IPM) and power IC technologies. Trench MOSFET switching speed has been greatly improved since 1999 in order to meet the requirement of high efficiency and high di/dt of Voltage Regulator Modules for CPUs. The details are described in Section III. Progress in IPM was already reviewed in Ref.[2]. Power IC technologies are classified into two categories. One is high voltage SOI power ICs[3] for monolithic DC motor control ICs and PDP flat panel display drivers. Figure 4

A.Nakagawa is with Semiconductor Company, Toshiba Corporation, 580-1, Horikawa-Cho, Saiwai-ku, Kawasaki, 212-8520, Japan. E-mail: akio.nakagawa@toshiba.co.jp

shows first 500V 1 and 3 ampere SOI one chip inverter ICs for DC motor control in 1994 and 1999[4]. The other is low voltage power ICs[5]. Figure 5 shows feature size trends in BCD power IC. Fine design rule is now often required in system power ICs for mobile equipments such as cell phones and automotive field.

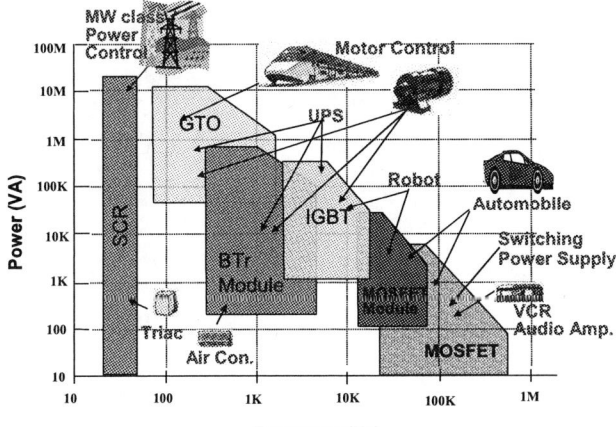

Fig.1 Application fields of power devices in 1997.

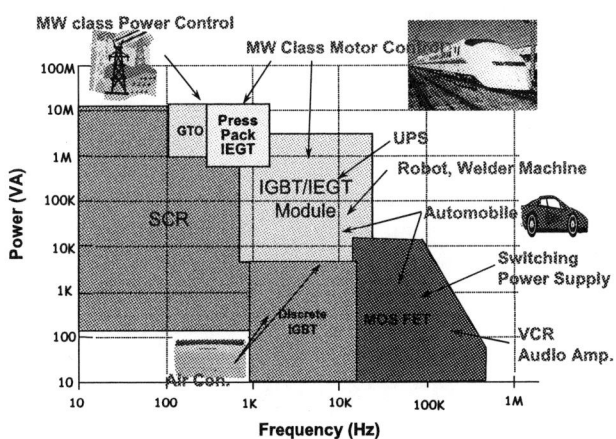

Fig.2 Application fields of power devices in 2005.

Recently, it is often pointed out that silicon devices face the material limit. It is important to make clear the limit characteristics of silicon devices and the future potential of silicon devices to be exploited. The author also proposes new FOM[6] in order to facilitate achieving the limit characteristics of high speed MOSFET.

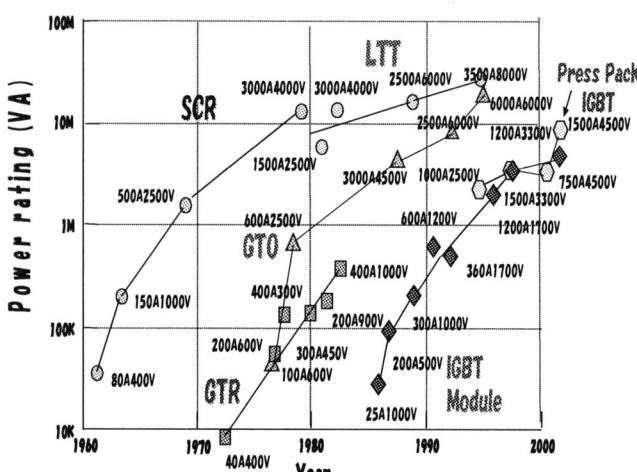

Fig.3 Evolution of power devices In Toshiba

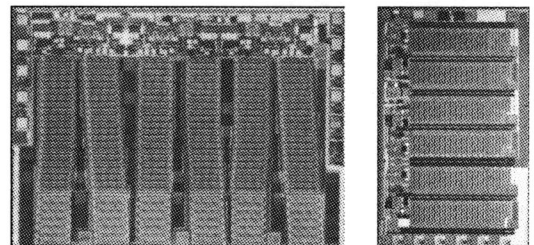

Fig.4 500V 3A&1A single chip inverter IC (After Ref.[4])

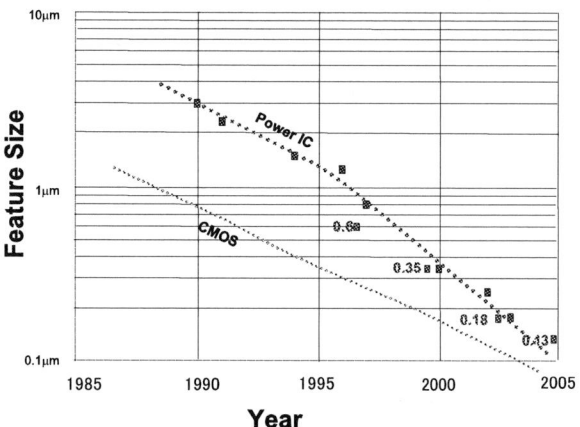

Fig.5 Technology roadmap for BCD power ICs

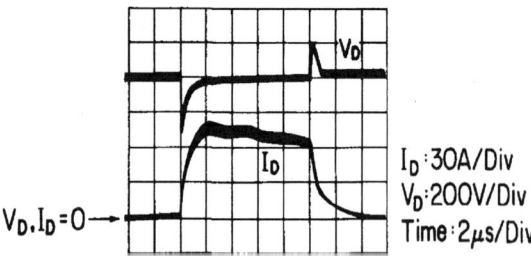

Fig.6 First demonstration of short-circuit withstanding capability of non-latch-up IGBT (after Ref.[12]).

IGBT History

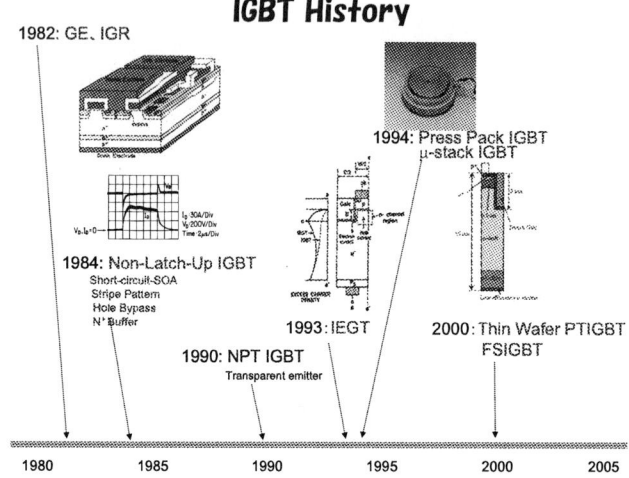

Fig.7 IGBT historical events.

II. IGBTs

A. Brief History of IGBT

Concept of IGBTs was first described in the patent by Becke et al[7]. The actual fabrication was reported by Baliga et al. in 1982[8]. Since then numerous papers were published to improve the device characteristics, such as first switching speed[9,10] and large current capability[11].

In early development stage, IGBTs suffered from the latch-up of the parasitic thyristors and the poor current capability. Non-latch-up IGBTs, satisfying the concept of Becke, were demonstrated for the first time, in 1985[12]. Figure 6 shows the first demonstration of short circuit capability of non-latch-up IGBTs.

IGBT electrical characteristics have been steadily improved. Figure 7 shows major technology achievements in IGBT history. However, there is no prediction of IGBT limit characteristics attributed to silicon material. The author, for the first time, predicts the silicon limit of IGBT.

B. Theory for silicon limit of IGBTs

This section proposes a theory to achieve the lowest forward voltage drop in IGBTs and proposes a new trench gate IEGT/IGBT, realizing the theoretical limit.

The adopted assumption is asymmetrical conduction: "all of the current flows by electrons." Holes contribute only to the conductivity modulation. From the assumption of no hole current flow, the following equations are valid under the high injection condition:

$$J_p = qD_p \frac{\partial p}{\partial x} - q\mu pE = 0 \quad \text{-----Eq.(1),}$$

$$J_n = J_{Total} = 2 \times qD_n \frac{\partial n}{\partial x} \quad \text{----Eq.(2)}$$

$$E = \frac{kT}{q} \frac{1}{n} \frac{\partial n}{\partial x} \quad \text{------Eq.(3)}$$

The situation is satisfied by assuming that the carrier density distribution is approximately a linearly decreasing function from cathode to anode. The current density(J) - voltage(V_F) relation of the proposed IGBT can be derived by integrating Eq.(2) with $D_n = \dfrac{a}{n+b}$, a=3.24e18, b=9.39e16 and appropriate boundary conditions[13].

$$V_F = \frac{2kT}{q}\ln[\frac{1}{n_i}\{(\sqrt{\frac{QJ}{qD_n}}+b)\exp(\frac{JW_n}{2qa})-b\}]$$
$$+ R_{ch}J \quad \ldots\ldots \qquad Eq.(4)$$

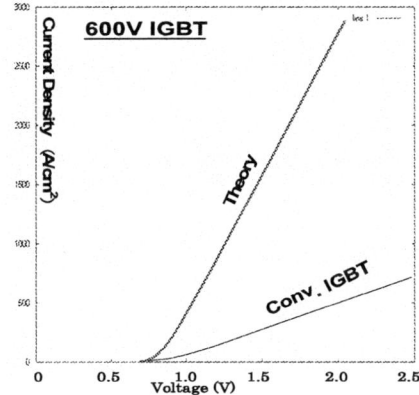

Fig.8   V-I curve comparison between proposed theory and conventional 600V IGBTs.

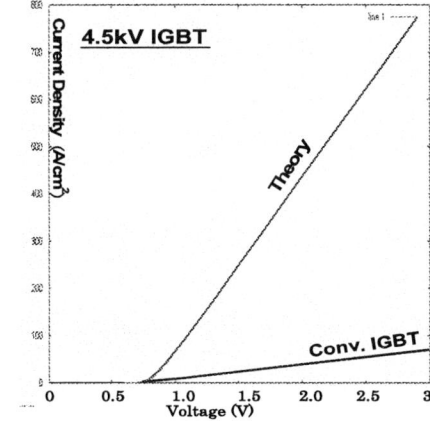

Fig.9   V-I curve comparison between proposed theory and conventional 4.5kV IGBTs.

Calculated V-J curves are shown in Figs. 8 and 9, comparing with those of conventional IGBTs. The proposed asymmetrical conduction IGBTs drastically improve IGBT characteristics.

The author proposes a new IGBT structure (ultimate IEGT) to realize a very high electron injection efficiency in MOS gate structure. If the trench to trench distance (mesa width) is as thin as the thickness of the inversion layer, the two inversion channel layers on the both trench side walls merge and constitute a high concentration N-type layer in the narrow mesa, serving as a barrier for holes. For example, if the mesa width is less than 40nm,

the induced electron density is greater than 1×10^{17}cm^{-3}, and effectively blocks the hole current flow, realizing electron injection efficiency of more than 0.9. The proposed narrow mesa IGBT realizes a low forward voltage even with the p-emitter of very low injection efficiency. The details will be published in ISPSD 2006[14].

Figure 10 compares the on-resistance of the proposed IGBT with state of the art devices. The proposed IGBT successfully reduces its on-resistance to below SiC limit for over 1.5kV.

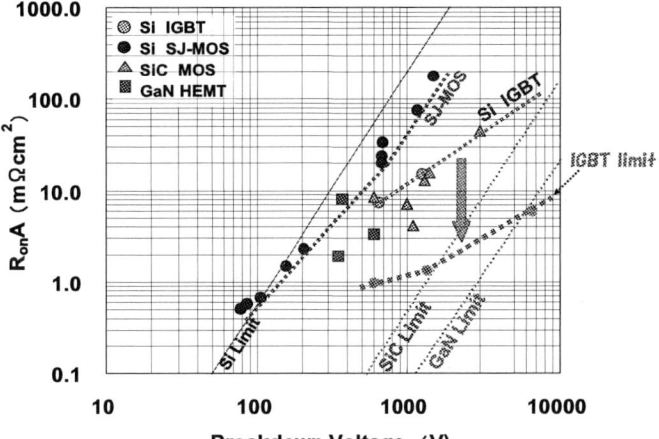

Fig.10 The proposed IGBT, denoted as "IGBT limit," is compared with state of the art devices. Predicted IGBT limit surpasses so called SiC limit for over 1.5kV range.

C. Design for Large Electrical Short-Circuit SOA

Achieving a large safe operating area is one of the big concerns for IGBT development. Short-circuit SOA is especially important for motor control application. In this section, the author shows a theoretical basis that IGBTs have a potential of infinitely large SOA. In fact, in 1996, Hagino et al. reported very high critical power density of 2MW/cm^2[15] for short-circuit withstanding capability. However, no theory has been presented, so far, how to design such large short-circuit SOA in IGBTs.

It is generally a good assumption in PTIGBT that the ratio of the hole current density(J_p) over the total current density(J) does not change throughout the high field region in the n-base. The ratio is equal to the anode efficiency γ if the high field reaches the n-fuffer.

In the present paper, the anode efficiency γ is defined as the ratio of the hole current over the total current at the n-base n-buffer junction, being identical to the product of p-emitter injection efficiency γ_{PE} and transport factor in the n-buffer α_T. The electron and hole densities can be calculated by the following equations in the high field region.

$\gamma = J_p/J$, $\quad$ p=J_p/qv_h, $\quad$ n=J_n/qv_e,

where p and n denote hole and electron densities, v_h and v_e denote hole and electron saturation velocities, respectively. The net charge in the high electric field region ρ is given

169

by Eq.(5) with the donor density N_D.

$$\rho = N_D + p - n = N_D + (\gamma/v_h + (\gamma-1)/v_e)J/q, \quad -----Eq.(5)$$

$$\gamma_c = v_h/(v_h + v_e) \quad \text{(high field case)} \quad -----Eq.(6)$$

If γ is lower than γ_c, the second term (mobile charge) in Eq.(5) is negative. The net charge ρ decreases as the current density J increases, and eventually changes its sign when J exceeds the critical current density J_C[16]:

$$J_C = qN_D/((1-\gamma)/v_e - \gamma/v_h) \quad -------Eq.(7)$$

Once the net charge becomes negative, the peak high electric field appears at the n-base n-buffer junction. It should be noted that the electric field is uniform in the n-base when $J=J_C$.

Figure 11 shows the electric field build-up in the n-base n-buffer junction in a 1200V IGBT as J increases. Avalanche breakdown will take place when the peak electric field in the n-base n-buffer junction exceeds the critical value E_C. This phenomenon is very similar to the second breakdown in npn bipolar transistors.

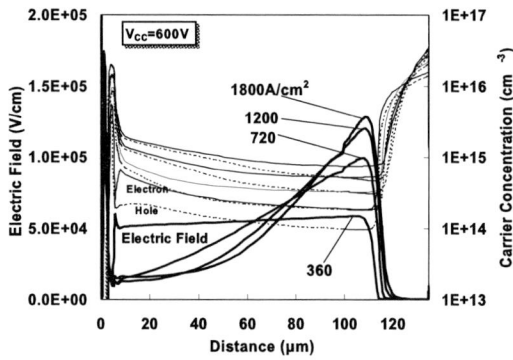

Fig.11 Simulated electric field distributions with current density as a parameter when forward voltage =600V. High electric field appears in n-base n-buffer junction.

SOA locus can be predicted by calculating the $p^+\pi n^+$ diode breakdown voltage, assuming that the impurity concentration of the π-region is the same as ρ given by Eq.(5). It is also assumed that the breakdown occurs at the critical peak field E_C of 1.8×10^5 V/cm. Figure 12 compares analytical results and TCAD results of 600V thin wafer PTIGBTs with a low γ. The arrows indicate the breakdown voltage points, caused by the high electric field at the n-base n-buffer junction. Good agreement is seen for high current density region. For low current density region, estimation of breakdown voltage based on the constant critical field, Ec, is not adequate.

It is predicted that high short-circuit withstanding voltage can be obtained if the saturated collector current is approximately the same as the critical current density, Jc, determined by Eq.(7). This is because the flat uniform electric field distribution is realized in the n-base, and the electric field magnitude can be minimized and the impact ionization can also be minimized.

The value of J_c simply increases as γ approaches γ_c.

Thus, analytically predicted short-circuit SOA increases enormously as γ approaches γ_c as shown in Fig.13. Electrical SOA can be sufficiently large if the device is designed so that the adequate γ is realized.

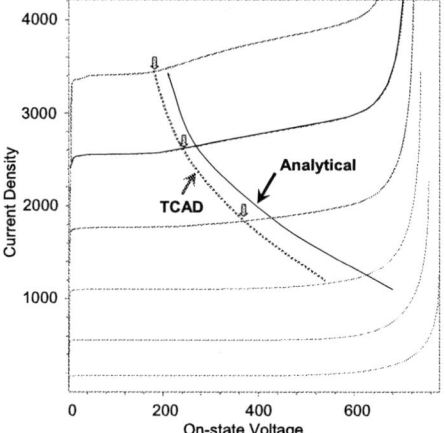

Fig.12 TCAD results are compared with analytical theory, shown by solid line. Dotted line indicates the locus, determined by TCAD, where significant impact ionization at n-base n-buffer junction starts to occur.

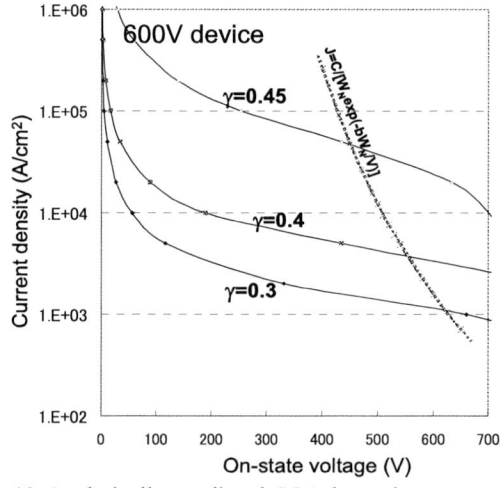

Fig.13 Analytically predicted SOA locus increases significantly as γ approaches γ_c. The dotted line shows the locus $(J=C/[W_N\exp(-bW_N/V)])$, under which no significant impact ionization occurs if γ is optimized.

Maximum SOA locus can be predicted also by calculating the impact ionization current under the assumption that an optimum γ is chosen and that the flat and uniform electric field is realized in the n-base. The maximum SOA is defined as the area where the impact ionization current density, J_{imp} is limited to below a constant value.

$$J_{imp} \cong \int J_e \alpha_\infty \exp(-\frac{b}{E}) \, dx = (1-\gamma)J\alpha_\infty W_N \exp(-\frac{bW_N}{V}) < Const,$$

where W_N denotes the n-base width and the impact ionization by hole current is ignored. If the electric field is constant and sufficiently small, the integral is easily

170

evaluated. The SOA boundary locus is expressed as:

$$J = C/[W_N \exp(-bW_N/V)], \qquad \text{-------Eq.(8)}$$

where W_N is the n-base width, C and b are constants.
In Fig.13, the dotted line shows the SOA locus given by Eq.(8) where J_{imp} is assumed to be approximately $200A/cm^2$, which corresponds to the generation rate of $2.5 \times 10^{23} cm^{-3}$ For example, a point of 500V and $10^4 A/cm^2$ is within the SOA locus and $5 \times 10^6 W/cm^2$ power dissipation can be allowed. More concrete device design based on TCAD will be presented in ISPSD2006[14].

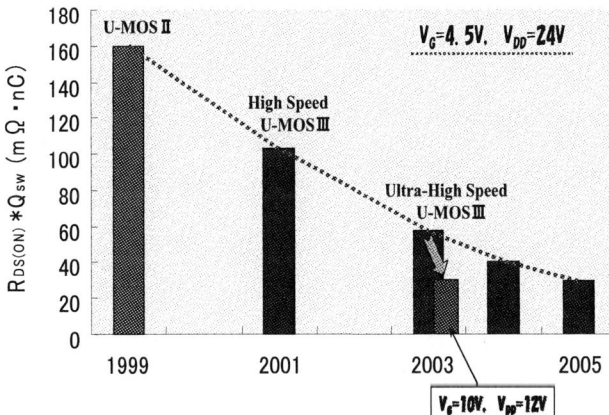

Fig.14 Toshiba's roadmap for high speed MOSFETs

III. POWER MOSFET

A. Recent advancement in MOSFET

Since 1999, switching speed of power MOSFETs has been greatly improved for the application of VRM (Voltage Regulator Module) for CPUs. The FOM of $R_{on}Q_{gd}$ is conventionally adopted for high speed MOSFETs as design guide.

Figure 14 shows the Toshiba's roadmap of 30V power MOSFET. $R_{on}Q_{gd}$ was improved from $160m\Omega nC$ in 1999 to $30m\Omega nC$ in 2005. The buck converter efficiency was improved from 85.5% in 2000 to 90% in 2004.

As the on-resistance of MOSFETs decreases, it is recognized that the package impedance itself occupy a large part of the total on-resistance. Recently, new packages adopt metal ribbons in place of bonding wires, shown in Fig.15.

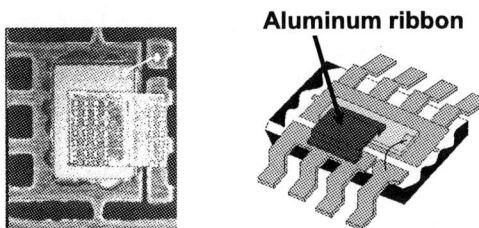

Fig.15 Low impedance package using Aluminum ribbon.

B. Multi-chip Module

If one try to pursue higher efficiency of synchronous buck converters, it is recognized that
(1) reduction in parasitic inductances of the power stage circuits, (2) prevention of self-turn-on of low-side MOSFET, and (3) dead-time optimization are equally important, as compared with the improvement of MOSFETs, themselves.

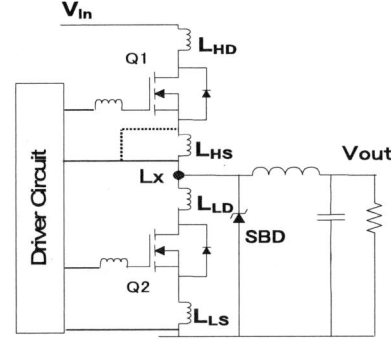

Fig.16 Analyzed circuit of buck converter

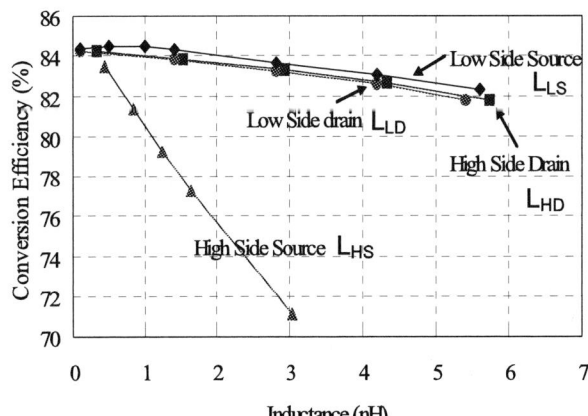

Fig.17 Influence of Parasitic inductances on converter efficiency

Figure 17 shows the influence of each parasitic inductance on buck converter efficiency[17]. The each parasitic inductance is defined in the circuit in Fig.16. The most influential one is the high side MOSFET source inductance, L_{HS}. If the MOSFET is turned-on, the drain current increase rate, dI_D/dt, induces the voltage drop in the parasitic inductance L_{HS}. The voltage applied by the gate driver circuit is the sum of the actually applied MOSFET gate voltage and the voltage drop in the inductance L_{HS}. The high dI_D/dt reduces the actually applied gate-source voltage, resulting in the delayed turn-on. In order to realize the first switching-on, the parasitic source inductance should be minimized. The dotted line in Fig.16 should be adopted for the gate driver ground connection.

The other parasitic inductances increase voltage spike in the switching transients of high side MOSFET and increases the power loss of the high side MOSFET.

The parasitic inductances include the ones inside the package and the ones in the PCB board. In order to reduce the parasitic inductances and resistances, multi-chip module was introduced. Figure 18 shows MCM, called DrMOS, proposed by Intel. Three chips of high-side and low-side MOSFETs and the driver circuit are mounted in the single package, thus minimizing the parasitic impedances. DrMOS improves converter efficiency by 2 or 3% as shown in Fig.19, even if the same rated MOSFET chips are used.

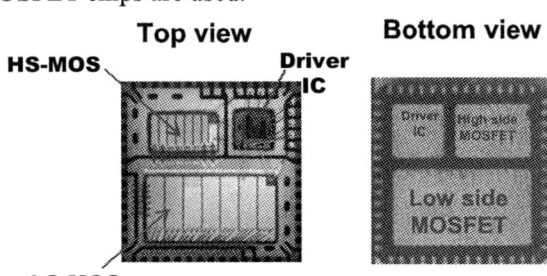

Fig.18 MCM (DrMOS)

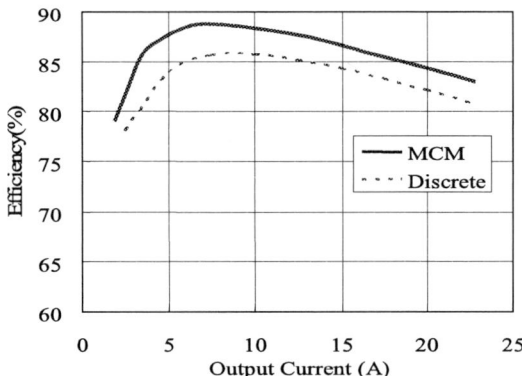

Fig.19 Comparison of converter efficiency between MCM and discrete solution

The parasitic inductances increase the possibility of self-turn-on of the low-side MOSFET. Self-turn-on of the low side MOSFET is closely related to the reverse recovery of the body diode. As shown in Fig.20, the forward voltage, V_{ds}, of the low side MOSFET does not increase immediately after the dV/dt is applied to the MOSFET, but the diode reverse current, I_{bd}, flows (time period t_a-t_b). Although the diode current stops in a short time period, the parasitic inductance tries to keep the current level, imposing a larger dV/dt to the low-side MOSFET (time period t_b-t_c). The actually applied dV/dt to the low side MOSFET is greater than that of originally imposed dV/dt to the node Lx. The dV/dt increases the gate voltage through the C_{gd} and charges the gate capacitance C_{gs}. Thus, it is often indicated that the ratio of C_{gd}/C_{gs} should be small to prevent the self-turn-on.

It should be noted that if the gate drive circuit impedance is sufficiently low, it can be expected that the gate circuit keeps the gate voltage below the threshold, and self-turn-on is prevented.

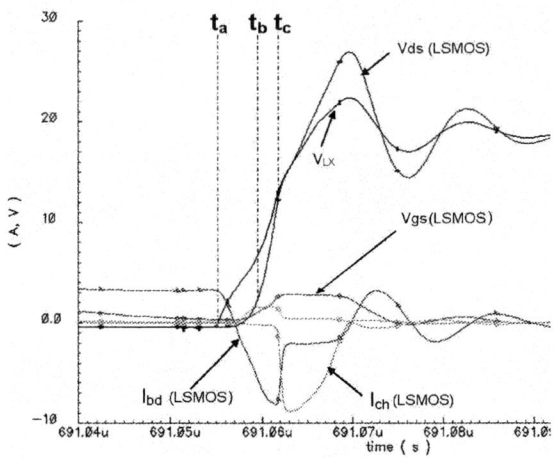

Fig.20 Waveforms explaining self-turn-on

C. Future technology for Multi-Chip-Module

In the conventional gate drive circuit, switching speed is determined by Q_{sw}/I_g.

$$P_{loss} = R_{on}I_D^2 + I_D V_D \frac{Q_{sw}}{I_g} f + \frac{1}{3} Q_{ds} V_D f + Q_G V_G f,$$

----- Eq.(9)

where 1st, 2nd, 3rd and 4th terms show on-state loss, switching loss, main junction capacitance loss and gate charge loss, respectively. The main junction capacitance loss, $Q_{ds}V_D/3$, is added. Here, Q_{ds} denotes output charge, Q_{oss}. The coefficient is 1/3 not 1/2. The reason is described in Appendix.

If the power loss is determined by the first two terms in Eq.(1), the product of R_{on} and Q_{sw} can be used reasonably as figure of merit (FOM) for MOSFETs.

If the gate drive circuit impedance is assumed to be very low and if the value of Q_{sw}/I_g is negligibly small, the 2nd term in Eq.(9) disappears and the switching loss is determined only by the main junction capacitance and the 3rd term expresses the switching-off loss as shown in Eq.(10).

$$P_{loss} = R_{on}I_D^2 + \frac{1}{3} Q_{ds} V_D f + Q_G V_G f$$ -------Eq.(10)

Again, if the gate loss assumed to be ignored compared with the first two terms, $R_{on}Q_{ds}$ is regarded reasonably as new FOM[6].

Figure 21 shows the MOSFET switching-off simulation results using the very low impedance gate drive. The gate and the source is shunted with a 1mΩ resistor. The turn-off time is only 2ns, which corresponds to the Q_{ds}/I_D value. There is no plateau in the gate voltage waveform, originating from Q_{gd} value. The switching-off loss can be minimized by the low impedance gate drive[18].

The low impedance gate drive is especially effective for MCM with low voltage MOSFETs or monolithic solution, because much faster switching of 2nsec will be realized by MCM or ICs. It should also be emphasized that the self-turn-on of the low side synchronous MOSFET can be prevented by the low impedance gate drive.

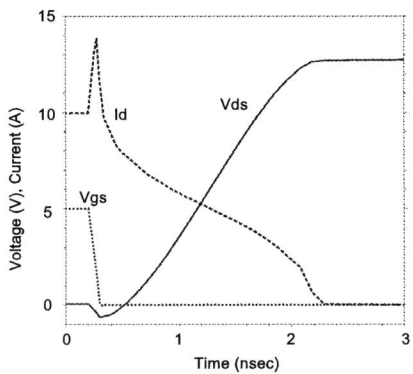

Fig.21 TCAD results of MOSFET Turn-off with ideal gate drive circuit. The turn-off time is 2 nsec.

The low impedance gate drive proposed in this section is explained in the following way: If the channel inversion layer still remains and conducts electron current after the depletion layer is formed in the main junction in the turn-off transient, joule loss occurs in the depletion layer. This makes the major switching loss and is expressed by the 2nd term in Eq.(9). The concept proposed in this section is that the MOS gate channel should cease before the main junction starts to recover so that no joule loss occurs in the switching-off transient but just the main junction capacitance is charged or the main junction recovers. The charged main junction capacitance is discharged and joule loss occurs in the turn-on transient.

It should be noted that the switching-off loss, $Q_{ds}V_D/3$, does not depend on the magnitude of the drain current. This is the distinguished difference from the conventional switching, whose switching time depends on the Q_{gd} value, which increases as the drain current increases.

It should be emphasized that it is difficult to reduce the turn-on loss even by the proposed gate drive method. This is because the drain current starts to flow through the main junction depletion layer immediately after the channel inversion layer is formed. This makes joule loss until the device forward voltage becomes low enough. The device forward voltage is determined by the outside circuit condition. Thus, the total power loss is expressed by Eq.(11).

$$P_{loss} = R_{on}I_D^2 + \frac{1}{3}\underbrace{Q_{ds}V_D f}_{joule\text{-}loss} + p_{turn\text{-}on} f + Q_G V_G f \quad \text{----Eq.(11)}$$

Figure 22 compares the efficiency of DCDC converters for the two cases, where conventional gate drive circuit and the extremely low impedance gate drive circuit are used. In the ideal gate drive condition, the efficiency will improve and achieve more than 90% at 20A output current even if the same MOSFETs are used. These results imply that efficiency in DC-DC converters is still expected to be improved in future. Thus, low voltage

trench MOSFETs will be still mainstream for these applications.

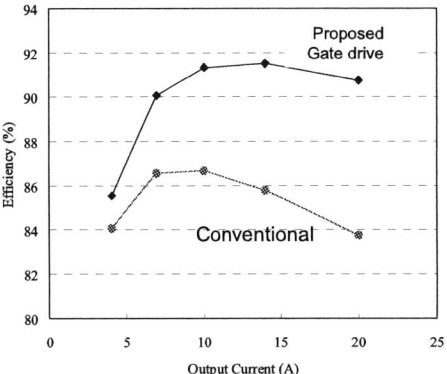

Fig.22 Prediction of converter efficiency using ideal gate drive

It is definitely important to reduce parasitic stray inductances in the power stage circuits and to adopt voltage clamping method in order to reduce voltage spikes caused by the high speed switching. The combination of low impedance gate drive and MCM or monolithic IC technology will provide the solution. A good method to realize the ideal gate drive circuit of very low impedance is to integrate the driver circuit within the power MOSFET chip itself. This can be easily realized using lateral MOSFET and BCD power IC technology. High speed switching can be easily realized in the integrated solution as seen in Fig.23.

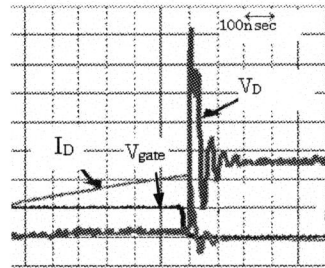

Fig.23 Integrated driver realizes high speed switching.

D. New FOM

In this section, we introduce a new figure of merit(NFOM) [6] based on the discussions in the previous section.

$$\text{NFOM} = R_{on}Q_{str} = T_{sw}V_F \quad \text{-------Eq.(12)}$$

Q_{str} is stored carrier quantity between the drain and the source. For MOSFETs, Q_{str} is equivalent to Q_{ds}. Switching time, T_{sw}, is expressed as follows:

$$T_{sw} = \frac{Q_{str}}{I_D}, \quad V_F = R_{on}I_D \quad \text{-------Eq.(13)}$$

Q_{str} and R_{on} are represented by the following equations, assuming ideal R_{on} and the applied voltage being near the breakdown voltage of the device:

$$Q_{str}=\varepsilon E_C \qquad\qquad ------Eq.(14)$$
$$Ron=4V_{BD}^2/\varepsilon\mu E_C^3 \qquad ------Eq.(15)$$
$$NFOM=4V_{BD}^2/\mu Ec^2 =4V_{BD}^2/BHFOM \quad ---Eq.(16)$$

It can be shown that NFOM is closely related to the BHFOM under special assumptions. NFOM can be defined specifically for each device including bipolar device by using the stored carriers, Q_{str}, and the on-resistance. This feature is the distinguished difference from the BHFOM.

It should be noted that Q_{str} depends on the operating condition just like Q_{gd}. Equation (14) assumes that the applied voltage is the same as the breakdown voltage.

As for 30V silicon MOSFET, device simulation shows that the turn-off time is expected to be 2 nsec by ideal gate drive. This value coincides with the value calculated from Q_{str}/I_D.

Figure 24 shows the comparison of NFOM among silicon devices and new material devices. NFOM of IGBTs depends on the design of the devices. The squares show the NFOM of the IGBTs described in Section II. The circles show high speed IGBTs, having flat carrier density distribution. Super-junction MOSFETs, shown by circles, show the ideal simulation results. SiC MOSFETs, shown by triangles, are plotted, using the product of the reported on-resistance and εE_c. NFOM values of GaN FETs, shown by squares, are calculated by multiplying the reported on-resistance and two-dimensional electron gas density of 1×10^{13}cm^{-2}.

For less then 100 V, NFOM of conventional silicon MOSFET is even superior to any devices shown in the figure and have the potential of fastest switching speed among the devices, although gate loss is not considered. The reported SiC MOSFET is not far better than silicon MOSFETs, because the currently available on-resistance is not sufficiently low from the view point of NFOM.

APPENDIX

The stored energy in the main junction, P_{loss}, in the turn-off transient can be estimated assuming step junction approximation and inductive load, where the drain current, I_D, keeps in the same level within the turn-off transient. Using the voltage, V, as a function of depletion layer width, d, the final result is easily derived in the following.

$$V=\frac{qN_D}{2\varepsilon}d^2, \quad I_D\delta t=qN_D\delta d, \quad V=\frac{I_D^2}{2\varepsilon qN_D}t^2, \quad V_D=\frac{I_D^2}{2\varepsilon qN_D}t_s^2,$$

where t_s denotes the switching time.

$$Q_{ds}=I_Dt_s=(2\varepsilon qN_DV_D)^{\frac{1}{2}} \quad P_{loss}=\int_0^{t_s}VI_Ddt=\frac{I_D^3}{6\varepsilon qN_D}t_s^3=\frac{1}{3}Q_{ds}V_D$$

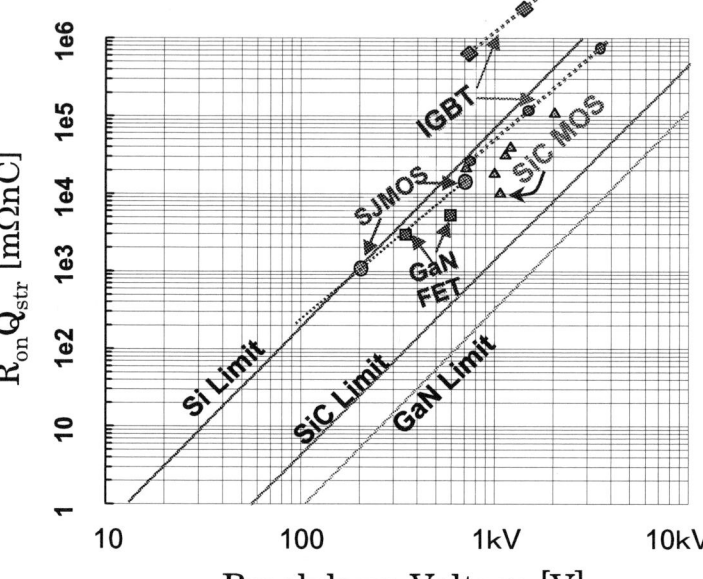

Fig.24 $R_{on}Q_{str}$ as function of breakdown voltage

ACKNOWLEDGEMENT

The author would like to thank Vice-President K. Tani, Technology Executives M. Hideshima, K. Murakami and Dr. K. Morizuka for their support for this work, and Mr. T. Kawano, Mr. K. Nishitani, Mr. M. Yamaguchi, Mr. N. Yasuhara, Mr. Y. Kawaguchi, Mr. K. Nakamura and Mr. S. Ono for their contribution to this article.

REFERENCES

[1] G. Deboy et. al. : IEDM Tech. Digest, p. 683, 1998
[2] G.Majumdar, 2004 PESC Record, p.10
[3] A.Nakagawa et. al., Proc. of ISPSD, p.97(1990)
[4] A.Nakagawa et. al., Proc. of ISPSD, pp.321-324.(1999)
[5] T. Efland, Proc. of ISPSD, p.2 (2003)
[6] A. Nakagawa et. al., IEEJ Journal,Vol.125, p.758(2005)
[7] H.W.Becke et. al., USP 4364073(1982)
[8] B.Baliga et. al., IEEE IEDM Tech. Digest, p.264(1982)
[9] J.P.Russel et al., IEEE EDL, p.63(1983)
    A.M.Goodman et. al., IEEE IEDM Tech. Digest, p.79(1983)
[10] M.F. Chang et. al., IEEE IEDM Tech. Digest,p. 83(1983)
[11] A.Nakagawa et. al., Ext. Abst. of SSDM, p.309(1984)
[12] A.Nakagawa et. al., IEEE IEDM Tech. Digest, p.150(1985)
[13] M.Naito et al., IEEE Trans. ED-28, p.231(1981)
[14] A. Nakagawa, to appear in Proc. of ISPSD'06
[15] H. Hagino et. al., IEEE Trans. ED-43, 490(1996)
[16] A. Nakagawa et. al., Proc. of ISPSD, p.103(2004)
[17] Y. Kawaguchi et. al. : Proc. of ISPSD 2005, p. 371
[18] M.Tsukuda et. al., Proc. of IPEC 2005, p.118

2006 25th International Conference on Microelectronics

High Speed Electro-Thermal Models for Inverter Simulations

P.A. Mawby, A.T. Bryant, P.R. Palmer, E. Santi, and J.L. Hudgins

Abstract— This paper presents the application of advanced compact models of the IGBT and PIN diode to the full electrothermal system simulation of a hybrid electric vehicle converter using a look-up table of device losses. The Fourier-based solution model is used, which takes account of features such as local lifetime control and field-stop technology. Device and circuit parameters are extracted from experimental waveforms and device structural data. Matching of the switching waveforms and the resulting generation of the look-up table is presented. An example of the use of the look-up tables in simulation of inverter device temperatures is also given, for a hypothetical electric vehicle subjected to an urban driving cycle.

Keywords- Compact models, IGBT, Diode, Thermal modeling, Power electronics.

I. INTRODUCTION

In power converter design, the devices should be specifically optimized to the application. The high power densities found in IGBT power modules subjects the power devices to high thermal stress. Knowledge of the transient device temperature profile is required to assist in the design of the devices and inverter.

Traditionally, transient simulations of device temperature have been achieved using physics-based compact device models. To this end, accurate physics-based electro-thermal compact device models have recently been developed for the IGBT and diode [1]–[3], using the Fourier-based solution of the ambipolar diffusion equation [4]. These may be used for circuit simulation and device and circuit optimization [5], [6], and are supported by an extensive parameter extraction process [7]. The accuracy of the models has been validated across a wide range of conditions and temperatures and for a range of device structures, including lifetime zoning and field-stop technology.

However, a number of difficulties arise with simulation of power converter systems. An important example is the hybrid electric vehicle. The device temperature varies as the load on the converter changes with the driving conditions.

P.A. Mawby and A.T. Bryant are with School of Engineering University of Warwick Coventry CV4 7AL, U.K., E-mail: P.A.Mawby@warwick.ac.uk

P.R. Palmer is with Department of Engineering, University of Cambridge, Trumpington Street, Cambridge CB2 1PZ, U.K.

E. Santi is with Department of Electrical Engineering, University of South Carolina, Columbia, SC 29208

J.L. Hudgins is with Department of Electrical Engineering University of Nebraska Lincoln, NE 68588-0511

Full simulation of every switching event within the load cycle is impractical since many thousands of switching events must be simulated.

This paper presents a solution to this problem, where a range of switching cycles under various currents and temperatures are pre-calculated using the detailed simulation. As the device conditions may not exactly coincide with any of the conditions used in the pre-calculation, it will be necessary to develop look-up tables to allow interpolation between values, for a particular device design. It will be shown that the look-up table developed is a rapid way of evaluating the switching performance of an IGBT/diode combination, as the values are used in a rapid simulation of the converter and drive [8]–[10]. In this manner, a range of IGBT and diode designs may be assessed, without the run times becoming excessive. Results for an implementation in Simulink will be presented for the hybrid electric vehicle drive.

II. DEVICE MODELS

The behaviour of conductivity modulated devices, such as PIN diodes and IGBTs, depends heavily on the excess carrier distribution in the wide lightly-doped drift region. The charge profile has a one-dimensional form over most of its volume; therefore a one-dimensional solution is adequate for the bulk of the device. Space-charge neutrality is maintained with the majority carrier profile closely matching the minority carrier profile (quasi-neutrality). Under these conditions, and assuming high-level injection, the charge dynamics are described by the ambipolar diffusion equation (ADE):

$$D \frac{\partial^2 p(x,t)}{\partial x^2} = \frac{p(x,t)}{\tau_{HL}} + \frac{\partial p(x,t)}{\partial t}. \qquad (1)$$

where $p(x,t)$ is the ambipolar (excess) carrier density, D is the ambipolar diffusivity and τ_{HL} is the high-level carrier lifetime.

A Fourier-based solution for this equation has been developed [4]. The carrier density $p(x,t)$ is represented as a sum of Fourier series components in space:

$$p(x,t) = \sum_{k=0}^{\infty} p_k(t) \cos\left(\frac{k\pi(x - x_1)}{x_2 - x_1} \right). \qquad (2)$$

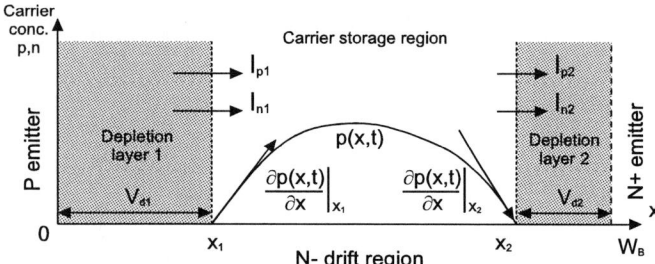

Figure 1. General arrangement of the carrier storage region and depletion layers in the drift (base) region, showing the excess carrier density p(x; t) and the depletion layers.

where k is the harmonic number. The boundaries x_1 and x_2 define the edges of the carrier storage region. Fig. 1 shows the general arrangement of the model. In the Fourier-based solution, the ADE (a second-order partial differential equation) is converted into a set of ordinary differential equations:

for $k > 0$:

$$\frac{2D}{(x_2 - x_1)}\left[\frac{\partial p}{\partial x}\bigg|_{x_2}(-1)^k - \frac{\partial p}{\partial x}\bigg|_{x_1}\right] = \frac{dp_k}{dt} + p_k\left[\frac{1}{\tau_{HL}} + \frac{D\pi^2 k^2}{(x_2 - x_1)^2}\right]$$

$$\cdot + \frac{2}{(x_2 - x_1)}\left(\sum_{\substack{n=1 \\ n\neq k}}^{\infty}\frac{n^2 p_n}{n^2 - k^2}\left[\frac{dx_1}{dt} - (-1)^{n+k}\frac{dx_2}{dt}\right] + \frac{p_k}{4}\left[\frac{dx_1}{dt} - \frac{dx_2}{dt}\right]\right). \quad (3)$$

for $k = 0$:

$$\frac{D}{(x_2 - x_1)}\left[\frac{\partial p}{\partial x}\bigg|_{x_2} - \frac{\partial p}{\partial x}\bigg|_{x_1}\right] = \frac{dp_0}{dt} + \frac{p_0}{\tau_{HL}}$$

$$+ \frac{1}{(x_2 - x_1)}\sum_{n=1}^{\infty}p_n\left[\frac{dx_1}{dt} - (-1)^n\frac{dx_2}{dt}\right]. \quad (4)$$

These are solved in the simulator, giving the excess carrier density across the undepleted drift region using equation (1).

This representation requires the following boundary conditions: the positions of the boundaries x_1 and x_2, and the hole and electron currents at the boundaries of the region (I_{p1}, I_{n1}, I_{p2}, I_{n2}). The currents, as in fig. 1, set the boundary carrier density gradients:

$$\frac{\partial p}{\partial x}\bigg|_{x_1} = \frac{1}{2qA}\left(\frac{I_{n1}}{D_n} - \frac{I_{p1}}{D_p}\right). \quad (5)$$

$$\frac{\partial p}{\partial x}\bigg|_{x_2} = \frac{1}{2qA}\left(\frac{I_{n2}}{D_n} - \frac{I_{p2}}{D_p}\right). \quad (6)$$

The boundary carrier densities calculated from the Fourier solution, p_{x1} and p_{x2} at x_1 and x_2 respectively, are fed

back to calculate the depletion layer voltages V_{d1} and V_{d2}. For positive values of p_{x1} (or p_{x2}), depletion layers do not exist, so V_{d1} (or V_{d2}) equal zero. Once p_{x1} (or p_{x2}) go negative, V_{d1} (or V_{d2}) become positive as the depletion layers form, and they are related to the boundary carrier densities by a constant factor K, e.g.,

$$V_{d1} = \begin{cases} 0 & \text{if } p_{x1} > 0 \\ -Kp_{x1} & \text{if } p_{x1} \leq 0 \end{cases} \quad (7)$$

This model can be used for both diodes and IGBTs provided appropriate relationships are used to calculate the boundary currents and the boundary positions. The following sections describe the model details specific to each device.

A. Diode Model

Local lifetime control is accounted for by adding extra terms to the ODEs in equations (5,6). This is described in full in [3]. The lifetime profile across the drift region is used to construct a look-up table used in the extended Fourier solution.

The boundary currents are set by the minority emitter recombination currents, I_{n1} and I_{p2} [1], e.g.,

$$I_{n1} = qAh_p p_{x1}^2 = I_{sne}\left(\frac{p_{x1}}{n_i}\right)^2. \quad (8)$$

where h_p is the recombination parameter [11] (this is a scaled equivalent of the minority carrier saturation current, I_{sne}), and p_{x1} is the boundary carrier density at x_1 calculated from the Fourier solution. These, the majority injected currents, I_{p1} and I_{n2}, and the boundary displacement currents, I_{disp1} and I_{disp2}, add up to equal the total anode current, I_A:

$$I_A = I_{n1} + I_{p1} + I_{disp1} = I_{n2} + I_{p2} + I_{disp2}. \quad (9)$$

Correct behaviour of the diode depletion layers must be modeled. Detailed device simulations show that depletion layers at both the PN- and N-N+ junctions form during reverse recovery (these must also be allowed to meet if punch-through occurs), while only that at the PN- junction exists in the off-state. This requires careful implementation, involving integration of the boundary displacement currents to obtain the electric fields, and is described in detail in [3].

The drift region voltage drop, V_B, is calculated numerically by splitting the carrier storage region (x_1 to x_2) into segments in order to integrate the electric field numerically [1]. Any variable N-doping that may be present throughout the drift region is also accounted for at this stage. V_B, the junction voltages, V_{j1} and V_{j2}, and the depletion layer voltages, V_{d1} and V_{d2}, are added to give the diode voltage V_{AK}:

$$V_{AK} = V_{j1} + V_{j2} + V_B - V_{d1} - V_{d2}. \quad (10)$$

B. IGBT Model

For NPT (non-punch-through) IGBTs, the emitter recombination current sets the boundary current at the P+ (anode) end of the IGBT, i.e. at x_1, given in equation (8). For PT (punch-through) and FS (field-stop) IGBTs, the hole current in the buffer layer is calculated to determine the conditions where the N buffer layer meets the N- drift region at x_1. This is based on the expressions in [2] but linearized to improve the convergence:

$$\frac{dQ_H}{dt} = I_C - Q_H \left[\frac{1}{\tau_{pH}} + \frac{2h_p N_H}{W_H} + \frac{2D_{pH}}{W_H{}^2} \right]$$

$$+ qA \left[h_p + \frac{2D_{pH}}{W_H N_H} \right] p_{x1}{}^2. \qquad (11)$$

$$I_{p1} = \frac{2Q_H D_{pH}}{W_H{}^2} - \frac{2qA D_{pH} p_{x1}{}^2}{W_H N_H}. \qquad (12)$$

where Q_H is the hole charge in the buffer layer, N_H and W_H are the doping and width of the buffer layer, D_{pH} and $\square_{pH}$ are the hole diffusivity and lifetime in the buffer layer, and h_p (equal to $I_{sne}/qAn_i{}^2$) sets the electron recombination from the buffer layer into the P+ emitter. For both NPT and PT/FS IGBTs, the boundary x_1 is always zero during the on-state or forward blocking and there is no depletion layer ($V_{d1}=0$).

At the cathode end of the IGBT, the electron current I_{n2} is set by the MOS channel current using the classic MOSFET equations [1]. The hole current I_{p2} is calculated from the P-well and gate displacement currents, I_{disp2} and I_{CG}, the electron current I_{n2} and the total IGBT current I_C:

$$I_{p2} = I_C - I_{n2} - I_{disp2} - I_{CG}. \qquad (13)$$

During forward operation there is always a depletion layer (voltage V_{d2}) at the PN junction, between the N- drift region and the P-well. This is approximately equal to the MOS channel voltage, and is calculated from the boundary carrier density p_{x2} using equation (7). The corresponding depletion layer width W_{d2} and boundary position x_2 are given by,

$$x_2 = W_B - W_{d2} = W_B - \sqrt{\frac{2\varepsilon V_{d2}}{qN_B + \dfrac{|I_C|}{Av_{sat}}}}. \qquad (14)$$

where W_B and N_B are the drift (base) region width and doping, ε is the silicon permittivity and v_{sat} is the carrier saturation velocity. W_{d2} and V_{d2} are also used to calculate

the displacement currents I_{disp2} and I_{CG}, further details are given in [1].

The drift region voltage drop, V_B, is calculated as for the diode [1] and used with the junction voltage drop V_{j1} (and V_{j0} in the PT/FS case [2]) to calculate the total IGBT voltage, V_{CE},

$$V_{CE} = V_{j1} + V_B + V_{d2} \left(+ V_{j0} \right). \qquad (15)$$

C. Model Implementation

The diode and IGBT models have been implemented in Pspice (using an equivalent R-C cell circuit to represent the Fourier series solution [1,4]) and MATLAB/Simulink [5]. Extensive validation has been performed over a wide range of temperatures (including cryogenic) [1], [12] and conditions, and for a variety of devices.

The models are implemented in Simulink here because this gives better integration with the inverter simulation, also implemented in MATLAB/Simulink [8]–[10]. The Fourier series solution is also easier to implement in Simulink since its signal vector and matrix capabilities can be used to make the model implementation more compact. It also allows the number of Fourier terms, M, to be changed easily. In both Simulink and Pspice, care must be taken with implementing differentiators (these are necessary in the displacement current calculations and in the Fourier series solution), achieved using frequency-limited differentiators of the form $G(s) = s/(1+s\tau)$. Fig. 2 shows the Simulink NPT IGBT model.

The models also include full temperature dependency. This is explained in more detail in [1], [12]. The parameters which are affected between 250-400 K are listed in table I. The necessary calculations are performed in a MATLAB script file run before the simulation.

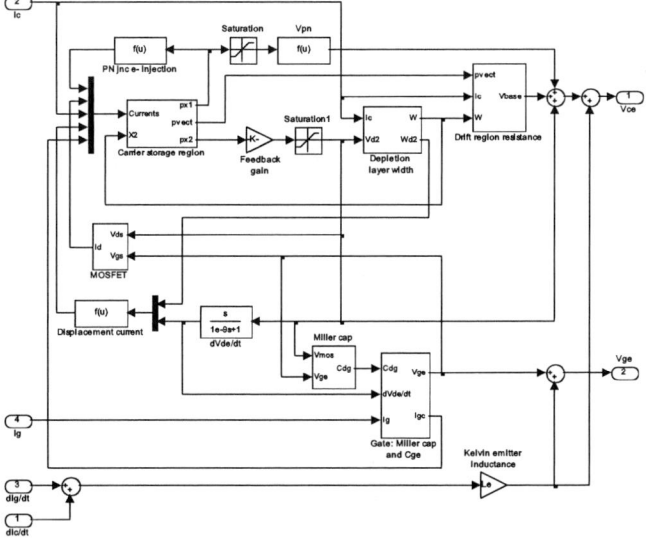

Figure 2. Simulink NPT IGBT model.

TABLE I.
TEMPERATURE DEPENDENT PARAMETERS

Parameter	Temperature dependency
μ_n, μ_p	$\mu_{n,p} = \mu_{n0,p0}\left(\dfrac{300}{T}\right)^{2.5}$
n_i	$n_i = \dfrac{3.88\times10^{16}\,T^{1.5}}{\exp(7000/T)}$
v_{sat}	$v_{sat} = \dfrac{2.4\times10^7}{1+0.8\exp(T/600)}$ [13]
h_p, I_{sne}	$h_p = h_{p0}\left(\dfrac{300}{T}\right)^{2.5} = \dfrac{I_{sne0}}{qAn_i^2}\dfrac{(T/300)^{0.5}}{\exp(14000(1/T-1/300))}$ [14]
τ_{HL}	$\tau_{HL} = \tau_{HL0}\left(\dfrac{T}{300}\right)^{1.5}$
V_{TH}	$V_{TH} = V_{TH0} - 9\times10^{-3}(T-300)$
K_p	$K_p = K_{p0}\left(\dfrac{300}{T}\right)^{0.8}$

TABLE II.
DEVICE PARAMETERS AND INITIAL ESTIMATION

Device	Parameters	Estimation Method
Diode	Area, A	Forward DC current, I_F Max. current density, J
	Lifetime, τ_{HL}	Forward DC current, I_F Reverse recovery charge, Q_{RR}
	Base width, W_B	Breakdown voltage, V_{BD} Ionisation coefficients, a, b
	Base doping, N_B	Estimated $\sim 1\times10^{14}$ cm^{-3}
	Recombination parameter, h_p	Estimated $1\text{-}10\times10^{-14}$ cm^{4}s^{-1}
IGBT	Area, A	Forward DC current, I_F Max. current density, J
	Lifetime, τ_{HL}	Decay rate of tail current
	Base width, W_B	Breakdown voltage, V_{BD}
	Base doping, N_B	Variation of capacitances C_{rss}, C_{oss} with voltage V_{CE}
	Recomb. parameter, h_p (NPT)	Initial tail current
	Intercell ratio, a_i	Capacitances C_{rss}, C_{oss}
	MOS conduatnce, K_p	Datasheet I-V plots
	Gate-emitter capacitance, C_{GE}	Datasheet value
	Oxide capacitance, C_{OX}	Capacitance C_{rss}
	Threshold voltage, V_{TH}	Datasheet value
	Buffer layer parameters (PT/FS): $W_H, N_H, \tau_{pH}, h_p(I_{sne})$	Tail current shape
Circuit	Stray inductance, L_S	V_{CE} overshoot and dI_C/dt at IGBT inductive turn-off
	Emitter inductance, L_E	Estimated 2-10 nHcm^2 Match at IGBT turn-on
	Diode inductance, L_D	Reverse recovery
	Gate inductance, L_G	Estimated 50-100 nH

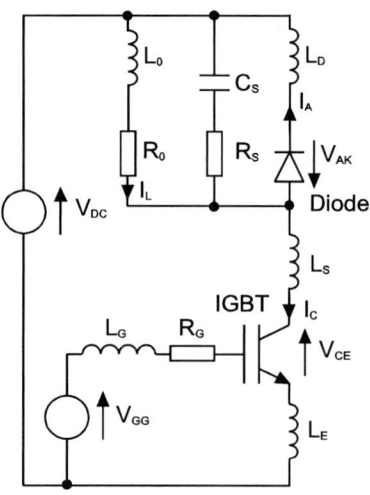

Figure 3. Chopper cell circuit used to simulate inductive switching. R$_0$;L$_0$: load, R$_S$;C$_S$: snubber, L$_S$: primary stray inductance, L$_D$: diode inductance, L$_E$: emitter Kelvin inductance, R$_G$;L$_G$: gate resistance and inductance.

D. Parameter Extraction of Devices and Circuit

The devices are simulated in a chopper cell circuit, shown in fig. 3, which includes all necessary circuit and parasitic components to capture inductive switching. The parameter extraction methods in [7] were used with experimental waveforms and device structural data to complete the modeling process. The parameter extraction methods are summarized in table II.

III. INVERTER SIMULATION

In [8]–[10], the system-level inverter electrical simulation uses a greatly simplified inverter, which generates the switching patterns required for electro-thermal device simulation. Simplified models of the heatsink, motor and drive are also used. The switching patterns and switching conditions are then used to access the look-up table of device losses. This method allows very large time steps to be made, so that the transient device temperature profile may be

quickly calculated throughout the entire load cycle. In the case of an electric vehicle this could be a standard driving cycle, e.g. the Federal Urban Driving Schedule (FUDS).

The switching conditions required in the look-up table are the duty ratio of the IGBT ρ, the load current I_L (this is assumed constant throughout one switching cycle due to the high load inductance), and the device junction temperature T_j. In [8]–[10], the look-up table was generated using datasheet and experimental values, while here it is generated using the device models in section II. The supply voltage, V_{DC}, and switching frequency, f_{SW}, are fixed, although in some cases they may not be, such as matrix converters or where "gear-changing" modulation is used.

The look-up table (of average power dissipation per switching cycle in this case) is then generated by running the simulation repeatedly through the range of switching conditions expected. This is achieved using MATLAB code to drive the Simulink model and to extract the necessary losses from the device waveforms.

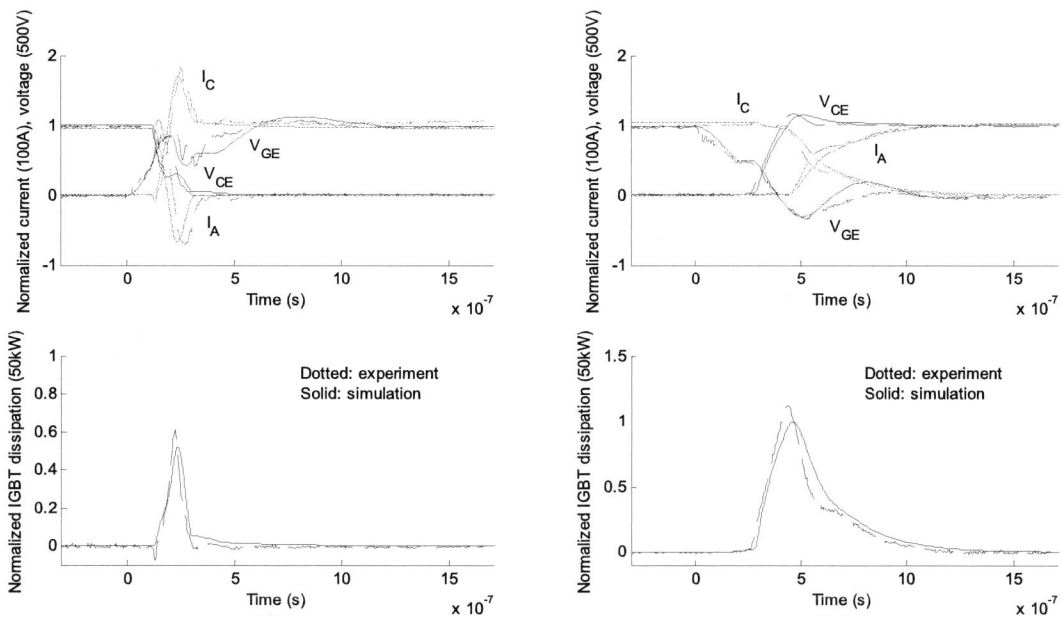

Figure 4. Matching of inductive switching waveforms for the IGBT and diode at 25 °C. The scaling for the IGBT gate voltage, V_{GE}, is 15V in the on-state.

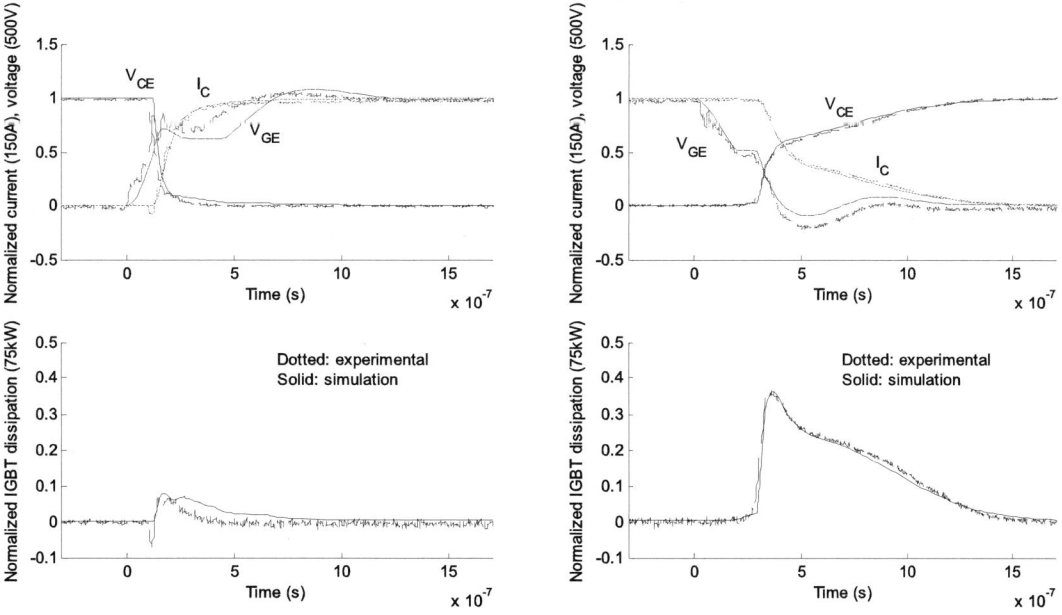

Figure 5. Matching of resistive switching waveforms for the IGBT and diode at 25 °C. The scaling for the IGBT gate voltage, V_{GE}, is 15V in the on-state.

IV. RESULTS

A. Device Simulation

Figs. 4 and 5 show the matching of the device switching waveforms for inductive and resistive switching respectively. The diode and IGBT use lifetime zoning and field-stop technology respectively to achieve the required trade-off between switching losses and forward voltage drop. Figs. 6 and 7 show the matching of the device forward (on-state) voltages for the diode and IGBT respectively.

B. Look-up Tables

The look-up tables generated from device simulations are shown in figs. 8 and 9 for the diode and IGBT respectively. The resulting points are shown as a three-dimensional scatter plot, with colour representing the average power dissipation (the switching frequency is 1 kHz) against the duty ratio, load current and junction temperature. Sample switching waveforms for the points in the look-up tables are given in figs. 9 and 10, showing clearly the effects of load current and temperature on the switching performance. The generation of the look-up table required approximately 5 minutes to complete.

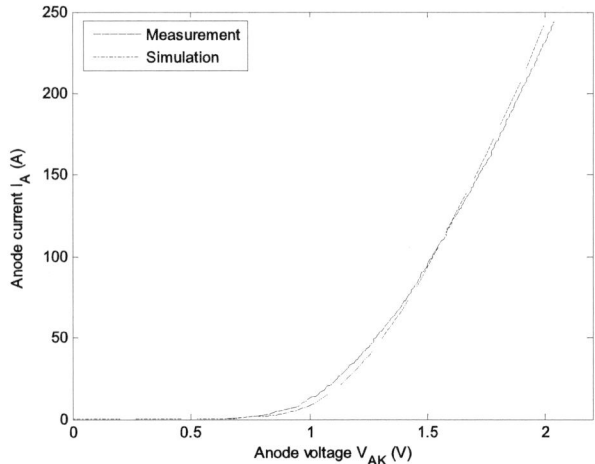

Figure 6. Matching of the diode forward voltage at 25 °C.

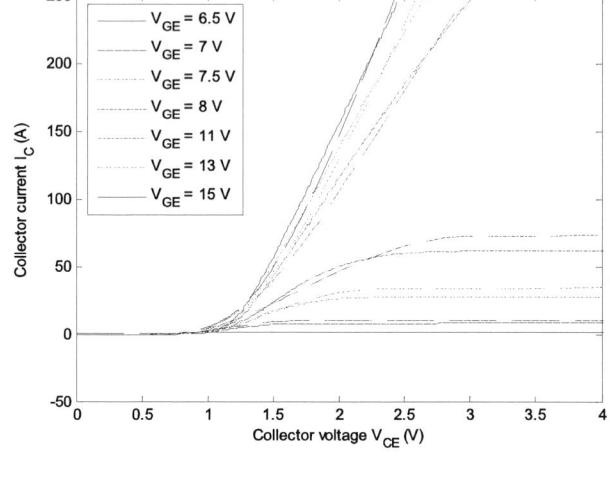

Figure 7. Matching of the IGBT forward voltage at 25 °C (dotted: simulation, solid: measurement).

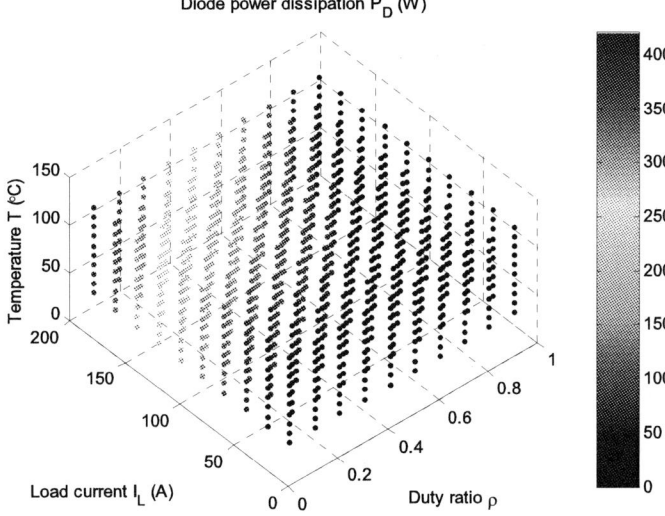

Figure 8. Average diode power losses plotted as a function of duty ratio, load current and temperature.

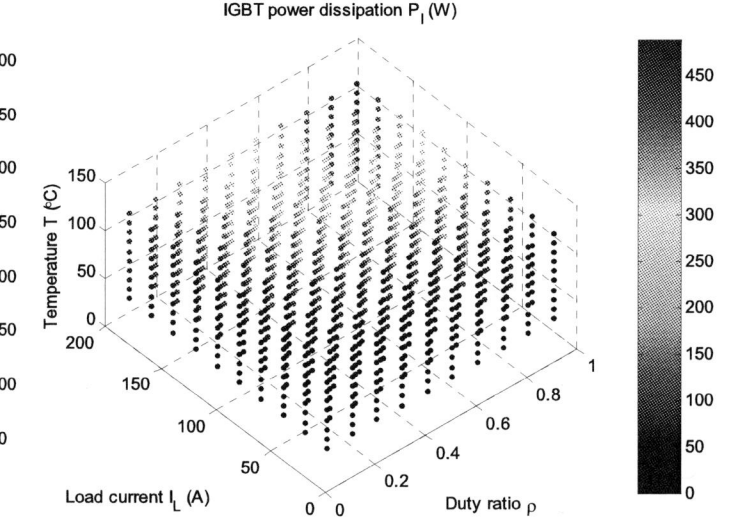

Figure 9. Average IGBT power losses plotted as a function of duty ratio, load current and temperature

C. Inverter Simulation

The look-up tables generated using the device simulations were then used in an inverter simulation. The inverter conditions were generated using a simple drive simulation of an electric vehicle subjected to the Federal Urban Driving Schedule (FUDS) [15], [16], and applied to the inverter model shown described in section III. The resulting ranges of the rms inverter output voltage and current were 100V and 56A respectively. The ambient temperature was set to 26 °C, with a simple model of a suitably-sized heatsink used to cool the devices.

The resulting transient temperature profiles for the devices are shown in figs 12-14. The complete inverter simulation required approximately 4½ minutes to complete using a Pentium IV processor.

V. DISCUSSION

The matching with experimental waveforms is good, particularly for the instantaneous IGBT power dissipation, indicating the accurate prediction of switching losses. The matching is also accurate for the devices' forward voltage drops, showing that the models are equally capable of simulating conduction losses.

The device temperature throughout the load cycle, fig. 12, increases significantly when the vehicle accelerates. Fig. 13 shows that the load conditions determine the balance between diode and IGBT losses, as there are occasions when the IGBT losses exceed those of the diode. In fig. 14 the instantaneous device temperature clearly varies during each modulation cycle, at the same frequency as the modulation (motor supply) frequency. This detail in the transient device temperature profile is useful in analyzing the effect on the

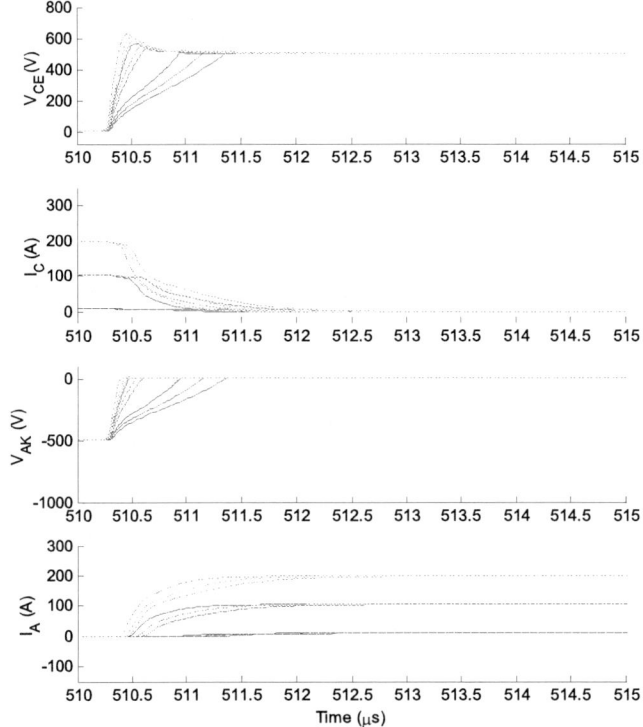

Figure 10.   IGBT and diode switching waveforms for different load currents and temperatures, used to generate the look-up tables. Waveforms shown are for IGBT turn-off/diode turn-on, referring to IGBT voltage, IGBT current, diode voltage and diode current respectively.

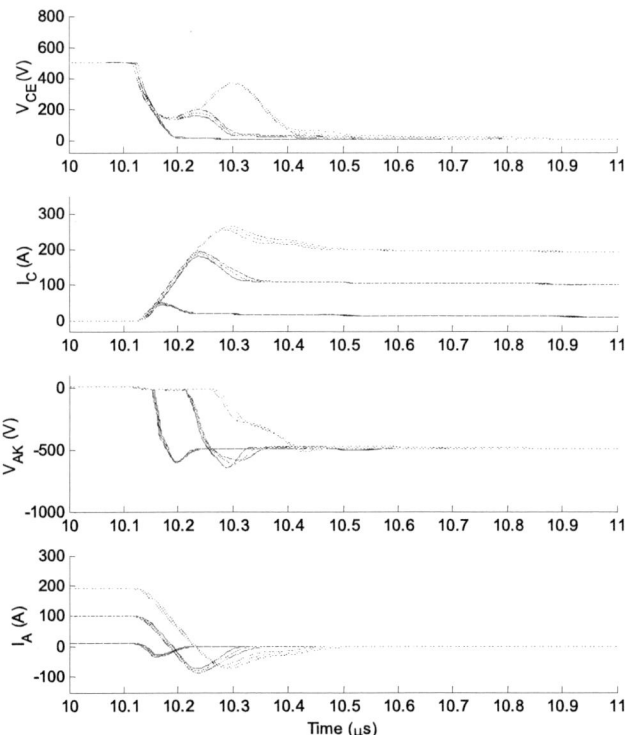

Figure 11.   IGBT and diode switching waveforms for different load currents and temperatures, used to generate the look-up tables. Waveforms shown are for IGBT turn-on/diode turn-off, referring to IGBT voltage, IGBT current, diode voltage and diode current respectively.

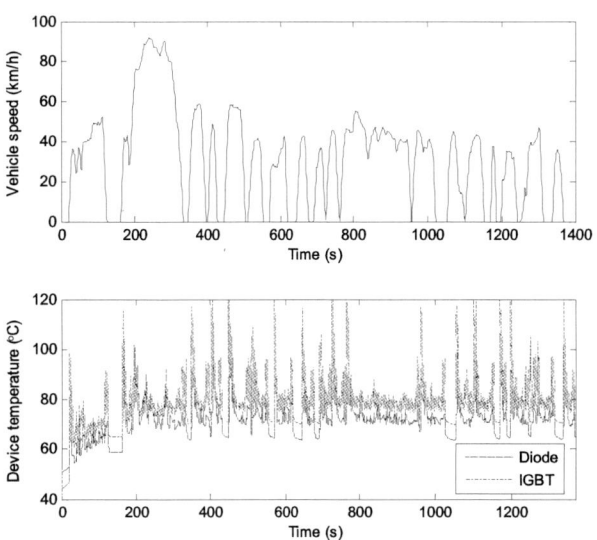

Figure 12.   Diode and IGBT junction temperatures during inverter simulation, shown in relation to the vehicle speed for the whole driving cycle.

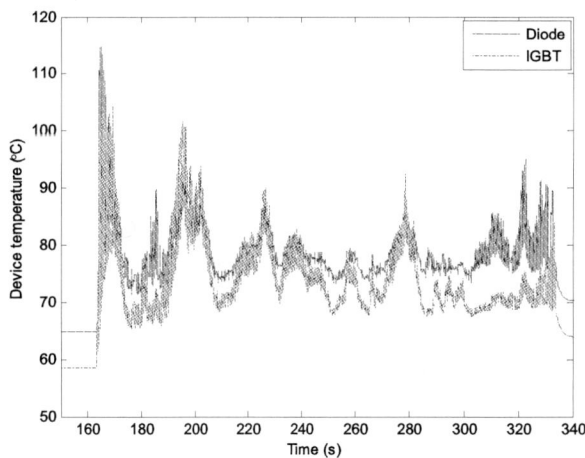

Figure 13.   Diode and IGBT junction temperatures during inverter simulation, shown for the high-speed section of the driving cycle.

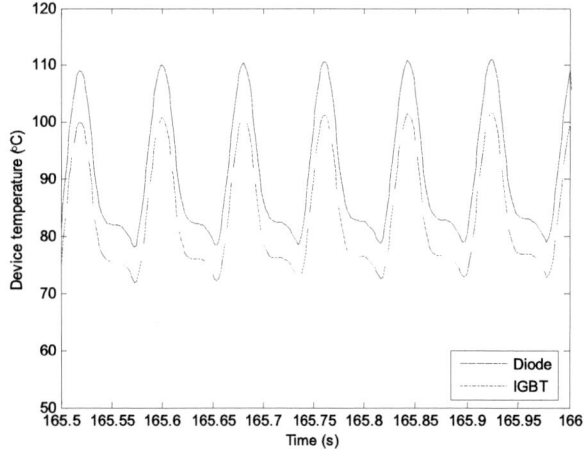

Figure 14.   Diode and IGBT junction temperatures during inverter simulation, shown at a low inverter modulation frequency during vehicle acceleration.

device reliability, for example the fatigue of bond wires and solder through thermal expansion [17]. This detail would not be accessible using just compact device simulation, since the simulation time would not be feasible (in the order of months to simulate every switching event in the load cycle). The simulation of the complete load cycle, including accurate simulation of device conduction and switching using the compact models, in only 10 minutes offers a powerful tool for device and inverter design, also making feasible its use in numerical optimization.

VI. CONCLUSIONS

The compact device models presented here, in conjunction with fast inverter simulation, have allowed for the first time quantitative and accurate estimates to be made of the trade-offs in converter design, such as size, cost and reliability.

Compact and accurate device models have been presented, with full temperature dependency and advanced features such as lifetime zoning and field-stop technology. Simulation waveforms obtained using the models match experimental measurements well for both switching and during the on-state. Thus the same models may be used with confidence for estimating both switching and conduction losses.

Look-up tables of device losses were generated as functions of duty ratio, load current and device temperature, for use in fast inverter simulations. Examples of transient device temperature profiles using the look-up tables have also been presented for the inverter of a hypothetical electric vehicle subjected to an urban driving cycle. The ability of the simulations to obtain detailed and accurate information regarding the device temperature has a potential use in the quantitative estimation of device reliability.

ACKNOWLEDGEMENTS

The authors would like to thank the INTRINSIC project for its kind financial assistance.

REFERENCES

[1] P.R. Palmer, E. Santi, J.L Hudgins, X. Kang, J.C. Joyce, and P.Y. Eng, "Circuit simulator models for the diode and IGBT with full temperature dependent features", IEEE Trans. Power Electronics, 18(5):1220–1229, September 2003.

[2] X. Kang, A. Caiafa, E. Santi, J.L. Hudgins, and P.R. Palmer, "Characterization and modeling of high-voltage field-stop IGBTs",

IEEE Trans. Industry Applications, 39(4):922–928, July/August 2003.

[3] A.T. Bryant, P.R. Palmer, E. Santi, and J.L. Hudgins, "A compact diode model for the simulation of fast power diodes including the effects of avalanche and carrier lifetime zoning", In PESC Conf. Rec., Recife, June 2005.

[4] Ph. Leturcq, "A study of distributed switching processes in IGBTs and other power bipolar devices", In PESC Conf. Rec., volume 1, pages 139– 147, St. Louis, 1997.

[5] P.R. Palmer, A.T. Bryant, J.L. Hudgins, and E. Santi, "Simulation and optimisation of diode and IGBT interaction in a chopper cell using MATLAB and Simulink", In IAS Conf. Rec., Pittsburgh, October 2002.

[6] A.T. Bryant, Y. Wang, S.J. Finney, T.C. Lim, and P.R. Palmer, "Numerical optimization of an active voltage controller for series IGBTs", In PESC Conf. Rec., Recife, June 2005.

[7] X. Kang, A. Caiafa, E. Santi, J. Hudgins, and P.R. Palmer, "Parameter extraction for a physics-based circuit simulator IGBT model", In APEC Conf. Rec., pages 946–952, Miami, February 2003.

[8] Z. Zhou, M. S. Khanniche, P. Igic, S. T. Kong, and P. A. Mawby, "Large time-scale electrothermal simulation model of power inverters for electrical vehicle applications", In Proc. International Power Electronics Conference, pages 1101–1106, Niigata, Japan, April 2005.

[9] Z. Zhou, M. S. Khanniche, P. Igic, S. T. Kong, M. Towers, and P. A. Mawby, "A fast power loss calculation method for long real time thermal simulation of IGBT modules for a three-phase inverter system", In EPE Conf. Rec., Dresden, September 2005.

[10] Z. Zhou, M.S. Khanniche, P. Igic, S.T. Kong, M. Towers, and P.A. Mawby, "A fast power loss calculation method for long real time thermal simulation of IGBT modules for a three-phase inverter system", Int. J. Numerical Modelling: Electronic Networks, Devices and Fields, 19:33–46, 2006.

[11] H. Schlangenotto and W. Gerlach, "On the effective carrier lifetime in p-s-n rectifiers at high injection levels", Solid-State Electronics, 12:267-275, 1969.

[12] A. Caiafa, X. Kang, E. Santi, J.L. Hudgins, and P.R. Palmer, "Cryogenic study and modeling of IGBTs", In PESC Conf. Rec., Acapulco, June 2003.

[13] Jacoboni, Canali, Ottaviani and Albergi Quaranta, "A Review of some Charge Transport Properties of Silicon", Solid-State Electronics, vol. 20, pp. 77-89, 1977.

[14] Hefner, "A Dynamic Electro-Thermal Model for the IGBT", IEEE Trans. Industry Applications, vol. 30, no. 2, pp. 394-405, March/April 1994.

[15] A.T. Bryant, A.R. Bradley, N.A. Parker-Allotey, and P.R. Palmer, "A hardware in the loop device testing system", In EPE Conf. Rec., Toulouse, September 2003.

[16] A.T. Bryant, N-A. Parker-Allotey, and P.R. Palmer, "The use of condition maps in the design and testing of power electronic circuits and devices", In IAS Conf. Rec., Seattle, October 2004.

[17] M. Held, P. Jacob, G. Nicoletti, P. Scacco, and M-H. Poech, "Fast power cycling test for insulated gate bipolar transistor modules in traction applications", Int. J. Electronics, 86(10):1193–1204, 1999.

2006 25th International Conference on Microelectronics

Influence of Small Emitter and Contact Non-Uniformities on the Current Filamentation in 3.3–kV $p^+ - n^- - n^+$ Silicon Diodes

<u>H.P. Felsl</u>, E. Falck, F.-J. Niedernostheide and J. Lutz.

Abstract—We investigate the current filamentation behavior during reverse recovery in high-voltage 3.3–kV silicon $p^+ - n^- - n^+$ diodes with transient S–shape negative differential resistance characteristics. The transient $I - U$–bistability occuring in the reverse recovery period leads to a non-uniform current distribution in the diodes when they are turned off with a high current rate di/dt. In this paper we compare the filamentation behavior of a homogeneous quasi one-dimensional structure with diodes providing small contact and emitter non-uniformities. The formation of these high-current domains is studied by means of isothermal device simulation and the different characteristics are explained by analyzing the transient electric-field and current-density distributions in the devices.

I. Introduction

When diodes are turned off with a high current rate di/dt the transient S-shape negative differential resistance characteristics leads to a non-uniform current distribution in these devices [1]. The basic process leading to the evolution of filaments is dynamic avalanche [2]. We analyzed a homogeneous structure and compared its filamentation behavior with diodes providing small contact non-uniformities and emitter non-uniformities. Characteristic differences concerning the destabilization of the current distribution and the subsequent development of current filaments were found. The results are discussed with regard to the safe operating area (SOA).

II. Device and Circuit Simulation

The analyzed diodes are shown in Figs. 2–6. They have a p^+–emitter with a depth of several μm, a large base width and a shallow n^+n^--emitter structure at the cathode side. The lateral dimension is $b = 1000 \ \mu m$.

Four structures with different non-uniformities in the contact and emitter regions were analyzed (Table I).

Diode $D1$ is equipped with homogeneous contact and emitter regions (Fig. 2). Diode $D2a$ was provided with a 1 μm broad non-uniformity of the anode contact at

H.P. Felsl, E. Falck and F.-J. Niedernostheide are with Infineon Technologies, Am Campeon 1-12, 85579 Munich/Neubiberg, Germany, Fax:+49-89-234-9552291, Phone:+49-89-234-26711, e-mail:Hans-Peter.Felsl@infineon.com.

J. Lutz is with the Chair for Power Electronics, Chemnitz University of Technology.

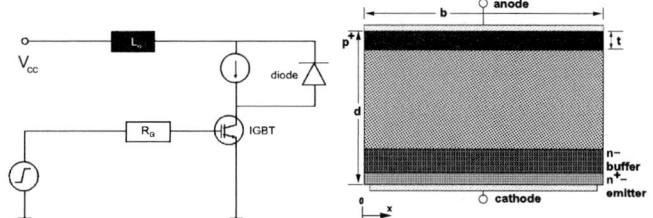

Fig. 1. Circuit for device simulation

Fig. 2. Diode $D1$ without any non-uniformity

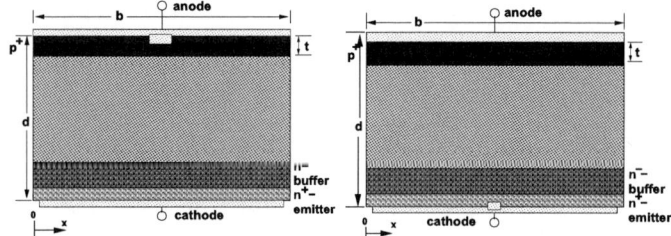

Fig. 3. Diode $D2a$ with a contact non-uniformity at the anode side

Fig. 4. Diode $D2b$ with a contact non-uniformity at the cathode side

the center of the anode (Fig. 3) and covers a part of the anode emitter, so that the anode emitter is reduced at this place. The depth of the non-uniformity in vertical direction was varied between 50 and 300 nm. In diode $D2b$ the same contact non-uniformities are placed at the cathode side of the diode (Fig. 4).

Small local doping variations in the p^+–emitter area were created in $D3$ by an additional gaussian acceptor profile (Fig. 5). The peak concentration was placed in a depth close to the $p^+ - n^-$-junction. Three values for the peak concentrations are assumed $1.0 \times 10^{14} cm^{-3}$, $1.0 \times 10^{15} cm^{-3}$ and $1.0 \times 10^{16} cm^{-3}$, respectively. The full width at half maximun of the added *Gauss* func-

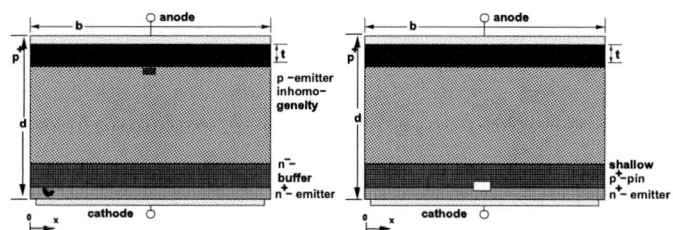

Fig. 5. Diode $D3$ with a doping non-uniformity at the anode side

Fig. 6. Diode $D4$ with a small p^+ area in the n^+n^- backside area

TABLE I
CONSIDERED HOMOGENEOUS AND INHOMOGENEOUS DIODE
STRUCTURES $D1$, $D2a$, $D2b$, $D3$ AND $D4$

	inhomogeniety at	value of varied parameter	type
$D1$	–	–	–
$D2a$	anode-side contact	$50 nm$ $150 nm$ $300 nm$	depth of metal contact
$D2b$	cathode-side contact	$50 nm$ $150 nm$ $300 nm$	depth of metal contact
$D3$	p^+–emitter	$1 \times 10^{14} cm^{-3}$ $1 \times 10^{15} cm^{-3}$ $1 \times 10^{16} cm^{-3}$	doping concentration
$D4$	$n^+ n^-$–area	$1 \times 10^{14} cm^{-3}$ $1 \times 10^{15} cm^{-3}$ $1 \times 10^{16} cm^{-3}$	doping concentration

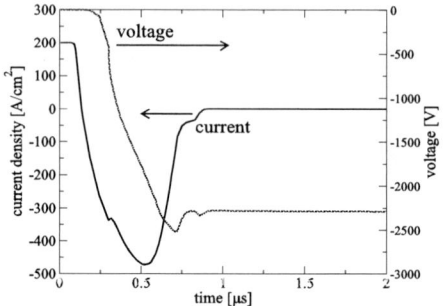

Fig. 7. Reverse recovery characteristic which is qualitatively similar for all diodes

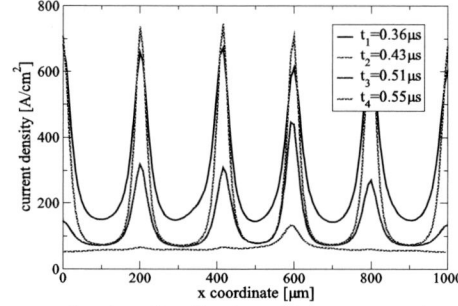

Fig. 8. Current density distribution $J(x, y = 5 \ \mu m)$ at $t = [0.36 \ \mu s, 0.43 \ \mu s, 0.51 \ \mu s, 0.55 \ \mu s]$ in $D1$

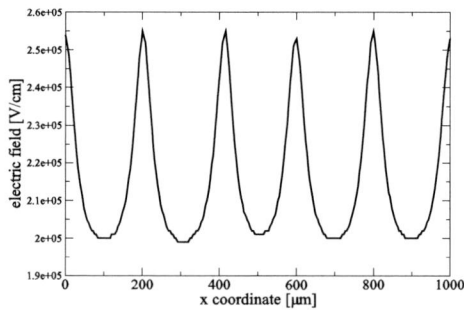

Fig. 9. Electric field distribution $E(x, y = 5 \ \mu m, t = 0.43 \ \mu s)$ in diode $D1$

tion is $1 \ \mu m$ for the lateral and vertical direction.

In diode $D4$ (Fig. 6) an additional gaussian acceptor profile was placed near the cathode side. It was placed in a shallow depth at the $n^+ n^-$-junction. The peak concentrations were $1.0 \times 10^{14} cm^{-3}$, $1.0 \times 10^{15} cm^{-3}$ and $1.0 \times 10^{16} cm^{-3}$, respectively. The added $Gauss$ function caused various small p^+-doped areas, which have equal dimensions in the lateral and vertical direction. Starting from a forward current density of $200 A/cm^2$ the diode was commutated by an IGBT switch (Fig. 1). The current rate of $di/dt \approx 2600 A/\mu s$ was defined by the circuit inductivity L_σ, the battery voltage U_{CC} and the switching behavior of the IGBT. As an example the reverse characteristic of $D2a$ is shown in Fig. 7. The reverse recovery characteristics are qualitatively equal for all conisidered diodes.

The simulations were carried out under isothermal conditions with the device simulator Medici [3].

III. SIMULATION RESULTS

In the diode $D1$ without any non-uniformity (Fig. 2) current filaments induced by dynamic avalanche may occur in the reverse recovery period as described, e.g., in [4] and [5]. Figs. 8 and 9 show lateral cross-sections of the current density distribution $J(x)$ and the electric field distribution $E(x)$, respectively, at a vertical position where the current density had its absolute maximum. The filaments envolving during the whole reverse recovery period are distributed uniformly along the contact area.

Neglecting thermal conduction during the reverse re-

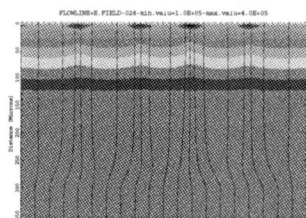

Fig. 10. Current flow lines in the diode $D1$ at $t = 0.43 \ \mu s$, showing the homogeneously distributed filaments

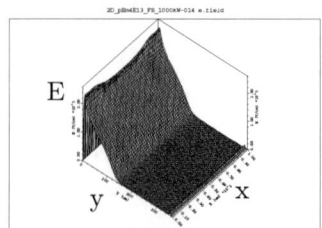

Fig. 11. Electric field distribution $E(x, y, t = 0.302 \ \mu s)$ in diode $D2a$

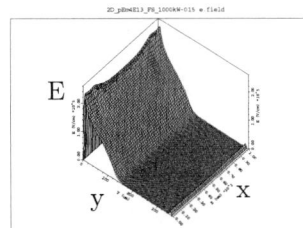

Fig. 12. Electric field distribution $E(x, y, t = 0.304\ \mu s)$ in diode $D2a$

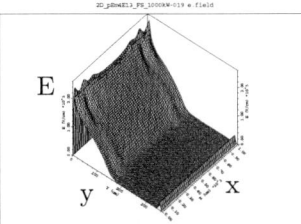

Fig. 13. Electric field distribution $E(x, y, t = 0.346\ \mu s)$ in diode $D2a$

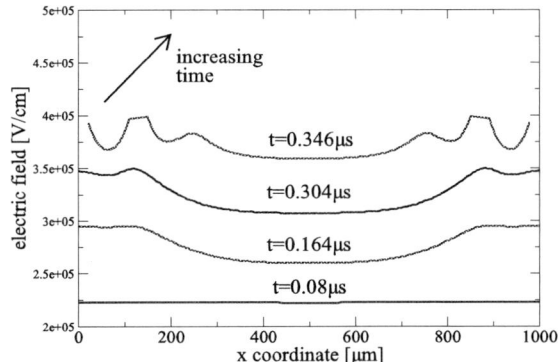

Fig. 14. Electric field distribution $E(x, y = 5.0\ \mu m)$ at different times in diode $D2a$ at the beginning of the reverse recovery phase, the waveforms are shifted by $0.5 \times 10^5\ V/cm$ for a better illustration

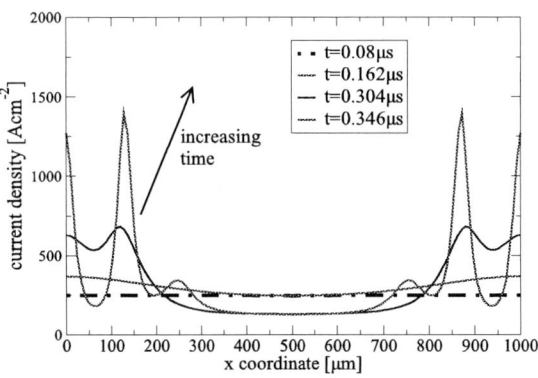

Fig. 15. Current density distribution $J(x, y = 0.350\ \mu m)$ at different times in diode $D2a$ at the beginning of the reverse recovery phase

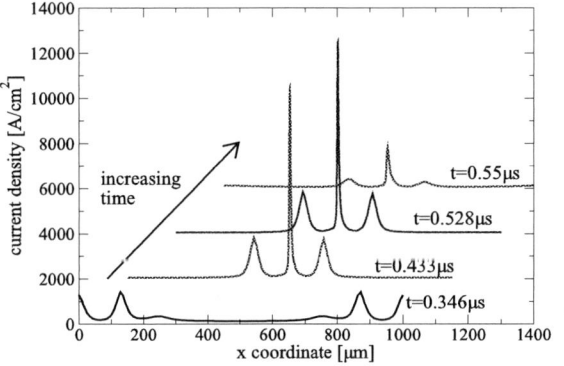

Fig. 16. Current density distribution $J(x, y = 5\ \mu m)$ at different times in diode $D2a$, the waveforms are shifted by $+2000 A/cm^2$ for better illustration

covery period, we can estimate the maximal temperature increase in a filament by using the relation:

$$c_{Si} \cdot \Delta T = P_H \cdot \Delta t \qquad (1)$$

where c_{Si} is the heat capacity of silicon, ΔT the heat increase, P_H the dissipated heat power which is correlated to $\vec{j} \cdot \vec{E}$, and Δt the heat duration. Assuming a duration of $0.3\ \mu s$ for the filamentation we obtain $\Delta T = 45\ K$.

The thermal limit is determined on the one hand by the intrinsic temperature T_i and on the other hand by the thermo-mechanical properties of the contacts. T_i of the low-doped base region is $T \approx 500K$. For the metal contacts we assumed a maximum temperature of $T = 700K$. These temperatures are not exceeded for initial device temperatures up to $T = 400\ K$.

In the diode $D2a$ provided with a contact non-uniformity at the anode side we found a different behavior. At the beginning of the recovery phase at $t = 0.346\ \mu s$ filament domains envolve close to the diode edges, but not directly in the center of the diode where the non-uniformity was located (Figs. 17-18).

The reason for this is that at the contact non-uniformity the extraction of the charge carrier plasma is slighly favoured at the beginning of the reverse recovery phase (Fig. 15). As a consequence, steepening of

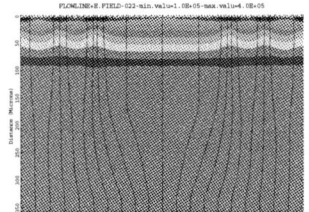

Fig. 17. Current flow lines in the diode $D2a$ at $t = 0.36\ \mu s$, showing the inhomogeneously distributed filaments

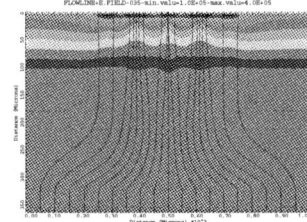

Fig. 18. Current flow lines in the diode $D2a$ at $t = 0.55\ \mu s$, showing the inhomogeneously distributed filaments

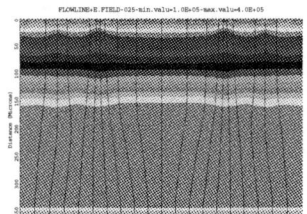

Fig. 19. Current flow lines in the diode $D2b$ at $t = 0.40\ \mu s$, showing the inhomogeneously distributed filaments at the beginning of the filamentation period

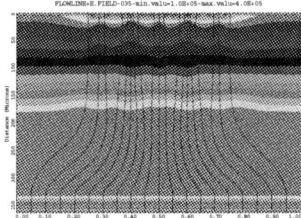

Fig. 20. Current flow lines in the diode $D2b$ at $t = 0.49\ \mu s$, showing the inhomogeneously distributed filaments at the end of the filamentation period

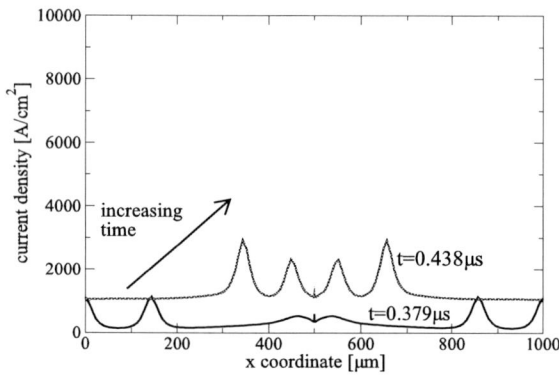

Fig. 21. Current density distribution $J(x, y = 5\ \mu m)$ at the beginning and the end of the filamentation phase in diode $D3$, waveform for $t = 0.438\ \mu s$ is shifted by $1000 A/cm^2$ for better illustration

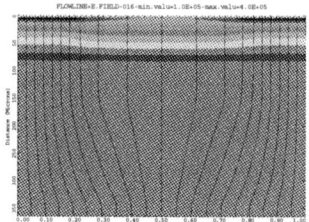 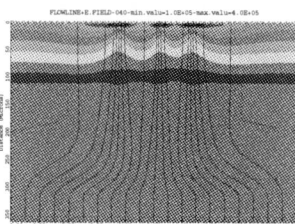

Fig. 22. Current flow lines in the diode $D3$ at $t = 0.31\ \mu s$, showing the inhomogeneously distributed filaments at the beginning of the filamentation, p^+-doping: $1.0 \times 10^{15} cm^{-3}$

Fig. 23. Current flow lines in the diode $D3$ at $t = 0.52\ \mu s$, showing the inhomogeneously distributed filaments at the end of the filamentation, p^+-doping: $1.0 \times 10^{15} cm^{-3}$

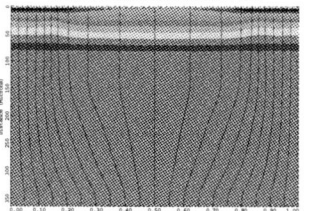 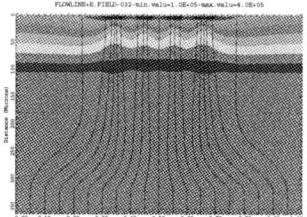

Fig. 24. Current flow lines in the diode $D3$ at $t = 0.31\ \mu s$, showing the inhomogeneously distributed filaments at the start of the filamentation, p^+-doping: $1.0 \times 10^{16} cm^{-3}$

Fig. 25. Current flow lines in the diode $D3$ at $t = 0.52\ \mu s$, showing the inhomogeneously distributed filaments at the end of the filamentation, p^+-doping: $1.0 \times 10^{16} cm^{-3}$

the electric field strength in the center region is lower than in the edge regions (Figs. 11, 12, 13, and 14).

Since the total current is distributed to a smaller number of filaments compared to the diode $D1$ the maximum current densities in these filaments in the edge regions in the diode $D2a$ were higher (Fig.8 and Fig.15).

The high current density, in turn causes a strong plasma extraction in the edge region, so that the filaments there disappear over time. Instead, filament formation in the center of the device is now favoured (Fig.16 and Fig.18). During this phase the contact non-uniformity acts similar to a point contact where the electric field strength is higher than in the surrounding. So filaments form in the center of the diode.

A similar scenario was found also for the two other cases with reduced depth of the additional local anode contact (150 and 50 nm). For a depth of 10 nm the behavior was comparable to that observed in diode $D1$. Since the maximum current density may exceed $\approx 1.0 \times 10^4 Acm^{-2}$ for a duration of $\Delta t = 0.1\ \mu s$, we conclude that contact non-uniformities with a depth higher than 50 nm may be critical for diode $D2a$.

If the non-uniformities are placed at the cathode side of the diode, we find a similar filamentation behavior (Figs. 19 and 20).

Diode $D3$ was provided with a doping non-uniformity in the p^+-emitter region. For a maximum acceptor concentration of $1.0 \times 10^{14} cm^{-3}$ inside the defection area we found a similar behavior as in diode $D1$. For maximum doping concentrations of $1.0 \times 10^{15} cm^{-3}$ and $1.0 \times 10^{16} cm^{-3}$ the filament formation begins close to the edge regions and ends in the center area, (Figs. 21-25).

The higher doping area in the p^+-emitter leads to a lowering of the electric field strength at the beginning of the reverse recovery phase and to an increase in the center region afterwards, when the filaments near the edge

disappeared due to the local strong plasma extraction. The behavior is qualitatively similar to that observed in the diodes with contact non-uniformities. However, since the maximum current density in the filament is lower than $J = 2000 A/cm^2$, these perturbations seem not to be critical with respect to the SOA.

For diode $D4$ with a buried p^+-acceptor profile in the $n^+ n^-$ emitter structure we found a filamentation behavior different from diode $D1$ only for a maximum doping concentration of $1.0 \times 10^{16} cm^{-3}$. For maximum concentrations of $1.0 \times 10^{14} cm^{-3}$ and $1.0 \times 10^{15} cm^{-3}$ the behavior is comparable to diode $D1$. It is worth noting that a maximum acceptor concentration of $1.0 \times 10^{15} cm^{-3}$ and $1.0 \times 10^{16} cm^{-3}$ causes an overcompensation of the defective area, resulting in a local p-type area inside the $n^+ n^-$-area.

In the diode with maximum doping concentration of $1.0 \times 10^{16} cm^{-3}$ in the burried p^+-zone filament formation also starts close to the edge regions. Overtime, a single filament appear in the center (Figs. 26-28). The high current density of $J = 7.0 \times 10^3 A/cm^2$ inside the filament leads to a temperature increase of $\Delta T \approx 150 K$ which could limit the SOA when the ambient temperature is high.

186

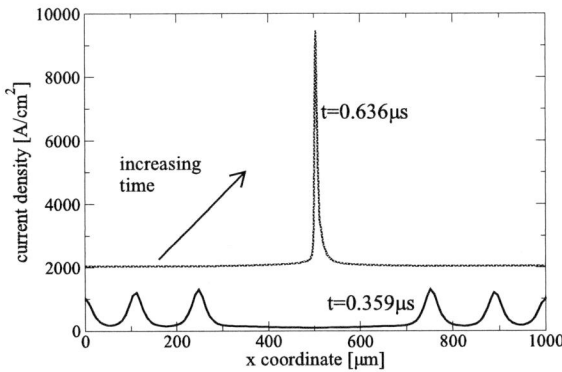

Fig. 26. Current density distribution $J(x, y = 5\ \mu m)$ at the beginning and the end of the filamentation phase in diode $D4$ with a maximum acceptor concentration of $1.0 \times 10^{16} cm^{-3}$ inside the defect area, waveform at $t = 0.636\ \mu s$ is shifted by $2000 A/cm^2$ for a better illustration

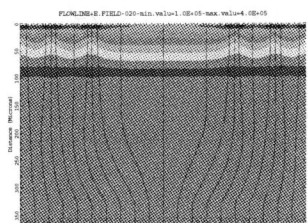

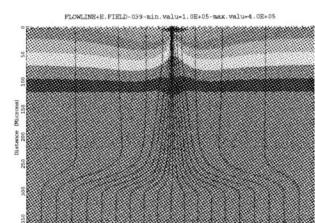

Fig. 27. Current flow lines in the diode $D4$ at $t = 0.36\ \mu s$, showing the inhomogeneously distributed filaments at the beginning of the filamentation, p^+-doping: $1.0 \times 10^{16} cm^{-3}$

Fig. 28. Current flow lines in the diode $D4$ at $t = 0.64\ \mu s$, showing the inhomogeneously distributed filaments at the end of the filamentation, p^+-doping: $1.0 \times 10^{16} cm^{-3}$

IV. CONCLUSIONS

We have analyzed the influence of various types of non-uniformities located in the contact and emitter regions on the formation of filaments during the reverse recovery phase in 3.3–kV diodes.

Our investigations have shown that filament formation near the defection area is less favoured in the beginning of the reverse recovery phase. However as soon as filaments have been found in other parts of the diode, strong local plasma extraction at this places result- together with the increasing diode voltage- in filament formation near the defective area.

Table II summarizes which inhomogeneities are critical ($-$), not critical ($+$) or may be critical (0) for the save operating area (SOA) of 3.3-kV diodes.

TABLE II

SOA OF THE CONSIDERED HOMOGENEOUS AND INHOMOGENEOUS DIODE STRUCTURES $D1$, $D2a$, $D2b$, $D3$ AND $D4$ ESTIMATED BY ISOTHERMAL DEVICE SIMULATION

	inhomogeniety at	value of varied parameter	type	SOA
$D1$	–	–	–	–
$D2a$	anode-side contact	$50nm$ $150nm$ $300nm$	depth of metal contact	0 0 0
$D2b$	cathode-side contact	$50nm$ $150nm$ $300nm$	depth of metal contact	0 0 0
$D3$	p^+–emitter	$1 \times 10^{14} cm^{-3}$ $1 \times 10^{15} cm^{-3}$ $1 \times 10^{16} cm^{-3}$	doping concentration	+ + +
$D4$	n^+n^-–area	$1 \times 10^{14} cm^{-3}$ $1 \times 10^{15} cm^{-3}$ $1 \times 10^{16} cm^{-3}$	doping concentration	+ + –

REFERENCES

[1] M.Domeij, J.Lutz, D.Silber, "On the Destruction Limit of Si Power Diodes During Reverse Recovery with Dynamic Avalanche, IEEE Trans. El. Dev., vol. 50 no. 2, 2003, pp.486-493.

[2] H.P. Felsl, B. Heinze and J. Lutz, "Avalanche and Post Avalanche Behavior of High Voltage Diodes", Proceedings of the 7th ISPS, Prague, September 2004, pp.83-88.

[3] $Medici^{TM}$ User manual, SYNOPSYS INC.

[4] F.-J.Niedernostheide, E.Falck, H.-J.Schulze, U.Kellner-Werdehausen, "Current-density patterns induced by avalanche injection phenomena in high-voltage diodes turn-off", Annalen der Physik, 2004, vol. 13, pp.418-426.

[5] F.-J.Niedernostheide, E.Falck, H.-J.Schulze, U.Keller-Werdehausen, "Avalanchen injection and current filaments in high-voltage diodes during turn-off", Proceedings of the ISPS 2004, Prague, pp.75-82.

2006 25th International Conference on Microelectronics

Schottky Contacts on Single-Crystal CVD Diamond

D. Doneddu, O.J. Guy, D. Twitchen, A. Tajani, M. Schwitters and P. Igić

Abstract – In this paper, a comparison of Gold, Nickel and Aluminium Schottky diodes fabricated on high-quality, Single-Crystal CVD Diamond is presented. Different metals, such as Gold, Nickel and Aluminium, have been deposited on the oxidised surface of an intrinsic diamond layer to serve as Schottky contacts, in order to investigate the physical properties of the different metal-semiconductor interfaces. A Cr-layer, followed by a subsequent Au deposition was used to form the Ohmic back contact for the Au-Schottky diodes, whereas a Cr-layer followed by a Ni deposition formed the Ohmic contact for the Ni and Al-Schottky contacts. Contacts have been characterized using I-V measurements. The gold Schottky contacts exhibited reverse leakage currents as low as $0.01\,\mu A$ at a reverse voltage of -600V, rising to $10\,\mu A$ at 1kV (without any periphery protection). Nickel and Aluminium contacts exhibited lower reverse leakage currents and higher average breakdown voltages, whilst giving poorer forward conduction.

I. INTRODUCTION

Diamond is a promising material for high-power, high-frequency and high temperature electronics applications, where its outstanding physical properties can be fully exploited. It exhibits the highest energy-gap, carrier mobilities, breakdown field strength, and thermal conductivity of any wide band-gap material [1,2]. It could therefore produce the fastest switching, highest power density, most efficient electronic devices obtainable with applications in the RF power, automotive and aerospace industries. Lightweight diamond devices, capable of high temperature operation in harsh environments, could also be used in radiation detectors and particle physics applications [1, 3, 4] where no other semiconductor devices would survive [5,6]. In the past, research could only have been focused on polycrystalline and nano-crystalline material, due to the high cost of natural diamond and the lack of large single crystal synthetic diamond. The recent development of Single-Crystal CVD (SC-CVD) Diamond [1,3] has opened up the possibility of producing diamond electronic devices capable of replacing current silicon, silicon carbide and gallium nitride technologies in certain niche high-power and

D. Doneddu, O.J. Guy and P. Igić are with the School of Engineering, University of Wales Swansea, Singleton Park, Swansea SA2 8PP, United Kingdom, E-mail: daniele.doneddu@gmail.com
D. Twitchen, A. Tajani and M. Schwitters are with Element Six Ltd., King's Ride Park, Ascot, Berkshire SL5 8BP, UK.

high-frequency markets. However, many physical and technological issues concerning this new material have yet to be addressed. Among them, the lack of a shallow n-type dopant at present had forced research to focus on unipolar p-type devices only, such as Schottky diodes and MESFETS, for RF [7,8] and high power applications. Successful n-type doping of diamond has recently been demonstrated [9], and bipolar devices on various crystal orientations have been reported to work at room temperature, whilst showing poor electronic properties [10,11].

II. EXPERIMENTAL WORK

An 18µm-thick, intrinsic SC-CVD diamond layer, homoepitaxially grown on a 300 µm-thick, heavily boron-doped – p-type – substrate has been used as a starting material. Prior to any further processing, the sample was cleaned using the following procedure: wipe with acetone; 10min solvent cleaning with isopropanol; 20 min refluxing in a saturated solution of sulphuric acid and potassium nitrate (1 g KNO_3 in 20 ml H_2SO_4), in order to Oxygen-terminate the surface. A 50 nm-thick Chromium layer, followed by a subsequent 200 nm-

Fig. 1. Schottky contacts on the i-p+ SCCVD Diamond Sample.

thick gold layer, were deposited on to the diamond sample, in an Edwards E306 coating system, under high vacuum (HV) conditions (1×10^{-6} mbar), to form the back (Ohmic) contact. The contact was then annealed in HV at 800°C for 10 min, at a pressure of 5×10^{-7} mbar, in order to form a layer of Chromium Carbide, which act as a more intimate contact, improving the specific

1-4244-0116-X/06/$20.00 ©2006 IEEE

contact resistance [12]. Several 300 nm Schottky Au contacts, with a diameter of 650 µm, were deposited on the intrinsic layer. Figures 1 and 2 show a image of the sample, together with some of the basic process steps. Due to the early stages of the SC-CVD technology, material is still not available in satisfactory volumes, therefore it has been necessary to re-use the sample for the three metallization. Fabrication and characterization of Al and Ni diodes has been therefore carried out as follows. The Au-Schottky contacts were etched away from the top intrinsic layer using diluted aqua regia. The same acid etchant was used for stripping the back Au/Cr layer off. A new 80 nm-thick Cr-layer, followed by a 200 nm-thick Ni-layer contact was then deposited in order to make the Ohmic contact.

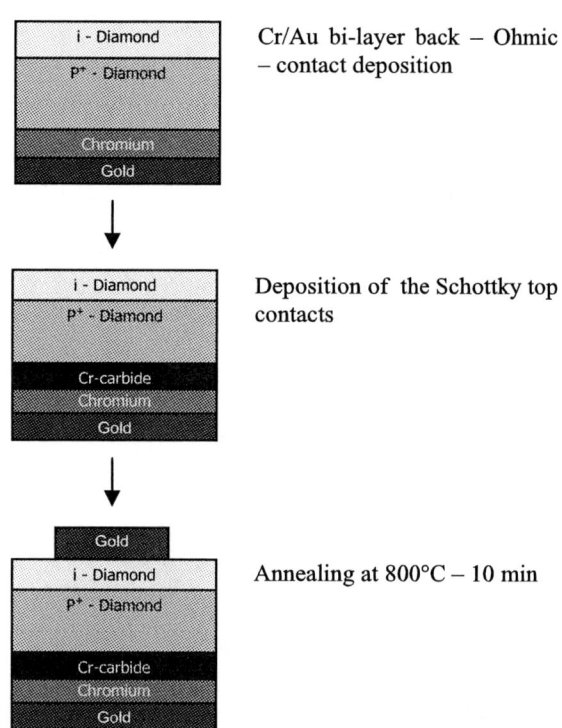

Cr/Au bi-layer back – Ohmic – contact deposition

Deposition of the Schottky top contacts

Annealing at 800°C – 10 min

Fig. 2. Technological processes for the fabrication of SCCVD Diamond Au-Schottky Diodes

Several Nickel Schottky top contacts were then deposited using the same metal evaporation equipment used for the Au-contacts deposition. Finally, Al-Schottky contacts have been fabricated. No contact periphery protection was used in these experiments. After fabrication, each set of diodes were characterized by I-V measurements, using a Karl Suss probe station connected to an HP4142b measurement kit.

III. RESULTS

A. Reverse Bias

The reverse I-V characteristics, performed at room temperature, of two diodes of each of the three different Schottky metal contacts sets are plotted in Figure 3. The total set of Au-Schottky diodes (25 diodes) exhibited quite a homogeneous behaviour in reverse bias, with all the diodes exceeding a reverse blocking voltage of 600V, and a yield at ~1000V close to 40%. The reverse leakage currents, at a given voltage, are quite widely spread over the 25 diodes within a range of 4 orders of magnitude, showing average values of about 2×10^{-6}A at -500V. Reverse leakage currents reached values as low as 10^{-7}A at 600 Volts, for a device active area of 1.33×10^{-2}cm^2. A reverse blocking voltage of more than 1kV was found for some devices.

Almost all Ni-Schottky diodes exceeded 1kV, which was the limit of our measurement kit, and did not show any insurgence of breakdown at this voltage. The average leakage current densities were in the range of 5×10^{-5}A/cm^2 at a given voltage of -500V.

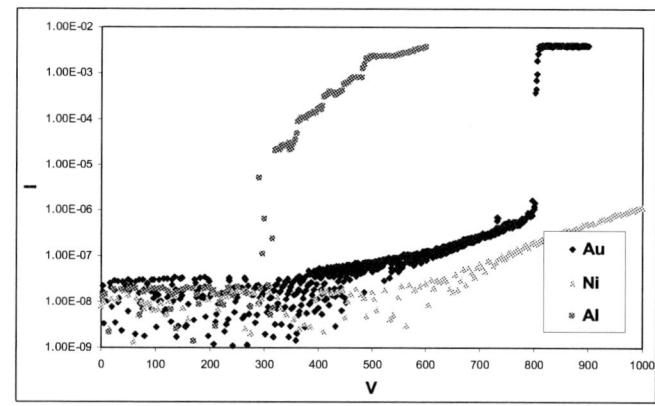

Fig. 3. Reverse characteristics of 3 fabricated Schottky diodes on single-crystal CVD diamond.

Al-Schottky diodes had an average leakage current density of about 1.5×10^{-4}A/cm^2 for the same reverse bias, but some diodes displayed an inhomogeneous behaviour at voltages higher than 500V, showing a sudden increase in the slope of the logarithmic plot of the current.

For all three metallization, current densities, at reverse bias approaching the reverse blocking capabilities of each diode, were quite high for a considerable percentage of the devices, and in the order of 5×10^{-2} A/cm^2.

B. Forward Bias

The forward characteristics, for the same diodes whose reverse current curves were previously plotted in Fig.3, are plotted in Fig. 4. As it is clear fromFig.4, diodes which had exhibited a reasonably good behaviour in reverse bias suffered from a poorer forward conduction. Though the Thermionic Emission theory does not strictly apply to these devices, a rough approximation of the Schottky Barrier Height - Φ_b – extracted from the semi-log I-V plot for the three metallizations is given. The better Au-Schottky diodes were able to reach a current density of 1A/cm^2, with a Φ_b of 0.98eV. Ni-Schottky diodes exhibit a lower current density, and have a comparable barrier height of

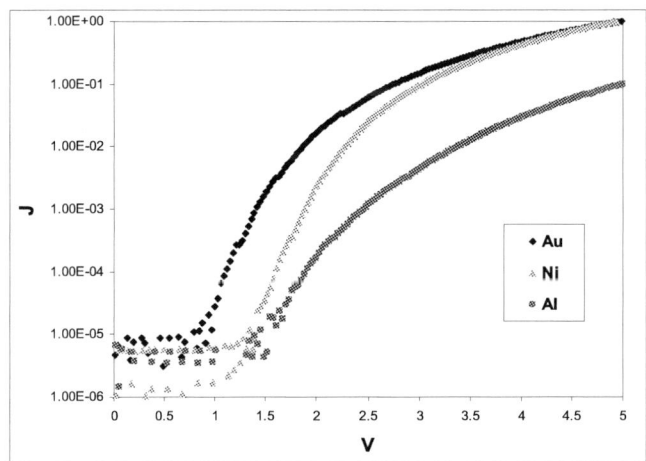

Fig. 4. Semi-log plot of the Forward current density of the diodes

1.03eV, whilst Al diodes had barriers which varied among the contacts in a range between 0.85eV and 1.07eV.

IV. DISCUSSION

Although metals with very different work functions have been used in this study, the related Schottky barrier heights Φ_b barely changed. This behaviour is probably due to the presence of an interfacial layer between the

TABLE I
SCHOTTKY BARRIER HEIGHT, Φ_b, FOR THE DIFFERENT SCHOTTKY DIODES

	Φ_b (eV)
Au	0.98
Ni	1.07
Al	1.16

top metal contact and the i-layer of the diamond sample. Oxygen from the cleaning procedure, or from the atmosphere, has created a semi-insulating layer, several atomic layers thick. It has been shown that this layer can be physically desorbed during annealing (further results will be reported on a future publication). This high density of surface states have the effect of "pinning" the barrier height, as explained by the Bardeen model (1947) [2].

The low current density levels in forward mode arise from the intrinsic – doping concentration $<10^{13}$cm^{-3} – 18μm-thick top layer, which gives rise to a very high specific on-resistance. The simple Mott-Gurney law:

$$J = \frac{9}{8}\varepsilon_r\varepsilon_0\mu\frac{V^2}{L^3} \qquad (1)$$

with J the current density, and L=18μm, theoretically gives 9 A/cm^2 for our structure, but it does not take in to account any carrier diffusion and the Ohmic losses. Ni and Al-Schottky diodes, fabricated using a Ni/Cr-layer to serve as the back contact, might also suffer from a higher specific contact resistance as a thin Ni-Chrome alloy might have formed upon deposition of the Cr/Ni back contact, giving a layer with higher resistivity. These assumptions need further investigation. The wide spread of values for the current, both in forward and in reverse, within the same batch of diodes can be explained by the presence of inhomogeneities in the Metal/Diamond contacts [13]. This phenomenon can be due to both the deposition process and the sample preparation itself. The cleaning procedure might also be partly responsible. The lack of any periphery protection for the Schottky contact is partly responsible for the increase in the leakage current, due to Electric field enhancement at the edge, which, in turn, lowers the barrier height, thus producing the increase in reverse current shown in Fig.3.

V. CONCLUSION

In this paper, vertical Schottky diodes on to a i-p+ Single-Crystal CVD Diamond sample, using different metals – Au, Ni, Al – as Schottky contacts, and different stacking layers – Cr/Au, Cr/Ni – for the back, Ohmic contact, have been fabricated and electrically characterized using I-V measurements. The uniformity of the barrier heights for different metallizations suggests the presence of an interfacial layer, whereas the rather high reverse leakage currents are suggested to be due to the lack of an edge termination. Current density has been shown to be mainly dependent on the resistivity of the intrinsic diamond layer.

ACKNOWLEDGEMENT

The authors gratefully appreciate financial support from Element Six Ltd.

REFERENCES

[1] D. Twitchen and A.J. Whitehead, Paper at 3rd European Conference on Diamond, Diamond-Like Materials, Carbon Nanotubes, Nitrides & Silicon Carbide, Sept. 2002, Granada, Spain

[2] Sze S. M. (ed) 1981 *Physics of Semiconductor Devices* 2nd edn. (New York: Wiley) p. 849

[3] J. Isberg, J. Hammersberg, E. Johansson, T. Wikstom, D.J. Twitchen, A.J. Whitehead, S.E. Coe, G.A. Scarsbrook, *Science*, 2002, 297, p. 1670

[4] P. Bergonzo, A. Brambilla, D. Tromson, C. Mer, B. Guizard, R.D. Marshall, F. Foulon, *Nuclear Instruments and Methods in Physics Research A*, 2002, 476, p. 694–700.

[5] W. Adam et al., *Nuclear Instruments and Methods in Physics Research A*, 2003, 514, p. 79

[6] J. Isberg , J. Hammersberg , D.J. Twitchen , A.J. Whitehead, *Diamond and Related Materials*, 2004, 13, p. 320

[7] M. Schwitters, M. P. Dixon, A. Tajani, D. Twitchen, S. E. Coe, H. El-Haji, M. Kubovic, M. Neuburger, A. Kaiser and E. Kohn, 'Diamond MESFET – Synthesis and Integration', 1st International Industrial Diamond Conference Barcelona October 20-21st 2005

[8] Yasar Gurbuz, Onur Esame, Ibrahim Tekin, Weng P. Kang and Jimmy L. Davidson, *Solid-State Electronics*, 2005, 49, Issue 7, p. 1055-1070

[9] S. Koizumi, K.Watanabe, M.Hasegawa, H.Kanda, Science 292 (2001) 1899.

[10] Satoshi Koizumi, Kenji Watanabe, Masataka Hasegawa and Hisao Kanda, *Diamond and Related Materials,* 2002, 11, Issues 3-6, p.307-311

[11] Toshiharu Makino, Hiromitsu Kato, Sung-Gi Ri, Yigang Chen and Hideyo Okushi, *Diamond and Related Materials,* 2005, 14, Issues 11-12 ,p. 1995-1998

[12] D. Doneddu, O.J. Guy, R.M. Baylis, L. Chen, P.R. Dunstan, P.A. Mawby, C.F. Pirri, S. Ferrero, D. Twitchen, A. Tajani and M. Schwitters, "SNOM investigation of surface morphology changes during Cr/Au contact fabrication on single-crystal CVD diamond", *Mater. Sci. For., in press.*

[13] R. T. Tung, Phys. Rev. B, vol 45, pp.13509-13523, 1992.

2006 25th International Conference on Microelectronics

Dynamic Thermal Simulation of Power Devices Operating with PWM Signals

Z. Zhou, M. S. Khanniche, P. Igic, S, N.Jankovic, S. G. Batcup, P. A. Mawby

Abstract - Fast power devices thermal simulation method based on averaging power losses over each cycle of PWM switching frequency is presented in this paper. For implementing a long real time dynamic thermal simulation of power devices, device power losses during transient process and static characteristics are defined as a function of device conduction current and junction temperature, and are represented by a lookup table. By carrying out the circuit electrical simulation, the device conduction current can be obtained. By combining the device conduction currents, global device temperature (GDT) and the data from the lookup table, the average power loss over each cycle of PWM switching frequency is then calculated for carrying out the thermal simulation. With the proposed method, a relative large simulation time step can be employed and simulation speed can be increased dramatically. The method is suitable for a long real time thermal simulation for complex power electronics systems.

I. INTRODUCTION

Inverter Power Modules (IPM) based on Insulated Gate Bipolar Transistors (IGBT) power modules, are more and more popular for automotive industry. However, the confined volumes, high temperature environment, high power density and huge power dissipation make the power package exposed to high thermal constraints, especially in system transients and very harsh electrical environments [1]. Since the reliability of the power module depends strongly on the device junction operating temperature, the correct information regarding effective power losses predicted by thermal simulation is essential for system designer.

An accurate approach to determine the power losses is to simulate a circuit by using fully physically based device model [2-4]; however, due to the complicate physics of the switching processes, the accurate prediction of power losses and thermal performance need a very detailed device model and very small simulation time steps, this will result in a computationally prohibitive computing time and memory requirement for modelling a complex power electronics system such as three phase inverter drive system. In addition, dynamic and unpredicted variation of load that results in variation of the power losses with time limits the use of the methods based on constant load assumption [8].

Z. Zhou, M. S. Khanniche, P. Igic, S, N.Jankovic, S. G. Batcup, P. A. Mawby with Electronics Systems Design Centre, School of Engineering, University of Wales Swansea, UK. Email: z.zhou@swansea.ac.uk

To overcome the difficulties for a long real time dynamic thermal simulation, we have developed fast power losses calculation method for a three-phase PWM inverter power module dynamic thermal simulation and modelling strategy and results are presented in this paper. Since the thermal time constant of device / heat-sink system is relatively larger than the PWM switching frequency, an average power loss over each cycle of PWM switching frequency is introduced for thermal simulation. This average power loss can be calculated according to the device electrical operating conditions by analyzing the device power loss characteristics corresponding to their operating conditions. As it is well known, semiconductor device power loss comprises two portions, e.g. on state conduction power loss and switching power loss. The on-state power loss is determined by device on-state V-I characteristics and duty period; the switching loss depends on the collector current (for IGBT) at which the device turns -on or -off. The key milestones for calculating the average power losses are: (1) determination of the device electrical operating conditions; (2) determination of the device power loss characteristics; and (3) determination of the duty period of each PWM switching frequency. An easy method would be based by using PSpice, Saber, etc conventional time-domain circuit simulation tools; however, the very small simulation time step makes it very difficult to simulate a long real time dynamic process (in range of minutes). The method proposed in this paper can be outlined as shown in Fig.1

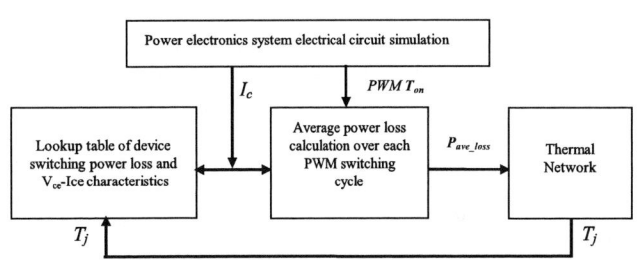

Fig. 1: Block diagram of the power device thermal simulation.

Here the key feature that allows a significant speed-up of the simulation CPU time is the simplified representation of the inverter during the system level modelling. We have taken the approach that the inverter can be described by a unity gain amplifier. This allows the system to work on effective voltages and currents, which are continuous quantities (rather than the actual PWM

1-4244-0116-X/06/$20.00 ©2006 IEEE

switching waveforms). By using the PWM reconstruction technique, reconstructed the device conduction current can be predicted from load current (according to its fundamental AC voltage), and then the accurate average power losses over each cycle of PWM switching frequency can be calculated and, finally, the dynamic thermal simulation can be carried out.

II AVERAGE POWER LOSS OF EACH PWM SWITCHING FREQENCY FOR THERMAL SIMULATION

Considering the time constant of the thermal transimution of semiconduction is relative large than the PWM switching period that used in most power eletronics circuit. So a average power loss of each PWM switching frequency is employed and to simulate the device global temerature. The average power loss of each PWM switching frequecy is calculated as follows

$$P_{ave\_loss}(k) = \frac{1}{T_{sw}}(E_{turn\_on}(k) + E_{cond}(k) + (E_{turn\_off}(k))) \quad (1)$$

where, Eturn_on, E cond and Eturn_off are the turn_on switching energy loss, steady on state energy loss and turn-off switching losses

$$P_{cond\_ave}(k) \approx \frac{1}{T_{sw}}V_{ce}(k) \cdot i_c(k) \cdot t_{on}(k)$$

$$E_{sw-on}(k) = f_{sw-on}(i_c(k), T_j(k)) \quad (2)$$

$$E_{sw-off}(k) = f_{sw-off}(i_c(k), T_j(k)) \quad (3)$$

where V_{ce}, i_c, represent the forward saturation voltage, collector current of an IGBT, respectively T_j represents the junction operating temperature. where t_{on}, represents the conduction duty period of the kth PWM switching cycle.

To illustrate the effectiveness of average power loss of each PWM switching frequency on the thermal simulation acuracy, two simulations have been carryed out on a resistive load single IGBT swiching circuit in Fig.3. (1) one is using the PWM based average power losses, (2) the second one is using the instantenous power loss that is implement based on a physically based compact model [10-11]. Fig.2 shows the instantanouse drain current of IGBT, and DC bus voltage is 500, load resistor 5Ω and the PWM switching frequency is 5kHz. Both simualtion use the same thermal model that for thermal simualtion purpose as shown in Fig.14. Fig. 3 shows the the device temperatures that obtained from the two simulations respectively. It can be found that using the avserage power loss of each PWM switchong frequency can get very acurate teperature performance.

A Physically based compact model [9-10] has been selected as a switching device in the inverter circuit. This model used the full ambipolar diffusion equation (ADE) solution and solving numerically the basic drift-diffusion equations. The power devices model uses one-dimensional approximation to the carrier distribution in the drift region. It includes conductivity modulation and non-quasistatic charge storage effect, which will describe correctly in static and dynamic behaviour of power bipolar devices. These models have been implemented in the SABER circuit simulator for today modelling environment.

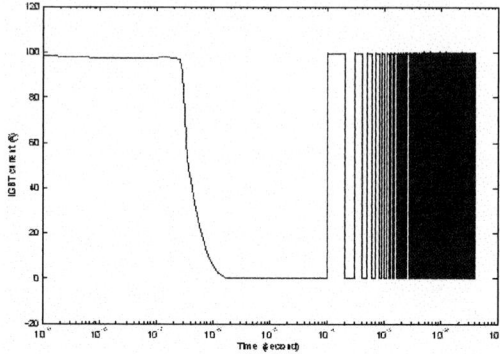

Fig. 2 PWM drain current of IGBT.

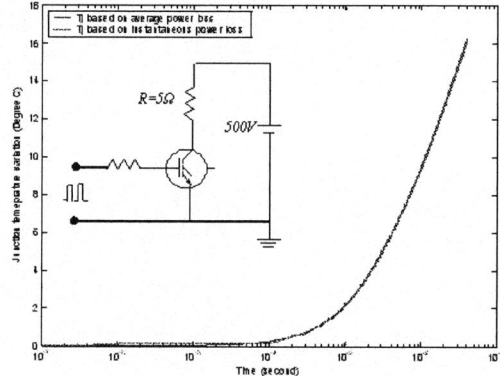

Fig.3 Simulated IGBT global temperature.

The behaviour of power devices are depends heavily on the excess carrier distribution in the low-doped drift region during the high-level injection mechanism. However, quasi-neutrality ($p \approx n$) is valid because the mobility will not undergo velocity saturation as a consequence of a high electric field. Under this condition, the ADE determines the carrier transport:

$$D_a\frac{\partial^2 p(x,t)}{\partial x^2} = \frac{p(x,t)}{\tau_A} + \frac{\partial p(x,t)}{\partial t} \quad (4)$$

where D_a represent the ambipolar diffusion constant and τ_A represent the ambipolar carrier lifetime. Our approach [10-11] is to model the ADE describing the internal plasma charge distribution using two exponential functions based on the ambipolar diffusion length.

$$p = Ae^{x/L} + Be^{-x/L} \quad (5)$$

where the diffusion length is given by $L = \sqrt{D\tau}$, and A and B are arbitrary constants determined by boundary conditions. These functions are fitted using weighting factors, which are adjusted using a numerical fitting algorithm. Although this approach accurately represents the evolution of the plasma with time, it requires considerable effort to ensure robustness of the model. Our approach does not make any assumptions about the shape of the plasma and so is more generic with a sound physical basis. Indeed compared with a full numerical model our approach shows excellent agreement over a wide range of operating conditions. The Punch-through IGBT structure and characteristics is shown in Figure 4.

In order to model heavily doped n^+ buffer layer existing in the IGBT device, the width of the undeleted region and the carrier current at the boundaries (Xl and Xr) of the region are required. Now, the boundary conditions for the n-base carrier storage region are defined by (6) and (7) as:

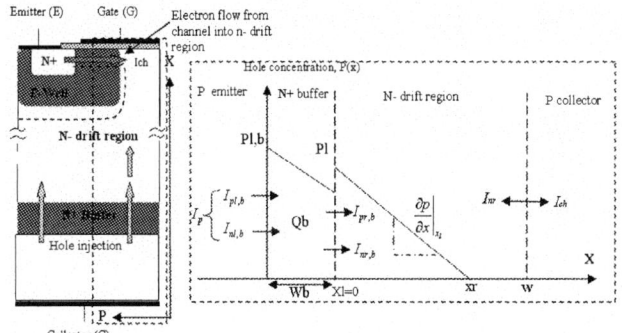

Figure 4 Punch-through IGBT structure and characteristics.

$$\left.\frac{\partial p}{\partial x}\right|_{x_l} = \frac{1}{2qA}\left(\frac{I_p - I_{pr,b}}{D_n} - \frac{I_{pr,b}}{D_p}\right) \qquad (6)$$

$$\left.\frac{\partial p}{\partial x}\right|_{x_r} = \frac{1}{2qA}\left(\frac{I_{ch}}{D_n} - \frac{I_p - I_{ch}}{D_p}\right) \qquad (7)$$

A is the cross sectional area of the device, D_p and D_n, the hole and electron diffusion coefficients, $I_{pr,b}$ is the minority carriers in the buffer layer and the hole current at the n-base end(x_l). I_{ch} The channel current in the inversion layer and I_p the total plasma current flow through the drift region.

III PWM AND DEVICE CURRENT RECONSTRUCTION FOR AVERAGE POWER LOSS CALCULATION FOR THREE PHASE INVERTERS.

By carrying an analytical system level electrical simulation, the effective motor voltage and current can be given; however, these waveforms are continuous quantities (rather than the actual PWM switching waveforms). To calculate the average power losses discussed previously, the voltage, current and the corresponding duty period are required. In this paper, we have proposed PWM reconstruction concept for calculating the on-state duty period of each PWM cycle can be reconstructed based on the PWM strategy using the fundamental motor voltage

vector. The device currents can also be reconstructed based on the inverter circuit topology and the statues of devices. Feeding this average power losses into the thermal network, the junction temperature can be calculated.

IV SIMULATION RESULTS

A three phase inverter power module consists of 12 IGBT modules (every arm contains two IGBT modules in parallel to share the current) and each of them is paired with an anti-parallel diode. The circuit connection is repeated in Fig. 5.

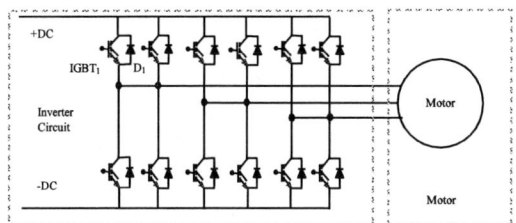

Fig. 5 -The three-phase inverter system

Using a curve fitting approach, the parameters *of each layer* can be extracted based on 3-D simulation or measurement. The corresponding equivalent thermal network of the inverter power module and heat-sink is shown in Fig. 6, where $T_a=50^\circ C$.

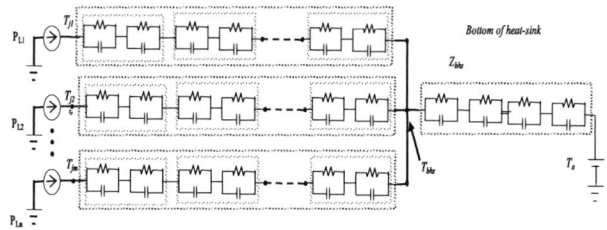

Fig. 6 - Eqivalent thermal network of the inverter power module and heat-sink

Simulation program is developed and implemented in Matlab/Simlink. Simulation of the power losses and thermal performance for the inverter drive system.

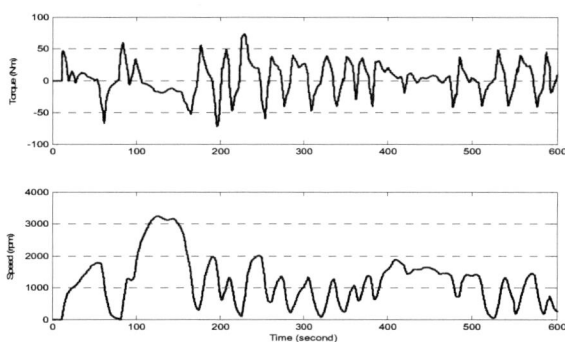

Fig.7 Motor torque (Nm) and speed (rpm)

The specified torque and speed profiles are shown in Fig. 7, the corresponding continuous effective motor stator currents and voltages are shown in Fig.18. The power losses in IGBT₁ and the corresponding junction temperature are shown in Fig 8-9 respectively. The temperature of inlet water is set to 50°C.

The PWM switching frequency is set to 2 kHz, and simulation time step is 0.5ms. The total simulation time takes about 1hr with a 3.12 GHz CPU, 1Gbyte memory PC.

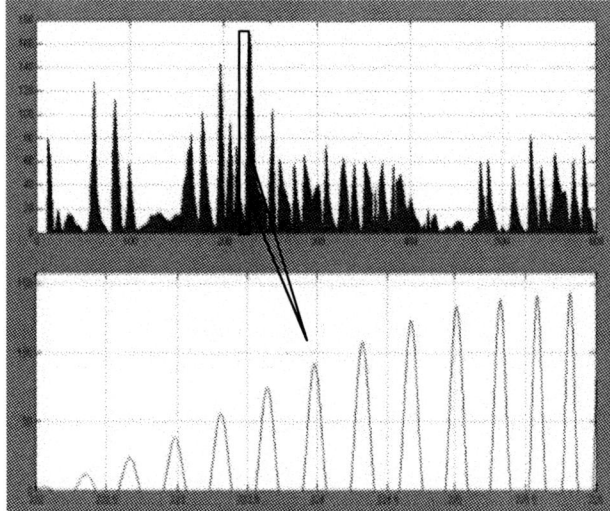

Time (sec.)
Figure 8 IGBT power losses (W).

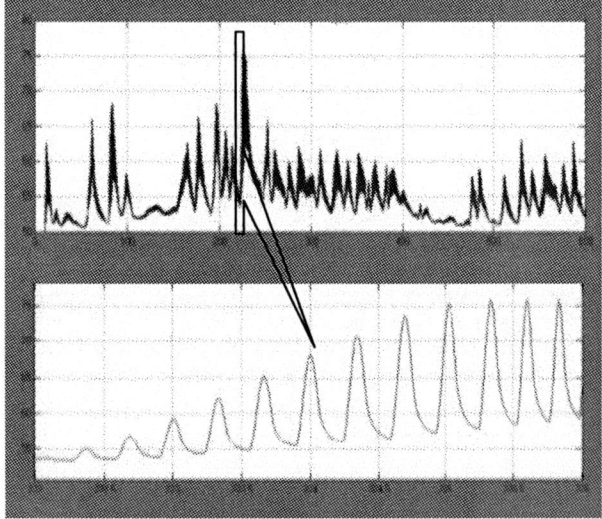

Time (sec.)
Figure 9 IGBT junction temperature (°C).

V. CONCLUSIONS

A fast power losses simulation model for a three-phase inverter power module (IPM) thermal simulation using PWM reconstruction technique is presented in this paper. This model is aimed for fast and efficient power loss and thermal analysis for inverter power devices in electric vehicle applications. This approach has a very fast simulation speed and produces the voltages and currents

associated with each of the semiconductor switches in real-time by using PWM reconstruction. This information is then used in conjunction with detailed device switching models, to describe the heat-source terms for a thermal solver, this allowing electro-thermal performance of the inverter to be predicted over long periods of real time (minutes of real time operation!). This simulation methodology brings together accurate models of the electrical systems performance, state of the art-device compact models and a realistic simulation of the thermal performance in a useable period of CPU time.

ACKNOWLEDGMENT

The authors would like to thank Toyota Motor Corporation for their financial support and technical advice.

REFERENCE

[1] Masayasu Ishiko, Masanori Usui, Takashi Ohuchi, Mikio Shirai, "Design Concept for Wire- Bonding Reliability Improvement by Optimizing Position in Power Devices", ISPS'04 7th International Conference on Power Semiconductors, Prague, 31 August – 3 September 2004, pp 39 44 (2004).

[2] Takashi Kojima, Yasushi Yamada, Mauro Ciappa, Marco Chiavarini and Wolfgang Fichtner, "A Novel Electro-thermal Simulation Approach of Power IGBT modules for automotive traction applications" Proceeding of 2004 international symposium on power semiconductor devices &ICs, Kitakyushu (2004).

[3] Alan Mantooth, Allen R.Hefner, " Electrothermal Simulation of an IGBT PWM inverter", IEEE Transaction on Power Electronics, vol. 12, no.3, May (1997).

[4] R. Hefner and D. L. Blackburn, "Thermal component models for electro-thermal network simulation," IEEE Transactions on Components Package. Manufacture. Technology, vol. 17, p. 413,(1994)

[5] A. D. Rajapakse, A. M. Gole, and P. L. Wilson," Electromagnetic Transient Simulation Models for Accurate Representation of Switching Losses and Thermal Performance in Power Electronic Systems", IEEE Transactions on Power Delivery, vol.20, no.1, January 2005, pp 319-327 (2005).

[6] Dewei Xu, Haiwei Lu, Lipei Hang, Satoshi Azuma, Masahiro Kimata, and Ryohel Uchida," Power Loss and Junction Temperature Analysis of Power Semiconductor Devices", IEEE Transaction on Industry Applications,Vol..38, No.5, pp, 1426-1431, September/October (2002).

[7] Introduction to the 600V ADD-A-pak ™ and INT-A-pak™ IGBT Modules, International Rectifier Application notes (1992).

[8] Alain Laprade and Ron H. Randal, "Numerical Method for Evaluating IGBT Losses", Application Note 7520 Rev. A1.

[9] P.M.Igic, P.A.Mawby, M.S.Towers and S.Batcup, "New Physically-based PiN diode compact model for circuit applications", in IEE Proc-Circuit Devices Syst. Vol.149,No.4, August , 2002.

[10]. P. M. Igic, P.A.Mawby, M.S.Towers and S.Batcup, " Physically based 2D Compact Model for Power Bipolar Devices" in International journal of Numerical Modelling Electronic Network. Model;17:397-405, 2004.

2006 25th International Conference on Microelectronics

Dynamic Characteristics of Novel Super-Junction LDMOS Switches under Charge Imbalance Conditions

K. Permthammasin, G. Wachutka, M. Schmitt, and H. Kapels

Abstract - The inductive switching performance of a smart super-junction (SJ) LDMOS switch is simulated and analysed, together with the reverse recovery properties of its internal diode. The SJ power switch under investigation differs from the conventional one in the so-called SJ structure which, in this case, consists of an array of p-type round pillars embedded in the n-type drift region. In view of the tolerances achievable in the present fabrication technology, the effect of slightly unbalanced doping charges in the SJ structure is of utmost importance and, hence, studied in detail. Two different SJ structures have been designed to tackle the effect of substrate-aided depletion on the blocking capability of the SJ switch. The 3D-simulation results reveal that a proper design of the SJ structure not only suppresses the substrate effect, but also enhances the switching capability of the SJ switch.

I. INTRODUCTION

In today's power switching circuits such as switched-mode power supplies, induction heating or motor drives, power MOSFETs are in extensive use because of their ultra-fast switching speed and high operating frequency. However, since the on-state resistance rapidly increases with the blocking voltage, the ratings of conventional power MOSFETs are limited to around 500 V [1]. Therefore, modern power MOSFETs contain a super-junction (SJ) structure in their drift region to form a novel class of high-voltage, low on-resistance transistors referred to as SJ devices [2]. An SJ device can handle much more power than a conventional one, but this property very sensitively depends on the exact charge compensation in the SJ structure [3].

This paper deals with the numerical analysis of the switching and reverse recovery behaviour of a smart SJ lateral-diffused MOS (LDMOS) power switch. Considering that perfect charge compensation in the SJ structure is hardly achievable in today's manufacturing technology, the transient analysis also includes the case of slightly unbalanced doping charges. Because the LDMOS structure is built on a bulk silicon substrate, the blocking capability of the SJ switch is affected by the non-uniformity of the vertical built-in electric field along the depleted portion of the

K. Permthammasin and G. Wachutka are with the Institute for Physics of Electrotechnology, Munich University of Technology, Arcisstrasse 21, 80290 Munich, Germany,
E-mail: komet@tep.ei.tum.de, wachutka@tep.ei.tum.de
M. Schmitt and H. Kapels are with Infineon Technologies AG, Power Semiconductor Development, Am Campeon 1-12, 85579 Neubiberg, Germany, E-mail:
Markus.Schmitt@infineon.com, Holger.Kapels@infineon.com

interface between substrate and drift zone [4]. In our study, two SJ structures have been considered: one is intended for counteracting the substrate effect, whereas the second serves as reference for comparison.

II. DEVICE STRUCTURE

Fig. 1 displays one unit cell of the SJ switches investigated, with a cell pitch of 10 μm. The design refers to a 50 μm long LDMOS manufacturable by an industrial smart power IC process.

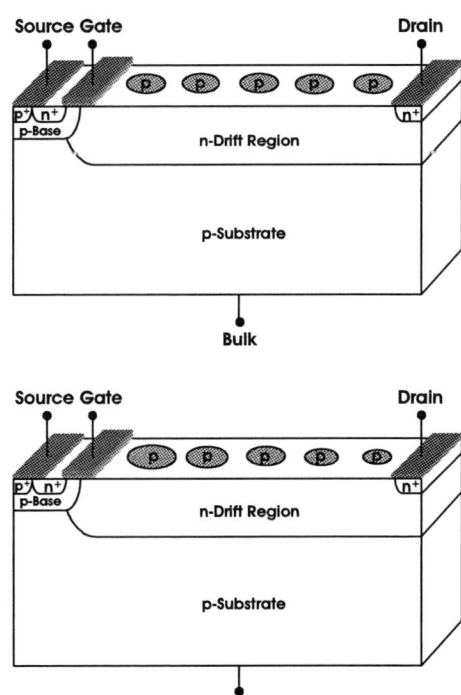

Fig. 1. Structure of two different super-junction switches.
Top: uniform SJ. Bottom: non-uniform SJ.

The essential feature of the structure is the lateral SJ drift zone with five p-type round pillars located at an equal distance to one another and embedded in a n-type layer of 15 μm thickness above the very thick p-type substrate. In order to counteract the substrate-aided depletion effect, which is commonly encountered in lateral power devices fabricated on bulk silicon, the size of the pillars is individually adjusted leading to the so-called non-uniform SJ structure: The pillar next to the source is thickest with a

1-4244-0116-X/06/$20.00 ©2006 IEEE   197

diameter of 7 μm, while the diameters of the following pillars decrease in steps of 0.5 μm along the drift region such that the thinnest pillar with a diameter of 5 μm is located next to the drain. As reference for comparison, we consider a second SJ structure denoted as uniform SJ where all the round pillars have equal diameters of 6 μm.

Optimising the unit cell design of the uniform SJ results in a gate threshold voltage $V_{GS(th)}$ = 3 V, a transconductance g_{fs} = 1.4×10^{-4} S, a specific on-state resistance $R_{DS(on)} \cdot A$ = 9.45 Ω·mm^2 and a breakdown voltage $V_{BR(DSS)}$ = 640 V. The optimum unit cell of the non-uniform SJ exhibits a gate threshold voltage $V_{GS(th)}$ = 3.3 V, a transconductance g_{fs} = 1.15×10^{-4} S, a specific on-state resistance $R_{DS(on)} \cdot A$ = 10.16 Ω·mm^2 and a breakdown voltage $V_{BR(DSS)}$ = 632 V, respectively. Its advantage is, however, a faster switching speed and a better robustness against dynamic avalanche breakdown as discussed in the following.

III. RESULTS AND DISCUSSION

Assuming that, with the present doping technique, the required optimum doping concentration $N_{A,opt}$ in the pillars can be adjusted within a tolerance better than 10%, i.e.

$$|\Delta N_A| = \left| \frac{N_A - N_{A,opt}}{N_{A,opt}} \right| \leq 10\% \qquad (1)$$

we considered the following three cases of charge imbalance in our transient simulations:
ΔN_A = –10%, ΔN_A = 0% and ΔN_A = +10%.

A. Gate Charge Characteristics

Since the basic function of a power MOS switch consists in controlling the drain current by the gate voltage, the relation between gate charge and gate voltage provides useful information for assessing the switching behaviour. Fig. 2 shows these characteristics, together with the drain voltages, for the two SJ structures at the optimum doping level.

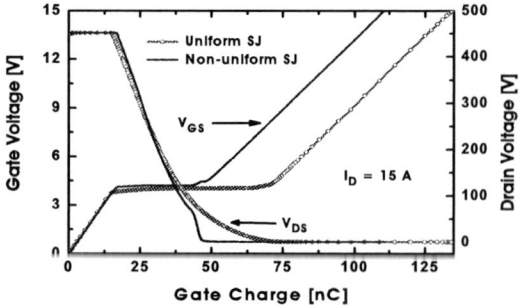

Fig. 2. Gate charge characteristics of two SJ switches in comparison.

The gate charge characteristics are extracted from the simulation of a test circuit with inductive load, in which the SJ switch is fed from a 450 V power supply. The load has a constant current of 15 A with a free-wheeling diode connected in parallel across it. From the initial slope of the gate charge characteristics it can be inferred that both device structures have comparable gate-to-source capacitances. Since the uniform SJ exhibits a lower threshold voltage and a higher transconductance, it reaches the Miller plateau region fairly faster. The non-uniform SJ consumes considerably less Miller charge (Q_{gd} = 47.5 nC) than the uniform SJ (Q_{gd} = 70.7 nC). This means that the non-uniform SJ can switch from the off-state to the on-state in a relatively short time, thereby reducing the switching losses. Furthermore, the total gate charge required for switching the non-uniform SJ is by far less than that of the uniform SJ. This low total gate charge (Q_g = 110.5 nC at V_G = 15 V) multiplied by the on-state resistance of the non-uniform SJ yields a figure of merit that is better by 10% compared to the uniform counterpart. A kink visible in the charge curve of the non-uniform SJ, shortly before the device leaves the Miller plateau, indicates that the value of the gate-to-drain capacitance has abruptly changed to such an extent that the rate of change of the drain voltage is also significantly raised.

B. Inductive Switching

Most electronic loads which are to be driven by SJ switches are inductive. The turn-on and turn-off switching transients of the two SJ structures connected to an inductive load are shown in Fig. 3 and Fig. 4, respectively, for the three charge imbalance conditions. In the test circuit, the SJ switch transfers power from a V_{DD} = 350 V power supply to an inductive load of L = 200 μH. A free-wheeling diode is connected in parallel across the load as protection against the voltage spikes induced on the drain side of the SJ switch during turn-off. The (virtually constant) current in the load is I_L = 20 A. A gate drive circuit with internal resistance R_G = 10 Ω provides a step voltage of 15 V to the gate terminal.

As expected, because of its lower Miller charge, the switching speed during turn-on and turn-off is faster with the non-uniform SJ, compared to the uniform SJ design. Generally, for the two SJ structures, the total turn-on time is shorter than the total turn-off time. Whereas a +10% charge imbalance in the pillars results in a faster switching time, a –10% charge deficiency leads to a reduction in switching speed. This is an obvious consequence of the fact that excessively doped round pillars provide less resistive paths for holes. It is interesting to compare the drain voltage waveforms during turn-on of the non-uniform SJ for ΔN_A = 0% and ΔN_A = +10%, respectively. There are obvious differences between the transients during the fall period, and the Miller plateau is finished before the switch enters the ohmic region. The cause of this untypical switching behaviour is the following:

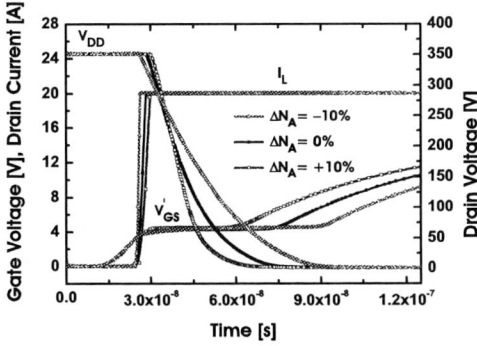

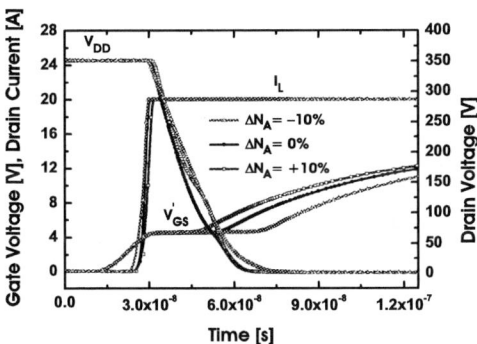

Fig. 3. Turn-on switching waveforms of the uniform SJ (top) and the non-uniform SJ (bottom).

At the beginning of the turn-on process a certain part of the pillars located in the vicinity of the p-base region is not completely depleted of holes. As the drain voltage falls, the holes in the undepleted section of the pillars start diffusing and, thereby, contribute to the variation of the parasitic capacitances in the SJ switch with the drain voltage.

C. Unclamped Inductive Switching

Robustness against harsh operating conditions is a challenge of increasing relevance. An example is unclamped inductive switching (UIS), where the power switch is driven into avalanche stress during the turn-off of an inductive load current in the absence of a free-wheeling diode across the load. We studied the behaviour of our two SJ LDMOS switches under UIS conditions by computer simulations, assuming that the SJ switches are fed from a $V_{DD} = 50$ V power supply with an unclamped inductive load of L = 25 μH and are controlled by a gate-drive unit with a voltage of $V_G = 15$ V and an internal resistance $R_G = 25$ Ω. Fig. 5 illustrates the capability of the SJ switches to endure the dissipated avalanche energy during UIS. The avalanche ener-gy E can be calculated from

$$ E = \frac{1}{2} L \cdot I_{AV}^2 \left[\frac{V_{(BR)eff}}{V_{(BR)eff} - V_{DD}} \right] \qquad (2) $$

where I_{AV} is the avalanche current and $V_{(BR)eff}$ is the breakdown voltage under avalanche conditions. For the two SJ

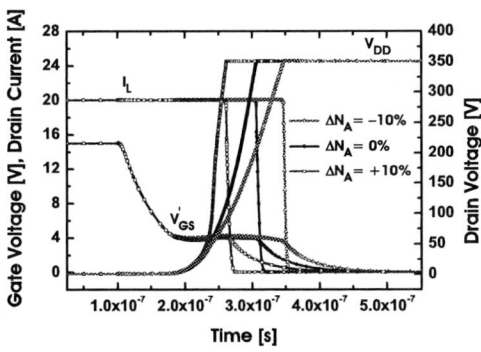

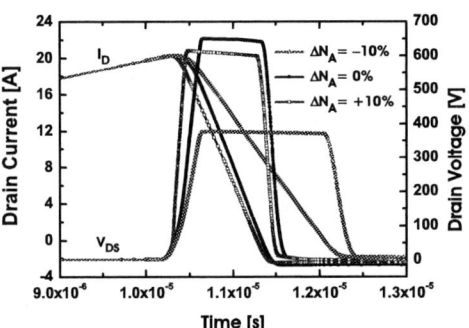

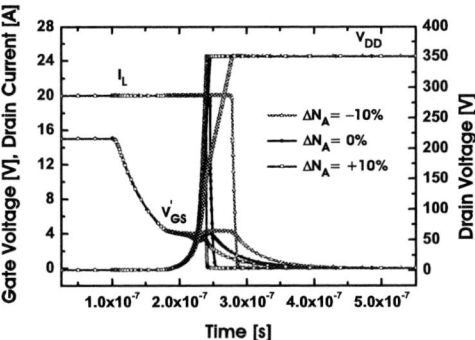

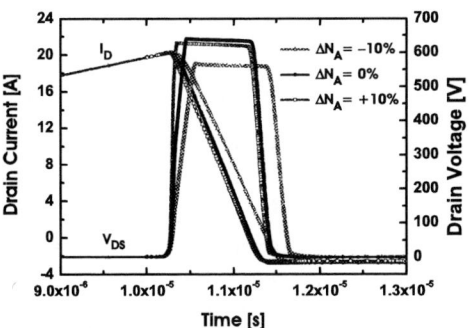

Fig. 4. Turn-off switching waveforms of the uniform SJ (top) and the non-uniform SJ (bottom).

Fig. 5. Unclamped inductive switching waveforms of the uniform SJ (top) and the non-uniform SJ (bottom).

structures, a +10% excess doping in the pillars has much less effect on the breakdown voltage during avalanche than a −10% charge deficiency. Since the peak avalanche current for each charge imbalance condition is nearly the same ($I_{AV} \approx 20$ A), while the breakdown voltage $V_{(BR)eff}$ is significantly lowered in the case of a −10% charge deficiency, we observe a relatively long avalanche time t_{AV} according to

$$t_{AV} = \frac{L \cdot I_{AV}}{V_{(BR)eff} - V_{DD}} \qquad (3)$$

If the avalanche current is high enough to trigger the parasitic bipolar transistor, such a long period of avalanche operation can lead to a rapid rise of the junction temperature above its critical value such that the SJ switch is eventually damaged by catastrophic thermal runaway [5]. All consi-dered, we find that the avalanche robustness of the non-uniform SJ is considerably less affected by charge imba-lance than that of the uniform SJ, and this implies the suppression of the substrate-aided depletion effect.

D. Reverse Recovery Properties

The SJ power switch, like conventional power MOS-FETs, has an inherently built-in "body diode". This is forward-biased when the source terminal has positive polarity with respect to the drain terminal. On the other hand, a positive voltage on the drain terminal will turn off the diode. The reverse-recovery waveforms, as shown in Fig. 6, describe how the forward-conducting diode of the two SJ structures recovers its reverse blocking capability during turn-off at a supply voltage of 350 V. The SJ device, in contrast to conventional power MOSFETs, cannot sustain a reverse-biased voltage until most of the excess carriers have been removed [6].

The initial fall rate of the forward diode current and the slope of the subsequent reverse current as well as the peak current value are comparable, since they are primarily determined by the parameters of the external circuits (i.e. I_F = 20 A and $|dI_{rr}/dt|$ = 100 A/μs). The non-uniform SJ exhibits "snappy recovery" for all three cases of charge imbalance. That is, after reaching the peak value, the reverse current falls off rapidly for a short while, during which the remaining excess carriers are completely removed. An abrupt change of the reverse current after reaching the peak value gives rise to high overvoltage and, subsequently, rapid oscillations in the current waveform. The same applies to the uniform SJ except for the case of $\Delta N_A = -10\%$.

For the uniform SJ with $\Delta N_A = -10\%$ the reverse current gradually decreases (soft recovery), suggesting that there are some excess carriers still left in the drift region and in the pillars at the time the reverse current reaches the peak value. In this case, one expects the gradual formation of depletion layers around each pn junction. Evidently, the soft recovery behaviour minimises the possibility of voltage overshoot which could damage the switch.

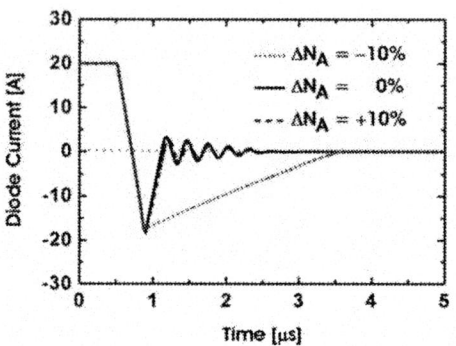

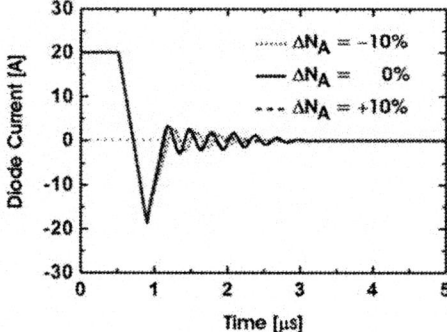

Fig. 6. Reverse-recovery current waveforms of the uniform SJ (top) and the non-uniform SJ (bottom).

IV. CONCLUSION

Using 3D device simulation, we studied two variants of a smart SJ power LDMOS switch, with the focus on the switching of inductive loads and the reverse recovery of the internal diode. In particular, we investigated the effects caused by charge imbalance in the SJ compensation structure. It turns out that the switch with the non-uniform SJ structure exhibits faster switching speed and enhanced avalanche robustness, compared with the uniform SJ design. The reverse recovery properties of the two SJ structures show, for most imbalance cases, no significant difference.

REFERENCES

[1] R. S. Ramshaw, *Power Electronics Semiconductor Switches*, London, Chapman & Hall, 1993.

[2] G. Deboy et. al., *Proc. IEDM*, pp. 683-685, 1998.

[3] P. M. Shenoy et. al., *Proc. ISPSD*, pp. 99-102, 1999.

[4] S. G. Nassif-Khalil et. al., *Proc. ISPSD*, pp. 81-84, 2002.

[5] K. Fischer et. al., *IEEE Trans. Electron. Dev.*, vol. 44, pp. 874-878, 1997.

[6] R. Ng et. al., *Proc. CAS*, pp. 461-464, 2001.

2006 25th International Conference on Microelectronics

Self-heating Eeffects in SOI NLDEMOS Power Devices

F. Dieudonné, S. Haendler, S. Chouteau, J. Rosa, P. Waltz, A. Perrotin, L. Boissonnet, B. Rauber, C. Schaffnit and C. Raynaud

Abstract – Self-heating is evaluated in high-voltage devices from a 0.13μm thin-film SOI CMOS technology. Our measurement procedure is described. The influence of different parameters such as layout variations is finally investigated, and main results are shown here. A strong linear correlation between the thermal resistance and the reverse of the active surface is demonstrated. Besides, a moderate impact of the metallization is observed.

I. INTRODUCTION

Integration of logic and power devices on the same chip has attracted a growing attention in recent years. Lateral Drain Extension (LDE) MOSFETs on SOI have been realized in a 130nm SOI Partially Depleted platform [1-2]. Due to the buried oxide and the thin silicon film thickness, the study of self-heating effects is of a great interest for this technology [3-4], and more particularly for NLDEMOS devices. In order to quantify these phenomena, experiments have been performed on dedicated NLDEMOS SOI structures for self-heating measurements.

This SOI HV device is based on a low doped unsalicided Drift region that sustains the drain bias (Figure 1).

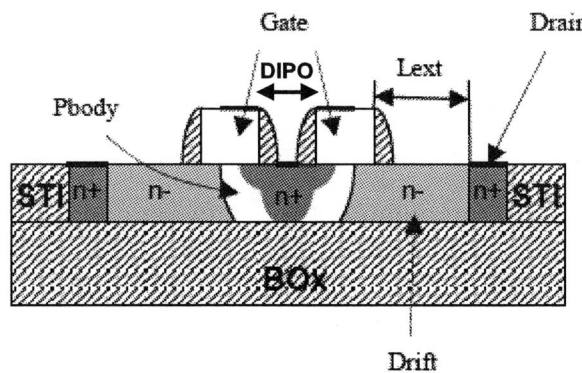

Figure 1. NLDEMOS device under test.

The initial SOI layer and the buried oxide are respectively 1600Å and 4000Å thick. High resistivity (>1 kΩ.cm) Unibond® SOI substrates have been used. Oxide thickness is 50Å. The body implant is self-aligned with the gate and allows targeting a 0.6V threshold voltage

F. Dieudonné, S. Haendler, S. Chouteau, J. Rosa, P. Waltz, A. Perrotin, L. Boissonnet, B. Rauber, C. Schaffnit and C. Raynaud belong to STMicroelectronics, Crolles R&D, 850 rue Jean Monnet, BP 16, 38926 Crolles, France.
E-mail: f.dieudonne-crolles@st.com.

independently of well implantations. The Drift region is protected from Silicidation whereas the Source & Body implants are short-circuited (the P+ Body implant – not shown in Figure 1 – is laterally realized [2]).

II. SELF-HEATING EVALUATION

A. Structures

The device under test is shown in Figure 1. Various dimensions (L_{ext}: drain extension length; DIPO: interpoly length; number of drain and source contacts...) have been implemented in order to analyze the impact of these variations on the thermal resistance.

For the standard NLDEMOS device (W/L=(2*5)/0.5μm and L_{ext}=0.6μm), the main electrical parameters are given in Table 1.

TABLE I
MAIN DC & RF PARAMETERS. V_T IS MEASURED WITH V_D=0.1V, I_{ON} WITH V_G=2.5 & V_D=5V, I_{OFF} AT V_D=5 & V_G=0V, AND F_T, F_{MAX} @V_D=4V.

V_t	$S.R_{on}$	BVds	I_{on}	I_{off}	F_t	F_{max}
V	mΩ.mm²	V	μA/μm	pA/μm	GHz	GHz
0.6	8.5	19.5	375	7.5	20	35

Good DC results are obtained with a reasonable I_{off} current level. Note that the I_{on} current is measured in DC conditions and thereby suffers from self-heating effects. RF measurements of F_t and F_{max} at V_d=4V also show some good performances similar to state-of-the-art results for bulk or SOI technologies [5-6].

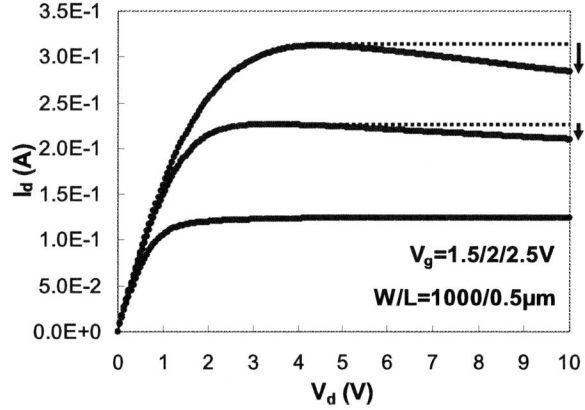

Figure 2. $I_d(V_d)$ of a W/L=1000/0.5μm device for different gate biases (V_g=1.5-2-2.5V).

1-4244-0116-X/06/$20.00 ©2006 IEEE 201

An example of $I_d(V_d)$ characteristic exhibiting self-heating is shown in Figure 2 for a large area device. Drain current reduction at high drain biases is clearly emphasized for V_g=2V & 2.5V in this case (see the arrows pointing out this phenomenon).

B. Measurement methodology

The thermal resistance R_{th} and the temperature rise ΔT_e are obtained by the measurement of the gate resistance using two gate contacts [3]. For each device, the gate resistance R_{goff} is measured at various temperatures to obtain a calibration curve $R_{goff}(\Delta T_e)$ (without self-heating, i.e. V_d=0V & V_g=2.5V). The following temperatures are chosen: 25, 50, 85 and 125°C. Then, measurements are performed for V_d=V_{dd}=5V & V_g=2.5V at room temperature to obtain R_{gon}.

Thanks to the measurement of R_{gon}, the temperature increase ΔT_e for a given test temperature is obtained as illustrated in Figure 3.

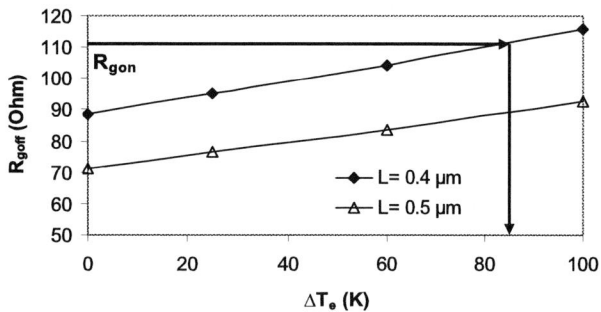

Figure 3. Calibration curves for two gate lengths with the extraction of ΔT_e.

Finally, the thermal resistance is calculated through the relation:

$$\Delta T_e = R_{th} \cdot P_{stat} \qquad (1)$$

where P_{stat} is the static dissipated power.

C. Self-heating results

Figure 4 shows the evolution of the temperature rise and the thermal resistance with the interpoly distance DIPO. A reduction of these parameters (respectively -8 and -20% between DIPO=0.42 and 1.2µm) is shown with the increase of the DIPO distance.

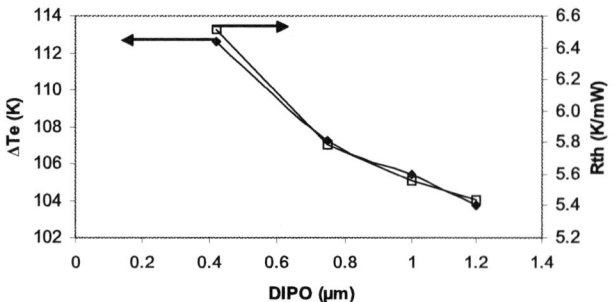

Figure 4. Evolution of ΔT_e & R_{th} with DIPO.

A better improvement for R_{th} (-20% compared to -8% for ΔT_e) is due to the I_{on} current increase with higher DIPO dimensions because of a reduction of the source series resistance.

The temperature rise and the thermal resistance variations with the extension dimension L_{ext} are then represented in Figure 5. Again, an improvement of ΔT_e and R_{th} is obtained by increasing L_{ext}. This time, a better reduction is found for ΔT_e due to the reduction of the power with L_{ext} (stronger drain resistance). Nevertheless, the increase of the L_{ext} dimension degrades the electrical performances of the devices (S.R_{on}, I_{on}, F_t and F_{max}…).

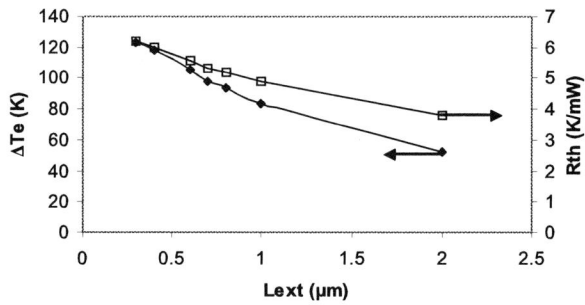

Figure 5. Evolution of ΔT_e & R_{th} with L_{ext}.

The influence of the active width W_a on R_{th} and ΔT_e is represented in Figure 6. As expected, R_{th} decreases when W_a increases due to the larger active area. However, ΔT_e increases because of the enhancement of the dissipated power.

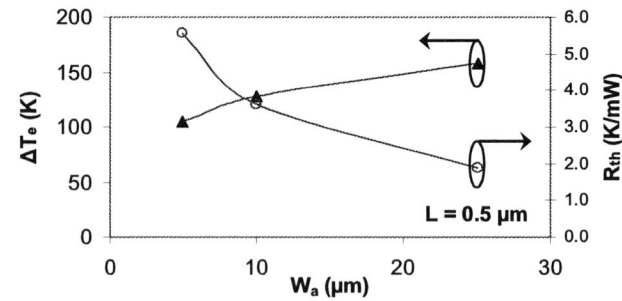

Figure 6. Evolution of ΔT_e & R_{th} with the active width.

A similar plot is done in Figure 7 for the impact of the channel length L_p on these parameters. Increasing L_p results in a decrease of both R_{th} and ΔT_e. Indeed, ΔT_e diminishes thanks to the R_{th} and P_{stat} reductions with L_p.

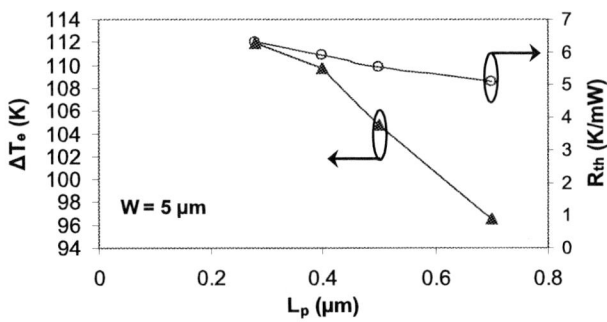

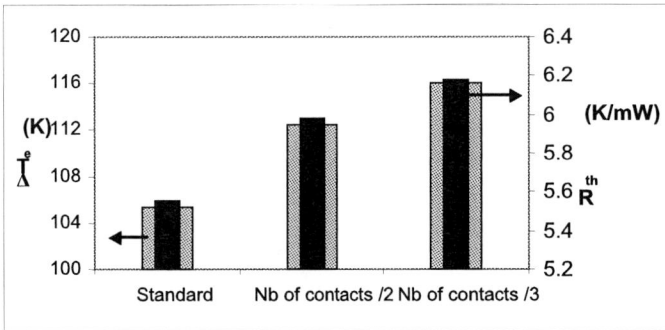

Figure 9. Impact of the active contact number on ΔT_e & R_{th}.

Figure 7. Evolution of ΔT_e & R_{th} with the channel length.

III. CONCLUSION

As shown in Figure 8, where $R_{th}(S_a)$ is plotted for various devices having different active surfaces, the thermal resistance is greatly proportional to the reverse of the active surface S_a. This emphasizes that heat is mainly evacuated through the buried oxide layer [4].

In this paper, self-heating measurements were carried out on NLDEMOS devices from a 130nm SOI PD technology platform. A DC method determining gate resistances was used in our case. Evolution of both temperature rise ΔT_e and thermal resistance R_{th} was shown varying some design rules or the device topology. Improvements were obtained increasing the DIPO & L_{ext} dimensions. A strong linear correlation between the thermal resistance and the reverse of the active surface was also demonstrated. Finally, a moderate impact of the metallization was observed.

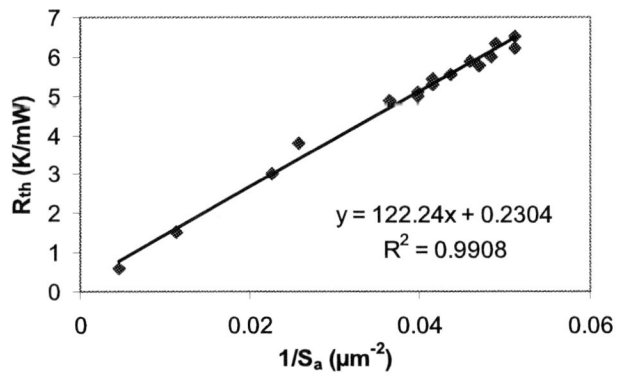

Figure 8. Evolution of R_{th} with the active surface.

REFERENCES

[1] C. Raynaud *et al.*, Is SOI CMOS a promising Technology for SOCs in High Frequency Range?, accepted to ECS'2005.

[2] O. Bon *et al.*, High voltage devices added to a 0.13µm high resistivity thin SOI CMOS process for mixed analog RF circuits, accepted to the 2005 IEEE Int. SOI Conf.

[3] C. Anghel *et al.*, Self-Heating Characterization and Extraction Method for Thermal Resistance and Capacitance in HV MOSFETs, IEEE EDL Vol. 25, N°3, pp. 141-143, March 2004.

[4] M. Reyboz *et al.*, Compact Modeling of the Self Heating Effect in 120nm Multifinger Body-Contacted SOI MOSFET for RF Circuits, 2004 IEEE Int. SOI Conf., pp. 159-161.

[5] S. Matsumoto *et al.*, RF Performance of a State-of-the-Art 0.5µm-Rule Thin-Film SOI Power MOSFET, IEEE TED, vol. 48, N°6, pp. 1251-1255, June 2001.

[6] B. Szelag *et al.*, Integration and Optimisation of a high performance RF Lateral DMOS in an advanced BiCMOS technology, ESSDERC 2003, pp. 39-42.

In Figure 9, the influence of the active contacts' number on R_{th} and ΔT_e is considered. It is shown that the reduction of the number of active contacts leads to a slight increase of the thermal resistance, and thus of the temperature rise. This is related to the reduced efficiency of heat evacuation through metallization. Nevertheless, this remains a second order effect compared to the active surface impact.

Poster Session
Power Devices and ICs

2006 25th International Conference on Microelectronics

An Alternative Process Architecture for CMOS Based High Side RESURF LDMOS Transistors

P.M. Holland and P.M. Igic

Abstract – An alternative CMOS manufacturing process architecture for implementing Power Integrated Circuits that may be used for applications requiring a bridge topology is presented. A RESURF N-LDMOS High-Side compatible power transistor was designed onto the new process using TSuprem4 and Medici TCAD software. Masks were designed using Cadence Virtuoso and the new structure was manufactured at X-Fab UK Ltd. The physical results show good transistor characteristics compatible for high-side applications. The Specific RDSon for the new device is 260mΩ.mm² and breakdown voltage for both high-side and low-side operation exceeds 100V.

I. INTRODUCTION

There is an increasing trend towards adding intelligence to power switching such that power management techniques are made as efficient as possible. This trend can only increase as legislation to control the emission of greenhouse gases is being implemented on a world-wide basis. Perhaps the best solution for incorporating intelligence into small to medium voltage applications is to use Power IC technologies. Here the analogue or digital control electronics and the power transistors are integrated onto the same chip [1].

One of the most common power switches used for Power IC technologies is the Lateral Double Diffused Metal Oxide Semiconductor (LDMOS) power transistor [2]. The LDMOS provides many benefits over other power transistor solutions such as ease of control, packing density, compatibility with foundry design libraries etc. that all lead to cost reductions in design and manufacturing [3,4]. However, the integration of LDMOS transistor technology with conventional CMOS circuitry presents some manufacturing processing and architecture problems that must be overcome to ensure correct operation and high manufacturing yields [5].

For many applications it is required to configure the power transistors on the IC in a bridge topology. This creates a problem for single chip solutions using LDMOS power transistors because for the high-side devices in the bridge during the switching cycle, the source may float to voltages close to the voltage supply [6].
This places a requirement on the LDMOS structure designer to electrically isolate the LDMOS source region

The authors are from the Electronics Systems Design Centre, University of Wales Swansea, Singleton Park, Swansea SA2 8PP, United Kingdom, E-mail: p.m.holland@swan.ac.uk

from the substrate such that other CMOS structures and LDMOS power transistors are not effected. To overcome this problem many manufacturers employ triple well architectures to provide the necessary isolation but this approach limits the achievable breakdown voltage due to punch-through between the P-Body of the lateral transistor and the substrate when operated in the high-side mode [7].

A cost effective solution to this problem is proposed with an architecture that adds a p-epi buffer layer between the p+ substrate and the n-epi layer.

II. DEVICE STRUCTURE

The structure, shown in figure 1. creates a higher breakdown voltage in the High-Side mode by allowing the depletion region caused by the Drain/Source to Substrate voltage to extend into the p- epitaxial silicon area. This delays the premature punch-through breakdown voltage from P-body to substrate to a value suitable for low to medium voltage applications (~100V). This structure allows the p-epi / n-epi doping and thickness values to be optimised for RESURF operation and thus creating the most favorable RDSon / Breakdown Voltage trade-off for Low-Side operation. Other advantages of the structure are that the n-epi is thin enough such that the p-sinker that is required for isolation and correct RESURF operation does not require expensive long and high temperature furnace operations to make electrical connection with the p-epi. Also a highly doped p+ substrate is still included that maintains the CMOS resistance to latch-up [8,9].

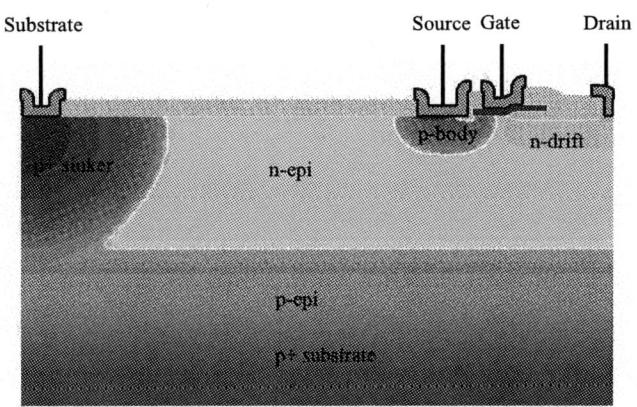

Fig. 1. Cross-Sectional diagram of LDMOS design.

1-4244-0116-X/06/$20.00 ©2006 IEEE   207

III. DEVICE DESIGN

The structure was modeled using T-Suprem4 and Medici. The n-type and p-type epitaxial silicon doping and thicknesses were varied to find the optimal values for High-Side and Low-Side breakdown and RDSon.

Figure 2 shows a 3-D plot of the electric field across the structure for simulated operation in the high-side mode. Here the drain, source and gate voltages have been increased until the device is just at the point of breakdown. The depletion region created by the potential difference between drain/source and substrate extends into the p-epi buffer region extending the breakdown voltage.

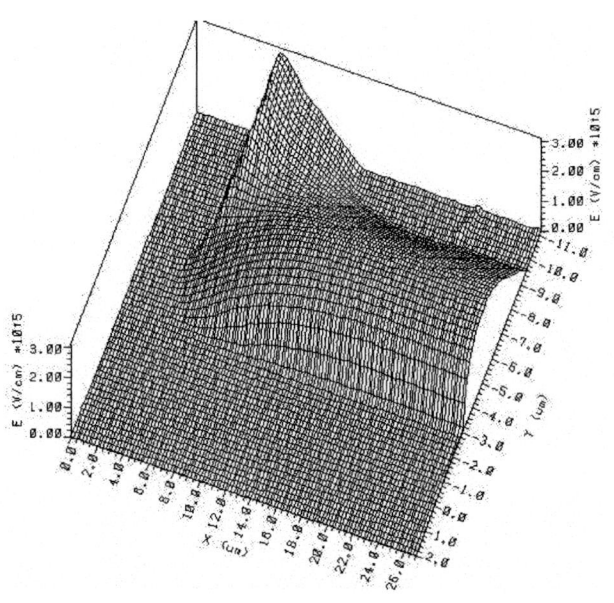

Fig. 2. Simulated Electric Field Distribution for High-Side Operation

IV. PHYSICAL LAYOUT

The LDMOS layout was designed using Cadence Virtuoso software and manufactured at X-Fab Plymouth. Figure 3. shows a photograph of the manufactured design used for test applications. Smaller versions were used for test purposes when experimenting with mask size and position to optimise performance. This structure was also used to calculate the RDSon for this type of design as it represents a typical design for a commercial application.

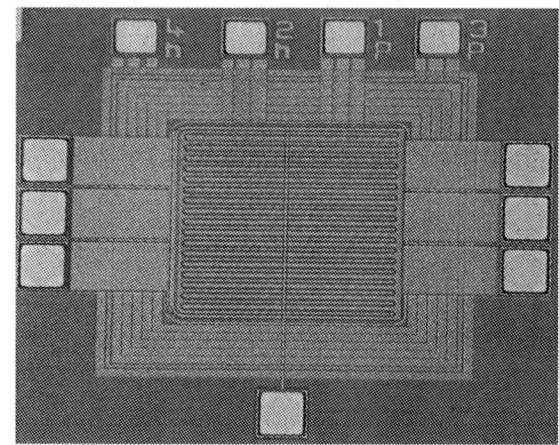

Fig. 3. Photograph of Interleaved LDMOSFET Test Structure

V. RESULTS

A. Measured Characteristics

The manufactured devices were probed and typical results for the transfer characteristic, output characteristic curves and breakdown characteristic are shown in figures 4, 5, and 6.

The transfer characteristic in figure 4. shows a threshold voltage (V_{TH}) of approximately 1V that is an optimal value for 5V Power IC CMOS process. This value may be changed if required by changing the p-body implant dose. This must be done carefully to prevent introducing a punch-through breakdown mechanism if the doping is reduced or significantly changing the effective length of the transistor channel and increasing RDSon.

Transfer Characteristic

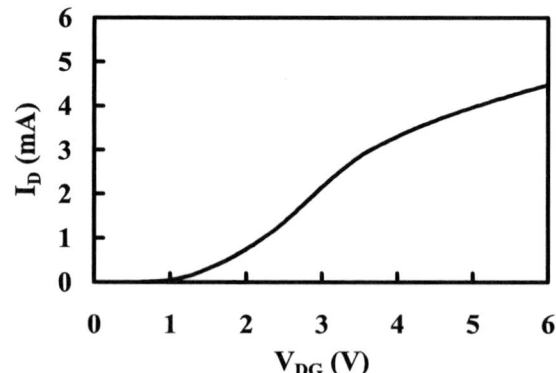

Fig. 4. Physical Transfer Characteristic

The experimental DC Characteristic Curves for the test structure are shown in figure 5. The 'square rule' often used to describe the saturation drain current dependency on the gate voltage in MOS designs does not apply to this LDMOS design [10,11]. This was attributed to electron mobility reduction due to the high gate field [12,13].

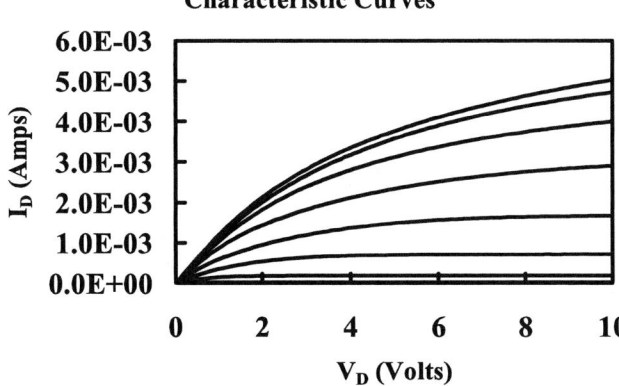

Fig. 5. Physical DC characteristic Curves. Gate Voltage from 1V – 5V step 0.5V

Figure 6 shows the experimental breakdown voltages for the conventional Drain to Source avalanche breakdown and for the punch-though breakdown between the p-body region and the p-epi/p+ substrate when the device is operated in the high-side mode. The two breakdown mechanisms were investigated by TCAD simulation and the maximum breakdown voltage values found by careful selection of the n-epitaxial / p-epitaxial silicon doping and thicknesses to use the RESURF effect and by careful selection of the poly-silicon gate on field oxide overlap.

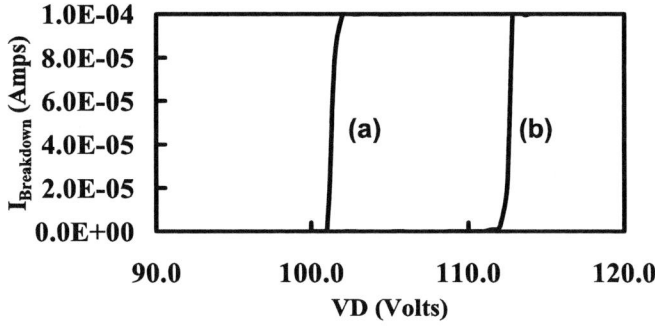

Fig. 6. Breakdown Characteristics for (a) Avalanche Breakdown from Drain to Source and (b) Punch-through Breakdown

B. 2-D RESURF Action

Test transistors were layed out that had varying overlaps of the poly-silicon gate over the field oxide. This was done to manipulate the electric field at the p-body / n-drift interface and to tune the RESURF effect. Figure 7 shows a plot of the Breakdown Voltage between Drain and Source as a function of the gate overlap. It can be seen that the optimal value is between 2 and 2.5µm where the breakdown voltage is greater than 100V.

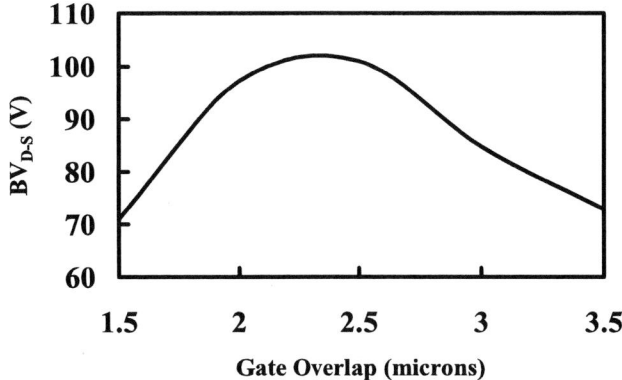

Fig. 7. Breakdown Voltage Verses Poly-Silicon Gate Overlap Onto Field Oxide.

V. CONCLUSION

The results show that a 2-D RESURF LDMOS device has been designed and implemented that may be used for high-side applications. The device was modeled using TSUPREM4 / MEDICI to achieve the desired breakdown performance, high-side application compatibility and RESURF operation. The mask-set to implement the device was designed using CADENCE Virtuoso and all the changes used were compatible with a standard CMOS process with only an additional n-type epitaxial layer and three additional implants required. The device was manufactured at X-Fab, Plymouth. The device gives performance of RDSon equal to $260m\Omega.mm^2$, the Breakdown Voltage exceeds 100V and the device is high-side compatible. By varying the poly-silicon gate on field oxide overlap the devices have been 'tuned' to display optimal RESURF operation

ACKNOWLEDGEMENT

The authors would like to thank the team at X-fab, Plymouth, United Kingdom. In particular Graham Chapman, John Ellis and Ian Daniels for their technical advice and help in the processing of the silicon wafers.

REFERENCES

[1] J.-L. Sanchez, "State of the Art and Trends in Power Integration", in *Technical Proceedings of the 1999 International Conference on Modelling & Simulations of Microsytems*, MSM1999, pp.20-29

[2] G. Charitat, M.A. Bouanane, P. Austin and P. Rossel "Modelling and improving the on-resistance of LDMOS RESURF devices", *Microelectronics Journal*, 1996, vol.27, pp. 181-190

[3] T. R. Efland "Integration of Power Devices in Advanced Mixed Signal Analog BiCMOS Technology", in *Proc. 13th International Symposium on Power Semiconductor Devices*, ISPSD 2000, pp. 39-44

[4] J.A. Appels and H.M.J. Vaes "High Voltage Thin Layer Devices (Resurf Devices)", IEDM-1979, p.238-241

[5] Adriaan W. Ludikhuize "A Review of RESURF Technology", in *Proc. 13th International Symposium on Power Semiconductor Devices* ISPSD' 2000, pp. 11-18

[6] B. Murari et al "Smart Power ICs", Springer 2002

[7] R. M. Forsyth "Technology and Design of Integrated Circuits for up to 50V applications"

[8] Y. Taur et al "Characterization and Modeling of a Latchup Free 1-μm CMOS Technology", IEDM-1984, p.398-402

[9] S. Wolf and R.N. Tauber "Silicon Processing for the VLSI Era", Vol 2, Lattice Press 1990

[10] S.M. Sze "Semiconductor Devices Physics and Technology", Wiley 2002

[11] B. Jayant Baliga "Modern Power Devices", Wiley 1987

[12] R. van Langevelde and F. M. Klaasen "Effect of Gate-Field Dependent Mobility Degradation on Distortion Analysis in MOSFET's", in *IEEE Transactions on Electron Devices*, November 1997, VOL. 44, NO 11, pp.2044-2053

[13] J. I. Lee et al "Mobility Reduction Parameter's in Short-Channel MOSFETs", in *Electronics Letters*, 25th May 1989, Vol. 25, No.11, pp. 753-754

2006 25th International Conference on Microelectronics

Comparison of measured and simulated characteristics of boron implanted 4H-SiC DiMOSFET

S. C. Kim, W. Bahng, K. H. Kim, I. H. Kang, S. J. Kim, N. K. Kim

Abstract – The vertical double implanted MOSFET (DiMOSFET) were fabricated on 4H-SiC(0001) and the device characteristics were investigated. The sheet resistance of the Ni/Ti metals on the P-well region were 41.7kΩ/□ and 12.7kΩ/□ for post ion implantation annealed at 1600°C and 1700°C, respectively. The DiMOSFET device had blocking capability of 980V and the threshold voltage of 0.5V. However abnormal characteristics were observed in the forward characteristics. In this paper, we focused on the narrowing effect of JFET region due to lateral diffusion of implanted boron ions. The structures with narrow or closed JFET at the bottom region showed drain voltage shift.

I. INTRODUCTION

Silicon carbide power devices have superior advantages compared to silicon power devices. SiC power devices are expected to be advantageous for saving energy consumption in electric power components. SiC MOSFET devices are very attractive devices for the high frequency and high power applications.

Among wide band-gap semiconductor materials, SiC hac the unique property that high quality thermal oxide can be easily grown. Furthermore, the high critical electric field (Ec) of SiC enables power MOSFETs using 10% drift layers thickness and 100 times the doping concentration as compared to silicon devices with comparable blocking voltage. This should result in a much smaller specific on-resistance $R_{on,sp}$. Therefore, SiC MOSFET is regarded as a leading candidate for the next-generation ultra-low-loss power device [1]. A number of vertical high-voltage SiC MOSFETs has already been demonstrated [2, 3]. However, the performances of the present-time experimental SiC MOSFETs are far from reaching the theoretical limit. One of the serious problems in SiC-based vertical MOSFET is the poor channel mobility because of the large number of interface traps existing at the SiO$_2$/SiC interface [4]. Much effort has been devoted to improve the SiC MOS channel mobility, but several difficulties still remain [5]. Generally, 4H-SiC is the preferred polytype since the bulk mobility is higher and more isotropic, compared to 6H-SiC. In this

paper, characteristics of fabricated 4H-SiC MOSFET are presented, and compared to simulation results.

II. MOSFET FABRICATION

The schematic diagram of fabricated 4H-SiC DiMOSFET is shown in Fig. 1. The DiMOSFET are made on 8° off axis Si face n+ substrates with 10μm thick n-type epilayer with doping concentration of 3.36×10^{15} cm^{-3}. The p-well regions were made by six subsequent boron ion implantations at 650 °C with the acceleration energies ranged in the range 30keV – 370keV and implantation doses ranging between 5.3×10^{11}cm^{-2} – 2.08×10^{13}cm^{-2} followed by post-implant activation for 30 min at 1600°C in argon atmosphere. As a result, P-well was formed with a surface concentration and junction depth of around 5×10^{17} cm^{-3} and 0.65μm. For ohmic contact with metal, the doping in the surface of p-well region was modified using three subsequent aluminum implantations at 650 °C using energies in the range 30keV – 140keV and implantation doses ranging between 1.5×10^{14}cm^{-2} – 5.0×10^{15}cm^{-2}. The final implantation depth and doping concentration are 0.2 μm and 5×10^{19}cm^{-3}, respectively. During high temperature activation, the surface was covered with a graphite cap layer in order to prevent sublimation and roughening of the surface.

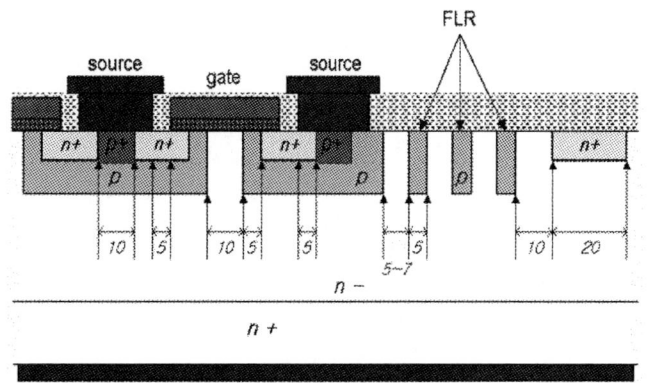

Fig. 1. Cross-section of simplified 4H-DiMOSFET.

For formation of deep p-well in the MOSFET, aluminum or boron implantation is generally used. Aluminum has a shallower acceptor level than boron and therefore its ionization is much higher and the resistance is

S. C. Kim, W. Bahng, K. H. Kim, I. H. Kang and N. K. Kim are with the Power Semiconductor Group, Korea Electrotechnology Research Institute, 28-1 Sungju-dong, Changwon-City, GyungNam, Rep. of Korea, E-mail: sckim@keri.re.kr

S. J. Kim is with the R&D Department, Itswell Co., Ltd, 9-4BL, Ochang Scientific Industrial Complex, 1115-4 Namchon-ri,, Oksan-myeon, Cheongwon-gun, Chungbuk, Rep. of Korea, E-mail: kseongjin@itswell.com

1-4244-0116-X/06/$20.00 ©2006 IEEE

much lower than boron at any given temperature [6]. However, aluminum ions are difficult to form deep p-well due to its heavy mass. And heavy ions implantation causes much damage in SiC lattice. Consequently, performance of the devices would be degraded. From the above reason, boron implantation was commonly used as a deep p-well formation.

The source region was formed by nitrogen ion implantation at 650°C. Using multiple implant energies and doses, a box profile of about 5×10^{19}cm^{-3} and a junction 0.1 μm was fabricated. Thereafter a sacrificial oxide with thickness of 250Å was grown in steam at 1100°C and removed in buffered HF solution. A PECVD oxide layer (0.5 μm thick) was then deposited as an inter-metallic dielectric layer, and the channel area was opened by wet etching. The field oxide deposition temperature was 300°C and deposition rate was 15nm/min, respectively. The gate oxide for the Si-face was then grown in 10 % N$_2$O ambient in a oxidation furnace at a temperature of 1175°C for one hour. The thickness of the gate oxide was measure to 20nm. A nickel and aluminum was used as the gate metal and nickel as the contact metal for source and drain. The field limiting ring (FLR) was fabricated for improve breakdown characteristics.

III. RESULT AND DISCUSSION

In the case of SiC devices, ohmic contact on the n-type substrate is relatively easy than on the p-type substrate. Therefore, many researchers would like to find different materials and different process for the ohmic contact. But this is obstacle for integration of the process in MOSFET. Accordingly, we measured the ohmic characteristics of Ni/Ti on the p-well formed aluminum implantation. To measure the specific resistance, we used linear and circular type TLM method. Fig. 2 shows the resistance characteristics of linear and circular pattern. From this figure, we find big difference depend on the annealed temperature between 1600°C and 1700°C. The transfer lengths were 2.6μm in the 1600°C annealed sample and 1.25μm in the 1700°C annealed sample, respectively. And also, the sheet resistances were 41.7kΩ/□ and 12.7kΩ/□, respectively. From the above result, we calculated the N$_a$-N$_d$ and activation ratios. The activation ratio of 1600 °C annealed sample was 10% and 60% for the 1700°C annealed sample, respectively. And the specific contact resistance between Ni/Ti and p-well region were 1.8×10^{-5}cm^2 and 5.6×10^{-5}cm^2, respectively.

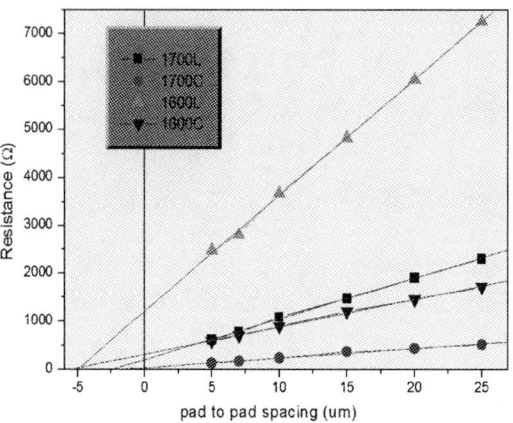

Fig. 2. Measured result of resistance depends on the annealing temperatures.

Fig. 3 shows the room temperature on-state characteristics of 4H-SiC DiMOSFET. The channel with and length of this device are 660 and 5μm, respectively. The threshold voltage extracted from saturation region is approximately 0.5V, which is significantly lower than the value reported up to now for the 1000V MOSFET devices. It was possible to low threshold voltage just using thin and high quality gate oxide layer. In the case of thin gate oxide, leakage current was increased through the thin gate oxide. And also, high voltage was applied to the source-drain and gate drain. Therefore, optimized termination structure is very important for the thin gate oxide. In this study, we optimized high voltage edge termination structure. The off-state characteristics are shown in Fig. 4. The device is normally-off, and showed stable avalanche behavior at a V$_{DS}$ of 980V.

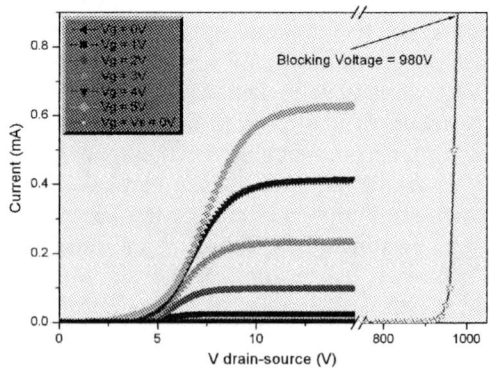

Fig. 3. Forward characteristics of fabricated 4H-SiC DiMOSFET.

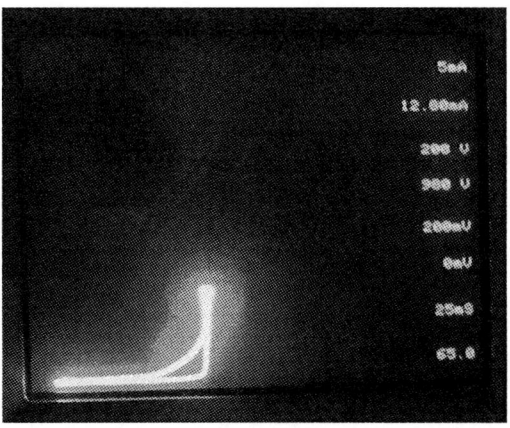

Fig. 4. Off-state characteristics of fabricated 4H-SiC power DiMOSFET.

The V_{DS}/V_{th} ratios of the fabricated 4H-SiC DiMOSFET are compared to those of published results in Table 1. From this table, most of the results of the V_{DS}/V_{th} are 100 to 400. I.e., high breakdown voltage causes the high threshold voltage in SiC MOSFET. However, V_{DS}/V_{th} ratio is 1960 in this study. This means that thin and high quality of gate oxide is very important factor for fabrication of SiC MOSFET.

TABLE 1.
COMPARISON OF V_{DS}/V_{th} RATIO IN THE PUBLISHED RESULT

V_{DS}/V_{th} ratio	V_{th} (V)	Breakdown Voltage (V)	reference	remark
-	4.6	?	Kiritani	4H-SiC
260 ~ 430	3 ~ 5	1300	Banerjee	Lateral, 6H-SiC
140 ~ 175	8 ~ 10	1400	Banerjee	Lateral, 6H-SiC
154	7.8	1200	Peters	4H-SiC
324	8.5	2750	Ryu	4H-SiC
267	3	800	Infineon SIPMOS	Silicon
1960	0.5	980	This work	4H-SiC

One of the un-expected point is that there is no drain current until drain voltage is reached to 5V. To explain this result, simulation for the devices with several possible p-well structures using TCAD was done. In generally, the diffusion of boron ions along to the lateral direction is faster than along to surface normal direction due to the anisotropic nature of 4H-SiC during high temperature activation process [7]. In many reports, they considered that boron diffusion along to the lateral direction is just 1 ~ 2μm. But, JFET region was almost closed in this study. Considering of our mask design, the distance of p-well was 10μm. Therefore, the lateral diffusion length is almost 5μm.

From this reason, the electron passed the channel region could not flow to drain contact at the low source-drain voltage. The considering DiMOSFET structure in the simulation is showed in Fig. 5. Because of retrograde doping profile, the concentration of the implanted ion of bottom side was higher than that of the surface. Therefore, diffused length in the bottom side of JFET was much longer than the surface. From this reason, the bottom of JFET region changed to intrinsic region. The simulation result of forward characteristics of considering structure is shown in Fig. 6. The dose dependencies of lateral diffusion during high temperature annealing were caused the JFET narrowing effect. This simulated forward characteristic was in good agreement with the measured one. As a result, current blocking layer was formed at the bottom side of JFET region due to lateral diffusion and retrograde doping profile of the boron ions. In generally, the diffusion of boron ions along to the lateral direction is faster than along to surface normal direction due to the anisotropic nature of 4H-SiC during high temperature activation process. But, the diffusion length to the lateral direction is much longer than we expected. Consequently, when we design and fabricate DiMOSFET using boron implantation, we have to consider lateral diffusion of boron ions.

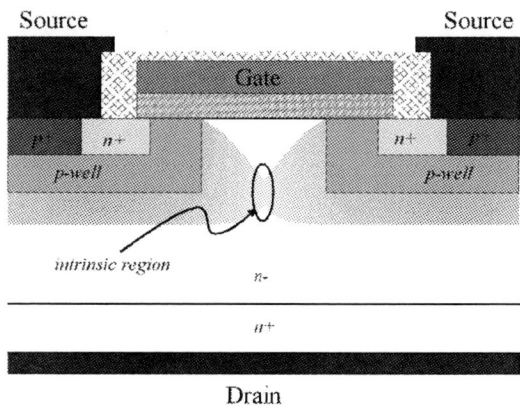

Fig. 5. Conception view of lateral diffusion in the SiC DiMOSFET.

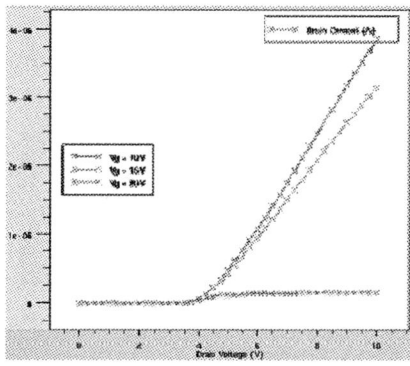

Fig. 6. Simulated forward characteristics of the considered DiMOSFET.

IV. SUMMARY

4H-SiC DiMOSFETs have been fabricated. The threshold voltage and avalanche breakdown voltage of the fabricated 4H-SiC DiMOSFET are 0.5V and 980V, respectively. However, abnormal forward characteristics of DiMOSFET devices have been observed because of the retrograde doping profiles in p-well region. To verify this result, we used the computer simulation. From the simulation result, current blocking layer was formed at the bottom side of JFET region due to lateral diffusion and retrograde doping profile of the boron ions. Consequently, when we design and fabricate 4H-SiC DiMOSFET using boron implantation, we have to consider lateral diffusion of boron ions.

ACKNOWLEDGEMENT

This work was done as a part of SiC Device Development Program (SICDDP) supported by MOCIE (Ministry of Commerce, Industry and Energy) Korea.

REFERENCES

[1] J.A. Cooper, M.R. Melloch, R. Singh, A.K. Agarwal, and J.W. Palmour: IEEE Electron Devices, Vol. 49 (2002), p. 658

[2] A.K. Agarwal, J.B. Casady, W.F. Valek, M.H. White, and C.D. Brandt: IEEE Electron Device Lett. Vol. 18 (1997), p. 586

[3] Y. Li, J.A. Cooper, and M.A. Capano: Materials Science Forum, Vol. 389-393(2002), p. 1191

[4] N.S. Saks, S.S. Mani, and A.K. Agarwal: Appl. Phys. Lett. Vol. 76 (2000), p. 2250

[5] M.K. Das: Materials Science Forum, Vols. 457-460 (2004), p. 1275

[6] W. J. Choyke, H. Matsunami, and G. Pensl, Silicon Carbide, Recent Major Advances, (Springer, 2003).

[7] M. K. Linnarsson, M. S. Janson, A. Shoner, A. Konstantinov and B. G. Svensson, Mat. Sci. Forum Vols. 457-460 (2003), p. 917.

2006 25th International Conference on Microelectronics

IGBT Behavioral PSPICE Model

Katya Asparuhova, Tsvetana Grigorova

Abstract - Behavioral IGBT macromodel for OrCad PSPICE simulator is presented. The model is parameterized and it is implemented as a subcircuit in the simulator. The nonlinear DC equations and voltage-controlled capacitors are precisely represented using ABM method, which is resulting in good accuracy. The static I-V characteristics offered by the behavioral model are shown. The tailing of the anode currents simulated by the behavioral model at a constant anode voltage-switching test are given. The good agreement between the simulation and experimental results can be observed in the paper.

I. INTRODUCTION

The IGBT is rapidly becoming as the preferred switching device in many power electronic circuits. As a consequence, several IGBT CAD models have been recently proposed, to describe the operating characteristics of the device [1]-[5]. Roughly speaking, the IGBT models available in literature can be subdivided as either behavioral or physics-based. The physics-based IGBT models proposed to date, are not easily implemented in circuit simulators, require heavy numerical computations and the knowledge of process parameters, which are not easy to extract from electrical measurements. A significant part of existing behavioral macromodels [5] are physical, based on the internal device structure presented in Fig.1. The others, considering the semi-empirical relations between terminal voltages and currents, have a poor accuracy and their results are valid only in a narrow range of operating conditions.

The paper presents IGBT PSPICE behavioral macromodel, which is built using the Hammerstein configuration. It consists of a nonlinear static block followed by a linear dynamic block. The proposed IGBT PSPICE behavioral macromodel is the improved Oh model given in [2] for Saber simulator.

The output characteristics of the IGBT appear to be similar to the BJT except that the controlling parameter is the input voltage rather than the input current. The transfer characteristic is identical to that of the power MOSFET. This curve is reasonably linear over the wide range of the collector current, becoming nonlinear only at low collector current where the gate-emitter voltage is approaching the

K. Asparuhova is with the Department of Electronics and Electronics Technologies, Faculty of Electronic Engineering and Technologies, Technical University - Sofia, 8 Kliment Ohridski blvd., 1000 Sofia, Bulgaria, e-mail: k_asparuhova@tu-sofia.bg

Tsv. Grigorova is with the Department of Electronics, Faculty of Electronics and Automation, Technical University - Sofia, Branch Plovdiv, 61 Sankt Petersburg blvd., 4000 Plovdiv, Bulgaria, e-mail: c_grigorova@abv.bg

threshold. The parameters and characteristics from datasheets are the starting points for determination the parameters for Spice simulations. Most of the model parameters are obtained from the output characteristics.

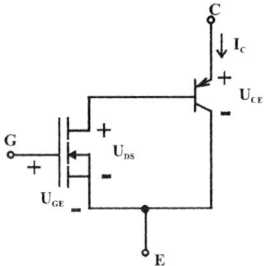

Fig.1 Equivalent circuit of the IGBT

II. DC MODEL OF IGBT

The DC part of the proposed model is based on the empirical formulas for the IGBT collector current given in [3]. This part combines the equations that describe the MOSFET in cutoff, the linear and saturation regions with the equations of a bipolar junction transistor operating in the active mode. The model accounts for high-level injection and the voltage drop in the extrinsic part of the IGBT.

A. IGBT output characteristics in the Cutoff region

The IGBT output characteristics are different from these of the MOSFET. The output current in the cutoff region is zero. Typical value of the output voltage is between 0,7V and 1V [3].

B. IGBT characteristics in the linear region

The linear region of the IGBT and MOSFET can be assumed to be the same. Only the transition from the linear into the saturation region must be corrected:

$$V_{DSsat} = V_{GE} - V_{th}, \qquad (1)$$

where V_{th} is the threshold voltage of the MOSFET.

From Fig.1 the saturation voltage of the output transistor can be derived as

$$V_{CEsat} = V_{GE} + V_D - V_{th}, \qquad (2)$$

where V_D is the voltage drop across the emitter-base junction.

In order to make the saturation voltage of MOSFET equal to the corresponding IGBT saturation voltage the

1-4244-0116-X/06/$20.00 ©2006 IEEE   215

correction function f_1, dependent on gate-emitter voltage, is introduced [3]:

$$f_1 V_{CEsat} = V_{GE} + V_D - V_{th}. \quad (3)$$

The correction function is approximated with an appropriate polynomial. The degree of this polynomial depends on the number of the accounted points and required accuracy:

$$f_1 = a_0 + a_1 V_{GE} + a_2 V_{GE}^2. \quad (4)$$

For $V_{DS} < V_{DSsat}$, the MOSFET drain current in the linear region is given by

$$I_D = k_p \left[(V_{GE} - V_{th}) V_{DS} - \frac{V_{DS}^2}{2} \right]. \quad (5)$$

The collector current in this region is expressed as

$$I_C = (1 + \beta) k_p \left[(V_{GE} - V_{th})(V_{CE} - V_D) - \frac{(V_{CE} - V_D)^2}{2} \right], (6)$$

for $V_{CE} < V_{GE} + V_D - V_{th}$; β is the current gain of the BJT.

C. IGBT characteristics in the saturation region

For $V_{DS} > V_{DSsat}$, the MOSFET drain current is saturated and is given by

$$I_D = k_p \frac{(V_{GE} - V_{th})^2}{2}, \quad (7)$$

and the collector current is expressed analogously as

$$I_C = (1 + \beta) k_p \frac{(V_{GE} - V_{th})^2}{2}, \quad (8)$$

for $V_{CE} > V_{GE} + V_D - V_{th}$.

The output current in the saturation region are fitting using the correction function f_2:

$$\frac{I_{Csat}}{f_2} = k \frac{(V_{GE} - V_{th})^2}{2}, \quad (9)$$

where $k = (1 + \beta) k_p$ and

$$f_2 = b_0 + b_1 V_{GE} + b_2 V_{GE}^2. \quad (10)$$

Finally the expression for a collector current becomes:

$$I_C = \begin{cases} 0, & \text{if} \quad V_{GE} \leq V_{th} \quad or \quad V_{CE} < V_D \\ k.f_2 \left[\begin{array}{c} (V_{GE} - V_{th})(f_1 V_{CE} - V_D) - \\ -\frac{(f_1 V_{CE} - V_D)^2}{2} \end{array} \right] \\ \text{if} \quad V_{CE} < V_{GE} + V_D - V_{th} \\ k.f_2 \frac{(V_{GE} - V_{th})^2}{2}, \\ \text{if} \quad V_{CE} > V_{GE} + V_D - V_{th} \end{cases} \quad (11)$$

There has to be noticed that the correction functions f_1 and f_2 are equal to unity for the low level

injection (it means for the small values of V_{GE}), which case is not observed in the practice.

The polynomial coefficients a_i and b_i of f_1 and f_2 are determined from the I_C-U_{CE} curves. Record the coordinates of the saturation points (U_{CEsat}, I_{Csat}) for three given gate voltages V_{GE}. The value of U_{CEsat} is read at the point where the tangent in saturation region is separated from the curve. Two systems with three equations are obtained substituting in Eqs. (3), (4) and (9). Using the systems' solutions the correction functions f_1 and f_2 can be calculated for each value of the gate voltage.

Thus the proposed functions (11) model more accurately the behavior of the dc collector current than those in [2]. The introduction of two-correction functions f_1 and f_2 allows the use of the more complex approach that relies on physical modeling [3].

III. DYNAMIC MODEL OF IGBT

The dynamic model is presented on Fig. 2.

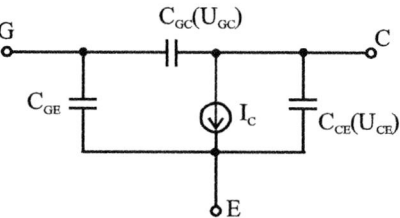

Fig.2 Dynamic behavioral model

From the datasheets, the following capacitance curves can be obtained:

$$C_{ies} = C_{GE} + C_{GC}, \quad C_{res} = C_{GC}, \quad C_{oes} = C_{CE} + C_{GC},$$

where C_{ies} is the input capacitance, C_{res} the reverse transfer capacitance, and C_{oes} the output capacitance. The values of the capacitors can be extracted from the dependence capacitance vs. V_{CE} for $V_{GE}=0$. To achieve an accurate description of IGBT's switching waveforms, it is necessary to develop a high precision voltage-controlled capacitance model that exhibit nonlinear variation of the corresponding voltages. The graphs from datasheet are digitizing using an appropriate program like GetData. In order to consider the dependence of the capacitors on the gate-emitter voltage their values are obtained by scaling on the fitting expression.

The process of the device switching is modeled using a voltage-controlled switch S with parameters: $V_{on} = \{V_{th}\}, V_{off} < \{V_{th}\}, R_{on} = \{V_{CE(sat)} / I_C\}$ and R_{off} - a very big value corresponding on turn-off transistor.

IV. IGBT PSPICE MODEL IMPLEMENTATION

The behavioral model described above has been developed using ABM method and implemented in the OrCad PSPICE simulator as a subcircuit. The DC part of the IGBT model includes the voltage control current sources GVALUE which implement Eqs. (11), using "IF-THEN-ELSE" operator in PSPICE. The correction functions f_1 and f_2 are realized using GPOLY source.

A voltage dependent capacitance can be specified by using a look-up table, or by using a polynomial. In this paper the look-up table in the ABM expression is used. The nonlinear capacitor in the model is replaced by a controlled current source G, which current is defined by

$$I = C(V)dV/dt \ . \qquad (13)$$

The time derivative, $dV(t)/dt$, is modeled by using the DDT function in PSPICE. This table contains (voltage, capacitance) pairs picked from points on the curve. The voltage input is nonlinearly mapped from the voltage values in the table to the capacitance values. Linear interpolation is used between table values. This voltage dependent capacitance is the multiplied by the time derivative of the voltage to obtain the output current.

The realization in Spice language is with GVALUE "look-up table" voltage control current source:

$$Table(V(\%IN+,\%IN-),voltage,capacitor)* \\ DDT(V(\%IN+,\%IN-)) \qquad (14)$$

C_{GC}, C_{CE} and C_{GE} are modeled in this manner.

An advantage of the proposed model it is parameterized. It allows easy implementations of the deferent IGBT types. A part of the preliminarily extracted parameters are included in Param operator – these are V_{th}, V_D, β, k_p. Other part is included in Table operator – these are the digitizing capacitance plot data. The parameters of the voltage-controlled switch S are parameterized too.

V. SIMULATION AND EXPERIMENTAL RESULTS

The type CM600-24H IGBT is chosen For the verification of the model. Fig. 3 shows the static I-V characteristics offered by the behavioral macromodel and Fig. 4 these received from the device implemented PSPICE model. In the calculation it was assumed that V_{th}=6V, β=100, k_p=1.01. The value of k_p is chosen from the data sheet. The simulations are made under following initiation conditions: $V_{CE} = 0 \div 12V$, $V_{GE} = 7 \div 14V$.

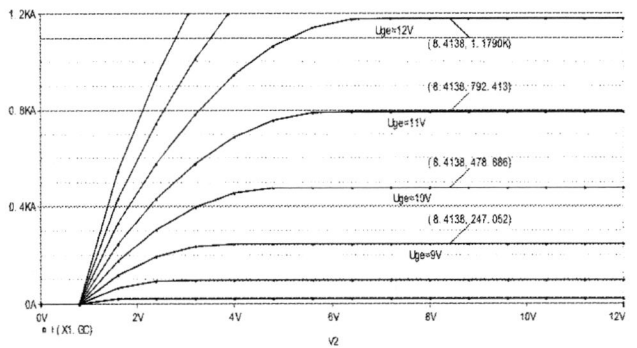

Fig.3. Behavioral macromodel simulated static I-V characteristics for the device CM600-24H

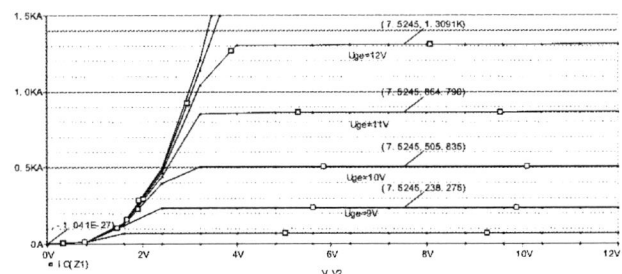

Fig.4. Implemented PSPICE model simulated static I-V characteristics for the device CM600-24H

The analyses results are summarized in Table I.

TABLE I

V_{GE},V	I_C, A Data Sheet	I_C, A Implemented PSPICE model		I_C, A Behavioral macromodel	
		Value	Error, %	Value	Error, %
9	236.269	238.275	0.85	247.052	4.56
10	484.855	505.835	4.33	478.886	1.23
11	795.855	864.790	8.66	792.413	0.43
12	1181.35	1309	10.8	1179	0.2

The average error between the theoretical predictions of the behavioral model and the experimental data is less 1% for V_{GE}>9V. The largest error occurs around V_{GE} = 8V and is equal to 8%.

Fig.5 shows one of the transient test circuits at resistor and inductor load to verify the transient predictions of the proposed model.

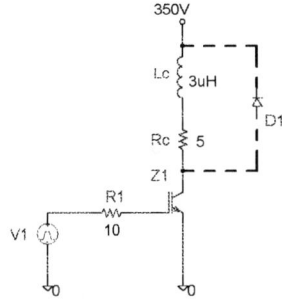

Fig.5. A test circuit at resistor and inductor load

The simulations results of the both models are given in the Fig.6.

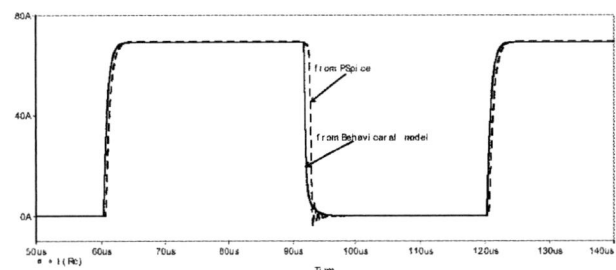

Fig.6 Simulated collector currents at a constant collector voltage-switching test

As seen a very good agreement between transient waveforms is achieved. The error is less than 2%.

PSPICE behavioral model of IGBT type HGTD10N40F1 is created too on the basis of the proposed macromodel [7].

Fig.7 shows the collector voltage at different R_1's, i.e. different gate driving conditions.

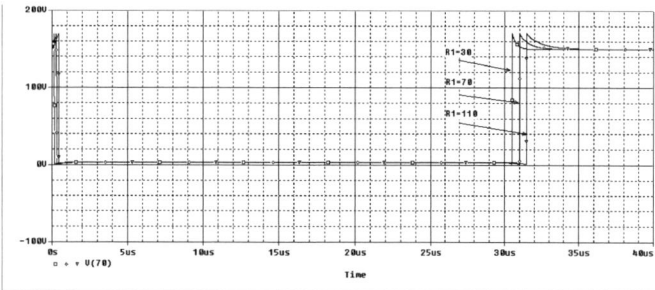

Fig.7 Simulated collector voltage at different R_1's

The simulation and experimental results of the IGBT HGTD10N40F1 connected in the test circuit are presented in Fig.8 and Fig.9. The results are given by the following conditions: the supply voltage V_{CC}=50V; a load -R_C=10Ω and L_C=10μH.

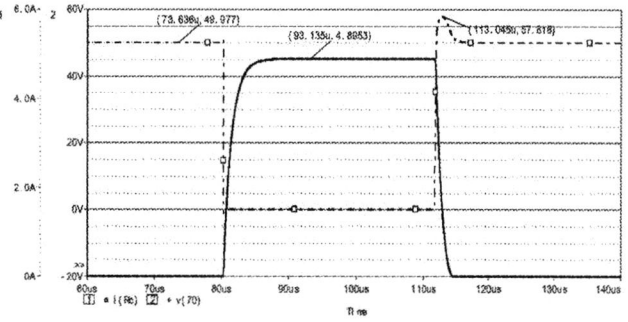

Fig. 8 Transient waveforms using inductive test circuit

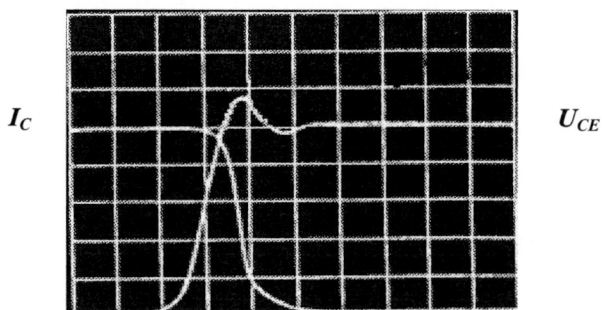

Fig. 9. Test circuit experimental results

The switching curves (U_{CE} and I_C) are given in Fig.9 at the resolution U_{CE}=10V/div and I_C=1A/div. The average error between simulation and experimental results is around 3%.

VI. CONCLUSIONS

The behavioral IGBT macromodel for OrCad PSPICE simulator is presented. The model is parameterized and it is implemented as a subcircuit in the simulator. The nonlinear DC equations and voltage-controlled capacitors are precisely represented using ABM method, which is resulting in good accuracy. The static I-V characteristics offered by the macromodel are shown. The tailing of the anode currents simulated by the behavioral model at a constant anode voltage-switching test are given. The good agreement between the simulation and experimental results can be observed in the paper as the average error is around 3%.

A future research direction could be the automatic extraction of the parameters from the datasheets and the inclusion of temperature effects in the IGBT model presented herein.

REFERENCES

[1] A. R. Hefner, D. M. Diebolt, "An experimentally verified IGBT model implemented in the Saber circuit simulator", *IEEE Transactions on Power Electronics*, vol. 9, No 5 pp. 532-542, Sep., 1994.

[2] H. S. Oh, M. El Nokali, "A new IGBT behavioral model", *Solid- State Electron.*, vol. 45, pp. 2069–2075, Nov. 2001.

[3] F. Mihalich et al., "IGBT SPICE model", *IEEE Transactions on Industrial Electronics*, vol.42, No1, pp. 98-105, Feb.1995.

[4] Z. Shen, T. P. Chow, "Modeling and characterization of the insulated gate bipolar transistor (IGBT) for SPICE simulation", *IEEE Conf. Rec. 5th Int. Symp. Power Semiconductor Devices and IC's*, pp. 165-170, 1993.

[5] A. Maxim, G. Maxim, "A novel analog behavioral IGBT Spice model", *IEEE Trans. on Power Electron.*, pp. 364-369, 1999.

[6] H. S. Kim et al., "Parameter extraction for the static and dynamic model of IGBT", *IEEE PESpConf.*, pp. 71-74, 1993.

[7] W. Kang et al., "A Parameter Extraction Algorithm for an IGBT Behavioral Mode", *IEEE Transactions on Power Electronics*, vol. 19, No6, Nov. 2004.

[8] Intersil Designer's Manuel, 1999.

2006 25th International Conference on Microelectronics

Numerical Analysis of a Trench Gate FLIMOSFET with No Quasi-Saturation, Improved Specific On-Resistance and Better Synchronous Rectifying Characteristics

R Vaid and N Padha

Abstract - In this paper, we report a novel Trench gate FLIMOSFET structure designed using the concept of "Opposite Doped Buried Regions (ODBR) and floating island (FLIMOSFET) along with Trench-gate technology. The proposed device structure exhibits quasi-saturation free output and transconductance characteristics over a wide range of voltages as well as gives reduced on-resistance when compared with the conventional FLIMOSFET for two trench depths 2.5 and 3 microns. The new device structure does not give any degradation towards the breakdown voltage which remains almost constant at 65 volts. It also exhibit better synchronous rectifying characteristics.

I. INTRODUCTION

The trade-off between specific on-resistance (R_{on}) and breakdown voltage (V_B) has been a key issue for researchers throughout the world. To address this issue, recently many power MOSFET configurations has been proposed and studied such as P^+ floating Island devices (FLIMOSFET) [1]-[6], Superjunction/COOLMOS^{TM} devices (SJ) [7]-[8], Opposite Doped Buried Regions (ODBR) MOSFET [9], and P-buried layer Schottky barrier diode [10]. The ODBR and FLIMOSFET devices are based on the concept that the triangular electric field distribution in the bulk is divided into several sections to decrease the magnitude of the peak electric field by inserting electrically oppositely doped floating buried layer in the drift region of the device. Thus the doping concentration of the bulk can be enhanced such that the maximum electric field kept less than the breakdown/critical field of the device (e.g. for silicon, breakdown electric field ~ $3*10^5$ V/cm), leading to reduce the on-resistance of the device proportionately. The FLIMOSFET structure has a distinct advantage over the SJ structure on boron implantation dose control for the P^+ buried floating layer. The SJ structure is designed with the RESURF (Reduced Surface Field) concept, where the implantation dose is to be severely controlled to compensate the space charge in the N^- drift region (i.e., the charge imbalance in N & P pillars) which is not the case in

R Vaid and N Padha are with the Department of Physics and Electronics, Faculty of Science, University of Jammu, Jammu-180006, Jammu & Kashmir, India, E-mail: rakeshvaid@gmail.com

FLIMOSFET. For voltage rating less than 180 V, the narrow SJ columns width of 4 µm or less is very difficult to fabricate and good charge balance between P and N columns cannot be maintained even if fabricateable. Until now, no SJ power MOSFET device has ever been made for voltage rating below 100 V [11].

It has been shown that well known conventional VDMOSFET Silicon Limit [$R_{on.S} = 8.33 \times 10^{-9}$ $V_B^{2.5}$ Ω-cm^2] can be overcome by using FLIMOSFET structure which appears to be one of the best Power MOSFET in low voltage applications. The FLIMOSFET limit has been given for one P-floating island by the equation as [$R_{on.S} = 5.87 \times 10^{-9}$ $V_B^{2.5}$ Ω-cm^2] [4]. Recently, the process flow mechanism required to fabricate FLIMOSFET structure using multi-epitaxial technology has been described [6], [10], which is less complex and less expansive than the SJ devices technology.

However, these FLIMOS devices exhibit a current limiting effect at high current levels similar to that of a conventional power MOSFET despite of its having low on-resistance known as the quasi-saturation. This effect takes place when both drain and gate voltages applied to the device are higher and the current carrying capability of the drift region becomes insufficient to remove the supply of electrons at the drain end. This basically happens due to the velocity saturation in the drift region. The quasi-saturation effect can be reduced by increasing the doping concentration of the drift region and by increasing the separation between adjacent MOS channels. However, these modifications cause the reduction in the forward blocking capability and also increase the total chip area of the device [13]. Trench gates have been used in the design of low on-resistance MOSFETs [14], [15] and removing quasi-saturation effect [16]. In this paper, we report a novel trench-gate FLIMOSFET by combining these two well established concepts i.e., of floating islands and trench gate together based on numerical device simulation tools [17]. The results obtained exhibit low on-resistance, reduced electric field distribution in the drift region along with the quasi-saturation free drain and transconductance characteristics over a wide range of voltages, making the proposed device very attractive for industrial applications.

1-4244-0116-X/06/$20.00 ©2006 IEEE   219

II. DEVICE STRUCTURE AND OPERATION

The structure of the proposed trench-gate FLIMOSFET is schematically shown in Fig.1, whereas Fig.2 shows the structure of the conventional FLIMOSFET along with their various dimensions in microns designed using ISE-TCAD MDRAW [17]. The doping densities used in the device are: N⁻ drift region equal to 5×10^{15} cm⁻³, N⁺ source equal to 5×10^{19} cm⁻³; P-base equal to 5×10^{19} cm⁻³; P⁺ floating island equal to 5×10^{19} cm⁻³; N⁺ substrate equal to 5×10^{19} cm⁻³. The 800.0 A° gate oxide and metallization has been extended into the trench-gate, which can be formed by RIE (reactive ion etching). The device has been studied for two trench depths TG1=2.5 μm and TG2=3.0 μm.

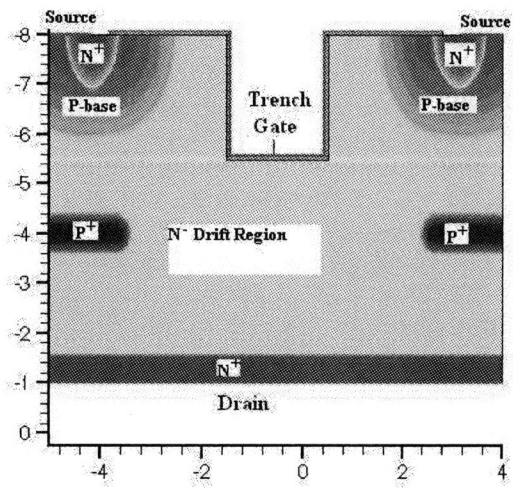

Fig. 1. Cross-section of the proposed Trench-Gate FLIMOSFET. All dimensions are in microns.

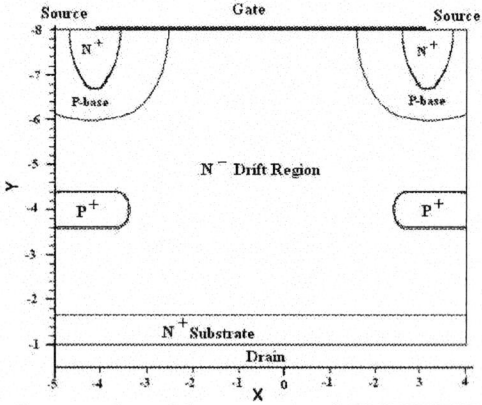

Fig. 2. Cross-section of the conventional FLIMOSFET structure.

The device simulations using ISE-DESSIS take into account the mobility models that are dependent on electric field, impurity concentration & impact ionization at a temperature of 300 degree Kelvin. The device structure is mapped onto a mesh which can be refined to follow closely the contours of the device. The various regions of the

device are then defined, doping profiles and concentrations specified, and electrodes placed at the appropriate positions. The proposed device could be fabricated using multi-epitaxial processes [6], [10] and deep trenches formations [11], [14], [15].

III. SIMULATION RESULTS AND DISCUSSIONS

To understand the behavior of the proposed device and its relative comparison with the conventional device, we have compared the electric field distribution in the drift

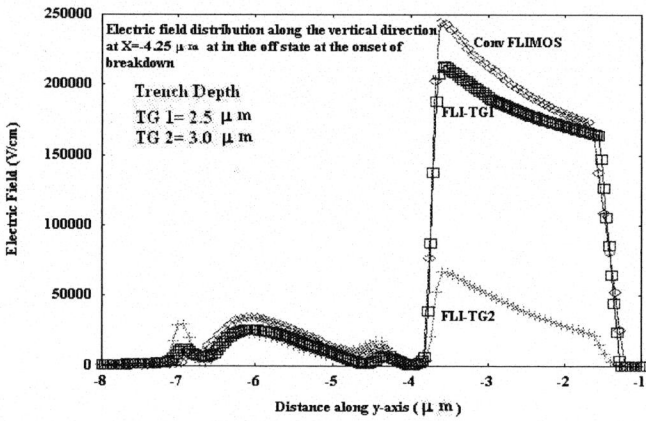

Fig. 3. Comparison of electric field distribution with and without Trench-Gate.

region at x = -4.25 μm with and without trench gate as shown in Fig. 3. The plot suggest that there is almost four fold reduction in the electric field distribution for a trench depth of 3.0 μm which enables the enhancement of the doping concentration of the p-substrate, thus reducing the overall device on-resistance, since the drift region contributes maximum in the total device on-resistance.

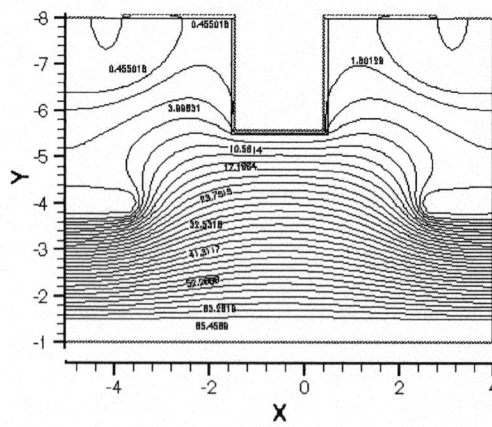

Fig. 4. Electrostatic potential contours with Trench-Gate.

Fig.4 and Fig.5 show the electrostatic potential contours with and without trench gate respectively when the drain bias is near avalanche breakdown. The plot clearly indicate that there are very little changes in the curvature of the potential contours in these two figures, which means there is no degradation in the breakdown voltage which is highly desired. Fig. 6 shows the forward I-V characteristics of the

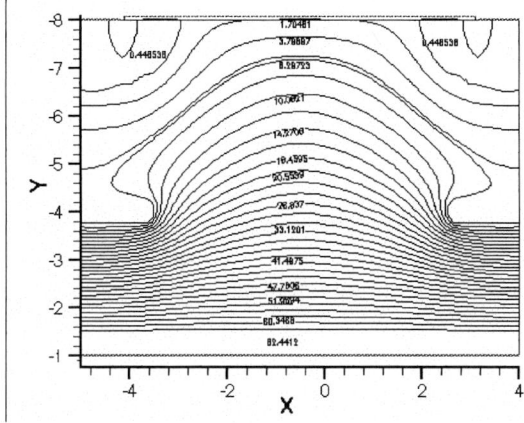

Fig. 5. Electrostatic potential contours without Trench-Gate (Conventional FLIMOSFET).

device at different gate biases. Characteristics of the equivalent conventional FLIMOSFET with no trench gate is also shown. It is clear that at gate voltage more than 3 V, the conventional device shows a current limitation with

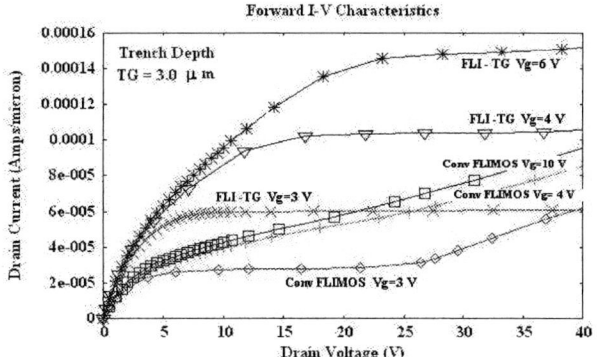

Fig. 6. Forward I-V characteristics

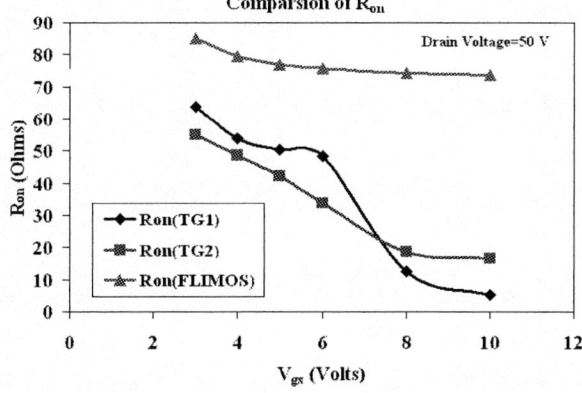

Fig. 7. Comparison of On-Resistance with and without Trench-Gate.

regard to the bias, but trench-gate device does not. In addition, the on-resistance of the new device is less than that of the conventional device as shown in Figure (7).

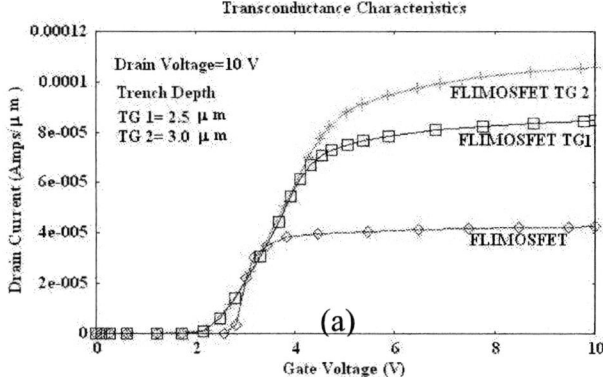

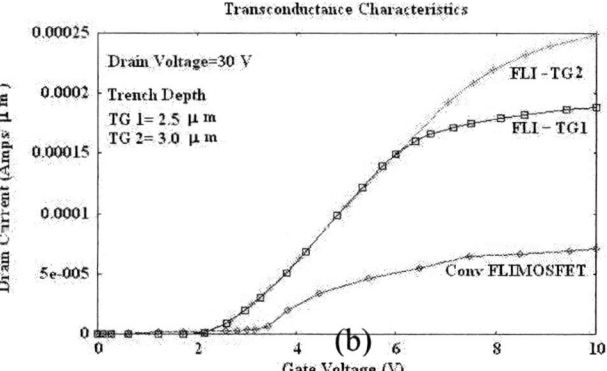

Fig. 8. Transconductance characteristics with trench depth for (a) drain bias=10 V and (b) drain bias =30 V.

Figures 8(a) and (b) present the transconductance characteristics of the proposed device structure with two trench depths TG1=2.5 μm and TG2=3.0μm at drain bias of 10 volts. The 3.0 μm trench gives a significant improvement in the quasi-saturation as compared to 2.5μm trench. The quasi-saturation is completely removed at the high gate bias of about 30 volts, thus making the proposed device structure very attractive and useful for many automotive/telecom applications within 50 volts breakdown. Figure (9) shows the synchronous rectifying characteristics with and without trench-gate. These characteristics are obtained if a positive gate voltage and negative drain voltage are applied to the device, an alternate path for current between the drain and the source, consisting only of the majority carriers, is created. If the on-resistance of the device is very low, the device will exhibit a very low forward voltage drop, significantly lower than for a P-N junction rectifier, and still retain its higher switching speed because no minority carriers are stored in the device.

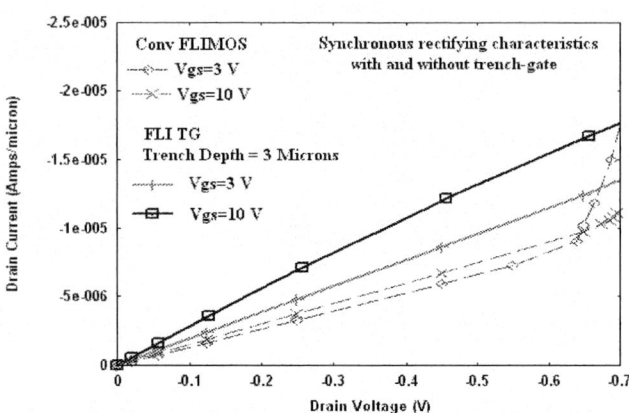

Fig. 9. The synchronous rectifying characteristics with and without Trench-Gate.

IV. CONCLUSION

A novel Trench gate FLIMOSFET structure designed using the concept of "Opposite Doped Buried Regions (ODBR) and floating island (FLIMOSFET) along with Trench-gate technology has been discussed. The proposed device structure exhibits quasi-saturation free output and transconductance characteristics over a wide range of voltages as well as gives reduced on-resistance when compared with the conventional FLIMOSFET for two trench depths 2.5 and 3 microns. The new device structure does not give any degradation towards the breakdown voltage which remains almost constant at 65 volts. It also exhibit better synchronous rectifying characteristics and almost four fold decrease in the electric field distribution in the drift region in off-state at the onset of breakdown voltage.

ACKNOWLEDGEMENT

One of the authors (R. Vaid) gratefully acknowledges Prof. V Ramgopal Rao, Dept. of Electrical Engineering, IIT Bombay, Mumbai for allowing the use of ISE-tools to carry out this work and the University Grants Commission (UGC), Govt. of India, India, for the award of Teacher Fellowship under the FIP scheme during the 10th plan period. The author also wishes to acknowledge Mr. K. Narasimhulu from dept. of Electrical Engineering, IIT Bombay, Mumbai for having very fruitful discussions regarding this work.

REFERENCES

[1] N. Cezac, P. Rossel, F. Morancho, H. Tranduc, A. Peyre-Lavigne and I. Pages, "A New Generation of Power Unipolar Devices Based on the Concept of the Floating Islands", in *Proc. 22nd International Conference on Microelectronics (MIEL 2000)*, NiS, Serbia, 2000, vol. 2, pp. 637-640.

[2] N. Cezac, F. Morancho, P. Rossel, H. Tranduc, and A. Peyre-Lavigne, "A New Generation of Power Devices: the Concept of the Floating Islands MOS Transistor (FLIMOST)", in *Proc. 12th International Symposium on Power Semiconductor Devices and ICs (ISPSD'2000)*, Toulouse, France, 2000, pp. 69-72.

[3] F. Morancho, N. Cezac, A. Galadi, M. Zitouni, P. Rossel and A. Peyre-Lavigne, "A New Generation of Power Lateral and Vertical Floating Islands MOS Structures", *Microelectronics Journal*, vol. 32, pp. 509-516, 2001.

[4] S. Alves, F. Morancho, J. M. Reynes and B. Lopes, "Vertical N-channel FLIMOSFETs for Future 12V/42V Dual Batteries Automotive Applications", in *Proc. 15th International Symposium on Power Semiconductor Devices and ICs (ISPSD'2003)*, Cambridge, UK, 2003, pp. 308-311.

[5] R. Vaid and N Padha, "Power VDMOSFETs with Vertical Floating Islands" in *Proc. International conference on Systemics, Cybernetics and Informatics (ICSCI'2004)*, vol. 1, Hyderabad, India, 2004, pp.42-45.

[6] R. Vaid and N. Padha, "Investigation of a Power FLIMOSFET Based on Two-Dimensional Numerical Simulations", *Indian Journal Engineering and Material Sc.*, (In press).

[7] T. Fujihara, "Theory of semiconductor superjunction devices", *Jpn. Journal Appl. Phys.*, vol. 36, pp. 6254-6262, 1997.

[8] X. B. Chen and J. K. O. Sin, "Optimization of the Specific On-Resistance of the COOLMOS[TM]", *IEEE Trans. Electron Devices*, vol. ED- 48, pp. 344-348, 2001.

[9] X. B. Chen, X. Wang and J. K. O. Sin, "A Novel High-Voltage Sustaining Structure with Buried Oppositely Doped Regions", *IEEE Trans. Electron Devices*, ED-47, pp. 1280-1285, 2000.

[10] W. Saito, I. Omura, K. Tokano, T. Ogura and H. Ohashi, "A Novel Low On-Resistance Schottky-Barrier Diode with p-Buried Floating Layer Structure", *IEEE Trans. Electron Devices*, vol. ED – 51, pp. 797-802, 2004.

[11] X. Yang, Y. C. Liang, G. S. Samudra and Y. Liu, "Tunable Trench gate Power MOSFET: A Feasible Superjunction Device and Process Technology", in *Proc. 30th Annual Conference of the IEEE Industrial Society*, Busan, Korea, 2004, pp. 729-733.

[12] R. Vaid and N. Padha, "Novel Power VDMOSFET Structure with Vertical Floating Islands and Trench Gate", *Indian Journal Pure and Appl. Phys.*, vol. 43, pp. 301-307, 2005.

[13] M. N. Darwish, "Study of the Quasi-saturation effect in VDMOS Transistors," *IEEE Trans. Electron Devices*, vol. ED-33, pp.1710-1716, 1986.

[14] D. Ueda, H. Takagi, and G. Kano, "Deep-trench MOSFET with a Ron Area Product of 160 milliohm-mm^2," in *IEEE IEDM Tech. Dig.*, 1986, pp. 638-641.

[15] H. R. Chang, R. D. Black, V. A. K. Temple, W. Tantraporn, and B. J. Baliga, "Ultra Low Specific On-Resistance UMOSFET", in *IEEE IEDM Tech. Dig.*, 1986, pp. 642-645.

[16] J. Zeng, P. A. Mawby, M. S. Towers and K. Board, "Numerical Analysis of a Trench VDMOST Structure with no Quasi-Saturation", *Solid-State Electronics*, vol. 38, 821-828, 1995.

[17] Integrated System Engineering, *ISE-TCAD Manuals*, AG, Zurich, Switzerland, 2002.

Session
Microsystem Technologies

2006 25th International Conference on Microelectronics

Wireless Ultra-Low Power Smart Data Acquisition System for Pressure Sensing in Medical Application

K. Arshak and E. Jafer

Abstract - The development of a wireless sensor microsystems containing all the components of data acquisition system, such as sensors, signal-conditioning circuits, analog-digital converter, embedded microcontroller (MCU), and RF communication modules has become now the focus of attention in many biomedical applications.

This paper discusses innovation circuits and system techniques for building advanced smart medical devices (SMD). Low power consumption and high reliability are among the main criteria that must be given priority when designing such wirelessly powered Microsystems. Different capacitive readout circuits used for pressure sensing will be described. An example for a low power wireless system developed for multi-sensors monitoring will be presented.

I. INTRODUCTION

The need for patient remote monitoring is vital to measure some biological signals. Any mobile medical system should consist of three main modules, these are: non-invasive technique to measure biological signals without doctor interfere, wireless system for transmitting data captured, and a user interface software to enable data acquisition. Pressure sensor can be used in a wide range of medical invasive measuring, some of these are:

- Blood pressure
- Respiration rate
- Gastrointestinal (GI) activity

This paper will concentrate on different aspects of designing smart wireless sensing systems that can be used for monitoring the GI activity.

A. Smart wireless interfacing

Figure 1 illustrates the most straightforward method in capturing data and transmitting it to the monitoring site. A sensor is placed in a process environment to provide relevant information, with the sole intention of being able to control the process as accurately as desired. The sensor is coupled to a sequence of stages, which each play their proper role in signal handling and conditioning. The first part is the analogue sensor interface circuit, which must be well adapted to the sensor and often requires careful design, in order not to jeopardize the usually small sensor output signals. Its role is to amplify the minute signals from

K. Arshak and E. Jafer are with the Department of Electronic and Computer Engineering, College of Informatics, University of Limerick, Limerick, Ireland, E-mail: Khalil. Arshak@ul.ie

the sensors up to acceptable levels for further treatment. In modern systems, an analogue-to-digital (A/D) conversion is carried out as soon as possible, to take advantage of the many possibilities that are offered these days by the ever-increasing mathematical and decisive power of digital electronics.

The transmitter stage is an important stage in remotely placed systems, since this stage will send the information over a longer distance to a matched receiver unit. This transmission can be carried out by a wireless system, which is most desirable in applications where monitoring has to occur in closed environments (e.g. vessels, but also in biomedical applications). After decoding the received signal, the data handling and storage unit will usually decide on the relevance of the signals, will store eventual interesting data, and will provide useful or important information to the outside, for example on a display, as shown in the Figure 1. The intention of the complete system is to give the operators a better control over the process they are responsible for. The displayed data, which should be corresponding to the sensor's output, is used as a decisive element to take the required measures for action and control of the process. Often, the control or action is taken on a fully automated base.

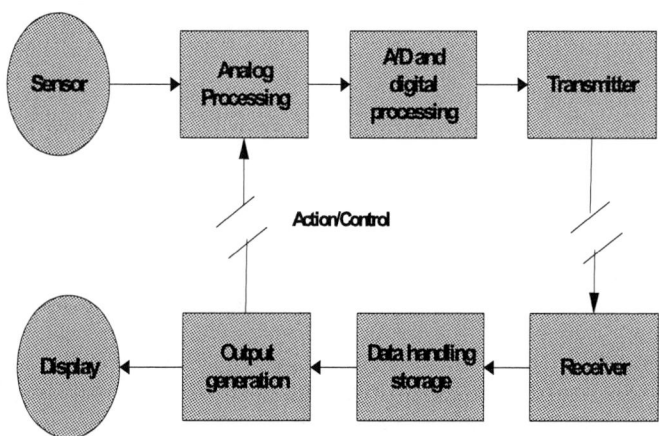

Fig. 1. The most general representation of sensing system and its use.

One of the most common problems that affects the system above is the stability of the sensor with time and that the interface circuits are able to follow the sensor signals at all times. No matter how desirable this situation is, in real life most sensors cannot fulfill this requirement. In particular, in long-term applications, sensor drift is a known problem.

1-4244-0116-X/06/$20.00 ©2006 IEEE   225

In many cases, the drift can become so important that the amplifier, which is coupled to the sensor, will not be able to follow the signal and will turn into saturation, with of course signal loss as a major consequence.

To tackle the problem of the lack of long-term stability of the sensing front-ends, the only way out is to provide a bi-directional communication link, instead of a single direction link as depicted in figure 1. Figure 2 illustrates the idea. The data transmission system has now been expanded with a second transmission stage, which allows one to communicate data to the device. By doing so, not only will the mere data captured by the sensors be transmitted, but also one can ask the device to go into a self-interrogating mode to verify the status of all the individual building blocks. After transmission of these data, the remote controlling station can adapt the settings of, for example, the offset of the amplifier, to cope for unwanted drift, or to re-program the built-in compression algorithms etc. The addition of such bi-directional communication has proven to be very advantageous, especially in novel applications, where the end user may not have had any experience with the possibility of measuring on such remote sites.

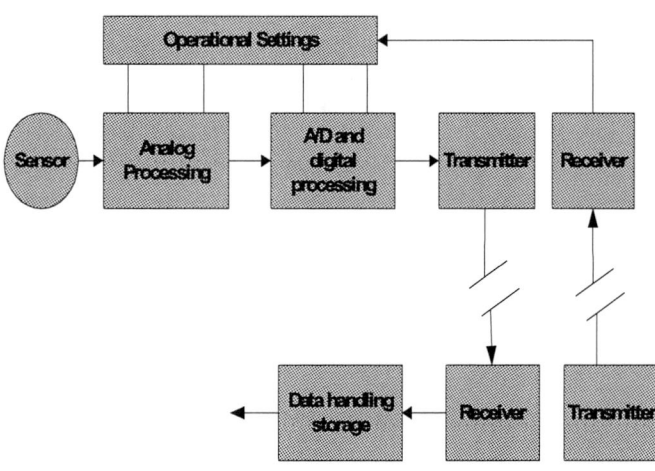

Fig. 2. Sensing system with reprogrammable operational setting points.

The paper is organized as follows: Basic types of telemetry systems are described in the next section. In section III, some of the main and recent capacitive analog circuitries are studied. Then, the fabrication process of biomedical compatible pressure sensors is given in section IV. Finally, an example of smart wireless system for multi-sensor monitoring developed by our research group has been presented.

II. TELEMETRY SYSTEMS

Wireless data gathering systems have the potential to restructure the instrumentation used in a variety of industries, including security, health care, and transportation.

Looking at wireless sensing technology, both active and passive telemetry are used. Active telemetry systems provide relatively long-range bidirectional sensor data transfer [1], but with increased size and decreased life when internal batteries are used. Passive telemetry significantly reduces transmission distances but allows the implementation of battery-free devices with indefinite lifetimes [2,3]. In the following sections, some basic aspects about the two telemetry systems are given:

A. Passive Telemetry System

Wireless passive telemetry systems based on the integration of MEMs and RF backscatter modulation has been realized for implantable pressure monitoring [4,5]. The block diagram of the passive telemetry system is depicted in Figure. 3. It consists of an external control unit (the base unit) and an implantable transponder. Wireless communication can then be established between the two units, based on an absorption modulation mechanism.

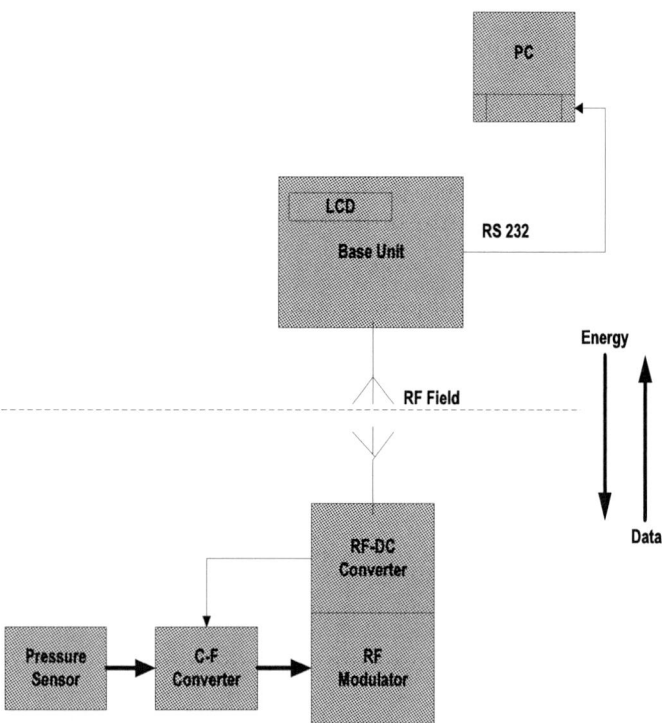

Fig. 3. Passive system comprised of a base and transponder unit.

The capacitance to frequency converter unit (C-F) will convert the signal read by the pressure sensor to a frequency tone to be transmitted later on through the RF-link to the base station. The transponder receives power and external control data through an RF field, while it can transmit data by modulating the absorption rate. The base unit, on the other hand, demodulates the transmitted data and processes it though (MCU) to convert the signal into a digital bit stream. The resulting data can then be sent to a

PC through a serial output or alternatively be displayed on an LCD on the base unit.

The performance of the RF-DC converter considered being crucial for the overall implant system performance. In this unit, received RF energy will be rectified and regulated to power the implant.

A typical RF-DC circuit is shown in Figure.4. A start up circuit is used to provide the bandgap reference with the rectified voltage on power up. Then the circuit will be switched to the regulated voltage when the latter reaches a sufficient level. The start up circuit compares V_{rec} and the high-regulated voltage and feeds the larger to the bandgap reference.

Low drop-out (LDO) linear regulators [6] are used for high and low voltage regulations. A step up switched capacitor circuit has been used to provide the 3.3V LDO with voltage since it leads to a better power efficiency. Temperature does not vary substantially, but the rectified voltage shows a strong ripple because a small tank capacitor must be used for at the input for size considerations.

It can be noticed that the above circuit is suitable for minimizing power consumption and can be further improved in the future.

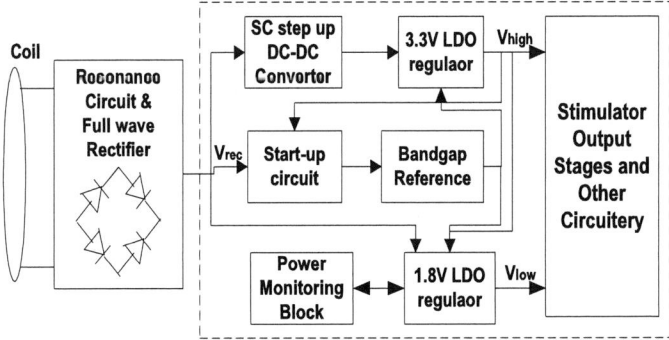

Fig. 4. Block diagram of the power recovery circuit.

B. Active Telemetry System

Power supply is needed internally for such system, which provides a long distance range for bi-directional sensor data flow. Because these portable systems will be battery powered, one of their key constraints is the overall power consumption, which must be minimized without scarifying performance.

Capacitive sensors have no intrinsic power dissipation and thus they considered being an attractive option for low power circuit techniques while offering high sensitivity and self-test capabilities [7,8]. The analog interface circuitries and development of the pressure capacitive sensors will be the issues of the coming sections.

The need for a generic system capable of reading out multiple sensors has widely increased. Furthermore, a generic interface circuit should provide standard communication link to the main system controller, support sensor self-test and self-calibration, support multi-ranging

within a single sensor, dissipate low power, and occupy very small size.

One of the common wireless multi-sensors system architecture is shown in Figure 5. The system here provides a highly modular framework with components that can be easily interchanged to meet a wide range of application-specific demands. The central control electronics manages microsystem operation, perform sensor signal processing including calibration and compensation, interface to front-end sensors and external systems, and implement system-level power management. These functions can be implemented using a commercial low power microcontroller, typically with 8b or 16b processor. A low power DSP could be used in application with high signal processing demands such as the Field Programmable Gate Array (FPGA) [9].

A variety of sensors busses have been used to interface sensors with microcontrollers including the IEEE P1451.2 standard. Several sensor busses are discussed and compared in [10], including the Intramodule Multielement Microsensor (IM^2) bus that was specifically designed to complement the system architecture shown in Figure.5. Key requirements for a sensor bus in modular low power multi-sensor systems include a physical bus, which minimizes hardware overhead, and supports power management features such as interrupt generation and multiple power supply signals.

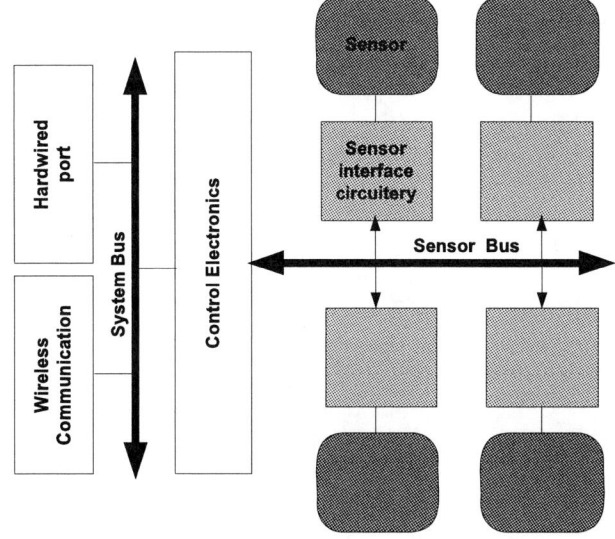

Fig. 5. Block diagram of a wireless modular sensor system.

C. Use of FPGA in wireless systems

FPGA have become a dominant technology for first-stage intermediate frequency (IF) in a variety of wideband wireless applications [11]. This is especially true in markets where "reprogrammability" and time-to-market concerns outweigh the costs benefits associated with application-specific integrated circuits (ASICs) in mass production. Table 1 compares between wireless smart systems based on the use of FPGA as a control unit and those use a microcontroller.

TABLE I
WIRELESS SYSTEMS COMPARISON USING FPGA AND MICROCONTROLLERS

FPGA	Microcontroller
Permits design upgrades with no hardware replacement.	Hardware is fixed and can't be upgraded.
Provides a good platform for a later on ASIC design.	ASIC is not possible.
Support a wide range of Digital Signal processing (DSP) operations.	Not efficient for implementing a complicated DSP.
Power consumption and chip size are still considerable.	Some processors now a day can be found with a very small size and low power.

III. CAPACITIVE READOUT CIRCUITS

Most developed pressure sensors have capacitive properties since they are highly sensitive. In general, two main approaches have been followed to design a capacitive readout circuitry. The measured output parameter will determine the approach type to be either capacitance-to-frequency (C-F) or capacitance-to-voltage (C-V). Types of circuits in each approach will be presented briefly as follows:

A. Capacitance to Frequency conversion

Switched-capacitor (SC) front-end (FE) is a good selection for the miniaturized capacitive readout circuits since the gain of this circuit is less sensitive to variations in input parasitic capacitance. A basic switched-capacitor relaxation oscillator used as capacitance transducer as shown in Figure 6 [12].

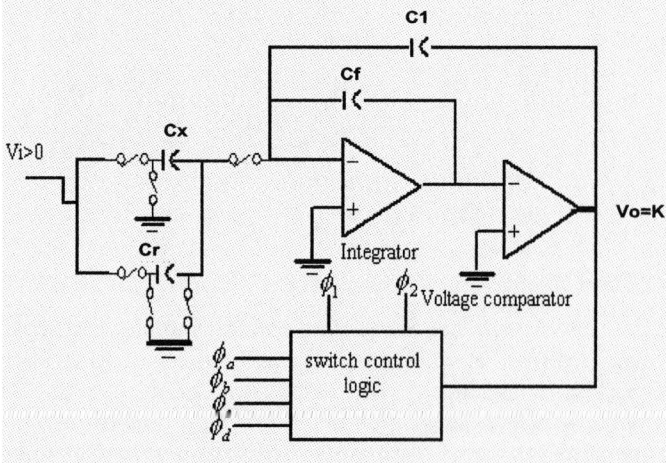

Fig. 6. Capacitive interface based on SC relaxation oscillator.

In this circuit, C_x denotes an unknown capacitance (capacitive sensor), C_r is the known reference capacitance and C_f is a non-critical value integration capacitance. The circuit consists of a standard switched-capacitor integrator, a voltage comparator (Schmitt trigger), and switch control logic. The MOS switches are controlled by a two-phase clock, Φ_1 and Φ_2 and switch control logic with logic signals Φ_a, Φ_b, Φ_c, and Φ_d as given by the following equation:

$$\Phi_A = \Phi_2, \ \Phi_B = \Phi_1, \Phi_C = \overline{K}\Phi_1, \ \Phi_D = \overline{K}\Phi_2 \tag{1}$$

The oscillation frequency is expressed as:

$$f \cong f_c \frac{(C_r - C_x)C_x}{2C_1C_r} \tag{2}$$

Where f_c is the clock frequency.

The above-mentioned circuit is sensitive to the offset voltage of the op-amp and to the clock feedthrough of the clock signals through the gate-source and gate drains parasitic capacitors of the MOS.

B. Capacitance to Voltage conversion

This type of capacitive interface circuitry is widely preferred in telemetry systems because of the frequency drift that occurs with the first type. SC circuits proved to be a good option when low power performance is required. A simple SC based C-V capacitive readout is shown in Figure.7.

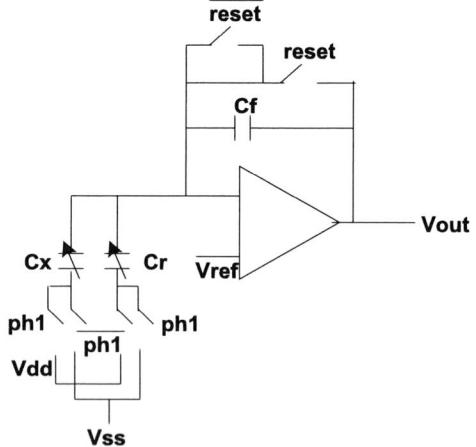

Fig. 7. SC capacitive readout circuit.

Cx is the sensor capacitor, and Cr is the reference capacitor. Both ph1 and ph2 are two non-overlapping clock signals. During ph1, the reset switch of the charge integrator is closed and Cs is charged through the charge integrator output to Cx (Vref-Vdd), and Cref is charged to C1 (Vref-Vss). Once ph1 goes low, charge stored in Cx changes to Cx (Vref-Vss) and that in Cr changes to Cr (Vref-Vdd). The net change in charge is (Cx-Cr)(Vdd-Vss). This change of charge is transferred to feedback

228

capacitor Cf. The magnitude of the output voltage will be equal to:

$$V_{out} = V_{ref} + (V_{dd} - V_{ss})(C_x - C_r)/C_f \quad (3)$$

The performance of the above design can be further improved by adding a second stage to form a two-stage Delta-sigma ($\Sigma\Delta$) modulator as shown in Figure 8. The feedback capacitor Cfb has been introduced to increase the charge transfer efficiency. Consequently the modulator exhibits a large sensitivity to drift and noise of the voltage sources.

In [13] another capacitive signal conditioning circuitry based on the principle of capacitance-frequency-voltage conversion has been developed. Here the sensor C_x and the reference C_r capacitances are converted first to a frequency tones Fx and Fr respectively using a low power CMOS timer IC's and then to a voltage values using a (Phase locked loop) PLL as presented by Figure 9. An analog switch, which is controlled by a processor unit, is used to output the measured sensor voltage value with respect to the reference voltage.

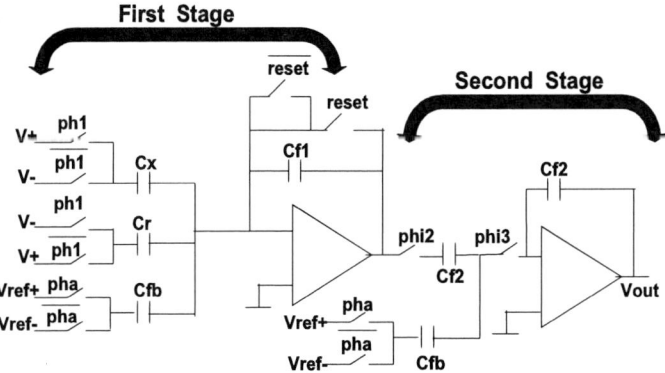

Fig. 8. Two stage $\Sigma\Delta$ modulator with sensor and reference connected to first stage.

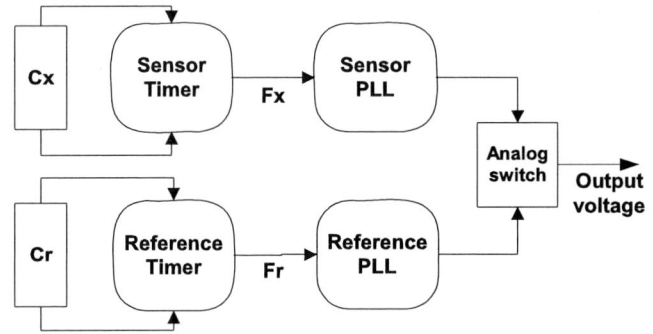

Fig. 9. Block diagram of the PLL capacitive interface.

The above design can be suitable for building a low power prototype if suitable units have been selected but it is not efficient an ASIC system. The main reason behind that is the non-linearity of the PLL unit, which has negative impact on system performance. It is found experimentally that the PLL design is sufficient for the pressure sensors

range developed as will be explained in the following section.

IV. PRESSURE SENSOR DEVELOPMENT

In this study, capacitive sensors were fabricated using interdigitated electrodes as they contain no moving parts, require one less process step than a sandwich structure and detects pressure/strain changes through the deformation of the dielectric layer [14]. Furthermore, the interdigitated arrangement is popular with designers as altering the length of the electrodes can easily change the structures capacitance. DuPont 4929 silver conductive paste was used to form the electrodes which were printed onto alumina and Melinex® substrates using a DEK RS 1202 automatic screen-printer. After printing the substrate were allowed to cure at 120 °C for 30 minutes. A Thelco Model 6 oven was used for this purpose.

The dielectric layer consists of a polymer thick film paste, prepared using polyvinylidene fluoride (PVDF) as the functional material. This was combined with 7 wt.% binder and 0.1 wt.% surfactant. Typically, commercial powders have a particle size of 30 μm or more and so mechanical milling is necessary to reduce this to between 0.5 μm and 5μm for the functional material and 0.2 μm and 2 μm for the binder. The binder used in this study is ethyl cellulose and lecithin was added to act as the surfactant. Finally, the solvent, Terpinol-α was used to form a paste of suitable consistency. Three layers of PVDF paste were deposited over the electrodes and then placed in the oven for curing.

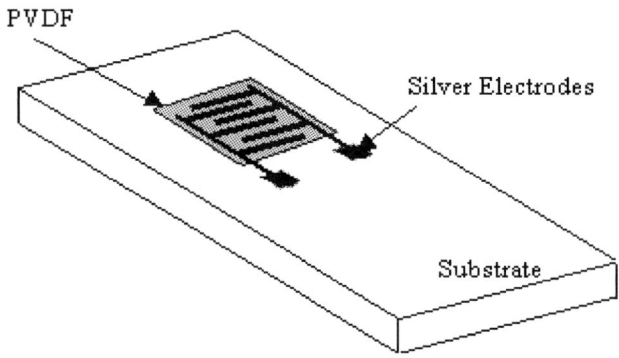

Fig. 10. Shows the structure of the interdigitated device.

A. Sensor Testing Mechanisms

After fabrication, the sensors on alumina substrates were placed in a cantilever beam arrangement so that the change in capacitance with applied strain could be measured. The experimental arrangement is shown in Figure 11. It can be seen from equation 4 that with knowledge of the beam geometry and displacement, the strain on the sensor can be calculated:

$$\varepsilon = \frac{3xyh}{2L^3} \quad , \quad (4)$$

where ε is the strain, x is the distance from the centre of the beam to the point where the load is applied, y is the displacement of the beam and L is the distance from the clamped end to the point where the load is applied. By measuring the change in capacitance, ΔC and the applied strain, the gauge factor (GF) can be calculated from equation 5, where C is base line capacitance. The change in capacitance was measured using a HP 4192A LF Impedance Analyser.

$$GF = \frac{\Delta C / C}{\varepsilon} \qquad (5)$$

To assess the suitability of the PVDF device for pressure sensing applications, sensors on flexible substrates were connected to the developed interface circuit and subjected to pressure in the range of 0 to 12 kPa. A flexible substrate was chosen, as there is a potential for these devices to be applied to irregularly shaped objects. The applied pressure causes a deformation of the dielectric layer, which leads to a change in the sensor capacitance.

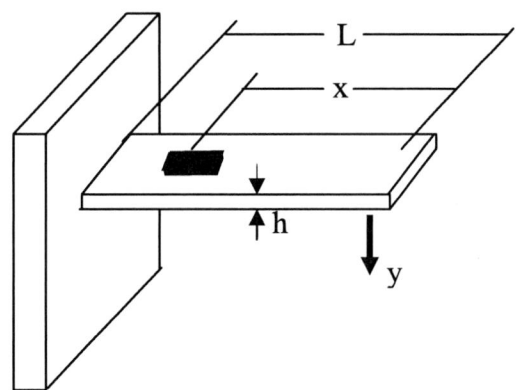

Fig.11. The cantilever beam arrangement

B. Testing the Strain Gauge Properties

The PVDF sensors were mounted in the strain gauge test rig, and the strain was increased from 0 to 500 μstrain. Previous results for a sandwich capacitor based on PVDF showed a gauge factor of 3.5 and linearity error of 5 %. In this study, a gauge factor of 6.2 was measured, which is higher than that measured for the sandwich structure. The sensors response showed a linearity error of 6 %, as shown in Figure 12.

V. DEVELOPMENT OF MULTI-SENSOR WIRELESS SYSTEM (ACTIVE TELEMETERY EXAMPLE)

An overview of the developed system main units is shown in Figure 13. The mote is configured to be for either resistive or capacitive measurements [15]. The output samples from the signal conditioning circuit will be processed and buffered by a microcontroller till the

transmitter becomes ready. The data will be sent to the Base Station over 433 MHz channel using FSK modulation

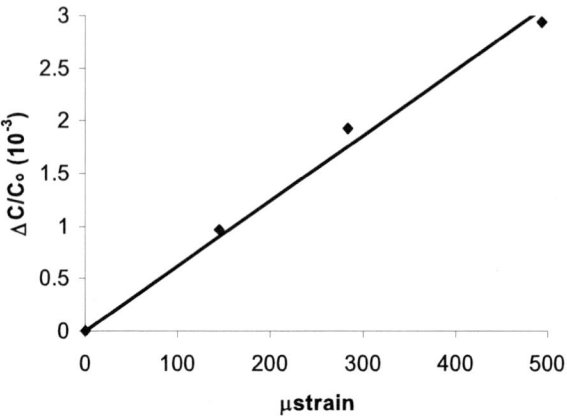

Fig. 12. Linearity of the PVDF sensor

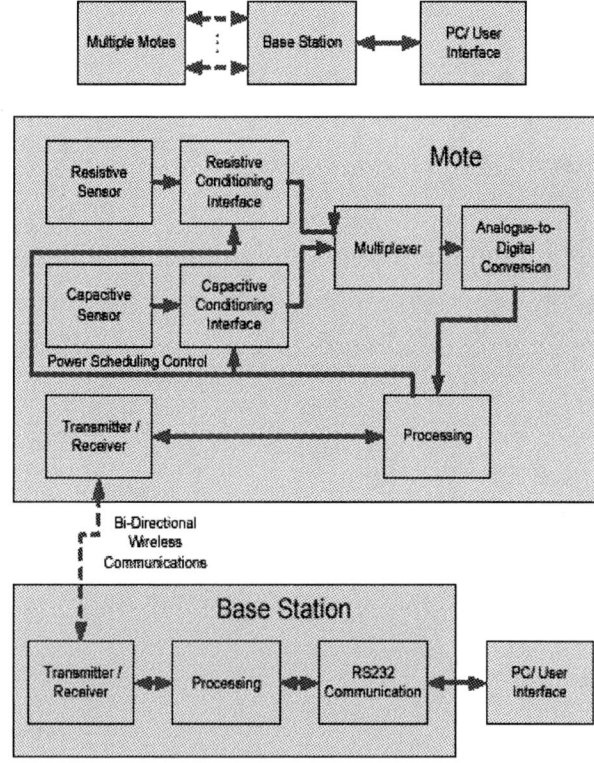

Fig. 12. System overview

type. The mote also has the ability to power off the appropriate conditioning interfaces if they are not being used between the sampling times of the two analog signals to save power.

The base station is able to communicate with multiple motes. This can be accomplished by using a form of time-division multiplexing, where the base station coordinates for each mote to start transmission. The data received from

230

all motes at the base station are then sent to a PC via RS232 serial communication.

In the following sections, a brief description of both capacitive and resistive analog interface circuitries employed in the system is given. More attention will be given to the techniques used to reduce overall power consumption.

A. Capacitive and Resistive interface circuitries

The capacitive readout circuit is based on Capacitance-to-Voltage conversion based on the use of the PLL unit as mentioned in section III. It is desired that the circuit can measure sensor of range between 0 to 40 pF. A PVDF developed pressure sensor and a commercial temperature sensor have been used for the capacitive and resistive interface circuitries respectively.
Anderson Loop [15] circuit topology has been selected for the resistive interface since it has a linear performance. The circuit has been designed to be configurable and able to read from sensors with different ranges.

B. Power management

The first Power scheduling has been designed carefully to control the different parts of the system since it is required to reduce power consumption on the mote side as much as possible. Figure 14 shows the power flow mechanism controlled mainly by the processor unit. The transceiver and MCU have a "sleep" mode that can be controlled by the embedded software.

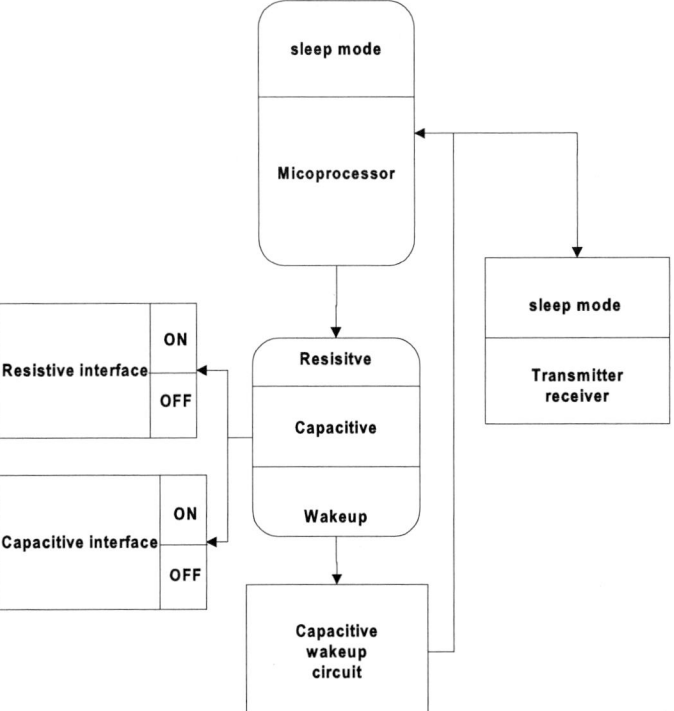

Fig.14. Power control of the mote system

As shown above, the sensor signal conditioning circuits can be switched either on or off by the control signals of the processor unit. A CMOS analog switch is used for this purpose. Table 2 clarifies the power scheduling options for the sensor interface circuits. In order to preserve the power, certain modules within the mote system are powered off when they are not in use.

TABLE 2. POWER SCEDUALING OF SENSOR INTERFACES

MCU	Resistive	Capacitive	Wakeup
Wakeup	ON	OFF	OFF
Wakeup	OFF	ON	OFF
Sleep	OFF	OFF	ON

A wakeup circuit has been introduced in the system to get further reduction in the power consumption. The idea came from the work done in [16] where a miniaturized low power wireless system has been described for remote environmental monitoring. The operation of the wakeup circuit is to put the MCU in sleep mode when the input signal is not changing. When the signal starts to give a significant change, an interrupt is generated that wakes up the MCU and it will start sampling again. The circuit is designed as shown in Figure 15.

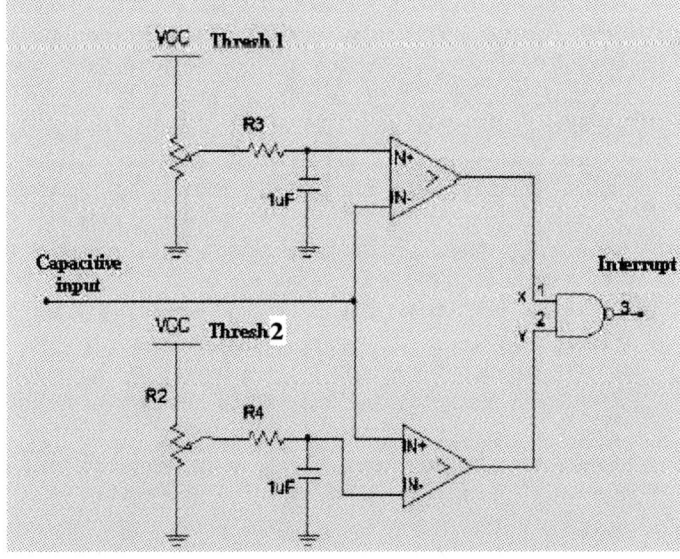

Fig.15. Circuit diagram of the wakeup circuit

The design is based on the principle of window detector, where two digitally controlled potentiometers are used to set the upper and lower trip points of the two comparators. At the beginning of operation, the MCU decides when the capacitive samples are not changing, then it will set the upper and lower limits of the window and powers up the wakeup circuit. The MCU will be in sleep mode when the sensor goes outside the limit. Than an interrupt signal will be generated to wakeup the MCU that will power down the wakeup circuit to save power.
The above design draws bout 700µA and a NAND gate that outputs zero when the sensor is within the window range

will generate the interrupt signal. If the sensor goes above the upper-trip or below the lower trip, the desired output is high, that will interrupt the MCU from the sleep mode.

It has been noticed that wakeup circuit is not suited for the resistive interface where a higher power is needed to keep the circuit operating constantly. The wakeup circuit must be on when the processor unit is in sleep mode.

VI. CONCLUSION

In this paper, the issue of building a smart low power wireless system has been considered. The focus of the paper was mainly on pressure sensing and the capacitive interface circuitries.

Passive telemetry systems are more efficient in term of power consumption but it only suits applications with a very short-range. On the other hand, active telemetry system can provide a long-range of bi-directional communications. According to this, active type requires more power to function properly but the level of consumption can be controlled with well-designed system power management.

It can be concluded that C-F interface circuitry is more suitable for passive telemetry systems where the transmission will not be affected much by any frequency drift. For the active telemetry systems C-V based interface circuitry is a good option to get a good and stable performance. The $\Sigma\Delta$ SC circuit type proved to have a high sensitivity to noise and drift and it can be incorporated within the ADC of the data acquisition system. Although the SC is designed for high-speed operation but there are a lot of special techniques that can be employed to reduce the supply of power. This should be done without affecting the ON resistance of the MOS switches.

As a case study, a multi-sensor wireless system has been presented as an example of development a smart active telemetry system. Two mechanisms have been adopted to reduce further the overall power consumption. These are power scheduling and wakeup circuit for the capacitive interface. The first one is implemented fully by the MCU in order to switch on/off all the system units. A wakeup circuit has been designed to interrupt the MCU when it is in the sleep mode and the capacitive samples are changing significantly. It was found that such design is more suited for the capacitive interface than the resistive one where the circuit consumes much less power even if it is kept working continuously.

ACKNOWLEDGEMENT

This work was supported by the Enterprise Ireland Commercialization Fund 2003, under technology development phase, as part of the MIAPS project, reference no. CFTD/03/425.

REFERENCES

[1] A. Leung,W. H.Ko, T.M. Spear, and J. A. Bettice, "Intracranial pressure telemetry system using semicustom

integrated circuits," *IEEE Trans.Biomed. Eng.*, vol. 33, pp. 386–395, Apr. 1986.

[2] A. DeHennis and K. D.Wise, "A double-sided single-chip wireless pressure sensor," in *Proc. IEEE Conf. Micro-Electromechanical Syst.*, Las Vegas, NV, Jan. 2002, pp. 252–255.

[3] M. A. Fonseca, J. M. English, M. V. Arx, and M. G. Allen, "Wireless micromachined ceramic pressure sensor for high-temperature application," *J. Microelectromech. Syst.*, vol. 11, pp. 337–343, Aug. 2002.

[4] S. Chatzandroulis, D. Tsoukalas, and P. A. Neukomm, "A miniature pressure system with a capacitive sensor and a passive telemetry link for use in implantable applications," *J. Microelectromech. Syst.*, vol. 9, pp. 18–23, Mar. 2000.

[5] K. Stangel, S. Kolnsberg, D. Hammerschmidt, B. J. Hosticka, H. K. Trieu, and W. Mokwa, "A programmable intraocular CMOS pressure sensor system implant," *IEEE J. Solid-State Circuits*, vol. 36, pp. 1094–1100, Jul. 2001.

[6] Y. Hu, M. Sawan, " A 900 mV 25µW High PSRR CMOS voltage reference dedicated to implantable micro-devices", IEEE-ISCAS, Bangkok, Vol.1, pp. 373-376, May 2003.

[7] A. Mason, N. Yazdi, K. Najafi, and K. Wise, "A low-power wireless microinstrumentation system for environmental monitoring", in: Di- gest Int. Conf. on Sensors and Actuators Transducer 95 Stockholm, Sweden, pp. 107–110, June1995.

[8] A. Mason, N. Yazdi, A. Chavan, K. Najafi, and K. Wise, "A generic multielement microsystem for portable wireless applications", *Proc. IEEE*, 86(8), pp.1733–1746, August 1998.

[9] J. Mendoza-Jasso, G. Ornelas-Vargas, R. Castaneda-Miranda, E. Ventura-ramos, Alfredo. Zepeda-Gerrido, and G. Herrera-Ruiz, "FPGA-based real-time remote monitoring system", *J. Computers and Electronics in Agriculture*, Vol 49, pp. 272-285, 2005

[10] J. Zhou and A. Mason, "Communication Buses and Protocols for Sensor Networks," *SENSORS, (ISSN: 1424-8220)*, vol. 2 (7), pp. 244-257, August 2002.

[11] A. Arshak, E. jafer, and D. McDonagh, "Modeling remote system for sensor monitoring using Verilog HDL and SIMULINK®. Co-simulation", *Proc IEEE BMAS*, pp. 64-69, Sept 22-23, 2005.

[12] A. Cichocki, and R. Unbehauen, R., "A Switched-Capacitor Interface for Capacitive Sensors Based on Relaxation scillators", *IEEE transactions on instrumentation and measurement*, VOL 39, pp. 797-799, Oct 1990.

[13] K.Arshak, E. Jafer, J. Orr, A. Arshak, D. Morris, O. Korostynska, D. McDonagh, J. D. Quartararo, H. Dämpfling, C. Y. Huang, "Design of a low power capacitive pressure sensor signal-conditioning interface using PLL", *IEEE conf on circuits and systems*, Tunisia, March 22-25, 2005.

[14] K. Arshak, D. Morris, A. Arshak, O. Korostynska, E. Jafer, D.Waldron, J. Harris," Developmet of polymer based sensor for integration into a wireless data acquisition system suitable for monitoring environmental and physiological processes", Presented in EMRS, France, 2005.

[15] K. Arshak, and E. Jafer, "Design of low power smart wireless system for Multi-syatem monitoring," Presented in IEEE sensors Conference, Irvine, CA (USA), Oct 2005.

[16] Y. K. Seok, G. Joonho, K. Jinbong, K. Hong-Jeong, K. Kyunghyun, P. Daesik, K. Myeung su, S. Hyungcheol, L. Kwyro, K. Juhyoun, Y. Euioik, "A miniaturized low power wireless remote environmental monitoring system based on electrochemical analysis," Sensors and Actuators, vol. 102, 2004, pp. 27-34.

2006 25th International Conference on Microelectronics

In-Line Concentration Measurement of Nanoliter Liquid Sample Using Low-Coherence Spectral Interferometry

Milos Tomic, Zoran Djinovic, and Aleksandar Vujanic

Abstract - A method for in-line measurement of the refraction index and the concentration of binary liquid mixture in a nanoliter volume is presented. Low-coherence spectral interferometric technique, based on fiber optic Mach-Zehnder interferometer, is applied for measuring the liquid refraction index, from which its volume fractions are found. The accuracy of volume fractions measurement, of about ±0.2%, was predominantly determined by the accuracy and resolution of reading the light spectrum. The data rate has been limited to 40 Hz by the time of the light spectrum capturing.

I INTRODUCTION

Great efforts are invested today in miniaturization of different analytical systems that belong to a rather broad application range in chemical and biochemical engineering, life science, etc. For instance, routine laboratory analyses are shrinking to the microliter, nanoliter, or even picoliter level, based on emerging microfluidic devices such as "Lab-on-a-chip" [1]. The result is a vast reduction in sample and reagent consumption, decreased waste generation, dramatically faster operation, and an incredible potential for automation and massive, parallel processing of laboratory procedures. These benefits come together with greater resolution in separations, exquisite control over mixing, and the capacity for expediting chemical reactions within highly controlled microenvironments performed by microchemical reactors [2].

Rapid development of miniature high-throughput analytical systems has also resulted in accelerated research in the sensing field. Typically, a sensing unit should provide an *in line* information extracted from the ultra small volume of probe of several nano or even pico liter range. Common analytical techniques, such as spectroscopy and liquid- or gas-chromatography are not suitable for a microchemical architecture with small overall dimensions of about 25x25x0.3 mm³.

Milos Tomic is with the Institut bezbednosti, Kraljice Ane BB, 11040 Belgrade, Serbia & Montenegro,
E-mail: milos.tomic@gmail.com
Zoran Djinovic is with the Institute of Sensor and Actuator Systems, University of Technology, Floragasse 7, Vienna, Austria
Aleksandar Vujanic is with the Integrated Microsystems Austria, Viktor Kaplan str.2/1, Wr. Neustadt, Austria

In addition, cross sectional dimensions of reaction channels are even smaller, typically 100x100 μm².

In this paper we propose a new fiber optic interferometric technique, applicable for measurement of the refraction index and the concentration of binary liquids in a nanoliter sample volume. The method is based on low-coherence spectral interferometry [3], performed by a fiber optic Mach-Zehnder interferometer. Our previous work, based on a similar interferometric configuration suffers from limited speed, determined by mechanical scanning [4].

II THEORY

The relationship between the concentration of components of a binary liquid (ϕ_1, ϕ_2) and the liquid mixture refraction index (n_{12}) is given by Lorenz-Lorentz equation [16]:

$$\frac{n_{12}^2 - 1}{n_{12}^2 + 2} = \phi_1 \frac{n_1^2 - 1}{n_1^2 + 2} + \phi_2 \frac{n_2^2 - 1}{n_2^2 + 2} \qquad (1)$$

Assuming that the Lorenz-Lorentz curve is not ambiguous, knowing the refraction indices of the pure components (n_1 and n_2) and measuring the refraction index of the mixture (n_{12}), we can find the volume fractions of the individual components in a binary liquid mixture through:

$$\phi_1 = \frac{n_{12}^2 - 1}{n_{12}^2 + 2} - \frac{n_2^2 - 1}{n_2^2 + 2}, \quad \phi_2 = 1 - \phi_1 \qquad (2)$$

In this way, the problem of measurement of concentration in binary liquids is converted to the refraction index measurement.

Mach-Zehnder interferometer (MZI) is a two- beams interferometer, containing two beam splitters and two completely separated optical paths. Fiber optic version of MZI is shown in Fig.1. Intensity of light at the interferometer output is described by classical interferometric relation [15]:

1-4244-0116-X/06/$20.00 ©2006 IEEE   233

$$I_D = I_1 + I_2 + 2\sqrt{I_1}\sqrt{I_2}\,\left|\gamma_{11}(\Delta L_{12})\right|\cos\left(\frac{2\pi}{\lambda}\Delta L_{12}\right) \quad (3)$$

where I_1 and I_2 are the irradiances of light beams in the two interferometric arms; ΔL_{12} is the optical path difference (OPD); $\gamma_{11}(\Delta L_{12})$ is the degree of coherence of interfering beams. OPD is the difference between the two separated light paths, from the splitting fiber optic coupler, throughout fibers and air gaps, till the second, combining fiber optic coupler. Air gap in one interferometer arm can be filled with a sample of the examined liquid; air gap in the other arm can be adjusted by some means. If the difference in fiber arms length is ΔL_{fib}, and the refraction index of liquid is n_{liquid}, the OPD is:

$$\Delta L_{12} = n_{LIQUID}\cdot L_{SAMPLEGAP} - L_{AIRGAP} + n_{GLASS}\cdot\Delta L_{FIB} \quad (4)$$

The interferometric term of the light intensity at the MZI output can be written as:

$$I_D\left(n_{liquid}, \lambda\right) \sim \left|\gamma_{11}(n_{liquid})\right|\cos\left(\frac{2\pi}{\lambda}\cdot\left[n_{LIQUID}\cdot L_{SAMPLEGAP} - \Delta L\right]\right) \quad (5)$$

where constant terms are omitted for clarity reasons and the interferometer path difference other than that of the sampling gap is denoted as ΔL. It can be seen in Eq. (5) that the light spectrum has a harmonic shape, whose periodicity linearly depends on the OPD and consequently, on the liquid refraction index. The reciprocal value of the period of spectrum is equal to:

$$\frac{1}{\Lambda} = \frac{n_{liquid}\cdot L_{SAMPLEGAP} - \Delta L - \lambda}{\lambda^2} \quad (6)$$

From Eq. (6), it is evident that the OPD and the index of refraction can be obtained by measuring the periodicity of the interferometer output light spectrum. Furthermore, it can also be noticed that frequency of the spectrum becomes equal or very near to zero for small overall OPDs. In these cases the spectrum period, being larger than source light spectrum width, cannot be measured. On the other hand, if the OPD is large, the degree of coherence becomes very small, causing vanishing of the interferometric term and periodicity in spectrum as well. The limiting factor in this case is, in fact, the spectral resolution of spectrometer [6]. The range of usable OPDs is therefore confined, usually between several tens of micrometers and several hundreds of micrometers. The length of the air gap L_{AIRGAP} in the reference interferometric arm can be used to adjust the measuring range, to optimize it for every specific case of binary liquid.

The light source employed in this experiment must have a relatively wide optical spectrum, to allow seeing clearly the spectrum periodicity. Such a source, at the same

time, has low coherence length, which yields a very fast dropping of the coherence degree term in Eq. (3) Moreover; it has also a low coupling efficiency, especially with single mode fibers. A good choice for the light source, similar as in other low coherence interferometric applications, is a superluminescent diode. These diodes have spectral width of several tens of nanometers and a relatively good coupling efficiency.

III EXPERIMENT

Central part of the experimental set-up, shown in Fig. 1, was a fiber-optic Mach-Zehnder interferometer, based on two single-mode fiber-optic couplers, 2x1 with 4/125 μm fibers (SMC1, SMC2). Low-coherence radiation (LCS) was generated by a superluminescent diode SLD-381, having FWHM spectral width of 25 nm (Superlum, Moscow). The light spectrum of interfering beams was detected by a fiber optic spectrometer (Ocean Optics S2000), then digitalized by 16 bit A/D converter (AD) and captured by a personal computer (PC).

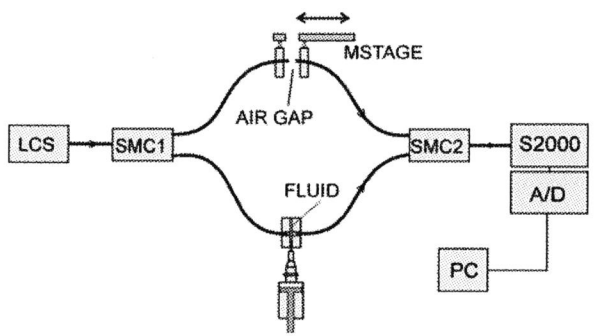

Fig. 1. Experimental set-up, LCS– low-coherence source, SMC$_1$, SMC$_2$-optical fiber couplers, MSTAGE-motorized stage, S2000-fiber-optic spectrometer, A/D-converter, PC-computer

The spectrometer S2000, having an asymmetric crossed Czerny-Turner design, is equipped with a 600 l/mm diffraction grating and Sony ILX511 linear array CCD with 2048 elements [7]. Maximum speed of spectrum capturing with this spectrometer was 3 ms, but averaging of 8 measurements was used, yielding the total acquisition time of 24 ms.

Binary mixture of liquids (water/glycerol in this case) was injected by a syringe in the sampling channel between the two fiber tips in the sensing arm of interferometer. The channel width of 300 μm and its cross-section of 200x200 μm^2 have defined the minimal probe volume of about 20 nl. Motorized stage has been used to adjust the range of OPDs in this specific case of water and glycerol index combination and to the light source employed.

The captured light spectrum was modulated in the wavelength space by a harmonic form, whose frequency is sensitive to the interferometer OPD. OPD was obtained from the spectrum and refraction index of the examined liquid (n) was easily calculated using $OPD=(n-1)*L$, where L is the sampling channel width. Partial concentrations of the sample constituents were calculated using Lorentz-Lorenz equation.

The method was proved by independent measurement of refraction indices of the same samples by Abbe refractometer.

IV RESULTS

Two captured light power spectra, for two different values of the interferometer OPDs are shown in Fig. 2. Harmonic modulation with a very good contrast can easily be seen in the spectrum, with OPD of 100 μm. There is a similar modulation in the 600 μm OPD spectrum, but with higher spatial frequency and with very small modulation depth. Shapes of the spectra at OPDs larger then 600 μm, as well as ones at very small OPDs are almost the same, as there is no interference. Between these two extremes, the frequency of harmonic modulation is increasing with the OPD increase, while, at the same time, "fringe" contrast is decreasing. Upper limit of the usable OPD is much larger then the light source coherence length in classical sense (defined by falling-off of $\gamma_{11}(OPD)$ to 0.5).

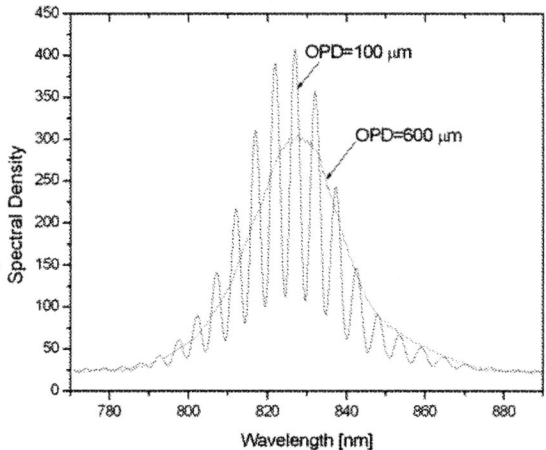

Fig. 2 Two spectrums, obtained by fiber optic spectrometer, for two OPD in fiber optic Mach-Zehnder interferometer.

Spatial spectra of the captured light spectra for several different OPDs, obtained by FFT analysis using MATLAB, are shown in Fig. 3. These results, with addition of many more measuring points in the 10-600 μm range, are used to make a calibration curve, showing the relationship between the peak position in the spatial frequency spectrum and the interferometer OPD.

Using results of the FFT analysis of captured spectra for several mixtures of water and glycerol, interferometers OPDs are obtained from the calibration curve. Thus, as the sampling gap width is known (500 μm), refraction indices of mixtures are easily calculated. These computations: FFT of the captured data looking at the calibration curve and calculations of the refractive index and the partial concentrations were performed off-line, using Matlab. That's why the time needed for signal processing, and thus the measurement speed, cannot easily be estimated. However, a custom-designed stand-alone signal-processing unit can certainly do these tasks in a period smaller then the spectrum capturing time. Therefore, the maximum data rate is capturing time limited to about 40 Hz.

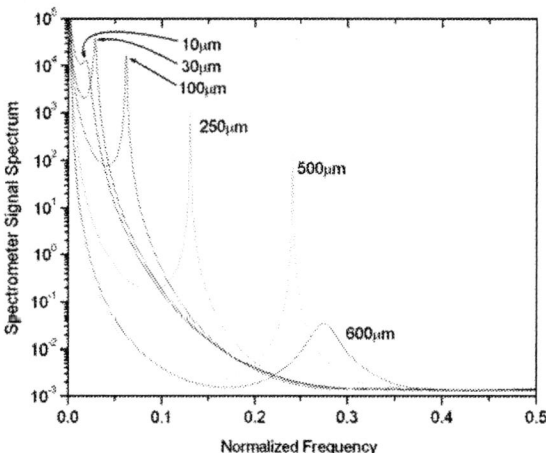

Fig. 3. Spatial spectra of light spectra for different OPDs, ranging from 10 μm to 600 μm.

The results of our measurement for various water/glycerol mixtures, together with measurements obtained by classical Abbe refractometer, are shown in Fig.4.

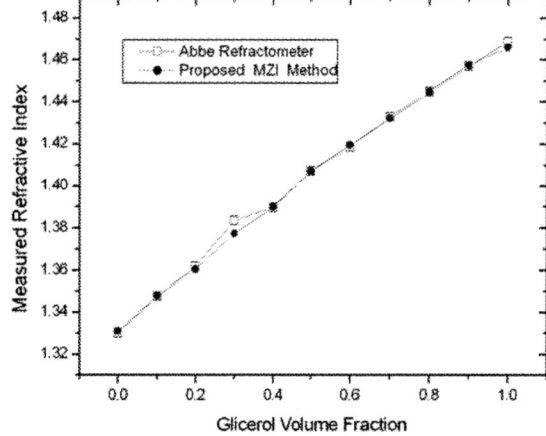

Fig. 4. Refractive indices for different water/glycerol mixtures, measured by the proposed method and by classical Abbe refractometer.

Very good agreement between the two methods can be seen. The RMS of its difference is 0.06 %, meaning that the measurement uncertainty of partial concentration is about 0.2%, according to the error analysis given in [4].

V CONCLUSION

In this paper we presented an optical interferometric method that can serve as a platform for building up a refractometer and further for concentration measurement of binary liquid constituents. Since the method is performed in a fiber-optic interrogation scheme, it is suitable for measurement of nanoliter scale volume of liquid samples in very confined spaces. We proved this capability by measuring the concentration of glycerol in various water/glycerol mixtures.

We obtained the measurement uncertanity of the glycerol concentration of about 0.2%, based on comparative measurement by standard laboratory Abbe refractometer.

REFERENCES

[1] K. Yamashita, Y. Yamaguchi, M. Miyazaki, H. Nakamura, H. Shimizu, and H. Maeda, *Lab on a chip*, 4(2004) pp.1-3

[2] W. Ehrfeld, V.Hessel, and H. Loewe, "*Microreactors: new technology for modern chemistry*", Wiley VCH Verlag, Weinheim, Germany, 2000

[3] L. Montgomery Smith, and C. Dobson, "Absolute displacement measurements using modulation of the spectrum of white light in a Michelson interferometer", *Appl.Opt.*, 28 (1989) pp.3339-3342

[4] M. Tomic, Z. Djinovic, and D. Vujanic, "Nanoliter volume liquid concentration measurement using fiber optic Mach-Zehnder low coherence interferometer", in *Proc. of XIX Eurosensors Conference*, Vol.I, MP72, Barcelona Sept. 2005

[5] Rita Mehra, "Application of refractive index mixing rules in binary systems of hexadecane and heptadecane with n-alkanols at different temperatures", *Proc. Indian Acad. Sci. (Chem. Sci.)*, Vol. 115, No.2, April 2003, pp 147–154

[6] U.Schnell, E.Zimmermann, and R. Dandliker, "Absolute distance measurement with synchronously sampled white light channelled spectrum interferometry", *Pure Appl.Opt.* ,4 (1995) pp.643-651

[7] Petr Hlubina, "Dispersive spectral/domain two/beam interference analyzed by a fibre-optic spectrometer", *Journal of Modern Optics*, 51 (2004) pp.537-547

Using Design of Experiment to Investigate the Effects of Conducting Polymer Composite Sensor Composition on the Response to an Homologous Series of Alcohols

K. I. Arshak, E. G. Moore, C. Cunniffe and L. M. Cavanagh

Abstract –Statistical design of experiment (DOE) is a critically important tool in the engineering, physical and chemical science worlds for improving existing products and also for the development of new products. This paper investigates the use of DOE for the design and characterisation of conducting polymer composite (CPC) vapour sensors. DOE has the advantage of providing a large amount of information on a process using a minimum amount of experimentation, equipment and time. This analysis tool is therefore ideal for the development of composite sensors, where more than one material determines the properties of the sensor. The effects of % plasticiser, polymer type, alcohol type and concentration on the response and response times of these materials were investigated. The results showed that plasticiser significantly improved the response but also increased the response time. PVB was the best polymer to use to achieve a high response but again it had a very long response time. The DOE analysis technique provided a large amount of information on the response of the chosen materials with three alcohols that using standard testing methods would have required a lot more experimental work.

I. INTRODUCTION

Much research has been carried out in recent years on the use of conducting polymer composites as vapour sensing devices. These materials consist of conductive particles embedded in an insulating polymer matrix. The composite material swells upon interaction with a vapour, altering its resistance. This change in electrical signal can be readily integrated with low cost, conservative signal conditioning circuitry. Arrays of these materials are used to produce a pattern for a particular vapour based on the changes in resistance of each sensor. A different pattern is produced for each vapour, which can be used as a means of vapour classification. These materials provide cost effective low powered vapour sensors that operate at room temperature which are ideal for use in electronic nose systems [1]. Chemical diversity can be achieved in these composites through combining a range of fillers, plasticisers and various other compounds. However, examining the effects of each substance in the mixture individually can require a lot

K. I. Arshak, E. G. Moore, C. Cunniffe and L. M. Cavanagh are with the Department of Electronic & Computer Engineering, College of Informatics & Electronics, University of Limerick, Plassey technological park, Limerick, Ireland. E-mail: Khalil.Arshak@ul.ie

of experimentation, time and equipment. Further tests are required to determine any interactions between the materials in the mixture. DOE can be used to overcome the time consuming experimentation required when using single component testing.

DOE is a statistical approach used to gain the maximum amount of information from a single experiment using the minimum amount of materials, time and equipment. [2]. The experimental array chosen for use in this work was an L9 Taguchi. The experiment consisted of nine runs with three replicates of each run carried out. Each factor had three levels. The factors examined were % plasticiser, polymer type, alcohol type and alcohol concentration as shown in TABLE I. This design was used as a method to investigate which factors have the largest effects on the response and response time. It also helps determine the level at which these influential factors should be set so as to obtain the desired value of the response. The information provided from this set of experiments can be used to further optimize the materials to improve their sensitivities to methanol, ethanol and propanol.

Plasticisers have been used in the literature to manipulate the optical and electrical properties of organic materials for the detection fof volatile organic compounds. Depending on the chemical make up of the plasticiser changes in the selectivity of the gas sensor can be achieved [3]. This paper presents the use of Monasil PCA for the plasticisation of three polymers and examines its effects on their gas sensing properties.

TABLE I
NINE RUNS CARRIED OUT IN L9 TAGUCHI EXPERIMENT

Run no.	% Plasti-ciser	Polymer Type	Alcohol	Alcohol Conc.
1	0	PVAc	Methanol	2500
2	0	PVB	Ethanol	5000
3	0	PS	Propanol	7500
4	10	PVAc	Ethanol	7500
5	10	PVB	Propanol	2500
6	10	PS	Methanol	5000
7	20	PVAc	Propanol	5000
8	20	PVB	Methanol	7500
9	20	PS	Ethanol	2500

II. EXPERIMENTAL

A. Sensor Preparation

The polymers used in the preparation of the composite gas sensors were polyvinyl butyral (PVB), polyvinyl acetate (PVAc) and polystyrene (PS), which were purchased from Sigma-Aldrich, BDH chemicals and Avocado Research Chemicals Ltd. respectively. Black Pearls 2000 carbon black was purchased from Cabot Carbon Ltd. Tetrahydrofuran (THF) was the solvent used to dissolve the polymers and was purchased from LABSCAN Ltd. Hypermer PS-3 surfactant was used to aid carbon black dispersion in the polymers and this was acquired from Uniqema. Monasil PCA was also acquired from Uniqema.

Nine samples were prepared as shown in Table 1. Each sample was prepared in a 60ml glass bottle with an airtight lid. The polymer was added to the 20ml of THF in the glass bottle and sonicated for an hour to encourage polymer dissolution into the THF. The required quantities of carbon black, surfactant and plasticiser were then added to each of the 9 polymer solutions prepared as outlined in TABLE II. Each bottle was then shear mixed for 300s at 16,000rpm using an Ultra-Turrax T25 basic shear mixer.

TABLE II
COMPOSITION OF EACH SAMPLE PREPARED FOR DOE EXPERIMENT

Sample no.	Polymer (mg)		Carbon black (mg)	Plasticiser (mg)	Surfactant (mg)
1	PVB	160	40	0	30
2	PVAc	160	40	0	30
3	PS	160	40	0	30
4	PVB	137	40	23	30
5	PVAc	137	40	23	30
6	PS	137	40	23	30
7	PVB	114	40	46	30
8	PVAc	114	40	46	30
9	PS	114	40	46	30

Three sensors were fabricated from each of the nine sample bottles prepared. 2µL's of each sensing material was drop coated onto Cu interdigitated electrodes on a PCB (printed circuit board) substrate using a microlitre pipette as shown in Figure 1. The interdigitated electrode pattern was designed using eagle software and then etched onto the PCB using standard etching techniques. The sensors were dried in air for 30 minutes and then placed in an oven at 60°C for 24 hours to remove any residing solvent.

Fig. 1 Surfactant Treated PVB \ CB sensor prepared on PCB substrate

B. Sensor Testing Apparatus

Each sensor was tested individually in the flask. The sensor was placed into the chamber and attached to a multimeter to record its resistance. The flask was closed and flushed with compressed air to remove any unwanted gases or vapours. The sensor baseline resistance was allowed to stabilize prior to the addition of the solvent. The required concentration of the solvent was injected into the flask through the septum using a microlitre syringe. The sensor was exposed to the solvent for 220s and then flushed with compressed air for 300s to remove the solvent and return the sensor to its baseline resistance. A hot plate was used to control the temperature and to vaporise the solvent in the flask while also preventing condensation of the solvent on its walls. The fan was also used to encourage homogenous distribution of the solvent vapour as shown in Fig. 2.

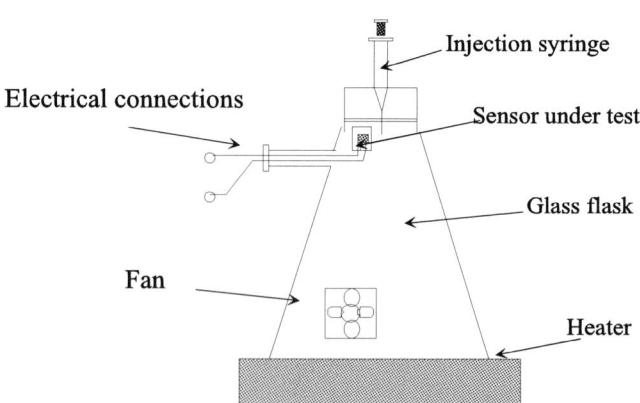

Fig. 2 Illustration of test chamber used to test sensors with alcohol vapours.

III. RESULTS & DISCUSSION

C. The Response curve of CPC sensor

The maximum relative differential response of each sensor upon exposure to a vapour was calculated using the following equation $((R_{Max} - R_B) / R_B)*100$ where R_{Max} is the maximum change in resisitance upon exposure to a vapour for 220s and R_B is the baseline resistance of the sensor prior to exposure. This value provides a dimensionless, normalized response that can compensate for inherently large or small signals [4]. The max time is the time it took the sensors to reach a maximum response when exposed to a vapour. The response curve of a PVAc\ CB \PS3 sensor containing 20% Monasil PCA when exposed to 7500ppm methanol is shown in Fig. 3. The response mechanism of CPC sensors is based on percolation theory. The vapour absorbs into the polymer in the composite causing it to swell. The increasing expansion of the material increases the distance between the conductive particles. This reduces the number of pathways available for the charge carriers and therefore increases the overall resistance of the material [5].

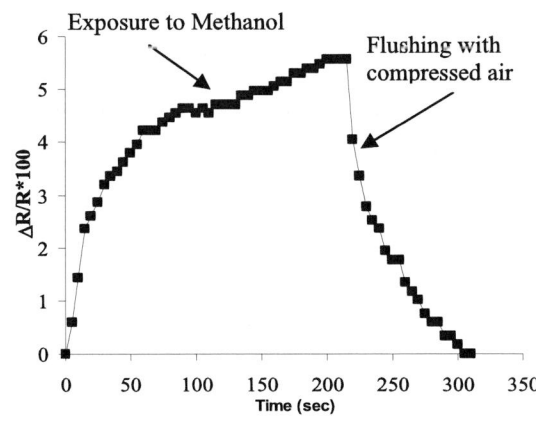

Fig. 3 The response curve obtained when PVAc\CB\PS3 sensor with 20% Monasil PCA was exposed to 7500ppm Methanol.

D. DOE software analysis of responses

27 runs were carried out based on the L9 Taguchi experiment designed using DOE Wisdom software. The results were placed into a DOE wisdom datasheet and statistical analysis of the results was carried out.

E. Main effects of the factors on the relative differential response

Fig. 4 shows the effects that each factor had on the relative differential response based on the results obtained from the DOE experiment. The objective of this study was to characterize the impact that each material had on the response of the CPC sensor to methanol, ethanol and propanol over a range of concentrations.

The main effects plot for the plasticiser ranging from 0 to 20% shows that increasing the plasticiser increases the response. The plasticiser contains both polar

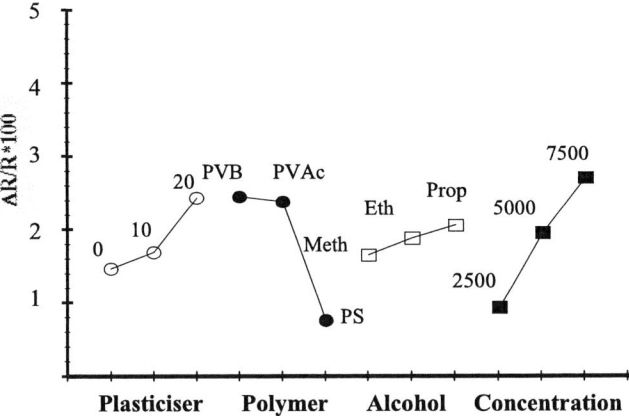

and non-polar groups. The non-polar groups ensure

Fig. 4 Main effects plot of each factor on the maximum relative differential response.

that the plasticiser remains in the polymer. The polar groups attenuate the attractive forces between the polymer chains to improve segmental chain mobility and free volume. This increase in free volume and segmental chain mobility allows increased absorption of vapour into the composite material, which breaks down more carrier pathways and increases the resistance change of the sensor [6]. The increase in the response on addition of plasticiser can also be contributed to the highly polar side group in Monasil PCA, which encourages the absorption of polar vapours like alcohols into the composite material [3].

TABLE III
PROPERTIES OF THE POLYMERS

Polymer	FH parameter	Tg	Mw
PVAc	1.034	30	160,000
PVB	1.577	67	65,000
PS	1.753	100	100,000

The effects of polymer type on the response of the sensors can be explained in terms of glass transition temperature (T_g), free volume, the Flory-Huggins interaction parameter and the molecular weight (Mw) of each polymer. The results in Fig. 1 show that PS had a much smaller effect on the response when compared to PVB and PVAc. PS has a very high glass transition temperature in comparison to the other polymers, which contributes to the poor response. Polymers with high T_g's tend to have very rigid glassy structures at room temperature that inhibit the absorption of vapours. PVB and PVAc on the other hand have rubbery structures, which encourage the absorption of vapours [7]. The Flory-Huggins interaction (FH) parameter as shown in Equation 1 is also important in determination of the response of a CPC sensor to a vapour [8]. The closer this value is to zero the better the attraction between the polymer and the vapour, which encourages a greater response from the sensor. The Flory – Huggins parameter for each polymer with methanol is given in Table III.

$$\chi_{12} = \frac{V_1^0}{R.T}(\delta_S - \delta_P)^2 \qquad (1)$$

where V_1^0 is the molar volume of the solvent. R is the gas constant. T is the temperature. δ_S is the solubility parameter of the solvent. δ_P is the solubility parameter of the polymer. PVAc has a lower Tg and FH parameter than PVB but PVB still has a bigger effect on the response. This is explained by the fact that the molecular weight (Mw) of PVB is much lower than that of PVAc as predicted by the DOE software. From the results it can be seen that polymers with low Mw, Low FH parameters and low Tg's give the best responses.

The effects of alcohol type on the response can be attributed to the Mw of the alcohol where increasing the molecular weight increases the response of the sensor [3]. The FH parameters for Methanol, ethanol and propanol are 1.577, 1.144 and 0.866 respectively. As discussed previously the closer this parameter is to zero, the higher the response.

F. Main effects of the factors on the relative differential response

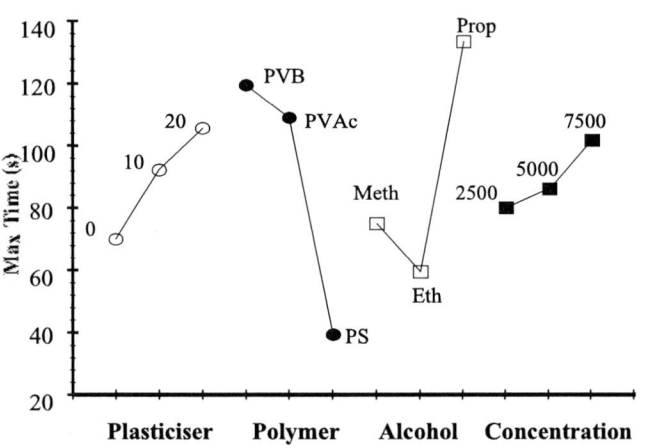

Fig. 5 Main effects plot of the Max response time.

The results shown in Fig.5 correspond with those obtained in section E. The increase in plasticiser improved vapour sorption into the CPC material as previously described. However this requires additional time for the vapours to percolate into the material. As a result more time is required for the material to reach equilibrium. This can also account for differences in response time for each polymer type. PVB and PVAc take longer to reach equilibrium when compared to PS because there is a larger quantity of the vapour trying to absorb into their structures.

Changes in the response time of the sensor to each of the alcohols investigated can be accounted for by considering their Mw. This will influence the rate of transport of vapours within the polymer matrix. Low Mw vapours, like methanol, can penetrate the polymer quickly. However larger Mw vapours increase the heat of sorption and effectively plasticise the polymer, allowing more molecules to permeate into the polymer. Therefore, in medium Mw vapours, for example ethanol, it is thought that both processes are present [6]. This results in an improved response time when compared to methanol and propanol. It can be seen from the results that the composite mixture, which achieves the best sensitivity takes the longest time to reach the steady state. The DOE analysis has shown that increasing the response of a CPC sensor prepared using the materials in this study will result in increasing the response time. Therefore a compromise has to be made between achieving a good response to a vapour and achieving that response in a short time period.

IV. CONCLUSION

In this work, statistical design of experiment was used to characterize three CPC sensors with a view to improving sensitivity, through the addition of Monasil PCA. It was found that the addition of a plasticiser increases sensitivity at the expense of the response time. It was also shown that low Mw polymers with low Tg's and FH parameters close to zero are required in order to maximize the response. It has been shown through this work that a large quantity of information can be obtained through the use of DOE with a minimum amount of experimentation. In addition, the effect of a combination of parameters can be investigated in order to understand and optimize device operation. The technique has been successfully applied to the analysis and improvisation of CPC sensors.

REFERENCES

[1] M. C. Lonergan et al. "Array-Based Vapor Sensing Using Chemically Sensitive, Carbon Black-Polymer Resistors," *Chem. Mater*, Vol. 8, pp.2298, 1996.
[2] D. C. Montgomery, Design and Analysis of Experiments, New York, John Wiley & Sons, Inc., 2001.
[3] M. E. Koscho, R. H. grubbs, N. S. Lewis, "Properties of Vapor Detector Arrays Formed through Plasticization of Carbon Black – Organic Polymer Composites", Anal. Chem., vol. 74, pp. 1307-1315, 2002.
[4] K. Arshak et al., "A Review of gas sensors employed in Electronic Nose Applications", vol. 24, pp. 181-198, 2004.
[5] K. Arshak et al., "Determination of the Electrical Behaviour of Surfactant Treated Polymer/Carbon Black Composite Gas Sensors", Composites: Part A, vol. 36, pp. 487-491, 2005.
[6] S. C. George, S. Thomas, "Transport Phenomena Through Polymeric Systems", Prog. Polym. Sci., vol. 26, pp. 985-1017, 2001.
[7] R. A. McGill, M. H. Abraham, "Choosing Polymer Coatings for Chemical Sensors", Chemtech, pp. 27-37, 1994.
[8] J. F. Feller, Y. Grohens, "Evolution of Electrical Properties of Some Conductive Polymer Composite Textiles with Organic Solvent Vapours Diffusion", Sensors and Actuators B, vol. 97, pp.231-242, 2004.

2006 25th International Conference on Microelectronics

IR Bimaterial Detectors Performance

Z. Djurić, D. Randjelović, J. Matović, J. Lamovec

Abstract - In this paper basic parameters of a bimaterial infrared thermal detector (BMD) – sensitivity, noise equivalent power, detectivity – and their dependence on relevant thermal and mechanical parameters are given. In order to find sensitivity and noise we introduce an equivalent "thermo mechanical" excitation force for microelctromechanical bimaterial oscillator and solved equation for this oscillator assuming only first mode of vibration. After the identification of all important noise mechanisms (temperature fluctuations, Brownian motion), we solved the appropriate Langevin stochastic equation and obtained the mean square deflection of the bimaterial cantilever oscillator. This enabled us to determine all of the important parameters for BMD.

I. INTRODUCTION

There is a considerable interest in 2D thermal detector arrays for low cost thermal imagers [1], whose moderate sensitivity can be compensated by a large number of elements. The appearance of MEMS technologies enabled development of a new generation of thermal detectors – the bolometer type. Today, bolometric focal plane arrays (FPA) with 320 x 240 elements are already in use [2]. The complexity of interconnections between pixels and of the read-out circuits is the main reason for high price which still represents the main obstacle for a wider use in commercial applications. Due to the electrical connection between each pixel and the circuit for signal analysis thermal isolation close to the radiation limit can not be achieved. Electronically based systems such as bolometers inherently generate 1/f and Johnsons noise which limit their detectivity far below radiation limit.

In recent years a number of papers appeared which suggest the use of opto-mechanical or capacitive-mechanical imaging systems in order to overcome the above stated problems [3-8]. Most of these papers are inspired by the fact that microcantilevers built of two different materials and optical detection of displacement as used in AFM systems can be applied for detection of small temperature variations. The deflection of a thermally sensitive bimaterial cantilever is measured with a laser optical lever and a position-sensitive silicon photodiode (displacements of the order 10^{-12} m can easily be measured [9]).

The first part of this paper presents a short theory of bimaterial effect and the equivalent "thermo-mechanical" force is introduced. After the identification of all pertinent noise mechanisms, the appropriate Langevin stochastic equation is solved. Using the results from the first part, the

Z. Djurić, D. Randjelović, J. Matović, and J. Lamovec are with the IHTM-IMTM, University of Belgrade, Njegoševa 12, 11000 Belgrade, Serbia & Montenegro,
E-mail: zdjuric@nanosys.ihtm.bg.ac.yu

detectivity of a BMD is obtained. In the third part, for a given structure of the BMD, its basic parameters (resonant frequency, Q-factor, thermal conductivity) are analysed as a function of gas pressure inside the housing. In the final part we apply our theory on a chosen BMD structure and give our calculation results of detector parameters.

II. BIMATERIAL EFFECT

Bimaterial effect is based on the bending of a microcantilever with a "sandwich" structure due to temperature increase induced by absorption of infrared radiation. The structure of a typical BMD is shown in Fig. 1. It consists of a relatively large absorbing area whose position is controlled by two active bimaterial cantilevers. The whole structure is connected to the support via two thermally isolating beams.

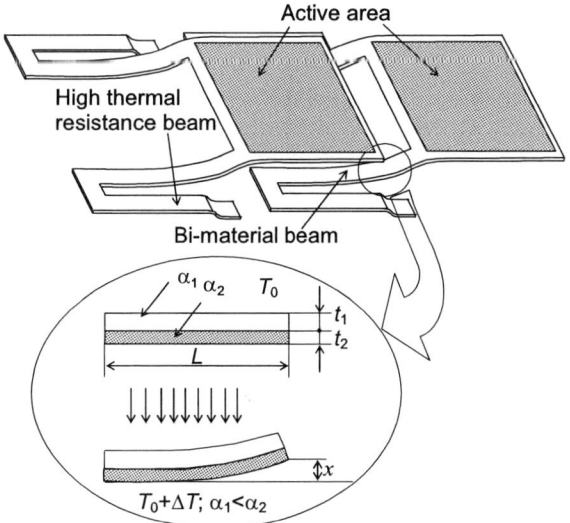

Fig. 1. IR detector based on the bimaterial concept.

By applying the theory of static deformation it can be shown that the deflection of the free end of a microcantilever at a constant temperature T is given by

$$x(L,T) = \frac{3L^2(\alpha_1 - \alpha_2)}{t_2} \frac{(1+n)(T-T_0)}{4+6n+4n^2+n^3\phi+1/n\phi} \quad (1)$$

where the following parameters were used: ratio of layer thicknesses, $n = t_1/t_2$ and ratio of their Young's moduli, $\phi = E_1/E_2$. It can be seen that the free end deflection is proportional to temperature difference, $\Delta T = T - T_0$, square of the cantilever length, L, and difference in thermal expansion coefficients of the materials used, $\alpha_1 - \alpha_2$.

1-4244-0116-X/06/$20.00 ©2006 IEEE   241

III. BIMATERIAL MICROCANTILEVER WITH EXTERNAL THERMAL DRIVE

Bimaterial microcantilevers belong to the class of movable micromechanical structures. In the first approximation these structures can be modeled as a mechanical oscillator with a certain damping. If we analyze bimaterial cantilever with a spring constant, k_{BM}, resonant frequency, ω_0, effective mass characteristic for the first harmonic, m_{eff}, mechanical resistance R ($R(\omega) = (\omega_0/Q)m_{eff}$) and quality factor Q, equation of vibrational motion can be written as

$$\frac{d^2x}{dt^2} + \frac{R}{m_{eff}}\frac{dx}{dt} + \frac{k_{BM}}{m_{eff}}x = \frac{F}{m_{eff}}, \quad (2)$$

where x is cantilever deflection. F is thermal driving force proportional with temperature difference, $F = K\,\Delta T$.

Taking into account that for all types of thermal detectors the increase of the temperature of the active area in the first approximation is given by

$$\Delta T = R_{th}\varepsilon P\left(1 + \omega^2\tau_{th}^2\right)^{-1/2}, \quad (3)$$

(where P is the absorbed optical power, ε is active area emissivity, $\tau_{th} = R_{th}C_{th}$ is thermal time constant equal to the product of thermal resistance and thermal capacitance) and assuming that both temperature and displacement of the cantilever are changing with frequency ω we obtain expression for cantilever displacement

$$x = \frac{K}{m_{eff}}\frac{R_{th}\varepsilon P}{\sqrt{1 + (\omega\tau_{th})^2}}\left(\left(\omega_0^2 - \omega^2\right)^2 + (\omega\omega_0/Q)^2\right)^{-1/2}. \quad (4)$$

BMD sensitivity is defined as variation of displacement due to variation of the power of the incident radiation

$$S = \frac{dx}{dP} = \frac{K\varepsilon R_{th}}{m_{eff}}\frac{1}{\sqrt{1 + (\omega\tau_{th})^2}}\left(\left(\omega_0^2 - \omega^2\right)^2 + (\omega\omega_0/Q)^2\right)^{-1/2} \quad (5)$$

On the other hand, it is well known that the force acting on an oscillating system is proportional to the displacement, $F = k_{BM}\, x(L)$. Combining (1) for the micro-cantilever displacement with the starting assumption, $F = K\,\Delta T$, the proportionality constant is obtained.

The spring constant for a bimaterial cantilever of width w_{BM} was calculated using [10]

$$k_{BM} = \frac{w_{BM}}{L^3}\left(E_1 t_1^3 + E_2 t_2^3\right) - \frac{3w_{BM}}{4L^3}\frac{\left(E_1 t_1^2 - E_2 t_2^2\right)^2}{E_1 t_1 + E_2 t_2}. \quad (6)$$

By substituting the expression for K in eqn. (5) and using the relation $\omega_0^2 = k_{BM}/m_{eff}$ the expression for sensitivity is easily obtained.

IV. NOISE IN BMD

Let us now analyze the cantilever displacement taking into account both temperature fluctuations, $F = K\Delta T$ and stochastic Brownian motion, F_{Braun}. Assuming that temperature and displacement are changing with frequency ω we can write the following expression for the spectral density of displacement

$$\overline{X^2(\omega)} = \frac{1}{\left(\omega_0^2 - \omega^2\right)^2 + \left(\frac{\omega_0\omega}{Q}\right)^2}\left(\frac{K^2\,\overline{\Delta T^2(\omega)}}{m_{eff}^2} + \frac{\overline{F_{Braun}^2}}{m_{eff}^2}\right). \quad (7)$$

The performance of BMDs is influenced by two basic noise generation mechanisms caused by temperature fluctuations [11]. The first one is causing spectral density of temperature fluctuations

$$\overline{\Delta T^2} = 4k_B T^2 R_{th}\big/\left(1 + \omega^2\tau_{th}^2\right) \quad (8)$$

and the second one is thermal mechanical noise resulting from thermo-mechanical force with a spectral distribution

$$\overline{F_{Braun}^2} = 4k_B T\left(\omega_0 m_{eff}\right)\big/Q. \quad (9)$$

Using these expressions the mean square value of displacement is obtained

$$\frac{\overline{X^2(\omega)}}{\Delta f} = \frac{1}{\left(\omega_0^2 - \omega^2\right)^2 + \left(\frac{\omega_0\omega}{Q}\right)^2}\left(\frac{K^2}{m_{eff}^2}\frac{4k_B T^2 R_{th}}{1 + \omega^2\tau_{th}^2} + \frac{4k_B T}{m_{eff}}\frac{\omega_0}{Q}\right) \quad (10)$$

Specific detectivity, D^* is given by relation

$$D^* = \sqrt{A\Delta f}\big/NEP = \sqrt{A\Delta f}\ S\big/\sqrt{\overline{X^2(\omega)}}. \quad (11)$$

Substituting the above derived expresions for the mean square value of displacement and sensitivity we finally obtain

$$D^* = \frac{\sqrt{A}}{\sqrt{\dfrac{4k_B T_d^2}{\varepsilon^2 R_{th}} + \dfrac{4k_B T_d\omega_0 m_{eff}}{\varepsilon^2 K^2 R_{th}^2 Q}(1 + \omega^2\tau_{th}^2)}}, \quad (12)$$

where R_{th} is the total thermal resistance of the detector comprising losses via conduction through supporting legs, G_{beam}, and surrounding gas G_{gas} and radiation losses G_{rad}:

$$1/R_{th} = G_{beam} + G_{gas} + 2\varepsilon A_{eff}\sigma k_B\left(T_d^3 + (T_0/T_d)^2 T_0^3\right). \quad (13)$$

The above expressions take into account the emissivity of the upper surface of active area, ε, and also the possibility that the detector temperature, T_d, differs from the background temperature, T_0. A_{eff} is the effective active area which takes into account the difference of emissivities of the front and back side. In our analysis of a BMD performance we assume $\varepsilon = 1$, and $T_d \approx T_0$.

242

It can be seen that specific detectivity depends both on thermal and mechanical properties of the detector. With a pressure decrease thermal resistance and Q-factor increase, therefore detectivity can be improved with pressure decrease and a proper choice of other bimaterial microcantilever parameters (increase of K). In this way, detectivity is approaching the background limited value, $D^* = 1,81 \cdot 10^{10}$ cmHz$^{1/2}$/W.

V. PRESSURE INFLUENCE ON BMD PERFORMANCE

As previously stated BMD can be regarded as a mechanical oscillator characterized by its resonant frequency, ω_0, and Q-factor [8]. On the other hand, its operation is based on temperature increase induced by heat absorption which depends on thermal conductance of the structure, G_{th}. The response time of a BMD depends both on thermal conductance and thermal capacitance (C_{th}) of the structure. All these basic parameters are dependent on the gas pressure inside the housing. Sensitivity and specific detectivity are functions of these parameters and are therefore also pressure dependent.

A typical pressure dependence of Q-factor of a microcantilever is shown in Fig. 2. (this curve is adapted from [12]). At low pressures (intrinsic region), below 1Pa, Q-factor is constant and has a maximum value determined by various physical mechanisms which cause intrinsic damping [8]. The largest change in quality factor occurs in the molecular region (1 Pa $\leq p \leq 10^3$ Pa), dominated by the interaction between the air molecules and the vibrating structure. In this region Q-factor is inversely proportional with pressure. For higher pressures, over 10^3 Pa, air should be considered as a viscous fluid. In this region Q-factor is again constant around 10^4 Pa and then drops with $p^{-1/2}$.

Resonant frequency of structures with bimaterial cantilevers is pressure dependent ([4], [12], [13]), but this effect will be neglected in our analysis.

In a BMD structure the absorbed heat is transferred via conduction through the bimaterial and the isolating beams, via conduction and convection through surrounding gas and by radiation mainly from the absorbing area.

For the structure shown in Fig. 1 the total thermal conductance of one beam is given by

$$G_{beam} = l_{BM}\left(\lambda_1 w_{BM} t_1 + \lambda_2 w_{BM} t_2\right)^{-1} + l_i\left(\lambda_i w_i t_i\right)^{-1} \quad (14)$$

where l_{BM} and w_{BM} are the length and width of bimaterial beam, $\lambda_{1,2}$ and $t_{1,2}$ are thermal conductance and thickness of materials composing the bimaterial beam, and with index i denoted are the same parameters for the isolating part of the beam.

Regarding the heat loss through the surrounding gas, it can be shown that for BMDs convection can be neglected. Thermal conductance via conduction is given by

$$G_{gas}^{cond} = \lambda_{gas} A/d , \quad (15)$$

where λ_{gas} is thermal conductivity of the gas inside the housing, A is absorber area and d is distance between the absorbing plate and the substrate. Thermal conductivity of the gas shows strong dependence on pressure. According to the model given by Eriksson et al. [13] λ_{gas} is given by

$$\lambda_{gas}^{-1} = \lambda_{hp}^{-1} + \left(\gamma_{lp} Pd\right)^{-1}. \quad (16)$$

Parameters λ_{hp} and γ_{lp}, characteristic for high and low pressure regions, respectively, are given by relations

$$\lambda_{hp} = 2c/\left(3\sigma_0\sqrt{\pi}\right)\sqrt{k_B TM} , \gamma_{lp} = c/3\sqrt{8M/\pi k_B T} \quad (17)$$

where c is specific heat, M is the molecule weight and σ_0 is the scattering cross section of the molecules.

VI. CALCULATION RESULTS

Theory of BMDs presented in this paper was applied on a structure designed and fabricated at ORNL (Fig. 3) [4]. This detector has an active area of 66 µm x 41 µm, and is made of silicon nitride 550 nm thick, covered with aluminum film of 100 nm thickness. Bimaterial beams are made of the same Al/SiN$_x$ combination and have width of 3 µm and length 101 µm. The thermal isolation parts of the beams are made of SiN$_x$ 250 nm thick, 30 µm long and 1.5 µm wide. The values of other relevant parameters used in calculation were taken from [6].

The analysis of thermal conductance was performed. The total thermal conductance of the two legs, including their bimaterial and isolation parts, is $G_{leg} = 3.85 \cdot 10^{-7}$ W/K, the radiative part is equal $G_{rad} = 1.01 \cdot 10^{-8}$ W/K and air conductance G_{gas} is pressure dependent. Summing all three components we obtained the pressure dependence of thermal conductance of BMD as shown in Fig. 2.

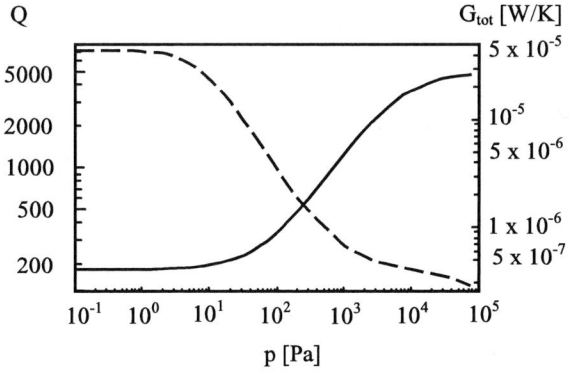

Fig. 2. Typical pressure dependence of Q-factor for cantilever (dashed line) and total thermal conductance of BMD (solid line).

Next we analyzed the influence of pressure on the detector thermal time constant. Taking into account the thermal capacitance of the active area ($C_A = 3.13 \cdot 10^{-9}$ J/K) and the total thermal conductance of the BMD structure we obtained that the time constant is decreasing with pressure.

In the pressure range of interest it drops from about 8 ms to nearly 0.1 ms.

Although for the analyzed detector a variation of resonant frequency with pressure was observed [4] in our calculations we assumed a constant resonant frequency of 9.3 kHz. Using this assumption and the above obtained results for other relevant parameters we calculated the sensitivity and the specific detectivity of a BMD for several oscillation frequencies. Pressure dependence of sensitivity is shown in Fig. 3.

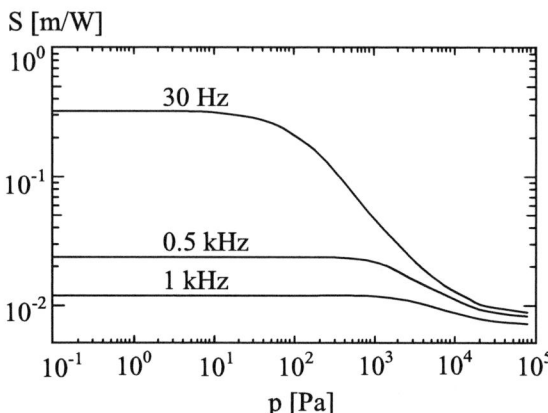

Fig. 3. Sensitivity of BMD as a function of pressure.

Fig. 4. shows the dependence of the specific detectivity of the BMD on the pressure in the detector housing. It can be concluded that high vacuum and low conduction heat losses (G_{leg}) are necessary to reach near-background limited detectivities (~$1.81 \cdot 10^{10}$ cmHz$^{1/2}$/W).

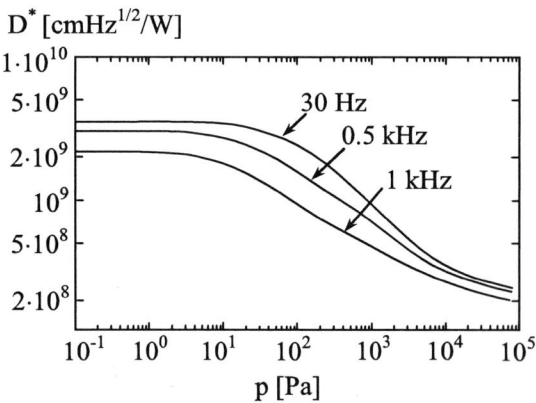

Fig. 4. Specific detectivity of BMD as a function of pressure in the detector housing.

VII. CONCLUSION

Using the theory presented in this paper, basic parameters of an IR BMD can be calculated. It was shown that detectivity of real structures can attain values only few times less than ideal ones.

ACKNOWLEDGEMENT

Authors would like to thank Dr. Slobodan Rajic, Dr. Panos G. Datskos and Dr. Dragoslav Grbovic for useful discussions and authorization for implementing our theory on a detector fabricated at Oak Ridge National Laboratory.

REFERENCES

[1] R. A. Wood, C. J. Hau, P. W. Kruse, "Integrated Uncooled Infrared Detector Imaging Arrays", in *Proc. IEEE Solid State Sensor and Actuator Workshop*, Hilton Head Island, June 1992 , pp. 132-135.

[2] N. Oda, Y. Tanaka, T. Sasaki, A. Ajisawa, A. Kawahara, S. Kurashina, "Performance of 320 x 240 Bolometer-Type Uncooled Infrared Detector", *NEC Research & Development* 2003/2 Paper 6, www.nec.co.jp/techrep/en/r_and_d/r03/r03-no2/rd06.pdf

[3] Y. Zhao, M. Mao, R. Horowitz, A. Majumdar, J. Varesi, P. Norton, J. Kitching, "Optomechanical Uncooled Infrared Imaging System: Design, Microfabrication, and Performance", *J. Microelectromech. Syst.*, Vol. 11, No. 2, pp. 136-146, April 2002.

[4] P. G. Datskos, N. V. Lavrik, S. Rajic, "Performance of uncooled microcantilever thermal detectors", *Rev. Sci. Instrum.*, Vol. 75, No. 4, pp. 1134-1148, April 2004

[5] L. R. Senesac, J. L. Corbeil, S. Rajic, N. V. Lavrik, P. G. Datskos, "IR imaging using uncooled microcantilever detectors", *Ultramicroscopy*, pp. 451-458, 2003.

[6] T. Ishizuya, J. Suzuki, K. Akagawa, T. Kazama, "160 × 120 pixels optically readable bimaterial infrared detector", *The Fifteenth IEEE International Conference on MEMS*, Las Vegas, USA, 20-24 January, pp. 578-581, 2002.

[7] S. Lim, J. Choi, R. Horowitz, A. Majumdar, "Design and Fabrication of a Novel Bimorph Micro-Opto-Mechanical Sensor", *J. Microelectromech. Syst.*, Vol. 14, 4, pp. 683-690, 2005.

[8] Z. Djurić, "New generation of Thermal Infrared Detectors", *Proc. 20th Int. Conf. on Microelectronics MIEL '95*, Vol 1, Niš, Serbia, 559-564, 1995.

[9] R. V. Jones, J. C. S. Richards, "Recording optical lever", *Journal of Scientific Instruments*, Vol 36, February 1959

[10] J. Hirai, R. Mori, H. Kikuta, N. Kato, K. Inoue, Y. Tanaka, „Resonance Characteristics of Micro Cantilever in Liquid", Jpn. J. Appl. Phys., Vol. 37, pp. 7064-7069, 1998.

[11] Z. Djurić, "Mechanisms of noise sources in microelectromechanical systems", *Microel. Reliability* 40, pp. 919-932, 2000.

[12] O. Brand, R. Lenggenhager, H. Baltes, "Influence of Air Pressure on Resonating and Thermoelectric Microstructures Realized with Standard IC Technologies", *Tech. Digest, Electron Dev. Meeting*, Washington, DC, USA, pp. 195-198, Dec. 1993.

[13] J. Mertens, F. Finot, T. Thundat, A. Fabre, M.-F. Nadal, V. Eyraud, E. Bourillot, "Effects of temperature and pressure on microcantilever resonance response", *Ultramicroscopy* 97, pp. 119-126, 2003.

2006 25th International Conference on Microelectronics

Dynamic Elastic Bending in Optically Driven Microcantilever: Surface Strain Effects

D. M. Todorović, T. Grozdić

Abstract – Dynamic thermoelastic (TE) and electronic deformation (ED) bending in a bimaterial rectangular microcantilever are investigated. The theoretical model for the dynamic elastic vibrations of the microcantileve, including the surface stress effects (the axial force) is given. The components of elastic bending of microcantilever, with dimensions typical for atomic force microscopy are calculated. These investigations are important for many sensors and actuators based on microcantilevers.

I. INTRODUCTION

A mass-production of inexpensive microcantilevers (MCLs) enables a revolution in the field of chemical and physical sensor design. A novel class of highly sensitive sensors is based on commercially available MCLs, such as those used in atomic force microscopy (AFM). The wide availability of AFM instrumentation generated substantial interest in MCLs as a platform for a variety of sensors.

Unlike many other types of transducers, MCLs are simple mechanical devices. When coated with a sensitizing overlayer, these MCLs show significant changes in two independent analyte-induced signals, resonance frequency and static bending, as the result of exposure to various chemical and physical phenomena. Resonance frequency shift has the particular advantage of being relatively insensitive to interference from external factors such as thermal drift [1].

A new approach for producing compact, light-weight, a highly sensitive micromechanical detector is provided by MCL technology, which functions based on the bending of a MCL upon absorption of optical energy. When a MCL is exposed to optical radiation, the temperature of the MCL increases due to absorption this optical energy. The bimaterial effect will cause the MCL to bend in response to this temperature variation, if MCLs are constructed from materials exhibiting dissimilar thermal expansion properties (such as silicon and thin gold film). The extent of bending is directly proportional, in first order, to the rate of energy absorption, which in turn is proportional to the radiation intensity.

The temperature sensitivity of the MCL arises due

to a differential expansion from a temperature gradient within the MCL profile (vertical depth) upon exposure to optical radiation which induces vertical deflection of the MCL or by the bimaterial effect when a dissimilar material is plated upon the MCL structure. In both situations the optical radiation flux upon the MCL was converted into a measurable MCL deflection.

In the past year techniques that drive the MCL directly, such as magnetic modulation [2] and thermal modulation with resistive heating [3] have been developed. The vibrations of the MCL can be generated by a modulated laser – photothermal (PT) modulation. Ratcliff *et al.*[4] reported that a PT excited MCL has a well-defined single resonance mode while a MCL mechanically driven by an external actuator generates an excitation spectrum showing many resonances.

The photoacoustic (PA) and photothermal (PT) science and technology extensively developed new methods in investigation of semiconductors and microelectronic structures during the last ten years. The PA and PT effects can be important as driven mechanisms for micromechanical structures, especially for MCLs [5]. The theoretical analysis of the thermoelastic (TE) and electronic deformation (ED) effects in micromechanical structures consists in modeling a complex system by simultaneous analysis of the coupled plasma, thermal and elastic wave equations. In previously published papers, Todorović et al. [6,7], the TE and ED effects contribution to the photoacoustic signal theoretically and experimentally analyzed. In literature exist small number of papers where the influence of the TE and ED effects to vibration of the MCL were investigated [6,8,9].

Surface stresses arising from the adsorption or deposition of material onto a surface can be very large. To investigate how the surface stress affects the dynamic vibrations of the MCL, the surface stress can be expressed as an equivalent force F_s and moment M_s acting at the free end of the MCL. In this way, the problem is converted to an axial force acting on the MCL similar to a string under compression and tension. With this model, the influence of the surface stress on the transverse vibrations of the MCL can be studied.

In this work the dynamic TE and ED bending in uncoated (Si) and coated (Au/Si) MCL are investigated including the surface stresses.

Authors are with Center for Multidisciplinary Studies, University of Belgrade, P.O.Box 33, 11030 Belgrade, Serbia and Montenegro, E-mail: dmtodor@afrodita.rcub.bg.ac.yu

1-4244-0116-X/06/$20.00 ©2006 IEEE   245

II. OPTICALLY DRIVEN MICROCANTILEVER

A number of MCL geometries were proposed since the invention of AFM. The most preferable among researchers are rectangular and triangular lever forms. The typical dimensions of a MCL for nanoscale resolution will be 100 μm long, 10 μm wide, and less than 2 μm thick. Hence it would have a spring constant of 0.1 - 10 N/m and a resonant frequency of 10 - 100 kHz. The most popular materials for MCL are monocrystalline Si and Si_3N_4. Si_3N_4 cantilevers

2.1 Thermal and electronic elastic deformation

The periodic generated excess carriers in a MCL produce heat due to carrier thermalization and recombination processes. The generated heat can produce elastic deformation in the sample and the TE strain, $\varepsilon^{TE}(r,\omega)$, that changes proportionally with temperature distribution, $\Delta T(r,\omega)=T(r,\omega)- T_o$:

$$\varepsilon^{TE}(\mathbf{r},\omega) = \alpha_T \Delta T(\mathbf{r},\omega), \qquad (1)$$

where r is the space coordinate, ω is the modulation frequency and α_T is the coefficient of linear expansion and T_o is the referent temperature. Also, as noted previously, the photogenerated plasma can directly produce a local strain, which then generates elastic waves in the semiconductor. The electronic strain, $\varepsilon^{ED}(r,\omega)$, changes linearly with excess carrier density, $n(r,\omega)$ and is given by:

$$\varepsilon^{ED}(\mathbf{r},\omega) = d_n\, n(\mathbf{r},\omega); \qquad d_n = \frac{1}{3}\left(\frac{\partial E_G}{\partial p}\right)_T \qquad (2)$$

where d_n is the coefficient of electronic elastic deformation (d_n denotes the pressure dependence on the band gap energy at a constant temperature. Since d_n is negative for silicon, it means that electronic strain and thermal expansion are opposite in sign; the generation of excess carriers causes a contraction of the material, while thermal heating results in an expansion. Than, the elastic displacement can be given as the sum of these two components:

$$\mathbf{u}(\mathbf{r},\omega) = \mathbf{u}^{TE}(\mathbf{r},\omega) + \mathbf{u}^{ED}(\mathbf{r},\omega), \qquad (3)$$

For these two components of elastic deformation it is possible to consider two types of elastic displacements: elastic expansion and bending. Our analysis showed that the elastic expansion is much smaller than the elastic bending and can be neglected in microstructure considered in this work [8].

2.2 Dynamic elastic bending of cantilever beam

The elastic bending calculation for a bimaterial thin rectangular beam, with a thickness h much smaller than the length l and width b ($h \ll l, b$) is given. One side of the bimaterial MCL is uniformly illuminated with an intensity-modulated homogeneous laser beam.

In accordance with the elastic theory for periodical excitation, including the surface stress effects, the dynamic elastic equation for MCL bending is

$$\left[\frac{\partial^4}{\partial x^4} - S\frac{\partial^2}{\partial x^2} - \underline{k}^4(\omega)\right]w(x,\omega) = \frac{\partial^2 \underline{m}}{\partial x^2}, \qquad (4)$$

$$S = \frac{sl}{B}, \qquad \underline{k}^4(\omega) = \frac{\rho_1 A_1 + \rho_2 A_2}{B}\omega^2$$

$$\underline{B} = 4\left[F_1 I_1 + F_2(I - I_1)\right], \qquad F_i = \frac{E_i}{1-v_i^2}, \qquad I_i = \frac{bh_i^3}{12},$$

where $w(x,\omega)$ is the displacement of the middle surface of the MCL, s is the surface stress (force per length), $\underline{k}(\omega)$ is the wavenumber, $\underline{m}$ is the elastic moment, B is the effective flexural rigidity of the bimaterial MCL, ρ_i is the density, A_i cross-section, E_i is Youngs's module, and v_i is Poisson's ratio of the layer i.

The dynamic elastic bending of MCL on the free end of MCL ($x = l$) including the thermal lateral flux is

$$w(l,\omega) = w_s(l,\omega)\, G(\omega), \qquad (5)$$

$$w_s(l,\omega) = -\frac{l^2}{4}\left[\frac{3}{2}\underline{m}_T(l,\omega) + \underline{m}_n(l,\omega)\right],$$

$$\underline{m}_T(x,\omega) = \left(\frac{x}{l}\right)^2 b\frac{F_1\alpha_{T1}m_{T1} + F_2\alpha_{T2}m_{T2}}{B},$$

$$m_{T1} = \int_0^{h_1} z\Delta T_1(x,z,\omega)dz, \qquad m_{T2} = \int_{h_1}^{h_1+h_2} z\Delta T_2(x,z,\omega)dz,$$

$$\underline{m}_n = b\frac{F_2 d_{n2} m_{n2}}{B}, \qquad m_{n2} = \int_{h_1}^{h_1+h_2} z\Delta n_2(x,z,t)dz,$$

where $w_s(l,\omega)$ is the quasi-static elastic bending and $\underline{G}(\omega)$ is the dimensionless dynamic frequency factor, $\underline{m}_T(x,t)$ and $\underline{m}_n(x,\omega)$ is the TE and ED moment generation term, i.e. the source term from the thermal and plasma waves.

III. ANALYSIS OF ELASTIC BENDING IN MICROCANTILEVER

The theoretical model, previously given, enables to calculate the elastic bending of the uncoated and coated MCLs. The dynamic vibration characteristics and TE and ED components were calculated, including the surface strain effects (the axial force). Fig.1 shows the influence of the surface strain effects (the axial force) on the dynamic elastic bending on the free edge of the coated Au / Si MCL. Elastic bending was calculated for typical parameters of Si (layer 2) and Au (layer 1) and dimensions: thickness of the Au film h_1 = 100 nm and Si substrata h_2 = 2 μm, length l = 200 μm and width b = 20 μm, typical for AFM. , The surface stress (force per length) is $s = 5*10^{-6}$ N/m. In this case, it is possible to see

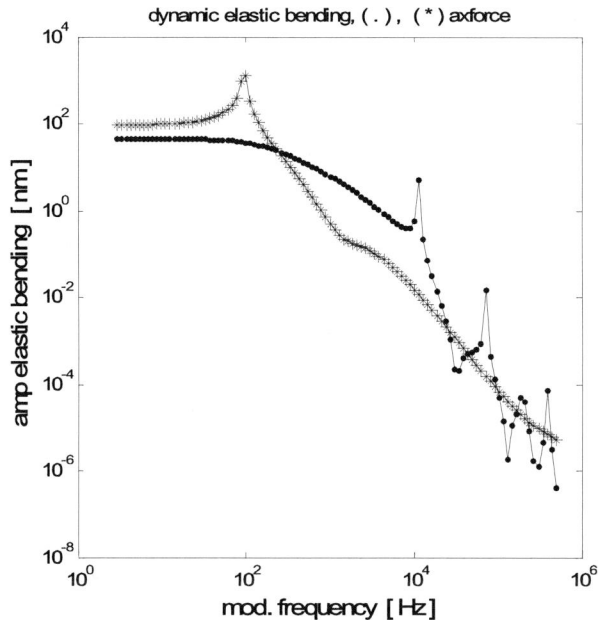

Fig.1 Elastic bending of coated Au / Si MCL: (•) without the surface strain effects (the axial forces); (*) with surface strain.

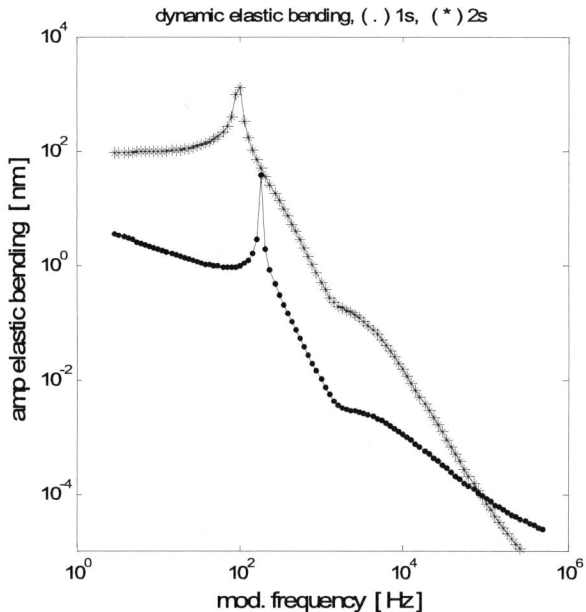

Fig.2 Comparing an uncoated (Si) and coateded (Au / Si) MCLs: (•) Si; (*) Au / Si .

that the surface strain drastically change the frequency charactcristic of MCL.

In Fig.2 the total dynamic elastic bending (TE + ED) of an uncoated (Si) and coateded (Au / Si) MCLs, including the surface strain effects, are compared. For both type of MCLs the elastic bending decrease with increasing the frequency. In the low frequency range, the total elastic bending of this uncoated MCL is smaller (about one order) then total elastic bending of coated MCL. On the other side, in the high frequency range the elastic bending for uncoated MCL begins higher than the elastic bending of coated MCL.

In Fig..3 the TE and ED components of dynamic elastic bending of coateded (Au / Si) MCLs, including the surface strain effects, are compared. It is possible to see that the ED component is much smaller then the TE component in the low frequency range. On the other side, in the high frequency range the ED component can be begin higher than the TE component of elastic bending of MCL.

3.1 Sensitivity

Sensitivvity optically excited microstructure is important characteristic for application as a sensor and it is defined as a ratio of the amplitude of the elastic displacement $|w(\omega)|$ and intensity of the optical excitation $I(\omega)$. There are at least three sources of noise which limit the sensitivity of the MCL detection system: (1) thermal vibrations of the MCL, (2) noise in the displacement sensor, and (3) noise generated in the oscillation control amplifier and other electronics. With a low-noise displacement sensor, thermal vibrations of the cantilever are the dominant noise source under most conditions.

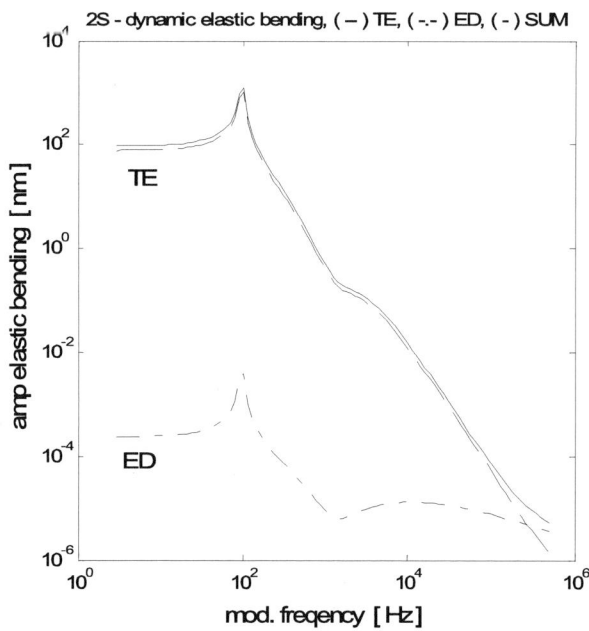

Fig.3 Components of elastic bending of coated Au / Si MCL with the surface strain effects (the axial forces) : (--) TE; (-.-) ED; and (-) sum of components (TE + ED).

3.2 Thermomechanical noise

To determine the ultimate sensitivity of the MCL as a sensor an estimate of the noise levels is required. The noise arises from external sources such as the electronics, mechanical vibrations, acoustic coupling, intensity fluctuations in our light source, and convection effects.

A fundamental noise limitation is that due to the thermal vibration noise of the sensor. By careful design of the MCL sensor, the noise from other sources can be minimized to the point where the thermal noise dominates. In the following analysis only the thermal noise contribution to the cantilever motion is considered.

In thermodynamic equilibrium the mean-square displacement of the MCL from its neutral position is described by

$$\sqrt{\langle w^2 \rangle} = \sqrt{\frac{k_B T}{k_L}}, \qquad (6)$$

where $\sqrt{\langle w^2 \rangle}$ is the mean-squared displacement of the MCL (w is the z-displacement of the free end of MCL), k_B, Boltzmann's constant, T the temperature of the surrounding heat bath, k_L the cantilever spring constant, and $\langle\rangle$ denotes the average in time. Note that this formula is independent of the cantilever geometry.

Usually employed MCLs possess force constants in the range of 0.01 - 100 N/m. The "soft" cantilevers which k_L is below 0.1 N/m are chosen mainly while using contact mode in order to affect the sample in minimal extent. The rigid ones with k_L >1 N/m are often used in non-contact or dynamic modes since they exhibit high resonant frequencies and small oscillation amplitudes of about several nanometers. It provides wide dynamic frequency range and substantially raises sensitivity. Typical values for these MCLs give a noise value of $(1-2) \times 10^{-2}$ nm/$\sqrt{\text{Hz}}$.

The quasistatic spring constant as obtained from a static finite element analysis (FEA) is k_L = 1.3 N/m (manufacturer's value: k_L = 1.0 N/m). From a variation of the thickness in the range from 0.1 to 5 μm, the empirical relation $k_L \approx 1.64 \cdot 10^{-3} \cdot h^3$ (with h in μm) was found to be a good approximation. Additionally, k_L is directly proportional to the material parameter Young's modulus E. Thus, already the simple analytic relation

$$k_L = \frac{Eb}{2}\left(\frac{h}{l}\right)^3 \qquad (7)$$

is a good approximation and differs only 5% from the values obtained by the numerical FEA. Then, for typical Si parameter E= 1.31·10⁷ N/m (b = 20 μm, h_2 = 2 μm, l = 200 μm), the quasistatic spring constant is $k_L \approx 1.31$ N/m.

The spectral components of the noise signal are not uniform because of the highly tuned MCL response. Thus, frequencies well below the natural resonant frequency ω_o of the sensor, the *rms* thermal noise amplitude in nm/$\sqrt{\text{Hz}}$ is

$$\sqrt{\langle w^2 \rangle} = \sqrt{\frac{4 k_B T B_\omega}{k_L \omega_o Q}} \qquad (8)$$

where B_ω the bandwidth of the measuring system, ω_o is the natural resonant frequency, and Q is the quality factor for the resonance and k_L is the effective spring constant of the MCL.

IV. CONCLUSION

The surface strain effects (the axial force) can have important influence to the dynamic elastic bending of uncoated (Si) MCL and coated (Au / Si) MCL. The analysis in this work showed that the surface strain can drastically change the frequency characteristic for both type of MCLs. For both type of MCLs the elastic bending decrease with increasing the frequency.

The TE and ED components of dynamic elastic bending of an uncoated (Si) and coateded (Au / Si) MCLs, including the surface strain effects, were compared. The analysis showed that the ED component is much smaller then the TE component in the low frequency range. On the other side, in the high frequency range the ED component can be begin higher than the TE component of elastic bending of MCL.

ACKNOWLEDGEMENT

This work was performed in the frame of the project "Micro and Nanosystems Technologies, Structures and Sensors", supported by grants from the Ministry for Science and Environmental Protection, Republic of Serbia, Grant No. TP – 6151 B.

REFERENCES

[1] E. A. Wachter and T. Thundat, "Micromechanical sensors for chemical and physical measurements", *Rev. Sci. Instrum.* , vol. 66(6), pp. 3662-3667, 1995

[2] W. Han, S. M. Lindsay, and T. Jing, *Appl. Phys. Lett.*, vol. 69, pp. 4111, 1996.

[3] A. C. Hiller and A. J. Bard, *Rev. Sci. Instrum.*, vol. 68, pp. 3083, 1997.

[4] G. C. Ratcliff, D. A. Erie, and R. Superfine, *Appl. Phys. Lett.*, vol. 72(15), pp. 1911, 1998.

[5] D. M. Todorović, "Photothermal and electronic elastic effects in microelectromechanical structures", *Rev. Sci.Instrum.*, vol. 74 (1), pp. 578-581, 2003.

[6] D.M.Todorovic, P.M.Nikolic, A.I.Bojicic, K.T.Radulovic, , "Thermoelastic and electronic strain contributions to the frequency transmission photoacoustic effect in semiconductors", *Phys.Rev.B*, vol. 55(23), pp.15631-15642, 1997.

[7] D.M.Todorovic, P.M.Nikolic, Ch. 9 in *Semiconductors and Electronic Materials* (A.Mandelis and P.Hess, Eds., SPIE Opt.Eng . Press, Belingham, Washington, 2000), p. 273-318.

[8] A.Prak, T.S.J.Lammerink, "Effect of electronic strain on the optically induced mechanical moment in silicon microstructures", *J.Appl.Phys.*, vol.71(10), pp.5242-5245, 1992.

[9] D. M. Todorović, "Photothermal dynamic elastic bending in microcantilever", *J. de Physique IV France*, vol 125, pp. 495-463, 2005.

2006 25th International Conference on Microelectronics

Real -Time Tracking of a Moving Object Inside a Tube

K. Arshak, F. Adepoju

Abstract – The need for accurate determination of location of objects in inaccessible positions has always presented a great challenge in the physical world. Although it has been successfully achieved in industrial robotics by using methods like optical triangulation, computer binary network, etc. Such techniques could not extend into some area of biomedicine due to the peculiarities of the human body organs. Most popular tracking methods employ radio frequency (RF) signals or ultrasound pulse waves to determine the location of the target object in 2 or 3-D. A round trip time-of-flight (TOF) data is recorded for all instances of pulse-echo transmission and by using the method of triangulation, the real-time location of the test object can be computed.

A modified method of acquiring the TOF data is employed in the work presented in this paper. A trajectory of the moving object was composed to within 5% accuracy using the TOF data obtained for the transmitter's motion in air and also with a layer of fluid (water) placed along the path of the transmitted pulse and the receiver.

I. INTRODUCTION

Ultrasonic sensors are commonly used to produce ultrasound signals in medical applications [1,2] and also in range tracking. Ultrasound waves are mechanical, that is, waves are generated as a result of high frequency vibration of crystals due to changes in incident voltage or current.

An incident ultrasound signal from a transmitter is directed into the object across an acoustic interface. Acoustic interface is the boundary between two material of different acoustic impedances. When sound strikes an acoustic interface at normal incidence, some amount of sound energy is reflected and some amount is transmitted across the boundary. The dB loss of energy on transmitting a signal from medium 1 into medium 2 is given by equation (1).

This signal suffers some attenuation [2,3] and diffraction but a large percentage is reflected back from the object through the medium of transmission back into the receiver. It is a fact that ultrasound attenuates as it progresses through a medium. Assuming no major reflections, there are three causes of attenuation. These are diffraction, scattering, and absorption. Attenuation results in loss of signal or a progressive reduction in signal strength as modeled by equation (1).

In range measurement, ultrasonic sensors are used for

K. Arshak and F. Adepoju are with the Electronic & Computer Engineering Dept., University of Limerick, Plassey Technological Park, Limerick, Ireland.
E-mail: khalil.arshak@ul.ie

measuring distances between the sensor and solid objects ahead of the sensor. Such arrangement consists of an acoustic transmitter and receiver. Measurements are based on the time-of-flight (TOF) of the ultrasonic wave from the moment of transmitting to the time of receiving an echo. Typically, transducers operating at ultra-sound frequencies are used to perform a single distance measurement [4] One crystal will transmit a burst of ultrasound, and a second crystal will receive this ultrasound signal. The elapsed time from transmission to reception is a direct and linear representation of the physical separation of the crystals [1]

The resulting transit time is easily converted to a distance if the speed of sound in the material being measured is known. Typically, in air at 25ºC, this speed is about 340 meters per second. In many practical cases, the ultrasonic energy is reflected in a diffuse manner. That is, regardless of the incident angle, it is reflected almost uniformly within a wide solid angle, which may approach 180º. The resolution of any ultrasound system depends on its ability to accurately detect the received ultrasound signal and on its ability to measure transit time. Advanced high-gain, low noise circuitry allows for the detection of the received ultrasound signal and precise tracking of the received signal regardless of its frequency. This allows for complete independence of crystal frequency and measurement resolution.

At the heart of many positioning methods is the idea of triangulation: Triangulation is essentially the use of the properties of triangles to calculate distances [5] In triangulation of points, if any two reference points are given, then, it is possible to calculate the distance from a reference point to an object with knowledge of the angles between both references and the object and also the distance between the references.

II. ARCHITECTURE

A generalized block diagram of a microprocessor-based ultrasonic position tracking system is illustrated in Fig. 1. This consists of the following elements:

(1) A dual element acoustic transducer for transmitting and receiving acoustic pulses.
(2) Signal amplifier and excitation stage to produce adequate voltage to excite the transducer.
(3) Band pass filter circuit to limit the range of input signal that the system will respond to.
(4) Pulse detection circuit and
(5) A microcontroller for logic, timing, interrupt and system control.

1-4244-0116-X/06/$20.00 ©2006 IEEE

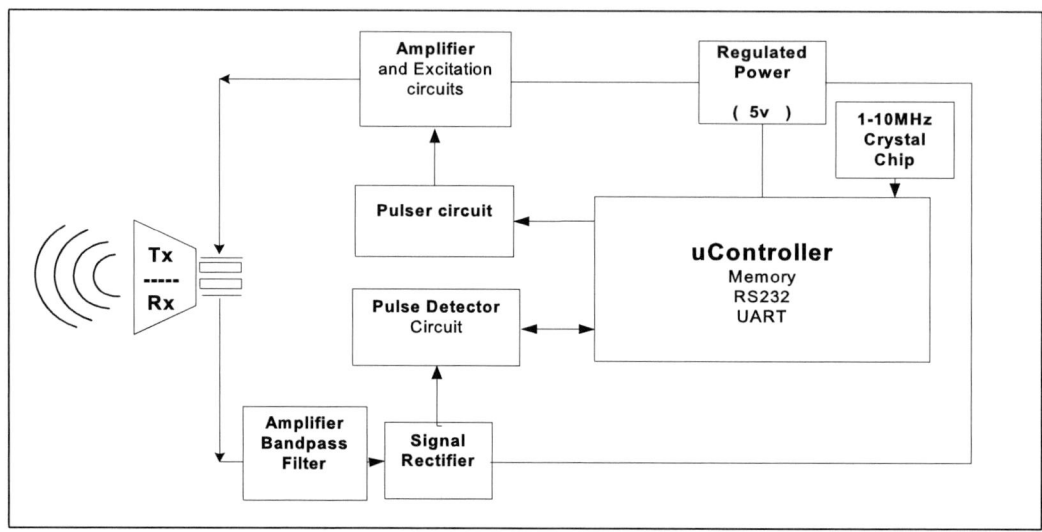

Fig. 1 Generalized circuit diagram

The use of integrated IC technology enables reduction of operational voltage to about 5V. Other necessary peripherals such as extra memory, RS232, etc. are connected to the microprocessor in order to facilitate storage of intermediate data to memory, and also to be able to communicate with a personal computer. The pulser, under the control of the microprocessor, provides an excitation pulse [6] to the transducer. The ultrasonic pulse generated by the transducer is coupled to the test piece. Echoes returning from the inside or the surface of the test piece are received by the transducer, converted to electrical signals, and fed to the receiver amplifier for processing. The microprocessor-based control and timing logic circuits both synchronize the pulser and selects the appropriate echo signals for the time interval measurement.

When echoes are detected, the timing circuit will precisely measure an interval corresponding to a round trip of the sound pulse in the test piece, and usually repeat this process several times to obtain the required position information as a function of time. The microprocessor then passes this time interval measurement, along with sound velocity and zero offset information to a PC running a triangulation algorithm in order to calculate and render the measurement information graphically.

Most ultrasonic measurements are made [7] with one of four types of transducer and appropriate electronics. These are contact, delay line, immersion, and dual element transducers. Although each type of transducer has its own advantages and limitations, dual element transducer has features that suit this application. Dual element transducers utilize separate transmitting and receiving elements mounted on delay lines. This configuration improves near surface resolution by eliminating main bang recovery problems to achieve better near-surface resolution. In addition, the crossed beam design nature of the dual element transducer provides a pseudo focus that enhances its sensitivity to returning echoes.

A   Circuit Description

The sensing is initiated by first creating an acoustic pulse at 40KHz frequency using the pulser circuit. This pulse train is passed on to the excitation circuits to produce the necessary voltage to excite the transducer. The circuit of Fig. 1 detects a reflected wave from the object after sending out ultrasonic pulses at a controlled interval. The detected pulse is converted to digital pulse by the rectifier circuit. If the detected signal measures up to the set threshold, then the micro controller is activated. By measuring the round trip time after emitting the ultrasonic pulse wave, a distance to the object is measured. This operation is executed for all object positions.

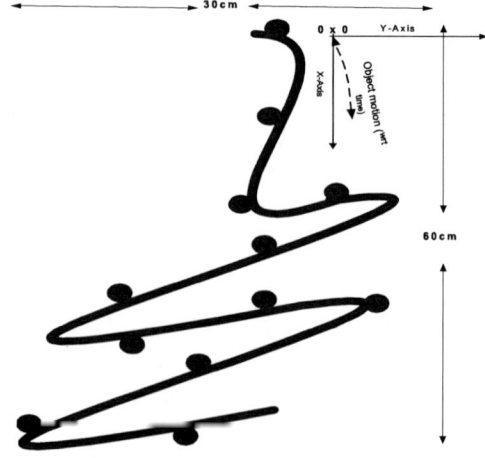

Fig. 2 Transmitter trajectory: Air Experiment

B Experimental

The receiver was taken through the trajectory shown in Fig. 2. First with air as transmission medium and then a packet of fluid (water) was placed in the signal path. At each object position along the trajectory, a pulse signal is transmitted and an echo received after a period of time corresponding to the round-trip time-of-flight (TOF). The returning echo is passed on to a rectification circuit then on to a comparator in order to determine the echo strength. If appropriate signal strength is attained, then the microprocessor forks an interrupt to determine the TOF. Such data are recorded for the three receivers. The sequence is repeated for the duration of object travel. The resulting data are uploaded onto the PC in order to compute the resultant location and length.

III. COMPUTATIONAL METHOD

The system presented in this paper employs a 3 - receiver network to measure TOF [8] of acoustic pulse emanating from the base station. One receiver/transmitter, the central station, is situated at x=0,y=0 (0,0), and the other two at known distances from 0,0. The location of the object (which has a transmitter/receiver/microcontroller attached to it) is found by plotting the locus of points where the signal could have originated for each receiver.

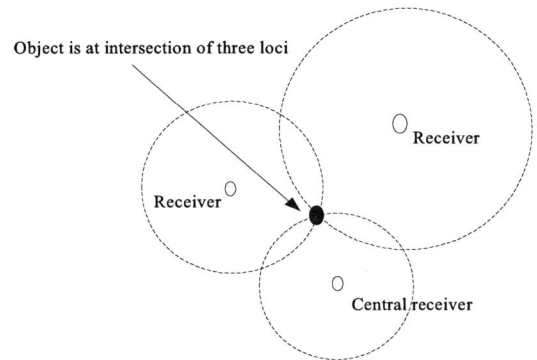

Fig. 3   TOF positioning method

As shown in Fig 3, the position of the object [9,10] is the point where all of the loci intersect. If the length prior to this position is known, then the current length is calculated using equations (2) and (3). As shown in Fig. 4, these two equations are combined with other intermediate equations to determine the length L at any successive object position O_n, with R_n as obtained from Fig. 3.

Accuracy of distance measurement using this method depends on how often the object position is sampled. That is, the limit as $\Delta x \to 0$. One downside of this system is that the transducers are vulnerable to interference from outside sources or, more importantly, multi-path fading caused by part of the signal bouncing off another object in the local area.

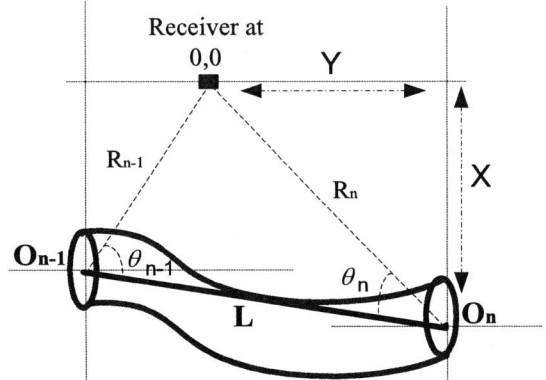

Fig. 4   Determination of length covered within the interval $t_{n-1} < t < t_n$

A List of Equations

$$dB\ loss = 10\ Log_{10}\ [4Z_1Z_2 / (Z_1 + Z_2)^2]\ [7] \qquad (1)$$

$$R = vt/2 \qquad (2)$$

$$Accumulated\ length = \sum_{n=0}^{k} \left(L_{n-1} \cdot L_n\right) \qquad (3)$$

where Z_i =Acoustic impedance of material i: i =1, 2.
R = distance from object to the base station,
v = speed of sound in medium (330m/s for air),
t = total round-trip time interval and
L_n = Length computed so far

IV. RESULTS

Fig. 5 shows a captured instance of a returning echo from the object. This is as result of a successful transmit and reflection of a pulse. The graph of Fig. 6 shows the objects relative positions with respect to time during experimentation. It was observed that for objects around adjacent curves, there is no big difference in TOF values. Also, the attenuation introduced by a packet of water placed along the signal path became more pronounced as distance increased.

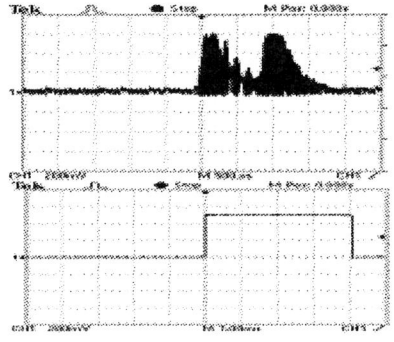

Fig. 5 A sample echo pulse

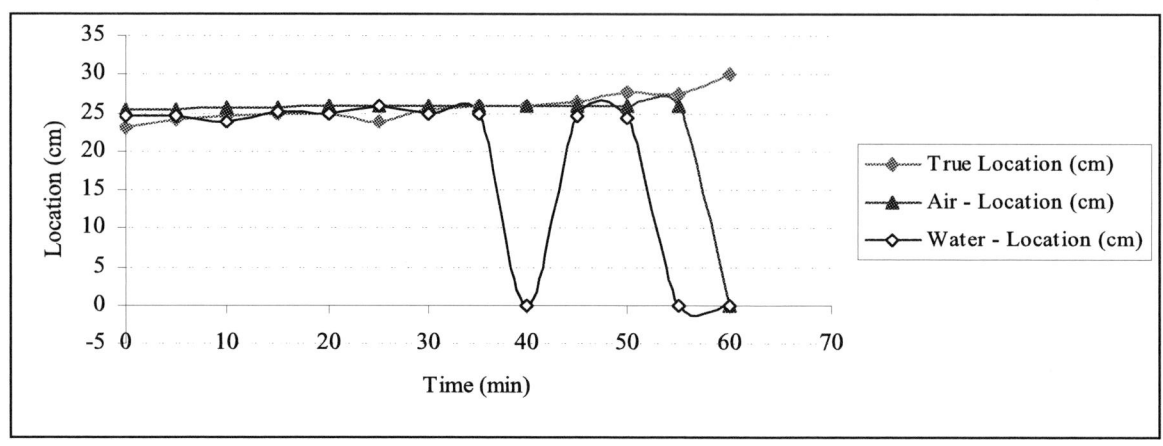

Fig 6: Graph of Object vs True length

V. CONCLUSION

Fig. 6 shows the real-time location of the object as it moved along the trajectory shown in Fig. 2. This trajectory was chosen in order to investigate how accurately the microprocessor would resolve objects at close proximity. It was confirmed however that within the experimental boundary of 60cm x 30cm, the resolution was fairly weak. Also, there was no way of confirming the direction of object travel. When the range was increased to 90cm, object detection became more problematic. Attenuation was introduced with a packet of water placed in the signal path. This made detection at 40KHz quite unreliable.

However, with increase in frequency (> 50kHz), echo was received in some instances. If this system is intended for medical instruments [10], then there is a need to up the frequency to a minimum of 1MHz in order to penetrate the human tissue.

Finally, with the right selection of amplifier, filter, pulser, detector and transducers, detection and real-time localization of objects for medical systems can be accomplished with minimum of costs.

ACKNOWLEDGEMENT

This work was supported by the Enterprise Ireland commercialization Fund 2003, under technology development phase, as part of the MIAPS project, reference no. CFTD/03/425. Funding was also received from the Irish Research Council for Science, Engineering and Technology: funded by the National Development Plan.

REFERENCES

[1]   US PATENT 2005/0556 (in press) Arshak K, Adepoju F 'A Tracking System'

[2]   DH Evans, WN McDicken.'Doppler Ultrasound – Physics, Instrumentation and Clinical Applications' John Wiley 1999 pp 29-35

[3]   W Hendricks D Hykes, D Starchman,'Ultrasound Physics And Instr, 3rd Ed Mosby 1999

[4]   http://www.sonometrics.com/index-p.html viewed on 15 Nov 2005

[5]   Lahanas, M., "Measurements", http://www .mlahanas.de/Greeks/Measurements.htm, visited on 18 May 2005.

[6]   JA Zagzebski Essentials of Ultrasound Physics Mosby publ, 1996 pp 12- 17, 29 –40

[7]   http://www.panametrics-ndt.com/ndt/ndt_technology /intro_ultrasonic_thickness.html viewed on 18 Oct 05

[8]   Togawa, T., T. Tamura, et al. (1997) *Biomedical Transducers And Instruments*, Crc Press Llc

[9]   TJ Brooks, H Baker, KA Mercer. 'A review of position tracking' 1st Int Conference on Sensing Technology Nov 21-23, 2005 Palmerston North, New Zealand

[10]   K. Arshak, F. Adepoju 'A Review of Object Tracking and Methods for Locating Telemetry Capsules' (unpublished)

Poster Session
Microsystem Technologies

2006 25th International Conference on Microelectronics

A Simplified Method for Analysis of MEMS Bimaterial Cantilever Elements

J. Matović

Abstract - This paper presents a simplified method for the calculation of the outputs available from a MEMS bimaterial cantilever. This includes the total work and force, the work against the inertial forces and concentrated forces. The approach is based on the fact that a bimaterial element performing the work against an external force, as in the case of an actuator, or against inertial forces (resonant sensor), has a single neutral plane. In such case, a bimaterial element can be studied as an equivalent mechanical element with an equivalent Young modulus E_0 and an equivalent thickness $h = h_1 + h_2$.

I. INTRODUCTION

Bimaterials were discovered in 1766 and since then they were used as sensors/actuators for the compensation of the ambient temperature variations in early chronometers. Theoretical analysis of a bimaterial element starts with the pioneering works of Timoshenko [1].

The application of bimaterial elements is considerable in MEMS, both for sensors and for micro actuators [2–5]. Alongside with an extensive application of bimaterial elements in MEMS devices, theoretical analysis of bimaterial elements response is continuing to develop, but at the same time the theories are becoming more complex [6 – 8]. This paper presents a simplified theory of the response of a bimaterial element.

II. THE EFFECTIVE YOUNG'S MODULUS OF A BIMATERIAL CANTILEVER

A bimaterial element is a sandwich structure composed of two materials which alter their volumes differently in response to an external stimulus. The stimuli used are numerous: temperature, magnetic and electrostatic field, chemical and photon absorption. Solid materials are practically incompressible. Thus bimaterial elements deform to conserve a constant volume of the structure. The stress distribution is homogeneous over the bimaterial surface, so the bimaterial plate reacts to the volume change by curving into an approximately spherical surface, and a narrow strip becomes a circular arc.

At first, an important property of a bimaterial element should be noted: when a bimaterial element

J. Matović is with the Institute of Microelectronic Technologies and Single Crystals, A Division of Institute of Chemistry, Technology and Metallurgy, Belgrade University, Serbia and Montenegro, E-mail: matjovan@eunet.yu

responds to a stimulus, each its constitutive layer has its own neutral plane. However, when a bimaterial element performs work against an external force, as in the case of an actuator, or against inertial forces (resonant sensor), then it has a single neutral plane. Then, instead of analyzing a bimaterial element with Young moduli E_1 , E_2 and layer thickness h_1, h_2 it is admissible to consider an equivalent mechanical element with an equivalent Young modulus E_0 and a thickness $h = h_1 + h_2$, Fig. 1.

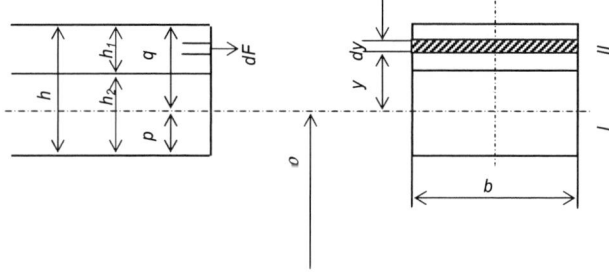

Fig. 1. Schematic of the bimaterial cantilever loaded with an external force.

where ρ is the radius of curvature of a bended cantilever, b is the cantilever width and p, q denote the position of the equivalent neutral plane. According to Fig. 1, the elementary force dF at the small area $dS = v\,dV$ and at the distance y is equal to

$$dF = \sigma b dy = \frac{Eybdy}{\rho} \qquad (1)$$

The coordinates of the neutral plane p and q can be calculated from the condition that the total sum of forces along the surface dF should be zero:

$$F = \int dF = \int_{p-h_1}^{p} \frac{E_1 b y dy}{\rho} + \int_{-q}^{-q+h_2} \frac{E_2 b y dy}{\rho} =$$
$$= \frac{b}{\rho}\left[E_1 h_1\left(p - \frac{h_1}{2}\right) - E_2 h_2\left(q - \frac{h_2}{2}\right)\right] = 0 \qquad (2)$$

and from (1), substituting $h = p + q$, the coordinates p and q are equal to:

1-4244-0116-X/06/$20.00 ©2006 IEEE

$$p = \frac{E_1 h_1^2 + E_2 h_2 (h + h_1)}{2(E_1 h_1 + E_2 h_2)}$$
$$q = \frac{E_2 h_2^2 + E_1 h_1 (h + h_2)}{2(E_1 h_1 + E_2 h_2)} \quad (3)$$

The bending moment of the internal forces M is

$$M = \frac{E_0 I_0}{\rho} = \int y dF = \int \sigma b y dy \quad (4)$$

Using the values for p and q from (3) and including the moment of inertia as $I_0 = (bh^3)/12$, the effective Young's modulus is [9]

$$E_0 = \frac{(E_1 h_1^2 - E_2 h_2^2)^2 + 4h_1 h_2 h^2 E_1 E_2}{(E_1 h_1 + E_2 h_2) h^3} \quad (5)$$

If one introduces the normalized units $m = h_1/h_2$ and $n = E_1/E_2$, the effective Young's modulus E_0 can be graphically presented as in Fig. 2.

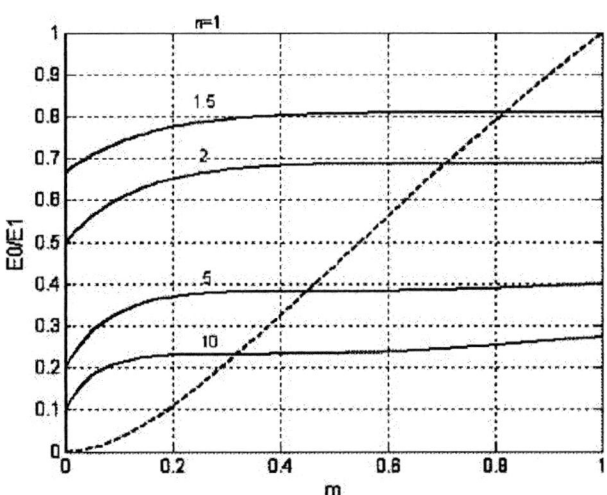

Fig. 2. The value of E_0 as function of units $m = h_1/h_2$ and $n = E_1/E_2$

III. THE OUTPUTS AVAIABLE FROM A THERMAL BIMATERIAL

A) Calculation of the work and force available from a thermal bimaterial element.

With E_0 calculated according to (5), a bimaterial element can be considered a homogeneous cantilever. Generally, the deformation energy of a small element is

$dV = b\ h\ dx$, regardless of the energy origin (thermal, magnetic, etc.). The total work of a deformed elastic beam under the constant moment along the z axis is equal to

$$A = \int_0^l dA = \frac{\sigma^2 dV}{2E_0} = \frac{M^2 dx}{2E_0 I_0} \quad (6)$$

where A is work, σ is stress, M is loading moment and I is modulus of rigidity. Considering that the bimaterial cantilever bends into a circular arc [1], and that this is an equivalent case to the bending of a cantilever loaded with a constant moment [9] , the bending radii of both are

$$\frac{M}{E_0 I_0} = \frac{1}{\rho} = \frac{2k\Delta T}{h} \rightarrow M = \frac{2E_0 I_0 k \Delta T}{h} \quad (7)$$

$$k = \frac{3(\alpha_1 - \alpha_2) h^2 h_1 h_2 E_1 E_2}{3h^2 h_1 h_2 E_1 E_2 + (E_1 h_1 + E_2 h_2)(E_1 h_1^3 + E_2 h_2^3)}$$

According to (6) and (7), the available total work from the thermal cantilever is simply

$$A = \int_0^l dA = \frac{2E_0 I_0 k^2 l \Delta T^2}{h^2} \quad (8)$$

where the $V = bhl$ is the total volume of cantilever. If one requires to calculate the partial work along the displacement Δy, it is equal to

$$A = \frac{2E_0 I_0 \Delta y^2}{l^3} = \frac{E_0 b h^3 \Delta T^2}{6l^3} = \\ = \frac{b^2 h^2 \rho_0^2 a^2 l^5}{40 E_0 I_0} = a \frac{3b\rho_0 l^5}{10h} \quad (9)$$

B) Work of a cantilever against the inertial forces

This case is important for the analysis of resonant sensors as well as for the calculation of noise of the microcantilevers. Here it is necessary to calculate first the total mass density of the bimaterial cantilever as $\rho_0 = (h_1\rho_1 + h_2\rho_2)/h$, where ρ_1 and ρ_2 are densities of the layers. Similarly as in the *Example A*, the work from inertial forces is

$$A = \int_0^l dA = a^2 \frac{b^2 h^2 \rho_0^2}{8 E_0 I_0} \int_0^l (l - x)^4 dx = \\ = \frac{b^2 h^2 \rho_0^2 a^2 l^5}{40 E_0 I_0} = a \frac{3b\rho_0 l^5}{10h} \quad (10)$$

256

C) Work from the concentrated force at the end of a bimaterial element

The work in this case is given as

$$A = \int_0^l dA = \frac{m^2 a^2}{2E_0 I_0} \int_0^l (l-x)^4 dx =$$

$$= a^2 \frac{m^2 l^3}{6E_0 I_0} = a^2 \frac{2m^2 l^3}{E_0 h^3}$$

(11)

where m is the concentrated mass. This case is important in infrared bimaterial sensors, which usually have a large plate fixed to the cantilever end. The response of the cantilever to other stimuli (magnetic, electrostatic, absorption, etc.) can be found in a similar manner.

IV. CONCLUSION

This paper presents a simple procedures for calculation of the force and work available of a bimaterial cantilever. The procedure also includes calculation of the effective mass and moment of inertia for a bimaterial cantilever.

ACKNOWLEDGEMENT

This work is partially supported by the Serbian Ministry of science, technology and development within the framework of the project IT.6151.B.

REFERENCES

[1] Timoshenko S 1925, *Analysis of bimetal termostat, J. Opt. Soc. Am.* 11 233–55W

[2] P. G. Datskos, T. Thundat, Nickolay V. Lavrik, *Micro and Nanocantilever Sensors* in Encyclopedia of Nanoscience and Nanotechnology, Edited by H. S. Nalwa, Volume X: pp. 1–10.

[3] Reithmuller, W. Benecke, Thermally Excited Sillicon Microactuators, *IEEE Tans. ED*, Vol 35 No 6, pp 758-762, June 1988

[4] J. Matović, J. Lamovec, Z. Djinović, Bimaterial Infrared Detector with Efficient Suppression of Interference from Ambient Temperature, *Proc. Miel* 2003

[5] P. I. Oden, E. A. Wachter, T. Thundat and R. J. Warmack, Optical and Infrared Detection Using Microcantilevers, SPIE Vol. 2744 (1966)

[6] G A Gehring, M D Cooke, I S Gregory, W J Karl and R Watts, Cantilever unified theory and optimization for sensors and actuators, Smart Mater. Struct. 9 (2000) 918–931.

[7] Arvi Kruusing, Analysis and optimization of loaded cantilever beam microactuators, Smart Mater. Struct. 9 (2000) 186–196.

[8] Wen-Hwa Ghut et al, Analysis of tip deflection and force of bimetallic cantilever microactuator, J.Microeng. 3 (1993) 5 - 7

[9] Timoshenko S., Theory of elasticity, McGraw.Hill Book Company, Inc., New York, 1934

258

2006 25th International Conference on Microelectronics

Design Parameters Optimization Using Process Variations of the Pull-In Voltage for MEMs

R. Voicu, C. Tibeica, M. Bazu

Abstract – Probability density function for the pull-in voltage of the electrostatically actuated micro-bridges due to variations of some dimensional parameters was estimated using Monte Carlo approach. Variations of the pull-in voltage were analyzed in order to find out optimum design parameters taking into account the given process variations in manufacturing micro-bridge structures.

I. INTRODUCTION

Pull-in voltage is an important parameter that characterizes the electrostatically actuated MEMS devices such as RF switches, or test structures for material properties measurements. The electrostatic actuation is the most used method for actuating MEMS devices due to its advantages vs. other methods (piezoelectric, magnetic, etc.).

The paper suggests a statistical scheme to estimate the effects of dimensional parameters variation on probability density function of pull-in voltage. Probabilistic simulations were performed in order to optimize the design parameters taking into account the process variations.

Probability density estimation is done using Monte Carlo method. In Monte Carlo approach each parameter is selected randomly according to its probability density function. Then, using these parameter values, simulation is run a number of times.

Our investigations are focused on how variations of the pull-in voltage can characterize the dimension of some bridge design parameters.

II. STRUCTURES MODELLING

For modelling a MEMS device we use the analytical approach [1][2]. We follow a model for a simple micro-bridge that is subject in few papers [1][3]. In our paper, we supposed that bridges are electrostatically actuated. We start our simulation from a set of gold micro-bridges having various lengths with the following design parameters [3]:

Bridges width, W = 40 μm

R. Voicu, C. Tibeica and M. Bazu are with the National Institute for R&D in Microtechnologies (IMT-Bucharest), 32B, Erou Iancu Nicolae street, 077190, PO-BOX 38-160, 023573, Bucharest, Romania, E-mail: rodicav@imt.ro, catalint@imt.ro, mbazu@imt.ro

Bridges length, L = from 160 up to 760 μm
Electrode length, L' = from 140 up to 740 μm
Bridges thickness, t = 1.3 μm
Air gap, g = 1.5 μm

The bridge thickness and air gap will be assumed to be nonuniform random variables.

The analytical pull-in voltage used for Monte Carlo simulation is given by a function of the gap (g) and of the thickness of the bridge (t) [1][3][6]:

$$V_{pull-in} = \sqrt{\left(8 \frac{K_{tot} \cdot g^3}{27 \cdot \varepsilon_0 \cdot L' \cdot W} \right)} \qquad (1)$$

where

$$K_{tot} = K_L + K_\sigma \qquad (2)$$

is the total stiffness of the bridge:

$$K_L = 32 \, EW \left(\frac{t}{L} \right)^3 \frac{1}{8 \left(\frac{L+L'}{2 \cdot L} \right)^3 - 20 \left(\frac{L+L'}{2 \cdot L} \right)^2 + 14 \left(\frac{L+L'}{2 \cdot L} \right) - 1} \qquad (3)$$

$$K_\sigma = 8 \, \sigma \left(1 - \nu \right) W \left(\frac{t}{L} \right) \frac{1}{3 - \left(\frac{L+L'}{L} \right)} \qquad (4)$$

where the structure is described in Fig. 1.

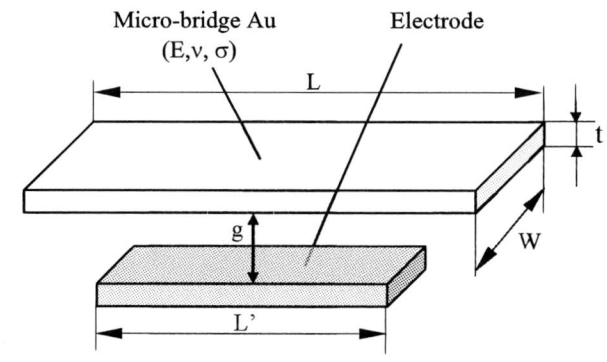

Fig. 1. The micro-bridge model.

III. Monte Carlo approach

We have used the Monte Carlo approach [4][5] to estimate the probability density for the pull-in voltage. The parameters gap and thickness were selected randomly according to their probability density function. For clarity, by Eq. 2, Eq. 3 and Eq. 4 we can conclude that K_{tot} is a function of variables t, L, L' and by Eq. 1 we can claim that $V_{pull-in}$=f (t, g).

A. Algorithm and Estimation for Probability Density of the Pull-in Voltage

First, we supposed that the parameters t and g follows normal density function and assumed that these input parameters are independent.

We have generated a 2-dimensional normal vector with independent components using C++ programming language. The algorithm that we have implemented is based on the following steps:

1) an independent standard normal vector, Z~N(0,I), using polar method has been generated
2) Cholesky matrix C has been computed for normal mean 1.3 μm and variance 0.13 μm (10% of mean), and for 1.5 μm and 0.15 μm, respectively
3) the normal arbitrary vector μ+CZ~N(μ,Σ) has been calculated.

We have run the program that generates mean and variance for pull-in voltage for a different number of entries, such as: 250X250, 500X500, 1000X1000, 5000X5000 and 10 000X10 000 random variables. We noticed that 5000X5000 gives a good approximation with minimum errors (differences) for mean and variance of outputs and vs. the duration of executions (see Fig. 2).

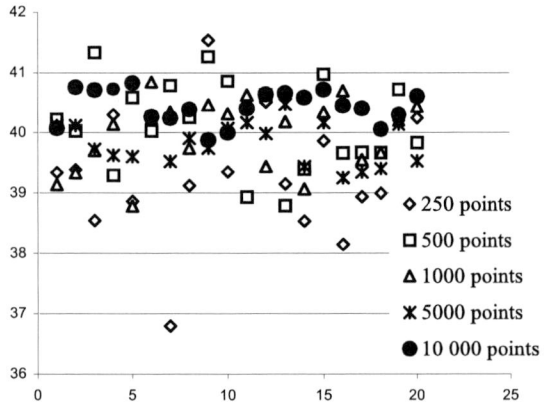

Fig. 2. Mean values for pull-in voltage for different numbers of entries.

The generator validation was made using the histogram representation in Matlab7.1 (see Fig. 3, left, for lognormal variable validation).

Then we have calculated the outputs of the pull-in function with Eq. 1, by using 5000X5000 normal random entries. One may notice in Fig. 2, right, that the probability density estimation for pull-in voltage is not a normal one for L=160 μm (and the same for L=760 μm). The histogram is not symmetric, suggesting a lognormal distribution.

The lognormal distribution is commonly used for general reliability analysis, cycles-to-failure in fatigue, material strengths and loading variables in probabilistic design. This distribution is positively skewed (most of the values being closer to the lower limit), like ~~(what)~~ we need, because we work with dimensions. A random variable is lognormally distributed if the logarithm of the random variable is normally distributed.

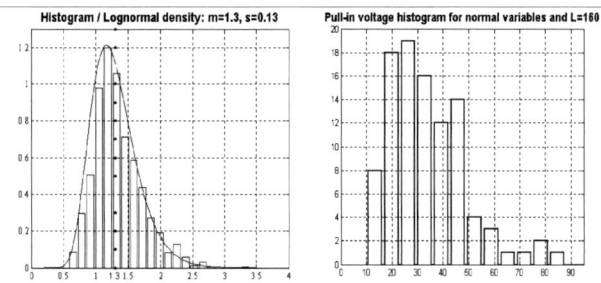

Fig. 3. The histogram and probability density function for lognormal validation (left) and for normal entries (right).

Consequently, we have used the first algorithm for a normal vector to generate independently a 2-dim lognormal vector with the lognormal variables following mean 1.3 μm and variance 0.13 μm (10% of mean value), and 1.5 μm and 0.15 μm, respectively.

The variations of gap and thickness produce slight variations of the pull-in voltage. In order to describe these variations we have computed the voltage functions with Eq. 1 for 5000X5000 entries and for two lengths: L=160 μm and L=760 μm, i.e. the shortest and the longest bridge. We noticed that the histogram of results [4] indicates for the pull-in voltage a lognormal distribution with mean and variance given by the points simulated using Monte Carlo method, as we can see in Fig. 4.

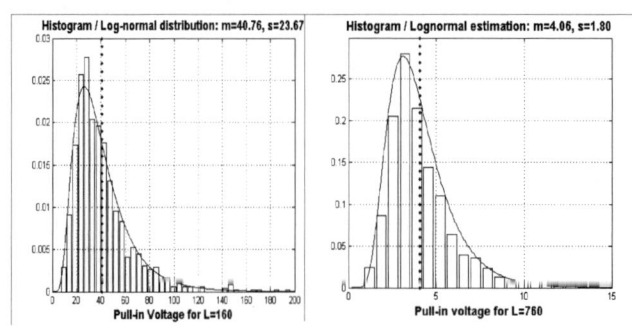

Fig. 4. The histogram and probability density function for lognormal entries (L=160 μm and L=760 μm).

260

B. Optimal Dimensions

Now we intend to find the behavior of thickness and gap such that the mean values of the pull-in voltage are close to a fixed value for the pull-in voltage (in particular, close to the values calculated from initial air gap and thickness).

For initial dimensions the pull-in voltage has been calculated with Eq. 1. A comparison between the results [3] can be seen in TABLE I.

TABLE I.
PULL-IN VOLTAGES DETERMINED BY ANALYTICAL EQUATIONS, FEM-BASED SIMULATIONS AND EXPERIMENTAL MEASUREMENTS

		Pull-in voltage [V]		
	Analytical	Monte Carlo	FEM simulation	Experimental
	$g=1.5\mu m$ and $t=1.3\ \mu m$	variance of g and t equal with 10%		
Short bridge	35.266	40.042	44.375	45.5
Long bridge	3.703	4.016	4.375	3.9

Then we have calculated different combinations for the variance of thickness and gap to observe the dependence between them and the mean value of the pull-in voltage calculated by Monte Carlo method.

TABLE II.
VARIATION OF PULL-IN VOLTAGE VS. VARIATION OF DIMENSIONS

Variance for g [% of mean]	0.5%	0.5%	10%	10%
Variance for t [% of mean]	0.5%	10%	0.5%	10%
Mean value of the Pull-in Voltage (short bridge)	35.58	37.78	37.70	40.04
Mean value of the Pull-in Voltage (long bridge)	3.71	3.93	3.79	4.01

In Table II, the extreme values of the pull-in voltages determined by a variation of the gap and thickness with variance taking values between 0.5% and 10% of mean are listed.

We can conclude that the pull-in voltages for short bridge increase with a slope of 2.20 if the variance of g is fixed and that of t is between 0.5% and 10% of mean and the slope is 2.12 if the variance of t is fixed and that of g is between 0.5% and 10% of mean. In this case the slopes are

approximately the same (Fig. 5 and Fig.7 left). For the long bridge one may notice a difference in slopes (Fig. 6 and Fig. 7 right).

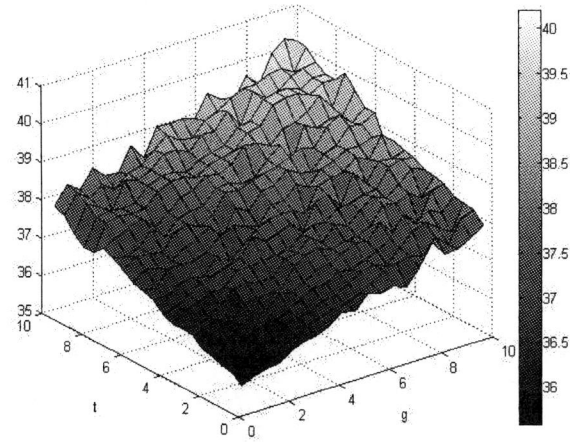

Fig. 5. Variations of t and g parameters and 3D approximation for pull-in voltage (L=160 μm bridge)

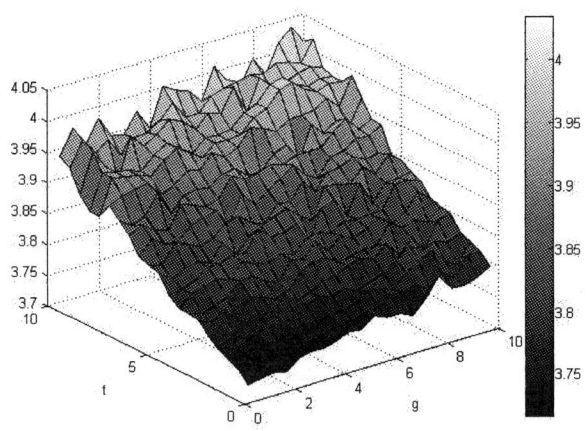

Fig. 6. Variations of t and g parameters and 3D approximation for pull-in voltage (L=760 μm bridge)

The pull-in voltages for long bridge increase with a slope of 0.22 if the variance of g is fixed and that of t is between 0.5% and 10% of mean and with a slope of 0.07 if the variance of t is fixed and that of g is between 0.5% and 10% of mean.

In Fig. 5, the behavior of the pull-in voltages vs. the variations of the variance of thickness and gap between 0.5% and 10% of the mean value is shown. So, if we establish a limit of variations of pull-in voltage, we can estimate a variation of variances for the gap and thickness and we can find a limit of variation for these dimensions.

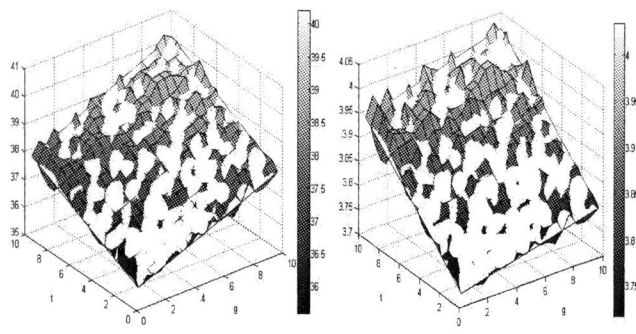

Fig. 7. Approximation by plane of the pull-in voltage function, $V_{pull-in} = f(g,t)$.

IV. CONCLUSION

The pull-in voltage of electrostatically actuated microbridges has been chosen as the target behavior of the designed structures. In order to assure the maximum acceptable variations of this parameter, it is possible to optimize the dimensional parameters affecting its values. By means of statistical simulations we have investigated the influence of process variations on the pull-in voltage. The analyses show that in the case of short bridges a variation of gap and thickness dimensions give both a strong variation for pull-in voltage. The long bridges seem to be more stable. A variation of gap dimension seems to have a smaller influence on pull-in voltage. On the other hand, the variation of thickness dimension proves to be more influential.

The results show that the Monte Carlo method is a significant and promising tool for studying the process of manufacturing electrostatically actuated micro-bridges.

REFERENCES

[1] L. X. Zhang and Y. P. Zhao, "Electromechanical model of RF MEMS switches", *Microsystem Technologies*, 9 (2003) 420-426

[2] P. M. Osterberg and S. D. Senturia, "M-test: A test chip for MEMS material property measurement using electrostatically actuated test structures", *Journal of Micromechanical Systems*, Vol. 6, No.2, pp. 107-118, 1997

[3] C. Tibeica, R. Voicu, M. Bazu, V.E. Ilian, D. Vasilache, L. Galateanu, P. Pons, K. Yacine, and F. Flourens, "Analytical vs. computational approach of the pull-in voltage of electro-statically actuated micro-bridges", *CAS 2005 Proceedings,* Vol 1, pp.179-182, Sinaia

[4] R. O. Topaloglu, "Double-Correlated Discrete Probability Propagation for Process Variation Characterization of NEMS Cantilevers for Molecular Diagnostics", *MAE*

[5] L. Schenato, Wei-Chung Wu, L. El. Ghaoui and K. Pister, "Process variation analysis for MEMS design", *SPIE Symposium on Smart Materials and MEMS*, Melbourne, Australia, December 2000.

[6] K. Yacine, F. Flourens, P. Pons, K. Grenier, D. Dubuc, R. Plana, M. Sartor and L. Dantas, "Mechanical characterization of metallic bridges used in RF MEMS", *Memswave2003 – 4th Workshop on MEMS for millimeter wave communications,* Toulouse, 2-4 July, 2003

2006 25th International Conference on Microelectronics

Development of a Portable Gamma Radiation Monitoring System

K. Arshak, D. Buckley, O. Korostynska

Abstract — This paper demonstrates the feasibility of a portable gamma radiation monitoring system, based on the Anderson Current Loop circuit, where Nickel Oxide (NiO) thick films are used as radiation-sensing elements. A cost effective prototype system was developed for remote ionising radiation monitoring that is required to minimise the necessity for human interaction and consequently prevent possible exposure.

I. INTRODUCTION

Ionising radiation is used extensively in areas such as environmental monitoring, food irradiation, diagnostic and therapeutic medical procedures, and aerospace and military applications, where accidents could pose serious health risks to those in the proximity of these radiation sources. Terrorist attacks are an ever-present and ever-increasing threat and as such we must be equipped to handle any type of possible attack, including exposure to ionising radiation. Successful and safe remediation of radioactive waste sites is dependant on accurate monitoring of radioactive plumes to ensure that they are not emitting radioactive material into the surrounding environment.

The ability to detect and measure radiation dose is of great importance, and necessitates the use of sensitive and accurate devices for these functions. It is therefore imperative to explore alternative detection strategies. This paper describes how the Anderson current loop [1] is used to measure the effect of Gamma radiation exposure on thick film Nickel Oxide radiation sensors [2]. This novel method is based on continuously measuring the resistance of the sensor as it is exposed to radiation and detecting a change in potential across the sensor as its resistance is increased due to absorbed dose.

II. SYSTEM DESCRIPTION

The diagram in Fig. 1 shows a simplified overview of the prototype system's setup. The ADuC812 micro-converter, the Anderson Loop circuit, and the sensor are shown separately even though they are amalgamated on one printed circuit board (PCB) in the final prototype. The System consists of the thick film radiation sensor connected to the resistive conditioning analogue front-end of the circuit, the Anderson Loop, who's output is connected to the Analogue to Digital Converter (ADC) of

K. Arshak, D Buckley, O. Korostynska are with the Electronic & Computer Engineering Department, University of Limerick, Limerick, Ireland, E-mail: khalil.arshak@ul.ie

the system's microcontroller, Analog Devices AduC812 micro-converter [3]. This microcontroller stores the output of the Anderson Loop in its internal memory and outputs the value to an on-board LCD display and to the serial port connection. This serial port connection can be (optionally) connected to a PC, using RS232 protocol, where a LabVIEW [4] graphical user interface (GUI) shown in Fig. 2 displays the voltage, dose and circuit temperature in real time as well as saving it to an Excel file (or to any other format the user wishes) for later use. It should also be noted that in Fig. 1 and in the final prototype the ADuC only receives readings from the Anderson Loop, however, two-way communication can be easily setup. This could allow the ADuC to access digital potentiometers in the Anderson loop circuit to allow control of the gains and other such configuration parameters through software (i.e. through the assembly code downloaded to the AduC812).

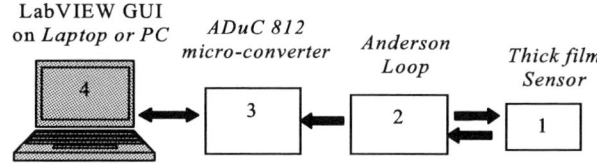

Fig.1. Prototype system overview.

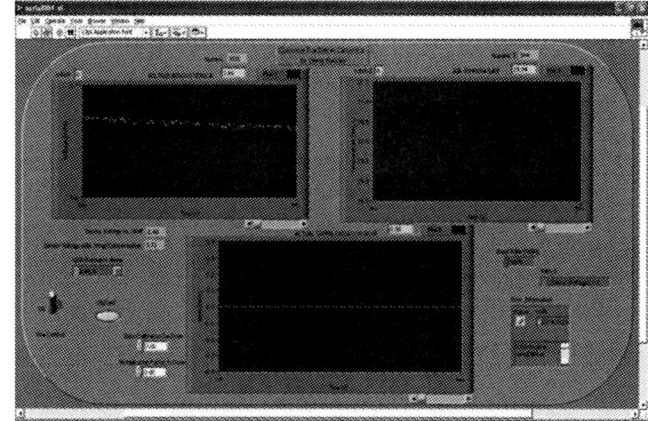

Fig. 2. LabVIEW graphical user-friendly interface, displaying voltage changes as a function of dose.

The serial port connection is also used to download readings stored in the ADuC812's internal memory that could have been previously taken at a remote location. This, along with the system's ability to run from the mains or battery power, allows the device to be portable or used directly with a PC. It can be worn on a person's body or clothing and used for personal radiation monitoring where

1-4244-0116-X/06/$20.00 ©2006 IEEE   263

the present dose can be easily read from the LCD display and will alarm if the dose level exceeds a preset safe level. It can also be left in a remote location taking measurements for up to 12 hours, a time easily increased, this was just for the prototype and only limited by memory capacity and battery life. It can then be retrieved, and information stored on it from radiation measurements, downloaded to a PC. The system also shuts off entirely when the battery potential falls below a certain level. This ensures that a false reading is not output to the user due to low battery power.

This system can easily be extended to a wireless format where data from many metres away, could be transmitted back to a receiver unit connected to a PC. Alternatively, up to eight sensor and Anderson Loop pairings (number only limited in the prototype by the number of ADC channels available on the microcontroller used) can be located with long wires at several locations and all eight sensors could relay information back to the ADuC812 and PC concurrently (every 300ns) as shown in Fig. 3. One of the applications for this could be the monitoring of radioactive plumes from radioactive waste storage sites. With the PC setup as a central monitoring station the system could detect and quantify radioactive plumes during remediation, verify the absence of radioactive materials and ensure that radioactive material is not contaminating the surrounding area.

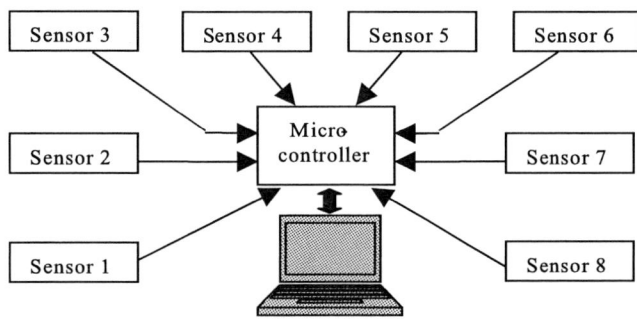

Fig. 3. Eight sensors relying radiation measurements back to PC.

A. Anderson loop

For resistive conditioning, this system uses an Anderson Loop arrangement. This was chosen for its low power consumption, and the accuracy that it achieves for its relatively simple design. The Anderson Loop concept was invented to deal with lead-wire resistance variation problems that occurred when connecting high temperature strain gages to conventional bridge signal conditioning in the Flight Loads Laboratory of the NASA Dryden Flight Research Centre. It was first described by Karl F. Anderson of NASA Dryden Centre in 1992 [1] and subsequently patented by NASA [5]. The Anderson Loop is a relatively simple, easy to design, reliable, cost effective, low power circuit with few components and can be easily calibrated for different sensors' resistance ranges.

The underlying concept of the Anderson Loop is a simple blending of continuous analog subtraction with Kelvin sensing as shown in Fig. 4. A current source provides the excitation for a reference resistor and a resistive sensor. The voltage differences induced across each resistor are subtracted from each other, and the output voltage is proportional to the difference in resistance between the two (when using a high-impedance device, like an instrumentation amplifier, to measure the reference and sensor resistances, such that very little current is drawn).

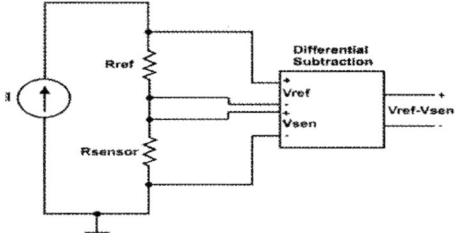

Fig. 4. The Anderson Current Loop concept showing active differential subtraction.

The output of the Anderson loop is simply a voltage that represents the differences in resistance of the sensor and reference resistor, as described by Eq. 6. These are initially setup to be of similar resistance and the output voltage can be set to null (or a predefined agreed "zero" point) and will remain like this until the resistance of the sensor is altered due to exposure to radiation dose. The Anderson Current loop automatically maintains a constant current across both the reference resistor and sensor and the output is then a change in potential. This potential or voltage change is relative to the change of the absorbed dose of the sensor as is shown in the Results section below.

By altering various variable resistors within the circuit the gain can be adjusted and essentially the output can be modified to suit any sensor or measurement range the user desires so essentially sensors manufactured to be more sensitive to particular radiation dose ranges can be multiplexed to give wide ranging sensitivity.

The basic Anderson Loop circuit configuration used is shown in Fig. 5. There were several additions made to this circuit, for example, filtering and necessary interfacing with the ADuC812 micro-converter some of which can be seen in the system block diagram in Fig. 6.

B. Thick film sensors

Gamma rays produce a change in the density of charge carriers in semiconductor material, which alters the material properties in measurable way. This change provides information on the dose absorbed by the material. It is believed that ionising radiation causes structural defects (called colour centres or oxygen vacancies in oxides) leading to a change in their density on exposure to γ-rays. In this work, thick film NiO sensors were used, whose electrical properties are altered by the presence of ionising radiation [2].

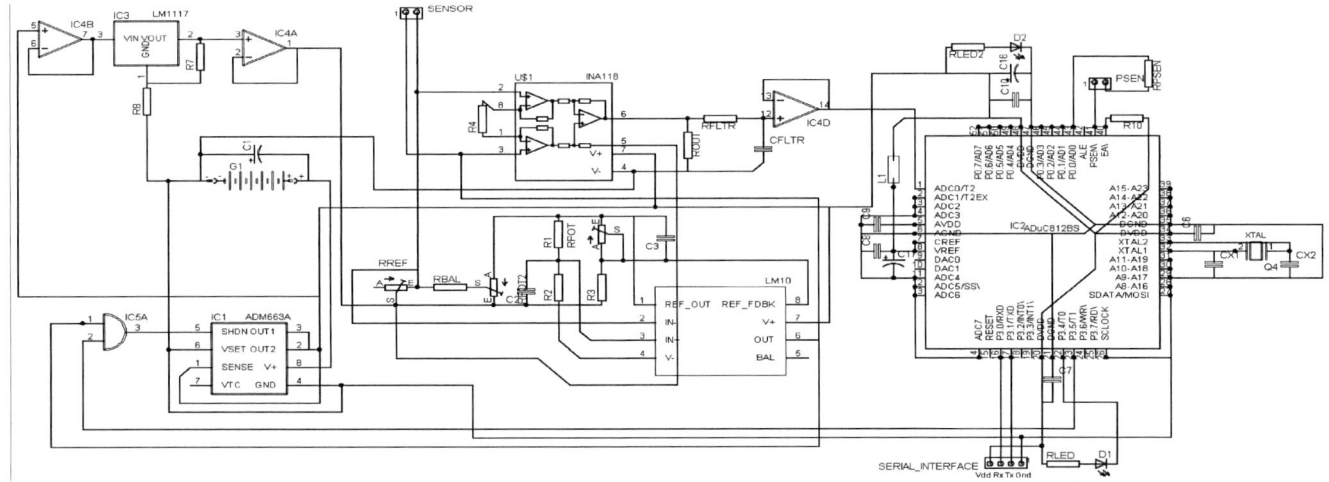

Fig. 6. The basic system block diagram used for the prototype.

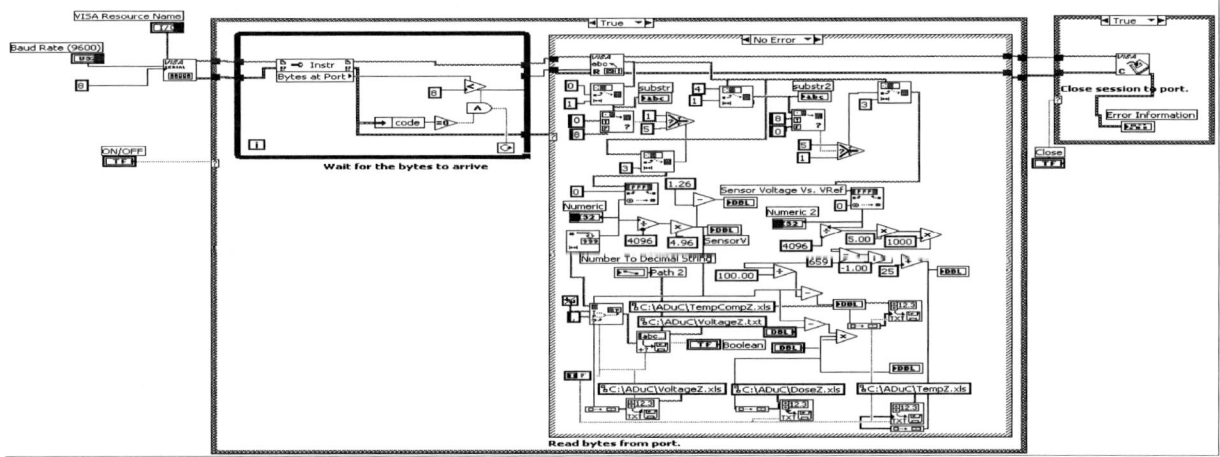

Fig. 7. The source diagram of the LabVIEW GUI shown in Fig. 2.

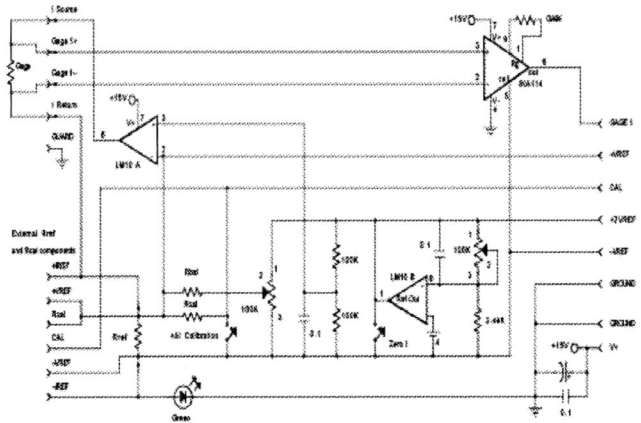

Fig. 5. Basic single-sensor Anderson Loop signal conditioner configuration used for the prototype circuit. [6] © IEEE 1998.

C. Software

The ADuC812 micro-converter required assembly code to be written and downloaded to its internal non-volatile memory. The code controlled the serial port settings, the ADC channels and conversion timing, as well as features such as the power supply monitor. Several code examples can be found on the Analog Devices web site [5]. This code and hence the settings can be altered for different system requirements depending on its application

National Instruments LabVIEW software was chosen due to its myriad of prewritten functions that allow simple manipulation of data and serial port communication. Its graphical programming environment makes program design intuitive and its use-friendly GUI minimises the user necessity for involved interaction. The source diagram for the GUI shown in Fig. 2 is shown below in Fig. 7. It shows how data received serially from the ADuC812 is dealt with. It is here that the byte streams from the ADuC812's ADC are separated, to be graphically displayed and saved into different locations. In this example, there are just two different blocks of information to separate: the temperature of the circuit and the sensor reading, each 4 bytes long, the first byte of each block being the channel ID and the next 3 bytes the actual measurement value. This can easily be extended to deal with all 8 ADC channels and hence 8 sensors.

III. EXPERIMENTAL PROCEDURE

A DEK RS 1202 automatic screen printer was used for thick film fabrication. A polymer paste was made of 92% NiO and 8% $C_8H_{18}O_3$ by weight with Diethylenglycolmonobutylether as a solvent. NiO polymer paste and commercial DuPont 4929 silver paste were used to fabricate the radiation sensitive material and contacts respectively. Pastes were printed on glass to form a sandwich Metal-Semiconductor-Metal structure with an active area of $1cm^2$ and a film thickness of 70μm.

^{137}Cs (0.662 MeV) disk-type source (provided by AEA Technology QSA GmbH as a standard reference γ-source) was used for exposing the samples to γ-radiation. The source was held at a distance of 1 cm from the NiO film at an angle of incidence of 0°.

IV. RESULTS

To validate the stability, reliability and repeatability of the system, extended tests with many different experimental setups were carried out. This included testing with: variable resistors in the place of the sensors; sensors that were not being exposed to radiation; sensors during exposure, exposing the circuit and sensors at the same time to radiation and also just the sensor on its own. All the results from these tests were similar and correlated with what was expected from the circuit.

To ensure the repeatability of the sensors results, multiple tests were performed on various sensors. Fig. 7, shows the output of one such test. A graph of change in voltage measured across the sensor, against accumulated gamma radiation dose. This real-time graph describes the potential change from the moment the sensor was exposed to the ^{137}Cs source until 10 hours later when the source was removed. This represented a total dose of 3420 μGy.

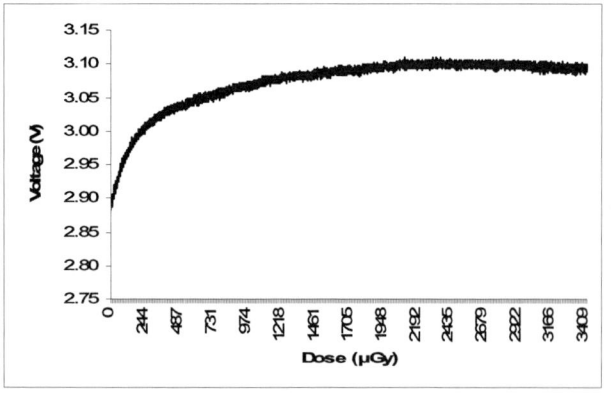

Fig. 7. Sensor voltage increase with dose.

It is important to monitor the response of both the sensor and the system as a whole during this period but it is the initial sharp rise slope that we are mostly interested in from a dosimeter perspective. This is shown in greater detail in Fig. 8, which shows the response of the system for the initial dose range from 0 to 223μGy.

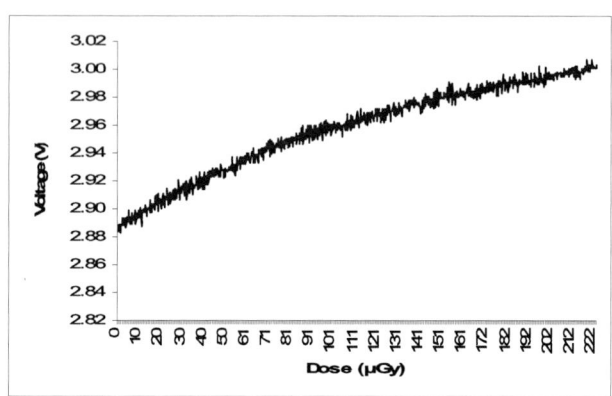

Fig. 8. Initial voltage response to radiation.

IV. CONCLUSION

A prototype real time portable gamma radiation monitoring system employing thick film NiO sensors was developed. The system employed the Anderson Loop for analogue resistive conditioning, which is a novel application of this concept. It is a cost effective circuit with very few components required. The system is portable, as it is battery operated and data can be saved locally in non-volatile memory for subsequent downloading and analysis. It is real-time in that it outputs its readings immediately to an LCD display or to a PC. The system described, provides a simple method of collecting, storing and displaying data from one or more radiation sensors. This monitoring system can be applied to a wide range of areas, such as homeland security tasks, environmental and personal radiation monitoring and nuclear waste control, where personnel's exposure to radiation can be minimised.

The system was exposed to a disc-type ^{137}Cs gamma source and the output voltage of the circuit was shown to monotonically increase with a corresponding increase in dose as expected. Repeatability, reliability and stability were demonstrated. All of which shows that this is a viable low cost alternative to existing Gamma radiation dosimeters and monitoring systems.

REFERENCES

[1] Anderson, K. F., 1992 "The Constant Current Loop: A New Paradigm for Resistance Signal Conditioning," NASA TM-104260.

[2] Arshak, K., Korostynska, O., and Harris, J., 2002. "γ-radiation dosimetry using screen printed nickel oxide thick films," in *Proc. 23rd. Int. Conf. MIEL.*, 1, 357–360.

[3] http://www.ni.com/labview/

[4] http://www.analog.com/en/prod/0,2877,ADUC812,00.html

[5] Anderson, K. F., 1994. "Constant Current Loop Impedance Measuring System That Is Immune to the Effects of Parasitic Impedances", U.S. Patent No. 5,731,469.

[6] Anderson, K. F., 1998. "NASA's Anderson Loop," IEEE Instrumentation and Measurement Magazine, vol. 1, Is. 1, pp. 5 – 15.

Thermal Characterization of the Micro Bonding Process Using a Hot Air Stream

D. Andrijasevic, I. Giouroudi, W. Smetana, W. Brenner, D. Esinenco

Abstract – The possible solution for bonding two different materials with intermediate layer of adhesive at relatively low temperatures in the micro domain is proposed in this paper. A stream of hot gas has been used for melting and softening different kinds of adhesives. After cooling down at room temperature, adhesives harden forming a stabile and strong bond. The gas is heated as it passes through the tube with the heater developed on it. The parameters which induce the heat transfer and the process performance generally are considered through various experiments and by numerical simulations. Some advantages of this process are applicability for different material combinations, cost and time efficiency, compact size of equipment etc.

I. INTRODUCTION

In the last decade, micro electro mechanical systems (MEMS) have made real revolution on the market of modern communications and technologies. It is almost not possible to find the area where MEMS cannot bring some benefits and improvements. Although MEMS are widely used, there are still some hindrances due to technological limitations and more important, due to very high production costs. The big stakes in total production costs is related on assembly and bonding techniques. This is the reason why researchers all around the world investigate this problem with aim to find better and chipper solutions.

A new concept of micro bonding of micro-parts with dimensions in the range of 50 μm to 300 μm is presented in this paper. The intermediate layer of adhesive is softened with the stream of gas which is heated while passing through the heating tube. Afterwards, a micro-part can be embossed in the softened glue or covered and shielded by it.

The use of adhesive in micro-system technique has been avoided regarding its instability and remarkable changes of properties during the time. But it has been used in some other applications. Optical fibers or micro-lenses can be made of adhesives. Electrical connections between components in flip chip configurations can be formed of electrically conductive adhesives [1]. In micro-fluidic

D. Andrijasevic, I. Giouroudi, W. Smetana, W. Brenner are with the Institute of Sensor and Actuator Systems, Vienna University of Technology, Floragasse 7/2 - MST, 1040 Wien, Austria, E-mail: Daniela.Andrijasevic@Tuwien,ac,at
D. Esinenco is with theThe National Institute for Research and Development in Microtechnologies, Erou Iancu Nicolae street, 077190, Bucharest, ROMANIA E-mail: dorine@imt.ro

systems, J. Simon, et al. have described the use of room-temperature-curing UV epoxy adhesive as micro-gaskets in an array of liquid-filled micro-relays [2].

The described principle is well-known in the macro-world, employed primarily for bonding thermoplastic foils in food industry. In the research presented in this paper, this concept is used for joining micro-components into complex, 3D structures. As an illustration of possible application, optical fibers have been positioned into V-grooves and firmly adhered to a basic substrate.

In order to minimize the time necessary for the optimization of processing parameters, finite element analyses using the ANSYS™ program package were performed. The temperature distribution along the set-up parts which participate in heat and gas transfer, on the substrate surface, as well as within the gas stream is evaluated and preliminary results are represented in the following discussion.

II. EXPERIMENTAL SET UP

To confirm the basic idea of the proposed principle, a special set up has to be developed. It comprises: an air reservoir (technically clean air), a compressor, a heating element on the metal tube, a nozzle, and additional pipes for connecting these elements. The heater was printed on a metal tube by a thick film technology. This provides small and compact dimensions of the complete set-up. On an austenitic stainless steel tube (inner diameter of 2 mm, outer diameter of 4 mm, length 120 mm), the heating element of a total length of 80 mm was developed in a way similar to that described in [3]. The maximum peak temperature on the surface of the heating tube was 600 °C and it is obtained within 5 seconds with heating power of 28 W. It is also possible to reach higher temperatures, but this is restricted with working temperatures of glues which connect thermocouples for the tube.

The schemas of the experimental set up and the heating tube with the developed heater on it are shown in Figure 1.

Metal nozzles (CrNiMo, inner diameter in the range of 150 μm and 300 μm, capillary length 11 mm), were installed on the bottom end of the heating tube in order to focus the gas stream on the requested position. The current temperature on the nozzle, on the substrate and on the heating tube were measured continuously with three K-type thermocouples made of NiCr-Ni wires, diameter 80 μm. A flow meter regulates the air flow according to the optimal values gained by experiments and simulations.

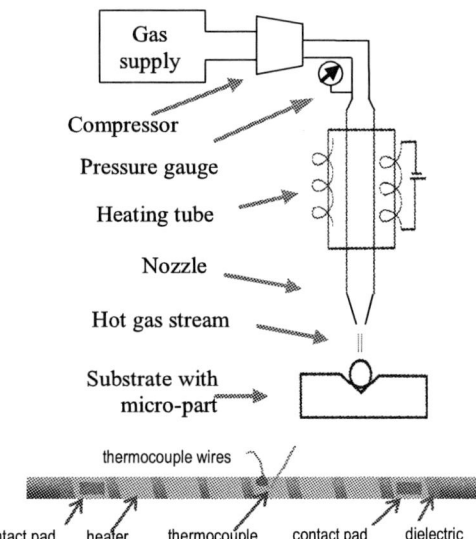

Fig. 1. Schema of experimental set-up and schema of the heating element with a thermocouple and electrical connections.

The basic substrate was placed on the micro-manipulation stage which has three degrees of freedom (3 translations) and whose movement is controlled by a LabVIEW program. The minimum step width of the step motors which drive the stage is 1 μm. The total range of movement is 6 cm in all three directions.

III. NUMERICAL SIMULATION

In order to shorten the time necessary for optimization of processing parameters, as well as to find the optimal geometry of the heating tube and the heater, many and various finite element analysis simulations in ANSYS™ program package were performed. The parameters included in the simulations were the temperature on the surface of the heating tube, on the nozzle's outlet, on the substrate, air pressure, air flow, etc.

Taking into account symmetrical structure of elements in the system, the simulation model was simplified to reduce the required computational power. Finally, a quarter of the cross-section of the tube and the nozzle was adopted as a simulation model and it is depicted on Figure 2.

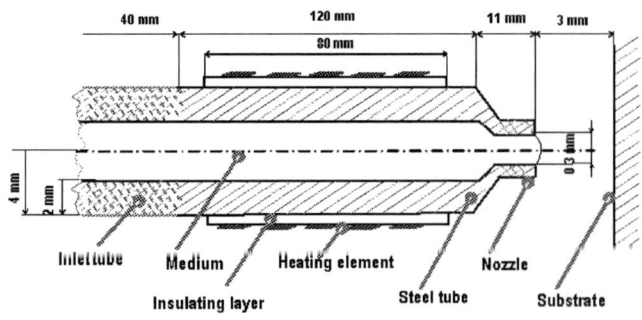

Fig. 2. Numerical simulation model.

The information about material properties considered in the simulations is summarized in Table 1.

TABLE I
MATERIAL PROPERTIES DATA [4], [5].

Parts	Material	Density (kg/m^3)	Thermal conductivity (W/mK)	Specific thermal capacity (J/kg K)
Inlet tube	silicon	1850	0.35	850
Heating tube and nozzle	austenitic steel (1.4301)	7900	16	500
Insulating layer and substrate	glass ceramic	2230	1.02	837
Medium	air	1. 229	0. 0252	1000

The air is injected into the heating tube at room temperature and under pressure of app. 1.25 bar. While it passes through the tube, in the region of the heater, the air temperature raises rapidly, depending on the actual heating power. Afterwards, in the region of the nozzle, it drops fast, because of the high thermal conductivity of the metal. The random temperature distribution gained by simulations is shown on Figure 3.

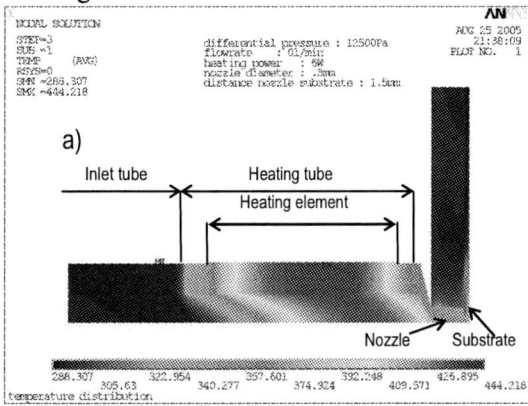

Fig. 3. The temperature distribution along the cross-section of the heating element, the nozzle, the air stream and on the substrate.

Based on the simulations results, it is concluded that the main parameter that influences the heat conductivity and the temperature reached on the outlet of the nozzle is the air flow. The distance between the substrate and the nozzle outlet influences the strongest the temperature on the substrate. Therefore, these two parameters, air flow and nozzle outlet-substrate distance must be optimized.

The heating-up characteristics of the temperature on the substrate's surface versus the air-flow and the heating element temperature derived by experiments and numerical simulations are shown on Figure 4. The nozzle diameter (300 μm) and the distance from the substrate (3 mm) were kept constant.

There is a small deviation between the experimental and numerical results due to the losses during interaction of the hot air with the surrounding. It was tried to eliminate

268

these losses as much as possible by placing the whole set-up into a Styrofoam isolation chamber and improving the sealing of the contact joins.

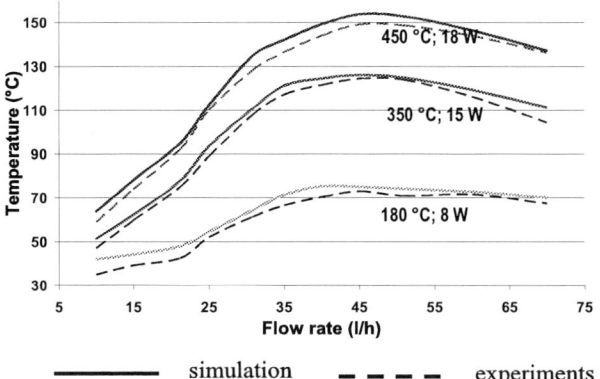

Fig. 4. Substrate surface temperature characteristics versus the air flow rate at different temperatures on the heating element.

Nevertheless, there is a good agreement between results gained by simulations and by measurement.

The distance between the substrate and the nozzle outlet influences the most the real temperature on the substrate. It is noticed that the temperature on the substrate is even higher than the temperature on the nozzle outlet, if the distance is shorter than 2 mm, and the maximal temperature is recorded for the distance of 1, 2 mm. This can be explained with different thermo-kinetic effects, but this region is very unstable and unpredictable. On the other hand, the distance is too small to provide conditions for normal work (handling and assembly).

Figure 5 shows the functionality "the substrate temperature versus nozzle-substrate distance", whereby the temperature on the nozzle's outlet varied.

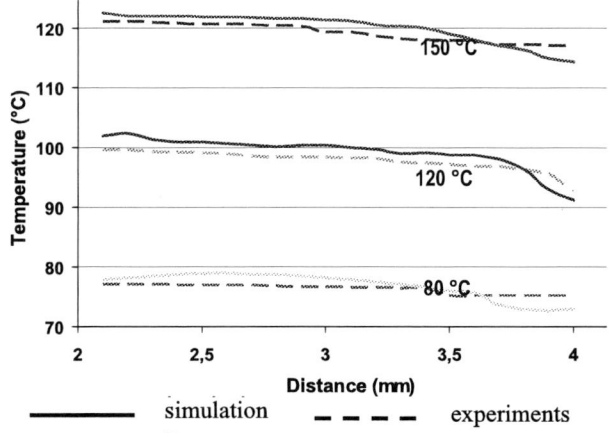

Fig. 5. Substrate surface temperature characteristics versus the distance between the substrate and the outlet of the nozzle, for different temperatures on the outlet of the nozzle.

The results obtained by the numerical simulations were proved by the experiments and the process parameters were optimized accordingly. An analytical model has not been developed because of the complexity of the phenomena which appeared during the process.

Another interesting aspect which influences directly the temperature distribution is the pressure distribution. In the tube, it depends mostly on the air flow; in the region between nozzle and substrate on the distance between them. Next figure illustrates the typical pressure distribution for the nozzle diameter 0.3 mm and the distance of 2.5 mm.

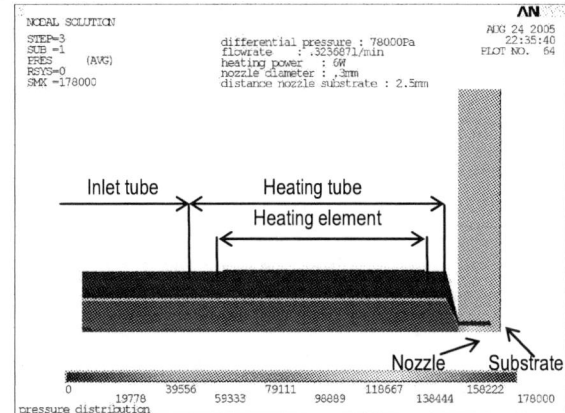

Fig.6. The pressure distribution along the cross-section of the heating element, the nozzle, the air stream and on the substrate.

IV. PROCESS OF JOINING

In order to confirm the advantages of the proposed bonding technique, numerous experiments were performed. As a micro-part, single mode glass optical fiber (core diameter 9 μm; cladding diameter 125 μm) was used for the simple evaluation of the bonding results. Optical fibers are positioned in V-grooves to achieve requested alignment and positioning precision. V-grooves are made by chemical etching of silicon in the "111" plane with the slope of 54, 7° [6]. The thick layer of SiO_2 (1.2 μm) was grown on the silicon layer by dry thermal deposition. The applied positive resist HPRD 505 and the oxide were removed in $HF:NH_3$ (1:6) with a speed of app. 1000Å/min (0,1 μm/min). Anisotropic etching of silicon was done in KOH (25 %) at a temperature of 85 °C. With these parameters the removing speed is app. 1 μm/min. The SEM photo presents the obtained structure.

Fig. 7. V-grooves obtained by KOH etching.

269

Up to date, two different kinds of glues were used. Their main properties are given in the following table.

TABLE II
ADHESIVE PROPERTIES

Type of adhesive	State	Softening point (°C)	Thickness (µm)	Deposition method	Manufacturer
Hot melt glue	solid	65	~ 12	spinning	2 SPI
Adhesive film	solid	72	50	lamination	Bemis Ass

The polyurethane adhesive film was laminated on the substrate, while the hot melt glue was applied by spinning (800 rpm, 40 s) after dissolution in PGMA ((1-Methoxy-2-Propyl)-Acetat) in mass-proportion of 55% of PGMA and 45% of glue. The hot melt glue softens at a temperature of around 65 °C and the fiber is embossed onto the high viscose composition.

The adhesive foil can be used for bonding in two ways. First, the foil is applied on the substrate, softened at 90 °C and the fiber is embossed in the soft material. Second, the fiber is positioned onto the V-grooves and covered with the foil. After the softened foil cools down, the fiber is firmly joined to the substrate.

The quality of the bond is evaluated by measuring a mechanical strength of bond. By increasing the softening temperature of the adhesive foil, the strength of bond increased significantly and maximum was 0,8 N/mm^2 for 102 °C. The strength of join made with hot melt glue dropped fast at the higher temperature, because adhesive evaporates, i.e. there is not enough adhesive to accomplish rigid bond. The maximum strength was 0,3 N/mm^2 achieved at the temperature of 65 °C. These results are comparable with other bonding techniques.

V. CONCLUSION

A thermal characterization of the bonding process using a hot air stream has been done in the work presented in this paper. As a proof for the applicability of investigated technique, glass fibers were positioned in V-grooves and successfully bonded on them.

In the future work, the selection of the most suitable adhesive for this application with regard to its thermal stability, mechanical strength and working life has to be done. The design and development of the new, integrative tool for pressing the fiber onto the requested position and supplying with focused air stream have to be considered. It should enable better positioning of the fiber and the decrease of the bonding time.

ACKNOWLEDGEMENT

This work was financially supported by FP6 Marie Curie Research Training Network "Advanced Methods and Tools for Handling and Assembly in Microtechnology - ASSEMIC", funded by the European Commision, Contract no. MRTN-CT-2003-504826 and it is the part of common work in the 4M Network of Excellence in Multy-Material Micro Manufacture.

REFERENCES

[1] F. Sarvar, D. A. Hutt, D. A. Whalley. "Application of Adhesives in MEMS and MOEMS Assembly: A Review", *IEEE Polytronic Conference*, Hungary, pp 29-34, 2002.

[2] J. Simon, L. S. Huang, B. Sridharan, C. J. Kim, "Micro-gasketing and Room Temperature Wafer Joining for Liquid-Filled MEMS Devices," *Proc. MEMS, ASME Int. Mechanical Engineering Congress and Exposition*, Dallas, USA. pp 29-34, 1997.

[3] D. Güleryüz, W. Smetana, H. Homolka, M. Mündlein, M. Unger. „Aspects Concerning the Optimization of a Calorimetric Flow Sensor Built Up in Thick-Film Technology", *28th International Spring Seminar on Electronics Technology*, pp 370 – 376, Austria, 2005.

[4] J. H. Keenan, F. G. Keyes, P. G. Hill, J. G. Moore. "Steam Tables : Thermodynamic Properties of Water Including Vapor, Liquid, and Solid Phases/With Charts (metric measurements)" Krieger Publishing Company; Rep/Charts edition, ISBN: 0471465011, 1992.

[5] K.Raznjevic. "Handbook of Thermodynamic Tables and Charts", Pub. McGraw-Hill; 2nd edition, ASIN: B000715DA4, 1976.

[6] D.F. Moorel, R.R.A. Syms. "Silicon technology for optical MEMS," Europhysics News, Vol. 34 No. 1, 2003, http://www.europhysicsnews.com/full/19/article1/article1.htl

2006 25th International Conference on Microelectronics

Designing and Modeling MEMS Resonator in VHF Range

M. Al_Khusheiny and B. Y. Majlis

Abstract - In this paper a new structure of Clamped-Clamped Mixed μresonator (CCMR) of resonance frequency (f_o) of 71 MHZ, motional resistance of 3.7 -KΩ and Q as high as 6230 is designed, modeled, and simulated using a two-layers-beam composed of Silicon Nitride and Cupper .*Intellisuite* simulator is used to model the mechanical properties of the new MEMS Resonator using a static displacement analysis, and helps to get the optimum values of the beams parameters: the effective spring constant and mass of the resonator were calculated. By calculating the effects due to mechanical non-linearity CCR, we can achieve and demonstrate a resonance frequency of 71 MHz in a Polysilicon surface micromachining technology which can be used to build a MEMS Filter.

I. INTRODUCTION

A Clamped-Clamped Resonators (CCR's) play an important role in realization the low power MEMS Filters, Oscillators and Mixers. CCR's can be used in communication because of their relatively high spring constant which enables large dynamic range and power handling, beside their ease of manufacturing . The μmechanical resonators offer the potential for the very high Q's in the context of conventional IC processes; as their Q determines the insertion loss and phase noise respectively. The resonance frequency, motional resistance and the Q-factor is the most characteristics for the CCR ,and these three characteristics is largely depend on the physical dimensions of the CCR .To be useful for direct insertion into present-day cellular and cordless phone applications, mechanical resonators used in IF filters must be capable of operating at frequencies from 70 to 250 MHz[1], while those aimed at RF filters must attain a range from 800 MHz to 1.8 GHz.

In this paper a MEMS CCMR of resonance frequency up to 71 MHZ and Q_{nom} of 6230 is modeled, and simulated. *Intellisuite* simulator is used to model the mechanical properties of the CCMR using a static displacement analysis; the effective spring constant and mass of the mixed resonator were calculated besides calculating the effects due to mechanical non-linearity CC-mixed resonator, by which one can achieve and demonstrate a high Q's and a resonance frequency in VHF range in a polysilicon surface micromachining technology.

M. Al_Khusheiny & B. Y. Majlis, senior member IEEE, are with the Institute of Microengineering and Nanoelectronic Universiti Kebangsaan Malaysia,43600 Bangi, Selangor ,Malaysia ,E-mail: mustafa1@vlsi.eng.ukm.my .

II. DEVICE STRUCTURE AND OPERATION

Figure 1 presents the perspective-view schematic of the lateral CC-mixed beam resonator in a typical measurement circuit. As shown in the figure the μresonator consists of two main parts; a CC-mixed beam, which is composed of two layers, the silicon-nitride layer of thickness (h_2) and the conductive (metallic) layer of thickness (h_1), this mixed beam is suspended to two anchors, and the metallic electrode beneath it, so by changing the dimensions of the resonator, a different resonance frequency for the resonator can be achieved. A dc bias voltage V_p is applied to the conductive layer of the mixed beam, while an ac excitation signal $v_i = V_i \cos(\omega_i t)$ is applied to the underlying electrode. In this configuration, a dominant force component is generated at ω_i , which drives the beam into mechanical resonance when $\omega_i = \omega_o$, creating a dc-biased (via V_p) time-varying capacitance between the electrode and resonator, and sourcing an output current $i_o = V_p(\partial C / \partial z)(\partial z / \partial t)$.

Also figure 1 shows the plot of i_o, v_o versus the frequency, traces out a bandpass biquad characteristic with a high-Q in vacuum ,which is very suitable for reference oscillators.

III. CLAMPED-CLAMPED MIXED-BEAM DESIGN

A. Resonance Frequency Design

In general ,the resonator beam width W_r is governed by transducer and length-to-width ratio design considerations, while its equivalent thickness $h_t = h_{t+} h_{2+...+} h_n$, where n is the number of the layers of the mixed beam. Thus, the length L_r becomes the main variable with which to set the overall resonance frequency.

For VHF designs, for which beam lengths begin to approach their thickness dimensions, the shear displacements and rotary inertias become more efficient. So to obtain accurate beam lengths for VHF μmechanical resonators, the Timoshenko design procedure is more appropriate. For a clamped–clamped mixed beam with uniform cross section, the resonance frequency of the CC-mixed beam (f_{nom}) can be obtained by solving the following equations [2]:

$$\tan\frac{\beta}{2} + \frac{\beta}{\alpha}\left(\frac{\alpha^2 + g^2(kG/E_t)}{\beta^2 - g^2(kG/E_t)}\right)\tanh(\alpha/2) = 0,$$

1-4244-0116-X/06/$20.00 ©2006 IEEE

$$\left.\begin{array}{c}\alpha^2\\\beta^2\end{array}\right\} = \frac{g^2}{2}\left[\mp\left(1+\frac{E_t}{kG}\right)+\sqrt{\left(1-\frac{E_t}{kG}\right)^2+\frac{4L_r^2 h_t W_r}{g^2 I_r}}\right] \quad (1)$$

Where $g^2 = \left[2\pi f_{nom}\right]^2 L_r^2\left(\frac{\rho_t}{E_t}\right)$

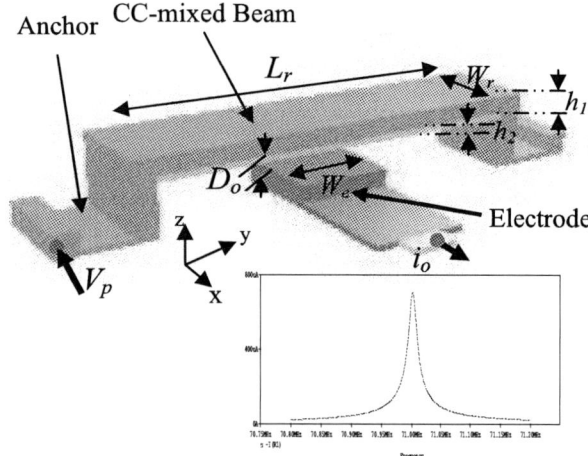

Fig. 1. Perspective-view schematic of a CC-mixed Beam ,with the output current of the resonator

and $I_r = \frac{W_r h_t^3}{12}$, and $G = \frac{E_t}{2(1+v_t)}$,while I_r is the bending moment of inertia, G is the shear modulus of elasticity, E_t is the equivalent Young's modulus of the mixed beam which can be calculated by equation (2),[3], v_t is equivalent Poisson's ratio which can be calculated by the same method of calculating the E_t, κ is the shear-deflection coefficient (for a rectangular cross section, κ is 2/3), ψ is the slope due to bending, and axis definitions are provided in Fig. 1.

$$E_t = \frac{\sum_n h_n E_n}{h_1 + h_1 + .. + h_n} \quad (2)$$

Where n is the number of the layers which compose the mixed beam.

When the support beams are designed as described above, the expression for resonance frequency f_o of the beam in Fig. 1 will actually be a:

$$f_o = f_{nom}\sqrt{1-\frac{k_e(y)}{k_m(y)}} \quad (3)$$

Where the f_{nom} is the resonance frequency without applying any voltage between the resonator beam and its electrode, and $k_e(y)$, $k_m(y)$ are the electrical and mechanical stiffness of the CC-mixed beam at the location y, and $k_e(y)/k_m(y)$ is a parameter representing the combined electrical-to-mechanical stiffness ratio integrated over the electrode width W_e ,and can be expressed as [4]:

$$\left\langle\frac{k_e(y)}{k_m(y)}\right\rangle = \int_{L_1}^{L_2}\frac{dk_e(y')}{k_m(y')} = \int_{L_1}^{L_2}\frac{(V_p^2 \varepsilon_o W_r}{[d(y)]^3 k_m(y)}dy \quad (4)$$

Where $L_1 = 0.5(L_r - W_e), and L_2 = 0.5(L_r + W_e)$ and L_r is the beam length .and $d(y)$ represent the deviation of the beam.

By applying the voltage between the beam and electrode the force which results will be as, $F = (V_p)^2 \partial C/\partial z$,where the $\partial C/\partial z = W_r W_e/(D_o)^2$, is the variance in the capacitance between the resonator beam and electrode.

B. Equivalent Lumped Parameter Mechanical Circuit

To conveniently model and simulate the Impedance behavior of this micro-mechanical resonator in an electro-mechanical circuit, an electrical equivalent circuit is needed [5], [6], [7]. As shown in Fig.3, both electrical and mechanical inputs and outputs are possible for this device, so the equivalent circuit must be able to model both. In addition, a circuit model that directly uses the lumped mechanical elements summarized by (6) is preferred. Fig. 3 presents the equivalent circuit used in this work, in which transformers model both electrical and mechanical

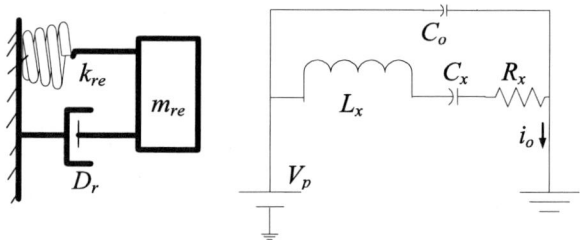

Fig. 3. The equivalent circuit for the μmechanical mixed resonator in mechanical and electrical domain.

couplings to and from the resonator, which itself is modeled by a core of LCR circuit with element values corresponding to actual values of mass, stiffness, and damping which is given by (5).

$$D_r(y) = \frac{\sqrt{k_m(y)m_r(y)}}{Q_{nom}} = \frac{k_m(y)}{2\pi f_{nom}Q_{nom}} \quad (5)$$

Where $D_r(y)$ is the damping factor at location y; and Q_{nom} is the quality factor of the CC-mixed beam without applying any voltage between beam and electrode.

The dynamic stiffness of the resonator is: $k_r(y) = k_m(y) - k_e(y)$,where $k_e(y)$ is the electrical spring constant, ,and $k_m(y)$ is the mechanical spring constant of the resonator without applying any external force on it.

Practically the transformed LCR circuit of the resonator elements values are given in the next equations:

$$C_x = \frac{\eta^2}{k_{re}} \qquad L_x = \frac{m_{re}}{\eta^2} \qquad, R_x = \frac{c_{re}}{\eta^2} \qquad, \eta = V_p\frac{\partial C}{\partial z} \quad (6)$$

The subscript e denotes the electrode location at $y = Lr/2$, R_x is the motional resistance seen a cross the electrical –to-resonator gap at resonance, and η is the electromechanical transformer turns ratio .

C. Pull-In Voltage V_{PI}

When the applied dc-bias voltage V_p is sufficiently large, the device will be broken down or failed due to either the breaking of the beam under the large force ,which is produced by the high dc voltage, or due to pulling down the resonator beam onto the electrode, and this leads to destruction of the device due to excessive current passing through the now shorted electrode-to-resonator path.

To calculate the pull-in voltage V_{pI} , the resonance frequency is set to be equal to zero in equation (3) and solve for finding V_p; but here unlike previous paper [8] ,

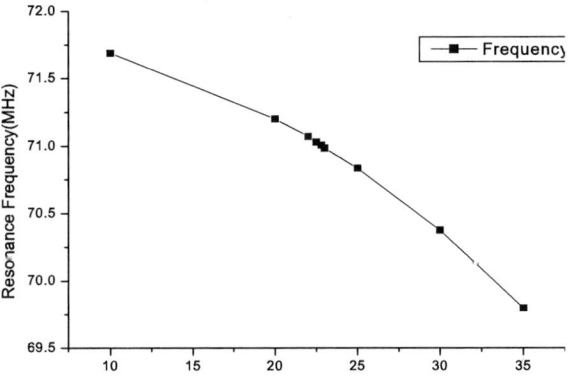

Fig. 5. the relation ship between the resonance frequencies of the mixed resonator and the dc-voltage V_P

the deviation of the beam is taking in our calculation; so while V_p is changed, the deviation of the beam $d(y)$ will be changed and as a result, the output of equation (4) will be changed, so in order to get the accurate value of the Pull-in-Voltage ,a certain closed loop in calculation should be used for each value of V_p to find the deviation $d(y)$, then solve for equation(3)=0.

IV. RESULTS AND ANALYSIS

Micromechanical CCMR is modeled and designed as detailed in Section II and III , using a polysilicon surface micromachining technology, the sacrificial oxide thickness in this process is 2 µm and the Silicon-nitride thickness is 0.12 µm ,the CC–mixed beam µmechanical resonator designed to be able to match 500 Å much more closely than [8] . Table I shows the dimensions of the reported device and its designed and simulated operating frequency by using *Intellisuite*, it is with 71 MHz resonance frequency. There were some variations from the simulated values which was gotten by *Intellisuite,* and the calculated values; we believe that much of the difference was caused by the residual stresses in the c-c mixed beam structure

which was not considered in our theoretical calculation of the designed frequency.

TABLE I
C-C-MIXED MMECHANICAL RESONATOR DESIGN
PARAMETERS AND PERFORMANCE SUMMARY

Parameter	Value		Unit
	Design	Input Simulated / *Intellsuitte*	
L_r	13	13	μm
Wr	8	8	μm
W_e	6	6	μm
h_1	2	2	μm
h_2	0.15	0.15	μm
D_o	500	500	$\mathring{A}$
Poisson Ratio ,v	0.27 Si$_4$N$_3$ 0.34 Cu	0.27 Si$_4$N$_3$ 0.34 Cu	……
Young Modulus, E	300 Si$_4$N$_3$ 122.5 Cu	300 Si$_4$N$_3$ 122.5 Cu	GPa
Density of polysilicon ,ρ	2784 Si$_4$N$_3$ 8916.5 Cu	2784 Si$_4$N$_3$ 8916.5 Cu	Kg/m^3

(A)

Parameter	Value		Unit
	Calculated	Simulated /*Intellsuitte*	
k_{re}(y=Lr/2)	4'/		kN/m
m_{re} (y=Lr/2)	0.232442805		pg
f_o	71.5	71.	MHz
V_p	30	30	$volts$
V_{PI}	132.5	129	$Volts$
Resonator Q,	6231 under 1 Torr	65 under atmosphere	……
C_0	8.5	9.2	fF
R_x	0.79		$k\Omega$

(B)

The loaded Mechanical quality factor Q_{load} of the CC-mixed beam, operating in atmospheric pressure was calculated using *Intellisuite*, which was around 65; while under low pressure of 1-Torr, Q_{load} was 6072; this difference is due to the increased damping caused by the air flow around the beam.

As the device is brought under vacuum, the Q and magnitude of the resonant peak increases ,so the resonators work well in air but a vacuum of less than 1 Torr is required for optimum performance [9].

Fig. 5 shows the relationship between the applied dc voltage V_p between the µresonator beam and its electrode, as shown, as the V_p increases the resonance frequency (f_o) decreases and this is due to the increasing in the deviation of the beam, until f_o becomes zero, that means breaking-down of the beam due to the large applied force on the beam at the Pull-In-Voltage V_{PI}.

Figures 6 shows the deviation of the mixed beam through vibration, along its length , using *Intellisuite* simulator,; from the figure, one can see the maximum deviation of the beam ,which is at the mid of the beam, is 5.66 Å.

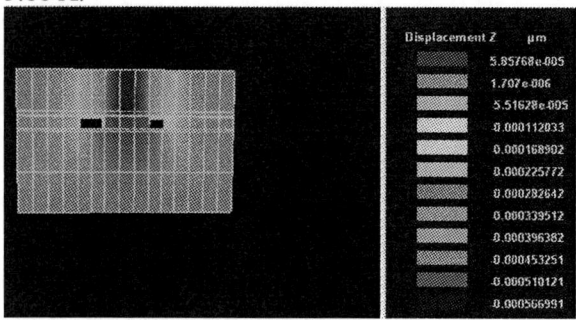

Fig. 6. Resonator displacement as along the mixed beam simulated by the *Intellisuite*.

Fig 7 plots the resonance frequency of the beam with conductive layer of different metals, versus the applied dc-voltage V_p .As shown, the beam with gold has the lowest resonance frequency, while the beam with aluminum has the highest resonance frequency of the beam and this is returned to the high value of the modulus and low mass density of the aluminum with respect to the Gold and

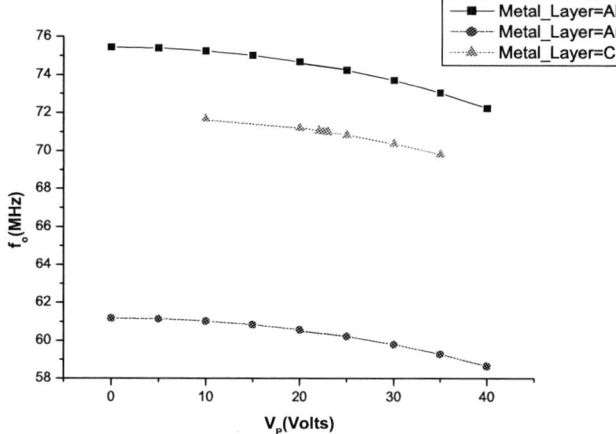

Fig. 7. Plots of both of motional resistance R_x and resonance frequency f_o with respect to the beam-to-electrode gap.

cupper. There is an important note here, the adhesion force between the two layers of the mixed beam is determined by the properties of the two layer's material, which will affect finally on the stability of the beam through vibrating, because through the vibration ,specially for high frequency, the metal layer may be separated a way from the high structural layer ,due to the attractive(vertical) force, between the electrode and beam, which may exceed the adhesion force between the two layers of the mixed beam; so it is important to choose the metal of low modulus with respect to structural material to prevent the layer's separation, and with low mass density to get a small mass for the mixed beam Thus, in our case, cupper was used due

its , relatively, low mass density compared with gold , and low Young's modulus compared with aluminum .

V. CONCLUSION

Some expected advantages of the C-C mixed beam presented in this paper are summarized: (I) a new design principle is proposed, by constructing a mixed resonator beam of two layers, one is an isolator- high structural layer (silicon-nitride) ,and the other is a conductive layer (cupper in our case) ,by this method, one can achieve higher resonance frequency with less motional resistance ,due the increasing in the effective resonator spring stiffness k_r ; while its effective mass m_r is les than that of single doped polysilicon, with the same resonance frequency.(II) the proposed mixed beam has higher Q than that of the doped polysilicon beam, with the same resonance frequency. (III) its easier to be fabricated than that of the domed polysilicon.

In general, the resonance frequency of this clamped-clamped mixed beam depends upon many factors, including geometry, structural material properties of its materials, stress, surface topography, and the magnitude of the applied dc-bias voltage V_P .

ACKNOWLEDGEMENT

This project is financially supported by the Ministry of Science, Technology and Environment of Malaysia under IRPA program: Development of MEMS Technology for Automotive Applications.

REFERENCES

[1] K. Wang, " VHF Free-Free Beam High-Q Micromech-anical Resonators " , *J. Microelectromechanical Systems*, pp.347-360 ,2000 .

[2] S. P. Timoshenko, *Vibration Problems in Engineering*, 1990.

[3] G. M. Rebeiz , *RF MEMS ,Theory ,Design, and, Technology* . Hoboken, New Jersy , John willy & Sons Inc.,2003.

[4] J. R. Clark "High-Q HF Microelectromechanical Filters,", *IEEE J. Solid-State Circuits*, pp. 512–526, 2000.

[5] R. T. Howe , "Resonant-Microbridge Vapor Sensor." *IEEE Trans. Electron Devices*, pp. 499–506, 1986.

[6] H. A. C. Tilmans, "Equivalent Circuit Representation Of Electromechanical Transducers: II ., " *J. Micromech. Microeng*, pp. 157–176, 1996.

[7] H. A. C. Tilmans and R. Legtenberg, "Electrostatically Driven Vacuumencapsulated Polysilicon Resonators, Part II: Theory and Performance," *Sens. Actuators A*, vol. 45, pp. 67–84, 1994.

[8] F. D. Bannon, "High-Q HF microelectromechanical filters," *IEEE J. Solid-State Circuits*,pp.512-526, 2000

[9] M. Dubey ,"Surface Micromachined Piezoelectric Resonant Beam Filters,"*U.S. Army Research Laboratory Adelphi, Maryland 20783-1197*

Session

Opto and Microwave Devices and ICs

Silicon Integrated Photonics for Microelectronics Evolution

Hei Wong, V. Filip, C. K. Wong, and P. S. Chung

Abstract—The advances in Si technology and the rapid growth of broadband communication via optical fiber allow silicon integrated photonics to begin revolutionizing the electronic devices, circuits, and systems. The pace of technological development has been recently speeded up. Using microfabrication technology we are now able to make waveguide structures and optical components of Si-based materials, such as silicon oxynitride or doped silica. Visible light can be obtained from Si-based materials such as Si quantum wire/dots and Si nanoclusters embedded in insulators. It is further demonstrated that lasing effect is also possible with the nanostructures. The impact of these moves will be revolutionary. Si integrated photonics will enable on-chip optical interconnects for future microprocessor and giga-scale circuits, chip-to-chip fiber interconnection and will greatly decrease the cost for fiber-to-home connection. This will be one of the major moves for the next technology revolution. The present article discusses some recent developments on these aspects.

I. INTRODUCTION

The microelectronic technology has now developed into the nanoscale range; the device size is approaching the size of atomic clusters, where the material properties differ dramatically from those of the bulk material. On one hand, this trend imposes constraints on the further downsizing of microelectronic devices [1-2], but, on the other hand, opens up various potential applications of quantum device structures [3-5]. One of these is the integration of microelectronics and photonics. Si integrated photonics has many merits. As the mainstream electronic devices are made of silicon, fabricating photonic devices on silicon is the most cost effective method for electro-optic system integration. With silicon technology, we can make any kind of geometry, e.g. optical cavities, 3D and motion structures, and nano structures. With nano geometries, even the quantum effects are possible in Si. On the other hand, silicon is the largest substrate available to date. It is particularly suited to large-scale integration and mass production, with excellent uniformity and reproducibility. In guided wave applications, the silicon-based optical devices have features of low insertion loss, polarization-independent behavior and efficient fiber pigtails. As silicon is a low-contrast material, a highly efficient fiber-chip coupling can be obtained [6-7].

The authors are with Department of Electronic Engineering, City University of Hong Kong, Tat Chee Avenue, Hong Kong, E-mail: eehwong@cityu.edu.hk

The successful development of micro-fabrication technology for nanoscale structures will also have a great impact on micro-optics. Optical systems in the past have been bulky with expensive optical components such as mirrors, filters, and beam splitters. These components require precise alignment in the micron scale. By employing microfabrication technology, optical micro-electro-mechanical systems (OMEMS) are possible and thus the size of present optical systems can be reduced greatly. Si-based optoelectronics is therefore low in cost, highly robust, and multifunctional. This leads to a bright future for inexpensive mass production of OMEMS and integrated photonics.

The revolution based on Si-integrated photonics is indeed not far away. Silicon materials can now also be used in light generation. Bulk silicon is not suitable for light emitting devices (LEDs) fabrication, as it is a non-direct band-gap material and present semiconductor lasers are based on low-dimensional compound semiconductors. This situation is changing. Quantum confinement of electrons and holes has been found to be possible in quantum dots and nanocluster silicon structures [8-9]. Strong luminescence can also be generated from radiative centers in silicon oxide and nitride. Intense research on the luminescence properties of silicon-based structures, including porous silicon (PS) [10-13], and silicon nanoclusters in amorphous SiO_2 [14-15], has been conducted. A wide range of luminescence, including the red-orange band and the blue band as observed [16]. A green line has also been found by using the silicon nitridestructure. This provides the possibility for fabricating full-color devices based on silicon technology [7]. Semiconducting nanowires have already been used to construct photodetectors [17] and light-emitting diodes (LEDs) [18].The potential for making silicon photonics and the possibility of making optical interconnect for future giga-scale microelectronics look very promising. The research is now spreading into industrial R&D. Intel has just announced its plan for integrating entire photonic devices onto future generations of microprocessors [19]. Here is their statement:

"Intel is focusing on ways to *siliconize* photonics and bring the benefits of Intel's volume manufacturing expertise to optical communications. Intel's goal is to make integrated, inexpensive photonic devices out of silicon instead of the exotic materials used today. By demonstrating how optical modulators can be made out of silicon using Intel's standard manufacturing processes in an existing fab, Intel researchers have removed a significant cost barrier in photonics. The next step is integrating entire photonic devices on a chip with digital intelligence. This should pave the way to produce photonics products based on silicon."

--- From Intel.com (2005)

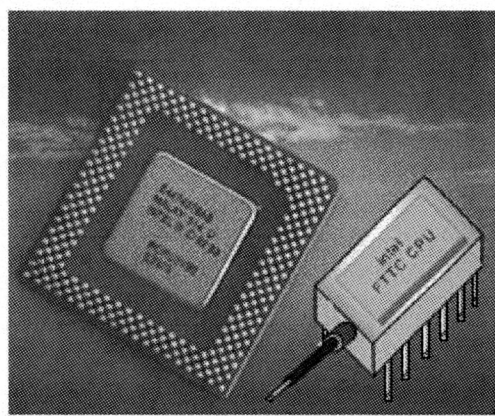

Fig. 1. Intel is focusing on ways to siliconize photonics. Future microprocessors may be optically connected to increase the speed and bandwidth. On-chip optical interconnection may also be introduced to resolve the interconnect complexity for giga-scale silicon microchips where hundreds of address lines, data lines and control lines between the functional blocks could then be replaced by a single optical waveguide.

II. LIGHTING FROM SILICON MATERIALS

The semiconductor laser is the key device for photonic applications. Bulk silicon is an indirect bandgap material and cannot be an efficient light emitter because the fast non-radiative recombination processes dominate the carrier transfer between the conduction and valence bands. The light sources currently available for optical communication are made of quantum well structures of compound materials. Because of a lattice constant larger than Si it is difficult to incorporate compound alloys into silicon microchips. As a result, the light emitters must be kept separate from the silicon-based processing devices. The separation of the light sources hinders the development of microelectronics-photonics integration as it is difficult to make precise alignment between the light sources and microelectronic circuits. In addition, the hybrid circuits are complicated and costly. The need for high capacity interconnect between the giga-scale integrated chips for teraflop power processing also calls for a Si-compatible optical emitter. Considerable effort has been devoted to developing efficient silicon-based light-emitters, and many significant achievements have been obtained. Recent results have shown that crystalline silicon can be an efficient light emitter with appropriate engineering. Many studies have found that light emission from silicon is possible in

low-dimensional Si structures (such as porous silicon [10-13], silicon nanocrystals, silicon/insulator superlattices [8-10], and silicon nano-pillars [20]) and in doped Si (with active impurities such as erbium) [21]. Fig. 2 depicts two example of visible light emitting from Si-based materials.

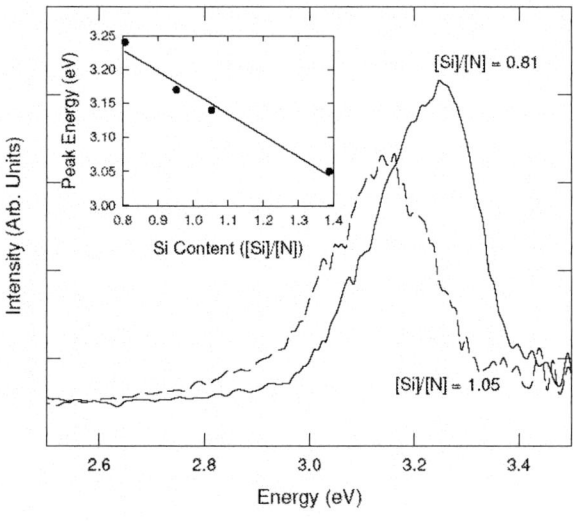

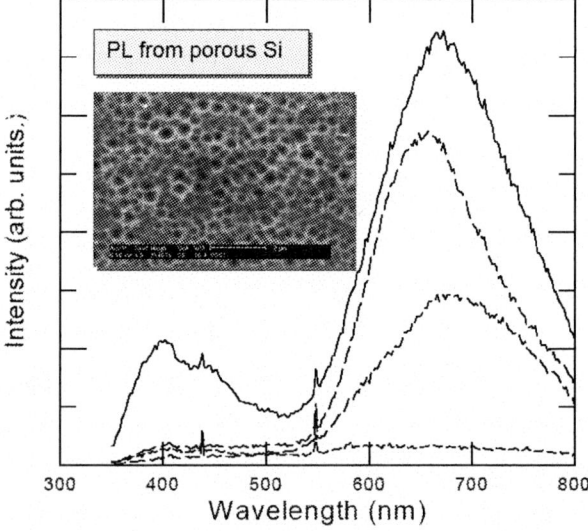

Fig. 2. Photoluminescence of silicon nitride embedded silicon nanoclusters (upper) and porous silicon (lower) can produce light in the UV and visible spectrum.

III. PROPOSED STRUCTURES FOR SI-LEDS

Several Si-based light-emitting structures have been recently proposed. Fig.3 gives some the light-emitting structures proposed for silicon-based materials. The light emission from a pn junction can be described by the generalized Planck's theory [22]. In this model, the electron and hole distributions are governed by quasi-Fermi levels. Photons with energy equal to the difference between the electrons' and holes' quasi-Fermi energies can be generated when electron-hole recombinations take place. To obtain a highly efficient light emission, the pn junction must have large carrier diffusion lengths, low surface recombination rates and negligible series resistance. Green and co-workers [23] recently reported a highly efficient Si-based LED with several novel features. In their design, inverted pyramids were etched on the top surface to enhance the absorptance and reduce reflection. The absorptance is further enhanced with the rear metal reflector. To reduce parasitic absorption, wafers are only moderately doped ($\sim 1.4 \times 10^{16}$ cm^{-3}). Meanwhile, parasitic recombination is minimized by fabricating the devices on float-zone grown wafers, which have large minority-carrier recombination lifetimes. The surfaces of the pn junction are passivated with high-quality thermal oxides and areas for metal electrodes are kept small. With this structure, the Si-based LED has a power conversion efficiency above 1%, which is close to the early compound semiconductor emitters.

Ng and co-workers [24] also described the fabrication of a silicon light-emitting diode (LED). In their proposed method, boron is implanted into silicon to dope the pn junction to introduce dislocation loops (see Fig.3(b)). With the dislocation loops, a local strain field in three dimensions is formed, which modifies the band structure and provides spatial confinement of the charge carriers in three dimensions and then electroluminescence. The dislocation loops (with diameters of about 80–100 nm) are spaced about 20 nm apart. A window was left in the large back AuSb contact for light emitting. Under forward bias, the device emits visible light.

By immersing Si in hydrofluoric acid, much of the material is etched away and a porous substance made up of a network of silicon nanowires is left behind. Canham [4] found that this porous silicon structure can emit visible light after laser illumination and explained this by the quantum confinement effect. Fauchet [25] even created a light-emitting diode (LED) using the porous Si. However the efficiency of electro-luminescence in PS-based devices is still too low (about 0.2%) to have any practical application. The main reasons for such a low power efficiency are the difficulties in injecting carriers from the contact into the PS and the poor transport properties in PS. As the blue PL has a measured lifetime of 1 ns [26], blue PS LEDs are not suitable for displays but are still useful in optical interconnects. Because the blue luminescence is believed to be due to defects in the oxide, luminescence efficiency can be improved with silicon oxide and silicon nitride structures [8]. In addition, good contacts and high current injection density can be achieved with silicon nitride based LED. Another problem blocking PS applications in LEDs is instability. The ageing phenomenon occurs when PS devices are exposed to the ambient. Indeed, the surface properties of porous silicon have profound effects on the characteristics of optoelectronic devices. As the surface layer is very unstable, the reliability of the devices is still one of the key problems. The quest for solutions to these issues is of great theoretical and practical importance. A

new method to produce porous polysilicon films (PPS) by plasma etching of the thermally oxidized polycrystalline silicon films was developed [27]. The major advantage of the plasma etching technique is that a native and stable silicon surface of the micro pores can be obtained.

Amplified stimulated emission and lasing have been reported for nanowires [28-32]. Duan et al. [28] investigated the feasibility of achieving electrically driven lasing from individual nanowires and demonstrated that a single nanowire can function as a standalone optical cavity and gain medium [32]. This approach can be readily extended to other nanowire materials such as Si. Pavesi [31] and his colleagues also produced quartz-embedded silicon nanocrystals (about 3 nm wide). They found that the nanocrystals do not only emit red light but also yield optical gain. This phenomenon opens a way for possible laser action in Si nanocrystals.

Quantum confinement is also possible in dielectric films with silicon dots/nanoclusters and it was suggested that Si nanocrystals embedded in silicon oxide are potential candidates for Si-based light emitting devices. Non-radiative recombination is always dominant in Si. Minimizing the carrier diffusion can be of help to enhance the radiative efficiency. Forming silicon clusters is thus a possible way for strong band-edge photoluminescence. The only achievable procedure is to incorporate these clusters inside large bandgap insulating materials. However, the insulating matrix prevents efficient carrier injection, making the devices difficult to be produced. The spatial fluctuation of the chemical composition in SiO$_x$ leads to the formation of potential wells where carrier confinement occurs. However, there are some constraints for the application of silicon oxide-based luminescence materials. According to the band structure, the field strength needed for electroluminescense in SiO$_x$ will be larger than 6 MV/cm for electron injection and more than 10 MV/cm for hole injection from Si into SiO$_2$. This field strength is very close to the breakdown field of silicon dioxide. This constraint can be resolved by employing the SiN$_x$ structure which has much smaller barriers at the Si-Si$_3$N$_4$ interface are 2.0 eV and 1.5 eV for electrons and holes, respectively. Fig.3(c) depicts the proposed tunneling injection Si/Nitride/low dimensional Si structure for light emitting devices based on this idea. The low dimensional Si could be silicon nanoclusters embedded in silicon oxide/nitride, silicon quantum dots or silicon quantum wires. This process is fully compatible with the existing MOS technology. In this tunneling injection LED structure, the low dimensional Si governs the photon generation efficiency and energy spectrum whereas the asymmetry of the barrier heights on the both sides (formed by the SiO$_2$ and Si$_3$N$_4$, respectively) provide high efficiency carrier injection based on direct tunneling and maximize the recombination events taking place in the low-dimensional silicon. To have a quantitative idea on the structural and functional parameters for the proposed device, numerical computations have been performed [33]. Figure 4 shows the computed recombination rate together with the total current density. The corresponding radiative power of the device was also estimated (lower inset of Fig. 4) by assuming 1% optical efficiency [23] which is governed by the optical properties of the nanostructure in the active region.

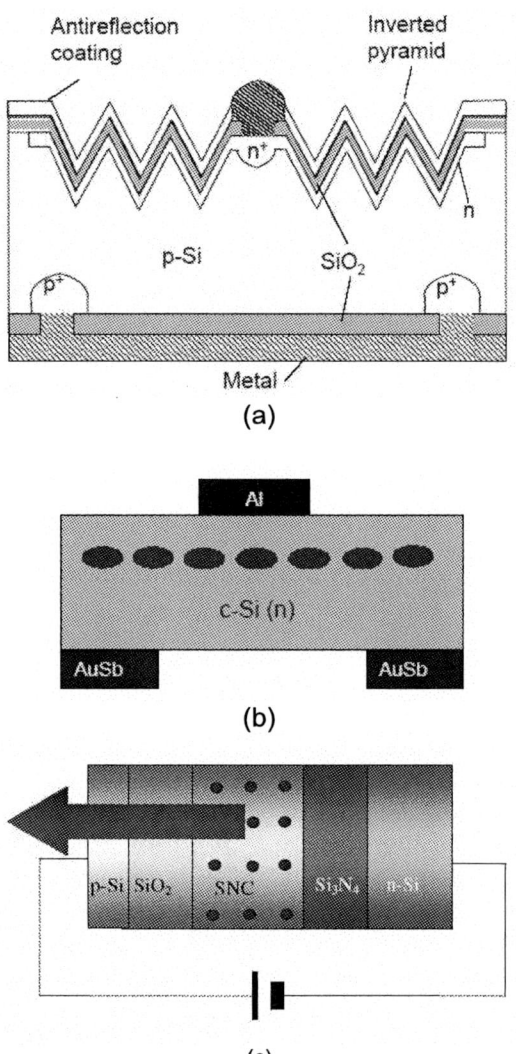

(a)

(b)

(c)

Fig. 3. (a) Careful design in device structure can enhance photon state occupation probability and minimize the non-radiative recombination [23]; (b) local strain fields are formed in the dislocation loops which modified the band structure and results in carrier confinement [24]; (c) tunneling carrier injection structure can enhance the injection efficient and the band-to-band recombination rate [33].

Further development of Si photonic devices is the silicon laser. The progress in this matter is also promising. Light amplification or stimulated emission is a necessary condition for laser action. However, stimulated emission is less efficient in bulk silicon because of the high free carrier

absorption efficiency and significant Auger saturation of the luminescence intensity at high power [34]. In addition, significant size-dependence of the radiative energies was found in Si nanostructures, resulting in large inhomogeneous broadening and optical losses [35]. The latest attractive development of Si-based laser was achieved by Intel's group [36]. Based on standard IC fabrication process, Rong et al demonstrated a compact all-silicon Raman laser on a single silicon chip using a 4.8 cm long S-shaped silicon waveguide which has an effective core area of about 1.6 mm^2. One of the waveguide facets was coated with a highly reflecting material to form the Raman laser optical cavity. Pumping the cavity with a pulse laser at a wavelength of 1536 nm, strong lasing at a wavelength of 1669.5 nm was observed [36]. Continuous wave Si laser is now under development and their ultimate goal for this project is to siliconize all the components of the transceiver for optical communication. Laser action in Si also possible by achieving stimulated emission in an efficient optical cavity. Micro-electro-mechanical systems (MEMS) can also be realized with the silicon microfabrication technology. With an adjustable cavity, using deformable Si_3N_4 membrane, a continuous range of tunable lasers can be obtained [3]. Therefore, the outlook of the progress in photonics-microelectronics integration is now quite optimistic.

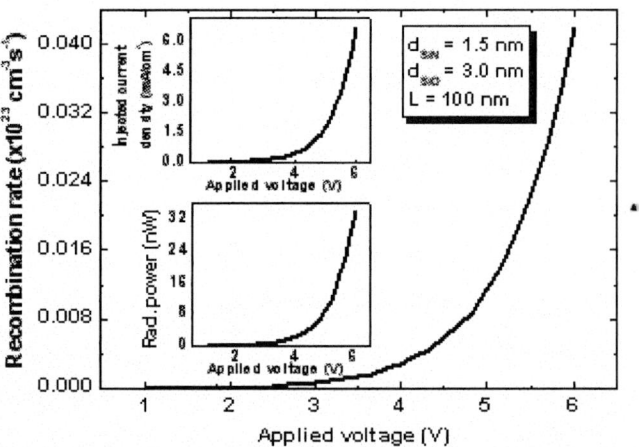

Fig. 4. Recombination rate, total current and radiative power for optimum values of the structural parameters of the proposed light emitting device. The radiative power was estimated for an assumed optical efficiency of 1 %.

IV. OXYNITRIDE WAVEGUIDE

Silicon oxynitride, having a wide range of tunable refractive index from 1.45 (SiO_2) to 2.0 (Si_3N_4) and being fully compatible with the existing mainstream CMOS technology, has been considered as the most promising material for integrated optical waveguides based on Si technology. The minimum allowable bending radius for this high-index-contrast material could be one order of magnitude smaller than that of silica (see Fig. 5) and is particularly suitable for compact integrated waveguide devices and optical interconnects. However, the propagation loss of this material is still a major concern. The optical loss in plasma enhanced chemical vapor deposition (PECVD) or LPCVD oxynitride is mainly due to the absorption loss (particularly in the 1460 – 1620 nm band) of inherent N-H bonds [3, 37]. Several methods for solving this drawback have been proposed. The hydrogen content of oxynitride film can be greatly reduced through thermal reoxidation of the as deposited silicon-rich silicon nitride or silicon oxynitride films [37, 38]. Figure 6 depicts the refractive index variation as a function of the NH_3 and N_2O flow rates for oxynitride films prepared with PECVD method. The refractive index changes from 1.65 to 1.48 as the N_2O flow rate decreases from 100 sccm to 500 sccm. The relationship between the refractive index and the N_2O flow rate is quite linear which enable an easy control of the process for a specific film [38]. Annealing conditions, both temperature and duration, have profound effects on the optical properties of the oxynitride film and can be used for fine tuning and process optimization for improving the waveguide properties. Figure 7 shows the annealing effect on the hydrogen content. Hydrogen bonds in oxynitride films do not only affect the refractive index but also are the major sources for absorption loss of optical transmission in the oxynitride waveguide. Gaussian decompositions of the Fourier transform infrared spectra reveal that the thermal annealing is particularly effective for removing the O-H related absorption bands. Annealing at 800 °C for 3 hr is already able to reduce the O-H bond density to the detectable limit. For N-H stretching vibration band a higher temperature (> 1000 °C) is required. The hydrogen content

structure for compact integrated waveguide devices and optical interconnects.
of the sample annealed at 1100 °C has been reduced down to 3.57×10^{20} cm^{-3} [39]. Preliminary measurements on the oxynitride channel waveguide found that the propagation loss is about 0.1 dB/cm.

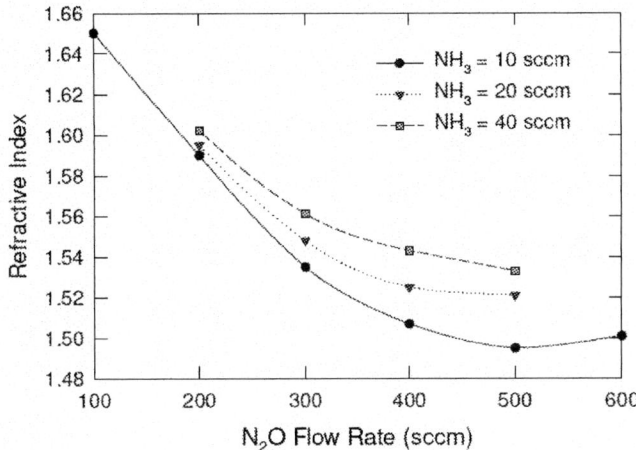

Fig. 6. The refractive index of oxynitride film prepared by the PECVD method can be readily tuned by varying the gas flow rate of nitric oxide and ammonia [39].

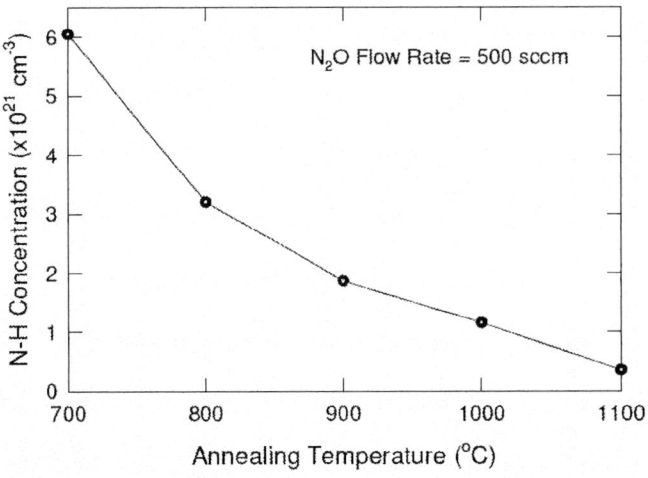

Fig. 7. Both N-H and O-H content of oxynitride waveguide can be greatly reduced by reoxidation of the as-deposited film at high temperatures, which provides a way for minimizing the propagation loss of the waveguide [39].

With the available processes for thickness, refractive index and hydrogen content control, low-loss channel oxynitride waveguides in the 1550-nm wavelength region were obtained. An example of ridge waveguide is shown in Fig.8. The cross-section and the refractive index of the core layer are 3×2.5 μm^2 and 1.55, respectively. The mode simulation shows that the ridge waveguide is able to support 9 modes for 1550-nm operation [39]. Since a large refractive index difference between the core and cladding

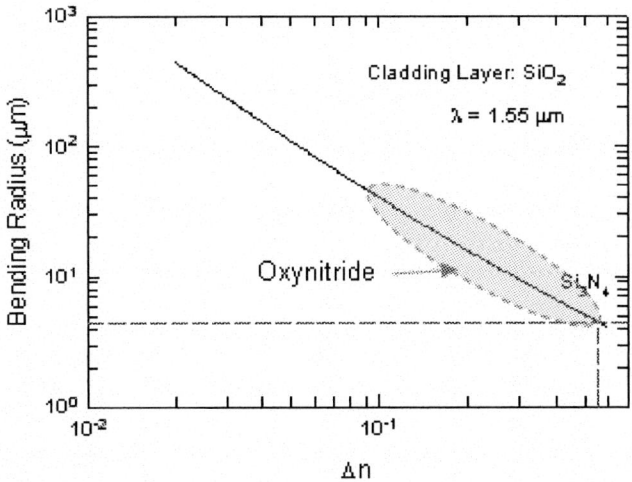

Fig. 5. Silicon oxynitride has a wide range of tunable refractive index from 1.45 (SiO$_2$) to 2.0 (Si$_3$N$_4$). The minimum allowable bending radius for this high-index-contrast material could be one order of magnitude smaller than of the silica and is an excellent

layers is achieved, this kind of waveguide devices has advantages of high contrast and compactness.

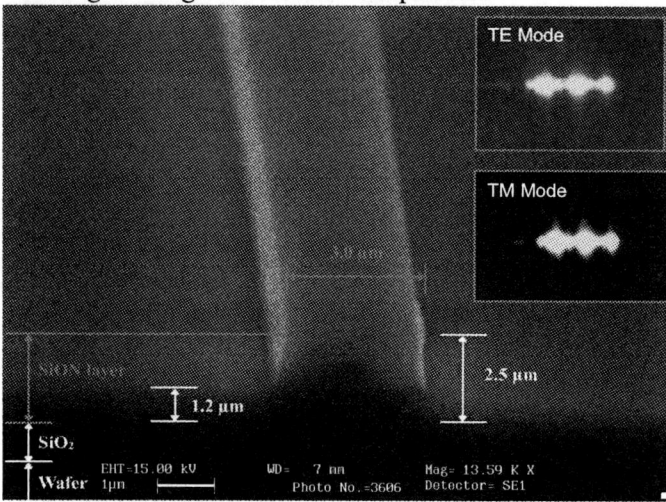

Fig. 8. A ridge oxynitride waveguide (without top cladding layer) with a large refractive index of 1.55 for the core layer. Both TE and TM modes transmission at 1550 nm were recorded. The large refractive index for the core layer makes the waveguide compact in size (smaller cross-section and bending radius) [39].

V. CONCLUDING REMARKS

Microelectronic technology has fueled the development of almost all technologies in the last four decades. It was predicted that the downsizing of microelectronic devices will end soon but the Moore's law of the chip density may still march forward for several decades because of the availability of new technological options for increasing transistor numbers and chip area without reducing the transistor size [1]. On-chip optical interconnection is the most attractive for future ultra-large scale Si microchips. Hundreds of address lines, data lines and control lines between functional blocks can be replaced by a single optical waveguide. An immediate impact is to provide low-cost Si-based wavelength filter for fiber-to-home applications. Si-based light emitting devices may be viable in several years' time. Other potential attractive applications include fiber-to-chip and light sources for fiber communications. Over the next 3-10 years, photonic components may be integrated with microelectronic circuits. By that time, the speed and bandwidth of computer and communication technologies will be revolutionized by the introduction of on-chip and inter-chip optical interconnects and the siliconized OEICs for optical communication.

Acknowledgement: The work described in this paper was support by grants from CityU (Project No. 7001513 and 7001704).

REFERENCES

[1] H. Wong and H. Iwai, "The road to miniaturization," *Physics World*, vol.18, No.9, pp.40-44, September 2005.

[2] H. Wong, "Silicon Integrated Photonics: Potentials and Promises (invited)," *11th IEEE International Symposium on Electron Devices for Microwave and Optoelectronic Application (EDMO 2003)*, Orlando, USA , 17-18 November 2003. pp.145-150

[3] H. Wong, "Recent Developments in Silicon Optoelectronic Devices," *Microelectron. Reliab.*, vol.42, pp.317-326, March 2002.

[4] L. T. Canham, "Silicon quantum wire array fabrication by electrochemical and chemical dissolution of wafers," *Appl. Phys. Lett.* Vol. 57, pp.1045-1048, 1990.

[5] L. Pavesi, "Silicon chips light up" *Physics World* vol.18, No.4, April pp25-26, 2005.

[6] A. Himeno, K. Kato, and T. Miya, "Silica-based planar lightwave circuits," *IEEE J Selected Topics Quantum Electron.*, Vol. 4, pp.913-924, 1998

[7] H. Hoffmann, P. Kopka, and E. Voges, "Low-loss fiber-matched low-temperature PECVD waveguides with small-core dimension for optical communication systems," *IEEE Photon. Technol. Lett.*, vol. 9, pp. 1238–1240, 1997.

[8] V. A. Gritsenko, K. S. Zhuravlev, A. D. Milov, H. Wong H, R. W. M. Kwok, J. B. Xu, "Silicon dots/clusters in silicon nitride: photoluminescence and electron spin resonance," *Thin Solid Films*, vol.353, pp.20-24, 1999.

[9] P. Mutti, G. Ghislotti, S. Bertoni, L. Bonoldi, G. F. Cerofoloni, L. Meda, E. Grilli and M. Guzzi, "Room-temperature visible luminiscence from silicon nanocrystals in silicon implanted SiO2 layers," *Appl. Phys. Lett.*, vol.66, pp.851-853, 1995.

[10] J. Piqueras, B. Méndez, R. Plugaru, G. Craciun, J. A. García, and A. Remón, "Cathodoluminescence from nano-crystalline silicon films and porous silicon," *Appl. Phys. Lett.*, vol.68, pp.329-331, 1999.

[11] H. Wong, P. G. Han, M. C. Poon, "Investigation of Silica Layer on the Porous Poly-Si Thin Films," *Microelectron. Reliab.*, vol. 41, pp.179-184, 2001.

[12] O. Bisi, S. Ossicini, and L. Pavesi, "Porous silicon: a quantum sponge structure for silicon based optoelectronics, *Surf. Sci. Rep.*, vol. 38, pp.1-126, 2000.

[13] G. D. Sanders and Y. C. Chang, "Theory of optical properties of quantum wires in porous silicon," Phys. Rev. B vol.45, pp.9202-9213, 1992.

[14] V. A. Gritsenko, Y. G. Shavalgin, P. A. Pundur, H. Wong, and W. M. Lau, "Catholuminescence and photoluminescence of amorphous silicon oxynitride," *Microelectron. Reliab.*, vol.39, pp.715-718, 1999.

[15] A. J. Kenyon, P. F. Trwoga, C. W. Pitt, G. Rehm, "The origin of photoluminiscence from films of silicon-rich silica," *J. Appl. Phys.*, vol.79: pp. 9291-9302, 1996.

[16] H. Wong and V. A. Gritsenko, "Defects in Oxynitride Gate Dielectric," *Microelectron. Reliab.*, vol.42, 2002, pp.597-605, 2002.

[17] J. F. Wang, M. S. Gudiksen, X. F. Duan, Y. Cui, and C. M. Lieber, "Highly polarized photoluminescence and photodetection from single indium phosphide nanowires," *Science*, vol. 293, pp.1455-1457, 2001.

[18] D. Leong, M. Harry, K. J. Reeson, K. P. Homewood, "A silicon/iron disilicide light-emitting diode operating at a wavelength of 1.5 nm," *Nature*, vol. 387, pp.686-688, 1997.

[19] A. G. Nassiopoulos, S. Grigoropoulos, and D. Papadimitriou, "Electroluminescent device based on silicon nanopillars," *Appl. Phys. Lett.*, vol. 69, pp.2267-2269, 1996.

[20] http://www.intel.com

[21] G. Franzo, F. Priolo, S. Coffa, A. Polman, and A. Carnera, "Room temperature electroluminescence from Er doped crystalline silicon," *Appl. Phys. Lett.*, vol.64, pp.2235-2237, 1994.

[22] P. Würfel, S. Finkbeiner, E. Daub, "Generalised Planck's radiation law for luminescence via indirect transitions," *Appl. Phys. A*, vol. 60, pp.67-70, 1995.

[23] M. A. Green J, Zhao, A. Wang, P. J. Reece and M. Gal, "Efficient silicon light-emitting diodes," *Nature*, vol. 412, pp.805-808, 2001.

[24] W. L. Ng, M. A. Lourencao, R. M. Gwilliam, S. Ledain, G. Shao and K. P. Homewood, "An efficient room-temperature silicon-based light-emitting diode," *Nature*, vol.410, pp.192-194, 2001.

[25] K. D.Hirschman, L.Tybekov, S. P. Duttagupta, and P.M. Fauchet, "Silicon-based visible light-emitting devices integrated into microelectronic circuits," *Nature*, vol.384, pp.338-341, 1996.

[26] L. Tsybeskov, J. V. Vandyshev and P. M. Fauchet, "Blue emission in porous silicon: oxygen-related photoluminescence," *Phys. Rev. B*, vol.49, pp.7821-7824, 1994.

[27] P. G. Han, H. Wong, A. H. P. Chan and M. C. Poon, "A novel approach for fabricating light-emitting porous polysilicon films" *Microelectron. Reliab.*, vol.42, pp. pp.929-933, 2002.

[28] X. F. Duan, Y. Huang, Y. Cui, J. F. Wang, and C. M. Lieber, "Indium phosphide nanowires as building blocks for nanoscale electronic and optoelectronic devices," *Nature*, vol. 409, pp. 66-69, 2001.

[29] M. S. Gudiksen, L. J. Lauhon, J. Wang, D. C. Smith and C. M. Lieber, "Growth of nanowire superlattice structures for nanoscale photonics and electronics," *Nature*, vol.415, pp.617-620, 2002.

[30] M. Kazes, D. Y. Lewis, Y. Ebenstein, T. Mokari, and U. Banin, "Lasing from semiconductor quantum rods in a cylindrical microcavity," *Adv. Mater.*, vol.14, pp.317-321, 2002.

[31] L. Pavesi, L. Dal Negro, C. Mazzoleni, G. Franzo and F. Priolo, "Optical gain in silicon nanocrystals, *Nature*, vol.408, pp.440-444, 2000.

[32] A. M. Morales, and C. M. Lieber, "A laser ablationmethod for the synthesis of crystalline semiconductor nanowires," *Science*, vol.279, pp.208-211, 1998.

[33] H. Wong, V. Filip, D. Nicolaescu, P.L. Chu, "High-efficiency light emitting device based on silicon nanostructures and tunneling carrier injection," *J. Vac. Sci. Techno. B*, vol. 23, pp.2449-2456, 2005.

[34] M. V. Wolkin, J. Jorne, P. M. Fauchet, G. Allan, and C. Delerue, "Electronic states and luminescence in porous silicon quantum dots: the role of oxygen," *Phys. Rev. Lett.*, vol. 82, pp.197-200, 1999.

[35] P. Blood, "On the dimensionality of optical absorption, gain and recombination in quantum-confined structures," *IEEE J. Quantum Electron.*, vol.36, pp.354-362, 2000.

[36] H. Rong, A. Liu, R. Jones, O. Cohen, D. Hak, R. Nicolaescu, A. Fang and M. Paniccia, "An all-silicon Raman laser," *Nature*, vol.433, pp.292-294, 2005.

[37] H. Wong, M. C. Poon, Y. Gao, T. C. W. Kok, "Preparation of thin dielectric film for non-volatile memory by thermal oxidation of Si-rich LPCVD nitride," *J. Electrochem. Soc.*, vol.148, pp.G275-278, 2001.

[38] C. K. Wong, H. Wong, C. W. Kok and M. Chan, "Silicon oxynitride prepared by chemical vapor deposition as optical waveguide materials," *J. Cryst. Growth*, in press.

[39] C. K. Wong, H. Wong, M. Chan, C. W. Kok and H. P. Chan, "Minimizing hydrogen content in silicon oxynitride by thermal oxidation of silicon rich silicon nitride," *Microelectron. Reliab.*, in press

2006 25th International Conference on Microelectronics

Diagnostics of Homogeneity of Individual Layers of Large-Area Silicon Solar Cells Using Local Irradiation

V. Benda, Z. Macháček and J. Salinger

Abstract - This paper deals with the possibility of checking the recombination rate distribution over the area of power (large-area) solar cells from measured values of open circuit voltage V_{OC} using local irradiation by monochromatic light of different wavelengths (LBIV – Light Beam Initiated Voltage). This method can provide information both about the distribution of the recombination centres in large-area solar cells and the surface recombination rate at the antireflection coating. From the V_{OC} distribution, also position and extent of local defects can also be determined. The method can be used to investigate the influence of technology on characteristics of solar cells as an in-process checking with the aim of increasing efficiency and reliability of solar cells.

I. INTRODUCTION

Important work is now being done on improving the production of solar cells and making them more effective, so that solar energy will be ale to compete with others energy sources. This is in line with the world-wide search for renewable energy sources

Low concentration of the recombination centres in the volume of crystalline silicon solar cells is very important for the fabrication of high-efficiency solar cells. Fast non-destructive diagnostics of solar cells can provide information both about both the input material and the quality of the technological operations.

This paper deals with the possibility of checking the recombination rate in the individual layers of a solar cell from measuring both short circuit current I_{SC} and open circuit voltage V_{OC} under conditions of local illumination by monochromatic light of a suitable wavelength (LBIC – Light Beam Initiated Current, LBIV – Light Beam Initiated Voltage). By positioning the illuminated spot, it is possible to obtain a map of I_{SC} or V_{OC} from which we can find the distribution of the recombination in the individual layers of the cell structure. Using infrared light, the method can give information about recombination centres distribution in large-area solar cells. From the I_{SC} or the V_{OC} distribution, the position and extent of local defects can also be determined. Shorter wavelengths can give information about recombination rate in surface layers. The method of solar cell diagnostics using local irradiation described in the paper presented is very simple and can be used to

V. Benda, Z. Macháček, and J. Salinger are with the Department of Electrotechnology, Faculty of Electrical Engineering, Czech Technical University in Prague, Technicka 2, 166 27 Praha 6, Czech Republic, E-mail: benda@fel.cvut.cz

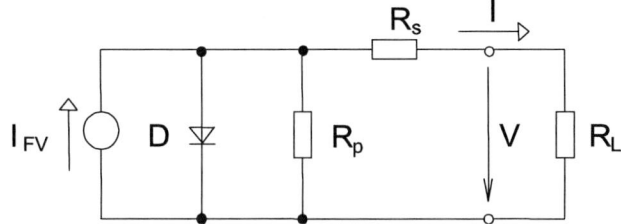

Fig.1. Eequivalent circuit of a solar cell

investigate the influence of technology on the homogeneity of solar cells and consequently to help to increase the efficiency and reliability of solar cells.

II. PRINCIPLE OF THE METHODS

Both LBIV and LBIC methods are based on measuring either open circuit voltage V_{OC} or short circuit current I_{SC} under conditions of local illumination by monochromatic light of a suitable wavelength. By positioning the illuminated spot it is possible to obtain a map of either V_{OC} or I_{SC}, from which we can find the distribution of the recombination rate in the individual layers of the cell structure.

Usually, solar cells are modelled by an equivalent circuit as shown in Fig.1. The characteristic of a solar cell with series resistance R_s and parallel resistance R_p and illuminated cell area A can be expressed as [1]:

$$I = AJ_{FV} - I_{01}\left[\exp\left(e\frac{V+R_sI}{kT}\right)-1\right] - I_{02}\left[\exp\left(e\frac{V+R_sI}{2kT}\right)-1\right] - \frac{V+R_sI}{R_p}$$
(1)

where I_{01} represents the diffusion component of the p-n junction reverse current, and I_{02} is the generation-recombination component of the p-n junction reverse current. J_{FV} is the density of current generated within the solar cell structure. For short circuit current (V = 0), it is possible to find out

$$I_{SC} = AJ_{FV} - I_{01}\left[\exp\left(e\frac{R_sI}{kT}\right)-1\right] - I_{02}\left[\exp\left(e\frac{R_sI}{2kT}\right)-1\right] - \frac{R_sI}{R_p} \cdot$$
(2)

For open circuit voltage it can be derived (supposing a high parallel resistance R_p)

$$V_{OC} = \frac{2kT}{e}\ln\left(\frac{-I_{02}+\sqrt{I_{02}{}^2+4I_{01}(I_{02}+I_{01}+AJ_{FV})}}{2I_{01}}\right).$$
(3)

1-4244-0116-X/06/$20.00 ©2006 IEEE

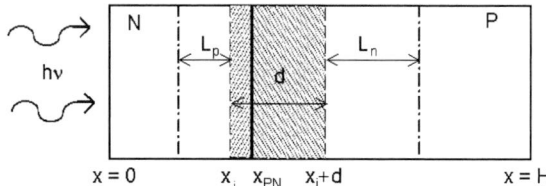

Fig. 2. The structure of a solar cell.

From that follows that both short circuit current I_{SC} and open circuit voltage V_{OC} increase with the current density J_{FV}; I_{SC} depends on linearly on the current density J_{FV} while V_{OC} depends logarithmically on the current density J_{FV}.

The solar cell structure of thickness H is shown in Fig. 2. If monochromatic light of the wavelength λ is applied, the current density J_{FV} is generated [3],

$$J_{FV}(\lambda) = e \int_0^H G(\lambda)dx - e \int_0^H \frac{\Delta n}{\tau}dx - J_{sr}(0) - J_{sr}(H) \cdot \quad (2)$$

The first term on the right hand side represents the current density generated by the incident light. The second term represents the recombination in the volume of the device, and $J_{sr}(0)$ and $J_{sr}(H)$ represent surface recombination. In the P-base of the c-silicon cell structure, the carrier lifetime is influenced practically only by recombination centre concentration N_t, $\tau \sim 1/N_t$. In the highly doped N^+ layer, the Auger recombination predominates and the carrier lifetime depends in practice only on the donor concentration N_D (for 300 K the Auger recombination coefficient in the N –type layer $C_{An}= 2.8 \times 10^{-43}$ $m^{-3}s^{-6}$ [6])

$$\tau \approx \frac{1}{C_{An}N_D^2} \cdot \quad (6)$$

Therefore, the recombination rate in the N^+-layer is strongly influenced by the both the donor concentration and junction depth x_j.

The current density generated by light J_{FV} consists of the hole current density J_p generated in the N^+-layer and the electron current density J_n generated in the P-base. Assuming that the carrier lifetime in the N^+-layer is mostly influenced by Auger recombination, the hole current density J_p can be influenced by both the recombination in the N^+-layer and the surface recombination. The electron current density J_n can be influenced practically only by the recombination on the local centres (the recombination rate at the back front contact can have an influence only for application of light with a wavelength over 1 µm).

Non-uniformity in either the generation or the recombination rate over the area of the solar cell results in a non-uniform distribution of J_{FV} and consequently, in non-uniform distribution of both V_{OC} and I_{SC} under conditions of local illumination [2].

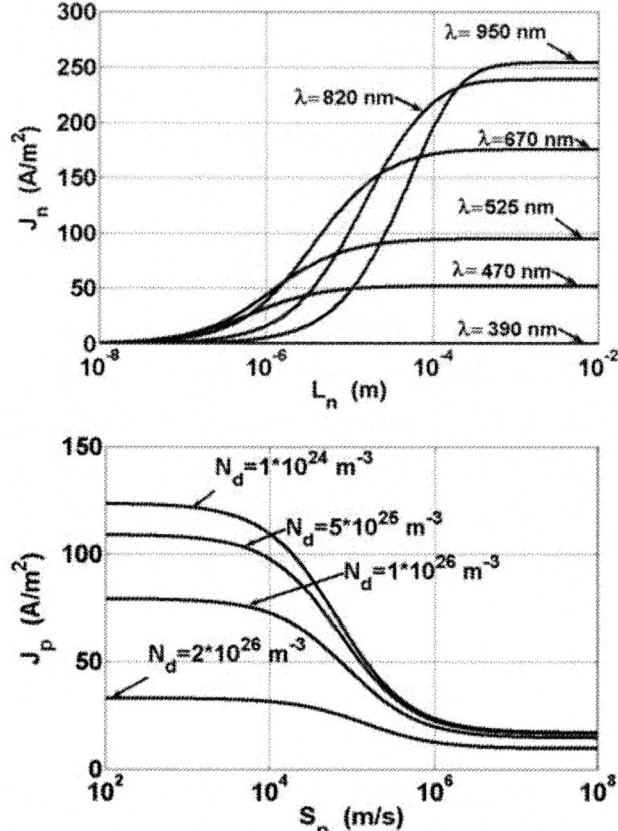

Fig. 3. Simulation results for a structure with $x_j = 0.5$ µm.
a) Current densities initiated by light of different wavelengths (950 nm, 820 nm, 670 nm, 525 nm, 470 nm and 390 nm). The influence of the electron diffusion length L_n in P base on the electron current density J_n
b) The influence of the hole surface recombination rate S_p and the donor concentration N_d in the N^+ layer on the hole current density J_p ($\lambda = 470$ nm).

As already shown in previous papers [4],[5], the measured distribution depends on the incident light wavelength. The use of different wavelengths of incident light allows different types of important information to be obtained about non-uniformity in the recombination rate in different parts of the cell structure. If infrared light is applied, the absorption length is relatively long (e.g. about 60 µm at λ = 950 nm) and the measured distribution of V_{OC}, or I_{SC} corresponds to the distribution of recombination centres in the bulk, and the influence of surface recombination and recombination in the thin high doped N^+-layer is nearly negligible. Infrared light of wavelength of 820 nm provides information about recombination rate at a lower distance (about 12 µm) form the surface. If visible light is applied, the absorption length is much shorter (about 3µm for λ=635 nm, about 1µm at λ=500 nm and about 0.5 µm in the case of λ=470 nm) and excess carriers are generated in the close vicinity of the p-n junction or in the N_+ layer. Recombination in the highly doped region and surface have a much greater influence than in the case of incident infrared light, and the influence of bulk recombination is recombination much lower. Ultraviolet light generates excess carriers close to the surface (about 0.1 μm at $\lambda = 390$ nm) and the resolution depends strongly on the p-n junction depth (as demonstrated in Fig..4.), donor concentration in the N^+-

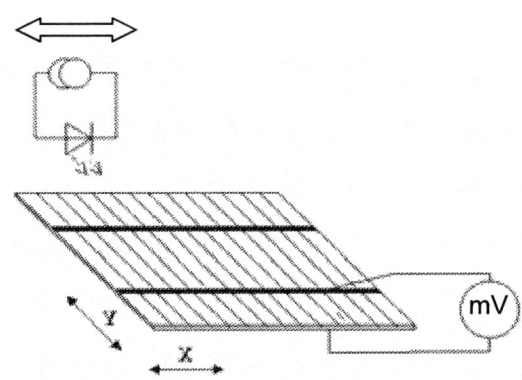

Fig. 5. Schematic diagram of the LBIV method.

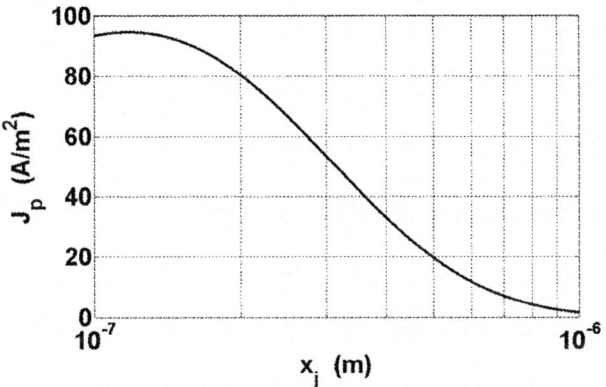

Fig.4. Dependence J_p on the junction depth x_j for $N_D = 10^{25} m^{-3}$ and $\lambda = 390$ nm.

layer and on the surface recombination rate. Therefore the maps of both V_{OC} and I_{SC} depend on the wavelength of the incident light.

III. EXPERIMENTAL RESULTS AND DISCUSSION

At our department, an equipment for measuring large-area solar cells at six different wavelengths has been constructed. The basic principle is shown in Fig.5. A PM8154 X-Y plotter was adapted to obtain the positioning system. The electrostatic paper holder was replaced by a table with an electrode system. The writing system of the plotter, i.e. namely holders of the writing pencil, were used as a system that holds and gradually moves the laser (or

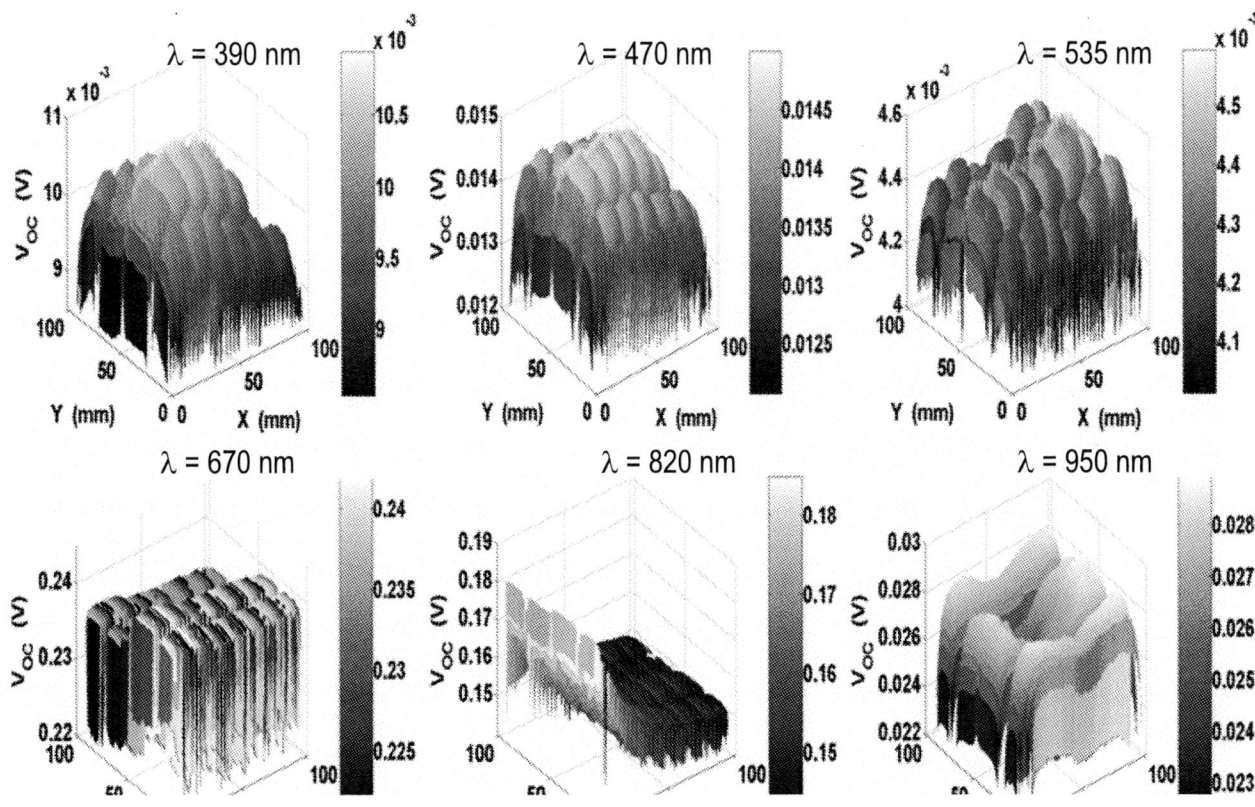

Fig. 6. An example of a set of LBIV maps obtained measuring a solar cell sample using six different wavelengths of incident light.

LED) diode which radiates light approximately 5mm in distance above the surface of a solar cell. During the measuring process, the plotter holds the laser (or LED) diode, which radiate an area with diameter of about 3 mmn on the cell surface. The holder moves gradually with the help of computer program from the bottom left beginning of a solar cell area to the last upper right side of the cell area in a programmable distance (e.g. 1 mm). The positioning accuracy is greater than 0.1 mm. The programme enables the source of light to be changed (maximum 6 different sources of light), so measurements using different wavelengths can be performed in exactly the same position. Since the digital voltmeter KEITHLEY is connected to the bus of the solar cell and then to the computer, it is easily possible to manage the system and measure the value of open circuit voltage V_{OC} at each point with in a relatively short time. The measured data are organised in a way that can be processed using EXCEL or MATLAB.

An example of distributions V_{OC} measured by the LBIV method forat one sample using different wavelengths of the light spot is shown in Fig.6. When short wavelengths are applied (ultraviolet, blue, green), the measured value of V_{OC} is strongly influenced by Auger recombination and the V_{OC} distribution is influenced mainly by inhomogeneity (in thickness or concentration) of the diffused N^+- layer (ultraviolet light can also provide information about the surface recombination rate). The results can be interpreted as a result of a temperature gradient in diffusion equipment during phosphorous diffusion (a thinner N^+-layer in the centre of the wafer).

Results obtained using longer wavelengths ($\lambda > 670$ nm) are more influenced by the recombination rate in the bulk of the cell within the region of thickness of the absorption depth of the light used. Using red light the sensitivity of the method (both LBIC and LBIV) is limited because carriers are generated close to the PN junction and gives resolution from a layer of less than 5 μm from the surface and about 15% of the current density is generated in the N^+ layer. In the case of infrared light of $\lambda = 820$ nm, the map obtained gives the resolution from a layer of about 20 μm from the surface. Therefore, application of incident light with a wavelength of 820 nm and less can provide information about diffusion length distribution in the P base close to the N^+P junction. As demonstrated in Fig. 6, the pattern obtained using $\lambda = 820$ nm can indicate a slightly damaged layer as a residuum after wafer fabrication (the surface damaged layer after cutting the rod was probably not fully etched off). On the other hand, the map obtained using $\lambda = 950$ nm reflects a carrier lifetime distribution deeper in the bulk of P base, because only 10%

of generated current density J_n can be influenced by the surface (about 5 μm in thickness) damaged layer and this map reflects the carrier lifetime distribution (striations [7]) in the original crystal.

This way, using different wavelengths of incident light we can determine uniformity of N^+-layer, defects in the area close to the P-N junction and the carrier lifetime distribution in the bulk of the cell. This may provide important information about the quality of the starting material and individual steps of the technological process.

IV. CONCLUSION

Light beam measuring techniques (LBIC, LBIV) are often used for mapping solar cell homogeneity. The resolution depends strongly on the wavelength of the incident light. Application of short wavelength incident light (ultraviolet, blue) provides information about the homogeneity of the N^+ layer. Maps obtained using a longer wavelength (600-800 nm) reflect the distribution of defects in the surrounding of the N^+P junction (a layer of several microns below the cell surface). Information about the carrier lifetime distribution in the bulk of the cell (the P base) can be obtained using an incident light wavelength longer than 900 nm. For a complex analysis to evaluate the technological process, the cell structure should be mapped using several wavelengths of incident light taking into account the part of the cell structure to be checked.

REFERENCES

[1] A. Goetzberger, J. Knobloch and B. Voss: *Crystalline silicon solar cells,* J.Wiley & Sons. (1998)

[2] S. M. Sze: *Physics of Semiconductor Devices, 2nd Edition,* John Wiley & Sons, 1981

[3] S. J. Fonash, *Solar Cell Device Physics,* Academic Press, 1981

[4] V. Benda and A. Asresahegn.: Diagnostics of Large-Area Solar Cell Homogeneity Using LBIV Method., Proc. *24th International Conference on Microelectronics.*MIEL´04, Niš, 2004, vol. 1, . 671-674.

[5] V. Benda: Diagnostics of Homogeneity of Recombination Rate in Individual Layers of Large-Area Silicon Solar Cells Using LBIV Method, Proc. *20th Europeand Photovoltaic Solar Energy Conference, Barcelona* 2005, s. 670-673.

[6] J. Dyiewior and W. Smith, Auger coefficients for highly dopped and highly excided silicon, Appl. Phys. Lett., 31 (5), pp.346-348, 1977

[7] K.V. Ravi: *Imperfections and impurities in semiconductor silicon,* J.Wiley & Sons, New York, 1981

2006 25th International Conference on Microelectronics

Improved Dual Grating-Assisted Directional Coupler for Silicon Nanophotonics

G. Z. Mashanovich, V. M. N. Passaro, G. J. Ensell, F. Y. Gardes, and G. T. Reed

Abstract - In this paper we present preliminary experimental results of an improved dual grating-assisted directional coupler. As the top SiON layer is thicker, the total insertion loss is reduced, and coupling efficiency significantly increased in the L-band wavelength region, compared to the previously reported results. The coupling efficiency is in the 42-56% range for both C- and L-band wavelength regions. Temperature dependence of the coupler has been also investigated.

I. INTRODUCTION

Silicon photonics is experiencing a dramatic increase in interest due to emerging applications areas and several high profile successes in device and technology development. Despite early work dating back to the mid 1980s, dramatic progress has been made in recent years [1-3]. Whilst many approaches to research have been developed, the striking difference between the work of the early to mid 1990s, and more recent work, is that the latter has been associated with a trend to reduce the cross sectional dimensions of the waveguides that form the devices. Whilst this reduction enhances the performance of the photonic circuit, it makes coupling of light to/from the circuit very difficult, particularly to/from standard optical fibres that typically have a core dimension of ~ 9 μm. Consequently several devices have been proposed for efficient coupling to/from the silicon photonic circuit [4-6] but none have yet produced satisfactory performance. We have previously presented theoretical work of an alternative device entitled the Dual Grating Assisted Directional Coupler (DGADC) [7] and recently experimental results from the first batch of devices that were fabricated in Silicon on Insulator (SOI) technology [8].

The light from a fibre is coupled to the thick silicon oxynitride waveguide (with refractive index close to that of the fibre), and via the first grating to the intermediate silicon nitride layer, and after that to the thin silicon waveguide using the second grating (Figure 1). Coupling lengths, periods, depths and duty cycles are generally different for the two gratings. The silicon nitride

G. Z. Mashanovich, F. Y. Gardes, and G. T. Reed are with the Advance Technology Institute, School of Electronics and Physical Sciences, University of Surrey, Guildford, Surrey, GU2 4EL, UK, E-mail: g.reed@surrey.ac.uk

V. M. Passaro is with the Dipartimento di Elettrotecnica ed Elettronica, Politecnico di Bari, Via Orabona 4, 70125, Bari, Italy

G. J. Ensell is with the School of Electronics and Computer Sciences, University of Southampton, SO17, 1BJ, UK.

waveguide is crucial for the operation of the coupler because it bridges the gap between the fibre and silicon refractive indices. The thick input waveguide and the two separation layers are fabricated in silicon oxynitride technology because of the possibility to control the refractive index over a broad range.

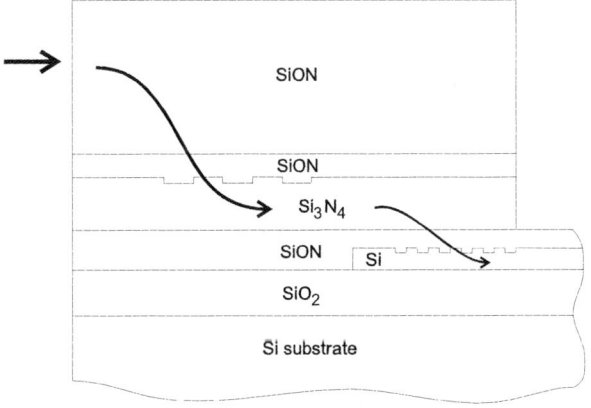

Fig. 1. Dual grating-assisted directional coupler in SOI technology.

Theoretical investigation of the coupler has shown that coupling efficiency in excess of 90% can be obtained [7], while measured efficiency for coupling to 230 nm thick silicon waveguide was 55% [8], which is the best result for grating-based coupling to such small semiconductor waveguides reported to date. In this paper we present preliminary experimental results of the second batch of devices with a thicker top SiON layer. The coupling efficiency has been improved by more than 10% at the wavelength of ~1.6 μm, compared to the first batch.

II. EXPERIMENTAL RESULTS

Unibond four inch wafers with a 3 μm buried oxide layer and a 230 nm silicon overlayer were used for the fabrication of the devices. Both 'single' (Fig. 1) and 'double' DGADCs [8] were fabricated. The former was less convenient for measurement of the coupling efficiency than the latter because of a highly divergent beam coming out from the thin silicon waveguide. ICP etching was used to define the silicon waveguide. Gratings with a period of 1.3 - 1.4 μm and 10 nm height were then patterned by plasma etching on the top of this waveguide. Plasma Enhanced Chemical Vapour Deposition (PECVD) was

1-4244-0116-X/06/$20.00 ©2006 IEEE   289

used for the fabrication of SiON layers. The films were deposited at 300°C with varying flow of nitrous oxide, and constant flows of 50 sccm (standard cubic centimeters) of ammonia and 155 sccm of 5% silane in nitrogen. Silane and ammonia were used as reactants for the Si_3N_4 deposition process; ammonia instead of nitrogen was chosen as a reactant gas because it provides better thickness and refractive index uniformity [9]. The pressure was 0.5 Torr, and silane gas flow five times as large as that in the SiON deposition. The grating on this layer has a larger period (≈ 5 µm) and height of 20 nm. The top three layers were then etched in CHF_3 plasma down to the second gap layer.

As in reference [8], the efficiency was determined by normalising to the light transmitted through a straight surface SiON waveguide, which is almost identical to the 'double' DGADC structure, on the same chip. It is, however, continuous, unlike the 'double' DGADC which has a silicon waveguide in the central region, and the straight waveguide does not have gratings so no coupling occurs. This approach was adopted because this excludes the reduced input coupling efficiency to the surface SiON waveguide, due to its non-optimal thickness, but still demonstrates the enhancement due to the DGADC. This approach has been also used extensively by other authors (e.g. in [4, 10]).

The thickness of the top SiON layer in the first batch was only 3.8 µm, hence limiting the maximum coupling efficiency to 60%. In this batch, the top SiON layer is ≈ 4.4 µm thick, which reduces insertion loss when coupling the waveguide with an optical fibre and increases maximum theoretical coupling efficiency to ~70% when the other layers have optimum thicknesses. Further improvement of the coupling efficiency can be expected for the SiON thickness of 5 µm.

Each sample contains a number of devices with different grating periods, enabling a shift of the resonant peak towards longer or shorter wavelengths still achieving high coupling efficiency. Preliminary results show that the maximum coupling efficiency in C-band is 56%. This maximum efficiency has been measured at the wavelength of 1560 nm, while previously efficiency of 55% was measured around 1540 nm. As for the L-band, the best measured efficiency was only 32% previously, and is 43% now, representing an improvement of 11% at a wavelength of ~1600nm. The fact that the improved efficiencies occur at longer wavelengths (1560 nm and ~1600 nm) rather than at the wavelength of 1550 nm, for which targeted layers' thicknesses were optimised, indicate that the silicon nitride thickness, for the samples examined so far, might be larger than optimal. Therefore, efficiencies in the C-band have not exceeded 56%, consistent with increased SiON waveguide thickness. Further investigation of the silicon nitride thickness will be carried out in order to confirm this assumption.

The temperature dependence of the resonant peak has been also measured (Fig. 2). Temperature of the samples was varied from 20 to 60°C. As the refractive index changes with temperature, a wavelength shift of the resonant peak can be expected. Consequently, coupling efficiency will also change. For a temperature change of $\Delta T=40°C$, the peak shifts by 5.6 nm. For $\Delta T=15°C$ the output at the resonant wavelength for $T=20°C$, drops by 3 dB. In other words, temperature variations of $\pm15°C$ results in output decrease of ≤ 3 dB at fixed wavelengths. To increase this temperature tolerance, the FWHM of the resonant peak has to be broadened. Work is underway to broaden the spectrum of the devices by grating chirping and by varying the duty cycle of the gratings.

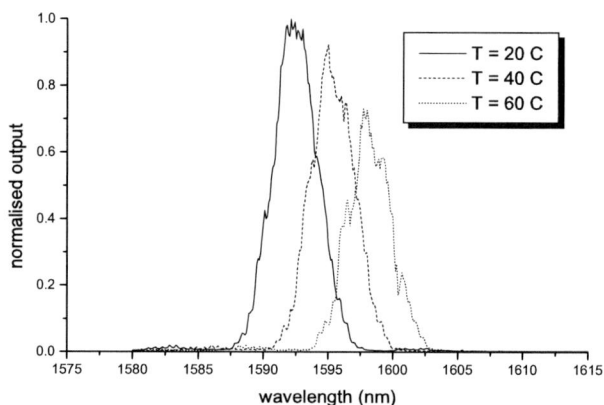

Fig. 2. Temperature dependence of the resonant peak.

We have already shown that the Transfer Matrix Method (TMM) can be used for design and analysis of the DGADC [7]. Using the TMM and the following thermo-optic coefficients:

$$\beta_{SiO2} = \beta_{SiON} = \beta_{Si3N4} = 10^{-5} \text{ K}^{-1} \qquad [11]$$

$$\beta_{Si} = 1.86 \times 10^{-4} \text{ K}^{-1} \qquad [11, 12]$$

the theoretical shift was estimated to be 6.9 nm. As for the normalised output, the experimental output at $T=60°C$ is 0.73 of the output at $T=20°C$, while the TMM calculation gives the value of 0.88. Therefore, both the experimental wavelength shift and the normalised output are in a good agreement with the theoretical predictions. Further measurements and analysis using the Floquet-Bloch theory (FBT) will be performed to additionally confirm these preliminary results.

III. CONCLUSION

DGADC devices with thicker SiON waveguides have been experimentally investigated. The thickness of 4.4 µm is closer to optimum thickness of 5 µm than our first batch of reported results, which gives maximum theoretical coupling efficiency of ~ 90% for optimal thickness. As the maximum theoretical efficiency is ~ 70% for the fabricated devices, and the previous experimental results were within

5% of the theoretical value, efficiency in excess of 60% was expected. However, maximum experimental efficiency is 56% for the devices that has been measured. The most significant improvement of 11%, compared to the previous experimental results, has been measured in L-band. These facts indicate that silicon nitride thickness is not optimum, and further investigation is needed to confirm this hypothesis.

Measurement of the temperature dependence of the resonant peak has been also performed. It has shown that temperature variations of ± 15 K result in a 3 dB reduction of the output signal at a fixed wavelength. To improve this temperature dependence, work is underway to broaden the spectrum of the devices by grating chirping and by varying the duty cycle of the gratings. Theoretical predictions, obtained by the TMM, and experimental results for both the peak wavelength shift and decrease of the output signal by changing the temperature in the 20-60°C temperature range are in a good agreement. Even better matching between the two results could be expected by using the FBT method.

ACKNOWLEDGEMENT

The authors would like to thank Tony Blackburn and Mike Josey of Innos Ltd for their contribution in the fabrication of the devices, and William Headley of the Advanced Technology Institute, University of Surrey for SEM analysis of the samples. The authors are grateful to UniSdirect and EPSRC for funding.

REFERENCES

[1] A. Liu, R. Jones, L. Liao, D. Samara-Rubio, D. Rubin, O. Cohen, R. Nicolaescu, and M. Paniccia, "A high-speed silicon optical modulator based on a metal-oxide-semiconductor capacitor," *Nature*, vol. 427, pp. 615-618, 2004.

[2] H. Rong, R. Jones, A. Liu, O. Cohen, D. Hak, A. Fang, and M. Paniccia, "A continuous-wave Raman silicon laser," *Nature*, vol. 433, pp. 725-728, 2005.

[3] V. R. Almeida, C. A. Barrios, R. R. Panepucci, and M. Lipson, "All-optical control of light on a silicon chip," *Nature*, vol. 431, pp. 1081-1084, 2004.

[4] V. R. Almeida, R. R. Panepucci, and M. Lipson, "Nanotaper for compact mode conversion", *Opt. Lett.*, vol. 28, pp. 1302-1304, 2003.

[5] A. Sure, T. Dillon, J. Murakowski, C. Lin, D. Pustai, and D. W. Prather, "Fabrication and characterization of three-dimensional silicon tapers," *Opt. Express*, vol. 11, pp. 3555-3561, 2003.

[6] Z. Lu and D. W. Prather, "TIR-Evanescent coupler for fiber to waveguide integration of planar optoelectronic devices," *Opt. Lett.*, vol. 29, pp. 1784-1750, 2004.

[7] G. Z. Masanovic, V. M. N. Passaro, and G. T. Reed, "Coupling to nanophotonic waveguides using a dual grating-assisted directional coupler," *IEE Proc. Optoelectronics*, vol. 152, pp. 41-48, 2005.

[8] G. Z. Masanovic, G. T. Reed, W. Headley, B. Timotijevic, V. M. N. Passaro, R. Atta, G. Ensell, and A. G. R. Evans, "A high efficiency input/output coupler for small silicon photonic devices," *Opt. Express*, vol. 13, pp. 7374-7379, 2005.

[9] V. S. Nguyen, S. Burton, and P. Pan, "The variation of physical properties of plasma-deposited silicon nitride and oxynitride with their compositions," *J. Electrochem. Soc.*, vol. 131, pp. 2348-2353, 1984.

[10] J. J. Fijol, E. E. Fike, P. B. Keating, D. Gilbody, J. J. LeBlanc, S. A. Jacobson, W. J. Kessler, and M. B. Frish, "Fabrication of silicon-on-insulator adiabatic tapers for low-loss optical interconnection of photonic devices," *Proc. SPIE*, vol. 4997, pp. 157-170, 2003.

[11] L. Eldada, "Advances in telecom and datacom optical components," *Opt. Eng.*, vol. 40, pp. 1165-1178, 2001.

[12] G. T. Reed and A. P. Knights, *Silicon Photonics: An Introduction*, Chichester, UK: Wiley, 2004.

2006 25th International Conference on Microelectronics

High Speed Modulator in Q-band Range on 4H-SiC p-i-n Diodes

R.D. Kakanakov, L.P. Kolaklieva, L.P. Romanov, A.V. Kirillov, A.A. Lebedev, M.S. Boltovets, K. Zekentes

Abstract – 4H-SiC p-i-n diodes with very good switching characteristics have been developed and used in high speed modulator in the Q-band range. The diode has a voltage drop at forward direction as low as 3.4 V at current density of 100 A/cm^2. Leakage currents less than 1 μA are measured at reverse voltages of 1200 V and 1700 V in air and a SF$_6$ atmosphere, respectively. The operating speed of the modulator is 5-7 ns which make it very perspective for applications in Q-band range. In the frequency range of 26-38 GHz, the modulator has insertion losses not more than 2 dB and isolation losses not less than 20 dB.

I. INTRODUCTION

Silicon carbide (SiC) has excellent electrical properties, which allow performance far beyond the capabilities of silicon in many applications. The unique combination of high breakdown voltage, short switching time and high operating temperature (up to 500°C) [1] makes the use of 4H-SiC p-i-n diodes very promising for applications in microwave devices working under extreme conditions.

A 4H-SiC p-i-n diode with high blocking voltage (up to 20 kV) and relatively low forward voltage drop (V$_F$) has been reported. [2] However, the commercialization of such devices is steel suffered from technological obstacles. Most of them relate to the quality of the i- layer, which should be lightly doped (~ 1x10^{15} cm^{-3}) and should combine a big thickness (6-10 μm) with structural perfection (very low micropipe and defect densities). Steps of the device processing such as low resistivity and thermally stable ohmic contacts and periphery protection are also critical.

In this work we present the results from the development and investigation of 4H-SiC p-i-n diodes with very good switching characteristics. Based on these diodes a high speed modulator in Q-band range is made and studied.

R.D. Kakanakov and L.P. Kolaklieva are with the Institute of Applied Physics, Bulgarian Academy of Sciences, 59, St. Petersburg Blvd., 4000 Plovdiv, Bulgaria, E-mail: ipfban@mbox.digsys.bg

L.P. Romanov, and A.V. Kirillov are Svetlana-Electropribor, St.Petersburg, 194156 Russia

A.A. Lebedev is with the Ioffe Physicotechnical Institute, Russian Academy of Sciences, St.-Petersburg, 194021, Russia

M.S. Boltovets is with the State Enterprise Research Institute "ORION", Kiev 03057, Ukraine

K. Zekentes is with MRG, IESL, Foundation for Research and Technology-Hellas, 71110 Heraklion, Crete, Greece

II. 4H-SiC P-I-N DIODE

A. 4H-SiC p-i-n diode device processing

The structure of the p-i-n diode is schematically presented in Fig. 1.

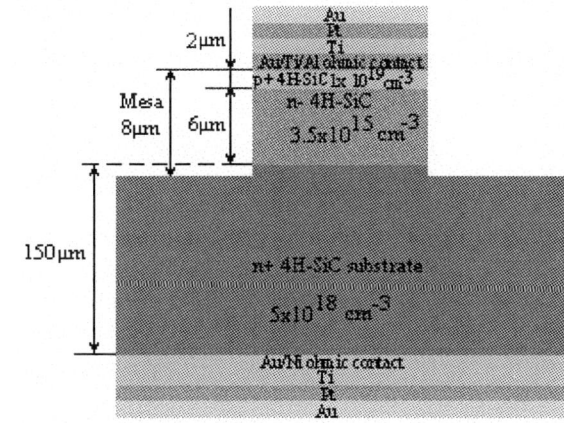

Fig. 1 Cross-section of the developed p-i-n- structure.

The 4H-SiC p-i-n diodes were made using a p$^+$-n$^-$-n$^+$- device structure. It was grown on commercially available (from Cree Co.) n$^+$- 4H-SiC substrates with thickness of 350 μm and doping concentration $N_d - N_a = 5 \times 10^{18}$ cm^{-3}. P$^+$- and n$^-$ epilayers were grown by sublimation epitaxy. The lightly doped n$^-$ - epilayer was grown on the "silicon" face of the substrate. It had a thickness of 6 μm and donor concentration as low as 3.5×10^{15} cm^{-3}, which corresponds to specific resistance of 20-30 Ω.cm. In order to lower the series resistance, the concentration of the uncompensated (i.e. electrically active impurities) of the p$^+$- type layer should exceed values of 1×10^{19} cm^{-3}. The high-quality of the metallurgical interface between the n- and p-type regions is very important for lowering the leakage current. The quality of the metallurgical boundary of the p$^+$–n$^-$ junction was improved by in situ pregrowth polishing sublimation etching of the initial layer, which was effected by changing the sign of the temperature gradient within the growth cell [3]. This technique allowed obtaining good metallurgical p-n junction and uniform distribution of the atomic concentration of Al in the p$^+$- type layer as it was observed by SIMS analysis (Fig.2). The concentration of the electrically active impurities determined by Hall and C-V measurements was found to be $N_P - N_A = 1 \times 10^{19}$ cm^{-3}.

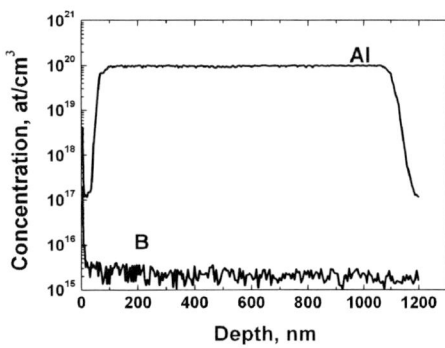

Fig. 2 Atomic concentration of the impurities in the p⁺-layer determined by SIMS analysis.

After the epitaxial growth of the p⁺-n⁻-n⁺ structure the substrate thickness was thinned up to 150 μm, which caused reducing the diode resistance.

The diodes were formed as mesa structures with diameter of 80 μm. The mesas were obtained by dry etching in mixture of SF_6 and 20% O_2 at working pressure of $2x10^{-3}$ torr. The etching rate of the p⁺- type layer was 40 Å/min, whereas that of the n-type layers was higher (50 Å/min). The periphery of the mesa was isolated by 100 nm thick thermal SiO_2. In addition, after the chip packaging it was protected by polyimide PI2525. The improved etching process and mesa protection allowed achieving breakdown voltage of 1200 V in air and 1700 V in a SF_6 atmosphere at reverse current less than 1 μA.

Low resistivity and thermally stable up to 600 °C ohmic contacts (Fig. 3) were developed for the p-i-n diode. [4] Ni/Au and Al/Ti/Au multilayers were consecutively deposited by magnetron sputtering, e-beam and thermal evaporation as ohmic contacts to the n⁺-substrate and p⁺-epilayer, respectively. Ohmic properties were formed by annealing at temperature of 900 °C in an inert ambient.

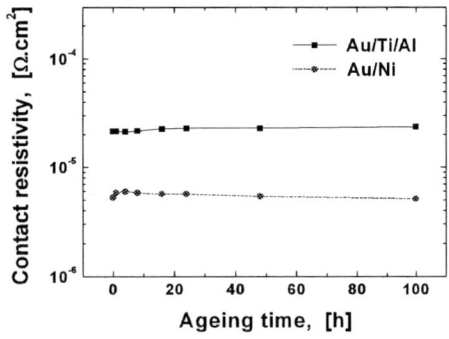

Fig. 3 Dependence of the resistivity of the ohmic contacts on the aging temperature of 600 °C.

Contact resistivity of $4.9x10^{-6}$ $\Omega.cm^2$ and $1.4x10^{-5}$ $\Omega.cm^2$ was obtained for n- and p-type contacts, respectively. After annealing Au/Pt/Ti multilayers were consecutively deposited over the ohmic contacts. For the subsequent device processing a thick Au film was plated. Both, low contact resistivity and thinned substrate contribute to reduction of the total diode resistance.

After the diode structures were formed, the wafer was cut into chips of 0.6x0.6 mm in size. The chips were mounted in a ceramic-metal packages type M15. The n⁺-substrate was glued to the package plate at 360°C in a N_2 atmosphere using an Au(88%)–Ge(12%) solder, while the contact of the p⁺-type region was realized by an Au wire of 25 μm in diameter by thermal compression. The principal scheme of a packaged p-i-n diode is presented in Fig. 4.

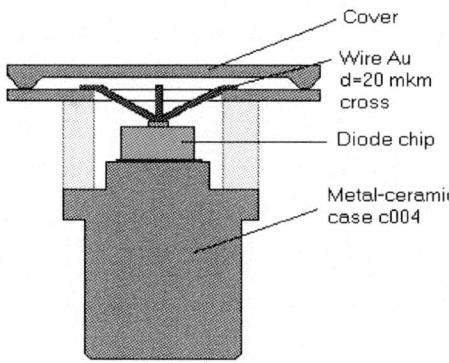

Fig. 4 Scheme of a package with a mounted p-i-n diode chip.

B. Characterization of the 4H-SiC p-i-n diode

The packaged p-i-n diode was electrically characterized at 25 °C by I-V and C-V measurements. The on-state performance of the 4H-SiC p-i-n diode with a 6 μm thick i-layer is shown in Fig. 5, where the current density is plotted versus the forward voltage drop. At room temperature, a forward voltage drop of 3.4 V was determined at current density of 100 A/cm^2. This result is indication for a further increase in the power handing capabilities. The turn-on voltage for the studied diode was found to be 2.8 V. This value compares favorably to the theoretical turn-on voltage for 4H-SiC rectifiers of about 2.6 V, which is approximately 75% of the bandgap of 4H-SiC. [5] In the forward direction, low on-resistance of 1.6 $m\Omega.cm^2$ was achieved, indicating effective conductivity modulation of the i-layer.

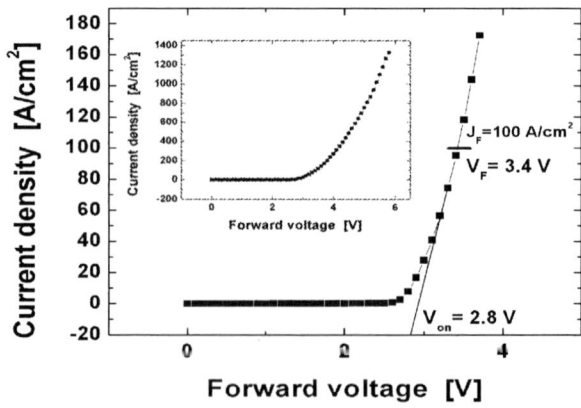

Fig. 5 Forward I-V characteristic of the developed 4H-SiC p-i-n diode.

The capability to handle very high current densities can be seen in the insert in Fig. 5. The forward I-V characteristic showed excellent on-state voltage drop of 5.8V at high current density of 1325 A/cm^2. This result allows, from an application point of view, to set nominal current density of 130 A/cm^2 and still being able to handle ten times higher currents as is required for short periods of time.

The static blocking characteristic up to 1.2 kV is shown in Fig. 6. The measurement of the leakage currents was carried out under d.c. conditions at room temperature in air. The leakage current was below 20 nA resolution limit up to 800 V reverse bias. The diode had a current leakage below 200 nA for reverse voltages up to 1.1 kV. As the reverse voltage was increased the leakage increased, but at 1.2 kV it remained steel lower 1μA. A catastrophic failure was observed in the breakdown characteristics at voltages over 1.2 kV and 1.7 V in air and a SF$_6$ atmosphere, respectively.

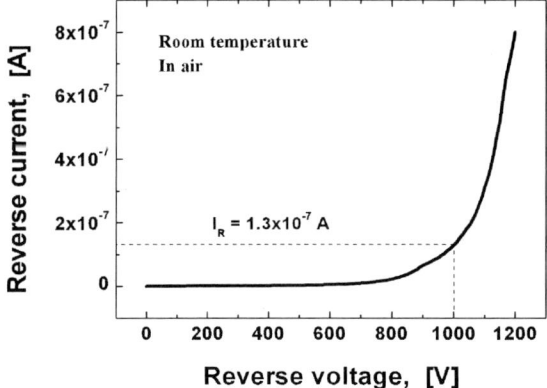

Fig. 6 Reverse current voltage characteristic of a 4H-SiC p-i-n diode.

The capacitance of the chip and the packaged diode obtained by C-V measurements is presented in Fig. 7.

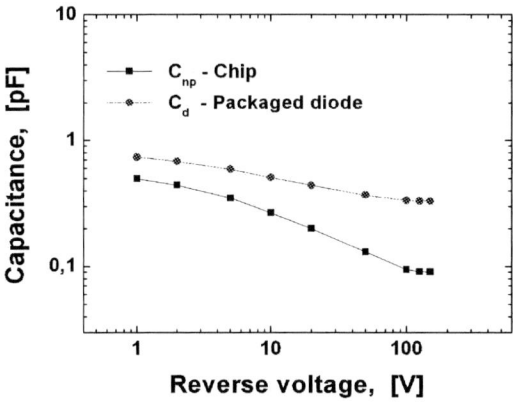

Fig. 7 Dependence of the capacitance of the chip and the packaged diode versus the reverse voltage.

II. MODULATOR ON 4H-SiC P-I-N DIODES

Diodes with minimal junction capacity and minimal differential resistance, 0.04÷0.052 pF and 2.1÷2.25 Ω, respectively, were selected for use in the modulator. The modulator was made as a part of fine-line with width of 6 mm and lowered height of 1.5 mm (the cross-section of the fin-line being 6.5x1.5 mm). The photography of the developed Q-band modulator is presented in Fig. 8.

Fig. 8. Design of the Q – band modulator made on the developed 4H-SiC p-i-n diodes.

Frequency characteristics of two modulators on the basis of a fin-line are shown in Fig. 9. The results show that characteristics of isolation have sufficient broadbandness. Namely, the modulator #1 at isolation on resonance frequency of 16.5 dB has an isolation band of 4 GHz at 15 dB, and the modulator #2 at isolation on resonant frequency of 15 dB has an isolation band of 4 GHz at 13.5 dB.

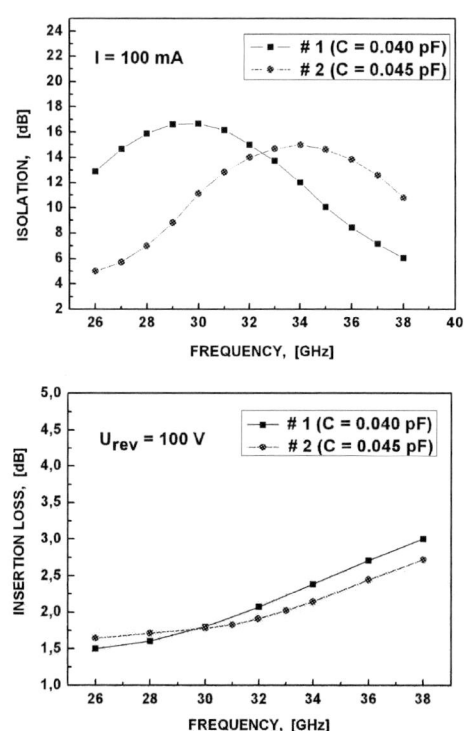

Fig. 9. Frequency characteristics of the modulators on the basis of a fin-line.

It was found that the insertion losses of the modulators increase by 2 times with frequency increase from 26 to 38 GHz due to the shunting action of a fin-line regular part by the diode capacity. To prevent this effect a two-cascade modulator (switch) was designed. It consists of two single cascades, located on length of a slot on distance $\lambda_w/4$, where λ_w is the length of a wave in the fin-line. As a result of the two-cascade construction, reflections from separate switches are mutually compensated, which reduces the insertion losses, especially at higher frequencies. Hence, the working frequency band of the modulator can be considerably expanded. The measurements in the frequency band ranged from 29 GHz to 34 GHz (15 %) determined isolation losses not less than 30 dB and insertion losses no more than 1.75 dB and in the frequency band 26÷38 GHz they were not less than 20 dB and no more than 2 dB, respectively (Fig. 10).

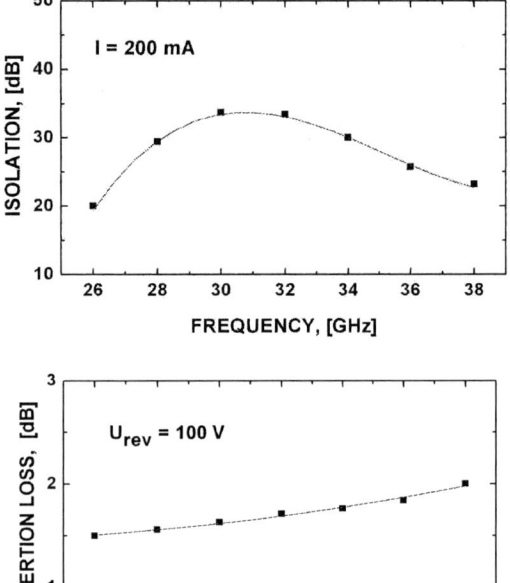

Fig. 10 Frequency characteristics of a two-cascade modulator.

The results obtained show that the two-cascade modulator can be used for modulations (or switching) practically in the whole range of a wave guide with section of 7.2 x 3.4 mm, which corresponds to the range of 26÷40 GHz.

Since the breaker speed of the modulator has a big impact for modulation applications, it was also evaluated. The estimation of breaker speed was carried out in a mode of low power 10 mW continuous signal modulation. The breaker speed was estimated on bending around the detected output power on levels of 0.1 and 0.9 of amplitudes (Mode of test: P = 10 mW, I_f = 200 mA, V_{rev}

=50 V). The transient period time and recovery time were practically identical and were between 5.0 and 7.0 ns with this mode test.

III. CONCLUSION

4H-SiC p-i-n diode has been developed for application in a high speed modulator in Q-band range. High quality metallurgical p^+-n^- junction lowering the leakage current is obtained by sublimation epitaxy. The 6 μm thick i-layer is grown with low doping concentration of $3.5 \times 10^{15} cm^{-3}$. The low resistivity of the ohmic contacts and the thinned substrate allow reducing the total diode resistance. The developed p-i-n diode has forward voltage drop of 3.4 V at current density of 100 A/cm^2 and turn-on voltage of 2.8 V, which is in good agreement with the theoretical value of 2.6 V for 4H-SiC. The achieved low on-resistance of 1.6 mΩ.cm^2 presupposes effective conductivity modulation of the i-layer.

On the base of the developed 4H-SiC p-i-n diode a high speed modulator is demonstrated. The insertion and isolation losses of the modulator are measured in the frequency range of 26-38 GHz. The insertion losses do not exceed 2 dB at value of the isolation losses not less than 20 dB. In the frequency interval 29-34 GHz the insertion losses are no more than 1.75 dB while the isolation losses are not less than 30 dB.

Preliminary analysis showed that decrease of the insertion losses up to 1 dB and increase of the isolation losses over 30 dB could be achieved by optimization of the diode and modulator parameters as well as by lowering the diode resistance.

The modulator has operating speed of 5-7 ns, which makes it very promising for applications in Q-band range.

ACKNOWLEDGEMENT

The support from the INTAS – 010603 project is gratefully acknowledged.

REFERENCES

[1] B.J. Baliga, *Power Semiconductor devices*, PWS Publishing Company, 1996.
[2] M.K. Das, *Latest Advantages in 4H-SiC p-i-n and MOS Power Devices*, International Semiconducor Device Research Symposium, Washington, DC, December 2003.
[3] N. S. Savkina, A. A. Lebedev, D. V. Davidov, *et al.*, *Mater. Sci. Eng. B*, vol. 61–62, p. 165, 1999.
[4] R. Kakanakov, L. Kasamakova Kolaklieva et al., *Materials Science Forum*, vols. 457-460, pp. 877-880, 2004.
[5] K. Rottner et al, *Appl. Mater. Sci. Eng. B*, vol. 61–62, pp. 330-338, 1999.

2006 25th International Conference on Microelectronics

Low-Frequency-Noise Characteristic of Quasi-Enhancement-Mode HEMT Using a Selectively Hydrogen-Pretreatment

I. H. Kang, S. C. Kim, W. Bahng, and N. K. Kim

Abstract – The DC, RF, and Low-frequency noise characteristics were investigated for a quasi-enhancement-mode (QE) HEMT using a selective hydrogen pretreatment (SHP). The QE-HEMT with SHP showed a large shift in threshold voltage without severe degradation of RF performances including cut-off frequency and maximum oscillation frequency, compared with those of HEMT without SHP. Moreover, the QE HEMT exhibited a reduction of low-frequency noise bulges compared with those of depletion-mode HEMT without an SHP, leading to an one-order smaller input noise spectral density at 100Hz, and offered a potential for application to a low phase noise oscillator.

I. INTRODUCTION

Many modern communication systems demand more available channels and more timing margin. It in turn puts more stringent requirements on the phase noise characteristics of the local oscillator and the clock generator [1]. From many previous works, the origin of the phase noise of a microwave oscillator as a building block of mobile communication systems has been known as a low-frequency noise (LFN) of microwave device that is up-converted through a mixing process into frequency fluctuation around the carrier signal [2]. Consequently, it is important to employ a device having a superior LFN characteristic, to achieve low phase-noise oscillator [1].

The low-frequency noise of microwave device is attributed to traps in the bulk, on the surface, or at the interface of the semiconductor materials [3]. These traps can be reduced by a proper hydrogen treatment and post annealing, due to a hydrogen passivation [4]. According to results of these pioneer works on the hydrogen treatment, the incorporation of hydrogen in the gate region of HEMT structure has considerable effects on device characteristics including drain current, threshold voltage, transconductance, gate leakage current, and low-frequency noise characteristics [5]. These changes are mostly related to the passivation of donor impurity and defects, and the change of surface stoichiometry [6].

In the previous work, we found that a selective hydrogen pretreatment (SHP) that caused the passivation of donor impurity and defects was a good candidate for

I. H Kang, S. C. Kim, W. Bahng, and N. K. Kim are with the Power Semiconductor Research Group, Korea Electrotechnology Research Institute, 28-1 Seongju-dong, Changwon-si, Gyeongsangnam-do, Korea, Email: ihkang@keri.re.kr

fabricating an enhancement-mode FET [7]. In this works, we implemented a QE-HEMT and investigated the DC, RF, and low-frequency noise characteristics. The quasi-enhancement-mode (QE-) and depletion-mode (D-) HEMTs showed a superior low frequency noise characteristic without severe degradation of transconductance and RF performances.

II. EXPERIMENT

Fig. 1 shows the structure of commercially available HEMT having an $Al_{0.22}Ga_{0.78}As$/ $In_{0.22}Ga_{0.78}As$ double heterostructure. An InGaAs channel layer and a double delta-doping scheme were employed to obtain a sufficient linearity and to prevent degradation in device performance due to the hydrogen passivation of donor.

Function	Material	Thickness [Å]	Doping [cm^{-3}]	
cap	GaAs	200	5e18	
cap	GaAs	200	1e18	
barrier	$Al_xGa_{1-x}As$	220	undoped	x=0.22
δ-doping			5e12	
spacer	$Al_xGa_{1-x}As$	50	undoped	x=0.22
channel	$In_xGa_{1-x}As$	120	undoped	x=0.22
spacer	$Al_xGa_{1-x}As$	40	undoped	x=0.22
δ-doping			1.5e12	
buffer	$Al_xGa_{1-x}As$	5000	undoped	
(100) semi-insulating GaAs substrate.				

Fig. 1. The structure of HEMT to fabricate QE-HEMT and D-HEMT.

Firstly, HEMTs were isolated by forming a mesa by means of wet-etching with phosphoric-based etchant, and then ohmic contacts were formed by evaporating Ni/Au/Ge/Ni/Au and annealing at 405°C for 20 sec. To reduce an over-etch effect, the gate recess was performed at low etch rate in a dilute citric-based etchant, concurrently monitoring the drain current. Before the hydrogen pretreatment, the reactive ion etching (RIE) chamber was cleaned with an oxygen plasma, since residual chemicals in the chamber can greatly influence the gate region during the RIE process. Subsequently the selective hydrogen pretreatment was carried out following two steps: hydrogen

1-4244-0116-X/06/$20.00 ©2006 IEEE

exposure using RIE (100W) and annealing at 470°C for 25 sec under N_2 atmosphere. During the hydrogen treatment, the samples were masked with photoresist except for the gate region ($1.5 \times 50 \mu m^2$) to prevent degradation of ohmic characteristics. After hydrogen treatment, Ti/Pt/Au gate metal ($1.5 \times 50 \mu m^2$) was deposited, followed by Si_xN_y device passivation and deposition of pad metal for probing.

The DC characteristics of p-HEMTs with and without the hydrogen-pretreatment were measured using an H4155B semiconductor parameter analyzer. The on-wafer s-parameter measurements of p-HEMTs were performed for frequencies between 1 and 40 GHz using an HP8510C Network Analyzer at room temperature. On-wafer low-frequency noise measurements were carried out for frequencies between 1 Hz and 1 MHz and temperatures between 180 K and 460 K. The temperature of the probe station was controlled by using a liquid nitrogen cooler and a heater operating in DC mode to prevent any possible low-frequency noise generated by the heater from affecting the measurement. The selectively hydrogen-pretreated HEMTs were biased in the common-source mode by using an HP 33150A bias network and low-noise wire-wound resistors to minimize the low-frequency noise contribution from the biasing circuitry. The output noise voltage from the DUT (device under test) was amplified by using an EG&G5185 low-noise amplifier and was measured by using an HP35670A dynamic signal analyzer.

III. RESULTS AND DISCUSSION

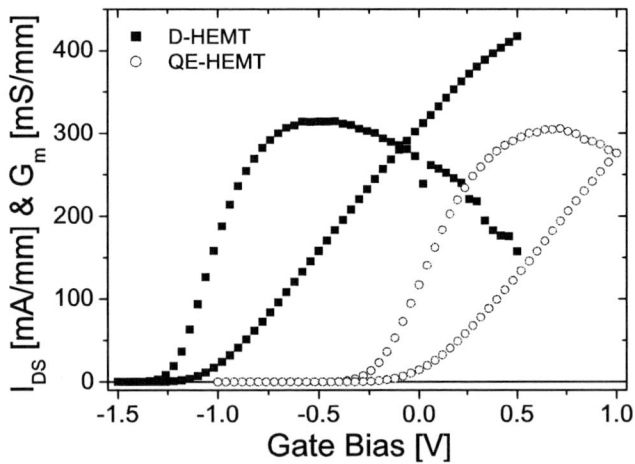

Fig. 1. Comparison of the drain current and the transconductance of QE/D-HEMTs biased at V_{ds}=2.5V

Figure 1 shows the drain current and the transconductance of QE/D-HEMTs biased at V_{ds}=2.5V. A large shift in the threshold voltage ($\sim$1V) was achieved using SHP. It is attributed to the partial passivation of donor, which dominantly adjusts the threshold voltage of

HEMT. However, despite of the large shift in the threshold voltage, the maximum transconductance was not seriously deteriorated.

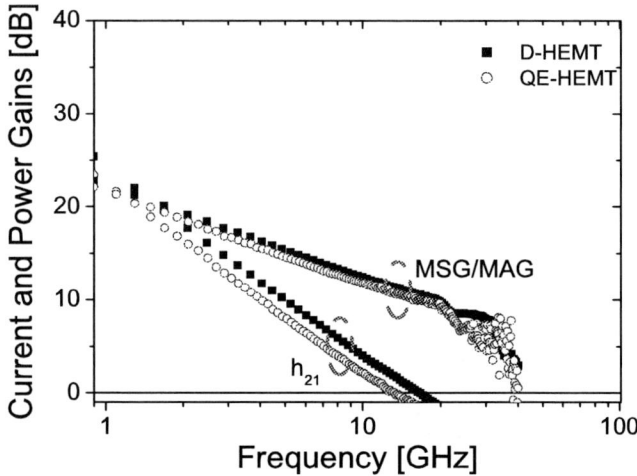

Fig. 2. Comparison of RF performances of QE/D-HEMTs biased at the gate voltage for the maximum transconductance and V_{ds}=2.5V

Figure 2 shows the comparison of the cut-off frequency and the maximum oscillation frequency of QE/D-HEMTs biased at the gate voltage of maximum transconductance (–0.5V for D-HEMT and 0.75V for QE-HEMT), and the drain voltage of 2.5V. The cut-off frequencies of QE/D-HEMTs were 13GHz and 17GHz, respectively. The maximum oscillation frequencies were 60GHz and 62GHz, respectively. No fatal degradation in the RF performances of QE/D-HEMTs was exhibited like the DC characteristics of QE/D-HEMTs. It is attributed to partial passivation of donor at the expense of the linearity of device as mentioned in previous work [7]. To analyze the low frequency noise characteristics of QE/D-HEMTs, their input noise spectral density S_{iv} was measured at the gate voltages for the peak transconductance, 0.75V and -0.5V by sweeping the DUT temperature from 180K to 460K.

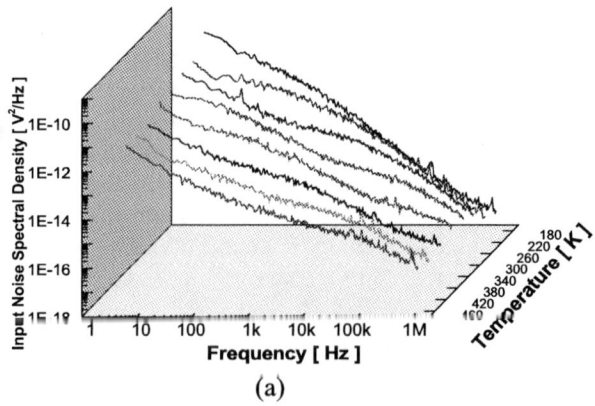

(a)

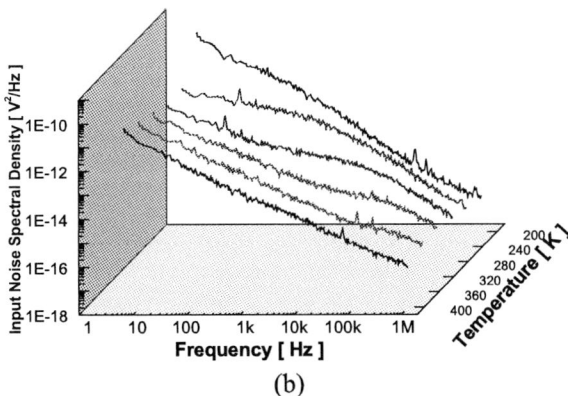

(b)

Fig. 3. Average input noise spectral density of (a) depletion-mode HEMT, and (b) quasi-enhancement-mode HEMT.

Fig. 3 shows the average input noise spectral density of QE/D-HEMTs. In these figures, two distictive G-R bulges are shown for D-HEMT in the temperature range from 260K to 300K, while QE-HEMT seems to have one distinctive noise bulge in the temperature range from 240K to 320K.

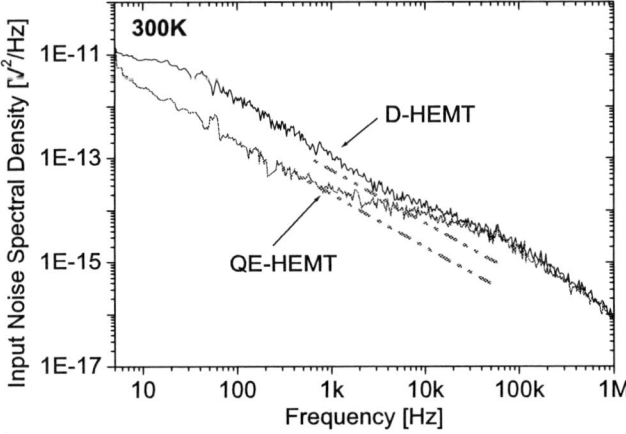

Fig. 4. Comparison of average input spectral density of QE/D-HEMTs measured at 100Hz

Figure 4 shows the average input noise spectral density of QE/D-HEMTs at 300K. The input noise of D-HEMT seems to deviate from an 1/f-slope and to have two noise bulges at f=~20Hz and ~40kHz. However, the input noise of QE-HEMT is lower than that of D-HEMT by an order below ~1kHz (For example, ~1.91E-13 V²/Hz for QE-HEMT and ~1.29E-12 V²/Hz for D-HEMT at 100Hz), due to no noise bulges (see fig.3, the input noise of QE-HEMT follows an 1/f-slope for f<1kHz), but increases with frequency to the same level of D-HEMT above 30kHz. This means that the low frequency noise characteristics of QE-HEMT can be improved by using SHP. In order to find out the detailed origin of this improvement, the Arrhenius plots of the $T^2\tau_c$ product as a function of the reciprocal temperature were obtained for QE/D-HEMTs using the following equation:

$$\omega_c = \frac{1}{\tau_c} \sim T^2 \exp\left(-\frac{E_a}{k_B T}\right)$$

where E_a is the trap activation energy, k_B is the Boltzmann constant, and T is the operating temperature.

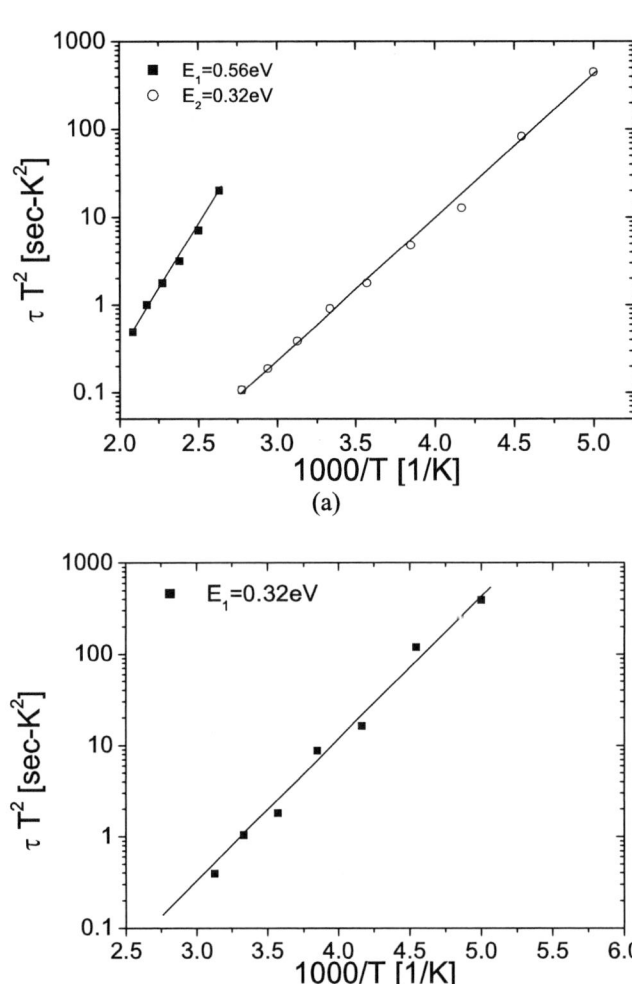

Fig. 5. Arrhenius plot of $\tau_c \cdot T^2$ product as a function of a reciprocal temperature for (a) D-HEMT and (b) QE-HEMT

In the figure 5, the QE-HEMT had a dominant trap at $E_1 \sim 0.32\text{eV}$, and the D-HEMT had two dominant trap at $E_1 \sim 0.56\text{eV}$ and $E_2 \sim 0.32\text{eV}$. Both noise bulges are ascribed to the DX center [8]. However, the noise bulge (~0.56eV) was not found for QE-HEMT compared with those of D-HEMT. It is believed that the hydrogen injected by SHP deactivates this noise bulge.

IV. CONCLUSION

We have fabricated the D-HEMT and the QE-HEMT by using selective hydrogen pretreatment composed of hydrogen exposure and RTA prior to the gate metallization, and characterized their DC, RF, and low frequency noise performances. Without severe degradation of the DC and RF performances, the QE-HEMT showed the superior low-frequency noise characteristics, and thus a potential for its application to the low-noise applications including a low-phase noise oscillator and a low-jitter timing generator.

ACKNOWLEDGEMENT

This work was supported by MOCIE in South Korea.

REFERENCES

[1] Ali Hajimiri, and Thomas H. Lee, "A general theory of phase noise in electrical oscillator", *IEEE Trans. Solid-State Circuits*, vol. 33, pp. 179-194, 1998.

[2] J. Verdier, O. Llopis, R. Plana, and J. Graffeuil, "Analysis of noise up-conversion in microwave field-effect transistor oscillators", *IEEE Trans. Microwave Theory and Techniques*, vol. 44, pp. 598-601, 1996.

[3] Y. J. Chan, and D. Pavlidis D, "Trap studies in GaInP/GaAs and AlGaAs/GaAs HEMT's by means of low-frequency noise and transconductance dispersion characterizations", *IEEE Trans. Elec. Dev.*, vol. 41, pp. 637-642, 1994.

[4] R. G. Pereira, M. Van Hove, M. de Potter, and M. Van Rossum, "Influence of CH4/H2 reactive ion etching on the deep levels of Si-doped AlxGa1-xAs (x=0.25)", *J. Vac. Sci. Technol. B*, vol. 14, pp. 1773-1779, 1996.

[5] I. H. Kang, J. H. Kim, H. J. Song, and J.-I. Song, "Effect of selective hydrogen pretreatment on the characteristics of AlGaAs/InGaAs p-HEMTs", *J. Korean Phys. Soc.*, vol. 42, pp. 281-284, 2003.

[6] T. Okumura, "Hydrogen-related issues in GaAs Schottky contacts", *Digest on Int. Conf. Compound Semiconductor Manufacturing Technology*, 1999.

[7] I. H. Kang, and J.-I. Song, "Enhancement-mode p-HEMT using selective hydrogen treatment", *Electronics Letters*, vol. 39, pp. 408-409, 2003.

[8] Adachi Sadao, *Properties of Aluminium Gallium Arsenide*, Short Run Press, 1993.

2006 25th International Conference on Microelectronics

Hyper Sound Amplification

S. V. Koshevaya, V. V. Grimalsky, M. Tecpoyotl-Torres, J. Escobedo-Alatorre, M. F. Díaz-Ayala and A. Garcia-B

Abstract - In this report, two basic mechanisms of the amplification of acoustic-electromagnetic waves are analyzed. The first mechanism is similar to mechanism of traveling wave tube in different materials due to piezo-effect, deformation potential and electrostriction. We analyze the second mechanism, which is due to the Gunn effect and the negative differential mobility in GaAs. We show device schemes for its use like filters, delay lines etc., in communication and control systems. It is demonstrated that this amplification is adequate for the obtaining of active filters and delay line so as it is strong. Other mechanism to obtain amplification is named deformation potential.

I. INTRODUCTION

The experimental investigation of the acoustic wave amplification has been carried out [1-3]. The application was very difficult because the hyper sound has big losses at high frequencies. From other hand, the usefulness of hyper sound ($f \succ 10GHz$) is very important for the application in communication and control systems. In this report, the full analysis of basic mechanisms of the amplification is presented. The possibility of strong amplification due to mechanism of resonance excitation and amplification of sound –space charge hybrid mode in GaAs is shown. As an example of second mechanism we demonstrate the amplification only using the piezoeffect but it is clear that the obtaining of frequencies bigger than 100GHz it is possible only in case of the use of potential deformation in GaAs.

II. BASIC MECHANISMS OF CLASSICAL APLIFICATION OF HYPER SOUND AND STRUCTURES

We had used the simplest models of crystals with piezoeffect, deformation potential and electrostriction [4]. For case of piezoeffect and electrostriction we use GaAs, considering the substrate with big value of *electric-mechanic coefficient of piezoeffect* $K^2 = \dfrac{\beta^2}{\varepsilon C}$ (non-

S. V. Koshevaya, M. Tecpoyotl-Torres, J. Escobedo-Alatorre and , M. F. Díaz-Ayala are with Autonomous State University of Morelos, CIICAp, Av. Universidad No. 1001, Z. P. 62209, Cuernavaca Mor., México, e-mail: svetlana@uaem.mx

V. V. Grimalsky and A. Garcia-B are with National Institute of Astrophysics, Optics and Electronics (INAOE), P. O. Box 51 & 216, Z. P. 72000, Puebla, México

dimensional), where β is piezomodule and C is the elastic module. All modulus are tensors.

We consider simplest case of the passing of longitudinal acoustic-electromagnetic mode (hybrid mode) in direction of the Z exes. The elastic theory equations are:

$$\rho \frac{\partial^2 U}{\partial t^2} = C \frac{\partial^2 U}{\partial z^2} + \beta \frac{\partial E}{\partial z} \qquad \text{a)}$$

$$D = \varepsilon_0 \varepsilon E - \beta \frac{\partial U}{\partial z} = 0 \qquad \text{b)} \qquad (1)$$

for case of pizoeffect [1]:

$$\rho \frac{\partial^2 U}{\partial t^2} = C \frac{\partial^2 U}{\partial z^2} - V_0 \varepsilon_0 \varepsilon \frac{\partial E}{\partial z} \qquad \text{a)}$$

$$D = \varepsilon_0 \varepsilon E - V_0 \varepsilon_0 \varepsilon \frac{\partial^2 U}{\partial z^2} = 0 \qquad \text{b)} \qquad (2)$$

for case of deformation potential and for electrostriction and electrostriction [4]:

$$\rho \frac{\partial^2 U}{\partial t^2} = C \frac{\partial^2 U}{\partial z^2} - a_0 E^2 \qquad \text{a)}$$

$$D = \varepsilon_0 \varepsilon E + \frac{1}{2} a_0 E_0 E = 0 \qquad \text{b)} \qquad (3)$$

where ρ is the density of material. $\varepsilon_0, \varepsilon$ are the dielectric constant of vacuum and relative constant of material respectively, U is the mechanical displacement, V_0 is the constant of deformation potential and a_0 is the constant of electrostriction (the ceramics with a very big value of the electrostriction is the material like $SrTiO_3 + BaTiO_3$ for room temperature [4]. All values without index 0 are variables here and later. The longitudinal electric constant and variable electric field, are in direction of Z axis. We must also use the following equations:

$$\frac{\partial D}{\partial z} = en \qquad \text{a)}$$

$$e \frac{\partial n}{\partial t} + \frac{\partial}{\partial z}(j_z) = 0 \qquad \text{b)}$$

$$j_z = env_z \qquad \text{c)}$$

$$v = v_z \approx \mu_{diff} E + D_{diff} \frac{\partial n}{\partial z} \qquad \text{d)}$$

$$D = \varepsilon_0 \varepsilon E \qquad \text{e)} \qquad (4)$$

A simple analysis of the space charges waves are used.

1-4244-0116-X/06/$20.00 ©2006 IEEE

The models of real structures for the filter and delay line are shown in Fig.1 a), b).

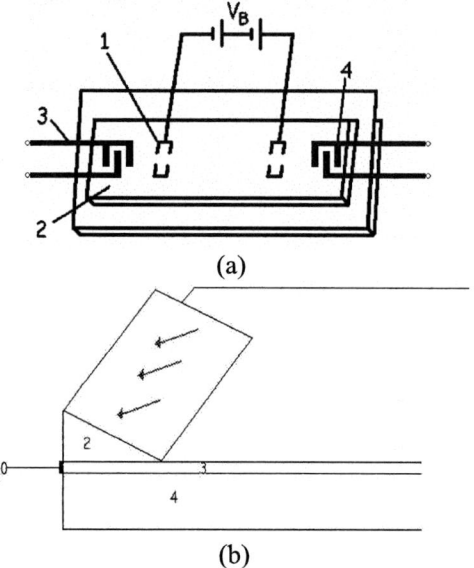

(a)

(b)

Figure 1. a) Top view: The contact 1 serves for the creation of the electron's beam, input and output antennas consist of contacts 2 and 3 (ohmic and Shottky, respectively), 4 is the output antenna. b) The input part of the complete device: The ohmic contact 1 it is used for deformation of the electron's beam. Input and output antennas are like the cuneiform element 2 of i-GaAs. Thin film 3 of n-GaAs is on 4, the substrate of i-GaAs.

For the analysis of the space charges waves, in the simplest model, considering the wave as $\propto \exp[i(\omega t - kz)]$ [5] –[7], we use (neglecting the sound):

$$\frac{\partial E}{\partial t} + v_o \frac{\partial E}{\partial z} + 2iD_{diff}k\frac{\partial^2 E}{\partial z^2} +$$

$$+(\omega_M + D_{diff}k^2)E - D_{diff}\frac{\partial^2 E}{\partial z^2} = 0 \qquad (5)$$

where μ_{diff}, is negative in case of electric constant field, which is more critical in case of GaAs, v_0 is the constant velocity of electrons in current j_z, n_0 is the material concentration, D_{diff} is the coefficient of diffusion, and $\omega_M = \frac{\sigma_{diff}}{\varepsilon_0 \varepsilon}$ is the Maxwell's relaxation frequency with differential negative conductivity $\sigma_{diff} \prec 0$.

Anatoliy Barybin was the first scientist in used the effect produced by the amplification of space charges [5]. The equations presented below are used in the analytical calculations and the simulation. Analytical calculations are presented for the two type of the amplification. The first mechanism of the amplification works like in traveling

wave tube; in second case, we use the Gunn Effect and the negative differential mobility.

III. ANALYSIS OF THE AMPLIFICATION WITH TRAVELING WAVE MECHANISM

For the case of the piezoeffect we use the system formed by equations 1a), b), and 4a), b), c), e) and describe the hybrid wave like $\propto \exp i[(\omega t - kz)]$. Dispersion equation is then:

$$k^2 = \frac{\omega^2}{s^2} - K^2 k^2 \frac{\omega - kv_0 - ik^2 D_{diff}}{\omega - kv_0 - i(k^2 D_{diff} + \omega_m)} \qquad (6)$$

Where: s is the sound velocity, $\omega_M = \frac{\sigma_0}{\varepsilon_0 \varepsilon}$ is the frequency of Maxwell's relaxation, which in this case, it is positive so as we take into account only positive case of mobility and conductivity μ_0, σ_0.

Taken into account that $k = k_0 + \Delta k(K^2)$ and $\Delta k = \Delta k' + i\Delta k''$ we find the solution for amplification, calculating at first to (with $K^2 \prec 1$ equal for piezoeffect):

$$\Delta k'' = -\frac{1}{2} \cdot \frac{\omega_M}{s} \cdot K^2 \cdot \frac{1 - \frac{v_0}{s}}{(1 - \frac{v_0}{s})^2 + (\frac{\omega^2}{\omega_D} + \omega_M)^2} \qquad (7)$$

where ω_M is positive (but $\omega_M = \frac{\sigma_{dif}}{\varepsilon_0 \varepsilon} < 0$, and

$\sigma_{dif} \prec 0$ if the electric field has critical values for GaAs, see below), v_0 is the constant velocity of electrons in current j_z, n_0 is the concentration of the material,

D_{diff} is the coefficient of diffusion, and $\omega_D = \frac{v_0^2}{D_{diff}}$ is the diffusion frequency.

We have classical traveling wave amplification in case of small value of $(s - v_0)s = \delta \prec 1$ like in traveling wave tube (the effect of Cherenkof-Vavilof). Experimental investigation has confirmed this result [2].

Similar calculations are obtained for case of deformation potential where the coefficient $\Delta k''$ is given by:

$$\Delta k'' - \frac{1}{2} \cdot \omega_M \cdot N^2 \frac{\omega^6}{s^6} \frac{1 - \frac{v_0}{s}}{\omega^2 (1 - \frac{v_0}{s})^2 + (\omega_M - \frac{\omega^2}{\omega_D})^2} \qquad (8)$$

302

where $N^2 = (\frac{V_0}{e})^2 \cdot \frac{\varepsilon_0 \varepsilon}{C}$ with units $cm^2 \cdot s^2$ (similar to K^2). The difference in amplification is the case of the deformation potential is explained as follows:

Amplification decreases with frequency so as $\frac{\text{Im}(k)_{def}}{\text{Im}(k)_{pies}} = \frac{N^2 \omega^4}{K^2 s^2}$, and in frequencies $\omega \succ \sqrt{\frac{K \cdot s}{N}}$ the mechanism of the deformation potential is dominated (this happen at frequencies of approximately $\omega \propto 2 \cdot 10^{11} s^{-1}$). For the case of GaAs, this is millimeter range so hyper sound has classical amplifications in case of the deformation potential. For case of ceramics with big values of electrostriction and conductivity, it is also obtained:

$$\Delta k'' = -\frac{1}{2} \cdot \omega_M \cdot M^2 \cdot \frac{\omega^2}{s} \cdot \frac{1 - \frac{v_0}{s}}{\omega^2 (1 - \frac{v_0}{s})^2 + (\omega_M + \frac{\omega^2}{\omega_D})^2} \quad (9)$$

where: the non dimensional coefficient $M^2 = \frac{a_0^2 E_0^2}{2 \varepsilon_0 \varepsilon C}$

(also similar to K^2). We have the possibility to change (to control) this mechanism of amplification by means of electric constant field E_0. This fact gives us the possibility to use a new technology, in a very useful frequency range for optoelectronics applications.

IV. SIMPLEST MODEL OF AMPLIFICATION OF HYPER SOUND DUE TO EFFECT GUNN

Considering a thin film of n-GaAs located between two mediums, the substrate from the same material and the transversal wave passed along Z axis (see Fig.1a) but without acoustic contact with environment. The component of displacement U is in direction of the Y axis. The film consists of a 2D gas with a high negative differential mobility. L is the length of the element (along axis OZ), h is the thickness of film (see Fig. 2). This film has a substrate with depth d of the same material. The hybrid wave is decrease in the substrate, then, all its energy is in the very thin film with 2D electron beam. The effective size of the acoustic waveguide is described by the parameter 2H=2(h+d/2).

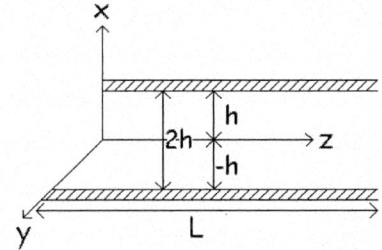

Fig.2. Geometry of the simplest model for simulation.

We analyze the simplest transversal acoustic mode using elastic theory equation with the piezoeffect. We use the variable electric field $E = -\frac{\partial \varphi}{\partial z}$ (in direction of axes Z) [8], [9] determined by potential φ:

$$\frac{1}{s^2} \frac{\partial^2 u}{\partial t^2} = \Delta u + \Gamma \Delta \frac{\partial u}{\partial t} - \frac{2\beta}{\rho s^2} \frac{\partial^2 \varphi}{\partial x \partial z} \quad \text{a)}$$

$$-\Delta \varphi + \frac{\beta}{\varepsilon_0 \varepsilon} \frac{\partial^2 \varphi}{\partial x \partial y} = 0 \quad \text{b)}$$

$$[\frac{\partial u}{\partial x} + -\frac{\beta}{\rho s^2} \cdot \frac{\partial \varphi}{\partial z}|x = +,-h/2] = 0 \quad \text{c)} \quad (10)$$

where we have the boundary condition (eq. 10c). In the case of free very thin film $\frac{h}{D/2 + h} \prec\prec 1$. We also considered to the viscosity coefficient Γ for the simulation. In the absence of current, the wave process in this element shows amplification if the synchronism between the sound transversal mode and the space charge wave exist:

$$k_0 \cong \sqrt{\frac{\omega_0}{s^2} - \frac{(2m+1)^2 \pi^2}{4H^2}} = \frac{\omega_0}{v_0} \quad (11)$$

where ω_0 is the frequency, s, v_0 is sound and space charges waves velocities, respectively; index m (1,2,3...) $\succ\succ 1$). H must have also an optimal value.

Resonance condition $\cos gH \cong 0$ must hold true. Parameters g and ω_{cr} are given by $g = \frac{\omega_{cr}}{s} = \frac{(2m+1)\pi}{2H}$ and $\omega_{cr}^2 = \frac{\pi^2 s^2}{4H^2}(2m+1)^2$.

These conditions for the amplification are better for the symmetric modes like:

$$u_n \propto B \cdot \sin gx \cdot e^{i(\omega t - kz)}, \quad \phi_n \propto \frac{ik\pi}{g} \cdot B \cdot \cos gx \cdot e^{i(\omega t - kz)}$$

Boundary conditions are taken in case of the absent of the acoustic contact of the film with the environment (eq. 10c). In the simulation, we use equations (5) and (10) and the resonance condition (11). We take into account to the 2D electron gas of n-GaAs thin film and the equations for slow changing amplitudes of acoustic-electromagnetic hybrid wave with potential φ.

IV. NUMERICAL SIMULATIONS

The numerical simulations show the effective amplification of hyper sound in the presence of the negative differential mobility (conductivity). The following parameters have been chosen: $n_0 \approx 10^{15} cm^{-3}$, the electron concentration in the film, $v_0 \approx 2 \cdot 10^7$ *cm/s* , the electron velocity, *L = 0.1 cm* , the length of the film, *H= 1 μm* ,the size of the wave-guide. The intensity of acoustic mode can reach value of *1 W/cm²* at microwave frequencies $\omega = 5 \cdot 10^{10} - 2 \cdot 10^{11}$ *s⁻¹*. In the Figs. 3 and 4, the amplification of the hyper sound (hybrid wave) are presented.

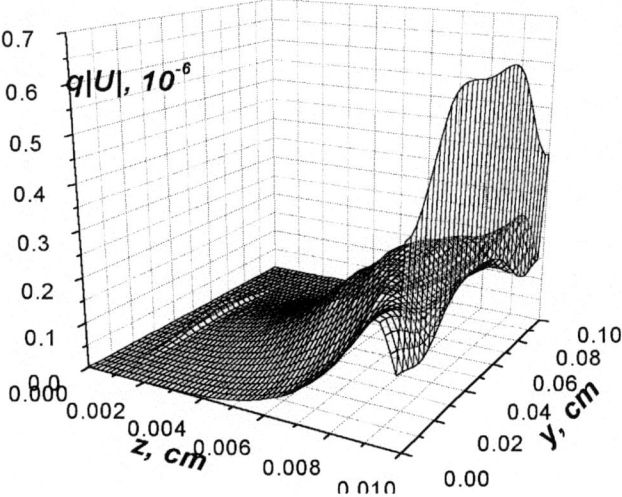

Fig. 3. Distribution of the amplitude of acoustic deformation $q|U|$ along the film for t = 4 ns. The resonant excitation is given at frequency $\omega = 10^{11}$ s⁻¹ of the hybrid mode, and the size of the wave-guide is $H = 1$ μ m.

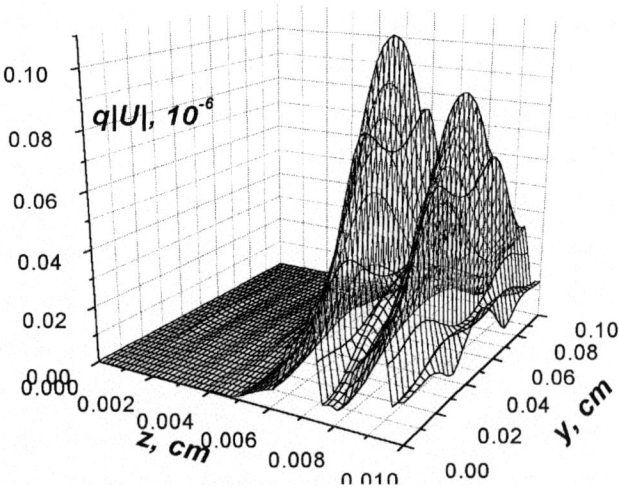

Fig.4 Distribution of the acoustic deformation amplitude along the film at t = 4 ns, with a size of wave guide H= 0.1 μ m.

Aditionally, in simulations were considered: $E^{\sim}(z,y,t) = E_{10}\exp(-((z-z_1)/z_0)^2 - ((y-y_1)/y_0)^2) \cdot \exp(-((t-t_1)/t_0)^2)$, the electric variable field distribution in input antenna, $z_1 = 0.001$ cm, $z_0 = 0.0001$ cm , $y_1 = 0.05$ cm, $y_0 = 0.02$ cm, $t_1 = 2.5$ ns, $t_0 = 1.25$ ns, $E_{10} = 200$ V/cm and $E_B = 4.56$ kV/cm, the drift field, $n_0 = 10^{11}$ cm⁻², the 2D concentration of electron gas, and q, the transverse wave number (index m) of the acoustic mode (approx. equal to 5).

V. CONCLUSIONS

The two mechanisms of the amplification of the hyper sound (traveling wave mechanism and the mechanism related with the Gunn effect, under negative differential mobility) and possible real structures for application like active filters and delay lines were shown here.

In the presence of current, an amplification of hybrid wave takes place due to the negative differential conductivity. This wave can excite to the acoustic modes of the film, due to piezoeffect.

The numerical simulations have demonstrated the effective excitation of the hyper sound in the presence of negative differential conductivity.

Also it is shown that the symmetric modes, emerging as transverse ones, interact more effectively with the space charge waves.

Another important result is the following: the dominating amplification mechanism was deformation potential at the frequencies $\omega \succ 2 \cdot 10^{11}$ s⁻¹.The mechanism of electrostriction will be very promising in future, with a new technology, which will be important for optoelectronics systems.

REFERENCES

[1] Kotsarenko N. Ya., Ostrovskiy I. V., "Generation of Ultra - Sound in the Piezosemiconductor Acoustic Wave Guide", Ukrainian Phizicheskiy Zhurnal, 15, N 9, 1565-1567, (1970).

[2] Kotsarenko N. Ya., Kucherov I. Ya.,Ostrovskiy I. V.,Protopopova L. F., Fedorchenko A. M., "Electrical Damping and Amplification of Lamb Waves in piezosemiconductors", Ukrainian Phizicheskiy Zhurnal, 16, N 10, 1707 - 1716, (1971).

[3] Kotsarenko Ya, S. V. Koshevaya, I. V. Ostrovskii, "Piezoelectric Interactions of the Transverse Normal Waves in 2 Layer Medium", Ukrainian Phizicheskiy Zhurnal, V. 18, No. 12, pp. 1937-1942, (1973).

[4] Levitsky S. M. and S. V. Koshevaya, "Vacuum and Solid State Electronics", "Vyshcha Shkola" Kiev publishing house, (1986), 271 pages, L2403000000-026/M211(04)-86 203-86.

[5] Barybin A. A., V. M. Prigorovsky, "The Waves in Thin Layers of Semiconductor with Negative Differential Mobility", Isvestiya VUZ, Fizika, V. 24, No. 18, pp. 28-41, (1981).

[6] Koshevaya S., J. Escobedo-A. , V.Grimalsky, M.Tecpoyotl-T., M.A. Basurto–P., "Superheterodyne Amplification of sub Millimeter Electromagnetic Waves in an n-GaAs Film", International Journal of Infrared and Millimeter Waves , V. 24(2), February, pp.201-209, (2003).

[7] Koshevaya S., V.Grimalsky, J. Escobedo-A. M.Tecpoyotl-T., "Superheterodyne Amplification of Sub Millimeter Electromagnetic Waves in an n-GaAs, Shur's Model", Microelectronics Journal, V. 34, pp. 172-177, (2003).

[8] Shur M., GaAs Devices and Circuits, Plenum Press, N.Y., (1989).

[9] Dieulesaint E. and D.Royer, Elastic Waves in Solids, Wiley, N.Y., (1980).

Poster Session
Opto and Microwave Devices and ICs

2006 25th International Conference on Microelectronics

Noise as a Diagnostic Tool for Quality of GaSb Laser Diodes

Z. Chobola, J. Vaněk, E. Hulicius, T. Šimeček

Abstract - Transport and noise characteristic of forward biased 2.3 μm CW GaSb laser diodes were measured in order to evaluate new technology. From the measurement results it follows that noise spectral density related to defects is of 1/f type and its magnitude was found to be proportional to the square of DC forward current at low injection levels.

I. INTRODUCION

Noise has been used for a long time as a diagnostic tool in device research [1]. Low frequency noise is a sensitive tool for degradation phenomena like electro-migration and short breakdown [2]. All types of noise, thermal, shot, generation-recombination and 1/f noise play different role in reliability analysis. The correlation between noise in a device, its reliability and the question why conduction noise, especially 1/f noise, is a quality indicator for devices is indicated in [3, 4]. Opportunities and limitations of the use of low/frequency noise as a diagnostic tool for device quality is discussed in Wandame [5].

II. EXPERIMENTAL RESULTS AND DISCUSSION

Fig. 1 shows U-I characteristic of forward-biased samples Nos.43712, 7672 and 7675. We can see excess current around $U_F = 0,3 - 0,4$ V.

The exponent β in $I = I_o \, e^{\beta \, U}$ plot equal between $\beta = 13,9$ V^{-1} to $\beta = 16,3$ V^{-1}.

The U-I characteristic give evidence of poor contact quality, with the contact resistance ranging from $R_S = 5.3$ Ω for specimen No. 7672 up to $R_S = 7$ Ω for specimen No. 7675.

Figs. 2 and 3 show the noise voltage power spectral density versus DC voltage plots. The noise voltage was measured across load resistance $R_L = 100$ Ω, $1 \, k\Omega$, $10 \, k\Omega$ and $100 \, k\Omega$ respectively. The pass band central frequency was 1 kHz, the band-width being equal to 20 Hz.

Z. Chobola and J. Vaněk are with the Department of Physics, Faculty of Civil Engineering, University of Technology Brno, Žižkova 17, 602 00 Brno, Česká Republika,
E-mail: chobola.z@fce.vutbr.cz
E. Hulicius and T, Šimeček are with the Institute of Physics, Academy of Sciences, Prague, Czech Republic.

The noise voltage spectral density for the specimens No. 7675. At DC voltage up to 0.2 V and load resistance $R_L = 100$ Ω and $R_L = 1 \, k\Omega$, the samples exhibit non-increasing S_U versus DC bias voltage plots, representing the thermal noise background. A marked excess current component can be observed at higher voltages, reaching maximum at the power match point, where the PN junction dynamic resistance equals the load resistance. This peak is shifting toward lower DC voltage if the load resistance is increased. In the noise spectral density vs. forward voltage plots, sample No. 7672, Fig. 3 has two peaks, whose positions appear to depend on the load resistance R_L.

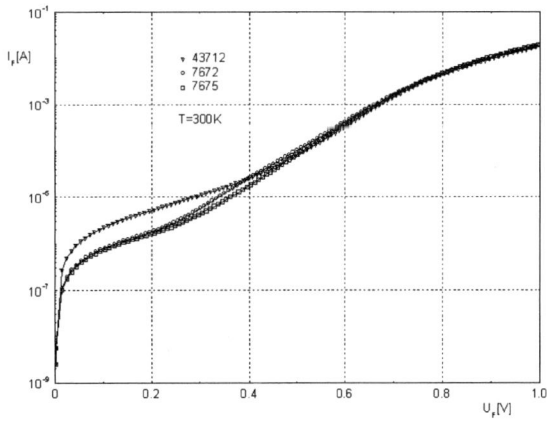

Fig. 1. I-V characteristic for laser diodes Nos. 43712, 7672 and 7675 under forward bias.

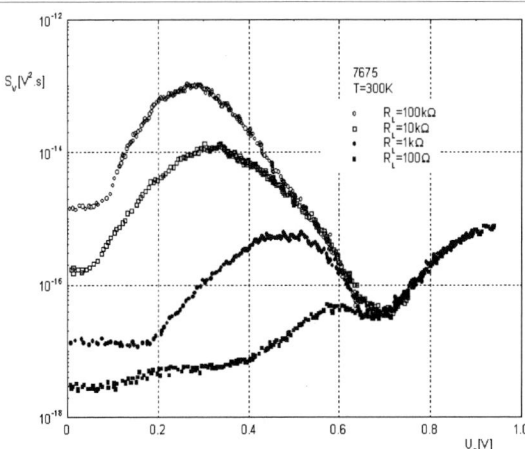

Fig. 2. The noise spectral density as a function of forward voltage for laser diode No.7675.

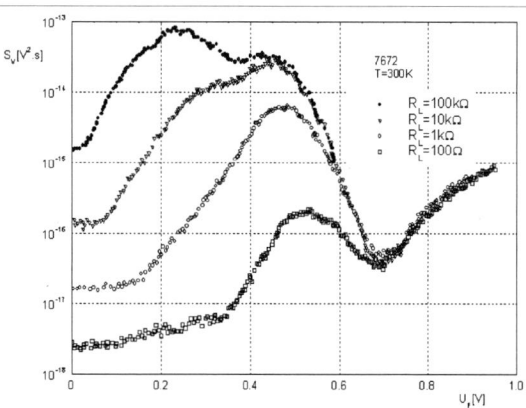

Fig. 3. The noise spectral density as a function of forward voltage for laser diode No.7672.

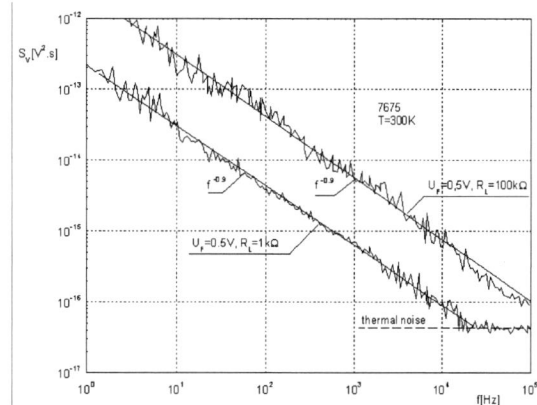

Fig. 4. The noise spectral density versus frequency for laser diode No.7675, $U_F = 0.5$ V, $R_L = 1$ kΩ and $R_L = 100$ kΩ.

For $R_L = 10$ kΩ, they correspond to voltage $U_{F1} = 0.3$ V and $U_{F2} = 0.45$ V, where as the corresponding peak voltages are $U_{F1} = 0.22$ V and $U_{F2} = 0.44$V for $R_L = 100$ kΩ.

It may be deduced that at least two separate structure defect related noise sources are present in the PN junction region.

The greatest excess noise we can see for sample No. 43712 when voltage $U_F = 0.3$ V Fig. 4. The spectral density S_U has maximum value $S_{UM} = 7.10^{-13}$ V^{2}s when $R_L = 100$ kΩ. Here we can see also two peaks.

At forward voltages exceeding $U_F = 0.7$ V, all samples show an increase in the excess noise component, which is characteristic of imperfect contacts, where resistance ranges from 5 Ω to 7 Ω.

Fig. 4 shows the noise voltage power spectral density versus frequency plots for the laser diode No. 7675. The noise voltage is measured across the load resistance $R_L = 1$ Ω and $R_L = 100$ kΩ, the DC forward bias voltage being $U_F = 0.5$ V and $U_F = 0.3$ V, i.e. in the power match region.

It may be seen that the excess noise component is of $1/f^\alpha$ type when α is very close to unity. This noise component is masked be thermal noise at higher frequencies.

III. CONCLUSION

Our experimental studies of GaSb laser diodes have shown that these diodes are of rather "inferior quality", which concerns both the PN junctions and the contacts as compared to wide-gap GaAs based lasers. The U-I characteristics have marked excess current component occurring at DC forward voltages below $U_F = 0.2$ V to 0.4 V. Also the contact resistance $R_S = 3$ Ω to 7 Ω should be improved.

These assumptions concerning the imperfections being present in the samples under study have been fully confirmed by our noise measurements, indicating $1/f^\alpha$ - type excess noise components for both the PN junction and the contact region.

Moreover sample No. LD 7672 contains at least two different types of defects , which were manifested by two separate local maxima in the spectral density versus frequency plots.

ACKNOWLEDGEMENT

This research has been supported by project of GACR No.102/04/0142.

REFERENCES

[1] M. S. Gusta, "Applications of Electrical Noise", in *Proceedings od IEEE*, vol. 63, no 7, 1975, pp. 996 – 1010.

[2] C. Cioti, B. Nevi, "Low – frequency noise measurements as a characterisation tool for degradation phenomena in solid-state devices". in *Journal of Physics – applied physics*, vol. 33, no. 21, 2000, pp. R199 – R216.

[3] L. K. L. Vandamme, "Noise as a Diagnostic Tool for Quality and Reliability of Electronic Devices", in *IEEE Transaction on Electron Devices*, vol. 41, no. 11, Now. 1994, pp. 2176 – 2187.

[4] Z. Chobola, "Noise as a tool for non-destructive testing of single-crystal silicon solar-cells", in *Microelectronics Reliability*, vol. 41, no. 12, Dec. 2001, pp. 1947 – 1952.

[5] L. K. L. Vandamme, "Opportunities and limitations to use low-frequency noise as a diagnostic tool for device quality", in *Proceedings of the 17th International conference ICNF 2003*, Prague 2003, pp. 735 – 748.

2006 25th International Conference on Microelectronics

Nonlinear Surface Ultrasonic Monopulses in Solid Film–Substrate System

V. Grimalsky, S. Koshevaya, E. Gutierrez- D., and A. Garcia-B.

Abstract - An excitation of ultra-high frequency (100 MHz – 1 GHz) nonlinear solitary acoustic waves, propagating along the interface between a solid film and a solid substrate, is theoretically analyzed. A quadratic elastic nonlinearity in both contacting media of the system is taken into account in the case of baseband waves. The Benjamin–Ono (B-O) type equation, due to the character of the linear wave dispersion, describes the baseband waves, but they are not true monopulse-like solitons and manifest as single shock-like pulses in the case of an excitation from the long initial rectangular pulse.

I. INTRODUCTION

During the last years, research of nonlinear wave phenomena in bounded solids in ultra-high frequency and microwave ranges has been of great interest [1-13]. Especially, nonlinear wave effects in structures, which include thin solid films, are perspective for experimental observations, due to a possibility to control the linear and nonlinear wave properties.

The surface ultrasonic waves (SAW) possess attracting properties for observing nonlinear phenomena. The surface acoustic waves are relatively slow and are localized at the interfaces. The integral power levels, which are necessary for a manifestation of nonlinearity, are quite low. The combination of materials of the film and the substrate makes possible to manage both the wave dispersion and the nonlinearity.

The nonlinear acoustic wave propagation in the solid layered systems has been considered in many papers [3-13]. The most results of simulations were obtained within an approximation of moderate nonlinearity, where the wave was presented as a superposition of partial harmonics with slowly varying amplitudes, but the transverse profiles of partial harmonics were assumed as weakly changed, in a comparison to the linear case. Those results can be summarized as follows. The baseband pulses can be excited experimentally [11,12,13] but they are not true solitons, because they do not preserve their shapes after collisions, as numerical simulations demonstrated [8,13].

V. Grimalsky, E. Gutierrez-D., and A.Garcia-B. are with National Institute for Astrophysics, Optics, and Electronics (INAOE), Puebla 72000, Pue., Mexico, E-mail: vgrim@inaoep.mx.
S. Koshevaya is with CIICAp, Autonomous University of Morelos (UAEM), Cuernavaca 62210, Mor., Mexico.

In the experiments, the initial acoustic pulses were quite short: 2 – 20 ns. The problem of formation of nonlinear baseband pulses from long initial ones was not considered.

The existence of envelope solitons has been pointed out in several papers [3,5,7,11]. The envelope solitons are governed by the nonlinear Schrödinger equation or its generalizations [11].

In this paper, a possibility of excitation of ultra-high frequency (100 MHz – 1 GHz) moderately nonlinear solitary surface acoustic waves propagating along the interface between a solid thin film and a substrate is under investigation. The quadratic elastic nonlinearity both in the film and in the substrate is taken into account in the case of baseband (monopulse) waves. When the exciting force is monopulse-like, then we can expect an occurrence of a baseband nonlinear wave; when the initial force possesses a certain carrier frequency, an envelope soliton can be excited. Note that baseband and envelope pulses lie in different frequency ranges. It has been shown that true baseband solitons are absent in such a system. Moreover, it seems impossible to approximate the nonlinear baseband wave dynamics by means of any fully integrable model.

II. BASIC EQUATIONS AND LINEAR DISPERSION RELATION

The surface acoustic wave propagation in the system a thin solid film – solid substrate is under investigation. Below, we consider isotropic contacting media without slipping. Usually, the Lagrangian coordinate frame is used within the solid media [1,2,4,14,15,16]. The dynamic equation for the mechanical displacement vector *u* in solids has the next form [1,15,16]:

$$\rho_0 \frac{\partial^2 u_i}{\partial t^2} = \frac{\partial \sigma_{ij}}{\partial x_j}; \sigma_{ij} = \sigma_{ij}{}^L + \sigma_{ij}{}^{NL} \qquad (1)$$

where ρ_0 is the unperturbed density of solid and σ_{ij} is the Piola – Kirchhoff tensor.

For an isotropic solid, the expression of σ_{ij} is [1,3]:

1-4244-0116-X/06/$20.00 ©2006 IEEE

$$\sigma_{ij} = \rho_0(s_l^2 - 2s_t^2)div\bar{u}\,\delta_{ij} + 2\rho_0 s_t^2 u_{ij} +$$

$$+ C(div\bar{u})^2\delta_{ij} + (\rho_0 s_t^2 + A/4)\left(\frac{\partial u_i}{\partial x_l}\frac{\partial u_j}{\partial x_l} + \frac{\partial u_i}{\partial x_l}\frac{\partial u_l}{\partial x_j} + \frac{\partial u_l}{\partial x_i}\frac{\partial u_l}{\partial x_j}\right) + \quad (2)$$

$$+ (\rho_0(s_l^2 - 2s_t^2) + B)\frac{\partial u_i}{\partial x_j}div\bar{u} + (A/4)\frac{\partial u_j}{\partial x_l}\frac{\partial u_l}{\partial x_i} + B\frac{\partial u_j}{\partial x_i}div\bar{u} +$$

$$+ (\rho_0(s_l^2 - 2s_t^2)/2 + B/2)(\frac{\partial u_l}{\partial x_m})^2\delta_{ij} + (B/2)\frac{\partial u_l}{\partial x_m}\frac{\partial u_m}{\partial x_l}\delta_{ij}$$

In such an expansion, the quadratic nonlinear terms are given. The coefficients in nonlinear terms are combinations of both material nonlinear modules (A,B,C) and linear ones, due to a presence of geometric nonlinearity [3]. Here $s_{l,t}$ are longitudinal and transverse acoustic velocities; u_i are the components of a mechanic displacement For a sake of simplicity, we have assumed in the figures presented below that the following condition is satisfied: $|C| >> |A|, |B|$. Thus, the nonlinear term $C(div u)^2 \cdot \delta_{ij}$ in the expansion (2) is assumed as dominating. For some glass materials, this inequality is valid [1]. In a general case, the coefficients of nonlinear equations are expressed through the combination of all material nonlinear and linear modules, but governing equations for slowly varying amplitudes preserve the same structure. The results of the simulation presented below are of qualitative character and do not depend essentially on the values of nonlinear modules. The wave dissipation due to viscosity, proportional to the square of frequency, is also taken into account, which is not presented in the Eq. (1).

The ordinary boundary conditions at the interface film - substrate are used; namely, the components of displacement vector u_i and the normal components of the tension force $\sigma_{ij}n_j$ are the same, where n is vector normal to the unperturbed interface (vector n is directed along OX-axis).

The geometry of the system under consideration is described here. The axis OZ is directed horizontally along the wave propagation in the interface plane, OX one is directed vertically upwards, see Fig. 1,a. The thickness of the film ($-h < x < 0$) is $h = const$. The solid substrate occupies the space $x < -h$. The surface $x = 0$ is assumed as free.

In this system, linear surface acoustic waves can propagate with the displacement components $u = (u_x, 0, u_z)$. The dispersion equation for the sound waves has been obtained from subjecting the partial solutions of the linearized dynamic Eq.(1) in each region to the boundary conditions, and it can be expressed in the matrix form.

The dispersion curve for the lowest acoustic mode is presented in Fig.1,b. The contact of the Si(Ge)O$_2$ film with Si substrate is considered. The following parameters are used: $h = 0.25 \ \mu m$, $c_t = 2.5 \cdot 10^5$ cm/s, $c_l = 4.0 \cdot 10^5$ cm/s, $\rho_1 = 3.0$ g/cm^3 (Si(Ge)O$_2$); $s_t = 5.0 \cdot 10^5$ cm/s, $s_l = 8.4 \cdot 10^5$ cm/s, $\rho_2 = 2.33$ g/cm^3 (Si), where $c_{l,t}, s_{l,t}$ are longitudinal and transverse bulk velocities in the film and the substrate, respectively; $\rho_{1,2}$ are densities. A crystalline anisotropy in Si is neglected here. These waves possess an essential

dispersion [3], see Fig.1,c. We note that $\partial^2 k/\partial \omega^2$ has different signs in the frequency regions $\omega < \omega_{crit}$ and $\omega > \omega_{crit}$, where ω_{crit} is the value of the frequency of the wave where the inflection of dispersion curve takes place: $\partial^2 k/\partial \omega^2 = 0$. The frequency spectrum of the baseband waves is in the frequency interval $\omega < \omega_{crit}$.

III. BASEBAND NONLINEAR WAVES

In our analysis, nonlinearity is assumed as moderate. This gives a possibility to use the spectral method for a consideration of the baseband wave dynamics. Namely, it means that the localized nonlinear wave can be presented as the set of harmonics with slowly varying amplitudes $A_j(z)$, but the transverse profiles $f_j(x)$ of the harmonics are almost the same as in the linear case [8-10]:

$$\frac{\partial \bar{u}}{\partial t} = \frac{c_t}{2}\sum_j A_j(z)\vec{f}_j(x)e^{i\omega_j\eta} + c.c. \quad (3)$$

where $\omega_j = (2\pi/T)j$, $(j=1,...,N)$, $\eta = t - z/v_0$ is the "running time", $v_0 = (\partial\omega/\partial k)|_{k=0}$. T is the temporal domain where the dynamics of localized nonlinear pulse occurs. Therefore, with a respect to η, periodical boundary conditions are applied. To derive the equations for the slowly varying amplitudes A_j (z), the following reciprocity relation has been used [16,17]:

$$\frac{\partial}{\partial t}(\rho\frac{\partial u_i^{(1)}}{\partial t}\frac{\partial u_i^{(2)*}}{\partial t} + c_{ijkl}u_{ij}^{(1)}u_{kl}^{(2)*}) = \frac{\partial}{\partial x_j}[(c_{ijkl}u_{kl}^{(1)} +$$

$$+ \sigma^{nl}_{ij})\frac{\partial u_i^{(2)*}}{\partial t} + c_{ijkl}u_{kl}^{(2)*}\frac{\partial u_i^{(1)}}{\partial t}] - \sigma^{nl}_{ij}\frac{\partial}{\partial t}\frac{\partial u_i^{(2)*}}{\partial x_j}; \quad (4)$$

$$\frac{\partial \bar{u}_j^{(1)}}{\partial t} = \frac{c_t}{2}A_j(z)\vec{f}_j(x)e^{i\omega_j\eta}; \quad \frac{\partial \bar{u}_j^{(2)*}}{\partial t} = c_t\vec{f}_j^{\,*}(x)e^{-i\omega_j\eta}$$

where $u_j^{(1)}$ are the positive frequency component of the mechanical displacement of the nonlinear wave of the frequency ω_j; $u_j^{(2)*}$ are the negative frequency component of the monochromatic linear wave; $c_{ijkl} = \rho_1 c_t^2 \cdot (\delta_{ik}\delta_{jl} + \delta_{il}\delta_{jk}) + \rho_1(c_l^2 - 2c_t^2)\cdot\delta_{ij}\delta_{kl}$ for the isotropic film (analogously for the substrate), σ^{nl}_{ij} are the components of the nonlinear part of the Piola – Kirchhoff tensor [1,3,15]. The normalization of the transverse profiles f_j has been chosen to the unity value of the flux of the energy of the mode. The integration of the Eq. (4) over the transverse coordinate x gives a possibility to obtain the orthogonality relations for the linear own modes and to derive the coupled equations for slowly varying amplitudes in the nonlinear case. The wave dynamics does not depend on the choice of the parameter T for the enough large values of T and for a number of used harmonics more than 200. Namely, several simulations with the same parameters have been provided with different values of T, and the results should be independent on the value of T. An alternative approach to derive the equations for amplitudes of harmonics is based on averaged Hamiltonian or Lagrangian function of the elastic field [4,13].

The propagation of baseband nonlinear wave is described by a set of coupled ordinary differential equations for harmonics [8], and the total number of harmonics is of about 256 - 1024. The linear wave dispersion law near $\omega = 0$ is:

$$k(\omega) = (\omega / v_0) - g\omega \mid \omega \mid . \qquad (5)$$

There exists the standard nonlinear Benjamin – Ono (B-O) equation [3,8,18]:

$$\frac{\partial U}{\partial z} + N \frac{\partial}{\partial \eta}(U^2) - \Gamma \frac{\partial^2 U}{\partial \eta^2} + \frac{g}{\pi} V.P. \frac{\partial^2}{\partial \eta^2} \int_{-\infty}^{+\infty} \frac{U(\eta',z)}{\eta'-\eta} d\eta' = 0$$

where $U(z, \eta)$ is slowly varying profile of the wave; $\eta = t - z/v_0$ is the "running time". The B-O equation describes the nonlinear propagation of waves in some systems possessing the linear dispersion law (5). The excitation of true baseband solitons within the framework of B-O equation was investigated in details in the literature [18]. A difference of the nonlinear baseband waves in the system film – substrate from the solitons of B-O equation is discussed below.

Both the B-O equation and one describing the dynamics of wave monopulse in the layered system film - substrate can be reduced to the following set of equations for amplitudes of harmonics:

$$\frac{dA_j}{dz} + i(k_j - jk_1)A_j + \Gamma_j A_j = i\omega_j (\sum_{m<j} P_{m,j}A_m A_{j-m} +$$

$$+ 2 \sum_{m>0} P_{m,m+j}A_{m+j}A_m^*) \qquad (6)$$

where $k_j \equiv k(\omega_j)$ is determined from linear dispersion equation; $\Gamma_j = j^2 \cdot \Gamma_1$ is the dissipation of the jth harmonic. In simulations devoted to the baseband waves it has been taken $\Gamma(\omega = 10^9 \text{ s}^{-1}) = 0.5 \text{ cm}^{-1}$. The quadratic nonlinearity is dominating there. The coefficients $P_{m,j}$ are expressed through the linear and nonlinear modules of two contacting media and, due to their complexity, they are not presented here. Using the relation (4), the nonlinear boundary conditions for stresses at the interfaces are taken into account simultaneously, and there is no necessity to separate the nonlinearity in the volume equations and in the boundary conditions here, in a distinction to [14].

In the case of the model described by the B-O equation, all the coefficients $P_{j,m}$ are equal. In our case, they increase with frequency growth. The numerical simulations show that the baseband nonlinear wave in this

system with above-pointed nonlinearity differs from the B-O solitons.

In our simulations, the generation of nonlinear baseband waves from initial almost rectangular pulse has been investigated (see Fig.2). Note that there exists the self-similar solution of the Eqs.(6), which describes the travelling wave with the constant velocity of propagation [8,13]. But, because the set of equations for the nonlinear acoustic surface wave is not fully integrable, it is impossible to state that such a solitary travelling wave must be generated from each initial data of enough large value [18]. In another words, it is impossible to separate the soliton and non-soliton parts of the nonlinear wave there, because of an absence of true solitons.

The nonlinear baseband wave possesses certain asymmetry under its propagation (see the Fig.2). The value of the nonlinear coefficient in both media has been taken as $C = -2 \times 10^{12}$ g·cm^{-1}·s^{-2} ($C/\rho_l c_t^2 = -10$). The polarity of the nonlinear pulse depends on the sign of quadratic nonlinear module. Stable true solitary wave is not formed during the evolution of input long rectangular pulses. The nonlinear baseband wave in such a system manifests rather like a shock-like pulse. When the input rectangular pulse is taken enough long (Fig.2, c,d), it is possible to obtain the nonlinear pulses of large peak values, where an approximation of moderate nonlinearity ceases to be valid. To generate the pulse with the duration ≤ 10 ns, it is necessary to use the structures with the thickness of the film $h < 1 \mu$m. The frequency range of the pulse under maximal compression ($z \sim 5 - 10$ cm) lies within the interval $0 < \omega < 10^9$ s^{-1}, namely, $\omega << \omega_{crit}$. Therefore, the approximation (5) is valid here. The similar dynamics of nonlinear wave excitation has been obtained for different values of nonlinear parameters A,B,C (Eq. 2), because the values of nonlinear coefficients $P_{m,j}$ are not constant for any set of nonlinear parameters of the system film – substrate. Note that in the case of fully integrable B-O system, it is possible to excite several solitons from the long initial pulse.

A formation of short acoustic monopulses (the duration ~ 1 ns, the longitudinal size $\sim 2 \mu$m) can be useful for a treatment and testing [9] of interfaces in microelectronic systems.

IV. CONCLUSIONS

Simulations of the baseband nonlinear wave dynamics in the layered system, which includes a film of a thickness of $0.2 - 2$ micrometers and half-infinite substrate, demonstrate a possibility of the formation of ultra-high frequency envelope solitons and collapsing pulses. The dynamics of baseband pulses has demonstrated the essential non-soliton behavior. The formation of ultrashort (~ 1 ns) acoustic shock-like pulses is important for signal processing and acoustic treatment of interfaces in microelectronics.

REFERENCES

[1] K. Naugolnykh, L. Ostrovsky, *Nonlinear Wave Processes in Acoustics*. Cambridge University Press, Cambridge, UK, 1998.

[2] D.F. Nelson, E*lectric, Optic, and Acoustic Interactions in Solids*. Wiley, N.Y., 1979.

[3] A.P. Mayer, "Surface acoustic Waves in Nonlinear Elastic Media", *Phys. Reports*, Vol. 256, May 1995, pp.237-366.

[4] M.F. Hamilton, Yu.A. Il'inskii, and E.A. Zabolotskaya, "Nonlinear Surface Acoustic Waves in Crystals", *J. Acoust. Soc. Am.*, Vol. 105, Febr. 1999, pp.639-651.

[5] A.V. Porubov and A.M. Samsonov, "Long Non-Linear Strain Waves in Layered Elastic Half-Space", *Intl. J. Non-Linear Mechanics*, Vol. 30, Nov. 1995, pp.861-877.

[6] Y.B. Fu and S.L.B. Hill, "Propagation of Steady Nonlinear Waves in a Coated Elastic Half-Space", *Wave Motion,* Vol. 34, June 2001, pp.109-129.

[7] C. Eckl, J. Schloemann, A.P. Mayer, A.S. Kovalev, and G.A. Maugin, "On the Stability of Surface Acoustic Pulse Trains in Coated Elastic Media", *Wave Motion*, Vol. 34, June 2001, pp.35-49.

[8] C. Eckl, A.S. Kovalev, and A.P. Mayer, "Do Surface Acoustic Solitons Exist?", *Phys. Rev. Lett.*, Vol. 81, Aug. 1998, pp.983-986.

[9] A.M. Lomonosov and P. Hess, "Impulsive Fracture of Silicon by Elastic Surface Pulses with Shocks", *Phys. Rev. Lett.*, Vol. 89, Aug. 2002, 095501, 4 pp.

[10] A.S. Kovalev, A.P. Mayer, C. Eckl, and G.A. Maugin, "Solitary Rayleigh Waves in the Presence of Surface Nonlinearities", *Phys. Rev. E*, Vol. 66, Sept. 2002, 036615, 15 pp.

[11] A.P. Mayer and A.S. Kovalev, "Envelope Solitons of Acoustic Plate Modes and Surface Waves", *Phys. Rev. E*, Vol. 67, June 2003, 066603, 14 pp.

[12] A.M. Lomonosov, P. Hess, and A.P. Mayer, "Observation of Solitary Elastic Surface Pulses", *Phys. Rev. Lett.* Vol. 88, Febr. 2002, 076104, 4 pp.

[13] C. Eckl, A.S. Kovalev, A.P. Mayer, A.M. Lomonosov, and P. Hess, "Solitary Surface Acoustic Waves", *Phys. Rev. E*, Vol. 70, Oct. 2004, 046604, 15 pp.

[14] Mingxi Deng, "Second-harmonic generation of Lamb modes in a solid layer supported by a semi-infinite substrate", *J.Phys.D: Appl. Phys.*, Vol. 37, May 2004, pp. 1385-1393.

[15] G.A. Maugin, *Nonlinear Waves in Elastic Crystals*. Oxford Univ. Press, N.Y., 1999.

[16] B.A. Auld, *Acoustic Fields and Waves in Solids, Vol.2*. Wiley, N.Y. 1973.

[17] D. Royer and E. Dieulesaint, *Elastic Waves in Solids 2: Generation, Acousto-Optic Interaction, Applications*. Springer-Verlag, N.Y., 1999.

[18] M. Ablowitz and H. Segur, *Solitons and the Inverse Scattering Transform*. SIAM, Philadelphia, 1981.

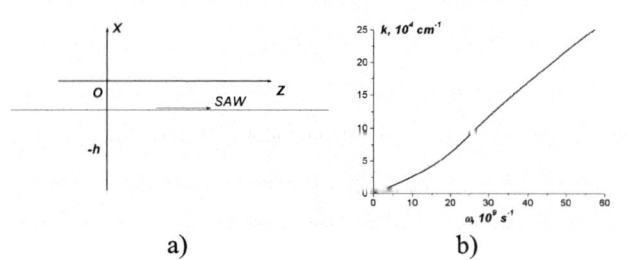

a)

b)

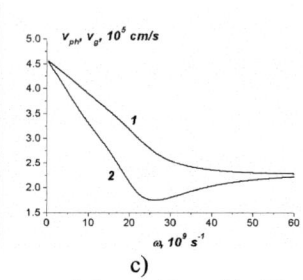

c)

Fig.1. The geometry of the problem (a). The film occupies the region $-h < x < 0$, the substrate is $x < -h$. The free surface is $x = 0$. Linear wave dispersion for surface acoustic wave in the layered solid system film – substrate where $\omega_{crit} = 2.62 \cdot 10^{10}\ s^{-1}$ for used parameters of Si(Ge)O$_2$ film – Si substrate (b). Corresponding dependencies of phase (curve 1) and group (curve 2) velocities on frequency (c). The thickness of the film is $h = 0.25\ \mu$m.

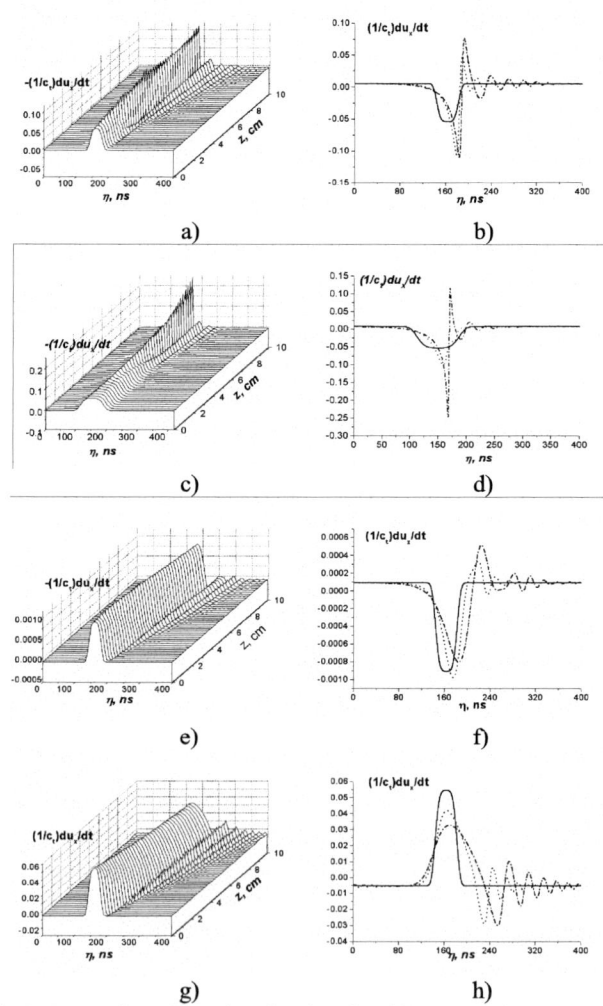

a)

b)

c)

d)

e)

f)

g)

h)

Fig.2. Dynamics of baseband pulses in Si(Ge)O$_2$ –Si structure: a) is spatial-temporal dynamics of the vertical component of velocity at the free surface of the film (note that $-V_x$ is presented); b) is the temporal distributions of the pulse at $z = 0$ (solid line), at $z = 5$ cm (dot line), and at $z = 10$ cm (dash-dot line); c) and d) are the same as a) and b), but for twice wider input pulse; e) and f) are the same as a) and b), but for smaller input pulse (linear regime); g) and h) are the same as a) and b), but for the opposite polarity of input pulse. The thickness of the film is $h = 0.25\ \mu$m.

2006 25th International Conference on Microelectronics

Accurate Noise Modeling of HEMT for Low-Noise Applications

I. H. Kang, S. C. Kim, W. Bahng, and N. K. Kim

Abstract – A novel physical device model was developed for designing a low-noise-application-oriented HEMT structure and modeling a high-frequency noise characteristic of HEMT. To calculate accurate noise figure as a figure-of-merit indicating the high-frequency noise characteristics, two factors were considered based on Ando's model: (1) The parasitic effects which are more dominant at higher frequency, including a gate-to-drain shunt feedback capacitance and a gate resistance, were taken into account. (2) The electron saturation velocity was assumed to depend on the gate length. Through the above two modifications, we calculated sheet carrier density, DC, RF parameters and noise figure (*NF*).

I. INTRODUCTION

The High Electron Mobility Transistor (HEMT) has been widely used for circuits in microwave and millimeter-wave frequency regions, due to its ultra-wide bandwidth and superior high-frequency noise characteristics. The analytic, CAD-oriented modeling of DC, RF, and noise characteristics of HEMT on the basis of semiconductor physics is more attractive for circuit simulators such as SPICE, ADS, and so on, because an accurate physical model can easily predict the device performances with respect to variation of the device geometry and structure [1]. Ando's analytical noise model for GaAs-based HEMT is very simple and easier to estimate noise performance of HEMT, compared with a numerical model based on the relaxation time approximation [2]. Although his model has shown ability to accurately predict noise figure of HEMT at relatively low frequency, it showed a serious discrepancy between experimental and estimated values as the frequency is increased. In this paper, a novel accurate noise model for HEMT will be presented based on Ando's model, by adding frequency-dependent parasitic effects, feed-back capacitance, C_{gd}, and gate-length-dependent electron saturation velocity.

II. THEORY

Ando's model was developed based on previous pioneer works on noise modeling of MESFET done by Van der Ziel, Pucel and Brookes [3]. However, two major modifications were added in his model to improve the

I. H Kang, S. C. Kim, W. Bahng, and N. K. Kim are with the Power Semiconductor Research Group, Korea Electrotechnology Research Institute, 28-1 Seongju-dong, Changwon-si, Gyeongsangnam-do, Korea, Email: ihkang@keri.re.kr

accuracy. First, a sheet electron concentration in the channel, $n_s(E_F)$ was approximated to the following equation to obtain accurate charge-control model :

$$E_F(x) = \gamma_f \cdot n_s^{\alpha} - E_{F0} \tag{1}$$

where E_F is the Fermi level, γ_f, α, and E_{F0} are fitting parameters. The fitting parameter, α is generally approximated to 2/3. A numerical self-consistent solution of Poisson and Schrödinger equations for HEMT structure is used to extract γ_f and E_{F0}. The extraction of the fitting parameter γ_f is important because it has much influence on the noise performance due to its influence on both the two-dimensional electron gas (2DEG) capacitance and the bias-dependent 2DEG equivalent thickness, which especially affects the noise generated in the saturation region. Although Ando's charge control model does not predict 2DEG saturation at low gate bias, this restriction is not much fatal in using this model since the HEMT in a low-noise application is biased typically below a gate bias for the maximum transconductance, g_m.

Second, a diffusion constant is modified. In many previous noise models, the diffusion constant was assumed to be independent of electric field when the high-field diffusion noise in the channel is calculated. However, the diffusion constant has a strong dependence on electric field in reality. In his model, the diffusion constant applicable even to high electric field was approximated to:

$$D_x = \frac{k_B T_0 v_{sat}}{q E_x} + D_h \tag{2}$$

where k_B, T_0, and v_{sat} are Boltzman constant, lattice temperature, and the saturation velocity respectively. E_x is the lateral electric field, and D_h is the diffusion constant at high electric field.

Basically, Ando's modeling follows the next steps: (1) a calculation of DC characteristics based on a charge control model, (2) an extraction of small-signal equivalent circuit parameters, (3) a determination of noise sources, and (4) finally a calculation of NF based on well-known noise theory.

Although Ando's noise model could estimate noise figure of HEMT accurately at low frequency, it showed larger deviation from measurements at higher frequency. Therefore, some modifications were added to increase the

1-4244-0116-X/06/$20.00 ©2006 IEEE 313

accuracy of the noise characteristics of HEMT at a high frequency, while following the above steps suggested in Ando's noise model. As the first step, a self-consistent analysis of Poisson and Schrödinger equations was carried out for the HEMT structures. The relationship between the 2DEG density, n_s and the Fermi level, E_F is shown in Fig. 1 for different device structures. Using this relationship, the fitting parameters, γ_f and E_{F0} are derived.

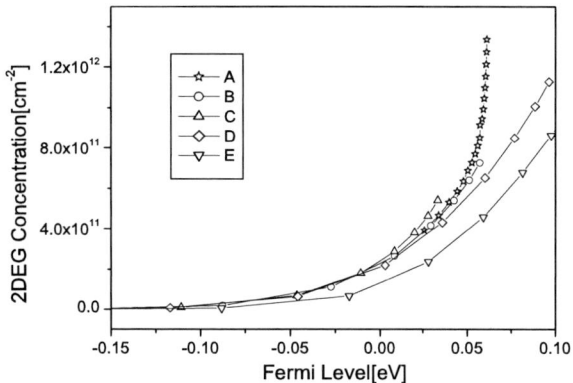

Fig.1. Relationship between 2DEG concentration versus Fermi level calculated from the 1-D self-consistent analysis for conventional AlGaAs/GaAs HEMT and AlGaAs/ InGaAs /GaAs p-HEMT structures.

It is observed that these parameters have dependence on device structure including doping concentration and thickness of barrier layer and conduction band offset. Table 1 shows the list of the extracted fitting parameters for different HEMT structures. The fitting parameters were used to estimate DC characteristics and small-signal equivalent circuit parameters of the transistors.

TABLE 1.

DEPENDENCE OF FITTING PARAMETERS ON DEVICE STRUCTURES INCLUDING CONDUCTION BAND OFFSET (ΔE_C), BARRIER LAYER THICKNESS (d) AND DOPING CONCENTRATION (N) FOR AlGaAs/GaAs (Σ) AND AlGaAs/InGaAs p-HEMT (Ω) STRUCTURES.

Type	A	B	C	D	E
Structure	Σ	Σ	Σ	Σ	Ω
ΔE_C	0.12	0.12	0.08	0.19	0.19
d [Å]	400	200	200	200	200
N [×10^{18}cm^{-3}]	2.5	4.0	4.0	4.0	4.0

The small-signal equivalent circuit used by Ando's noise model along with the modified one is shown in Fig. 2. The accuracy of the noise model at high frequency was improved by addition of a frequency-dependent gate resistance and a gate-to-drain feedback capacitance. In Ando's noise model, the variation of gate resistance as a function of frequency was ignored.

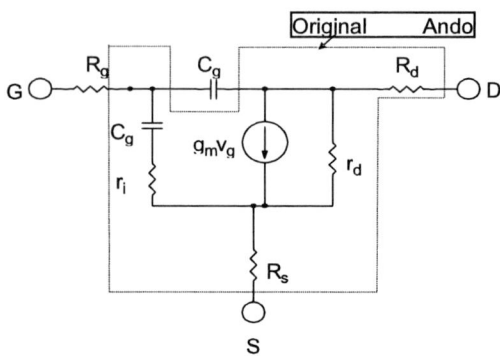

Fig. 2. Small-signal equivalent circuit for Ando's and the modified Ando's model.

However, as the frequency is increased, the gate resistance increases due to a decrease in skin depth, resulting in an increase in NF. The parasitic gate resistance is given by the following equation:

$$R_g = \frac{\rho Z_g}{3m^2 h L_g} \qquad (3)$$

where ρ, m, and h are a resistivity of gate metal, number of gate finger, and an effective thickness of gate metal respectively, and Z_g is a total gate width. The factor 1/3 is used to consider the distributed feature of the gate at high frequency [4]. In Eq. 3, h is not constant, but is expressed as

$$h = \begin{cases} \sqrt{1/(\pi f \mu \sigma)} \equiv \delta \\ h_0 \end{cases} \qquad (4)$$

where μ and σ are a permeability and a conductivity of gate metal respectively, and h_0 is a real thickness of gate metal.

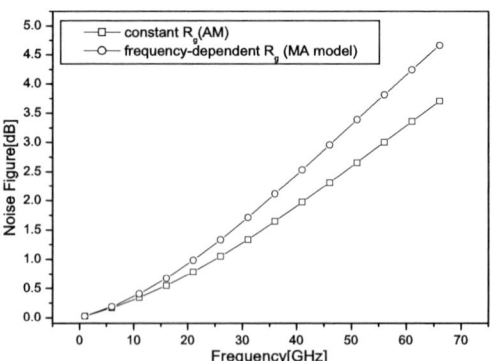

Fig. 3. Effects of R_g as a function of frequency on NF, compared with NF predicted by Ando model assuming constant R_g. They were calculated from AlGaAs/GaAs HEMT which was biased at V_{dd}=3V and I_{ds} for the optimum NF.

314

h is replaced by a skin depth (δ) at higher frequency, where the skin depth becomes smaller than h_0. Fig. 3 shows the effect of adding frequency-dependent gate resistance on the noise performance of HEMT. As shown in Fig. 3, the Ando's noise model using constant gate resistance underestimates the NF of device significantly at high frequency.

An inclusion of a gate-drain feedback capacitance C_{gd} is essential to improve the accuracy of the noise model at high frequency. The feedback capacitance can be approximated to:

$$C_{gd} \approx \frac{\varepsilon^2 Z_g^2 L_g}{2d_i C_{gs} - 1.5\varepsilon Z_g L_g} \tag{5}$$

where d_i is a gate insulator thickness [5].

After the small-signal equivalent circuit is constructed, the noise sources are calculated. Another modification considered in this work is an electron saturation velocity v_s. As the device is scaled down to submicron, the non-stationary transport effects become significant and it must be considered. In addition, v_s also has a large impact on DC characteristics and thus on both small-signal parameters and NF in Ando's model, as addressed briefly in [2]. The equivalent v_s therefore is considered and reflected in this model by means of the empirical model suggested by Matthias *et al.* [5], and is given as

$$v_s = \frac{v_{s0}}{\sqrt{L_g}} \tag{6}$$

To include the parasitic effects mentioned above, an equivalent noise resistance R_n, an equivalent noise conductance G_n, and a correlation admittance Y_{cor} should be transformed into the following equations respectively:

$$R_n' = R_n |n_{11} + n_{12}Y_{cor}|^2 + G_n |n_{12}|^2 \tag{7}$$

$$G_n' = \frac{G_n R_n}{R_n'} |n_{11}n_{22} - n_{12}n_{21}|^2 \tag{8}$$

$$Y_{cor}' = \frac{R_n}{R_n'}(n_{21} + n_{22}Y_{cor})(n_{11}^* + n_{12}^* Y_{cor}^*) + \frac{G_n}{R_n'}n_{22}n_{12}^* \tag{9}$$

where the transform matrix parameters for the shunt feedback are given as

$$n_{11} = \frac{S_{21}C_2'}{S_{21}C_2' + S_{21}'C_2} \tag{10}$$

$$n_{12} = 0 \tag{11}$$

$$n_{21} = \frac{S_{21}P - S_{21}'Q}{S_{21}C_2' + S_{21}'C_2} \tag{12}$$

$$n_{22} = 1 \tag{13}$$

In Eq. 9-12, [S] and [S'] are the s-parameter extracted from the simple Ando's small signal model and the s-parameter extracted from a shunt circuit composed of only gate-drain capacitance itself respectively. P, Q, C_2 and C_2' are given as follows.

$$P = (1 - S_{11}')(1 - S_{22}') + S_{12}'S_{21}' \tag{14}$$

$$Q = (1 - S_{11})(1 - S_{22}) + S_{12}S_{21} \tag{15}$$

$$C_2 = (1 + S_{11})(1 + S_{22}) - S_{12}S_{21} \tag{16}$$

$$C_2' = (1 + S_{11}')(1 + S_{22}') - S_{12}'S_{21}' \tag{17}$$

A final transformation to more common noise-parameter based on the Ando's model can be obtained through complicated substitution between Eq. 7 through 17, as described by Vendelin [6].

$$F_{min} = 1 + 2R_n'(G_{cor}' + G_{0n}') \tag{18}$$

$$R_n = R_n' \tag{19}$$

$$G_{0n}' = \sqrt{\frac{G_n'}{R_n'} + G_{cor}'^2} \tag{20}$$

III. RESULTS AND DISCUSSION

To check the validity of modified Ando's model (MA), we used the same experimental data referenced in [2] for convenience. Fig. 4 shows the modified Ando's model agrees well with the measured NF of AlGaAs/GaAs HEMT with respect to the drain current. The HEMT was assumed to be biased at V_{dd}=3V and operated at f=12GHz.

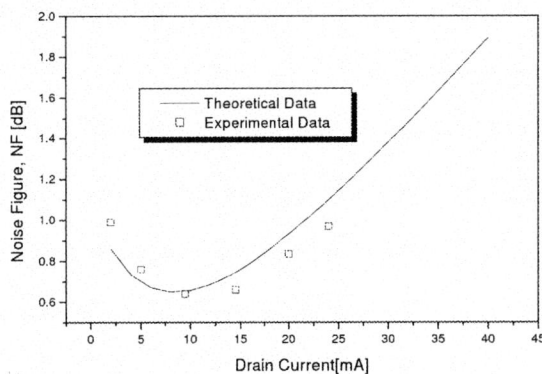

Fig. 4. Comparison of the measured NF with the calculated NF as a function of the drain current for AlGaAs/GaAs HEMT.

Fig. 5 is comparison of measured minimum NF (NF$_{min}$) with NF$_{min}$ of the original Ando's model and NF$_{min}$ of the MA model, and shows the dependency trend of each model about the frequency. For the original Ando's model, the calculated NF$_{min}$ tends to overestimate experimental data at higher frequency because Ando's model didn't consider the drain-to-gate feedback effects together with variation in parasitic gate resistance due to the skin effect. However the MA model relatively follows the trace of measured NF$_{min}$ particularly at higher frequency. Moreover the NF$_{min}$ predicted by the MA model exhibits a linear-dependency on frequency, if it is expressed in decibels. It also agrees well with previous researches reported by Brian, that the minimum noise figure of a FET in decibels increases approximately linearly with frequency as frequency is approaching the FET's maximum oscillation frequency f_{max} [7]. The Ando's model however does not prove linear-dependence, because it did not include the feedback effects generally improving linearity.

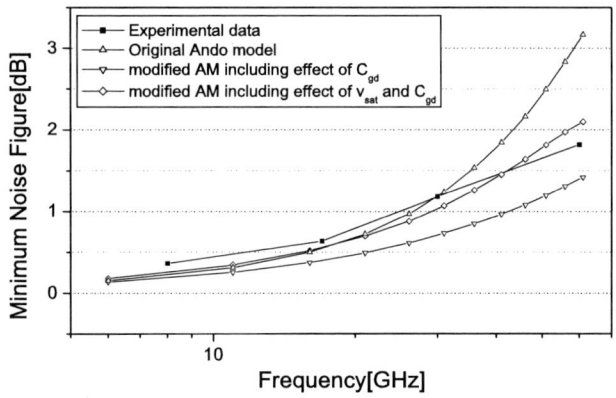

Fig. 5. Comparison of the measured NF$_{min}$ with the NF$_{min}$'s calculated by both AM and MA as a function of frequency for AlGaAs/GaAs HEMT. (all operation conditions are the same as those suggested in [2])

IV. CONCLUSION

An analytical noise model for GaAs-based HEMT was developed based on Ando's noise model and accurate self-consistent simulation along with the parasitic effects at higher frequency and the gate-length-dependent electron velocity. It has been verified for the conventional HEMT, resulting in coincidence with experimental data especially at higher frequency.

ACKNOWLEDGEMENT

This work was supported by MOCIE in South Korea.

REFERENCES

[1] F.Bonani, G.Ghione, C.U.Naldi *et al.*, "HEMT short-gate noise modelling and parametric analysis of NF performance limits", *IEDM*, pp.581, 1992.

[2] Yuji Ando and Tomohiro Itoh, "DC, small-signal and noise modeling for two-dimensional electron gas field-effect transistors based on accurate charge-control characteristics", *IEEE Trans. Electron Devices*, Vol. 37, No. 1, pp.67, January, 1990.

[3] Robert A. Pucel, Daniel J. Massé, Charles F. Krumm, "Noise performance of gallium arsenide field-effect transistors", *IEEE J. Sold-State Circuits*, vol. SC-11, pp.243-255, April, 1976.

[4] Peter H. Ladbrooke, "MMIC Design GaAS FETs and HEMTs", Artech House, 1989.

[5] Matthias Weiss and Dimitris Pavlidis, "The influence of device physical parameters on HEMT large-signal characteristics", *IEEE Trans. Microwave Theory Tech.*, vol.36, no.2, pp.239, Feburary, 1988.

[6] George D. Vendelin, Anthony M. Pavio, and Ulrich L. Rohde, "Microwave Circuit Design using Linear and Nonlinear Techniques", John Wiley & Sons, 1990.

[7] Brian Hughes, "A linear dependence of F$_{min}$ on frequency for FETs", *IEEE Trans. Microwave Theory Tech.*, vol.41, no.6/7, pp.979, June/July, 1993.

2006 25th International Conference on Microelectronics

Modeling and Constituent Design of SPR Based and/or Waveguide Type Sensors

N.L. Dmitruk, O.I. Mayeva, O.V.Korovin, M.V. Sosnova, O.S. Lytvyn, V.I.Min'ko

Abstract – This work is motivated by interest to the waveguide geometry which is attractive because it allows the surface polariton (SP) sensors to be incorporated into the integrated optical system. Inclusion of additional covering and/or (intermediate) dielectric (SiO_x; x=2) and/or conducting (ITO) (tuning) layers in geometry with diffraction grating (DG) as input - coupling component to excite SPs is useful in modifying the SP resonant (SPR) characteristics and creating the potential for universal devices with multifunctional sensing capability (optochemical sensors [1], polarization – and wavelength (angle) – selective photodetectors etc.[2]). Properties (reflectance/ transmittance) of a multilayer stack with DG as optical waveguide (or a leaky quasi-waveguide) are described. Particularly emphasis is placed on modeling the spectral (angular) resonance characteristics of the device modified by inclusion of a heavily doped oxide (ITO) layer. The adequate optical model approaches for the dispersion of the optical constants of ITO films were proposed to full description of the photoresponse and to optimization of future device design. In this paper the possibility of implementing the SPR sensing principle in optical (quasi)waveguide structures designed was demonstrated. We explored constituent design considerations to show how variation structural parameters affect SPR based sensor (detector) response.

I. INTRODUCTION

The effect of the resonant excitation of a surface plasmon or guided wave in optical multilayer structures is being intensively studied for applications in opto(bio)chemical and environmental sensors. In addition to sensor application SP devices have proved useful as photodetectors. Current technology permits us to employ noble metals (Au, Ag) or Al films embedded between low refractive index dielectric layers (SiO_x) or conducting films (ITO) integrated in a multilayered stack with diffraction grating as optical waveguide structure. The properties of ITO are very sensitive to deposition conditions and post-deposition annealing and can be changed in a complicated manner. The importance of considering the processing conditions effects becomes clear if one takes into account that ITO commonly grows with a graded microstructure which introduces grading into the film optical properties (refractive index and extinction coefficient).

In this paper we deal with the post-deposition characterization of multilayer (quasi)waveguide structures

N.L. Dmitruk, O.I. Mayeva, O.V.Korovin, M.V. Sosnova, O.S. Lytvyn, V.I.Min'ko are with the Institute for Physics of Semiconductors, NAS of Ukraine, 41 prospect Nauki, Kyiv, Ukraine, E-mail: dmitruk@kiev.isp.ua

by spectroscopic ellipsometry (VASE) and related optical techniques (the transmittance and the reflectance measurements). The peculiarities of ITO optical properties may have definitive effects on the efficiency of input – coupling light into waveguide systems and on a designed device performance, accordingly. The variation of the refractive index and extinction coefficient are assumed to be only the direction of the film normal. Ellipsometric ψ and Δ data were acquired at multiple angles of incidence over the wide wavelength range. Transmission data for ITO films was also acquired along with the VASE at normal incidence. Particular attention is devoted to theoretical analysis of light reflectance/transmittance (angular and spectral) characteristics of multilayer (quasi)waveguide structures, using the differential formalism [3]. Prior to theoretical analysis of the proposed sensor structures it was necessary to determine optical constants of the deposited layers. Au, Ag, Al and bilayer Au/Ag were chosen as a plasmon-carrying metals because of their outlook for polaritonic optoelectronics device application due to the optical properties of theirs (when the important requirement for a SP excitation occurs). The basic properties of the main approaches to the determination of the optical constants of metal films are discussed in [1] for Au and in [2] for Al. Optical parameters for Ag were taken from [4]. One of the aims of this paper is to propose the optical modeling approaches resulted in the dispersion of the optical constants of ITO films, adequately represented. Multiple angles and wavelengths data for ITO films were analyzed and fitted simultaneously along with the transmission spectra in the two optical models: (i) quasi-continuous medium (point-to-point fit provide a means for determining the approximate film thickness and refractive index), (ii) the film is divided into the two sublayers (more exact fit was achieved by modeling the film with a graded profile and using Lorentz oscillators (three oscillators) dispersion models).

II. SAMPLES DESIGN AND METHODS

One-dimensional DG interfaces were received by using holographic etching [1,2]. Manufacturing layered structures were produced by thermal evaporation thin metal (Au, Ag, Al layers or Au/Ag bilayers and dielectric layers (SiO_x)) as well as by d.c. magnetron sputtering (ITO) on the processed GaAs semiconductor substrates at RT. Then ITO films were annealed at 350°, 30 min. Prior to evaporation of metal layer, deposition of ultrathin metal layer of Cr (~3nm) was used to promote metal adhesion to substrate. A conductive (ITO) and dielectric (SiO$_x$) film were deposited onto flat both glass and GaAs substrates and processed semiconductor substrates (grating) in one deposition run.

Measurements of the morphological and statistical features of the DGs were performed with the help of processing AFM – images (Fig.1). Surface profile was obtained by sectioning of the DG surface perpendicular to grooves. Intensity spectrum of various profile harmonics was calculated from the Fourier transform of the DG profile: the two periods (d) are seen – the fundamental (0.85 µm) and second-harmonic (0.43 µm) one.

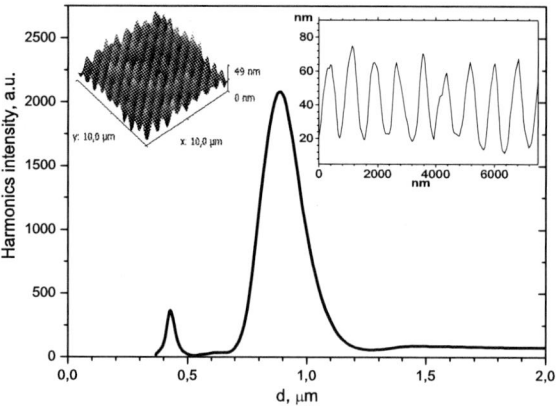

Fig.1. Intensity spectrum of various harmonics calculated from the Fourier transform of the DG profile (insert: AFM-images of the DG on GaAs - on the left hand; section analysis of this profile – on the right hand).

A. Optical Properties of ITO Films

For the characterization of the ITO layers VASE technique was employed. It is based on measurement of the polarization state of a light beam reflected from the sample surface as function of known angle of incidence and wavelength. Ellipsometric measurements are expressed in terms of ellipsometric angles ψ and Δ defined as

$$\rho = \frac{r_p}{r_s} = tg\,\psi\,e^{i\Delta} \tag{1}$$

where r_p, r_s are the Fresnel reflection coefficients for p- and s- polarized light, respectively.
Then the complex dielectric constant ε of the film can be determined by

$$\varepsilon = \varepsilon_1 + i\varepsilon_2 = \sin^2\theta\big[(1-\rho)/(1+\rho)\big]^2 tg^2\theta \tag{2}$$

where ϑ is the angle of incidence of the probing light; $\varepsilon_1 = n^2 - k^2$, $\varepsilon_2 = 2nk$, n and k are the real and imaginary parts, respectively, of the complex index of refraction N, given as: $N = n - ik$.

In order to relate the VASE – measured parameters with actual characteristics of the layer, we constructed a model from which the Fresnel reflection coefficients are calculated. Then spectroscopic optical function used in the model must be selected and fitting procedure to determine the fitted parameters must be included. The results depend on the amount of information available in the experimental data. Ellipsometric ψ and Δ data were acquired at three angles of incidence (65°, 70°, and 75°) over the spectral range 250-1000 nm in steps of 10 nm and were combined with the intensity transmission data also acquired at normal incidence over the same spectral range. As a first attempt we treated ITO layer of unknown thickness as quasi-continuous medium. This model is useful for obtaining an approximate film thickness and refractive index n, because ITO is nearly transparent over the visible spectrum (k=0). Grading and absorption (if it occurs) were not included in this initial model, so the (point-to-point) fitting are not perfect and the results are only approximate. But this fitting can be used as a starting point for construction of more complicated model. In spectral ellipsometry modeling, if optical constants and thickness of the films must be determined, a parametric representation of the spectra is often used [1,2]. The spectra (Vis and near – IR) in the optical constants of the ITO films can be represented using a harmonic oscillator approximation as:

$$\varepsilon(\omega) = \varepsilon_\infty + \sum_i \frac{A_i}{E_i^2 - (\hbar\omega + i\Gamma_i/2)^2} \tag{3}$$

where ε is defined above, $\hbar\omega$ – photons energy, i denotes the oscillator number (i=1,2,3,...), ε_∞ is the dielectric function of infinite energy; A_i, E_i, and Γ_i are the amplitude, center energy and broadening of each oscillator, respectively. Keeping in mind the graded structure of ITO films, we divided the total film thickness into two sublayers in order to use two sets of Lorentz oscillators (three oscillators) to model the ITO refractive index at the top and bottom the film. The results of the analysis are presented in Table I.

TABLE I
SET OF THE FIT PARAMETERS OBTAINED FROM THE ANALYSIS BASED ON LORENTZ OSCILLATOR MODEL FOR THE ITO SUBLAYERS ONTO GaAs SUBSTRATE

Sub-layer	Thickness nm	Amplitude A_i	Energy E_i	Broadening Γ_i	ε_∞
the top	121,27	24,036	1,176	1,069	2,13
		0,001	0,001	1,713	
		8,912	0,548	4,359	
the bottom	82,40	0,104	0,471	8,124	1,18
		0,913	0,001	0,671	
		1,592	0,000	4,824	

Fig. 2a,b shows ellipsometric angles ψ and Δ with that generated from the model over the spectral range

250-1000 nm. The investigated configuration is shown in Fig.2 (insert). Reference optical data [5] were used for the GaAs substrate and the thin native GaAs oxide layer. The thickness of the oxide layer was fixed in the analysis. A good agreement between calculated and experimental data in the Vis is seen. The best-fit optical constants at the top and bottom of the film are shown in Fig.3. It is seen that the extinction coefficient k of ITO film is lower at the bottom film that is the film is more conductive near the surface. Some studies [6] have reported that the conductivity of ITO is improved by a post-annealing in non-oxidizing atmosphere, because of forming oxygen vacancies in the film. The oxygen vacancies are difficult to form at a depth far from the surface during annealing.

B. Modeling and Measurements optical characteristics of Multilayer structures

To test and analyze the properties of the light reflectance/transmittance (angular and spectral) characteristics the differential formalism [3] was applied to multilayer (quasi)waveguide structures using the measured n and k for ITO film and the reference data for SiO_x [7]. Shown in Figs.4 are results of computer calculation of light reflection (R_p, R_s) from- and transmission (T_p - T_{p0}) through a multilayer DG into a semiconductor substrate, which are resonant relative to angle (a) and wavelength (b) also. Electric field vector perpendicular (p – polarized light) or parallel (s – polarized light) to the grating grooves.

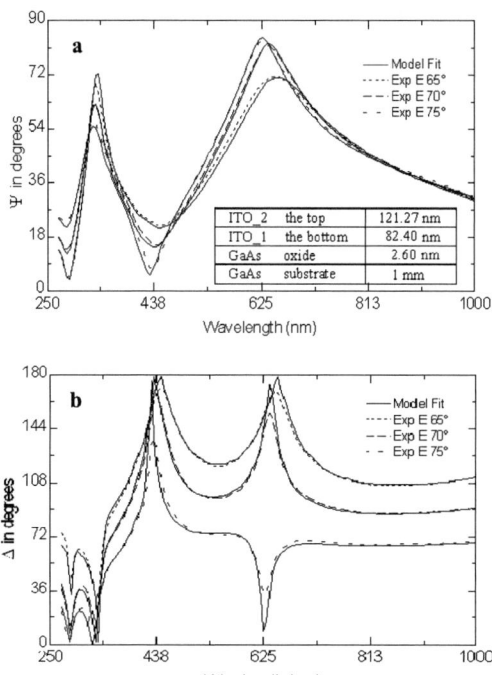

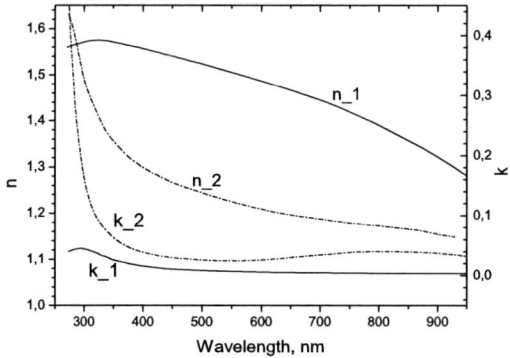

Fig.2. Ellipsometric angle ψ (a) and Δ (b) data for ITO thin film on flat GaAs substrate. Experimental and best-fit calculated data and the configuration under investigation (in the insert) are shown.

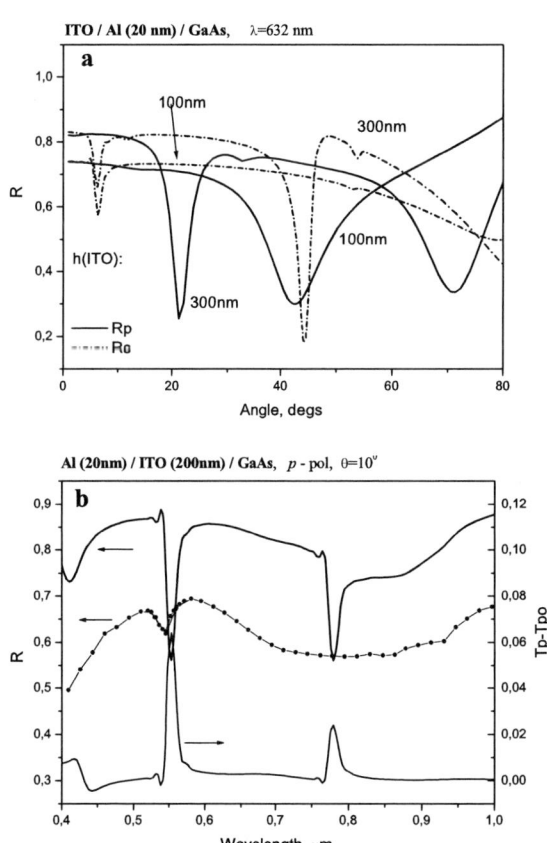

Fig.4. The angular (a) and spectral (b) dependencies of the reflectance R and the differential transmittance enhancement (Tp -Tpo) of multilayer DG on the thickness h of covering and metal layers, where Tp is the transmittance of the structure with DG, Tpo is the same for the flat multilayer structure. Points are experimental data, solid lines are generated from theory with DG parameters: period d = 0.7μm and groove depth H = 35 nm.

The main feature of these results a strong dependence of the reflectance/transmittance on the thickness as well as the refractive index of the active layers. It is may be connected with the change of the polarization state of the light beam reflected from multilayer system (ITO/Al/GaAs): the metal surface with over – (or

Fig.3. Optical constants n and k obtained from analysis of VASE data for ITO_2 and ITO_1.

319

intermediate) layer may allow the operation range of the sensor to be shifted because of the effective refractive index of a surface plasma waves propagating along the metal/dielectric interface is changed. In this manner the "tuning effect" can be achieved in order to obtain maximal sensitivity at the characteristic spectral region. And when designing sensor structures (constituents) for maximum sensitivity these characteristics must be taken into account first of all. For example, when a photoelectric mechanism of sensor signals generating, a basic principle underlaying the device operation is a resonant enhancement of the photoresponse generated under SPR conditions in Schottky junction with the DG at the interface. The photoelectric signal maximum will correspond to a peak of light transmission through a multilayer DG into a semiconductor substrate.

As for a structure of resonant metallic film based on bimetallic Ag/Au layers, it combines advantages of both Au and Ag resonant layers.

It is known resonant guided modes exists in a dielectric layer with a thickness greater than $\sim \lambda/4n$.

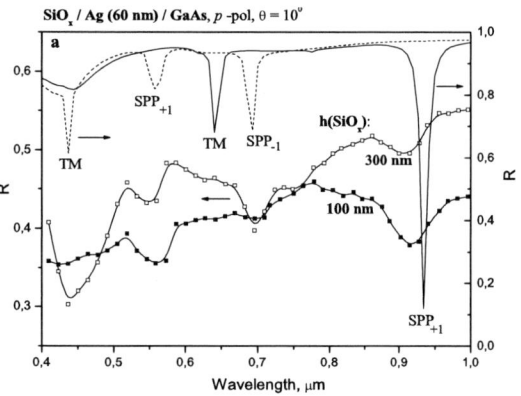

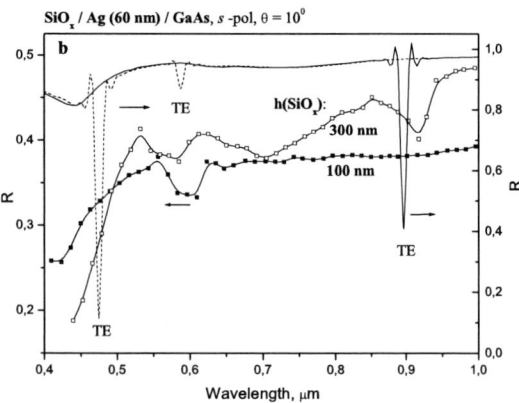

Fig.5. Influence of the dielectric thickness on the intensity of coupling to SPP and waveguide modes at both the fundamental (solid lines) modes and second-harmonic (dotted) ones for *p*-polarized light (a) and *s*-polarized light (b). Points are experimental data, solid and dotted lines are generated from theory with DG parameters: period d' = 0.85µm, d"=0.43µm, depth H' = 35 nm, H"=7 nm and h(SiO$_x$)=300nm.

In the structures of similar type (SiO$_x$/Ag/GaAs) the waveguides (TE, TM) modes may be observed depending on the measurement conditions (*p*- or *s*- polarization of the incident light) (Fig.5). In similar systems these modes will propagate with less loss than SPPs as their associated fields are contained mainly in the non-absorbing dielectric rather than the metal. Slight absorption (interface enriched by Si) and porosity of dielectric (SiO$_x$) cause these modes to be broader than was expected (Fig.5, experimental points).

III. CONCLUSION

Optical modeling performed at successive stages of the processing has demonstrated the general trend in the behavior of the (quasi) waveguided modes in the multilayer structures with diffraction gratings. Owing to grating – coupling techniques by proper selection of intermediate, covering and metal layers optical properties and the geometrical parameters of gratings as well polaritonic sensor (detector) may be tuned in a wide operating range. Reasonable agreement the calculation results and the experimental ones is found.

Angle dependences of R$_p$ for structures of interest and fitting procedure for whole set of the measurements to reveal the optical parameters of the design constituents are in progress.

REFERENCES

[1].N.L. Dmitruk, O.I. Mayeva, S.V. Mamykin, O.B. Yastrubchak, and M. Klopfleisch. "Characterization and application of multilayer diffraction gratings as optochemical sensors". *Sensors and Actuators A88*, 2001, pp.52-57.

[2].N.L. Dmitruk, M. Klopfleisch, O.I. Mayeva, S.V. Mamykin, E.F. Venger, and O.B. Yastrubchak. "Multilayer diffraction gratings Al/GaAs as polaritonic photodetectors". *Phys. Stat. Sol.(a) 184, №1*, 2001, pp.165-174.

[3].J. Chandezon, M.T. Dupius, G. Cornet, and D. Maystre, "Multicoated gratings: a differential formalism applicable in the entire optical region". *J.Opt.Soc.Amer., 72*, 1982, pp.839-846.

[4].P.B. Johnson and R.W.Christy, "Optical Constants of the Noble Metals". *Phys.Rev.B,* vol. 6, 1972, pp. 4370-4379.

[5].D.E.Aspnes and A.A. Studna. "Dielectric functions and optical parameters of Si, Ge, GaAs, GaSb, InP, InAs, and InSb from 1,5 to 6,0 eV". *Phys Rew B27, 2*, 1983, p.985

[6].J.A. Woollam, W.A. McGahan and B. Johs. "Spectroscopic ellipsometry studies of indium tin oxide and other flat panel display multilayer materials". *Thin Solid Films, 241*, pp.44-46, 1994.

[7].E.D. Palik (Ed.), Handbook of Optical Constants of Solids, Academic Press, Orlando (Fl.), 1985.

2006 25th International Conference on Microelectronics

Effect of Interface Microrelief on Optical, Electrical and Photoelectric Characteristics of Heteroepitaxial Al$_x$Ga$_{1-x}$As/GaAs Structures

N.L. Dmitruk, O.Yu. Borkovskaya, R.V. Konakova, A.V. Korovin, I.B. Mamontova, O.S. Kondratenko

Abstract – We have investigated the dependence of optical, photoelectric and electrical characteristics of heteroepitaxial Al$_x$Ga$_{1-x}$As/GaAs structures on active interface microrelief morphology. Experimental study of heterostructures included analysis of surface microrelief parameters by AFM techniques, measurements of specular and diffuse reflection spectra, the dark and light current-voltage (*I-V*) characteristics and spectra of the short-circuit photocurrent. Investigation showed that advantages of optical and recombination properties of an anisotropically etched GaAs surface may be remained in Al$_x$Ga$_{1-x}$As/GaAs heterostructures at a suitable choice of processing regimes. To explain reduction of the open circuit voltage (*V$_{oc}$*) in some heterostructures with textured interface a theoretical analysis has been made for V$_{oc}$ dependence on the nonuniformity of doping with impurity atoms caused by interface roughness.

I. Introduction

Semiconductor surface microrelief of appointed morphology is known as a promising means to reduce optical losses and to increase the photodetector response and photoconverter (solar cells, SC) efficiency [1]. To enhance efficiency of such well elaborated photoconverters as p^+-Al$_x$Ga$_{1-x}$As/p^+-n-n^+-GaAs heterostructure [2] with the help of microtextured active interface it is necessary to ensure the epitaxial growth of thin (submicron), and in the same time continuous and homogeneous, highly doped layers of p^+-GaAs and p^+-Al$_x$Ga$_{1-x}$As on the microrelief substrate. However, the process of epitaxy is accompanied with smoothing surface microrelief that first of all manifests itself in the change of optical properties and photocurrent values. On the other hand, the preservation of microrelief on the active interface of hetero- or p/n-junction may cause the lateral nonuniformity (fluctuations) of emitter layer doping and also can distort fields of built-in potential barrier, increase electron/hole losses by recombination, and so change the mechanism of current flow (diffusion, generation/recombination, tunnel).

The purpose of this paper is to investigate the

N.L. Dmitruk, O.Yu. Borkovskaya, R.V. Konakova, A.V. Korovin, I.B. Mamontova, O.S. Kondratenko are with the Institute for Physics of Semiconductors, National Academy of Sciences of Ukraine, 45 Nauki Prospect, Kyiv 03028, Ukraine, E-mail: dmitruk@isp.kiev.ua

mechanism of changes of optical, electrical and photoelectric characteristics of Al$_x$Ga$_{1-x}$As/GaAs heterostructure depending on active interface microrelief parameters and on their transformation induced by the process of epitaxy and doping impurity diffusion.

II. Investigated Structure Processing

Heterostructures with textured interface were fabricated by different modification of liquid phase epitaxy (LPE) or vapor phase epitaxy (VPE) techniques on n- or n-n^+-GaAs substrates anisotropically etched to obtain surface microrelief of dendrite or quasigrating type [3, 4].

The geometric statistical parameters of the surface microrelief were varied by the etching regime (temperature of etchants, duration) and were investigated by AFM techniques. Transformation of GaAs surface microrelief morphology as a result of VPE heteroepitaxy is shown in Fig. 1. Unlike the LPE, VPE technique allows to fabricate thin films heteroepitaxial structures with remaining the substrate microrelief character, although certain smoothing of it takes place too. Parameters of Al$_x$Ga$_{1-x}$As layer (x value, thickness) were determined by X-ray diffraction methods.

II. Experimental Results

Experimental study of heterostructures included measurements of specular and diffuse reflection spectra, the dark and light *I-V* characteristics, spectra of the short-circuit photocurrent. Fig. 2 demonstrates the optical properties of textured GaAs substrates and of heterostructures fabricated on them by VPE: spectra of the specular, diffuse and total reflection. To get a quantitative description of the efficiency of micro(nano)relief as light trap, we calculated (from the optical studies of structures obtained using different technologies) the effective reflectance coefficients ($\overline{R}$) averaged over the solar AM0 spectrum in the λ = 0.4–0.9 μm spectral range:

$$\overline{R} = \int_{0.4\,\mu m}^{0.9\,\mu m} R(\lambda)N(\lambda)d\lambda \bigg/ \int_{0.4\,\mu m}^{0.9\,\mu m} N(\lambda)d\lambda . \tag{1}$$

1-4244-0116-X/06/$20.00 ©2006 IEEE

The results are presented in Table I.

TABLE I
(QGL – QUASI-GRATING-LIKE, DL – DENDRITE-LIKE,
F – FLAT)

Samples	$\overline{R}$, %	Growth technology	Structure
Su (F) #7 Su (DL) #8 Su (QGL) #9	37.44 16.14 19.9	VP diffusion of zinc (550 °C, 30 min.)	p^+-n^+-GaAs
MC-1 (QGL) MC-2 (DL) MC-3 (F)	47.88 26.61 45.06	VPE	p^+-GaAs/ p^+-Al$_x$Ga$_{1-x}$As/ p^+-n-GaAs/ n^+-GaAs
MO-10 (F) MO-12 (QGL) MO-13 (DL)	36.69 28.99 15.9	VPE and diffusion of carbon (600-750 °C)	p^+-GaAs/ n-n^+-GaAs
Ka (F) #23 Ka (DL) #23 Ka (F) #25 Ka (DL) #25	39.55 38.25 27.32 14.02	LPE of n-GaAs + VP diffusion of zinc + heteroepitaxy	p^+-Al$_x$Ga$_{1-x}$As/ p-n^M-GaAs/ n^+-GaAs

It was found that the smallest total reflectance coefficient $\overline{R}$ is obtained at VP diffusion for a dendrite-like microrelief (15.9% for the MO-13 sample).

Our investigations of the optical reflection spectra of p-n junctions and heteroepitaxial structures (obtained on flat and microtextured n^+-GaAs substrates using different technologies) were aimed at (i) choosing of the most efficient technology that would retain the advantages of microtextured surface in reduction of optical losses in SC, and (ii) determination of the value and spectral dependence of optical losses for calculation of the spectrum of internal quantum efficiency and estimation of recombination parameters of the structures studied.

From an analysis of the results of our investigations one can draw the following conclusions:

1. If the heteroepitaxial layer thickness is big (≥ 3 μm) as compared to the microrelief depth, then the front surface microrelief is smoothed out. This manifests itself in reduction (or even leveling) of the microtexturing effect on $\overline{R}$ (the samples Ka-23fl and Ka-23d). This effect is also displayed at VPE, though in a lesser degree (the MC samples).

2. The smallest $\overline{R}$ values were obtained in the cases of dendrite-like reliefs (MO 13, Ka 25d) and more developed quasi-grating-like relief (MO-12), as compared with the flat structures made using the same technology.

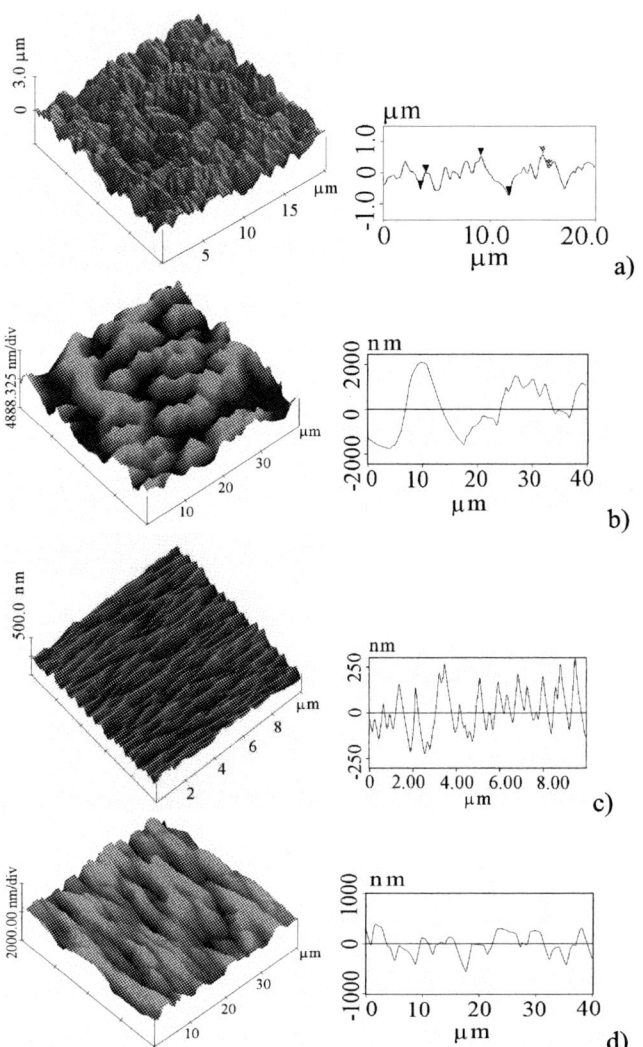

Fig. 1. AFM images and section analysis of the GaAs (a, c) and Al$_x$Ga$_{1-x}$As/GaAs (b, d) surfaces with dendrite- (a, b) and quasigrating-like (c, d) microrelief.

Some smoothing out of relief occurs also in the course of VP diffusion of zinc if it proceeds at a temperature ≥ 800 °C during a rather long time (60÷90 min.). This leads in increase of $\overline{R}$ values for a textured surface.

It is seen that, both in the case of LPE and VPE, the least total reflection is obtained for structures with dendrite and more developed quasigrating microreliefs. They cause greater values of external quantum efficiency (Fig. 3) and of the short circuit photocurrent of the light I-V dependencies, correspondingly (Fig. 4a). However the open-circuit voltage (V_{oc}) of such structures may be reduced due to greater value of the forward dark current, as it may be conclude comparing light (Fig. 4a) and dark (Fig. 4b) I-V curves. Since the V_{oc} value essentially depends on the electronic and recombination parameters of p/n interface and on the mechanism of the current flow, so theoretical analysis of these dependencies has been made with taking into account the fluctuation of p-layer doping impurity concentration, caused by interface microrelief.

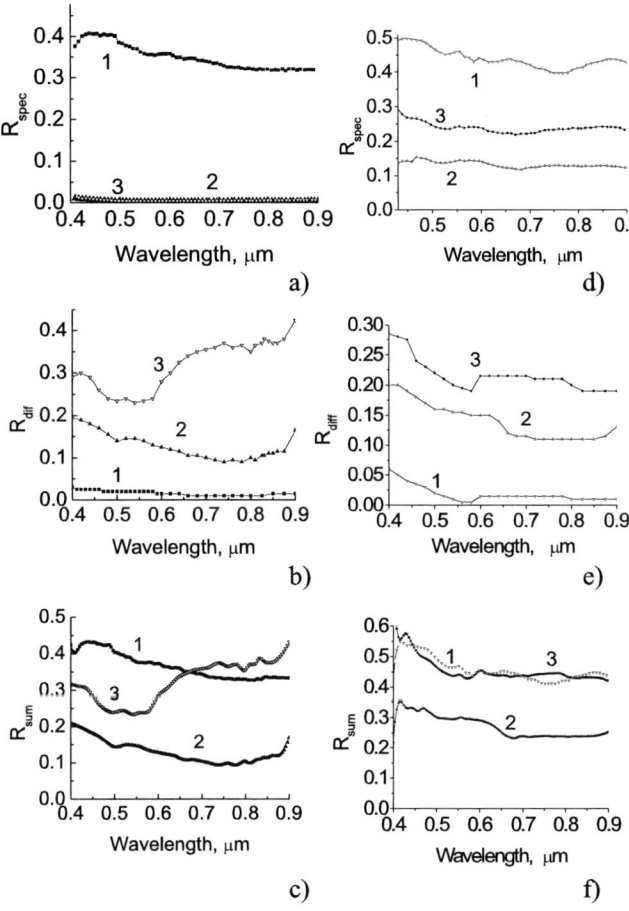

Fig. 2. Spectra of specular (a, d), diffuse (b, e) and total (c, f) reflectance for n^+-GaAs surfaces (a, b, c) and p^+-GaAs/p^+-Al$_x$Ga$_{1-x}$As/p-n-GaAs n^+-GaAs (texturized substrate) (d, e, f): chemically polished (1) and with surface microrelief of dendritic (2) and quasigrating (3) type.

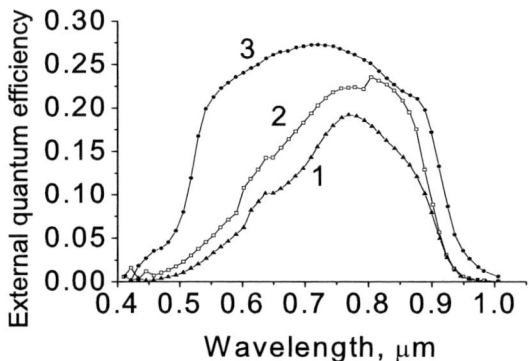

Fig. 3. Experimental spectral characteristics of the external quantum efficiency for VPE p^+-GaAs/p^+-Al$_x$Ga$_{1-x}$As/p^+-n-n^+-GaAs/n^+-GaAs heterostructures with flat (1) and microrelief of quasigrating (2) and dendrite (3) type interfaces (c).

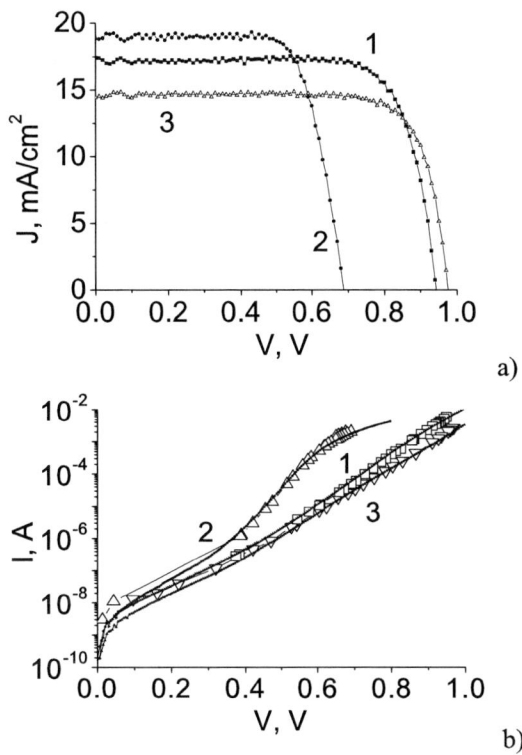

Fig. 4. Light (a) and the forward branches of dark (b) I–V curves for p^+-Al$_x$Ga$_{1-x}$As/p^+-n-GaAs structures, fabricated by LPE with flat (1) and microrelief of quasigrating type (2, 3) interfaces.

III. THEORETICAL ANALYSIS OF V$_{OC}$ DEPENDENCE ON DOPING FLUCTUATION IN P-N JUNCTION WITH MICRORELIEF INTERFACE

The model of in parallel connected elementary p^+-n diodes with the Gaussian distribution of barrier height, caused by fluctuation of doping in microrelief structure, was analyzed.

Average open circuit voltage $\overline{V}_{oc}$ was computed from the condition of equality of the short circuit photocurrent (I_{sc}) to averaged total dark current at $V = \overline{V}_{oc}$ by partial averaging of the diffusion (I_d), recombination (I_r) and tunnel (I_t) components of the current flow:

$$
I_{sc} = \left\langle I_d(\overline{p}, x) \right\rangle \left[\exp\left(\frac{q\overline{V}_{OC}}{kT} \right) - 1 \right] +
$$
$$
+ \left\langle I_r(\overline{p}, x) \right\rangle \left[\exp\left(\frac{q\overline{V}_{OC}}{2kT} \right) - 1 \right] +
$$
$$
+ \left\langle I_t \right\rangle \left[\exp\left(\frac{q\overline{V}_{OC}}{\varepsilon} \right) - 1 \right] + \frac{\overline{V}_{OC}}{R_{sh}}
\qquad (2)
$$

where $\overline{p}$ is mean concentration of holes in the p^+-layer of p^+-n-junction, $x = \ln(p/\overline{p})$, p – is a local concentration of holes with Gaussian character of their distribution and with distribution width σ:

$$f(x,\sigma) = \frac{e^{-\frac{x^2}{2\sigma^2}}}{\sqrt{2\pi}\sigma}, \qquad (3)$$

ε is a characteristics energy of tunneling, R_{sh} is an ohmic shunting resistance.

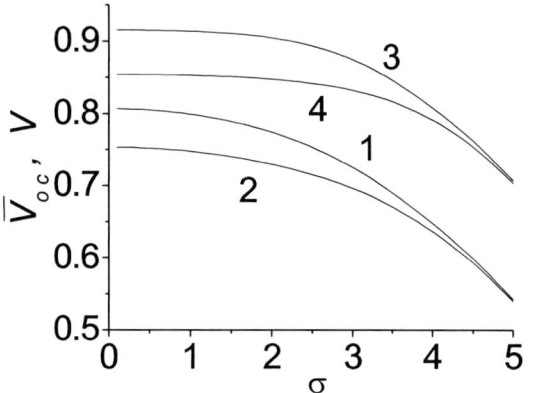

Fig. 5. Averaged open-circuit voltage vs dispersion at current averaging in the case of the diffusion, recombination and tunnel components of the total current. The hole concentration $\overline{p}$ [cm^{-3}] in the emitter: $2 \cdot 10^{15}$ (1, 2), $2 \cdot 10^{18}$ (3, 4). The lifetime τ_p [s]: $2 \cdot 10^{-8}$ (1, 3), $2 \cdot 10^{-9}$ (2, 4) the front surface recombination velocity. $S = 10^4$ cm/s.

Expressions for I_d and I_r are taken from [5], and for I_t from [6]. An example of V_{oc} vs σ dependencies for various $\overline{p}$ and recombination parameters of junction is shown in Fig. 5. Computer simulation of dark I-V dependencies taking into consideration the main components of current flow (diffusion, recombination, tunnel and ohmic shunting) and their fitting to experimental one allowed to determine the change of barrier parameters and current flow mechanisms caused by interface microrelief of heterojunction. The recombination mechanism of the current flow was shown to dominante in LPE and VPE Al$_x$Ga$_{1-x}$As/GaAs heterostructures (with flat and microrelief interface) which have maximal value of V_{oc} and $\sigma \leq 2$. The diffusion mechanism with $\sigma \geq 4$ (at greater V) or tunneling through dislocation crossing p/n junction dominate for heterostructures with microrelief interface having minimal $\overline{V}_{oc}$. As it is seen from Fig.4, the Al$_x$Ga$_{1-x}$As/GaAs heterostructures with microrelief

interface, specifically of quasigrating type, may be so perfect to ensure the V_{oc} value comparable to or even exceeding that of flat structures.

IV. Conclusions

Our investigation has shown that suitable choice of the fabrication technology of p^+-Al$_x$Ga$_{1-x}$As/p^+-n-n^+-GaAs heterostructures with microrelief interface, specifically of anisotropic etching and heteroepitaxy regimes, allows to retain advantages of optical properties of microtextured surface and to obtain corresponding increase of the photocurrent. Also the optimal choice of processing regimes allows to ensure sufficient perfectness of the interface for obtaining the open-circuit voltage value not less or even greater than that of the flat structure.

Theoretical analysis of the $\overline{V}_{oc}$ dependence on the fluctuation of doping impurity concentration at various mechanisms of the current flow and comparison of the experimental I-V dependencies with simulated ones allowed to characterize electrical properties of heterojunction and their dependence on the technological regimes.

References

[1] M. A. Green, *High efficiency silicon solar cells*, Trans. Tech. Publ., Brookfield, VT, 1987.

[2] V. M. Andreev. *Photovoltaic and Photoactive Materials– Properties, Technology and Application.* Ed. by Y. M. Marshall and D. Dimova-Malinovska, Ser. II: Mathematics, Physics and Chemistry, Kluwer Acad. Publ., London, v.80, 2002.

[3] N. L. Dmitruk, O. Yu. Borkovskaya, I. N. Dmitruk, I. B. Mamontova, "Analysis of thin film surface barrier solar cells with a microrelief interface", *Solar Energy Materials & Solar Cells*, 2003, vol. 76, pp. 625-635.

[4] N. L. Dmitruk, O. Yu. Borkovskaya, V. P. Kladko, R. V. Konakova, Ya. Ya. Kudryk, O. S. Lytvyn, V. V. Milenin, "Effect of Surface Roughness on the Properties of Ohmic Contacts to GaAs", in *Proc. 24th International Conference on Microelectronics, MIEL 2004*, Niš, Serbia and Montenegro, vol. 2, pp. 499-502.

[5] O. Yu. Borkovskaya, N. L. Dmitruk, V. G. Lyapin, A. V. Sachenko, "Computer simulation of the photocurrent collection coefficient in solar cells based on the textured thin-film Al$_x$Ga$_{1-x}$As-GaAs heterostructure", *Thin Solid Films*, 2004, vol. 451-452, pp. 402-407.

[6] V. V. Evstropov, M. Dzhumaeva, Yu. V. Zhilyaev, N. Nazarov, A. A. Sitnikova, and L. M. Fedorov, "The dislocation origin and model of excess tunnel current in GaP p–n structures", *Semiconductors*, vol. 34, issue 11, pp. 1305-1310.

Session
Physics and Modeling

2006 25th International Conference on Microelectronics

a-Si$_{1-x}$C$_x$:H TFTs: Fabrication and Modeling

M. Estrada, A. Cerdeira, R. García and B. Iñiguez

Abstract - In this paper we resume results obtained in the fabrication and characterization of first a-Si$_{1-x}$C$_x$:H TFTs. The behavior with temperature of these devices was studied using a unified model and extraction parameter method, UMEM, previously developed by us and applied to amorphous, polycrystalline and nanocrystalline Si TFTs as well as to organic TFTs. The behavior and variation with temperature of the transfer and output characteristics, of mobility and saturation parameter, as well as other features of these TFTs with active layer containing both Si and C are shown and compared to those of typical a-Si:H TFTs. Localized traps density distribution and their activation energies were estimated. Instability after voltage stress is also shown. First results after polarization of the amorphous TFts by laser annealing are presented.

I. INTRODUCTION

Due to its high thermal conductivity, absorption characteristic and emission spectra, silicon carbide has been of great interest as optoelectronic material, as well as for operation at high voltage and temperature. At present, many devices are fabricated with crystalline silicon carbide for power and photovoltaic or optoelectronic application, while a-Si$_{1-x}$C$_x$:H devices have also found applications in solar cells and optoelectronics [1].

Since low-cost glass or polymeric substrates are desirable for large area and low cost electronic applications, the characteristics of amorphous silicon carbide films deposited at temperatures below 300 °C by PECVD have been studied by several authors as function of gas composition and other process parameters [2-3]. However, in spite of the potential application of this material to TFTs, first reports on a-Si$_{1-x}$C$_x$:H TFTs seems to be in [4]. In this paper we summarize the electrical characteristics of TFTs fabricated with a-Si$_{1-x}$C$_x$:H films deposited by PECVD, with carbon composition x=0.23.

We also show the possibility of fabrication devices that can work at low voltage range up to 10 V, as well as at higher voltages, up to 90 V.

Analytical modeling and electrical simulations complemented measurements in order to represent differences in their behavior with respect to a-Si:H TFTs and to estimate the localized state energy distribution

M. Estrada and A. Cerdeira are with Departmento de Ingeniería Eléctrica, Sección de Electrónica del Estado Sólido, CINVESTAV-IPN, Av. IPN No. 2508 San Pedro Zacatenco, CP 07360, Mexico D.F., Mexico, E-mail: mestrada@cinvestav.mx
R. García and B. Iñiguez, are with the Dept. Enginyeria Electronica Electrica i Automàtica, Universitat Rovira i Virgili, Avinguda Paisos Catalans, 26, 43007 Tarragona, Spain.

present in the TFT active layer.

The behavior of device characteristics with temperature was analyzed. Preliminary studies on device instabilities are also shown. Finally, first results on polycrystalline SiC TFTs, after laser annealing of the a-Si$_{1-x}$C$_x$:H TFTs are presented.

II. EXPERIMENTAL PART

a-Si$_{1-x}$C$_x$:H films from 150 to 250 nm thickness were deposited by PECVD using a gas mixture of pure SiH$_4$, CH$_4$ and H$_2$. Carbon concentration of 0.23 was obtained varying pure SiH$_4$ and CH$_4$ flows at 300 °C, 1 Torr and 65 mW/cm^2. Since it is difficult to obtain efficiently n-doped a-Si$_{1-x}$C$_x$:H material, [5], we used 50 nm thick a-Si:H n$^+$ layers as drain and source regions. The deposition of a-Si$_{1-x}$C$_x$:H and n$^+$ type a-Si:H layers was done in sequence without turning off the plasma, by turning off the CH$_4$ and introducing PH$_3$ diluted in H$_2$. The SiO$_2$ gate dielectric was deposited from N$_2$O and SiH$_4$, obtaining thicknesses from 120 to 600 nm. Aluminum was used for gate, drain and source contacts; final annealing was performed at 350 °C for 30 min in H$_2$. Drain and source regions, gate, and metal contacts were defined by photolithography. The basic steps of the fabrication process for TFTs with carbon content x=0.23 are summarized in Table 1.

TABLE I
FABRICATION STEPS FOR a-Si$_{1-x}$C$_x$:H TFTs

Step	Fabrication Process for x=0.23
1	PECVD deposition of a-Si$_{1-x}$C$_x$:H layer T=300 °C, 1 Torr; 65 mW /cm^2; t=25 min Gas flows: SiH$_4$=3.9 sccm; CH$_4$=12.5 sccm; H$_2$=80 sccm.
	Deposition of n$^+$ a-Si:H layer by PECVD at 300 °C; 1 Torr; 65 mW /cm^2; t=5 min Gas flows: SiH$_4$=9.9 sccm; H$_2$=140 sccm; 60 sccm of 1% PH$_3$ in H$_2$,
2	Photolithography of Si N$^+$ layer
3	PECVD deposition of SiO$_2$ layer at 300 °C; 0.5 Torr; 190 mW/cm^2; SiH$_4$=2 sccm, N$_2$O=2.75 sccm, t=30 min
4	Photolithography of SiO$_2$ layer
6	Al deposition
7	Photolithography of Al
8	Annealing in H$_2$ at 350 °C 30 min

1-4244-0116-X/06/$20.00 ©2006 IEEE

Fabrication process for TFTs with carbon content x=0.17 was also done, following the same processing steps, just increasing the SiH₄ flow from 3.9 to 39 sccm. Since a-Si$_{1-x}$C$_x$:H films are expected to be more resistive than a-Si:H, TFTs with upper gate were used.

Fig. 1 shows the measured linear and saturation transfer characteristics for an a-Si$_{1-x}$C$_x$:H TFT with x=23 and x=17.

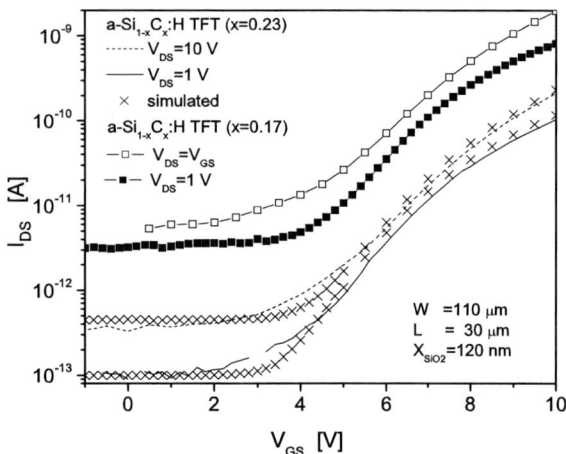

Fig. 1. Comparison between experimental linear and saturation transfer curves for two devices with different carbon concentration, modeled curves using UMEM and curves simulated in ATLAS using the using the following localized trap distribution: g_{Ad}=4x10^{18} cm^{-3}, g_{Dd}=4x10^{18} cm^{-3}, g_{At}=g_{Dt}=4x10^{21} cm^{-3}, E_{Ad}=0.176 eV, E_{Dd}=0.256 eV, E_{At}=0.05 eV and E_{Dt}=0.107 eV.

A feature of a-Si$_{1-x}$C$_x$ TFTs is that in subthreshold, the drain current can be very low. At the same time, gate SiO$_2$ is deposited by PECVD with current density expected to be around 10^{-8}- 10^{-7} cm^{-2}. For these reasons, depending on the transistor geometry, the off current can be of the same order as the current flowing through the gate and the advantage of having very low off or subthreshold current may turn into a problem as the carbon concentration is increased toward stoichiometry.

Device parameters were extracted using the Unified Model and Parameter Extraction Method, UMEM, previously developed by us, which can be applied to a-Si:H [6,7], polySi [8], nanocrystalline [9] and organic TFTs [10]. The effective mobility in above threshold regime is modeled through standard parameters μ_0, γ_a and V_{aa} first used only for a-Si:H TFTs, where:

$$\mu_{FET} = \mu_0 \cdot \frac{\left(V_{GS} - V_T\right)^{\gamma_a}}{\left(V_{aa}\right)^{\gamma_a}}. \quad (1)$$

The drain current is calculated as:

$$I_{DS} = K \frac{\mu_{FET}}{f_R} \frac{V_{GT} V_{DS} (1 + \lambda V_{DS})}{\left[1 + \left(\dfrac{V_{DS}}{V_{DSsat}}\right)^m\right]^{1/m}}, \quad (2)$$

where

$$K = \frac{W}{L} C_i, \quad (3)$$

$$V_{GT} = V_{GS} - V_T, \quad (4)$$

$$f_R = 1 + K \cdot R \cdot \mu_{FET} \cdot V_{GT}, \quad (5)$$

and $$V_{DSsat} = \alpha_s \cdot \left(V_{GS} - V_T\right). \quad (6)$$

We extracted model parameters for each measured temperature to study their behavior with temperature, as can be seen in Table 2.

TABLE II
EXTRACTED VALUES OF MODEL PARAMETERS AT DIFFERENT TEMPERATURES

T [°C]	V$_T$ [V]	γ_a	α_s	λ [V^{-1}]	μ_{FET} [cm^2/Vs] V$_{GS}$=10 V
30	3.7	2.2	0.26	-6x10^{-4}	3x10^{-4}
40	3.6	2.2	0.24	-2x10^{-4}	4.7x10^{-4}
50	3.5	2.1	0.27	-9x10^{-3}	5.6x10^{-4}
60	3.3	1.9	0.28	-1x10^{-2}	1.1x10^{-3}
70	3.2	1.6	0.32	-9x10^{-4}	1.210^{-3}
80	2.8	1.3	0.3	+0.015	3.3x10^{-3}
90	2.6	0.8	0.3	+0.025	6x10^{-3}

V$_T$ decreases linearly with temperature and can be modeled as:

$$V_T(T) = V_{T300} + B \cdot \left(\frac{T - 300}{300}\right), \quad (7)$$

where V_{T300} is the threshold voltage value at 300 K. The reduction of V$_T$ with temperature was in general somewhat smaller than the value reported for a-Si:H TFTs of -36 mV/K, [11]. For the device shown in Table 2, the reduction was around -25 mV/K.

Parameter α_s varied from 0.32 to 0.24 for carbon content x=0.23, which is around half the typical value for a-Si:H. For this reason, output characteristics show the saturation region starting from smaller values of V$_{DS}$ compared to a-Si:H TFTs with similar V$_T$ and geometry. This is desirable for devices working in switching regime. This parameter stays practically constant with temperature, although the saturation voltage V_{DSsat} increases, basically due to the reduction of V_T. As carbon content decreases, α_s increases.

In Fig. 1 and 2 we show the excellent agreement of modeled output curves, with experimental ones at 30 °C. In general, if model parameters are extracted at 300 K, and the variation law with temperature of each parameter is known or determined for a given fabrication process, transfer and output characteristics for different temperatures can be modeled using a modification of UMEM, where each model parameter is multiplied by the function describing its variation with temperature.

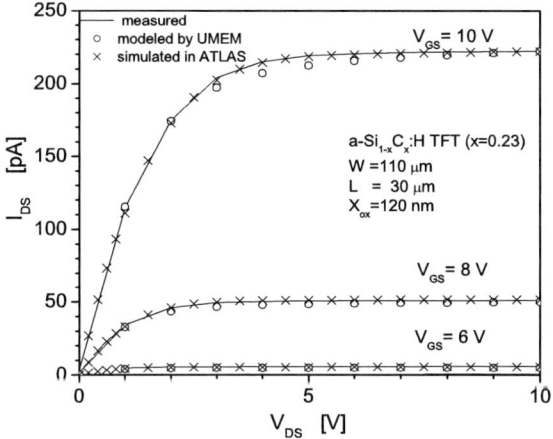

Fig. 2 Comparison between measured output characteristics of an a-Si1-xCx:H TFT with x=0.23, curves modeled by UMEM and curves simulated in ATLAS using the same localized trap distribution indicated in Fig. 1.

Simulation in ATLAS[1] using different localized state energy distribution was used to determine the effect of variations of the activation energy and state density on the output and transfer curves. Fig. 1 and 2 also show the good correspondence of simulated in ATLAS transfer and output curves with measured ones at 30 °C, for devices with x=0.23, for a localized state distribution in which acceptor and donor deep states are considered $g_{Ad}=4 \times 10^{18}$ cm^{-3}, $g_{Dd}=4 \times 10^{18}$ cm^{-3}, respectively similarly as reported for a-Si:H in [12]; acceptor and donor tail density were both considered equal to $g_{At}=g_{Dt}= 4 \times 10^{21}$ cm^{-3}; the activation energy for acceptor and donor deep states was increased to $E_{Ad}=0.176$ eV, $E_{Dd}=0.256$ eV, respectively, while the activation energy for acceptor and donor tail states was increased to $E_{At}=0.05$ eV and $E_{Dt}=0.1068$ eV.

Devices were tested up to 240 °C. Fig. 3 shows measured linear transfer characteristic between T=50 and 150 °C. As can be seen I_{DS} increases while V_T decreases with temperature. Up to 90 °C, the value of S is not altered and the I_{on}/I_{off} ratio increased in almost 2 orders. At 150 °C, however, this ratio reduces since I_{off} increases very

significantly with the conductivity of the film. The ratio further reduces as T is increased.

The I_{on}/I_{off} ratio of the transfer characteristic in saturation reduces with T, see Fig. 4. Although I_{on} increased, I_{off} increased more significantly with temperature. In all cases, after cooling, devices needed several hours to return to previous characteristics measured at room temperature before heating, suggesting that at least part of the increase in I_{DS} is associated with the localized traps.

On some devices, after T=80 °C an abrupt kink appeared, although this was not a normal behavior. Curves displaced several volts toward negative values of gate voltage, corresponding to the decrease of V_T.

An Arrhenius plot of the drain current for $V_{GS}=10$ V gives an activation energy around 0.4 eV for both $V_{DS}=10$ V and, $V_{DS}=1$ V, corresponding to an increase in I_{DS} of one order per 50 °C increase in temperature, see Fig. 5.

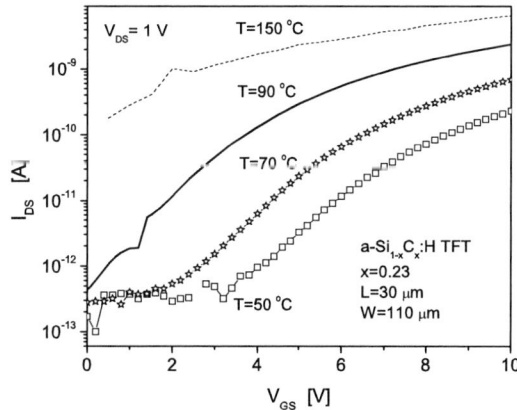

Fig. 3. Behavior with temperature of the linear transfer curves for an a-Si$_{1-x}$C$_x$:H TFT with x=0.23

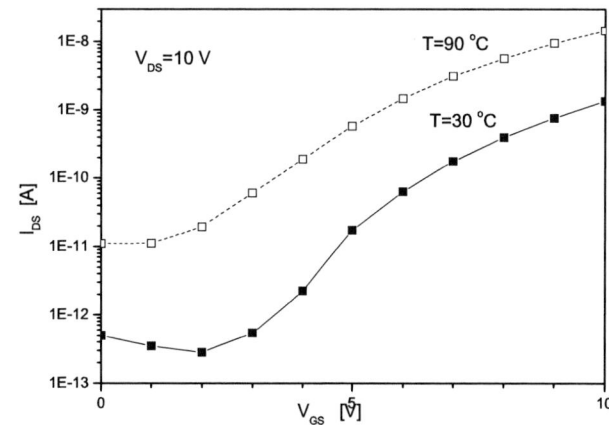

Fig. 4. Behavior with temperature of the saturation transfer curves of an a-Si$_{1-x}$C$_x$:H TFT with x=0.23.

---

[1] ATLAS is a product of Silvaco International.

As temperature was increased, the output characteristics showed, for low gate voltages, a rapid increase of the drain current that appears when V_{DS} is greater that a critical value. This effect seems to be due to band to band tunneling. As V_{GS} decreases, this critical value of V_{DS} decreases. The effect can be simulated in ATLAS using the field dependent term in the generation expression represented as [13]:

$$G = C1 \cdot E^{C2} \cdot \exp\left(-\frac{C3}{E}\right), \qquad (8)$$

where C1, C2 and C3 are adjustment parameters. Modeling this effect with default values of C1=4x10^{14} V^{-2}s^{-1}cm^{-1}, C2=2.5 and C3=1.9x10^7 V/cm, included in ATLAS simulator show the increase in current at V_{DS}= 20 V for V_{GS}>30 V at 90 °C. If C3 is taken equal to 3x10^6 V/cm, the increase in I_{DS} starts for V_{DS}<10 V, which corresponds to experiment for the low voltage device shown in Fig. 6. This effect appears when the longitudinal electric field near the drain is greater than 1x10^5 V/cm.

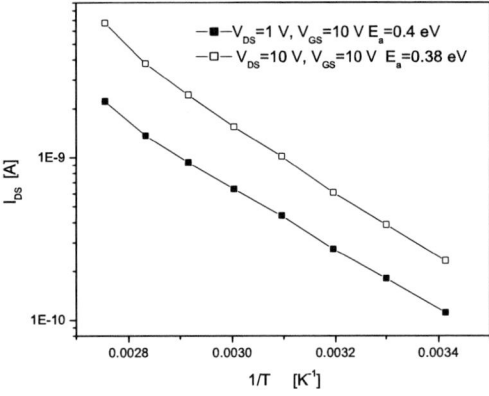

Fig. 5 Variation of drain current with temperature at V_{GS}=10 V, for V_{DS}=1V and V_{DS}=10 V.

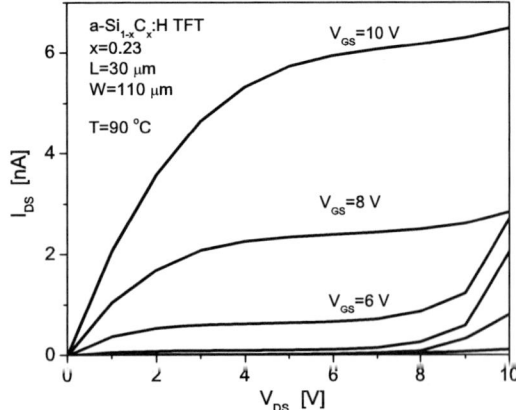

Fig. 6 Effect of band to band tunneling in the output characteristics of an a-Si$_{1-x}$C$_x$:H TFT with x=0.23, at 90 °C.

High voltage devices were fabricated that worked up to 90 V, see Fig. 7. However, the increase in I_{DS} already mentioned for drain voltages above a critical value, may limit the maximum value of V_{DS} that is possible to apply.

Finally devices were tested for instability and degradation, which is known to be a major issue in a-Si:H devices.

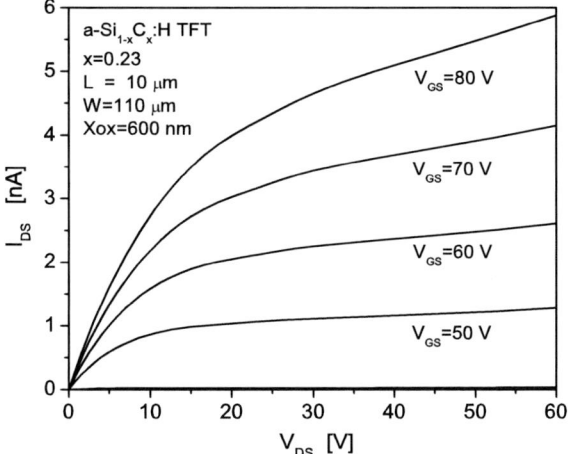

Fig. 7. Output characteristic of a high voltage a-Si$_{1-x}$C$_x$:H TFT with x=0.23.

After stressing an a-Si$_{1-x}$C$_x$:H TFT at V_{GS}=20 V and V_{DS}=1 V linear transfer curves slightly displace to the left, the drain current in subthreshold slightly increases, while it practically does not change in above V_T, as shown in Fig. 8.

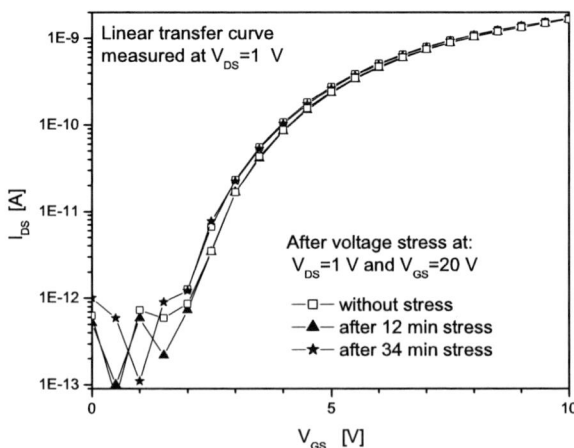

Fig. 8. Instability in the linear transfer curve at V_{DS}=1 V after voltage stress of V_{GS}=20 V and V_{DS}=1 during 12 and 34 min.

Under this stress condition, the longitudinal electric field is small and the variation in the drain current an V_T are not much different from that observed in a-Si:H. This effect is attributed to the increased number of electrons in the channel due to the transversal electric field sufficiently high to interact with weak Si-H bonds, releasing H and creating positively charged localized states along the interface with the dielectric or near-by it. The appearance of positive charge is responsible for the reduction of V_T [14].

Since there is practically no deformation of the curves after this stress, the generated traps are expected to be located in the lower part of the middle of the gap. In our case the effect becomes more significant when the stress is done at V_{GS}>20 V, when the transversal electric field is higher than 10^5 V/cm and the surface carrier concentration is above 10^{16} cm^{-2} along most of the channel, extending from the drain.

Some of the a-Si$_{1-x}$C$_x$:H TFTs were polycrystallized, using a KrF laser excimer as described in [4]. In Fig. 9 output curves for V_{GS}=10 V are plotted for an a-Si$_{1-x}$C$_x$:TFT with x=0.17 before and after polycrystallization. The output curve for the polycristallized TFT at 100 ^oC is also plotted. It is clearly seen the increase in drain current of the crystallized TFT with respect to the amorphous at both temperatures.

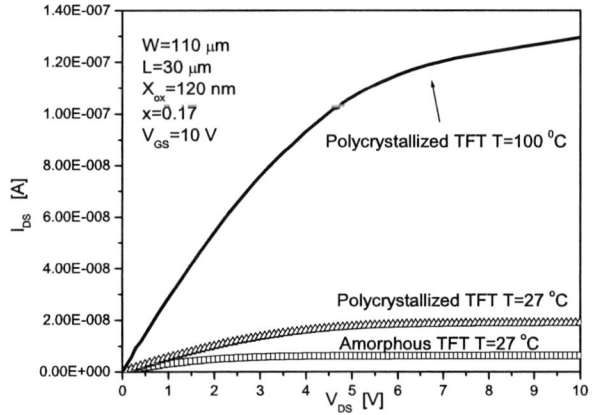

Fig. 9 Comparison of output characteristics for an a-Si$_{1-x}$C$_x$:H TFT before and after polycrystallization and for the polycrystallized TFT at T=100 ^oC.

IV. CONCLUSIONS

a-Si$_{1-x}$C$_x$:H TFTs were first fabricated by PECVD deposition with carbon content of x=0.23 and x=0.17, corresponding to a band gap of 2.1 and 1.9 respectively. Transfer curves show a subthreshold swing S of 2V/decade, with I_{on}/I_{off} greater than 200. I_{DS} in a-Si$_{1-x}$C$_x$:H TFTs increases around one decade per 50 ^oC.

Mobility shows a stronger dependence on gate voltage than that for a-Si:H TFTs. which reduces as temperature is increased. V_T decreased with temperature at a rate of -25 mV/K, which is less than for a-Si:H. The saturation factor α_s is around half the value typically observed for a-Si:H TFTs, resulting in an advantage for devices working in the switching regime.

It was demonstrated the possibility of fabricating a-Si$_{1-x}$C$_x$:H TFTs that work at voltages, up to 90 V.

Localized trap distribution was estimated. Deep and tail trap density are similar to values reported for a-Si:H devices, but the activation energy for both of them is hiagher. The effect of voltage stress at V_{DS}=1 V and V_{GS}=20 V is similar to the observed in a-Si:H TFTs.

First results of polycrystallization of the amorphous TFTs show a significant increase in drain current for similar conditions. Further work most be done regarding polycrystallization to reduce surface damage after laser annealing.

ACKNOWLEDGEMENT

We acknowledge Olga Gallegos, Edmundo Rodriguez, Benito Nepomuceno and Enriqueta Aguilar for device fabrication, K.F. Albertin and L. Resendiz for electrical measurements with temperature and B. García for all the work related to polycrystalline TFTs. This work was supported by CONACYT project 39708 and program ACI2002-34 of the Generalitat de Catalunya in Spain.

REFERENCES

[1] J. Kanicki in *Amorphous and Microcrystalline Semiconductor Devices*, Vol 1, Boston. MA., Artech House 1991.

[2] I. Pereyra and M. N. P. Carreño, "Wide Gap a-Si$_{1-x}$C$_x$:H Thin Films Obtained Under Starving Plasma Deposition conditions", *J Non-Crystalline Solids*, Vol. 201, pp. 110-118, 1996.

[3] S. E. Hicks, A. G. Fitzgerald and S. H. Baker, "The Structural, Chemical and Compositional Nature of Amorphous Silicon Carbide Film", *Phil. Mag. B,* vol. 62, pp. 193-212, 1990.

[4] B. Garcia, M. Estrada, K. F. Albertin, M. N. P. Carreño, I. Pereyra and L. Resendiz, "Amorphous and Excimer Laser Annealed SiC Films for TFT Fabrication", available on line, to be published in *Solid State Electronics* 2006.

[5] M. N. P. Carreño, I. Pereyra and H. E. M. Peres, "N-type Doping in PECVD of a-Si$_{1-x}$C$_x$:H Obtained Under Starving Plasma Conditions", *J Non-Crystaline Solids* vol. 227-230, pp. 483-487, 1998.

[6] A. Cerdeira, M. Estrada, R. García, A. Ortiz-Conde and F. J. García Sanchez, "New Procedure for the Extraction of Basic a-Si:H TFT Model Parameters in the Linear and Saturation Regions", *Solid-State Electronics*, vol. 45pp. 1077-1080, 2001.

[7] L. Reséndiz, M. Estrada and A. Cerdeira, "New Procedure for the Extraction of a-Si:H TFTs Model Parameters in the Subthreshold Region", *Solid State Electronics*, vol. 47 pp. 1351-1358, 2003.

[8] M. Estrada, A. Cerdeira, A. Ortiz-Conde, F. J. García and B. Iñiguez, "Extraction Method for Polycrystalline TFT Above and Below Threshold Model Parameters", *Solid-State Electronics*, vol. 46, pp. 2295-2300, 2002.

[9] A. Cerdeira, M. Estrada, B. Iñiguez, J. Pallares and L. F. Marsal, "Modeling and Parameter Extraction Procedure for

Nanocrystalline TFTs", *Solid-State Electronics*, vol. 48, pp. 103-109, 2004.

[10] M. Estrada, A. Cerdeira, J. Puigdollers, L. Resendiz , J. Pallares, L. F. Marsal, C. Voz and B. Iñiguez, "Accurate Modeling and Parameter Extraction Method for Organic TFTs", *Solid-State Electronics,* vol. 49, pp. 1069-1016, 2005.

[11] L. Wang, T. Fjeldly, B. Iñiguez, H. Slade and M. Shur, "Self Heating and Kink Effects in a-Si:H Thin Film Transistors", *IEEE Trans. Electron Devices*, vol ED-47, pp. 387-397, 2000.

[12] Y-T. Tsai, K-D. Hong, and Y-L. Yuan, "An Efficient Method for Calculating Trapped Charge in Amorphous Silicon", *IEEE Tran on Computer Aided Design of Integrated Circuits* vol. CADIC-13, pp. 725-728, 1994.

[13] G. A. M. Hurkx, D. B;. Klassen and M. P. G. Knuvers, "A new Recombination Model for Device Simulation Including Tunneling", *IEEE Trans. Electron Devices*, vol ED-39, pp. 331-338, 1992.

[14] W. B. Jackson, J. M. Marshal and M. D. Moyer, "Role of Hydrogen in the Formation of Metastable Defects in Hydrogenated Amorphous Silicon", *Phys Rev B*, vol. 39, pp. 1164-1179, 1989.

2006 25th International Conference on Microelectronics

Boron Redistribution During SOI Wafers Thermal Oxidation

Zoran Đurić, Milče M. Smiljanić, Katarina Radulović, Žarko Lazić

Abstract – We fabricated Silicon-On-Insulator (SOI) pressure sensors intended for operation in a wide temperature range. After the thermal oxidation process, the measured sheet resistance of the silicon active layer was higher than expected for the case when the dopant distribution in the active layer is uniform and the layer thickness reduced. The reason was the redistribution of dopant (boron) in the active layer during the high temperature thermal oxidation processes. We developed a model to calculate the boron redistribution in an SOI wafer which was initially uniformly doped. The temperature dependence of hole mobility (and, based on it, the sheet resistance and its temperature dependence) was determined for a calculated impurity profile after oxidation in wet O_2. The experimental temperature dependence of sheet resistance was obtained by measuring the temperature dependence of piezoresistor resistance. The best match between the theoretical and experimental TCR (temperature coefficient of resistance) was achieved by slightly modifying the Arora's model for hole mobility.

I. INTRODUCTION

Diffused piezoresistors of IHTM-CMTM piezoresistive pressure sensor consist of *p*-type regions in an *n*-type substrate. The previously designed pressure sensor had a limited range of operating temperatures because the leakage currents of *pn* junction became unacceptably high at temperatures exceeding about 120 ^{0}C. A novel sensor structure for a wide range of operating temperatures has been fabricated by the standard planar technology and micromachining technology on SOI wafers (Fig.1a). The new SOI pressure sensors have piezoresistors dielectrically isolated by SiO_2 from each other and from the substrate [1], Fig. 1b.

The boron concentration profile in the active layer and the boron surface concentration on the Si-SiO₂ surface determine the sheet resistance and the temperature behaviour of the piezoresistor, respectively. A hole surface concentration of 2-3·10^{18} cm^{-3} in piezoresistors enables an "intrinsic" temperature compensation of pressure sensor sensitivity [2], [3].

In SOI bulk micromachining, it is necessary to protect the silicon active layer from wet etchants by thermal oxide. During thermal oxidation boron impurities move from

silicon to silicon dioxide through the Si-SiO₂ surface. This effect causes depletion and redistribution of boron impurities in the SOI active layer.

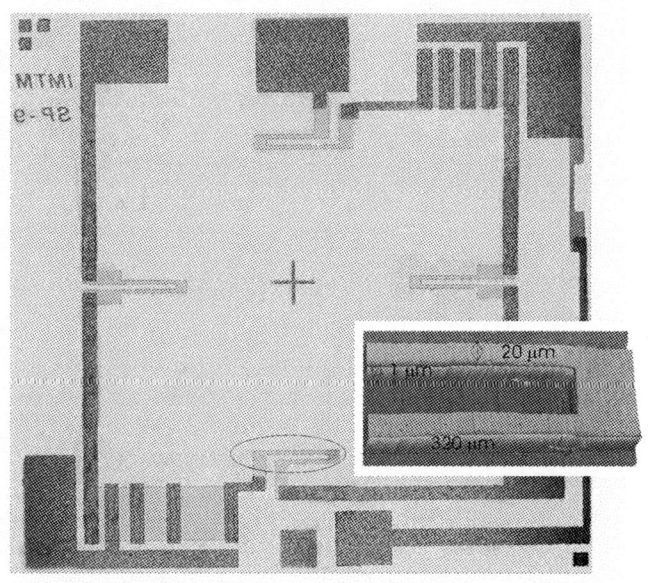

Fig. 1a IHTM-CMTM SOI piezoresisitive pressure sensor for wide operation temperature range. The inset represents the AFM scan of the piezoresistor.

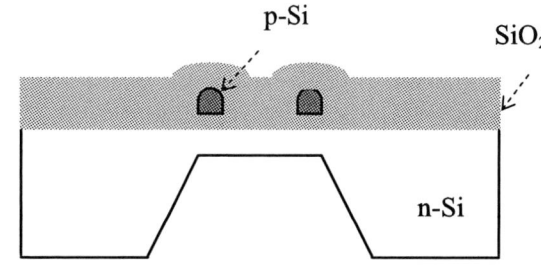

Fig. 1b The schematic cross section of the piezoresistor of IHTM-CMTM SOI piezoresisitive pressure sensor.

II. THEORETICAL MODEL

The redistribution process during thermal oxidation was studied by Grove et al. [4] for uniformly doped semi-infinite bulk silicon and by Huang et al [5] for a Gaussian

---

Zoran Đurić, Milče M. Smiljanić, Katarina Radulović and Žarko Lazić are with the IHTM – Institute of Microelectronic Technologies and Single Crystals, Njegoševa 12, 11000 Belgrade, Yugoslavia

Fax +381 11 182 995, e-mail zdjuric@nanosys.ihtm.bg.ac.yu

1-4244-0116-X/06/$20.00 ©2006 IEEE   333

profile of dopant distribution. Analogously with the model given by Huang et al., we developed a model of boron redistribution during thermal oxidation in a uniformly doped active layer of SOI with a finite thickness (l) and an initial boron concentration N_B. Because the Huang's model is valid for bulk Si, we modified the boundary conditions for SOI wafer.

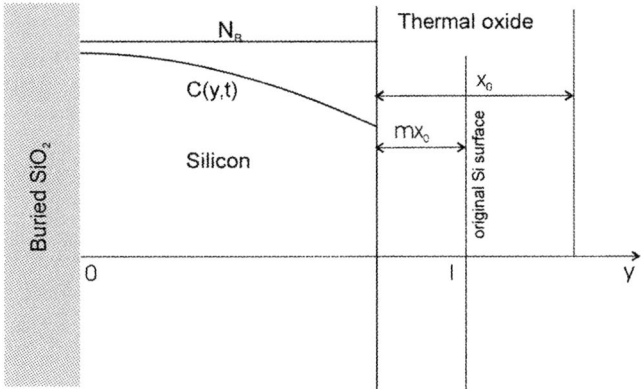

Fig. 2 The shematic model of boron redistribution in SOI wafer during thermal oxidation

Fig. 2 shows the redistribution model and the used analytical axes. The origin is at the Si-buried SiO$_2$ interface. The depth in the active p-Si layer is y, where for $t=0$ the thermally grown SiO$_2$-Si interface is at $y=l$. In the model, parabolic thermal oxide growth is assumed. In that case the thickness of the thermally grown SiO$_2$ layer is given by $x_o = \sqrt{Bt}$, where B is the growth rate constant. mx_0 corresponds to the active layer thickness consumed due to the oxidation.

The resulting boron distribution within the active layer can be represented as the sum of two parts [5].

First, it is considered that there is no flow of impurity atoms into the thermal oxide from silicon, and the original Si surface is impermeable. Oxidation will cause an amount of impurity contained within mx_o to be absorbed by the oxide. The impurity distribution, in the case of uniformly doped Si layer, is then given by $C_1(y,t)=N_B$.

The conditions for the second part (C_2) of the solution are:

$$\bullet \quad C_2(y,0) = 0 ; \text{ initial condition} \tag{1}$$

$$\bullet \quad D\frac{\partial C_2}{\partial y}\bigg|_{y=0} = 0 ; \text{ at Si-buried SiO}_2 \text{ interface} \tag{2}$$

where D is the boron diffusion constant in Si.

There is a net flow of impurities across the Si-thermal SiO$_2$ interface. The rate of flow is obtained under consideration that in the small time interval dt, a layer of silicon mdx_o is converted into oxide of thickness dx_o. The amount of impurity in the original silicon is $Cs(l-mx_o, t)\cdot mdx_o$, and the amount in the oxide is

$$Cs'(l - mx_o,t)dx_o = kCs(l - mx_o,t)dx_o ,$$

where k is the segregation coefficient, and Cs and Cs' are the surface concentrations in the Si and SiO$_2$ layer, respectively. The rate of flow of impurities across the Si-thermal SiO$_2$ interface is then given by

$$\bullet \quad D\frac{\partial C_2}{\partial y}\bigg|_{y=l-mx_o} = (k-m)Cs(l-mx_o,t)\frac{dx_o}{dt} = \frac{(k-m)Cs(l-mx_o,t)}{2}\sqrt{\frac{B}{t}} \tag{3}$$

According to these conditions, the second part of the solution $C_2(y,t)$ is given by

$$C_2(y,t) = A\cdot\left[erfc\left(\frac{l-m\sqrt{Bt}+y}{2\sqrt{Dt}}\right) + erfc\left(\frac{l-m\sqrt{Bt}-y}{2\sqrt{Dt}}\right)\right] \tag{4}$$

Equations (3) and (4) can be combined to give

$$C_2(y,t) = \frac{k-m}{2}\sqrt{\frac{B\pi}{D}}Cs(l-mx_o,t)\cdot \frac{erfc\left(\frac{l-m\sqrt{Bt}+y}{2\sqrt{Dt}}\right) + erfc\left(\frac{l-m\sqrt{Bt}-y}{2\sqrt{Dt}}\right)}{1-\exp\left(-\left(\frac{l-m\sqrt{Bt}}{\sqrt{Dt}}\right)^2\right)} \tag{5}$$

The resulting impurity distribution, $C(y,t)$ is obtained by adding $C_1(y,t)$ and $C_2(y,t)$. By setting $y = l - mx_o = l - m\sqrt{Bt}$ in relation for $C(y,t)$, the surface concentration $Cs(l-mx_o, t)$ is determined

$$Cs(l - mx_o,t) = \frac{N_B\left(1-\exp\left(-\left(\frac{l-m\sqrt{Bt}}{\sqrt{Dt}}\right)^2\right)\right)}{1-\exp\left(-\left(\frac{l-m\sqrt{Bt}}{\sqrt{Dt}}\right)^2\right) + \frac{k-m}{2}\sqrt{\frac{B\pi}{D}}\left(1+erfc\left(\frac{l-m\sqrt{Bt}}{\sqrt{Dt}}\right)\right)} \tag{6}$$

Finally, the concentration profile is given by

$$\frac{C(y,t)}{N_B} = 1 - \frac{k-m}{2}\sqrt{\frac{B\pi}{D}}\;\frac{erfc\left(\frac{l-m\sqrt{Bt}-y}{2\sqrt{Dt}}\right)+erfc\left(\frac{l-m\sqrt{Bt}+y}{2\sqrt{Dt}}\right)}{1-e^{\left(\frac{l-m\sqrt{Bt}}{\sqrt{Dt}}\right)^2}+\frac{k-m}{2}\sqrt{\frac{B\pi}{D}}\left(1+erfc\left(\frac{l-m\sqrt{Bt}}{\sqrt{Dt}}\right)\right)} \qquad (7)$$

In Fig. 3 the calculated impurity profile after the oxidation processes in wet O_2 at 1100°C for 9 h 15 min is given. The theoretical profile of the boron distribution is given by (7) for which the values of the diffusion and oxidation parameters, corresponding to the oxidation temperature are given in Table 1 [5].

The sheet resistance and its temperature dependence are calculated using equation

$$R_{sheet}(y,T) = 1/\left(q\int_{o}^{l-mx_o}C(y,T)\mu_p(y,T)dy\right) \qquad (9)$$

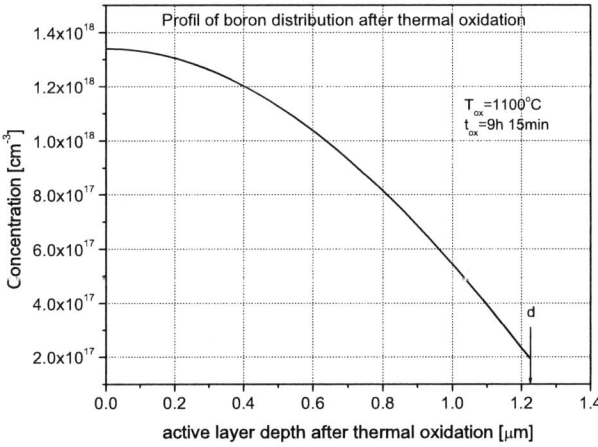

Fig. 3 Calculated boron distribution in active silicon layer of SOI after thermal oxidation at 1100°C for 9h 15 min

Fig. 4 Hole mobility μ_p obtained by Arora's model for dopant profile given in Fig. 3

III. EXPERIMENTAL RESULTS AND DISCUSSION

We used SOI wafers with a p-type silicon active layer to omit process of boron diffusion and simplify fabrication of the new pressure sensor. The thickness of the active layer was ~2.1 μm and its boron concentration approximately $5 \cdot 10^{18}$ cm^{-3} (ρ~0.0195 Ωcm)

In order to obtain the pressure sensor diaphragm and dielectrically isolated piezoresistors, wet oxidation was performed at 1100°C for 9h 15min. During this process, the active layer thickness was reduced from l~2.1 μm to d~1.21μm.

The experimental temperature dependence of the sheet resistance of the active layer after the oxidation process was obtained by measuring the piezoresistors resistance change with temperature. We measured values of piezoresistor resistance in a temperature chamber at temperatures of -20, 0, 20, 40 and 60 ^{0}C. The values of experimentally obtained resistance are given in Table 2.

TABLE I

THE VALUES OF THE (OXIDATION) PARAMETERS USED IN THEORETICAL CALCULATION [5]

Oxidation temperature, T	1100^{0}C
Oxidation conditions	wet O_2
Diffusivity [cm^2/s], D	3.5 x 10^{-13}
Growth constant [cm^2/s], B	1.55 x 10^{-12}
Initial boron concentration [cm^{-3}], N_B	5 x 10^{18}
Segregation coefficient, k	9

The hole mobility $\mu_p(y,T)$ profiles in the active layer for the various temperatures from -20 to 250°C given in Fig.4 are obtained by Arora's empirical relation [6]

$$\mu_p = 54.3\left(\frac{T}{300}\right)^{-0.57} + \frac{407\left(\frac{T}{300}\right)^{-2.23}}{1+C/\left(2.67e17\left(\frac{T}{300}\right)^{2.546}\right)} \qquad (8)$$

The sheet resistance obtained based on these experimental values ($R_{sheet}=R/n_{\square}$, $n_{\square}\sim33$) is higher than expected $R_{sheet}>\rho/d$ for the case when the dopant distribution in the active layer is uniform and the layer thickness reduced. The calculated and experimental temperature dependences of sheet resistance are given in Fig. 5.

TABLE II

MEASURED VALUES OF PIEZORESISTORS RESISTANCES

T [^{0}C]	R_1 [kΩ]	R_2 [kΩ]	R_3 [kΩ]	R_4 [kΩ]
-19.33	13.82	14.93	16.43	14.6
1.8	14.24	15.39	16.96	15.05
22.9	14.92	16.61	17.65	16
41.78	15.53	16.81	18.57	16.44
61.47	16.21	17.55	19.39	17.16

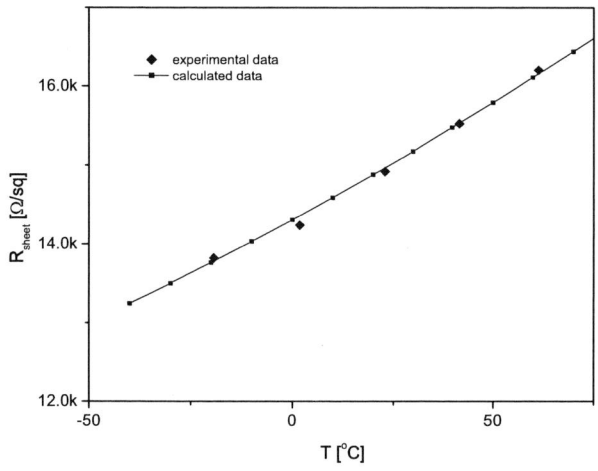

Fig. 5 The calculated and experimental temperature dependences of the sheet resistance

The best agreement between the theoretical TCR (temperature coefficient of resistance) and the experimentally obtained one was achieved by modifying the value of the parameter 2.67 in the denominator of (8) to 2.37. The reason is that Arora's model is valid for bulk silicon, while in SOI there is the hole mobility degradation caused by SiO_2-Si surface roughness scattering, which is temperature dependent.

IV. CONCLUSIONS

The theoretically calculated and experimentally measured values of the sheet resistance after high temperature thermal SOI wafer oxidation in wet oxygen exceed the values calculated using data sheet values specified for the active SOI layer. The reason is the redistribution of dopant (boron) in the active layer during the thermal oxidation processes. A consideration of the results given in Fig. 5 indicates a good agreement between theory and experiment, thus validating the diffusion model for boron redistribution during thermal oxidation as proposed in this work.

ACKNOWLEDGEMENTS

The authors wish to thank Dr Zoran Jakšić for helpful suggestions.

This work has been partially supported by the Serbian Ministry of Science, Technologies and Development within the framework of the project TR 6151.

REFERENCES

[1] Jyrki Kühämäki, "Fabrication of SOI micromechanical devices", VTT Technical Research Centre of Finland, VTT Publication 559, 2005.
[2] В.И.Ваганов, Интегральные тензопреобразователи, Москва энергоатомиздат 1983
[3] Semiconducror Sensors, Edited by S.M.Sze, John Wiley & Sons Inc., 1983.
[4] A. S. Grove, O. Leistiko, Jr., and C. T. Sah, "Redistribution of Acceptor and Donor Impurities during Thermal Oxidation of Silicon", *Journal of Applied Physics*, 1964, Vol.35, No.9, pp. 2695-2701.
[5] J. S. T. Huang and L. C. Welliver, "On the Redistribution of Boron in the Diffused Layer during Thermal Oxidation", *J.Electrochem.Soc.: Solid State Science*, 1970, Vol. 117, No.12, pp. 1577-1580.
[6] VLSI Handbook, Edited by Norman G. Einspruch, Academic Press, Inc., 1985.

2006 25th International Conference on Microelectronics

Transport of Non-Equilibrium Charge Carriers in Bipolar Semiconductor Materials

Yuri G. Gurevich, and J.E. Velázquez-Pérez

Abstract – In this work we present results concerning the fundamental equations of charge carrier transport in semiconductor structures. We discuss the modeling of the recombination term in the charge transport equations for each type of carriers that respects the charge conservation law. It was obtained that under stationary conditions and equal generation rates of electrons and holes the recombination rates of both kinds of carriers must be matched. Under low excitation conditions (linear regime) the recombination rate can be expressed as a linear combination of the variation of the carrier concentrations δn and δp. Explicit calculation of the charge variation has been carried out in the framework of the Shockley-Read-Hall statistics. The study of the space charge in non-equilibrium has been addressed. It was found that the quasi-neutrality condition contradicts the usual assumption $\delta n = \delta p$. Finally, boundary conditions for the Poisson equation have been presented.

I. INTRODUCTION

The charge carrier transport is at the base of the electrical behaviour of applications based on any semiconductor device. Despite the efforts to correctly modelling the transport in semiconductors across the years, many questions remain open in the study of transport phenomena: for instance the recombination processes have to be described in a way that do not contradicts the Maxwell equations [1]. The boundary conditions (BC) commonly used when solving the transport equations need to be carefully reviewed (the ones currently used are only valid for devices operating in open-circuit conditions [2]). In this paper we will show that also a crucial concept as it is the quasi-neutrality (QN), that is at the base of the models used to explain the operation of most of the semiconductor devices, [3-5] need to be revised.

To describe the transport of the non-equilibrium charge carriers we used the conventional transport equations for the non-equilibrium charge carriers and the Poisson's equation [6], however the common expressions to describe the recombination was carefully modified in order to ensure that the law of conservation of charge is satisfied. The details of the model can be found in [7, 8].

One of the main conclusions of this paper is that when

Yuri G. Gurevich is on sabbatical leave at the Universidad de Salamanca.. Permanent address: Departamento de Física, CINVESTAV-IPN, Apdo. Postal 14-740, 07000 México, D.F. México. E-mail: gurevich@fis.cinvestav.mx.

J.E. Velázquez-Pérez, is with the Departmento de Física Aplicada, Universidad de Salamanca, Pza. de la Merced, s/n, E-37008 Salamanca, Spain, E-mail: js@usal.es.

in a device a contact between a p-type semiconductor and a metal is present this contact, and consequently the whole device, cannot be considered as being unipolar. And this results is independent of the ohmic or Schottky nature of the contact.

II. PHYSICAL MODEL AND RECOMBINATION

Otherwise indicated and without loss of generality, we will refer in this paper to a semiconductor that contains an impurity with a single energy level able of trapping electrons.

The macroscopic description of the charge transport is done by the continuity equations for the electrons (j_n) and hole (j_p) current densities and the Poisson equation:

$$\frac{\partial n}{\partial t} = g_n + \frac{1}{e} div j_n - R_n \tag{1}$$

$$\frac{\partial p}{\partial t} = g_p - \frac{1}{e} div j_p - R_p \tag{2}$$

$$div E = \frac{4\pi}{\varepsilon} \rho \tag{3}$$

where, n and p are the local electron and hole concentrations (n_0 and p_0 are the equilibrium values), g_n and g_p are the electron and hole external generation ratio, E is the electric field, ρ is the bulk electrical charge, e is the hole charge, ε is the permittivity, and R_n and R_p are the electron and hole recombination rates.

Subtracting the Eq. 1 from the Eq. 2 we obtain:

$$e(g_n - g_p) + div(j_n + j_p) - \\ -e(R_n - R_p) = e \frac{\partial(n-p)}{\partial t} \tag{4}$$

The charge conservation can be written as:

$$div(j_n + j_p) = div j = e \frac{\partial}{\partial t}(n - p + n_t) \tag{5}$$

where j is the total current and n_t the concentration of electrons captured by the impurities. From the Eqs. 4 and 5 we obtain the following relationship:

1-4244-0116-X/06/$20.00 ©2006 IEEE

$$g_n - g_p = R_n - R_p - \frac{\partial n_t}{\partial t} \qquad (6)$$

In general, the variation of the concentration of the electrons trapped in the impurity level (δn_t) is function of deviations of the electron and hole concentrations from their equilibrium values (δn, δp) and of g_n, g_p and t.

Under stationary conditions if the generation rates g_n and g_p are the same (this includes the trivial case of both g_n and g_p being null) from the Eq. 6 it follows that both recombination rates must also match:

$$R_n = R_p = R \qquad (7)$$

It can readily be shown that, [1], that the recombination centre follows the Shockley-Read-Hall statistics we can write from Eq. 7:

$$\delta n_t = \frac{1}{\alpha_n(n_0 + n_1^0) + \alpha_p(p_0 + p_1^0)} \qquad (8)$$

$$\left[\alpha_n(N_t - n_t^0)\delta n - \alpha_p n_t^0 \delta p\right]$$

where α_n (α_p) is the electron (hole) capture coefficient, N_t is the impurity concentration, n_t^0, n_1^0 and p_1^0 are the equilibrium values of the respective magnitudes (n_1 (p_1) are the electron (hole) concentration when the Fermi-level matches the activation energy of the impurity) and, in the linear approximation, $n_t = n_t^0 + \delta n_t$.

Also in linear regime, R can be written as a linear combination of δn and δp (this expression can be extended also to the case of band to band recombination):

$$R_n = R_p = R = \frac{\delta n}{\tau_n} + \frac{\delta p}{\tau_p} \qquad (9)$$

Under the same conditions the following relationship holds:

$$\frac{\tau_n}{\tau_p} = \frac{n_0}{p_0} \qquad (10)$$

It has been demonstrated, [1], that despite the time dimensions of τ_n and τ_p, these parameters cannot be straightforwardly identified with the lifetimes of the non-equilibrium carriers, contrary to what is widely used in semiconductor modelling [9].

In the particular case of an intrinsic semiconductor, the above conditions are trivially fulfilled ($n_t = 0$ and the only generation and recombination mechanism available is band to band); nevertheless, even in this special situation, in general, the inequality $\delta n \neq \delta p$ holds as the possible existence of a local electric field cannot, in principle, be ruled out.

II. RESULTS AND DISCUSSION

From the above, it follows that assuming linear conditions (low injection), uniform temperature ($T=T_0$) across the sample, stationary conditions and absence of generation processes in the region under study, the system of Eqs. 1, 2 and 3 can be re-written as:

$$\mathrm{div}\,\boldsymbol{j}_n = eR \qquad (11)$$

$$\mathrm{div}\,\boldsymbol{j}_p = -eR \qquad (12)$$

$$\mathrm{div}\,\delta\boldsymbol{E} = 4\pi\delta\rho = 4\pi e(\delta n - \delta p + \delta n_t) = \\ = A\delta n + B\delta p \qquad (13)$$

For simplicity we assumed a value for the permittivity equal to the unit. In the second member of Eq. 13 we write $\delta\rho$ as $A\delta n + B\delta p$ in agreement with Eq. 8. It should be noted that in Eq. 13 we use the variation of magnitudes, not the magnitudes themselves. In this way we can separate and remove the equilibrium contributions; for instance, the electric field can be written as $\boldsymbol{E} = \boldsymbol{E}_0 + \delta\boldsymbol{E}$, $\boldsymbol{E}_0$ contains the built-in electric-field and cancels with the contribution to the equilibrium electric charge (ρ_0) in the second member.

From the last equation, it follows that quasi-neutrality (QN) does not necessarily implies $\delta n \approx \delta p$. Whereas the latter is commonly accepted, it is a wrong assumption in most of the cases. Now we must recall that quasi-neutrality means $\rho = 0$. QN condition actually means that $A\delta n + B\delta p \approx 0$.

Let us come back to the system of Eqs. 1-3. In a general case, δn and δp can be written in terms of the variations of their respective chemical potentials as:

$$\delta n = \frac{n_0}{T_0}\delta\mu_n \qquad (14)$$

$$\delta p = \frac{p_0}{T_0}\delta\mu_p \qquad (15)$$

Let us consider, for the sake of simplicity, that he current's flux only takes place in the x-dimension. The current densities can be calculated as:

$$j_n = -\sigma_n\frac{d\widetilde{\delta\varphi}_n}{dx} \qquad (16)$$

$$j_p = -\sigma_p\frac{d\widetilde{\delta\varphi}_p}{dx} \qquad (17)$$

where σ_n (σ_p) is the electrical conductivity of electrons (holes) and $\widetilde{\delta\varphi}_n$ ($\widetilde{\delta\varphi}_p$) the variation of the electrochemical potential, or quasi-Fermi levels, of electrons (holes):

$$\delta\widetilde{\varphi}_n = \delta\varphi - \frac{\delta\mu_n}{e} \qquad (18)$$

$$\delta\widetilde{\varphi}_p = \delta\varphi + \frac{\delta\mu_p}{e} \qquad (19)$$

where φ is the electrical potential. Using Eqs. 9, 11, 12, 16 and 17 we obtain:

$$-\sigma_n \frac{d^2\delta\widetilde{\varphi}_n}{dx^2} = e\left(\frac{\delta n}{\tau_n} + \frac{\delta p}{\tau_p}\right) \qquad (20)$$

$$-\sigma_p \frac{d^2\delta\widetilde{\varphi}_p}{dx^2} = -e\left(\frac{\delta n}{\tau_n} + \frac{\delta p}{\tau_p}\right) \qquad (21)$$

By using the Eqs. 14, 15 and 10 the above two equations become:

$$-\sigma_n \frac{d^2\delta\widetilde{\varphi}_n}{dx^2} = e\left(\frac{n_0\delta\mu_n}{T_0\tau_n} + \frac{p_0\delta\mu_p}{T_0\tau_p}\right) =$$
$$= \frac{en_0}{T_0\iota_n}\left(\delta\mu_n + \delta\mu_p\right) \qquad (22)$$

$$-\sigma_p \frac{d^2\delta\widetilde{\varphi}_p}{dx^2} = -e\left(\frac{n_0\delta\mu_n}{T_0\tau_n} + \frac{p_0\delta\mu_p}{T_0\tau_p}\right) =$$
$$= -\frac{en_0}{T_0\tau_n}\left(\delta\mu_n + \delta\mu_p\right) \qquad (23)$$

These two equations can be conveniently rewritten in a more compact form:

$$\frac{d^2\delta\widetilde{\varphi}_n}{dx^2} = \frac{\tau_M^n}{\tau_n l_n^2}\left(\delta\widetilde{\varphi}_n - \delta\widetilde{\varphi}_p\right) \qquad (24)$$

$$\frac{d^2\delta\widetilde{\varphi}_p}{dx^2} = \frac{\tau_M^p}{\tau_n l_n^2}\left(-\delta\widetilde{\varphi}_n + \delta\widetilde{\varphi}_p\right) \qquad (25)$$

where l_n is the electron Debye length, τ_M^n and τ_M^p are the Maxwell relaxation times for electrons and holes respectively. Eqs. 24 and 25 do not depend explicitly on the electric potential. This means that in the system of equations 1, 2 and 3 we have uncoupled the equation of Poisson from the two equations that only involve the quasi-Fermi levels. The intimate physical reason underlying the uncoupling of the Poisson equation from the transport ones is that space charge do not influence the charge carrier transport.

We must emphasize the importance of a proper calculation of the recombination rates as they are basic to model the transport in semiconductor devices [10]. Quite

commonly, the simulation of ultra short CMOS [11] is carried out neglecting the carrier recombination across the sample since the device's simulated length is several orders of magnitude lower than the diffusion one and/or the devices are unipolar. Concerning this later point, we must realize that this approximation is not fully correct. If fails, for instance, in unipolar p-type devices as they involve metal contacts (and therefore electron-hole recombination will necessarily take place) and, evidently, in devices like MOSFET that are not truly unipolar devices. In order to illustrate the impact of the recombination on the electric characteristics of a sample [7-8] let us recall the expression for the resistance of a semiconductor slab with length "2a" and section equal to the unit in linear approximation of the electric field:

$$R_s = \frac{2a}{\sigma_n + \sigma_p}\left[1 + \frac{\sigma_p/\sigma_n}{\lambda a\coth(\lambda a) + \lambda^2\tau a S}\right] \qquad (26)$$

Where λ is the inverse of the diffusion length, τ is calculated as $\tau^{-1} = \tau_n^{-1} + \tau_p^{-1}$ in the absence of trapping centres ($n_t=0$) and S the surface recombination velocity. The term outside the brackets in Eq. 26 is the classical value of the ohmic resistance of the sample, whereas the term in the brackets is associated to bulk and surface recombination. Assuming that no recombination exists in the structure ($\tau\to0$, $S\to0$) leads to:

$$R_s \approx \frac{2a}{\sigma_n} \qquad (27)$$

This result fails to describe the expected value of a p-doped resistor: the reason for that is that the hypothesis that neglects recombination in a p-type semiconductor/metal heterojunction is wrong and leads to unphysical results. Consequently, the study of Schottky and ohmic contacts made on p-type semiconductors are different from those made on n-type semiconductors. The first ones must be considered like a bipolar semiconductor and recombination cannot be neglected.

Finally, to conclude this section let us discuss briefly the issue of the boundary conditions (BC) that must be used when solving the system of Eqs. 1 to 3 and to obtain Eq. 27. For the transport equations (Eq. 1 and 2) the BC were already presented and discussed in [2]. For the Poisson equation (Eq. 3) the issue was not previously discussed. Nevertheless it is easy to show that in the absence of double space charge at the interface (charge of opposite sign at each side of the semiconductor-metal interface) the BC is the continuity of the electric potential. If a double space charge is present (like in the junction of two metals) the electric potential will exhibit a discontinuity at the interface that in general is unknown, nevertheless since we are under QN conditions the equation: $A\delta n + B\delta p \approx 0$ holds and we will only need Eqs. 24 and 25 to study the charge transport. Therefore the Poisson equation should not be solved and consequently to know its BC is pointless.

IV. CONCLUSIONS

In this work we present results concerning the fundamental equations of charge carrier transport in semiconductor structures. We discuss the modeling of the recombination term in the charge transport equations for each type of carriers that respects the charge conservation law. It was obtained that under stationary conditions and equal generation rates of electrons and holes the recombination rates of both kinds of carriers must be matched. Under low excitation conditions (linear regime) the recombination rate can be expressed as a linear combination of the variation of the carrier concentrations δn and δp. Explicit calculation of the charge variation has been carried out in the framework of the Shockley-Read-Hall statistics. The study of the space charge in non-equilibrium has been addressed. It was found that the quasi-neutrality condition contradicts the usual assumption $\delta n = \delta p$. Finally, boundary conditions for the Poisson equation have been presented.

ACKNOWLEDGEMENTS

Yu. G. Gurevich wants to thank CONACYT for financial support and Ministerio de Educación y Ciencia of Spain and Universidad de Salamnca (España) for hospitality under grant number (SAB2004-0184). J.E. Velázquez thanks for financial support to Junta de Castilla y León (SA072A05) and MEC-FEDER (TEC2005-02719/MIC).

REFERENCES

[1] I.N. Volovichev, G.N. Logvinov, O.Yu. Titov, and Yu.G. Gurevich, "Recombination and lifetimes of charge carriers in semiconductors", *J. Appl. Phys.*, vol. 95, pp. 4494-4496, 2004.

[2] O. Yu. Titov, J. Giraldo, and Yu. G. Gurevich, "Boundary conditions in an electric current contact", *Appl. Phys. Lett.*, vol. 80, pp. 3108-3110, 2002.

[3] S.R. in't Hout, "Quasineutrality in semiconductors", *J. Appl. Phys.*, vol. 79, pp. 8435-8444, 1996.

[4] W. van Roesbroeck, "Current-carrier transport with space charge in semiconductors", Phys. Rev., vol. 123, pp. 474-490, 1961.

[5] Y. Moreau, J.-C. Manifacier, H.K. Henisch, "Minority carrier injection into relaxation semiconductors", *J. Appl. Phys.*, vol. 60, pp. 2904-4496, 19864.

[6] J. Singh, *Semiconductor Devices: Basic Principles*, New York, Wiley, 2001.

[7] Yu.G. Gurevich, G.N. Logvinov, G. Espejo, O.Yu. Titov, and A. Meriuts, "The role of non-equilibrium carriers in linear charge transport (Ohm's Law)", *Semiconductors*, vol. 34, pp. 755-758, 2000.

[8] Yu.G. Gurevich, G.N. Logvinov, I.N. Volovichev, G. Espejo, O.Yu. Titov, and A. Meriuts, "The role of non-equilibrium carriers in the formation of thermo-EMF in bipolar semiconductors", *phys. stat. sol. b*, vol. 231, pp. 278-293, 2002.

[9] P.T. Landsberg, *Recombination in Semiconductors*, Cambridge, Cambridge University, 1991.

[10] M. Reisch, *High-Frequency Bipolar Transistors*, Berlin, Springer-Verlag, 2003.

[11] P. Dollfus, A. Bournel, S. Galdin-Retailleau, S. Barraud, P. Hesto, "Effect of discrete impurities on electron transport in ultrashort MOSFET using 3D MC simulation", IEEE Trans. Electron Dev., vol. ED-51, pp. 749-756, 2004

2006 25th International Conference on Microelectronics

A High-Frequency Extension of a Surface-Potential-Based Substrate Model for Noise Coupling Analysis

Nebojša Simić, Fredrik Ingvarson, Simon Kristiansson, Marinel Zgrda and Kjell O. Jeppson

Abstract – In this paper we present a high-frequency extension of our surface-potential-based substrate model for predicting substrate noise coupling in integrated circuits. The model handles an arbitrary number of aggressor and victim devices on a multi-layered substrate with either biased or floating backside. We show that the dielectric properties of the substrate are easily included in this model for providing a more accurate description above GHz frequencies. Finite element calculations are used for validating the model.

I. INTRODUCTION

In system-on-chip solutions, where both digital and analog circuitry is present on the same silicon substrate, it has become a serious problem that noise created by the digital parts affects the analog parts [1]. The switching noise distributed through the substrate has been identified as an important issue already in the early 1990s [2]. Existing IC design tools offer possibilities to analyze the amount of substrate-coupled noise, but usually only in the post-layout design phase. However, there is also a need for substrate models for use already in the beginning of the design flow. Such models must be simple, yet accurate, and should assist designers in analyzing and minimizing substrate noise during the early phases of the design flow.

The model presented in our earlier work [3] is valid up to frequencies in the low GHz region for practical substrate resistivities. The purpose of this paper is to extend this model to include the dielectric properties of the substrate for a more accurate high-frequency description. Magnetic characteristics are not included.

This paper is organized as follows. Our substrate model presented in [3] is outlined in Section II and the high-frequency extension is presented in Section III together with a validation against finite element calculations. Section IV concludes this paper.

II. CURRENT SUBSTRATE MODEL

The voltages V and the currents I of N devices in an IC interacting through the substrate (see Fig. 1) can be related through a coupling matrix Z as

$$V = ZI. \qquad (1)$$

Authors are with Solid-State Electronics Laboratory, Dept. of Microtechnology and Nanoscience, MC2, Chalmers University of Technology, SE-412 96, Göteborg, Sweden, E-mail: simon.kristiansson@mc2.chalmers.se, simicn@yahoo.com

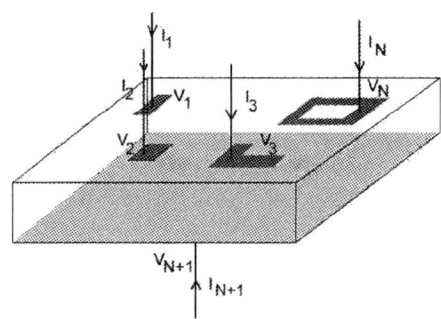

Fig. 1. Illustration of an IC with N devices interacting through the substrate.

The voltage of each contact (each device is modeled as a surface contact) is given with reference to the substrate potential V_{ref} far from all devices. The $N \times N$ substrate coupling elements z_{ij} of the Z matrix depend on the substrate doping profile and geometries and locations of the contacts and are generally difficult to determine. However, for circular contacts on a multi-layer substrate, as shown in Fig. 2, the z_{ij} elements can be calculated by

$$z_{ij} = \frac{\rho_1}{2\pi a_j} \int_0^\infty A_1(u) \frac{\sin(a_j u)}{u} J_0(d_{ij}u)du, \qquad i \neq j$$

$$z_{ij} = \frac{\rho_1}{\pi a_j^2} \int_0^\infty A_1(u) \frac{\sin(a_j u)}{u^2} J_1(a_j u)du, \qquad i = j$$

$$(2)$$

where ρ_1 is the resistivity of the surface layer, a_j is the radius of contact j, d_{ij} is the distance between the centers of contacts i and j, and J_0 and J_1 are Bessel functions of the first kind of order 0 and 1, respectively.

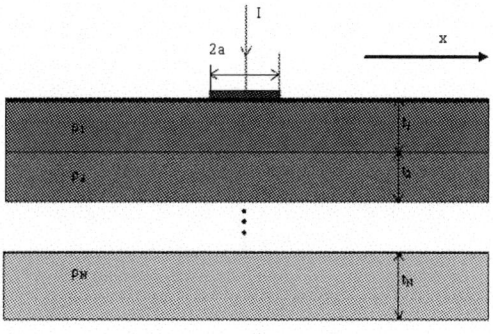

Fig. 2. Cross-section of a multi-layer substrate.

The function $A_1(u)$ captures information about the resistivities and thicknesses of the different substrate layers through

$$A_i = \frac{1 - k_i e^{-2t_i u}}{1 + k_i e^{-2t_i u}} \quad \text{and} \quad k_i = \frac{\rho_i - \rho_{i+1} A_{i+1}}{\rho_i + \rho_{i+1} A_{i+1}} \tag{3}$$

Here t_i and ρ_i are the thickness and resistivity of layer I, respectively. For the last layer $A_i = 1$.

In Fig. 3, the modeled surface potential induced by current injection into a circular contact of radius 1 μm is compared with finite element calculations using FEMLAB [4] for the four silicon substrate profiles shown in Table I. As seen in Fig. 3, the model is very accurate compared with 3-D FEM calculations.

The surface potential at any point induced by N contacts is calculated using superposition of all potential contributions. This is utilized for modeling non-circular contacts, which are found in ICs. Such contacts are approximated by many small circular contacts as shown in Fig. 4. An example of the surface potential close to the contact (biased to 1 V) in Fig. 4 is shown in Fig. 5.

TABLE I
SUBSTRATE PROFILES.

Substrate # of layers	a 1	b 2	c 2	d 3
ρ_1 (Ωcm)	10	0.1	10	1
t_1 (μm)	700	0.4	4	1
ρ_2 (Ωcm)	-	10	0.001	15
t_2 (μm)	-	700	700	4
ρ_3 (Ωcm)	-	-	-	0.001
t_3 (μm)	-	-	-	700

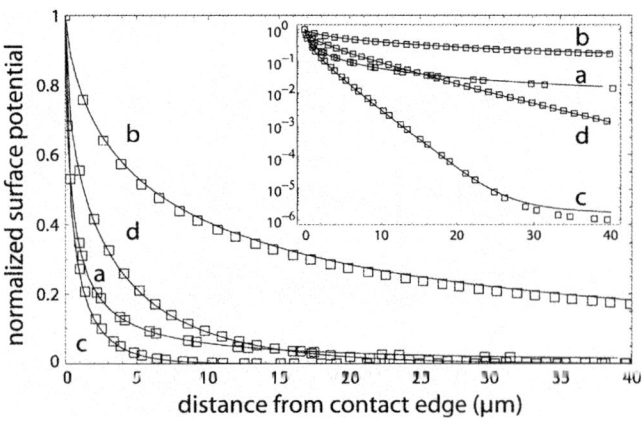

Fig. 3. The modeled surface potential (lines) matches 3-D finite element results (symbols).

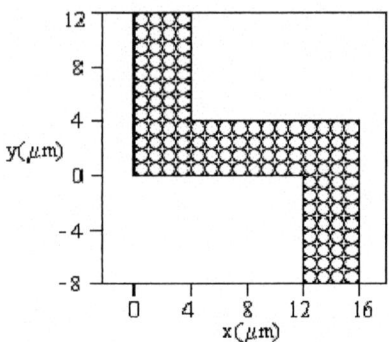

Fig. 4. A non-circular contact modeled by many small circular contacts.

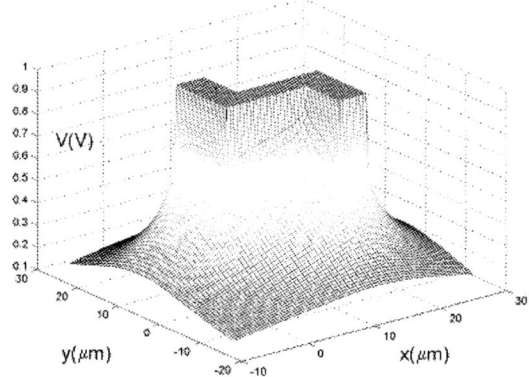

Fig. 5. Surface potential close to the contact shown in Fig. 4.

III. EXTENDED SUBSTRATE MODEL

At high frequencies the capacitive effects of the substrate become pronounced and should be included in the substrate model. In the following calculations a very wide frequency range, 0 Hz to 10^{18} Hz, is utilized. It should be noted that these extremely high frequencies are not of practical interest but are used only for evaluation purposes; the model is not expected to be accurate for such high frequencies. It is expected to be accurate up to tens of GHz.

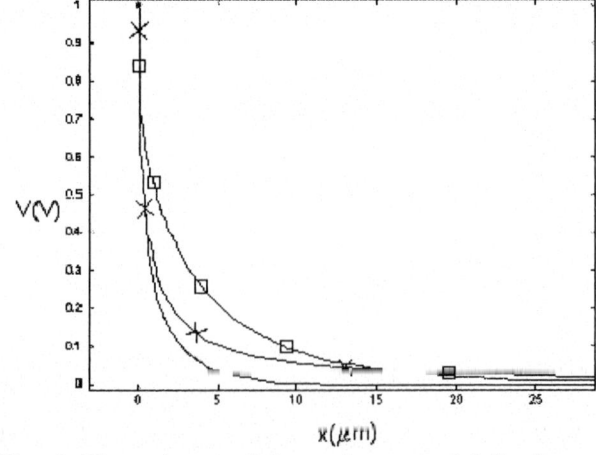

Fig. 6. The real part of the surface potential for frequencies 10^6 Hz (squares), 10^{12} Hz (no symbol), and 10^{18} Hz (crosses).

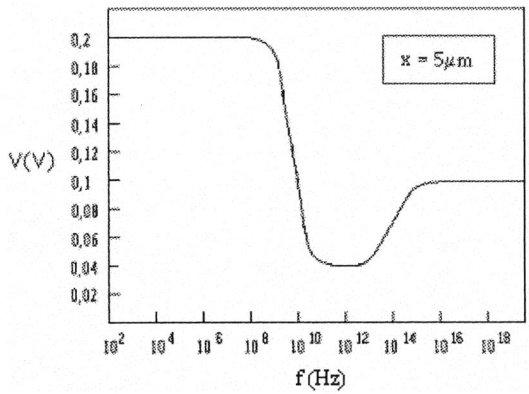

Fig. 7. Real part of the surface potential at the point x=5 μm.

The influence of the capacitive effects on the surface potential is illustrated in Fig. 6 showing the real part of the surface potential. FEMLAB was used to produce surface potential data for a single 1 μm contact injecting current at three different frequencies. The substrate is type d) in Table I. A quasi-static approach where induced currents were neglected was employed in FEMLAB.

Fig. 7 shows the real part of the surface potential 5 μm away from the contact edge for different frequencies. The observed step-like changes in the real part of the surface potential is due to the capacitive effects of the different layers becoming pronounced. The frequency at which a layer exhibits pronounced capacitive effects is roughly given by $1/(2\pi\rho_i\varepsilon_i)$ where ε_i is the permittivity of substrate layer i.

To account for the behavior observed in Fig. 7 using equivalent RC networks different networks are commonly used in the different frequency ranges [5], making that approach somewhat cumbersome. Since our substrate model is not based on equivalent networks, such frequency-controlled switching between networks is not required.

To account for the permittivity of substrate layer i we use the generalized complex resistivity Γ_i, as did Brandtner and Weigel [6],

$$\Gamma_i = \frac{1}{\dfrac{1}{\rho_i} + j\omega\varepsilon_i} \tag{4}$$

where $\omega = 2\pi f$ is the angular frequency. Next we show that replacing ρ_i in (2) and (3) with Γ_i is a feasible approach for extending our substrate model to include capacitive effects.

The real and imaginary parts of the surface potential for a single 1 μm circular contact on a substrate of type d) calculated using 3D FEMLAB and the model are shown in Figs. 8-13 for frequencies 10^6, 10^{12}, and 10^{18} Hz. As seen in the figures, the model and FEM results agree thus showing that the capacitive effects can easily be included in our substrate model using the simple approach presented above. Again, the extremely high frequencies used are for evaluation purposes only under the given conditions (quasi-static conditions and no induced currents).

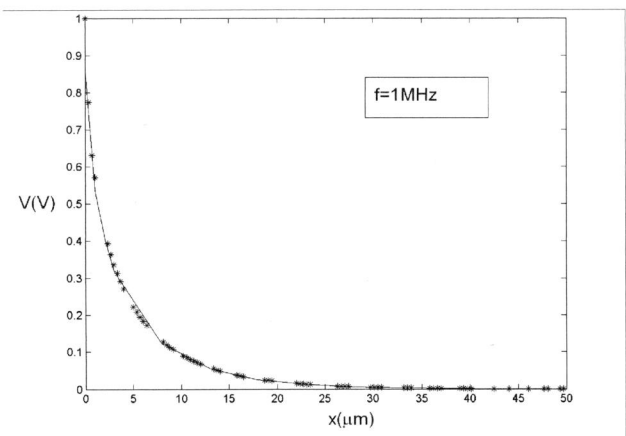

Fig. 8. Real part of the surface potential. Line=model, symbols=FEM.

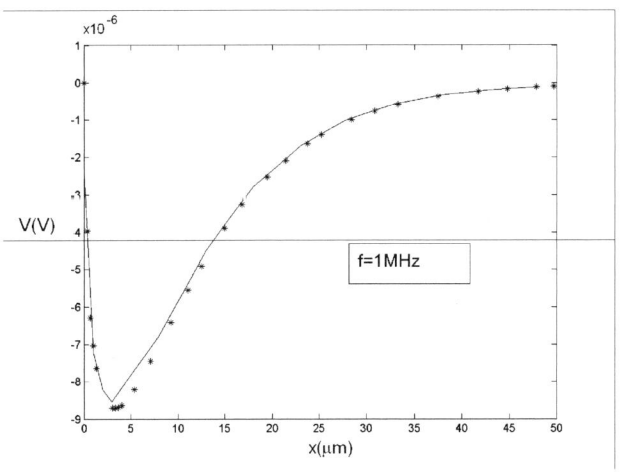

Fig 9. Imaginary part of the surface potential. Line=model, symbols=FEM.

Silicon-on-Insulator (SOI) substrates are becoming more widely used in the semiconductor industry. By including the dielectric properties of the substrate layers, the coupling through SOI substrates is easily simulated using our model. The buried oxide layer is simply a layer in the multilayer structure with the two parameters $\rho_{oxide}=\infty$ and ε_{oxide}.

Fig. 14 shows a cross-section of an SOI-based test structure. The structure consists of two rectangular contacts 100 μm apart situated on top of the oxide. The bulk substrate is grounded on the backside. The rectangular contacts are modeled using many small circular contacts as described previously. The transmission parameter S_{21} was simulated using the model and Agilent's Momentum [7]. A good agreement between the two simulations is achieved as seen in Fig. 14. The case of having no buried oxide is also included in the figure.

343

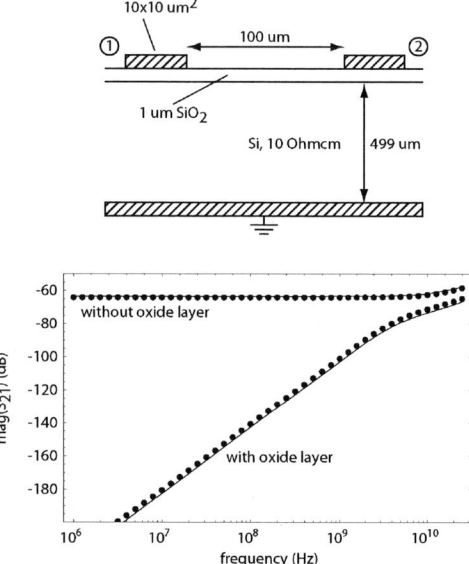

Fig. 14. Cross-section (top) and magnitude of S_{21} with and without buried oxide (bottom). Symbols=Agilent Momentum, lines=model.

IV. CONCLUSIONS

The continuous increase of the number of the devices in modern ICs requires efficient simulation algorithms for predicting noise coupled through the substrate. The extended substrate model presented in this paper include, apart from the resistive coupling, the dielectric behavior of the silicon substrate for modeling high-frequency conditions. The dielectric behavior is accounted for by including a complex resistivity into an existing spreading resistance model. The accuracy of the model is shown to be very good compared with three dimensional finite element calculations.

REFERENCES

[1] Min Xu, David K. Su, Derek K. Shaeffer, Thomas H. Lee and Bruce A. Wooley, "Measuring and Modeling the Effects of Substrate Noise on the LNA for a CMOS GPS Receiver, *IEEE J. Solid-State Circuits*, vol. 36, no. 3, March 2001.

[2] D. K. Su, M. J. Loinaz, S. Masui and B. A. Wooley, "Experimental results and modeling techniques for substrate noise in mixed-signal integrated circuits", *IEEE J. Solid-State Circuits*, vol.28, no 4. pp. 420-430, Apr. 1993.

[3] Simon Kristiansson, Fredrik Ingvarson, Shiva Prasad Kagganti, Nebojsa Simic, Marinel Zgrda, and Kjell O. Jeppson, "A Surface Potential Model for Predicting Substrate Noise Coupling in Integrated Circuits", *IEEE J. Solid-State Circuits*, vol. 40, no. 9, September 2005.

[4] Femlab 3.0 Comsol. Available: http://www.comsol.com/

[5] Chonggang Xu, Terri Fiez and Karti Mayaram, "High Frequency Lumped Element for Substrate Noise Coupling", *IEEE, Behavioral Modeling and Simulation, 2003, BMAS 2003. Proceedings of the 2003 International Workshop*, Oct. 7-8, 2003

[6] Thomas Brandtner and Robert Weigel, "SubCALM: A Program for Hierarchical Substrate Coupling Simulation on Floorplan Level", *IEEE, Proceedings of the Design, Automation and Test in Europe Conference and Exhibition(Date'04)*

[7] Momentum. Agilent Technologies http://www.eesof.tm.agilent.com/

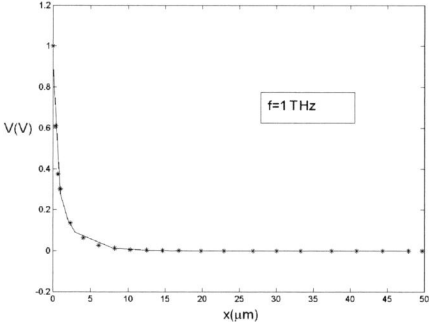

Fig.10. Real part of the surface potential. Line=model, symbols=FEM.

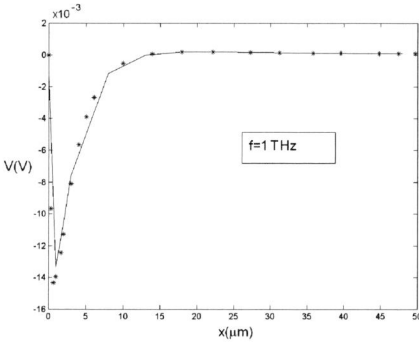

Fig. 11. Imaginary part of the surface potential. Line=model, symbols=FEM.

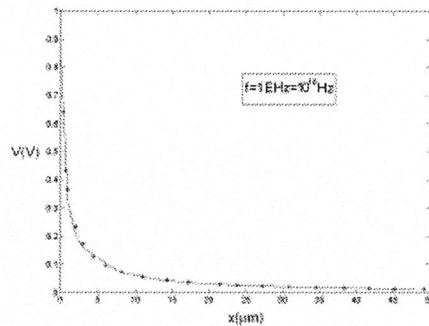

Fig.12. Real part of the surface potential. Line=model, symbols=FEM.

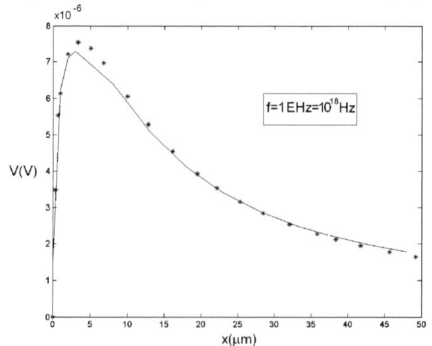

Fig. 13. Imaginary part of the surface potential. Line=model, symbols=FEM.

2006 25th International Conference on Microelectronics

Analytical Modeling of the Triggering Drain Voltage at the Onset of the Kink Effect for PD SOI NMOS

Milija Sarajlić and Rifat Ramović

Abstract - Here we give a new approach for calculating triggering drain bias at the onset of the kink effect utilizing electron drift properties in the channel. This approach directly relates electron mobility in the channel of the PD SOI NMOS devices to the kink effect and gives possibility for determining mobility from the kink voltage V_{kink}. We compare our theory to the previously published experimental results and based on this match we predict behaviour of the kink effect for PD SOI NMOS components for various technology parameters. This theory is applicable to the PD SOI NMOS devices with effective channel length below 600 nm. Theory could be extended to the prediction of the breakdown drain-to-source bias at the PD SOI NMOS devices.

I. INTRODUCTION

Silicon-on-insulator (SOI) CMOS offers a 20–35% performance gain over bulk CMOS [1]. Some of the recent applications of SOI are in high-end microprocessors and its upcoming uses are in low-power, radio-frequency (rf) CMOS and embedded DRAM (EDRAM), to name a few. As we move to the 0.1 μm generation and beyond, SOI is expected to be the technology of choice for system-on-a-chip applications which require high-performance CMOS, low-power, embedded memory, and bipolar devices [1].

The primary feature of MOS in SOI is that the local substrate ("body") of the device floats electrically, and therefore the substrate–source bias voltage, V_{BS}, is not fixed. This is especially prominent for Partially Depleted (PD) SOI MOS devices where active layer is not totally depleted but it leaves an island of non-depleted region close to the buried oxide, Fig. 1.

As V_{BS} changes, the device threshold voltage, V_T, will change. This "instability" in V_T is what has made SOI device design very challenging. One manifestation of the threshold variation is the "kink effect," or increase in the output conductance of the device near drain-to-source bias, V_{DS}, of 1 V, the band gap of silicon, even though kink effect could appear for the drain-to-source bias below band gap of silicon, inset of Fig. 1, [2]. This is caused by the impact-ionisation induced increase in V_{BS} with increasing

M. Sarajlić is with the Center for Microelectronic Technologies and Single Crystals, IHTM, Njegoševa 12, 11000 Belgrade, Serbia&Montenegro, E-mail: milijas@nanosys.ihtm.bg.ac.yu

R. Ramović is with Department for Microelectronics, School of Electrical Engineering, Belgrade University, Bulevar Kralja Aleksandra 73, 11000 Beelgrade, E-mail: ramovic@etf.bg.ac.yu

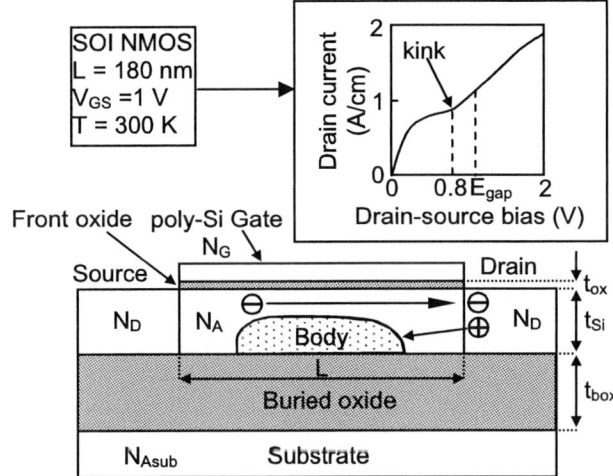

Fig. 1. Schematic representation of the Partially Depleted SOI NMOS device. Mechanism of impact ionisation and kink is sketched. At the inset, typical drain current characteristic is shown [2].

V_{DS}, and the resulting reduction of V_T; when V_{DS} becomes large enough, impact ionisation current (holes) flows to the non-depleted body, increasing the body charge and V_{BS}, resulting in a decrease in V_T. This is much more pronounced at the SOI NMOS devices than at the SOI PMOS devices because effective cross-section for impact ionisation is much higher for electrons than for holes. For this reason we will restrict our consideration to the NMOS devices although the same approach could be implemented for PMOS.

II. THEORY

In this consideration we will employ classical mechanics. First question arises: could we properly model electron drift in the channel by classical mechanics? At the Fig. 2 we give comparison of the electron velocity in the channel as calculated by classical mechanics and Monte-Carlo method [3]. Monte-Carlo method is more accurate because it takes into account single collisions of electron with phonons in the lattice. As can be seen, Fig. 2, there is apparent difference between these two approaches for the electron velocity along the channel, but maximum velocity is the same in both cases. Device parameters for the simulated component at the Fig. 2 are: front oxide

1-4244-0116-X/06/$20.00 ©2006 IEEE

thickness $t_{ox} = 1$nm, active layer thickness $t_{Si} = 15$nm, buried oxide thickness $t_{box} = 400$nm, doping within drain/source regions $N_D = 1\times10^{20}$cm^{-3}, active layer doping $N_A = 1\times10^{16}$ cm^{-3}, substrate bias $V_{BS} = 0$ V, temperature $T = 300$ K, channel length $L = 40$ nm, gate bias above threshold and drain bias V_{GS} - $V_T = V_{DS} = 1$ V. We have assumed constant longitudinal electric field and constant mobility along the channel. What matters to us in estimating kink effect is maximum electron velocity which electron has at the drain interface.

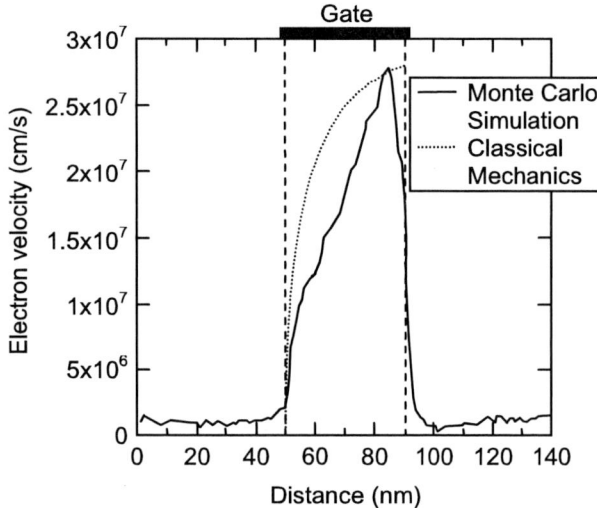

Fig. 2. Electron velocity vs. distance in the channel. Comparison between Monte Carlo simulation [3] and Classical Mechanics. $L_{eff} = 40$ nm, V_{GS} - $V_T = V_{DS} = 1$ V, $N_A = 1\times10^{16}$ cm^{-3}, estimated effective mobility $\mu_{eff} = 120$ cm^2/Vs.

At the Fig. 3 we give electron velocity vs. distance along the channel for different channel lengths and for the same drain-to-source bias $V_{DS} = 1$V. Electron mobility is assumed to be equal to the intrinsic mobility, $\mu = 1400$ cm^2/Vs. We will see reason for this lately. Velocity is derived from differential equation for electron drift:

$$mv(x)\frac{dv(x)}{dx} = q\frac{V_{DS}}{L} - \frac{q}{\mu}v(x) \quad (1)$$

where m is the electron rest mass, x is the distance along the channel, $v(x)$ is the electron drift velocity along the channel, q is the electron charge, V_{DS} is the drain-to-source bias, L is the channel length and μ is the electron mobility in the channel.

Solution to this differential equation is:

$$v(x) = \frac{\mu V_{DS}}{L}\left(1 + ProductLog\left(-e^{-1-\frac{qL}{m\mu^2 V_{DS}}x}\right)\right) \quad (2)$$

where *ProductLog[z]* gives the principal solution for w in $z = we^w$.

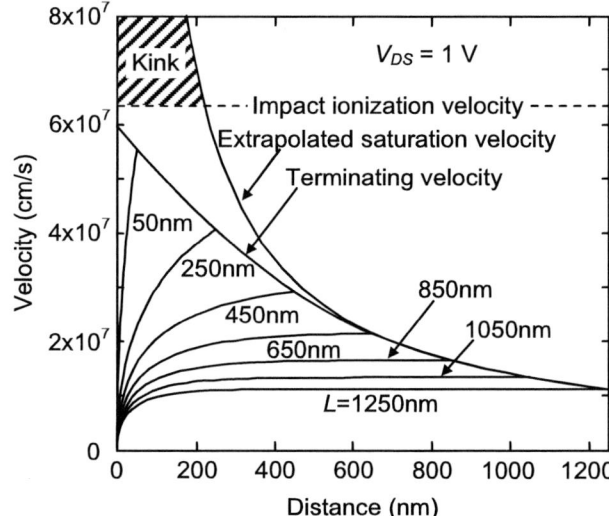

Fig. 3. Velocity in the channel vs. distance for different channel lenghts. Terminating velocity at the drain interface and extrapolated saturation velocity are also depicted. Shaded area represents the onset of the kink. Drain-to-soure bias is 1V.

As can be seen on Fig. 3 there are two distinct regions of the electron drift. First region belongs to the acceleration, where electron is accelerated up to the saturation velocity v_{sat}, and after that region it travels with constant velocity v_{sat}. On Fig. 3 it is shown that for sufficiently short channel, electron will not attain saturation velocity at the point where it reach drain interface; this is pointed out by two different curves that depict saturation velocity and terminating velocity at the drain interface. For channel lengths below 600 nm, we need to extrapolate electron drift in order to find saturation velocity. Below 600 nm there is apparent difference between terminating velocity and extrapolated saturation velocity, Fig. 3. We will assume that in acceleration region electron will suffer no recombinations. In that manner, same electrons that emerge at the source will immerge at the drain; this means that we could model situation at the drain interface knowing situation at the source. This is not applicable for the longer channel because electron will suffer multiple recombinations and subsequent generations before it reaches drain interface. This assumption is reminiscent of the famous Shockley theory of "lucky electron" where "lucky electron" will travel all channel without collisions with phonons or it could even receive more phonons than it emits, so final velocity will be above average. Here we only expect electron not to suffer recombination during flight; but we consider no ballistic electrons. Taking all this into account we will restrict our theory only to the electrons where saturation and terminating velocity do not equal.

346

Now we have a possibility to formulate our basic model. Impact ionization in the drain region will occur if saturation energy is equal to the silicon band-gap energy:

$$E_{gap} = \frac{1}{2}mv_{sat}^2 = \frac{1}{2}m\left(\frac{\mu V_{kink}}{L}\right)^2 \qquad (3)$$

where V_{kink} is the drain-to-source bias at the onset of the kink effect. It is straightforward from here to express kink voltage:

$$V_{kink} = \frac{L}{\mu}\sqrt{\frac{2E_{gap}}{m}} \qquad (4)$$

Very special issue in this approach is modelling of the channel mobility. There is no universal model that depends on geometry and technology parameters of the SOI device. In this situation we will employ two successive stages in mobility modelling. First, we find mobility for doped Si with same doping that active layer is made. We use semi-empirical formula [4]:

$$\mu(N_A) = 200\frac{cm^2}{Vs}\left(1 + \frac{2\times10^{18}cm^{-3}}{N_A}\right). \qquad (5)$$

This value will be used as maximum mobility in the channel of the SOI NMOS, while electric filed dependence of the mobility is estimated in the following manner:

$$\mu_{eff} = \frac{2\sqrt{b}\,\mu(N_A)\left(E_s(V_{GS}) - E_s(V_T) + \sqrt{b}\right)}{b + \left(E_s(V_{GS}) - E_s(V_T) + \sqrt{b}\right)^2} \qquad (6)$$

where μ_{eff} is the effective electron mobility in the channel, E_s is the transverse electric field in the front channel, V_{GS} is the gate-to-source bias, V_T is the threshold voltage, $\mu(N_A)$ is the maximum mobility as calculated from (5), b is the fitting parameter. $E_s(V_{GS})$ represents dependence of transverse electric field in the front channel on the gate-to-source bias. We do not take into account mobility dependence on the lateral electric field in the channel because that would prevent us for obtaining sufficiently high electron velocity for the impact ionisation to occur. In (6) it is envisioned that maximum mobility correspond to the threshold bias V_T. Therefore, we need to find threshold for a given device technology. We will utilize one-dimensional formula obtained through full depletion approximation [5]. We do not take into account Short Channel Effects. Threshold formula read as [5]:

$$V_T = V_{FB} - 2\phi_f +$$
$$\frac{t_{ox}}{t_{box} + \gamma t_{Si}}\left(-2\phi_f - V_{sub} + V_{FBsub} + \frac{qN_A t_{Si}(2t_{box} + \gamma t_{Si})}{2\gamma\varepsilon_{Si}}\right) \qquad (7)$$

$$V_{FB} = \frac{kT}{q}\ln\left(\frac{N_A N_G}{n_i^2}\right) \qquad (8)$$

$$\phi_f = \frac{kT}{q}\ln\left(\frac{N_A}{n_i^2}\right) \qquad (9)$$

$$\gamma = \frac{\varepsilon_{Si}}{\varepsilon_{ox}} \qquad (10)$$

$$V_{FBsub} = \frac{kT}{q}\ln\left(\frac{N_{Asub}}{N_A}\right) \qquad (11)$$

where V_{FB} is the flat-band voltage in the front channel, k is the Bolzman constant, q is the electron charge, N_G is the doping of the polisilicon gate, n_i is the intrinsic carrier density, ϕ_f is the Fermi level, γ is the dimensionless constant, ε_{Si} is the dielectric permittivity of silicon, ε_{ox} is the dielectric permittivity of SiO2, V_{FBsub} is the flat-band voltage in the back channel, V_{sub} is the substrate bias, N_{Asub} is the doping of the substrate.

Now we can find electric field in the front channel E_s:

$$E_s = \left(-\frac{\partial\phi}{\partial x}\right)_{x=0} = \gamma\frac{V_{GS} - V_{FB} - \phi_s}{t_{ox}} \qquad (12)$$

where ϕ_s is the front channel potential. Expressions for $\phi_s(V_{GS})$ and $E_s(V_{GS})$ are:

$$\phi_s(V_{GS}) = V_{GS} - V_{FB} -$$
$$\frac{t_{ox}}{t_{box} + \gamma t_{Si}}\left(-V_{sub} + V_{FBsub} + \frac{qN_A t_{Si}(2t_{box} + \gamma t_{Si})}{2\gamma\varepsilon_{Si}}\right) \qquad (13)$$

$$E_s(V_{GS}) =$$
$$\frac{\gamma\left(\phi_s(V_{GS}) - V_{sub} + V_{FBsub} + \frac{qN_A t_{Si}(2t_{box} + \gamma t_{Si})}{2\gamma\varepsilon_{Si}}\right)}{t_{box} + \gamma t_{Si}}$$
$$\qquad (14)$$

347

III. RESULTS

On Fig. 4 we compare experimental and theoretical results for SOI NMOS device with technology parameters: $t_{ox} = 4$ nm, $t_{Si} = 126$ nm, $t_{box} = 360$ nm, $N_A = 2 \times 10^{18}$ cm^{-3}, [6]. As the inset of the Fig. 4 we give original drain-to source current characteristics from which kink data is derived [6].

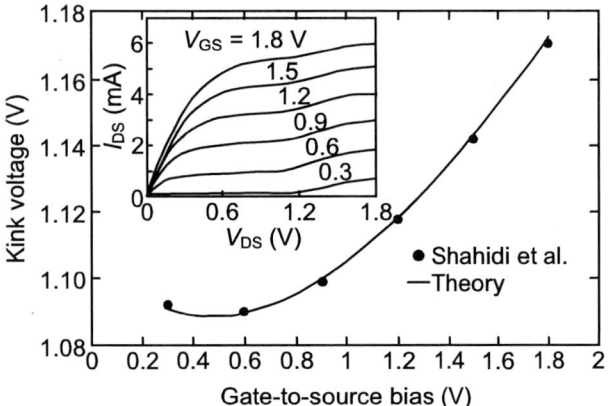

Fig. 4. Kink voltage vs. gate-to-source bias. Comparison with experimental data from Shahidi et al. [6]. As the inset we give original drain-to source current characteristics from which kink data is derived [6].

At Fig. 5 we give prediction for the kink voltage vs. gate-to-source bias for different channel doping. At Fig. 6 we give prediction for the kink voltage vs. channel length for different gate-to-source bias.

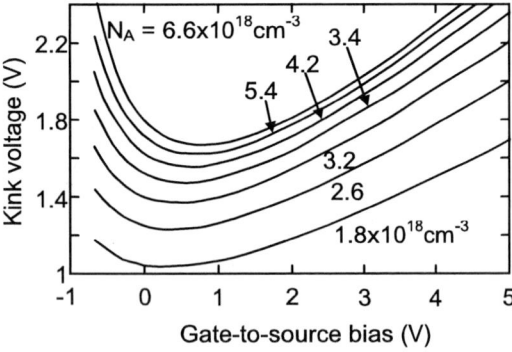

Fig. 5. Kink voltage vs. gate-to-source bias for different channel doping. Device data: $t_{ox} = 5$ nm, $t_{Si} = 50$ nm, $t_{box} = 500$ nm.

IV. CONCLUSION

We have shown that it is possible to directly relate electron mobility in the channel at the PD SOI NMOS devices to the kink voltage i.e. triggering drain-to-source bias at the onset of the kink effect. This approach could be

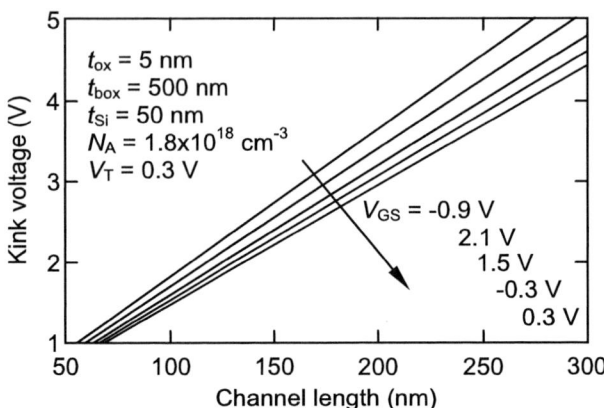

Fig. 6. Kink voltage vs. channel length. Lowest slope is for the device at the treshold.

used in two manners: first, to give prediction for behaviour of the kink voltage for various technology parameters; this can be utilized in modelling of the PD SOI NMOS devices or circuitry simulations; second, it could be new method for measuring electron mobility in the front channel of these devices. It is important to notice that theory is applicable for the devices with channel lengths below 600 nm. This theory could be extended to the prediction of the breakdown drain-to-source bias at the SOI NMOS devices by equalling terminating electron energy to the Si band-gap. This is the subject of the subsequent work.

ACKNOWLEDGEMENT

This work has been partially supported by the Serbian Ministry of Science, Technologies and Development within the framework of the project IT.6151.B.

REFERENCES

[1] G. G. Shahidi, "SOI technology for the GHz era", *IBM J. RES. & DEV.* VOL. 46, NO. 2/3 MARCH/MAY 2002.

[2] Shahidi, et al. IBM DAMOCLES tutorial. Available at: www.research.ibm.com/DAMOCLES/html_file/segii.html

[3] S. Barraud, L. Clavelier, T. Ernst, "Electron ransport in thin SOI, strained-SOI and GeOI MOSFET by Monte Carlo simulation", *Solid State Electronics*, 49 (2005) 1090-1097.

[4] Siegfried Selberherr, *Analysis and Simulation of Semiconductor Devices*, pp. 88, Wien, Springer-Verlag, 1984.

[5] Milija Sarajlić, Rifat Ramović, "Modification of the Quasi 2D Model for Short-Channel SOI MOSFET", *Proc. ETRAN*, Budva, Serbia and Montenegro, June , 2005.

[6] G. G. Shahidi, J. D. Warnock, J. Comfort, S. Fischer, P. A. McFarland, A. Acovic, T. L. Chappell, B. A. Chappell, T. H. Ning, C. J. Anderson, R. H. Dennard, J. Y.-C. Sun, M. R. Polcari, B. Davari, " CMOS scaling in the 0.1 μm, 1 .X-volt regime for high-performance applications", *IBM J. RES. & DEV.* VOL. 39, NO. 1/2, JANUARY/MART 1995.

AUTHOR INDEX

Abdullah, S.H.	605	Cherkaoui, K.	55, 379
Abelein, U.	127, 131	Chobola, Z.	307, 501
Adepoju, F.	249	Chouteau, S.	201
Agarwal, R.	671	Chung, P.S.	277
Ahmad, I.	605	Claeys, C.	67
Ahmadi, A.	655	Cunniffe, C.	123, 237
Al Khusheiny, M.	271		
Aleksić, O.	479, 619	Dakhel, A.A.	115
Alexiou, G.	517	d'Alessandro, V.	483
Alvarado, J.	491	Danković, D.	639, 645
Anderson, D.	601	Davidović, V.	639, 645
Andjelković, B.	659	De Leonardis, F.	137, 141
Andrejević, M.	437	De Mey, G.	529
Andrijašević, D.	267	De Paola, F.	483
Arora, V.K.	17	de Souza, M.	509
Arpatzanis, N.	513	Delides, C.G.	391
Arshak, K.	123, 225, 237, 249, 263	Diaz-Ayala, M.	301
Asparuhova, K.	215	Dieudonne, F.	201
Atanassova, E.	47, 581, 585	Dimitriadis, C.A.	513
Axelevitch, A.	361	Dimitrijev, S.	557
Azhniuk, Yu.M.	111	Ding, P.W.	487
		Dmitruk, N.	317, 321
Baling, W.	211, 297, 313	Dojčinović, I.P.	145, 149
Batcup, S.G.	193	Doneddu, D.	189
Batyrev, I.	89	Dzhagan, V.M.	111
Bauer, A.J.	589		
Bazu, M.	259	Djinović, Z.	233
Belaroussi, M.T.	459	Djordjević, G.	697
Bellis, S.J.	671	Djordjević, S.	447
Benda, V.	285	Djorić-Veljković, S.	639, 645
Bhuwalka, K.K.	127, 131	Djurić, Z.	11, 103, 241, 333
Blyzniuk, M.	413, 525		
Boissonnet, L.	201, 593	Eisele, I.	127, 131
Boltovets, M.	293	Eneman, G.	67
Borejko, T.	679	Eng, Y.-C.	521
Borkovskaya, O.	321	Ensell, G.	289
Born, M.	127, 131	Escobedo-Alatorre, J.	301
Boselli, G.	429	Esinenco, D.	267
Bouzerara, L.	459	Estrada, M.	327
Bravaix, A.	593	Exarchos, M.	597
Brenner, W.	267		
Bryant, A.T.	175	Falck, E.	183
Buckley, D.	263	Felsl, H.P.	183
		Feng, H.	409
Capizzo, M.	497	Filip, V.	277
Cavanagh, L.	123, 237	Flandre, D.	491, 509
Cerdeira, A.	327, 491, 509	Fleetwood, D.	89
Chan, M.	383	Fobelets, K.	487
Chatterjee, P	3	Fragiadakis, D.	119

Frantlović, M.	103	Iniguez, B.	327
		Inkman, B.	601
Gallagher, C.	577	Itoh, K.	77
Garcia, R.	327		
Garcia-B, A.	301, 309, 541	Jablonski, G.	529
Gardes, F.	289	Jaćimovski, S.	533
Ghibaudo, G.	551	Jafer, E.	225
Giouroudi, I.	267	Jakšić, A.	577
Gocek, P.	529	Jakšić, O.	103
Goguenheim, D.	565	Jakšić, Z.	107, 153
Golan, G.	361	Jalar, A.	605
Golubović, S.	639, 645	Janicki, M.	529
Gonda, V.	369	Janković, N.	193
Gorbunov, M.	545	Januszkiewicz, P.	529
Gorenstein, B.	361	Jeppson, K.	341
Gorobchuk, A.	537	Jevtić, M.	569, 573, 623, 627
Grasser, T.	475	Jia, X.	387
Grigorova, T.	215	Johnson, B.	601
Grigoryev, Y.	537	Jokić, I.	103
Grimalsky, V.	301, 309, 541	Jomaah, J.	551
Grmela, L.	501	Jovanović, D.	153
Grozdić, T.	245, 611	Jovanović, G.	667
Guan, X.	409	Jović, V.	611
Guermaz, M.B.	459	Jutman, A.	679
Gunnar Malm, B.	25		
Gurevich, Y.	337	Kakanakov, R.	293
Gutierrez-D., E.	309, 541	Kamarinos, G.	513
Guy, O.J.	189	Kanapitsas, A.	391
		Kang, I.H.	211, 297, 313
Hadži-Vuković, J.	569	Kapels, H.	197
Haendler, S.	201	Khanniche, M.S.	193
Hallstedt, J.	25	Kilshytska, V.	491
Han, J.	557	Kim, E.D.	211
Hatzopoulos, A.A.	513	Kim, K.H.	211
Hatzopoulos, A.T.	513	Kim, N.K.	211, 297, 313
Heinzl, R.	475	Kim, S.C.	211, 297, 313
Hellstrom, P.-E.	25	Kirillov, A.	293
Hold, L.	557	Kok, C.W.	383
Holland, P.M.	207	Kolaklieva, L.	293
Holzer, S.	465	Konakova, R.	321
Hruska, P.	501	Kondratenko, O.	321
Hu, S.-F.	157	Kong, F.	557
Huang, K.-D.	157, 373	Konofaos, N.	517
Hughes, G.	379	Kontou, E.	391
Hughes, P.J.	577	Korostynska, O.	263
Huidgins, J.L.	175	Korovin, A.	321
Hulicius, E.	307	Korovin, O.	317
Hurley, P.	55, 379	Koshevaya, S.	301, 309, 541
		Kouvatsos, D.	597
Igić, P.	189, 193, 207	Kovac, J.	357
Ilić, D.	533	Kristiansson, S.	341
Ingvarson, F.	341	Kuchmii, S.Ya.	111

Kuo, J.B.	61	McDonnell, S.	379	
Kuraica, M.M.	145, 149	Michalas, L.	597	
		Mijalković, S.	471	
La Spina, L.	365	Milijić, M.	705	
Lachenal, D.	593	Milosavljević, M.	685	
Lai, P.T.	561	Milovanović, B.	705	
Lamovec, J.	241	Min'ko, V.	317	
Lapsker, I.	361	Mitić, D.	667	
Lau, K.M.	561	Mitrović, M.	149	
Lazić, Ž.	333	Modreanu, M.	379	
Lebedev, A.	293	Mohan, D	3	
Lebedev, E.V.	391	Moore, E.	123, 237	
Lee, T.-Y.	521			
Lehouidj, B.	459	Nakagawa, A.	167	
Lemberger, M.	589	Nanver, L.	365, 369	
Lempinen, J.	649	Napieralski, A.	529	
Liberali, V.	429	Negara, A.	379	
Lin, C.H.	61	Nenadović, N.	365	
Lin, J.-T.	157, 373, 521	Niedernostheide, F.-J.	183	
Lin, K.-C.	521	Nikolić, M.	701	
Lin, L.	561	Novkovski, N.	585	
Lin, S.-T.	373	Novotny, I.	357	
Litovski, V.	437, 659	Nowakowski, J.	483	
Logakis, E.	119, 391			
Lončar, B.	631, 693	O'Flynn, B.	671	
Lorito, G.	369	Ogier, J.-L.	565	
Lu, H.	601	O'Keeffe, C.	671	
Lui, S.	369	Osmokrović, P.	631, 693	
Lukić, P.	505	Ostling, M.	25	
Lupan, O.	161			
Lutz, J.	183	Padha, N.	219	
Lytvyn, O.	317	Palmer, P.R.	175	
		Pandis, C.	391	
Machacek, Z.	285	Pantelides, S.T.	89	
Maguire, P.	39	Panwar, N.S.	455	
Mahony, C.	39	Papaioannou, G.J.	597	
Majlis, B.Y.	271	Pappas, I.	513	
Maksimović, M.	107, 153	Paskaleva, A.	47, 581, 589	
Malović, G.	39	Passaro, V.M.N.	137, 141, 289	
Mamontova, I.	321	Paszkowski-Rogacz, M.	529	
Mamunya, Y.P.	391	Pauč, N.	39	
Manevych, V.	361	Pavanello, M.	509	
Manić, I.	639, 645	Permthammasin, K.	197	
Mao, L.F.	635	Perrotin, A.	201	
Marano, I.	417	Persano Adorno, D.	497	
Marić, D.	39	Pershenkov, V.S.	689	
Marić, V.	479	Pešić, B.	395	
Mashanovich, G.Z.	137, 141, 289	Petković, M.	443	
Matić, M.	145, 149	Petković, P.	447, 701	
Matović, J.	241, 255	Petrović, Z.	39	
Mawby, P.A.	175, 193	Pic, D.	565	
Mayeva, O.	317	Pissis, P.	119, 391	

Pleskacz, W.A.	679	Smiljanić, M.	611
Poole, K.	3	Smiljanić, M.M.	333
Popa, C.	425, 451	Sokolović, M.	701
Popovici, E.M.	671	Sosnova, M.	317
Poriazis, S.	433	Spassov, D.	581
Prijić, A.	395	Spevak, M.	475
Prijić, Z.	395	Stamenković, Z.	401
Protić, D.	685	Stanimirović, I.	623, 627
Purić, J.	145, 149	Stanimirović, Z.	623, 627
		Stanković, S.	693
Radmilović-Radjenović, M.	39	Stanković, T.	697
Radojčić, B.	619	Stanković, Z.	705
Radulović, K.	333	Stathis, J.	83
Raevskaya, A.E.	111	Stojadinović, N.	639, 645
Ramović, R.	345, 505, 619	Stojčev, M.	667, 697
Randjelović, D.	145, 149, 241	Stroyuk, A.L.	111
Rauber, B.	201	Sulima, T.	127, 131
Raynaud, C.	201	Sung, C.-L.	157
Razvalyaev, A.U.	689	Sutta, P.	357
Reed, G.	289		
Rey-Tauriac, Y.	593	Šašić, R.	505, 631
Rinaldi, N.	417, 483	Šetrajčić, J.	533
Rodgers, K.	577	Šimeček, T.	307
Rodgers, M.P.	89		
Romanov, L.	293	Tadić, N.	421
Rosa, J.	201	Tajani, A.	189
		Talmi, M.	675
Sajfert, V.	533	Tamigi, F.	417
Salinger, J.	285	Tanasković, D.	107
Saniter, J.	675	Tanner, P.	557
Santi, E.	175	Tassis, D.H.	513
Sarajlić, M.	107, 153, 345	Tecpoyotl-Torres, M.	301
Satarić, M.V.	99	Tibeica, C.	259
Schaffnit, C.	201	Todorović, D.M.	245, 611
Schellevis, H.	365, 369	Tomić, M.	233
Schindler, M.	127, 131	Tošić, B.	533
Schmidt, M.	127, 131, 197	Tremasov, A.D.	689
Scholtes, T.L.	369	Trucco, G.	429
Schrimpf, R.D.	89	Tsetseris, L.	89
Schwaha, P.	475	Tsonos, C.	391
Schwitters, M.	189	Tuominen, A.	649
Selberherr, S.	465	Tvarozek, V.	357
Shishiyanu, S.	161	Twitchen, D.	189
Shishiyanu, T.	161		
Shtereva, K.	357	Ubar, R.	679
Simić, N.	341		
Simoen, E.	67	Vaid, R.	219
Singh, A.P.	455	Valakh, M.Ya.	111
Singh, R.	3, 161	Vanek, J.	307
Siskos, S.	513	Vaseashta, A.	31
Slimane, A.	459	Vasić, A.	631, 693
Smetana, W.	267	Vasiljević-Radović, D.	107, 153

Velazquez-Perez, J.E.	337, 487	Yang, L.	409
Venca, A.	417	You, H.	387
Venkateshan, A.	3	Yukhymchuk, V.O.	111
Videnović-Mišić, M.	573		
Vincze, A.	357	Zarcone, M.	497
Voicu, R.	259	Zebrev, G.	545
Von Haartman, M.	25	Zekentes, K.	293
Voss, S.-H.	675	Zerbe, V.	659
Voutsas, A.	597	Zgrda, M.	341
Vučenović, S.	533	Zhang, S.	25
Vujanić, A.	233	Zhang, Z.	25
		Zhou, J.	601
Wachutka, G.	197	Zhou, M.	601
Waltz, P.	201	Zhou, X.J.	89
Wang, A.	409	Zhou, Z.	193
Wang, L-M.	615	Zimmermann, H.	421
Wang, S.	89, 337	Zirmi, R.	459
Wang, Z.O.	635	Zlatković, V.	663
Wong, C.K.	277, 383	Zwolinski, M.	437, 655
Wong, H.	277, 383		
		Živanov, Lj.	479

2006 25th International Conference on Microelectronics

Serbia and Montenegro
14-17 May 2006

Volume 2 of 2

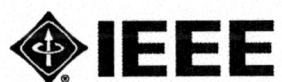

IEEE Catalog Number: 06TH8868
ISBN: 1-4244-0116-X

Copyright © 2006 by The Institute of Electrical and Electronics Engineers, Inc.
All Rights Reserved

Copyright and Reprint Permissions: Abstracting is permitted with credit to the source. Libraries are permitted to photocopy beyond the limit of U.S. copyright law for private use of patrons those articles in this volume that carry a code at the bottom of the first page, provided the per-copy fee indicated in the code is paid through Copyright Clearance Center, 222 Rosewood Drive, Danvers, MA 01923.

For other copying, reprint or republications permission, write to IEEE Copyrights Manager, IEEE Operations Center, 445 Hoes Lane, Piscataway, New Jersey USA 08854. All rights reserved.

IEEE Catalog Number:	06TH8868
ISBN:	1-4244-0116-X
LOC:	2005938573

Additional Copies of This Publication Are Available from:

IEEE Service Center
445 Hoes Lane
Piscataway, NJ 08854
IEEE Service Center
445 Hoes Lane
Piscataway, NJ 08854
Phone: (800) 678-IEEE
                    (732) 981-1393
Fax: (732) 981-9667
E-mail: customer-service@ieee.org

MIEL 2004 Best Paper Award Winners

Left to right: F. De Paola, J. Millan, N. Stojadinović (Conference Chairman),
A. Paskaleva, H. Hein

MIEL 2004 Best Oral Paper Award

X. Perpina, X. Jorda, P. Godignon, J. Millan, H. Von Kiedrowski,
J. Vobecky, N. Mestres

for

Direct Measurement of Self-Heating Effects at the Drift Region of 600V PT-IGBTs

J. Millan accepts the award from N. Stojadinović

MIEL 2004 Best Poster Paper Award

P. Baureis, H. Hein, M. Peter, F. Oehler

for

A Fully Integrated 0.35 μm CMOS MMIC Amplifier for Short Range 443 MHz ISM Band Transceiver Applications

H. Hein accepts the award from N. Stojadinović

MIEL 2004 Best Student Paper Award

F. De Paola, V. Aliberti, B. Rajaei, N. Rinaldi, J. Burghartz

for

A Scalable Physical Model for Coplanar Waveguide Transition in Flip-Chip Applications

F. De Paola accepts the award from N. Stojadinović

MIEL 2004 Best Paper Award
on behalf of
Microelectronics Reliability

A. Paskaleva, M. Lemberger, S. Zurcher, A.J. Bauer

for

Electrical Properties and Conduction Mechanisms in $Hf_xTi_ySi_zO$ Films
Obtained from Novel MOCVD Precursors

A. Paskaleva accepts the award from N. Stojadinović

STEERING COMMITTEE

Asia & Pacific

C. Y. Chang, National Chiao-Tung University, Taiwan
Y.-I. Choi, Ajou University, Korea
S. Dimitrijev, Griffith University, Australia
H. Iwai, Tokyo Institute of Technology, Japan
C. Jagadish, Australian National University, Australia
B. Z. Li, Fudan University, China
T. Ohmi, Tohoku University, Japan
M. K. Radhakrishnan, Philips Electronics, Singapore
H. Wong, City University of Hong Kong, Hong Kong

Europe

H. Detter, Technical University of Vienna, Austria
G. Ghibaudo, ENSERG, France
G. Golan, ATCT Ltd., Israel
V. Liberali, University of Milan, Italy
M. Ostling, Royal Institute of Technology, Sweden
V. Pershenkov, Moscow Physics Engineering Institute, Russia
R. Popović, EPFL, Switzerland
S. Selberherr, Technical University of Vienna, Austria
N. Stojadinović, University of Niš, Serbia and Montenegro

America

V. Arora, Wilkes University, PA, USA
J.-P. Colinge, University of California at Davis, CA, USA
M. Estrada, CINVESTAV, Mexico
J. Liou, University of Central Florida, FL, USA
A. Nathan, University of Waterloo, Canada
V. Oklobdžija, University of California, CA, USA
A. Ortiz-Conde, University Simon Bolivar, Venezuela
K. Shenai, University of Illinois at Chicago, IL, USA
R. Singh, Clemson University, SC, USA

PROGRAMME COMMITTEE

Chairman:
N. Stojadinović, University of Niš, Serbia and Montenegro

Vice-Chairmen:
S. Dimitrijev, Griffith University, Australia
H. Iwai, Tokyo Institute of Technology, Japan
S. Selberherr, Technical University of Vienna, Austria
J. Liou, University of Central Florida, FL, USA

Scientific Secretary:
I. Manić, University of Niš, Serbia and Montenegro

Organization Secretary:
T. Pešić, University of Niš, Serbia and Montenegro

Corresponding Members:

I. Adesida, University of Illinois at Urbana-Champaign, USA
K. Arshak, University of Limerick, Ireland
E. Atanassova, Bulgarian Academy of Sciences, Bulgaria
V. Benda, Technical University of Prague, Czech Republic
M. Blyzniuk, Ivan Franko National University of L'viv, Ukraine
T. Brožek, PDF Solutions, Inc., USA
J. Burghartz, IMS Chips, Germany
I. De Wolf, IMEC, Belgium
A. Dziedzic, Technical University of Wroclaw, Poland
Z. Đurić, IHTM-CMTM, Serbia and Montenegro
D. Fleetwood, Vanderbilt University, USA
D. Flores, National Centre of Microelectronics, Spain
G. Ghibaudo, ENSERG, France
P. Hagouel, Aristoteles University of Thessaloniki, Greece
S. Haque, Philips, USA
S. Horita, Japan Advanced Institute of Science and Technology, Japan
K. Itoh, Hitachi, Ltd., Japan
H. Ishiwara, Tokyo Institute of Technology, Japan
N. Janković, University of Niš, Serbia and Montenegro
Z. Jakšić, IHTM-CMTM, Serbia and Montenegro
M. Jevtić, Institute of Physics, Serbia and Montenegro
E.D. Kim, Korea Electrotechnology Research Institute, Korea
K. Kim, Samsung, Korea
S. Koshevaya, National Institute of Astrophysics, Mexico
D. Kouvatsos, NCSR Demokritos, Greece
J. Kuo, National Taiwan University, Taiwan
K. Lee, KAIST, Korea
V. Litovski, University of Niš, Serbia and Montenegro
J. Lutz, Technical University of Chemnitz, Germany
P. Mawby, University of Warwick, United Kingdom
S. Mijalković, Delft University of Technology, The Netherlands
E. Miranda, University of Buenos Aires, Argentina
T. Mouthaan, University of Twente, The Netherlands
A. Napieralski, Technical University of Lodz, Poland
J. Nicolics, Vienna University of Technology, Austria
P. Nikolić, University of Belgrade, Serbia and Montenegro
N. Novkovski, Faculty of Natural Science and Mathematics, Macedonia
M.K. Radhakrishnan, Philips Electronics, Singapore
R. Ramović, University of Belgrade, Serbia and Montenegro
G. Reeves, Royal Melbourne Institute of Technology, Australia
N. Rinaldi, University of Naples, Italy
A. Rusu, Technical University of Bucharest, Romania
N. Shammas, Staffordshire University, United Kingdom
E.M. Shankar, De Montfort University, United Kingdom
D. Tjapkin, University of Belgrade, Serbia and Montenegro
V. Tvarožek, Slovak Technical University, Slovak Republic
R. Ubar, Tallin Technical University, Estonia
V. Vashchenko, National Semiconductor Corporation, USA
J. Vobecky, Czech Technical University in Prague, Czech Republic
G. Wachutka, Technical University of Munchen, Germany
P. Yu, University of California, San Diego, USA
M. Zwolinski, University of Southampton, United Kingdom
Lj. Živanov, University of Novi Sad, Serbia and Montenegro

CONTENTS

VOLUME 1

Workshop
Nanotechnologies

Invited Keynote Papers

Semiconductor Manufacturing in the Nanotechnology World of the 21st Century
R. Singh, A. Venkateshan, K. Poole, D. Mohan, P. Chatterjee 3

Nanoscience and Nanotechnologies in Serbia
Z. Djurić 11

Failure of Ohm's Law: Its Implications on the Design of Nanoelectronic Devices and Circuits
V.K. Arora 17

Device Integration Issues Towards 10 nm MOSFETs
M. Ostling, B. Gunnar Malm, M. Von Haartman, J. Hallstedt, Z. Zhang, P.-E. Hellstrom, S. Zhang 25

Nanoscale Materials, Devices, and Systems for Sensing, Detection, and Environmental Pollution Monitoring and Mitigation
A. Vaseashta 31

The Role of Non-Equilibrium Plasmas and Micro-Discharges in Top Down Nanotechnologies and Selforganized Assembly of Nanostructures
Z. Petrović, G. Malović, M. Radmilović-Radjenović, N. Pauč, D. Marić, P. Maguire, C. Mahony 39

Challenges of Ta$_2$O$_5$ as high-k Dielectric for Nanoscale DRAMs
E. Atanassova, A. Paskaleva 47

Electrically Active Defects at the Interface between (100)Si and Hafnium Dioxide Thin Films
P. Hurley, K. Cherkaoui 55

Capacitance Behavior of Nanometer FD SOI CMOS Devices with HfO$_2$ High-K Gate Dielectric Considering Gate Tunneling Leakage Current
J.B. Kuo, C.H. Lin 61

Plenary Sessions

Invited Keynote Papers

Defect Engineering and Stress Control in Advanced Devices on High-Mobility Substrates
C. Claeys, G. Eneman, E. Simoen 67

Reviews and Prospects of Low-Voltage Nano-Scale Embedded RAMs
K. Itoh 77

Gate Oxide Reliability for Nano-Scale CMOS
J. Stathis 83

Effects of Device Aging on Microelectronics Radiation Response and Reliability
D. Fleetwood, M.P. Rodgers, L. Tsetseris, X.J. Zhou, I. Batyrev, S. Wang, R.D. Schrimpf,
S.T. Pantelides 89

Session
Nanotechnologies

Nonlinear Model of Microtubule Dynamics and Its Impact on Kinesin Motion
M.V. Satarić 99

Adsorbed Mass Fluctuations of a Micro/Nanoresonator Surrounded by an Arbitrary Gas Mixture
Z. Djurić, O. Jakšić, I. Jokić, M. Frantlović 103

Scanning Probe-Shaped Nanohole Arrays with Extraordinary Optical Transmission as Platform for Enhanced Surface Plasmon-Based Biosensing
Z. Jakšić, M. Maksimović, D. Vasiljević-Radović, D. Tanasković, M. Sarajlić 107

Synthesis and Optical Properties of CdSe Nanocrystals Obtained from $CdCl_2$ and Na_2SeSO_3 Aqueous Solutions in the Presence of Gelatine
A.E. Raevskaya, A.L. Stroyuk, S.Ya. Kuchmii, Yu.M. Azhniuk, V.M. Dzhagan,
V.O. Yukhymchuk, M.Ya. Valakh 111

Growth and ac-Properties of Lanthanum-Manganese Oxide Films on Si Substrates
A.A. Dakhel 115

Low - k Polyimide/Silica Nanocomposites for Microelectronics Applications
E. Logakis, D. Fragiadakis, P. Pissis 119

Investigation of Electrode Patterns Suitable for Nano-Litre Drop Coated Conducting Polymer Composite Sensors
K. Arshak, C. Cunniffe, E. Moore, L. Cavanagh 123

A Novel Vertical Impact Ionisation MOSFET (I-MOS) Concept
U. Abelein, M. Born, K.K. Bhuwalka, M. Schindler, M. Schmidt, T. Sulima, I. Eisele 127

Tunnel FET: A CMOS Device for High Temperature Applications
M. Born, K. Bhuwalka, M. Schindler, U. Abelein, M. Schmidt, T. Sulima, I. Eisele 131

Poster Session
Nanotechnologies

Analysis of Nonlinear Effects in Nanometer-Scale Silicon-on-Insulator Rib Waveguides
V.M.N. Passaro, F. De Leonardis, G.Z. Mashanovich 137

Analysis of Pulsed Excitation in Small Silicon-On-Insulator Microring Resonators
F. De Leonardis, V.M.N. Passaro, G.Z. Mashanovich 141

Silicon Surface Exfoliation Under Compression Plasma Flow Action
I.P. Dojčinović, M.M. Kuraica, D. Randjelović, M. Matić, J. Purić 145

Magnetic Field Influence on Silicon Surface Periodic Structures Obtained by Plasma Flow Action
I.P. Dojčinović, M.M. Kuraica, M. Mitrović, D. Randjelović, M. Matić, J. Purić 149

A Consideration of Transparent Metal Structures for Subwavelength Diffraction Management
Z. Jakšić, M. Sarajlić, M. Maksimović, D. Vasiljević-Radović, D. Jovanović 153

The Fabrication of Single Electron Transistor by Polysilicon Thin Film and Point-Contact Lithography
K.-D. Huang, J.-T. Lin, S.-F. Hu, C.-L. Sung 157

Impact of Rapid Photothermal Processing on Properties of ZnO Nanostructures for Solar Cell Applications
S. Shishiyanu, R. Singh, T. Shishiyanu, O. Lupan 161

Session
Power Devices and ICs

Invited Keynote Paper

Evolution of Silicon Power Devices and Challenges to Material Limit
A. Nakagawa 167

Invited Keynote Paper

High Speed Electro-Thermal Models for Inverter Simulations
P.A. Mawby, A.T. Bryant, P.R. Palmer, E. Santi, J.L. Huidgins 175

Influence of Small Emitter and Contact Non-Uniformities on the Current Filamentation in 3.3-kV p^+-n^--n^+ Silicon Diodes
H.P. Felsl, E. Falck, F.-J. Niedernostheide, J. Lutz 183

Schottky Contacts on Single-Crystal CVD Diamond
D. Doneddu, O.J. Guy, D. Twitchen, A. Tajani, M. Schwitters, P. Igić 189

Dynamic Thermal Simulation of Power Devices Operating With PWM Signals
Z. Zhou, M.S. Khanniche, P. Igić, N. Janković, S.G. Batcup, P.A. Mawby 193

Dynamic Characteristics of Novel Super-Junction LDMOS Switches under Charge Imbalance Conditions
K. Permthammasin, G. Wachutka, M. Schmitt, H. Kapels 197

Self-Heating Effects in SOI NLDEMOS Power Devices
F. Dieudonne, S. Haendler, S. Chouteau, J. Rosa, P. Waltz, A. Perrotin, L. Boissonnet, B. Rauber,
C. Schaffnit, C. Raynaud 201

xii

Poster Session
Power Devices and ICs

An Alternative Process Architecture for CMOS Based High Side RESURF LDMOS Transistors
P.M. Holland, P. Igić 207

Comparison of Measured and Simulated Characteristics of Boron Implanted 4H-SiC DiMOSFET
S.C. Kim, W. Bahng, K.H. Kim, I.H. Kang, S. J. Kim, N.K. Kim 211

IGBT Behavioral PSPICE Model
K. Asparuhova, T. Grigorova 215

Numerical Analysis of a Trench Gate FLIMOSFET with No Quasi-Saturation, Improved Specific On-Resistance and Better Synchronous Rectifying Characteristics
R. Vaid, N. Padha 219

Session
Microsystem Technologies

Invited Keynote Paper

Wireless Ultra-Low Power Smart Data Acquisition System for Pressure Sensing in Medical Application
K. Arshak, E. Jafer 225

In-Line Concentration Measurement of Nanoliter Liquid Sample using Low-Coherence Spectral Interferometry
M. Tomić, Z. Djinović, A. Vujanić 233

Using Design of Experiment to Investigate the Effects of Conducting Polymer Composite Sensor Composition on the Response to an Homologous Series of Alcohols
K. Arshak, E. Moore, C. Cunniffe, L. Cavanagh 237

IR Bimaterial Detectors Performance
Z. Djurić, D. Randjelović, J. Matović, J. Lamovec 241

Dynamic Elastic Bending in Optically Driven Microcantilever: Surface Strain Effects
D.M. Todorović, T. Grozdić 245

Real-Time Tracking of a Moving Object Inside a Tube
K. Arshak, F. Adepoju 249

xiii

Poster Session
Microsystem Technologies

A Simplified Method of Analysis of MEMS Bimaterial Cantilever Element
J. Matović 255

Design Parameters Optimization Using Process Variations of the Pull-In Voltage for MEMS
R. Voicu, C. Tibeica, M. Bazu 259

Development of a Portable Gamma Radiation Monitoring System
K. Arshak, D. Buckley, O. Korostynska 263

Thermal Characterization of the Micro Bonding Process Using a Hot Air Stream
D. Andrijašević, I. Giouroudi, W. Smetana, W. Brenner, D. Esinenco 267

Designing and Modeling MEMS Resonator in VHF Range
M. Al Khusheiny, B.Y. Majlis 271

Session
Opto and Microwave Devices and ICs

Invited Keynote Paper

Silicon Integrated Photonics for Microelectronics Evolution
H. Wong, V. Filip, C.K. Wong, P.S. Chung 277

Diagnostics of Homogeneity of Individual Layers of Large-Area Silicon Solar Cells Using Local Irradiation
V. Benda, Z. Machacek, J. Salinger 285

Improved Dual Grating-Assisted Directional Coupler for Silicon Nanophotonics
G. Mashanovich, V. Passaro, G. Ensell, F. Gardes, G. Reed 289

High Speed Modulator in Q-Band Range on 4H-SiC p-i-n Diodes
R. Kakanakov, L. Kolaklieva, L. Romanov, A. Kirillov, A. Lebedev, M. Boltovets, K. Zekentes 293

Low-Frequency-Noise Characteristic of Quasi-Enhancement-Mode HEMT Using a Selectively Hydrogen-Pretreatment
I.H. Kang, S.C. Kim, W. Bahng, N.K. Kim 297

Hyper Sound Amplification
S. Koshevaya, V. Grimalsky, M. Tecpoyotl-Torres, J. Escobedo-Alatorre, M. Diaz-Ayala, A. Garcia-B 301

xiv

Poster Session
Opto and Microwave Devices and ICs

Noise as a Diagnostic Tool for Quality of GaSb Laser Diodes
Z. Chobola, J. Vanek, E. Hulicius, T. Šimeček 307

Nonlinear Surface Ultrasonic Monopulses in Solid Film-Substrate System
V. Grimalsky, S. Koshevaya, E. Gutierrez-D., A. Garcia-B. 309

Accurate Noise Modeling of HEMT for Low-Noise Applications
I.H. Kang, S.C. Kim, W. Bahng, N.K. Kim 313

Modeling and Constituent Design of SPR Based and/or Waveguide Type Sensors
N. Dmitruk, O. Mayeva, O. Korovin, M. Sosnova, O. Lytvyn, V. Min'ko 317

Effect of Interface Microrelief on Optical, Electrical and Photoelectric Characteristics of Heteroepitaxial $Al_xGa_{1-x}As/GaAs$ Structures
N. Dmitruk, O. Borkovskaya, R. Konakova, A. Korovin, I. Mamontova, O. Kondratenko 321

Session
Physics and Modeling

Invited Keynote Paper

a-$Si_{1-x}C_x$:H TFTs: Fabrication and Modeling
M. Estrada, A. Cerdeira, R. Garcia, B. Iniguez 327

Boron Redistribution During SOI Wafers Thermal Oxidation
Z. Djurić, M.M. Smiljanić, K. Radulović, Ž. Lazić 333

Transport of Non-Equilibrium Charge Carriers in Bipolar Semiconductors Materials
Y. Gurevich, J.E. Velazquez-Perez 337

A High-Frequency Extension of a Surface-Potential-Based Substrate Model for Noise Coupling Analysis
N. Simić, F. Ingvarson, S. Kristiansson, M. Zgrda, K. Jeppson 341

Analytical Modeling of the Triggering Drain Voltage at the Onset of the Kink Effect for PD SOI NMOS
M. Sarajlić, R. Ramović 345

Author Index 349

VOLUME 2

Session
Processes and Technologies

P-Type Conduction in Sputtered ZnO Thin Films Doped by Nitrogen
K. Shtereva, V. Tvarozek, I. Novotny, J. Kovac, P. Sutta, A. Vincze ... 357

Investigation of Metal-Polycrystalline Silicon Carbide Bonding While Metallization
G. Golan, V. Manevych, I. Lapsker, B. Gorenstein, A. Axelevitch ... 361

PVD Aluminium Nitride as Heat Spreader in Silicon-on-Glass Technology
L. La Spina, H. Schellevis, N. Nenadović, L. Nanver ... 365

Reliability Issues Related to Laser-Annealed Implanted Back-Wafer Contacts in Bipolar Silicon-on-Glass Processes
G. Lorito, V. Gonda, S. Lui, T.L. Scholtes, H. Schellevis, L. Nanver ... 369

A Novel Bottom Gate Polysilicon Thin-Film Transistor with Smart Body Tie
J.-T. Lin, K.-D. Huang, S.-T. Lin ... 373

Poster Session
Processes and Technologies

Electrical Properties of HfO_2 Films Formed by Ion Assisted Deposition
K. Cherkaoui, A. Negara, S. McDonnell, G. Hughes, M. Modreanu, P. Hurley ... 379

PECVD Growth of Thick Silicon Oxynitride for On-Chip Optical Interconnects Applications
C.K. Wong, H. Wong, C.W. Kok, M. Chan ... 383

The Characterization and Optimization of the Thermal Oxidation Process Equipment Using Experimental Design and Data Transformation
H. You, X. Jia, S. Wang ... 387

PTC Effect and Structure of Polymer Composites Based on Polypropylene/Co-Polyamide Blend Filled with Dispersed Iron
A. Kanapitsas, C. Tsonos, E. Logakis, C. Pandis, P. Pissis, E. Kontou, Y.P. Mamunya, E.V. Lebedev, C.G. Delides ... 391

A New Method of Evaluation of Liquidus Temperatures of Ternary Alloys
A. Prijić, Z. Prijić, B. Pešić ... 395

Session
Circuit Design

Invited Keynote Paper

System-on-Chip Design: Engineering of Art
Z. Stamenković 401

Invited Keynote Paper

A New Circuit Model for Designig Fully Integrated Class-A Power Amplifier
X. Guan, H. Feng, A. Wang, L. Yang 409

Gate Layout Improvement Aimed at Testability
M. Blyzniuk 413

An Improved Write Driver for Miniaturized Hard Disk Drives
F. Tamigi, I. Marano, A. Venca, N. Rinaldi 417

Optical Receiver with Voltage-Controlled Transimpedance in BiCMOS Technology
N. Tadić, H. Zimmermann 421

Improved Linearity Active Resistor Using Equivalent FGMOS Devices
C. Popa 425

A Stochastic Approach to Crosstalk Analysis in Mixed-Signal ICs
G. Boselli, G. Trucco, V. Liberali 429

The 4-Phase Frame Partitioning Circuit
S. Poriazis 433

Fault Diagnosis in Digital Part of Mixed-Mode Circuit
M. Andrejević, V. Litovski, M. Zwolinski 437

Poster Session
Circuit Design

Simulation of Crosstalk between Several Interconnection Lines in CMOS Integrated Circuits
M. Petković 443

Reordering in Topology Decision Diagram Method for Symbolic Circuit Analysis
S. Djordjević, P.M. Petković 447

An Improved Performance FGMOS Voltage Comparator for Data Acquisition Systems
C. Popa 451

On Silicon Timing Validation of Digital Logic Gate "A Study of Two Generic Methods"
A.P. Singh, N.S. Panwar 455

High Speed Low Power CMOS Comparator for Pipeline ADCs
M.B. Guermaz, L. Bouzerara, A. Slimane, M.T. Belaroussi, B. Lehouidj, R. Zirmi 459

xvii

Session
Modeling and Simulation

Invited Keynote Paper

Optimization Issue in Interconnect Analysis
S. Holzer, S. Selberherr 465

Invited Keynote Paper

Advanced Circuit and Device Modeling with Verilog-A
S. Mijalković 471

Process and Device Simulation With a Generic Scientific Simulation Environment
M. Spevak, R. Heinzl, P. Schwaha, T. Grasser 475

Design and Simulation of Thick Film Thermistors Using Commercial Simulation Tools
V. Marić, O. Aleksić, Lj. Živanov 479

Fully Automated Electrothermal Simulation using Standard CAD Tools
F. De Paola, J. Nowakowski, V. d'Alessandro, N. Rinaldi 483

A Novel 3D Embedded Gate Field Effect Transistor: Device Concept and Modelling
K. Fobelets, P.W. Ding, J.E. Velazquez-Perez 487

A Modified EKV PDSOI MOSFETs Model
J. Alvarado, A. Cerdeira, V. Kilshytska, D. Flandre 491

Poster Session
Modeling and Simulation

Monte Carlo Analysis of Voltage-Current Characteristic Nonlinearity and Harmonic Generation in Submicron Semiconductor Structures
D. Persano Adorno, M. Capizzo, M. Zarcone 497

Diode I-U Curve Fitting with Lambert W Function
P. Hruska, Z. Chobola, L. Grmela 501

A New Treshold Voltage Analytical Model of Strained Si/SiGe MOSFET
P. Lukić, R. Ramović, R. Šašić 505

Graded-Channel SOI nMOSFET Model Valid for Harmonic Distortion Evaluation
M. de Souza, M. Pavanello, A. Cerdeira, D. Flandre 509

A Simple Polysilicon Thin-Film Transistor SPICE Model
I. Pappas, A.T. Hatzopoulos, D.H. Tassis, N. Arpatzanis, S. Siskos, A.A. Hatzopoulos,
C.A. Dimitriadis, G. Kamarinos 513

Modeling and Simulation of Submicron MOSFETs with Alternative Gate Dielectrics for DRAM Cells
N. Konofaos, G. Alexiou 517

Investigation of the Novel Attributes of a Vertical MOSFET with Internal Block Layer (bVMOS): 2-D Simulation Study
J.-T. Lin, K.-C. Lin, T.-Y. Lee, Y.-C. Eng 521

Filling of Learning Process in Electronics/Microelectronics Studies by Teaching for Understanding Properties Using Open Training CAD Tools
M. Blyzniuk 525

Distributed Framework for Teaching Fundamentals of Heat Transfer in Electronic Circuits
M. Paszkowski-Rogacz, P. Januszkiewicz, P. Gocek, G. Jablonski, M. Janicki,
G. De Mey, A. Napieralski 529

Charge Carriers Distributions in Rectangular Quantum Rod
J. Šetrajčić, B. Tošić, V. Sajfert, D. Ilić, S. Jaćimovski, S. Vučenović 533

Numerical Investigation of O_2 Adsorption Effect in $Si-CF_4/O_2$ System
Y. Grigoryev, A. Gorobchuk 537

Dispersion Relation for Two-Valley Quasi-Hydrodynamic Models in SCWs Propagation in n-GaAs Thin Films
A. Garcia-B., V. Grimalsky, E. Gutierrez-D., S. Koshevaya 541

Diffusion-Drift Model of Fully-Depleted SOI MOSFET
G. Zebrev, M. Gorbunov 545

Session
Reliability Physics

Invited Keynote Paper

Low Frequency Noise and Fluctuations in Sub 0.1 μm Bulk and SOI CMOS Technologies
G. Ghibaudo, J. Jomaah 551

Analysis of Surface Generation Mechanisms in MOS Capacitors
F. Kong, P. Tanner, L. Hold, J. Han, S. Dimitrijev 557

Effects of NO Annealing on the Characterizations of GaN MIS Capacitor
L. Lin, P.T. Lai, K.M. Lau 561

Long Range Statistical Lifetime Prediction of Ultra-Thin SiO_2 Oxides: Influence of Accelerated Ageing Methods and Extrapolation Models
D. Pic, D. Goguenheim, J.-L. Ogier 565

Low Frequency Noise of GaAs MESFET Degraded in ESD Test
M. Jevtić, J. Hadži-Vuković 569

Dependence of DGMOSFET 1/f Noise on Transistor Geometry and Technology Parameters
M. Videnović-Mišić, M.M. Jevtić 573

xix

Use of RADFETs for Quality Assurance of Radiation Cancer Treatments
A. Jakšić, K. Rodgers, C. Gallagher, P.J. Hughes 577

Effect of the Metal Electrode on the Characteristics of Ta_2O_5 Capacitors for DRAM Applications
E. Atanassova, D. Spassov, A. Paskaleva 581

Reliability Properties of Ta_2O_5 Films Grown on N_2O Plasma Nitrided Silicon
N. Novkovski, E. Atanassova 585

Stress Induced Leakage Currents and Charge Trapping in Thin Zr- and Hf-Silicate Layers
A. Paskaleva, M. Lemberger, A.J. Bauer 589

Reliability Investigation of NLDEMOS in 0.13µm SOI CMOS Technology
D. Lachenal, Y. Rey-Tauriac, L. Boissonnet, A. Bravaix 593

Physics and Electrical Characterization of Excimer Laser Crystallized Polysilicon TFTs
L. Michalas, M. Exarchos, G.J. Papaioannou, D. Kouvatsos, A. Voutsas 597

Reliability Assessment of a RF PA Assembly With Embedded Coin Construction
J. Zhou, H. Lu, M. Zhou, B. Inkman, D. Anderson, B. Johnson 601

The Effect of Number of Zincation in Electroless Nickel Immersion Gold (ENIG) Under Bump Metallurgy (UBM) on Reliability in Microelectronics Packaging
S.H. Abdullah, I. Ahmad, A. Jalar 605

Poster Session
Reliability Physics

Investigation of the Ion Defect States by Photoacoustic Spectroscopy
D.M. Todorović, V. Jović, M. Smiljanić, T. Grozdić 611

Relationship between Intrinsic Breakdown Field and Bandgap of Materials
L-M. Wang 615

Electrode Effect on NTC Planar Thermistor Volume Resistivity
O. Aleksić, B. Radojčić, R. Ramović 619

Performances of Conventional Thick-Film Resistors After Multiple High-Voltage Pulse Stressing
I. Stanimirović, M.M. Jevtić, Z. Stanimirović 623

Influence of Simultaneous Mechanical and Electrical Straining on Conventional Thick-Film Resistors
Z. Stanimirović, M.M. Jevtić, I. Stanimirović 627

Influence of Electrode Materials and the Manner of Electrode Surface Processing on Gas-Filled Surge Arresters Relevant Characteristics
B. Lončar, P. Osmokrović, A. Vasić, R. Šašić 631

New Insights into Tunneling Current through Oxynitride/Oxide Stack for MOSFETs
L.F. Mao, Z.O. Wang..635

Spontaneous Recovery in DC Gate Bias Stressed Power VDMOSFETs
I. Manić, S. Djorić-Veljković, V. Davidović, D. Danković, S. Golubović, N. Stojadinović................639

Lifetime Estimation in NBT Stressed P-Channel Power VDMOSFETs
D. Danković, I. Manić, S. Djorić-Veljković, V. Davidović, S. Golubović, N. Stojadinović................645

Evaluation of Reflow Ovens for Lead-Free Soldering
J. Lempinen, A. Tuominen..649

Session
System Design

Word-Length Oriented Multiobjective Optimization of Area and Power Consumption in DSP Algorithm Implementation
A. Ahmadi, M. Zwolinski..655

A Mission Level Design Language Based on AleC++
B. Andjelković, V. Litovski, V. Zerbe..659

Clocking Challenges in High Speed Source Synchronous Interfaces
V. Zlatković..663

An Adaptive Pulse-Width Control Loop
G. Jovanović, D. Mitić, M. Stojčev..667

Low Power Computing for Secure and Reliable Sensor Networks
R. Agarwal, E.M. Popovici, C. O'Keeffe, B. O'Flynn, S.J. Bellis..671

Design and Implementation of an FPGA Based High-Speed Data Buffer for Optical Interconnects
S.-H. Voss, M. Talmi, J. Saniter..675

DefSim: Measurement Environment for CMOS Defects
T. Borejko, A. Jutman, W.A. Pleskacz, R. Ubar..679

Poster Session
System Design

NNARX Model of Speech Signal Generating System: Test Error Subject to Modeling Mode Selection
D. Protić, M. Milosavljević..685

The Electronic and Program Control System of X-Ray Ion Mobility Spectrometer
V.S. Pershenkov, A.U. Razvalyaev, A.D. Tremasov..689

The Innovative Method for Determining Characteristics of Over-Voltage Protection Elements
P. Osmokrović, B. Lončar, S. Stanković, A. Vasić..693

Approach to Partially Self-Checking Finite State Machine Design
G. Djordjević, T. Stanković, M. Stojčev ... 697

Laboratory ADC Tester Based on NI-6251 Acquisition Card
M. Nikolić, M. Sokolović, P. Petković ... 701

Late Submission Paper

Hybrid Empirical-Neural Model of the Loaded Microwave Cavity Applicators
Z. Stanković, B. Milovanović, M. Milijić ... 705

Author Index ... 709

Session

Processes and Technologies

P-Type Conduction in Sputtered ZnO Thin Films Doped by Nitrogen

K. Shtereva, V. Tvarozek, I. Novotny, J. Kovac, P. Sutta, and A. Vincze

Abstract - Nitrogen doped zinc oxide (ZnO:N) thin films were prepared by RF diode sputtering from ZnO target in different ratio of Ar/N_2 gas mixture. The p-type features of ZnO:N thin films have been caused by the incorporation of the nitrogen acceptor NO into ZnO what was proven by Second Ion Mass Spectroscopy (SIMS) analysis. The minimum value of resistivity of 790 Ωcm, a Hall mobility of 22 $cm^2V^{-1}s^{-1}$ and the carrier concentration of $3.6 \times 10^{14} cm^{-3}$ were yielded at 75 % N_2. X-ray diffraction measurements (XRD) showed that ZnO:N films had the preferential orientation of (002) plane at 25 % N_2 and of (100) plane for higher N_2 concentrations. The average grain size was from 7 to 42 nm for all Ar/N_2 ratios. ZnO:N films exhibit relatively high microstrains (10×10^{-3}).

I. INTRODUCTION

Zinc oxide (ZnO) is transparent, wide band gap II-VI semiconductor, with direct band gap of 3.37 eV at room temperature, large exiton binding energy of 60 meV, and high refractivity (melting point of 1975 °C) and chemical stability. To realize the light emitting devices, an important issue is the fabrication of p-type ZnO with a high hole concentration and a low resistance [1].

ZnO films can be grown by variety of methods such as magnetron sputtering, evaporation, metalorganic chemical vapor deposition, (MOCVD), molecular beam epitaxy (MBE), sol-gel process, spray pyrolysis, plasma enhanced chemical vapor deposition (PECVD), pulsed laser deposition (PLD), atomic layer deposition (ALD) and filtered cathodic vacuum arc technique. Undoped zinc oxide exhibits intrinsic n-type conductivity and it turned out to be more difficult to fabricate low-resistivity p-type ZnO due to high activation energy of acceptors, low solubility of acceptor dopants and self-compensating process on acceptor doping. ZnO can be doped intrinsically, through native defects, and extrinsically, through impurity doping with group III elements (Al, Ga or In), group I elements (Li, Na and K) or group V elements (N, P and As).

K. Shtereva is with the Department of Electronics, University of Rousse, Studentska 8, BG - 7017 Rousse, Bulgaria, E-mail: kshtereva@ecs.ru.acad.bg

V. Tvarozek, I. Novotny, J. Kovac and A. Vincze are with the Department of Microelectronics, Slovak University of Technology, Ilkovicova 3, SK-812 19 Bratislava, Slovakia, E-mail: vladimir.tvarozek@stuba.sk

P. Sutta is with the West Bohemia University, New technologies - Research Center, Univerzitni 8, 306 14 Plzen, Czech Republic, E-mail: sutta@ntc.zcu.cz

Nitrogen doping, which has been successful in fabricating p-type ZnSe, is considered an effective method to realize p-type ZnO thin films [2]. As known are few articles reporting p-type conductivity in nitrogen doped ZnO films and one of the most quoted is Joseph et al. [3]. These studies point that the key elements for successive synthesis of p-type ZnO thin films were N dopant source and the substrate temperature. Novelty of our approach is in use of plasma assisted deposition method - RF diode sputtering - which allows to perform direct action of ions, ion complexes and energetic electrons on growing film with the aim to increase the substrate temperature as well as to form suitable nitrogen acceptors.

II. EXPERIMENTAL

Thin films were prepared in a planar RF sputtering diode system Perkin Elmer 2400/8L. ZnO:N films were deposited on Corning glass substrates from ZnO target in Ar/N_2 atmosphere. Pre-sputtering for 10 minutes cleaned the target. The vacuum chamber was evacuated to a base pressure of 2×10^{-5} Pa. High purity nitrogen (N_2) acted as a doping source and the content of N_2 in sputtering gas varied from 0 % to 100 %. The total gas pressure of 1.33 Pa was maintained constant during the sputtering process. The sputtering power was 500 W and the sputtering time varied from 30 to 60 minutes and accordingly thicknesses of ZnO films were in range from 430 nm to 870 nm.

The film thickness was measured by Talystep instrument. The crystal orientation and microstrains of the films were investigated by an automatic powder X-ray diffractometer AXS Bruker D8 with Eulerian cradle and 2D detector ($CoK\alpha$, $\lambda = 0.179$ nm). The depth profile of the films was measured with secondary ion mass spectroscopy (SIMS) using TOF-SIMS IV analyzer from ION TOF GmbH, Muenster. Hall measurements were carried out at room temperature and at following measurement parameters: currents 10 nA - 200 nA, temperature 300 K, magnetic field 0.385 T.

III. RESULTS AND DISCUSSION

The XRD diffraction lines of ZnO (100), (002), (101), and (110) were observed for all studied samples. The set of two-dimensional diffraction patterns of ZnO:N films deposited at 75 % N_2 in sputtering gas displays three Debye rings of ZnO Bragg reflections (Fig.1). The white circle is a Laue spot diffracted by substrate material (c-Si) due to the continuous X-ray radiation. The diffraction lines of

Fig. 1. Two dimensional diffraction patterns of ZnO films deposited at 75% N_2 in sputtering gas.

ZnO (100), (002), (101), and (110) were observed for all studied samples. ZnO:N films demonstrated the preferential orientation of (002) plane at 25 % N_2, which is attributed to undoped ZnO thin films. Highly N_2 (50 % N_2

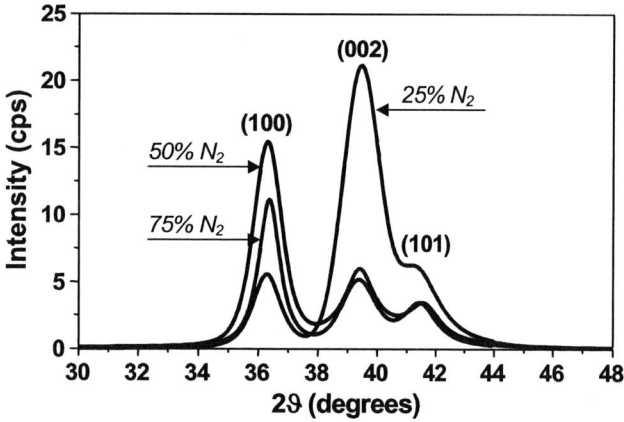

Fig. 2. XRD patterns for ZnO films doped with different percentage of N_2 in sputtering gas

and 75 % N_2) doped showed the preferential orientation of (100) plane (Fig. 2). Other authors observed the same change of prefer orientation with the increase of the dopant concentration [4, 5]. A decrease in the intensity of (002) peak of ZnO and the increase of the (100) peak with an increase of N_2 content might be due to the higher concentration of NO and the decreasing of oxygen vacancies (V_O) into the films, which are detrimental to the crystallinity of ZnO films. Corresponding parameters of real structure ZnO:N: the average grain size was in the range 7 – 42 nm for all Ar/N_2 ratios; average microstrains were relatively high (10×10^{-3}).

SIMS depth profile of as grown ZnO:N thin films, deposited in 75% N_2 in Ar/N_2 sputtering gas shows NO and NO_2 incorporated into the films (Fig. 3). It is clear that the

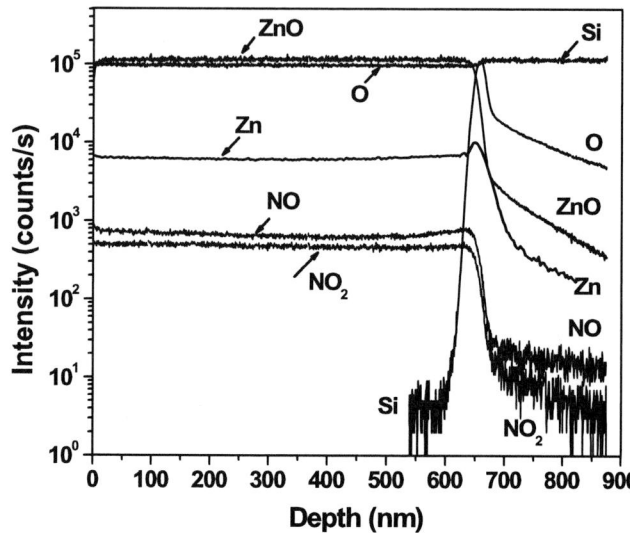

Fig. 3. SIMS depth profiles of ZnO:N grown in 75 % N_2 in the sputtering gas

layer is doped with N as intended, although we cannot get the absolute value of the concentration of N. The p-type conductivity of our ZnO thin films can be attributed to the presence of NO molecules. The calculations of Lee et al. show that NO molecules will introduce low-formation energy N_O, which is an acceptor in ZnO [2].

The electrical properties of N-doped ZnO thin films as a function of nitrogen percentage in Ar/N_2 sputtering gas are shown in Table 1 and Fig. 4a, b.

Results of Hall measurements of films prepared at different dopant concentration show a strong dependence of the electrical properties on dopant concentration. The undoped ZnO thin film (0 % N_2 in sputtering gas) was not measurable and ZnO films deposited in 10 % N in the sputtering gas showed a high resistivity (5.4×10^4 Ωcm). The p-type conductivity was recorded at ZnO:N samples deposited in 25 %, 75 % and 100 % nitrogen in the sputtering gas. The conductivity of the sample deposited at 50 % nitrogen was rather controversial and statistically was determined to be n-type.

The film's resistivity decreases with the increase of the percentage of nitrogen from 25 % to 75 %. The further increase of nitrogen percentage in the sputtering gas from 75 % to 100 % leads to the increase of the resistivity. p-type ZnO films show the lowest resistivity of 7.9×10^2 Ωcm, the highest Hall mobility of 22 $cm^2V^{-1}s^{-1}$ and the carrier concentration of P ~ 3.6×10^{14} cm^{-3}. The lowest resistivity is a result of the high mobility. The highest carrier concentration P ~ 2.6×10^{15} cm^{-3} is recorded for 25 % N_2 in the sputtering gas.

TABLE I.

ELECTRICAL RESISTIVITY ρ, HALL CONSTANT R_H, HALL MOBILITY μ_H, CARRIER CONCENTRATION P OR N, GRAIN SIZE D AND MICROSTRAIN ε OF AS GROWN ZnO:N FILMS DEPOSITED ON CORNING GLASS SUBSTRATE

ex. no	H [nm]	N$_2$ [%]	ρ [Ωcm]	R_H [cm^{3}C^{-1}]	μ_H [cm^{2}V^{-1}s^{-1}]	P, N [cm^{-3}]	D [nm]	ε x 10^2 [-]
8	600	0	---	---	---	---	42	1.01
9	870	10	5.4 x 10^4	---	---	---	41	1.40
10	840	25	1.5 x 10^3	2.4 x 10^3	2	P ~ 2.6 x 10^{15}	9	0.77
11	710	50	7.0 x 10^2	4.7 x 10^3	7	N ~ 1.3 x 10^{15}	12	0.82
12	660	75	7.9 x 10^2	1.7 x 10^4	22	P ~ 3.6 x 10^{14}	10	0.69
13	580	100	2.8 x 10^3	3.5 x 10^3	1	P ~ 1.8 x 10^{15}	11	0.23

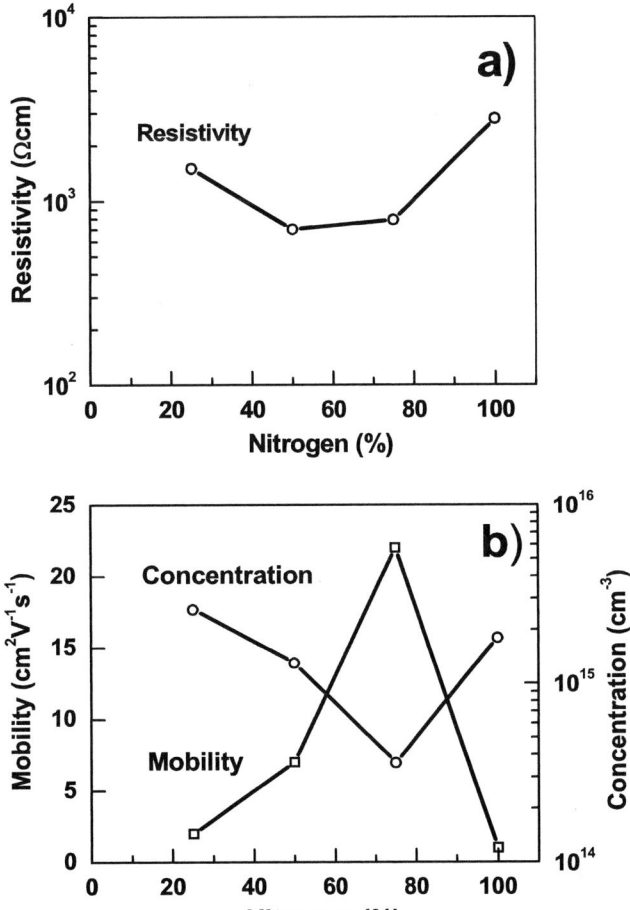

Fig. 4 Dependence of a) resistivity, b) Hall mobility and carrier concentration on nitrogen percentage for as grown ZnO:N films

The strong dependence of the electrical properties of sputtered ZnO:N films on dopant content can be explained with the different scattering mechanism of free carriers [6]. Films with small dopant content exhibit good crystal structure in terms of grain size and grain orientation. The dopant atoms are effectively incorporated in the ZnO lattice but their concentration is low. Thus ionized impurity scattering becomes dominant in comparison to grain boundary scattering. Hence a low carrier mobility and a low hole concentration were obtained at samples deposited with 25% N$_2$ in the sputtering gas. The increasing of the dopan content in ZnO films introduce furthers scattering centers and thus the decrease of carrier concentration at high dopant levels. The increase of the carrier concentration in films deposited at 100% N$_2$ in the sputtering gas can be a result of the high concentration of NO incorporated into the films. The compensation effect by N$_2$ molecules at O sites ((N$_2$)$_O$) and N$_O$ - (N$_2$)$_O$ complexes can be responsible for non-stable conductivity types of the film deposited at 50% N$_2$ in the sputtering gas [4].

Josef et al. reported p-type ZnO:N with a hole concentration of only 2 x 10^{10} cm^{-3} for ZnO:N films [3]. Nitrogen doped p-type ZnO films obtained from Z.-Z Ye et al. [7] had the lowest resistivity of 1.2 x 10^3 Ωcm, Hall mobility of 84.9 cm^{2}V^{-1}s^{-1} and the carrier concentration of 6.0 x 10^{13} cm^{-3}. These results are ascribed to the low concentration of NO. Thus only low concentration p-type ZnO can be achieved [8]. The electrical parameters of our ZnO:N films (the lowest resistivity of 7.9 x 10^2 Ωcm, and the highest carrier concentration P ~ 2.6 x 10^{15} cm^{-3}) confirm the incorporation of NO and p-type features of ZnO thin films prepared by RF diode sputtering.

IV. CONCLUSION

We synthesized p-type ZnO thin films on Corning glass substrate using nitrogen as a dopant by RF diode sputtering. ZnO films were prepared in Ar/N$_2$ sputtering gas and the percentage of N$_2$ varied from 0 % to 100 %. The hole carrier concentration, Hall mobility and resistivity are in the range of 10^{14} - 10^{15} cm^{-3}, 1 – 22 cm^{2}V^{-1}s^{-1}, and 7 x 10^2 Ωcm –2.8 x 10^3 Ωcm. SIMS analysis proved the incorporation of N$_O$ acceptor in as grown ZnO thin films. RF diode sputtering is deposition process conditions of which significantly differ from thermodynamic balance. Therefore we can expect that p-type properties of RF sputtered ZnO:N films will be metastable and highly dependent on nitrogen content in sputtering gas. In order to improve p-type properties of ZnO thin films will be necessary to combine nitrogen doping with other co-dopants.

ACKNOWLEDGEMENT

Presented work was supported by the SK Grant VEGA 1/3098/06 of the Slovak Grant Agency and in part by PPP Programme project DAAD 5/2005.

REFERENCES

[1] G. Xiong, J. Wilkinson, B. Mischuck, S. Tuzemen, K.B. Ucer, and R.T. Williams, "Control of p- and n-type conductivity in sputter deposition of undoped ZnO", *Applied Physics Letters*, 2002, vol. 80, pp. 1195-1198.

[2] E.Ch. Lee, Y.S. Kim, Y.G. Jin, and K.J. Chang, "First principles study of the compensation mechanism in N-doped ZnO", *Physica B*, 2001, vol. 308-310, pp. 912-915.

[3] M. Joseph, H. Tabata, H. Saeki, K. Ueda, and T. Kawai, "Fabrication of the low-resistive p-type ZnO by codoping method", *Physica B*, 2001, vol. 302–303, pp. 140–148.

[4] C. Zhang, X. Li, J. Bian, W. Yu, and X. Gao, "Structural and electrical properties of nitrogen and aluminum codoped p-type ZnO", *Solid State Communications*, 2004, vol. 132, pp. 75-78.

[5] F. Zhuge, L.P. Zhu, Z.Z. Ye, J.G. Lu, B.H. Zhao, J.Y. Huang, L. Wang, Z.H. Zhang, and Z.G. Ji, "Effects of growth ambient on electrical properties of Al–N co-doped p-type ZnO films", *Thin Solid Films*, 2005, vol. 476, pp. 272– 275.

[6] B. Szyszka, S. Jäger, J. Szczyrbowski, and G. Bräuer, "Compariso of transparent conductive oxide thin films prepared by a.c. and d.c. reactive magnetron sputtering", *Surface and coatings Technology*, 1998, vol. 98, pp. 1304-1314.

[7] Z.Z. Ye, F.Z. Ge, J.G. Lu, Z.H. Zhang, L.P. Zhu, B.H. Zhao, and J.Y. Huang, "Preparation of p-type ZnO films by Al+N co-doping method", *Journal of Crystal Growth*, 2004, vol. 265, pp. 127–132.

[8] S.B. Zhang, S.H. Wei, and Y. Yan, "The thermodynamics of codoping:how does it work?", *Physica B*, 2001, vol. 302-303, pp. 135-139.

2006 25th International Conference on Microelectronics

Investigation of Metal-Polycrystalline Silicon Carbide Bonding While Metallization

G. Golan, V. Manevych, I. Lapsker, B. Gorenstein, and A. Axelevitch

Abstract - Polycrystalline silicon carbide heaters o rheated substrates are widely used within the semiconductor industry. The problem of making reliable contacts between such SiC and various metals is most relevant. The main goal of our investigation was an experimental study of molten metals (Fe, Cu, Cr) behavior on top of surfaces of polycrystalline silicon carbide SiC. The mechanism of melt-polycrystalline SiC interaction was found and reported. Non-wetting metal in a liquid phase penetrates into the micro and macro volumes in the poly-crystalline SiC surface and holds there due to the residual stresses originated by the difference in the linear expansion coefficients. Metal layers obtained on the poly-crystalline SiC surfaces by the described method were durable and stable.

I. INTRODUCTION

Polycrystalline silicon carbide is widely used in the semiconductor industry. A major condition for silicon carbide application is a reliable contact between the SiC and electrodes plug on it within the electronic circuit. It is well known that welding and bonding the electrodes will ensure an effective contact. Therefore, silicon carbide surface behavior in a contact with a molten metal is most interesting.

The main goal of our work was to study the joint behavior of molten metals (Fe, Cu, Cr) on top of polycrystalline silicon carbide heater or substrate as part of the connection process.

II. EXPERIMENTAL DETAILS

The Polycrystalline SiC for our experiments were prepared by vacuum sintering of SiC powder pressed with bakelite with surplus silicon. This technology is faultless by separating the free surplus Si on the surfaces of SiC crystallites. Experimental samples were of (3-5)×(3-5)×30 mm^3 and had a density of ~2.5-2.7 g/cm^3. The metals (Fe, Cu, Cr) were applied as powders and melt on the SiC surface in air and in vacuum.

Heating of metal powders was provided up to melting with a heating rate of 2-50 ^{0}C/sec in air and 500-1000 ^{0}C/sec in vacuum of ~5×10^{-5} Torr. Figure 1 illustrates the laboratory setup used for our experiments. The samples were cooled down and annealed by in free air; The

G. Golan, V. Manevych, I. Lapsker, B. Gorenstein, and A. Axelevitch are with the Department of Electrical and Electronics Engineering, Holon Academic Institute of Technology, 52 Golomb St., P.O.Box 305, 58102 Holon, Israel, E-mail: golan@hait.ac.il

temperature of the Cu sample at the first second was of about 170 ^{0}C.

Surfaces of the samples after the treatment were studied using an optical microscope "Optihot-100" and a scanning electron microscope (SEM) "Stereoscan 430" equipped with the energy-dispersive analyzer EDX.

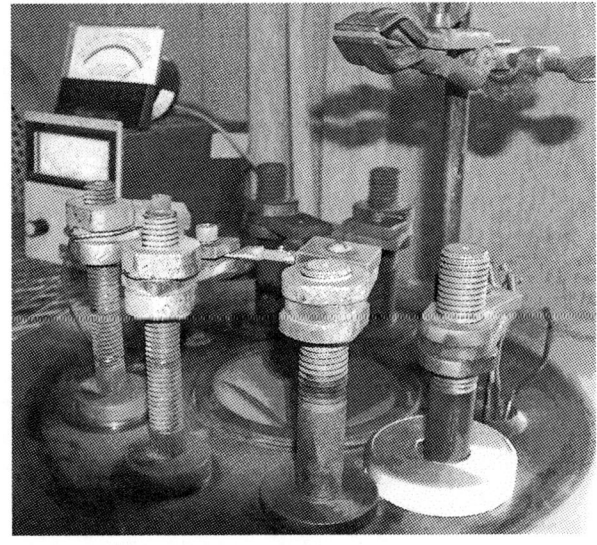

Fig. 1. Internal view of the laboratory vacuum setup.

The chemical composition of the samples was studied using EDX in a depth of 3 μm. Figure 2 presents a schematic view of the sample cleaving system to examinee the SiC – metal adhesion forces. The samples were cleaved and the cleaved samples were studied using SEM and EDX.

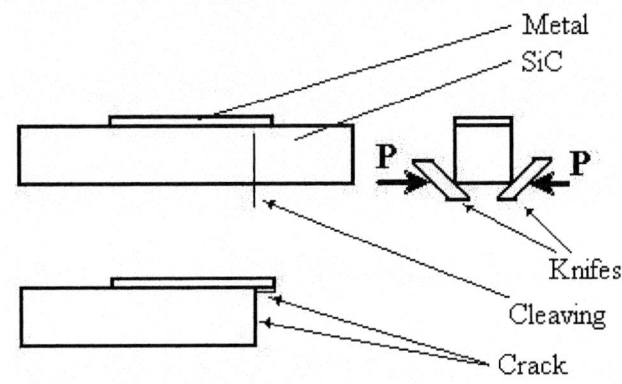

Fig. 2. Schematic view of the sample cleaving system.

III. EXPERIMENTAL RESULTS AND DISCUSSIONS

Figure 3 presents an external view of the polycrystalline silicon carbide sample. One can see that the sample consists of micro-crystals of about 3-40 μm. Micro-relief of the surface sample had ledges and cavities formed by micro-crystals.

Fig. 3. Polycrystalline silicon carbide, an upper view.

Metal powder in contact with SiC begins to melt while heating. This heat is transferred from the SiC to the metal. Figure 4 shows a broken surface microphotography of iron powder heated on top of a SiC surface up to the agglomeration.

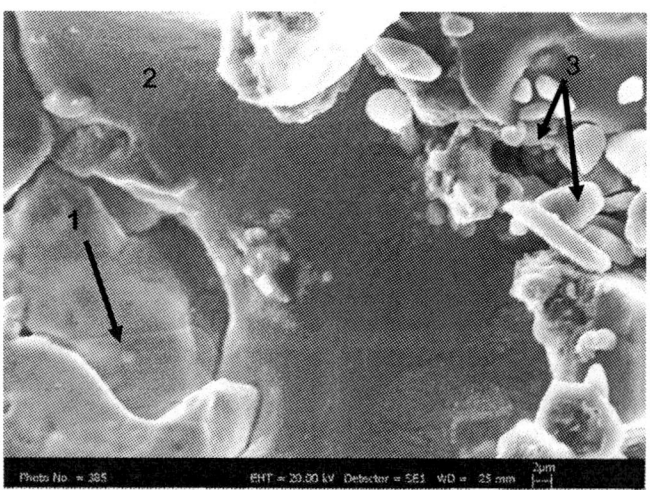

Fig. 4. Broken surface microstructure: (1) Metal powder particles, (2) Liquid phase of metal, (3) Silicon carbide.

The liquid phase (2) appeared between the SiC surface (3) and the Fe powder (1) in a thickness of 10-20 μm. The iron particles placed along the surface are compacted and deformed due to heating. The melt does not wet the SiC surface and does not form surface bonding between two materials. However, the melt holds on its surface due to wetting of the own solid phase (powder). Thus, the melt is not coagulate in the vicinity of the SiC. Overheating of 50-100⁰C above the melting point of a metal, results in coagulation as shown in figure 5.

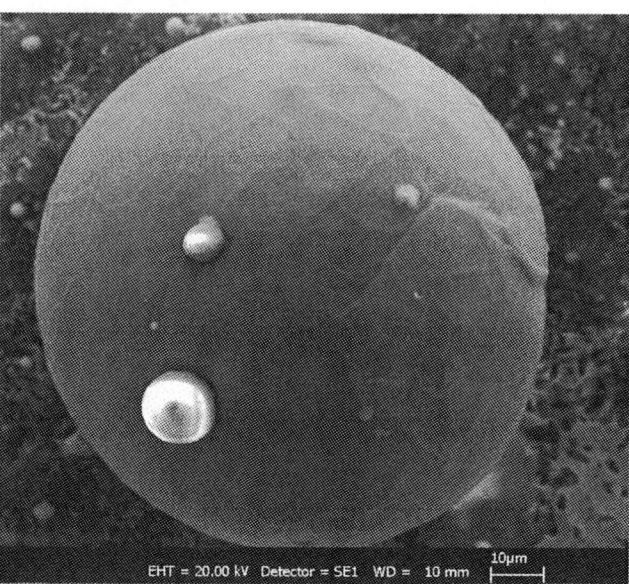

Fig. 5. Coagulated Fe melt on a surface of SiC.

The coagulated drops were of various dimensions ranging from ~0.1mm up to 1-2 mm as shown in figure 5. On the spherical surface of the coagulated Fe drop there were found additional small drops of more fusible compounds. EDX analysis showed the follows composition of these small drops: Fe-1.6%, Si-53%, Ca-23%, Ti-0.4%. Analysis of the wetting angle between the drops and the SiC surface showed an angle of 120-160⁰. Thus, coagulation is a result of internal bonding forces within a metal exceeding the bonding forces between two different materials. This wetting angle depends on the formation time and may be much more than the value estimated using the Young's equation [1] :

$$\cos\theta = (\sigma_{sv} - \sigma_{ms})/\sigma_{mv} , \qquad (1)$$

here θ is the wetting angle, σ is an interfacial tension of forces, and m, s and v represent the metal, substrate (SiC), and vapor, respectively.

The melted metal in the first stage of heating holds on the SiC surface due to wetting of self solid phase. Following a complete melting, a liquid metal begins to coagulate in a drop form and moves on to the SiC substrate, having surface tension forces. These drops penetrate in the SiC pores and fill them. After the heat stops, a surface joining process begins. Cooling starts from the substrate, thus the liquid metal nearest the SiC begins to crystallize and the viscosity of the liquid increases from the substrate side. Due to this high viscosity, the metal drops penetrate into the SiC surface and hold inside surface cavities. Also, cooling results in residual tensions. Two basic mechanisms are causing this residual tensions: metal volume decrease due to crystallization (shrinkage) and linear expansion coefficients difference for metal and SiC. For example, the densities of liquid and solid phases of Fe are 7 g/cm^3 and 7.85 g/cm^3, consequently [2]. The shrinkage leads to compressive stresses within the SiC substrate and tensile stresses within the metal layer. The compressive stress relates to the lower linear expansion coefficient [3]. Both these two mechanisms contribute to the stresses at the same direction. Thus, the tensile forces may reach up to significant values and even breaking point, for specific metals. Figure 6 presents a metal surface covered by cracks net. These cracks demonstrate that the tensile forces are greater than the breaking point for the Cr layer (300 MPa) [4].

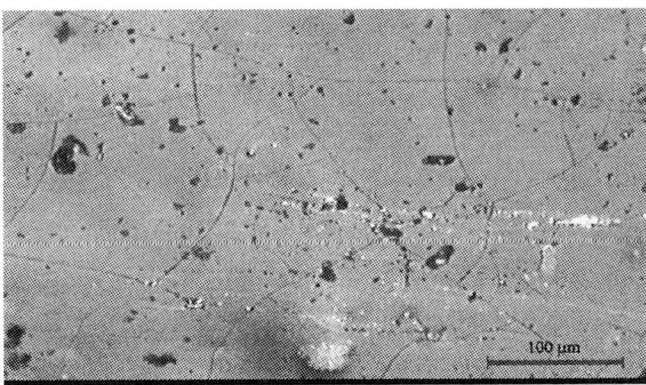

Fig. 6. Cracked metal layer.

Figure 7 presents a sample of metallized SiC. This photography shows the fragile type of at a destruction pont.

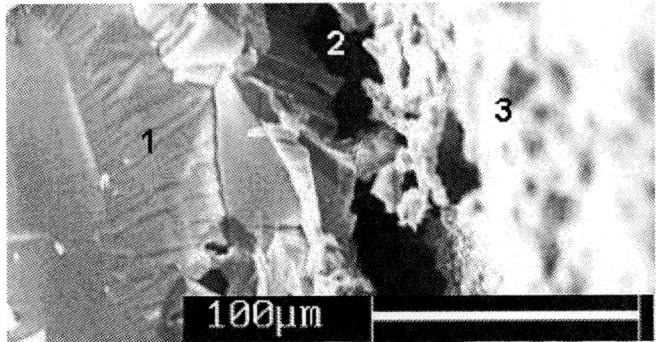

Fig. 7. A metallized SiC sample: (1) broken surface, (2) finite dimensions fracture, (3) residual part of the broken substrate.

A broken surface presents a specific type of fragile breaking: material "rivers" [5] exist in this surface; they flow and join in the direction of fracture expansion. Figure 8 illustrates the origin of a fracture. The fracture begins from the knife and expands to the interface between the SiC substrate and the metal layer. Near the interface, the

fracture meets a compressive stress fields which complicate the fracture evolution. The fracture changes the expansion direction and forms several branches of cracks. Part of these branches form finite dimensions fractures, and a residuary part, which develops in the short side, comes out.

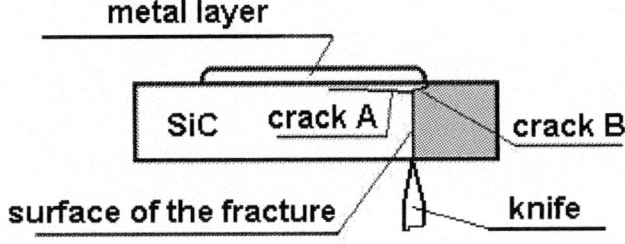

Fig. 8. A fracture origin diagram.

Figure 9 presents a chip surface coated with metal layer after breaking. Comparing this picture with the cracking diagram in Fig. 8 shows the surface of the fracture and the upper part of the fracture adjacent to the metal coating. One can see there the SiC crystallites on these surfaces.

Fig. 9. SiC sample with a metal layer after breaking.

A break develops inside the silicon carbide. The adhesion forces between SiC and metal layer are greater than the internal bonds of silicon carbide.

Two regions A and B are marked out in Fig. 9. Enlarged micro-photography of these regions is presented in figures 10 (a, b). These pictures present microcrystallites of SiC at various magnifications. This confirm our thoughts on excess adhesion forces between the metal layer and SiC versus internal bonding forces within the silicon carbide.

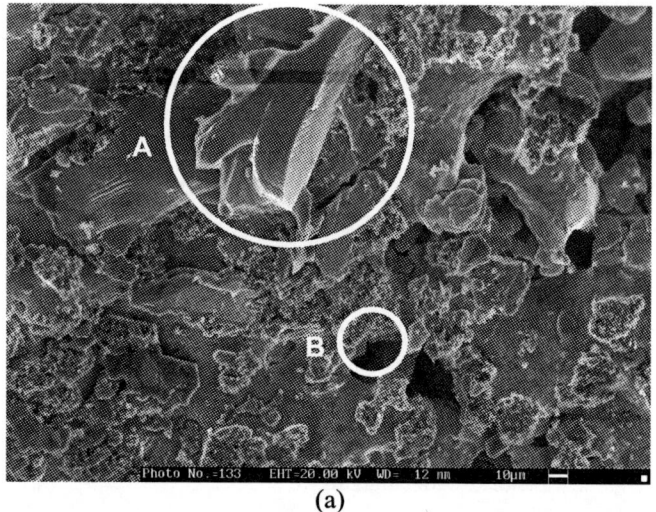

(a)

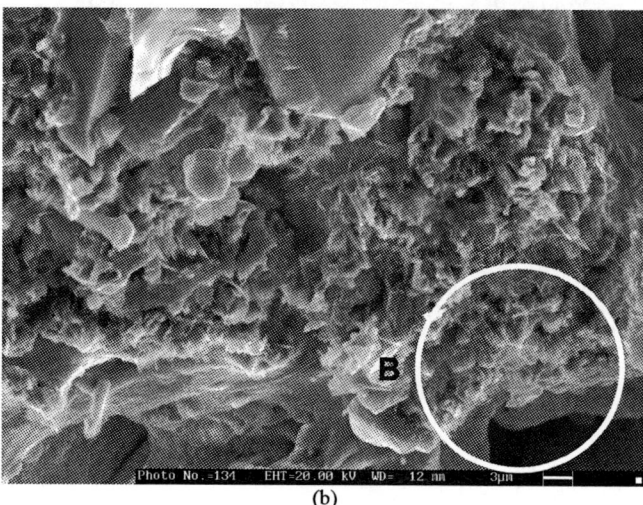

(b)

Fig. 10 (a, b) present the breaking surfaces of the sample with various magnifications. The marked regions A and B connect these photos with Fig. 9.

High adhesion forces between the metal layer and the SiC substrate are defined by residual stresses fields. These fields are originated by variations between the linear expansion coefficients in metals and silicon carbide. This difference leads to compressive stresses within the SiC substrate and tensile stresses in the metal layer. The joining action of these forces results in strong and reliable connection between the metal coating and the silicon carbide substrate.

IV. CONCLUSION

Experimental study of melted metals behavior in polycrystalline silicon carbide surface showed the follows:
1. Iron melt reacts with free surplus silicon to form FeSi (a silicide) alloy;
2. FeSi coagulates on the SiC surfaces in a macro- and micro-volumes due to nonwettability;
3. Iron penetrates into the inter-crystalline voids of the SiC at depth of 1-2 crystallites of middle dimensions;
4. The mechanical connection of the metal layer with the silicon carbide surface occurs due to residual stresses fields originated due to differences in the linear expansion coefficients of the joining materials;
5. Iron coating has the highest adhesion to the silicon carbide surface. Sample destruction while cleaving occurs via the bulk material.

REFERENCES

[1] M. Ohring, *Material Science of Thin Films, Deposition and Structure*, San Diego, Academic Press, 2002.
[2] V.A. Kudrin, *Steel Metallurgy*, Moscou, Metallurgy, 1989 (in Russian).
[3] S. Timoshenko, *Strength of Materials: part 1, elementary theory and problems*, New York, Krieger, 1976.
[4] *Handbook of Chemistry and Physics*, R.C. Weast and S.M. Selby, Eds., Cleveland, The Chemical Rubber Co., 1967.
[5] D. Broek, *Elementary engineering fracture mechanics*, Leyden, Noordhoff International Publishing, 1974.

2006 25th International Conference on Microelectronics

PVD Aluminium Nitride as Heat Spreader in Silicon-on-Glass Technology

L. La Spina, H. Schellevis, N. Nenadović, and L. K. Nanver

Abstract - Physical-vapor-deposited aluminium nitride has been integrated in a silicon-on-glass NPN BJT process. Deposition conditions have been developed for which suitable electrical, mechanical and thermal properties are achieved. Electrothermal device characterization is used to demonstrate the effective heat spreading of this thin-film material.

I. INTRODUCTION

To enhance the speed performance of modern silicon devices and circuits, different electrical insulating techniques are being used in combination with aggressive device size reduction. The silicon-on-glass technology [1] represents a radical example of this approach, where transistors are fabricated in very small silicon islands fully surrounded by electrical insulators, like glass and silicon oxide, as can be seen in Fig. 1. Unfortunately, such devices are subject to very strong electrothermal feedback [2], [3] due to poor thermal conductivity of the surrounding materials. The inability to remove the heat from the active components leads to effects such as thermal breakdown, which is recognized to be one of the major limiting factors for performance of high frequency (HF) silicon devices [4]. A possible way to reduce electrothermal effects is to integrate materials that are at the same time good thermal conductors and electrical insulators. Due to its electrical and thermal properties [5], aluminium nitride (AlN) is a candidate for heat spreading in IC applications. It is neither contaminating nor poisonous, unlike other materials, such as e.g., beryllia, and can be both deposited and etched by conventional silicon processing techniques.

This paper examines the viability of using physical-vapor-deposited (PVD) AlN as an integrated heat spreader in silicon-on-glass technology. The main compatibility issues are shown to be related to the material stress, electrical properties, etch rates and selectivity. The material properties of the AlN in thin film form have been investigated through appropriate analysis and electrical measurements. The compatibility with silicon IC processing is demonstrated through successful integration in an in-house bipolar process. The NPN transistor characteristics clearly demonstrate the effectiveness of the

L. La Spina, H. Schellevis, and L. K. Nanver are with the Laboratory of ECTM, DIMES, Delft University of Technology, Feldmannweg 17, P.O. Box 5053, 2600 GB Delft, The Netherlands. E-mail: l.laspina@tudelft.nl

N. Nenadović was with ECTM, DIMES, Delft University of Technology. He is currently with Philips Semiconductors CTO, 6534 AE Nijmegen, The Netherlands.

AlN heat spreaders. In fact, due to the high sensitivity of the silicon-on-glass devices respect to the electrothermal feedback, they represent an ideal test vehicle for the heat spreading capability of this material.

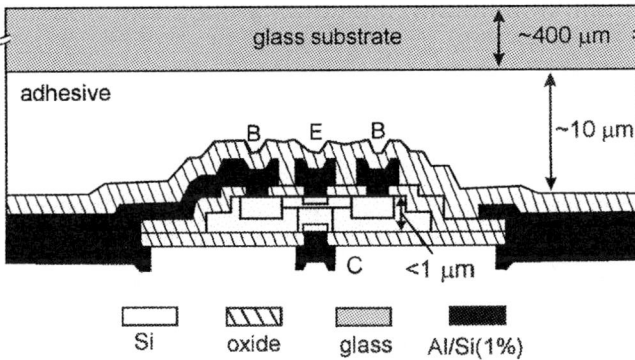

Fig. 1. Schematic cross section of the reference silicon-on-glass NPN BJT.

II. FABRICATION PROCESS

Thin films of AlN are deposited at a substrate temperature of 300 °C and pressure of 5 mtorr, using a pure Al target. Nitrogen incorporation is achieved with a N_2 flow of 75 standard cubic centimeters per minute (sccm) and argon flow of 38 sccm. The stress exerted from the AlN layers can be manipulated by varying different parameters. For layers of less than 1 μm thick, a pulsed-DC power of 2 kW with a pulse width of 1616 ns is used. For a thickness of 800 nm, a tensile stress of about 400 MPa will result. With the same processing parameters and the addition of a 20 W RF biasing, a compressive stress of about −500 MPa is obtained, as reported in the Table I. Thus, to reduce the stress, the AlN layers thicker than 1 μm are built up of alternating 0.2-μm-thick layers, the one deposited by pulsed DC-sputtering and the other with the addition of RF biasing. With this technique it is possible to deposit up to 4-μm-thick almost free-stress AlN layers on fully processed wafers. The processing is completed with a 400 °C 90 min long anneal to give the necessary outgassing of the material and hence avoid bubble formation, which can be very detrimental for the bonding process used in the silicon-on-glass technology.

A wide range of etchants has been tested on the deposited AlN layers. Some typical etch rates of wet etchants are listed in Table II. However, for the dimensional control of window etching in AlN layers, dry etching in Cl_2/HBr chemistry is preferred. Thick resist

1-4244-0116-X/06/$20.00 ©2006 IEEE

masking layers should be used since the etch rates for resist and AlN are respectively 306 nm/min and 188 nm/min, i.e., the selectivity is 1.6:1.

TABLE I
DEPOSITION PARAMETERS AND RESULTING MATERIAL PROPERTIES OF TWO DIFFERENT PVD AlN LAYERS.

	Layer A	Layer B
RF Power [W]	0	20
Freq. [kHz]	250	250
Pulse Width [ns]	1616	1616
Stress [MPa]	+407	-510
d_{AlN} [nm]	864	841
n_f (λ=365 nm)	2.17	2.20
n_f (λ=633 nm)	2.12	2.13

TABLE II
ETCH RATES OF AlN FOR DIFFERENT WET ETCHANTS.

Solution	Temp [°C]	Etch rate [nm/min]
0.55% HF	20	5.6
BHF 1:7	20	13.5
CrO_3/H_3PO_4	70	60
TMAH 25%	85	4100
H_3PO_4 4.25%	80	93.9

III. MATERIAL PROPERTIES

The electrical properties of bulk AlN are typically those of an insulating material. The electrical resistivity for the AlN substrates is, depending on the process technology, between $4 \times 10^{11} \ \Omega \cdot cm$ [6] and $5 \times 10^{13} \ \Omega \cdot cm$ [7].

The deposited AlN thin film investigated in this work, has clearly a granular structure, as seen in the TEM image shown in Fig. 2. The electrical properties are strongly influenced by the grain structure and can have directional dependencies. In order to measure the vertical and lateral electrical resistivity of the deposited AlN, appropriate structures were designed. For extraction of the vertical resistivity, several Al-AlN-Al capacitors, with different aluminium nitride thickness, were fabricated and characterized. The measured electrical resistivity is in the order of $10^{12} - 10^{14} \ \Omega \cdot cm$, depending on the thickness of the material.

To extract the lateral resistivity of the AlN layers, multifinger metal structures with a wide range of pitches and metal thicknesses were fabricated and covered with AlN layers of a thickness up to 0.8 μm. To avoid hillocks formation during thermal processing of the first Al layer, this layer was capped by a 100-nm-thick TiN layer. The lateral resistivity was extracted using the four probes method and found to be higher than $10^{11} \ \Omega \cdot cm$.

The measured dielectric constant of the AlN layers is about 8 at 1 MHz. As comparison, note that the dielectric constant is 3.9 for SiO_2, which is one of the most common insulating materials employed in microelectronics field, and it is 6.5 and 7.9 for SiN and Al_2O_3 respectively [8].

The refractive index n_f, measured at two different wavelengths λ on the Sopra ES4G ellipsometer, is given in Table I for two AlN layers, deposited in different conditions, and it is seen to be about 2.1 - 2.2. These high values indicate a good crystalline rather than amorphous structuring of the layers [9], and they are in agreement with the results given in [10] and [11].

The thermal conductivity k_{TH} of bulk materials can be very different from that of the corresponding thin film. The AlN theoretically has a thermal conductivity of $320 \ Wm^{-1}K^{-1}$ [12], while the maximum experimental value reported for the bulk material is $270 \ Wm^{-1}K^{-1}$ [13]. However, the k_{TH} of AlN film is much lower, in the range of $0.4 - 26 \ Wm^{-1}K^{-1}$ [14], [15], since it depends on the deposition process details, grain size and shape, film thickness and impurity concentration. The lateral thermal conductivity of the AlN film is extracted from a dedicated test structure [16], and it is around $11 \ Wm^{-1}K^{-1}$. This value is high compared to other materials used in IC technology. For example, the silicon dioxide SiO_2 has a k_{TH} of about $1.4 \ Wm^{-1}K^{-1}$, and silicon nitride films have a value in the order of $1.5 - 2 \ Wm^{-1}K^{-1}$ [17], [18].

Fig. 2. TEM image of a 2-μm-thick AlN layer deposited on 100 nm thermal oxide.

IV. INTEGRATION IN SILICON-ON-GLASS PROCESS

The PVD AlN layers are integrated in the silicon-on-glass bipolar process in the manner shown in Fig. 3:

(a) The processing starts on silicon-on-insulator wafers with a 0.94 μm n-doped top silicon layer. The active device areas are isolated by means of trenches etched to the 0.4-μm-thick buried oxide. A pedestal collector region is defined by a phosphorus implant and extrinsic base regions by low-dose deep boron implants that compensate the n-epi. The emitter and base are processed and contacted by applying standard processing procedures for the front side of the wafer (front-wafer). This is followed by a deposition of the first 0.8-μm-thick layer of PVD AlN that thus has direct contact to aluminium metallization. In principle,

366

since no further patterning of this front-wafer AlN layer is necessary, any layer thickness with tolerable stress levels could be used. The front-wafer processing is completed by depositing a 1-μm-thick PECVD SiO_2 layer that is necessary to guarantee the adhesion of the subsequent substrate transfer gluing process.

(b) After gluing, the silicon substrate is removed, followed by the processing of the backside of the wafer (back-wafer). Front- to back-wafer high-precision waferstepper alignment [19] is used to align minimum-dimension contact windows on the back-wafer to the front-wafer. Low-ohmic contacts to the collector are formed on the back-wafer by arsenic implantation, dopant activation by laser annealing and metallization [1]. The front-wafer bondpads are then opened from the back-wafer and metallized.

(c) Next a second 0.8-μm-thick layer of AlN is deposited directly on the back-wafer metal and patterned to allow contacting to the bondpads.

(d) To further improve the heat-sinking capabilities, a 4-μm-thick copper layer is electroplated on the back-wafer.

V. ELECTROTHERMAL DEVICE BEHAVIOR

The resulting devices summarized in Table III are used to experimentally monitor the heat spreading and sinking capabilities of the AlN. Emitter-current-controlled I_C-V_{BE} measurements for a fixed V_{CB} are performed as shown in Fig. 4. The starting point of negative differential resistance represents the onset of thermal instability, which can be used for accurate calculation of the thermal resistance R_{TH} [20]. From Fig. 4 and R_{TH} values from Table III, it is clear that even very thin AlN layers are effective in spreading the heat away from the active device areas. In particular, this is the case for the device from Fig. 3(d1) where very good vertical conduction of AlN is combined with the heat sinking capability of Cu, which is electroplated directly below the active device area. Moreover, the lateral thermal conductivity of deposited layers is investigated by splitting the Cu heat sink in two parts and moving them away from the active device area. Experiments show that, even when the distance L between the two Cu parts is 34 μm (device Fig. 3(d2)L34), a reduction of the 50% in R_{TH} compared to the reference device (see Table III) is obtained. Thus, unlike SiO_2 layers [21], the AlN is very effective in transferring the heat to the Cu heat sink.

VI. CONCLUSION

PVD aluminium nitride layers have been developed. The deposition conditions are tuned to guarantee good properties of the layers. For 0.8-μm-thick layers, the electrical resistivity is found to be higher than $10^{13} \, \Omega \cdot cm$, the dielectric constant is about 8, and the lateral thermal conductivity is around 11 $Wm^{-1}K^{-1}$. The high value of the refractive index (about 2.1-2.2) measured at different wavelengths indicates a good poly-crystalline quality of the layers. The mechanical stress has been prevented and the AlN film has been integrated in fully processed silicon-on-glass NPNs. The results demonstrate that thin-film layers of PVD AlN can be used with profit for heat spreading in silicon integration processes.

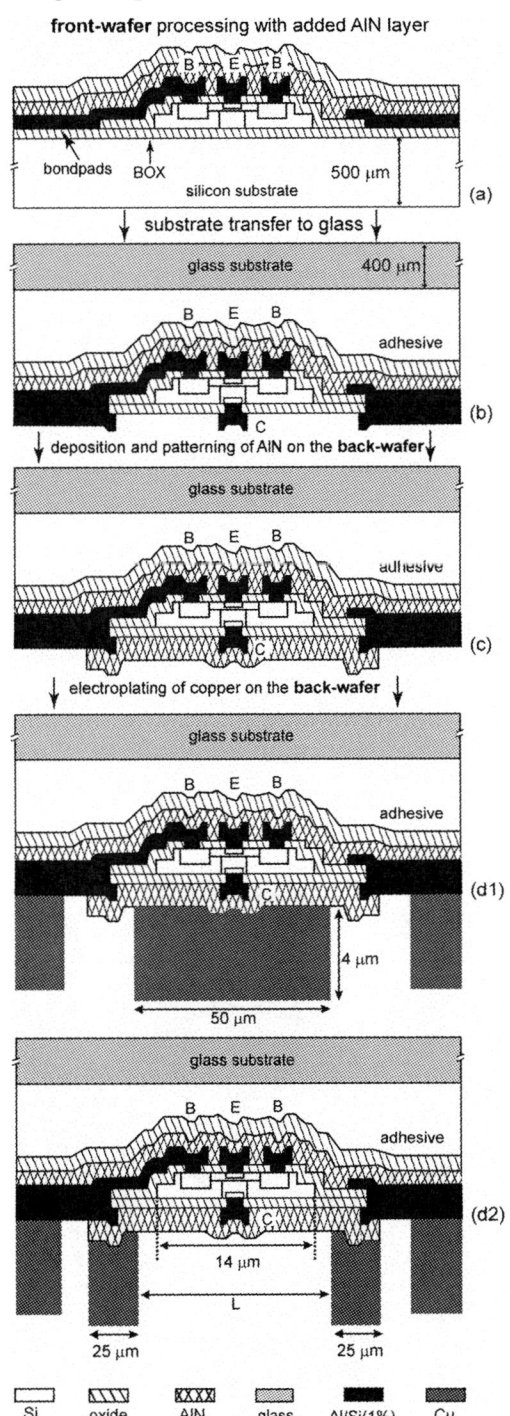

Fig. 3. Schematic of the process flow for the fabrication of the silicon-on-glass NPN's with AlN depositions on the front- and back-wafer as well as Cu-plating of the final AlN layer.

TABLE III
DIFFERENT TEST NPNS AND CORRESPONDING R_{TH}

Device	AlN on front-wafer [μm]	AlN on back-wafer [μm]	L [μm]	R_{TH} [K/W]
Fig. 1	0	0	no Cu	12600
Fig. 3(b)	0.8	0	no Cu	8800
Fig. 3(c)	0.8	0.8	no Cu	7550
Fig. 3(d1)	0.8	0.8	0	3600
Fig. 3(d2)L18	0.8	0.8	18	5600
Fig. 3(d2)L24	0.8	0.8	24	6500
Fig. 3(d2)L34	0.8	0.8	34	6530

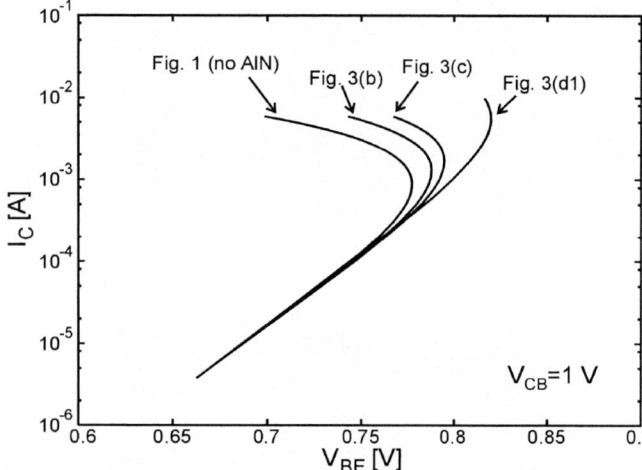

Fig. 4. Emitter-current-controlled Gummel plots of devices with different heat spreaders.

REFERENCES

[1] L. K. Nanver, N. Nenadović, V. d'Alessandro, H. Schellevis, H. W. Van Zeijl, R. Dekker, D. B. De Mooij, V. Zieren, and J. W. Slootboom, "A Back-Wafer Contacted Silicon-On-Glass Integrated Bipolar Process – Part I: The Conflict Electrical Versus Thermal Isolation," *IEEE Trans. Electron Devices*, 51 (1): 42-50, 2004.

[2] K. Lu and C. M. Snowden, "Analysis of Thermal Instability in Multi-Finger Power AlGaAs/GaAs HBT's," *IEEE Trans. Electron Devices*, 43 (11): 1799-1805, 1996.

[3] J.-S. Rieh, J. Johnson, S. Furkay, D. Greenberg, G. Freeman, and S. Subbanna, "Structural Dependence of the Thermal Resistance of Trench-Isolated Bipolar Transistors," in *Proc. IEEE BCTM*, pp. 100-103, 2002.

[4] N. Nenadović, L. K. Nanver, and J. W. Slotboom, "Electrothermal limitations on the current density of high-frequency bipolar transistors," *IEEE Trans. Electron Devices*, 51 (12): 2175-2180, 2004.

[5] S. Strite and H. Morkoç, "GaN, AlN, and InN: A review," *J. Vac. Sci. Technol. B*, 10 (4): 1237-1266, 1992.

[6] W. Werdecker and F. Aldinger, "Aluminum Nitride-An Alternative Ceramic Substrate for High Power Applications in Microcircuits," *IEEE Trans. Comp., Hybrids, Manufact.*, 7 (4), 1984, pp. 399-404.

[7] Y. Kurokawa, K. Utsumi, H. Takamizawa, T. Kamata, and S. Noguchi, "AlN Substrates with High Thermal Conductivity," *IEEE Trans. Comp., Hybr., Manifact., Technol.*, Vol. Chmt-8, No. 2, 1985, pp. 247-252.

[8] K.-H. Allers, P. Brenner, M. Shrenk, "Dielectric reliability and material properties of Al_2O_3 in metal insulator metal capacitors (MIMCAP) for RF bipolar technologies in comparison to SiO_2, SiN and Ta_2O_5," in *Proc. IEEE BCTM*, 2003, pp. 35-38.

[9] Properties of group III Nitrides, J. H. Edgard (Ed.), London, INSREC, 1994.

[10] F. Engelmark, G. Fuentes, I. V. Katardjiev, A. Harsta, U. Smith, and S. Berg, "Synthesis of highly oriented piezoelectric AlN films by reactive sputter deposition," *J. Vac. Sci. Technol. A*, 18 (4), 2000, pp. 1609-1612.

[11] V. Mortet, M. Nesladek, K. Haenen, A. Morel, M. D'Olieslaeger, M. Vanecek, "Physical properties of polycrystalline aluminium nitride films deposited by magnetron sputtering," *Diamond and related materials*, 13, 2004, pp. 1120-1124.

[12] G. A. Slack, "Nonmetallic crystals with high thermal conductivity," *J. Phys. Chem. Solids*, 1973, Vol. 34, pp. 321-335.

[13] K. Watari and S. L. Shinde, *MRS Bulletin*, June 2001, pp. 440-441.

[14] P. K. Kuo, G. W. Auner, Z. L. Wu, "Microstructure and thermal conductivity of epitaxial AlN thin films," *Thin Solid Films*, 253, 1994, pp. 223-227.

[15] J. W. Lee, J. J. Cuomo, Y. S. Cho, R. L. Keusseyan, "Aluminum Nitride Thin Films on an LTCC Substrate," *J. Am. Ceram. Soc.*, 88 (7), 2005, pp. 1977-1980.

[16] L. La Spina, N. Nenadović, A. W. van Herwaarden, H. Schellevis, W. H. A. Wien, and L. K. Nanver, "MEMS test structure for measuring thermal conductivity of thin films," *Proc. IEEE ICMTS*, 2006.

[17] A. Irace, P. M. Sarro, "Measurement of thermal conductivity and diffusivity of single and multi-layer membrane," *Sensors and Actuators*, Vol. A76, No. 1-3, 1999, pp. 323-328.

[18] M. von Arx, O. Paul, and H. Baltes, "Process-Dependent Thin-Film Thermal Conductivities for Thermal CMOS MEMS," *J. Microelectromech. Syst.*, Vol. 9, No. 1, 2000, pp. 136-145.

[19] H. W. van Zeijl and J. Slabbekoorn, "Front- to back-wafer overlay accuracy in substrate transfer technologies," in *Proc. ISTC*, Vol. 17, pp. 356-367, 2001.

[20] N. Nenadović, V. d'Alessandro, L. K. Nanver, F. Tamigi, N. Rinaldi, and J. W. Slotboom, "A Back-Wafer Contacted Silicon-On-Glass Integrated Bipolar Process – Part II: A Novel Analysis of Thermal Breakdown," *IEEE Trans. Electron Devices*, 51 (1): 51-62, 2004.

[21] N. Nenadović, V. Cuoco, S. J. C. H. Theeuwen, H. Schellevis, G. Spierings, A. Griffo, M. Pelk, L. K. Nanver, R. F. F. Jos, and J. W. Slotboom, "RF Power Silicon-On-Glass VDMOSFETs," *IEEE Electron Device Letters*, 25 (6): 424-426, 2004.

2006 25th International Conference on Microelectronics

Reliability Issues Related to Laser-Annealed Implanted Back-Wafer Contacts in Bipolar Silicon-on-Glass Processes

G. Lorito, V. Gonda, S. Liu, T. L. M. Scholtes, H. Schellevis and L. K. Nanver

Abstract – Silicon-on-glass vertical NPN's and PNP's with collector contacts on the back of the wafer directly under the emitter are investigated in relationship to the collector contacting method. Increased base-leakage and impact-ionization currents were found when the collector contacts were implanted. This effect is related to the residual implantation damage at a distance from the contact that cannot be thermal annealed during the low-temperature back-wafer processing.

I. INTRODUCTION

Substrate transfer technologies are gaining acceptance as promising processes for the fabrication of high-performance radio-frequency (RF) integrated circuits [1]. The silicon-on-glass back-wafer contacted bipolar process developed in DIMES is a good example of the advanced device design that these technologies enable [2]. In fact, they offer much more design flexibility than the comparable bulk silicon technology. Consequently, better compromises for the common performance trade-offs can be achieved and several parasitic effects, that usually degrade the device operation in bulk processes, can be easily minimized. Moreover, the fabrication in the same process of both n- and p-type high-performance bipolar transistors becomes straightforward [3] as well as the integration of high-quality passive components [4].

In the DIMES silicon-on-glass bipolar process low-ohmic back-wafer contacts are formed by high-dose, low-energy implantations in the contact windows followed by high-energy excimer laser annealing to activate the dopants [5]. The clear advantage of such back-wafer contacts is the elimination of the usually large series resistance and capacitance associated with the conventional bulk buried layer contacts. Nevertheless, in this innovative design it is not possible to resort to a high-temperature thermal process to repair the implantation-induced damage associated with the back-wafer contact doping. This damage consists of Si interstitials, which are injected into the silicon during the implantation [6]. Since they can travel easily in silicon, they are not confined to the actual implanted contact region. The interstitials are not directly annealed by the

G. Lorito, V. Gonda, S. Lui, T. L. M. Scholtes, H. Schellevis and Prof. L. K. Nanver are with the Laboratory of ECTM, Dept. of Microelectronics, DIMES, Delft University of Technology, Feldmannweg 17, 2628 CT Delft, The Netherlands. E-mail: g.lorito@ewi.tudelft.nl

laser annealing, which only recrystallizes a few tens of nanometers thin region at the surface [7].

Previous work on silicon-on-glass varactors has shown that in these devices the arsenic-implanted back-wafer contact induces defects at least as far as 0.2 μm from the contact [8]. A comparable result for BF_2^+ implantation has been reported in [9] where the defect distribution created by such a process is analyzed by positron annihilation Doppler broadening technique. Moreover, in the above case of silicon-on-glass varactors, it has been observed that the breakdown voltage decreases if the diode depletion region extends into the defect region.

In this paper the focus is on the influence of the back-wafer contacts in a silicon-on-glass bipolar process containing both high-performance vertical NPN's (VNPN's) and PNP's (VPNP's). In particular, alternative back-wafer contact designs are presented that reduce the damage created in the area surrounding the back-wafer contact. It is experimentally demonstrated that this results in a reduction of the device leakage and impact-ionization currents, which can be detrimental to the device reliability.

II. EXPERIMENTAL MATERIAL

The DIMES silicon-on-glass back-wafer contacted complementary bipolar process is illustrated in Fig. 1. The starting material is a SOI wafer with 0.2 μm intrinsic mono-crystalline silicon on a 0.4 μm buried oxide (BOX) layer. The top Si layer thickness is increased by epitaxy to 0.94 μm and trench etching is used to define islands electrically-isolated by SiO_2. In such silicon islands the fully-implanted transistors are fabricated. The low-doped collector region and the emitter/base regions are implanted during the processing of the front side of the wafer (front-wafer) and the emitter/base are contacted by the front-wafer metallization. For the investigated devices the emitter contact area is 40×1 μm^2. A 400 °C alloy in forming gas hydrogen-passivates all oxide/silicon interfaces. The front-wafer is glued to a glass substrate and then the bulk silicon is removed by selective wet etching using the BOX as etch stop layer. In the back-wafer processing the ohmic contacts to the collector are made by implantation of As^+ and BF_2^+ for NPN's and PNP's, respectively, into contact windows opened directly below the emitter regions. The implantations are performed at 5 keV with a dose of 10^{15} cm^{-2}. A high-power excimer laser

1-4244-0116-X/06/$20.00 ©2006 IEEE

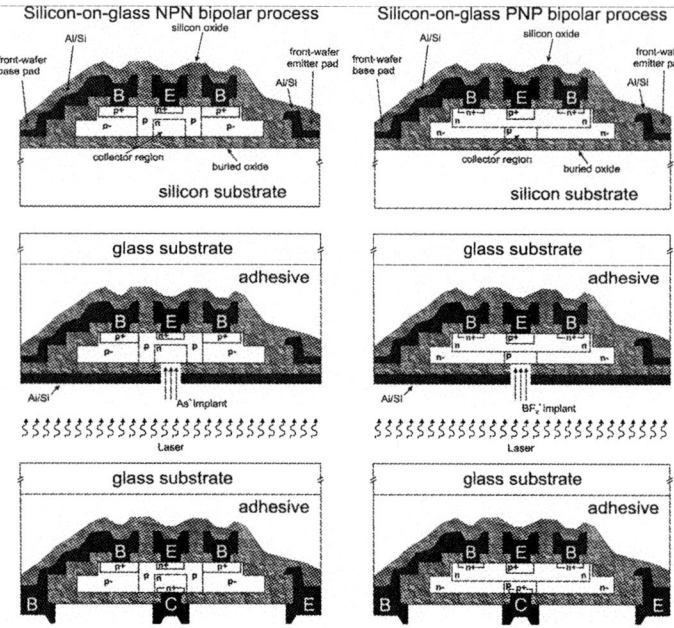

Fig. 1. Schematic process flow of the DIMES silicon-on-glass back-wafer-contacted complementary bipolar technology.

annealing process with an energy density of 1000 mJ/cm² is used to activate the dopants. The back-wafer contact metallization is realized by a standard Al/Si(1%) PVD followed by a 3 hours alloying step at 300 °C. This is the maximum temperature tolerated by the glue.

In the investigated devices three different collector contact designs are implemented. The laser-annealed back-wafer contact implantation is performed either with a 7° or a 30° tilt angle. Moreover, in some devices the implantation is omitted and thus the collector is contacted by a Schottky junction [3].

III. RESULTS AND DISCUSSION

The doping profiles of the NPN and PNP active device regions are shown in Fig. 2. In the depletion regions over the junctions the presence of interstitials from the collector contact implantation will give an extra source of generation-recombination (g-r) centers. The associated leakage currents are a monitor of the amount of residual damage that has been created. Moreover, the impact ionization generated when the collector current and the electric field over the C-B junction is high enough will also be enhanced by the presence of extra g-r centers. These parasitic currents are therefore investigated here in relationship to the type of transistor and the three methods of collector contact formation.

A. Base-leakage currents

Base-leakage currents that could be related to the method of collector contacting were very prominent in the PNP's. This is shown in Figs. 3 and 4 for forward and

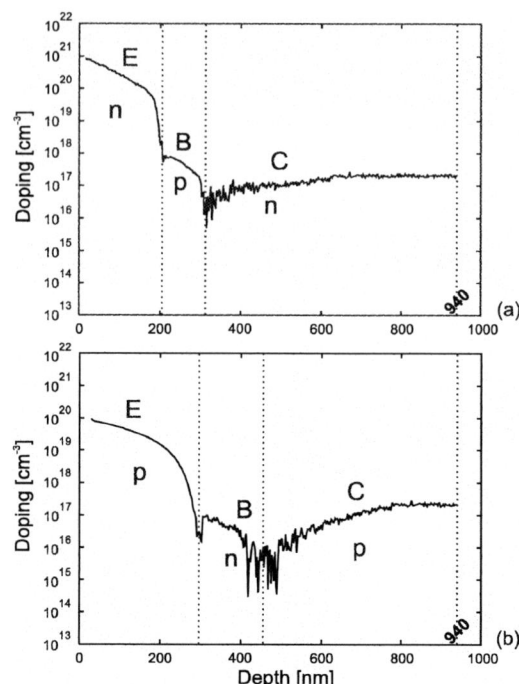

Fig. 2. SIMS of the doping profiles in the active E-B-C device regions for (a) NPN's and (b) PNP's after implantation and the 950 °C thermal anneal. The ultra-shallow collector contact implantation is not included.

reverse operation mode, respectively.

In the forward Gummel plots, the reduced slope of the base current at low V_{EB} is due to g-r centers in the E-B junction depletion region. In the case of 7° tilted collector contact implant, an ideality factor of n = 1.42 is found in the forward-mode low-voltage region, while n = 1.26 is found for the 30° tilted implant. This is in accordance with the fact that the 30° implant channels the Si interstitials in a direction away from the emitter. For the PNP's with Schottky collector contact the base current is near-ideal with n = 1.01 over the whole low- to medium-current range, which confirms that the collector contact implantation is the origin of the extra damage.

As a further support of this conclusion, the distribution of the g-r centers over the E-B junction is evaluated by comparing the bulk to the perimeter leakage. The base-leakage current, $I_{B,leakage}$, is extracted by extrapolating the near-ideal medium-current level to the low-current region and subtracting it from the measured base current. Devices with emitter and collector contact areas from 40×1 to 40×10 μm² are measured and the perimeter-to-area analysis is performed by plotting the relation

$$\frac{I_{B,leakage}}{A} = K_P \frac{P}{A} + K_A \qquad (1)$$

where P is the perimeter length, A the area, K_P the current per unit perimeter length and K_A the current per unit area.

370

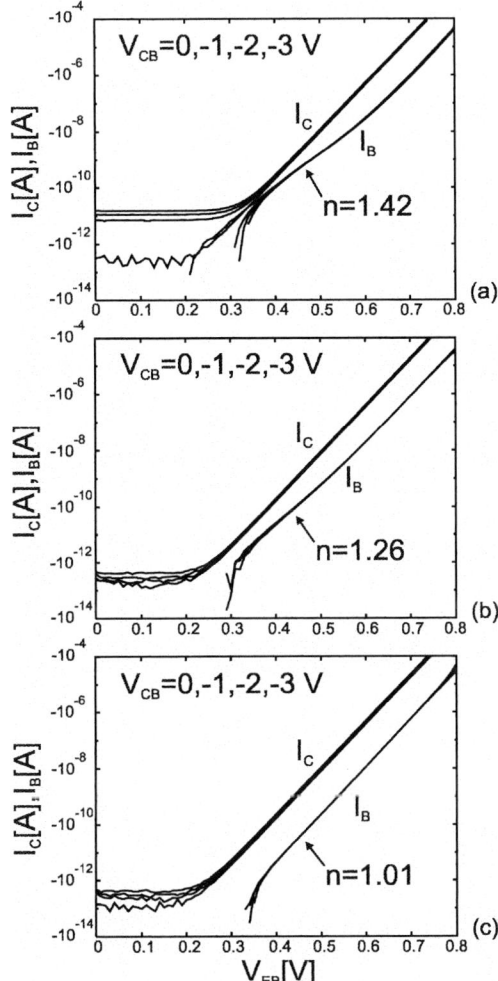

Fig. 3. Measured forward Gummel plots of PNP's with the back-wafer collector contact formed by (a) a 7° and (b) a 30° tilted implantation, and (c) a Schottky contact. The emitter contact area is $40 \times 1 \ \mu m^2$.

The results are shown in Fig. 5 and clearly indicate that the base leakage is not a pure perimeter effect. The latter would be the case if interface states at the Si-SiO$_2$ boundary were the main source of g-r. In the PNP's the damage is clearly distributed over the whole emitter area, as it would be expected if it were caused by the collector contact implant.

In reverse operation the implantation-induced base leakage is even more evident as it can be seen in Fig. 4. The C-B junction that now is forward biased is much closer to the collector contact and the depletion region is also much wider than that one over the E-B junction. As in forward mode, the reduction in damage when switching from a 7° to a 30° implant is overly evident (see (a) and (b) lines in Fig. 4). This could be related to the fact that the Si-SiO$_2$ surface acts as a sink for interstitials and so for the 30° implant the total defect concentration in the C-B depletion region is expected to be lower due to the trapping of interstitials at the bottom Si-SiO$_2$ interface. The latter effect also explains the reduction of the C-B junction

leakage itself in the case of 30° contact implant, which can be seen from the collector current in forward mode at very low current regimes where the C-B junction leakage dominates (see (a) and (b) in Fig. 3).

In the NPN transistors no significant base leakage was detected in connection with the collector contact implantation. This is understandable since the E-B junction is further away from the collector contact and also the depletion region is much narrower than in the PNP's.

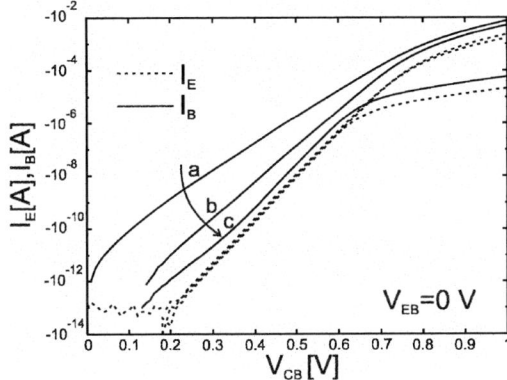

Fig. 4. Measured reverse Gummel plots of the PNP transistors referred to in Fig. 3.

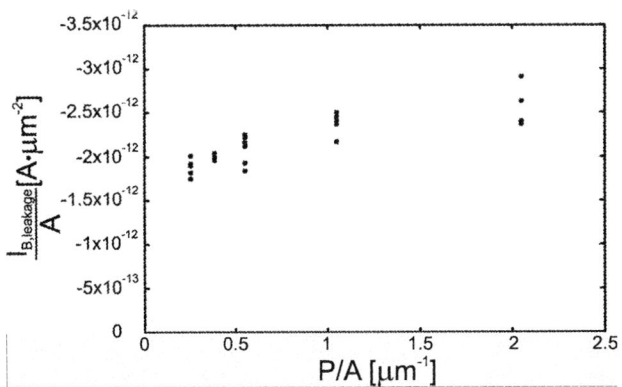

Fig. 5. Base-leakage current at $V_{EB} = 0.4$ V versus emitter contact perimeter-area ratio for PNP's with the collector contact formed by a 7° tilted implantation. The investigated devices have emitter contact sizes of 40×1, 40×2, 40×4, 40×6 and $40 \times 10 \ \mu m^2$.

B. Impact-ionization current

The influence of the collector contact formation on the collector-current-induced impact-ionization current is investigated in the NPN's, for which the E-C breakdown is an avalanching mechanism. The PNP's are not suitable for such a study since the lower base doping results in breakdown by a punch-through mechanism.

For the NPN's analyzed here the actual E-C breakdown voltage is not a correct monitor of the impact ionization level because the available devices were not provided with heat-sinks [10]. The selfheating of the transistors will therefore cause an excessive increase in I_C at high-level current regimes and dominate the breakdown

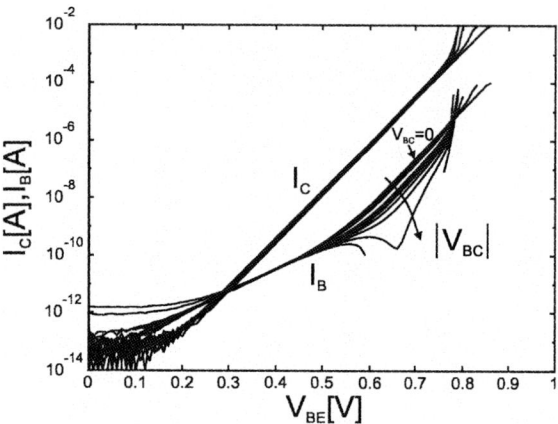

Fig. 6. Measured forward Gummel plots of NPN's for increasing values of base-collector reverse voltage. The emitter contact area is 40×1 µm².

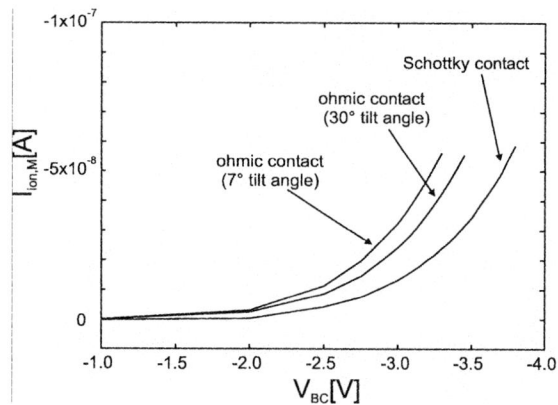

Fig. 7. Impact-ionization current versus base-collector reverse voltage at $V_{BE} = 0.66$ V for NPN's with different collector contacting and an emitter area of 40×1 µm². The investigated devices have the same forward current gain.

mechanism.

However, the impact-ionization current that leads to avalanching can be measured accurately at low- and medium currents where the selfheating is negligible. An example of forward Gummel plots for increasing V_{CB} is shown in Fig. 6. In forward mode the base current at low- and medium-level current regimes can be described as:

$$I_B(V_{BE},V_{BC}) = I_{p,B-E}(V_{BE}) + I_{B,leakage}(V_{BE}) - I_{ion}(V_{BE},V_{BC}) \quad (2)$$

where $I_{p,B-E}$ is the biasing hole current from the base to the emitter, also referred to as the ideal base current, and I_{ion} is the hole current to the base as a result of g-r processes in the C-B depletion region. To have a measure of I_{ion} we can define

$$I_{ion,M}(V_{BE},V_{BC}) = I_B(V_{BE},V_{BC} = 0) - I_B(V_{BE},V_{BC}) \quad (3)$$

which is valid since the base-leakage current is only dependent on V_{BE}. The $I_{ion,M}$ is plotted in Fig. 7 for transistors with collector contacting by the three different

methods. To ensure a correct comparison the devices have the same forward current gain. As it can be seen, for each value of V_{BC} the impact-ionization current is low when the expected damage from the collector-contact implant is low.

For the Schottky collector-contacted devices it must be noted that the effective reverse voltage over the C-B junction is lower than the applied voltage due to the voltage drop on the Schottky barrier. This decreases the electric field over the C-B depletion region and thus also contributes to reduce the impact ionization in this case.

IV. CONCLUSION

This work demonstrates that the implantation damage induced by the back-wafer implanted and laser-annealed contacting of the collector can lead to device degradation both in terms of increased base leakage and reduced E-C breakdown voltage. In advanced high-frequency bipolar (mono- and heterojunction) transistor designs the C-B junction is brought close to the collector contact because a very low collector resistance is required. The present method of back-wafer contacting, made possible by the silicon-on-glass substrate transfer, can provide extremely low collector resistance. However, the degradation issues related to the collector-contact implantation will increase with decreasing vertical device dimensions. It is shown here that the detrimental effects can be reduced by using tilted implants. The use of a Schottky collector contact completely eliminates the contact implantation and the associated damage, but instead other speed versus collector-contact-resistance trade-offs are introduced.

REFERENCES

[1] R. Dekker et al., "Substrate transfer for RF technologies," IEEE Trans. Electron Devices, vol. 50, 2003, pp. 747-757.
[2] L. K. Nanver et al., "A back-wafer contacted silicon-on-glass integrated bipolar process – Part I: The conflict electrical versus thermal isolation," IEEE Trans. Electron Devices, vol. 51, Jan. 2004, pp. 42 – 50.
[3] G. Lorito et al., "Offset voltage of Schottky-collector silicon-on-glass vertical PNP's," in Proc. BCTM, 2005, pp. 22-25.
[4] K. Buisman et al., "Distortion-Free Varactor Diode Topologies for RF Adaptivity," in Proc. IEEE IMS 2005.
[5] L. K. Nanver et al., "Electrical characterization of silicon diodes formed by laser annealing of implanted dopants," in Proc. 23rd ECS 2003, vol. 14, pp. 119-30.
[6] E. C. Jones et al., "Shallow doping technologies for ULSI," Materials Science and Engineering R, 24, pp. 1-80, 1998.
[7] K. S. Jones et al., "Transient enhanced diffusion after laser thermal processing of ion implanted silicon," Appl. Phys. Lett., 75, pp. 3659-3661, 1999.
[8] K. Buisman et al., "High-Performance Varactor Diodes Integrated in a Silicon-on-Glass Technology," in Proc. ESSDERC 2005.
[9] A. Burtsev et al., "Surface morphologies of excimer-laser annealed BF_2^+ implanted Si diodes," in Proc. E-MRS Symposium B, 2004.
[10] L. La Spina et al., "PVD Aluminum Nitride as Heat Spreader in Silicon-On-Glass Technology," accepted for presentation at IEEE MIEL 2006.

2006 25th International Conference on Microelectronics

A Novel Bottom Gate Polysilicon Thin-Film Transistor with Smart Body Tie

Jyi-Tsong Lin, Kuo-Dong Huang and Shih-Tsong Lin

Abstract –In this paper, a novel bottom gate polysilicon thin-film transistor (TFT) with smart body tie is presented. Besides, several body-recessed structures are firstly presented and studied. Comparing with the corresponding conventional bottom gate TFT, the local-thinned-body structure device with the smart body tie has many advantages including the lower sub-threshold swing, the higher current turn on/off ratio and the less DIBL.

I. INTRODUCTION

Thin-film transistors (TFTs) fabricated on polysilicon are generally used in active-matrix LCDs (AMLCDs) as pixel switches, drivers, and peripheral analogue circuits [1]-[2]. Ultra-thin body polysilicon TFTs are reported to have higher current drive compared to their thicker film counterparts [3]. However, conventional polysilicon TFTs present several undesirable effects including large off-current [4], kink-effect [5], and hot-carrier instabilities [6]-[7]. The usual approach to reduce this effect is to limit the impact ionization contribution by decreasing the electric field at the drain junction. One of the solutions is forcing a fixed potential to the device body region using any of several body-tied structures [8]-[11], but all generally suffer from adverse costs such as area penalty and/or process complexity.

In this paper, several novel structures with bottom gate polysilicon TFTs, in which the body is locally thinned with smart body tie (UTB_SBT), are firstly investigated. Meanwhile, to reduce the PN junction area and PN parasitic capacitance, for the first time, we dig the over PN junction out and allow the source metal electrode to contact with the channel body directly from the top side. Therefore, the passivation oxide can replace the major part area of the PN junction between the source and channel body, which was dug out in advance. As a consequence, the PN junction leakage can be reduced greatly. Utilizing this UTB_SBT TFT can overcome the short channel effect resulting from the scaling down, relieve the self-heating effect due to the conventional fully dielectric isolation of buried oxide, suppress the kink effect attributed to the impact ionization, and enhance the breakdown voltage of the device. The experiments and results are investigated in the following sections.

Jyi-Tsong Lin, Kuo-Dong Huang and Shih-Tsong Lin are with the Department of Electrical Engineering, National Sun Yat-Sen Univ., 70 Lien-hai Rd, Kaohsiung 804, Taiwan ROC, E-mail: jtlin@ee.nsysu.edu.tw, d933010002@student.nsysu.edu.tw, m933010140@student.nsysu.edu.tw

II. PROCESS SIMULATION

For the comparison, several structures, bulk_BG (bulk structure with local thinned body and without smart body tie), one-side_SBT (only the left or right side PN junction area is dug out), two-side_SBT (both the left and right PN junction area are dug out) and UTB_SBT (thinned body with smart body tie), are prepared, the fabricating and simulating profiles are shown in Figure 1.

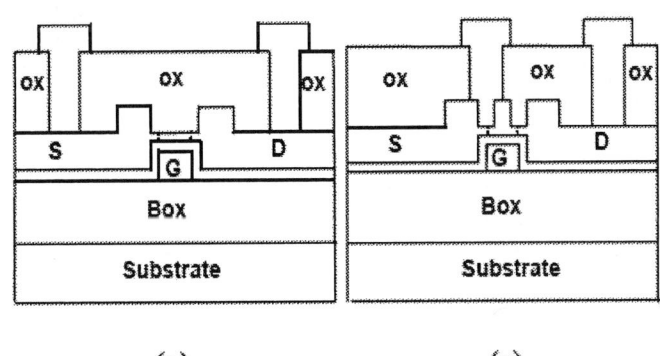

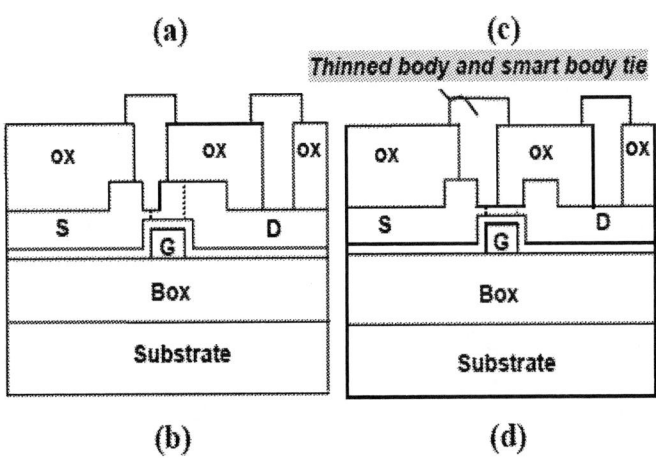

Fig. 1. Device structures designed, (a) Bulk_BG, (b) One-side_SBT, (c) Two-side_SBT and (d) UTB_SBT.

The fabrication steps can be itemized with ISE TCAD simulator. The silicon wafers with 200 nm of thermally grown oxide are used as starting substrates. Poly silicon film of 250 nm deposited at 550 °C by low-pressure chemical vapor deposition (LPCVD) is used as the bottom gate material. The poly film is heavily implanted with phosphorus, and subsequently patterned. It is worthwhile noting that multi-layer structure of the gate that can also be allowed for the process. For example: The associations of a

1-4244-0116-X/06/$20.00 ©2006 IEEE   373

polysilicon and metal silicide can be considered. In addition, the gate layer need proper high concentration dopant and can be replaced by single or multi-layer of the different materials for the threshold voltage optimization. Next, a 100 nm gate oxide is formed by low temperature oxide (LTO) deposition. A 250 nm polysilicon film is then deposited by LPCVD. After that RIE technology and photolithography process are utilized to pattern the middle thin body region of the device. It is also worthwhile noting that only near the S/D regions of four devices were dug out after deposition, so there is enough thickness for reducing the series resistance in the S/D regions. Besides, it improves the Miller parasitic capacitances between the self-raising S/D regions and the gate. Then, a 50 nm screen layer is also deposited by LPCVD. Next, we define PR to protect the gate and channel body. The poly silicon film is then doped with phosphorus by implantation to form the source and drain regions. The implantation dose and energy are $5e^{21}$ cm^{-2} and 47 keV respectively. In order to reduce the series resistance of the source and drain effectively, the energy and the dose of the ion implantation need to be adjusted properly. Then thermal annealing at high temperature is done to form the most proper PN junction between the source/drain and the channel body. Finally, the back-end steps including contact-hole etch, metal deposition, and metal pattering steps were performed to complete the process.

III. RESULTS AND DISCUSSION

In this paper, the devices are designed by the ISE TCAD. The fabrication process and electrical characteristics of the devices are simulated.

Because the body is grounded, the heating energy produced by the electron-hole pairs generated from the impact ionization of high electric field near the drain edge can be driven away effectively. The hot carrier tunneling effect is neglected in this case.

In Figure 2, comparison of the impact ionization rate near the drain junction area among the four structures shown in Figure 1 is presented. The impact ionization rate resulted from the high electric field in the Bulk_BG device is greater than the body tie device. The reasonable explain is that the potential gradient near the drain junction in the body tie devices is sharper than in the Bulk_BG device, so the ionized charges are rapidly recombined in the grounded source. On the other hand, the conventional device has no path to release the excess ionized charges, so the floating body effect becomes serious.

In Figure 3, the output current characteristic of the four structures with the gate length of 400 nm and the thickness of 120 nm is presented. The kink effect is clearly observed in the conventional TFT (Bulk_BG) device. The other three body tie devices show smooth output current characteristics implying that there is no kink effect being observed. It can be considered that the body of the conventional TFT (Bulk_BG) device accumulated the

charges created from the impact ionization near the drain high field region thus the body potential is rose up and induces a lower threshold voltage and the kink is observed. Besides, the non-thinned body tie TFT (SBT) in Figure 3 has the highest output current because its body thickness is thicker than others. The kink effect only occurs in the partially depleted body of the SOI/TFT without the body tie. In our results, if the body thickness is thinned below 100 nm, the kink effect is disappeared.

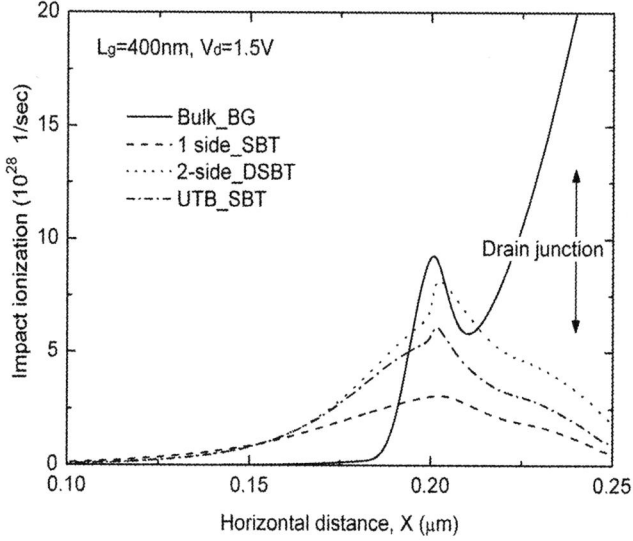

Fig. 2. The simulated impact ionization characteristics of the body tie structures compared with that of the conventional one. The impact ionization rate is more serious in the Bulk_BG device.

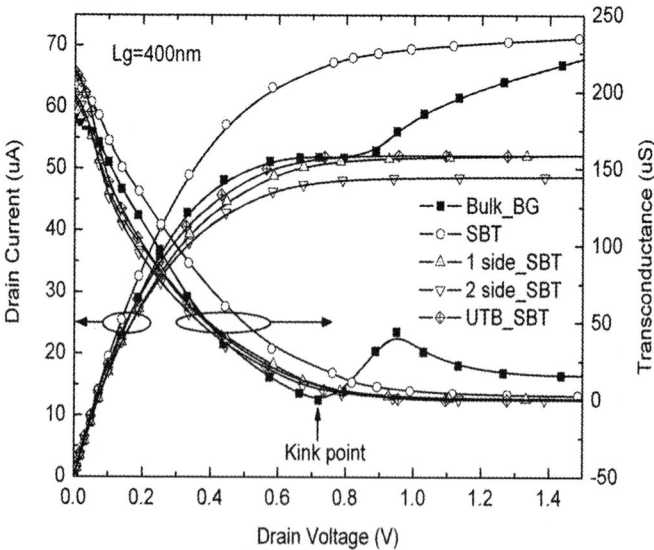

Fig. 3. The simulated *Id-Vd* characteristics of the body tie structures compared with the conventional one, at *Lg*=400 nm, *Tsi*=120 nm *Vgt*=1 V. The kink effect is obviously suppressed.

Figure 4 shows the comparison of the breakdown voltage of the four structures under different gate overdrive voltage (V_{gt}), 0.5 V and 1 V. The body tie devices have much higher breakdown voltage than conventional one. In addition, the output curves in the saturation region of the body tie devices have smaller output conductance than that of the bulk_BG, which indicates that they have higher output impedance thereby suppressing the short channel effect.

Figure 5 shows the sub-threshold characteristic of the four structures. Due to p^+ polysilicon be used as the gate material, the threshold voltage of four structures are of reasonable positive values. The body tie structure thus present the lower leakage current and higher sub-threshold slope than conventional one. It is considered that the thinned recessed body and the body tie structure suppresses the depletion region controlled by the drain bias encroaching into the body.

The sub-threshold characteristics of the n-channel UTB_SBT with different recessed film thickness (100 nm ~ 250 nm) in the channel body region are shown in Figure 6. The device performance is considerably improved when the channel body thickness is thinned from 250 nm to 100 nm. The reason is that the junction capacitances of the drain/source are reduced. For example, the minimum off-current and the sub-threshold factor are reduced from 1.54×10^{-9} A to 1.79×10^{-16} A, and 0.86 V/dec to 0.39 V/dec respectively. The current turned on/off ratio is increased from 7.12×10^4 to 1.85×10^{10}. The great performance improvement in the device with the thinner channel body region is due to the reduced amount of body depletion zone and the PN junction area.

The other interesting electrical characteristics are shown in Table 1. It is interesting that the threshold voltage of body tie TFT is greater than the conventional one. With a concept of charge sharing effect, the threshold voltage of a device is depended on the depletion region controlled by the gate bias. Without the body tie, the depletion area of the drain/source junction, especially of the drain junction, will share the induced charge from the body channel which is originally controlled by gate. Thus, the threshold voltage will roll off as the device is scaled down. On the other hand, the grounded body tie keeps the depletion area from encroaching on the thinned body so that the induced charge will not be shared and the threshold voltage is higher than the conventional one. Beside, the sub-threshold swing is better than the conventional device that is also shown in Table I. The current leakage and the turned on/off ratio of the device shown great improvements and the DIBL is reduced about 4~7 time.

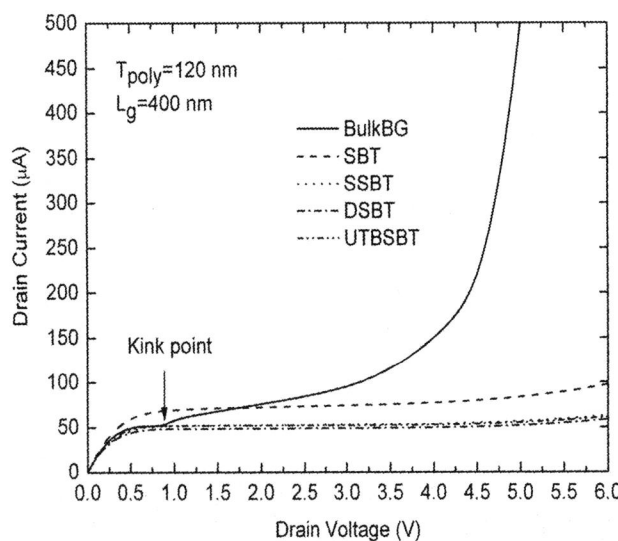

Fig. 4. The simulated *Id-Vd* characteristics of the body tie structures compared with the conventional one, at *Lg*=400nm, *Tsi*=120nm *Vgt*=1V. The breakdown voltage of the bulk BG device is the worst (the smallest) among the structures. The body tie device has great improvement breakdown voltage compared with the conventional one.

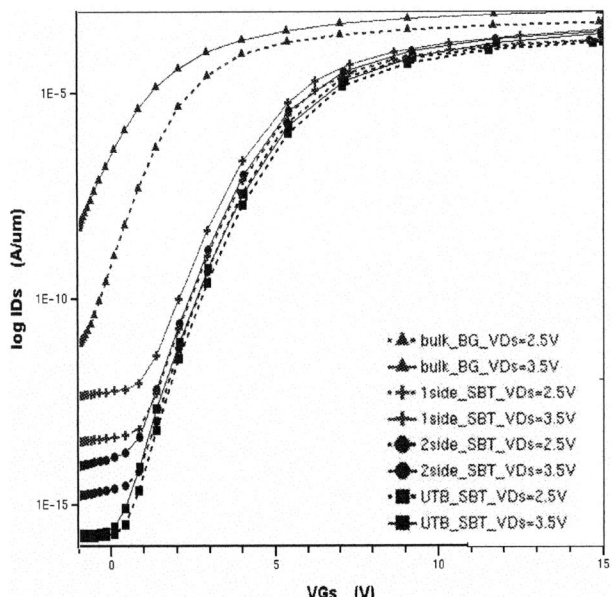

Fig. 5. The simulated *Id - Vg* characteristics of the four structures, at *Lg* = 250nm, *Tsi* = 120nm, *Vds* = 2.5 and 3.5V. The turned off leakage on the body tie devices is much smaller than the conventional one.

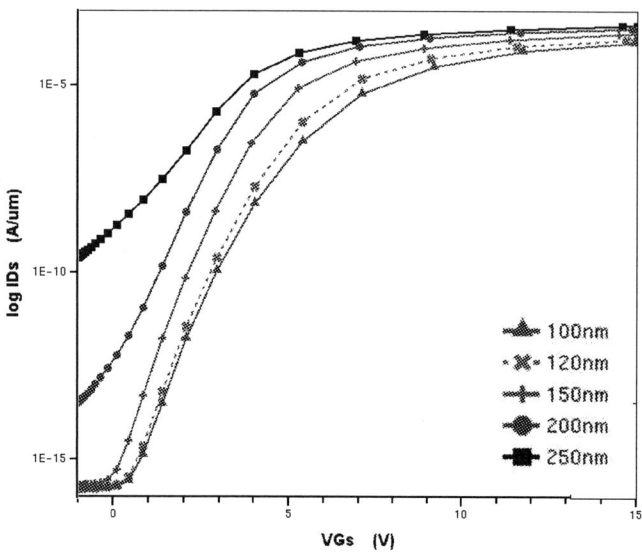

Fig. 6. The simulated sub-threshold characteristics of the UTB_SBT with different channel body thickness (*Tsi*), at *Vds* = 2.5 V, *Lg* = 250 nm.

TABLE I

THE SUMMARY OF THE CHARACTERISTICS OF THE FOUR STRUCTURES AT *Vds*=2.5 V, *Lg* = 250 nm, *Tsi* = 120 nm.

Structure/ Parameter	V_t (V)	S (V/dec)	I_{off} (A/um)	I_{on}/I_{off}	DIBL (V/dec)
bulk_BG	2.86	0.54	4.9×10^{-10}	4.3×10^5	1.26
1side_SBT	6.11	0.48	3.8×10^{-14}	2.6×10^8	0.34
2side_SBT	6.13	0.46	1.6×10^{-15}	3.9×10^9	0.28
UTB_SBT	6.69	0.43	2.0×10^{-16}	1.5×10^{11}	0.17

IV. CONCLUSION

In conclusion, the novel four thinned body TFT structures with self-raising S/D and the smart body-tie have been studied and simultaneously formed, so that the device drive current is increased, the self-heating and the floating body effect which is resulting from the hot carriers and the impact ionization is also suppressed. The performances of the smart body-tied bottom gate TFTs has been shown greatly improved compared with those of the conventional bottom gate TFTs.

REFERENCES

[1] A. G. Lewis, D. D. Lee, and R. H. Bruce, "Polysilicon TFT circuit design and performance," *IEEE J. Solid State Circuits*, vol. 27, p1833–1842, Dec. 1992.

[2] H. G. Yang, S. Fluxman, C. Reita, and P. Migliorato, "Design, measurement and analysis of CMOS polysilicon TFT operational amplifiers," *IEEE J. Solid State Circuits*, vol. 29, p727–732, June 1994.

[3] T. Naguchi, H. Hayashi, and T. Oshima, "Low temperature polysilicon super-thin-film transistor (LSFT)," *Jpn. J. Appl. Phys.*, vol. 25, no. 2, pL121, 1986.

[4] S. D. Brotherton, J. R. Ayres, and M. J. Trainor, "Control and analysis of leakage currents in poly-Si thin-film transistors," *J. Appl. Phys.*, vol. 79, p895–904, 1996.

[5] M. Valdinoci, L. Colalongo, G. Baccarani, G. Fortunato, A. Pecora, and I. Policicchio, "Floating body effects in polysilicon thin-film transistors," *IEEE Trans. Electron Devices*, vol. 44, p2234–2241, Dec. 1997.

[6] J. R. Ayres, S. D. Brotherton, D. J. McCulloch, and M. J. Trainor, "Analysis of drain field and hot-carrier stability in polysilicon thin film transistors," *Jpn. J. Appl. Phys.*, vol. 37, p1801–1808, 1998.

[7] S. Yamada, S. Yokoyama, and M. Koyanagi, "Two-dimensional device simulation for avalanche induced short channel effect in Poly-Si TFT," *IEDM Tech. Dig.*, p859–862, 1990.

[8] M. Matloubian, "Smart body contact for SOI MOSFETs", *Proceedings of IEEE SOI Conference*, p128-129, 1989.

[9] B. W. Min, M. Mendicino, "Effective body contact in SOI devices by partial trench isolated body-tied (PTIBT) structure", *IEEE Semiconductor Device Research Symposium*, p469-472, Dec. 2001.

[10] Y. H. Koh, J. H. Choi, M.H. Nam, J. W. Yang, "Body-contacted SOI MOSFET structure with fully bulk CMOS compatible layout and process", *IEEE Electron Devices letters*, vol. 18, p2102-104, Mar. 1997.

[11] P. Y. Kuo, T. S. Chao, and T.F. Lei, "Suppression of the floating-body effect in poly-Si thin-film transistors with self-aligned Schottky barrier source and ohmic body contact structure", *IEEE Electron Devices Letters*, vol. 25, p634-636, Sep. 2004.

Poster Session
Processes and Technologies

2006 25th International Conference on Microelectronics

Electrical Properties of HfO₂ Films Formed by Ion Assisted Deposition

K. Cherkaoui, A. Negara, S. McDonnell, G. Hughes, M. Modreanu, P. K. Hurley

Abstract –In this paper, the electrical and structural analysis of HfO₂ thin films formed by Ion Assisted Deposition are reported. The electrical results show excellent layer uniformity and very good reproducibility before post deposition annealing. The influence of the Oxygen flow during the growth upon the electrical properties of the film has also been investigated. Forming gas annealing removed the hysteresis observed in the capacitance measurements. However, the electrical results and TEM measurements show the presence of an important SiO₂ interfacial layer.

I. INTRODUCTION

In order to sustain the fast increase of performance undergone by the CMOS integrated circuits in the last decades, the semiconductor industry will very soon need to incorporate new materials at the device level, including the gate oxide, the gate metal and possibly the semiconductor substrate. For the gate oxide high dielectric constant (high k) metal oxides such as HfO₂ offer a possible alternative to nitrided SiO₂ [1]. These high k materials will increase the gate capacitance for the same SiO₂ thickness and hence increase the transistor drive current without increasing the gate leakage current. HfO₂ films have already been formed using different growth techniques such as Metalorganic Chemical Vapour Deposition (MOCVD)[2], Atomic Layer Deposition (ALD) [3-4] and Pulsed Laser Deposition (PLD) [5]. Ion Assisted Deposition (IAD) is a novel technique that has been employed to produce optical coatings or Bragg mirrors. This growth technique presents many advantages over other techniques, such as: formation from solid oxide sources, low growth temperature (25-300°C) and film densification by argon ions during the growth. The argon ions are accelerated towards the substrate and transfer their momentum to the condensing films, increasing the film packing density.

The aim of this work was to assess if the benefits of film densification, inherent in this technique, could be transferred to the formation of thin (2.5-5nm) film on Si (100) substrates with a view to high k gate applications.

K. Cherkaoui, A. Negara, M. Modreanu and P. K. Hurley are with the Tyndall National Institute, University College Cork, Lee Maltings, Cork Ireland, E-mail: karimc@tyndall.ie

S. McDonnell and G. Hughes are with the Physics Department, Dublin City University, Ireland, E-mail: stephen.mcdonnell2@mail.dcu.ie

II. EXPERIMENTAL DETAILS

The HfO₂ films were deposited on 4 inch Si (100) wafers in an ion assist ebeam evaporator (Leybold LAB600). Both n and p type Si wafers were cleaned using a standard SC1/SC2 clean followed by a native oxide removal in HF. The deposition was performed at 150°C by electron beam heating of a solid source consisting of monoclinic HfO₂ pellets. Argon ions were generated in a plasma to assist the deposition, the Ar flow rate was 4 SCCM. In addition, O₂ was introduced in the chamber during the deposition. For this study the O₂ flow rate ranged from 0 to 7 SCCM. The base pressure in the chamber prior to Ar and O₂ flow was below 1×10^{-6} mBar and during deposition the pressure increased to $\sim 1 \times 10^{-4}$ mBar. The ebeam power was controlled to achieve a deposition rate around 0.5Å/s. The electrical test structures consisted of MOS capacitors of different sizes. The metal areas were defined using lithography and lift-off process after Ebeam evaporation, in a separate e-beam evaporator.

In the present study, two sets of samples with two HfO₂ target thicknesses of 2.5 nm and 5 nm were produced.

III. RESULTS AND DISCUSSION

The initial step for this study was to select a suitable metal for the MOS capacitor electrical characterisation. After investigating several metals and metal alloys (Ni, TiN, Al, TiAlN, W, NiSi) the lowest leakage current was measured on the Ni gate structures. It is worth noticing that among the metals screened in the current study Ni presents the highest work function (~5.1eV). All the results presented in this paper were obtained on Ni gate structures.

Figure 1 shows a typical high frequency Capacitance Voltage (CV) characteristic of a MOS capacitor fabricated on a 5nm nominal thickness HfO₂ sample. The CV shows a significant hysteresis effect when sweeping the gate voltage from inversion to accumulation and back. The negative CV curve shift observed could arise from hole injection in the insulator indicating the presence of hole traps in the HfO₂ layer. The SiO₂ equivalent oxide thickness from the measured Capacitance in accumulation (CET) is estimated to about 42Å. Complementary analysis is required to explain such a high CET.

1-4244-0116-X/06/$20.00 ©2006 IEEE

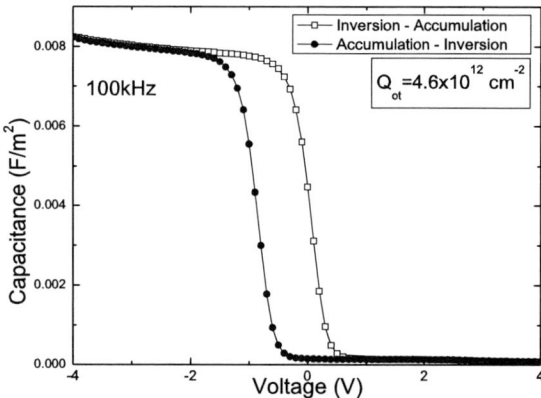

Fig. 1. Capacitance Voltage Hysteresis characteristic for un-annealed Ni/HfO$_2$/Si(100) structures.

This sample was subsequently annealed in an open furnace in forming gas ambient (5%H$_2$/95%N$_2$) at 400°C for 1hour. Figure 2 shows the hysteresis CV characteristic after the FGA. Under these annealing conditions no shift was observed probably as result of defect passivation in the oxide. In addition there was no significant variation of the accumulation capacitance. This indicates that the dielectric constant and CET were not affected under these annealing conditions.

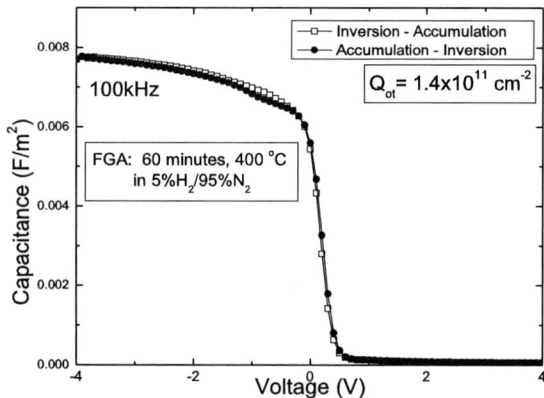

Fig. 2. Capacitance Voltage Hysteresis characteristic of Ni/HfO$_2$/Si(100) after FGA.

The high CET value obtained for the 5nm HfO$_2$ capacitors is confirmed in the high resolution TEM performed on theses samples. Figure 3 shows clearly the presence of an important SiO$_2$ interfacial layer (3nm) responsible for the high CET. The TEM results combined with the CET value yield a k value around 19 for HfO$_2$, assuming dielectric constant of 3.9 for the interfacial SiO$_2$.

Within the current study another set of HfO$_2$ samples was produced. The target thickness was reduced to 2.5nm. Several deposition runs were performed under different O$_2$ flow rates (0, 2.5, 5and 7 SCCM) to understand the role of this extra source of Oxygen on the films properties.

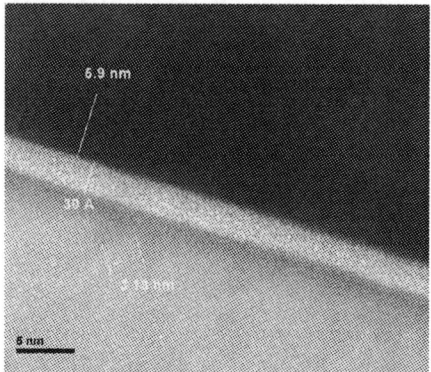

Fig. 3. High Resolution TEM cross Section

Figures 4 and 5 show typical Current Density versus Voltage (JV) characteristics measured on the 2.5nm HfO$_2$ nominal thickness wafers. The electrical measurement presented very good uniformity across the wafer (figure 4). The JV repeatability on the same device is shown in Figure 5. Successive JV sweeps below breakdown didn't induce any excess gate leakage current. The CET values for this set of samples was reduced (32-35Å) as compared to the previous one, however, they are still higher than expected probably due to a significant interfacial SiO$_2$. It is nevertheless interesting to note that these good electrical properties were obtained on wafers that didn't undergo any post deposition annealing.

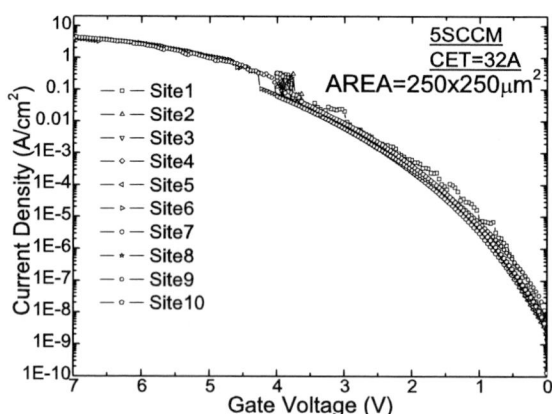

Fig. 4. Current Voltage characteristics for several devices across an n-type Si wafer

The previous remarks are valid for all the samples with HfO$_2$ layers deposited with O$_2$ flow rates (2.5, 5 and 7 SCCM). A comparison of the JVs from samples formed under different flow rates (figure 6) shows that 7 SCCM gives the best results, the 2.5 and 5 SCCM flow rates present similar results while the samples formed with no O_2 flow exhibit very high gate leakage current.

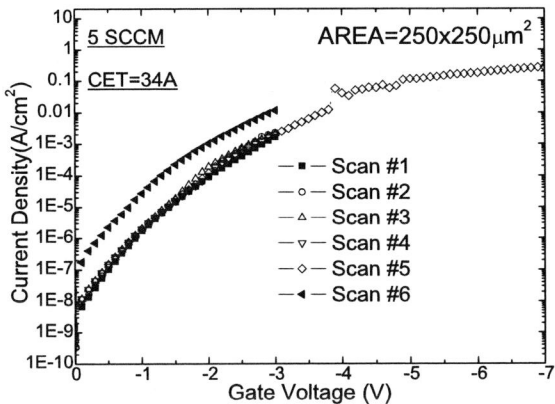

Fig. 5. Gate leakage current density evolution after several voltage sweeps (p-type Si substrate)

XPS measurements were carried out on the same set of samples in order to understand their electrical behaviour. Figure 7 shows the Hf 4f XPS spectra for the different O_2 flow rates. All of the binding energies presented in this study have been referenced to the Silicon substrate at 99.3 eV to account for the surface charging effects charging [6]. The Hafnium oxide 4f 7/2 peak shifts to lower energy (from 18 eV to 17.4 eV) as the flow rate is reduced this is consistent with a slightly sub stoichiometric oxide.

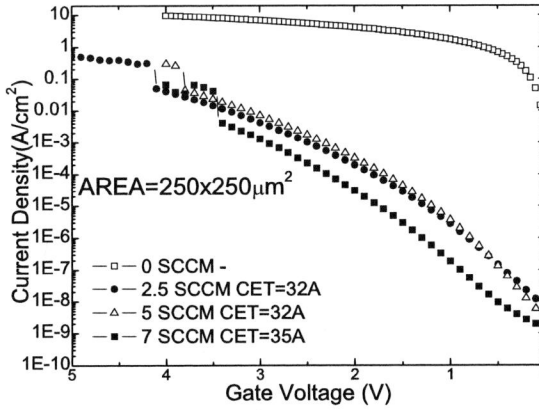

Fig 6. Gate leakage current density for HfO_2 films formed with different O_2 flow rates (n-type Si Substrate)

The presence of metallic Hafnium or Hafnium Silicide being responsible for the high leakage current density on the film formed under 0 SCCM O_2 flow rate can

be ruled out as the characteristic binding energy for the metallic Hf or Hafnium Silicide 4f 7/2 electron is ~14.3 eV [6,7]. It is noted that during sputtered depth profiling of the HfO_2 films the XPS Hf or Hf Silicide signal at 14.3 eV is detected. However this is due to HfO_2 dissociation during the sputtering process. This has been confirmed on samples analysed in this work where it is possible to detect the substrate Si 2p signal prior to profiling and metallic Hf or Hf Silicide were not detected. The 14.3 eV binding energy signal only emerges upon Ar bombardment confirming this is an artefact of the sputtering process.

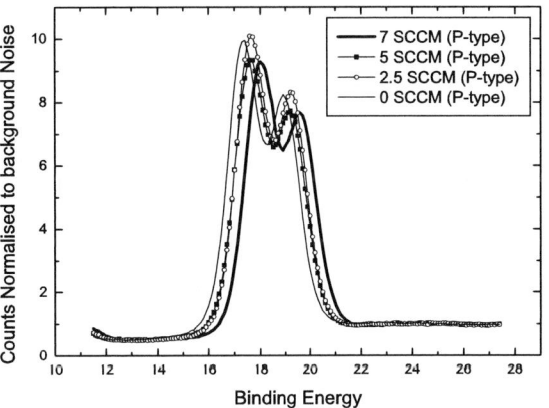

Fig 7. Hf 4f XPS spectra for different O_2 flow rates.

Figure 8 shows the Si 2p XPS spectra for the same set of samples. The low energy peak corresponds to the silicon substrate and the high energy one corresponds to Si in an oxide. The relative intensity of the oxide to the substrate peak depends on the thickness of the SiO_2 interfacial layer.

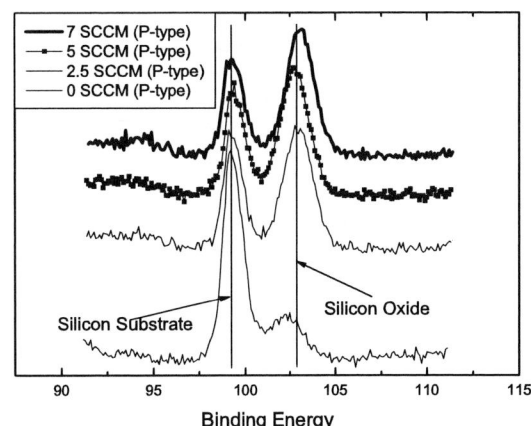

Fig 8. Si 2p XPS spectra for different O_2 flow rates.

The most significant result is the reduction of the SiO_2 XPS peak for the film formed with no O_2 flow. This indicates that under these conditions the SiO_2 interfacial layer is significantly reduced. Therefore we can conclude that the interfacial layer is probably grown due to the Si wafer exposure to the plasma and O_2 flow prior to and during the HfO_2 deposition. It still unclear whether the poor electrical characteristics are due to the absence of the SiO_2 layer or due to a lower quality HfO_2 film resulting from the lack of O_2 supply.

III. CONCLUSION

This study showed the possibility of forming uniform thin films of HfO_2 using Ion Assisted Deposition. The uniformity of the deposited film was confirmed by the reproducibility of the current density versus voltage characteristics across the wafer. It was also shown that CV hysteresis was removed after FGA at 400°C for one hour. However, the presence of Oxygen in plasma prior to and during the deposition is responsible for the relatively thick SiO_2 interfacial layer (3nm) increasing the CET value. Future work will be aimed at reducing the interfacial layer and understanding the Hafnium oxide bulk defects.

ACKNOWLEDGEMENT

The authors gratefully acknowledge Science Foundation Ireland for funding this work, Mr. Dan O'Connell for the film deposition and Robert Dunne and Stephen Cosgrove from Intel Ireland for the TEM analysis.

REFERENCES

[1] G. D. Wilk, R.M. Wallace, J. M. Anthony, "High-k gate dielectrics: Current status and materials properties considerations", *J. Appl. Phys.* Vol. 89 (10) pp.5243-5275, 2001.

[2] C. Dubourdieu, H. Roussel, C. Jimenez, M. Audier, J. P. Senateur, S. Lhostis, L. Auvray, F. Ducroquet, B.J. O'Sullivan, P.K. Hurley, S. Rushworth, L. Hubert-Pfalzgraf, "Pulsed liquid-injection MOCVD of high-K oxides for advanced semiconductor technologies" *Mater. Sci. and Eng.* Vol. B 118, pp. 105-111, 2005.

[3] A. C. Jones, H.C. Aspinall, P.R. Chalker, R.J. Potter, K. Kukli, A. Rahtu, M. Ritala, M. Leskela, "Recent developments in the MOCVD and ALD of rare earth oxides and silicates" *Mater. Sci. Eng.* Vol. B 118 (1-3) pp. 97-104, 2005.

[4] D.H. Triyoso, M. Ramon, R.I. Hegde, D. Roan, R. Garcia, J. Baker, X.D. Wang, P. Fejes, B.E. White, P.J. Tobin, "Physical and electrical characteristics of HfO2 gate dielectrics deposited by ALD and MOCVD" *J. Electrochem. Soc.* Vol. 152 (3): pp.G203-G209, 2005.

[5] H. Ikeda, S. Goto, K. Honda, M. Sakashita, A. Sakai, S. Zaima, Y. Yasuda, "Structural and electrical characteristics of HfO2 films fabricated by pulsed laser deposition" *Jap. J. Appl. Phys.* Vol. 41 (4B) pp. 2476-2479, 2002.

[6] J.F. Moulder, W.F. Stickle, P.E. Sobol, K.D. Bomben, *Handbook of X-ray Photoelectron Spectroscopy* (J. Chastain, editor), Perkin Elmer Corporation, 1992.

[7] H.T. Johnson-Steigelman, A.V. Brinck, S.S. Parihar, P.F. Lyman, "Hafnium Silicide Formation on Si(100)" *Phys. Rev. B* Vol. 69 (23) Art. No. 235322, 2004.

PECVD Growth of Thick Silicon Oxynitride for On-Chip Optical Interconnects Applications

C. K. Wong, H. Wong, C. W. Kok and M. Chan

Abstract—Thick oxynitride films were prepared by plasma enhanced chemical vapor deposition (PECVD) with N_2O, NH_3 and SiH_4 precursors. The composition and the bonding structure of the oxynitride films were investigated with Fourier transform infrared spectroscopy (FTIR) and x-ray photoelectron spectroscopy (XPS). Results showed that the silicon oxynitride deposited with gas flow rates of $NH_4/N_2O/SiH_4 = 10/400/10$ (sccm) has favorable properties for integrated waveguide applications. The refractive index of this layer is about 1.5 and the layer has comparative low densities of O-H and N-H bonds. The O-H bonds can be readily eliminated with high temperature annealing of the as-deposited film in nitrogen ambient. Annealing at temperature of 1000 °C or above which can significantly suppress both the N-H bonds and O-H bonds is preferred. Simple ridge type waveguide with cross-section of 3 μm$\times$2.5 μm for the core layer (n = 1.57) was fabricated. This waveguide is able to transmit signal in either TE or TM mode and the number of mode is eight and the bending radius of the waveguide can be reduced to about 6 μm.

I. INTRODUCTION

Recently, the possibility of developing on-chip optical interconnects technology has drawn significant attentions of researchers from both microelectronics and optoelectronics [1-6]. The optical interconnects would result in substantial changes, including shorter interconnect delays, higher bandwidth, smaller power consumption and EMI immunity, in the signal transmission within the ultra large scale microelectronic chip. This work explores the technology for preparing low propagation loss silicon oxynitride film for on-chip optical interconnects applications using CMOS compatible technology.

It was reported that silicon oxynitride has readily tunable optical properties such as refractive index as well as the reduction of absorption loss in spectral regions of interest

The work described in this paper was fully supported by a project (Project No. 7001513) funded by City University of Hong Kong.

C. K. Wong and H. Wong with the Department of Electronic Engineering, City University of Hong Kong, Tat Chee Avenue, Kowloon, Hong Kong, E-mail: eehwong@cityu.edu.hk

C. W. Kok and M. Chan are with the Department of Electrical and Electronic Engineering, The Hong Kong University of Science and Technology, Clear Water Bay, Kowloon, Hong Kong.

[2]. Compared to the conventional silica materials, oxynitride films offer much larger refractive index for the core layer of optical waveguide and are able to lower minimum allowable bending radius for waveguide designs. This is particular important for on-chip optical interconnect in microelectronic technology. It was reported recently that the radii of curvature in a ring-resonator filter could be as small as a few hundred micrometers by using the silicon oxynitride material [3]. However, the optical loss is still the major concern for integrated oxynitride based waveguide devices. High hydrogen content was found with oxynitride films prepared by chemical deposition method. Hydrogen in oxynitride films does not only affect the refractive index but also the major source for absorption loss of light transmission in the oxynitride waveguide [2, 7-9]. The objective of this work is to explore the processing effects on the material and optical properties of thick oxynitride films prepared using plasma enhanced chemical vapor deposition (PECVD) technique.

II. EXPERIMENTAL

Silicon oxynitride films were deposited using a STS 310 PECVD machine operated at 187.5 kHz with 60W plasma power. The chamber pressure and substrate temperature are maintained at 600 mTorr and 300 °C, respectively, during the deposition. Different film compositions were obtained by changing the flow rates of silane gas, ammonia and nitrous oxide. To remove the hydrogen content from the films, the silicon oxynitride layers were annealed in a furnace at several different temperatures for 3 hr. Ellipsometry and Fourier transform infrared (FTIR) spectroscopy characterization were conducted to study the effects of thermal annealing. The thickness and refractive index of the films was characterized by Rudolph Auto EL II Ellipsometer with 632.8 nm wavelength He-Ne laser light source. Physical Electronics PHI5600 X-ray Photoelectron spectroscopy (XPS) with an AlK_{α} X-ray source was used to probe the composition profile and bonding features of the deposited and annealed oxynitride films. The take off angle was 45 degree. A Bio-Rad FTIR spectrometer FTS 6000 was used to make analysis on the compositional as well as structural properties of the material. The apparatus has a detection range of 2 cm^{-1} resolution.

1-4244-0116-X/06/$20.00 ©2006 IEEE

III. RESULTS AND DISCUSSION

Fig. 1 depicts the atomic force microscopic (AFM) results on the thickness uniformity and surface smoothness of a typical ridge oxynitride waveguide. It is clear that using typical microelectronics dry etching process is able to produce a well-defined and smooth-surface oxynitride layer. The typical surface roughness is about 4.5 nm which is small enough when compared to the typical dimension of waveguide cross-section. This small surface roughness should not be the major source of optical loss. Major optical loss is due to the hydrogen-related absorption band of the oxynitride films. Particular measure has to be employed to reduce the hydrogen content. High temperature annealing was found to be an effective technique.

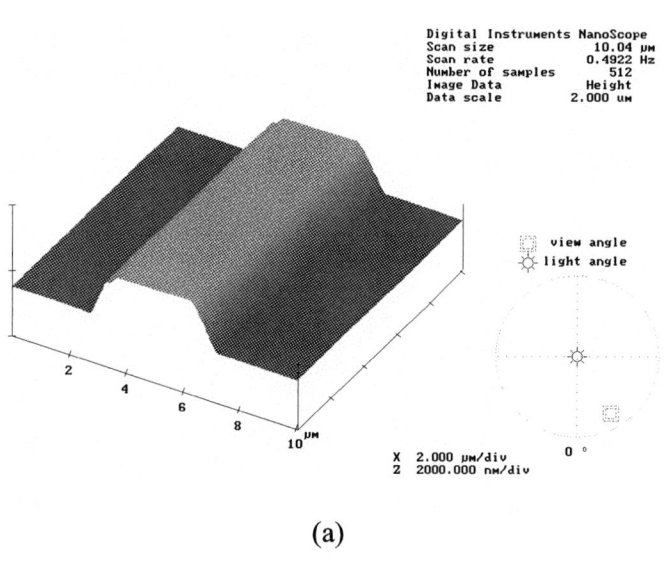

(a)

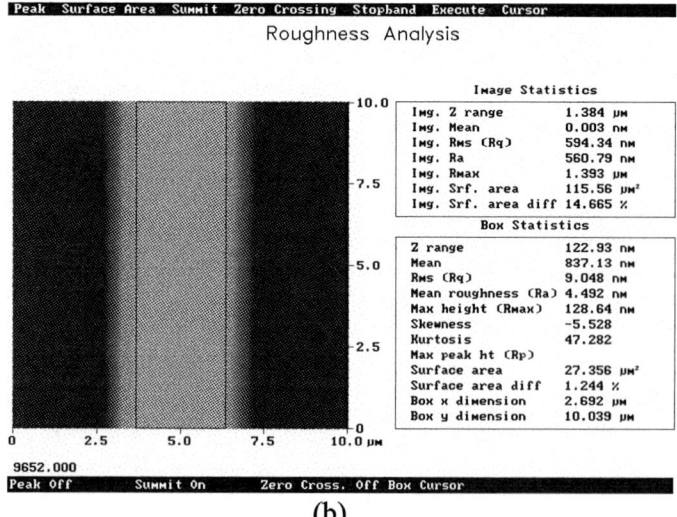

(b)

Fig. 1. AFM studies of the uniformity and smoothness of the prepared ridge oxynitride waveguide.

Fig.2 plots the refractive index and film thickness as a function of the annealing temperature. It is found that a high temperature annealing can significantly improve the material and optical properties of as-deposited oxynitride films. The annealing was conducted in nitrogen ambient and the annealing duration was 3 hr. The film thickness decreases and significantly, and almost linearly as the annealing temperature increases. This observation can be attributed to the densification effects involving both removal of microvoids and hydrogen in the film [9].

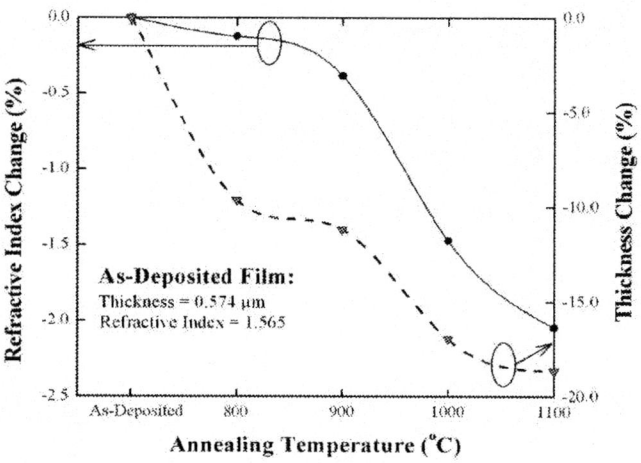

Fig. 2. Plot of the refractive index as a function of annealing temperature. The annealing duration is 3 hr.

Fig. 3(a) depicts the nitrogen 1s XPS spectra of the as-deposited oxynitride films and films with annealing at different temperatures for three hours. It can be seen that the N 1s for non-annealed sample has peak at energy of 398.6 eV. It indicates that the nitrogen atoms in the oxynitride are essentially in the form of Si-N bonding with high hydrogen content [10]. Si-N bond in stoichiometric Si_3N_4 film has a feature peak at around 397.4 eV. The excess silicon and hydrogen will results in the shift of the N 1s to higher energy side. In NH_3 nitrided oxide, Bhat et al [11] found that the N 1s peak is 398.6 eV which is attributed to the Si-N=H_2. For samples with 3 hr annealing at temperatures larger than 800 °C, the intensity of the N 1s peak reduced pronounced and its location shifts to even higher energy side to about 399 eV. This change cannot be explained with the removal of hydrogen which should have an opposite effect. High energy shift of N 1s peak should be due to the decomposition of Si-N bonds. With fewer N neighbours, the nitrogen atoms in SiN_4 phases at the interface are more electronegativity and have a large binding energy. The allegation is confirmed with the concentration data. The nitrogen content in the oxynitride film is smaller as the film was annealed at higher temperature and longer duration.

Fig. 3(b) shows the effect of annealing temperature on the

384

Si 2p feature. The Si 2p has a peak at energy of 103.3 to 103.9 eV. This data indicate that the samples are essentially a silicon oxide layer. Amorphous SiO_2 has a Si 2p peak of about 104 which Si_3N_4 has a peak of about 102 eV. A bulk oxynitride film prepared by LPCVD method that there is only one broaden peak (in the range of 102 to 104 eV) in the XPS spectra for different chemical composition. The different location of Si 2p is also governed by the hydrogen content. As will be shown later, the oxynitride prepared by PECVD have higher content of hydrogen whereas in annealed films low hydrogen content can be reduced quite significantly. Reduction of hydrogen would cause more positive charges on Si.

band ranging from 3200 to 3800 cm^{-1} was found for all samples (see Fig.4). This band is the main cause of the optical absorption at 1.55 micrometer [2]. Annealing has profound effect on this band. With 3 hr annealing at 1100 °C, the absorption band reduced below the detectable range (see Fig.5). Gaussian deconvolution [12] of this absorption band indicated that the major this band is mainly composed with N-H and O-H bands. The O-H band can be effectively suppressed with thermal annealing at temperature higher than 800 °C. To remove the N-H band, higher temperature (> 1000 °C) annealing is required.

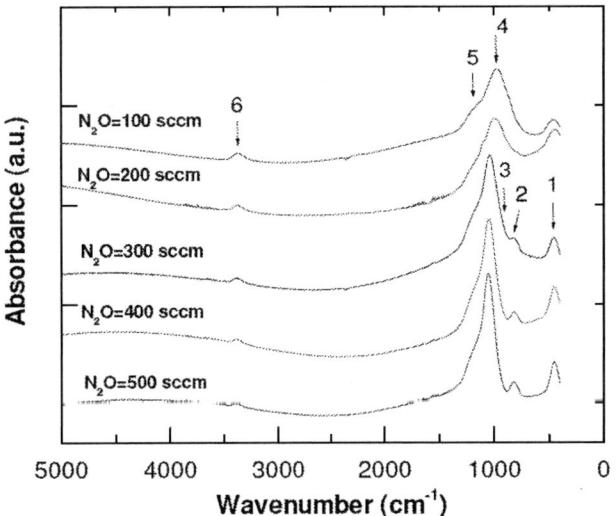

Fig. 4. Infrared absorbance of silicon oxynitride films deposited with 10 sccm NH_3 flow rate and various N_2O flow rates. Numbers indicate the major features of the absorbance spectra. 1:Si-O rocking, 2:Si-O bending, 3:Si-N stretching, 4: Si-O symmetric stretching, 5: Si-O asymmetric stretching, and 6: band related to H bonds.

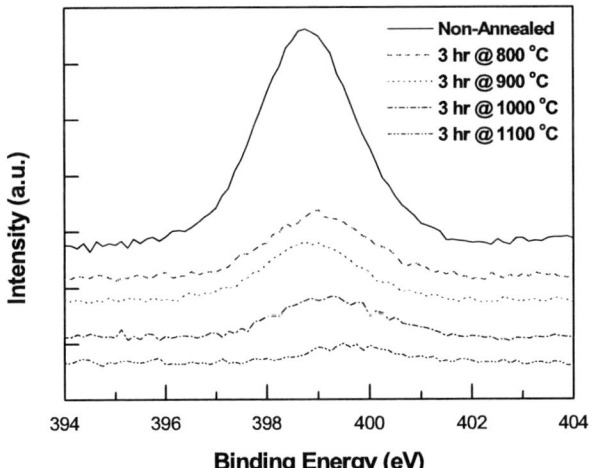

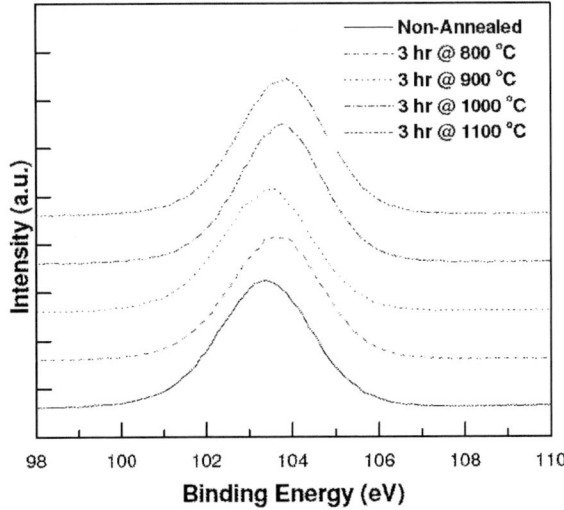

Fig. 3. Nitrogen 1s (upper) XPS and Si 2p (lower) sputtering spectra for silicon oxynitride with different annealing conditions.

To study the processing and annealing effects on the oxynitride films FTIR measurements of the samples were also conducted. Clear N-H and O-H stretching absorption

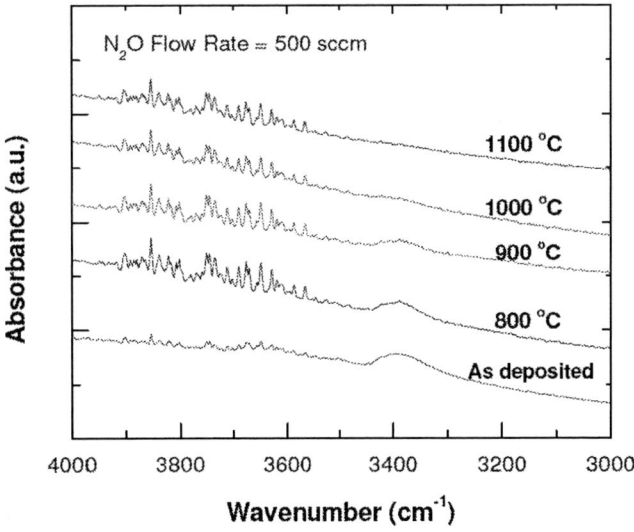

Fig. 5. Effect of annealing temperature on residual the O-H and N-H concentration in the oxynitride films.

With systematic optimization processes, we found that

the silicon oxynitride deposited with gas flow rates of $NH_4/N_2O/SiH_4 = 10/400/10$ (sccm) has favorable properties for integrated waveguide applications. The refractive index of this layer is about 1.5 and the layer has comparative low densities of O-H and N-H bonds with are the major loss for silica or oxynitride waveguide operated at C+L band (1520-1610 nm). Annealing the as-deposited samples at temperature of 1000 °C or above is preferred in order to suppress both the N-H bonds and O-H bonds. A simple ridge type waveguide with cross-section of 3 μm×2.5 μm for the core layer (n = 1.57) (see Fig.1) was fabricated. This waveguide is able to transmit signal in either TE or TM mode and the number of mode is eight. The propagation loss at 1550 nm is less than 0.5 dB/cm and the bending radius of the waveguide can be reduced to about 6 μm. The loss can be reduced further with optimal design and with a proper top layer cladding.

IV. Conclusions

Detailed study on the processing effects on material and optical properties of PECVD silicon oxynitride films has been conducted. The composition and the bonding structure of the oxynitride films were investigated with Fourier transform infrared spectroscopy (FTIR) and x-ray photoelectron spectroscopy (XPS). Results showed that the silicon oxynitride deposited with gas flow rates of $NH_4/N_2O/SiH_4 = 10/400/10$ (sccm) has favorable properties for integrated waveguide applications. The refractive index

of this layer is about 1.5 and the layer has comparative low densities of O-H and N-H bonds. The high content of O-H bond can be readily eliminated with high temperature annealing of the as-deposited film in nitrogen ambient. Annealing at temperature of 1000 °C or above which can significantly suppress both the N-H bonds and O-H bonds is preferred.

REFERENCES

[1] H. Wong 11th IEEE Int'l Sympo. Electron Devices for Microwave and Optoelectronic Application (EDMO 2003), Orlando, USA, November 2003. p.145.

[2] H. Wong, Microelectron. Reliab. 42 (2002) 317

[3] A. Melloni, R. Costa, P. Monguzzi, M. Martinelli. Optics Lett., 28 (2003) 1567

[4] R.M. de Ridder, K. Worhoff, A. Driessen, P.V. Lambeck, H. Albers, IEEE J. Select. Topics Quantum Electron. 4 (1998) 930.

[5] C. David, D. Wiesmann, R. Germann, F, Horst, B.J. Offerin, R. Beyeler, H.W.M. Salemink, G.L. Bona, Microelectron. Eng., 57-58 (2001) 713.

[6] G.-L. Bona, R. Germann, B.J. Offerin, IBM J. Res. Dev. 47 (2003) 239.

[7] H. Wong, M. C. Poon, Y. Gao, T. C. W. Kok, J. Electrochem. Soc. 148 (2001) G275.

[8] F. Ay and A. Aydinli, Optical Matter. 26 (2004) 33.

[9] C.K. Wong, H. Wong, C. W. Kok and M. Chan, J. Cryst. Growth, in press

[10] M. C. Poon, C. W. Kok, H. Wong and P. J. Chan, Thin Solid Films, 462 (2004) 42.

[11] M. Bhat, G. W. Yoon, J. Kim, D. L. Kwong, M. Arendt, and J. M.White, Appl. Phys. Lett. 64 (1994) 2116.

[12] W.A. Landford, M.J. Rand, J. Appl. Phys. 49 (1978) 2473.

2006 25th International Conference on Microelectronics

The Characterization and Optimization of the Thermal Oxidation Process Equipment Using Experimental Design and Data Transformation

YOU Hailong, JIA Xinzhang, and Wang Shaoxi

Abstract -Using design of experiment and data transformation, the model of thermal oxidation furnace is developed. The model involves six input factors and two responses. Only 16 run experiments are needed through the fractional factorial experimental design. The Box-Cox transformation method is used to transform the experimental data, and the optimum transformation style is determined. The responses are transformed so that the residuals meet the assumption of ANOVA and the model can explore more information here. After responses transformation the figure of merit for fitting, Adjust-R^2, is improved from 93.54% to 98.64%. The models are subsequently used to optimize the thermal oxidation process. The non-uniformity of film is improved from 0.2% to 0.08% with the other specification satisfied.

I. MANUSCRIPT SUBMISSION

Developing new processes, decreasing the discreteness of processes, and improving yields and the product's quality depend on the characterization and optimization of the process equipment in semiconductor manufacturing. Therefore the model of process equipments is needed. Physical model can only be used in the principle research of the process and the design of procedure [1] , and not be used for characterizing the process variability, such as the uniformity of the oxidation film discussed in this paper. It is essential that the statistical model of process equipments be founded. In order to build the statistical model, the process output/input relation is developed via design of experiment such as fractional factorial experimental design.

There are many researchers who use the experimental design method to characterize the process. Gary S. May etc develop the plasma model using the sequential experimental design [1]; Kyeong K. Lee etc characterize the molecular beam epitaxy (MBE) process using experimental design and neural networks [2]. And Byungwhan Kim etc, construct the model of the plasma etching using full factorial experiment and neural networks [3].

In this paper the modeling of microcircuit process equipments, thermal oxidation furnace, is investigated using fractional factorial experimental design and data transformation. The fractional factional experimental design is used to generate design matrix for the experiment

You Hailong, Jia Xinzhang, Wang shaoxi are with the Department of Microelectronics, Xidian University, Xi'an 710071,China , E-mail: youhailongl@126.com

of six input factors. The target values include the film thickness and uniformity measured from the spectrophotometer. The process models are founded by the analysis of variance (ANOVA). In order to make the models meet the assumption of ANOVA and explore more data information, the experimental data is transformed using Box-Cox transformation.

II. EXPERIMENTAL DESIGN AND RESPONSE TRANSFORMATION

A. Fractional factorial experimental design

Oxidation process is one of the basic process in the microcircuit manufacturing. The thermal oxidation process is widely used and is the main method to grow the film of SiO_2. The film is grown inside the oxidation furnace using the technology of H_2、 O_2 composition. The furnace is a traditional structure tube [4]. The main target values in the process include the oxidation thickness film (signed by D) and the uniformity (signed by U) within the wafer. They are defined as follows:

$$D = (x1 + x2 + x3 + x4 + x5)/5 \qquad (1)$$

$$U = \frac{\sqrt{\frac{1}{4}[(x1-D)^2 + (x2-D)^2 + (x3-D)^2 + (x4-D)^2 + (x5-D)^2]}}{(x1+x2+x3+x4+x5)/5} \qquad (2)$$

Where x1, x2, x3, x4, x5 represent the oxidation thickness film of the fixed points in the wafer.

In order to characterize the thermal oxidation process equipment, the input factors of interest and their level in the experiment are summarized in Table 1.

TABLE I

OXIDATION PROCESS INPUT FACTORS AND LEVEL

Factor	Level/Unit
Crystal orientation	(<111>,<100>)
Oxidation Temperature	(950,1100)/°C
H2 Flow	(5.5,6.5)/slpm
O_2 Flow	(4.5,5.5)/slpm
Oxidation time	(30,130)/min
Location	(-1 , +1)[1]

[1] The level of −1 in the location factor represents the gas input port of the furnace, and level of −1 represents the other port.

1-4244-0116-X/06/$20.00 ©2006 IEEE

The thermal oxidation process involves six input factors with two qualitative factors and four quantitative factors, and also include six two-factor interactions of interest, i.e. Oxidation Temperature*H_2 Flow, Oxidation Temperature*O_2 Flow 、 H_2 Flow*O_2 Flow 、 H_2 Flow*Location 、 O_2 Flow*Location and Oxidation* Temperature Location. Thus there are total 12 effects that should be estimated. Response surface Method (RSM) techniques are most effective when the number of input factor is limited to six or fewer [5]. The Factional designs are widely used in experiments involving several factors where it is necessary to study the joint effect of the factors on a response. The 2^K design provides the smallest number of runs with which K factors can be studied in a complete factorial design for the experiment of K factors [6]. Therefore we choose the 2^K factional designs.

The thermal oxidation process involves six input factors, and each factor at two levels. A full 2^K factorial experiment to determine all effects and interactions for six factors would require 2^6, or 64 experimental runs. In order to reduce the experimental cost, the effects of two-factor interaction of non-interest and higher order interactions are neglected. Therefore a 2^{6-2} fractional factorial design requiring only 16 runs is performed. This design uses a Resolution VI format which prevents main effects from being aliased with any other main effects as well as two and three-factor interactions. But some two-factor interactions are aliased with each other. We can choose the matrix to prevent the effects two-factor interactions of interest from being aliased with each other, so the effects of two-factor interactions of interest can be estimated.

The experimental sequence was randomized in order to avoid biases due to equipment aging during the experiment. In order to make sure that the experimental results can represent the actual process equipment, it is important to bring and maintain the operation under statistical process control (SPC).

B. Box-Cox transformation [6]

Because of scale effects and the nature of the measurements, the assumption of normality and homogeneity for residuals is often violated, and it results in a loss of information. The loss in information is usually reflected by a loss of power of the statistical test. The goal of data transformation is to change the scale by which the data are analyzed so that the residuals meet the assumption homogeneity (or violated lesser degree) and the model can explore more information.

Box-Cox transformation is used to transformed data. Box and Cox (1964) have shown how the transformation parameter λ in $y^* = y^{\lambda}$ may be estimated simultaneously with the other model parameters (overall mean and treatment effects) using the method of maximum

likelihood. The procedure consists of performing a standard analysis of variance on Eqn (3) for various values of λ

$$y^{(\lambda)} = \begin{cases} \dfrac{y^{\lambda}-1}{\lambda \dot{y}^{\lambda-1}}, & \lambda \neq 0, \\ \dot{y} \ln y & \lambda = 0, \end{cases} \quad (3)$$

where $\dot{y} = \ln^{-1}[(1/n)\sum \ln y]$ is the geometric mean of the observations.

The parameter λ may be determined using the method of the maximum likelihood estimate. Through derivation, Equation 3 that can be used to determines the value of λ is found as follows

$$\ln L(\lambda) = -\frac{n}{2}\ln SS_E(\lambda) \quad (4)$$

where n is the number of data, SSE(λ) represents the error sum of squares. The maximum likelihood estimate of λ is the value for which the value of LnL (λ) is maximum, or the error sum of squares, SSE(λ), is a minimum. The value of λ is usually found by plotting the graph of the likelihood estimate of λ (LnL(λ)) versus λ and reading the value of λ that maximizes LnL(λ) from the graph. In the paper the Response Transformation graph of LnL(λ) versus λ in SAS shown in Fig 1 is used to determine if the style of the response transform is appropriate.

$$\ln L(\lambda)^* = \ln L(\lambda)(1+\frac{t^2_{\alpha/2,v}}{v}) \quad (5)$$

In addition we recommend that the experimenter use simple choices for λ. An approximate $100(1-\alpha)$ percent confidence interval for λ can be found by computing Equation (3) and then by reading the corresponding confidence limits directly form the graph, as shown in Fig 1(B).

where v is the number of errors degrees of freedom. If this confidence interval includes the value $\lambda =1$, this implies that the data do not support the need for transformation.

The responses of the film thickness and uniformity are analyzed by the Box-Cox transformation. With the method mentioned above, we can get the Box-Cox analysis graph of the film thickness and uniformity, such as Fig 1. Through Fig 1 the optimum transformation is determined: the parameter λ for the film thickness is 0.2, but for uniformity is 1. Thus the film thickness D should be transformed to $\sqrt[5]{D}$, and the uniformity need not be transformed.

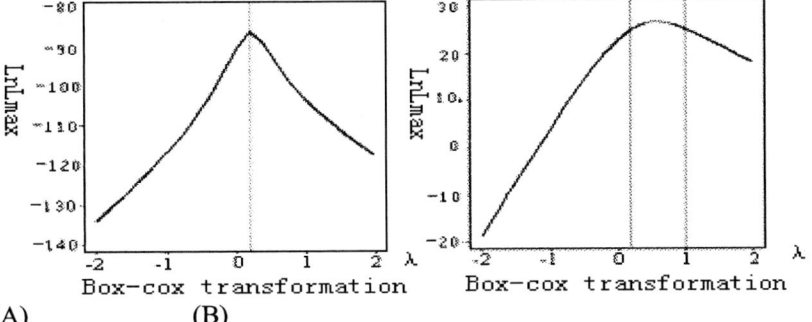

(A)   (B)

Fig. 1. (A)Film thickness Box-Cox analysis and (B)Uniformity Box-Cox analysis

III RESULTS AND DISCUSS

A. Responses model

The regression model used to fit the experimental data for the 26 - 2 fractional factorial design is as follows

$$y = \beta_0 + \sum \beta_i x_i + \sum \beta_{ij} x_i x_j + \varepsilon \qquad (6)$$

where y represents the response ,xi represents the input factors, β_i represents regression coefficients. The regression coefficients are estimated using the method of the least squares. The model only includes the factors and two-factor interactions that significantly impact on the responses. The P-value is used to determine which effects of factors are statistically significant. If the p-value is less than α =0.005(95%), then the effects of the input factors and two-factor interactions is of significance for the response with 95% confidence. Therefore we can get the models of $\sqrt[5]{D}$ and U.

Through the inverse transformation of $\sqrt[5]{D}$, we can find the value of the film thickness D.

R-square and adjusted R-square values are usually used to judge the adequacy of the regression model:

$$R^2 = \frac{\sum_{i=1}^{n}(\hat{y}_j - \overline{Y})^2}{\sum_{i=1}^{n}(y_j - \overline{Y})^2} \qquad (7)$$

$$R_{Adj}^2 = 1 - (\frac{n-1}{n-m})(1-R^2) \qquad (8)$$

where $\hat{y}_j$, $\overline{Y}$ and y_j are the value estimated by the model ,the observed value and the mean value of the response respectively. The number of runs is denoted by n, and m is the number of terms in the model. Because R^2

always increases as we add terms to the model, some regression model builders prefer to use an adjusted R^2 statistic .In general, the adjusted R^2 statistic will not always increase as variables are added to the model .In fact, if unnecessary terms are added, the value of the adjusted R^2 will often decrease. The closer to unity R_{Adj}^2 is, the more accurately the model fits the data. Adjust-R^2 value that represents the proportion of variation in the response explained by the model is 93.54% for D, 98.64% for $\sqrt[5]{D}$, and 69.82% for U. It is shown that the overall fit capability of the models is satisfactory.

B. Comparison analysis of the model precision before transformation and after transformation

From Adjust-R^2, it is shown that after the response transformation the model provide the better explanation of variation in the thickness film. Fig 2 shows the results of the measured film thickness versus the predicted film thickness. As can be seen, the model for the transformed data can fit to the measured values better than the model for the no-transformed data.

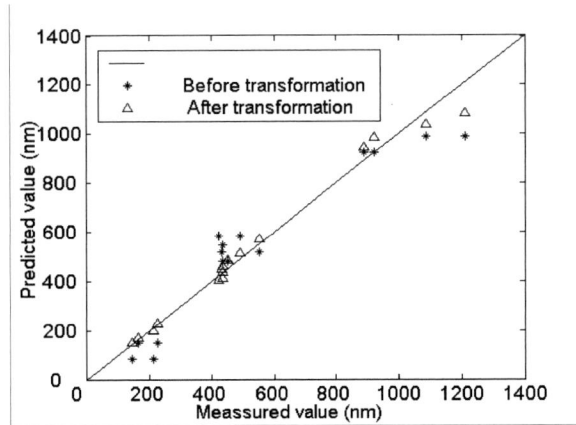

Fig. 2. Measured versus predicted the film thickness

The reasons can be found by the statistical significance analysis. The effects of the input factors Crystal orientation ,H_2 Flow and two-factor interaction O_2

Flow*Location on the response film thickness become significant after the response transformation. Only the effect of two-factor interaction H_2 Flow*Location becomes no-significant. Thus the model for the transformed response includes more items and explores more information

C. Process optimization using the equipment models

The models of the film thickness and uniformity have been used to design a new oxidation recipe, which exhibits improvement in the response uniformity. The recipe is designed to minimizing non-uniformity while meeting the special film thickness goal. The optimum recipe was determined using the Matlab Optimization Toolbox. After the optimum recipe is determined, an experiment was undertaken to confirm the improvement and the non-uniformity of film is improved from 0.2% to 0.08% with the other specification satisfied.

IV CONCLUSION

An economical 2^{6-2} fractional factorial experiment has been designed and conducted to characterize the oxidation film thickness, uniformity. In order to explore more information from the experiment data and improve the accuracy of the model to fit the data, the response is transformed using the Box-Cox transformation. In addition,

the models are used to optimize the process. Based on the results, the method of characterizing microcircuit process equipment using experimental design and the response transformation can be used to reduce the development cost and improve the model accuracy of semiconductor manufacturing.

REFERENCES

[1] Gary S.May , Jiahua Huang,Costas J.Spanos, Statistical Experimental Design in Plasma Etch Model[J] , IEEE Transactions on semiconductor manufacturing 1991.VOL.4.NO.2 83-98.

[2] Kyeong K.Lee, R.Bicknell-Tassius, G.dagnall, A.Brown, etc Statistical Experimental Design for MBE Process Characterization[J],IEEE/CPMT Int'l Electronics Manufacturing Technology Symposium (1996),378-385

[3] Byungwhan Kim, Kyungyoung Park, Modeling plasma etching process using a radial basis function network[J], Microelectronic engineering,(2004 article in press)

[4] Stephen A.Campbell, The science and engineering of microelectronic fabrication(second edition), Oxford University Press ,February 1, 1996.

[5] M.W.Jenkins,M.t.Mocella,K.D.Allen,and H.H.Sawin, Modeling pasma etching processed using response surface methodology [J],Solid State Tech., Apr. 1986

[6]Montgomery DC. , Design and analysis of experiments New York:Wiley ,1997.

PTC Effect and Structure of Polymer Composites Based on Polypropylene/Co-Polyamide Blend Filled with Dispersed Iron

A. Kanapitsas, C. Tsonos, E. Logakis, C. Pandis, P. Pissis,
E. Kontou, Y.P. Mamunya, E.V. Lebedev, C.G. Delides

Abstract – The temperature dependence of resistivity, structure and thermal expansion of composites based on polymer blend polypropylene/co-polyamide (PP/CPA) filled with dispersed iron (Fe) has been studied. The dependence of conductivity on filler content shows percolation behavior with the values of the percolation thresholds equal to 5 vol.% for the filled blend PP/CPA-Fe with two steps character of the conductivity curve. The evolution of structure of the composite PP/CPA-Fe demonstrates transition from polymer matrix CPA-Fe with inclusions of PP through co-continuous phases of both CPA-Fe and PP to PP matrix with inclusions of CPA-Fe. Such a structure occurs due to localisation of the filler only in one of the polymer phases, namely in CPA. The PP/CPA-Fe composites demonstrate PTC effect with the presence of a resistivity plateau. In such a system the PTC is caused by the break of the conductive structure of the filler inside the CPA-Fe phase due to the morphological changes in the vicinity of the melting temperature of CPA.

I. INTRODUCTION

A specific feature of composites with conductive dispersed fillers is the presence of the so-called percolation threshold, i.e. of the value of the filler content where a sharp transition conductive/insulating state occurs. One of the peculiar characteristics of the filled semi-crystalline polymers is the presence of the positive temperature coefficient of resistance (PTC) which has reversible character [1, 2]. High performance of PTC effect for the polymer blend filled with metal particles (low initial resistivity, high intensity of PTC effect, high reproductibility) makes such composites proper for

A. Kanapitsas and C. Tsonos are with the Technological Educational Institute of Lamia, Department of Electronics, 3rd km. Old National Road Lamia – Athens, 35100 Lamia, Greece, E-mail: kanapitsas@teilam.gr

A. Kanapitsas, E. Logakis, C. Pandis, P. Pissis are with the National Technical University of Athens, Department of Physics, Zografou Campus, 157 73 Athens, Greece.

Y.P. Mamunya, E.V. Lebedev are with the Institute of Macromolecular Chemistry of National Academy of Sciences of Ukraine, 48, Kharkivske chausse, 02160 Kyiv, Ukraine.

E. Kontou is with the National Technical University of Athens, Department of Mechanics, Zografou Campus, 157 80 Athens, Greece.

C.G. Delides is with the Technological Educational Institute of Kozani, Laboratories of Physics and Materials Technology, 50100 Kila, Kozani, Greece.

applications as the material for temperature sensors, thermistors, self-regulating heaters, current-limiting devices and disruptors.

II. EXPERIMENTAL

For the preparation of the composites the following materials were used: *co-polyamide* (CPA) PA6/PA6.6/PA6.10 with melt flow index (MFI) 11.9 g/10 min, *polypropylene* (PP), with MFI = 2.0gr/10 min, *iron powder* (Fe) with average size of particles is 3μm and shape of the particles close to spherical one. The composites based on the polymer blend were prepared in two stages. First a master batch CPA-Fe with 35 vol. % of the iron powder was prepared, then on the second stage the ground master batch was mixed with pure PP in the needed relation and the composites were subjected to the homogenisation in the extruder. Composites based on PP and CPA were named PP-Fe and CPA-Fe respectively, composites based on the polymer blend were named PP/CPA-Fe. Details on preparation and experimental methods used have been given elsewhere [1].

III. RESULTS AND DISCUSSION

Structure of the composites

During the second stage of the composite preparation, owing to the significant difference of the viscosities of PP and CPA (MFI of CPA is much higher than that of PP), the filler localizes in the polymer phase with lower viscosity, namely in CPA. A change of the composite composition leads to a dramatic alteration of the structure. The composite PP/CPA-Fe with content of Fe equal to 30 %, close to that of the master batch (35 vol. %) and accordingly with small amount of PP (14 vol. %), demonstrates the structure of the matrix of filled CPA with isolated inclusions of PP (Fig. 1-f). Increase of PP content in the composite results in incorporation of the PP particles and creation of extended inclusions of the PP phase (Fig. 1-e). Further increase of the PP content leads to merging of the PE inclusions and to creation of interpenetrating networks of two phases: the PP phase and the CPA-Fe filled phase (Figs. 1-c and 1-d). In this case conductive and non-conductive phase are co-continuous. Further increase of PP content results in the collapse of the CPA-Fe network and the composite exhibits the structure of a matrix of PP with the isolated inclusions of CPA-Fe, i.e. the island structure is formed (Fig. 1-b). In this case the conductive PP phase is in a form of separated inclusion in a conductive phase matrix. High predominance of PP in the composite (that corresponds to low content of Fe, for the composition of composites see the legend of Fig. 1) decreases the size of the inclusions of CPA filled with Fe (Fig 1-a, b). Such an evolution of structure demonstrates as a result a phase inversion, namely the CPA-Fe matrix with inclusions of PP transforms into the PP matrix with inclusions of CPA-Fe (compare Figs. 1-a and 1-e with each other). Due to such a structure of the composite PP/CPA-Fe, the filler interacts only with CPA, whereas PP interacts with the filled phase of CPA-Fe. An interpenetrating network of PP and CPA-Fe occurs in the interval of the filler content 7-10 vol.%. Such peculiarity of the structure of the composites PP/CPA-Fe allows us to accept a structure model with two co-existing phases, namely pure PP and filled CPA-Fe. In the latter one the local content of filler is identical to that of the master batch (35 vol.%) and is not changed during the process of mixing with pure PP at any ratio.

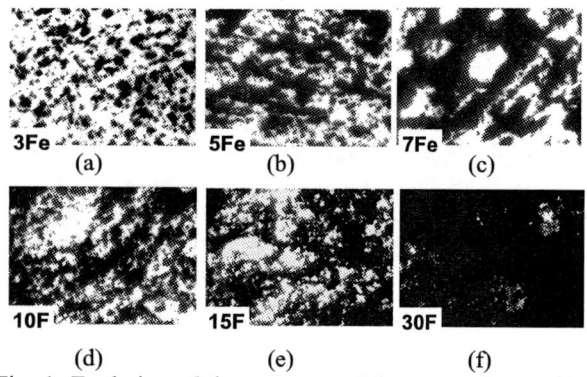

3Fe (a)   5Fe (b)   7Fe (c)

10F (d)   15F (e)   30F (f)

Fig. 1. Evolution of the structure of the composites PP/CPA-Fe with change of their composition: a - 91/6-3, b – 86/9-5, c – 80/13-7, d – 71/19-10, e – 57/28-15, f – 14/56-30.

Electrical conductivity of the composites

The dependence of conductivity of the composites on the volume filler content φ is shown in Fig. 2. The values of the percolation threshold φ_c for the composites PP-Fe, CPA-Fe and PP/CPA-Fe are 0.12, 0.26 and 0.05 respectively. In this case the conductivity can exist under the conditions of double percolation [3, 4], when first the continuity of the conductive filled polymer phase is necessary and second the conductive filler inside the polymer component should create the conductive network. The structure model assumes the maintenance of constant value of the filler content (namely 35 vol. %) in the CPA-Fe phase during diluting of the master batch by pure PP. Hence, the percolation threshold in the composite PP/CPA-Fe is related to the process when the conductive filled phase of CPA-Fe appears/disappears with change of the composite composition. The dependence of conductivity on the filler content for the composites PP/CPA-Fe is not monotonous, showing a plateau after the jump of conductivity at the percolation threshold. Such "two-step" rise of conductivity with increase of the filler content was earlier described in ref. [4] for the filled polymer blends, when the filler is located in one polymer phase only or at the boundary of two polymer phases. In our case such a plateau exists between 6 and 10 vol.% of filler, when the structure of the composite exhibits co-continuous networks of PP and CPA-Fe. With increase of Fe content, when the conductive phase CPA-Fe transforms into the matrix, the conductivity rises again owing to the increase of the amount of the conductive component. In the case of composite systems scaling low may used [1, 5]:

$$\frac{\sigma - \sigma_c}{\sigma_m - \sigma_c} = \left(\frac{\varphi - \varphi_c}{F - \varphi_c} \right)^t \quad (1)$$

where σ_c is the conductivity of the composite at the percolation threshold, σ_m is the maximal value of the composite conductivity, F is the value of packing-factor (equal to maximal content of the filler in the volume of the composite).

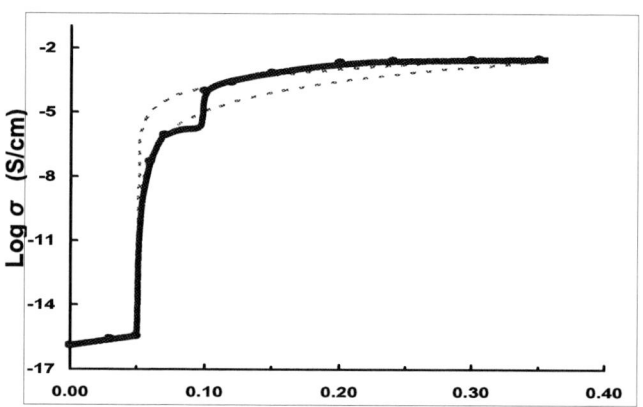

Fig. 2. Dependence of the conductivity on filler content for the composites PP/CPA-Fe.

From the calculations of the theoretical curves (dotted lines in Fig. 2) we found the following parameters:

PP/CPA-Fe	t	φ_c	F	$\log\sigma_c$	$\log\sigma_m$
1st step	3.2	5	35	-15.5	-2.48
2nd step	2.1	5	35	-15.5	-2.32

Dynamic Mechanical Analysis measurements

The temperature dependence of storage modulus E', loss modulus E'' and $\tan\delta$ ($=E'/E''$) of PP, PP-Fe, CPA, CPA-Fe and PP/CPA-Fe composites reveals three transition regions observed in the plots, not shown here. The low temperature $\tan\delta$ peaks for CPA and CPA-Fe composites located at -144^0C and -75^0C for γ and β relaxations respectively, as well as α-relaxation peak observed at 20^0C, does not affected by the presence of filler content in the composites. The $\tan\delta$ spectra of PP/CPA and PP/CPA-Fe composites dominated by the presence of two secondary relaxations at -140 and -75^0C which are more intense than those of PP system. At temperature region 20 – 40⁰C a relaxation peak was recorded as the contribution of PP and CPA α transitions. A significant increase in α peak magnitude with increasing filler content, as well as, no any temperature position changes of PP/CPA-Fe α peaks was observed [6].

Dielectric Relaxation Spectroscopy Measurements (DRS)

Fig. 3 shows results obtained by DRS: ε' and ε'' versus the temperature for the 95CPA-5Fe composite at three different constant frequencies. The data have been recorded isothermally by frequency scanning and have been repotted here to facilitate a comparison with DMA and DSC data. Two dipolar relaxations, a secondary γ relaxation at T = -120⁰C and a secondary β relaxation at T = -50⁰C can be observed, in agreement with DMA measurements The peaks shifted to higher temperatures with increasing frequency. At higher temperatures conductivity effects dominate $\varepsilon''(T)$ behavior. This conductivity effect masks the α relaxation peak.

Thermomechanical Analysis measurements (TMA)

In many references, the temperature dependence of the resistance of the filled polymer systems is explained in terms of the thermal expansion of the polymer matrix, leading to gradual destruction of the conductive chains of filler during the heating of the composite and, consequently, to increase of the resistance. This effect is reversible, because the created gaps between conductive particles as a result of thermal expansion vanish after cooling and the conductive chains are restored. However, other explanations of the PTC effect have been proposed as well by complicated changes of the polymer morphology under melting that disturbs the initial distribution of the conductive particles [1, 7, 8].

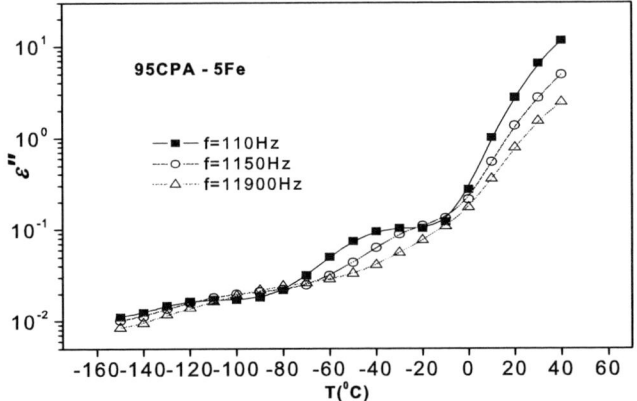

Fig. 3. Dielectric losses ε'' at different frequencies, against temperature for 95CPA-5Fe composite.

The curves of thermal expansion of the composites PP/CPA-Fe with various contents of components are shown in Fig. 4. The curves show a close to linear dependence in the temperature interval 30 – 100⁰C with the slope corresponding to the coefficient of linear thermal expansion *a*. It can be seen that the change of the

composite composition does not influence on the value of α (α_{PP} = 1.8x10⁻⁴ ⁰C⁻¹, α_{CPA-Fe} = 1. 78x10⁻⁴ ⁰C). When the temperature reaches the polymer melting region (T_{mCPA}=120⁰C, T_{mPP}=160⁰C) the expansion is substituted by the deformation of the sample under loading. In such a way the melting temperatures of the pure polymers can be defined, in good agreement with DSC data. For the filled composites the CPA phase melting temperature is shifted by 10⁰C to higher values as compared to pure CPA, whereas the PP phase melting temperature shifts to lower values for filler content 15 – 30 vol %. In the region of intermediate compositions (7 – 10 vol.% Fe) two peaks (or shoulders) are present on the curves, caused by the melting of two polymer phases in the composite. When the filler content is less than 5 vol.% the polymer matrix is PP and when the filler content is more than 15 vol. % the polymer matrix is filled CPA-Fe. Hence, the conductive continuous structure of the CPA-Fe phase appears in the range of phase inversion and the composite can change its conductive properties when this structure transforms into the CPA-Fe matrix with inclusions of PP.

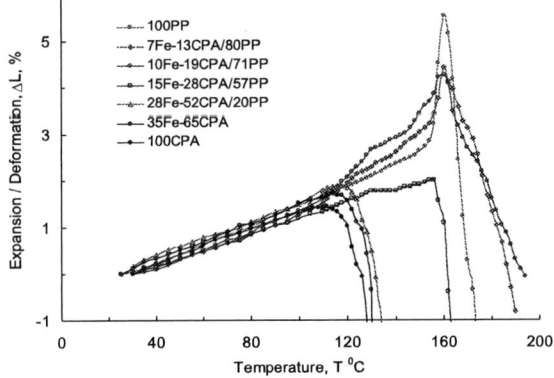

Fig. 4. Thermal expansion of the composites PP/CPA-Fe with various relationships of the components.

Differential Scanning Calorimeter measurements (DSC)

Fig. 5 shows DSC thermograms of the 31PP/45CPA-24Fe composite at temperature range -50 to 180⁰C. The two endothermic peaks observed define the melting range of CPA phase at T= 120 – 130 ⁰C and the peak at T=160⁰C the melting of crystalline PP phase. The observed peaks provides evidence for the biphasic character of the composite. No any temperature shifting of PP and CPA crystalline phase melting temperatures was recorded with filler content. It is necessary to have in view that during first heating/cooling circle recrystallization takes place, as a result a melting range 120 – 130 ⁰C appears instead of melting point at 130⁰C for CPA which is a copolymer with complex crystalline phase. The observed enthalpy jumps at T=30 and 40⁰C may be contributed with the glass transition temperatures of PP and CPA amorphous phases respectively.

Temperature – electrical characteristics of the composites

The temperature dependence of the resistivity for the

composites is given in Fig.6. The curves recorded during the second run as evidently a stabilization of the composite structure takes place during first heating/cooling cycle. The curves show great growth of resistivity under heating (PTC effect), which related with the particular structure of the conductive phase, in the temperature region of 80-100^{0}C for the composites with filler content higher than percolation threshold (> 5 vol% Fe). In contrary, the resistivity of the pure polymers decrease with increasing temperature. The master batch CPA-35Fe shows resistivity maximum at T=110^{0}C. After introduction of PP into the master batch the peak shifts to lower temperatures.

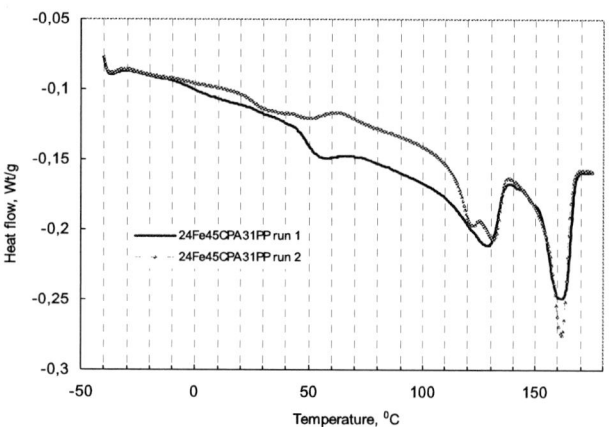

Fig. 5. DSC thermograms of 31PP/45CPA-24Fe composite.

The resistivity values of PP/CPA-Fe composites with filler content more than 15vol% coincide with those of pure CPA at temperature range 100-160^{0}C, whereas those of composites with filler less than 6 vol% are close to pure PP resistivity values. The composites with filler content 6 -10 vol.%, in which their morphology are in the form of two co-continuous networks of PP and CPA-Fe, shows resistivity values between those of master batch and pure PP. The presence of the plateau (or sligthly decreasing of resistivity region) at resistivity values observed at temperature region of 100 to 160^{0}C (i.e. between the melting temperatures of CPA and PP crystalline phases respectively) reduces the undesirable NTC effect which occurs at T > 160^{0}C. The NTC effect is a result of the increase of conductivity of the pure polymer which can be caused by charge release from traps, increase of the charge mobility and presence of ionic conductivity. The appearance of PTC effect is caused by break of the conductive structure of the filler inside the CPA-Fe phase due to the morphological changes in the vicinity of the melting temperature of CPA and causes the appearence of the peak at T_p. The creation of additional amorphous phase during melting of the crystalline regions leads to interaction between the polymer melt and the Fe conductive phase and promotes the formation of additional gaps between the filler particles, as well as the agglomeration of the particles that gives additional contribution to increase of resistivity [3].

IV. CONCLUSIONS

A transition from the structure of nonconductive matrix with conductive inclusions to the structure of conductive matrix with nonconductive inclusions is realized through phase inversion with co-continuous conductive and nonconductive phases. The co-continuous structure of composites leads to an appearance of a plateau on the conductivity curves. The reason of such differences is the rheology of polymer blend: the higher is the difference between viscous of polymer phases, the more narrow is the region of phase inversion and more shifted to the smaller content of low viscous polymer phase.

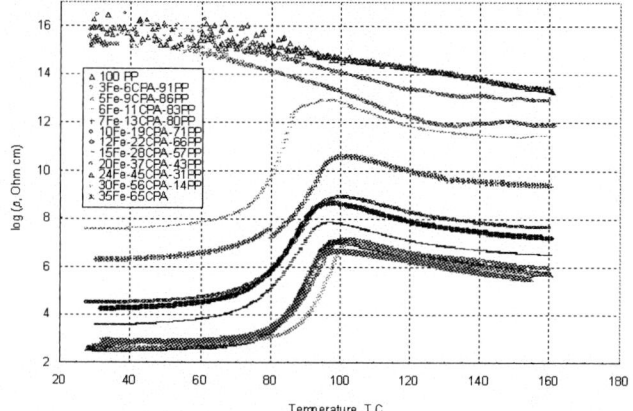

Fig. 6. Values of log resistivity vs temperature for the PP, CPA-Fe and PP/CPA-Fe composites.

The composites displays PTC effect, with the presence of a plateau at resistivity values within the temperature region between melting points of CPA and PP phases, caused by break of the conductive structure of the filler inside CPA-Fe phase due to the morphological changes in the vicinity of CPA melting temperature.

REFERENCES

[1] Y. Mamunya et.al. *"PTC effect and structure of polymer composites based on polyethylene/ polyoxymethylene blend filled with dispersed iron"*, Journal of Appl. Polym.Sci., submitted for publication, (2005).

[2] M. *Narkis*, S. Srivastava, R. Tchoudakov, O. Breuer, Synthetic *Metals*, **113**, 29 (2000).

[3] M. *Sumita*, K. Sakata, S. Haykawa, K. Asai, K. Miyasaka, M. Tanemura, *Colloid Polym. Sci.*, **270**, 134 (1992).

[4] M.Q. Zhang, G. Yu, H.M. Zeng, H.B. Zhang, Y.H. Hou, *Macromolecules*, **31**, 6724 (1998).

[5] Y.Mamunya et.al, *Polym.Eng.Sci.*, 42, 90 (2002)

[6] R.Kotsilkova, D.Fragiadakis, P.Pissis, *J.Appl. Polym. Sci., Part B, Polym.Phys.*, **43**, 522-33 (2005)

[7] J.F. Feller, I. Linossier, G. Levesque, *Polym. Adv. Technol.*, **13**, 714 (2002).

[8] J-F. Zhang, Q. Zheng, Y-Q. Yang, X-S. Yi, *J. Appl. Polym. Sci.* **83**, 3117 (2002).

This work was fulfilled under the financial support of the programm "Archimedes II" (financed 75% by the European Commission and 25% by the Greek State.

2006 25th International Conference on Microelectronics

A New Method of Evaluation of Liquidus Temperatures of Ternary Alloys

A. Prijić, Z. Prijić, B. Pešić

Abstract – This paper presents results of liquidus surface estimation of ternary Sn-Pb-Bi and Sn-Pb-In alloys applying a method which is based on the property of 3D ternary phase diagrams to form smooth surfaces in specific composition and temperature range. Standard program for solid modeling is applied for construction of liquidus surface of an alloy using phase diagrams of three subsystem binary alloys and only a few experimental data. Advantage of presented method is the possibility of 3D graphical representation and determination of liquidus temperature of arbitrary composition of considered ternary alloy.

I. INTRODUCTION

Surface mount assembly processes for products that does not experience harsh temperature environments, as well as production of electronic devices whose operation is based on their temperature response, such as thermal cutoff fuses, often use low-temperature solders as substantial materials. For that purpose, various low-melting (fusible) alloys are used, since one single alloy may not be appropriate as a universal solution. These alloys generally melt below 250°C and are multicomponent, usually composed of tin, lead, silver and one of low melting element such as bismuth, indium, cadmium or gallium.

In the design, development, processing and understanding of multicomponent alloys, their phase diagrams (visual representation of the state of a material as a function of composition, temperature and pressure) play an important role. Since experimental determination of multicomponent phase diagrams can be time-consuming, expensive and difficult, theoretical evaluation based on a far less number of experimental data is substantial. While the graphical form of binary phase diagrams at constant pressure is well-known, for ternary systems a three-dimensional representation is needed. Therefore, liquidus and solidus surfaces of ternary systems are usually projected to a 2-D base by a series of isotherm and cotectic lines. Determination of these lines and overall phase diagrams of ternary systems have been one of the main interest of the CALPHAD (CALculation of PHAse Diagrams) group [1] during past 3 decades.

In conventional methods, for the calculation of phase equilibria in multicomponent systems, minimization of the total Gibbs energy G of the system is necessary [2]. Contribution of the Gibbs energy of each phase φ that takes part

A. Prijić, Z. Prijić, and B. Pešić are with the Department of Microelectronics and Microsystems, Faculty of Electronic Engineering, University of Niš, Aleksandra Medvedeva 14, 18000 Niš, Serbia & Montenegro, E-mail: aneta@elfak.ni.ac.yu

in the system equilibrium - G^φ is determined by the mole number n_i, thus function:

$$G = \sum_{i=1}^{p} n_i \cdot G_i^\varphi = \min \qquad (1)$$

has to be resolved. For this calculation appropriate thermodynamic function that describes each phase has to be assigned. The CALPHAD method employs variety of models to identify the Gibbs energy dependence on temperature, pressure and concentration of various phases. On the basis of thermodynamic database (collection of the thermodynamic properties of the phases in multicomponent system) and appropriate calculation software, by this method an optimized set of parameters for adopted free-energy model is obtained [2].

Frequently used software packages for phase diagram calculations are *FactSage* (previously *ChemSage*), *Thermo-Calc*, *MTDATA*, so-called *Lucas* programs, *Parrot* program [2] and newly released *PANDAT* software [3]. Also, well established thermodynamic databases are Scientific Group Thermodata Europe (SGTE) [4] and FACT [5] data banks. All these software and databases are commercial, as well as, most of the published results of thermodynamic assessments of ternary systems obtained by CALPHAD method.

In this paper a new approach in estimation of ternary phase diagrams (particularly liquidus surfaces) on the basis of phase diagrams of three subsystem binary alloys and a few experimental data is presented. The new method is used for determination of liquidus surfaces of low-melting Sn-Pb-Bi and Sn-Pb-In alloys and results are compared with those available in the literature.

II. THEORETICAL BASIS

State of the material consisting of three components as a function of composition and temperature generally can be graphically presented by 3-D surfaces. Geometrically, ternary phase diagram is bounded by a regular equal three-sided prism, formed when three binary phase diagrams of subsystems are joined together. Height of the prism is determined by temperature axis. Liquidus surface of the system consists of one or more smooth 3-D surfaces bounded by liquidus lines of binary phase diagrams.

Standard programs for 3-D surface modeling incorporate NURBSas mathematical tool for construction of guided (or bounded) surfaces. Thus, complex surfaces can be created on a basis of predefined guiding lines and so-called keypoints. This enables one to apply these programs

1-4244-0116-X/06/$20.00 ©2006 IEEE

for construction of ternary liquidus surface using binary liquidus lines as guiding ones and experimentally determined ternary liquidus temperatures of specific alloy compositions as controlling points.

Number of liquidus subsurfaces, their cutting lines (cotectic lines) and eutectic or peritectic points is governed by phase diagrams of constituent (isomorphus or eutectic) subsystems.

III. LIQUIDUS SURFACE CONSTRUCTION

Proposed alternative method is used for determination of liquidus surfaces of Sn-Pb-Bi and Sn-Pb-In alloys. These are low-melting alloys and knowledge of liquidus temperatures of their specific predefined compositions is of importance in solder metallurgy. As a tool, almost any of standard programs for solid modeling could be employed.

A. Sn-Pb-Bi Alloy

For construction of the liquidus surface binary phase diagrams of subsystems (Sn-Pb, Sn-Bi, Pb-Bi) are obtained from the literature [4], [6]. Graphics of liquidus lines are used as bounding lines, while keypoints are selected from collection of measured data given by the world known manufacturer of low-melting alloys [7]. In Table I values of composition and liquidus temperatures of selected alloys are summarized.

TABLE I
MEASURED LIQUIDUS TEMPERATURES OF Sn-Pb-Bi ALLOYS

Alloy number	mass % Sn	mass % Pb	mass % Bi	T liquidus (°C)
1	60	14,5	25,5	180
2	46	46	8	173
3	27	51,5	21,5	170
4	37	42	21	152
5	30,8	38,4	30,8	139
6	22	22	56	104
7	22	28	50	100
8	15,50	32	52,5	95

Constructed 3-D liquidus surface graphically is presented in Fig. 1. Since this system consists of eutectic subsystems, where Pb-Bi binary alloy forms incongruently melting compound, in 3-D graph four subsurfaces can be distinguished. These subsurfaces intersect each other along cotectic lines which meet at eutectic and peritectic points.

2-D representation of liquidus surface is given by isothermal (at 300°C, 250°C, 200°C and 150°C) and cotectic lines projections as shown in Fig. 2. In this figure keypoints from Table I are denoted, and comparison is made with calculated liquidus projection from [8] presented by black bolded lines. Good agreement between two diagrams exists for all isothermal lines above 150°C. Also, determined eutectic points are close. Discrepancies can be observed for

isothermal lines at 150°C. It is evident that reference point 4 from Table I (for 152°C) is very close to isotherm determined by the new method. On the other hand, reference point 5 (for 139°C) is close to isotherm depicted from [8]. This implies conclusion that our 150°C isotherm is more realistic one. Peritectic points also diverge as a consequence of differences between isotherms. They are 139,6°C (29%Sn-46%Pb-25%Bi) by the new method and 135,7°C (21,9%Sn-47,6%Pb-30,5%Bi) from [8].

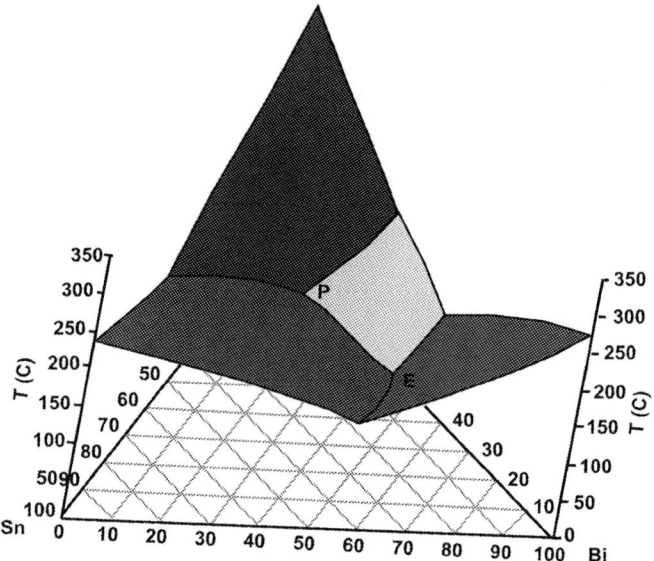

Fig. 1. Liquidus surface of ternary Sn-Pb-Bi alloy.

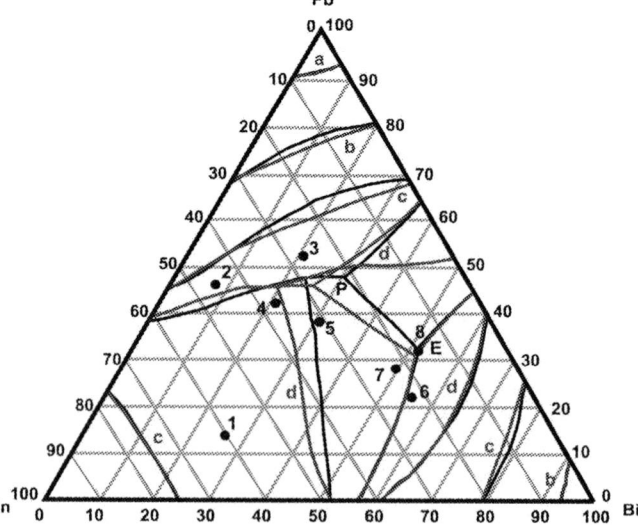

Fig. 2. Comparison of ternary phase diagram of Sn-Pb-Bi alloy evaluated by the new method and from [8]. Denoted isothermal lines are: a) 300°C, b) 250°C, c) 200°C, d) 150°C; keypoints are from Table I.

396

B. Sn-Pb-In Alloy

Phase diagram and thermodynamic description of this system are rare in the literature [9], [10]. For description of three binary subsystems (Sn-Pb, Sn-In, Pb-In) liquidus lines from [5] are used. Measured liquidus temperatures of 7 ternary alloys, summarized in Table II, are taken from specification of Indium Corporation [7] (alloys 1-4) and as referred in [10] (alloys 5-7). These experimental data are used as keypoints for construction of liquidus surface along with binary bounding lines.

TABLE II
MEASURED LIQUIDUS TEMPERATURES OF Sn-Pb-In ALLOYS

Alloy number	mass % Sn	mass % Pb	mass % In	T liquidus (°C)
1	37,5	37,5	25	181
2	70	18	12	167
3	54	26	20	154
4	40	20	40	130
5	60,5	31,5	8	170
6	57	20	24	147
7	50,5	31,5	44	124

Liquidus surface of Sn-Pb-In system in 3-D representation is given in Fig. 3. There are three main subsurfaces which intersect along cotectic lines and determine one peritectic point.

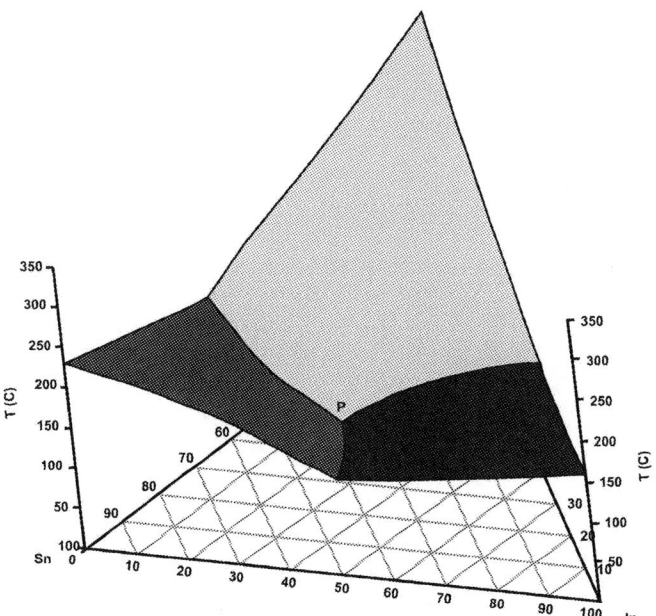

Fig. 3. Liquidus surface of ternary Sn-Pb-In alloy.

Evaluated ternary phase diagram is presented in 2-D by projection of cotectic and isothermal lines in Fig. 4. It is compared with calculated cotectic lines from [10], which are presented by bold black lines. Differences between cotectic lines, are based on different choice of referenced experimental data. From the same reason, peritectic points determined by the new method - 123,2°C (42,5%Sn-17,5%Pb-40%In) and from [10] - 134,8°C (47,5%Sn-27,6%Pb-24,9%In) diverge considerably. Namely, results for isothermal cut in Sn-Pb-In system at 172°C from [10] do not satisfy condition of liquidus temperature for alloy No. 2 from Table II, which is used as keypoint in this work. Since this point is taken from collection of measured data of the world leading manufacturer of indium alloys [7], here presented liquidus surface is taken as correct for adopted set of experimental points. Moreover, from Fig. 4 it is noticeable that isothermal lines for 250°C and 200°C are smooth, while two lines for 150°C have distortions as a result of liquidus surface adjustment to measured points. Overall, it is evident that lack of reliable experimental data imposes some uncertainties in evaluation of liquidus temperatures for any chosen methodology.

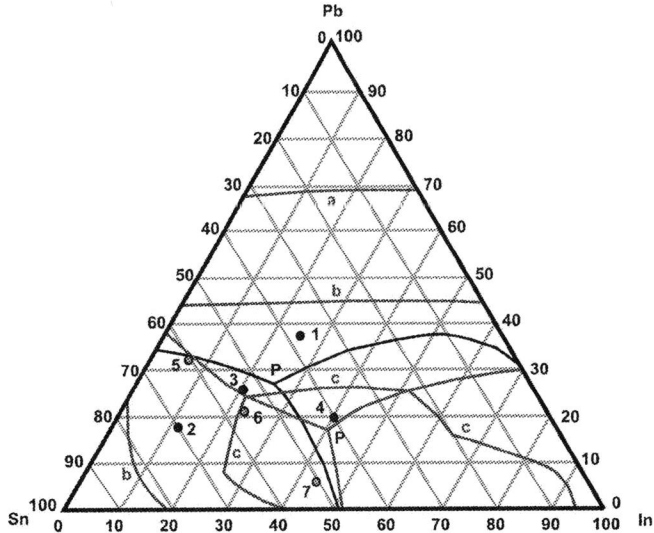

Fig. 4. Comparison of ternary phase diagram of Sn-Pb-In alloy evaluated by the new method and cotectic lines from [10]. Denoted isothermal lines are: a) 250°C, b) 200°C, c) 150°C; keypoints are from Table II.

Liquidus surfaces obtained by the proposed method enable determination of liquidus temperature for an alloy of any arbitrary composition. Namely, intersection of liquidus surface and the line that goes from desired alloy composition normally onto 2-D plane, gives point whose coordinate on temperature axis determines liquidus temperature.

IV. CONCLUSION

On the basis of standard 3-D surface modeling principles, new method of liquidus surface evaluation of ternary systems at constant pressure is proposed. Method is applicable for any ternary system whose phase diagrams of binary subsystems and a few appropriate experimental points are known. Liquidus surfaces of two low-melting ternary alloys are constructed as illustration of the proposed method. For any arbitrary composition of ternary system liquidus temperature can be easily obtained from 3-D surface by geometrical rules. It is shown that critical point in any methodology for thermodynamic assessment of multi-component systems is determination of reliable experimental data.

ACKNOWLEDGEMENT

This work was supported by grant of the Serbian Ministry of Science and Environmental Protection under contract No. TR6140.

REFERENCES

[1] CALPHAD, New York: Pergamon, 1977-2005.

[2] U. Kattner, "The Thermodynamic Modeling of Multicomponent Phase Equilibria ", *Journal of Materials, 1997,* Vol. 49, No. 12, pp. 14-19.

[3] S. L. Chen, F. Zhang, S. Daniel, F. Y. Xie, X. Y. Yan, Y. A. Chang, R. Schmid-Fetzer, and W. A. Oates, "Calculating Phase Diagrams Using PANDAT and PanEngine", *Journal of Materials*, 2003, No. 12, pp.48-51.

[4] http://www.sgte.org

[5] http://www.crct.polymtl.ca/fact/documentation

[6] http://www.metallurgy.nist.gov/phase/solder/bipbsn.html

[7] http://www.indium.com/products/tableofalloys.xls

[8] S. W. Yoon and H. M. Lee, "A Thermodynamic Study of Phase Equilibria. in the Sn-Bi-Pb Solder System", *CALPHAD*, 1998, vol. 22, No. 2, pp. 167–178.

[9] Z. Mei, H. Holder, and H. Vander Plas, "Low-Temperature Solders ", *Hewlett-Packard Journal*, 1996, No. 8, Article 10.

[10] N. David, K. El Aissaoui, J. M. Fiorani, J. Hertz, and M. Vilasi, "Thermodynamic Optimization of the In-Pb-Sn System Based on New Evaluations of the Binary Borders In-Pb and In-Sn", *Thermochimica Acta*, 2004, No. 413, pp. 127-137.

Session
Circuit Design

400

System-on-Chip Design: Engineering or Art

Z. Stamenković

Abstract - The paper presents a specific approach to SoC design, aimed to provide a library of configurable, extensible, and reusable modules (processors, various hardware cores, memories, etc.) for wireless applications. It familiarises you with the concept of library and its hierarchy. It also describes how to specify, synthesise, layout, verify, and reuse the library modules.

I. INTRODUCTION

Today's emerging sub-micron silicon technologies bring enormous gains in respect of chip complexity, but they also dramatically increase design and verification complexity. A System-on-Chip (SoC) integrates processor cores, memory blocks, and custom logic into a closed system that must be assembled and verified within a short timeframe [1]-[7].

The industry as we know it today relies on two ways to build systems. One camp gets their product to market quickly, but without much differentiation, by utilising Application Specific Standard Products (ASSPs) [8]. The other camp builds proprietary and differentiated systems on silicon by utilising Application Specific ICs (ASICs) [9]. The tradeoffs and benefits of each approach are well understood by most engineers.

The future promises a different breed of products that offer the time-to-market advantages of an ASSP, and the proprietary differentiation of an ASIC; behold the ASPP, the Application Specific Programmable Part. The specifics of this new breed of devices will unfold in the next chapter of our industry: Moore's Law begins to defeat the EDA industry's ability to inhabit real estate on silicon but fails to overcome the ability of creative and relentless engineers to utilise silicon smartly [10].

A SoC design has to deal with a wide range: it starts with a functional description on system level, where major function blocks are defined and no timing information is given. The other end of the spectrum is the result of the design process, where all functionalities described before are mapped to hardware and all hardware is defined down to Register-Transfer Language (RTL) level. At that point in time a cycle accurate model exists, which is ready for production. Fig. 1 depicts the domains and levels of abstraction in SoC design. The goal of SoC design paradigm is to make proper decisions in the hardware-software co-design. This is only possible by following a well-defined flow of design steps.

Z. Stamenković is with the IHP – Innovations for High Performance microelectronics, Im Technologiepark 25, 15236 Frankfurt (Oder), Germany, E-mail: stamenko@ihp-microelectronics.com

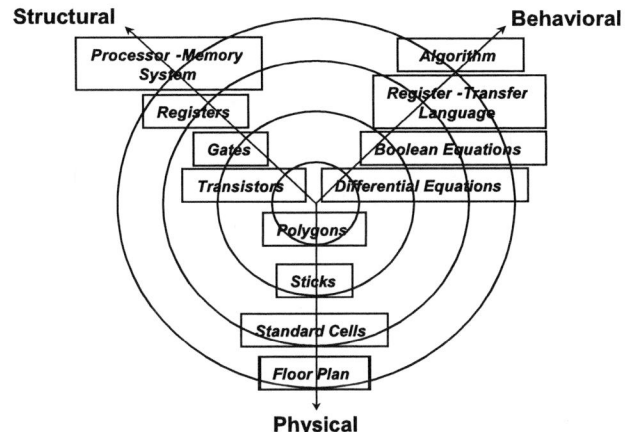

Fig. 1. Abstraction levels in SoC design [11]

In the next sections we describe main components of a system-on-chip and corresponding design flow. Section II is dedicated to configurable, customizable and extensible processors, and processor generators. Section III deals with standard cell libraries and memory generators. In Section IV we describe SoC design flow. Finally, we conclude the paper in Section V.

II. PROCESSORS

Processors have long been playing a role in embedded systems design, but fundamental changes in the economics, technology and complexity of systems are making processors even more central. Especially in wireless applications, embedded protocols and data-types are becoming more sophisticated. Paradoxically, each new processor must support these new algorithms and functions, yet must generally be radically smaller and more power efficient than previous solutions. Resolving this paradox justifies a new approach to processors – automatic generation of system-specific processors.

A. Configurable Processors

A configurable processor [12]-[14] can be configured according to the requirements of various applications and tasks. Configurations include one or more coprocessors, one or more caches, scratchpad memory, on-chip trace memory, on-chip buses, etc. Usually, the highest attention is paid to configuring caches. The configuration process supposes making decision on the size, associativity and organization of the instruction and data cache. To select appropriate configurations and choose the power-optimal one for specific application is a difficult task [15]-[22].

The configurable processor core usually implements the architecture with Memory Management Unit (MMU) interfacing between Execution Unit and Cache Controller. MMU can implement a simple Fixed Mapping Translation (FMT) or more complex Translation Lookaside Buffer (TLB) mechanism. It performs virtual to physical address translation and provides attributes for the mapped regions. The block diagram of the configurable core MIPS32 4KEp [12] is shown in Fig. 2. The core is divided into *required* and *optional* blocks.

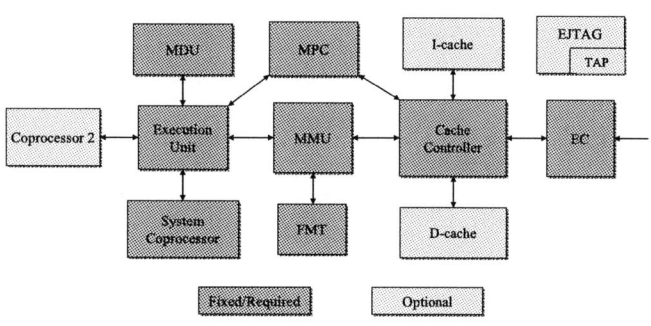

Fig. 2. MIPS32 4KEp core block diagram [12]

Master Pipeline Control (MPC) is responsible for recognizing and managing dependencies in the pipeline. This involves decoding each instruction and checking to see if there are interlocks that need to block the issue of the next instruction. Execution Unit implements fundamental 32-bit integer data manipulation functions. In addition to data calculations, it contains logic for branch determination, branch target calculation, and the load aligner. To conserve area, MIPS32 4KEp core uses an area-efficient iterative Multiply-Divide Unit (MDU). Multiplier performs an iterative one-bit shift-add every cycle. Multiplies are completed in 32 clock cycles. Multiply accumulates and multiply subtracts take additional two cycles to do the final accumulate step. Divides are also iterative, but do not have the data-dependent early-in optimization. System Coprocessor supports the virtual memory system and exception handling.

EC interface contains the logic to drive the external interface signals. Additionally, it contains implementation of the 32-byte collapsing write buffer. The purpose of this buffer is to store and combine the write transactions before issuing them at the external interface.

Instruction and data caches are fully configurable from 0 - 64 Kbytes in size. In addition, each cache can be organized as direct-mapped or 2-way, 3-way, or 4-way set associative. Load and fetch cache misses only block until the critical word becomes available. The pipeline resumes execution while the remaining words are being written to the cache. Both caches are virtually indexed and physically tagged to allow them to be accessed in the same clock that the address is translated. An optional Enhanced JTAG (EJTAG) block allows for single-stepping of the processor as well as instruction and data virtual address breakpoints.

Reusing the configurable processor MIPS32 4KEp, AMBA bus [23], memory controller and several system peripherals [24], we have designed a high performance, low power SoC for wireless applications [25] in short time with minimal resources. System architecture is presented in Fig. 3 and integrates both instruction and data scratchpad memories (ISPRAM and DSPRAM) of 8 KB each. A memory controller is attached to the AMBA advanced high performance bus (AHB). It provides an interface to an external flash memory and static RAMs. The slower AMBA advanced peripheral bus (APB) is attached to AHB via the bridge AHB-to-APB. Two UARTs and GPIO are connected to APB. The chip photo is shown in Fig. 4.

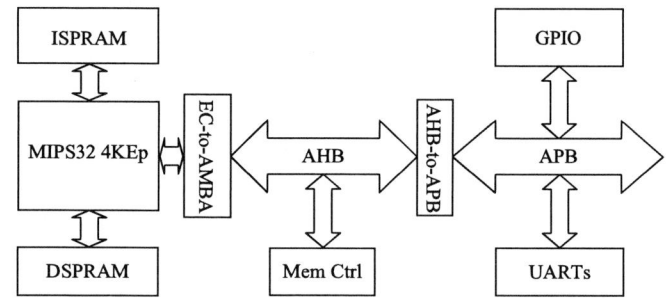

Fig. 3. System architecture [25]

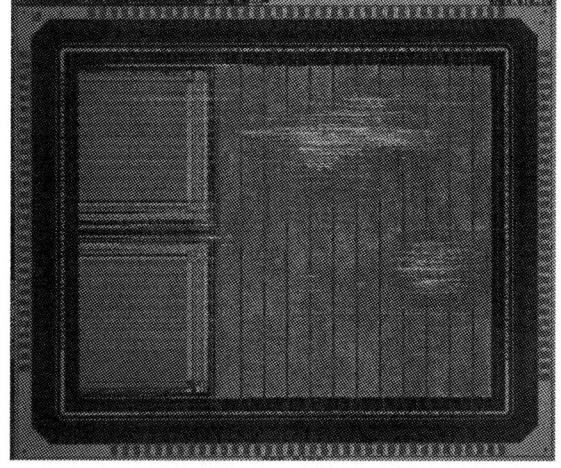

Fig. 4. Layout of the system

B. Customizable Processors

The basic motivations for the Von Neumann architecture developed in 1945 (e.g. hardware is expensive, applications are scientific computation, power is not an issue) are no longer relevant for the embedded systems being implemented in system-on-chip methodology. An approach which better exploits the characteristics of modern hardware technology to meet present day application and flexibility requirements can achieve efficiency gains of orders of magnitude in area and power over software solutions. For example, it is now possible to put down 1000's of multipliers on a chip, so the critical

402

task for computation intensive applications is to determine how to harness this potential parallel computation. The efficient exploitation of many parallel computation units is one way to achieve orders of magnitude reduction in area and energy consumption over a conventional architecture.

One way to do this is to develop low power domain specific architectures (Multiprocessors, Very Long Instruction Word (VLIW) processors or Superscalar processors) that explicitly capture and implement all the potential instruction-level parallelism. We present a processor customized for IEEE 802.11a MAC layer operations [26]. The implemented architecture exploits dedicated hardware for timing critical tasks.

The IEEE 802.11 MAC layer provides reliable data delivery for the upper layers over the wireless medium with data rate up to 54 Mbit/sec. It specifies how a computer on the wireless network gains access to receive and transmit data, and once communication is established how it is maintained. The basic medium access control technique used in the IEEE 802.11 standard is Carrier Sense Multiple Access with Collision Avoidance (CSMA/CA).

The first step of processor customization was to establish a detailed simulation model of the IEEE 802.11 protocol using Specification and Description Language (SDL). This model was used to verify the functional correctness of our design. On the other hand, it served as a basis for performance investigations to identify those parts of the protocol that require either software optimization (hand-optimized C code) or implementation in hardware in order to meet the real-time requirements. The top level structure of the SDL simulation model is shown in Fig. 5.

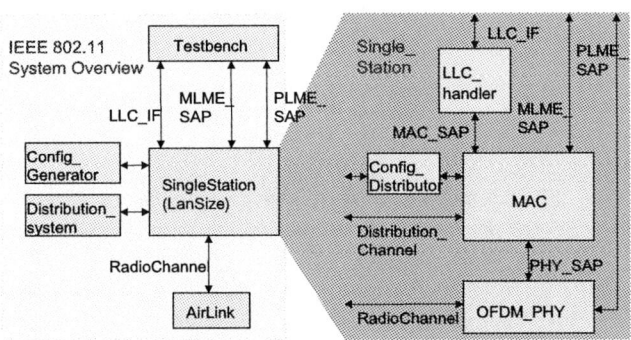

Fig. 5. SDL model of IEEE 802.11 WLAN [26]

We have generated the C model of the protocol and compiled it for the target MIPS processor. During simulations on MIPS32 4KEp, it turned out that excessive invocations of the SDL run-time system cause heavy overheads when executing the software. Therefore, we have replaced some of the automatically generated C code with hand-optimized code. It has helped us to speed up the execution, but due to the tight timing requirements of the protocol, the processor executing the software would have to be clocked at nearly 1 GHz in order to comply with the standard. This would, of course, lead to high power consumption, which is not feasible for operation in mobile

devices. Therefore, we have decided to implement some of the timing critical MAC functionalities into dedicated hardware. Based on profiling information from the C implementation and analysis of the real-time requirements specified in the standard, we have conducted hardware-software partitioning of our model. Among the timing critical algorithms implemented in hardware are CRC calculation, RC4 encryption and decryption as well as a large part of the frame reception process including address filtering and generation of the acknowledgment. All other functionalities that are not timing critical are executed in software.

The system architecture (Fig. 6) is based on MIPS32 4KEp that performs most of the MAC functionality. The core communicates with the rest of the system through the X-bus. Two targets are connected to the X-bus: General Purpose Input/Output (GPIO) and Peripheral Bus Controller (PBC). The interface to the physical layer is provided through EPP port implemented via GPIO. EPP represents a standard 8-bit parallel port specified by the IEEE 1284 standard. Fig. 7 shows the chip photo.

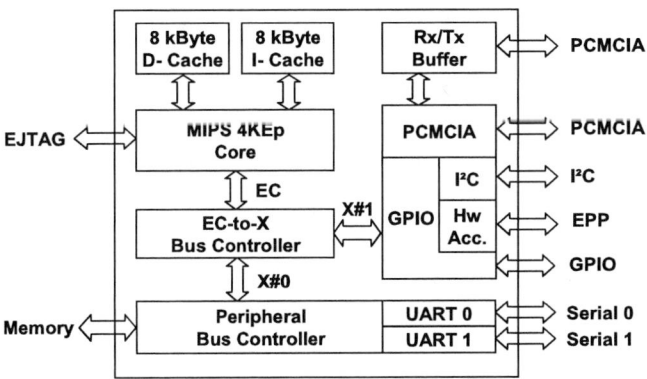

Fig. 6. Architecture of IEEE 802.11a MAC layer processor

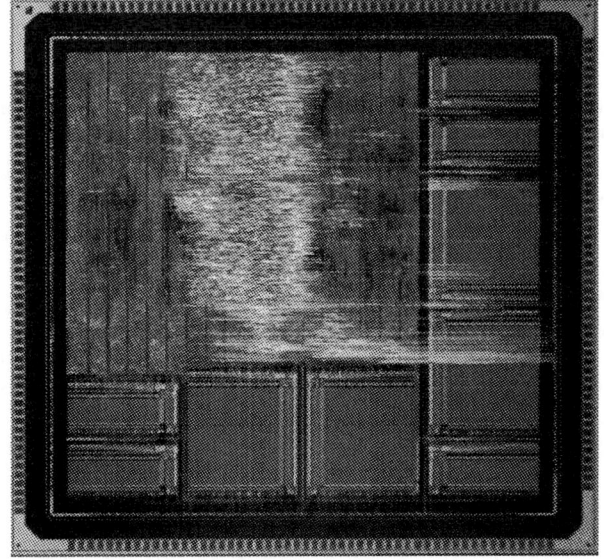

Fig. 7. Layout of IEEE 802.11a MAC layer processor

403

The timing critical MAC functionality is implemented into a hardware accelerator. This unit acts as a data pipe between the processor and IEEE 802.11 physical layer connected via EPP port. An interface to other external components has been provided through I²C bus and two serial ports. I²C bus controller is implemented as a part of GPIO. PCMCIA is connected to the processor via GPIO too. The serial ports are controlled by two UARTs connected via PBC. PBC also provides an interface to external SRAM, flash and other memory-like peripherals.

C. Processor Generators

Automatic processor generators allow chip architects to reliably discover and quickly describe the key instruction, memory, peripheral and interface functions required by their application. The generators produce a complete hardware design, verification and software development environment for that custom processor in minutes. The resulting designs consistently execute applications faster, dissipate less power and use less code than previous processors.

The real impact of automatic processor generation is improvement in the tradeoffs between silicon optimality and product development flexibility. Traditionally, designers have turned to hardwired designs whenever their problem was well-understood and unlikely to change. The complexity explosion, however, mandates that more of the system remain flexible, to accommodate fast design cycles, design bugs and market changes. Designers can now extend the processor with instruction sets unique to the data-types and computations of their application, yet use standard C/C++ programming to exploit these proprietary extensions in flexible ways. So the answer to the question "Software or Hardware" is this: hardware and software really can work together in new ways for system-on-chip design.

Generation of extensible processors [27], [28], and [29] addresses three architectural levels: the instruction set extensions (the designer specifies their functionalities), inclusion/exclusion of predefined blocks (scaleable register files, special registers, memories, interfaces, peripherals, selectable DSP coprocessors, built-in self-test logic, etc.), and parameterization (clock gating, cache size, size of almost every data path, number and type of execution units, number and type of peripherals, number and size of load/store units and ports into memory, size of instructions, number of operations encoded in each instruction, etc.). To customize an extensible processor to a specific application, one should start with profiling the application on the basis of a high-level language, probably C language. The profiling tools determine bottlenecks of the application. After the designer defined a frame within extensible instructions, predefined blocks and parameter settings are available, the optimal instruction set, used blocks and parameters with respect of the considered application are automatically calculated, and, the adapted compilers and

evaluation tools are generated (Fig. 8). Customizations are evaluated using these tools.

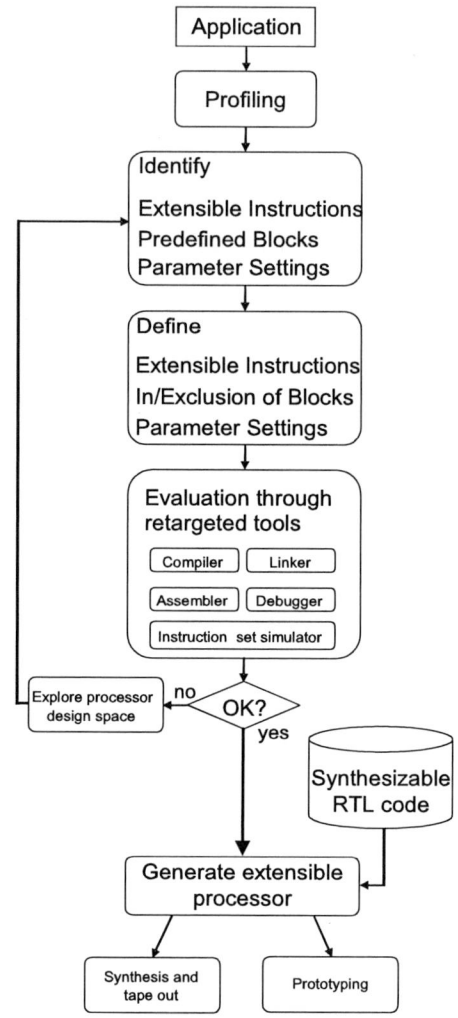

Fig. 8. Design flow of an extensible processor [27]

III. STANDARD CELLS AND MEMORIES

SoC design requires small area and high speed, but also low power consumption. Careful design of standard cells and memories (as the basic building blocks of a SoC) is crucial to obtain SoCs which are fast, compact, and easily testable out of the overall synthesis process. As one can imagine, criteria used to design standard cell libraries and memories optimized for area and speed differ from the ones applied to minimize power dissipation.

A. Standard Cell Libraries

A standard cell library is a collection of electronic devices implementing basic logic functions in a given technology. Characteristics of the library and its cells play a fundamental role in the process of automatic synthesis of digital circuits [30]-[35].

Building a standard cell library requires development of the physical layout for all the leaf cells ranging from simple gates to latches and flip-flops (Fig. 9). These can then be used to build arithmetic blocks like adders and multipliers. However, when the standard cell library must be ported to a new fabrication process, the physical layout of all the cells needs to be changed [36], [37]. Automation helps the process, but eventually expert designers and layout engineers create process-optimized layout that takes advantage of special layers, local interconnect, salicide and many other features unique to a given fabrication process.

Each memory generator is a set of various, parameterized generators: Layout generator generates an array of custom, pitch-matched leaf cells (Fig. 10); Schematic generator and Net-lister extracts a net-list used for both layout vs. schematic and functional verification; Function and Timing model generators create models for gate level simulation, dynamic/static timing analysis and synthesis; Symbol generator generates schematic; and special purpose generators such as Critical Path generator are used for both circuit design and timing characterization.

Fig. 9. Physical layout of a latch

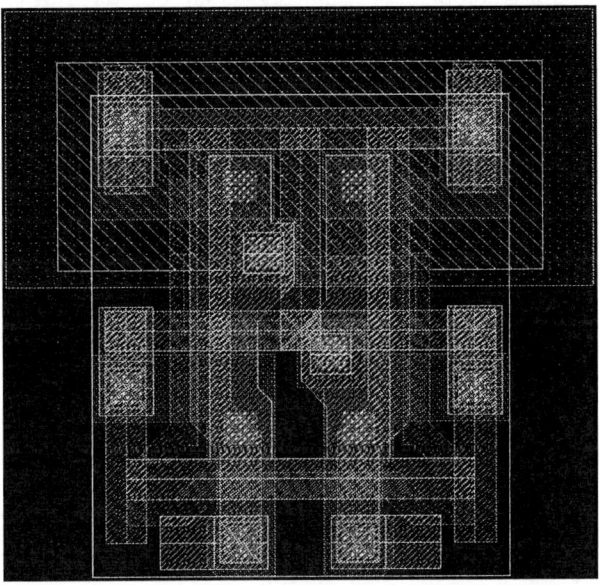

Fig. 10. Physical layout of a 6-transistor SRAM cell

IV. SoC Design Flow

Each standard cell contains necessary port and connectivity information required for simulation and synthesis. Additionally, data on typical rise and fall delays, load characteristics, maximum clock frequency and other specifications are available in the form of simulation and synthesis library interface files which can be used by industry standard simulation and synthesis tools.

B. Memory Generators

A memory generator [38]-[43] is a tool which can create memories (RAM or ROM blocks) in a range of sizes as needed. The customer usually wants a particular number of words (depth) and bits (width) for each memory ordered. Each of the final building blocks (physical layout) will be implemented as a stand-alone, densely packed, pitch-matched array. Using complex layout generators and adopting state-of-the-art logic and circuit design technique, today's embedded memories can reach extreme density and performance [44], [45]. Designers can choose the memory aspect ratio according to the requirements of SoC level layout.

A SoC design methodology assumes getting a design from the architectural level or RTL level to the layout. It should provide the designer with a working starting point for each stage of the design process. We describe such a methodology that relies on a flexible and extensible library of reusable ASIC modules and satisfies the unique needs of custom applications [46].

A. Modular Library

A library comprises of modules described in RTL, flat or hierarchical net-list format, Library Exchange Format (LEF), Timing Library Format (TLF) and Standard Delay Format (SDF), and test benches. This subsection describes the hierarchical structure of the library, module formats, and test benches.

Having written a design specification, the next job is to describe the design in computer readable form using the client's Hardware Description Language (HDL) of choice (VHDL or Verilog). The aim is to produce RTL code that clearly exhibits the functionality prescribed in the specification, whilst meeting the constraints placed upon it by the target technology and standard cell library. The skill and experience of the designer is the most important factor in this process.

To create a module database for the library, we synthesize flat and hierarchical Verilog net-lists. The module net-lists are compiled, saved and added to the library database. The basic elements of a net-list are global design data, components, and nets. Global design data is described in terms of special keywords. Examples are a design name and comments specifying the design engineer, revision numbers, etc. Components are described in terms of instance names associated with names of standard cells. Nets are described in terms of groups of standard cell pins names.

The first LEF file must include the technology information and a complete description of standard cells, I/O pads, and macro blocks. You can *incrementally* add data from other LEF files to the library in memory. A single macro block may have multiple LEF files (logical, physical and timing). The description within LEF files must follow the structure and syntax rules defined by the layout tool. If you plan to use a timing-driven placement, a TLF file with timing information is necessary.

Timing of a library module is contained in a corresponding SDF file. This file can be generated after logic and/or layout synthesis and is necessary for simulation.

A test bench is written in order to prove correctness of the RTL model against the specification. The aim of a good test bench (or test benches) is to:

- Prove that the device functions as described in the specification reporting any discrepancies to the transcript in order that they may be fixed;
- Fully exercise every line of RTL code;
- Allow re-simulation of the post-layout net-list. In many cases, it is only possible to perform a subset of the total test suite at gate level due massive increase in computing power required;

Test benches are sometimes used to produce *cycle based* test vectors for final electrical testing during manufacture. We provide configurable and hierarchically linked test benches. The hierarchy of a top test bench follows the hierarchy of the library. Configuration of test benches is possible using an in-house prepared *Tcl interface*.

B. Design Flow

Design flow starts with the creation of a HDL file describing the system and its components as black boxes. Most of components are configurable functional building modules, which are automatically, after choosing the parameters, described by generation of the net-list and physical layout. Configuration of functional modules is possible using an in-house prepared *Tcl interface*. In addition, area and average power dissipation estimates are

also generated to support exploration of design alternatives. This is done using the hierarchical structure of the library.

Fig. 11 shows all main steps of the design flow including both logical and physical stage. Logic synthesis tools are used to synthesize a net-list representation of the design [47]. RTL and net-list (without and with timing information) simulations are performed by HDL simulation tools [48], [49]. Layout tools can do the floor planning, placement, clock tree generation, routing (including the delay optimization), and verification of layout [50].

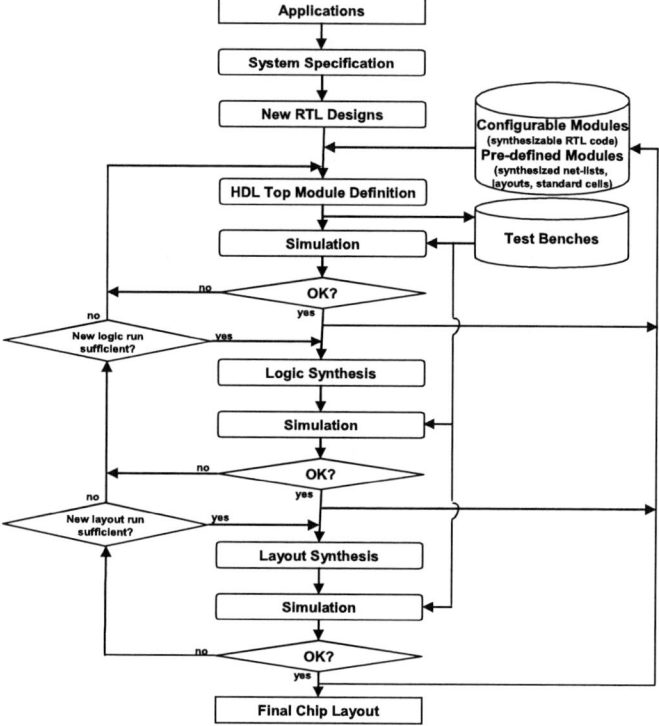

Fig. 11. Overview of IHP's SoC design flow [46]

Standard SoC design flow usually starts with an RTL description of the design. Then synthesis tools take the design to produce a gate-level equivalent design with specified timing constraints, i.e. to synthesize it into target standard cell library. At the end of this synthesizing step, a Verilog net-list and a SDF file are generated, and the design is simulated. To shorten the time necessary for synthesis, we have carefully prepared the synthesis *scripts with design constraints*.

Layout tools provide a foundation for the deep sub-micron technology design with three or more metal layers. We use First Encounter® [51], which is a tool for hierarchical designs that have timing critical blocks. As the core of the Cadence® Encounter™ digital IC design platform, First Encounter quickly produces a silicon virtual prototype of the physical design, which provides both rapid feedback on chip performance and a fully functional, physically feasible layout. With this physical prototype, our front-end designers quickly explore the impact of their implementation choices on chip performance and physical feasibility. The back-end designers produce a floor plan and placement optimized for rapid, reliable design closure. We do first the floor planning including the power planning, and then placement, clock tree generation and routing using our *own scripts*.

First Encounter is intended for a *hierarchical* floor planning. Floor planner uses an interactive approach, combining automatic functions and interactive editing to plan locations for blocks and core rows. It can predict and assess the effects of physical layout before you place and route the design. This speeds up the design process by minimizing costly iterations. The primary goal of floor planning is to create rows where the placer can place cells. Each row is assigned to a site type, limiting the number of cells than can be placed in a row. The floor planner estimates the required layout area (die size), defines the layout domain (shape, aspect ratio), and plans and modifies the row configuration. The row configuration implies the position, spacing, orientation and row type. You can also align rows, adjust the channel space and modify the cell distribution within rows. Finally, it is possible to edit groups and group constraints, resize the layout area and modify the global routing grid. After the floor plan is ready and read in, the next step is to create power paths in the design. Then power rings and power stripes can be added.

Generally, I/O pads (or pins) are placed before blocks. If there are some predetermined I/O pad (or pin) placement requirements, an I/O constraint file should be used to place I/O pads. Otherwise I/O pads and cells can be placed at the same time. After this, we can analyze the timing at different points in the design with ideal clock, as there are no clock trees yet. First Encounter is capable to perform in-place optimization resizing gates (including flip-flops), and inserting buffers and inverters to correct timing and electrical violations.

Clock distribution and skew are controlled using an automatic clock tree generator. The clock buffer space and clock net must be defined. The clock tree generator automatically constructs an optimized clock tree, minimizes skew and clock tree min/max insertion delay, and uses buffer/inverter selection control. First Encounter can perform a *hierarchical* clock tree synthesis.

At this point, we have a fully placed design and ready to route it. In general, the routing of a design is done in three stages: the routing of power nets, the routing of clock trees and the routing of the remaining nets. We use Cadence® NanoRoute Ultra™ [52] to perform the global and detail routing. There is a critical synergistic relationship between placement and routing for all aspects of design closure, including timing, power, cross talk and congestion. That is why a post-routing optimization step is necessary.

The routed design is checked out against connectivity and geometry violations. At this phase, we extract parasitic capacitances from the layout and generate a Verilog net-list, a SDF file, and a LEF file of the design. This step ends with the design simulation (including the generated SDF file) or timing analysis to verify the design performance. The operation is known as a back-annotation.

We use the HDL simulator to verify that all RTL, logically synthesized net-list and physically synthesized net-list have the functionality as expected. The design is simulated with a specific hierarchical test bench. Iteration may be needed to get the final layout and, at the end, a new hard-core library module (as a LEF file).

V. Conclusion

We have described main hardware components of modern SoCs for wireless communications. Most of the components are extensible or configurable functional building modules, which are automatically, after choosing the parameters, described by generation of the net-list and physical layout. We have also developed the specific SoC design flow based on the flexible library of ASIC modules.

Acknowledgement

My best acknowledgements go to my colleagues at IHP H. Frankenfeldt, U. Jagdhold, M. Krstić, G. Panić, and K. Tittelbach-Helmrich for putting enormous efforts in design, implementation, and verification of presented SoCs.

References

[1] W. Wolf, *Modern VLSI design: system-on-chip design (3rd Edition)*, Englewood Cliffs, New Jersey: Prentice Hall, 2002.

[2] S. Sarkar, S. G. Chandar, and S. Shinde, "Effective IP reuse for high quality SoC design", *Proc. IEEE International SOC Conference*, Washington, 2005, pp. 215-224.

[3] A. Hekmatpour, K. Goodnow, and S. Hemen, "Standards-compliant IP-based ASIC and SoC design", *Proc. IEEE International SOC Conference*, Washington, 2005, pp. 322-323.

[4] M.-A. Dziri, W. Cesario, F. R. Wagner, and A. A. Jerraya, "Unified component integration flow for multi-processor SoC design and validation", *Proc. Design, Automation and Test in Europe Conference*, Paris, 2004, pp. 1132-1137.

[5] M. Bocchi, C. Brunelli, C. De Bartolomeis, L. Magagni, and F. Campi, "A system level IP integration methodology for fast SoC design", *Proc. International Symposium on System-on-Chip*, Tampere, 2003, pp. 127-130.

[6] S. Nugent, D. S. Wills, and J. D. Meindl, "A hierarchical block-based modeling methodology for SoC in GENESYS", *Proc. 15th IEEE International ASIC/SOC Conference*, Rochester, 2002, pp. 239-243.

[7] S. J. E. Wilton and R. Saleh, "Programmable logic IP cores in SoC design: opportunities and challenges", *Proc. IEEE Conference on Custom Integrated Circuits*, San Diego, 2001, pp. 63-66.

[8] P. G. Paulin, "Chips of the future: soft, crunchy or hard?", *Proc. Design, Automation and Test in Europe Conference*, Paris, 2004, pp. 844-849.

[9] J. Koehl, D. E. Lackey, and G. Doerre, "IBM's 50 million gate ASICs", *Proc. 8th Asia and South Pacific Design Automation Conference*, Kitakyushu, 2003, pp. 628-634.

[10] B. Dipert, S. Rawat, and S. Tam, "Future systems-on-chip: software or hardware design?", *Proc. 37th Design Automation Conference*, Los Angeles, 2000, pp. 336-336.

[11] D. D. Gajski, *High-level synthesis: introduction to chip and system design*, Boston, Massachusetts: Kluwer Academic Publishers, 1992.

[12] http://www.mips.com/content/Products/

[13] http://www.arc.com/configurablecores/

[14] http://www.gaisler.com/Processors

[15] P. R. Panda, N. Dutt, and A. Nicolau, *Memory issues in embedded systems-on-chip: optimizations and exploration*, Boston, Massachusetts: Kluwer Academic Publishers, 1999.

[16] D. H. Albonesi, "Selective cache ways: on-demand cache resource allocation", *Proc. 32nd Annual International Symposium on Microarchitecture*, Haifa, 1999, pp. 248-259.

[17] W. T. Shiue and C. Chakrabarti, "Memory exploration for low power embedded systems", *Proc. 36th Design Automation Conference*, New Orleans, 1999, pp. 140-145.

[18] A. Malik, B. Moyer, and D. Cermak, "A low power unified cache architecture providing power and performance flexibility", *Proc. International Symposium on Low Power Electronics and Design*, Rapallo, 2000, pp. 241-243.

[19] T. Givargis, F. Vahid, and J. Henkel, "System-level exploration for Pareto-optimal configurations in parameterized SoC", *IEEE Trans on VLSI Systems*, vol. 10, pp. 416-422, 2002.

[20] R. Banakar, S. Steinke, B.-S. Lee, M. Balakrishnan, and P. Marwedel, "Scratchpad memory: a design alternative for cache on-chip memory in embedded systems", *Proc. 10th International Symposium on Hardware/Software Codesign*, Estes Park, 2002, pp. 73-78.

[21] Z. Stamenkovic, F. Vater, and Z. Dyka, "A framework for selection of cache configurations for low power", *Proc. 4th International Workshop on IP-Based System-on-Chip Design*, Grenoble, 2003, pp. 137-140.

[22] P. Kalla, X. S. Hu, and J. Henkel, "A flexible framework for communication evaluation in SoC design", *Proc. 10th Asia and South Pacific Design Automation Conference*, Shanghai, 2005, pp. 956-959.

[23] http://www.synopsys.com/cgi-bin/designware/amba

[24] http://www.synopsys.com/products/designware

[25] G. Panic, Z. Stamenkovic, K. Tittelbach-Helmrich, J. Lehmann, and G. Schoof, "Design of wireless systems utilizing scratchpad memories", *6th International Workshop on IP-Based System-on-Chip Design*, Grenoble, 2005, pp. 221-226.

[26] G. Panic, D. Dieterle, Z. Stamenkovic, and K. Tittelbach-Helmrich, "A system-on-chip implementation of the IEEE 802.11a MAC layer", *Proc. 3rd EUROMICRO Symposium on Digital System Design*, Antalya, 2003, pp. 319-324.

[27] J. Henkel, "Closing the SoC design gap", *Computer*, vol. 36, pp. 119-121, 2003.

[28] C. Rowen, *Engineering the complex SoC: fast, flexible design with configurable processors*, Englewood Cliffs, New Jersey: Prentice Hall, 2004.

[29] http://www.tensilica.com/products/xtensa_LX.htm

[30] E. Macii and M. Poncino, "The influence of cell library characteristics on power consumption of CMOS circuits", *Proc. 7th Mediterranean Electrotechnical Conference*, Antalya, 1994, pp. 537-540.

[31] K. Scott and K. Keutzer, "Improving cell libraries for synthesis", *Proc. IEEE Custom Integrated Circuits Conference*, San Diego, 1994, pp. 128-131.

[32] M. Hashimoto, K. Fujimori, and H. Onodera, "Standard cell libraries with various driving strength cells for 0.13, 0.18 and 0.35 μm technologies", *Proc. 8th Asia and South Pacific Design Automation Conference*, Kitakyushu, 2003, pp. 589-590.

[33] B. Hu, Y. Watanabe, and M. Marek-Sadowska, "Gain-based technology mapping for discrete-size cell libraries", *Proc. 40th Design Automation Conference*, Anaheim, 2003, pp. 574-579.

[34] R. Aitken and S. Becker, "Cell library techniques using advanced transistor structures", *Proc. International Conference on Integrated Circuit Design and Technology*, Austin, 2004, pp. 199-204.

[35] S. Chandrasekar, G. K. Varshney, and V. Visvanathan, "A comprehensive methodology for noise characterization of ASIC cell libraries", *Proc. 6th International Symposium on Quality of Electronic Design*, San Jose, 2005, pp. 530-535.

[36] http://www.artisan.com/products/standard_cell.html

[37] http://www.prolificinc.com/products.html

[38] J. C. Tou, P. Gee, J. Duh, and R. Eesley, "A submicrometer CMOS embedded SRAM compiler", *IEEE J. Solid State Circuits*, vol. 27, pp. 417-424, 1992.

[39] D. Donnelly, "Memory generator method for sizing transistors in RAM/ROM blocks", *Proc. IEEE International Workshop on Memory Technology, Design and Testing*, Los Alamitos, 1998, pp. 10-11.

[40] H. Lim, A. Shubat, V. Duvalyan, S. Dandamudi, S. Raviv, and A. Kablanian, "A widely configurable EPROM memory compiler for embedded applications", *Proc. IEEE International Workshop on Memory Technology, Design and Testing*, Los Alamitos, 1998, pp. 12-16.

[41] T. Tsang and R. Thukral, "An area-efficient 0.25 μm memory compiler designed for 780 MHz operations", *Proc. IEEE Canadian Conference on Electrical and Computer Engineering*, Edmonton, 1999, pp. 533-537.

[42] R. Rajsuman, "Design and test of large embedded memories: an overview", *IEEE Design and Test of Computers*, vol. 18, pp. 16-27, 2001.

[43] R.-F. Huang, L.-M. Denq, C.-W. Wu, and J.-F. Li, "A testability-driven optimizer and wrapper generator for embedded memories", *Proc. IEEE International Workshop on Memory Technology, Design and Testing*, San Jose, 2003, pp. 53-56.

[44] http://www.viragelogic.com/Memory

[45] http://www.dolphin-ic.com/Memory.html

[46] Z. Stamenkovic, G. Panic, U. Jagdhold, H. Frankenfeldt, K. Tittelbach-Helmrich, G. Schoof, and R. Kraemer, "Modular Processor: A Flexible Library of ASIC Modules", *Proc. IASTED International Conference on Applied Simulation and Modelling*, Rhodes, 2004, pp. 428-432.

[47] http://www.synopsys.com/products/logic

[48] http://www.model.com/products

[49] http://www.cadence.com/products/functional_ver

[50] http://www.cadence.com/products/digital_ic

[51] http://www.cadence.com/products/digital_ic/first_encounter

[52] http://www.cadence.com/products/digital_ic/nanoroute_ultra

2006 25th International Conference on Microelectronics

A New Circuit Model for Designing Fully Integrated Class-A Power Amplifier

Xiaokang Guan, Haigang Feng, Albert Wang, Liwu Yang

Abstract—This paper presents a new circuit model for Class-A power amplifier (PA), which can be used to guide design of fully-integrated PA by accurately estimating the optimum load resistance (Ropt), hence to achieve optimized PA performance, e.g., gain, output power (Po), power-added efficiency (PAE) and harmonics, etc. A new figure-of-merit (F) is introduced as a design criterion in practical PA design to ensure optimum overall circuit performance. This new model is verified using a two-stage fully-integrated 2.4GHz Class-A PA designed and implemented in a commercial 0.35μm SiGe BiCMOS technology.

I. INTRODUCTION

SiGe BiCMOS technology has become a major contender to III-V technologies in RF power amplifier (PA) design. This is because SiGe HBT has many attractive features such as: high f_t and f_{max}, considerable current gain, good substrate thermal conductivity and high breakdown voltage [1-2]. Additionally, due to its full compatibility to standard silicon process, SiGe BiCMOS technology enables IC designers to easily integrate PA with other modules to realize real single-chip RF transceivers, resulting in better chip performance and lower system costs. Several power amplifiers using SiGe HBT technology have been reported recently [3-5].

Class-A power amplifier is widely used in communication systems with variable envelop modulation scheme, despite its relatively low efficiency. One of the most challenging tasks in Class-A PA design is to determine the optimal load resistance R_{opt}, which ensures that maximum output power can be achieved. Conventionally, this work can be done by using an off-chip commercial load-pull system [6]. Once R_{opt} is determined, an output impedance matching network is built with off-chip components or transmission lines so that a 50 ohm load is transformed to R_{opt}. However, although it can provide post fabrication tuning ability to optimize circuit performance, this design method is time-consuming and costly, and cannot be used in designing a fully-integrated PA [7-8]. Consequently, how to effectively estimate the R_{opt} to ensure optimized circuit performance, e.g., signal gain, P_o, PAE and harmonics, etc., becomes a challenging issue in designing a fully-integrated PA. This paper presents an improved Class-A PA circuit model and a new

X. Guan, H. Feng, and A. Wang are with Department of ECE, Illinois Institute of Technology, Chicago, IL-60616, USA, E-mail: awang@ece.iit.edu

L. Yang is with RF Integrated Corp., Irvine, CA-92618, USA.

figure-of-merit to address this design problem. The new design method is verified by a two-stage fully-integrated 2.4GHz Class-A PA implemented in a commercial 0.35μm SiGe BiCMOS technology.

II. NEW CIRCUIT MODEL FOR CLASS-A PA

An HBT Class-A power amplifier, along with its ideal circuit model and waveforms are shown in Fig. 1, where V_{AC} and I_{AC} are amplitudes of the collector voltage and current, respectively.

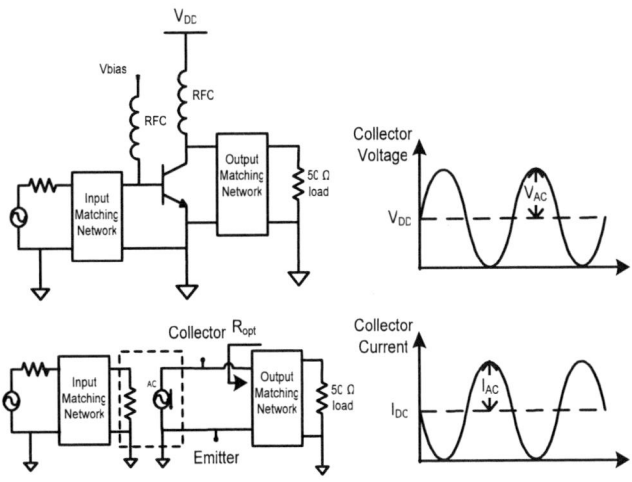

Fig. 1. Class-A PA with its ideal circuit model and waveforms.

Based on the ideal PA model, the optimal load resistance R_{opt} can be written as:

$$R_{opt} = V_{DD}/I_{DC}. \qquad (1)$$

Considering the saturation voltage V_{CEsat} of the HBT and the resistance associated with the radio frequency chock (denoted as R_{RFC}), we can modify (1) as:

$$R_{opt} = (V_{DD} - I_{DC}R_{RFC} - V_{CEsat})/I_{DC}. \qquad (2)$$

The output power P_o can be written as a function of the transformed load resistance R as:

$$P_o = 0.5I_{DC}^2R, \quad R \le R_{opt}, \qquad (3)$$

1-4244-0116-X/06/$20.00 ©2006 IEEE

$$P_o = 0.5\left(V_{DD} - I_{DC}R_{RFC} - V_{CEsat}\right)^2 / R, \ R > R_{opt}. \quad (4)$$

This traditional PA circuit model treats the transistor as an ideal controlled current source. From (4), if transformed load resistance $R > R_{opt}$, the collector voltage swing could reach its maximum value, that is, $2(V_{DD} - I_{DC}R_{RFC} - V_{CEsat})$. However, this will never happen in real PA circuit where the output power could be much less than that calculated by (4). In order to describe this phenomenon and to provide a better way to predict output power, we propose a new PA circuit model as shown in Fig. 2, where a new parameter R_{int} is introduced to model the inevitable internal resistance of the controlled current source for the transistor. R_{int} can be determined from S-parameter measurement of the PA transistor at its actual working bias, e.g. R_{int}= Re (S_{22}).

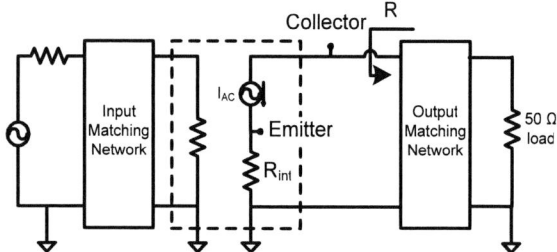

Fig. 2. Proposed new circuit model of Class-A PA.

From Fig. 2, one can easily derive the following equations when assuming $I_{AC} < I_{DC}$:

$$V_{CEmax} = (V_{DD} - I_{DC}R_{RFC}) + I_{AC}(R + R_{int}) \leq BV_{ceo}, \quad (5)$$

$$V_{CEmin} = (V_{DD} - I_{DC}R_{RFC}) - I_{AC}(R + R_{int}) \geq V_{CEsat}, \quad (6)$$

$$V_{AC} = I_{AC}R, \quad (7)$$

$$P_o = 0.5I_{AC}^2 R, \quad (8)$$

where V_{CEmax} and V_{CEmin} are maximum and minimum collector-emitter voltage, respectively. $V_{DD} - I_{DC}R_{RFC}$ is the collector-emitter voltage at DC condition, and BV_{ceo} is breakdown voltage of HBT. From (5) and (6), it is readily observed that only the sum of voltage swings across R_{int} and R can approach $BV_{ceo} - V_{CEsat}$, while the voltage swing at collector, which is equal to $2I_{AC}R$, is always smaller than the maximum value $BV_{ceo} - V_{CEsat}$ as in real PA operation.

To verify this improved PA circuit model, we provide following detailed analysis of the output stage of the PA designed in this work. Fig. 3a shows the $P_o \sim R$ curves of the output stage obtained using the traditional and new circuit model as well as from SpectreRF simulation. Clearly, the old model deviates from the circuit behavior significantly, while the new model describes the output power fairly well. The slight deviation in the low-R region is attributed to the nonlinear distortion generated as the transistor approaches cut-off region, which cannot be

addressed by the model yet. Further, Fig. 3b, c & d show the small-signal gain, PAE and 2nd harmonic of the output stage, all are critical parameters in practical design.

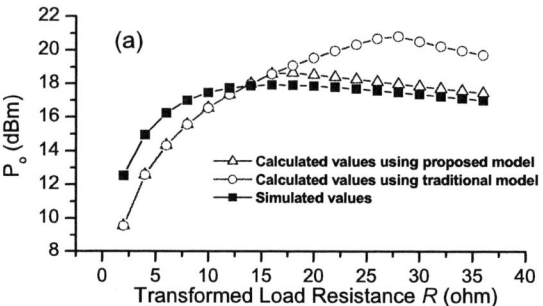

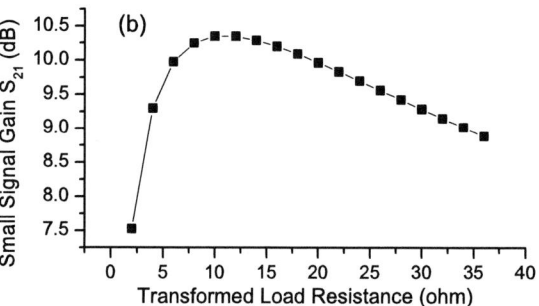

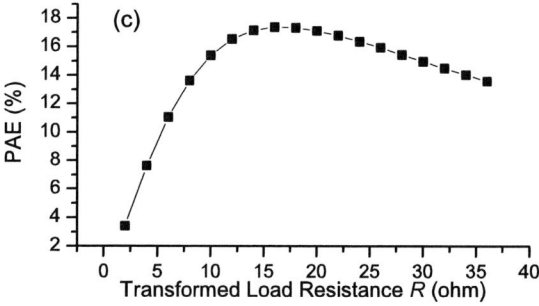

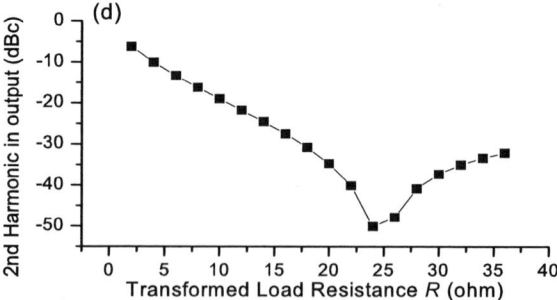

Fig. 3. Critical PA specifications: (a) $P_o \sim R$; (b) small signal gain $\sim R$; (c) PAE $\sim R$; (d) 2nd harmonic $\sim R$; (calculation and simulation conditions: V_{DD}=3.3 V, I_{DC}=95 mA, BV_{ceo}=6 V, V_{CEsat}=0.2 V, R_{RFC}=5.26 ohm, R_{int}=10.8 ohm, P_{in}=8.5 dBm).

Traditionally, one may use the old model in practical PA design to estimate the required R_{opt}, hence to achieve the maximum output power P_o. However, the so-selected R_{opt}, using the old model from Fig. 3a would be ~28Ω, which is quite different from the one that might be suggested from the new model and circuit simulation. This indicates that the traditional method would be misleading in PA design optimization. More importantly, in selecting the real optimum R_{opt} for PA performance optimization, one should not only consider the output power (P_o), but also, and most importantly, consider the overall PA circuit specifications, e.g., gain, PAE, 2nd harmonic, etc, in order to strike the right design trade-off balance and achieve the overall PA performance optimization. However, Fig. 3 clearly shows conflicting indications from different specifications. To address this matter, we introduce a new figure-of-merit, denoted by F, to describe the overall PA performance. The F is defined below in (9), which readily suggests that, for overall optimum PA performance, a large F is preferred. Hence, the real optimal R_{opt} should be determined against maximum F as expressed in (10):

$$F = 20\log(\frac{P_o \times PAE \times Gain}{P_{2nd,harmonic}}), \qquad (9)$$

$$R_{opt} = R(F_{max}). \qquad (10)$$

Following this new rule, the F~R curve is obtained in Fig. 4 that allows one to easily determine the real R_{opt}, which is ~24Ω as opposed to ~28Ω suggested by the old model. This new R_{opt} was used in designing the output stage of the 2.4GHz PA in this work.

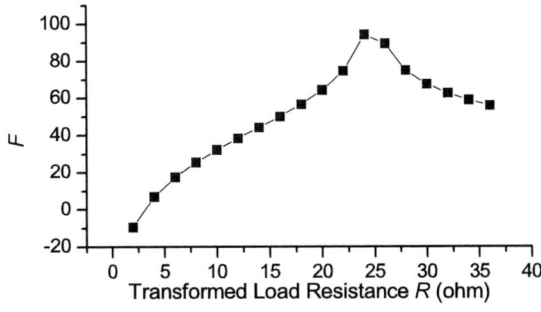

Fig. 4. New F~R curve to determine the real R_{opt}.

III. EXPERIMENTAL VERIFICATION

In order to validate the new PA circuit model and its usefulness in practical single-chip PA design verification, a 2-stage 2.4GHz fully-integrated PA was designed in this work using the new method and fabricated in a 0.35μm SiGe BiCMOS technology. Fig. 5 shows the complete PA schematics where an integrated diode linearizer technique [9] is used in biasing to offset the gain compression effect.

The f_t of the HBTs used is 27 GHz. The total emitter areas for the gain and power stage are 146μm² and 730μm², respectively. Fig. 6 shows the die photo of the PA circuit that has a die size of only 1.2mm x 1mm including biasing and probing pads, a good proof of size advantage of full integration design.

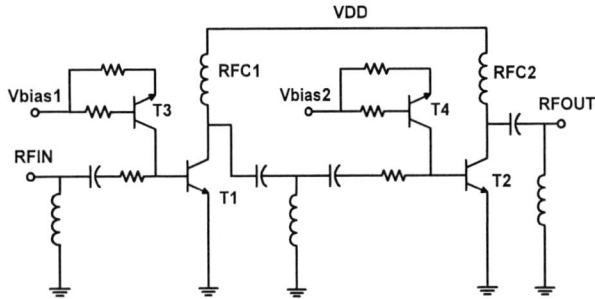

Fig. 5. Complete schematics of the 2-stage Class-A PA.

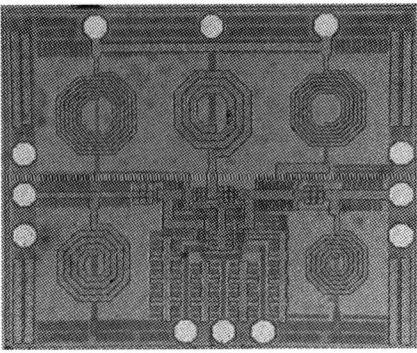

Fig. 6. Die photo of the 2-stage fully integrated PA.

Fig. 7 gives the S-parameter results of the power amplifier, showing a small signal gain S_{21} of 27.8 dB at 2.45 GHz and return loss of S_{11} = -19.3dB and S_{22} = -8.3 dB, respectively. A good input and output VSWR is achieved.

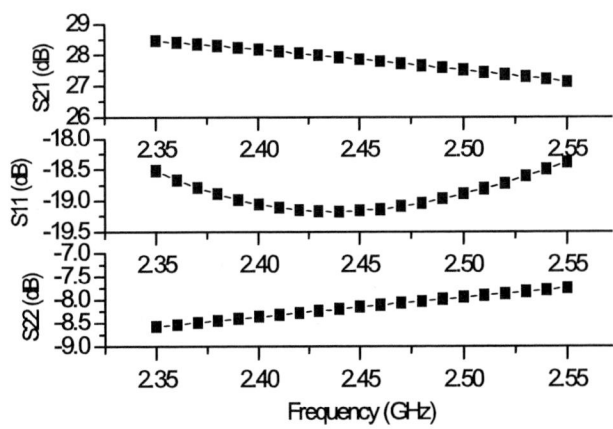

Fig. 7. S-parameter results of the 2.4 GHz PA.

Fig. 8 shows the output power, signal gain and PAE data obtained by periodic steady state (PSS) analysis. The maximum linear output power is 18.87 dBm with a PAE of 18.5%. A 1-dB compression point is reached when the input power is -8 dBm. The relatively low output power and PAE are mainly due to the large internal resistance R_{int} of the SiGe HBT and poor Q-factor of the on-chip RFC available in the SiGe process used.

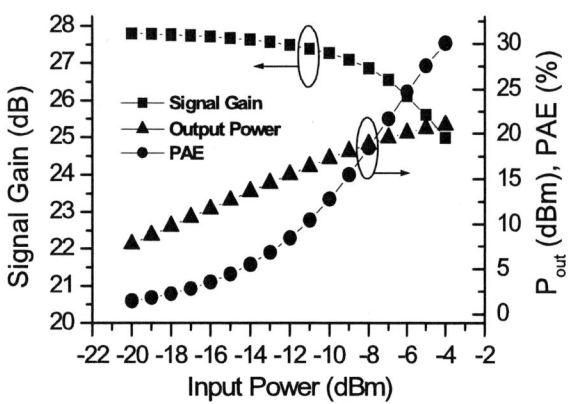

Fig. 8. Small signal gain, output power P_o and PAE of the PA at 2.45 GHz.

Table I compares the PA specs with R_{opt}=24Ω & 28Ω, which clearly shows the design optimized by the new model achieves better overall performance.

TABLE I.

PA SPECS COMPARISON WITH NEW AND TRADITIONAL CIRCUIT MODEL (INPUT POWER: -8 DBM).

		New Model	Old Model
Design Parameters	R_{opt} (Ω)	24	28
Critical Specs of the PA	P_o (dBm)	18.87	18.67
	2nd Harmonic (dBc)	-41.12	-39
	PAE	18.5%	17.67%

III. CONCLUSION

In summary, this paper reports a new circuit model for Class-A PA, which can be used to optimize fully-integrated PA design by accurately estimating the real R_{opt} for optimum PA performance. A new F parameter is introduced to address the overall PA performance and a R_{opt} against the maximum F is suggested in practical PA design optimization to achieve better gain, P_o, PAE and harmonics, etc. The new model is verified using a 2-stage fully-integrated 2.4GHz PA designed in a 0.35μm SiGe BiCMOS.

REFERENCES

[1] J. Cressler, "SiGe HBT technology: a new contender for Si-based RF and microwave circuit applications," IEEE Trans. Microwave Theory and Techniques, Vol. 46, No. 5, May 1998, pp.572–589.

[2] Inoue, et al, "The maximum operating region in SiGe HBTs for RF power amplifiers," Digest IEEE MTT-S Symp., Vol. 2, 2002, pp.1023–1026.

[3] R. Gotzfried, et al, "Design of RF integrated circuits using SiGe bipolar technology," IEEE Journal of Solid-State Circuits, Vol. 33, No. 9, Sept. 1998, pp.1417–1422.

[4] A. Raghavan, et al, "A 2.4 GHz high efficiency SiGe HBT power amplifier with high-Q LTCC harmonic suppression filter," Digest IEEE MTT-S Symp., Vol. 2, 2002, pp.1019–1022.

[5] J. Johnson, et al, "SiGe BiCMOS technologies for power amplifier applications," Digest GaAs IC Symp., 2003, pp.179–182.

[6] S. Cripps, RF Power Amplifiers for Wireless Communicatins, Norwood, MA, Artech House, 1999.

[7] I. Rippke, et al, "A fully integrated, single-chip handset power amplifier in SiGe BiCMOS for W-CDMA applications," Digest IEEE MTT-S Symp., Vol. 1, 2003, pp.153- 156.

[8] W. Bakalski, et al, "A Fully Integrated 5.3-GHz 2.4-V 0.3-W SiGe Bipolar Power Amplifier With 50 Ohm Output," IEEE Journal of Solid-State Circuits, Vol. 39, No. 7, July 2004, pp.1006–1014.

[9] T. Yoshimasu, et al, "An HBT MMIC power amplifier with an integrated diode linearizer for low-voltage portable phone applications," IEEE J. Solid-State Circuits, Vol. 33, No. 9, Sept. 1998, pp.1290–1296.

2006 25th International Conference on Microelectronics
Gate Layout Improvement Aimed at Testability

Mykola Blyzniuk

Abstract - In the presented paper the improvement of the layout of complex standard gates from the industrial cell library aimed at decreasing the probability of occurrence of undetectable faults is considered. Such improvement allows us to determine the defect coverage table correctly and as a result to estimate properly the optimal sequence of input test pattern for defects detection. The ability of gate layout improvement is based on the results of defects probabilities determination and identification of functional faults caused by these defects. The results are obtained by *FIESTA-Extra* software tool.

I. INTRODUCTION

The design for testability (DfT) includes the development of the design techniques which are used to make the testing of a circuit more efective. Most designs for testability have been done at the logic level with the goal of increasing the controllability and observability of the circuit. As researchers realize the importance of realistic physical faults, the need for a level of a physical design for testability (PDfT) has become evident. This level deals with physical layout-level faults and implies the changing of the layout of the circuit to make it more testable [1-3].

One of the approaches in which the testability of circuits may be increased is the improvement of layout of complex gates from the industrial cell library in order to make more testable cells. The main reason for doing this is to take advantage of the reuse of cells within a circuit and through many designs. By investing effort in a cell design, a large improvement in testability can be gained as the cells will be used many times [3].

This work focuses on improving the testability of complex gates from the industrial cell library. Such improvement is provided by the developed approach and software tool for more precise and more detailed defect/fault analysis of standard cells.

I. TEST-BASED DEFECT/FAULT CHARACTERISATION OF COMPLEX GATES

The proposed approach is developed for test-based defect/fault characterisation of complex gates from the industrial cell library. Special software tool named *FIESTA–EXTRA (Faults Identification and EStimation of TestAbility by EXTRAction of faults probabilities, kinds of faults and usefulness of test patterns for faults detection)*

Mykola Blyzniuk is with Ivan Franko National University of L'viv, 1, Universytetska St., 79001, L'viv, UKRAINE, E-mail: MB_IK@org.lviv.net.

was developed for automation of defect/fault analysis process [4-7].

The aim of defect/fault characterisation is the realistic representation of physical defects influence on gate behaviour in fault models. The characterisation process is focused [4,5,7]:

- on the precise analysis of gate layout geometry, defect size distribution, and density of physical defects;
- on careful identification of real faulty function from actual behaviour of failure circuit,
- and on test-based fault characterisation for finding the best sequence of test patterns for all faults detecting.

The process of test-based defect/fault characterisation includes the solution of the next tasks [4,7]:

1) Formation of its own model of the conductive layers of gate layout: to do this the output text files generated by Cadence Layout Editor (files containing the information about conductive layers of gate layout) are processed.

2) Estimation of probability of occurrence of shorts between nodes and opens in certain branches of a circuit graph: to do this the computational experiment with the use of critical area model of conductive layer of gate layout is carried out for the estimation of probability of shorts/opens.

3) Identification of types of faulty functions resulting from probable defects of short/open type in a gate: to do this the analogue simulation of a gate with introduced short or open defect is used.

4) Estimation of probability of different faulty functions resulting from defects: the vector of probable kinds of faulty functions resulting from defects of short/open type and the vector of these faults probabilities are formed. The list of defects, types of faulty functions caused by these defects, and probabilities of occurrence of these faults are main results of probabilistic analysis.

5) Estimation of the effectiveness of input test patterns for detection of faulty functions caused by physical defects: to do this the defect coverage table is determined.

6) Estimation of optimal sequence of input test pattern for defects detection: to do this the usefulness of the test pattern is estimated as test pattern possibility to detect the greatest number of defects, the sum of probabilities of occurrence of which is the highest.

The estimation of usefulness of the test pattern for defect detection is the main result of defect/fault characterisation of complex gates by *FIESTA-Extra* software tool [5]. Such estimation of the usefulness of test vector components allows to improve the work on the development and generation of test cycles and to provide high quality of tests. The high quality of the test and

1-4244-0116-X/06/$20.00 ©2006 IEEE   413

improvement of work of the test developers is provided by finding the best sequence of patterns for the detection of all faults and by reducing the average length of the sequence of patterns to a minimum. For example, the diagram in Figure 1 shows the obtained optimal sequence of test patterns for the complex gates from the cell library in 0.8 μm CMOS technology [4].

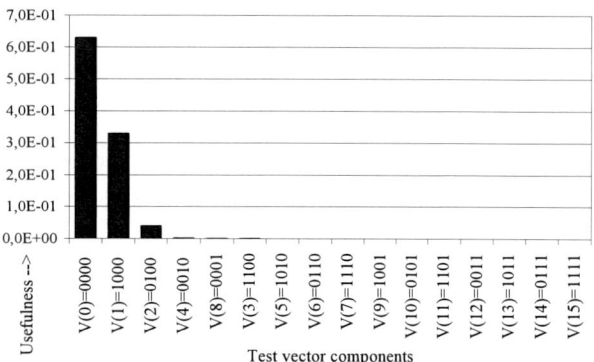

Fig.1. Optimal sequence of test patterns for physical defects detection in And2_Nor3 complex gate.

The main problem in complex gate characterization process is to represent correctly the realistic behaviour of a faulty circuit in a fault model. It is the problem of identification of kinds of functional faults resulting from defects in gate layout. The results of faults identification during test-based characterisation of complex gates from the industrial cell library show that defects may cause different kinds of functional faults including undetectable or difficult-to-detect faults with significant probabilities. These kinds of functional faults do not allow to determine defect coverage table correctly and as a result to estimate properly the optimal sequence of input test pattern for defects detection. Such situation forces the improvement of the layout of standard gates in order to decrease the probability of occurrence of undetectable faults.

II. UNDETECTABLE FUNCTIONAL FAULTS: IDENTIFICATION

The behaviour of the digital circuit with physical defects can be correctly represented when we treat this circuit in the same way as the analogue circuit [7]. In spite of the purely digital function of a gate we should carry out the analogue simulation of the circuit (with the introduced defect, for example short/open) in the time domain. This analogue circuit simulation is performed in time domain using SPICE or similar circuit simulator. During the transient analysis of the circuit with the introduced defect full combination of logic levels of inputs (*logic-0* and *logic-1*) are provided by time-varying input voltage sources. Processing the inputs/outputs waveforms obtained from such simulation we determine the actual faulty function by the automatic formation of the truth tables and by the logic function extraction from them.

The process of faults identification shows that complex gates may perform different faulty functions resulting from defect. For example for defect of short type:

a) result of short is "stuck-at-1" at the output Q of a gate (SA1) or result of short is "stuck-at-0" at the output Q of a gate (SA0);

b) result of short modifies the gate own logic function into another logic function. For instance: $Q=not(A+B+C+D)$ for the Nor4 gate (Figure 2 shows the logic diagram of gate) may become $Q=not(A*B+C+D)$, $Q=A*(not(B+C+D))$, $Q=(((C+D)*not(A+B))+(C*D))$, $Q=(((A+B)*not(C+D))+(A*B))$, and so on [4];

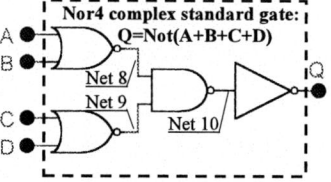

Fig. 2. Logic diagram of Nor4 complex gate.

c) result of short modifies the gate logic function into the analogue function, like, for example, oscillation on the output of complex gate:

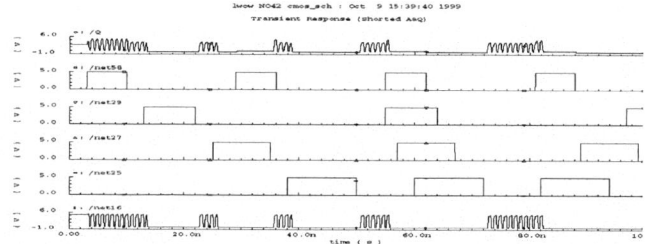

Fig. 3. Results of transient analysis of the faulty Nor4 complex gate with introduced short between A and Q nets.

d) result of short modifies the gate logic function into another undetectable and illogical function:

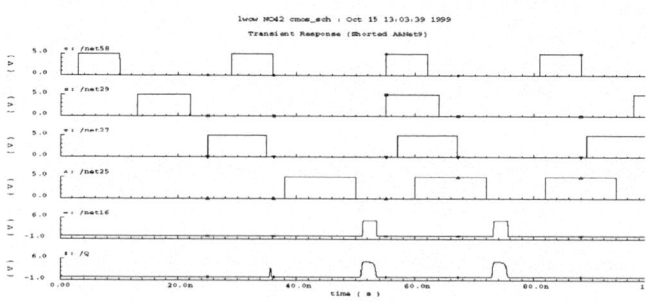

Fig. 4. Results of transient analysis of the faulty Nor4 gate with introduced short between A and internal Net 9.

f) or result of short does not change the logic function of a circuit (for example, shorted internal Net 8 and Net 9 for Nor4 complex gate) but decreases the reliability by increased current densities and temperature caused by increased power dissipation.

The examples *c,d,f* namely show the "not desired" types of functional faults which do not allow to determine clearly the defect coverage table and estimate properly the usefulness of the test pattern, because there is no logic test that guarantees the functional faults shown above will be detected. The negative influence of such faults on circuit testability will depend on their probabilities: if the probability of occurrence of such faults is significant we will not be able to guarantee the quality of the test. Let us consider the results of identification of functional faults caused by defects of short type in Metal 1 layer of Nor4 gate layout (see Table 1). *"Unlikely"* in Table 1 indicates that identification of functional faults caused by defects is formed by taking into account the probability of short between certain nodes of the gate. If the probability of a short between the given nodes equals zero or if it is not significant then a faulty function caused by this short is not identified [6]. In the given table there are 3 types of difficult-to-detect faults. Table 2 shows the estimated probabilities of identified functional faults. The results indicate that the value of probabilities of hard-to-detect faults is significant. So in order to guarantee the complex gate testability we should make these faults unlikely to occur.

TABLE 1
IDENTIFIED TYPES OF FUNCTIONAL FAULTS CAUSED BY PROBABALE DEFECTS OF SHORT TYPE IN NOR4 COMPLEX GATE.

	Net[1]=A	Net[2]=B	Net[3]=D	Net[4]=C	Net[5]=Net8	Net[6]=Net9	Net[7]=Net10	Net[8]=Q	Net[9]=Vdd	Net[10]=Vss
Net[1]=A		not((A*B)+C+D)	Unlikely	Unlikely	A*(not(C+B+D))	Unlikely	Unlikely	Unlikely	Unlikely	Unlikely
Net[2]=B			not((D*B)+C+A)	Unlikely	Unlikely	Unlikely	Unlikely	Unlikely	Unlikely	Unlikely
Net[3]=D				not((C*D)+A+B)	Difficult-to-detect	Unlikely	Unlikely	Unlikely	Unlikely	Unlikely
Net[4]=C					Difficult-to-detect	C*(not(A+B+D))	Unlikely	Unlikely	Unlikely	Unlikely
Net[5]=Net8						Difficult-to-detect	(((A+B)*not(C+D))+(A*B))	Unlikely	not(C+D)	SA0_for_Q
Net[6]=Net9							(((C+D)*not(A+B))+(C*D))	Unlikely	Unlikely	SA0_for_Q
Net[7]=Net10								(A+B+D+C)	SA0_for_Q	SA1_for_Q
Net[8]=Q									SA1_for_Q	SA0_for_Q
Net[9]=Vdd										Unlikely
Net[10]=Vss										

TABLE 2
IDENTIFIED TYPES OF FUNCTIONAL FAULTS IN NOR4 GATE AND THEIR PROBABILITIES

Defects of short type (shorted Net_i & Net_j)	Type of identified functional faults caused by defects	Probabilities of faults
Net8&Vdd	not(C+D)	2,83E-09
A&Vdd; D&Vdd; C&Vdd; Q&Net8; Net8&Vss; Q&Net9; Net9&Vss; Net10&Vdd	SA0_for_Q	1.98E-09
Net9&Net10	(((C+D)*not(A+B))+(C*D))	1.88E-09
C&Net9	C*(not(A+B+D))	1.68E-09
Net8&Net10	(((A+B)*not(C+D))+(A*B))	1.52E-09
Q&Vdd; Net10&Vss	SA1_for_Q	1.25E-09
Q&Net10	(A+B+D+C)	1.20E-09
C&D	not((C*D)+A+B)	1.07E-09
A&Net8	A*(not(C+B+D))	9.58E-10
C&Net8; D&Net8; Net8&Net9	Difficult-to-detect	5.54E-10
B&D	not((D*B)+C+A)	4.41E-10
A&B	not((A*B)+C+D)	3.53E-10
Net9&Vdd	not(A+B)	2.21E-10
B&Net8	B*(not(A+C+D))	2.14E-10
A&D	not((A*D)+C+B)	2.00E-10
B&C	not((C*B)+A+D)	1.43E-10
D&Net9	D*(not(A+B+C))	1.04E-10
C&Vss	not(A+B+D)	2.47E-11
B&Vss	not(A+C+D)	1.39E-11
A&C	not((A*C)+B+D)	1.11E-11

III. GATE LAYOUT IMPROVEMENT

We can do goal-directed modifications of the gate layout by decreasing the critical area for shorts/opens, which cause undetectable types of functional faults. Our software tool *FIESTA-Extra* make it possible for designers to determine the certain shapes of gate layout which belong to corresponding nets of gate schematic and to determine the critical area between these nets, i.e. the probability of defect occurrence. Figure 5 visually demonstrates it.

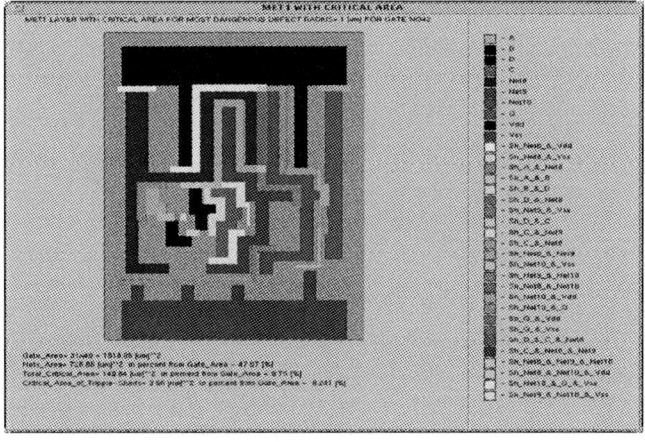

Fig.5. Conductive layer of Nor4 gate layout with extracted critical area for shorts between nets for most dangerous defect radius.

The gate layout improvement can be directed to increase the distance between shapes of corresponding nets which cause undetectable faults, or to avoid placing certain shapes in adjacent by replacing with others if it is possible.

Obviously, if the gate layout was designed optimally, the probability of detectable faults will increase in this case. In general, we may reassign the probabilities between the detectable and not detectable faults by goal-directed modifications of gate layout on condition that the total probability of faults is still constant. In other words a designer can provoke the "desired" (detectable) types of functional faults.

Figures 6 and 7 demonstrate the fragments of Metal 1 layer of Nor4 gate layout before and after modification correspondingly. The light layout modification allows to make difficult-to-detect faults caused by shorted Net 8 and D unlikely to occur.

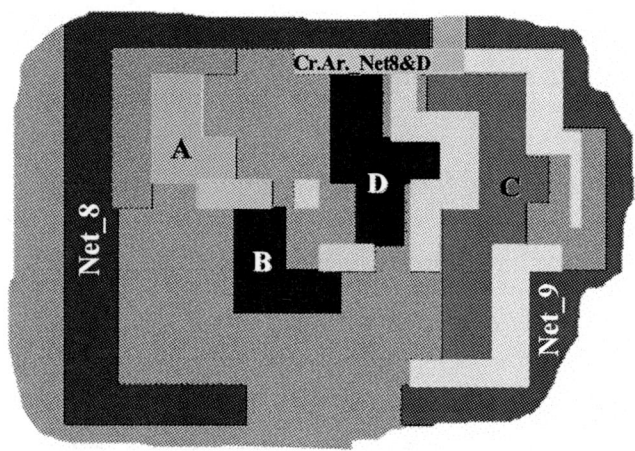

Fig.6. Fragment of layout with critical area before modification.

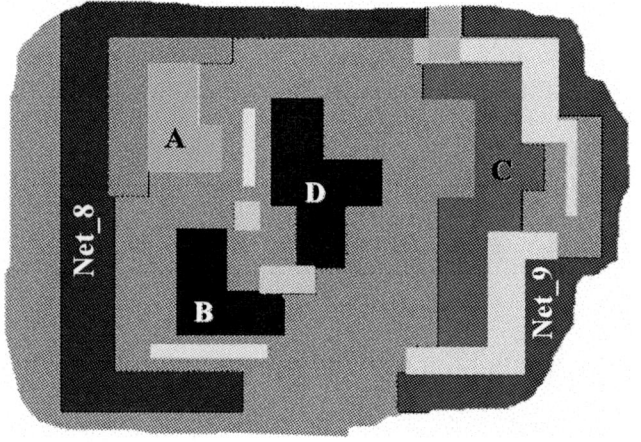

Fig.7. Fragment of layout with critical area after modification.

The results of gate layout improvement aimed at elimination of hard-to-detect faults allow to increase complex gate testability and to determine properly optimal sequence of input test pattern for defects detection. The estimated usefulness of the test vector components in defects detecting for Nor4 gate is shown in Figure 8.

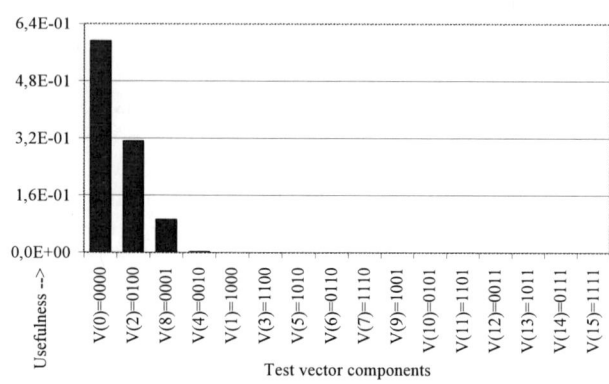

Fig.8. Usefulness of test vector components for physical defects detection for Nor4 complex gate.

IV. Conclusions

Defect/fault characterisation of complex gates from the industrial cell library by *FIESTA-Extra* software tool which is based on precise analysis of gate layout geometry, defect size distribution, density of physical defects and careful identification of real faulty function from actual behaviour of failure circuit allows designers to make layout improvement in order to increase standard complex gates testability and as a result the testability of the whole VLSI circuit.

References

[1] J.A Prieto, A Rueda, I Grout, E Peralias, J L Huertas & A M Richardson, "An Approach to Realistic Fault Prediction and Layout Design for Testability in Analogue Circuits". // *Proc. of the Design, Automation & Test in Europe Conference*, Paris, France, Feb. 23rd-26th 1998, pp.905-912.

[2] R. McGowen and F. J. Ferguson, "Elimination of undetectable shorts during channel routing". // *Proc. of 12th IEEE VLSI Test Symposium*, , IEEE, 1994, pp. 402-407.

[3] Jee, F.J. Ferguson. "A methodolgy for characterizing cell testability". // *Proc. of 15th IEEE VLSI Test Symposium* (VTS'97), 1997, pp. 384-390.

[4] M. Blyzniuk, I. Kazymyra "Probabilistic-based Defect/Fault Characterisation of Complex Gates from Standard Cell Library" // *Journal "Electronics and Electrical Engineering"*. – Kaunas: Technologija, 2004. – No. 3(52). – P. 76-81.

[5] Mykola Blyzniuk and Irena Kazymyra "FIESTA-Extra: Cell-Oriented Software for the Defect/Fault Analysis in VLSI Circuits". // *Proc. of IEEE 24rd International Conference on Microelectronics (MIEL 2004)*, Vol.2, Nis, Serbia and Montenegro, May, 2004, pp. 793-796.

[6] M. Blyzniuk, I. Kazymyra "Approach to prune the list of defects by neglecting statistically insignificant shorts in standard cells defect/fault analysis". // *In Proceedings of the 8th Biennial Conference on Electronics and Microsystem Technology (BEC 2002)*, Tallinn, Estonia, October, 2002, pp. 279-282.

[7] Blyzniuk M, Kazymyra I., Kuzmicz W, Pleskacz W, Raik J, Ubar R, "Probabilistic analysis of CMOS physical defects in VLSI circuits for test coverage improvement", *Microelectronics Reliability*, 2001, vol. 41/12, pp 2023-2040.

2006 25th International Conference on Microelectronics

An Improved Write Driver for Miniaturized Hard Disk Drives

F. Tamigi, I. Marano, A. Venca, and N. Rinaldi

Abstract – The present work proposes the design of an enhanced driver for magnetic storage write heads to be employed in mini- and micro-drive applications. It is intended for running from a reduced-voltage supply (as commonly found in portable devices) with low power consumption, yet retaining an excellent write speed performance, and is thoroughly programmable. Although originally developed in BiCMOS technology, the design is compatible with a full CMOS process.

I. INTRODUCTION

A challenging request from the mobile equipment market is an ever larger storage capability on phones, music players and handheld PCs with more and more reduced weight, size, power consumption and cost. In order to face this growing demand, new technologies have to be adopted in magnetic storage along with a deeper integration of the electronic subsystems.

One of the most troublesome issues to cope with in miniaturized hard disk drives is the availability of only a low-voltage single-ended power supply (typically, a battery), which urges for new circuit solutions to be sought after, especially in those sections where relatively high currents are required. This applies therefore to the write driver, as being the front-end directly deputed to drive large currents into the inductive head in order to generate the adequate magnetic excitation.

Several preamplifier designs were lately proposed [1]-[3], along with the earlier [4], [5], all of which realized either in bipolar or BiCMOS technology, working at 5 V or higher and devised for the best speed performance at the expense of power consumption. A first approach to a full CMOS write driver was undertaken in [6]; again, power saving is less than a priority issue. In less recent times, a low-power preamplifier in bipolar technology [7] was demonstrated for a 3.3-V minimum supply, but the specifications are by far outdated in view of modern designs. Only the latest [8] explicitly addresses the low-voltage low-power case.

The inductive-head driver design presented in this work is specifically targeted to mini- and micro-drives and supersedes former circuits under several aspects. Based on ST Microelectronics patents [9], [10], with a bunch of core improvements aimed to make it suitable for mobile appliances, it runs from as low as a 3-V single-ended supply with reduced power consumption. The write speed is raised up to a maximum of 500 Mbit/s, which is well beyond the requirements of the ultra-portable range.

The driver is fully programmable as to both the overshoot and the dc current levels in order to comply with the widest span of manufacturers specifications on the write head and the magnetic medium. Also, the overshoot duration can be made adjustable by endowing the circuit with an appropriate delay generator.

This design was developed on the basis of the ST Microelectronics 0.35-µm BiCMOS technology design kit (although straightforwardly portable to full CMOS) and is to be implemented in an upcoming preamplifier test chip.

II. CIRCUIT DESCRIPTION

The read/write subsystem schematic layout is represented in Fig. 1. It is clearly visible that the heads are connected to the preamplifier chip via a transmission line, namely a flexible (or suspension) interconnect, which is responsible for a number of inconveniences encountered in engineering the chip. Being made up of heterogeneous segments with bends and corners, in fact, the flex – as it is commonly termed – exhibits an irregular shape and consequently an impedance profile that cannot be even nearly taken as uniform or easily matched.

F. Tamigi, I. Marano and N. Rinaldi are with the Department of Electronics and Telecommunications Engineering, University of Naples Federico II, via Claudio 21, 80125 Naples, Italy. Email: fatamigi@unina.it

A. Venca is with ST Microelectronics, Data Storage Division, via Remo De Feo 1, 80022 Arzano, Naples, Italy.

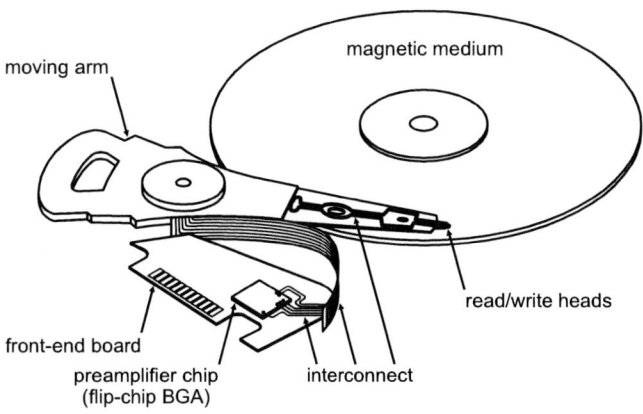

Fig. 1. Read/write subsystem in a miniaturized hard disk drive.

1-4244-0116-X/06/$20.00 ©2006 IEEE

The proposed design is based upon a voltage-boosted architecture to the purpose of deriving the large overshoot current I_{OS} required for writing onto the medium (some more than 100 mA through the head) from the low 3-V power supply. Power management considerations lead to discard the solution of definitely raising the supply voltage by means of an always-on converter, rather choosing to boost on the fly only when needed.

Fig. 2 displays the blocks making up one quarter of the overshoot circuitry. As can be seen, the interconnect is tied to the inner nodes of the H-bridge MN_B, MP_B (the transistors on the right hand side not being shown): a reference bias V_{REF} is applied to the gates such that the switches stay turned off slightly under threshold, unless the voltage values on their sources fall out of the supply rails, that is, the related branches of the bridge are activated at pre-charge release.

A programmable current-mirror reference (or DAC) formed by MN_{REF} and the logarithmic scale MN_{DAC0}, ..., $MN_{DAC(n-1)}$ is employed to set the overshoot current, followed by a separator (a low output-impedance buffer) in order to properly drive the bridge without affecting the DAC's output voltage. An unbalanced logic inverter MN_{SW}, MP_{SW} connects the reference level to the gate of MP_{OS} acting as a multiplying current source. The latter is to be endowed with a very large aspect ratio W/L because of the considerable mirroring factor, even though this may be expensive in terms of area.

Voltage boosting is implemented as follows. Via the diode and MN_{PC}, capacitor C_{BOOST} is pre-charged to V_{CC} (and similarly is its opponent on the other side) while not in use, that is, when current in the head is flowing the

opposite way. At overshoot time both capacitors are switched so to swing the voltage across the bridge symmetrically out of the supply rails, theoretically tripling it (actually, peak voltage is only approximately doubled due to drops). Boost capacitors are chosen as result of a trade-off between power consumption, silicon usage, charging speed and pulse sustain capability: an acceptable compromise value sits around 50 pF.

The differential structure yields a common mode fixed at $V_{CC}/2$ at all times. It is worth noting that this layout also guarantees a prompt response when a write is initiated since all the capacitors are always kept charged, thus working out the issue of latencies.

As for the dc current, this is sourced by a second DAC made up of a logarithmic-scaled resistor ladder for I_{DC} programmability (see the left half of it in Fig. 2): each resistance value is either inserted or detached by means of a three-way selector. Interconnect matching, i.e., holding a constant network impedance with varying bias conditions, is not pursued since several simulations proved it to be of no advantage to head current dynamics (because of the reasons discussed in the opening of this Section). The balanced design achieved by a switching scheme with two different voltage sources V_{CC} and V_{CM} again ensures the proper common mode level even for null I_{DC} without any drawback on consumption.

Eventually, some amount of support logic is added to handle the signals within the circuit itself. All the control signals and the programming codes, except write data, are to be exchanged with the host controller via a syncronous serial interface (neglected at this stage) and stored in a file of registers.

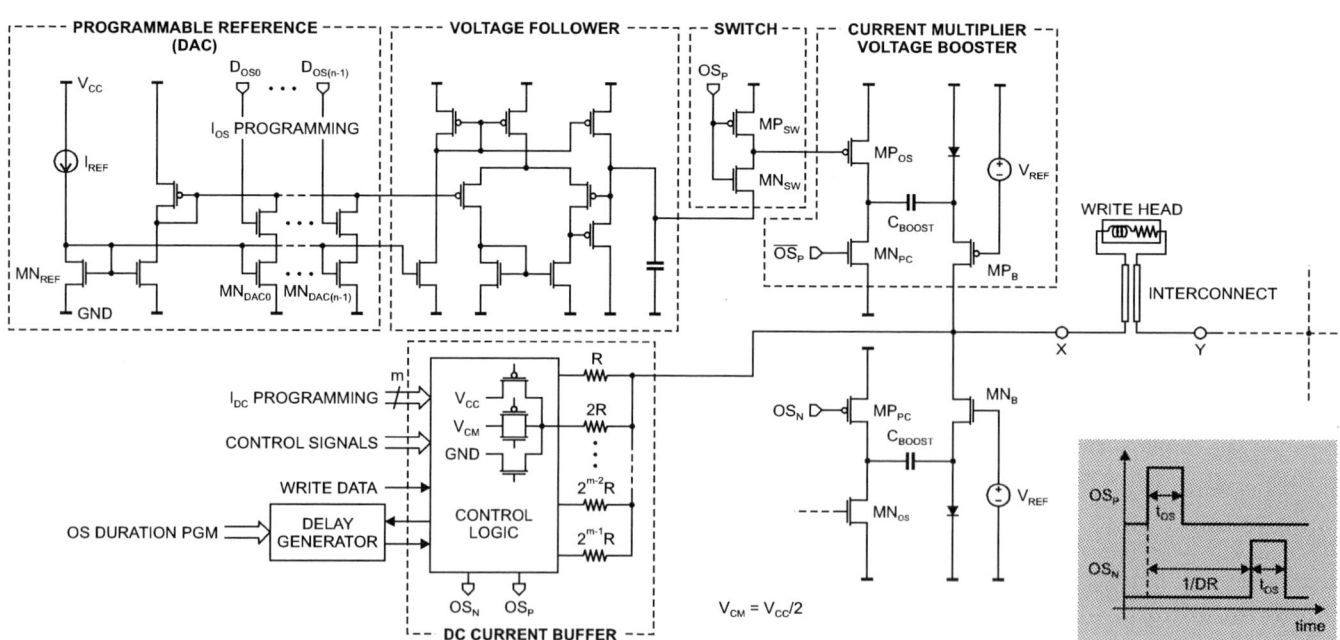

Fig. 2. Circuit schematic of the write driver with main blocks in evidence. See full-rate timing diagram on bottom right (DR = data rate).

418

III. LOW-POWER ARCHITECTURE

As apparent from Fig. 2, the presented driver is made up of a sum of subcircuits, of which the bridge is but one part, although the most power-consuming. However, the bridge draws a variable current and only when a write cycle is in progress, whereas the other sections constantly dissipate, even when no write is being accomplished. This quantity can well reach a considerable share of the overall consumption; therefore, careful design is demanded in order to reduce the amount of power required to supply those parts.

In the present Section, the low-power enhancements adopted in the driver are described; these mark the actual improvement upon the aforementioned references, which are either unconcerned with power optimization or do not account for it but within the bridge.

A first note is dealing with the currents drawn by each single branch along power rails. These quantities are optimally assigned the lowest values adequate for the task, that is, about a few hundred microamps, with the sole exception of the DAC's most significant bits. This is especially true of the reference value I_{REF}, being set at as low as 700 µA and subsequently elevated by a factor of one hundred through the bridge.

As previously evidenced, voltage boosting is not permanent; instead, it is performed on write request only, along with current multiplication. Morover, the arranged scheme allows for capacitors to be kept constantly on charge while not in use without causing any steady-state power consumption.

Noteworthy is also that the DAC circuit is not entirely reproduced four times alike: only one specimen is needed, together with just a bunch of current mirrors to drive the remaining branches of the bridge, which favors further reducing supply current.

The most important device introduced is probably the dc network being devised for a null overhead. The resistor ladder draws exactly as much as the provided dc current, which saves considerable power since I_{DC} is to be seamlessly sourced during a write.

IV. SIMULATION RESULTS

The results of circuit simulation are shown in Fig. 3: the boosted voltages and the write head current with varying codes are depicted at 250 Mbit/s (the rate of interest, also considering the interconnect features) for a given input string. It is apparent that the waveforms exhibit very neat shapes and peaks; only the overshoot leading edge appears somewhat fuzzy and distorted in dynamics due to the presence of the interconnect. The graph also evidences the excellent linearity of both the overshoot and the dc currents versus the settings.

The used simulation engine is Mentor Graphics' Eldo; the development was thoroughly performed under Cadence environment. A database model extracted from EM finite-element simulations of a reference commercial sample was adopted for the interconnect.

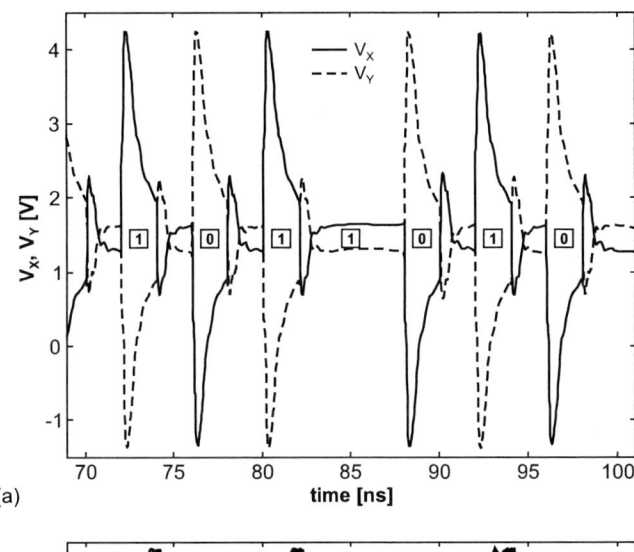

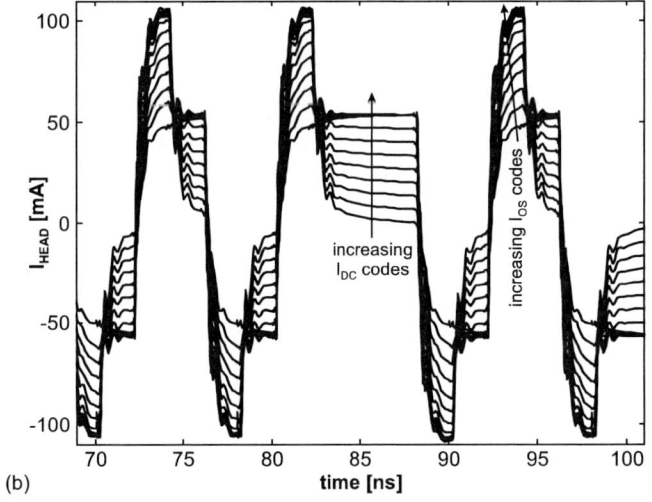

Fig. 3. Overshoot voltages (a) and write head current with varying programming codes (b). Note write data in the middle of (a).

In Table I the mean dissipated power values are reported as a function of overshoot duration, while Fig. 4 displays the waveform with both I_{OS} and I_{DC} set at their maximum, evidencing the effects of a variable overshoot duration on the write head current.

TABLE I
POWER DISSIPATION WITH VARYING OVERSHOOT DURATION
(I_{OS} = 100 mA, I_{DC} = 50 mA)

DUR [ns]	1	2	3	4
P_D [mW]	303	444	554	605

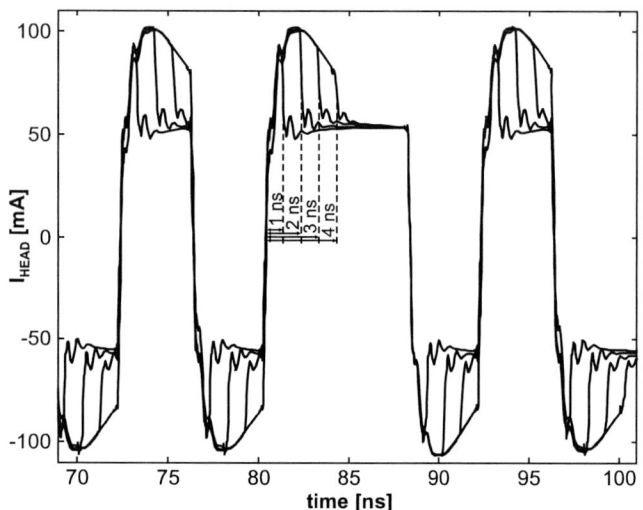

Fig. 4. Current in write head with varying overshoot duration (maximum current levels; same write pattern as Fig. 3).

V. CONCLUSION

In this work, an enhanced low-voltage low-power inductive write head driver targeted to mobile hard disk drive applications has been demonstrated. A voltage-boosted architecture was proposed for operation from a 3-V single-ended supply, such as a battery, and accurate power optimization was performed, resulting in an average consumption around 440 mW at 250 Mbit/s data rate. Moreover, the circuit was designed to be thoroughly programmable as to the overshoot and the dc currents (up to 100 and 50 mA respectively) as well as the overshoot duration (in the span 1 ÷ 4 ns) for commercial compliance purposes. The write driver was developed using 0.35-μm BiCMOS technology; however, effortless portability to full CMOS was pursued at design stage.

REFERENCES

[1] H. Veenstra, J. Mulder, Luan Le, and G. Grillo, "A 1 Gb/s read/write-preamplifier for hard-disk-drive applications," in *IEEE International Solid-State Circuits Conference Digest of Technical Papers*, pp. 188-189, 445, Feb. 2001.

[2] N. Fujii, M. Kuraishi, T. Mochizuki, S. Irikuraz, and T. Hirose, "A SOI-BiCMOS 800 Mbps write driver for hard disk drives," in *Proc. IEEE Conference on Custom Integrated Circuits*, pp. 451-454, May 2001.

[3] H. Yoshizawa, Y. Kobayashi, M. Yoshinaga, Y. Ookuma, K. Maio, and K. Irikura, "A 1.2 Gbps SOI-BiCMOS write driver for hard disk drives," in *Proc. IEEE Custom Integrated Circuits Conference*, pp. 349-352, May 2002.

[4] R.J. Reay, K.B. Klaassen, and C.S. Nomura, "A resonant switching write driver for magnetic recording," *IEEE Journal of Solid-State Circuits*, Vol. 32, No. 2, Feb. 1997, pp. 267-269.

[5] L. Le, E. Pieraerts, and G. de Jong, "A 500 MHz write amplifier for hard disk drives with low output impedance," in *Proc. European Solid-State Circuits Conference*, pp. 62-64, Sept. 1999.

[6] S. Lamb, L. Cheng, and D. Young, "A 550 Mb/s GMR read/write amplifier using 0.5 μm 5 V CMOS process," in *IEEE International Solid-State Circuits Conference Digest of Technical Papers*, pp. 358-359, 470, Feb. 2000.

[7] T. Ngo, C. Brannon, and J. Shier, "A low-power 3 V-5.5 V read/write preamplifier for rigid-disk drives," in *IEEE International Solid-State Circuits Conference Digest of Technical Papers*, pp. 286-287, Feb. 1994.

[8] T. Kawashimo, H. Yamashita, M. Yagyu, and F. Yuki, "A Low-Power Write Driver for Hard Disk Drives," in *Symposium on VLSI Circuits Digest of Technical Papers*, pp. 226-229, June 2005.

[9] A. Venca, B. Posat, R. Alini, "Power-optimized impedance-matched write driver," *US patent application*, ST Microelectronics, 2003.

[10] A. Venca, B. Posat, R. Alini, "Boosting technique to improve the output voltage swing in a write driver for a disk drive device," *US patent application*, ST Microelectronics, 2003.

2006 25th International Conference on Microelectronics

Optical Receiver with Voltage-Controlled Transimpedance in BiCMOS Technology

Nikša Tadić and Horst Zimmermann

Abstract - An optical receiver in 0.6 μm BiCMOS technology with voltage-controlled resistor-based variable transimpedance is presented. The frequency bandwidth is independent of the photodiode capacitance. A linearity error smaller than 2.8 %, a sensitivity dynamic range of 76.3 (37.7 dB), an offset voltage smaller than 0.47 mV, a bandwidth up to 204 MHz, and a maximum power consumption of less than 4.05 mW have been achieved.

I. INTRODUCTION

In order to meet the requirements for permanent increase of data rate and decrease of power of laser light, optoelectronic integrated circuits (OEICs) should include high-speed and high-sensitivity both integrated photodiodes and transimpedance amplifiers (TIAs). In addition, due to the different operating conditions in optical storage systems (different wavelengths of the laser light used, read and write operation, and multilayer discs), TIAs should have variable gain. The examples of such OEICs are [1] and [2] with switchable gain based on digitally-controlled pass-transistor arrays and feedback polysilicon resistors, [3] and [4] with a two-stage current amplifier followed by a classical TIA stage, [5] based on a Gilbert cell, and [6] and [7] with non-linear transfer characteristics using single-MOSFET voltage-controlled resistors (VCRs). On the other hand, it is desirable to use photodiodes with large sensitive area to ensure that all the incident light is coupled into it [1]. However, large photodiode diameters cause large PN junction capacitances, which limit both the speed and the sensitivity of standard OEIC [8].

The proposed optical receiver satisfies the demands for high-speed and high-sensitivity OEICs with variable gain. Moreover, the frequency bandwidth of the proposed optical receiver is independent of the photodiode capacitance. The design of the optical receiver with voltage-controlled transimpedance in BiCMOS technology is based on a mixed current-mode and voltage-mode approach by using the first-generation current conveyor (CCI) and a voltage amplifier. This OEIC includes an

Nikša Tadić is with the University of Montenegro, Faculty of Electrical Engineering, Cetinjski put bb, 81000 Podgorica, Montenegro, Serbia and Montenegro, E-mail: niksa_tadic@yahoo.com.
Horst Zimmermann is with the Vienna University of Technology, Faculty of Electrical Engineering and Information Technology, Institute for Electrical Measurements and Circuit Design, Gusshausstrasse 25/354, 1040 Vienna, Austria, E-mail: horst.zimmermann@tuwien.ac.at.

integrated PIN photodiode and a variable-gain TIA. The transimpedance gain is directly proportional to the equivalent resistance of the VCR, which is continuously changeable by varying the control voltage.

II. BASIC APPROACH

The simplified circuit schematic of the proposed optical receiver is shown in Fig. 1. It consists of the integrated PIN photodiode [9] and the variable gain TIA including CCI [10], [11], the non-grounded VCR R_{VCR} [12], the reference voltage V_{REF} for reverse photodiode biasing, and the non-inverting voltage amplifier. If operational amplifier's (OA's) transfer characteristic is approximated by the dominant pole ω_p and given by $A(j\omega)=A_0/(1+j\omega/\omega_p)$, the DC open-loop gain is $A_0 \gg 1+R_4/R_3$, and the output resistance R_Z of the CCI's terminal Z is much larger than the largest equivalent resistance $R_{VCRmax}\sim100$ kΩ, the transimpedance T of the optical receiver is

$$T(j\omega) \approx \frac{(1+R_4/R_3)R_{VCR}}{(1+j\omega/\omega_{p1})(1+j\omega/\omega_{p2})(1+j\omega/\omega_{p3})} \quad (1)$$

where $\omega_{p1}\approx1/(R_{VCR}C_{ZT})$, $\omega_{p2}=A_0\omega_p/(1+R_4/R_3)$, $\omega_{p3}=1/(R_X C_{XT})$, $C_{ZT}=C_Z+C_{OA}+C_{VCR}$, $C_{XT}=C_X+C_{PD}$, C_Z is the output capacitance at the CCI's terminal Z, C_{OA} is the input capacitance of the OA, C_{VCR} is the input capacitance of the VCR, C_X is the input capacitance at the CCI's terminal X, C_{PD} is the photodiode capacitance, and R_X is the input resistance at the CCI's terminal X. Because all poles are real and different, there is no gain peaking in the frequency response of the optical receiver, and there is no stability problem. One can see the following features of the frequency response of the proposed optical receiver with voltage-controlled transimpedance in BiCMOS technology:

- The larger the equivalent resistance R_{VCR}, and/or the larger the gain $1+R_4/R_3$ of the voltage amplifier, the smaller the frequency bandwidth f_{-3dB}, and vice-versa;
- The bandwidth f_{-3dB} can be made independent of the photodiode capacitance C_{PD} by a careful design of the CCI with small enough input resistance R_X;
- Gain peaking in the frequency response of the optical receiver cannot be caused by the dominant pole of the OA configuring the voltage amplifier.

The last two features are unique in optical receiver design, and cannot be achieved using a voltage-mode approach only.

1-4244-0116-X/06/$20.00 ©2006 IEEE

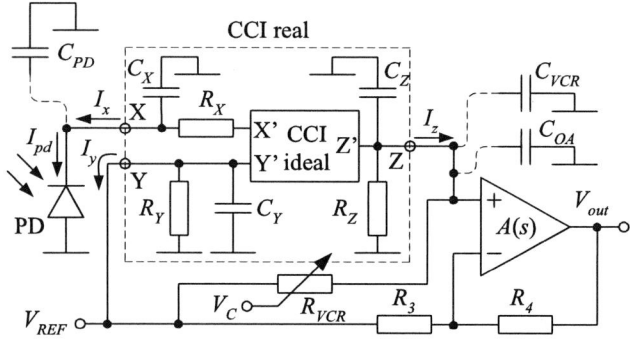

Fig. 1. Simplified circuit schematic of the proposed optical receiver.

III. CIRCUIT DESCRIPTION

The complete circuit schematic of the optical receiver with voltage-controlled transimpedance in BiCMOS technology is shown in Fig. 2.

The integrated $n^+n^-p^-p^+$ PIN photodiode used has the same design as that in [9].

The CCI [10] in BiCMOS technology consists of the BJTs Q_1 and Q_2 and MOSFETs M_3-M_8. Assuming ideal matching of the BJTs Q_1 and Q_2, and MOSFETs M_3 and M_4, and neglecting the base currents of the BJTs Q_1 and Q_2, the input resistance R_X of the CCI shown in Fig. 2 becomes $R_X=0$ Ω. So, the pole $\omega_{p3}=1/(R_X C_{XT})$ is infinite. However, because of the BJTs and MOSFETs matching limitations and finite current gains β_1 and β_2 of the BJTs Q_1 and Q_2, the input resistance R_X of the CCI can be estimated to be less than a few tens of ohms for all values of the input current I_x (photodiode current I_{pd}). Because the capacitance C_{XT} is smaller than a few pF (in the worst case), the pole $\omega_{p3}=1/(R_X C_{XT})$ is larger than the bandwidth of the photodiode, and the bandwidth f_{-3dB} of the optical receiver is insensitive to the photodiode capacitance C_{PD}.

The VCR using bisection of the input voltage [12] consists of the MOSFETs M_1 and M_2, the differential pair made by the BJTs Q_5 and Q_6 with active loads made by the MOSFETs M_{15}, M_{16}, M_{18} and M_{19}, biased by the MOSFET M_{21} and the voltage V_{B1}, and the resistive voltage divider with $R_1=R_2$ biased by the emitter follower with the BJT Q_8. The control voltage V_{C1} is presented by the source-to-gate voltage $V_{SG2}=V_{C1}$ of the saturated p-channel MOSFET M_2 carrying the control current I_C. It is designed as the voltage-controlled current source using the control voltage V_{C2}, the resistor R_5, and the CCI made by the MOSFETs M_9-M_{14}, BJTs Q_3 and Q_4, and biasing voltage V_{B2}.

To obtain the largest output voltage variation of the optical receiver $V_{outmax}-V_{outmin}=1$ V with a large enough resistance dynamic range R_{VCRmax}/R_{VCRmin}, the voltage variation V_{ds1} across the VCR should be amplified by the voltage amplifier shown in Fig. 1. The voltage amplifier consists of the differential input stage made by BJTs Q_5 and Q_7 with active loads made by MOSFETs M_{15}, M_{17}, M_{18} and M_{20}, biased by the MOSFET M_{21} and the voltage V_{B1},

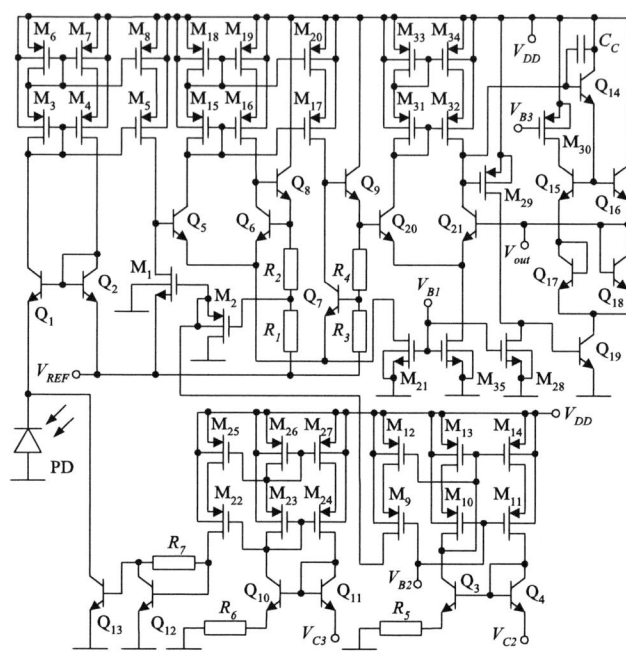

Fig. 2. Complete circuit schematic of the optical receiver with voltage-controlled transimpedance in BiCMOS technology.

and the feedback resistors R_3 and R_4 biased by the emitter follower with the BJT Q_9. In this way, the voltage across the VCR V_{ds1} is amplified by the gain $A=1+R_4/R_3$.

The class AB output stage with all npn BJTs is a modified version of the output stage presented in [13]. The core of this output stage consists of the BJTs Q_{15}-Q_{18}. The MOSFET M_{30} biased by V_{B3} acts as a simple current source to produce the constant sum of the base-to-emitter voltages of the BJTs Q_{15} and Q_{17} as well as the BJTs Q_{16} and Q_{18} $V_{BE15}+V_{BE17}=V_{BE16}+V_{BE18}=const.$ In order to increase the frequency bandwidth of the optical receiver by increasing the pole $\omega_{p2}=A_0\omega_p/(1+R_4/R_3)$, the voltage follower consisting of the differential pair made by the BJTs Q_{20} and Q_{21} and the active loads made by the MOSFETs M_{31}-M_{34}, biased by the MOSFET M_{35} and the voltage V_{B1}, is inserted between the output of the voltage amplifier and the class AB output stage.

The offset voltage V_{OFF} is compensated by using the cancellation current source designed as the voltage-controlled peaking current source [13]. It is realized by the BJTs Q_{12} and Q_{13}, the resistor R_7, and the voltage-controlled current source using the compensation control voltage V_{C3}, the resistor R_6, and the CCI made by the MOSFETs M_{22}-M_{27}, and BJTs Q_{10} and Q_{11}.

IV. EXPERIMENTAL AND SIMULATION RESULTS

The optical receiver with voltage-controlled transimpedance was fabricated in 0.6 μm BiCMOS technology. The chip photo is shown in Fig. 3. It occupies an active area of 330 μm x 190 μm. The chip has been glued onto a printed circuit board and wire-bonded to it.

422

The light source was a laser diode with an optical wavelength of $\lambda=660$ nm modulated via a bias tee. The laser light was coupled into the photodiode via an optical multimode fibre adjusted on a wafer prober. An octagonal PIN photodiode with a diameter of 50 μm and anti-reflection coating has been used. This photodiode has a responsivity of $R=0.51$ A/W at a wavelength of $\lambda=660$ nm, which corresponds to a quantum efficiency of $\eta=96.4$ % [9]. For all measurements the supply voltage V_{DD}, the reference voltage V_{REF}, the control voltages V_{C2} and V_{C3}, the biasing voltages V_{B1}, V_{B2}, and V_{B3}, and the output voltage V_{out} have the following values: $V_{DD}=5$ V, $V_{REF}=2.1$ V, 50 mV$<V_{C2}<$500 mV, $V_{C3}=93$ mV, $V_{B1}=1.2$ V, $V_{B2}=3.5$ V, $V_{B3}=3.8$ V, and 2.1 V$<V_{out}<$ 3.1 V. The resistors R_1-R_7 have the following values: $R_1=R_2=R_3=1.25$ kΩ, and $R_4=R_5=R_6=R_7=10$ kΩ. The resistors' tolerances do not affect the optical receiver operation because only the ratios R_2/R_1 and R_4/R_3 are important. The results are summarized in Table I.

Post-layout simulated characteristics of the output voltage V_{OUT} versus photodiode current I_{PD} of the optical receiver, for different values of the control voltage V_{C2}, are shown in Fig. 4. The simulations have been performed by using CADENCE software tools. The maximum linearity error E_l for the output voltage range 2.1 V$<V_{OUT}<$3.1 V is $E_{lmax}=2.8$ %. The achieved linearity is much higher in a much larger output voltage range compared to a simple MOSFET working as feedback resistor in a TIA. Using a single non-saturated MOSFET instead of the VCR [12], the linearity error rises up to $E_l=27.7$ % for the same operating conditions. The range of the post-layout simulated transimpedance T of the optical receiver for different values of the control voltage V_{C2} is 11.13 kΩ$<T<$846.7 kΩ. The transimpedance dynamic range is $T_{max}/T_{min}=76.07$.

The range of the measured sensitivity S for different values of the control voltage V_{C2} is 4.781 mV/μW$<S<$364.89 mV/μW. The measured sensitivity dynamic range is $S_{max}/S_{min}=76.32$, which is in very good agreement with the simulated transimpedance dynamic range T_{max}/T_{min}.

After offset compensation performed for only one arbitrary control voltage V_{C2} by adjusting the voltage-controlled peaking current-source, the measured offset voltages became $|V_{OFF}|<0.47$ mV for all sensitivities. The compensation control voltage was $V_{C3}=93$ mV. Achieved offset voltages are much smaller than 17.8 mV and 11.6 mV reported in [2] and [4], respectively.

The range of the measured maximum power consumption P_{max} for different values of the control voltage V_{C2} is 1.7 mW$<P_{max}<$4.05 mW. The power measurements have been performed without any load impedance at the output of the optical receiver. Achieved maximum power consumptions are much smaller than 8.3 mW, 37.5 mW, and 20.2 mW reported in [2], [3], and [4], respectively.

The measured frequency response of the optical receiver with voltage-controlled transimpedance in BiCMOS technology is shown in Fig. 5. The range of the

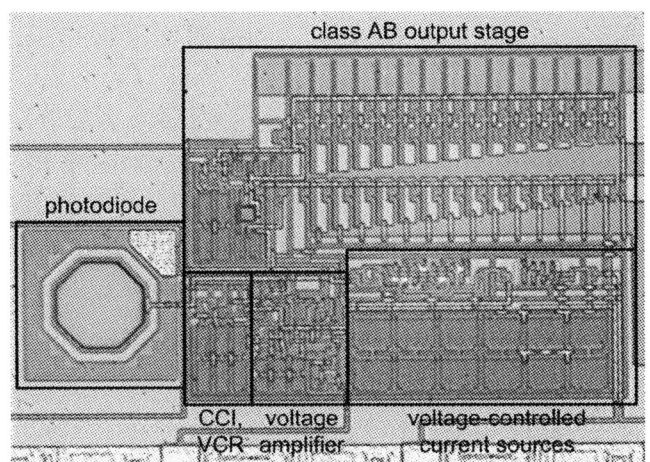

Fig. 3. Microphotograph of the optical receiver.

TABLE I
EXPERIMENTAL AND SIMULATION* RESULTS OF THE OPTICAL RECEIVER

V_{C2} |mV|	T* |kΩ|	E_l* |%|	S |mV/μW|	V_{OFF} |mV|	P_{max} |mW|	f_{-3dB} |MHz|
50	846.7	2.80	364.89	0.47	1.7	14.44
70	161.8	0.74	81.16	-0.28	1.92	56.55
100	56.02	0.42	27.89	-0.42	2.16	147.52
150	28.17	0.50	13.58	-0.42	2.63	177.71
200	19.95	0.47	9.420	-0.42	2.94	195.48
250	16.13	0.59	7.493	-0.42	3.18	202.32
300	14.15	0.64	6.457	-0.41	3.52	203.13
350	12.86	0.65	5.600	-0.42	3.65	203.28
400	11.93	0.63	5.173	-0.41	3.82	204.02
450	11.37	0.82	4.940	-0.41	3.93	203.63
500	11.13	0.95	4.781	-0.41	4.05	204.24

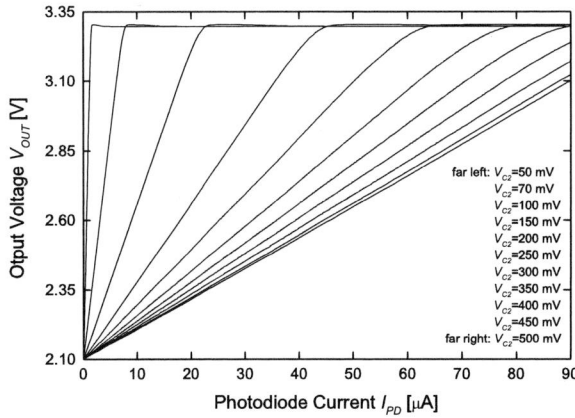

Fig. 4. Post-layout simulated characteristics of the output voltage versus photodiode current of the optical receiver.

measured frequency bandwidth f_{-3dB} for different values of the control voltage V_{C2} is 14.44 MHz$<f_{-3dB}<$204.24 MHz. The external load of (8.2 pF ∥ 270 Ω) has been added at the output of the voltage amplifier. The frequency bandwidth of the proposed optical receiver for $V_{C2}=100$ mV is 2.45 times higher at a 1.55 times lower

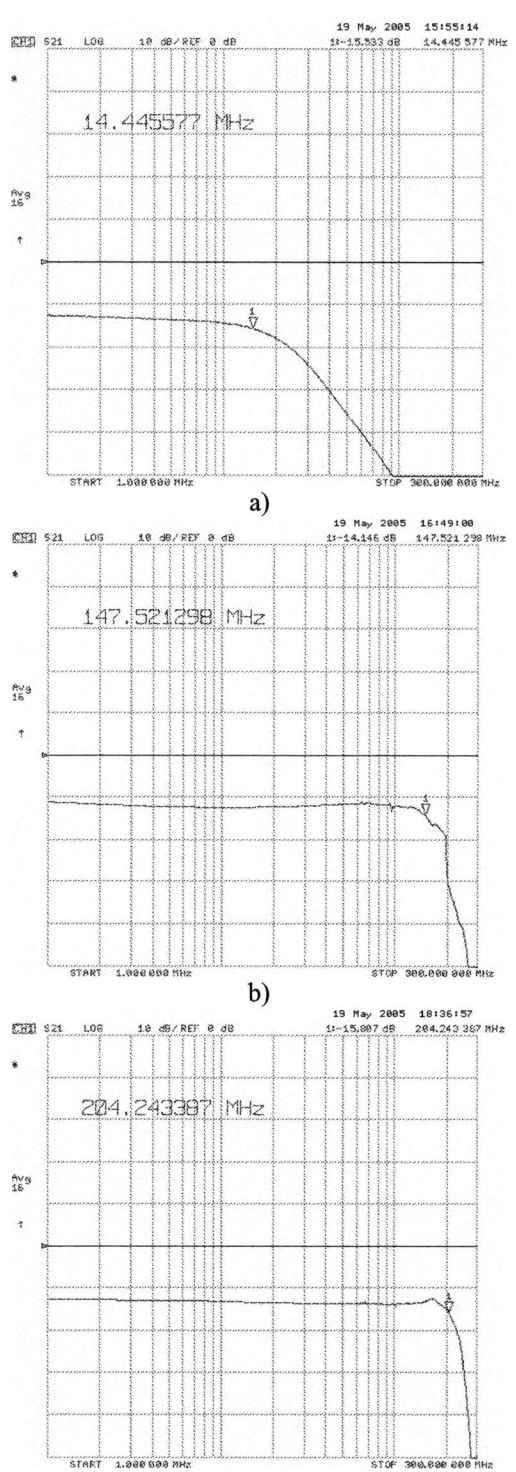

Fig. 5. Measured frequency response of the optical receiver: a) V_{C2}=50 mV, b) V_{C2}=100 mV, c) V_{C2}=500 mV.

sensitivity, compared to the fast channel in [2]. Compared to the sensitive channel in [2] with a bandwidth of 20.8 MHz and a sensitivity of 72 mV/µW, the bandwidth of the proposed optical receiver for V_{C2}=70 mV is 2.72 times higher at a 1.13 times larger sensitivity.

V. CONCLUSION

An optical receiver with voltage-controlled transimpedance in 0.6 µm BiCMOS technology is considered. It is based on a mixed current- and voltage-mode signal processing. In addition to very good linearity, wide sensitivity dynamic range, very small offset voltage, and low power consumption, the frequency bandwidth is independent of the photodiode capacitance. This unique feature will enable employment of large-area photodiodes in OEICs with a high transimpedance-bandwidth product.

REFERENCES

[1] K. Phang, D. A. Johns, "A CMOS optical preamplifier for wireless infrared communications", *IEEE Transactions on Circuits and Systems, Part II: Analog and Digital Signal Processing*, vol. 46, pp. 852-859, July 1999.

[2] K. Kieschnick, and H. Zimmermann, "High-sensitivity BiCMOS OEIC for optical storage systems," *IEEE J. Solid-State Circuits*, vol. 38, pp. 579-584, April 2003.

[3] J. Sturm, M. Leifhelm, H, Schatzmayr, S. Groiss, and H. Zimmermann, "Optical receiver IC for CD/DVD/blue-laser application," *IEEE J. Solid-State Circuits*, vol. 40, pp. 1406-1413, July 2005.

[4] C. Seidl, H. Schatzmayr, J. Sturm, S. Groiss, M. Leifhelm, D. Spitzer, H. Schaunig, and H. Zimmerman, "A programmable OEIC for laser applications in the range from 405 nm to 780 nm", *Proc. of European Solid-State Circuits Conference (ESSCIRC)*, pp. 439-442, 12 - 16 September 2005, Grenoble, France.

[5] T. Ruotsalainen, P. Palojarvi, J. Kostamovaara, "A current-mode gain-control scheme with constant bandwidth and propagation delay for a transimpedance preamplifier", *IEEE J. Solid-State Circuits*, vol. 34, pp. 253-258, Feb. 1999.

[6] R. G. Meyer, and W. D. Mack, "A wideband low-noise variable-gain BiCMOS transimpedance amplifier," *IEEE J. Solid-State Circuits*, vol. 29, pp. 701-706, June 1994.

[7] R. G. Meyer, and W. D. Mack, "Monolithic AGC loop for a 160Mb/s transimpedance amplifier," *IEEE J. Solid-State Circuits*, vol. 31, pp. 1331-1335, Sept. 1996.

[8] H. Zimmermann, *Silicon Optoelectronic Integrated Circuits*, Springer, Berlin, Heidelberg, Germany, 2004.

[9] M. Förtsch, H. Zimmermann, W. Einbrodt, K, Bach, and H. Pless, "Integrated PIN photodiode in high-performance BiCMOS technology," *IEEE International Electron Devices Meeting*, Tech. Digest, pp. 801-804, 2002.

[10] A. S. Sedra and G. Roberts, "Current conveyor theory and practice," in *Analogue IC design: the current-mode approach*, C. Toumazou, F. J. Lidgey, and D. G. Haigh (eds.), Stevenage, U. K.: Peter Peregrinus, 1990, chapter 3, pp. 93-126.

[11] B. Wilson, "Performance analysis of current conveyors", *Electronics Letters*, vol. 25, pp. 1596-1598, November 1989.

[12] N. Tadić and D. Gobović, "A voltage-controlled resistor in CMOS technology using bisection of the voltage range", *IEEE Transactions on Instrumentation and Measurement*, vol. 50, pp. 1704-1710, December 2001.

[13] P. R. Gray, R. G. Meyer, *Analysis and Design of Analog Integrated Circuits*, 3rd edition, New York: John Wiley & Sons, 1993.

2006 25th International Conference on Microelectronics

Improved Linearity Active Resistor Using Equivalent FGMOS Devices

C. Popa

Abstract – An original active resistor circuit will be presented in this paper. The main advantages of the new proposed implementations are the improved linearity, the small area consumption and the improved frequency response. An original technique for linearizing the $I(V)$ characteristic of the active resistor will be proposed, based on a parallel connection of two quasi-identical circuits opposite excited and different polarized having the result of improving the circuit linearity with about an order of magnitude. The errors introduced by the second-order effects will be also strongly reduced, while the advantage of using FGMOS devices is achieved maintaining the compatibility with classical technologies. The frequency response of the circuit is very good as a result of operating all MOS transistors in the saturation region. The circuit is implemented in $0.35 \mu m$ CMOS technology, the SPICE simulation confirming the theoretical estimated results and showing a linearity error under a percent for an extended input range $(\pm 500 mV)$ and a small value of the supply voltage $(\pm 3V)$.

Key words: active resistor, linearity error, squaring and square-root circuits, complementary functions

I. INTRODUCTION

CMOS active resistors are very important blocks in VLSI analog designs, mainly used for replacing the large value passive resistors, with the great advantage of a much smaller area occupied on silicon. Their utilisation domains includes amplitude control in low distortion oscillators, voltage controlled amplifiers and active RC filters. These important applications for programmable floating resistors have motivated a significant research effort for linearising their current-voltage characteristic.

The first generation of MOS active resistors [1], [2] used MOS transistors working in the linear region. The main disadvantage is that the realised active resistor is inherently nonlinear and the distortion components were complex functions on MOS technological parameters.

A better design of CMOS active resistors is based on MOS transistors working in saturation [3], [4]. Because of the quadratic characteristic of the MOS transistor, some linearisation techniques were developed in order to minimize the nonlinear terms from the current-voltage characteristic of the active resistor. Usually, the resulting linearisation of the $I-V$ characteristic is obtained by a first-order analysis. However, the second-order effects

which affect the MOS transistor operation (mobility degradation, bulk effect and short-channel effect) limits the circuit linearity introducing odd and even-order distortions, as shown in [4]. For this reason, an improved linearisation technique has to be design [5], [6] to compensate also the nonlinearities introduced by the second-order effects.

II. THEORETICAL ANALYSIS

The original idea for implementing a linear current-voltage characteristic of the active resistor, similar to the characteristic of a classical passive resistor is to use two complementary functions, the desired linearity being obtained by a mutual compensation of their nonlinearities. Because of the requirements for a good frequency response, only MOS transistors working in saturation could be used. For this reason, the original choose is to use the quadratic and the square-root functions as complementary functions. So, square and square-root circuits must be designed based exclusively on strong-inverted MOS devices (more exactly on MOS transistors working in saturation). In order to improve the circuit performances by using the FGMOS device and to maintain, also, the compatibility with classical CMOS technologies, an original equivalent FGMOS device will be proposed.

A. The block diagram of the FGMOS active resistor

The proposed active resistor is based on three important blocks: a voltage-current squarer, a current square-root circuit and a current-pass circuit, named x^2, $\sqrt{}$ and I, respectively on the block diagram from Fig. 1.

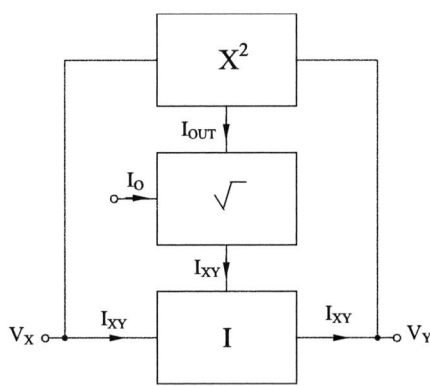

Fig. 1. The block diagram of the active resistor

C. Popa is with The Faculty of Electronics, Telecommunications and Information Technology Bucharest, 1-3 Iuliu Maniu, Bucharest, Romania, E-mail: cosmin_popa@yahoo.com.

1-4244-0116-X/06/$20.00 ©2006 IEEE   425

The I_{XY} current is proportional to the square-root of I_{OUT} and I_O currents, while I_{OUT} current is proportional to the square of the differential input voltage, $V_X - V_Y$. So, the result will be a linear relation between the differential voltage across the two pins, $V_X - V_Y$ of the active resistor and the current passing through it, I_{XY}. This linearity is valuable for a first-order analysis, that is without taking into account the second-order effects that affect the MOS transistor operation. Considering these undesired effects, a small error will affect the circuit linearity, quantitative evaluated by a total harmonic distortion coefficient that will be further determined.

B. The equivalent FGMOS transistor

The FGMOS transistor is a MOS transistor whose gate is floating (Fig. 2a), while the symbolical representation of this device is shown in Fig. 2b. The first silicon layer over the channel represents the floating-gate and the second polysilicon layer, located over the floating-gate implements the multiple input gates. This floating-gate is capacitive coupled to the multiple input gates.

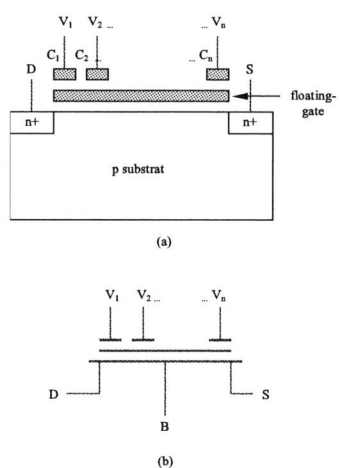

Fig. 2. (a) The basic structure of a n-channel FGMOS transistor; (b) symbolic representation

The drain current of a FGMOS transistor with n-input gates in the saturation region is given by the following relation:

$$I_D = \frac{K}{2}\left[\sum_{i=1}^{n} k_i(V_i - V_S) - V_T\right]^2 \qquad (1)$$

where $K = \mu_n C_{ox}(W/L)$ is the transconductance parameter of the transistor, μ_n is the electron mobility, C_{ox} is the gate oxide capacitance, W/L is the transistor aspect ratio, $k_i, i = 1,...,n$ are the capacitive coupling

ratios, V_i is the i-th input voltage, V_S is the source voltage and V_T is the threshold voltage of the transistor.

The great limitation of using FGMOS devices is that they are available only in few CMOS technologies, restricting the area of utilization of the circuits based on FGMOS transistors. The original idea for replacing the classical FGMOS device with two inputs by five MOS transistors is presented in Fig. 3.

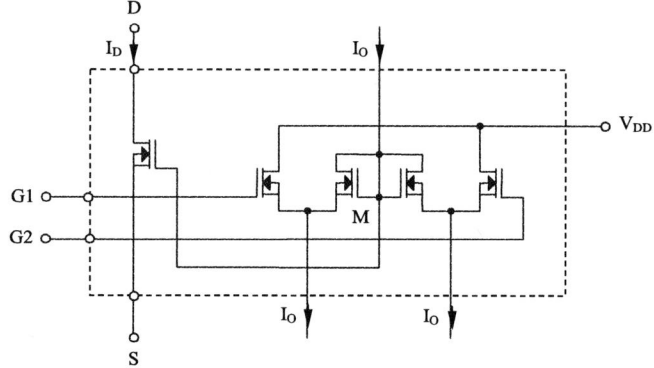

Fig. 3. The equivalent circuit of the FGMOS transistor

The four pins of the equivalent FGMOS device are, respectively: D (drain), S (source), $G1$ and $G2$ (gates). Because of the connection of the four MOS transistors from the right part of the circuit, the M potential is equal to the arithmetic mean of the gates' potentials, $V_M = (V_{G1} + V_{G2})/2$. For this reason, the drain current of the entire equivalent FGMOS device from Fig. 3 will have the following expression:

$$I_D = \frac{K}{2}\left(\frac{V_{G1} + V_{G2}}{2} - V_S - V_T\right)^2 \qquad (2)$$

similar to the equation which characterizes the operation of the classical FGMOS transistor.

The original proposed implementation presents the great advantage of avoiding the use of any resistor for computing the arithmetic mean, with the result of reducing the circuit area and of improving the achieved accuracy.

C. The voltage-current squarer

The new proposed squarer is based on the perfect symmetrical structure presented in Fig. 4.

The output current expression has a linear dependence on the drain currents of T_X, T_Y and T_Z transistors:

$$I_{OUT} = I_X + I_Y - I_Z \qquad (3)$$

Considering a saturation operation of all MOS devices from Fig. 3 and supposing that the area of T_Z is twice that

426

the T_X and T_Y areas, the previous currents will have the following expressions:

$$I_X = \frac{K}{2}\left(V_X - V_Y - V_T\right)^2 \qquad (4)$$

$$I_Y = \frac{K}{2}\left(V_Y - V_X - V_T\right)^2 \qquad (5)$$

$$I_Z = KV_T^2 \qquad (6)$$

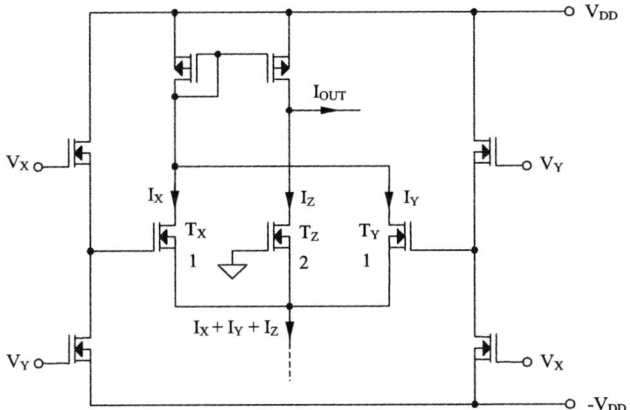

Fig. 4. The voltage-current squarer

From the previous relations it results a quadratic dependence of the output current I_{OUT} on the differential input voltage $V_X - V_Y$:

$$I_{OUT} = K\left(V_X - V_Y\right)^2 \qquad (7)$$

D. The square-root circuit

The square-root circuit represents also a perfect symmetrical structure (Fig. 5), using MOS transistors and an equivalent FGMOS device ($T_1 - T_5$) working in saturation for improving the circuit frequency response.

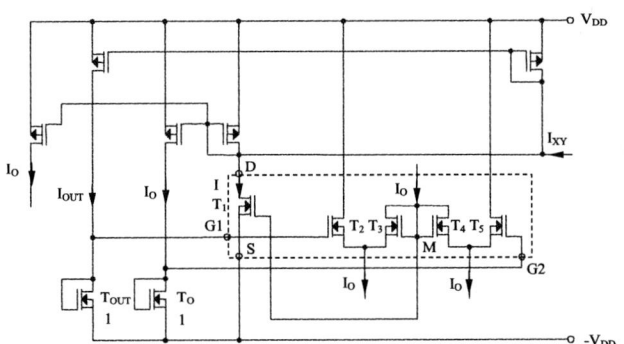

Fig. 5. The square-root circuit

Because the potential M is equal to the algebraic mean of $G1$ and $G2$ potentials, the drain current I of transistor T_1 (having an aspect ratio fourth time greater than the other devices) will be equal to:

$$I = \frac{4K}{2}\left(\frac{V_{GS_{OUT}} + V_{GS_O}}{2} - V_T\right)^2 \qquad (8)$$

where $V_{GS_{OUT}}$ and V_{GS_O} represents the gate-source voltages of T_{OUT} and T_O transistors. It results the expression of the output current of the square-root circuit from Fig. 4:

$$I_{XY} = \left(\sqrt{I_{OUT}} + \sqrt{I_O}\right)^2 - I_{OUT} - I_O \qquad (9)$$

equivalent to a square-root dependence of the output current on the two input currents:

$$I_{XY} = 2\sqrt{I_{OUT}I_O} \qquad (10)$$

E. The current-pass circuit

The necessity of designing this circuit is derived from the requirement that the same current to pass between the two output pins, X and Y. The implementation in CMOS technology of this circuit is very simple, consisting in a simple and a multiple current mirrors (Fig. 6).

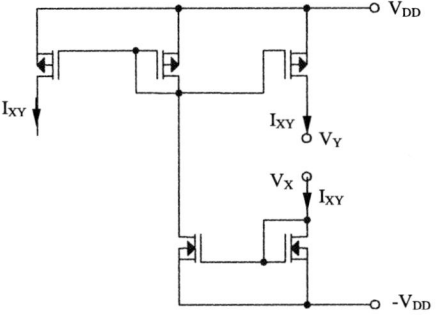

Fig. 6. The current-pass circuit

F. The linear characteristic of the active resistor

Because of the complementary characteristics (7) and (10) of the squaring and square-root circuits, the current-voltage characteristic $I_{XY}(V_X - V_Y)$ of the active resistor will be, in a first-order approximation, perfectly linear.

$$I_{XY} = 2\sqrt{KI_O}\left(V_X - V_Y\right) \qquad (11)$$

For this reason, it is possible to define an equivalent resistance between X and Y pins as follows:

$$R_{ECH.} = \frac{V_X - V_Y}{I_{XY}} = \frac{1}{2\sqrt{KI_O}} \quad (12)$$

The great advantage of the previous presented circuit is that the value of the equivalent active resistance could be very easily controlled by modifying the reference current I_O. For usual values of K and I_O parameters, the value of $R_{ECH.}$ resistance covers about three decades $(1k\Omega - 1M\Omega)$, equivalent with an important reduction of the silicon occupied area, especially for large values of the active resistance.

G. The second-order effects

The linearity (11) of the current-voltage characteristic for the active resistor circuit having the block diagram presented in Fig. 1 is slightly affected by the second-order effects that affect the MOS transistor operation. Considering that the design condition $\lambda = \theta_D$ is fulfilled, the gate-source voltage of a MOS transistor working in saturation at a drain current I_D will be:

$$V_{GS} = V_T + \sqrt{\frac{2I_D}{K}} + \theta_G \frac{I_D}{K} \quad (13)$$

The last term represents the error which affects the quadratic characteristic of the MOS transistor operated in saturation, caused by the previous presented second-order effects. The result will be a small accuracy degradation of the entire circuit linearity, quantitative evaluated by the superior-order terms in the $I(V)$ characteristic of the active resistor:

$$I_{XY} = \sum_{k=1}^{\infty} a_k (V_X - V_Y)^k \quad (14)$$

Because of the circuit symmetry, the odd terms from the previous relation are usually cancel out, so the main circuit nonlinearity caused by the second-order effects could be approximated by the third-order term from the $I(V)$ characteristic of the active resistor.

III. Simulated Results

The low-power CMOS active resistor was implemented in $0.35\mu m$ CMOS technology. The maximum nonlinearity error of the active resistor for limited input voltage range $(|V_X - V_Y| \leq 500mV)$ is less than a percent.

IV. Conclusions

A new active resistor circuit has been presented in this paper. The main advantages of the original proposed implementations are the improved linearity, the small area consumption and the improved frequency response. An original technique for linearizing the $I(V)$ characteristic of the active resistor has been proposed, based on a parallel connection of two quasi-identical circuits opposite excited and different polarized, having the result of improving the circuit linearity with about an order of magnitude. The errors introduced by the second-order effects will be also strongly reduced, while the advantage of using FGMOS devices was achieved maintaining the compatibility with classical technologies (the classical FGMOS device was replacing by an original equivalent circuit using exclusively classical MOS devices). The circuit frequency response of the circuit is very good as a result of operating all MOS transistors in the saturation region. The circuit is implemented in $0.35\mu m$ CMOS technology, the SPICE simulation confirming the theoretical estimated results and showing a linearity error under a percent for an extended input range $(\pm 500mV)$ and a small value of the supply voltage $(\pm 3V)$. A small linearity degradation could appear as a result of considering the parameters mismatches.

Acknowledgement

This work was supported by the Research Project 15.05.14 – UEFISCSU ET 2970.

References

[1] Z. Wang, "Current-controlled Linear MOS Earthed and Floating Resistors and Application", in *IEEE Proceedings on Circuits, Devices and Systems*, Dec. 1990, pp. 479-481.
[2] L. Sellami, "Linear Bilateral CMOS Resistor for Neural-type Circuits", in *Proceedings of the 40th Midwest Symposium on Circuits and Systems*, Ian. 1997, vol. 2, pp. 1330-1333.
[3] S.P. Singh, J.V. Hansom, and J. Vlach, "A New Floating Resistor for CMOS Technology", in *IEEE Transactions on Circuits and Systems*, Sept. 1989, pp. 1217-1220.
[4] S. Sakurai, and M. Ismail, "A CMOS Square-law Programmable Floating Resistor", in *IEEE International Symposium on Circuits and Systems*, June 1993, pp. 1184-1187.
[5] H. Huanzhang, E.K.F. Lee, "Low voltage technique for active RC filter", *Electronics Letters*, 1998, Volume 34, Issue 15, pp. 1479 – 1480.
[6] X.D. Jia, R.M.M. Chen, "Switched capacitor nonlinear circuits derived from Chua's circuit", in *Circuits and Systems, 1996. ISCAS '96*, 1996, pp. 213 – 216, vol.3 12-15.

2006 25th International Conference on Microelectronics

A Stochastic Approach to Crosstalk Analysis in Mixed-Signal ICs

Giorgio Boselli, Gabriella Trucco, and Valentino Liberali

Abstract— **In this paper we present a new approach to the analysis of digital switching noise, by modeling it as a stochastic process. As well known, a pre-layout estimation of digital switching noise is a very difficult task. Therefore, mixed-signal IC design would benefit from a simple model that uses few parameters and allows to estimate digital switching noise with an acceptable level of approximation.**

I. Introduction

The integration of complete systems on a single chip requires a careful modeling of crosstalk between analog and digital parts, as disturbances injected by the large digital processor can severely degrade analog block operation, thus affecting overall system performance [1].

To analyze a large mixed-signal circuit, we propose to use a stochastic model to describe the amplitude density of current pulses drawn by a switching logic circuit. When switching instants in a logic circuit are uniformly distributed in time, we can model them as a Poisson process, and describe digital switching currents as a shot noise process.

It is well known that, at microscopic level, currents vary in unpredictable ways; this unpredictable variation is called noise. In general, shot noise is due to the corpuscular nature of transport. In electronic, shot noise is always associated with direct current flow, which results from the motion of discrete and independent charged particles. Shot noise was studied to model the carriers passing through a p-n junction in a conductor. The passage of each carrier across the depletion region of the junction can be considered as a random event; since the events are random and independent, Poisson statistics describe this process. Sect. II is intended as a brief reminder to the basic concepts of Poisson process.

At macroscopic level, the same approach can be used to model the overall switching current in a large digital circuit (Sect. III).

Finally, in Sect. IV, we compare results from theoretical calculations with circuit simulations.

II. Poisson Process

The Poisson process is an example of continuous-time random process. Considering events occurring at random instants of time at an average rate of λ events per second, let us define $n(t)$ as the number of event

The authors are with Department of Information Technologies, University of Milan, Via Bramante 65, 26013 Crema, Italy, Email: boselli@dti.unimi.it

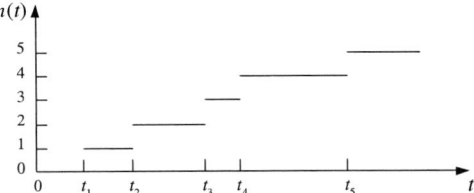

Fig. 1. A sample path of the Poisson counting process

occurrences in the time interval $[0, t]$. Then $n(t)$ is a nondecreasing, integer-valued, continuous-time random process. Fig. 1 shows a sample function of $n(t)$. Let us suppose that the interval $[0, t]$ is divided into m subintervals of very short duration $\Delta t = t/m$, and assume that: (i) the probability of more than one event occurrence in a subinterval is negligible compared to the probability of observing one or zero events; and (ii) whether or not an event occurs in a subinterval is independent of the outcomes in other subintervals. If the probability of an event occurrence in each subinterval is p, then the expected number of event occurrences in the interval $[0, t]$ is mp. Since events occur at a rate of λ events per second, the average number of events in the interval $[0, t]$ is λt. Then we have:

$$\lambda t = mp.$$

If $m \to \infty$ and $p \to 0$ while $\lambda t = mp$ is fixed, then the number of event occurrences $n(t)$ in the interval $[0, t]$ has a Poisson distribution with mean λt:

$$\Pr[n(t) = k] = e^{-\lambda t} \frac{(\lambda t)^k}{k!} \text{ for } k = 0, 1, 2, \dots \quad (1)$$

The process $n(t)$ is called *Poisson process* [2]. Fig. 2 shows the Poisson probability density function (p.d.f.)

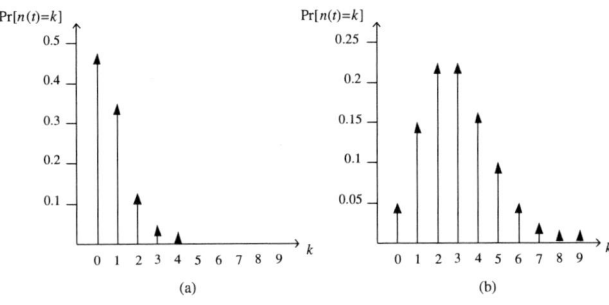

Fig. 2. Probability density functions of a Poisson random process

1-4244-0116-X/06/$20.00 ©2006 IEEE   429

for two different values of λt. For $\lambda t < 1$, $\Pr[n(t) = k]$ is maximum at $k = 0$ (Fig. 2(a)); if λt is a positive integer, $\Pr[n(t) = k]$ is maximum at $k = \lambda t$ and at $k = \lambda t - 1$ (Fig. 2(b)).

III. STOCHASTIC MODEL OF DIGITAL SWITCHING NOISE

Let us consider an asynchronous digital network, made up with identical logic cells driving equal capacitive loads. This assumption is useful to derive a simplified model. Later on, we will refine the model by removing some restrictive hypotheses. If all switching instants of input signals applied to logic gates are independent and randomly distributed in a uniform manner, the digital switching noise could be well approximated with a *shot noise* process. The instants when the input signals of logic gates switch can be considered as Poisson points. Given the time interval $[0, t]$, we define the random variable n as the number of transitions of signals at logic gate inputs. The probability to have exactly $n = k$ events in the considered time interval is given by (1), and the number of Poisson points in an interval of length t is a Poisson distributed random variable with parameter $a = \lambda t$, where λ is the density of the points [2].

Let us consider the input transition of a logic gate as a Dirac impulse. Then we have a train of impulses, each of which taken at random instants, to model the behavior of the circuit. Then the stochastic process $X(t)$, that in our case represents the input transitions of logic gates, is:

$$X(t) = \sum_i \delta(t - t_i). \qquad (2)$$

If the logic gates of the circuit are all equal, in the sense that all gates have the same i_{DD} and i_{SS} current absorption, the convolution between the train of impulses and the current absorbed by the single logic gate is equal to the total current absorbed by the entire digital circuit:

$$Y(t) = h(t) * X(t), \qquad (3)$$

where $h(t)$ is the impulse response, representing the current absorbed by a single gate in one logic transition. Fig. 3 illustrates the filtering of an impulse train with an LTI system. The filter output $Y(t)$ is the stochastic process representing the total current absorbed by the digital circuit.

The described process, called shot noise, is based on the statistical independence of the considered events [2], which are, in our case, the transitions of logic gates.

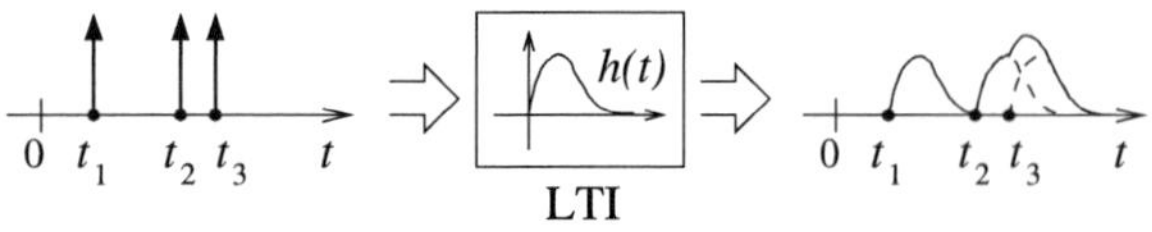

Fig. 3. Train of impulses filtered through an LTI system

Then, in an asynchronous system where all gates have equal current absorption, crosstalk noise is equivalent to a shot noise process. If the impulse density λ is constant over time, the process is stationary.

By substituting (2) in (3), we obtain:

$$Y(t) = \sum_i h(t - t_i). \qquad (4)$$

The p.d.f. of the process $Y(t)$ can be calculated from the p.d.f. of the single current pulse $f_H(x)$. In the following calculations, we assume that the LTI system is a finite impulse response filter, i.e., the current pulse has a finite duration τ. This approximation leads to an impulse at the origin in the p.d.f. [3].

Let us consider an arbitrary time instant t_1. The total current at time t_1, $Y(t_1)$, is a random variable, whose p.d.f. depends on both the number of Poisson impulses falling in the interval $[t_1 - \tau, t_1]$ and the p.d.f. of the single current pulse $f_H(x)$, as follows.

If no impulse occurs in $[t_1 - \tau, t_1]$, then $Y(t_1) = 0$, and the p.d.f. is $\delta(x)$, with probability $\Pr[n = 0]$.

If a single impulse occurs in $[t_1 - \tau, t_1]$, then $Y(t_1)$ has a p.d.f. $f_H(x)$, with probability $\Pr[n = 1]$.

If two impulses occur in $[t_1 - \tau, t_1]$, then the p.d.f. of the sum of two current pulses is given by the convolution between the two p.d.f. of the pulses $f_H(x) * f_H(x)$, with probability $\Pr[n = 2]$.

If exactly k impulses occur in $[t_1 - \tau, t_1]$, then the p.d.f. of the total current is:

$$\underbrace{f_H(x) * f_H(x) * \ldots * f_H(x)}_{k \text{ factors}},$$

with probability $\Pr[n = k]$.

Of course, $\sum_{k=0}^{\infty} \Pr[n = k] = 1$. The p.d.f. of the total current, $f(x)$, can be calculated by taking the average sum:

$$
\begin{aligned}
f(x) = {}& \delta(x) \Pr[n = 0] + f_H(x) \Pr[n = 1] + \ldots + \\
&+ \underbrace{f_H(x) * f_H(x) * \ldots * f_H(x)}_{k \text{ factors}} \Pr[n = k] + \ldots = \\
= {}& \sum_{k=0}^{\infty} f_k(x) \Pr[n = k],
\end{aligned}
$$
$$(5)$$

where

$$
\begin{aligned}
f_0(x) &= \delta(x) \\
f_1(x) &= f_H(x) \\
f_2(x) &= f_H(x) * f_H(x) \\
f_k(x) &= \underbrace{f_H(x) * f_H(x) * \ldots * f_H(x)}_{k \text{ factors}}
\end{aligned}
$$

By using the Poisson probability (1) in (5), we obtain:

$$f(x) = \sum_{k=0}^{\infty} f_k(x) e^{-\lambda \tau} \frac{(\lambda \tau)^k}{k!} \qquad (6)$$

If $\lambda\tau < 1$, i.e. the duration of current pulses is small compared to the average interval between Poisson impulses, then we have the *low-density shot noise*, and the p.d.f. of the total current can be obtained by adding just a few terms of the series (6), since the general term vanishes quickly as k increases. If $\lambda\tau > 1$, then we have the *high-density shot noise*, and the p.d.f. of the total current tends to be gaussian.

Now, let us change the initial hypothesis and suppose that the digital gates are not all equal, but they have different capacitive loads. In this case, the absorbed currents due to logic transitions are different, and we can model this situation by considering different amplitudes of Dirac impulses, proportional to the current absorbed by logic gates. Therefore we have a *generalized Poisson process* [2].

Moreover, let us suppose that the analyzed circuit is not an asynchronous net, but a pseudo-synchronous net, in which the digital switching activity occurs after the clock switching impulse. Then, the Dirac impulses of the input signal $X(t)$ are not uniformly distributed in time domain, but they are concentrated in some limited time intervals. Under the above assumptions, the digital switching noise can be modeled as a Poisson process passed through a time-windowing function, the *rectangular pulse train* $w(t)$ with period T and duty-cycle α, shown in Fig. 4. The Dirac impulses of $X_w(t)$, calculated as $X_w(t) = X(t) \cdot w(t)$, are localized in the time intervals where $w(t) = 1$; then the signal $X_w(t)$ is time-windowed.

The total current absorbed by the entire circuit is equal to:

$$Y(t) = (X(t) \cdot w(t)) * h(t) = X_w(t) * h(t),$$

and the process can be modeled as shown in Fig. 5. Since the impulse rate λ of the windowed process $X_w(t)$ is not constant, $Y(t)$ is not a stationary process. However, $Y(t)$ is cyclostationary, with period T.

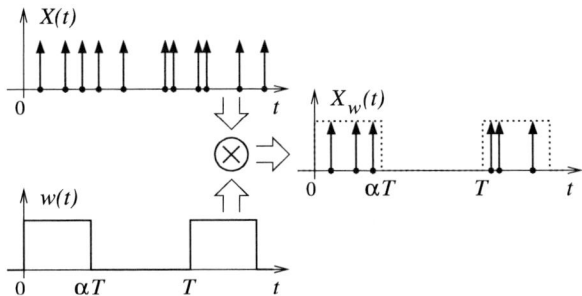

Fig. 4.   Windowed pulse train

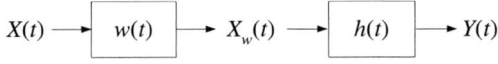

Fig. 5.   Switching noise with time-windowing function

IV. An Example

The theory presented in Sect. III has been applied to a practical circuit, to compare theoretical calculations with circuit simulations.

We have analyzed the amplitude density functions of the digital switching currents of the two-phase clock generator shown in Fig. 6. This circuit was extensively simulated at a device level, both with a commercial simulator (SPECTRE) [4], and with a dedicated tool implementing an algorithm for fast calculation of switching currents [5]. Fig. 7 shows the simulated current waveforms in time domain.

This circuit is a pseudo-synchronous net, since it is driven by an external clock signal. Therefore, current pulses in digital cells are not independent, as the output logic value of each cell switches in response to the corresponding input switching. However, we point out that the switching instants of logic gates are spread over a finite time, due to propagation delays introduced by capacitive loads and interconnection elements.

Moreover, logic gates and their capacitive loads are different. Therefore, current peaks are different, and this effect could be modeled with a generalized Poisson process.

In order to apply the theoretical formula (6), we assume that each current pulse has a triangular shape in time domain, and different logic gates have different values for the peak current, uniformly distributed in an interval $[i_1, i_2]$. The resulting probability density function of a single current pulse is shown in Fig. 8.

Fig. 9 shows the probability density function of the

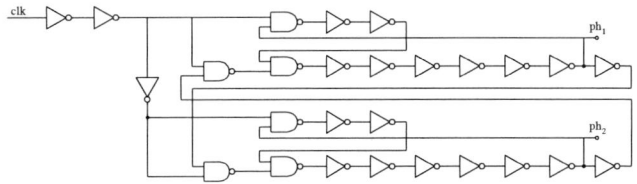

Fig. 6.   Circuit of the two-phase clock generator

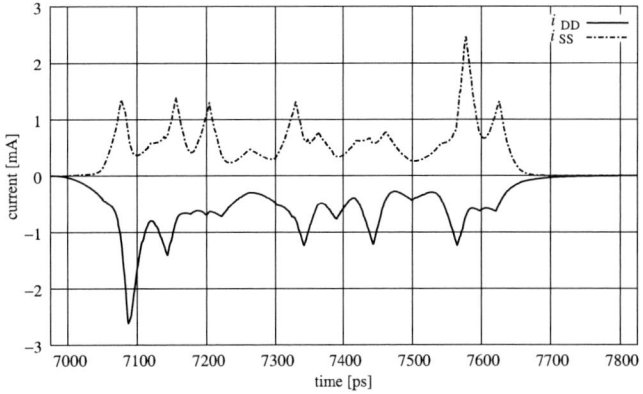

Fig. 7.   Switching currents of the two-phase clock generator

431

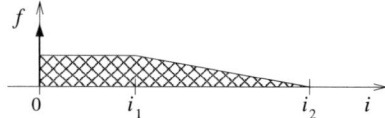

Fig. 8. Probability density function of a single current pulse

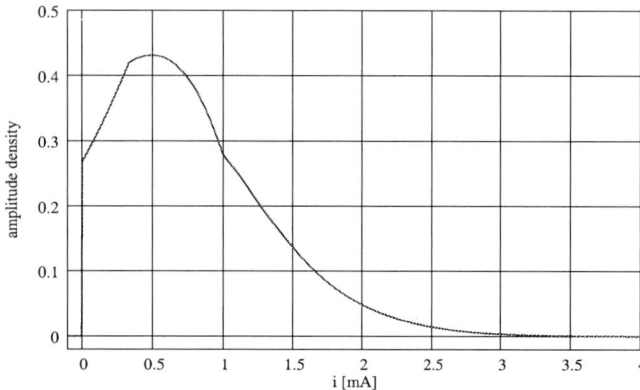

Fig. 9. Probability density function of the digital switching currents, calculated using eq. (6)

digital switching currents, calculated using (6), with the following parameters: $i_1 = 0.3$ mA, $i_2 = 1$ mA, $\lambda\tau = 10$.

Histograms in Figs. 10 and 11 were derived from simulated waveforms. We can observe a good agreement with the amplitude density in Fig. 9. The main deviations between theoretical curve and simulated results can be explained by the following considerations.

The probability density function of the single current pulse in Fig. 8 has an impulse at zero. This happens because our model considers current pulses having finite duration. In a real logic gate, however, current waveforms exponentially decrease towards zero value, and the probability of zero amplitude is negligible. This explains the difference between the theoretical curve and the histograms, for low current values.

Histograms in Figs. 10 and 11 were obtained with a time interval including the beginning and the end of the transition activity in Fig. 7. This results in a high amplitude density at zero current.

The simulated circuit has a low number of logic gates. In a larger circuit, the averaging effect due to the huge number of switching gates will lead to a better agreement with theoretical analysis.

V. Conclusion

Digital switching noise can be modeled as a stochastic process. By considering switching activity of logic gates as a random process, with activation instants randomly distributed in time, we can model digital switching currents as a time-windowed shot noise process. The amplitude density of switching currents can be derived from the probability density function of the single

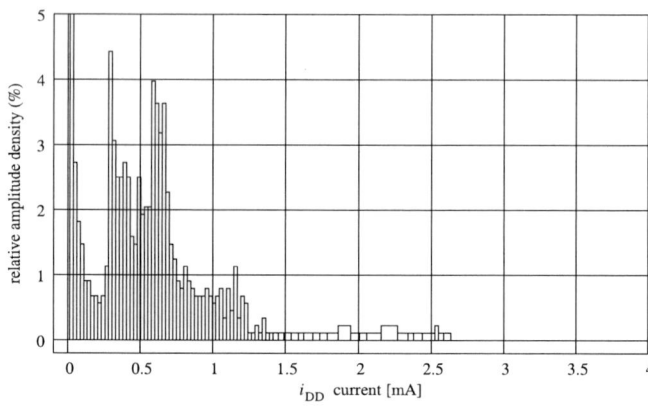

Fig. 10. Histogram of the relative amplitude density of the digital switching current i_{DD}

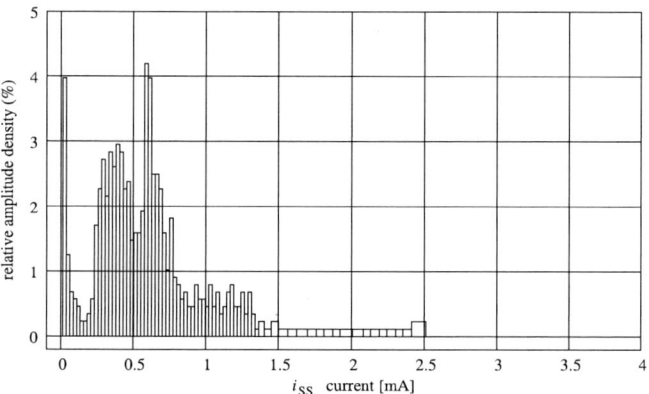

Fig. 11. Histogram of the relative amplitude density of the digital switching current i_{SS}

current pulse.

Theoretical results are in good agreement with circuit simulations.

To speed up the simulation process for a large mixed-signal circuit, the digital section can be modeled with a current generator having the same amplitude distribution function and the same spectral characteristics. The obvious advantage of a stochastic approach is that we can obtain a faster estimation of digital switching noise, using few parameters.

References

[1] S. Donnay and G. Gielen (Eds.). *Substrate Noise Coupling in Mixed-Signal ASICs*. Kluwer Academic Publishers, Boston, MA, USA, 2003

[2] A. Papoulis and S. U. Pillai. *Probability, Random Variables and Stochastic Processes*. McGraw-Hill, 4th ed., 2002

[3] O.-C. Yue, R. Lugannani, and S. Rice, 'Series approximation for the amplitude distribution and density of shot processes,' *IEEE Trans. Communications*, vol. 26, pp. 45–54, Jan. 1978

[4] G. Trucco, G. Boselli, and V. Liberali, 'Simulation of crosstalk through bonding and package in mixed-signal CMOS ICs,' in *Proc. Midwest Symp. on Circ. and Syst.*, Hiroshima, Japan, July 2004, volume I, pp. 121–124

[5] G. Trucco, G. Boselli, and V. Liberali, 'An analysis of current waveforms in CMOS logic cells for RF mixed circuits,' in *Proc. IEEE Int. Conf. on Microelectronics (MIEL)*, Niš, Serbia, May 2004, pp. 563–566

The 4-Phase Frame Partitioning Circuit

S. Poriazis

Abstract−The behavior of the 4-Phase Frame Partitioning (FPT4) circuit is analyzed. The circuit performs the phase partitioning of the input clock signal into one out of fifteen available partitions for a given four-phase frame. Each partition is being selectable by a 4-bit control word. A phase difference equal to the half period of the clock signal is used internally to achieve the correct pulse timing of the output signals. The VHDL description of the FPT4 cell is given and the simulation and synthesis results are presented.

Index Terms−clock, circuit design, frame, partitioning, phase, VHDL.

I. INTRODUCTION

Automatic test-pattern generation (ATPG) and design for testability (DFT) techniques are based on clock partitioning and selective clock freezing [1] in order to break the global feedback loops and to generate clock waves to test a sequential circuit with self-loops. Clock partitioning increases the testability of the sequential circuit. Algorithms can be run on the circuit with partitioned clocks to identify combinationally and sequentially untestable faults. A model with two time-frames is utilized, where a vector is generated in the first one and then the changed values of internal FFs are propagated to the QFI inputs of the next time-frame before a new fault is targeted; these values are frozen for ATPG.

Considering the Power Reduction aspect of FSM design, this can be achieved through clock gating and disabling the primary inputs to the sub-FSMs not active [2]. Clock-gating is an effective approach to reduce power consumption of Finite State Machines (FSMs) which have plenty self-loop events [3]. A low-power asynchronous communication controller is proposed by [4] for the interaction between the sub-FSMs. These are controlling the gating of the global clock. They utilize a circuit called Clock Controller Block (CCB) and the decomposed FSM consists of: (1) a number of sub-FSMs (partitions), (2) an equally large number of asynchronous CCBs, (3) nand-gates for gating the local clocks, and (4) one inverter for the global clock signal. The number of CCBs is equal to the number of partitions.

Instead of adapting techniques of gating the clock, we consider in this paper a dedicated circuit that can produce

Dr. S. Poriazis is with the R&D Department of Phasetronic Laboratories at http://www.phasetroniclab.com
Address: 6 Depasta Street., N.Smirni, GR-17122, Athens, Greece (email: serafim.poriazis@phasetroniclab.com).

sets of periodic patterns of clock phases under the control of an external binary word. In particular, we analyze the behavior of a 4-Phase Frame Partitioning (FPT4) circuit, which generates specific patterns of clock pulses for each of the available fifteen clock phase partitions for a time-frame of length 4, under the control of a 4-bit word. The subject circuit represents a reliable solution to the challenging problem of synchronizing the individual modules of a multiphase model [5], whose operation adopts a 4-phase timing pattern. The present work is based on the principles of operation of the Two-Phase Twisted Ring Counter (2P-TRC) circuit [6] and is targeting data streaming applications. The unit phase duration that is used at the outputs is equal to the period of the input clock signal, while internally the half period of the clock is used for the generation of a set of phased signals. The VHDL description of the circuit is given. The simulation of the FPT4 and the synthesis results are presented. The EXOR operator is used to express the timing relationships of the input clock signal, the internal phased signals and the output signals of the circuit.

II. THE FPT4 CELL

A. The basic cell operation

The fundamental cell, which partitions the four phases within a frame of the clock signal CLK of frequency *f*, is called the 4-Phase Frame Partitioning (FPT4) circuit

Table I. The phase assignment of the partitioning operation

index	Binary Control Word	Phase Partitions	Output line utilization
1	0000	1234	y_1
2	0001	1/234	
3	0010	2/134	
4	0011	3/124	
5	0100	4/123	y_2/y_3
6	0101	12/34	
7	0110	13/24	
8	0111	14/23	
9	1000	1/2/34	
10	1001	1/3/24	
11	1010	1/4/23	
12	1011	2/3/14	$y_2/y_3/y_4$
13	1100	2/4/13	
14	1101	3/4/12	
15	1110	1/2/3/4	$y_1/y_2/y_3/y_4$
16	1111	1234	y_1

```vhdl
library IEEE;
use ieee.std_logic_1164.all;

entity FPT4 is
    port (CLK , RESET : in std_logic;
          CTL : in std_logic_vector(4 downto 1);
          RSTFLAG : out std_logic;
          FPCLK : out std_logic_vector(4 downto 1));
end FPT4;

architecture behavioral of FPT4 is
  type validcodewords is array(1 to 16) of
        std_logic_vector(8 downto 1);
  constant phased_output : validcodewords :=
      ( ( '0', '0', '0', '0', '0', '0', '0' ,'0'),
        ( '0', '0', '0', '0', '0', '0', '0', '1'),
        ( '0', '0', '0', '0', '0', '0', '1', '1'),
        ( '0', '0', '0', '0', '0', '1', '1', '1'),
        ( '0', '0', '0', '0', '1', '1', '1', '1'),
        ( '0', '0', '0', '1', '1', '1', '1', '1'),
        ( '0', '0', '1', '1', '1', '1', '1', '1'),
        ( '0', '1', '1', '1', '1', '1', '1', '1'),
        ( '1', '1', '1', '1', '1', '1', '1', '1'),
        ( '1', '1', '1', '1', '1', '1', '1', '0'),
        ( '1', '1', '1', '1', '1', '1', '0', '0'),
        ( '1', '1', '1', '1', '1', '0', '0', '0'),
        ( '1', '1', '1', '1', '0', '0', '0', '0'),
        ( '1', '1', '1', '0', '0', '0', '0', '0'),
        ( '1', '1', '0', '0', '0', '0', '0', '0'),
        ( '1', '0', '0', '0', '0', '0', '0', '0') );
  signal index : integer := 0;
  signal present_state1, present_state2, next_state, pclk:
        std_logic_vector(8 downto 1);
  signal invalidcode_flag : std_logic := '0';
begin
reg1 : process (CLK, RESET)
begin
  if RESET = '1' then present_state1 <=
     phased_output(1);
  elsif (CLK='1' and CLK'event ) then present_state1 <=
     next_state ;
  end if;
end process;
reg2 : process (CLK, RESET)
  begin
  if RESET = '1' then present_state2 <=
     phased_output(9);
  elsif (CLK='0' and CLK'event ) then present_state2 <=
     next_state ;
  end if;
end process;
next_state_logic : process (CLK, RESET)
begin
case CLK is
when '1' => if RESET = '1' then index <= 1; else
              index <= index + 1; end if;
when '0' => if RESET = '1' then index <= 9; else
              index <= index + 1; end if;
when others  => null;
end case;
if index < 16 then next_state <=
              phased_output(index + 1);
else next_state <= phased_output(1); index <= 1;
end if;
for i in 1 to 16 loop
  if next_state = phased_output(i) then invalidcode_flag
     <= '0'; exit;
  else invalidcode_flag <= '1'; end if;
end loop;
end process;
output_logic : process (index, present_state1,
              present_state2)
begin
case CLK is
when '1' => pclk <= present_state1;
  if RESET = '1' then RSTFLAG <= '1'; else
              RSTFLAG <= invalidcode_flag; end if;
when '0' => pclk <= present_state2;
  if RESET = '1' then RSTFLAG <= '1'; else
              RSTFLAG <= invalidcode_flag; end if;
when others  => null;
end case;
case CTL is
when "0000" => FPCLK(1) <= (pclk(1) xor pclk(2)) or
  (pclk(3) xor pclk(4)) or (pclk(5)xor pclk(6)) or
  (pclk(7) xor pclk(8));
  FPCLK(2) <= '0'; FPCLK(3) <= '0'; FPCLK(4) <= '0';
when "0001" => FPCLK(1) <= '0';
FPCLK(2) <= pclk(1) xor pclk(2);
FPCLK(3) <= (pclk(3) xor pclk(4)) or (pclk(5) xor
              pclk(6)) or (pclk(7) xor pclk(8));
  FPCLK(4) <= '0';
when "0010" => FPCLK(1) <= '0';
  FPCLK(2) <= pclk(3) xor pclk(4);
  FPCLK(3) <= (pclk(1) xor pclk(2)) or (pclk(5) xor
              pclk(6)) or (pclk(7) xor pclk(8));
  FPCLK(4) <= '0';
when "0011" => FPCLK(1) <= '0';
  FPCLK(2) <= pclk(5) xor pclk(6);
  FPCLK(3) <= (pclk(1) xor pclk(2)) or (pclk(3) xor
              pclk(4)) or (pclk(7) xor pclk(8));
  FPCLK(4) <= '0';
when "0100" => FPCLK(1) <= '0';
  FPCLK(2) <= pclk(7) xor pclk(8);
  FPCLK(3) <= (pclk(1) xor pclk(2)) or (pclk(3) xor
              pclk(4)) or (pclk(5) xor pclk(6));
  FPCLK(4) <= '0';
when "0101" => FPCLK(1) <= '0';
  FPCLK(2) <= (pclk(1) xor pclk(2)) or (pclk(3) xor
              pclk(4));
  FPCLK(3) <= (pclk(5) xor pclk(6)) or (pclk(7) xor
              pclk(8));
  FPCLK(4) <= '0';
when "0110" => FPCLK(1) <= '0';
  FPCLK(2) <= (pclk(1) xor pclk(2)) or (pclk(5) xor
              pclk(6));
  FPCLK(3) <= (pclk(3) xor pclk(4)) or (pclk(7) xor
              pclk(8));
  FPCLK(4) <= '0';
when "0111" => FPCLK(1) <= '0';
  FPCLK(2) <= (pclk(1) xor pclk(2)) or (pclk(7) xor
              pclk(8));
  FPCLK(3) <= (pclk(3) xor pclk(4)) or (pclk(5) xor
              pclk(6));
  FPCLK(4) <= '0';
when "1000" => FPCLK(1) <= '0';
  FPCLK(2) <= pclk(1) xor pclk(2);
  FPCLK(3) <= pclk(3) xor pclk(4);
  FPCLK(4) <= (pclk(5) xor pclk(6)) or (pclk(7) xor
              pclk(8));
when "1001" => FPCLK(1) <= '0';
  FPCLK(2) <= pclk(1) xor pclk(2);
  FPCLK(3) <= pclk(5) xor pclk(6);
  FPCLK(4) <= (pclk(3) xor pclk(4)) or (pclk(7) xor
              pclk(8));
when "1010" => FPCLK(1) <= '0';
  FPCLK(2) <= pclk(1) xor pclk(2);
  FPCLK(3) <= pclk(7) xor pclk(8);
  FPCLK(4) <= (pclk(3) xor pclk(4)) or (pclk(5) xor
              pclk(6));
when "1011" => FPCLK(1) <= '0';
  FPCLK(2) <= pclk(3) xor pclk(4);
  FPCLK(3) <= pclk(5) xor pclk(6);
  FPCLK(4) <= (pclk(1) xor pclk(2)) or (pclk(7) xor
              pclk(8));
when "1100" => FPCLK(1) <= '0';
  FPCLK(2) <= pclk(3) xor pclk(4);
  FPCLK(3) <= pclk(7) xor pclk(8);
  FPCLK(4) <= (pclk(1) xor pclk(2)) or (pclk(5) xor
              pclk(6));
when "1101" => FPCLK(1) <= '0';
  FPCLK(2) <= pclk(5) xor pclk(6);
  FPCLK(3) <= pclk(7) xor pclk(8);
  FPCLK(4) <= (pclk(1) xor pclk(2)) or (pclk(3) xor
              pclk(4));
when "1110" => FPCLK(1) <= pclk(1) xor pclk(2);
  FPCLK(2) <= pclk(3) xor pclk(4);
  FPCLK(3) <= pclk(5) xor pclk(6);
  FPCLK(4) <= pclk(7) xor pclk(8);
when "1111" => FPCLK(1) <= (pclk(1) xor pclk(2)) or
  (pclk(3) xor pclk(4)) or (pclk(5)
  xor pclk(6)) or (pclk(7) xor pclk(8));
  FPCLK(2) <= '0'; FPCLK(3) <= '0'; FPCLK(4) <= '0';
when others  => null;
end case;
end process;

end behavioral;
```

Fig. 1. The VHDL description of the FPT4 cell

and its behavior is shown in Table I. For a set of four phases we have fifteen possible phase partitions and each one can be selected by a 4-bit control word. The content of each block of a partition is assigned phases of the frame of the clock, e.g. for the binary word "1001" we have a three block partition where the first block holds the phase-1 of the frame, the second block holds the phase-3 of the frame, and the third block holds both the phase-2 and the phase-4 of the frame of the input clock signal. For the above (line 10 of table 1), the first block of the partition is routed at the output line signal y_2, the second block is routed at y_3, and the third block at y_4. Similarly holds for each of the remaining partitions. We note that each phase has a duration equal to the period T of the input clock signal with a duty cycle of 50 percent, thus having a pulse of logic-1 or logic-0 value for a half period.

B. The algorithm aspects of cell operation

The VHDL description of the FPT4 cell is given in Figure 1. The entity section has an input port CLK on which the clock signal of frequency f is applied, an input port RESET on which a reset flag is applied, and an input port CTL[4..1] on which a 4-bit control word is applied. The output signal FPCLK[4..1] carry the phase information of each block of the partition of the frame of signal CLK according to Table 1 specification, and we have the notation as follows: FPCLK(1)=y_1, FPCLK(2)=y_2, FPCLK(3)=y_3 and FPCLK(4)=y_4. Each output signal can carry phase information only under the control of input CTL, that is, during a frame we have the selected phase partition present at the assigned output lines of FPCLK. The above ports are shown in Figure 2 on the block diagram of FPT4.

The architecture section is of type "behavioral" and utilizes a state machine model, where two internal registers are being used, reg1 and reg2, one for the present state named "present_state1" clocked by the rising edge of the clock and the other for the present state named "present_state2" clocked by the falling edge of the clock, respectively. The next state logic block and the output logic block of the model are specified by the corresponding processes "next_state_logic" and "output_logic". The internal set of sixteen valid codewords of the circuit is stored in an indexed array of size 16*8=128 bits that is represented by the constant named "phased_output". The index of the above array cycles through the integer values 1 to 16 specifying the valid codeword entry for the next state signal. The phase information at each output line of FPCLK[4..1] is derived from the internal phased signals for each value of the 4-bit control word CTL by applying the EXOR and the OR operators on the signal lines of $pclk$[8..1].

C. The valid codeword sequence

Internally, the eight overlapping phased signals $pclk_1$, $pclk_2$, $pclk_3$, ... , $pclk_8$ have frequency equal to $f/8$ (period of each $pclk_i$, i=1,2,3,...,8 equals 8·T, where T the period of CLK) with signal $pclk_1$ leading $pclk_2$, $pclk_2$ leading $pclk_3$, ... , $pclk_7$ leading $pclk_8$ by a T/2 time difference. The logic-'1' or logic-'0' pulse width of each of the above phased signals is equal to 8·T/2. Consecutive changes of logic value at each signal $pclk_i$, for i=1,2,3,...,8 occur at the rising or falling edges of CLK at a distance of 8·T/2.

When the clock signal CLK is applied to the circuit, the following cyclic sequence of codewords is presented at the internal phased signals: $pclk_1,pclk_2,pclk_3$, ... , $pclk_8$= 00000000→10000000→11000000→11100000→1111000 0→11111000→11111100→11111110→11111111→0111 1111→00111111→00011111→00001111→00000111→0 0000011→00000001, which is considered as being the normal internal cell operation. Each codeword remains stable for the state time of the circuit, that is T/2, and the above sequence is repeated internally throughout the operation of the cell. Thus the cycle time for the $pclk$ pattern is defined by the sixteen-tuple of codewords of length equal to 8·T. This duration forms the period of each phased signal. The additional 240 codewords out of the total 256 possible codewords that are not included in the above internal cyclic sequence should be considered during the design of the circuit for achieving reliable operation of the FPT4 cell. If the circuit reaches any of these 240 invalid codewords, then an invalid codeword flag is set. This flag maintains the proper initializing behavior until a valid codeword appears internally in the cell.

D. The VHDL simulation and synthesis

The VHDL testbench simulation results for the FPT4 cell are given in Figure 3. The duration of this simulation is defined by the value of the signal "done". Each of the signals "next_state", "present_state1", "present_state2" and $pclk$ have each a width of 8 bits, and CTL and FPCLK of 4 bits. The output port FPCLK is analyzed into four individual output signals with waveforms that verify the correct operation of the circuit (for demonstration purposes only the control words "0000", "0001", "0101", "1001" and "1110" are shown, each for a duration of two frames). The logic value changes of the internal phased signals $pclk$ occur at each rising and at each falling edge of the input signal CLK. The synthesis of the FPT4 cell targeting an FPGA device was successfully performed giving us the following results:

- flip flops with asynchronous reset = 8
- flip flops with asynchronous preset = 8

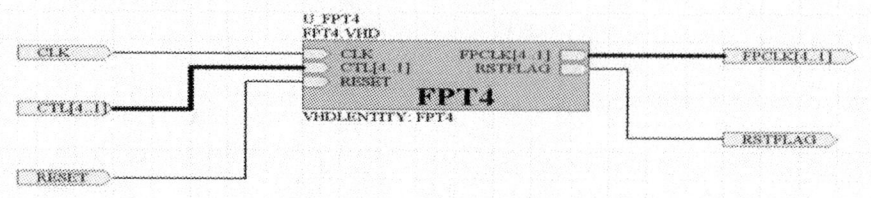

Fig. 2. The FPT4 block diagram

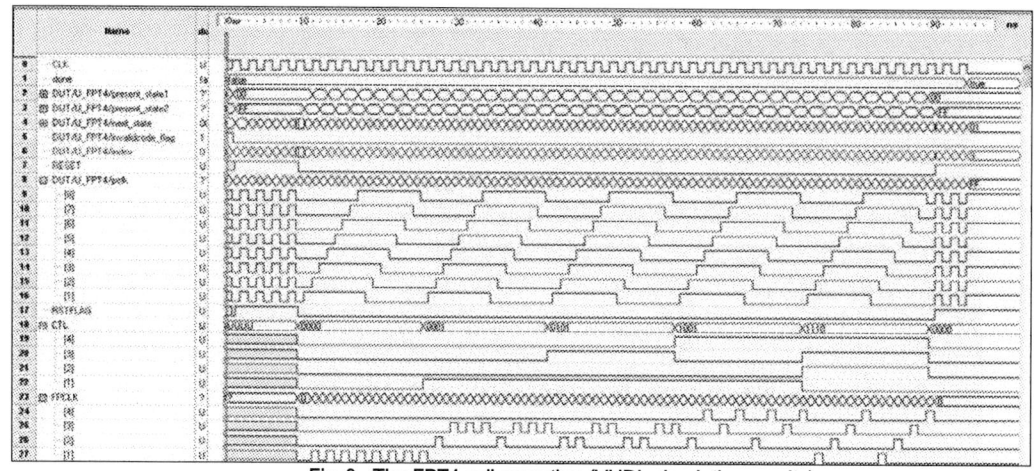

Fig. 3. The FPT4 cell operation (VHDL simulation results)

- combinational feedback paths = 41
- combinational logic area estimate = 171 LUTs

III. THE PHASE ASSOCIATIONS OF THE FPT4 SIGNALS

We examine the associations between the input clock signal CLK, the internal phased signals $pclk$, and the output signals $FPCLK$. We have the EXOR function that can be utilized to express the following relationships:

$$CLK = pclk_1 \oplus pclk_2 \oplus pclk_3 \oplus ... \oplus pclk_8$$
$$pclk_1 = CLK \oplus pclk_2 \oplus pclk_3 \oplus ... \oplus pclk_8 \qquad (1)$$
$$...$$
$$pclk_8 = CLK \oplus pclk_1 \oplus pclk_2 \oplus ... \oplus pclk_7$$

For the output signals, and for the binary control word "1001" as an example, we have the following relationships:

$$FPCLK_2 = pclk_1 \oplus pclk_2$$
$$FPCLK_3 = pclk_5 \oplus pclk_6 \qquad (2)$$
$$FPCLK_4 = (pclk_3 \oplus pclk_4) + (pclk_7 \oplus pclk_8)$$

IV. CONCLUSION

The design aspects of the 4-Phase Frame Partitioning (FPT4) circuit are being considered in this paper. The operation of this circuit is based on a set of internal phased signals that is used to generate timing patterns, which correspond to the blocks of each of the available fifteen phase partitions under the control of a 4-bit word and for a time-frame of length 4. The VHDL description of the FPT4 cell is given. The corresponding simulation results verify the proper circuit operation while the internal phased signals $pclk[8..1]$ and the output signals FPCLK[4..1] maintain the specified phase associations with the input clock signal CLK. The synthesis results are given targeting an FPGA device.

REFERENCES

[1] M. Abramovici, Y. Xiaoming, and E.M. Rudnick, "Low-cost sequential ATPG with clock-control DFT", in *Proceedings of the 39th Design Automation Conference, DAC 2002*, New Orleans, Louisiana, USA, 10-14 June 2002, pp. 243-248.

[2] C. Cao, and B. Oelmann, "Mixed synchronous/asynchronous state memory for low power FSM design", in *Proceedings of the Euromicro Symposium on Digital System Design, DSD 2004*, 31 Aug. – 3 Sept. 2004., pp. 363-370.

[3] W-K. Lee, and C-Y. Tsui, "Finite state machine partitioning for low power", in *Proceedings of the 1999 IEEE International Symposium on Circuits and Systems, ISCAS'99*, Orlando, FL, USA, Vol. 1, 30 May – 2 June 1999, pp. 306-309.

[4] B. Oelmann, Oapos, and M. Nils, "Asynchronous control of low-power gated-clock finite-state-machines", in *Proceedings of the 6th IEEE International Conference on Electronics, Circuits and Systems, ICECS'99*, Pafos, Cyprus, Vol. 2, 5-8 Sep. 1999, pp. 915-918.

[5] S. Poriazis, *Logic Design of Multiphase Finite State Machines*, Monograph, Phasetronic Laboratories, http://www.phasetroniclab.com Athens, Greece, 2001.

[6] S. Poriazis, "The Two-Phase Twisted-Ring Counter Circuit", in *Proceedings of the 2002 IEEE International Symposium on Circuits and Systems, ISCAS'02*, Phoenix, Arizona, USA, vol. IV, 26-29 May 2002, pp. 858-861.

2006 25th International Conference on Microelectronics

Fault Diagnosis in Digital Part of Mixed-Mode Circuit

Miona Andrejević, Vančo Litovski and Mark Zwolinski

Abstract – In this paper artificial neural networks (ANNs) are applied to diagnosis of catastrophic defects in the digital part of a nonlinear mixed-mode circuit. The approach is demonstrated on the example of a relatively complex sigma-delta modulator. A set of faults is selected first. Then, *fault dictionary* is created, by simulation, using the response of the circuit to an input ramp signal. It is represented in a form of a look-up table. Artificial neural network is then trained for modeling (memorizing) the look-up table. The diagnosis is performed so that the ANN is excited by faulty responses in order to present the fault codes at its output. There were no errors in identifying the faults during diagnosis.

I. INTRODUCTION

Every complex system is liable to faults or failures. In most general terms a fault is any change in a system that prevents it from operating in the proper manner. We define diagnosis as the task of identifying the cause of a fault that is manifested by some observed behavior. Then some method of determining what fault has occurred is required. This is most often considered to be a two-stage process: firstly the fact that fault has occurred must be recognized – what is referred to as fault detection. Secondly, the nature should be determined such that appropriate remedial action may be initiated.

The explosion of integrated circuit technology has brought with it some difficult testing problems. The recent growth of mixed analogue and digital circuits complicates the testing problem even further. It becomes more complicated to determine a set of input test signals and output measurements that will provide a high degree of fault coverage. There is also a timing problem of testing the circuits even on the fastest automated equipment.

In this paper we will show that feed-forward ANN may be applied to the diagnosis of non-linear dynamic electronic circuits that are mixed with digital ones. In order to make the explanation easier to understand, only a reduced set of faults will be used i.e. catastrophic defects in the digital part of the converter. Only single faults are considered.

The simulation before test concept was adopted. This means that after choosing the set of faults of interest (say the most probable ones), repetitive simulation is performed in order to create the system response for every fault. Codes are associated to the responses and used as part of

Miona Andrejević and Vančo Litovski are with the Department of Electronics, Faculty of Electronic Engineering, University of Niš, Aleksandra Medvedeva 14, 18000 Niš, Serbia & Montenegro, E-mail: (miona, vanco)@elfak.ni.ac.yu

Mark Zwolinski is with University of Southampton, Southampton SO17 1BJ, England, E-mail: mz@ecs.soton.ac.uk

the fault dictionary that, in addition, contains the faulty responses themselves. Of cause, the responses are represented in a form that is easy to manipulate.

The ANN is first trained for modeling the look-up table. This means that faulty responses are repeatedly brought to the input, while the ANN is forced to present the fault codes at its output. Then, the ANN running with the given vector of stimuli (measured output signals of a faulty or, possibly, fault free system) may be viewed as search of the look-up table. The ANN response, if the network properly trained, will immediately find the fault and produce the fault code at its output.

The procedure applied is reminiscent to the one implemented to analog circuits in [1]. To our knowledge this is the first application of ANNs to diagnosis of mixed signal circuit.

II. CONCEPTS OF DIAGNOSIS

Besides the human expert that is usually performing the diagnostic project, one needs tools that will help, and what is most desired, will perform diagnosis automatically. Such tools are a great challenge to design engineers that pertains to the fact that generally the diagnostic problem is indeterminate. In addition, it is a deductive process with one set of data creating, in general, unlimited number of hypotheses among which one should try to find the solution. This is why permanent attention of the research community is attracted by this problem [2].

During the life-cycle of a product, testing is implemented in both the production phase and the implementation phase. We claim, however, that the sustainability of a product is strongly influenced by the design phase. So, to make a sustainable product, one should design the test procedure and synthesize test signals early in the design phase.

It is frequently possible to perform functional verification of the system. That, most frequently, happens when a small number of input/output terminals is present. In the majority of cases however, full functional testing becomes time consuming and is not acceptable. So, one applies defect-oriented (structural) testing, as will be discussed in more detail as follows.

We consider testing to be: the selection of a set of defects regarded as the most probable, the description of a set of measurements, the selection of a set testing points (or output signals) and most importantly, the synthesis of optimal testing signals that will be applied at the system inputs allowing for detectability and observability of the listed fault effects. Here, optimality means that one test signal covers as many faults as possible.

1-4244-0116-X/06/$20.00 ©2006 IEEE

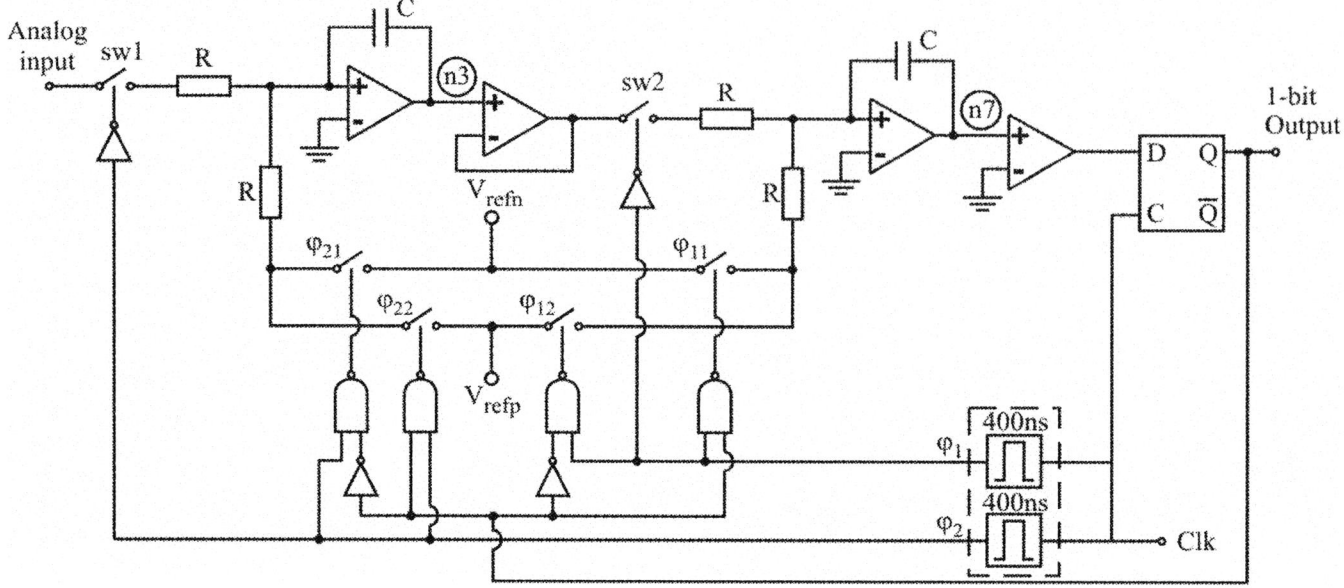

Fig. 1. Sigma-delta modulator architecture.

Selection of the type of measurements and testing points is specific to the circuit. One should stick to those measurements that are prescribed for functional verification. Specific measurements such as supply current monitoring are frequently adopted, too. Separate test points may be added in order to improve detectability or observability. Specific design for testability concepts can be applied.

After selection of test signals, the fault coverage has to be evaluated. To do that, as many replicas of the original circuit as the number of predicted faults have to be created. For large complex systems containing mechanical, analogue and digital parts, the number of replicas becomes huge. Each replica has one fault inserted. The fault coverage is evaluated after simulation of the faulty systems by comparing the results thus obtained with the response of the fault-free system. If these two differ, the fault is covered and the corresponding entry in the fault list can be removed. To reduce the computational effort, algorithms have been proposed to simulate multiple faulty circuits concurrently in both the analogue and the digital domains but not in mixed signal, and mixed description systems.

III. SIGMA-DELTA MODULATOR ARCHITECTURE

As an example of a complex non-linear dynamic electronic circuit with mixed signals, the architecture of sigma-delta modulator is chosen.

Sigma-delta modulators are very attractive for design low frequency high-resolution analog-to-digital converters. Sigma-delta modulators trade speed for resolution. They employ coarse quantization in one or more feedback loops. By sampling at a frequency that is much greater than the signal bandwidth, it is possible for the feedback loops to

shape the quantization noise so that most of the noise power is shifted out of the signal band. The out of band noise can then be attenuated with a digital filter. The degree to which the quantization noise can be attenuated depends on the order of the noise shaping and the oversampling ratio [3].

In addition to their tolerance for circuit nonidealities, oversampled A/D converters simplify system integration by reducing the burden on the supporting analog circuitry. Because they sample the analog input signal at well above the Nyquist rate, precision sample-and-hold circuitry is unnecessary. Also, the burden of analog antialiasing filter is considerably reduced. Much of its function is transferred to the digital decimation filter, which can be designed and manufactured to precise specifications, including a linear phase characteristic.

IV. FAULTS IN THE SPECIFIC MODULATOR DESIGN

As an example of a complex circuit, the sigma-delta modulator in Fig. 1 is chosen [2]. This is a mixed-signal circuit, having both analogue and digital elements. Switches in the circuit are modeled as truly ideal switches, with zero resistance for closed switch and infinite resistance for open switch [4].

The integrator charging time is invariable with respect to clock rate in order to keep the gain constant. This means that the analog switch must be turned on for fixed time duration regardless of clock rate. This is achieved by using monostable multivibrator as a fixed-width pulse generator in the circuit. The monostable multivibrator between the clock input and switch control block functions as a pulse generator to produce control signals of fixed time duration. Fig. 2 shows reaction of the system when the input is excited by a sinusoidal signal.

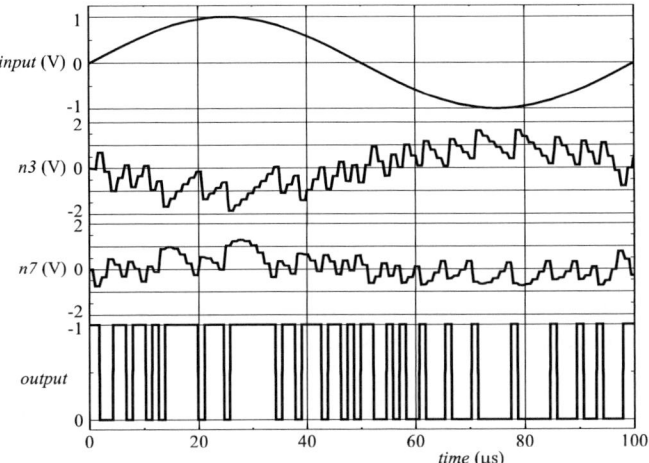

Fig. 2. Simulation results for linear sinusoidal excitation

In this paper only the faults in the digital part of the circuit will be considered. Digital signal can be "stuck-at-1" or "stuck-at-0". In the circuit in Fig. 1, analogue switches are controlled by digital signals, so there are pairs of the same fault effects, such as: the effect is the same when the switch is stuck at ON (OFF) and the logic circuit's output is "stuck-at-1" ("stuck-at-0"). So, we will consider hard faults (which refer to the analogue part of the circuit) as stuck switches.

TABLE I
FAULT DICTIONARY

Type of fault	Signature	Fault code
FF	C9CA9	0
sw_1OFF	99999	1
$\varphi_{12}OFF$	38E38	2
$\varphi_{21}OFF$	00000	3
$\varphi_{22}OFF$	FFFFF	4
sw_1ON	63655	5
$\varphi_{11}OFF$	1F07C	6
sw_2ON	E0F83	7
sw_2OFF	AAAAA	8

Fault dictionary is created using the response of the circuit to an input ramp signal. We published one approach to fault dictionary creation, where output signals of the fault-free and of the faulty circuits are transformed using the Fast Fourier Transformation, in [5].

In the alternative approach given here, the circuit output value is registered after every clock period, so these output digital values form the output signature. These are then represented in more compact hexadecimal presentation. Accordingly, fault dictionary is created and shown in Table 1. In the first column of Table 1, eight selected faults are named. FF stands for the fault free circuit. The cases when switches in the feedback loop (φ_{11}, φ_{12}, φ_{21}, φ_{22}) are permanently closed are excluded, because voltage references V_{refp} and V_{refn} would be shorted in such cases. The second column contains the signature seen at the output. The signature is then coded as shown in the third column. In the coding procedure we had in mind that similar responses must not have similar fault codes (for example response 99999 and AAAAA (hexadecimal value A is presented to the neural network as decimal value 10, or F as decimal value 15)).

TABLE II
ANN WEIGHTS AND THRESHOLDS

weight (1,1)(2,1)	120.812
weight (1,2)(2,1)	-71.5911
weight (1,3)(2,1)	-170.517
weight (1,4)(2,1)	145.099
weight (1,5)(2,1)	10.6883
weight (1,1)(2,2)	104.461
weight (1,2)(2,2)	-85.9051
weight (1,3)(2,2)	-181.814
weight (1,4)(2,2)	142.592
weight (1,5)(2,2)	5.02798
weight (1,1)(2,3)	118.426
weight (1,2)(2,3)	-80.095
weight (1,3)(2,3)	-166.541
weight (1,4)(2,3)	139.481
weight (1,5)(2,3)	-8.45216
weight (2,1)(3,1)	14.4496
weight (2,2)(3,1)	10.0822
weight (2,3)(3,1)	-14.6015
threshold (2,1)	-18.3932
threshold (2,2)	5.31276
threshold (2,3)	0.747092
threshold (3,1)	0.25

Artificial neural network was trained for modeling the look-up table. It is a feed-forward neural network with one hidden layer. The structure of the network is shown in Figure 3. The signatures are inputs to the network, and the fault code is network output to be learned. It means that the neural network has five input (one input per hexadecimal digit) and one output neuron. Hexadecimal values are presented as decimal when they are inputs to the network. After learning was completed, the number of hidden neurons in the resulting ANN was three, what was found by trial and error after several iterations starting with an estimation based on [6] and [7]. Parameters of the obtained network, its weights and thresholds, are presented in Table 2.

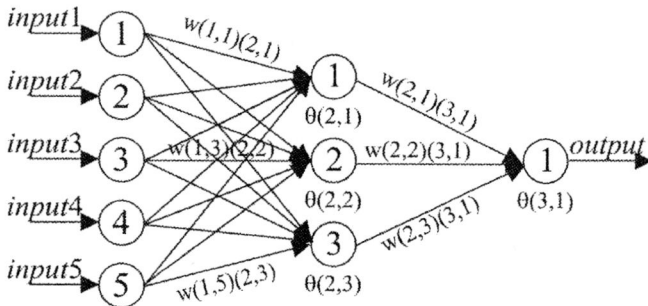

Fig. 3. The structure of the ANN used for diagnosis of digital faults in the circuit in Fig.1.

TABLE III
ANN OUTPUT CODES

Type of fault	Fault code	ANN output
FF	0	0.00754392
sw_1OFF	1	1.00436
φ_{12}OFF	2	2
φ_{21}OFF	3	2.99988
φ_{22}OFF	4	4.00265
sw_1ON	5	5.00078
φ_{11}OFF	6	5.9999
sw_2ON	7	7.0013
sw_2OFF	8	8.00416

The structure and the parameters of the obtained ANN are verified by exciting the ANN with faulty inputs. Responses of the ANN show that there were no errors in identifying the faults what is presented in Table 3. Negligible discrepancies may be observed (less than 0.5%).

The diagnosis was successful. Although only the catastrophic defects were diagnosed in this example, soft faults can be easily introduced. Accordingly, we may conclude that ANNs are convenient and powerful means for diagnosis, and, what is important, realisable as a hardware that may be as fast as necessary to follow the changes of the system's response in real time.

V. CONCLUSION

ANN approach is applied here, for the first time, to diagnosis of catastrophic defects in a digital part of nonlinear mixed-mode circuit. We consider this result as a full success. Although only the catastrophic defects were diagnosed in this example, soft faults can be easily introduced. Accordingly, we may conclude that ANNs are convenient and powerful means for diagnosis, and, what is important, realisable as a hardware that may be as fast as necessary to follow the changes of the system's response in real time.

Our future work will be devoted to implemetation of this idea to complete set of faults that include faults in analog and the digital part. Catastrophic as well as soft faults are intended to be introduced.

REFERENCES

[1] Vančo Litovski, Miona Andrejević, Mark Zwolinski, "Analogue Electronic Circuit Diagnosis Based on ANNs", *Microelectronics Reliability*, 2006, in printing.
[2] Xu, X., and Lucas, M. S. P., "Variable-Sampling-Rate Sigma-Delta Modulator for Instrumentation and Measurement", IEEE Transactions on Instrumentation and Measurement, Vol. 44, No. 5, October 1995, pp. 929-932.
[3] Candy, J., Temes, G., "Oversampling methods for A/D and D/A conversion", in *Oversampling Delta-Sigma Data Converters*. New York: IEEE Press, 1992, pp. 1-29.
[4] Mrčarica, Ž., Ilić, T., and Litovski, V.B., "Time domain analysis of nonlinear switched networks with internally controlled switches", IEEE Trans. on Circuits and Systems – I Fundamental Theory and Applications, Vol. 46, 1999, pp. 373-378.
[5] Miona Andrejević, Milan Savić, Miljan Nikolić, "Fault effects in sigma-delta modulator", Proceedings of the ETRAN, Budva, Montenegro, June, 2005, pp. 86-89.
[6] Masters, T., "Practical Neural Network Recipes in C++", Academic Press, San Diego, 1993.
[7] Baum, E. B., and Haussler, D., "What size net gives valid generalization", Neural Computing, Vol. 1, 1989, pp. 151-60.

Poster Session
Circuit Design

2006 25th International Conference on Microelectronics

Simulation of Crosstalk between Several Interconnection Lines in CMOS Integrated Circuits

M. Petković

Abstract - This work presents simulation of the crosstalk introduced by wiring in CMOS integrated circuits. Interconnections are modelled as lines with distributed parameters. Simulation results are shown on practical example. An electomagnetic simulator was used for calculation of coupled capacitances between conductors followed by the OrCad PSpice simulation to obtain the waveforms of output voltages on each line.

I. INTRODUCTION

Interconnection wiring is gaining a significant importance in speed of modern VLSI circuits. One of the reasons interconnections limit their performance are delay time and crosstalk effects [1]. One of the most challenging issues for semiconductor circuit design is how to overcome RC delays and crosstalk in the interconnect layers. Since the wiring may cover up to eighty percent of nowdays chip area, special care must be devoted to this problem. With larger dimensions and complexity of the chip, number of interconnections is increasing and also possibility of non-negligible crosstalk between them. Therefore, when designing for high speed circuits, designers have to pay special attention to the signal propagation through the wires.

Of course, a reliable and valuable simulation can help to the great extent to one's insight into the circuit behavior [2]. A special case of two interconnection lines is considered in [3]. Results presented in this paper generalize these results and are applicable to any number of lines. We gave the method for computing maximal deviation of signal caused by crosstalk effect.

This paper is organized as follows: Section 2 deals with the capacitance model of interconnections. An electromagnetic simulator is used for calculation of the coupled capacitances between conductors. Section 3 describes an electrical model used in the circuit simulator. Presented results show that, as it has been expected, dominant crosstalk is between the neighbor conductors.

II. MODEL OF INTERCONNECTION LINES FOR DELAY TIME AND CROSSTALK SIMULATION

Thin film technology which is commonly used in modern CMOS circuits requires dealing with intercon-

Marko Petković is undergraduate student of Faculty of Electronic Engineering, University of Niš, A. Medvedeva 14, 18000 Niš, Serbia & Montenegro. E-mail: dexter_of_nis@neobee.net.

nection lines with distributed parameters. These lines are commonly treated as RC lines [1,2,4]. On higher frequencies the inductance of interconnections is not negligible and must be included in the model. Also in the case of more than two coupled interconnections, there exists capacitance between each pair of interconnections, i.e. there are $N \cdot (N-1)/2$ coupled capacitances. As we will see, this number can be drastically reduced because many of these capacitances can be neglected. Figure 1 describes capacitance model of interconnection lines.

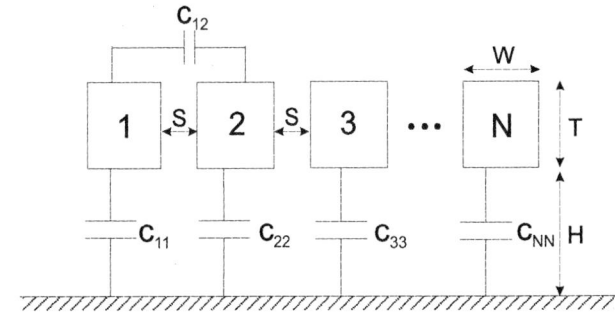

Fig. 1. Capacitance model of interconnections

For the cases $N = 2$ and $N = 3$ there are accurate analytic formulas for distributed capacitances c_{ij} between conductors [5]. For higher values of N there are no such formulas, so we have to use an electromagnetic simulator to calculate distributed capacitances. Simulator we used, Maxwell Student Version [6], is able to compute required capacitances directly form the model. In our approach, we used the same dimensions of conductors as used in [1]: $W = 1.5 \, \mu m$, $T = 0.35 \, \mu m$, $S = 2.5 \, \mu m$, $H = 1 \, \mu m$. We obtained results for different values of N (5, 7, 9, 11 and 15).

Calculated coupled capacitances c_{ij} for the case $N = 7$ are presented in Table I. All values are in pF/m.

It can be noticed that values of c_{ij} for $N \geq 7$ almost do not depend on N. Also, for $|i - j| > 2$ capacitance c_{ij} is less than 1% of c_{ii}, so they can be neglected. It means that every conductor influences two neighbor conductors on each side (for example, 4th conductor influences on 2nd, 3rd, 5th and 6th). So, we just need to consider three classes of

1-4244-0116-X/06/$20.00 ©2006 IEEE   443

capacitances: c_{ii}, $c_{i,i+1}$ and $c_{i,i+2}$. In the first class, for the sake of symmetry there should hold:

$$c_{3,3} \approx c_{4,4} \approx \ldots \approx c_{N-2,N-2}$$

but $c_{1,1} = c_{N,N}$ and $c_{2,2} = c_{N-1,N-1}$ should be slightly different due to the boundary effects. Our calculation shows that maximum relative difference between capacitances $c_{i,i}$ for $i = 3,4,\ldots,N-2$ is less than 1%, $c_{2,2} = c_{N-1,N-1}$ is 5% greater and $c_{1,1} = c_{N,N}$ is about 16% greater. So for the description of first class capacitances, we require three values.

TABLE I
CALCULATED CAPACITANCES FOR $N = 7$ CONDUCTORS

$c_{i,j}$ [pF/m]	1	2	3	4	5	6	7
1	130	14.5	2.33	0	0	0	0
2	14.5	117	14.1	2.29	0	0	0
3	2.33	14.1	116	13.9	2.29	0	0
4	0	2.29	13.9	114	13.9	2.29	0
5	0	0	2.29	13.9	116	14.1	2.33
6	0	0	0	2.29	14.1	117	14.5
7	0	0	0	0	2.33	14.5	130

Situation is similar in the second and third class. There holds

$$c_{2,3} \approx c_{3,4} \approx \ldots \approx c_{N-2,N-1}$$

and $c_{1,2}$ is about 3% greater. In the third class, it is sufficient to consider just one value of capacitance and as we will see later, this class can be also neglected.

Finally for the circuit simulation we require just six values of capacitances. Calculated values in our example are:

$$c_{11} = 1.325 \times 10^{-10} \text{ F/m}, \quad c_{22} = 1.20 \times 10^{-10} \text{ F/m},$$
$$c_{ii} = 1.14 \times 10^{-10} \text{ F/m}$$
$$c_{12} = 1.43 \times 10^{-11} \text{ F/m}, \quad c_{i,i+1} = 1.39 \times 10^{-11} \text{ F/m},$$
$$c_{i,i+2} = 2.29 \times 10^{-12} \text{ F/m}$$

III. CROSSTALK SIMULATION IN TIME DOMAIN

To simulate crosstalk, we used OrCad PSpice simulator. System of interconnection lines is modeled as cascade connection of multiport sections. An electrical circuit representing one section is shown on Figure 2.

Complete model of interconnections is formed by cascade connection of $k = 15$ sections. Total length of all interconnections in our example is $l = 1300 \, \mu\text{m}$. Capacitances C_{ij} are calculated using the formula

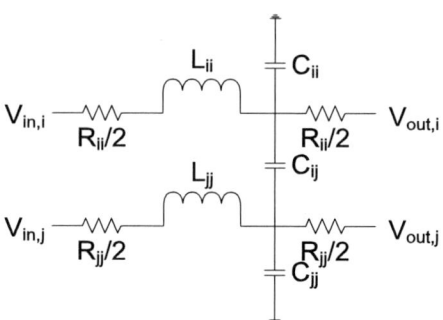

Fig. 2. Section of electrical model for crosstalk simulation between i^{th} and j^{th} line.

$C_{ij} = \dfrac{c_{ij} \cdot l}{k}$. Inductive and resistive parameters ($L = L_{ii}$ and $R = R_{ii}$) were calculated using formulas from [2]. Obtained values are: $L = 157.79 \text{ pH}$ and $R = 5.2 \, \Omega$.

We considered $N = 7$ lines made of aluminum placed on the same distance $d = 2.5 \, \mu\text{m}$. Equivalent circuit is shown in Figure 3.

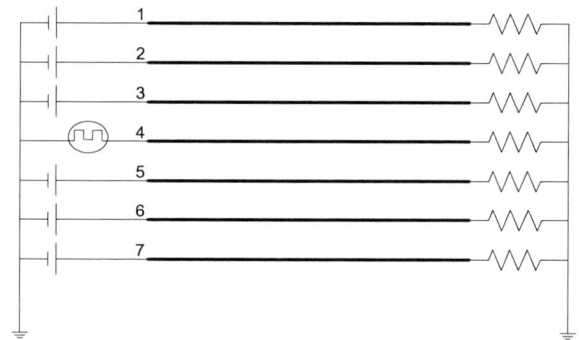

Fig. 3. Equivalent electrical scheme of the simulated circuit for $N=7$ conductors

All resistances are equal to $R_p = 1 \text{k}\Omega$ and all input DC sources have the same voltage $V_{in} = 5 \text{ V}$. Signal on the input of 4th line has periodical trapezoidal waveform with the frequency of 25 MHz and rise and fall times equal to 0.01 ns. Waveforms of output signals on lines 4, 5 and 6 are presented on Figures 4, 5 and 6, respectively.

As can be seen from the simulation results, the crosstalk effect can be noted at switching moments. Small graphs on each figure show the waveform of the impulse around switching moments (from 39 ns to 41 ns). Maximal deviation of signal at lines 5 and 6 due to the crosstalk are 700 mV and 110 mV respectively. Note that the ratio of these two values is almost the same as the ratio of the coupled capacitances between 4th and 5th line (c_{45}) and between 4th and 6th line (c_{46}). If we apply pulse signal to the first line, similar values of maximal deviation are obtained (750 mV and 120 mV).

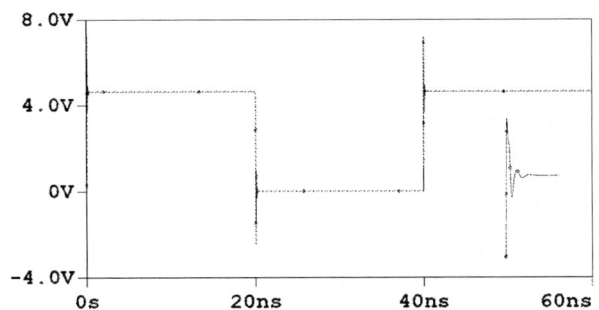

Fig. 4. Output signal on 4th line

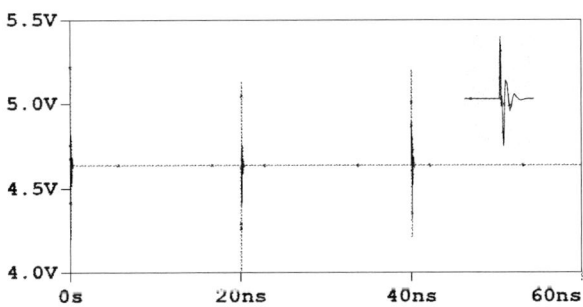

Fig. 5. Output signal on 5th line

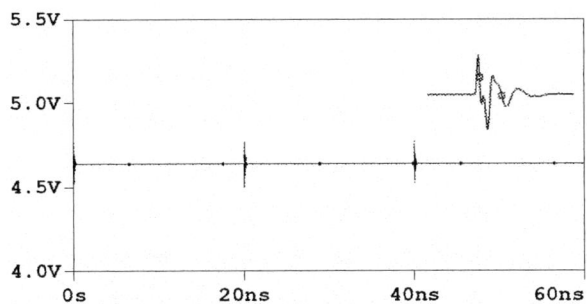

Fig. 6. Output signal on 6th line

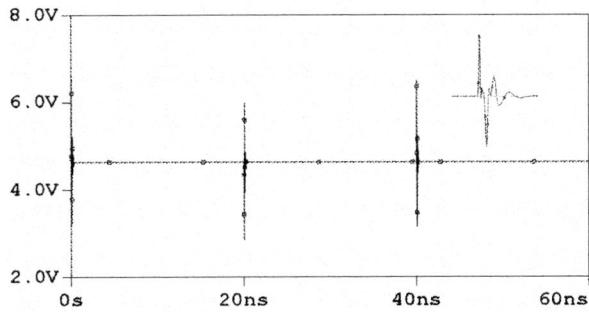

Fig. 7. Output signal on 4th line when input signals on 2nd, 3rd, 5th and 6th lines are pulse

Now we will try to obtain maximal possible deviation due to the crosstalk. Let us modify the circuit, such that input signal on 2nd, 3rd, 5th and 6th line is pulse, and on 4th line is constant (5 V, as in the previous case). Waveform of the output signal on line 4 is shown on Figure 7. Maximal deviation of the signal is now 1.9 V.

This value represents the maximum deviation of signal due to the crosstalk effect and should be compared with the noise margins of logical elements in order to verify proper design.

III. CONCLUSION

In this work we considered crosstalk between several interconnections in modern CMOS VLSI circuits. The interconnections are represented as lines with distributed parameters over the entire length. First we used Maxwell SV electromagnetic simulator to compute coupled capacitances, then we presented an electrical model of interconnections and finally we used OrCad PSpice simulator to obtain output waveforms on each line. We calculated the deviations of the signal for different combination of the input voltages. Maximal signal deviation due to the crosstalk can be up to 41% and it determines lower bound for the noise margins of the logical elements. Crosstalk effects can be reduced by increasing the distance between conductors (d).

ACKNOWLEDGEMENT

The author wishes to thank Professor Vančo Litovski, head of the Laboratory for Electronic Design Automation (LEDA) at the Faculty of Electronic Engineering, University of Niš for the oportunity of this research given, and to Milan Savić for usefull discussions.

REFERENCES

[1] A. Sheikholeslami, C. Heitzinger, S. Selberherr, F. Badrieh and H. Puchner, "Capacitances in the backend of a 100nm CMOS process and their predictive simulation", proceedings of the conference Mikroelektronik 2003, pp. 481-486, Vienna, October 1-2, 2003.

[2] R. Bauer and S. Selberherr: "Calculating Coupling Capacitances of Three-Dimensional Interconnections", ICSICT '92, pp. 697-702, Beijing, China, 1992.

[3] Ž. Mrčarica, V. Litovski, V. Živković: "Delay Time and Crosstalk Simulation in CMOS Integrated Curcuits", TELSIKS'97, pp. 122-125, Niš 1997.

[4] J. Rubinstein, P. Penfield JR, and M. Horowitz, "Signal Delay in RC Tree Networks", IEEE Trans. CAD/CAS., vol CAD-2, no. 3 July 1983.

[5] N. Delorme, M. Belleville and J. Chilo: "Inductance and capacitance analytic formulas for VLSI Interconnect", IEEE Electronic Letters vol. 32, no. 11, 23rd May 1996.

[6] http://www.ansoft.com/maxwellsv/.

446

2006 25th International Conference on Microelectronics

Reordering in Topology Decision Diagram Method for Symbolic Circuit Analysis

S. Djordjević and P. M. Petković

Abstract –This paper introduces reordering method in Topology Decision Diagram (TDD) in order to enhance symbolic analysis method for RLCg$_m$ network function generation in nested form. The improvement is obtained in circuit function compression and the execution time. An example of *n*-th order ladder network illustrates the method.

I. INTRODUCTION

Requirements for memory and CPU time consumption put the major limitations in symbolic analysis of the large circuits.

The technique of hierarchical decomposition results with compact symbolic expressions in nested form. Different methods for large circuits symbolic analysis mainly relay on hierarchical decomposition and can be classified as graph based [1, 2] and matrix based [3-7] techniques.

The method for symbolic analysis presented in this paper is topology oriented and represents a modification of Topology Decision Diagram method (TDD) developed by the same authors [8]. It generates exact symbolic network function in nested form. The proposed procedure represents symbolic network function by a diagram, like in DDD based algorithm [6, 7].

Graphic representation of the expression allows more efficient symbolic manipulation, derivation and evaluation. Instead of matrix entries in DDD, vertices in TDD are admittances or transconductances. The TDD method will be explained briefly in the next section.

II. TOPOLOGY DECISION DIAGRAM

We consider linear, time invariant, lumped RLCg$_m$ circuits characterized by network function in form of rational functions in the complex frequency s and the circuit parameters p that can be presented by (1) .

$$H(s) = \frac{N(s)}{D(s)} = \frac{\sum_i \left(\prod_j \left(p_{i,j} \cdot s^{n_j} \right) \right)}{\sum_k \left(\prod_l \left(p_{k,l} \cdot s^{n_l} \right) \right)} \tag{1}$$

Authors are with Department of Electronics, Faculty of Electronic Engineering, University of Niš, Aleksandra Medvedeva 14, 18000 Niš, Serbia & Montenegro, E-mail: srdjan@elfak.ni.ac.yu, predrag@elfak.ni.ac.yu.

where n_i is power of complex frequency $n_i \in \{0, 1\}$, while p_{ij} and p_{kl} represent circuit parameters.

Factorization of the network function in arbitrary circuit parameter p_k can be expressed as follows.

$$H_{0pk} = \lim_{p_k \to 0} H = \frac{N(p_k = 0)}{D(p_k = 0)} = \frac{N_{0pk}}{D_{0pk}} \tag{2}$$

$$H_{\infty pk} = \lim_{p_k \to \infty} H = \frac{N(p_k \to \infty)}{D(p_k \to \infty)} = \frac{N_{\infty p_k}}{D_{\infty p_k}} = \frac{p_k \cdot \dfrac{dN}{dp_k}}{p_k \cdot \dfrac{dD}{dp_k}} \tag{3}$$

$$H = \frac{N_{0p_k} + p_k \cdot N_{\infty p_k}}{D_{0p_k} + p_k \cdot D_{\infty p_k}} \tag{4}$$

Numerator and denominator are obtained separately. The expressions of the numerator and denominator of the circuit function are not known in advance, but one can easily determine the corresponding topology reducing circuit by setting $p_k \to 0$ and $p_k \to \infty$. The former case corresponds to elimination of parameter p_k, while $p_k \to \infty$ corresponds to parameter extraction [9]. From the aspect of the circuit topology, the parameter elimination can be treated as a branch reduction and parameter extraction as a node reduction .

During the circuit reduction every two-port device is treated as an admittance. When RLCg$_m$ circuits (CMOS circuits) are in scope, the only other type of devices besides the admittances is voltage controlled current source (VCCS).

When an admittance is the considering parameter, then $p_k \to 0$ corresponds to removed parameter from the circuit. Oppositely, the admittance is replaced by short for $p_k \to \infty$.

If p_k is a transconductance of a VCCS, the circuit topology remains unchanged except in two cases:
1. the current is controlled by the voltage across the same branch and VCCS operates as an admittance;
2. two or more VCCSs, mutually control each other and all transconductances, g_m, are extracted ($g_m = p_k \to \infty$); this implies that nodes of all these VCCSs are connected into one node.

During this recursive process, a tree-like topology decision diagram (TDD) is formed.

Every step in TDD construction results in new circuit topology, strongly related to the former one. During this process, some of the branches are eliminated and some of the circuit nodes are merged into one. Eventually, TDD outcomes with a graphical representation of the circuit function in nested form.

During the generation of TDD, each vertex in the diagram represents a circuit. The procedure for new vertex generation (child vertex), from previously generated vertex (parent vertex) in TDD can be:

- parameter elimination,
- parameter extraction or
- combination of both.

During the every step of TDD generation, symbolic expressions representing numerator or denominator of the transfer function are divided into two addendums. Both of these expressions do not contain the circuit parameter. They correspond to two new vertices of TDD and simultaneously, to two subcircuits. The first represents the part of the circuit that was independent of that parameter – actually the part from which the parameter was eliminated (branch reduction). The second is obtained after the parameter was extracted (node reduction). Elimination and extraction are always applied together during TDD generation, as graph in Figure 1 shows. These two procedures can be seen as a circuit device suppression with respect to the circuit topology reduction.

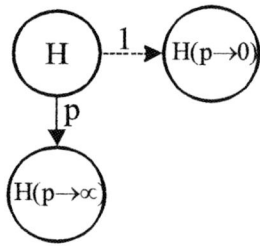

Figure 1. Suppression of a parameter.

The number of product terms (admittance order) in each addendum is the same and it is equal to the number of independent nodes in the network. Elimination of one circuit parameter results in a new vertex at the same level with the outgoing edge value equal to 1. After extraction of a single parameter, the number of nodes is decreased for 1 and admittance order is reduced for one. This corresponds to the lower level node in TDD. Outgoing edge is directed to the child node on the lower level of TDD, with weight p_k^l for k-th parameter extracted at l-th level, $k=1,...,m_l$), as shown in Figure 2.

The number of addenda at l-th level is equal to the number of eliminated parameters, while admittance order is equal to the number of levels. Actually, the subcircuit obtained at l-th level where m_l parameters are eliminated, is characterized by sum of m_l products each consisted of l multipliers.

Set of l parameters in every of m_l paths from leaf up to the root, represents one addendum in circuit function. Simultaneously, from the scope of circuit topology, it

represents the set of branches connecting all nodes in the circuit [10].

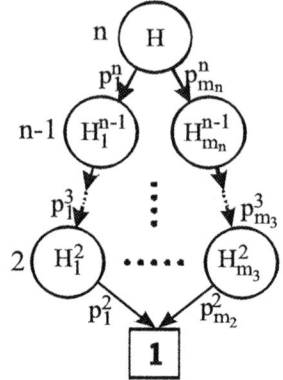

Figure 2. Construction of TDD.

It follows that less TDD nodes will be generated, if one extracts subsequently parameters from the branches connected to common node. Therefore, the fewest parameters set that can be extracted corresponds to the circuit node connecting the smallest number of branches.

In conclusion, the construction of the circuit function expression in symbolic form starts from the lives of TDD and proceeds up to the root, while vertices are represented by symbolic expressions.

The previous analysis indicates that proper ordering during construction of TDD has significant affect on efficiency of TDD method. Therefore the next section describes ordering method that guaranties more compact symbolic expression form.

III. REORDERING

Any part of the circuit can be generally represented as an *n*-port network, where *n* is the number of boundary nodes between this subcircuit and the rest of the circuit. The subcircuit is described in terms of the corresponding *n*-port network parameters . The nodes of the whole circuit can be divided, for the considered subcircuit, into three disjoint groups:

- internal,
- boundary and
- remainder.

We will consider the simplest case where circuit H is divided into subcircuits H_1 and H_2, by tearing node T, as illustrated in Figure 3.

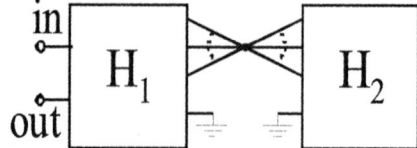

Figure 3. Two subcircuits with one common node.

Suppose that H_1 and H_2 have n_1 and n_2 nodes, respectively. If the controlled generators do not transmit

448

signal between subcircuits, H_2 can be treated as a two port. In this case, it can be presented as an admittance, according to Figure 4.

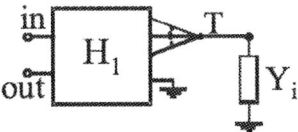

Figure 4. Modeling of subcircuit H_2 by an admittance.

According to (4) the resulting expression of the circuit function is:

$$H = \frac{N_0 + N_1 Y}{D_0 + D_1 Y} = \frac{N_0 N_Y + N_1 N_Y}{D_0 D_Y + D_1 D_Y}, \qquad (5)$$

where:

$$Y = \frac{N_Y}{D_Y} \qquad (6)$$

It follows that the transfer function is determined by six analytic expressions. Namely, N_0, D_0, N_1, D_1, N_Y, and D_Y. Expressions N_Y and D_Y are obtained from H_2. The reordering method relay on fact that these two expressions share many subexpressions. As will be explained, this can save a lot of computing time and to results in more compact circuit expression in symbolic form.

Let suppress H_2 first. After extraction of n_2 parameters successively (n_2 being the number of internal nodes), topologies obtained as lives in TDD are almost identical. In order to exploit the similarity between circuits, it is advised to proceed with admittance Y_i ($i=1,\ldots,m_{n1}$) that is connected to the terminal nodes, as indicated in Figure 4.

The number of nodes in the whole circuit is n_1+n_2 where n_1 is the number of nodes in H_1. After n_2 node reductions the remaining number of nodes is equal n_1.

The parameter extraction and elimination from all subcircuits that represent vertex on n_2-th level in TDD, result in two identical circuits. Accordingly, all outgoing edges are directed only to two vertices. The part of the TDD related to this step, is illustrated in Figure 5a and the corresponding circuits are shown in Figure 5b

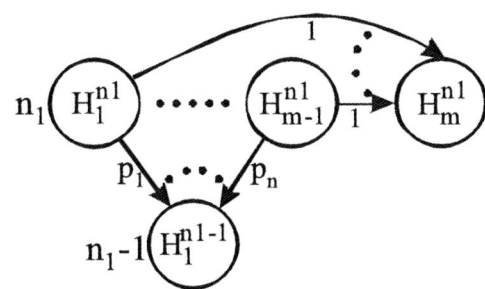

Figure 5. a. Reduction into two vertices.

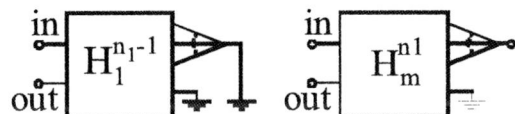

Figure 5. b. Two corresponding circuits.

In the subsequent step, two mutually independent sub-diagrams that correspond to these two circuits are formed. The proposed vertex ordering exploit shearing among expressions N_y and D_y in (6). Instead of two independent subdiagrams, only one is sufficient to provide generation of both expressions.

The order of parameter extraction can be reversed. Namely, it is possible to start with analysis of subcircuit H_1 instead of H_2. More compact expression will be gained if the larger subcircuit is analyzed first because the shearing of common expressions is exploited in more efficient manner.

Generally, bipartiotioning contains more than two tearing nodes. The circuit reduction obtained by extraction/elimination of parameters that belong to one subcircuit will result in many vertices with the identical circuit topology. The difference between circuits that correspond to these vertices are exploited in branches connected between terminal nodes. Extraction and elimination of these parameter will give identical circuits.

The number of levels in TDD corresponds (is equal) to the number of terminal nodes between two subcircuits. Simultaneously, the number of circuits corresponds (is equal) to the combination of the shortcuts and opens between every pair of terminal nodes.

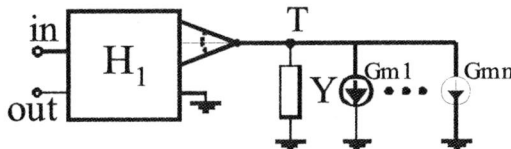

Figure 6. Modeling of a subcircuit by VCCS and admittance.

The procedure can be spread to the subcircuits that contain VCCS controlled by voltages from the rest of the circuit. The necessary condition is that the controlled generators transmit signal in one direction. Namely, from the rest of the circuit to the subcircuit. This case is illustrated in Figure 6.

IV. EXAMPLE

The efficiency of the circuit parameter reordering is illustrated on the ladder network illustrated in Figure 7.

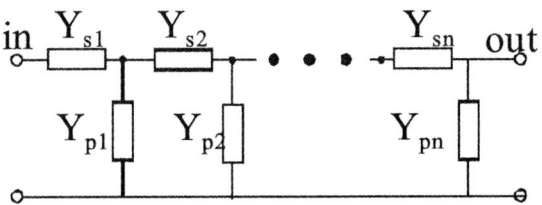

Figure 7. The ladder network.

Table I presents number of multiplications, additions, and intermediate expressions for different number of sections n. Obviously, reordering of the circuit

449

parameter extraction in TDD gives significantly compacted results in terms of number of operations.

TABLE I

n	without oredering			with ordering		
# of sections	#mul	#add	#expr.	#mul	#add	#expr.
5	18	10	33	17	10	15
10	51	34	90	37	24	36
15	94	70	165	62	40	60
20	134	107	246	82	54	81
30	258	212	459	127	84	126
40	332	261	594	172	114	171
50	481	379	870	217	144	216
100	1218	948	2140	442	294	441

Tabele II gives the simulation time for the same example. Simulation is performed on Pentium III processor at 551 MHZ with 256 MB of RAM.

TABLE II

n # of sections	time [mS]	
	without ordering	with ordering
5	13,3	9,7
10	53,57	35,25
15	180	90,8
20	230	157
30	625	401
40	1670	916
50	2070	1350
100	14000	11200

V. CONCLUSION

The primary goal of the symbolic analysis, which result in nested form of network function, is to generate expressions as compact as possible. Hierarchical decomposition is imposed as a natural component of this procedure. Optimization goal for the proposed method is minimization of the vertices number in assigned TDD. The proposed reordering of circuit parameter extraction can significantly reduce the size of TDD that describes circuit function.

Symbolic circuit analysis based on the TDD generation is very suitable for circuit partitioning. Every new TDD node represents a circuit with reduced topology. This paper considers a ladder example simple enough to serve for method explanation but sufficiently complex to explore benefits of the reordering method. The proposed procedure can be spread to the bipartioning with more tearing nodes.

The compactness of symbolic expression is achieved by shearing common expressions between subcircuits having same topology.

ACKNOWLEDGEMENT

This work was partially supported by the Serbian Ministry of Science and Environment Protection through project No. TR 006108.B.

REFERENCES

[1] A. Konczykowska and J. Strzyk, "Computer analysis of large signal flowgraphs by hierarchical decomposition method," in *Proc. European Conf. Circuit Theory Design*, (Warsaw, Poland) , 1980, pp. 408-413.

[2] Marwan M. Hassoun, Kevin S. McCarville "Symbolic analysis of large-scale networks using a hierarchical signal flowgraph approach" *J.Analog VLSI Signal Process.*, vol. 3, pp. 31-42, Jan. 1993.

[3] S.J. Jou, M. F. Perng, C. C. Su and C. K. Wang, "Hierarchical Techniques for symbolic analysis of large electronic circuits" *IEEE Inter. Symp. Circuits and Systems*, pp.21-24, June 1994.

[4] Marwan M. Hassoun, Pen-Min Lin "A hierarchical network approach to symbolic analysis of large-scale networks" *IEEE Trans.Circuits Syst.*, vol. 42, pp. 201-211, April 1995.

[5] S. Đorđević, P. Petković, "A hierarchical approach to large circuit symbolic simulation", *Microelectronics Reliability*, vol.41 , pp. 2041-2049, 2001.

[6] X.-D.Tan and C.-J.Shi, "Hierarchical symbolic analysis of analog integrated circuits via determinant decision diagrams," *IEEE Trans.Computer-Aided Design*, vol. 19, pp. 401-412, Apr. 2000.

[7] C.-J.Richard Shi, and Xiang-Dong Tan, "Compact Representation and Efficient Generation of s-Expanded Symbolic Network Functions for Computer-Aided Analog Circuit Design," *IEEE Trans. Computer-Aided Design*, vol. 20, pp. 813-827, Jul. 2001.S. Đorđević and P. M. Petković, "Generation of Factorized Symbolic Network Function by Circuit Topology Reduction", *Proceedings of MIEL'04*, Niš, 2004 pp. 773-776.

[8] Đorđević, S., Petković, P., "A Modified Method For Symbolic Network Function Extraction Based On Circuit Topology Decision Diagram", *Proc. of the XLIX Conf. of ETRAN, ETRAN 2005*, Vol. I, pp. 103-106, June 2005, Budva, in Serbian.

[9] Đorđević, S., Petković, P, "Generation of Factorized Symbolic Network Function by Circuit Topology Reduction", Proceedings of MIEL'04, Niš, 2004 pp. 773-776

[10] L. O. Chua and P. M. Lin, *Computer Aided Analysis of Electronic Circuits-Algorithms and Computational Techniques.* Englewood Cliffs, NJ: Prentice-Hall, 1975.

2006 25th International Conference on Microelectronics

An Improved Performances FGMOS Voltage Comparator for Data Acquisition Systems

C. Popa

Abstract - A new voltage comparator will be presented, using exclusively MOS transistors working in the saturation region for improving the circuit speed. In order to increase the controllability of the proposed voltage comparator and to achieve the possibility of comparing differential voltages, FGMOS (Floating Gate MOS) transistors will be used to replace classical MOS active devices. An original method for reducing the linear region of the comparator will be proposed, based on the biasing in the neighborhood of the subthreshold region of the MOS active devices.

I. INTRODUCTION

The differential amplifier is an important stage of a very large area of applications, including high-performances analog/mixed ICs, such as operational amplifiers, voltage comparators, voltage regulators, video amplifiers, modulators and demodulators or A/D and D/A converters. Replacing bipolar transistors with MOS transistors, it was solved the problem of relatively large values of the input bias and input offset currents and of the small value of the input impedance, with the disadvantage of reducing the voltage gain due to the quadratic characteristic of the MOS transistor working in saturation. Besides all these parameters, the linearity of the circuit still remains poor because of the fundamental nonlinear characteristic of both bipolar and MOS transistors, resulting the possibility of achieving a relatively good linearity only for a restricted input voltage range (the amplitude of the input voltage for the classic differential amplifier using MOS transistors in saturation have to be below a few hundreds of mV). In conclusion, it results the necessity of implementing a linearization technique for decreasing the superior-order nonlinearities of the MOS differential stage and for increasing the available range for the input voltage amplitudes. It exists in literature many circuit techniques used to improve the MOS differential amplifier linearity. It was presented in [1], [2] a third and fifth-order harmonics cancellation with good results and a relatively simple circuit implementation. A constant-sum of the gate-source voltages circuit connection was described in [3] and it allows an important reduction of the total harmonic distortions coefficient of the circuit. In [4], it was presented and implemented in CMOS technology a simple technique based on square-root circuits for improving the

CMOS differential stage linearity, which compensates the quadratic characteristic of the MOS transistor in saturation. An immediate application of the CMOS differential amplifier is represented by the voltage comparator. The new proposed circuit is based on the replacing of the classical MOS transistor by a FGMOS device, having the important advantages of an increased controllability and of the possibility of comparing differential voltages.

II. THEORETICAL ANALYSIS

A. The FGMOS transistor

The multiple-input floating-gate transistor is an ordinary MOS transistor whose gate is floating. The basic structure of a n-channel floating-gate MOS transistor is shown in Fig. 1a. The floating-gate is formed by the first silicon layer over the channel while the multiple input gates are formed by the second polysilicon layer which is located over the floating-gate. This floating-gate is capacitively coupled to the multiple input gates. The symbolical representation of such devices is shown in Fig. 1b. The drain current of a FGMOS transistor with n-input gates in the saturation region is given by the following equation:

$$I_D = \frac{K}{2}\left[\sum_{i=1}^{n} k_i (V_i - V_S) - V_T\right]^2 \qquad (1)$$

where $K = \mu_n C_{ox}(W/L)$ is the transconductance parameter of the transistor, μ_n is the electron mobility, C_{ox} is the gate oxide capacitance, W/L is the transistor aspect ratio, $k_i, i = 1,...,n$ are the capacitive coupling ratios, V_i is the i-th input voltage, V_S is the source voltage and V_T is the threshold voltage of the transistor. The capacitive coupling ratio is defined as:

$$k_i = \frac{C_i}{\sum\limits_{i=1}^{n} C_i + C_{GS}} \qquad (2)$$

C. Popa is with The Faculty of Electronics, Telecommunications and Information Technology Bucharest, 1-3 Iuliu Maniu, Bucharest, Romania, E-mail: cosmin_popa@yahoo.com.

1-4244-0116-X/06/$20.00 ©2006 IEEE

where C_i are the input capacitances between the floating-gate and each of the i-th input and C_{ox} is the gate-source capacitance which is equal to $(2/3)C_{ox}$ for operation in the saturation region. All the overlap capacitances are assumed to be considerably smaller than capacitances summation $\sum_{i=1}^{n} C_i + C_{GS}$. Equation (1) shows that the FGMOS transistor drain current in saturation is proportional to the square of the weighted sum of the input signals, where the weight of each input signal is determined by the capacitive coupling ratio of the input.

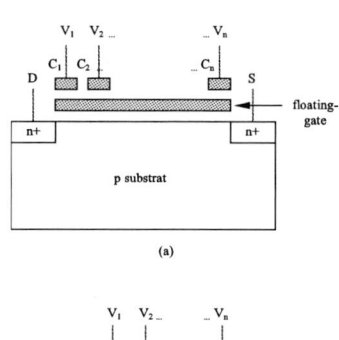

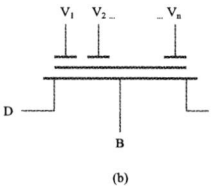

Fig. 1. (a) The basic structure of a n-channel FGMOS transistor; (b) symbolic representation

B. The classical MOS voltage comparator

The structure of the classical voltage comparator is presented in Fig. 2.

Considering a biasing in saturation of all MOS transistors from Fig. 2, the drain currents of the circuit will have the following expressions:

$$I_1 = \frac{K}{2}\left(v_1 - v_A - V_T\right)^2 \qquad (3)$$

and:

$$I_2 = \frac{K}{2}\left(v_2 - v_A - V_T\right)^2 \qquad (4)$$

resulting:

$$v_{id} = v_1 - v_2 = \sqrt{\frac{2I_1}{K}} - \sqrt{\frac{2I_2}{K}} \qquad (5)$$

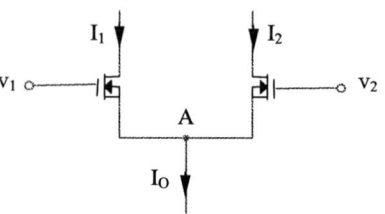

Fig. 2. The classical MOS voltage comparator

After some computations, the expression of the drain currents I_1 and I_2 with respect to the differential input voltage v_{id} will be:

$$I_1 = \frac{I_O}{2} + \frac{\sqrt{KI_O}}{2} v_{id}\sqrt{1 - \frac{Kv_{id}^2}{4I_O}} \qquad (6)$$

and:

$$I_2 = \frac{I_O}{2} - \frac{\sqrt{KI_O}}{2} v_{id}\sqrt{1 - \frac{Kv_{id}^2}{4I_O}} \qquad (7)$$

It is possible to demonstrate that the linear region of the classical CMOS voltage comparator is included in the $\left(-\sqrt{2KI_O}; \sqrt{2KI_O}\right)$ range. An important disadvantage of the classical voltage comparator is that only two input potentials (referred to the ground) could be compared. In order to compare two differential voltages and to increase the circuit controllability, FGMOS transistors will replace the classical MOS devices.

C. The FGMOS voltage comparator

The original proposed FGMOS voltage comparator is presented in Fig. 3.

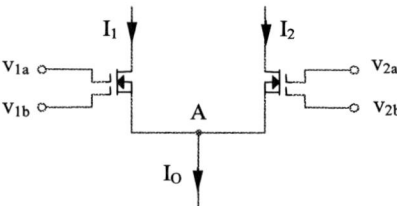

Fig. 3. The FGMOS voltage comparator

Considering a biasing in saturation of all FGMOS transistors from Fig. 3, the drain currents of the circuit will have the following expressions:

$$I_1 = \frac{K}{2}\left(\frac{v_{1a} - 2V_A + v_{1b}}{2} - V_T\right)^2 \qquad (8)$$

and:

$$I_2 = \frac{K}{2}\left(\frac{v_{2a} - 2V_A + v_{2b}}{2} - V_T\right)^2 \qquad (9)$$

resulting:

$$\frac{(v_{1a} + v_{1b}) - (v_{2a} + v_{2b})}{2} = \sqrt{\frac{2I_1}{K}} - \sqrt{\frac{2I_2}{K}} \qquad (10)$$

Using the following notations:

$$v_{ida} = v_{1a} - v_{2a} \qquad (11)$$

$$v_{idb} = v_{1b} - v_{2b} \qquad (12)$$

and:

$$v_{id} = \frac{v_{ida} + v_{idb}}{2} \qquad (13)$$

it results:

$$v_{id} = \sqrt{\frac{2I_1}{K}} - \sqrt{\frac{2I_2}{K}} \qquad (14)$$

and the drain currents of the FGMOS voltage comparator having the following expressions:

$$I_1 = \frac{I_O}{2} + \frac{\sqrt{KI_O}}{2} v_{id}\sqrt{1 - \frac{Kv_{id}^2}{4I_O}} \qquad (15)$$

and:

$$I_2 = \frac{I_O}{2} - \frac{\sqrt{KI_O}}{2} v_{id}\sqrt{1 - \frac{Kv_{id}^2}{4I_O}} \qquad (16)$$

The SPICE simulation of the graphical dependence $I_{1,2}(v_{id})$ is represented in Fig. 4.

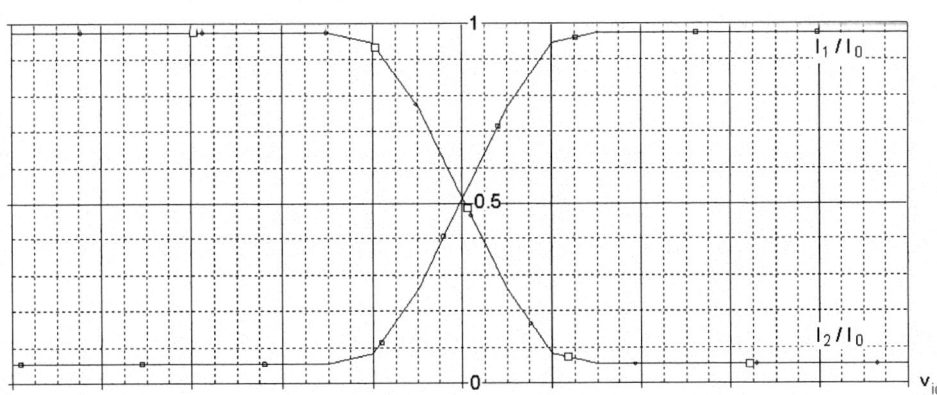

Fig. 4. The graphical dependence $I_{1,2}(v_{id})$

There are two important regions on the previous graphic:

a.

$$v_{id} > \sqrt{\frac{2I_O}{K}} \qquad (17)$$

equivalent to:

$$v_{ida} + v_{idb} > 2\sqrt{\frac{2I_O}{K}} \qquad (18)$$

In this case, it results:

$$I_1 - I_2 = I_O \qquad (19)$$

b.

$$v_{id} < -\sqrt{\frac{2I_O}{K}} \qquad (20)$$

equivalent to:

$$v_{ida} + v_{idb} < -2\sqrt{\frac{2I_O}{K}} \qquad (21)$$

In this case, it results:

$$I_1 - I_2 = -I_O \qquad (22)$$

In conclusion, the graphical dependence $v_{idb}(v_{ida})$ is represented in Fig. 5.

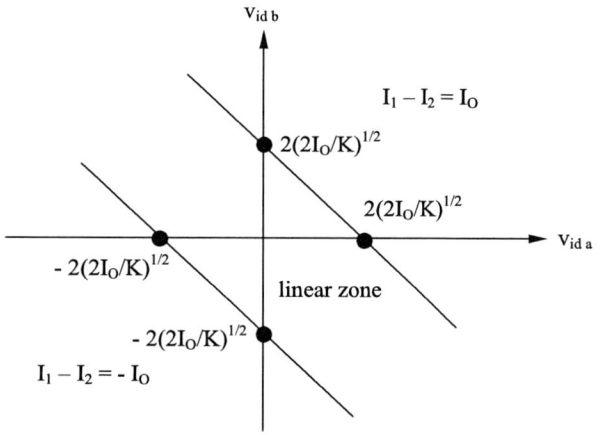

Fig. 5. The graphical dependence $v_{idb}(v_{ida})$

The original method for reducing the linear region from Fig. 4 is based on the decreasing of the value for the polarization current I_O toward the subthreshold operation limit.

III. CONCLUSIONS

A new voltage comparator has been presented, using exclusively MOS transistors working in the saturation region for improving the circuit speed. In order to increase the controllability of the proposed voltage comparator and to achieve the possibility of comparing differential voltages, FGMOS (Floating Gate MOS) transistors have been replaced classical MOS active devices.

An original method for reducing the linear region of the comparator has been proposed, based on the biasing in the neighborhood of the subthreshold region of the MOS active devices. With respect to the previous reported voltage comparators, the proposed original implementation presents the advantage of a reduced simplicity and of a reduced width of the linear region.

ACKNOWLEDGEMENT

This work was supported by the Research Project 15.05.14 – UEFISCSU ET 2970.

REFERENCES

[1] Popa C., "Linearity Improvement Design Technique for a CMOS Differential Amplifier", in *Scientific Bulletin, University "Politehnica" of Bucharest,* 2000, volume 62, number 4, series C, pp. 51-60.

[2] Popa C., "Linear Rail-to-rail CMOS Input Stage", in *The 13th International Conference on Control System and Computer Science,* University "Politehnica" of Bucharest, 2001, pp. 536-539.

[3] Hung C., Ismail M., Halonen K. and Porra V., "Low-voltage Rail-to-rail CMOS Differential Difference Amplifier", in *IEEE Proceedings of International Symposium on Circuits and Systems,* 1997, pp. 145-148.

[4] Hyogo A., Fukutomi Y. and Sekine K., "Low Voltage Four-quadrant Analog Multiplier Using Square-root Circuit Based on CMOS Pair", in *IEEE Proceedings of International Symposium on Circuits and Systems,* 1999, pp. 274-276.

2006 25th International Conference on Microelectronics

On Silicon Timing Validation of Digital Logic Gates "A Study of Two Generic Methods"

A.P. Singh, N.S. Panwar

Abstract - The world of electronic industry, is working with Nano-Seconds domain of Timings, so it's always a challenge for the electronic designers to know the exact, if not exact, then at least 99% accurate of on-silicon-delay values of their design components. By component, means, the smallest possible element for circuit designing. These components are a part of standard library, known as Standard Cell Library. The Technology trends are almost following the Moore's Law and thus every year we see the technology shrinks by roughly a factor of 1.5. This trend will go on as predicted by the experts. With the increasing complexity in designs, the need for a better silicon evaluation of our building blocks is also increasing. The ASIC Design flows use characterized-data for sign off, so to be aware of its accuracy is a must. As Flip-Flops are very important part of ASIC Design flow, the characterization of a Flip Flop of the ASIC Library on silicon with a good accuracy is a challenge. Being a sequential element, FF's delays and power consumption are very important for ASIC designers.

This study work will be dealing with the basic understanding of these logic gates. The "two-methods" used for showing the techniques which can be readily used in the electronics industry for measuring/validating the silicon-timings for the digital gates. The methods are **Dummy Path method** and **Ring-oscillator method.** The experimental work is performed to study the behavior of various sequential when observed under these two methods.

I. INTRODUCTION

Electronic Circuits which combine digital signals according to the Boolean algebra, are referred to as logic gates; gates because they control the flow of information. Positive logic is an electronic representation in which the true state is at a higher voltage, while negative logic has the true state at a lower voltage. We can use the positive logic type or the negative logic type. Two main categories for Boolean logic expressions are, namely: *Combinational and Sequential.*

Combinational Logic: Any combination of logic, such as for two inputs (AB or A+B or A!+B!), and even further combination of these terms can be generated under combinational logic. Truth table helps in generating the optimized combinational Boolean expression for final circuit design. The components like AND, INV, OR etc are used to realize the Boolean combinational expression. Figure 1 shows an example of combinational circuit.

A.P. Singh is with the University School of Information Technology, GGSIPU New Delhi (India)

N.S. Panwar is with the M.Tech (IT), School of Information Technology, GGSIPU New Delhi, India,

E-mail: nitinsingh77@gmail.com

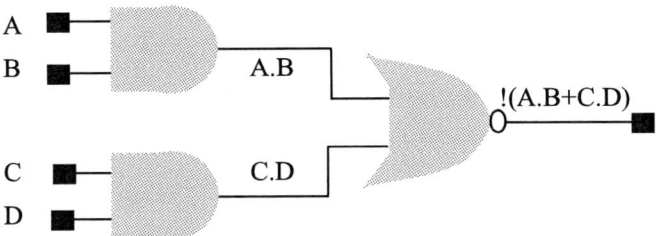

Fig. 1. AND-OR-INV combinational logic

Sequential Logic: FlipFlop, a sequential logic component, is a very important element of any ASIC standard cell library. Latches and Flip Flops are very important in today's designing world. A Latch changes its state on the positive or negative level, and a flip flop does so on the basis of positive or negative edge of the a signal generally called as clock, CLK. A particular sequence of states can be observed using the flip flop designing. Two latches can be connected back to back for implementing the flip flop circuit.

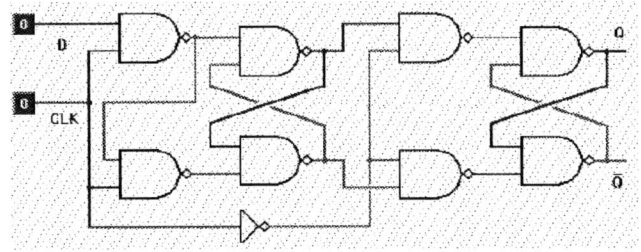

Fig. 2. D-Type Flip-Flop using NAND logic gates.

Other types of flip flops include JK-flip-flops etc.

II. UNDERSTANDING DIGITAL CIRCUIT TIMINGS

Timing Parameters for Combinational/Sequential Logic

When implemented physically, combinational circuits, such as AND and OR gates, exhibit certain timing characteristics. When a binary value (0 or 1) is applied at the input to a combinational circuit, the change at the circuit output is not instantaneous due to electrical constraints. Circuit input-to-output delay in combinational circuits can be expressed with two parameters, *tpd* and *tcd*, defined as follows: **Propagation delay (*tpd*)** - This value indicates the amount of time needed for a change in a logic input to result in a permanent change at an output.

1-4244-0116-X/06/$20.00 ©2006 IEEE   455

Combinational logic is guaranteed not to show any further output changes in response to an input change after *tpd* time units have passed. In case of Sequential logic, this value indicates the amount of time needed for a change in the flip flop-clock input (e.g. rising edge) to result in a **permanent** change at the flip-flop output (Q). When the clock edge arrives, the D input value is transferred to output Q. Note from Figure, that the output of the flip-flop may be at an intermediate value for a while (indicated by the cross-hatched area) before the final output value is created. After *tClk-Q*, the output is guaranteed not to change value again until next clock edge trigger (e.g. rise) arrives.

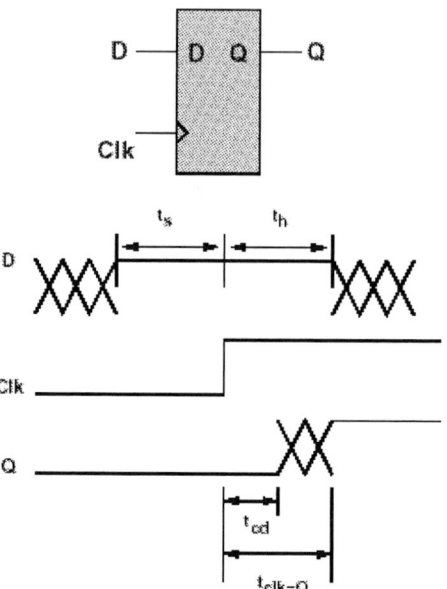

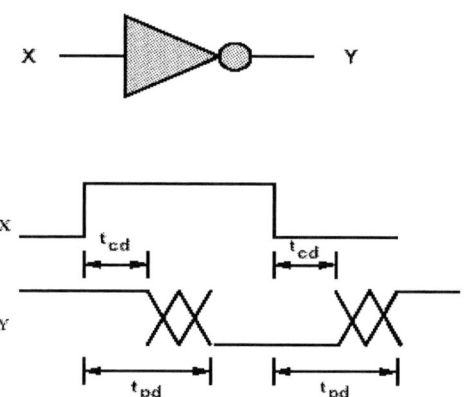

Fig. 3. Combinational Propagation and Contamination Delay

Contamination delay (*tcd*) - This value indicates the amount of time needed for a change in a logic input to result in an initial change at an output. Combinational logic is guaranteed not to show any output change in response to an input change before *tcd* time units have passed. This value indicates the amount of time needed for a change in the flip-flop clock input to result in the **initial** change at the flip-flop output (Q). Note from Figure, that the output of the flip-flop maintains its initial value until time *tcd* has passed. The flip-flop is guaranteed not to show any output change in response to an input change until after *tcd* has passed.

Setup time (*ts*) - This value indicates the amount of time **before** the clock edge that data input D **must** be stable. As shown in Figure below, D is stable *ts* time units before the rising clock edge.

Hold time (*th*) - This value indicates the amount of time **after** the clock edge that data input D **must** be held stable. As shown in Figure below, the hold time is always measured from the rising clock edge (for positive edge-triggered) to a point after the edge.

Silicon Timing Validation for Logic Gates
The term "Validation" is used here for silicon-proof of propagation delays, setup times etc, timing parameters. We will be discussing more on two methods used for study.

Fig. 4. Setup and hold time for Sequential Cells

Dummy Path Method: This method is not accurate but if very easy to understand, implement and is very close to the exact values. In fact it should be called as approximate value dummy path method, "appValueDummyPath". Dummy path says, "*First time, pass your signal through a path containing chain of cells, and second time, pass your signal through a path without chain of cells. Pen down the difference observed in two operations, and divide it by number of cells used in chain. This will give you the delay per cell*". The method involves a lot of good tactics while PnR-ing (Placement and Routing) the block, and also involved a bit of inaccuracy because of dummy path delay plus ATE tools limitations.

The following calculation can be made for measuring the delays of a logic cell:

> ***Cell Delay= (d1 - d2) / n***
> where   n   = number of cells in chain
> d1 = delay calculated through chain
> d2 = delay calculated through dummy path

Figure 5 depicts the actual circuit, which will be used to perform this delay-path measurement. Here the block used is NAND cell, and is shown here just for illustrating the dummy path method. The timing diagrams are also shown, which depicts how can we measure the delay.

Ring Oscillator Method: This method is always the first priority for validating the cell timings because of its easy and simple approach. Only by measuring the frequency of oscillations, one can find the delay per cell. The ring oscillator method says, "Design *a ring of cell connected in ripple fashion, apply a triggering pulse if required, and*

456

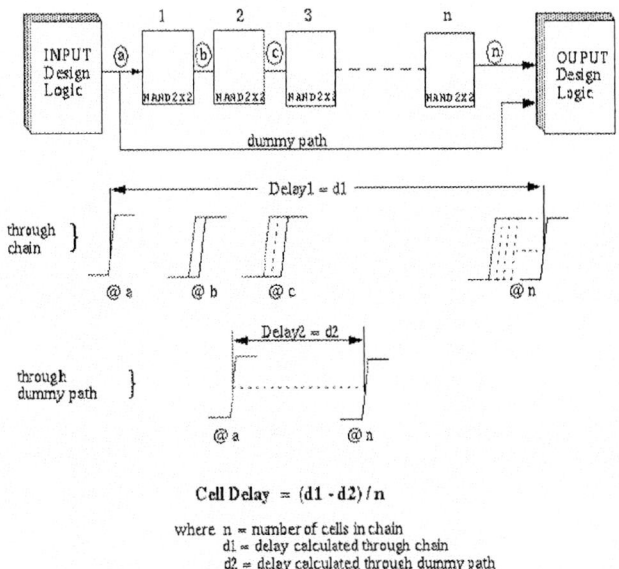

Cell Delay = (d1 - d2) / n

where n = number of cells in chain
d1 = delay calculated through chain
d2 = delay calculated through dummy path

Fig. 5. Dummy path method illustration

observe the frequency of oscillations. The inverse of frequency will help you in calculating the delay of cell used in ring". This frequency depends on number of cells used, process and the output load seen by each cell in ring with its driving strength. The following figure can illustrate the concept more clearly. Please note that cells to be connected in odd number to make the oscillations possible. Take care we should have odd numbers of cell connected in chain-fashion.

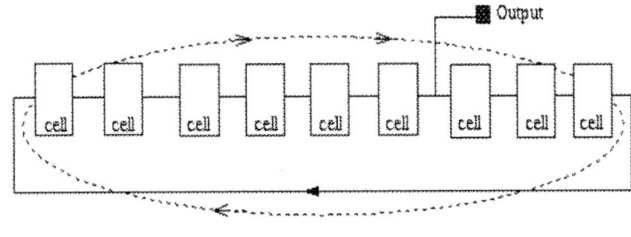

Frequency measured at Output = f Hz, (ATE dependent).

Delay per cell observed = $\dfrac{1}{2 * f * N}$

where :

"f" is the frequency of oscillations

N is the number of cells in ring.

Fig. 6. Diagram to illustrate ring oscilator method

This method is very accurate as by only measuring the frequency, one can calculate the delay of a single element of the ring and also the power consumed at that frequency can be measured (to calculate the powerXdealy product)

III. CIRCUIT IMPLEMENTATION

Circuit implementation for all the standard cells, is done in Cadence's environment for Schematic Designing. Cadence Design System Inc. has majority of the world design industries as the customers who work on design platform from Cadence. The real tool schematics/circuits are designed in Cadence's Composer environment, which was used to perform two methods for silicon validation of sequential cell propagation delays. This platform helps in easy drawing of schematic design, and to perform different checks and test in a some easy steps, and can be integrated with 3rd party tools, e.g. for design verification, netlisting etc.

Circuits are designed based on some standard cells from the library available for the latest technology, and are not produced here as for confidential database. The circuits are similar to the approach mentioned in above paragraphs.

IV. RESULTS

TABLE I
RESULT AND OBSERVATIONS, BASED ON THE TOP LEVEL CIRCUIT SIMULATION.

Serial Number	Sub_Block_Name	BLOCK	Observed Delay (ns)
1	NAND2X2	RingOsc	0.053
2	XOR2X2	RingOsc	0.113
3	JKFFSRX2	RingOsc	0.219
4	DFFSX2	RingOsc	0.151
5	NAND2X2	Dummy Path	0.055
6	XOR2X2	Dummy Path	0.120
7	DFFRX2	Dummy Path	0.151
8	DFFSX2	Dummy Path	0.156

TABLE II
COMPARING ALL THE VALUES FOR STANDARD CELLS TIMING SIMULATION, OBSERVED UNDER DIFFERENT CONFIGURATION.

S. No.	Cell Name	A	B	C	D	E
1	NAND2X2	0.054	0.0556	0.0545	0.055	0.053
2	XOR2X2	0.110	0.119	0.111	0.120	0.113
3	DFFSX2	0.148	0.1556	0.156	0.156	0.151
4	JKFFSRX2	0.209	x	0.215	x	0.219
5	DFFRX2	0.142	0.150	x	0.151	x

In Table above:

A : Actual single cell delay in nano seconds after simulation

B : Dummy path dealy (individual cell) in nano seconds

C : ring oscillator delay (nano seconds)
D : Top Level DP delay (nano seconds)
E : Top Level RO delay (nano seconds)

The delays are approximately matched with the actual delay using the simulation method. This illustrates the close approximation "time-validating" methods for sequential cells.

V. CONCLUSIONS

The basics of all digital gates, Combinational and Sequential cells, which resides in the heart of almost all chip designing today, were studied with their importance in today's electronic industry. We had performed two methods for "on-silicon" timing measurements: Dummy path method, Ring Oscillators. Both the methods have given a very good approach of validating *time* on silicon for these cells (logic gates).

All the block are integrated under one TOP Block and again the analysis is done to measure the delays on chip level. The result is observed and tabulated, found to be in approximate to the individual block level observations. This study is carried out as a "Test-Chip-Design" approach for the digital logic gates, in which both of the methods are used for "propagation delay measurement", along with some extra circuitry for input and output pins minimization. The further approach can be to study the "measurement of setup time and hold time" for sequential circuit, and to measure the timings at top level.

REFERENCES

[1] Walker, R., A Monolithic High-Speed Voltage Controlled Ring Oscillator , Proceedings of the Hewlett-Packard 1987 VLSI Design Technology Conference, May 18-20, 1987, S10.6.1-5

[2] Walker, R. C., A Fully Integrated High-Speed Voltage Controlled Ring Oscillator, U.S. Patent #4884041, Nov. 28, 1989

[3] "Design Challenges for High-Performance SOI Digital CMOS VLSI," C. T. Chuang, Invited Review/Tutorial Paper, Proc. of Tech. Papers, 1999 International Symposium on VLSI Technology, Systems, and Applications, Taipei, Taiwan, R. O. C., June 8-10, 1999, pp. 270-273.

[4] *G.M. Blair*; *"CMOS Buffer Tapering with Interconnect Capacitances"*, IEE Electronics Letters, 32, No 21, pp. 1984-1985, 10th Oct 1996.

[5] "SOI Digital CMOS VLSI - A Design Perspective," C. T. Chuang and R. Puri, Invited Tutorial Paper, Proc. 36th Design Automation Conference, New Orleans, LA, June 21-25, 1999, pp. 709-714.

[6] G.M. Blair; "Self-generating clocks using an augmented distribution network", Proc. IEE Circuits, Devices and Systems, 144, No 4, pp. 219-222, Aug 1997

[7] G. Morris, G.M. Blair; "Local generation of the falling clock edge for silicon resource sharing", IEE Electronics Letters, 34, No 5, pp. 436-7, Mar 1998.

[8] Sung-Mo Kang, Leblebici "CMOS Digital Integrated Circuits- Analysis and Design Second Ed"

[9] Y. Leblebici and A. Dervisoglu, "Computer-aided circuit analysis using modified nodal equations," Proc. Electrical Energy Symposium, Istanbul, October 1984.

[10] The Design and Analysis of VLSI Circuits (The VLSI systems series) by Lance A. Glasser, Daniel Dobberpuhl

[11] S. Ward and R. Halstead. Computation Structures. McGraw-Hill, Boston, Ma, 1991

[12] Semicondcutors: Infineon Technologies's Technical Book on Semiconductors 2nd revised edition 2004.

[13] N.H. Weste "Principles of VLSI design" 2nd edition

[14] VLSI Systems Design for Digital Signal Processing; by Bowen

[15] Basic VLSI Design; by Douglus A. Pucknell and Kamran Eshrighian, Prentice Hall

[16] "Physical Design of a Fourth-Generation POWER GHz Microprocessor," Carl Anderson et al., ISSCC 2001 Digest of Technical Papers, February, 2001, p. 232.

[17] Dubey P, ST-CRnD, Noida Paper on "Characterizing On Silicon A Flip- Flop In An ASIC Library"

[18] Randall L. Geiger, Phillip E. Allen "VLSI Design Techniques for Analog and Digital Circuits"

[19] Malvino, Leach "Digital Principles and Applications – 4th Ed".

[20] Microelectronic Circuit Design (McGraw-Hill Series in Electrical and Computer Engineering. Electronics and VLSI Circuits.) by Richard C. Jaeger, Travis N. Blalock

[21] "Introduction to VLSI Sysems": by Carver Mead, Lynn Conway.

[22] Y. Leblebici, "CMOS digital circuit design guidelines for improved long-term reliability," Proc. 1995 European Conference on Circuit Theory and Design, pp. 167-170, August 1995.

[23] Hspice tool manual: Users guide for simulation using Hspice (Synopsys Inc. USA)

[24] Cadence's Openbook (Cadence Design Systems Inc. USA)

2006 25th International Conference on Microelectronics

High Speed Low Power CMOS Comparator for Pipeline ADCs

M. B. Guermaz, L. Bouzerara, A. Slimane, M. T. Belaroussi, B. Lehouidj and R. Zirmi

Abstract - This paper describes and analyzes a low power and high speed differential comparator. The designed comparator is intended to be implemented in a 10bit 20MHz pipeline Analog-to-Digital Converter dedicated to RF WLAN applications. This comparator is based on the switched capacitor network using a two-phase nonoverlapping clock. The offset voltage of the designed comparator has been reduced by means of an active positive feedback. The analyses and simulation results which have been obtained using $0.8\mu m$ CMOS AMS process parameters, with a power supply voltage of 5V and an input common mode of $2-3V$, show that this comparator exhibits a propagation delay of $17.3ns$, a good accuracy and a low power consumption of about $0.8mW$.

I. INTRODUCTION

Low power and high speed Analog to Digital Converters (ADCs) are the main building blocks in the front-end of a radio-frequency receiver in most of the modern telecommunication systems. The ever-growing application of portable devices makes the power consumption a very critical constraint for circuit designers. In terms of power and speed performances, the pipeline architecture is considered as the most interesting one compared to other converter topologies in the telecommunication applications [1], [2], [3], [4] (Fig. 1). This arrangement consists in placing several stages in cascade of low bit resolution per stage, and thus very fast. Typically this resolution is 1.5 or 2.5 bit/stage [5].

The ith stage provides two outputs, the first q_i is a coarse resolution digital representation of the input while the second r_i represents the residual voltage obtained by measuring the difference between the input and the voltage expected by q_i. This voltage will be measured by a gain factor before being sent to the next stage and the code q_i will be sent to the digital error correction. The subsequent

M.B. Guermaz, L. Bouzerara, A. Slimane and M.T. Belaroussi are with the Microelectronics and Nonotechnologies Division, Centre de Développement des Technologies Avancées, cité 20 Août 1956 BP.17, Baba Hassen, 16303, Algiers, Algeria. E-mail: guermaz@hotmail.com

B. Lehouidj is with the University of Sciences and Technology Houari Boumediene, Systems Engineering Laboratory, Algiers, Algeria, E-mail: b_lehouidj@hotmail.com

R. Zirmi is with FGEI of the Univerity Mouloud Maameri of Tizi-Ouzou, Algeria, E-mail: r_zirmi_meln@hotmail.com

stages will try to improve the final code by quantifying each time the residual voltage [2], [6]. All the stages are synchronized by the same clock. Once the first stage has produced q_i and r_i, the second stage will start to quantify r_i while the first one processes the next sample of the input.

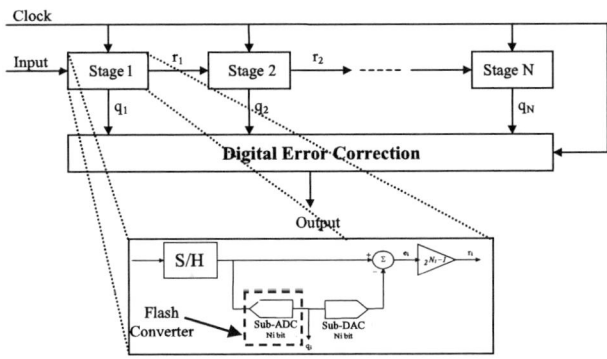

Fig. 1. General pipeline ADC Architecture with its typical stage

These kind of converters display a low power consumption, with a small area of integration [1]. These considerations make their implementation in CMOS technology feasible and more attractive [2].

After the initial latency of the pipeline converter, the output code will be available at each clock cycle. The stages are synchronized alternately with a clock [7]. Under these conditions, each component of a stage should have a settling time which is equal to the half of the clock period [2]. This pipeline converter uses several small Flash converters combined in time with this specific period (Fig. 1) [7].

As the comparator is one of the blocks which limits the speed of the converter, its optimization is of utmost importance. The choice of the technique 1.5bit/stage relaxes considerably the optimization constraints in this design. In these considerations, the Flash sub-converter will have only two comparators. The choice of this latter has been inspired from [2]. This comparator is characterized by its differential configuration and consists of a clocked comparator with a switched capacitor network.

II. CLOCKED COMPARATOR CIRCUIT

The clocked comparator consists of a preamlification

stage followed by a positive feedback stage forming the latch. This approach uses a dynamic latch, contrary to a static latch, in order to reduce the current consumption in the static regime [5].

The Fig. 2 shows the circuit topology of the clocked comparator. According to the decision of the preamplifier stage, the transistors M_4 (M_5) and M_{15} (M_{14}) determine the digital output. This combination allows a good output signal excursion. All the transistors constituting this comparator must be maintained in the saturation region [2]. The transistors M_2, M_3, M_4 and M_5 form the decision circuit. The transistors M_2 and M_3 are connected in diode to maintain the gate voltage of the transistors M_4, M_5, M_6 and M_7 at a constant value.

This arrangement allows to use a buffer, which consists of inverters at the output. This approach enabled us to get a gain lower than $20dB$, which is the maximum gain required for the preamplifier used in such a design, and consequently an increase in the bandwidth. The gain expression of the preamplifier is given by :

$$A_{preamp} = A_0 \frac{g_{m13}}{g_{m13} - g_{mx}} \qquad (1)$$

with

$$A_0 = \frac{1}{\alpha_B} \frac{g_{m1} g_{m15}}{g_{m4} g_{m13}} \qquad (2)$$

where A_0 represents the gain of the preamplifier without the transistors M_x et M_y and α_B is the gain factor due to the substrate effect.

The buffer plays an important role which consists in preloading and memorizing the decision in order to deliver it at the output. This latter is formed by the inverters (M_6, M_{17}) and (M_7, M_{18}) in order to restore the full logic levels. This buffer is controlled by a clock signal PHI1_P ($\overline{LATCH}$) with the switches M_8, M_9 and M_{21}. The output will be locked when the $\overline{LATCH}$ will be at high level. The memorization is carried out at level of the transistors M_{19}, M_{20} and M_{21}, to be delivered when the $\overline{LATCH}$ will be at low level. This particular design has a precharged output that will erase memory from the previous comparison [2].

When the comparator is not active, the output is brought back at high level, this allows to M_{17} and M_{18} to have minimal dimensions [2]. The other transistors of the buffer represent current mirrors to provide a current of $20\mu A$.

This architecture suffers considerably from the problem of the offset voltage, which should be maintained as low as $125mV$, which is the maximum value permitted, corresponding to $V_{ref}/4$ [2], [7]. For this purpose, the

transistors M_x and M_y are added in the preamplifier stage, which are connected in a positive feedback [8], [9], in order to further reduce this offset voltage with an increase in the gain of the preamplifier by maintaining it lower than $20dB$. The offset voltage without the transistors M_x, M_y by considering $f(x_i, x_j) = f(x_i) - f(x_j)$, is given by the following expression:

$$V_{os} \cong \sqrt{\frac{2I_{ds_{12,13}}}{\beta_{0,1}}} \left[\frac{\Delta W_{12,13}}{2W_{12,13}} + \frac{\Delta L_{12,13}}{2L_{12,13}} \right] + \Delta V_{th_{0,1}} - \sqrt{\frac{\beta_{12,13}}{\beta_{0,1}}} \Delta V_{th_{12,13}} \qquad (3)$$

with $$\beta_{i,j} = \mu_{i,j} C_{ox_{i,j}} W_{i,j}/L_{i,j}$$

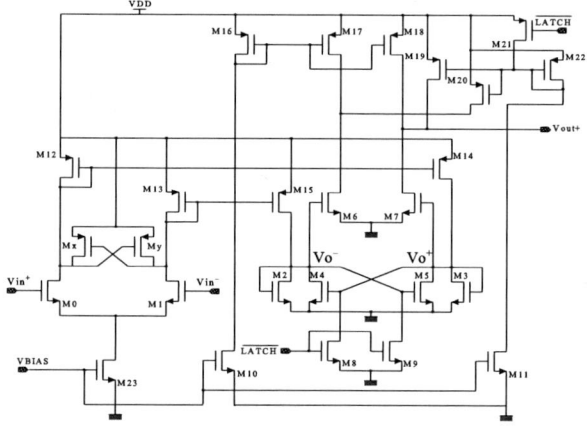

Fig. 2. Clocked comparator circuit

Thus, the offset voltage with the transistors M_x and M_y is given by the same expression, where $f'(x', y', z', ...) = f(x, y, z, ...)$. We can easily notice a decrease of the offset voltage since $I_{ds'_{12,13}} = I_{ds_{12,13}} - I_{ds_{x,y}} < I_{ds_{12,13}}$, and the drain voltages of the transistors M_0 et M_1 have increased, leading to a reduction of $\Delta W_{12,13}$, $\Delta L_{12,13}$, $\Delta V_{th_{0,1}}$ and $\Delta V_{th_{12,13}}$. The dimension ratio of the transistors M_x and M_y is the half with respect to the transistors M_{12} and M_{13}.

The speed of conversion of this comparator is given and determined by its propagation delay. This time is calculated by appraising the capacitors at the output nodes of each stage constituting the circuit. This time depends, in a crucial way, on the dimensions of the transistors M_{14} (M_{15}), M_4 (M_5), M_7 and M_{18} (M_6) and M_{17}, M_{19} as well as on the dimensions of the transistors M_8 and M_9 used as switches. The reduction of this time requires the minimization of the dimensions of these transistors, on the one hand, and a sufficient current to increase the slew rate, on the other hand.

III. DIFFERENTIAL COMPARATOR TOPOLOGY

The sub-ADC of each stage forming the pipeline architecture, contains two differential comparators (Fig. 3). In a 1.5 bit/stage architecture, the sub-ADC has two thresholds at respectively $+V_{ref}/4$ and $-V_{ref}/4$ [2], [6]. The threshold $-V_{ref}/4$ is obtained by inverting the connections $-V_{ref}$ and $+V_{ref}$ for the second comparator with respect to the former. In this design, the two references $-V_{ref}$ and $+V_{ref}$ have been fixed respectively at $2V$ and $3V$. The switched capacitor network operates with the two-phase nonoverlapping clock, using the four signals Φ_1, Φ_2, Φ_1' and Φ_2' (Fig. 4), where t_{nov} represents the time of nonoverlapping and t_{lag} is the time from falling edge of Φ_1' to the falling edge of Φ_1 (latency time).

Four capacitors are used to get the thresholds of the comparator and are sized with a ratio of $1/3$ to divide the reference voltage V_{ref} by 4 (Fig. 3) [6]. This network, which consists of a differential configuration and exploits the bottom plate technique, makes a sample on the capacitor C during the phase Φ_2 whereas the input of the capacitor 3C is short-circuited, thus giving a zero differential. During the phase Φ_1 the input signal V_i is applied to the inputs of the two capacitors then producing a differential voltage which is proportional to ($V_i - V_{ref}/4$) at the input of the clocked comparator. At the end of the phase Φ_1, the latch is locked to make the comparison and to produce the logical level at the output. The capacitor C has been fixed at a minimal value of $50fF$, which is the optimal value for this differential comparator topology.

The conservation of the charge, giving the deduced input appearing at the output of the preamplifier, is ensured by maintaining the nodes of the preamplifier at the same voltage after the addition of the switched capacitor network [7]. As brought earlier, in this differential comparator configuration, the offset voltage has been minimized by adding the transistors M_x and M_y which increase moderately its gain.

IV. SIMULATION RESULTS

The predicted performance is verified by analyses and simulations using PSPICE tool, based on AMS $0.8\mu m$ CMOS process ($V_{thn}=0.78V$, $V_{thp}=-0.83V$ and $T_{ox}=15.5nm$). The designed voltage comparator which is intended to be used in a 10bit 20MHz pipeline analog-to-digital converter dedicated to radio-frequency wireless local area network (RF WLAN) applications, has been optimized for a settling time higher than the double of the required clock period of a frequency of $20MHz$. A design optimization has been carried out, and it has been noticed that an increase in the speed of the comparator can degrade its accuracy characteristic. The output is valid after $17.3ns$, implying the use of a clock at a frequency of $57.8MHz$, which is largely higher than that required. The settling time of the comparator is only about $4ns$.

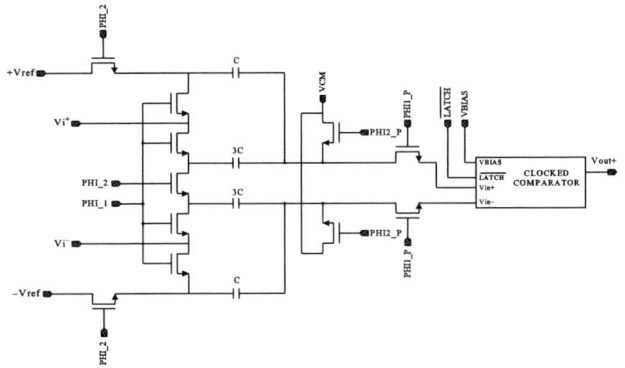

Fig. 3. Differential Comparator

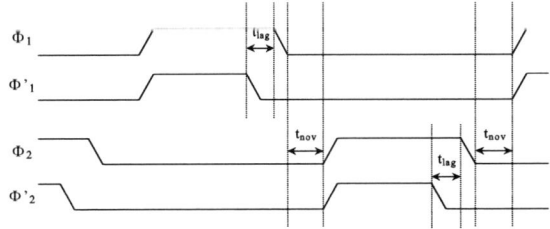

Fig. 4. Responses of two-phase nonoverlapping clock

The clock used in this comparator topology has been designed to work at a frequency of $40MHz$ and this fulfills the synchronization conditions, namely t_{nov} and t_{lag} which are respectively equal to $768ps$ and $320ps$. The designed comparator presents an offset voltage of $77.3mV$ which is lower than the maximum value allowable ($125mV$), in such a comparator design. This is clearly illustrated in Fig. 5, and exhibits a low power consumption of about $0,8mW$. The Fig. 6 depicts the performances of the clocked comparator simulated at a clock frequency of $20MHz$. We can easily notice that when the $\overline{\text{LATCH}}$ is at low level, the output is locked and the decision is made along the phase Φ_2'. In addition, we notice the good accuracy of the comparator to make the decision and maintaining it in spite of the threshold shift from their nominal positions.

After the addition of the switched capacitor network, the performances of the differential comparator have been simulated at the two threshold voltages with a clock operating at a frequency of $40MHz$. The Fig. 7, summarizes these performance, where the mention A represents the characteristics of the differential comparator with the threshold $+V_{ref}/4$ while the mention B represents these characteristics with the threshold $-V_{ref}/4$.

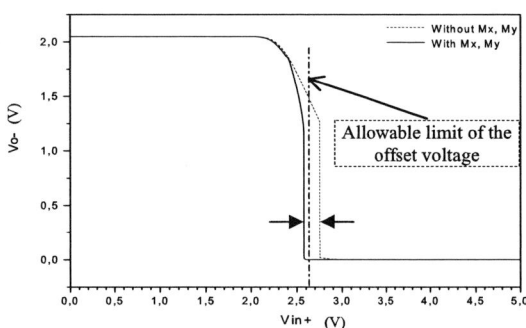

Fig. 5. DC transfer characteristic of the preamplifier

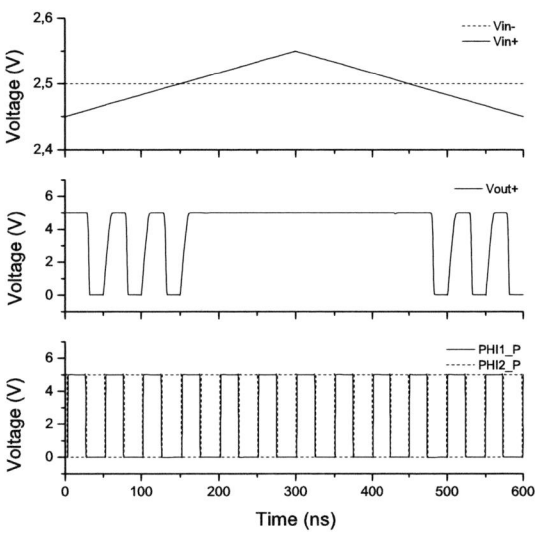

Fig. 6. Performances of the clocked comparator

We notice that this differential comparator exhibits good performances at this clock frequency. Furthermore, we can see that when the $\overline{\text{LATCH}}$ is at low level, the input is locked for the comparison whereas the output is valid on the phase Φ_2'.

V. CONCLUSION

The paper has presented and analyzed a $0.8\mu m$ CMOS clocked comparator. The designed comparator, which is intended to be used in a 10bit 20MHz pipeline analog-to-digital converter, presents promising and good optimized performances in terms of speed and power consumption. The simulation results obtained by considering an input range of $2-3V$, show that this comparator displays an acceptable offset voltage by using

the active positive feedback approach. The main advantage of such a comparator architecture resides at the very low power consumption feature combined with a high speed of conversion.

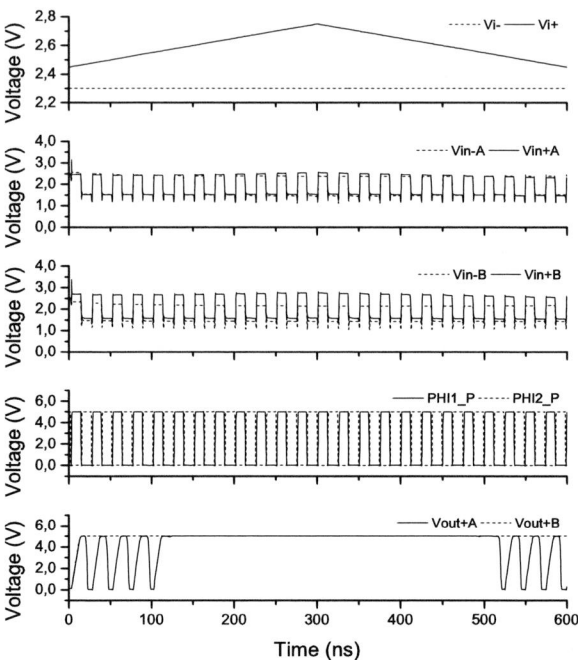

Fig. 7. Performances of the differential comparator

VI. REFERENCES

[1] M. Uster, "Current-Mode Analog-to-Converter for Array Implementation." Doctor of Technical Sciences, Swiss Federal Institute of Technology, Zurich, 2003.

[2] K. Sockalingam, "Error Compensation in Pipeline ADC." Master of Sciences in Electrical Engineering, B.S. University of Maine, 2000.

[3] T. Cho, P.R. Gray, "A 10b 20MS/s, 35mW Pipeline A/D Converters.", IEEE J. Solid-State Circuits, Vol. 30, pp. 166-172, Mar. 1995.

[4] Y. Okaniwa, H. Tamura, M. Kibune, D. Yamasaki, TS Cheung, J. Ogawa, N. Tzartzanis, W.W. walker and T. Kuroda, " A 0.11μm CMOS Clocked Comparator for High-Speed Serial Communication.", Symposium on VLSI Circuits 2004, Digest of Technical Papers, pp. 198-201, June 2004

[5] P. Amaral, J. Goes, N. Paulino and A. Steiger-Garção, "An Improved Low-Voltage Low-Power CMOS Comparator to be Used in High Speed Pipeline ADCs." IEEE International Symposium on Circuits and Systems, ISCAS 2002, pp. V-141-V-144, May 2002.

[6] A.M. Abo, "Design of Reliability of Low-Voltage Switched-Capacitor Circuits.", Thesis of Doctor of Philosophy in Engineering, Electrical Engineering and Computer Sciences, University of California, Berkeley, May 1999.

[7] A.M. Abo and P.R. Gray, "A 1.5V, 10bit, 14.3MS/s CMOS Pipeline Analog-to-Digital Converter.", IEEE Journal of Solid-State Circuits, Vol.34, No.5, pp. 599-606, May 1999.

[8] R. Wang, R. Harjani, "Partial Positive Feedback fot Gain Enhancement of Low Power CMOS OTA's.", Analog Integrated Circuits and Signal Processing, pp. 21-34, 1995, Kluwer Academic Publishers.

[9] L. Bouzerara, M.T. Belaroussi, "Low-Voltage, Low-Power and High Gain CMOS Operational Transconductance Amplifier.", IEEE International Symposium on Circuits and Systems, ISCAS 2002, pp. I-325-I-328, May 2002.

Session
Modeling and Simulation

2006 25th International Conference on Microelectronics

Optimization Issue in Interconnect Analysis

Stefan Holzer and Siegfried Selberherr

Abstract— State-of-the-art semiconductor devices with feature sizes in the deca nanometer regime require extremely shrinked geometries for their interconnect structures. Since various physics-based limits are already reached or, at least, rapidly approached, new materials are considered. The arrangements of these new materials can often not yet be rigorously described due to limited knowledge and limited resources, such as time and money. For practical applications, the uncertainty of material parameters and the limited knowledge of material interactions can be compensated by the introduction of parameterized models to describe the global behaviour sufficiently. In this work, such models are used to calibrate and optimize complete interconnect structures for enhanced applications.

I. INTRODUCTION

New integrated circuits' designs require high quality analysis for their sophisticated interconnect structures as well as for the devices. Therefore, new models and methods need to be developed to allow a sufficiently accurate description of the observed behavior. Interconnect structures have already reached a level of complexity where the behavior of the included materials and their interactions cannot be rigorously described by simple and basic equations because of limited knowledge of fundamental material parameters or by limited time. Therefore, parameterized models are required which allow a sufficient description of the observed behavior with reasonable computational effort in the desired range of interest and within a reasonable amount of time. Using sophisticated simulation and optimization tools new information can be obtained to develop new models and to adjust the parameters by calibration and optimization methods for appropriate needs. To exemplify this procedure we consider a complex fusing structure consisting of several polycrystalline interconnect materials. In the following section, the principle for obtaining uncertain or unknown parameters is presented. Afterwards the simulation models are sketched and the necessary equations are explained, where we introduce the adoptions for our simulation tools to obtain the temperature-dependent equation systems. Further-

more, for the investigated fusing structure a comparison between measurements and simulation results is carried out.

II. OPTIMIZATION AND CALIBRATION TOOLS

State-of-the-art simulation and optimization frameworks [1–4] offer a wide range of optimization strategies and interfaces to various simulation tools. The *Simulation Environment for Semiconductor Technology Analysis* (SIESTA) [4] provides numerous optimizers and interfaces to simulators which can be appropriately chosen for a particular problem. An overview of the data flow in SIESTA for parameter extraction is shown in Fig. 1. The core part for parameter extraction is the optimizer and a tool which compares the simulation result with a reference under certain user-defined constraints in order to calculate a score value for the quality of the currently available simulation result. With the calculated score value the optimizer tries to improve the simulation result by varying the unknown parameters in order to optimize the score value using different strategies. The other parts of the data processing blocks which are located in the upper half of Fig. 1 are related to the evaluation of the models and their parameters required to obtain the corresponding simulation result.

For achieving best results for the optimization of uncertain and unknown parameters, different optimization strategies can be chosen from gradient-based [5, 6], genetic, evolutionary [7, 8], and heuristic approaches [9–11]. If there are good initial guesses available, the gradient-based algorithms work best compared to genetic algorithms in terms of time efficiency.

Stefan Holzer is with the Christian Doppler Laboratory for TCAD in Microelectronics, Faculty of Electrical Engineering and Information Technology, TU Wien, Gußhausstraße 27-29 / E360, 1040 Wien, Austria, E-mail: Holzer@iue.tuwien.ac.at

Siegfried Selberherr is with the Institute for Microelectronics, Faculty of Electrical Engineering and Information Technology, TU Wien, Gußhausstraße 27-29 / E360, 1040 Wien, Austria, E-mail: Selberherr@iue.tuwien.ac.at

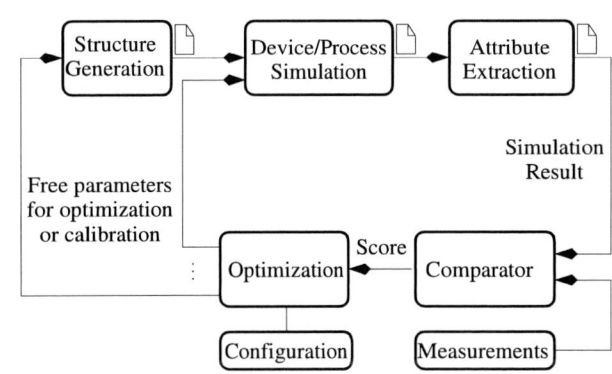

Fig. 1. Data flow for the optimization and calibration of SIESTA.

1-4244-0116-X/06/$20.00 ©2006 IEEE

The blocks in Fig. 1 symbolically depict important parts of typical simulation steps: structure generation which means geometry composition and mesh generation to generate the interconnect device structure, the actual underlying device and process simulation, and the necessary post processing steps to extract attributes which can be compared to measurements or to other reference data.

For our example device the shown steps have to be particularly adopted. The following section presents the corresponding models and mathematical equations, which we selected.

III. Simulation Models

For the simulation of the transient temperature evolution the three-dimensional interconnect simulator STAP [12,13] has been used. STAP calculates Joule self-heating by solving Euler's equation (1) and the heat conduction equation (2) which is coupled with the power loss equation through the heat source term p:

$$\nabla \cdot (\sigma \nabla \varphi) = 0 \tag{1}$$

$$c_\mathrm{p}\, \rho_\mathrm{m}\, \frac{\partial T}{\partial t} - p = \nabla \cdot (\lambda \nabla T) \tag{2}$$

$$p = \sigma\, (\nabla \varphi)^2 \tag{3}$$

Here, φ denotes the electrical scalar potential and σ and λ the electrical and the thermal conductivities. c_p represents the specific heat capacitance, ρ_m the mass density, and T the temperature. The heat source p is assumed to be the Joule power loss [12] due to negligible time changes in carrier concentrations as well as negligible inter-material heating effects [14,15].

For our investigation we have assumed that the material parameters of the thermal and the electrical conductivity are temperature-dependent and follow the polynomial models

$$\sigma(T) = \frac{\sigma_0}{1 + \alpha_\sigma(T - T_0) + \beta_\sigma(T - T_0)^2} \tag{4}$$

$$\lambda(T) = \frac{\lambda_0}{1 + \alpha_\lambda(T - T_0)}, \tag{5}$$

where σ_0 and λ_0 are the conductivities at a certain reference temperature T_0 and α_σ, β_σ, and α_λ are the corresponding first- and second-order temperature coefficients.

The temperature-dependence of the heat capacitance c_p is modeled with the Shomate equation [16]

$$c_\mathrm{p}(\tau) = A + B\tau + C\tau^2 + D\tau^3 + \frac{E}{\tau^2}, \tag{6}$$

where $\tau = T/1000\,\mathrm{K}$ represents the normalized temperature.

As a solution from the transient electro-thermal simulation we obtain the evolution of the resistance of the fusing structure due to the internal temperature distribution.

IV. Example Device

For processing technologies with feature sizes smaller than 350 nm, the fuses made of polycrystalline interconnect materials became an interesting option for programmable memory cells and one-time programmable switches. In technology nodes with larger feature size fusing can cause damage to the passivation layers and it is thus rated as critical [17,18]. Arrangements of integrated fuses can be used as one-time programmable memory blocks in a range of several kilobits. Thus, they provide a cheap and efficient alternative to standard non-volatile programmable memory cells, because

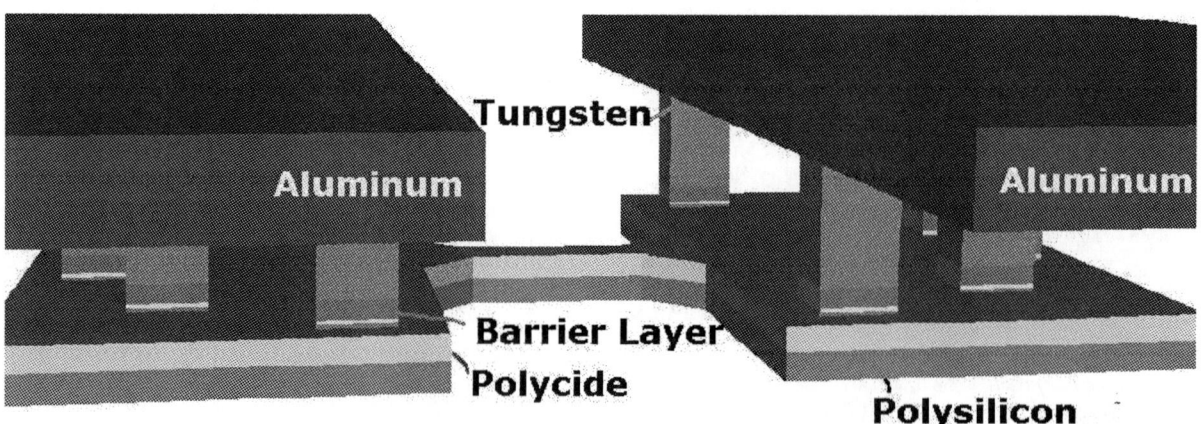

Fig. 2. An overview of the fusing structure showing the variety of different interconnect materials.

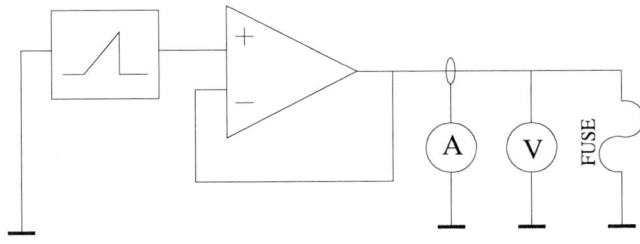

Fig. 3. Schematic of the test circuit for the poly crystalline fuse.

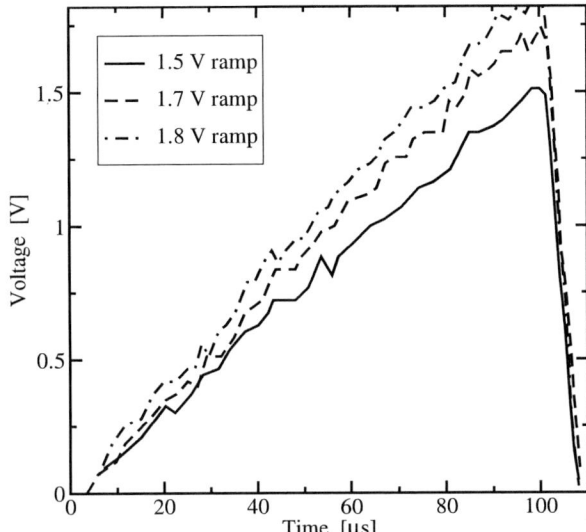

Fig. 4. Different voltage ramps applied to the fuse.

the additional process costs are very low [17]. Moreover, approaches have been reported to increase the memory density by using multi-layered tri-state fuses [19]. Another important application type is to use these fuses in field programmable gate arrays for trimming circuits to obtain a certain analog or digital performance [20]. These fuse applications can be used as elements for trimable resistors and capacitor arrays [21]. Furthermore, the fuses can also act as classic protective elements for critical circuit components [22].

The structure of a typical interconnect fusing device is shown in Fig. 2 which also displays the complex material composition. This fusing structure is used in a surrounding silicon dioxide layer [17].

The dual-layered rod in the region between the two aluminum pads is the actual fusing region which consists of stacked polycide and polysilicon layers. Compared to the polysilicon layer, the polycide layer has a high electrical conductivity. Hence, this layer is intended to be the hottest region and to be removed first.

The fuse is programmed by sending a current pulse through the fuse at an appropriate bias. This performs that the polycide layer is removed by electromigration [18]. Hence, the remaining polycrystalline silicon film is opened due to thermal second-breakdown. The final transition takes place when parts of the polycrystalline silicon layer reach the melting point. The molten silicon is transported from the negative biased side to the positive side through the drift of ions [23].

In terms of power and area consumption, fuses made of available interconnect materials are attractive compared to hybrid technologies [24] with other materials. Moreover, the programming is more time-efficient than that of laser fuses which also require much more additional space on the chip and additional processing steps during the fabrication [25].

However, scaling down to smaller feature sizes requires also a decreased supply voltage [26]. This constraint demands a careful design and a rigorous optimization of the fusing structure to ensure the reliability of the programming mechanism [27] and to minimize the power consumption during the fusing process.

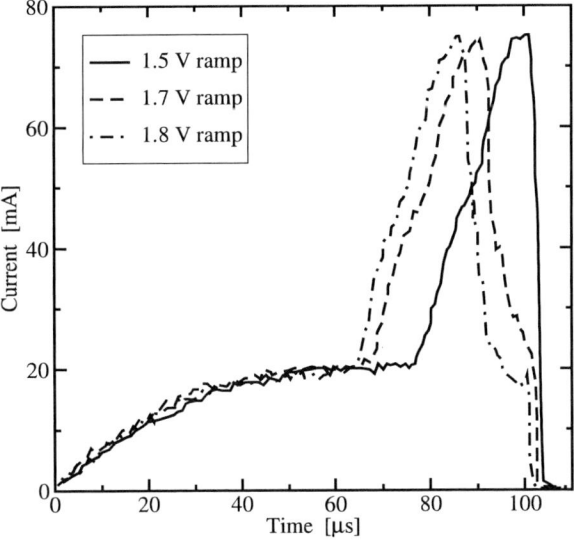

Fig. 5. Measurements of the currents due to the different applied voltages.

Since the fusing mechanism takes place within a couple of 10 ns for a voltage step and several micro seconds for a voltage ramp, direct measurements are hard to obtain [17]. Previous work [28] already brought some new insights into the physics of the fusing mechanism. In addition, an optimization of the fusing structure is required for a fast fusing process and to ensure the reliability. Obtaining these required measurements was only possible through costly experiments using test chips.

We focus on achieving a better insight into the electrical and thermal characteristics of the materials used in the structure shown in Fig. 2. In particular, we are

467

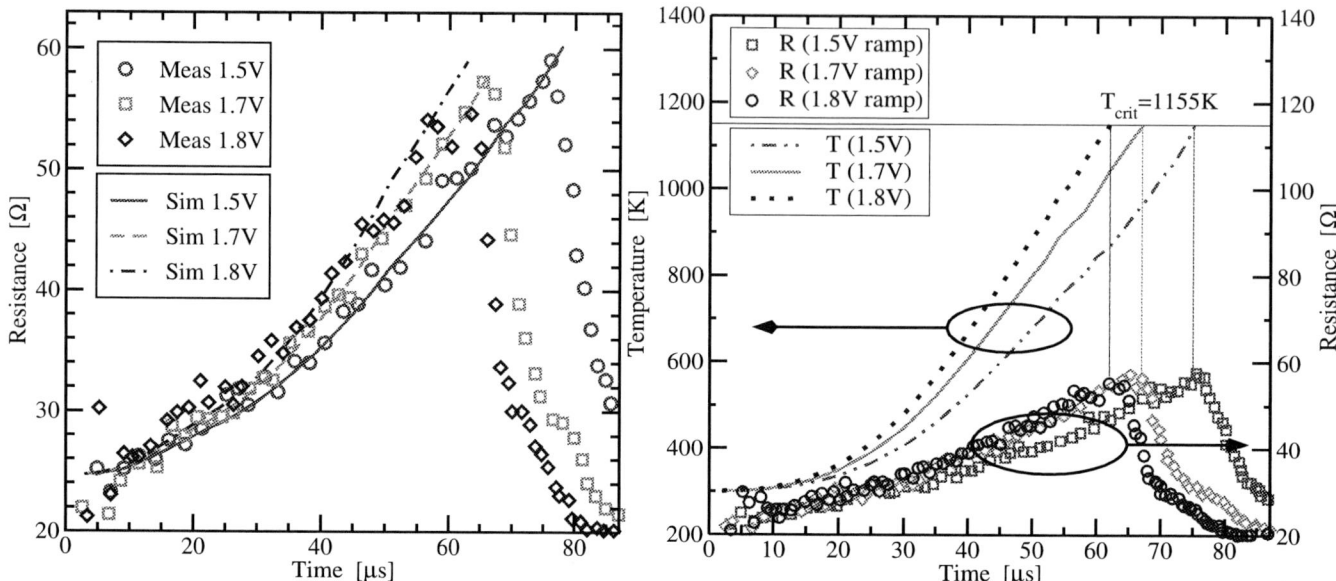

Fig. 6. The simulated fusing resistance compared with the measurements for the different applied voltages.

Fig. 7. Comparison of the simulation results with the measurements.

interested in the temperature dependence of the thermal and the electrical conductivity of the key materials polysilicon and the polycide (WSi$_x$). Better knowledge of these parameters allows to perform a layout optimization for higher reliability of the fusing procedure and a faster fusing operation. With the parameter identification procedure shown in Fig. 1 and the given mathematical models we are able to identify values for certain models within user-defined constraints. In order to obtain reasonable results from the simulation we need accurate information on the test circuit of the fusing device [29] which is depicted in Fig. 3.

Since the measurements of the fusing mechanism have to be carried out within a couple of 10 ns for several μs we prolongated the fusing time by applying a voltage ramp with a rising slope period of 100 μs, which allows to measure the fusing current with reasonable accuracy (cf. Fig. 4 and 5).

These measurement results serve as reference data for the parameter identification procedure. Fig. 5 shows the non-linear behaviour of the fuse current during the heat-up period of the fuse. After a certain time the current jumps to a value which is only constrained by the parasitics of the fuse and the test circuit. Therefore, this point will confine our currently available capability for the simulation and prediction with the introduced models (4) and (5).

For the parameter identification, initial values for the thermal and electrical conductivity of polysilicon and polycide obtained from literature [30–32] were used. The gradient-based optimization algorithm DONLP [5] and the heuristic approach using variants of the simulated annealing approach [10, 33] served well to improve

the initial values. Since the constraints were set appropriately to our particular problem, the convergence of the optimization was quite fast. Hence, we obtained very good agreement with the reference data from the measurements with fairly small computational effort.

V. RESULTS AND DISCUSSION

The temperature coefficients of the electrical and thermal conductivities of polysilicon and polycide have been computed by minimizing the difference between the reference data obtained from the measurements and the simulation results. In order to achieve that in reasonable time, we had to include several consistency checks to ensure that also the intermediate simulation data are physically reasonable.

After the completion of the optimization (minimization) task we obtained excellent agreement with the measurements as shown in Fig. 6 and 7. The different voltage ramps shown (cf. Fig. 4) result in different points of time, where the thermal run-away starts, visible as discontinuities of the currents in Fig. 5. Moreover, reducing the programming voltages may result in lower programming reliability. According to Fig. 5 the corresponding current for the voltage ramp with 1.5 V seems to be the worst case for a successful programming of this fusing structure. Lowering the programming voltage results in structures not completely shortened, because the heat-up of the fusing region is not sufficient to cause the thermal run-away.

Despite of the different applied voltages, and since the measurements were not averaged, the observed discrepancies between measurement and simulation

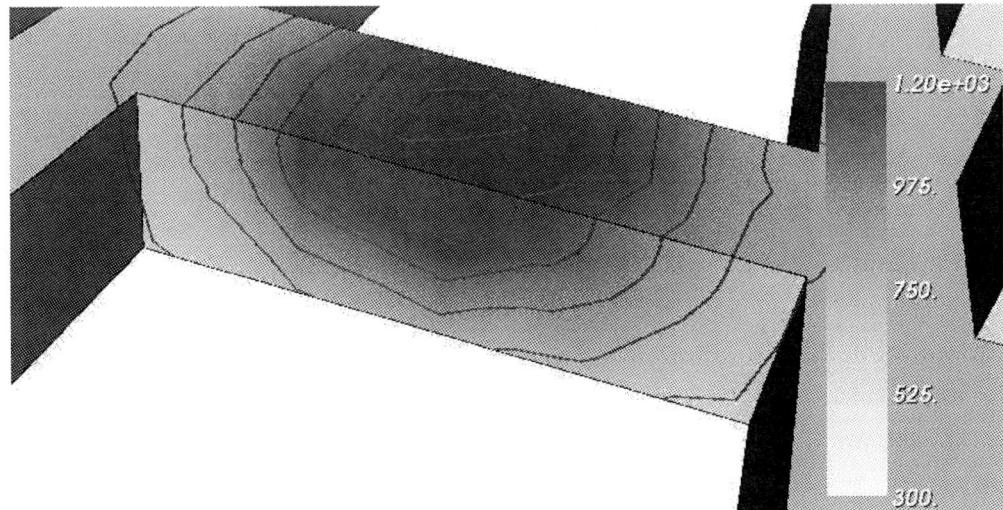

Fig. 8. The temperature distribution [K] at the hottest spot in the fusing area.

in Fig. 6 can be partially traced back to the different measurement methods for obtaining the reference data. However, the overall agreement is still within the accuracy of the simulation results.

The corresponding model coefficients are shown in Tab. I and Tab. II where the extracted and optimized values are compared to data found in the literature [31, 32, 34, 35]. In different technologies, polysilicon is used with different doping concentrations. Therefore, available literature data for the coefficients of the polysilicon model highly depend on the technology used.

Another interesting outcome of our investigation was that the temperature $T_{\text{crit}} = 1155\,\text{K}$, at which the resistance drops, is the same for all applied voltage ramps. Therefore, we can assume that this particular temperature corresponds to a material-specific phenomenon which is related to the thermal run-away due to the starting electromigration process of the polycide layer [18]. When the electromigration process in the polycide region has formed a void, the high conductivity path is disconnected and the whole current flows through the polysilicon layer. Therefore, the polysilicon layer starts to heat up very rapidly towards its melting point. As expected, the area with the highest local temperature is located in the polycide layer of the fusing area in between the two interconnect pads as shown in Fig. 8.

The extracted parameters can be used for investigation of local temperature distributions and self-heating effects in other interconnect structures with similar materials.

VI. Conclusion

We presented an optimization application to obtain important electrical and thermal material parameters of a complex interconnect structure only from electrical measurements. For this purpose, we have tailored our simulation and optimization environment SIESTA [3, 4]

to achieve fast and automatic parameter adaptations following the chosen gradient-based and heuristic optimization strategies.

These thereby identified parameters have been used to describe the programming operation of fusing structures until the electromigration process and melting of the materials take place in the fusing region of the investigated device.

Further investigations of optimizing the geometry of such fusing structures have been recently shown in [18] to increase the reliability of the electromigration processes. In particular, a lifetime study for polyfuses in $0.35\,\mu\text{m}$ CMOS process has been presented in [36], which shows the lifetime drift and yield as functions of the applied programming conditions. The obtained tempera-

TABLE I

EXTRACTED PARAMETERS FOR POLYSILICON COMPARED TO LITERATURE.

Quantity		Poly Si	Literature
σ_0	$[1/\mu\Omega\text{m}]$	0.12	-
α_σ	$[1/\text{K}]$	9.1×10^{-4}	10^{-3}
β_σ	$[1/\text{K}^2]$	7.9×10^{-7}	-
λ_0	$[\text{W}/\text{Km}]$	45.4	40
α_λ	$[1/\text{K}]$	2×10^{-2}	10^{-2}

TABLE II

EXTRACTED PARAMETERS FOR POLYCIDE COMPARED TO LITERATURE.

Quantity		Polycide	Literature
σ_0	$[1/\mu\Omega\text{m}]$	1.25	$0.1 - 18.8$
α_σ	$[1/\text{K}]$	8.9×10^{-4}	$5 - 10 \times 10^{-3}$
β_σ	$[1/\text{K}^2]$	8.1×10^{-7}	3.5×10^{-7}
λ_0	$[\text{W}/\text{Km}]$	119.4	$100 - 179$
α_λ	$[1/\text{K}]$	2.98×10^{-2}	-

ture coefficients of the electrical and thermal conductivities are used to investigate fusing structures for faster and more reliable programming processes under different operational conditions. These are two major design issues for the possible use of such fusing structures in existing semiconductor device applications. The newly obtained information is also used to determine the electrical and thermal behavior of more complex interconnect systems to support thermo-mechanical investigations for stress analysis [37] which can be used for electromigration analysis [38] in critical interconnect components.

REFERENCES

[1] Synopsys, *Taurus Work Bench User Manual*, Synopsys, 2003.

[2] ISE Integrated Systems Engineering, *ISE TCAD Manuals*, Integrated Systems Engineering, 2003.

[3] S. Holzer, A. Sheikoleslami, S. Wagner, C. Heitzinger, T. Grasser, and S. Selberherr, "Optimization and Inverse Modeling for TCAD Applications", in *SNDT 2004, Symposium on Nano Devices Technology 2004*, Hsinchu, Taiwan, May 2004, pp. 113–116.

[4] Institute for Microelectronics, *SIESTA – The Simulation Environment for Semiconductor Technology Analysis, Version 1.1std*, Technische Universität Wien, Austria, 2003.

[5] P. Spellucci, "An SQP Method For General Nonlinear Programs Using Only Equality Constrained Subproblems", *Mathematical Programming*, vol. 82, no. 3, pp. 413–448, 1998.

[6] R. Plasun, *Optimization of VLSI Semiconductor Devices*, Dissertation, Technische Universität Wien, 1999, http://www.iue.tuwien.ac.at.

[7] J. Holland, "Adaption in Natural and Artificial Systems", *University of Michigan Press, Ann Arbor, MI*, 1975.

[8] D. E. Goldberg, *Genetic Algorithms in Search and Optimization*, Addison-Wesley, 1989.

[9] V. Černy, "Thermodynamical Approach to the Traveling Salesman Problem: an Efficient Simulation Algorithm", *J. Opt. Theory Appl.*, vol. 45, pp. 41–45, 1985.

[10] L. Ingber, "Very Fast Simulated Re-Annealing", *Mathematical Computer Modelling*, vol. 12, pp. 967–973, 1989, http://www.ingber.com/asa89_vfsr.ps.gz.

[11] L. Ingber, "Genetic Algorithms and Very Fast Simulated Re-Annealing: A Comparision", *Mathematical and Computer Modelling*, vol. 16, pp. 87–100, 1992, http://www.ingber.com/asa92_saga.ps.gz.

[12] R. Sabelka and S. Selberherr, "A Finite Element Simulator for Three-Dimensional Analysis of Interconnect Structures", *Microelectronics Journal*, vol. 32, pp. 163–171, 2001.

[13] Institute for Microelectronics, *The Smart Analysis Programs*, Technische Universität Wien, Austria, 2003.

[14] G.K. Wachutka, "Rigorous Thermodynamic Treatment of Heat Generation and Conduction in Semiconductor Device Modeling", *IEEE Trans.Computer-Aided Design*, vol. 9, no. 11, pp. 1141–1149, Nov. 1990.

[15] R. Sabelka, *Dreidimensionale Finite Elemente Simulation von Verdrahtungsstrukturen auf Integrierten Schaltungen*, Dissertation, Technische Universität Wien, 2001.

[16] A. Cezairliyan, *Specific heat of solids*, Hemisphere Publishing Corp., 1988.

[17] R. Minixhofer, S. Holzer, C. Heitzinger, J. Fellner, T. Grasser, and S. Selberherr, "Optimization of Electrothermal Material Parameters Using Inverse Modeling", in *Proc. 33rd European Solid-State Device Research Conference (ESSDERC 2003)*, José Franca and Paulo Freitas, Eds., Estoril, Portugal, Sept. 2003, pp. 363–366, IEEE.

[18] W. R. Tonti, J. A. Fifield, J. Higgins, W. H. Guthrie, W. Berry, and C. Narayan, "Reliability and Design Qualification of A Sub-Micron Tungsten Silicide E-Fuse", in

Proc. 42nd Annual Intl. Reliability Physics Symposium, Apr. 2004, pp. 152–156.

[19] A. Doyle, "A Thick Polysilicon Three-State Fuse", *Motorola Technical Developements 3*, pp. 31–32, 1993.

[20] O. Kim, "CMOS Trimming Circuit Based on Polysilicon Fusing", *Electr.Lett.*, vol. 34, no. 4, pp. 355–356, 1998.

[21] D. J. Nickel, "Element Trimmable Fusible Link", *IBM Technical Disclosure Bulletin*, vol. 26, no. 8, pp. 4415, 1984.

[22] J. R. Lloyd and M. R. Polcari, "Polysilicon Fuse", *IBM Technical Disclosure Bulletin*, vol. 24, no. 7A, pp. 3442, 1981.

[23] D. W. Greve, "Programming Mechanism of Polysilicon Resistor Fuses", *IEEE Trans.Electron Devices*, vol. 29, no. 4, pp. 719–724, 1982.

[24] Y. Fukada, S. Kohda, K. Masuda, and Y. Kitano, "A New Fusible-Type Programmable Element Composed of Aluminum and Polysilicon", *IEEE Trans.Electron Devices*, vol. 33, no. 2, pp. 250–253, 1986.

[25] K. Arndt, C. Narayan, A. Brintzinger, W. Guthrie, D. Lachtrupp, J. Mauger, D. Glimmer, S. Lawn, B. Dinkel, and A. Mitwalsky, "Reliability of Laser Activated Metal Fuses in DRAMs", in *Proc. Twenty-Fourth IEEE/CPMT Electronics Manufacturing Technology Symposium*, Oct. 1999, pp. 389–394.

[26] A. Kalnitsky, I. Saadat, A. Bergemont, and P. Francis, "CoSi$_2$ Integrated Fuses on Poly Silicon for Low Voltage 0.18 μm CMOS Applications", in *Proc.IEDM Tech.Dig*, 1999, pp. 765–768.

[27] D. W. Greve, "Programming Mechanism of Polysilicon Fuse Links", *IEEE Trans.Electron Devices*, vol. 17, no. 2, pp. 349–354, 1982.

[28] S. Das and S.K. Lahiri, "Transient Response of Polysilicon Fuse-Links for Programmable Memories and Circuits", in *SPIE's Semiconductor Devices*, May 1996, vol. 2733, pp. 232–234.

[29] S. Holzer, R. Minixhofer, C. Heitzinger, J. Fellner, T. Grasser, and S. Selberherr, "Extraction of Material Parameters Based on Inverse Modeling of Three-Dimensional Interconnect Fusing Structures", *Microelectronics Journal*, vol. 35, no. 10, pp. 805–810, 2004.

[30] W. Harth, *Halbleitertechnologie*, Teubner, 1981.

[31] C. Vahlas, P.-Y. Chevalier, and E. Blanquet, "A Thermodynamic Evaluation of Four Si-M (M = Mo, Ta, Ti, W) Binary Systems", *Computer Coupling of Phase Diagrams and Thermochemistry*, vol. 13, no. 3, pp. 273–292, 1989.

[32] A. D. McConnell, S. Uma, and K. E. Goodson, "Thermal Conductivity of Doped Polysilicon Layers", in *Proc. of the Intl. Conference on Heat Transfer and Transport Phenomena in Microscale Structures*, G. P. Celata *et al.*, Ed., New York, 2000, pp. 413–419, Begell House.

[33] M. M. Ali, A. Torn, and S. Viitanen, "A Direct Search Variant of the Simulated Annealing Algorithm for Global Optimization Involving Continuous Variables", *Computers and Operations Research*, vol. 29, no. 1, pp. 87–102, 2002.

[34] K. Kells, *General Electrothermal Semiconductor Device Simulation*, Hartung-Gorre Verlag, Konstanz, 1994.

[35] D. C. Katsis and J. D. van Wyk, "Experimental Measurements and Simulation of Thermal Performance Due to Aging in Power Semiconductor Devices", in *Proc. Industry Applications Conference, 2002, 37th IAS Annual Meeting*, Pittsburgh, USA, 2002, pp. 1746–1751.

[36] J. Fellner, P. Boesmueller, and H. Reiter, "Lifetime Study for a Poly Fuse in a 0.35μm Polycide CMOS Process", in *Proc. 43rd Annual Intl. Reliability Physics Symposium*, Apr. 2005, pp. 446–449.

[37] C. Hollauer, S. Holzer, H. Ceric, S. Wagner, T. Grasser, and S. Selberherr, "Investigation of Thermo-Mechanical Stress in Modern Interconnect Layouts", in *Sixth International Congress on Thermal Stresses*, Wien, Austria, May 2005, pp. 637–640.

[38] H. Ceric, C. Hollauer, S. Holzer, T. Grasser, and S. Selberherr, "Comprehensive Analysis of Vacancy Dynamics Due to Electromigration", *Proceedings of the 13th European Symposium on Reliability of Electron Devices, Failure Physics and Analysis*, pp. 100–103, June 2005.

2006 25th International Conference on Microelectronics

Advanced Circuit and Device Modeling with Verilog-A

Slobodan Mijalković

Abstract— The hardware description language Verilog-A is today *de facto* the standard language for compact circuit and device modeling and it is implemented in almost all major commercial SPICE-like circuit simulators. The goal of this paper is to present the principles and techniques for implementation of various circuit and device modeling constructs in Verilog-A. Practical examples include a mixed lumped element and network function based modeling of a bipolar transistor as well as a physical, PDE based ,modeling of a PIN diode.

I. Introduction

The last three decades of EDA history are particularly recognized by the development and widespread use of Hardware Description Languages (HDLs). HDLs are offering a whole range of advantages in specification, documentation, design, simulation, formal verification and synthesis of digital and analog systems in comparison to traditional approaches. The large number of proposed HDLs initiated also standardization efforts in this field. It has started with digital HDL formulations and extended latter with analog language constructs. In the course of time, two dominant and competitive standard mixed signal HDL concepts have been created: Verilog-AMS [1] (standardized by Accelera consortium [2]) and Vhdl-AMS [3] (standardized by IEEE [4]). Both HDL standards have quite comparable capabilities and power of expressiveness [5].

Apart from other AMS languages, Verilog-AMS HDL has been developed keeping also its continuous only, or analog only, language subset Verilog-A [6] as the separate language entity. The initial expectations that the Verilog-A language will become obsolete with the widespread utilization of the Verilog-AMS language in simulation tools has appeared to be wrong. Due to its compatibility with the SPICE modeling philosophy, the Verilog-A language has been very early recognized as an indispensable environment for compact model development. While the implementation of full AMS languages has been mainly restricted to single engine mixed-signal simulators, Verilog-A can be easily supported by any SPICE-like simulator. To this end, variety of commercial and open source [7] Verilog-A compilers have been developed. They automatically supply SPICE-like simulators with executable models having almost the same

S. Mijalković is with the Faculty of Electrical Engineering, Mathematics and Computer Science, Delft University of Technology, Mekelweg 4, 2628 CD Delft, The Netherlands, E-mail: slobodan@ieee.org

efficiency as traditionally hand written C-code models. On the other hand, the size of a Verilog-A model source code is typically ten times smaller in comparison to the corresponding C-code implementation. There is continuously growing interest today in writing and distribution compact circuit models exclusively in Verilog-A language. Even the standardization activity has been turned into the direction of proposing various Verilog-A language extensions dedicated to compact modeling [2].

The language constructs of the Verilog-A language are by no means restricted to lumped SPICE-like compact modeling descriptions. Verilog-A also supports various behavioral modeling approaches based on signal flow diagrams, network functions in the frequency domain, noise, look-up tables as well as incorporation of quite general systems of differential-algebraic equations. The latest could originates from the models based on partial differential equations (PDEs). The main objective of this paper is to give an overview of the various advanced circuit and device modeling constructs in the Verilog-A language.

II. Verilog-A Modeling Concepts

A. Nodes, Ports, Branches and Access Functions

The basic structure for the model building in the Verilog-A language is the abstract graph, composed from a set of *nodes*

$$N = \{x_1, x_2, \ldots, x_n\} \qquad (1)$$

and their connectivity defined by a set of *branches*

$$B = \{b_1, b_2, \ldots, b_r\} \qquad (2)$$
$$b_i = \{x_{ik}, x_{il}\} \quad i = 1, r \qquad (3)$$

where $x_{ik}, x_{il} \in N$ and $k \neq l$. Some of the nodes from N are offered for external connectivity and referred to as *ports* while the remaining nodes are internal to the model description.

A system state is defined in Verilog-A at every instance of time in terms of *potential* and *flow* quantities associated with the nodes N and branches B, respectively. The nodes (and the corresponding branches) may belong to different physical disciplines (electrical, thermal, mechanical, etc.) with different meaning of the potential and flow. In the sequel, without losing the generality, we will restrict ourself to electrical discipline having electrical potential at nodes N and current

1-4244-0116-X/06/$20.00 ©2006 IEEE

through the branches B as potential and flow quantities of the system.

The numerical values of the potential differences and currents can be accessed by *potential and flow access functions* $V(\cdot)$ and $I(\cdot)$ using nodes and branches as their arguments. Applied to a single internal node, the access functions return the node potential and current with respect to the ground node while for a conservative port they return the port potential and current in the external port branch. The non-conservative ports are limited to signal flow accessible by $V()$ function.

B. Conservative Modeling

The conservative modeling is imposed in Verilog-A by the requirement that the sum of the currents leaving any node must be equal to zero at any instance of time. In that respect Verilog-A fully resembles the electric circuit *Kirchhoff's Current Law* (KCL).

Let the first $p \leq n$ nodes in N represent port nodes being all conservative. Given the vector of currents in the external port branches at time t as

$$I_P(t) = (I(x_1), I(x_2) \ldots, I(x_p))^T,\qquad(4)$$

the vector of the branch currents

$$I_B(t) = (I(b_1), I(b_2) \ldots, I(b_r))^T\qquad(5)$$

has to satisfy at any instance of time satisfy KCL equation

$$A^T \cdot I_B(t) + \begin{pmatrix} I_P(t) \\ 0 \end{pmatrix} = 0\qquad(6)$$

where $A \in \mathbb{N}^{r \times n}$ represent the branch incident matrix having nonzero elements $A(i,k) = 1$ and $A(i,l) = -1$ if $b_i = \{x_{ik}, x_{il}\}$ for $k, l = 1, n$ and $i = 1, r$.

Since the conservation KCL equations are implicitly imposed In Verilog-A, the model is practically defined specifying for each branch b_i, $i = 1, r$ the potential or current branch constitutive equation

$$V(b_i) | I(b_i) <+ f(V_B | I_B(t), P)\qquad(7)$$

where

$$V_B | I_B(t) = (V(b_1) | I(b_1), V(b_2) | I(b_2), \ldots, V(b_r) | V(b_r))^T$$

is the vector of branch potentials (or currents) and P is the vector of static model parameters. The contribution operator $<+$ is summing multiple contributions to the branch current (or voltage) as a function of all branch voltages (or branch currents). The branch constitutive equations could be implicit if the left hand side branch potential or current appears also on the right-hand side of the contribution statement. The function f in the assign statement (7) is practically realized in Verilog-A as the sequence of mathematical expressions given in a form of a programming code similar to C-language.

Equations containing time derivatives (and time integrals) are also supported, so that a system of ordinary differential and algebraic equations (DAE) can be easily specified in Verilog-A language.

C. Signal Flow Modeling

Verilog-A provides also possibility for the creation of control block diagrams similar to those used in the Matlab/Simulink (Scilab/Scicos) environments. It is based on non-conservative ports (nodes) having only potential nature. The current flow in the non-conservative port branches is not defined.

The signal flow models may be written as a pure input-output control diagrams where the potential at the output ports are purely function of the potential at the input ports without taking the flow into account. It is also possible to mix conservative and non-conservative ports. An example is introduction of control current or voltage sources into a circuit.

D. Transfer Function Modeling

The Linear Time-Invariant (LTI) models are often specified in a form of continuous transfer functions

$$H(s) = \frac{\sum\limits_{k=0}^{M} n_k s^k}{\sum\limits_{k=0}^{N} d_k s^k}\qquad(8)$$

where s is complex frequency while $n_i, i = 1, M$ and $d_i, i = 1, M$ are constant real numbers. Although Verilog-A language does not provide tools for the transformation between the time and frequency domains, the branches and complete modules described by the transfer functions are equally well operational in both domains. The network transfer function modeling construct can be specified in Verilog-A language using the *Laplace transform filter constructs* [2] offering besides the rational polynomial formulation (8) also the equivalent pole-zero formulation.

E. Noise Modeling

The behavior of the circuit is often analyzed under small-signal stimulus. In that case it also important to analyze the impact of the device-generated noise in a circuit or the circuit sensitivity to input noise. To this end, Verilog-A offers stimulus functions for the white noise, the stochastic processes whose spectral density does not depend on frequency, for the flicker noise and also the possibility to define the spectral noise density as piecewise linear function of frequency. Each of the noise functions is uncorrelated with the noise generated by other functions. Sharing the output of the single noise function could be used to introduce perfectly or partially correlated noise in the model description.

The noise stimulus functions take the expressions of the spectral noise power as the input argument. The expression is evaluated at each frequency point in the frequency sweep during frequency simulation. It is not required for the spectral power to be static and it can refer to values of signals, variables and parameters in the analog model description. During computations in time domain all noise source quantities are constrained to be zero.

F. Look-up Table Modeling

The look-up table based models describe the behavior of a system by interpolating between data points that are samples of that systems behavior. To build such a model, the user provides a set of N sample behavior points $(x_{i1}, x_{i2}, \ldots, x_{in}, y_i)$, $i = 1, N$ so that

$$g(x_{i1}, x_{i2}, \ldots, x_{in}) = y_i \qquad (9)$$

where g is the model function. Using interpolation techniques, one can get the model value at any point in the domain of the sample points approximately. In the case when the evaluated point is outside the domain of the sample points, extrapolation techniques can be used to calculate the model value at that point. However, extrapolation is usually unreliable when the evaluated point is far from the domain of the sample points. The Verilog-A provides at the moment only piecewise-linear interpolation for the look-up table models using either internal vectors or external files as data source.

III. EXAMPLES

A. BJT with Distributed Substrate Network

The existing bipolar junction transistor (BJT) compact models have either no or oversimplified representation of the substrate coupling effects. It could be a problem in modeling devices operating at RF/Microwave frequencies when the semiconductor region between the intrinsic substrate-collector junction and the substrate contact act as a distributed network with significant influence on overall transistor performance.

A solution could be a mixed modeling concept shown in Fig. 1 that combines a compact BJT model (Mextram 504) with four ports ($CBES_i$), where S_i denotes internal substrate port, and transfer function block $H(s)$ that represent distributed substrate coupling network. The BJT model parameters has been extracted in the standard way from measured CV, DC and fT characteristics of SiGe HBT test device. On the other hand, the transfer function coefficients are identified from the S-parameter off-state measurements of the substrate impedance. The order of the transfer function is $M = N = 6$ [8].

The effect of the substrate on the BJT characteristics is best visible in the frequency dependence of the transistor Y_{22} parameter shown in Fig. 2 [8]. It should

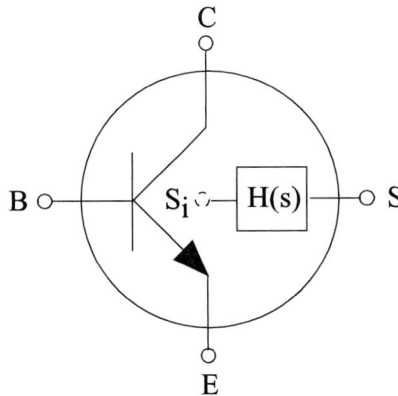

Fig. 1. Mixed compact and transfer function bipolar transistor modeling approach

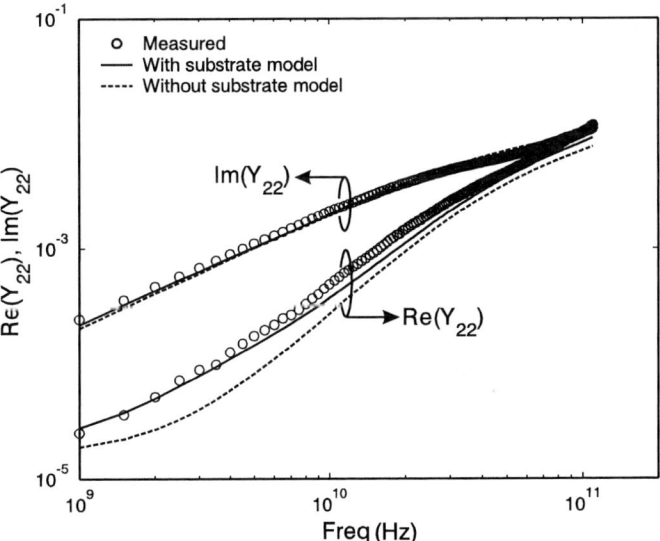

Fig. 2. Measured and simulated Y_{22} parameter of the bipolar transistor with and without substrate coupling transfer function modeling.

be noted that the mixed compact and transfer function bipolar transistor significantly improves the accuracy in modeling experimental data.

B. PDE Based Modeling

The behavioral and compact modeling approach is often insufficient to describe all necessary detail included in the modern semiconductor devices. In that case it is essential to directly employ physical models based on the underlying partial differential equations (PDEs). It requires to represent a discretized PDE in a form of the Verilog-A node-branch graph.

As an example of handling differential-algebraic models originating from discretization of PDEs, let us consider a distributed model of the PIN diode shown in Fig. 3. The one-dimensional grid with $N = 100$ space points is introduced as basis for discrete representation of the electric potential and concentration of electrons

473

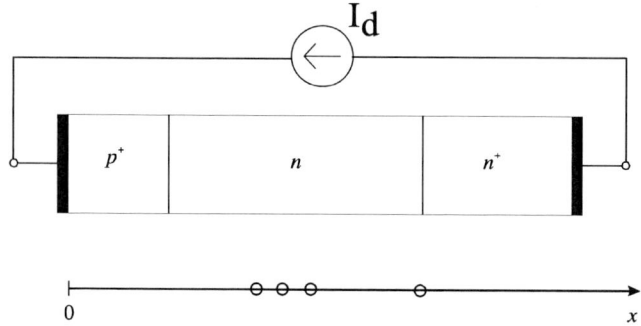

Fig. 3. The PIN diode structure

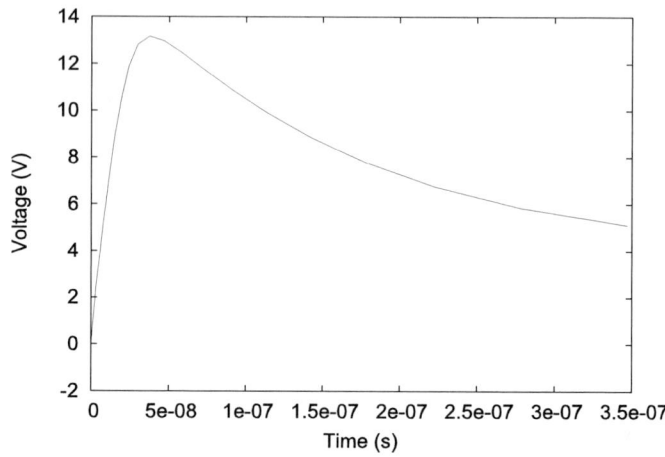

Fig. 5. The time response of the PIN diode voltage to linearly ramped forced diode current

and holes. The governing semiconductor equations are discretized using finite-volume method. The single discretization cell is implemented in Verilog-A as a node-branch structure shown in Fig. 4, where I_n and I_p are

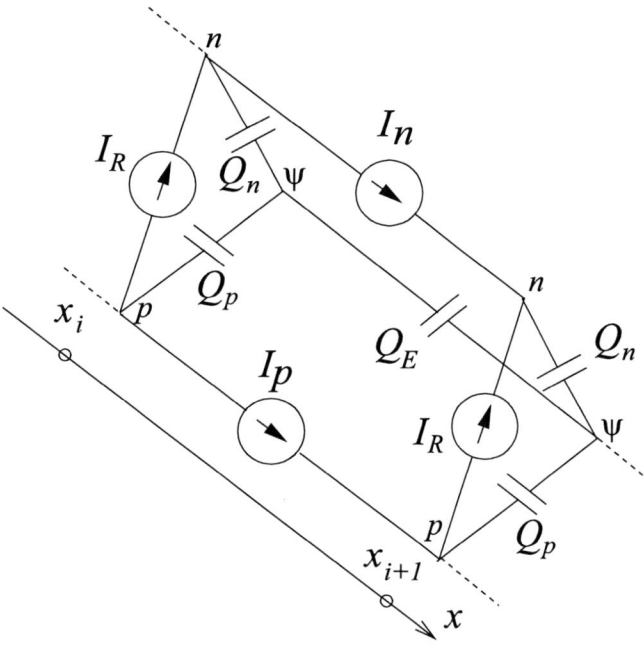

Fig. 4. The node-branch representation of the semiconductor equations

electron and hole current sources, $Q_E = \epsilon E$ is displacement charge, E is electric field, while Q_n and Q_p represent the diffusion charges.

Figure 5 shows the time dependence of the voltage across the PIN diode in response to the forced ramped current. The PIN diode voltage exhibits a characteristic overshoot phenomena.

IV. SUMMARY

Verilog-A is a subset of the standard mixed signal hardware description language Verilog-AMS supplying its continuous signal functionality. However, Verilog-A has found also its own way of existence as a standard language for compact modeling in SPICE-like cir-

cuit simulators. Today, Verilog-A is implemented in almost all major commercial circuit simulators and has its own influence on the Verilog-AMS standardization procedure.

The Verilog-A language modeling concepts are quite general and provide framework for implementation of mixed time and frequency domain modeling approaches as well as incorporation of quite general differential algebraic equations originating from the discretization of PDE-based models. With this features Verilog-A language represents an excellent environment for rapid development and verification of the compact and behavioral modeling ideas in the commercial circuit simulator. Two practical simulation examples demonstrate combined compact and network function modeling of the bipolar transistor operation at ultra-high frequencies and PDE based model of a PIN diode.

REFERENCES

[1] K.S. Kunderth and O. Zinke, *The Designer's Guide to Verilog-AMS*, Boston: Kluwer, 2004.
[2] Accelera [Online]. Available: http://www.acellera.org
[3] P.J. Ashenden, G.D. Peterson and D.A. Teegarden, *The System Designer's Guide to VHDL-AMS*, San Francisco: Morgan Kaufmann Publishers, 2003.
[4] 1076.1-1999 *IEEE Standard VHDL Analog and Mixed-Signal Extensions Language Reference Manual.*
[5] F. Pecheux, C. Laaaement, and A. Vachoux, "VHDL-AMS and Verilog-AMS as Alternative Hardware Desription Languages for Efficient Modeling of Multidiscipline Systems," *IEEE Tran. Computer-Aided Design of Integrated Circuit and Systems*, Vol. 24, No. 2, Feb., 2005.
[6] D. Fitzpatrick and I. Miller, *Analog Behavioral Modeling with the Verilog-A Language*, Norwel, MA: Kluwer, 1998.
[7] L. Lemaitre and C. NcAndrew, "An Open-Source Software Tool for Compact Modeling Applications," *IEEE Circuit and Device Magazine*, pp. 6-9, March/April, 2005.
[8] H-C. Wu, S. Mijalković, J. G. Macias, J. Burghartz, "Mixed Compact and Behavior Modeling Using AHDL Verilog-A," *Proc. 2003 International Behavioraral Modeling and Simulation Workshop (BMAS 2003)*, San Jose (CA), pp. 139–143, 2003.

2006 25th International Conference on Microelectronics

Process and Device Simulation with a Generic Scientific Simulation Environment

Michael Spevak, René Heinzl, Philipp Schwaha, Tibor Grasser

Abstract—**We present a high performance environment for scientific simulation applications (GSSE). This environment is based on the three orthogonal concepts of topology, geometry, and quantities. Lambda calculus is used in order to assemble various partial differential equations for TCAD and other physical equations. We present examples from device and process simulation which show the applicability of our environment. Despite the high abstraction level we can archieve high performance.**

I. INTRODUCTION

A general environment for TCAD equations has to provide methods for the solution of different physical phenomena such as carrier transport, diffusion, electromagnetic wave propagation, heat transfer, mechanical deformation, fluid flow, and quantum effects. Due to the wide range of applications it is not trivial to develop an environment which is capable of handling all equations within a homogenous environment. In the field of TCAD coupled partial differential equations and multi quantity equations are often employed. For the solution of these equations we use discretization schemes such as the finite element method, the finite volume method, the finite difference method or the boundary element method.

Each of these schemes has its merits and shortcomings and is therefore more or less suited for different classes of equations. All of these methods have in common that they require a proper tessellation and adaption of the simulation domain [1], [2]. Due to the diversity of the mathematical structures themselves, combined with efficiency considerations, in particular in three dimensions, the development of simulation software is quite difficult. Supporting libraries for numerical simulations exist, but no complete environment for scientific computing to tackle the following issues:
- Support for different geometries and topologies
- Complete and tested discretization schemes
- Support for mathematical modeling
- High performance calculation
- (Real-time) visualization

Hence, the simulation tools are typically written by experts specialized in other fields. In the extreme case all

M. Spevak is with the Institute for Microelectronics, Technical University Vienna, E-mail: spevak@iue.tuwien.ac.at

R. Heinzl, P. Schwaha, and T.Grasser are with the Christian Doppler Laboratory for TCAD in Microelectronics at the Institute for Microelectronics, Technical University Vienna, E-mail: {heinzl|schwaha|grasser}@iue.tuwien.ac.at

areas of simulator development, like software design, programming, testing, and evaluation are done by one person. In the last decade our institute has developed different simulators and libraries, like SAP [3], Wafer-State-Server [4], and Minimos-NT [5]. However, none of them has shown to be perfect for the rapid progress in scientific software development. Even the reuse of simple code parts is difficult, due to the non-generic library approach.

Therefore a scientific environment with high flexibility and adaptivity of meshes combined with great flexibility in numerical treatment and discretization schemes in all dimensions is called for. It should be possible to use a common code base which is easily adaptable to special requirements but does not require specialized features for different discretization schemes such as element matrices like many other specialized finite element simulation environments.

On that account, we have extracted the main concepts from all of our simulation tools and developed the generic and lightweight environment GSSE which suits scientific requirements. On the one side, generic library means that each part of GSSE can be used separately. The complete GSSE is based on header files only and therefore the required mechanisms can be included without incurring additional dependencies. On the other hand, generic means that all data types are parameterized and can be exchanged easily, for instance the numerical data type for quantity storage.

GSSE is designed for rapid development of simulation software. One of the most important facts is that errors can be easily found and even prevented. As errors are often not obvious to detect it may already take a lot of experience to decide if the result from a simulation is erroneous or not. If the result is not correct, it might be reduced to a programming bug, a logical error in the program flow, or a badly chosen parameter. Due to this reason it is necessary to locate an error quickly. Each of the data structures which we provide can itself be tested before compound data structures are tested. Thus the development effort for the final code can be reduced enormously.

In the following we will present some examples of our simulations from the field of semiconductor device and process simulation. First we solve a simple device simulation example. Then we benchmark a finite element example with the electromagnetic simulation tool SAP.

1-4244-0116-X/06/$20.00 ©2006 IEEE

Both examples will show that our environment can perform fast calculations.

II. THE GENERIC SCIENTIFIC SIMULATION ENVIRONMENT

Due to the large variety of available models and differential equations, discretization schemes, and simulation domains we had to extract the base concepts of a simulation environment. The main aspect of software design is orthogonality as well as modularity. Each component should be usable without any dependences to another. If a higher structure combines two base structures (e.g. quantities on a topology) it does not depend on the lower structures but it only relies on the concept.

Topology. Within a scientific simulation environment it is crucial to have neighborhood information of vertices, edges, faces, and cells available within a constant time. For this reason we have implemented a data structure which covers only topological information of vertex cell incidence and cell vertex incidence. Even though storing one of these incidence functions is redundant it is necessary to guarantee constant time for traversal. Based on this incidence information we generate inter-dimensional objects such as edges and faces. The incidence information of edges and faces does not need to be stored explicitly because it can be derived from the base traversal functions and the archetype information.

The archetype concept implies that each cell of the tessellation has the same topological shape or very few different shapes. Therefore we need not store all edge and face information but we can derive it from an archetype. For this reason the unstructured topology is highly flexible. We can use archetypes of different dimensions and shapes.

Geometry. Even though we have a convenient method for describing the topology of a simulation domain this does not imply that we store the coordinate information on the vertices and cells. Topology tells us for instance if two vertices are connected by an edge. It however does not provide any information about the real geometry of the curve. This is the task of the geometry concept. The basic geometry concept is the point list. The point list contains the geometrical point coordinates associated to the topological vertices. From this information we can derive the geometry of all edges, faces and cells. We can perform orientation tests, volume measurements, and the calculation of the voronoi information.

Quantity. Quantities are numerical attributes which can be stored on all topological elements using their handles. Using an associative storage concept we can store values with respect to an associative key of the topology and a quantity descriptor key (which might be a string). The value types of the quantities can be chosen differently. The simplest case is a scalar floating point value. We also provide vector and tensor quantities as well as string quantities. The quantity data type can be parameterized on the data type so that it is possible to use any type as data type for the quantity.

Based on these core concepts, a mathematical function layer was implemented to provide easy development of all different modeling issues on the one side and a high performance computing on the other side. This layer includes all necessary functions for a convenient numerical analysis as well as accumulation functions which will be discussed later on in the example section. We even have the possibility to work with linearized functions which provide direct access to the system matrix for line-wise entry as well as finite element stencils.

As we have parameterized data types for numerical calculations it is possible to introduce abstract matrix data types with lazy evaluation concepts which can reduce the execution time as well as the number of temporary objects. For this reason we use expression templates [6], [7] as well as lambda expressions [8].

A solver interface is integrated in this environment for the use of all different kinds of state-of-the-art solvers called TRILINOS [9]. For the important visualization purpose within scientific computing, IBM's data explorer [10] is integrated with a few modifications to make a real-time visualization possible.

III. DEVICE SIMULATION

The field of device simulation requires the use of many different numerical techniques. Macroscopic transport [11] models are among the most widely employed calculations. These models can be derived by applying the moment method on the Boltzmann equation. Together with the Poisson equation they form a system of partial differential equations which are capable of describing the behavior of semiconductor devices. We use the simplest macroscopic model, the drift diffusion transport equations. In the following we show how the equations can be discretized in our environment by the means of the finite volume method. The discretization formula for the Poisson equation yields

$$\sum_{\text{edge vertex}} (\Psi_j - \Psi_i) \frac{A}{d} = V (n - p + N_{\text{A}} - N_{\text{D}}) q \quad (1)$$

Based on this discretization formula we obtain the final discretization routine (1) we can write the following statement in the GSSE.

```
eqn = (sum<vertex_edge>
[diff<edge_vertex>(0.0)[potential] * A / d]
 - q * volume * (n - p + N_A - N_D))(v);
```

The same assembly routine can be applied to the current relations using the Scharfetter-Gummel discretization [12] (generation and recombination rates are omitted here).

$$J_w = \frac{1}{\Lambda} \left(n_j \mathcal{B}\left(\Lambda\right) - n_i \mathcal{B}\left(-\Lambda\right) \right) , \qquad (2)$$

where $\mathcal{B}$ is the Bernoulli function. Using the method of finite volumes we have discretize the current integral on the boundary of the control volumes. If we consider the stationary case as well as zero recombination we obtain the following code snippet for our discretization scheme.

```
eqn = sum<vertex_edge>(0.0)
 [area / distance * diff<edge_vertex>()
 [n * bern(d_psi/u_th)]
 [n * bern(-d_psi/u_th)]]
(v);
```

Both discretization terms need quantities which are located on topological elements (Fig. 1). The potential and charge terms as well as the box volume are stored on the vertices. The distance (d) and area (a) as well as the potential difference d_ψ are stored on the edges. The sum as well as the difference operations change the locality. The expression within the brackets gives the kind of traversal; `vertex_edge` iterates over all edges which cover a vertex, whereas `edge_vertex` iterates over all vertices which are covered by an edge. The formula in the square brackets are evaluated on all elements of the traversal and accumulated. `bern` denotes the Bernoulli function $\mathcal{B}$ which is used in the Scharfetter-Gummel current relation. From this specification the Jacobian

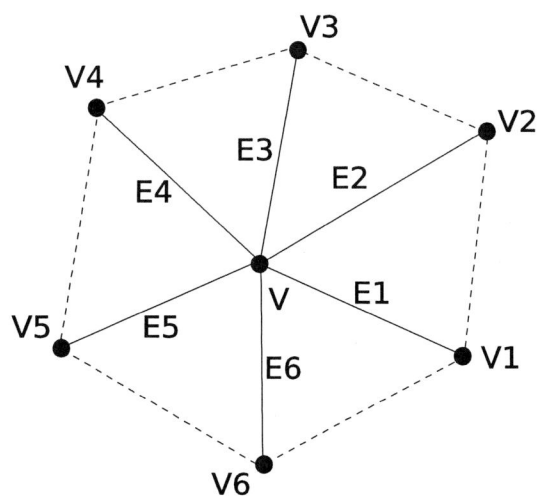

Fig. 1. The local patch of a vertex. We outlined the vertices as well as the edges.

matrix and the right hand side are assembled automatically. An automatic derivation of the linearized function, which is very error prone if done by hand, can be achieved by using an associative data structure. This data structure contains matrix entries as well as a

right hand side value. For structures which are called linearized equations, we have implemented basic operations such as addition and multiplication as well as other numerical functions such as the already demonstrated Bernoulli function. This method can be used in order to assemble linear as well as nonlinear equations which include discrete couplings between single topological elements. The complete application does not need more than 100 lines of code, the core is only about 25 of these lines.

IV. PROCESS SIMULATION

In general, three-dimensional process simulation steps need special surface treatment and must provide the ability to handle surface elements of arbitrary complexity containing degenerated or even faulty elements.

For the solution of problems in process simulation as well as interconnect simulation finite element methods are commonly used. In the following example we apply the finite element method to the Laplace equation in order to calculate capacitances as well as resistances of interconnect structures. In the following we will calculate two simple structures in order to evaluate the correctness of our simulation and to show the performance of the environment. The example is a single interconnect line. Even though it is very simple it shows the applicability of the GSSE as well as the performance.

V. RUNTIME EFFICIENCY

To compare the runtime efficiency of the generic scientific environment we used the fastest (Poisson only) simulation tools from our institute (SAP) and run different application benchmarks with an automatic benchmark system.

As our environment is on a high semantic level and also does not impose a high abstraction penalty it is easy to make special optimizations if simpler partial differential equations have to be discretized:

- There is only one solution quantity
- We only apply one kind of differential equation

Even though these conditions are not always met we can gain performance if any of them is fulfilled. The genericity of our environment allows us to use these simplifications for all kinds of differential equations.

For the testing of the finite element code we divide the simulation time into the following parts. Each of the parts is measured independently in order to show the differences.

- Preparation (I/O, mamory allocation)
- Assembly (Jacobian matrix and RHS)
- Solution

Our performance test example has to be very simple because we have to eliminate implication which result from problems with complicated geometries. Our test

477

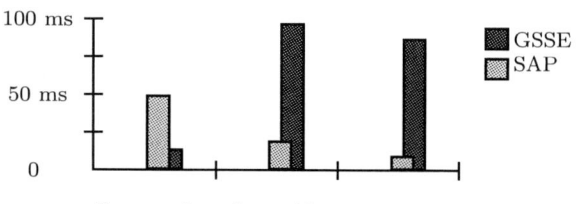

Fig. 2. Structure with 8.700 tetrahedra and 2.300 vertices.

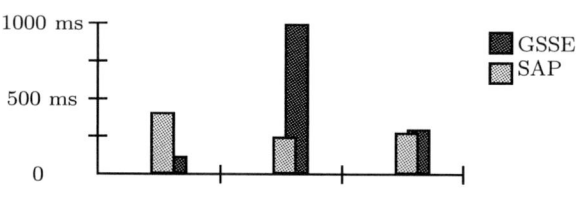

Fig. 3. Structure with 64.000 tetrahedra and 14.000 vertices.

sample is a simple via line with a length of 10μm and a square cross-section of 1μm^2. For the test case we apply a potential difference of 1V. For both examples we use meshes which have been generated with **vgmodeler** [1] with 8.700, 14.000 and 97.000 tetrahedra. The comparison of the run time on an Intel Pentium 4 with 2.80GHz shows the following results (Fig. 2, Fig. 3, Fig. 4):

The benchmarks (Fig. 2, Fig. 3, Fig. 4) show that the GSSE does not have a high time consumption for the preparation of the quantities. However in assembly time as well as in solution time the highly optimized code is faster but within the same order of magnitude. The solution time shows that the TRILINOS solver package is well designed for large matrices.

VI. CONCLUSION

A generic environment for scientific computing has been presented. It can handle a large variety of differential equations which can be specified with different discretization schemes such as finite elements, finite volumes and finite differences. We have shown that a high semantic level does not necessarily imply an abstraction penalty so that the performance is comparable to highly optimized programs.

Even though we have shown the applicability of our environment on very simple structures it is possible to extend the features very conveniently. First the simulation domain can be taken from any meshing output. The partial differential equations can be extended to more complex models using a C++ embedded language.

Using this environment it is possible for scientists to formulate PDE problems with a full topological and geometrical support for the development of applications with minimal in-depth knowledge of internal data structures.

REFERENCES

[1] R. Heinzl and T.Grasser, in *International Conference on the Simulation of Semiconductor Processes and Devices (SIS-PAD), Kobe* (2005), pp. 211–214.

[2] P. Schwaha, R. Heinzl, M. Spevak, and T.Grasser, in *International Conference on the Simulation of Semiconductor Processes and Devices (SISPAD), Kobe* (2005), pp. 235–238.

[3] R. Sabelka and S. Selberherr, in *Proc. Intl. Interconnect Technology Conference* (1998), pp. 250–252.

[4] A. Hössinger, R. Minixhofer, and S. Selberherr, in *International Conference on the Simulation of Semiconductor Processes and Devices (SISPAD)* (), pp. 129–132.

[5] IμE, *MINIMOS-NT 2.0 User's Guide*, Institut für Mikroelektronik, Technische Universität Wien, Austria, 2002, http://www.iue.tuwien.ac.at/software/minimos-nt.

[6] T. L. Veldhuizen, in *Proceedings of PEPM'99, The ACM SIGPLAN Workshop on Partial Evaluation and Semantics-Based Program Manipulation* (University of Aarhus, Dept. of Computer Science, 1999), pp. 13–18.

[7] J. G. Siek and A. Lumsdaine, in *ECOOP Workshops* (1998), pp. 466–467.

[8] in *Lambda-Calculus and Computer Science Theory*, Vol. 37 of *Lecture Notes in Computer Science*, edited by C. Böhm (Springer, 1975).

[9] M. A. Heroux *et al.*, ACM Transactions on Mathematical Software , for TOMS special issue on the ACTS Collection.

[10] *IBM visualization Data Explorer*, 3rd ed., 1993.

[11] T. Grasser, T. Tang, H. Kosina, and S. Selberherr, Proceedings of the IEEE **91**, 251 (2003).

[12] D. Scharfetter and H. Gummel, IEEE Transaction on Electron Devices **16**, 64 (1969).

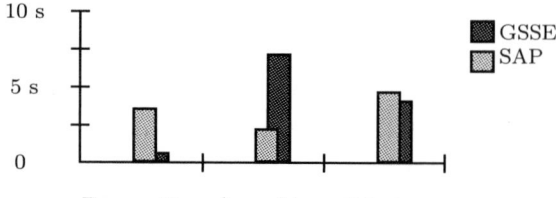

Fig. 4. Structure with 520.000 tetrahedra and 97.000 vertices.

2006 25th International Conference on Microelectronics

Design and Simulation of Thick Film Thermistors Using Commercial Simulation Tools

Viktor Marić, *Member, IEEE*, Obrad Aleksić, and Ljiljana Živanov, *Member, IEEE*

Abstract — Methods for predicting and analyzing the behavior of NTC thermistors have remained the same and they typically include accurate measurements of resistance and temperature, followed by curve-fitting methodologies for determining the relationship of temperature to resistance for similar thermistor devices. In this paper, new approach based on using commercial software tools was applied on the thick film segmented thermistor models. Obtained simulation results were compared with experimental results and validity of the adopted approach verified. Realized segmented thermistors were made of a low temperature NTC thick film paste NTC 3K3 95/2 with nanometer particle powder printed on alumina.

I. INTRODUCTION

NTC thermistors have been in use for a long time in wide range of temperature-related measurement and control applications. This was due to their good characteristics such as low cost, large change in resistance vs. temperature resulting in better measurement resolution, fast thermal response, small size and thus potential of use in different types of assemblies. Some of the typical applications include temperature measurement, temperature alarm, temperature control, liquid level sensing, fluid flow sensing (air sensors), gas analysis using thermistors in self heat mode (thermal conductivity gas analysis instruments), time dependency applications, time delay, surge suppression, inrush current limiting, temperature compensation and voltage regulation [1, 2].

Although thermistors have been in practice over a long time, methods for predicting the behavior of thermistors have remained the same. Typically, accurate measurements of resistance and temperature are made, and then curve-fitting methodologies are applied to relate temperature to resistance for similar devices.

This method however does not describe the thermistor resistance dependency on complex thermistor structure parameters such as geometric dimensions (size, shape, layer thickness) and material characteristics (ε_r-relative permittivity, σ-bulk conductivity, δ-loss tangent), which are very basic parameters to consider during a thermistor design process. The approach used in the past was to manufacture series of thermistors with these parameters as variables, perform the measurements and extrapolate the required characteristics, which could then be used for future designs of similar

V. Dj. Marić and Lj. D. Živanov are with the Faculty of Technical Sciences, University of Novi Sad, Trg Dositeja Obradovica 6, 21000 Novi Sad, Serbia & Montenegro, E-mail: vmaric2000@yahoo.com, lilaziv@uns.ns.ac.yu

O. S.-Aleksić is with the Faculty of Electrical Engineering, University of Belgrade, Bulevar Revolucije 73, 11000 Belgrade, Serbia & Montenegro, (email: obradal@yahoo.com).

devices. This is obviously not the most efficient design process with numerous inconveniences involved.

II. EXPERIMENTAL RESULTS

Thermistor device chosen to verify the validity of the proposed approach is a segmented thermistor, manufactured with a low temperature NTC thick film thermistor paste 3K3-95 / 2 EI Iritel + Ferrites [3, 4] on alumina substrate. This thermistor type was characterized by EDS, XRD, SEM, FIR and PA spectroscopy [5-8]. Different thick film thermistor geometries were printed with the same paste and characterized electrically [9-10]. Some of the test geometries were applied in custom design thermistors as well as in volume air flow sensor comprised of thick film segmented thermistors. The construction of this segmented thermistor enables integral temperature measurements on external electrodes and differential temperature (or temperature gradient) measurements on internal electrodes. It comprises of 8 cells (17 internal electrodes) and has a room temperature value of 1.5kOhm.

It integrates 16 smaller disc/sandwich thermistors in series which amount to about 90 Ohm each. The thickness of the final thermistor was 36µm and in-plane electrode spacing was 1mm.

Measurements were done on two different samples which had the same dimensions and were developed with the same manufacturing procedure, but their resistances at room temperature differed (sample 2 has a resistance of 1320Ohm at room temperature). This difference is due to the nature of the thick film technology. Measured resistance vs. temperature change for the same segmented thermistor was done on sample 1 and is given in Fig. 1.

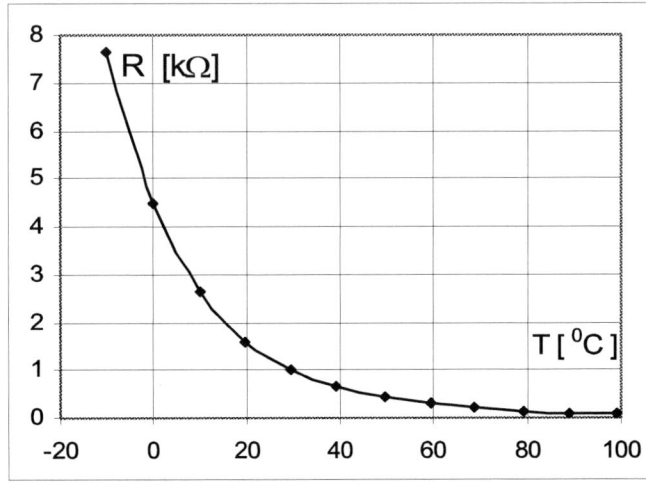

Fig. 1. NTC behavior of the segmented thermistor printed of NTC 3K3 paste (sample 1.)

1-4244-0116-X/06/$20.00 ©2006 IEEE
479

The same segmented thermistor exhibits a capacitance of 80-100pF at low frequencies such as 100Hz, 1kHz and 10kHz, but has a low Q factor because of the low distributed resistance in parallel to capacitance. This type of planar integrated thermistors has a parasitic resistance across the in–plane electrodes, but it is much higher and depends on the distance clearance between the electrodes (spacing distance). In tested case of 1mm electrode clearance, this resistance amounts to few kOhm and contributes few percents to overall cell resistance. At higher frequencies, total impedance of segmented thermistor depends on RC cell impedance. The normalized impedance of sample 2 was measured by network analyzer HP 8566B and is given in Fig. 2.

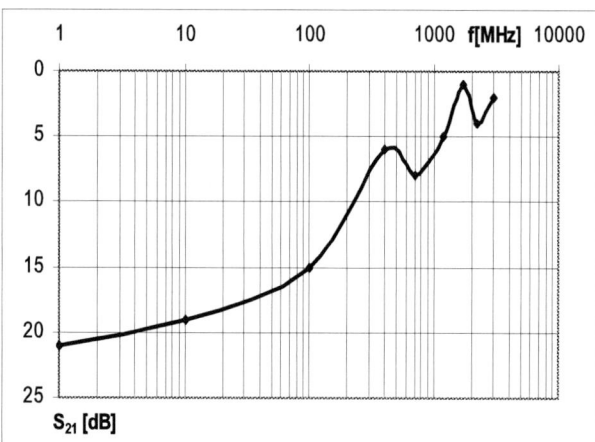

Fig. 2. S_{21} frequency response of the thick film segmented thermistor (sample 2.)

The diagram demonstrates a large attenuation at a couple of hundreds of MHz due to capacitance distributed over thermistor electrodes. The insertion loss of the same segmented thermistor was measured by the network analyzer and is given in Fig. 3.

The diagram exhibits a plateau of approximately -20 dB up to 100 MHz and then a gentle slope up to few dB in GHz region. The device actually acts as a high-pass filter with distributed R/C parameters.

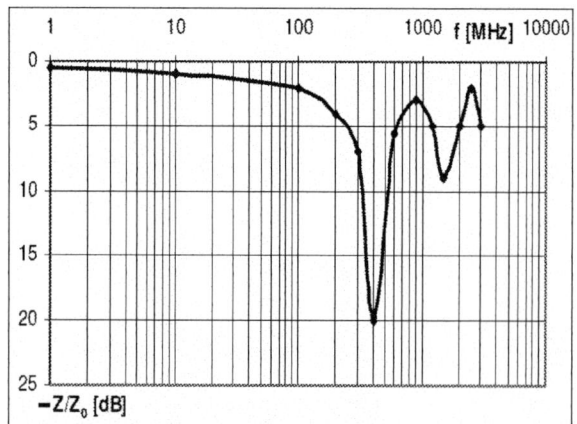

Fig. 3. Normalized impedance behavior Z/Z_0 for the thick film segmented thermistor (sample 2.)

III. PROPOSED APPROACH

Difficulties of experiment-based development have made us look for commercial software tools, which could aid the design process. Electromagnetic (EM) simulators use Maxwell.s equations to simulate planar 3D structures containing multiple metallization and dielectric layers, extracting the response of a structure from its physical geometry [11]. They are used mainly for radio frequency and microwave circuit simulation and modeling. Using EM simulators for the design of complex thermistor structures is a novel idea evaluated and presented in this paper.

Electromagnetic 3D structure is created by defining layers (their material characteristics: relative permittivity, bulk conductivity, loss tangent and layer thickness), adding conductors (shapes and conductivity), interconnecting them with vias (if connections between different layers are needed) and defining ports (i.e. measurement terminals). In solution process, conductors are initially discretized on a uniform rectangular mesh grid, then the amplitudes of basis functions (required for calculation of currents on conductors) for each individual grid element are calculated, and finally summed together over all discrete grid elements resulting in a good approximation of the current on conductors. To solve the basis functions required for calculating current distribution, so called moment matrix, which contains elements representing electromagnetic coupling between two basis functions, must be solved. This is possible either with direct or iterative matrix solvers, which differ in algorithms used, and thus in time and accuracy required for solving the moment matrix [11].

EM simulators typically do not support thermal behavior needed for thermistor design process, which was as well the case with our tool of choice, method-of-moments based software tool developed by Applied Wave Research. In its latest version however, it offers a scripting WinAPI interface (similar to Visual Basic for Applications), which gave us a possibility to extend its capabilities to address our problem of thermistor design. Validity of the idea was first checked against simple thermistor structures such as standard disc/sandwich or rectangle thick film thermistors seen in [2].

The 3D model with real-life physical dimensions (in μm) of a segmented thermistor used in MWO is shown in Fig. 4. It represents accurately the realized thermistor structure with the addition of the bottom and top air layers required for defining the boundaries of the electromagnetic enclosure.

Additional input parameters relevant for the electromagnetic simulation were material characteristics of the thermistor paste: relative permittivity (ε_r=7), bulk conductivity (σ=0.01745S/m at room temperature) and loss tangent (δ=0.001). Alumina substrate layer thick d=0.63mm was modeled by relative permittivity (ε_r=9.6), bulk conductivity (σ=0S/m) and loss tangent (δ=0.006).

Electrode conductors were modeled as palladium-silver conductors with conductivity of $2*10^6$ S/m and thickness of 10μm. Both top and bottom boundaries were modeled as "approximately open".

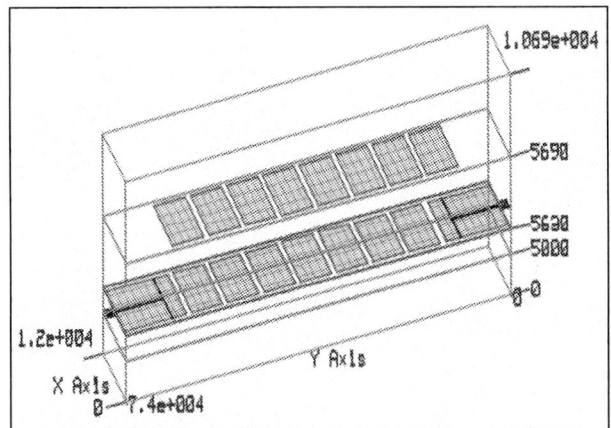

Fig. 4. Segmented thermistor structure as modeled in MWO

The thermal coefficient B can be calculated by the known approximate equation R(T)=A*exp(B/T). For two temperatures T_1 and T_2 and respective resistances R_1 and R_2, we have a relationship:

$$R_1 = R_2 \exp \left[B \left(\frac{1}{T_1} - \frac{1}{T_2} \right) \right]. \qquad (1)$$

Beta coefficient B, after solving the equation above, can be expressed as:

$$B = \frac{T_1 T_2}{T_2 - T_1} \ln \frac{R_1}{R_2} . \qquad (2)$$

By analogy to this relationship, we can assume that the bulk conductivity of the thermistor layer σ as main variable parameter of the electromagnetic simulation in MWO has the same exponential relationship to temperature:

$$\sigma(T) = C \cdot \exp(D / T). \qquad (3)$$

Using the experimental data of NTC behavior R(T) as given in Fig. 1, and choosing T as 273K and 303K and their respective resistances (4.5kOhm and 1kOhm), we obtained B= 4151K. To fit R(273K)=4.5kOhm in MWO, the bulk conductivity needs to be σ(273K)=0.00436S/m, and for R(303K)=1kOhm, σ(303K)= 0.02046S/m, resulting in C=26556.87S/m and D= -4267.24K. This was determined through iteration procedure within MWO tool Incorporating these results into the equation (3), and running the MWO simulation with its scripting interface, the simulated NTC R(T) behavior is obtained and shown in Fig. 5.

The same figure shows the calculated impedance based on the experimental data and equations (1) and (2). It can be seen that the results obtained from experimental impedance results correspond to the results interpolated from the bulk conductivity simulation values. This indicates that by knowing bulk conductivity values, it is possible to determine the impedance behaviour of more complex thermistor structures. The match is particularly close in the two-point defined range for which the beta values have been determined. By analogy, a three-point calibration based on Steinhart-Hart equation is possible which can give the closest match over whole temperature range of interest.

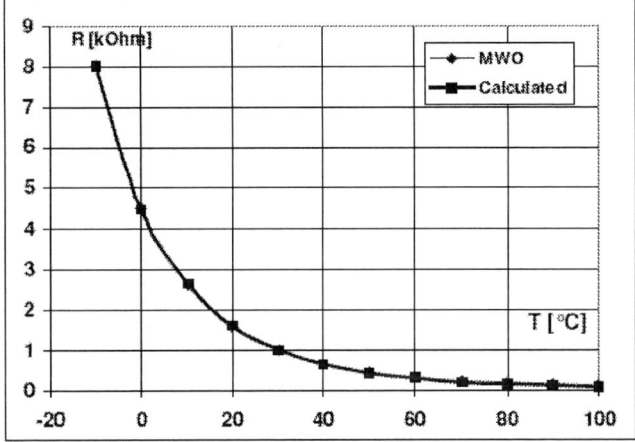

Fig. 5. R(T), MWO vs. calculated

In order to complete the comparison of the simulated with measured results for this hybrid component, two more plots, Z(f) and S_{21}(f) were done, shown in Fig. 6. and 7.

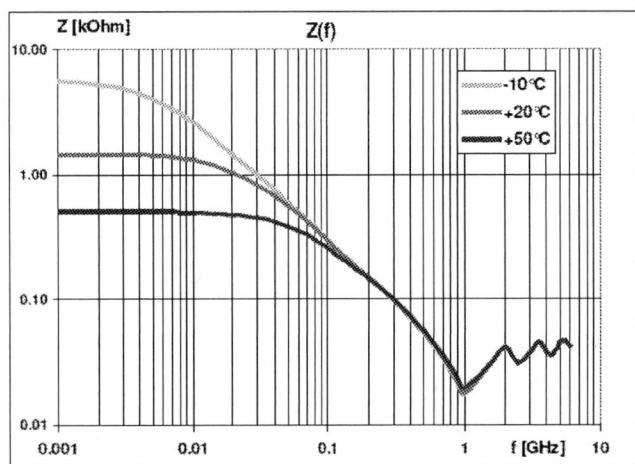

Fig. 6. NTC thick film segmented thermistor impedance Z, simulated by method of moments in MWO

Both of these plots display results for three temperature points, -10°C, 20°C and 50°C. Bulk conductivity for these temperatures was determined using equation (3). Room temperature simulations (20°C) correspond to a great degree with the experimental results shown in Fig. 2 and Fig. 3, indicating the validity of the obtained results.

At low frequencies, the impedance decreases with temperature, which translates into lower insertion loss S_{21}. The cut-off frequency for the high pass filter behavior differs for different temperatures, being around 100MHz

for room temperature, 200MHz for 50°C and 20MHz for
– 10°C.

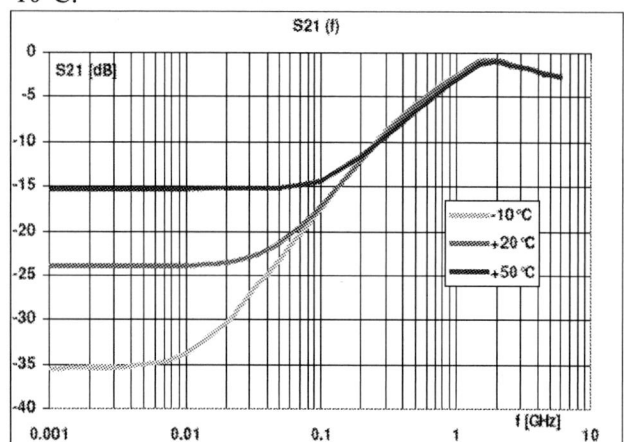

Fig. 7. NTC thick film segmented thermistor S_{21}, simulated by
method of moments in MWO

IV. DISCUSSION

Simulation results obtained by method-of-moments
based electromagnetic analysis in MWO tool (Fig. 5-7),
exhibit results fairly close to experimental, which
demonstrates a possibility of using MoM-based simulation
tools for more complex construction structures where a
development of an equivalent circuit schematic could be a
difficult task. This would be the case with custom
thermistor structures with miscellaneous shapes, where we
expect wider application of similar simulation tools.
Figures 6-7 indicate that measurements at low or high
temperatures, which would be technically very difficult to
perform, can be replaced by simulation results.

With the approach shown in this paper, thermistor
design process with commercial simulation tools becomes
based on temperature characterization of thermistor paste
in terms of bulk conductivity and relative permittivity. The
paste can be quickly and easily analyzed on simple disc
thermistor structures and these results exploited for more
complex structures.

MoM-based simulation approach has longer
simulation times than SPICE-similar simulations, but its
main input parameters being material and geometric
properties, it is closer to real life situation, and it gives
quick hints on how to improve process parameters in order
to optimize the structure performance. Another advantage
which simulation tools such as MWO offer is the
possibility of integration of individual electromagnetic
structures into higher-level schematics and models, thus
enabling an easy simulation of the whole system.

V. CONCLUSION

The advances in the area of development of
electromagnetic simulation software have allowed
uncommon application of such software tools in area of
thermistor development process. It was demonstrated that
such tools gave the possibility to study and successfully
evaluate complex thermistor structures (with different
shapes, multiple layers, different materials). The NTC
characteristics and high frequency behavior have been
shown, followed by the comparison of the experimental
and simulated data.

The results shown here imply that novel complex
thermistor structures could be fairly quickly evaluated for
feasibility without a need of producing them first, by
determining the effect of their parameters, thus speeding up
the whole development process. Since the future evolution
of custom design thermistors to miscellaneous shapes and
sizes is expected, the appropriate simulation tools
supporting this development should find its place in this
process as well.

ACKNOWLEDGMENT

The authors would like to acknowledge AWR, Applied
Wave Research Ltd. UK, for providing its powerful
software tools for this project.

In Memoriam: Vladan Desnica (1972-2005), who
contributed greatly to this paper before his tragic death.

REFERENCES

[1] Maclean E.D., *"THERMISTORS"*, Electrochem. Pub., Glasgow,
1979, pp. 5-11

[2] O.S.Aleksić, P.M.Nikolić, L.Lukić, "Analysis of thick film NTC
thermistor properties and their applications", NTB, Vol. 1, 2003,
pp. 3-16.

[3] Vakiv M., Shpotyuk O., Mrooz O., Hadzaman I., " Controlled
thermistor effect in system C_xNi_{1-x-y} Co_{2y} Mn_{2-y} O_4", *Journal
European Ceramic Society*, Vol. 21., 2001, pp.1783-5.

[4] Altenburg H., Mrooz O., Plewa J., Shpotyuk O., Vakiv M.,
"Semiconductor ceramics for NTC thermistors: the reliability
aspects, *Journal European Ceramic Society*, Vol. 21., 2001, pp.
1987-91.

[5] Aleksić S.O., Nikolić M.P., Simić N.M., Pejović Ž.V.,
Vasiljević-Radović G.D., "Resistivity vs. geometry relation in
bulk sintered and thick film Mn,Co,Fe–oxide thermistors",
Advance Science and Technology of Sintering, Kluwer
Academic/ Plenum Publisher 1999, pp. 425-30.

[6] Aleksić S.O., Nikolić M.P., Vasiljević Radović G.D., Luković
D. M., Djurić S., Pejović Ž.V., Radulović T.K., Luković T.D.,
Vujošević Lj.D., "Properties of thick film NTC thermistor layers
based on nanometric Mn, Co, Fe-oxide powder mixture", X
World Round Table Science of Sintering, .Belgrade 3-6
September 2002, pp 427-433.

[7] Aleksić S.O., Nikolić.M.P., Luković T.D., Radulović.T.
K.,Vasiljević-Radović G.D., Savić M. S., "Diffusivity for thick
film NTC layers by photoacustic technique" Microelectronics
Int., Emerald UK, Vol. 21, 2004, pp. 10-14.

[8] O.S.Aleksić, P.M.Nikolić, M.N.Simić, V.Ž.Pejović, D.G.
Vasiljević-Radović, "Resistivity versus geometry relation in
bulk-sintered and thick film Mn, Co, Fe–oxide thermistors",
Edited in Advanced Science and Technology of Sintering,
Plenum Pub, 1999, pp. 425-30.

[9] O.S.Aleksic, et al:"Analysis of thick film thermistor
geometries", Proc. of MIEL, 1997 pp. 431-434.

[10] O.S.Aleksić, P.M.Nikolić, D. Luković, S.Savić, V.Z.Pejović,
B.M. Radojčić, "Thick film NTC fhermistor air flow sensor",
Proc. 24th International Conference on Microelectronics MIEL-
IEEE 2004, Vol 1, Niš, Serbia, May 16-19, pp. 185-188, 2004

[11] *»MWO/VSS Getting started guide"*, Version 6.53, July 2005,
Applied Wave Research, Inc.

2006 25th International Conference on Microelectronics

Fully Automated Electrothermal Simulation Using Standard CAD Tools

F. M. De Paola, J. P. Nowakowski, V. d'Alessandro, and N. Rinaldi

Abstract— This contribution presents a novel simulation tool for the electrothermal analysis of solid-state devices and circuits in both static and transient conditions. The code relies on an effective analytical approach to describe the 3-D thermal process, and exploits the engine of the commercial software ADS (Advanced Design System) to consistently solve the electrical and thermal networks. Features like model accuracy, lack of convergence problems, low time/memory requirements, flexibility and user friendliness make the proposed software a good alternative to SPICE-like and fully numerical tools for the optimization of both reliability and performance of state-of-the-art devices/circuits.

I. INTRODUCTION

Electrothermal effects can significantly affect both reliability and performance of multifinger bipolar transistors [1] as well as integrated analog circuits [2]. The need of minimizing their impact through device/circuit optimization has long motivated research efforts to conceive and develop suitable electrothermal simulation programs. A widely diffused approach lies in resorting to SPICE-like tools, in which the macromodeling technique is employed to account for the electrothermal feedback, temperature sensitivity of key electrical parameters, and unique transistor phenomena [3], [4]. Unfortunately, such a technique is computationally viable only for circuits where the number of individual active devices is relatively low due to the large number of additional components often needed for the description of the elementary transistor. On the other hand, numerical procedures such as finite differences or thermal networks [5] usually require a – not effortlessly determinable – choice of the discretization level, as a proper trade-off between accuracy and time/memory requirements. Thus, alternative strategies are being currently sought. Conveniently, some tools adopted in the IC CAD arena – such as Agilent ADS [6] – nowadays incorporate recently developed BJT/HBT models (e.g., Mextram 504 [7]), which are provided with a supplementary terminal, namely a thermal node, and include a default value for the self-heating thermal resistance. Hence, the temperature increase above ambient can be evaluated from the dissipated power and considered, in turn, as an additional input that modifies the temperature-sensitive parameters, thereby allowing the description of self-heating (SH) effects. However,

mutual thermal interactions between active devices integrated in the same chip are not accounted for, which might lead to an unavoidable simulation inaccuracy, especially in the case of multifinger structures and high-density ICs. Besides, other built-in parts (e.g., resistances, diodes, and FETs) are not equipped with a thermal node.

In our communication, we present an extended version of the software ADS devised for the electrothermal simulation of ICs, as a valuable attempt to overcome the aforementioned limitations. The developed tool is user-friendly and provides appreciable accuracy with low time/memory requirements and no convergence problems. Both steady-state and transient analyses are allowed and nonlinear thermal effects can be enabled. Illustrative simulations are performed on GaAs-based multifinger HBTs and bipolar current mirrors with thermally-insulating substrates.

II. THE ELECTROTHERMAL SIMULATION CODE

As described in Fig. 1, the program is based on the ADS solving engine and is subdivided into three basic blocks, i.e., a pre-processing routine, a main core, and a post-processing code. At the pre-processing stage, the standard – purely electrical – ADS schematic is translated into an intermediate file format (.iff) that can be efficiently modified to account for SH and mutual thermal interactions. The self-heating and mutual 3-D thermal impedances Z_{THij} are evaluated from the layout (.dsn) file associated to the circuit as follows. First, the user provides information about depth and thickness of the heat sources along with the thermal properties of the substrate. Afterwards, the code scans the .dsn layout file for detecting the coordinates of the emitter windows (i.e., the projections of the heat sources on the top surface). Lastly, the thermal impedance matrix is computed through the closed-form analytical formulations proposed in [8] and [9] for the steady-state and transient cases, respectively. Such a process is fully automated, and allows avoiding troublesome discretization aspects that are required when dealing with numerical approaches. The thermal impedance matrix can be also "directly" provided by the user. This is needed when analyzing domains with complex geometries (as e.g., high-speed BJTs with manipulated substrates [10]), in which the thermal process is troublesome – or even impossible – to be analytically modeled. In this case, one

Department of Electronics and Telecommunications Engineering, University of Naples "Federico II", via Claudio 21, 80125 Naples, Italy, E-mail:fdepaola@unina.it

1-4244-0116-X/06/$20.00 ©2006 IEEE

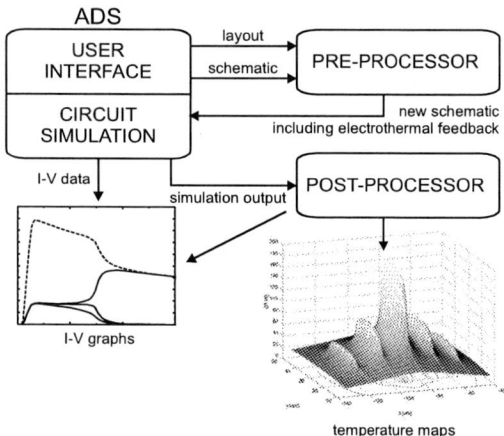

Fig. 1. Schematic *block diagram* of the proposed ADS-based electrothermal simulation tool.

can preventively extract the thermal impedance values by means of numerical 3-D thermal tools (as e.g., Femlab [11]) or experimental techniques [12]. As a second step, the code (i) creates a Verilog-A block including the thermal impedance matrix and (ii) automatically connects it to the active devices of the original (i.e., isothermal) schematic. Such a block is treated as an embedded ADS element and is employed to calculate the powers dissipated by all transistors (treated as currents) and the corresponding junction temperature increases above ambient (reviewed as voltages) during the electrothermal ADS-handled simulation. The connection procedure and the electrothermal feedback (EF) action of the Verilog-A component are illustrated in Fig. 2 for the case of a bipolar circuit and can be described as follows. After identifying all the transistors embedded in the original schematic (a), the code separates the collector terminal C_i of each device into a pair of nodes C_{ia}, C_{ib} (b) that are tied to two input terminals of the Verilog-A block and electrically shortened within the block itself to make the collector current I_{Ci} accessible (c). An additive input terminal is connected to the emitter node in order to sense the collector-to-emitter voltage V_{CEi}. Starting from I_{Ci} and V_{CEi}, the block evaluates the dissipated power as

$$P_i = V_{CEi} I_{Ci} \qquad (1)$$

and the temperature increase above ambient as

$$\Delta T_i = \sum_{j=1}^{N} Z_{THij} P_j \qquad (2)$$

which represents the output of the Verilog-A block, and is, in turn, fed back to the input thermal node of the i-th device. Thus, the EF effect is included into a new "nonisothermal" .iff file that is imported into ADS. The corresponding schematic can be displayed and solved by the main core via the embedded ADS engine. Afterwards, all currents and voltages in the electrothermal circuit are available. The post-processing stage is

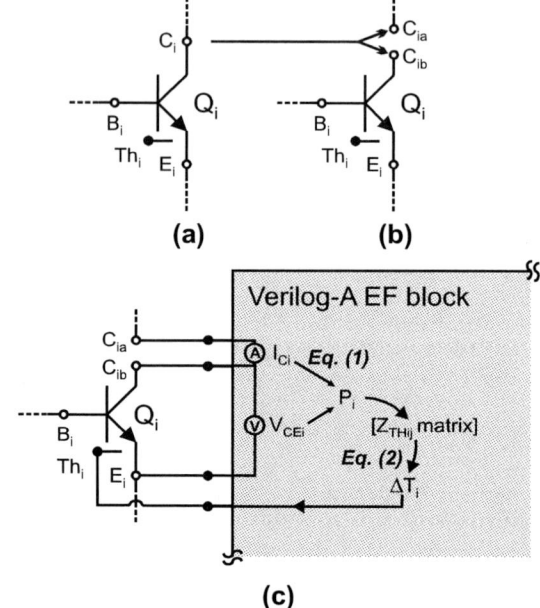

Fig. 2. Automated connection procedure of the "electrothermal feedback" block to the original isothermal schematic.

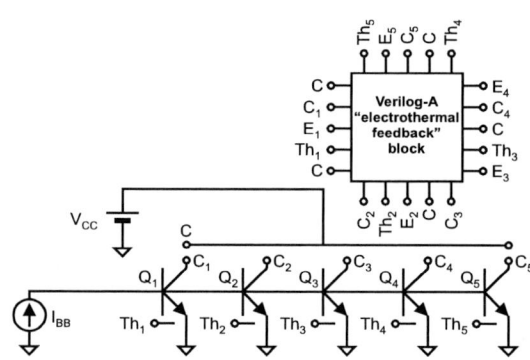

Fig. 3. Nonisothermal ADS schematic of a 5-finger GaAs-based HBT. The Verilog-A block handling the electrothermal feedback is also represented.

charged to both ADS and another in-house code, which handles the simulation results and evaluates the temperature field under assigned bias conditions over a chosen grid. The Verilog-A analog behavioral language has been chosen due to its high flexibility toward the analytical description of components, and its adoption has allowed also (1) extending the models of ADS components not intrinsically equipped with a thermal node, and (2) defining other custom active/passive device models to be incorporated into the program.

III. SIMULATION RESULTS AND DISCUSSION

As a first application, the electrothermal behavior of a GaAs-based 5-finger HBT was analyzed. The corresponding nonisothermal ADS schematic with the Verilog-A EF component is shown in Fig. 3. The built-in Agilent model equipped with thermal node was adopted for the description of the individual fingers.

484

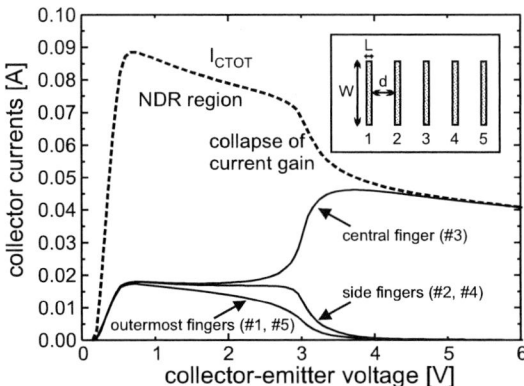

Fig. 4. Individual (solid lines) and total (dashed) collector currents for the 5-finger HBT as evaluated by the proposed code. The schematic layout of the device under test is shown in the inset ($L = 1$ μm, $W = 20$ μm, and $d = 15$ μm). As can be seen, for voltages beyond $V_{CE} = 2.5$ V an uneven current distribution arises and the NDR (Negative Differential Resistance) region is replaced by the "gain collapse" one [1].

Simulations were performed by applying a total base current $I_B = 1$ mA and increasing V_{CE}. Results are illustrated in Fig. 4, along with the device layout (reported in the inset); it is shown that at $V_{CE} > 2.5$ V the electrothermal interactions give rise to a nonuniform current (and temperature) distribution among fingers, which leads to the "collapse of current gain" [1]. Fig. 5 shows the temperature increase above ambient over the top surface of the device as evaluated by the post-processing custom code at $V_{CE} = 2$ (a), 3 (b), and 4 V (c). Inspection of Figs. 4 and 5 clarifies that for $V_{CE} > 3.5$ V the central finger bears the whole current, with a consequent temperature focusing (hot spot). Secondly, the influence of the spacing d between fingers upon the electrothermal behavior of a 3-finger structure has been investigated. In this case, all simulations were carried out by applying a total base current I_B amounting to 300 μA. Results are depicted in Fig. 6, where both the individual and the total collector currents are reported as a function of V_{CE}. The analysis evidences the stabilizing effect of a spacing reduction (which actually corresponds to an increase in mutual thermal resistances); it is indeed shown that the uneven current distribution and the consequent current gain collapse occur at higher V_{CE} values for structures with lower spacing.

Lastly, the electrothermal behavior of bipolar current mirrors fabricated in silicon-on-glass (SOG) technology has been analyzed. As exhaustively detailed in recently published papers [4], [10], [12], [13], SOG transistors are characterized by very high thermal resistances due to the low thermal conductivity of the high-resistivity glass substrate and all other materials surrounding the active silicon area. The self-heating thermal resistance R_{TH} of transistors Q_1 and Q_2 (see inset of Fig. 7) as well as the mutual coupling resistance R_{THM} have been evaluated through the ac measurement technique pro-

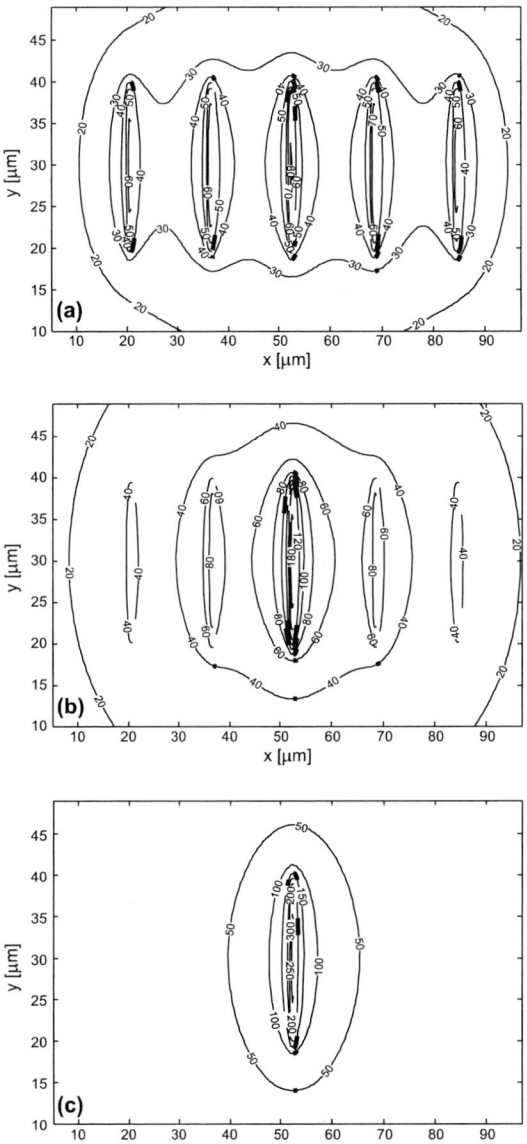

Fig. 5. Contour plots representing the top surface distribution of the temperature increase above ambient for the 5-finger HBT device; cases $V_{CE} = 2$ (a), 3 (b), and 4 V (c). The temperature focusing over the central finger in the "collapse" region at 4 V is apparent.

posed in [12] for all the fabricated structures. Concerning the configuration chosen, R_{TH} and R_{THM} equate 15100 and 4200 K/W, respectively. All simulations were performed by resorting to a simple, yet accurate in-house transistor model [15] accounting for all the key physical parameters influencing the electrothermal device behavior. Such a model has been first tuned on the basis of data measured under isothermal conditions and then imported into the ADS environment through the Verilog-A approach. Fig. 7 details the electrothermal I_{OUT} vs. V_{OUT} characteristics of transistor Q_2 for various I_{REF} values spanning from 0.5 to 3.0 mA. As can be seen, electrothermal effects lead not only to a poor mirroring behavior, but also to an undesirable flyback

485

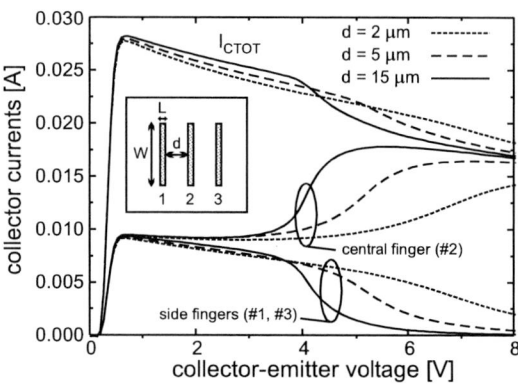

Fig. 6. Individual and total collector currents for the 3-finger HBT as calculated by the ADS-based tool for various values of the spacing d between fingers. The schematic layout of the device is shown in the inset ($L = 1$ μm, $W = 20$ μm).

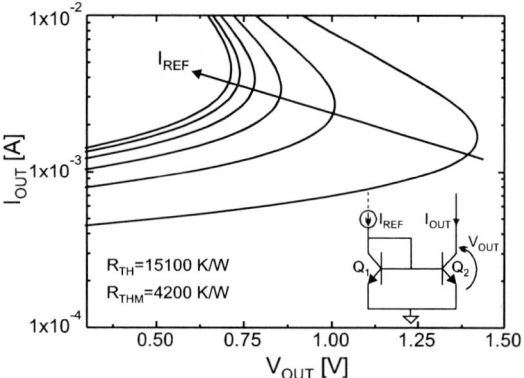

Fig. 7. Calculated I_{OUT} vs. V_{OUT} curves for $I_{REF} = 0.5, 1.0, 1.5, 2.0, 2.5,$ and 3.0 mA. The scheme of the current mirror analyzed is reported in the inset.

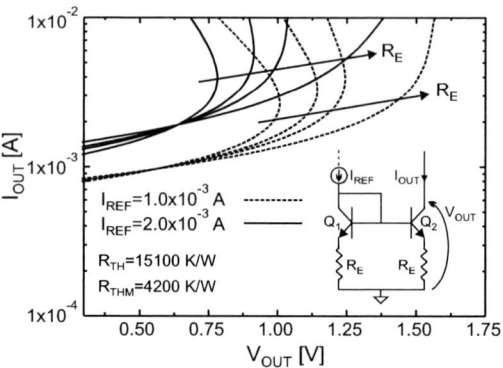

Fig. 8. Calculated I_{OUT} vs. V_{OUT} curves for $I_{REF} = 1.0$ mA (dashed curves) and 2.0 mA (solid): effect of the emitter ballasting resistors ($R_E = 0, 3, 7,$ and 10 Ω).

pared to other approaches. Furthermore, it is manageable in all platforms that support the ADS software.

occurrence [13], [14], well within the limits imposed by the avalanche mechanism. Fig. 8 illustrates the weakening action on the electrothermal feedback due to ballasting resistors tied to the emitters of Q_1 and Q_2 (see inset) for I_{REF} amounting to 1.0 and 2.0 mA, respectively. In particular, it is seen that the minimum R_E value needed to avoid the occurrence of unstable negative resistance branches equates 10 Ω for both I_{REF} cases.

IV. CONCLUSION

In this paper, a novel strategy for the electrothermal simulation of discrete and integrated devices/circuits has been addressed, which is based on the fully automated employment of the commercial ADS software and the Verilog-A behavioral language. The developed code makes use of an effective analytical approach for the evaluation of the 3-D thermal resistance matrix in both steady-state and transient conditions, and ADS embedded and Verilog-A custom in-house models for the description of active/passive devices. The program has been demonstrated to be user-friendly, flexible and accurate, yet requiring low computational effort com-

REFERENCES

[1] W. Liu et al., 'Current gain collapse in microwave multifinger heterojunction bipolar transistors operated at very high power densities,' *IEEE Trans. Electron Devices*, vol. 40, pp. 1917–1927, 1993.

[2] R. M. Fox et al., 'The effects of BJT self-heating on circuit behavior,' *IEEE J. Solid State Circuits*, vol. 28, pp. 678–685, 1993.

[3] A. Maxim and G. Maxim, 'Electrothermal SPICE macro-modeling of the power bipolar transistor including the avalanche and secondary breakdowns,' *Proc. IEEE IECON*, 1, pp. 348–352, 1998.

[4] V. d'Alessandro et al., 'A novel SPICE macromodel of BJTs including the temperature dependence of high-injection effects,' *Proc. IEEE MIEL*, 1, pp. 253–256, 2004.

[5] M. Latif and P. R. Bryant, 'Network analysis approach to multidimensional modeling of transistors including thermal effects,' *IEEE Trans. on Computer-Aided Design of ICs and Systems*, vol.1, pp. 94–101, 1982.

[6] ADS user's manual, Agilent, 2004.

[7] J.C.J. Paasschens et al., 'Mextram (level 504). The Philips model for bipolar transistors,' *Proc. FSA Modeling Workshop*, 2002.

[8] N. Rinaldi, 'Thermal analysis of solid-state devices and circuits: An analytical approach,' *Solid-State Electronics*, vol. 44, pp. 1789–1798, 2000.

[9] N. Rinaldi, 'On the modeling of the transient thermal behavior of semiconductor devices,' *IEEE Trans. Electron Devices*, vol. 48, pp. 2796–2802, 2001.

[10] L.K. Nanver et al., 'A back-wafer contacted silicon-on-glass integrated bipolar process – Part I: The conflict electrical versus thermal isolation,' *IEEE Trans. Electron Devices*, vol. 51, pp. 42–50, 2004.

[11] Femlab reference manual, Comsol AB, 2003.

[12] N. Nenadović et al., 'Extraction and modeling of self-heating and mutual thermal coupling impedance of bipolar transistors' *IEEE J. Solid-State Circuits*, vol. 39, pp. 1764–1772, 2004.

[13] N. Nenadović et al., 'A back-wafer contacted silicon-on-glass integrated bipolar process – Part II: A novel analysis of thermal breakdown,' *IEEE Trans. Electron Devices*, vol. 51, pp. 51–62, 2004.

[14] N. Rinaldi and V. d'Alessandro, 'Theory of electrothermal behavior of bipolar transistors: Part I – Single-finger devices,' *IEEE Trans. Electron Devices*, vol. 52, pp. 2009–2021, 2005.

[15] N. Nenadović et al., 'Restabilizing mechanisms after the onset of thermal instability in bipolar transistors,' *IEEE Trans. Electron Devices*, 2006 (accepted for publication).

2006 25th International Conference on Microelectronics

A Novel 3D Embedded Gate Field Effect Transistor: Device Concept and Modelling

K. Fobelets, MIEEE, P.W. Ding, and J.E. Velazquez-Perez

Abstract - A novel 3D Field Effect Transistor on SOI – the screen grid FET (SGFET) – for ultra-low power applications is proposed and TCAD analysis of the device is presented. The device is designed with the aim of decoupling the need for aggressive scaling of the gate oxide thickness when reducing the channel length. Other scaling objectives are: retaining low doping in the channel, maintaining the drain conductance and optimizing the low power/low voltage device behaviour. The simulation results show that these objectives are fulfilled: oxide thickness and channel doping have a reduced influence on the threshold voltage and do not need to be scaled aggressively to reduce the short channel effects. Finally, we show that the device performance for low-power/low-voltage applications is excellent.

I. INTRODUCTION

One of the objectives in semiconductor industry is higher operation speed and lower power consumption of the devices used in circuits and systems. To that aim, in field effect transistor (FET) technology, gate lengths are aggressively scaled to nanometer dimensions and gate oxide's thickness are currently no thicker than 2 nm. One of the problems associated with this downscaling is the increased leakage current which has a detrimental impact on power consumption for ULSI circuits. From the point of view of the analog circuit designer two main problems arise: poor output conductance and increased matching difficulties.

In order to solve the different problems occurring with downscaling, the next generation of FETs will use hetero-structures [1] and/or different gate topologies [2,3]. These new approaches, based on progress in nano-technology, introduce 2D/3D gating configurations, which require in general difficult or labour intensive processing.

The proposed embedded-gate screen grid FET (SGFET) in this manuscript is based on similar principles as the permeable base transistor (PBT) [4], but with drastically reduced processing complexity, compatible with current SOI technology and shows excellent improvements in DC behaviour compared to the other FET structures.

K.Fobelets, and P.W. Ding are with the Department of Electrical and Electronic Engineering, Imperial College London, Exhibition Road, London SW7 2BT, UK, E-mail: k.fobelets@imperial.ac.uk.

J.E. Velázquez-Pérez, is with the Department of Applied Physics, Universidad de Salamanca, Pza. de la Merced, s/n, E-37008 Salamanca, Spain, E-mail: js@usal.es.

II. DEVICE STRUCTURE LAYOUT AND FUNCTIONING

A schematic drawing of a proposed SGFET is given in Fig.1. The SGFET depicted has two rows consisting each of three gate cylinders, this distribution is by no means exclusive.

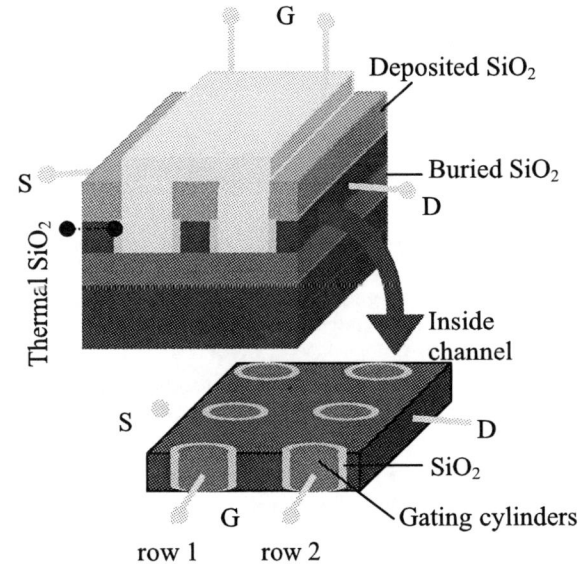

Fig. 1. Schematic configuration of a screen-grid FET with two gating rows, each consisting of 3 gate cylinders. Top: 3D side view. Bottom: channel region only. S, D indicate the start of the of the source and drain region resp., G is the gate.

The SGFET is fabricated in Silicon-on-insulator (SOI) and can be equally fabricated in the more advanced SSOI (strained-Si on insulator) technology [1] to benefit from the improved characteristics of the strained-Si channel. In the SGFET the deposited top oxide inhibits top surface gating and thus the source-drain conduction is only controlled by the gate cylinders *inside* the channel region.

The gating action for a unipolar SGFET is schematically illustrated in Fig.2 in 2D, the rectangle represents the Si channel, the dotted region the gate metal, the grey area the gate oxide and the shaded region the depletion region. The source and drain regions (not drawn) are heavily doped and, in the simulations, have the same width as the channel.

The gating occurring between two gate cylinders in the same row is consistent with a double gating action. At $V_{GS}=0V$, the depletion width extending from the gate into

1-4244-0116-X/06/$20.00 ©2006 IEEE

the channel is determined by the workfunction difference between the gate metal and the channel. In Fig.2 top, this is insufficient to deplete the channel region between the two gate cylinders and a current determined by the distance "a" will flow. Increasing the depletion widths by the gate potential pinches off the channel – a=0nm – and switches the current off (Fig.2 bottom). The choice of the gate metal allows the control of the absolute value and the sign of the threshold voltage (V_{th}).

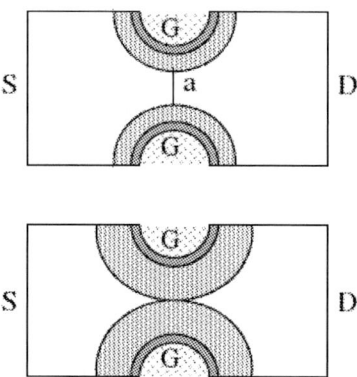

Fig. 2. Illustration of gating action between gate cylinders. Top: when V_{GS}=0V, bottom: when $V_{GS} \approx V_{th}$. Dotted region gate contact, grey region: gate oxide, shaded region: space charge region., white region: channel (Si).

Introducing the SGFET gate topology as shown in Figs.1 and 2 implies that the control of the gate and the source-to-drain (SD) distance is decoupled. In a conventional MOSFET both are related because when the SD distance is reduced (in order to decrease the transit time, or to increase its inverse, the cut-off frequency f_T) the gate length has to shorten. This scaling leads to a poorer control of the channel by the gate (the channel conductance g_d increases) and whilst increasing f_T the low-frequency voltage-gain unfortunately decreases (in spite of the every effort made in the scaling: particularly the reduction of the gate oxide thickness). This is a trivial result: unless one changes the topology of the gate, one cannot change the low-frequency gain-bandwidth product. The SGFET can vary this product due to the novel gate topology of which we can also change the distribution in order to optimize its operation.

II. PHYSICS BASED STUDY OF THE DEVICE

In this work we present a preliminary study of the device operation. Our main goal here is to understand the operation principles of the transistor and to verify that the aims in the design are fulfilled. These objectives are: to decouple the need for aggressive scaling of the gate oxide thickness, t_{ox} when reducing the channel length L_g – this will prevent current leakage, to maintain a low value of the doping in the channel, N_D when scaling – this will prevent mobility degradation, to avoid the degradation of the drain

conductance g_d when the channel length is shortened – control drain induced barrier lowering (DIBL), and to optimize the device behaviour for low power/low voltage applications – thus minimizing the subthreshold slope, S.

Bearing the above in mind, the most suitable method to study the electrical behaviour of the device is the drift-diffusion one (we used the Taurus/Medici 2D simulator from Synopsys [5]). It is well known that this method fails to correctly describe the electron transport in deep submicron devices since the carriers are intensely heated in the channel. However, in this paper we will limit ourselves to the study of relatively long channels (source to drain distances of about 0.8μm). Moreover, as the carriers are not confined in a quantum well (and therefore subject to the action of the perpendicular component of the electric field) their mobility becomes less dependent on the local value of the electrical field. As the device is 3D, we assume that the channel thickness, doping and electrostatic effects in the third dimension (from top surface to BOX) are uniform, deviations will require 3D simulations to study their impact.

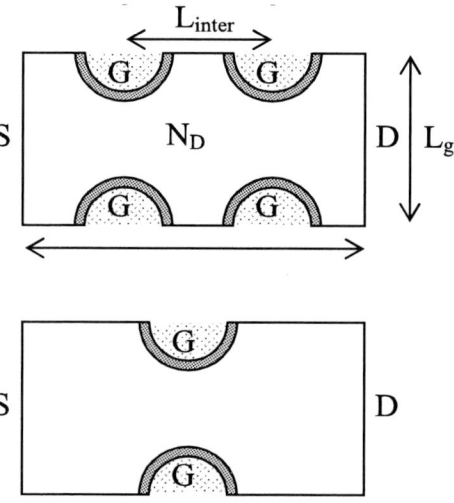

Fig. 3. Top view of a SGFET's section. Top: Section of a 2-row configuration. Bottom: Section of a 1-row configuration. The S and D regions are not shown. Grey ring-shaped regions: gate oxide. Grey dotted regions: Al. The total width of the device is NL_g, N being the total number of sections.

For normalization we assume the Si channel thickness to be 1μm. We will only consider here the n-channel case, but p-channel and complementary channels can be, in principle, built. Unless otherwise indicated, the parameters for the simulations are N_D=10^{15}cm^{-3} (active region doping), t_{ox}=2nm (gate oxide thickness), L_{SD}=800nm (source to drain distance), the finger's diameter is 120nm (for t_{ox}=2nm), L_g=280nm and L_{inter}=240nm (in the 2-row structure, see Fig. 3 top), The gate metal is chosen to be Al (workfunction φ=4.1 eV). Fig. 3 shows a schematic drawing of a unit section of the SGFET. Dependent on the needed current drive unit

488

sections can be attached. Simulations have shown that the current I_{tot} through N unit sections is equal to $N \times I_N$, with I_N the current through 1 unit section.

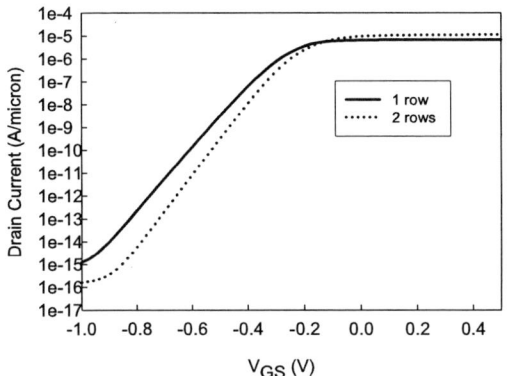

Fig. 4. Transfer characteristics of the devices in Fig. 3. V_{DS}=100mV.

The height of the source to channel barrier (SCB) controls the injection of carriers into the channel. In short channels the drain potential influences the SCB height. SCB lowering via the drain increases the current and is also responsible for the increase of the output conductance (g_d) as the gate length is shortened (DIBL).

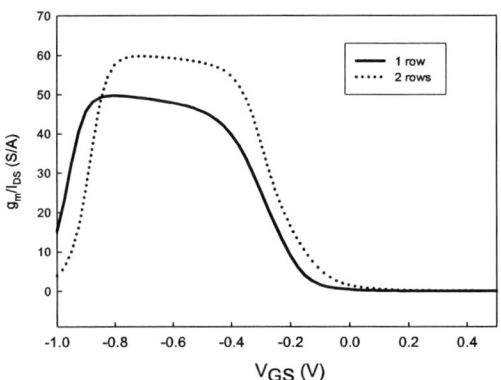

Fig. 5. Efficiency ratio (g_m/I_{DS}) of the devices in Fig. 3. V_{DS}=100mV.

The way to mitigate DIBL is to improve screening of the drain potential. This task can be effectively accomplished in the SGFET by using two rows of gate electrodes: 2-row configuration given in Fig.3 top rather than the 1-row configuration as given in Fig.3 bottom. The role of this second row is to screen the drain potential in order to keep the SCB height independent of V_{DS}. In principle, it is possible to work in a multigate configuration by applying different voltages to each row. For instance, the second row can be kept at as constant bias such as in a cascode gain stage. Of course, the position of the two rows modifies the device's performance. In this work we will

limit ourselves to the qualitative study of the influence of the most significant parameters.

Fig. 4 plots the transfer characteristics for the 1 and 2 row configurations. In this figure we can confirm that by adding the second row the device's ability for low-power/low-voltage applications is improved: S is lowered from 71 to 66 mV/decade, and the current range in which the subthreshold current is exponential extended.

TABLE I
INFLUENCE OF THE GATE OXIDE THICKNESS FOR L_g=294nm

t_{ox} (nm)	2	4	6	8	10
S (mV/dec.)	63.0	63.0	64.2	64.9	65.6
V_{th} (mV)	-250	-250	-259	-268	-277

In Fig. 5 the efficiency of the transconductance given by the ratio g_m/I_{DS} as a function of the gate voltage is plotted. This figure of merit measures the ability of a transistor to deliver a high gain independent of drain current sourcing. The 1-row structure exhibits a plateau value greater than 40S/A, whereas the 2-rows one is close to 60S/A. These results clearly show that the second row suppresses the parasitic drain control on the channel and helps to recover gate control.

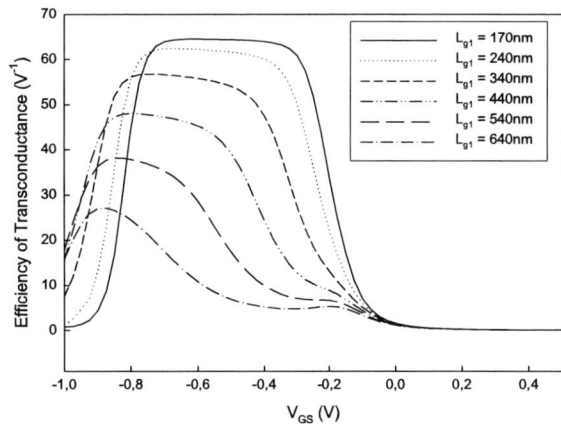

Fig. 6. Efficiency of the transconductance (g_m/I_{DS}). V_{DS}=100mV for different values of the gate length (L_g).

In agreement with the above, from the calculations of the output characteristics, we obtain at V_{GS}=0V and V_{DS}=2.5V a value of g_d=2.24μS/μm for the 1-row configuration and g_d=1.42μS/μm for the 2-rows one.

In Table I the influence of the oxide thickness on S and V_{th} is studied in the 2-row configuration. The data shows that the oxide thickness plays a marginal role in determining the value of V_{th} and S. This removes the need for aggressive oxide scaling as discussed above. Moreover, as the influence of this parameter on S and V_{th} is slight, it could at least partially lift the stringent control of t_{ox} needed in order to improve the matching. The simulations also show that N_D does not modify V_{th} noticeably as far as it is kept below $5.10^{17} cm^{-3}$. This alleviates the problem of the control of the channel doping to adjust V_{th}, and allows the use of undoped channels in order to improve the mobility.

For a given value of the finger's diameter (this diameter will be essentially limited by the processing), the threshold voltage will be essentially determined by L_g and the gate metal. L_g will become the key parameter of the device operation as it will strongly influence the performance of the transistor.

Fig. 6 plots the efficiency of the transconductance in the 2-row configuration with L_g as a parameter. The shift to the negative values of V_{GS} as the channel opening increases clearly shows the great influence of L_g on V_{th}. It is also noticeable that as L_g decreases the performance is improved – in contrast to traditional short channel effects. In contrast to the huge impact of L_g scaling on threshold voltage and efficiency of transconductance, the inter-gate row distance L_{inter} proves to have a minor influence on the performance of the device. This has important consequences for the reduction of the source-drain distance: When reducing the SD distance from 825nm to 425nm while keeping the L_{inter} value, the DIBL is reduced from 20.5mV/V to 6.4mV/V.

III. CONCLUSIONS

A novel 3D Field Effect Transistor on SOI – the screen grid FET (SGFET) – for ultra-low power applications is proposed and TCAD analysis of the device is presented. The simulations show that the definition of the gate cylinders inside the channel allow 3D-gating for excellent carrier control and creates sufficient flexibility to control short channel effects without the need for aggressive gate-oxide and doping scaling. The introduction of a second row of gate cylinders near the drain contact controls the influence of DIBL and this double gating row configuration leads to near-ideal values for subthreshold slope, and improves overall characteristics with gate length reduction: improved transconduction efficiency without increased leakage currents due to DIBL.

ACKNOWLEDGEMENTS

J.E. Velázquez thanks for financial support to Junta de Castilla y León (SA072A05) and MEC-FEDER (TEC2005-02719/MIC).

REFERENCES

[1] T.A. Langdo, M.T. Currie MT, Z.Y.Cheng, J.G. Fiorenza, M. Erdtmann, G. Braithwaite, C.W. Leitz, C. Vineis, J.A. Carlin , A. Lochtefeld, M.T. Bulsara, I. Lauer, D.A. Antoniadis, M. Somerville, "Strained Si on insulator technology: from materials to devices", Solid State Electron., vol. 48 (8), pp.1357 (2004).

[2] K. De Meyer, M. Caymax, N. Collaert, R. Loo, P. Verheyen, "The vertical heterojunction MOSFET", *Thin Solid Films*, vol. 336 (1-2), pp. 299-305, 1998.

[3] D. Hisamoto, W.C. Lee, J. Kedzierski, H. Takeuchi, K. Asano, C. Kuo, E. Anderson, T.J. King, J. Bokor, C.M. Hu, "FinFET - A self-aligned double-gate MOSFET scalable to 20 nm", *IEEE Trans. Electron. Devices*, vol. TED-47, pp. 2320-2325, 2000.

[4] A. Schuppen, L. Vescan, M. Marso, A. v.d. Hart, H. Luth and H. Beneking, "Submicrometer silicon permeable base transistors with buried CoSi2 gates", *Electon.Lett*. vol. 29(2), pp. 215-17. 1993.

[5] Taurus Medici User Guide, Synopsys, Mountain View, CA, USA, 2005.

2006 25th International Conference on Microelectronics

A Modified EKV PDSOI MOSFETs Model

J. Alvarado, A. Cerdeira, V. Kilshytska and D. Flandre

Abstract — The modeling of MOSFET I-V curves for distortion analysis in analog circuit design requires compact models for both long and short channel devices, which describe the transistor behavior with high precision based on the physics of the device. In the present paper, to achieve such precision, modifications of the EKV model equations are presented, while using the same parameters. A comparison for PD SOI MOSFETs of different channel lengths shows a very good agreement between experimental and modeled data. This enables the capability for studying, at circuit CAD level, the nonlinearity present in these devices thanks to the accuracy obtained in high derivatives of the drain current by our modified EKV model.

I. INTRODUCTION

There are three different types of MOSFET models. The first and more extended model is based on the threshold voltage [1], which gives good accuracy for digital applications. The second type concerns models based on the surface potential determination [2], which gives good coincidence with the experiment in all regions of operations, but has a large number of parameters, around 160, complicated expressions and parameter extraction procedure. Finally the charge sheet models as EKV [3] have more simple analytical expressions, less parameters, simplified extraction procedure and present good coincidence with experimental data.

In this paper we introduce modifications to the EKV model in order to obtain better results in the fitting of current-voltage characteristics and their derivatives. The analysis is applied to Partially Depleted (PD) SOI MOSFETs with body contacts, which present similar behavior to bulk MOSFETs. The proposed modifications are: a new interpolation function for the normalized inverted charge, the channel vertical field calculation, change in the mobility model, considerations of the velocity saturation, the channel length modulation, the series resistance and DIBL.

The main focus of this work is to improve the high order derivatives of the I-V characteristics modeled by EKV. The inadequacy of the basic EKV equations is related to the logarithm term used for the inversion charge expression. In order to model only the first derivatives, EKV uses another function: the transconductance-current

J. Alvarado and A. Cerdeira are with the Solid-State Electronics Section (SEES), CINVESTAV-IPN, México D.F., México. E-mail: jalvarado@gap.sees.cinvestav.mx

V. Kilshytska and D. Flandre are with the Microelectronics Laboratory, Université Catholique de Louvain, Louvain-la-Neuve, Belgium. E-mail: flandre@dice.ucl.ac.be

ratio g_m/I_D, which can be used to calculate the transconductances (i.e. g_m, g_{ms} and g_{md}). Another proposed improvement holds for the negative drain voltages in the linear regime, where EKV fails to yield fair agreement with measured data.

II. MODEL

A. Inverted Charge

The EKV model is based on the linearization of the inversion charge, which is given by the following interpolation function [3]:

$$q_{if(r)} = \sqrt{\frac{1}{4} + i_{f(r)}} - \frac{1}{2}, \quad (1)$$

$$i_{f(r)} = q_{if(r)}^2 + q_{if(r)} = \ln\left[1 + \exp\left(\frac{V_G - V_t - n \cdot V_{S(D)}}{2 \cdot n \cdot Ut}\right)\right]^2, \quad (2)$$

where q_{if} and q_{ir} are the source (forward) and drain (reverse) normalized inversion charges, respectively. V_t is the threshold voltage, U_t is the thermal voltage, n is the subthreshold ideality factor, given by:

$$n = 1 + \frac{\gamma}{2\sqrt{\Psi_p}}, \quad (3)$$

and Ψ_p is the pinch-off potential, defined as [4]:

$$\cdot \Psi_P = V_G - V_{FB} - \gamma\left(\sqrt{\frac{V_G - V_{FB}}{\gamma^2} + \frac{1}{4}} - \frac{1}{2}\right) \quad (4)$$

Equation (2) gives the normalized forward and reverse currents, depending on source and drain voltages, respectively. The drain current can also be represented in terms of the normalized charge and the specific current I_S, as:

$$I_D = I_S\left(q_{if}^2 + q_{if} - q_{ir}^2 - q_{ir}\right), \quad (5)$$

$$I_S = 2 \cdot n_q \cdot C_{ox} \cdot \mu_o \cdot U_t^2 \cdot W/L, \quad (6)$$

where C_{ox} is the oxide capacitance, μ_o is the mobility for low vertical field, $n_q = 1 + \gamma/\left(\sqrt{2\Phi f} + \sqrt{\Psi_P}\right)$ is the slope factor for the inverted charge W and L are the channel width and length respectively. The extraction procedure for I_S is described in [5].

In order to obtain better coincidence with the experimental data in the case of the PD SOI transistors analyzed in this paper, some changes were introduced. The normalized charge was found using as interpolation function, the Lambert`s W function [6]:

$$q_{if(r)} = \frac{n}{2.29} LambertW\left(\exp\left(\frac{V_G - V_t - n \cdot V_{S(D)}}{n \cdot Ut}\right)\right) \quad (7)$$

1-4244-0116-X/06/$20.00 ©2006 IEEE

Figure 1 shows the logarithmic drain current in order to emphasize the transition between weak and strong inversion with $V_D = 50$ mV and V_G varied from 0 to 1.8 V. The drain current is also depicted in figure 1 with linear vertical axes. Excellent accuracy is obtained by the modified model, using the same parameters as extracted by the direct parameter extraction method proposed in [5]. The experimental data was obtained for the analog PD SOI MOSFETs from a 0.12μm SOI CMOS process with 150 nm silicon layer, 5 nm gate oxide, 2.5 μm channel width and channel length from 0.32 μm to 10 μm. EKV 3.0 [7] model was implemented in Mathematica.

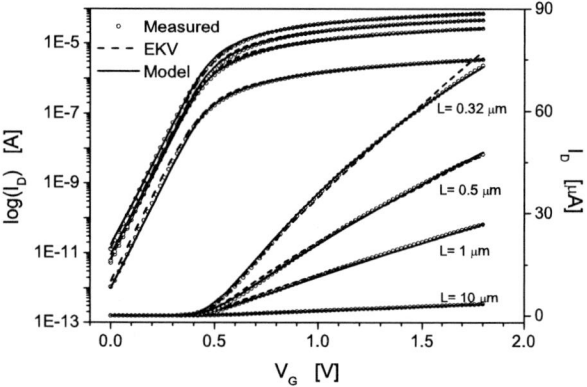

Fig 1. Transfer characteristics for different channel lengths at $V_D = 50$mV, $V_S = 0$V

B. Vertical field and mobility model

In [7], the vertical field is written in terms of the inversion and depleted charge as:

$$E_\perp = \frac{Q_B' + 1/2\, Q_I'}{\varepsilon_{si}},\qquad (8)$$

where ε_{si} is the silicon permittivity, Q_B' is the depleted charge per unit area and Q_I' is the inverted charge per unit area, given by:

$$Q_B' = \gamma \cdot C_{ox} \cdot \sqrt{\Psi_p/U_t} + Q_I' \cdot \left(\frac{n_q - 1}{n_q}\right),\qquad (9)$$

$$Q_I' = 2 \cdot n_q \cdot U_t \cdot C_{ox} \cdot q_i,\qquad (10)$$

$$q_i = \sqrt{1/4 + (i_f - i_r)} - 1/2,\qquad (11)$$

A more accurate expression for the vertical field [8], taking into account the inverted and depleted charges near the drain and source terminals, gives better accuracy in high orders derivative, and in the linear region, since it takes better into account both the source and drain influences which are either omitted or non symmetrical in (9-10):

$$E_\perp = \frac{\left(Q_{Bf}' + Q_{Br}'\right) + 1/2\left(Q_{If}' + Q_{Ir}'\right)}{2\varepsilon_{si}},\qquad (12)$$

$$Q_{Bf(r)}' = \gamma \cdot C_{ox} \cdot \sqrt{\Psi_p/U_t} + Q_{If(r)}' \cdot \left(\frac{n_q - 1}{n_q}\right),\qquad (13)$$

$$Q_{If(r)}' = 2 \cdot n_q \cdot U_t \cdot C_{ox} \cdot q_{if(r)} \cdot\qquad (14)$$

The mobility model takes into account the three main scattering effects, i.e Coulomb, phonon and surface roughness, as:

$$\mu_{eff} = \frac{\mu_o}{\mu_C + \mu_{Ph} \cdot E_\perp^{1/3} + \mu_{RS} \cdot E_\perp^2},\qquad (15)$$

where μ_C is the mobility for the Coulomb scattering, μ_{Ph} for the phonon scattering and μ_{SR} for the surface roughness scattering. EKV accounts for the same scattering mechanisms but instead of a local model like (15), uses a global model given by the integration of the mobility for the scattering mechanisms with respect of the channel as [9]:

$$\mu_{eff} = \mu_o \cdot \left[\frac{1}{L}\int_0^L\left[\frac{1}{\mu_C} + \frac{1}{\mu_{Ph}}E_\perp^{1/3} + \frac{1}{\mu_{SR}}E_\perp^2\right]dx\right]^{-1}.\qquad (16)$$

The integration for each scattering mechanism is given from an expression which includes dependences with some parameters like the channel doping, temperature and others. A complex analytic expression for this integration is implemented in the simulator. In our improved model we consider μ_C, μ_{Ph} and μ_{SR} from (15) as fitting parameters.

The total series resistance R_{ds} can be included in (15) as part of the effective mobility:

$$\mu_R = \frac{\mu_{eff}}{1 + \frac{W}{L}R_{ds}C_{ox}\mu_{eff}\left(V_G - V_T - \frac{1}{2}n_q \cdot V_D\right)}.\qquad (17)$$

Improvements of our model versus the EKV model, in linear regime, are clearly depicted in Figure 2. We can see that the rigorous first derivative from the expression of the drain current of the EKV model does not present good accuracy (dashed lines), but using a relation proposed in [7] where $g_m = (g_{ms} - g_{md})/n$, $g_{ms} = (I_S/U_t) q_{if}$ and $g_{md} = (I_S/U_t) q_{ir}$, we get a better approximation at low field (full symbols). This however fails after the maximum g_m, where the surface roughness scattering dominates. We can see that the first derivative of our modified model presents excellent accuracy (solid lines).

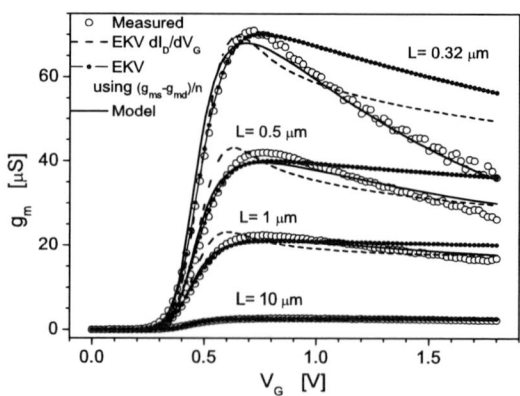

Fig 2. g_m characteristics for different channel lengths at $V_D = 50$mV.

492

C. Short channel effects

Drift velocity, as a function of the longitudinal field, is usually introduced as [10]:

$$v_d = v_{sat} \frac{E_{\parallel}/E_C}{\left[1 + \left(E_{\parallel}/E_C\right)^m\right]^{1/m}}, \qquad (18)$$

where v_{sat} is the carrier velocity saturation, $E_{\parallel}$ is the longitudinal field and E_C is the critical longitudinal field. In order to obtain good accuracy in high order derivatives we used the next function, proposed by [9,11], which is based on (18):

$$v_d = v_{sat} \frac{E_{\parallel}/E_C}{\sqrt{1 + \frac{\left[2 \cdot (2-d) \cdot \left(E_{\parallel}/E_C\right)\right]^2}{0.025 + \left|2 \cdot (2-d) \cdot \left(E_{\parallel}/E_C\right)\right|} + \left(E_{\parallel}/E_C\right)^2}}, \qquad (19)$$

where d is a fitting parameter and has values between 1 and 2. Note that if $d = 2$, (19) becomes (18).

Using the relation between the mobility and the saturation velocity and solving the integration along the channel we found the next mobility expression [9]:

$$\mu = \frac{\mu_R}{\sqrt{1 + \frac{\left[4 \cdot E_{\perp} \cdot (2-d)\left(q_{if} - q_{ir}\right)\right]^2}{0.025 + \left|4 \cdot E_{\perp} \cdot (2-d)\left(q_{if} - q_{ir}\right)\right|} + \left[4 \cdot E_{\perp} \cdot \left(q_{if} - q_{ir}\right)\right]^2}} \qquad (20)$$

Equation (20) is valid only when the saturated velocity is not reached. We can separate the channel in two parts, the first part near the source where the velocity saturation effect is not considered, and the other, near the drain with the velocity saturation effect. For this part the drain voltage is smoothened by the following expression [12]:

$$V_{dseff} = V_{Dsat} + \frac{1}{2}\left[V_D - V_{Dsat} + \Delta V - \sqrt{\left(V_D - V_{Dsat} + \Delta V\right)^2 + 4\Delta V \cdot V_{Dsat}}\right], (21)$$

where $\Delta V = \frac{U_t}{d} \cdot \frac{q_{irsat}}{q_{if}}$. q_{irsat} is the small normalized charge present when the condition of saturation velocity near the drain junction is reached. q_{irsat} is calculated by equating the saturated drain current $I_{Dsat} = W \cdot Q'_{Ir} \cdot v_{sat}$ to the drain current expression given by (5), (6) and (20), where μ is used instead of μ_o. The reverse normalized inversion charge q_{ir} obtained solving these equations, corresponds to q_{irsat}.

The expression for the saturation voltage is:

$$V_{Dsat} = U_t\left[2\left(q_{if} - q_{irsat}\right) + \ln\left(\frac{q_{if}}{q_{irsat}}\right)\right]. \qquad (22)$$

Equation (21) can substitute V_D in all equations. The estimation of the channel length reduction, ΔL, considering V_{Dsat} is equal to [7]:

$$\Delta L = \lambda \cdot L_C \cdot \ln\left(1 + \frac{V_D - V_{Dsat}}{L_C \cdot \frac{v_{sat}}{\mu_o}}\right), \qquad (23)$$

where λ is the channel length model parameter, and $L_C = \left(\varepsilon_{si}X_j/C_{ox}\right)^{1/2}$ with X_j the silicon layer thickness in SOI. The threshold voltage expression must include the DIBL effect, i.e. from [13]:

$$V_t = V_{to} - 2 \cdot \frac{\varepsilon_{si}}{\varepsilon_{ox}} \cdot \frac{t_{ox}}{L} \cdot \left[\left(\Phi f + V_S\right) + \sigma \cdot V_D\right], \qquad (24)$$

where V_{to} is the drain bias independent threshold voltage, ε_{ox} the oxide permittivity, Φf the bulk Fermi potential and σ is the DIBL parameter.

The transfer characteristics in saturation, for $V_D = 1$ V, and the corresponding g_m are analyzed in Fig. 3 where the improved model presents better coincidence for all L. The output characteristics for four lengths at fixed $V_G = 1.5$ V, in Fig. 4, show very good coincidence with the experiment in all regions. The extracted parameters are shown in Table I, including d=1.8, λ=0.27 and σ=1.1 for the four lengths.

TABLE I.
EXTRACTED PARAMETERS

L [μm]	V_{to} [mV]	μ_C [cm²/Vs]	μ_{Ph} [cm²/Vs]	μ_{SR} [cm²/Vs]	R_{ds} [Ω]	v_{sat} [cm/s]
10	400	1.0E-16	7.2E-3	01E-15	200	1E7
1	405	1.0E-16	7.2E-3	17E-15	200	1E7
0.5	420	1.0E-16	7.1E-3	11E-15	200	1E7
0.32	423	1.0 E-16	5.6E-3	29E-15	200	1E7

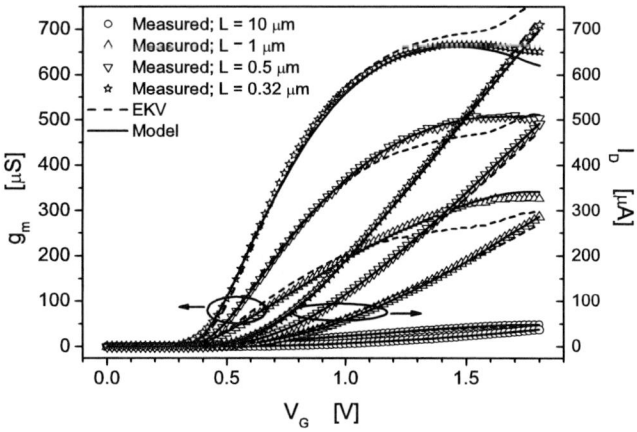

Fig 3. Transfer characteristics in saturation and g_m characteristics for different channel lengths at $V_D = 1V$, $V_S = 0V$.

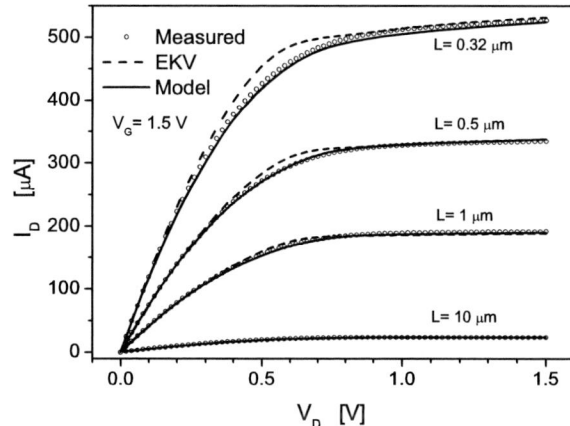

Fig 4. Output characteristics for different channel lengths.

Few models have the capability to accurately simulate the MOSFET current derivatives in the quasi-linear region for fixed V_G and V_D varied from negative to positive values. This region is very important for its applications in e.g. continuous-time filters of analog systems.

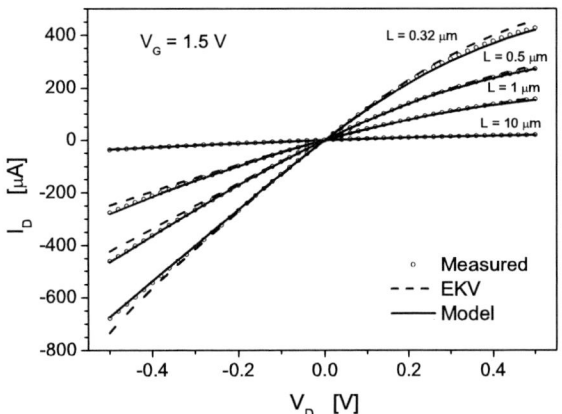

Fig 5. Linear I-V for different channel lengths with $V_G = 1.5$ V.

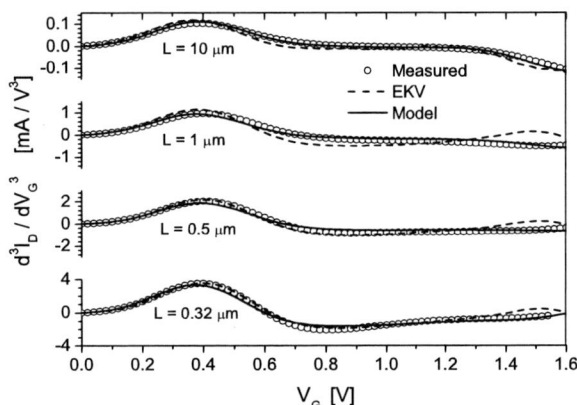

Fig 6. Third derivative for the transfer characteristics in saturation (fig. 3), with $V_D = 1$V and $V_S = 0$ V.

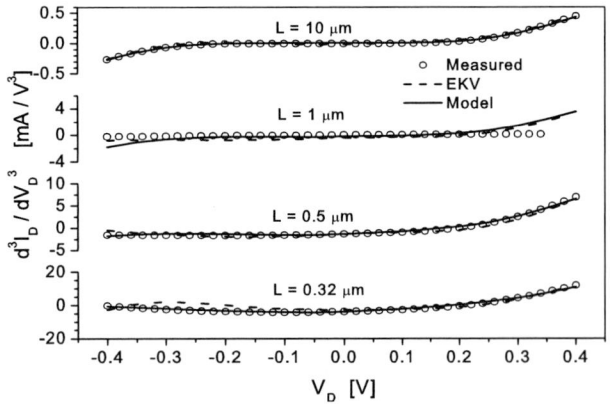

Fig 7. Third derivative for the quasi-linear characteristics (fig. 5) at $V_G = 1.5$ V.

The obtained results are plotted in Fig. 5, where the improved model shows better coincidence again, in particular for the third order derivative (i.e. the dominant term in differential circuits) of both the transfer saturation characteristics, Fig. 6, and the quasi-linear characteristics, Fig. 7.

III. CONCLUSIONS

This work presents a modified model based on the EKV 3.0 model. The parameters are the same as in EKV and can be extracted using the same procedures, while several internal equations have been modified to allow better simulation of analog behaviors such as the current derivatives to the third order. The model was validated by its excellent agreement with the experimental characteristics of PD SOI MOSFETs with short to long channel lengths of interest for analog design: transfer characteristics in linear and saturation regions, the corresponding transconductances, the output characteristics, the quasi-linear characteristics and there third order derivatives of the transconductance in saturation and quasi-linear characteristics.

ACKNOWLEDGEMENTS
This work was supported by CONACYT project 39708.

REFERENCES

[1] W. Liu, "MOSFET Models for SPICE Simulation Including BSIM3v3 and BSIM4". New York: Wiley, 2001.
[2] Klaassen DBM, et al., *The MOS model, level 1101*, Philips Research Laboratories.
[3] C.C. Enz et al., "An Analytical MOS Transistor Model Valid in All Regions of Operation and Dedicated to Low-Voltage and Low-Current Applications", Analog Int. Circuits and Signal Proc. Journ., Vol. 8, pp. 83-114, 1995.
[4] J.-M. Sallese et al, "Inversion charge linearization in MOSFET modeling and rigorous derivation of the EKV compact model ", Solid State Electronics, Vol. 47, pp. 677-683, 2003.
[5] M. Bucher et al., "An Efficient Parameter Extraction Methodology for the EKV MOST Model", IEEE Int. CMTS, Vol.9, pp. 145-150, Trento, Italy, March, 19.
[6] R. M. Corless et al., "On the Lambert W Function", Advances in Computational Mathematics, Vol. 5, pp. 329-359, 1996.
[7] M. Bucher et al., "The EKV 3.0 Compact MOS Transistor Model: Accounting for Deep-Submicron Aspects", MSM'2002, pp. 670-673, 2002.
[8] B. Iñiguez et al., "Accurate Compact MOSFET Modeling Scheme for Harmonic Distortion Analysis", Journal of Semi. Tech. and Science, Vol.4, No. 3, 2004.
[9] M. Bucher, "Analytical MOS Transistor Modeling for Analog Circuit Simulation", PhD Thesis, École Polytechnique Fédérale de Lausanne, 1999.
[10] D. M. Caughey et al., "Carrier Mobilities in Silicon Empirically Related to Doping and Field", Proc. IEEE (Letters), pp. 2192-2193, 1967.
[11] S. M. Sze, *Physics of Semiconductor Devices*, John Wiley, 1981.
[12] Y. Cheng et al., *BSIM3v3 Manual*, CA: UC Berkeley, 1995-1996.
[13] Y. Tsividis, *Operation and Modeling of the MOS Transistor*, WCB/McGraw-Hill, 1999.

Poster Session
Modeling and Simulation

2006 25th International Conference on Microelectronics

Monte Carlo Analysis of Voltage-Current Characteristic Nonlinearity and Harmonic Generation in Submicron Semiconductor Structures

D. Persano Adorno, M.C. Capizzo and M. Zarcone

Abstract – Using a multiparticles Monte Carlo technique, we investigate the dependence of the nonlinear carrier dynamics in GaAs n+nn+ structures operating under very intense sub-terahertz signals from some process parameters as: i) the frequency and the intensity of the excitation signal and ii) the length of the n region.

I. INTRODUCTION

The use of semiconductor systems for broad-band telecommunication and computer devices stimulates a more accurate knowledge of both their low and high frequency electric response properties. Recent advances in electronics pushes the devices to achieve higher output power and efficiency at very high frequencies around the THz region. Therefore, active devices usually operate under large-signal and time-periodic conditions. The miniaturization of integrated circuits implies that even at moderate applied voltages these systems are typically exposed to very intense electric fields.

For these reasons, nonlinear processes involving dynamical effects and high-order harmonic generation in semiconductor structures subject to intense radiation field are attracting increasing attention. Furthemore, the understanding of such processes in semiconductors exposed to far-infrared radiation can be fruitfully exploited for implementing coherent sources in the THz region. Both experimental and theoretical analysis have shown high conversion efficiency for the third and fifth harmonics in low-doped Si, GaAs and InP crystals in the frequency range 30-500 GHz and temperature range 80-400 K due to the nonlinearity of the velocity-field relation [1-5].

The problem of nonlinear processes in the sub-terahertz region recently has been investigated by considering the nonlinearity of the I-V characteristic, the harmonic generation and the electronic noise behavior in a nanometric n+n metal GaAs Schottky-barrier diode [6]. In this case, if also a static voltage is applied, both even and odd harmonics of the radiation field can be generated. The high-order harmonic intensity spectra is similar to that exhibited by the bulk material. The hysteresis-like behavior

D. Persano Adorno, M.C. Capizzo and M. Zarcone are with the Dipartimento di Fisica e Tecnologie Relative dell'Università di Palermo, Viale delle Scienze, Edificio 18, 90128 Palermo, Italy, E-mail: dpersano@unipa.it

of the curve <I(V)> is accompanied by a rapid increase with the frequency of the amplitude of the high order harmonics.

Under nearly similar conditions, submicron heavily doped GaAs Schottky-barrier diodes provide an environment more favorable to that of bulk semiconductors for extracting high order harmonics above the intrinsic noise level [6].

The aim of this work is to present and discuss the dependence of the voltage-current hysteresis cycle and the high-order harmonic efficiency in GaAs n+nn+ structures operating under sub-terahertz signals by the frequency and the intensity of the excitation signal. These very simple structures have been chosen because they form the basis for various high-frequency semiconductor devices.

This paper is organized as follows. In section II we describe the physical model used in the simulations and the procedure for the numerical calculations of J-V characteristic and harmonic generation under cyclostationary conditions. In section III we report the main results of our simulations and in Section IV we report our conclusions.

II. PHYSICAL MODEL AND CALCULATIONS

Using a multiparticles Monte Carlo (MC) code, self-consistently coupled with a one-dimensional Poisson solver, we simulate electronic transport in the structure with an applied periodic voltage V of amplitude V_0 and frequency f, given by $V(t)=V_0\sin(2\pi ft)$. With the aim to compare the results with those obtained in a previous analysis of the inertia effects in GaAs bulk [7], we have first analysed the voltage-current characteristic nonlinearity and the harmonic generation of a symmetric GaAs n+nn+ structure with doping levels of n+=10^{17} cm^{-3} and n=10^{15} cm^{-3}, n+ region length ln+=0.15 µm and n region length ln+=1 µm (structure 1). Then, we have calculated the dependence of the nonlinear carrier dynamics from: i) the frequency of the excitation signal and ii) the length of the n region, in a different GaAs n+nn+ structure having doping level of n+=10^{17} cm^{-3} and n= 10^{15} cm^{-3}, lengths of 0.15 µm and 0.30 µm for the cathode and anode n+ regions, respectively, and length varying from 0.4 µm to 1 µm for the n region (structure 2). In all case our simulations are performed at lattice temperature T=300 K. To solve the

Poisson equation the self-consistent electric field is updated every 10 fs and the structure is divided into meshes of 10^{-8} m each. The total simulated history duration is greater than 100 periods of the frequency of the applied voltage and it is obtained by using 10^3 particles for the whole diode.

The algorithm of MC simulation of the electron motion in the alternating electric field used involves the nonparabolicity of the band structure and the intervalley and intravalley scattering of electrons in multiple energy valleys. Electron scatterings due to ionized impurities, acoustic and polar optical phonons in each valley as well as all intervalley transitions between the equivalent and non-equivalent valleys are accounted for. Since the far-infrared frequencies are below the absorption threshold, in our model we consider the electrons in the conduction band as the only source of nonlinearity. We assume field-independent scattering probabilities; accordingly, the influence of the external fields is only indirect through the field-modified electron velocities. The conduction bands of GaAs are represented by the Γ valley, the four equivalent L-valleys and the three equivalent X-valleys. The parameters of the band structure and scattering mechanisms are taken from Ref. [3].

As concerns the spectral components of the current response, since no static voltage is applied to the structure, only odd harmonics are excited, whose number and amplitude depends from the amplitude and the frequency of the signal. The theory for the calculation of the harmonic generation efficiency has been derived in the paper [3].

III. NUMERICAL RESULTS

By studying the "structure 1" behaviour we have found that, at fixed frequency f=200 GHz, when a small oscillating voltage is applied to the structure, the response is quite similar to the static case. But, increasing the value of the applied voltage, the structure is not capable of following the excitation signal. The J-V characteristic broaden and the current density reaches a value different from that corresponding to the d.c. case, which is dephased with the maximum excitation voltage. The effect increases with the magnitude of the applied voltage. This result is similar to that found by Perez and Gonzalez in the analysis of voltage noise in sub-micrometer semiconductor structures under large-signal regime [8].

To analyse the carrier inertia effects, for a fixed value of V_0, we have studied the J-V characteristic as a function of the frequency f in the range from 10 GHz to 1 THz. In fact, at sufficiently low frequencies, dynamical relations are expected to recover those of the static case. However, increasing the applied signal frequency the inertia of carrier transport, the heating/cooling processes and the possible cut-off of some scattering mechanisms can significantly modify the static J-V relation.

Figure 1 reports the instantaneous total current density <J> flowing through the n+nn+ junction as a function of instantaneous values of the applied voltage

$V(t) = V_0 \sin(2\pi ft)$ during a period of V(t) oscillations with $V_0 = 4$ V, and frequencies f=10, 20, 50, 62.5 GHz (curves 1 to 4, respectively). Here brackets <...> denote averaging over a large number of successive periods simulated by MC runs. The inertia of the current response is nearly absent up to frequency f≈20 GHz. In this case, the <J(V)> diagram follows practically the static J-V relation. At f>20 GHz, the instantaneous <J(V)> characteristic begins to be different from that of the static case.

In figure 2 we report the harmonics efficiency spectra of the current density obtained for the values of frequency f= 20, 50, 62.5 and 100 GHz. A strong reduction of harmonic emission is clearly evident for f=62.5 GHz, which is the frequency value in which the J-V curve changes its shape and begins to be totally different with respect to the d.c. case.

Therefore, for f>100 GHz, the hysteresis-like behaviour of <J(V)> produces a similar increasing of the amplitude of the high order harmonics with f observed in bulk samples and in n+n metal GaAs Schottky-barrier diode. This increasing ends only for very high values of frequency (f≥400 GHz). But in the n+nn+ structure some peculiar mechanism is present and produces a significant reduction in the harmonic emission rate in a very low frequency range.

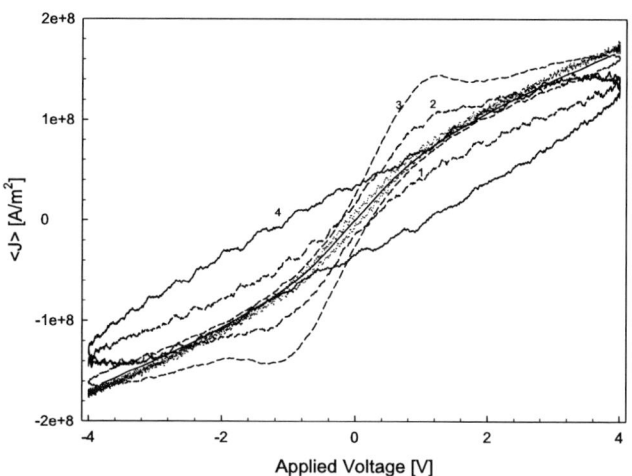

Fig.1 Instantaneous total current density <J> as a function of the instantaneous periodic voltage V(t) applied to the n+nn+ junction and frequencies f=10, 20, 50, 62.5 GHz (curves 1 to 4, respectively). The thick solid line shows the static behaviour of the structure, drawn for comparison.

To have some additional information about both the spatial and the frequency dependence of nonlinearities we have computed the J-V characteristic and the spectral component of the current density in the "structure 2". In this case we have studied the nonlinear processes in the presence of a periodic voltage V(t)=V_0 sin(2πft) applied to the structure, with V_0=2 V, and f ranging from 100 GHz to 1 THz.

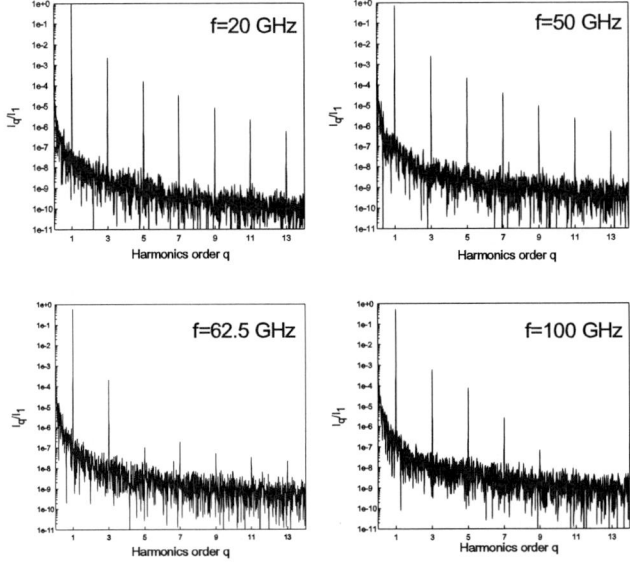

Fig.2 Harmonics generation efficiency versus their order. V_0= 4 V and f= 20, 50, 62.5 and 100 GHz.

Figure 3 reports the harmonics generation efficiency obtained in the structure 2 with length of the n region l_n=0.6 μm and different frequencies f=200, 400, 800 GHz (spectra (a) to (c), respectively). For f≥ 400 GHz the spectra show a reduction of harmonic emission. As in bulk, this effect can be related to the cut-off of the intervalley scattering mechanism. Another interesting result is the change of the electronic noise shape. In fact, for increasing values of frequency, a peak appears in the noise spectra. This result is quite similar to that found in the analysis of electronic noise in GaAs bulk, where a resonant peak with the excitation frequency is expected for f≥ 400 GHz [9].

At fixed frequency values we have calculated the nonlinear carrier dynamics dependence by the length of the n region, with l_n varying from 0.4 to 1 μm. As example, Figure 4 shows the harmonics efficiency spectra obtained in the structure at fixed frequency f=400 GHz and l_n= 0.4, 0.6, 0.8 μm (spectra (a) to (c), respectively). As expected, the current density amplitude decreases with the increasing of the length of the n region, since the resistance is directly related to the n-region spatial length. If we look at the harmonic spectra we note that the efficiency too decreases with the increasing of the length of the n region.

At very high values of frequency (f≥800 GHz) by increasing the length of the n region, we have not harmonic emission at all. This result depends from the fact that in this case: (i) the electric fields swing in the n region have an amplitude high enough to generate nonlinear processes and (ii) for very high frequency values some cut-off in the mechanisms scattering are present [9].

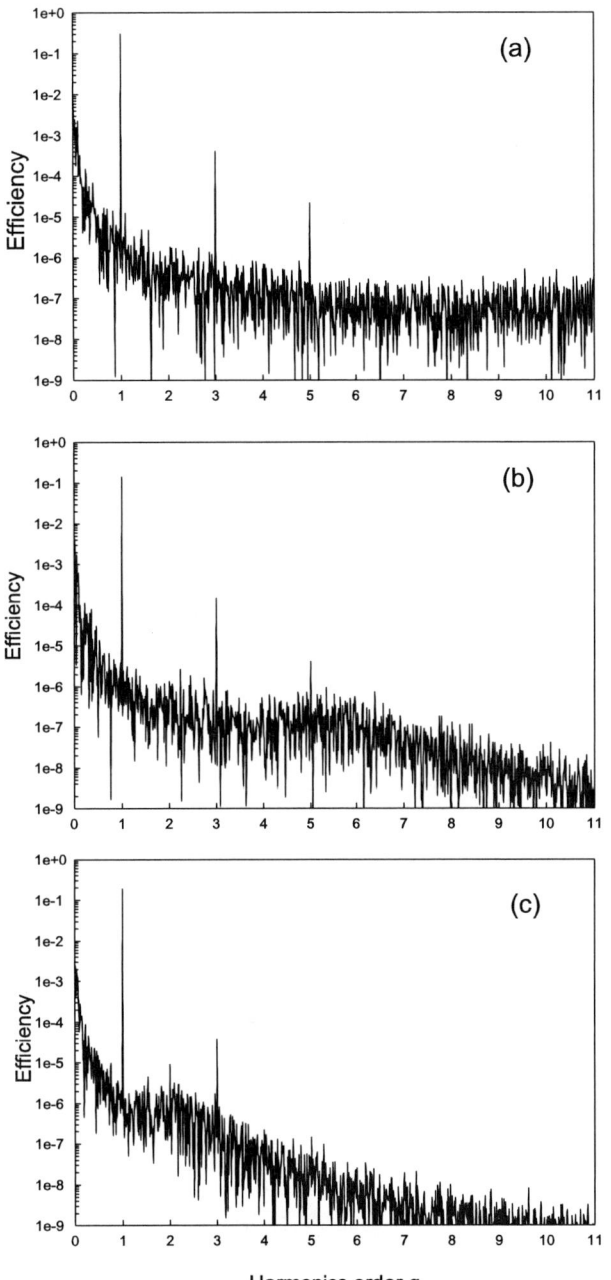

Fig.3 Harmonics generation efficiency versus their order. V_0= 2 V and l_n= 0.6 μm. (a) f= 200 GHz; (b) f= 400 GHz; (c) f= 800 GHz

III. CONCLUSION

We have reported the preliminary results of our study of the nonlinear behaviour of a n+nn+ junction driven by an applied voltage with frequency in the sub-terahertz range. If we compare our results with those obtained in previous analysis of GaAs bulk and of n+n metal GaAs Schottky-barrier diode we can conclude that: (i) as in bulk and in the Schottky diode, for f>100 GHz, the hysteresis-like behavior of $\langle J(V)\rangle$ produces an increasing with the frequency of the amplitude of the high order harmonics and this increasing ends only for very high value of frequency

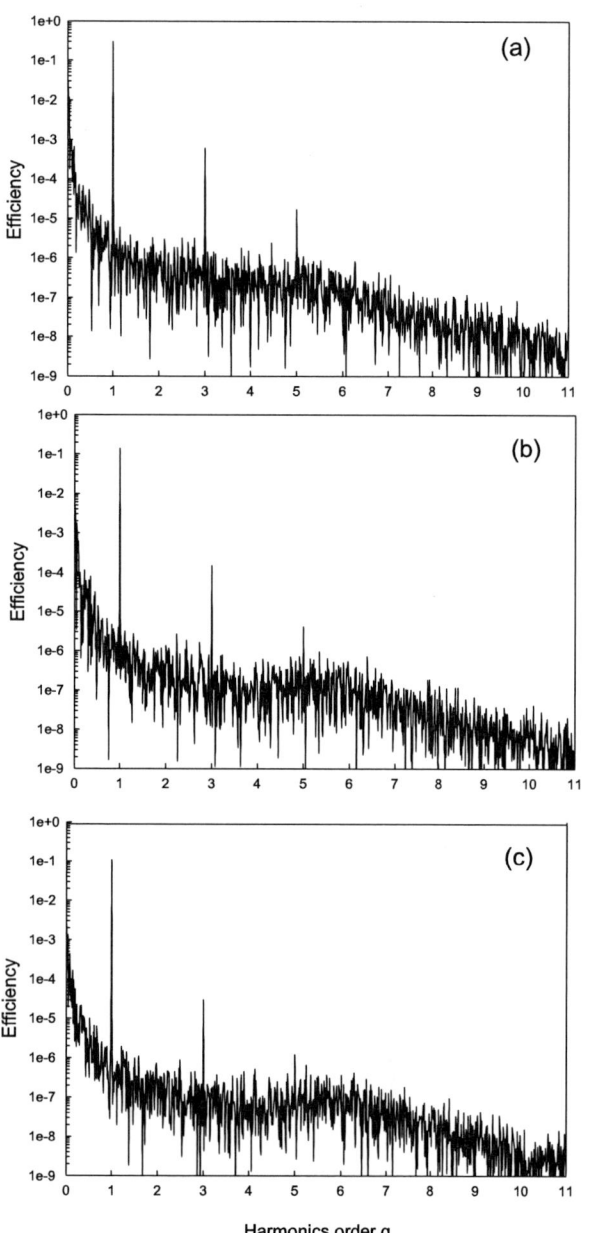

Fig.4 Harmonics generation efficiency versus their order. $V_0 = 2$ V and f= 400 GHz. (a) l_n= 0.4 μm; (b) l_n= 0.6 μm; (c) l_n= 0.8 μm.

(f≥400 GHz); (ii) at very low frequency values of the applied voltage, in the n+nn+ diode it is present some peculiar mechanism that produces a significant reduction in the harmonic rate emission; (iii) the harmonic emission efficiency is strictly related to the n-region length. Our future work will be oriented to the comprehension of these dependencies and to the extension of this type of analysis to more complex structures. A relevant part will be devoted to the analysis of the electronic noise and the signal-noise ratio.

REFERENCES

[1] M. Urban, Ch. Nieswand, M. R. Siegrist, and F. Keilmann, "Intensity dependence of the third-harmonic generation efficiency for high-power far-infrared radiation in n-silicon", *J. Appl. Phys*, 1995, vol. 77, pp 981-984.

[2] R. Brazis, R. Ragoutis, and M. R. Siegrist, " Suitabilty of drift nonlinearity in Si, GaAs, and InP for high-power frequency converters with a 1 THz radiation output", *J. Appl. Phys.*, 1998, vol. 84, pp. 3474-3482.

[3] D. Persano Adorno, M. Zarcone, and G. Ferrante, "Far-Infrared Harmonic Generation in Semiconductors:A Monte Carlo Simulation",Laser Physics, 2000, vol. 10, pp. 310-315.

[4] D. Persano Adorno, M. Zarcone, and G. Ferrante, "High Order Harmonic Generation Efficiency in n-Type Silicon and InP", *Laser Physics*, 2001, vol. 11, pp. 291-295.

[5] P. Shiktorov, E. Starikov, V. Gružinskis, M. Zarcone, D. Persano Adorno, G. Ferrante,L. Reggiani, L. Varani, and J. C. Vaissière, "Monte Carlo Analysis of the Efficiency of Tera-Hertz Harmonic Generation in Semiconductor Nitrides", *Phys. stat. sol. (a)*, 2002, vol. 190, pp. 271-279.

[6] D. Persano Adorno, M. Zarcone, G. Ferrante, P. Shiktorov, E. Starikov, V. Gružinskis, S. Perez, T. Gonzalez, L. Reggiani, L. Varani, and J. C. Vaissière, "Monte Carlo simulation of high-order harmonics generation in bulk semiconductors and submicron structures", *Phys. stat. sol. (c)*, 2004, vol. 1, pp. 1367-1376.

[7] D. Persano Adorno, PhD Thesis, Palermo University (2005)

[8] S. Perez and T. Gonzalez, "Monte Carlo analysis of voltage noise in sub-micrometre semiconductor structures under large-signal regime", *Semicond. Sci. Technol.*, 2002, vol. 17, pp. 696-700

[9] P. Shiktorov, E. Starikov, V. Gružinskis, L. Reggiani, L. Varani, and J. C. Vaissière, "Monte Carlo calculation of electronic noise under high-order harmonic generation", *Appl. Phys. Lett.*, 2002, vol 27, pp 4759-4761.

Diode I-U Curve Fitting
with Lambert W Function

P. Hruska, Z. Chobola, L. Grmela

Abstract – Application of the Lambert W function in diode forward *i-u* curve fitting for the purpose of diode parameters determination is described. CFTOOL of Matlab7 is utilized. Fitting and its results are discussed. Semiconductor PN junction solar cells are used as the diode samples. Brief Lambert W function appendix is attached.

I. INTRODUCTION

Knowledge of semiconductor diode parameter values is the basic prerequisite for technology modification and improvements. Many various methods of the parameters measurement have been developed, both direct and indirect. The method submitted in the paper is based on dc *i-u* curve fitting.

Several relevant parameters are included in the diode forward characteristic *i-u* curve. Extraction of the parameters from experimental curve requires fitting a model – an explicit function *i* of *u* - to experimental data. Simple functional dependence is usually written as

$$i = i_0 \left(\exp\left(\beta_0 (u - ir) \right) - 1 \right) \qquad (1)$$

where i is the diode current, u voltage across diode terminals, i_0 is the saturation current, r diode series resistance, $\beta_0 = \dfrac{e}{nkT}$, e elementary charge, T absolute temperature, k Boltzmann constant and n ideality factor, ranging between 1 and 2.

The product ir is the potential drop that has to be deducted from net voltage u in order to obtain PN junction voltage u_{pn}. It applies for higher current values. Effect of the series resistance r within the range (0.2 Ω, 1.2 Ω) and i_0 within (0.2 nA, 50 nA) can be seen in Fig.1.

P. Hruska, L. Grmela are with the Department of Physics, Faculty of Electrical Engineering and Communication, Brno University of Technology, Technicka 8, 616 00 Brno, Czech Republic, e-mail: hruskap@feec.vutbr.cz

Z. Chobola is with the Department of Physics, Faculty of Civil Engineering, Brno University of Technology, Zizkova 17, 602 00 Brno, Czech Republic, Chobola@dp.fce.vutbr.cz

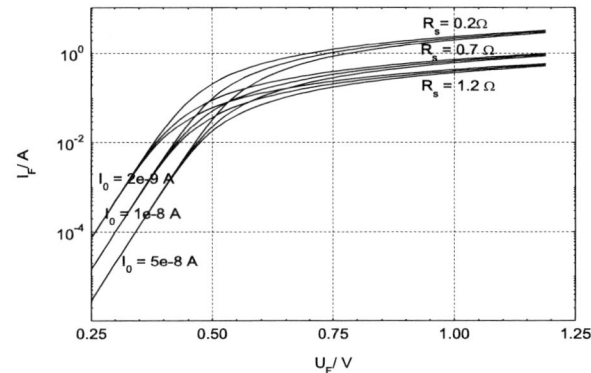

Fig. 1. Effect of r on diode forward current. Parameters: r and i_0. Asymptotic value of the current is fully determined by the series resistance.

The function (1) is an implicit function that is not possible to convert into explicit form in terms of basic mathematical functions. Introduction of Lambert's W function [3] and its support by programs Maple and Matlab offers a solution to the problem. It enables to convert (1) into explicit form and to perform requested fit. Application of Lambert's W function is the main contribution of the paper presented. The explicit form reads

$$i = 1 / \beta_0 / r \times$$
$$\text{lambertw}\left(\beta_0 r i_0 \exp\left(\beta_0 (u + r i_0) \right) \right) - i_0 \qquad (2)$$

For more exact analysis we have to consider generation-recombination and diffusion components of current, [1, 2]. General formula reads

$$i = i_{gr} + i_d, \qquad (3)$$

where the generation-recombination current i_{gr} resp. diffusion (Shockley) current are

$$i_{gr} = B \sqrt{u_{bi} - u_{pn}} \left(\exp\left(\beta_1 u_{pn} \right) - 1 \right) \text{ resp.} \qquad (3a)$$

$$i_d = i_2 \left(\exp\left(\beta_2 u_{pn} \right) - 1 \right) \qquad (3b)$$

Quantities B and i_2 depend on material parameters, such as impurity and trap concentrations, lifetimes, diffusion lengths. Quantity v_{bi} (built-in potential) depends on donor and acceptor concentrations, see e.g. Sze [1], Colinge [2],

$$\beta_1 = \frac{e}{2kT}, \ \beta_2 = 2\beta_1 \ ,$$

u_{pn} is the voltage across PN junction.

II. CHARACTERISTIC CURVE FITTING

A. Model function

Fitting to model (3) requires substitution of experimental value of diode voltage u into (3),

$$u = u_{pn} + ir.$$

Conversion of such formula into implicit function shows to be beyond power of Maple10 (the most advanced tool in Lambert's W function treatment) and cannot be done in full extent. Extensive numerical analysis indicates that the substitution is indispensable in the diffusion current component (3b), that describes higher currents. The substitution can be omitted in the case of (3a), since the currents here are low and potential drop rather negligible. Similar analysis shows that the unity term in (3b) is negligible.

Thus, the implicit model function can be written as

$$i = B\sqrt{u_{vbi} - u}\left(\exp\left(\beta_1 u\right) - 1\right) \\ + i_2 \exp\left(\beta_2 u - ir\right) \quad (4)$$

Its explicit form, in terms of Lambert's W function, reads

$$i = B\sqrt{v_{bi} - x}\exp(19.6x) - 39.2rB\sqrt{v_{bi} - x} \\ + \text{lambertw}\left(39.2ri_2 \exp\left(\theta\right)\right)\frac{1}{39.2r}, \quad (5)$$

with

$$\theta = 39.2\left(1 - rB\sqrt{v_{bi} - x}\left(\exp(19.6x) - 1\right)\right)$$

Substitution of β_1, β_2 values corresponds to room temperature of 295.77 kelvin (22.61°C).

B. Curve fitting procedure

Matlab Curve Fitting Tool is a large and powerfull program. It accepts x-y data from the Matlab operation memory (workspace) and asks for the model function. The model function could be polynomial, Fourier, rational, Gaussian, power series or exponential function of x. It accepts also a custom function – a function according to user's (reasonable) wish. It can use only selected intervals of experimental data – to be chosen in menu 'exclusion rules'. It can, among other features, compare results of a number of fits. It can generate a M-function for a given fitting problem. It is based on the Matlab functions FIT, OPTION.

For a large number of similar fitting, one can construct an 'ad hoc' time and memory saving M-function. Such M- functions could be two- or multiple-step [4]. They evaluate one parameter in a selected data interval and, consequently, input it into a model function for the next step of fitting. It allows to print required parameters into txt or xls files, to save figures into jpg and fig files, to mention possible outputs

III. RESULTS OF FITTING

The fitting procedure was applied to a set of solar cells k300 manufactured by an European manufacturer A. The actual fitting procedure was performed using Matlab7 CFTOOL . The model function (5) was used. A result of fitting one of the cell (sample k312) is presented in Fig.2.

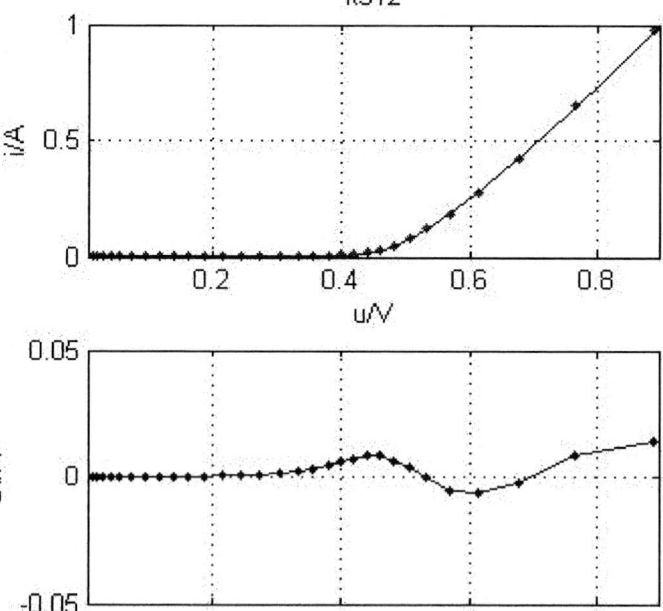

Fig.2. Results of fitting sample k312.
Upper part – experimental curve (dots) and fit (solid).
Lower part - residuals

The fit particulars, including diode parameters are listed in Tab.1.

TAB. 1.
FIT PARTICULARS

fittedmodel1 =
  General model:
  fittedmodel1(x) = (1.0000*r*B*(vbi-
1.*x)^(1/2)*exp(19.600*x)-1.0000*r*B*(vbi-1
    .*x)^(1/2)+.25510e-
1*lambertw(39.200*r*i2*exp(-39.200*r*B*(vbi-1

.*x)^(1/2)*exp(19.600*x)+39.200*r*B*(vbi-
1.*x)^(1/2)+
    39.200*r*i2+39.200*x))-1.0000*r*i2)/r

Coefficients (with 95% confidence bounds):

  B = 2.236e-007
  i2 = 3.207e-010
  r =   0.479 (0.4722, 0.4858)
  vbi = 0.8747 (0.8698, 0.8796)

  rsquare = 0.9975

FitOption: NonlinearLeastSquares
Algorithm: Trust-Region Reflective Newton

Experimental *i-u* curves of the k300 samples are visualized in Fig.3

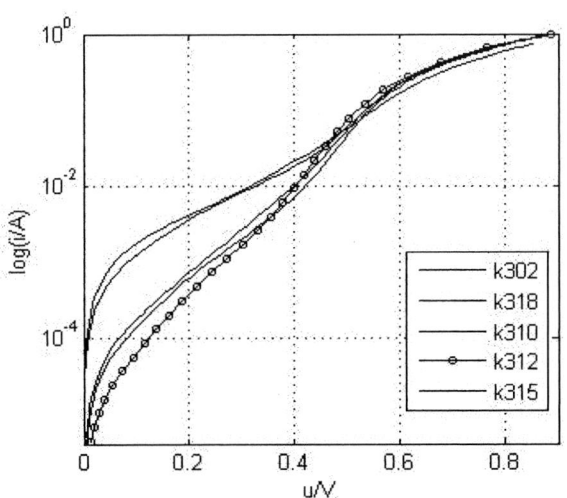

Fig. 3. Experimental *i-u* curves of five k300 samples in semi logarithmic co-ordinates.

Results of fitting of the five samples of k300 solar cells set are given in Tab.2.

TAB. 2.
DIODE PARAMETERS OBTAINED BY FITTING

	B/ V$^{1/2}$	i2/ A	r/ Ω	vbi /V
k302	2.09e-7	3.24e-10	0.481	0.880
k310	1.47e-7	3.16e-10	0.358	0.891
k312	2.24e-7	3.21e-10	0.479	0.875
k315	2.77e-7	2.53e-10	0.379	0.885
k318	1.52e-7	3.53e-10	0.373	0.895

IV. APPENDIX. LAMBERT'S W FUNCTION FEATURES

Lambert's W solves the equation

$$w \exp(w) = x \qquad (6)$$

for w as a function of x.
The function is real-valued for *x* in the range exp(-1) to infinity.
Notation of Lambert's function in Matlab is
$w(x) = \text{lambertw}(x)$
Derivative of the lambertw(*x*) is

$$\frac{d}{dx}\big(w(x)\big) = \frac{w(x)}{(1+w(x))x} \qquad (7)$$

Simple graphs of Lambert's W function are shown in Fig.4:
lambertw(x), lambertw(2x), *d* (lambertw(x))/*d* x

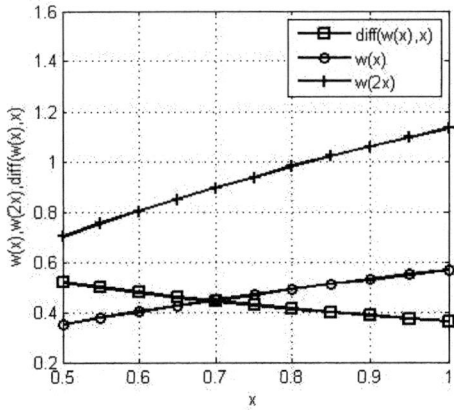

Fig. 4.. Lambert's W function of x, 2x and its derivative.

One can find a comprehensive discussion of Lambert's W function features in paper of R. M. Corrles et all [3]. Brief definition and few examples of expressions using the function are in Maple and Matlab Help pages. Selected function values are listed in table 3.

TAB. 3
NUMERICAL VALUES

x	w(x)	$d\,w(x)/d\,x$
0	0	NaN
0.2000	0.1689	0.7225
0.4000	0.2972	0.5727
0.6000	0.4016	0.4775
0.8000	0.4901	0.4111
1.0000	0.5671	0.3619

W = lambertw(K,X) is the K-th branch of this multi-valued function.

V. CONCLUSION

Authors show how to convert implicit diode i-u curves (1), (3) into explicit form using Lambert's W function. Explicit function form is used as the model function in i-u curves fitting with Matlab7 CFTOOL. The output of the paper are diode parameters: series resistance r, PN junction build-in potential v_{bi} and quantities B, i_2, that are related to semiconductor material parameters.

ACKNOWLEDGEMENT

This research has been supported by the Czech Ministry of Education in the frame of MSM 0021630503 Research Intention MIKROSYN New Trends in Microelectronic System and Nanotechnologies, project of GACR No.102/04/0142 and project MZP VaV/300/01/03.

REFERENCES:

[1] S. M. Sze: *Physics of Semiconductor Devices*, 2nd edition, J.Wiley, New York 1981

[2] J. P. Colinge, C. A. Colinge: *Physics of semiconductor devices*, Kluwer Academic Publishers, Boston 2002

[3] R.M. Corless, G.H. Gonnet, D.E.G. Hare, D.J. Jeffrey, and D.E. Knuth. "On The Lambert W Function". *Advances in Computational Mathematics 5* (1996): 329-359.

[4] HRUŠKA, P., CHOBOLA, Z., GRMELA, L. CF toolbox in solar cell diagnostics In *Proc.of International Conference "Technical Computing Prague 2005"*. Praha, November 15, 2005, pp. 49 - 54, ISBN 80-7080-577-3

[5] P. Hruska, Z. Chobola and J. Sikula: Transport reliability indicators for solar cells, *2nd ELEN Workshop*, October 25-27, 1995, Grenoble, France.

A New Threshold Voltage Analytical Model of Strained Si/SiGe MOSFET

P. M. Lukić, R. M. Ramović and R. M. Šašić

Abstract – In this paper a new analytical threshold voltage model of a strained Si/SiGe MOSFET is presented. Developed model includes all relevant parameters and it is very precise. Besides the previously mentioned fact and the fact that exposed model describes complex physical processes, the model is relatively simple and easily applicable. Presented model is modular, thus it can be easily observed, tested and eventually improved. This model can be used for strained Si/SiGe MOSFET parameters optimization. By using the proposed model, simulations were performed. Obtained results are in very good agreement with the already known ones.

I. INTRODUCTION

The semiconductor device world can be generally divided up based on the type of material used. Each material has its own device advantages and applications. Silicon with its MOS (Metal Oxide Semiconductor) based technology, InP with its optical uses and microwave devices, GaAs with its microwave applications etc [1 - 8]. However, the most widely used semiconductor material for fabrication of electron components was, and currently is, silicon.

It is the trend in the silicon and compound microelectronic technology to continuously develop semiconductor devices which have to be faster, smaller and consume less power. To achieve this, new concepts are investigated [1 – 8].

Metal Oxide Semiconductor Field Effect Transistor (MOSFET) is one of the most commonly used semiconductor component [1 - 3]. Much of what is referred to as modern electronics today would not be possible if it were not this device.

Strained Si/SiGe MOSFET has some superior characteristics comparing with standard silicon MOSFET [4 – 6].

P. M. Lukić is with the Department of Physics and Electrical Engineering, Faculty of Mechanical Engineering, University of Belgrade, Kraljice Marije 16, 11000 Belgrade, Serbia and Montenegro, E-mail: plukic@mas.bg.ac.yu

R. M. Ramović is with the Department of Microelectronics, Faculty of Electrical Engineering, University of Belgrade, Bulevar kralja Aleksandra 73, 11000 Belgrade, Serbia and Montenegro, E-mail: ramovic@etf.bg.ac.yu

R. M. Šašić is with the Department of Physics, Faculty of Technology and Metallurgy, University of Belgrade, Karnegijeva 4, 11000 Belgrade, Serbia and Montenegro, E-mail: plukic@mas.bg.ac.yu

Carriers mobility in the inversion layer of a Si MOSFET is significantly smaller than in the bulk of the same device. This fact is troublesome for Si pMOS devices since CMOS device performance has been limited by the lower intrinsic holes mobility. By using SiGe layer as a MOSFET channel, hole mobility can be significantly increased .

One of the basic characteristics of MOSFET is threshold voltage. Threshold voltage is closely correlated with channel carriers (hole) density. In this paper, the new analytical threshold voltage model of a strained Si/SiGe MOSFET is proposed

II. PROPOSED ANALYTICAL MODEL

Threshold voltage is defined as the gate voltage at which the internal maximum inversion carrier concentration equals substrate doping, implying the onset of heavy inversion inside the device.

The difference between conventional MOSFET and strained Si/SiGe MOSFET is that in strained Si/SiGe MOSFET two threshold voltages exist: V_{TH} which produces strong inversion on the Si/SiGe heterojunction (SiGe channel) and V_{TS} which produces strong inversion on the SiO2/Si junction (Si surface channel).

In Fig. 1. schematic illustration of a strained Si/SiGe MOSFET, with given type and thickness of each layer, is shown.

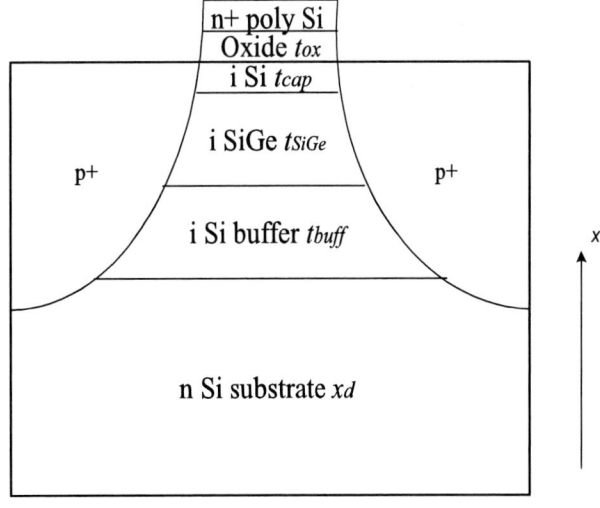

Fig. 1. Cross sectional view of a strained Si/SiGe MOSFET.

The corresponding Poisson's equations in the MOSFET layers are:

- substrate - n Si:

$$\frac{d^2\phi}{dx^2} = -\frac{q_e \cdot N_b}{\varepsilon_{Si}};$$ (1)

- buffer - undoped Si:

$$\frac{d^2\phi}{dx^2} = 0 ;$$ (2)

- channel – undoped SiGe:

$$\frac{d^2\phi}{dx^2} = \begin{cases} -\dfrac{q_e \cdot p(x)}{\varepsilon_{Si}} & |V_G| > |V_{TH}| \\ 0 & |V_G| \le |V_{TH}| \end{cases};$$ (3)

- undoped Si:

$$\frac{d^2\phi}{dx^2} = 0 ;$$ (4)

- oxide:

$$\frac{d^2\phi}{dx^2} = 0 .$$ (5)

In equations (1) – (5) ϕ is the potential, x is the vertical direction from bottom to top, q_e is the electron charge, N_b is the dopant concentration in substrate, $p(x)$ is the hole concentration, V_G is the gate voltage, V_{TH} is the threshold voltage.

Assuming that layers are depleted of free carriers, the set of equations (1) – (5) can be solved:

- substrate – n Si:

$$\phi_S(x) = -\frac{q_e \cdot N_b}{2\varepsilon_{Si}} \cdot x^2; \quad 0 \le x \le x_d ;$$ (6)

- buffer – undoped Si:

$$\phi_B = \phi_S(x_d) - \frac{q_e \cdot N_b}{\varepsilon_{Si}} \cdot x \cdot x_d; \quad 0 \le x \le x_{buff} ;$$ (7)

- channel – undoped Si/SiGe:

$$\phi_C = \phi_B(t_{buff}) - \frac{q_e \cdot N_b}{\varepsilon_{Si}} \cdot x \cdot x_d; \quad 0 \le x \le t_{SiGe} ;$$ (8)

- undoped Si:

$$\phi_U = \phi_C(t_{SiGe}) - \frac{q_e \cdot N_b}{\varepsilon_{Si}} \cdot x \cdot x_d; \quad 0 \le x \le t_{cap} ;$$ (9)

- oxide:

$$\phi_O = \phi_U - \frac{q_e \cdot N_b}{\varepsilon_{Si}} \cdot x \cdot x_d; \quad 0 \le x \le t_{ox} .$$ (10)

Assuming that Fermi level is constant and equal to its value in the bulk, it can be written:

$$\phi_F = -\frac{kT}{q_e} \cdot \ln\left(\frac{N_b}{n_i}\right),$$ (11)

where ϕ_F is the Fermi potential, k is the Boltzmann constant, T is the temperature and n_i is the intrinsic concentration of carriers.

Now, surface potential on the Si/SiGe interface can be defined as:

$$\phi_{TH} = 2\phi_F + \frac{\Delta E_V}{q_e},$$ (12)

and surface potential on the SiO$_2$/Si interface as:

$$\phi_{TS} = 2\phi_F .$$ (13)

In equation (12) ΔE_V is the valence band offset between Si and strained SiGe.

Heavy inversion takes place when surface potentials become equal to ϕ_{TH} and ϕ_{TS}:

$$\phi_C(t_{SiGe}) = \phi_{TH} ,$$ (14)

$$\phi_C(t_{cap}) = \phi_{TS} .$$ (15)

By using solutions of Poisson's equations (6) – (8), it is obtained:

$$\phi_C(t_{SiGe}) = -\frac{q_e \cdot N_b}{2\varepsilon_{Si}} \cdot \left(x_d^2 + 2x_d \cdot t_{buff} + 2x_d \cdot t_{SiGe}\right).$$ (16)

Combining (7) and (8) with (14) maximum depletion layer thickness x_d can be determined.

$$x_d = -\left(t_{buff} + t_{SiGe}\right) + \sqrt{\left(t_{buff} + t_{SiGe}\right)^2 - \frac{2\varepsilon_{Si} \cdot \phi_{TH}}{q_e \cdot N_b}}$$ (17)

In the moment when heavy inversion starts in the SiGe channel, surface potential on SiO$_2$/Si junction is:

$$\left|\phi_U(t_{cap})\right| = \left|\phi_{TH} - \frac{q_e \cdot N_b}{\varepsilon_{Si}} \cdot x_d \cdot t_{cap}\right| .$$ (18)

For proper transistor operating, it is necessary that heavy inversion occurs in the SiGe channel first:

$$\left|\phi_U(t_{cap})\right| < \left|\phi_{TS}\right| .$$ (19)

Now, for a given substrate doping concentration and Ge mole fraction, maximum value of Si cap thickness t_{cap} can be derived:

$$t_{cap} \le \frac{\varepsilon_{Si} \cdot \Delta E_V}{q_e^2 \cdot N_b \cdot x_d} .$$ (20)

If Si cap thickness t_{cap} exceeds maximum value defined by equation (20), carriers will gather on the SiO$_2$/Si interface and gate electric field will be completely compensated. These holes cover SiGe channel, thus inversion will not occur in it. In that case, device can be recognized as a conventional MOSFET with no improvements.

Heavy inversion occurs when gate voltage reaches threshold voltage:

$$\phi_O(t_{ox}) = V_{TH} - V_{FB} .$$ (21)

In equation (21) V_{FB} is flat band voltage. Combining (10), (21) and boundary conditions, it can be calculated:

$$\phi_O(t_{ox}) = \phi_{TH} - q_e \cdot N_b \cdot x_d \cdot \left(\frac{t_{cap}}{\varepsilon_{Si}} + \frac{t_{ox}}{\varepsilon_{ox}}\right) .$$ (22)

Using equations (21) and (22), threshold voltage V_{TH}, which produces heavy inversion on the Si/SiGe heterojunction, can be derived:

$$V_{TH} = V_{FB} + \phi_{TH} - q_e \cdot N_b \cdot x_d \cdot \left(\frac{t_{cap}}{\varepsilon_{Si}} + \frac{t_{ox}}{\varepsilon_{ox}} \right). \quad (23)$$

If gate voltage is further increased, hole concentration in SiGe channel also increases, thus electric field between upper Si/SiGe heterojunction and SiO_2/Si surface raises. Surface voltage becomes more negative and that causes strong inversion on the SiO_2/Si surface. In this case, hole density in SiGe channel is saturated and surface voltage is higher than its threshold value (for a few thermal voltages V_T).

$$\phi_{TH1} = \phi_{TH} - \frac{k \cdot T}{q_e} \cdot \ln \frac{N_v}{N_b}, \quad (24)$$

where N_v is the density of states in valence band.

If the impact of free carriers is neglected, electric field on the upper heterojunction can be written as:

$$E_U = \frac{\phi_{TH1} - \phi_{TS}}{t_{cap}}. \quad . \quad (25)$$

Now, electric field in gate oxide on the SiO_2/Si surface can be easily calculated.

Threshold voltage, which produces heavy inversion on the SiO_2/Si junction is:

$$V_{TS} = V_{FR} + \phi_{TS} - \frac{\varepsilon_{Si}}{\varepsilon_{ox}} \cdot \frac{t_{ox}}{t_{cap}} \cdot \left(\frac{\Delta E_v}{q_e} - \frac{k \cdot T}{q_e} \cdot \ln \frac{N_v}{N_b} \right) \quad (26)$$

III. RESULTS AND DISCUSSION

Based on the presented model, corresponding simulation algorithms were made. The following parameter values were used: $\varepsilon_{Si}=1.054 \times 10^{-10}$F/m; $\varepsilon_{ox}=3.45 \times 10^{-11}$F/m; $t_{buff}=10$nm; $t_{SiGe}=10$nm; $t_{cap}=5$nm; tox=10nm; $n_i=1.45 \times 10^{16}$/cm^3; $V_{FB}=-0.173$V. Achieved results are presented graphically in Figs. 2 – 5.

Threshold voltages V_{TH} and V_{TS} dependences on thickness of undoped silicon t_{cap}, are shown in Fig. 2. (for dopant concentration in substrate is taken $N_b=10^{17}$/cm^3). The same dependences are shown in Fig. 3., but for dopant concentration in substrate $N_b=10^{16}$/cm^3.

It can be observed that V_{TH} increases together with the increase of Si cap thickness t_{cap}. The obtained dependence also shows clear saturating character for bigger values of t_{cap}. At the same time, V_{TS} slowly decreases with the constant slope (Figs. 2. and 3.).

Threshold voltages V_{TH} and V_{TS} dependences on thickness of oxide t_{ox}, are shown in Fig. 4. (for dopant concentration in substrate is taken $N_b=10^{17}$/cm^3). The same dependences are shown in Fig. 5., but for dopant concentration in substrate $N_b=10^{16}$/cm^3. Both threshold voltages decrease with a constant slope versus oxide thickness.

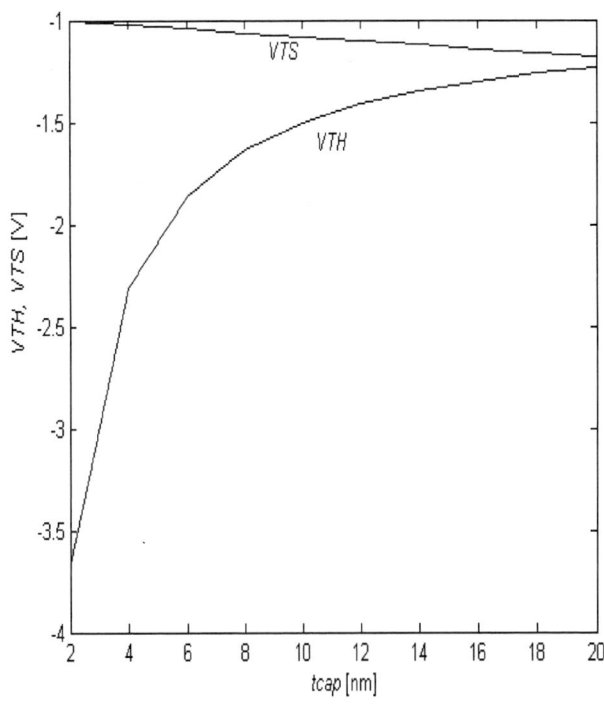

Fig. 2. Threshold voltage vs. thickness of undoped silicon ($N_b=10^{17}$/cm^3)

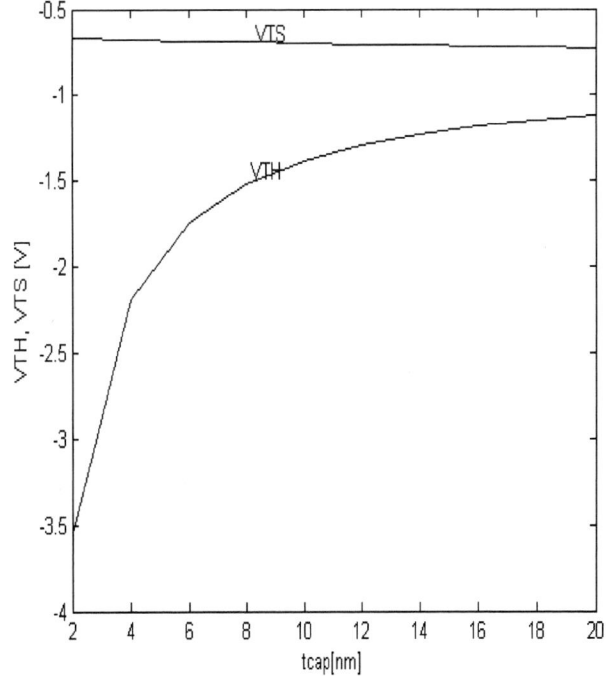

Fig. 3. Threshold voltage vs. thickness of undoped silicon ($N_b=10^{16}$/cm^3).

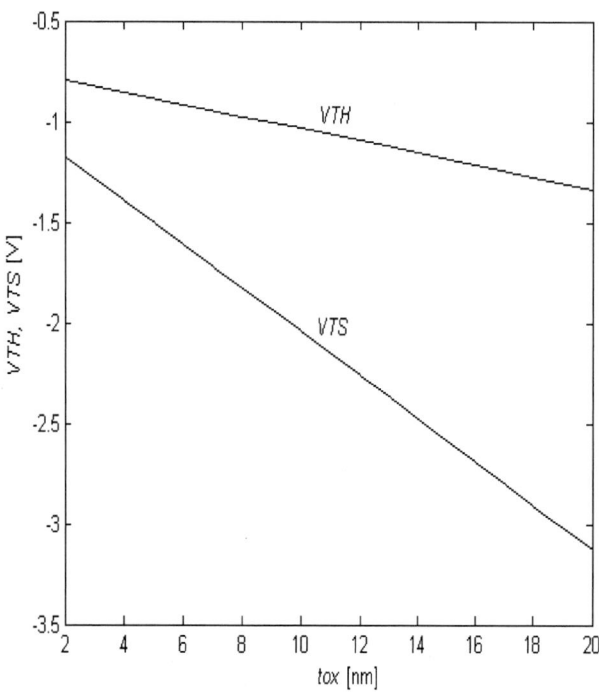

Fig. 4. Threshold voltage vs. thickness of oxide ($N_b=10^{17}/cm^3$).

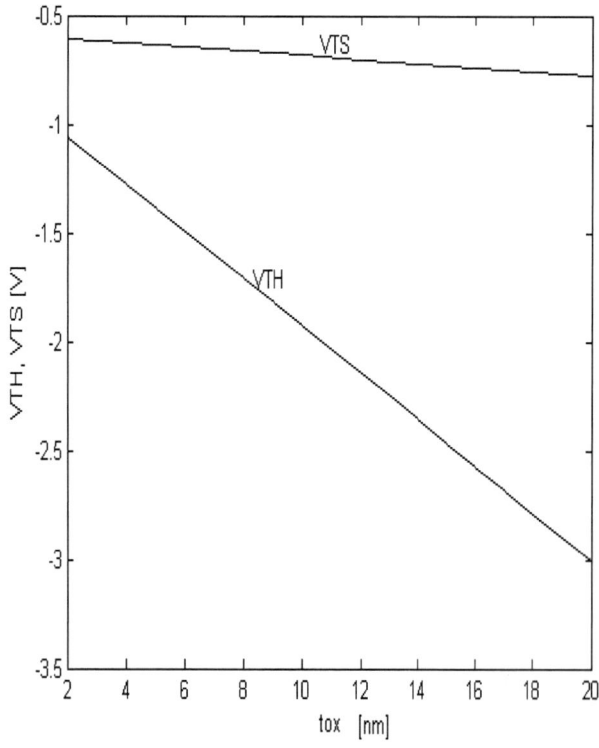

Fig. 5. Threshold voltage vs. thickness of oxide ($N_b=10^{16}/cm^3$).

IV. CONCLUSION

In this paper a new analytical threshold voltage model for strained Si/SiGe MOSFET has been exposed.

The developed model includes all relevant parameters and shows significant degree of accuracy. Therefore, it can be used for parameter extraction and optimization.

Besides the fact that it describes complex physical processes, its application is very simple.

This model is modular, so it can be easily examined and eventually changed – improved.

The results of the simulations performed according to this model are in very good agreement with those available in the literature.

REFERENCES

[1] Rifat Ramović and Rajko Šašić: *Analyze and Modeling of Unipolar Transistors with Small Dimensions* (In Serbian), DINEX, Belgrade, 1999.

[2] Petar M. Lukić: *New Analytical Models of Heterostructure Unipolar Transistors* (In Serbian), PhD Dissertation, Faculty of Electrical Engineering, University of Belgrade, 2005.

[3] Rifat Ramović, Sadeta Krijestorac and Petar Lukić: "Three-Dimensional Potential Distribution Model in Channel of Small Geometry MOSFET with Gauss Impurity Distribution", Proceedings of the 24th International Conference on Microelectronics MIEL 2004., Vol. 1, pp. 307. – 310., May 2004.

[4] Weimin Zhang and Jerry G. Fossum: "On the Threshold Voltage of Strained Si-SiGe MOSFETs", IEEE Transactions on Electron Devices, Vol. 52, No 2, February 2005.

[5] Yee-Chia Yeo, Vivek Subramanian, Jakub Kedzierski, Peiqi Xuan, Tsu-Jae King, Jeffrey Bokor and Chenming Hu: "Design and Fabrication of 50-nm Thin-Body p-MOSFETs with a SiGe Heterostructure Channel", IEEE Transactions on Electron Devices, Vol. 49, No 2, February 2002.

[6] J. Liu, H. J. Kim, O. Hul'ko, Y. H. Xie, S. Sahni, P. Bandary and E. Yablanovitch: "Ge Films Grown on Si Substrates by Molecular Beam Epitaxy Below 450^{0}C", Journal of Applied Physics, Vol. 96, No 1, July 2004.

[7] Petar M. Lukić, Rifat M. Ramović: "The New Analytical Model of SiC-Based MOSFET", Proceedings of the 27th International Convention MIPRO 2004, Conference Microelectronics, electronics and electronic technologies MEET, pp. 53. – 58., Croatia, Opatija, May 24. – 28. 2004.

[8] P. M. Lukić, R. M. Ramović: "Threshold voltage model of SiC MOS structure" (In Serbian), CD Proceedings of the XI Serbia and Montenegro Physicist Congress, pp. 8-103. – 8-106., Petrovac, June 3. – 5. 2004.

2006 25th International Conference on Microelectronics

Graded-Channel SOI nMOSFET Model Valid for Harmonic Distortion Evaluation

M. de Souza, M. A. Pavanello, A. Cerdeira and D. Flandre

Abstract – In this paper an evaluation of the harmonic distortion of graded-channel SOI nMOSFETs is performed. The analysis is carried out by comparing an analytical continuous model and experimental results. The total harmonic distortion, as well as the third and second order terms are used as figures of merit in this comparison. It is shown that GC SOI devices present better gain and linearity behavior than conventional devices and that these advantages are well described by the proposed analytical model. The results show that the proposed set of equations is able to describe the linearity behavior of GC devices, indicating its potential to be used in analog circuit simulation and design.

I. INTRODUCTION

Harmonic distortion is a key parameter for the analog performance of a MOS transistor. Therefore, an adequate modeling of the MOSFET very non-linear behavior, for analog circuit design, should not only provide continuity of the current-voltage characteristics and its derivatives, but also describe accurately its harmonic distortion.

The graded-channel SOI nMOSFET (GC) is an asymmetric channel device that has been proposed and demonstrated to improve the SOI MOSFET analog characteristics such as in operational transconductance amplifiers (OTA) and current mirrors, thanks to its significantly reduced output conductance [1]. In this device, the threshold voltage ion implantation is performed at the source side only, giving rise to a lightly doped region near the drain, reducing the effective channel length ($L_{eff} = L - L_{LD}$, L being the mask channel length and L_{LD} the length of the lightly doped region). By reducing the doping concentration near the drain, the electric field is substantially diminished, and as a result of the reduced impact ionization, the breakdown voltage is increased. Previous works reported several GC devices advantages

M. de Souza and M. A. Pavanello are with Laboratory of Integrated Systems of Polytechnic School, University of São Paulo. Av. Prof. Luciano Gualberto, trav. 3, n. 158, 05508-900, São Paulo, Brazil. M. A. Pavanello is also with Centro Universitário da FEI, São Bernardo do Campo, Brazil. E-mail: michelly@lsi.usp.br (M. de Souza), pavanello@fei.edu.br (M. A. Pavanello).

A. Cerdeira is with the Section of Solid State Electronics, Department of Electrical Engineering, CINVESTAV, Mexico. E-mail: cerdeira@cinvestav.mx

D. Flandre is with the Laboratoire de Microélectronique, Université Catholique de Louvain, Louvain-la-Neuve, Belgium. E-mail: flandre@dice.ucl.ac.be

over conventional ones for analog applications. Among these advantages it has been shown that GC SOI structure also provides appreciable reduction in the harmonic distortion when compared to conventional SOI devices [2].

Recently, aiming to explore the potential of GC devices for analog circuit design, an analytical charge-based continuous model has been proposed for analog simulation of the DC characteristics of submicron GC SOI devices [3].

In this work we evaluate the capability of this model for harmonic distortion description by comparing modeled and measured results. Total (THD), second order (HD2) and third order harmonic distortions (HD3) have been adopted as figures of merit in this comparison. These parameters have been extracted using the Integral Function Method (IFM) [4]. In this method the distortion characteristics of the transistor are obtained by using only DC measurements instead of Fourier-based methods that require complicated AC characterization.

II. MODEL DESCRIPTION

In the present model, the GC drain current (I_{DS}) is obtained computing that of a conventional (uniformly doped from source to drain) SOI MOSFET [5] corresponding to the highly doped part of the channel (eqn. 1) which includes short-channel effects such as mobility reduction, channel length modulation and carrier velocity saturation. The highly doped region acts as a "main" transistor, whose effective drain voltage (V_{DSE}) is a fraction of the drain bias applied to the GC structure:

$$I_{DS} = \frac{W}{L_{eff}} \frac{\mu_n}{1 + V_{DSE} \frac{\mu_n}{L_{eff} v_{sat}}} \left[v_T \left(Q_{nf,D} - Q_{nf,S} \right) - \frac{Q_{nf,D}^2 - Q_{nf,S}^2}{2 n C_{oxf}} \right] \quad (1)$$

where μ_n is the inversion layer mobility (dependent on the electric field both in perpendicular and parallel directions, as in [5]), C_{oxf} is the gate oxide capacitance per unit of area, v_T is the thermal voltage, v_{sat} is the saturation velocity, n is the body factor and $Q_{nf,D}$ and $Q_{nf,S}$ are the inversion charge densities at the drain and source edges of the highly doped region, given by [5]

$$Q_{nf} = C_{oxf} n v_T \left(1 - \sqrt{1 + \frac{4 Q_{nf2}^2}{\left(C_{oxf} n v_T \right)^2}} \right) \quad (2)$$

1-4244-0116-X/06/$20.00 ©2006 IEEE   509

where

$$Q_{nf2} = -C_{oxf}nv_T S_{NT} \cdot ln\left[1 + \sqrt{\frac{-Q_0/(2C_{oxf})}{nv_T S_{NT}^2}}\right.$$
$$\times exp\left(\frac{V_{GF} - nV_{thfi} - nV(y)}{2nv_T}\right) + exp\left(\frac{V_{GF} - V_{thf} - nV(y)}{2nv_T S_{NT}}\right)\right] \qquad (3)$$

with V(y) equal to V_{DSE} and V_S, respectively, at y = L-L_{LD} and 0, V_{GF} being the applied front gate voltage, V_{thf} and V_{thfi} the equivalent threshold voltages in strong and weak inversion regimes; Q_0 the inversion charge density at V_{GF} = V_{thfi} [5] and S_{NT} (<1) a fitting parameter that controls the transition between weak and strong inversion regimes.

The V_{DSE} voltage, which corresponds to the drain voltage that effectively reaches the virtual drain of the highly doped part of the channel, can be obtained through eqn. (4)

$$V_{DSE} = V_{DS,SAT} - V_{DS,SAT}\frac{ln\left[1 + exp\left(A_{TS}\left(1 - \frac{V_{D,HD}}{V_{DS,SAT}}\right)\right)\right]}{ln[1 + exp(A_{TS})]} \qquad (4)$$

A_{TS} being a fitting parameter that controls the transition from triode to saturation regions, $V_{DS,SAT}$ the saturation voltage (see ref. [5]) and $V_{D,HD}$ the potential drop on the highly doped region obtained from eqn. (5)

$$V_{D,HD} = V_{D,HD,SAT} - V_{D,HD,SAT}\frac{ln\left[1 + exp\left(B_{TS}\left(1 - \frac{V_{D,HD,LIN}}{V_{D,HD,SAT}}\right)\right)\right]}{ln[1 + exp(B_{TS})]} \qquad (5)$$

where B_{TS} is a fitting parameter that controls the transition from triode to saturation and $V_{D,HD,SAT}$ and $V_{D,HD,LIN}$ are the drain voltages in the saturation and linear regions accounting the effect of the lightly doped part of the channel, respectively. $V_{D,HD,LIN}$ is given by

$$V_{D,HD,LIN} = \frac{-B_T + \sqrt{B_T^2 - 4A_T C_T}}{2A_T} \qquad (6)$$

where $A_T = -\frac{\mu_{nHD} \cdot n}{2 \cdot (L - L_{LD})} + \frac{\mu_{nLD}}{L_{LD}}\left(\frac{n}{2} - 1\right),$

$B_T = \frac{\mu_{nHD} \cdot V_{GTHD}}{(L - L_{LD})} + \frac{\mu_{nLD} \cdot [V_{GTLD} + V_D(1-n)]}{L_{LD}}$

and $C_T = -\frac{\mu_{nLD}}{L_{LD}}\left[V_{GT,LD} \cdot V_D - \frac{nV_D^2}{2}\right];$

and $V_{D,HD,SAT}$ is given by

$$V_{D,HD,SAT} = V_{INTERM,SAT} + V_{D,HD0} \qquad (7)$$

where

$$V_{INTERM,SAT} = \frac{1}{n}\left(\frac{v_{sat LD}}{v_{satHD}}\frac{Q_{nfD,LD}}{C_{oxf}} + V_{GT,HD}\right) \qquad (8)$$

and

$$V_{D,HD0} = V_{D,HD,LIN}\big|_{V_D(V_{DS,LD} = V_{DS,SAT,LD})} \qquad (9)$$

being $V_D\left(V_{DS,LD} = V_{DS,SAT,LD}\right) = \frac{-B_S + \sqrt{B_S^2 - 4A_S C_S}}{2A_S}$ with

$$A_S = \frac{\mu_{nLD}}{2v_{sat,LD}v_T S_{NT}}\frac{\mu_{nHD}V_{GT,HD}L_{LD} + S_{NT}\mu_{nLD}V_{GT,LD}(L - L_{LD})}{S_{NT}v_T[\mu_{nHD}V_{GT,HD}L_{LD} + S_{NT}\mu_{nLD}V_{GT,LD}(L - L_{LD})]}$$

$$B_S = \frac{\mu_{nHD}V_{GT,HD}L_{LD} + S_{NT}\mu_{nLD}V_{GT,LD}(L - L_{LD})}{S_{NT}v_T[\mu_{nHD}V_{GT,HD}L_{LD} + S_{NT}\mu_{nLD}V_{GT,LD}(L - L_{LD})]} \quad \text{and}$$

$$C_S = -log\left(\frac{-Q_0/C_{oxf}}{nv_T S_{NT}}\right) - \frac{V_{GF} - V_{thfi,LD}}{nv_T}.$$

In the above expressions, V_D is the overall drain voltage applied to the GC structure, $V_{GT} = V_{GF} - V_{thfx}$ is the gate voltage overdrive, V_{thfx} being the threshold voltage (where x = HD for the highly doped and x = LD for the lightly doped region of the channel), $Q_{nfD,LD}$ is the inversion charge density at the drain of the lightly doped region, and V_{thfix} is the equivalent threshold voltage in weak inversion.

III. DEVICES CHARACTERISTICS

GC and conventional SOI devices have been fabricated according to the process described in [1]. Starting from a SOI wafer with doping concentration of 10^{15} cm^{-3} and buried oxide thickness of 390nm, devices were fabricated with a 30nm thick gate oxide in a silicon layer with final thickness of 80nm. The threshold voltage ion implantation led to a body concentration level of about 10^{17} cm^{-3}. The measured GC devices are W/L=18/0.5 (W is the channel width) with three different L_{LD}/L ratios: 0.16 (GC A), 0.29 (GC B) and 0.53 (GC C), resulting in effective channel length of 0.42, 0.35 and 0.24µm, respectively. A conventional SOI device (Conv.) has also been measured for comparison purposes.

The proposed GC model has been used to simulate devices with the same dimensions and characteristics as the measured ones. For all performed comparisons, model parameters were obtained as presented in ref.[3].

The measured and modeled I_{DS} vs. V_{GT} curves obtained at V_{DS} = 1.5V are shown in Figure1 (A). As expected, there is an increase of I_{DS} level as the length L_{LD} increases, due to the effective channel length reduction. The transconductance over drain current ratio (g_m/I_{DS}) curves, which is of high importance for analog performance, are also shown in Figure 1.

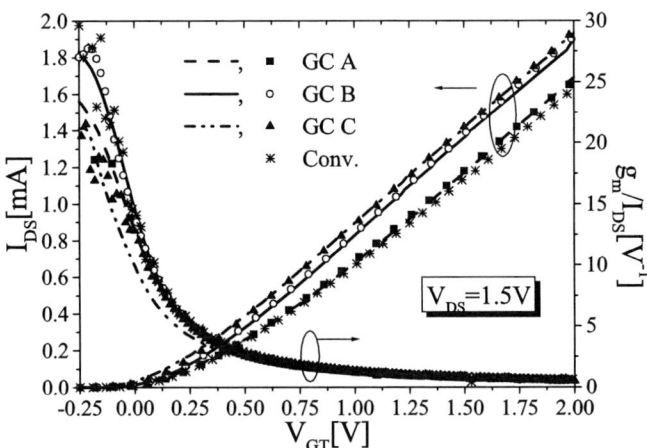

Figure 1 - Measured (symbols) and modeled (lines) I_{DS} and g_m/I_{DS} vs curves (V_{DS}=1.5V) for 0.5μm long GC devices.

Using the experimental and modeled I_{DS} *vs.* V_{DS} curves obtained at V_{GT}=200mV the output conductance ($g_D=\partial I_{DS}/\partial V_{DS}$) has been extracted and the low-frequency open-loop gain has been calculated ($A_V=g_m/g_D$) at V_{DS}=1.5V. The results are shown in Table 1, where can be noted the good agreement between measured and modeled results. From the presented data, one can see the improvement in A_V provided by the GC structure in comparison with the conventional device, which is in the order of 12dB in the worst case (GC A).

TABLE 1
EXPERIMENTAL AND MODELED DC OPEN-LOOP GAIN FOR THE STUDIED DEVICES BIASED AT V_{GT}=200mV AND V_{DS}= 1.5V.

L_{LD}/L	Measured [dB]	Modeled [dB]
Conv.	24.40	-
GC A	36.22	36.03
GC B	39.67	39.69
GC C	36.88	35.60

IV. HARMONIC DISTORTION ANALYSIS AND DISCUSSION

The input DC transfer characteristic required by the Integral Function Method in order to carry out the distortion analysis is the same as presented in Figure 1(A). IFM considers the input signal as composed by a bias voltage source (Vo, which coincides with the gate voltage overdrive, V_{GT}) associated to a sinusoidal signal with amplitude Va. The analysis of harmonic distortion has been performed by evaluating the total harmonic distortion (THD), as well as the third (HD3) and second order (HD2) terms.

The first analysis was performed considering a fixed input signal amplitude of 50mV (100mV peak-to-peak), superposed to the bias voltage, which has been varied from 0 to 1.5V.

The THD is an important parameter for evaluating the distortion properties in amplifiers. For the distortion analysis we will consider the device under test as part of an amplifier used in feedback configuration. The amplifier THD is then dominated by the THD of the active device divided by the gain of the feedback loop. This is equal to the amplifier low-frequency open-loop gain (A_V) for a closed-loop unity gain configuration.

In order to dissociate the linearity and the gain, taking into account the differences in the gain presented in Table 1, the harmonic distortion figures have been normalized by A_V. Besides, aiming to relate the current consumption to the linearity, we have plotted the distortion as a function of g_m/I_{DS}. This way, the THD/A_V *vs.* g_m/I_{DS} curves can be treated as a figure-of-merit of the drain current needed to attain a certain linearity level for a given gain-bandwidth product. The comparison between modeled and measured THD/A_V is shown in Figure 2(A), presenting an excellent matching. Figure 2(B) shows the comparison between HD2/A_V and HD3/A_V extracted from measurements and modeled results. The analysis has been performed in strong inversion only, i.e. for g_m/I_{DS} smaller than 10 V^{-1}. For GC SOI transistors the THD/A_V is virtually constant with g_m/I_{DS}; however for g_m/I_{DS} smaller than 1 or 2V^{-1} the devices come into the triode region, which is not of interest for amplifier operation.

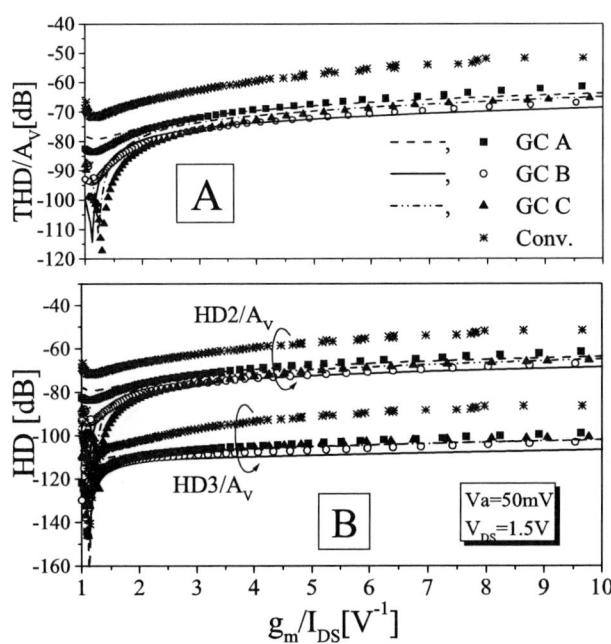

Figure 2 - Comparison between THD/A_V (A) and HD2/A_V and HD3/A_V (B) obtained from measured (symbols) and modeled curves (lines) at input voltage amplitude of 50mV.

The device distortion is being dominated by HD2 as clearly stated in figure 2 independently of the channel architecture. In all analyzed figures of merit the GC SOI presents improved behavior compared to conventional SOI. The THD/A_V increases with the effective channel length increase: THD/A_V of GC A is about 3-5 dB larger than in GC B and C at g_m/I_{DS}=5 V^{-1} (close to the largely used V_{GT}

of 200 mV), as given in Table 2. Also devices GC B and C can provide a reduction of about 13 dB in THD/A_V with respect to conventional SOI with similar total channel length. Even the presence of a short lightly doped region formation (as in the case of GC A) can provide improvements on the gain and 9 dB on THD/A_V over the conventional SOI. Among the GC SOI the HD3/A_V varies about 10 dB depending on the L_{LD} configuration but this figure is always better than in conventional SOI.

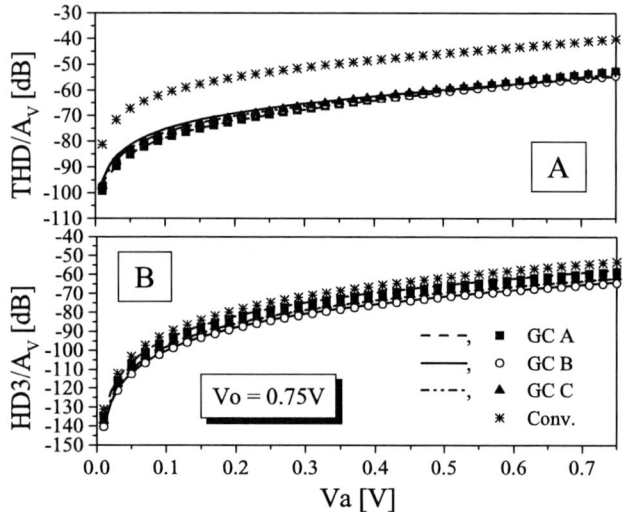

Figure 3 – THD/A_V (A) and HD3/A_V (B) obtained from measured (symbols) and modeled (lines) data varying Va at Vo=0.75.

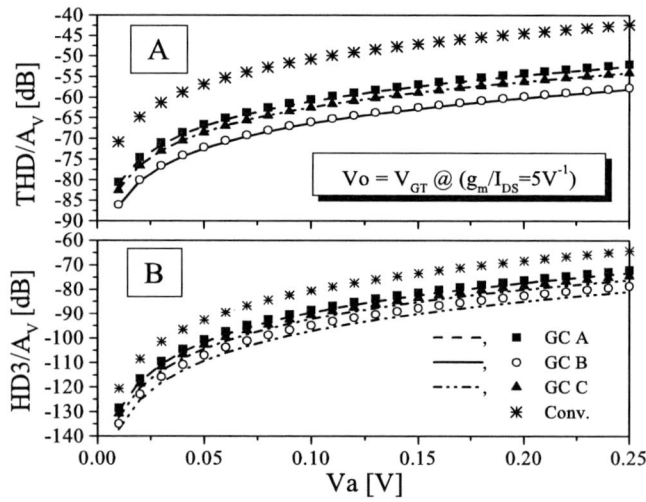

Figure 4 – THD/A_V (A) and HD3/A_V (B) obtained from measured (symbols) and modeled (lines) data varying Va at g_m/I_{DS}=5V^{-1}.

Another important feature that can be analyzed for the linearity of analog circuits is the distortion as a function of the input signal amplitude. In this case, THD/A_V and HD3/A_V were obtained at a fixed bias point Vo=0.75V and the input signal amplitude varied from 10mV to 0.75V. The obtained results are presented in Figure 3. The distortion figures were also obtained at a fixed value of g_m/I_{DS} (5V^{-1}), varying Va from 10mV to 250mV (figure 4). From the

presented curves one can see that the coincidence between results obtained from modeled and measured curves is very good and the behavior of the distortion figures is well described by the proposed set of equations as well as the THD reduction provided by the GC structure in comparison to conventional devices.

TABLE 2
MEASURED AND MODELED THD/A_V EXTRACTED AT G_M/I_{DS}=5V^{-1}.

L_{LD}/L	Measured [dB]	Modeled [dB]
Conv.	-57.09	-
GC A	-66.95	-67.72
GC B	-71.73	-72.14
GC C	-70.10	-69.46

V. CONCLUSION

This work presented the analysis of the linearity of SOI devices with asymmetric channel configuration. It is shown that GC SOI devices tend to attain better gain and distortion behavior in comparison to conventional SOI, as the L_{LD}/L ratio increases. The harmonic distortion figures were obtained through experimental and modeled DC curves, by using the Integral Function Method. From the obtained results, it is clearly seen that the proposed GC SOI model can describe the harmonic distortion figures. It indicates that the model is suitable to be used in the analog circuit simulation and design, allowing the prediction of the harmonic distortion of GC SOI devices.

ACKNOWLEDGEMENT

M. A. Pavanello acknowledges CNPq for the financial support. D. Flandre is Honorary senior research associate of FNRS, Belgium.

REFERENCES

[1] M. A. Pavanello, J. A. Martino and D. Flandre. "Graded-Channel Fully Depleted Silicon-On-Insulator nMOSFET for Reducing Parasitic Bipolar Effects". *Solid-State Electronics*, 2000, vol. 44, pp. 917-922.

[2] A. Cerdeira, M. Alemán, M. Pavanello, J. A. Martino, L. Vancaillie and D. Flandre. "Advantages of the Graded-Channel SOI FD MOSFET for Application as a Quasi-Linear Resistor". *IEEE Transactions on Electron Devices*, 2005, v. 52, pp. 967-972.

[3] M. de Souza, M. A. Pavanello, B. Iñiguez and D. Flandre. "A Charge-Based Continuous Model for Submicron Graded-Channel nMOSFET for Analog Circuit Simulation". *Solid-State Electronics*, 2005, v. 49, pp. 1683-1692.

[4] A. Cerdeira, M. Alemán, M. Estrada and D. Flandre. "Integral function method for determination of nonlinear harmonic distortion". *Solid-State Electronics*, 2004, v. 48, pp. 2225-2234.

[5] B. Iñiguez , L. F. Ferreira, B. Gentinne B and D. Flandre D. "A Physically-Based C_∞-Continuous Fully-Depleted SOI MOSFET Model for Analog Applications". *IEEE Transactions on Electron Devices*, 1996, v. 43, pp. 568-575.

2006 25th International Conference on Microelectronics

A Simple Polysilicon Thin-Film Transistor SPICE Model

I. Pappas, A. T. Hatzopoulos, D. H. Tassis, N. Arpatzanis, S. Siskos,
A.A. Hatzopoulos, C. A. Dimitriadis and G. Kamarinos

Abstract – A simple current-voltage model for polysilicon thin-film transistors (TFTs) is proposed, including the sixth-order polynomial function coefficients fitted to the effective mobility versus gate voltage data, the channel length modulation and the impact ionization effect. The model possesses continuity of current in the transfer characteristics from weak to strong inversion and in the output characteristics throughout the linear and saturation regions of operation. It has been applied in a number of long and short channel TFTs and the statistical distributions of the model parameters involved have been derived. The new model was adapted in the simulation program AIM-SPICE, with the extracted parameters used as input parameters of the new polysilicon TFT model.

I. INTRODUCTION

Polysilicon thin-film transistor (poly-Si TFT) technology is emerging strongly in large area electronics [1], such as display technology, memories and scanners. The efficient and economic design of polysilicon TFT integrated circuits depends to a great extent on the simplicity and accuracy of the model available for the TFT devices involved. In most of the proposed models, several parameters are needed to reproduce the experimental output characteristics or long computing time is required [2]-[7]. Relatively, there is a great interest for the development of analytical models suitable for circuit simulation programs, such as SPICE.

Recently, a simple analytical on-state drain current model for polysilicon TFTs has been developed, based on the carrier transport through latitudinal and longitudinal grain boundaries [8]. From the experimental transfer characteristics in the linear region, the experimental data of the effective mobility versus gate voltage can be extracted. By employing these data, the output characteristics can be successfully reproduced over wide range of bias voltages in devices with different gate length [9]. In this work, the current-voltage model has been applied in a number of long channel and short channel polysilicon TFTs and the

I. Pappas, A. T. Hatzopoulos, D. H. Tassis, N. Arpatzanis, S.Siskos and C. A. Dimitriadis are with the Department of Physics, Aristotle University of Thessaloniki, Thessaloniki, 54124 Greece, E-mail: ilpap@auth.gr

A. A. Hatzopoulos is with the Department of Electrical & Computer Engineering, Aristotle University of Thessaloniki, Thessaloniki, Greece

G. Kamarinos is with IMEP, ENSERG, 23 rue des Martyrs, 38016 Grenoble, Cedex 1, France

statistical distribution of the model parameters are derived.

The new polysilicon TFT model was adapted in the circuits simulation program AIM-SPICE. The extracted parameters are used as input parameters and their measured values are used as default values of the new TFT model. In order to verify the feasibility of the new model in the AIM-SPICE, the input and output characteristics of the fabricated TFTs were reproduced through simulation.

II. POLYSILICON TFT I-V MODEL

The drain current in the linear region of a polysilicon TFT can be expressed by the equation:

$$I_d = \frac{W}{L} \mu_{eff} C_{ox} \left(V_g - V_{inv} \right) V_d \qquad (1)$$

where W is the channel width, L is the channel length, C_{ox} is the gate capacitance per unit area, V_g is the gate voltage, V_d is the drain voltage and V_{inv} is the charge inversion voltage corresponding to the gate voltage at which the onset of the channel conductance is observed in the transfer characteristic. Considering carrier transport through latitudinal and longitudinal grain boundaries, the effective carrier mobility μ_{eff} is [8]:

$$\mu_{eff} = \frac{\mu_{gi}}{1 + \frac{w}{L_g - w} \exp\left(\frac{qV_b}{kT}\right)} + \frac{L_{gb}}{L_g} \mu_{gb//} \qquad (2)$$

where μ_{gi} is the mobility for a carrier within the grain region, w is the width of the depletion region at the grain boundary which is dependent on the grain boundary potential barrier height V_b, L_{gb} is the effective grain boundary width fixed at the value 2 nm, L_g is the average grain size of the polysilicon layer and $\mu_{gb//}$ is the mobility for a carrier passing along a grain boundary. The dependence of the potential barrier height at the grain boundary V_b on V_g is modeled as [8]:

$$qV_b = \frac{0.56}{1 + \left(\frac{q\left(V_g - V_{inv}\right)}{E_t} \right)} \qquad (3)$$

1-4244-0116-X/06/$20.00 ©2006 IEEE

The parameter E_t is related to the quality of the polysilicon material. For smaller parameter E_t, qV_b decreases more rapidly with V_g, thus the smaller value of E_t corresponds to better quality of the active polysilicon layer. Fit of the experimental I_d versus V_g data with Eqs. (1)-(3) enables the determination of the parameters L_g, E_t, $\mu_{gb//}$, μ_{gi} and V_{inv}. Thus, from analysis of the transfer characteristics, the experimental data of μ_{eff} versus V_g can be extracted. The use of the charge inversion voltage (V_{inv}) instead of the more commonly used threshold voltage, enables well correlation to the experimental data in the gate voltage region from weak to strong inversion.

In polysilicon TFTs, the experimental data of μ_{eff} versus V_{eff} can be represented by the nth-order polynomial function [9, 10]:

$$\mu_{eff} = \sum_{i=0}^{n} c_i \left(V_{eff}\right)^i \quad \text{for } V_{eff} > 0 \qquad (4)$$

where $V_{eff} = V_g - V_{inv}$ and $\{c_i\}$ are the polynomial coefficients. Considering ideal saturation region and using Eq. (4), the output characteristics can be described by the expression [8, 9]:

$$I_{di} = C_{ox} \frac{W}{L} \sum_{i=0}^{n} \left[\frac{c_i}{i+2} \left[\left(V_{eff}\right)^{i+2} - \left(V_{eff} - V_d\right)^{i+2} \right] \right]$$

$$\text{for } V_d \leq V_{eff}$$

$$I_{di} = C_{ox} \frac{W}{L} \sum_{i=0}^{n} \left[\frac{c_i}{i+2} \left(V_{eff}\right)^{i+2} \right]$$

$$\text{for } V_d > V_{eff} \qquad (5)$$

A more realistic description of the output characteristics is obtained by taking into account the channel length modulation effect associated with the channel length reduction due to the finite extent of the saturated part of the channel near the drain and the impact ionization effect due to the high electric field near the drain region. When these effects are included, the transistor characteristics can be described by equation:

$$I_d = I_{di} \times F_m \times F_i \qquad (6)$$

The channel length modulation effect is described properly by the function F_m:

$$F_m = \left(1 + \lambda V_d\right) \qquad (7)$$

The gate voltage dependence of the fitting parameter λ is empirically described by:

$$\lambda = \lambda_o - \lambda_1 V_{eff} \qquad (8)$$

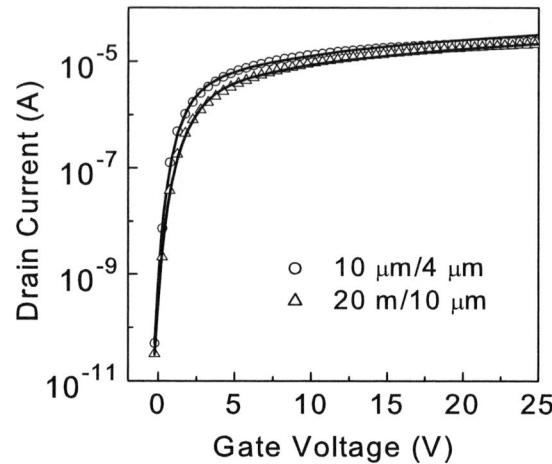

Fig. 1. Measured and simulated input characteristics of long and short channel TFTs and drain voltage V_d=0.1V.

where λ_o and λ_1 are model parameters. The current increase due to the impact ionization near the drain is described by the factor F_i :

$$F_i = 1 + K_n \upsilon_{sat} \exp\left(-\frac{B_n L}{V_d}\right) \qquad (9)$$

where u_{sat} is the carrier saturation velocity ($u_{sat} = 10^7$ cm/s) and K_n, B_n are empirical parameters. By employing the coefficients $\{c_i\}$ of the μ_{eff} versus V_{eff} polynomial function and using λ_o, λ_1, K_n, B_n as fitting parameters, the output characteristics can be reproduced.

III. EXPERIMENTAL APPLICATION

A. Input Characteristics

The above I – V model has been applied successfully in a number of TFTs, with channel lengths 4 μm and 10μm, fabricated on solid-phase crystallized polysilicon layers, irradiated by XeCl excimer laser with energy density 435 mJ/cm^2.

Fig. 1 shows the transfer characteristics of polysilicon TFTs with different channel dimensions, measured at drain voltage $V_d = 0.1$ V. The symbols correspond to the experimental data and the solid lines to the model results using the parameters: (a) $\mu_{gi} = 75.63$ cm^2/Vs, $\mu_{gb//} = 0.174$ cm^2/Vs, $E_t = 0.4$ eV, $L_g = 0.168$ μm, $V_{inv}= -0.698$ V for the TFT with L = 4 μm and (b) $\mu_{gi} = 64.56$ cm^2/Vs, $\mu_{gb//} = 0.102$ cm^2/Vs, $E_t = 0.4$ eV, $L_g = 0.075$ μm, $V_{inv}= -0.76$ V for the TFT with L = 10 μm. It is worth to mention that, independently on the channel length of the device, the experimental transfer characteristics are reproduced with the model using the parameter $E_t = 0.4$ eV, indicating that the quality of the polysilicon layer is similar in all the investigated devices. In the investigated transistors, the experimental I_d versus V_g data in the on-current region can be reproduced from Eqs. (1)-(3), using the three fitting parameters L_g, μ_{gi} and V_{inv}. These three parameters will be used as input parameters of AIM-SPICE.

B. Output Characteristics

For accurate reproduction of the experimental $I_d - V_d$ curves using Eqs. (5)-(9), it was found that the lower-order polynomial representation for μ_{eff} (V_{eff}) is a sixth-order polynomial function.

Fig. 2 shows the reproduced output characteristics (a) of long channel TFTs and (b) of short channel TFTs. In long channel TFTs (L= 10 μm), the experimental output characteristics can be reproduced with the model using the μ_{eff} (V_{eff}) sixth-order polynomial function coefficients and the impact ionization parameters K_n and B_n. The measured $I_d - V_d$ curves (open cycles) and the simulated curves (continuous lines) show a good agreement.

In short channel TFTs (L = 4 μm), the output characteristics can be reproduced with the model using the μ_{eff} (V_{eff}) sixth-order polynomial function coefficients and the channel length modulation fitting parameters λ_o and λ_1 described in Eqs. (7) and (8). A good agreement between measured data (open cycles) and calculated $I_d - V_d$ curves (solid lines) is obtained, showing that the channel length modulation effect is sufficient to explain the output characteristics of the investigated short channel polysilicon TFTs.

IV. SPICE IMPLEMENTATION

SPICE is the most commonly used analog circuit simulator today and is enormously important for electronics industry. AIM-SPICE is a version of circuits simulation program SPICE. The default poly-Si TFT model included in AIM-SPICE is the PSIA2 level 16 poly-Si TFT model. Our purpose was to adapt the new polysilicon TFT model into AIM-SPICE. In order to achieve this, the source code of the new model was written with the use of C language. The source code of the model includes the input parameters of the model and its functions. The source code was compiled with AIM-SPICE, so that the compatibility of the new model to be insured. This means that we were able to run all the routines of simulation (DC, AC and transient analyses) with the new model. The input parameters of the new model are the μ_{eff} (V_{eff}) sixth-order polynomial function coefficients, the impact ionization parameters K_n and B_n and the channel length modulation fitting parameters λ_o and λ_1.

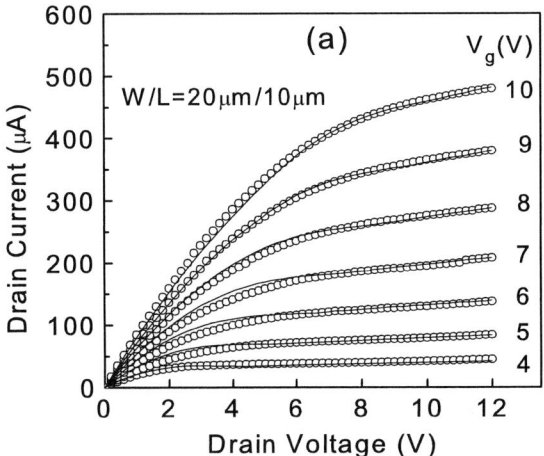

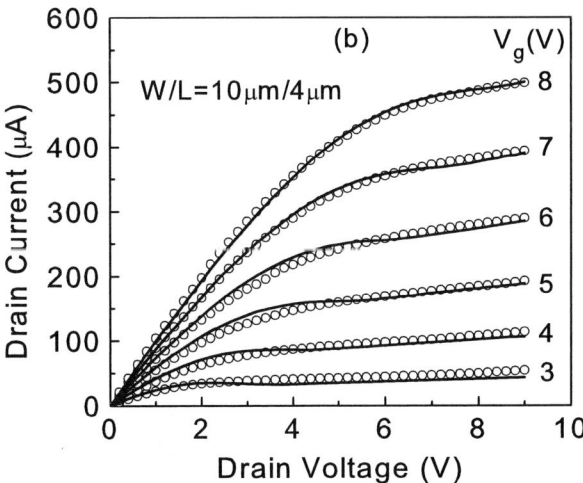

Fig. 2 Measured and simulated output characteristics: (a) for long channel TFT using the impact ionization parameters $K_n = 1 \times 10^{-7}$ s/cm, $B_n = 1.55 \times 10^4$ V/cm and μ_{eff} (V_{eff}) coefficients : $\{c_0,c_1,c_2,c_3,c_4,c_5,c_6\} = \{-17.56044, -1.002, 8.987922, -2.1117, 0.208307, -0.00963, 1.71 \cdot 10^{-4}\}$ and (b) for short channel TFTs using the channel length modulation parameters $\lambda_o = 0.111$ V^{-1}, $\lambda_1 = 8 \times 10^{-3}$ V^{-2} and μ_{eff} (V_{eff}) coefficients: $\{c_0,c_1,c_2,c_3,c_4,c_5,c_6\} = \{-88.178, 73.9, 13.7274, 1.279789, -0.06002, 0.001191, -3.76 \cdot 10^{-6}\}$

The default values of the inputs parameters are the statistical values obtained from the investigated devices and they are presented in Table 1. The convenience of the new model is that there is only one polynomial drain current equation for all the regions of operation which leads to more easy deviations and calculations for the source code of the new model.

Fig. 3 shows the simulated output characteristics of the new poly-Si model for short channel TFTs, reproduced with AIM-SPICE. As it can be seen from Fig. 3, there is a very good correlation between the simulated output and the measured characteristics. This shows that it is possible to perform realistic simulations with AIM-SPICE using the proposed TFT model and thus demonstrate the feasibility of circuits designed using poly-Si TFTs.

TABLE I
DEFAULT VALUES OF THE INPUT PARAMETERS

Input Parameters	Description	Default Values
V_{inv}	Charge inversion voltage	-0.61 ± 0.15 V
λ_o	Channel length modulation parameter	0.111 V^{-1}
λ_1	Channel length modulation parameter	$8 \cdot 10^{-3} \pm 8.6 \cdot 10^{-4}$ V^{-2}
B_n	Impact ionization parameter	$1.55 \cdot 10^4 \pm 2.9 \cdot 10^3$ V/cm
K_n	Impact ionization parameter	$1 \cdot 10^7$ s/cm
L_g	Average grain size of the polysilicon layer	0.104 ± 0.025 µm
μ_{gi}	Carrier mobility within the grain region	70 ± 5.56 cm^2/Vs

V. CONCLUSION

A simple and continuous $I - V$ model for the output characteristics of polysilicon TFTs is proposed, including the sixth-order polynomial coefficients fitted to the μ_{eff} versus V_{eff} data, the channel length modulation effect and the impact ionization effect. The model has been applied in TFTs fabricated on SPC polysilicon layers, irradiated by XeCl excimer laser with energy density 435 mJ/cm^2. This new model was adapted in the circuits simulation program AIM-SPICE and the measured model parameters were set as the default values of the input parameters for the AIM-SPICE. The output characteristics were reproduced by simulations with AIM-SPICE and they have shown very good correlation with the measured ones.

REFERENCES

[1] W. G. Hawkins, "Polycrystalline-silicon device technology for large area electronics", in *IEEE Trans. Electron Devices*, vol. 33, pp. 447.

[2] M. Valdinoci, L. Colalongo, G. Baccarani, A. Pecora, I. Policicchio, G. Fortunato, F. Plais, P. Legagneux, "Analysis of electrical characteristics of polycrystalline silicon thin-film transistors under static and dynamic conditions", *Solid State Electron.*, vol. 41, pp. 1363-1369, 1997.

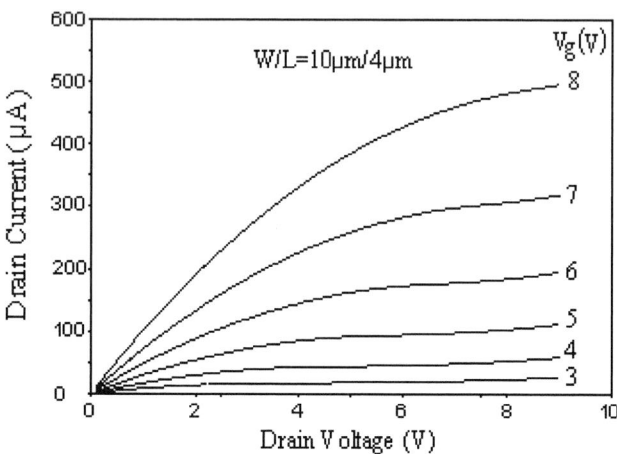

Fig. 3. Simulated output characteristics reproduced with AIM-SPICE of the short channel polysilicon TFT.

[3] L. Mariucci, A. Pecora, S. Giovannini, R. Carluccio, F. Massusi, and G. Fortunato, "Hot carrier effects in polycrystalline silicon thin-film transistors: analysis of electrical characteristics and noise performance modifications", *Microelectronics Reliability*, vol. 39, pp. 45-52, 1999.

[4] M. D. Jacunski, M. S. Shur, A. A. Owusu, T. Ytterdal, M. Hack, and B. Iniguez, "A short-channel DC SPICE model for polysilicon thin-film transistors including temperature effects," *IEEE Trans. Electron Devices*, vol. 46, pp. 1146-1158, 1999.

[5] M. Estrada, A Cerdeira, A. Ortiz-Conde, F. J. Garcia Sanchez, and B. Iniguez, "Extraction method for polycrystalline TFT above and below threshold model parameters," *Solid State Electron*, vol. 46, pp. 2295-2300, 2002.

[6] A Sehgal, T. Mangla, M. Gupta, and R. S. Gupta, "Temperature dependence on electrical characteristics of short geometry poly-crystalline silicon thin film transistor," *Solid State Electron.*, vol. 49, pp. 301-309, 2005.

[7] S. Bindra, S. Haldar, and R. S Gupta, "A semi-empirical approach to study a high performance poly-Si TFT with selectively floating a:Si layer, " *Solid State Electron.*, vol. 49, pp. 558-561, 2005.

[8] A. T. Hatzopoulos, D. H. Tassis, N. A. Hastas, C. A. Dimitriadis, and G. Kamarinos, "On-state current modelling of large-grain polycrystalline silicon thin-film transistors based on carrier transport through latitudinal and longitudinal grain boundaries ," *IEEE Trans. Electron Devices*, vol. 52, pp. 1727-1733, 2005.

[9] R. L. Hoffman, "A closed-form DC model for long-channel thin-film transistors with gate voltage-dependent mobility characteristics," *Solid State Electron*, vol. 49, pp. 648-653, 2005.

[10] G. Y. Yang, S. H. Hur, and C. H. Han, "A physical-based analytical turn-on model of polysilicon thin-film transistors for circuit simulation," *IEEE Trans. Electron Devices*, vol. 46, pp. 165-172, 1999.

2006 25th International Conference on Microelectronics

Modeling and Simulation of Submicron MOSFETs with Alternative Gate Dielectrics for DRAM Cells.

N. Konofaos *Senior Member IEEE* and G.Ph. Alexiou *member IEEE*

Abstract. In this study, we designed and simulated MOS devices having alternative gate dielectrics and examined their electrical behaviour as well as their effectiveness on the performance of a DRAM cell containing them. The devices under test had gate dielectrics of $SrTiO_3$, $(Ba_{1-x}Sr_x)TiO_3$, TiO_2, Ta_2O_5, Y_2O_3 and $(Pb_1La_x)TiO_3$. Equivalent oxide thicknesses were derived between 0.95 and 4 nm. Among the candidate materials, $SrTiO_3$ was the most promising when data retention is required, while Y_2O_3 depicted the best I-V behavior. An increased cell capacitance resulted to higher memory refreshing times, improving substantially the device reliability.

I. INTRODUCTION

In order to continue decreasing the size of the Metal-Oxide-Semiconductor Field Effect Transistors (MOSFETs), high-k dielectrics have been examined as potential candidates to replace SiO_2 as the gate material. The new sub-micron MOSFETs with the alternative gate dielectrics have emerged as a new technology for use in high-performance logic or low power memory circuits [1-3]. Many materials have been examined so far as potential candidates for the gate dielectric [4], but none of them has prevailed for being the optimum material, due to many problems related to leakage currents, electron mobility issues and metalization. Moreover, apart from the material science point of view, new challenges have emerged regarding the design, simulation and testing of the new devices and the circuits made of them [5,6,7]. In order to successfully design a MOS device with a new gate dielectric, several parameters need to be taken into consideration such as: the dielectric constant (ε), the threshold voltage (V_t), the leakage currents (J_g), process related phenomena while secondary effects such as the presence of interface states, electron/hole mobility and the metal contacts compatibility add extra challenges to the task [5,6,7].

In this paper, we examine MOSFETs with different gate dielectrics and test their suitability for use in Dynamic-Random-Access Memories (DRAMs) cells. We present a comparative study involving MOSFETs containing different high-k dielectrics, together with design

The authors are with the Computer Engineering & Informatics Department, University of Patras, GR-26500 Patras, Greece, & G.Ph.Alexiou is also with Research Academic Computer Technology Institute, GR-26500, Patras, Greece. email: nkonofao@ceid.upatras.gr, alexiou@cti.gr.

and performance evaluation. DRAM circuits have different performance requirements than logic ones and the materials under test for replacing SiO_2 at the MOSFET gate are different [2,3]t. The aim of this work is to address issues regarding the use of high-k dielectrics on the construction of MOSFETs and derive some useful values regarding the design and performance evaluation of these submicron new generation MOSFETs.

II. MODELING AND SIMULATION ISSUES

In previous work, we had reported a stepwise procedure to implement the use of high-k dielectrics into the MOS device design [6,7]. In these articles, we considered some of the secondary effects that may degrade the device performance such as the presence of interface states and leakage currents, but we did not consider two important issues that are to be addressed in the following paragraphs: a) The need to model the carrier mobility correctly and b) the effect of the leakage current to the correct estimation of the Equivalent Oxide Thickness (EOT).

For the design, simulation and performance evaluation of both the MOSFETs and the DRAM cell, the 90 nm, 6 metal CMOS technology was used with 1.0V source voltages, while the 1T-1C architecture was used for the construction of the memory cell. At our disposal for this work we had the latest Microwind 3.0 EDA tool. For the simulation we used the latest BSIM4 SPICE model, available within the Microwind 3.0 in a form of 20 parameters [8,9].

The devices under test contained the following high-k dielectrics: $SrTiO_3$, $(Ba_{1-x}Sr_x)TiO_3$, TiO_2, Ta_2O_5, Y_2O_3 and $(Pb_1La_x)TiO_3$.

The leakage currents were introduced into the design of the MOSFET using the standard current equations of the BSIM4 model. The geometrical factors and parasitic capacitance related effects were introduced after calculations using literature reports on structures involving the various high-k dielectrics.

Mobility of carriers is the most crucial parameter for MOS modeling. The mobility model in presence of high-k dielectric, trapped charges, vertical electric field, and surface roughness, should be selected properly. Otherwise, the estimation of the parameter:

1-4244-0116-X/06/$20.00 ©2006 IEEE

$$K_{n.p} = \frac{1}{2}\mu_{n,p}C_{ox}\left(\frac{W}{L}\right)$$

where μ is the carrier mobility (indexes n and p refer to electrons and holes respectively), W is the channel width and L is the channel length, may be in error in that situation making the results doubtful.

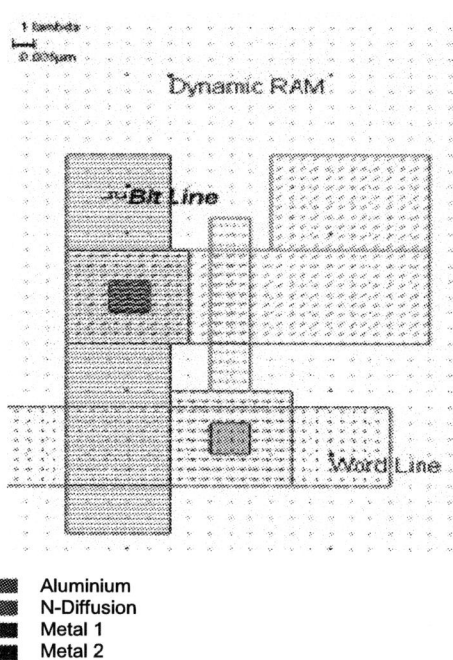

Aluminium
N-Diffusion
Metal 1
Metal 2

Figure 1. The designed DRAM cell.

Moreover, in a first approximation of the EOT value, one may neglect the effect of the leakage current. However, if precise values are to be used, then extra calculations are needed. In particular, the equivalent oxide thickness (EOT) has to be calculated considering not the same value of leakage current. The vertical electric field within the oxide and in the interface gets increased for the same potential difference between the gate and the substrate. This field influences the mobility of carriers, and has to be accounted for.

In our case, the following procedure was applied: It is well known that in the BSIM4 model, the carrier mobility is modeled via the parameter U0 which may have one of three different expressions denoted by the parameter MODMOB. Moreover, other parameters such as UA, UB and UC are used in order to model secondary effects correctly. The basis of this modeling is concerned with the form of the expression providing the mobility value and the model concerns the effect of the normal electric field while the lateral one is considered as constant. In the presence of a dielectric different than the SiO_2, the default approximation which is inherited in standard SPICE BSIM4 routines does not apply and all values in the basic

expression have to be recalculated. Hence, numerical calculations had to take place in order to replace every

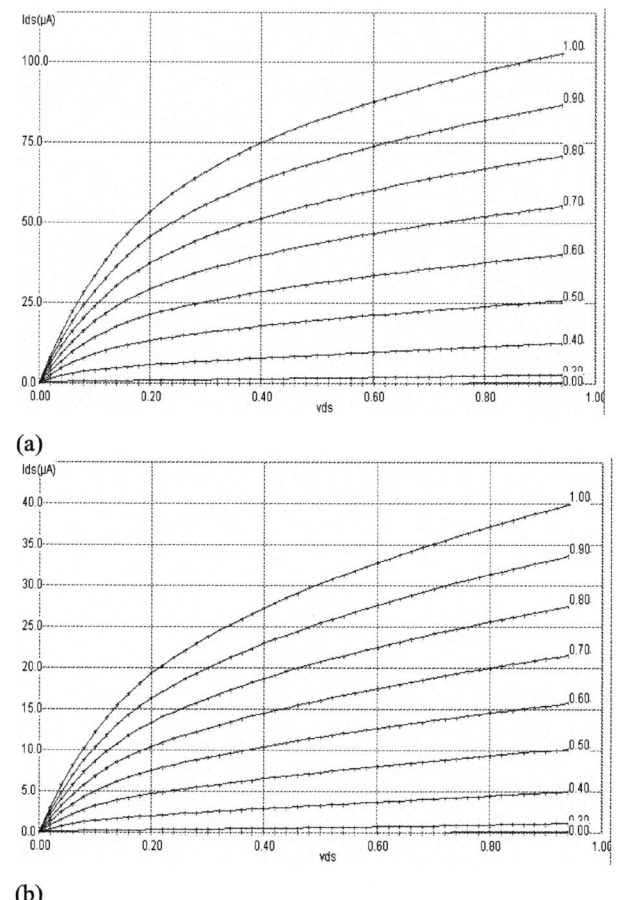

Figure 2. The I_d-V_{ds} curves derived by simulation of a NMOS device having a TiO_2 (a) and a Y_2O_3 (b) gate dielectric.

value with respect to the existence of the new gate dielectric, using a methodology used previously for MOSFETs with very thin oxides [10]. The simulation was made for a steady temperature. The same equations used for numerical evaluation of the mobility values that replaced the default ones in SPICE were used as numerical validation in conjunction with a MATLAB environment. Hence, the Spice simulation was checked via an analytical equation analysis running in parallel with the basic MOSFET equations in use [8,10].

For values of the extra charges affecting the correct device performance and for leakage currents, the values used were derived from literature reports on MOS devices with the same dielectrics [4 and references therein]. Despite these efforts, a lot of work is still needed in order to correctly model the mobility of carriers in the channel, since many other parameters have to be taken into consideration such as short-channel effects, collisions with the interface, non zero lateral field etc. These issues are left for future work.

Finally, as reported in previous papers [6,7] and in the references within, the presence of interface states affects parameters such as the threshold voltage (V_t) and the parasitic capacitances. In the case of a new dielectric, it is important to know the quality of the interface and also the band alignment at the interface while predicting the performance of the MOS device. For this exercise, a continuum distribution of states was considered allowing a smooth interface simulation and calculation of the parasitic values. To the best of our knowledge there are no reports up to date in the literature indicating otherwise.

Figure 1 depicts the designed device under test using the standard rules of the 90nm design based in a lambda scale. The use of Al as the metal contact has been shown to work effectively in real MOS devices with the materials used here as reported in the literature [11,12]. It has to be noted that for all cases examined in this work, both the gate dielectric and the DRAM capacitor dielectric were considered to be of the same material in each case.

III. RESULTS AND DISCUSSION

The MOSFET BSIM4 model was used in full for the whole design, simulation and performance evaluation of the devices. Geometrical factors reported in the literature and derived by experimental techniques were considered for the devices under test. Equivalent oxide thicknesses were derived between 0.95 and 4 nm and the numerical results are summarized in table 1.

Analytically derived values were also used for validation, using analytical equations of the standard MOS modeling, in a MATLAB programming environment. The method proposed by Seekamp et al [13] and the corrections applied to the basic equations as derived by Ohata [14] were used in these calculations.

Figure 2 depicts the calculated I_d-V_{ds} curves derived by simulation of a NMOS device having a TiO_2 (a) and a Y_2O_3 (b) gate dielectric. The values reported in table 1 are used. The depicted curves show that the leakage current reduces the final current output significantly.

The performance of the designed MOS devices was also tested when parts of a DRAM cell with the 1T-1C architecture. One important requirement for such a task is the calculation of the minimum capacitance value that is critical for the correct functioning of the DRAM [1,3,9]. The calculated values for the corresponding materials are also depicted in table 1.

As explained and presented in [7], a test of the DRAM cell can be made as follows: In order to examine the DRAM characteristics, a simulation of the Read/Store (R/S) process can be made. The simulation is initiated by the requirement of a single bit R/S process. For example, the test procedure made for the Store process is the following: First the Bit Line of the DRAM is set equal to '1'. When the Word Line is selected (is set to '1') the value

of the Bit Line is stored into the DRAM cell. Then the Data signal takes the value of '1' which means that the cell stored a value of '1'. This value must be kept until a new Word Line signal is set to '1' driving the DRAM cell to store a '0' because the Bit Line signal is then set to '0'.

As a result the DRAM performance appears in figure 3 for the selective MOSFET with the use of $Ba_{1-x}Sr_xTiO_3$. The time scaling of the pulses and their amplitude for testing the DRAM circuit were chosen by a relevant data sheet and subsequent implementation of the values following standard simulation rules [7] for the Read/Store process.

As shown in figure 3 we can observe that the value of '1' is being held for 25 ns. This value is high, since for

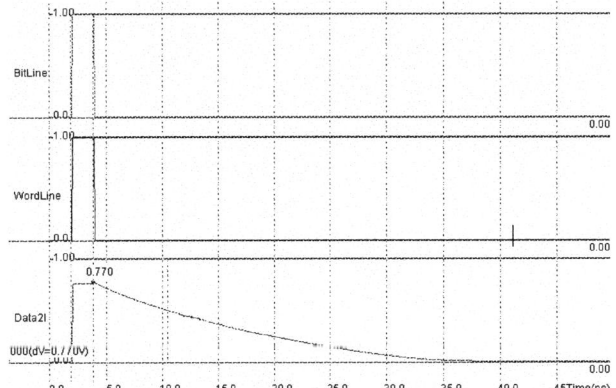

Figure 3. The timing diagram of the DRAM cell write operation for data '1' and data '0'. The used material is $Ba_{1-x}Sr_xTiO_3$ (see table 1).

conventional circuits with SiO_2 NMOS and capacitor, values of around 12 ns have been reported [7,9]. This is an important observation revealing that the newly constructed cell discharges twice as fast as the conventional ones. As a consequence, an important conclusion concerning operational DRAM characteristics is obtained. The above result is directly connected to the fact that at DRAM circuits a periodic refresh operation is always required to compensate for the steady leakage of charge from the storage nodes and to ensure the conservation of data. In this case, the refresh operation needed to recharge the DRAM cell capacitor of the cell made with conventional SiO_2 MOSFET should be completed in shorter time than the operation to refresh the cells made of MOSFETSs with high-k dielectrics. This is an advantage of the circuits made with the use of the new NMOS devices, and their operation requires longer refresh rates.

IV. CONCLUSIONS

Judging from the results presented in the previous paragraphs, modeling of a single MOS device requires a careful study of some physical parameters that are standard for the SiO_2 technology (i.e. the mobility), while it should take into account previously measured values of some electrical parameters which are reported in the literature. The choice of the high-k dielectric is important since prior knowledge of the physical properties of the material is required in order to cope with operational requirements.

The study showed that the designed MOSFETs, for the cases examined by simulation and analytical solutions of the MOSFET equations, revealed very good Current-Voltage (I-V) characteristics. The DRAM cells constructed by such devices showed better refresh rates than those reported to be made of conventional SiO_2 devices. An increased cell capacitance resulted to higher memory refreshing times, improving substantially the device reliability. Final evaluation of the circuit performance will be related to the independent MOSFET behaviour and proper suggestions will be made in order to improve performance and compatibility.

Open matters remain though regarding the use of the new dielectrics. Short-channel effects and a detailed modelling of the mobility are still open to improvement. The compatibility with existing logic technology is still an open matter and has to be addressed.

TABLE I. RESULTS.

Dielectric	k	t_{OX} nm	C_{crit} $fF\mu m^{-2}$	EOT nm	max I_{DS} μA
$SrTiO_3$	230	53	55	0.9	280
$(Ba_{1-x}Sr_x)TiO_3$	320	70	41	0.85	285
Ta_2O_5	25	100	14	15.6	53
TiO_2	40	80	9	7.8	95
Y_2O_3	17	100	5	22.9	37
$(Pb_{1-}La_x)TiO_3$	1400	500	25	1.4	275

REFERENCES.

[1] D.J.Frank, R.H.Dennard, E.Nowak, P.M.Solomon, Y.Taur, H.S.P. Wong, "Device scaling limits of Si MOSFETs and their application dependencies", Proc. IEEE 89(3), pp. 259-288, 2001.

[2] K.Kim, C.G.Hwang, J.G.Lee., "DRAM technology perspective for gigabit era", IEEE Trans. Electr. Dev., 45, pp. 598-608, 1998.

[3] Y.Kakagome, M.Horiguchi, T.Kawahara, K.Itoh, "Review and future prospects of low-voltage RAM circuits", IBM J. Res. & Dev., 47, pp. 525-552, 2003.

[4] S.Ezhilvalavan, T.Y.Tseng, "Progress in the developments of (Ba,Sr)TiO3 (BST) thin films for Gigabit era DRAMs", *Mater. Phys. Chem.*, 65, pp. 227-248, 2000.

[5] B.Cheng, M.Cao, P.Vande Voorde, W.Greene, H.Stork, Z.Yu, J.C.S.Woo, "Design considerations of high-κ gate dielectrics for sub-0.1-μm MOSFET's", IEEE Trans Electr. Dev., 46(1), pp. 261 – 262, 1999.

[6] N.Konofaos and G.Ph.Alexiou, "New challenges emerging on the design of VLSI circuits made of MOSFETs using new gate dielectric materials", proceedings of 5th International Symposium on Quality Electronic Design (ISQED04), San Jose CA, pp.92-95, 2004.

[7] N.Konofaos, Th.Voilas, G.Ph.Alexiou, "Design and simulation of an embedded DRAM cell made up with MOSFETs having alternative gate dielectrics", proceedings of SPIE Int. Ser, vol. 5837, pp. 598-604, 2005.

[8] W. Liu *MOSFET Models for SPICE simulation including Bsim3v3 and BSIM4"* (New York, Wiley & Sons, 2001).

[9] S-M Kang, Y.Leblebici, 'CMOS Digital Integrated Circuits, Analysis and Design' (McGraw-Hill N.Y., 3rd Edition, 2003).

[10] M.Liang, J.Choi, P.Ko and C.Hu, "Inversion-layer capacitance and mobility of very thin oxide MOSFETs", IEEE Trans. Electr. Devices, 33, pp. 409-416, 1986.

[11] Maiti C.K., Chatterjee S., Dalapati G.K., Samanta S.K., "Electrical properties of Ta_2O_5 gate dielectric on strained-Si", Electron Lett, 39, pp. 497-499, 2003.

[12] N.Konofaos, E.K.Evangelou, Z. Wang and U.Helmersson, "Characterisation of $Al/SrTiO_3/ITO$ capacitors for microelectronic applications". IEEE Trans. Electr. Dev., 51, pp. 1202-1205, 2004.

[13] Seekamp A., Avellan A., Schwantes S. and Krautschneider W., "Simple estimation of the effect of hot-carrier degradation on scaled nMOSFETs", Int. J. Electr, 90, pp. 607–612, 2003.

[14] A.Ohata, "Evaluation of performance degradation factors for high-k gate dielectrics in n-channel MOSFETs", Solid-State Electr., 48, pp. 345-349, 2004.

2006 25th International Conference on Microelectronics

Investigation of the Novel Attributes of a Vertical MOSFET with Internal Block Layer (bVMOS): 2-D Simulation Study

Jyi-Tsong Lin, Kao-Cheng Lin, Tai-Yi Lee, and Yi-Chuen Eng

Abstract – In this paper, a vertical n-channel enhancement-type MOSFET with internal block layer (bVMOS) is investigated theoretically. In the proposed structure, the internal block layer comprises a buried block layer and a sidewall block layer. We also test three blocking materials (ex. doped Si, nitride and oxide) for performance comparisons. That is, the p-n junction region between the substrate and drain is isolated by the buried block layer thereby reducing the p-n junction leakage current and the parasitic capacitance. Similarly, the electrical field between the body and drain is blocked or shielded by the sidewall block layer; hence the intolerable ultra-short-channel effects, such as drain-induced barrier lowering (DIBL), hot-carrier effect, source/drain (S/D) punchthrough, and charge-sharing effect, are ameliorated tellingly. Owing to the suppression of the ultra-short-channel effects, excellent subthreshold swing is also successfully achieved by the nano-scale regime. Moreover, the proposed vertical structure has a path between the body and the substrate, the generated hole current by impact ionization and generated heat in channel can be banished from this pass way. Thus, both the floating-body effects and the self-heating effects are avoided synchronously.

I. INTRODUCTION

The features of the silicon-on-insulator (SOI) devices, such as high-performance and low-power, have been proved when compared to the bulk Si. However, the floating-body effects in partially depleted (PD) SOI MOSFETs cause a strong bipolar action resulting in threshold voltage reduction and noise overshoot [1-2]. For diminished ultra-short-channel effects (USCEs), the thin-film thickness of SOI MOSFET must be reduced. As a result of the ultra-thin body (UTB) is considered, the fully depleted (FD) SOI MOSFET is, however, persecuted with serious self-heating effects [3] and high series resistance [4]. Furthermore, the conventional lithography is very difficult to achieve the nano-scale regime. Some vertical transistors were thus presented to solve the issues of lithography [5-7]. Because of the disposed position of the source and the drain is different, the USCEs are effectively suppressed when compared to the planer devices (see Fig. 1). For the excellent USCEs control, the vertical bulk Si needs ultra-shallow S/D extension depths. The vertical SOI MOSFET also needs UTB to offer better electrical

Jyi-Tsong Lin, Kao-Cheng Lin, Tai-Yi Lee, and Yi-Chuen Eng are with the Department of Electrical Engineering (EE), National Sun Yat-Sen University (NSYSU), 70 Lien-hai Rd. Kaohsiung 804, Taiwan ROC, E-mail: jtlin@ee.nsysu.edu.tw

characteristics. Additionally, the issues of p-n junction leakage current, ultra-short-channel effects, floating-body effects, and self-heating effects cannot be ameliorated or eliminated simultaneously. This work investigates a new vertical MOSFET device with internal block layer (bVMOS) theoretically and compares it with the other vertical devices. The 2-D numerical simulation studies demonstrate that the proposed vertical structure results in reduced drain-induced barrier lowering (DIBL), higher drain output resistance, excellent subthreshold swing (SS), kink free output characteristic, and increased drain breakdown voltage. In addition, the body of new architecture maintains solid ties with the substrate, the generated heat in bVMOS's channel is dissipated and the self-heating effects can be overcome.

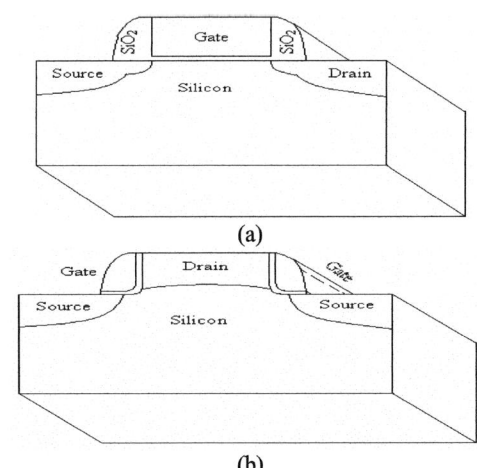

Fig. 1: Devices of (a) planar bulk Si and (b) vertical bulk Si.

II. DEVICE FABRICATION

The process flow for the vertical MOSFET with internal block layer (bVMOS) is outlined in Fig. 2. Firstly, both 50-nm thick oxide and nitride layers are deposited on the Si wafer, and then the conventional lithography is applied for the definition of the buried block layer (see Fig. 2(a)). Approximate 45-nm of oxide is deposited with a chemical vapor deposition (CVD) (see Fig. 2(b)). The sidewall spacer formation technique is applied here to form the sidewall block layer. After that, the nitride layer is cleaned by a wet chemical process (see Fig. 2(c)). Secondly, the thick α-Si is deposited and polished. The body implantation is performed with boron, $1.2e13/cm^2$, 11 KeV.

1-4244-0116-X/06/$20.00 ©2006 IEEE   521

Then, conventional lithography is applied for the definition of the mesa sidewall. The optional sacrificial oxide (2 minute, 900 degC, 97% oxygen, 3% HCl) is used to investigate the influence of Si/SiO2 interface on electrical characteristics and to remove etch damage from the mesa sidewall as shown in Fig. 2(d). The α-Si is then re-crystallized by using solid phase crystallization (SPC). After silicon S/D is realized, the conventional shallow trench isolation (STI) technology is carried out for isolation between the active regions. Thirdly, the gate oxide is growth and then the polysilicon is deposited. After that, conventional lithography is applied to remove most of the CVD polysilicon leading behind the thicker polysilicon on the sidewalls of the silicon S/D. To carry on this process, the screen oxide deposition before the phosphorus implant (2.2e13/cm^2, 15 KeV) is used to form the n+ S/D and gate. Finally, the contact formation is formed by metal silicide contacts on all the active areas of silicon S/D and polysilicon gate; a 25-nm thick titanium metal is deposited on the wafer surface and the two RTP rapid annealing (5 second, 450 degC and 6 second, 650 degC) are undergone to form the Ti-silicide.

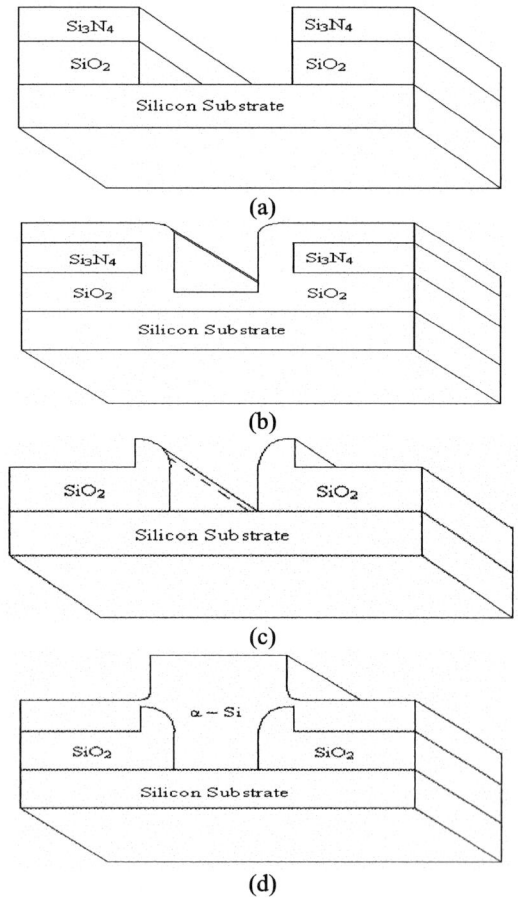

Fig. 2: Process steps for fabricating bVMOS; (a) formation of the buried block layer, (b) CVD oxide film deposition, (c) formation of sidewall block layer, (d) α-Si film deposition, CMP process, and smoothing the α-Si pillar with oxidation.

III. RESULTS AND DISCUSSION

Fig. 3 shows the all vertical MOSFETs, viz. bulk Si, bVMOS, PUD PiFET (partially insulating oxide (PiOX) under drain (PUD) partially insulated field-effect transistor (PiFET)) [8] and SOI MOSFET. To compare the blocking performances, we used different materials to form the internal block layer such as nitride or silicon with high dopant (boron, 1e18/cm^3). Fig. 4 shows the effective channel length for the all structures under the same process parameters.

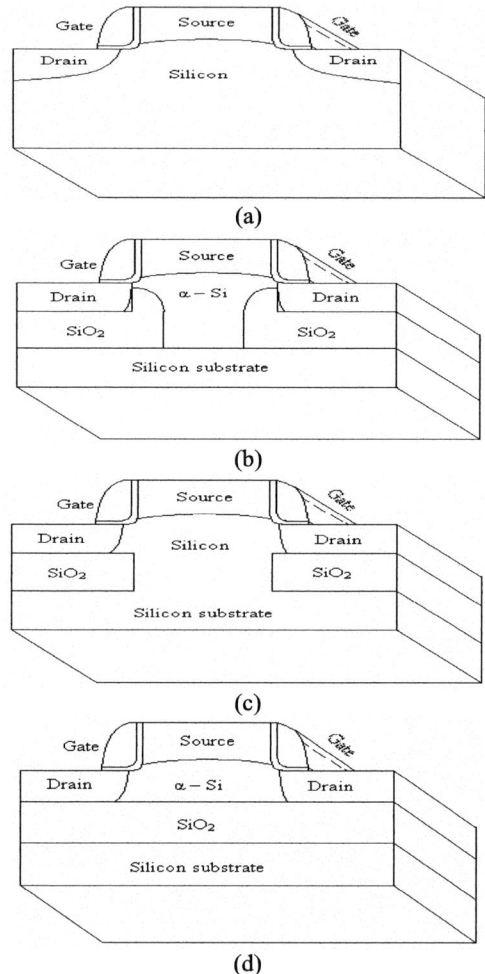

Fig. 3: Vertical transistors of (a) bulk Si, (b) bVMOS, (c) PUD PiFET [8], and (d) SOI MOSFET.

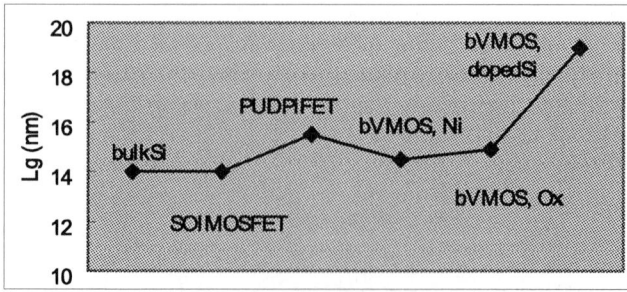

Fig. 4: The effective channel length of different structures.

A. Reducing the junction leakage current

For planar devices, such as enhancement-type MOSFET, the body forms p-n junctions with the S/D regions; the larger depleted region leakage current and bigger junction parasitic capacitance are easily produced thus the circuit performance is influenced (see Fig. 1(a)). If the drain voltage is increased, this phenomenon is significant to the device. The vertical bulk Si also faces with this issue. Whereas, in the proposed structure, the internal block layer can isolate the most parts of the p-n junction between the body and the drain thereby reducing the leakage current of the device (see Fig. 5(a)). On the other hand, the leakage current of the PUD PiFET and vertical SOI MOSFET are similarly higher than that of the bVMOS owing to the numerous p-n junctions between the body and S/D are "exposed" (see Fig. 3(c)-(d)).

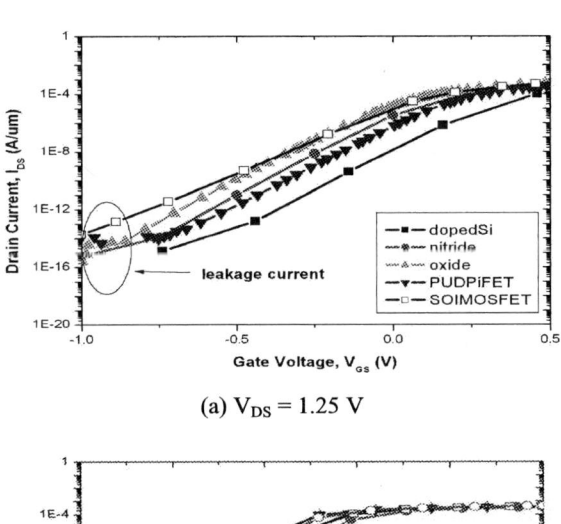

(a) V_{DS} = 1.25 V

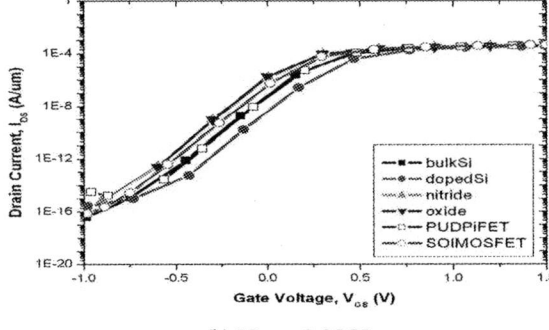

(b) V_{DS} = 0.05 V

Fig. 5: I_{DS}-V_{GS} characteristics of different structures.

B. Ultra-short-channel effects (USCEs)

The ability of suppressed *USCEs* in vertical devices is better than that of the planar devices due to the disposed direction between the source and the drain is not the same direction. However, for excellent USCEs control, the ultra-shallow junction depth or UTB must be applied to enhance the electrical properties. Because the sidewall block layer blocks out the drain depletion region to the channel, the proposed structures exhibit lower DIBL characteristics when compared to the other structures (see Fig. 6). That

means the USCEs are suppressed in our proposed structures. Additionally, the DIBL of bVMOS with block oxide and doped-silicon are similar; our explanation is that the block doped-silicon served as hallo scheme, so the lower DIBL characteristics can be obtained. On the other hand, the DIBL characteristics among the bVMOS with block nitride, the PUD PiFET, and the SOI MOSFET are likeness due to the PUD PiFET and SOI MOSFET are in similar structure. The permittivity of the silicon nitride is higher than the silicon oxide thus the blocking ability is slightly worse than the silicon oxide and similar to the PUD PiFET and SOI MOSFET.

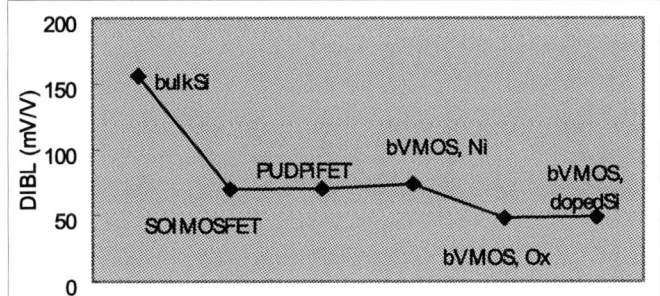

Fig. 6: The DIBL of different structures (V_{DS}= 0.05 V and 1.25 V).

The gate can improve its controllability of the channel depletion region thereby causing excellent subthreshold swing (SS) as shown in Fig. 7. By the same token, the above-mentioned concept can be used to explain the results of SS. Although the off-state current and the threshold voltage of our proposed structures are slightly worse and smaller than the other structures (see Fig. 8 and Fig. 9), the work function engineering developed can be applied to obtain the suitable threshold voltage and the low off-state current.

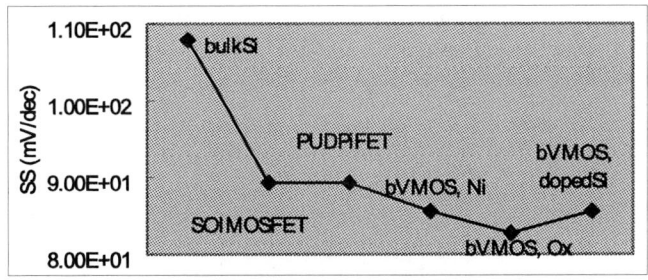

Fig. 7: The subthreshold swing of different structures (V_{DS} = 0.05 V).

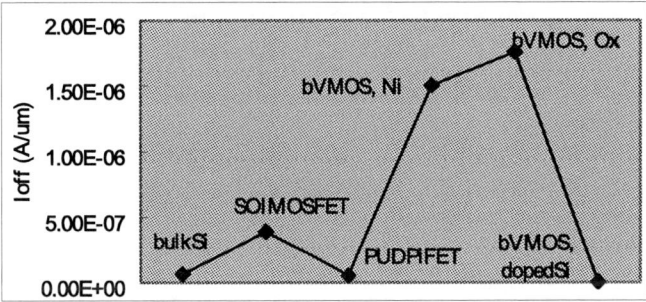

Fig. 8: The off-current of different structures (V_{DS} = 0.05 V).

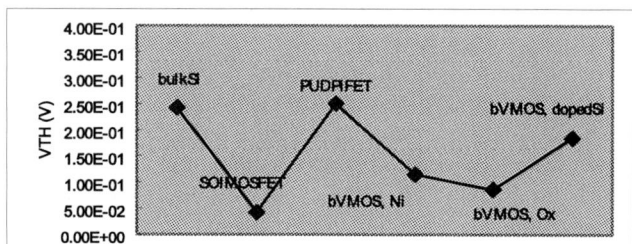

Fig. 9: The threshold voltage of different structures (V_{DS}= 0.05 V).

C. A path between the body and substrate

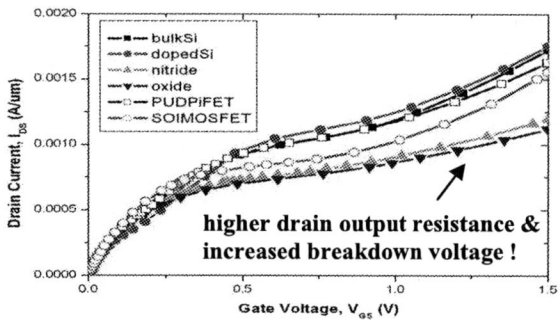

Fig. 10: I_{DS}-V_{DS} characteristics of different structures.

In the proposed structures, the kink effects and the self-heating effects are totally removed due to the body is tied to the substrate (see Fig. 10). In short, the path between the body and substrate play a role to eliminate the generated holes current by impact ionization as well as the generated heat in device's channel is rapidly dissipated. In addition, the bVMOS is based on the bulk Si wafer; the cost can be reduced when compared to the SOI wafer. In the same bias condition, our proposed structures have the corresponding drain current to the other structures (see Fig. 11). But, we do not worry about the trade-off between the driving current speed and our proposed structures if the advanced re-crystallization technique and high-k dielectric materials are introduced. This problem of bVMOS can be ameliorated dramatically.

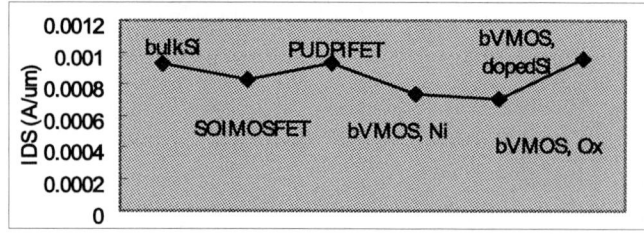

Fig. 11: The drain current of different structures (V_{GT} = V_{DS} = 0.5 V).

Deservedly, the bVMOS also meet other problems, such as the gate to S/D overlap, the random fluctuation of dopant doses, and the practical fabrication, are still surrounded by the bVMOS. We try to overcome these issues. The modified bVMOS and undoped Si channel will be focus in the on-going projects.

IV. CONCLUSION

We are for the first time to present a new vertical MOSFET with internal block layer (bVMOS) for reduced p-n junction leakage current and suppressed ultra-short-channel effects. The internal block layer comprises a buried blocking layer that can eliminate the most parts of the p-n junction between the body and the drain thus reducing the junction leakage current and parasitic capacitances. In addition, the sidewall block layer can isolate the electrical field between the body and drain. Therefore, it can improve the undesirable ultra-short-channel phenomena in depth. Moreover, the floating-body effects and the self-heating effects are also avoided in our proposed structures due to the body is tied to the substrate. If thick enough buried block layer and sidewall block layer are applied, bVMOS is believed to show excellent electrical performances. Besides, high-k dielectric materials, advanced gate engineering and re-crystallization techniques can be simultaneously introduced to bVMOS so that the on-state drain current can thus be further improved. Thus, the advantages of bVMOS provide stimulation to further experimental exploration.

REFERENCES

[1] M. J. Kumar and Vikram Verma, "Elimination of Bipolar Induced Drain Breakdown and Single Transistor Latch in Submicron PD SOI MOSFET", *IEEE Trans. Reliability*, vol. 51, pp. 367-370, Sep. 2002.

[2] Jerry G. Fossum, Mario M. Pelella, and Shrinath Krishnan, "Scalable PD/SOI CMOS with Floating Bodies", *IEEE Electron Devices Lett.*, vol. 19, pp. 414-416, Nov. 1998.

[3] Sun Zimin, Liu Litian, and Li Zhijian, "Self-Heating Effect in SOI MOSFET's", in *Proc. IEEE Int. ICSICT Conf.*, pp. 572-574, Oct. 1998.

[4] Zhikuan Zhang, Shengdong Zhang, and Mansun Chan, "Self-Align Recessed Source Drain Ultrathin Body SOI MOSFET", *IEEE Devices Lett.*, vol. 25, pp. 740-742, Nov. 2004.

[5] Thomas Schulz, Wolfgang Rösner, Lothar Risch, Adam Korbel, and Ulrich Langmann, "Short-Channel Vertical Sidewall MOSFETs", *IEEE Trans. Electron Devices*, vol. 48, pp. 1783-1788, Aug. 2001.

[6] M. Kittler, F. Schwierz and D. Schipanski, "SCALING OF VERTICAL AND LATERAL MOSFETS IN THE DEEP SUBMICROMETER RANGE", in *Proc. IEEE Int. ICCDCS Conf.*, pp. D58/1-D58/6, Mar. 2000.

[7] Meishoku Masahara, Yongxun Liu, Shinichi Hosokawa, Takashi Matsukawa, Kenichi Ishii, Hisao Tanoue, Kunihiro Sakamoto, Toshihiro Sekigawa, Hiromi Yamauchi, Seigo Kanemaru, and Eiichi Suzuki, "Ultrathin Channel Vertical DG MOSFET Fabricated by Using Ion-Bombardment-Retarded Etching", *IEEE Trans. Electron Devices*, vol. 51, pp. 2078-2085, Dec. 2004.

[8] Kyoung Hwan Yeo, Chang Woo Oh, Sung Min Kim, Min Sang Kim, Chang Sub Lee, Sung Young Lee, Sang Yeon Han, Eun Jung Yoon, Hye Jin Cho, Doo Youl Lee, Byung Moon Yoon, Hwa Sung Rhee, Byung Chan Lee, Jeong Dong Choe, Ilsub Chung, Donggun Park, and Kinam Kim "A Partially Insulated Field-Effect Transistor (PiFET) as a Candidate for Scaled Transistors", *IEEE Electron Devices Lett.*, vol. 25, pp. 387-389, Jun. 2004.

Filling the Learning Process in Electronics/Microelectronics Studies by Teaching for Understanding Properties Using Open Training CAD Tools

Mykola Blyzniuk

"What I hear, I forget.
What I see, I remember.
What I do, I understand."
- Confucius -

Abstract - The approach for filling the learning process in electronics and microelectronics studies by teaching for understanding properties using open training CAD tools is presented. Teaching for understanding in electronics and microelectronics is rather pedagogic art than academic work. Significant efforts and inventiveness are required to adhere the teaching for understanding concept in these subjects. Inestimable helper for the teachers in this work is the development of the open training program systems. Such systems allow to realise main concepts of teaching for understanding much easier.

I. TEACHING FOR UNDERSTANDING BY OPEN TRAINING CAD SYSTEMS

In learning process of electronics/microelectronics studies a teaching staff of technical universities should put teaching for understanding (TfU) concept up front. Teaching for understanding in such technical subjects is rather pedagogic art than academic work. The main problem is how to present the complex material of a high level of abstraction in a simple and understandable form [1,4]. That is why a teaching staff looks for the ways of developing approaches which allow to present the material with better understanding, to explain clearly, to find the analogies in order to clarify, etc. [1-3,5].

According to [2] understanding is "…a matter of being able to do a variety of thought-demanding things with a topic – like explaining, finding evidence and examples, generalising, applying, analogising, and representing the topic in a new way". There are four key concepts of TfU that give teachers tactics and strategy for enhancing their efforts to teach for greater understanding [2,7]:

- Generative topics. Generative topic should be central to the discipline, accessible to students, and connectable to diverse topics inside and outside the discipline. If generativity of topics is insufficient the teacher should give the topic more generative cast by adding a theme or a perspective.
- Understanding goals. A few specific understanding goals should be found from many different understandings which can be developed from each topic. These goals should be clear and should give the answer on "what students will understand".

Mykola Blyzniuk is with Ivan Franko National University of L'viv, 1, Universytetska St., 79001, L'viv, UKRAINE, E-mail: MB_IK@org.lviv.net.

- Understanding performances are defined as the heart of developing understanding [2]. Most students' activities during learning process are not performances that demonstrate understanding. These performances are too routine to be understandable, they build knowledge, routine skills but they do not build understanding. The teachers need to design understanding performances that support the understanding goals, so that the students should be involved into activities that make them to generalise, find new example, carry out applications, find analogies and represent topic problems in a new way, from another point of view.
- Ongoing assessment. The process of assessment should be more than just evaluation. The students need criteria, feedback, and opportunities for reflection from the beginning of and throughout any sequence of instruction [2]. The criteria should be clear, relevant and public and the feedback should be provided in order to inform students and teachers about both what students currently understand and how to proceed with the subsequent teaching and learning [7].

Together with keeping TfU concepts in our own teaching activities we pursue a goal of raising the students interest in electronics and microelectronics studies using additional concepts [4,6]: a) instilling the confidence into students that they will be able to understand very complex modern electronics and microelectronics; b) avoiding the fear of students before complex theoretical material; c) presenting complex material of a high level of abstraction in a simple and understandable form; d) finding and using the analogies for the explanation of different complex phenomena; f) keeping the students interest by teaching training material in an active way.

Our teaching experience shows that significant efforts, inventiveness, and pedagogic art are required to adhere the teaching for understanding concept in electronics and microelectronics education. Inestimable helper for teachers in this work is the development of the open CAD tools for education and training in electronics and microelectronics.

The modern CAD systems, which are being successfully applied both in industry and in training, unfortunately are so called "black boxes" for student. They are completely locked. It should be noted that the students cannot be trained on the basis of inaccessible and locked up "black boxes". In fact the students cannot gain access to knowledge and experience concentrated in modern CAD systems. This is to some extent a philosophical problem. If the approach does not change our students will be like the characters in the famous science fiction story by I.Asimov[1]. They got their knowledge artificially, they did not know the source of it and therefore they were not able to deepen their knowledge and to perfect skills.

We consider it is necessary to have some open systems filled with training components. Being given systems like these the students will be able to access to knowledge and experience concentrated in these tools, to study the peculiarities of the software development of such systems (including algorithms, methods, models, etc).

We suggest that open training system that teaches circuit designers has to agree with the following principles [8,9]:

1. *Openness principle* - completely open access to all the system components.

2. *Correspondence principle* - training system has to be brought into line with similar industrial systems. That principle includes: a) functional; b) linguistic correspondence.

3. *Teaching principles* are the main principles for training system. They include: a) *visual teaching principle* - training system should demonstrate how the circuit mathematical model is formed, how the input data are processed, how the algorithms work, etc.; b) *choice principle* - training system should contain: library of different algorithms, methods and programs for solving one problem; different models of the circuit elements. There should exist a possibility to choose one of these models, algorithms, techniques, etc.; c) *comparative evaluation principle* - when different algorithms are parallel or serial executed there should be a possibility, firstly, to compare their efficiencies; and secondly, to compare the accuracy of different models; d) *investigation principle* - training system should enable students to carry out step-by-step investigation of the algorithms work and programs execution; f) *viewing principle* - when solving one of the tasks the possibility should be provided to look at the algorithms flow charts, to look through the source texts of programs realising algorithms that are being currently used.

The use of the above mentioned principles allows us to develop open training software tools successfully. Teaching experience shows that use of such tools in the teaching process provides better understanding of complex material. These tools are very popular with students. The circuit design system *"Micro-PC"* presented in this paper is an example of the open training system. This system is destined for the teaching courses "Circuit Design and Analysis", "Circuit Theory", "CAD Systems for Circuit Design". Training system *"Micro-PC"* was designed as MS Windows application with graphical user interface (GUI). Figure 1 visualises the concept of GUI development that realises the dialogue with the system in such a way that it is quite understandable for the user [8]. *"Micro-PC"* includes solving the problems of circuit simulation, parametric optimization, tolerances assignment, development of circuit simplified models, providing the manufacturability, etc.

---
[1] I.Asimov. Profession. Astounding Science Fiction, July, 1957, pp. 8-56

A special attention in the developed system is paid to the circuit analysis because considerable experience has been gained in this area; various methods, models, algorithms and programs have been developed. There are different algorithms for solving the same problems, e.g. integration algorithms used in transient analysis, different transistor models, etc. The specific character of the design object (electronic circuit) has resulted in emergence of some original techniques, programs, adaptive algorithms. For example, sparse matrices technique with its different algorithms of optimal node renumbering, algorithms of *LU*-decomposition, programming of sparse matrices should be considered as the classic illustration of adaptive algorithm taking into account the specific features of the object of design i.e. the structure of electronic circuit. Including in open system existing different methods, algorithms, models gives students a possibility to study the basics of circuit analysis and design and to use the acquired knowledge in the field of electronics and microelectronics or even in other fields by analogy.

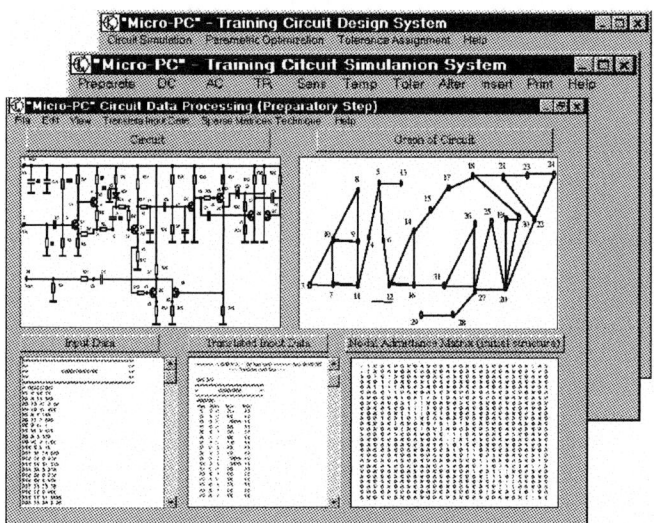

Fig. 1. GUI of *"Micro-PC"* training system.

The developed circuit design training system *"Micro-PC"* agrees with the proposed principles. According to the 1st principle this training system solves all typical tasks of circuit design. For example, during circuit simulation students may perform next typical analyses: DC analysis; AC analysis; transient analysis; sensitivity analysis; analysis of the influence of the temperature of environment; worst case analysis; insertion or alteration of a circuit element; etc. The circuit description, control statements and the form of input data presentation are just traditional and similar to those used in P-Spice. Together with standard control commands .DC, AC, .TRAN, .TEMP, etc, which are used in industrial programs, *"Micro-PC"* has special set of training control operating statements. The example of set of traditional and training control statements of the *"Micro-PC"* circuit design training system is presented in Figure 2.

526

In accordance with the 2nd principle *"Micro-PC"* is a completely open system. Graphical user interface of *"Micro-PC"* is developed to conform to the principles 3.a-3.f which are considered as the most important for training system development. According to the principle 3.a special attention is paid to the formation of circuit mathematical model. Special control statement *.PREPARE* allows to visualise and to investigate the circuit data processing, the work of program realising sparse matrices technique, the peculiarities of programming of sparse matrices, the process of forming nodal admittance matrix and the structure of this matrix, the work of the algorithm of optimal node renumbering.

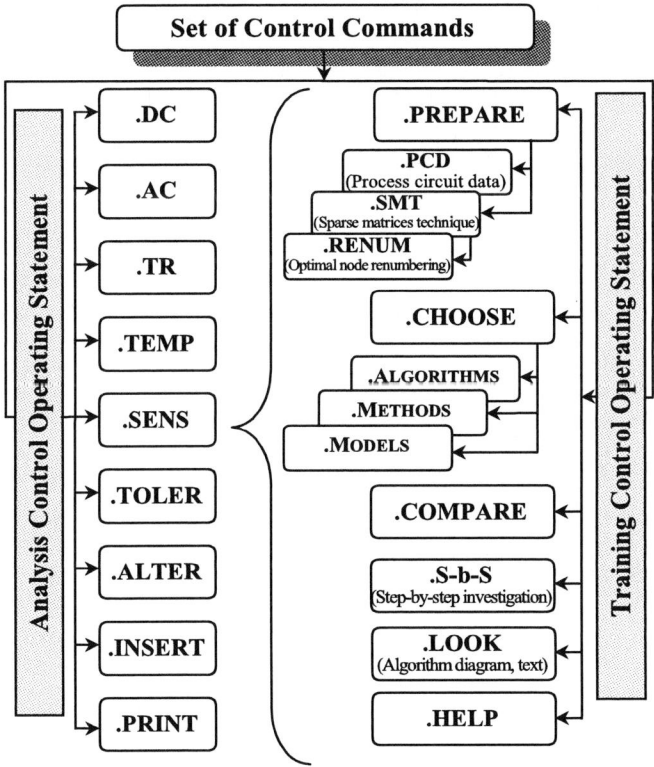

Fig. 2. Example of the set of control statements in the circuit design training system *"Micro-PC"*.

There is a possibility of choosing different algorithms, namely control statement *.CHOOSE*, for solving one and the same task with their parallel or serial execution. It is foreseen at the stage of circuit data processing as well as at all other stages of circuit simulation. For example, a student can choose different algorithms for optimal node reordering at the stage of forming of the circuit model, or different algorithms of integration for transient analysis. The comparison of the effectiveness of the algorithms, methods and models is performed with the help of control statement *.COMPARE.* This statement automatically fulfils comparative evaluation as to the time expenses and convergence of the algorithms, as to the models accuracy and so on. Control statement *.S-B-S (Step-by-step investigation)* allows to investigate the work of the algorithm or the program realising this algorithm in step-

by-step mode. The source texts of the programs could be viewed using control statement *.LOOK*. This control statement can be used also for looking at the algorithms flow charts or at the companion models associated with circuit elements for different methods.

The use of the open training systems like *"Micro-PC"* in learning process makes easy the work of teachers that try to implement TfU concepts in their pedagogic practice.

II. FILLING THE LEARNING PROCESS BY TFU PROPERTIES USING THE TRAINING CAD TOOLS

When an open training system is used in learning process we are able to give a more generative cast to a topic. For example, topic *"Circuit Theory"* for present-day students seems to be merely "dry" and unattractive. The theme *"Circuit Theory: computational approach"* will be more interesting for the students. And more generative, accessible and interesting the topic *"Circuit Theory with Circuit Simulator Developers Eyes"* will be because it implies the learning of circuit theory on the base of the existing circuit simulators with the possibility to develop heir own simulator or some part of the simulator, with the possibility to animate "dry" theory in an alive software tool. Obviously it is necessary to have the open training simulator for such topic. This software tool should be open for the students, and the students should have the possibility to change, to improve or to develop some parts of the simulator (to realise certain kind of analysis), to recompile the system and to check it by themselves.

When planning understanding goals we stick to the opinion that understanding of the nature of processes proceeding in the circuits is impossible without understanding of the mathematical tools describing these processes. Understanding of not only the nature of circuits behaviour but also the mathematical basics of circuit theory will guarantee that the students will become highly qualified specialists. We maintain this opinion by expressive formula of G.W.Leibniz *"Cum Deos culculat, fit mundus"*[2] (it means that in accordance with God's calculation, the world is designed). If we paraphrase this sentence for our case, we may say that "how the mathematical models of the circuit is analysed, so the circuit operates" or in other words, "if the circuit is hard for analysis, the alive circuit will be heavy in functioning". For instance, we consider that the main advantage of circuit nodal analysis is that it reflects the nature of circuit behaviour: if the circuit is correctly designed, in general we have no problem with its analysis; if we have convergence troubles - you should look for the gaps in a circuit. Obviously the better understanding of mathematical tools of circuit theory will be provided if the students have the possibility to look inside the process of mathematical model formulation, to look and to investigate the performed algorithms of analyses. Thereto, better understanding will be provided if the students have the possibility to realise certain method or algorithm, model of

---

[2] Kline M. *"MATHEMATICS: The Loss of Certainty"*. Oxford Univ. Press, 1980.

elements, to include it into open circuit simulator and to check it by themselves. *"Micro-PC"* allows to do all these things. In our learning process we pursue a goal to teach the students not only to analyse the existing circuits, but also to design circuits (to perform structural and parametric synthesis of a new circuit) by their own hands. To do this the students should understand the nature of circuit behavior and the mathematical tools for the description of processes inside the circuit.

The open training system gives the best possibilities for the design of understanding performances. Such an open system with full access (with access up to the source text of programs, with possibility to add some parts, to introduce changes and to recompile) allows to involve students into activities where they may create something new (a new circuit, new models of devices, circuit macromodels, improved algorithm or methods, etc.) by generalising, expanding and reshaping the information given, and apply what they already know. Students analyse the circuits by using the set of training control commands that provide the possibility to look inside the mathematical process of circuit simulation and design. At the same time they can introduce any change in models, methods, algorithms, etc. Students also try to design a new non-standard (special purposes) analogue circuit.

Students are involved into such activities as the development circuit simplified models on the basis of factor experiment design with the full circuit model [10]. In our learning process we try to teach students to develop the circuits simplified models which can answer the questions put in the considered problem but do not mimic the real circuit behaviour. The motto of such activities is a quotation given by A.Einstein that *"everything should be made as simple as possible but not simpler"*. We try to explain that sometimes very complex non-linear processes may proceed inside the circuit, and the circuit will be too complex for simulation, but circuit output parameters of interest can be successfully approximated by simple polynomial dependencies. When the students have developed simplified models of circuit, they should make decision about models adequacy and about applicability of these models in the process of circuit design by themselves. Training system *"Micro-PC"* (which includes the subsystems *"Factor"* for development of circuits simplified models on the basis of factor experiment design [10]) allows to do it. Such activities give the students understanding that correct interpretation of the results of factorial experiments is a guarantee of successful application of circuit factorial models.

Other understanding performances directed to solve the tasks of circuit design process are developed for the students, and their performing is provided by *"Micro-PC"* open training system. For example, they are: evaluation and providing of circuit manufacturability and electro-thermal compatibility, tolerance analysis and assignment, sensitivity analysis and estimation of circuit vulnerability to local temperature influence, optimization and synthesis, etc.

In the course of learning process the students investigate and try to improve a few standard complex analogue circuits and develop a simple non-standard

analogue circuit with special purpose. The results of simulation of the investigated and designed circuit by *"Micro-PC"* are compared with simulation results obtained by P-Spice simulator. Criteria for ongoing assessments are the comparison of the obtained results and degree of their matching. The designed simple non-standard analogue circuit is usually breadboard because only practical experience may be more useful criteria of truth. Students share the results of their design with one another for feedback and critique.

III. CONCLUSION

In conclusion it should be noted that intuitively every good teacher tries to teach for understanding. But for some subjects the teacher needs auxiliary tools in order to realize concepts of TfU. That is why open training CAD tools should be developed. It is just a matter of electronics and microelectronics studies and other technical subjects that are considered to be unglamorous and too difficult.

REFERENCES

[1] M. Blyzniuk, "Computer-Based Training Systems for Teaching for Understanding in Electronics and Microelectronics Studies". *In Proceedings of 2nd International Conference "Theoretical and Applied Aspects of Program Systems Development" (TAAPSD'2005),* Kyiv, Ukraine, 2005, pp. 10-16.

[2] Perkins, David and Tina Blythe, "Putting Understanding Up Front". *Educational Leadership.* 51 (5), 4-7, 1994.

[3] Chris Unger. "What Teaching for Understanding Looks Like". *Educational Leadership.* 51 (5), 8-10, 1994.

[4] M. Blyzniuk, I. Kazymyra, "Teaching for Understanding in Electronics/Microelectronics by Training CAD Tools". *Journal "Electronics and Electrical Engineering".* – Kaunas: Technologija, 2004. – No. 6(55). – pp. 14-19.

[5] Elliot M. Rothkopf, "Teaching for Understanding-Analogies for Learning in Electrical Technology". *ASEE/IEEE Frontiers in Education '95.*

[6] M. Blyzniuk, I. Kazymyra, "How to raise schoolchildren's interest to microelectronics studies". *In Proceedings of 9th International Conference "Mixed Design of Integrated Circuits and Systems" (MIXDES'2002).* Wroclaw, Poland, 2002, pp.129-133.

[7] Tina Blythe and Associates. "The Teaching for Understanding Guide", Jossey-Bass, San Francisco, 1998.

[8] I. Kazymyra, M. Blyzniuk, V.Shcherbacov "Training-Research System of Circuit Design "Micro-PC" as Educational Tool for Teaching CAD Systems Developers in the Field of Microelectronics". *Journal "Radio-Electronics and Telecommunications",* Lviv Polytechnic State University, Lviv, Ukraine, Vol. 352, 1998, pp. 61-71.

[9] I. Kazymyra, M. Blyzniuk, M. Lobur "Circuit Simulation Program for Training Specialists in the Field of Circuit Design Software Development". *Proc. of the 4th International Workshop "Mixed Design of Integrated Circuits and Systems".* Poznan, Poland, 1997. – pp. 673-678.

[10] Koval V.A., Blyzniuk M.B., Kazymyra I.Y. "Simplified Models of IC's for the Acceleration of Circuit Design". *Book "Mixed Design of Integrated Circuits and Systems",* Kluwer Academic Publishers, Boston/Dordrecht/London, 1998. – pp.149-155.

2006 25th International Conference on Microelectronics

Distributed Framework for Teaching Fundamentals of Heat Transfer in Electronic Circuits

[1,2] M. Paszkowski-Rogacz, [1,2] P. Januszkiewicz, [1,2] P. Gocek, [1] G. Jablonski,
[1] M. Janicki, [2] G. De Mey, and [1] A. Napieralski

Abstract - This paper presents a student project aimed at the development of a distributed framework for thermal simulations of electronic circuits. The architecture of the framework is well suited for e-learning or teaching through Internet. When creating the user interface, particular attention has been paid to the issues of security. Owing to the use of the latest Internet technologies, the framework allows running simulations without the necessity of revealing the computing engine code to a user. The input files for simulators can be created using the interface on any computer through an Internet browser. Additionally, the software renders possible for a network administrator to control the data traffic, the management of user groups and the edition of their particular access rights. Currently, the thermal simulations can be performed using a numerical FDM simulator and a semi-numerical simulator employing the thermal influence coefficient method.

I. INTRODUCTION

Nowadays, thermal modeling is an indispensable stage in the design of electronic systems and the thermal issues should be introduced into the curricula for all the students of electronics. The work presented here was focused on the creation of the distributed software for thermal simulation. The software, integrated into a WWW application, is made available to the entire Internet community. The employed solution is particularly suitable for e-learning purposes and may be used in virtual laboratories for students.

The next section is devoted to the detailed description of the overall architecture of the created framework. Then, the two thermal simulators incorporated into the framework are introduced briefly. This is followed by the presentation of exemplary simulation results. Finally, some important conclusions and indications for future developments of the framework are provided.

II. FRAMEWORK ARCHITECTURE

Due to the huge CPU and memory requirements of the simulation algorithms, placing the computing engine on the same machine as the web server is not advisable and they should be implemented on different computers. Moreover, the separation of the engine from the user interface would allow multiple users to take advantage of the computing

[1] The Department of Microelectronics and Computer Science, Technical University of Lodz, Al. Politechniki 11, 93-590 Lodz, Poland, E-mail: janicki@dmcs.p.lodz.pl.

[2] The Department of Electronics & Information Systems, Ghent University, St. Pietersnieuwstraat 41, 9000 Ghent, Belgium.

application at the same time. Another aspect of spreading the software is to allow users with less efficient machines to run computing tasks on other, more powerful machines. Thus, the authors adopted for the framework the three-part architecture, shown in Fig. 1, comprising: the user interface which can be loaded into the client browser, the computing engine residing on a server of the hosting institution and the proxy running on the web server connecting the client and the computing machine. The engine performs thermal computations, whereas the interface provides the necessary user interaction. The third framework component serves as a proxy assuring data transfer between the former two parts. All the components are run separately on different machines, but work together as one system.

The user interface has been implemented in the Java Applet technology. The applet is loaded into a user browser upon visiting a specific Web site. The main functions of the interface are: collecting input data, visualizing simulation results and transferring data between the user interface and the computing engine. Additionally, the interface provides real-time validation for user input data through client-side data pre-processing. Thus, a user is immediately informed about incorrect input and prevented from making mistakes in the input file. Additional advantage of such an approach is the reduction of data exchange between the client and the

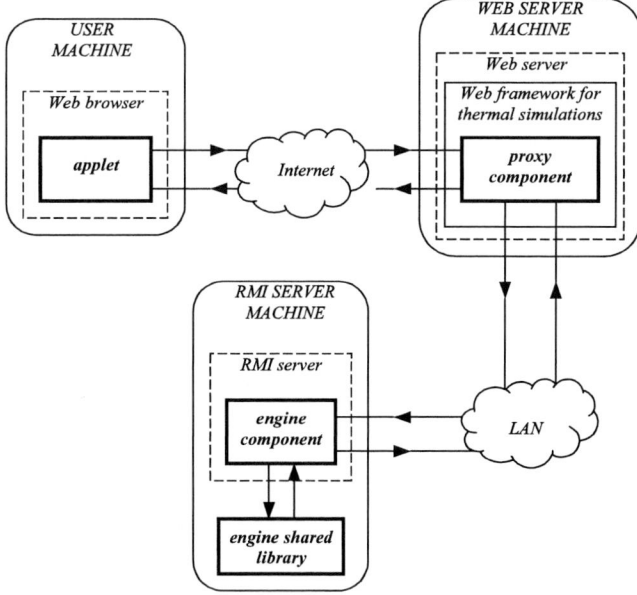

Fig. 1. Components of the distributed framework.

1-4244-0116-X/06/$20.00 ©2006 IEEE

server because there is no need to send any error messages from the server. The interface contains also a convenient interactive system for the processing and the visualization of thermal simulation results. The interface is independent from the operating system installed on a user computer and the data between the interface and the computing engine are transmitted through an open HTTP port. The third part of the framework is the proxy running on a WWW server. The main task of the proxy is handling the communication between the applet and the computing engine implemented in C++. Java supports network communication employing the following distributed object programming technologies: Common Object Request Broker Architecture (CORBA) and Remote Method Invocation (RMI) [1].

CORBA allows the integration between programming languages, thus it could directly bind the computing engine written in C++ and the proxy written using Java language. On the other hand, RMI uses object serialization and allows passing the local objects to remote object methods, which simplifies the process of object exchange. Thus, taking into account that the data exchanged between the user interface and the computing engine are fairly complex and contain internal relations, the communication between the server and the computing machine has been implemented using the Java RMI technology. Besides the above mentioned technologies the framework employs the Apache Tomcat as a servlet container, the Struts as an MVC framework, and the MySQL Database Server for keeping user activity records and their data. The detailed architecture adopted for the final version of the distributed framework is presented in Figure 2, which shows the programming technologies implemented for each particular framework component as well as their internal communication schemes [2].

III. THERMAL SIMULATORS

This section will introduce the partial differential heat equation describing the temperature distribution and two different methods for the solution of the equation employed in the computing engines implemented in the framework

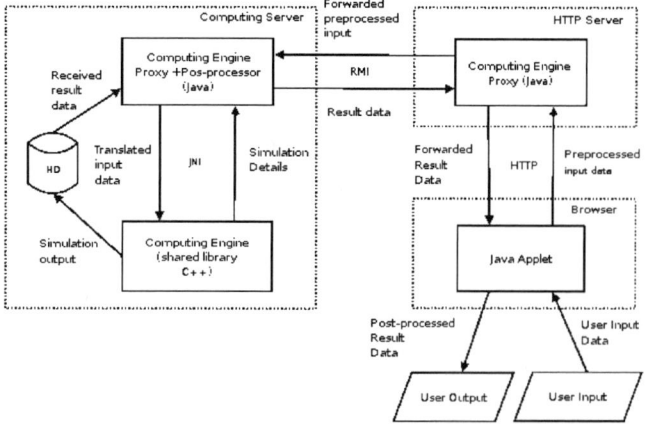

Fig. 2. Detailed architecture of the framework.

A. Heat Equation

The temperature distribution in any structure obeying the Fourier law is described by the following second order partial differential heat equation[3]:

$$\frac{\partial T}{\partial \tau} = \frac{\lambda}{c_p \rho} \nabla^2 T + \frac{g_v}{c_p \rho} \tag{1}$$

where: T – temperature, τ – time, λ – thermal conductivity, ρ – density, c_p – specific heat, g_v- generated heat density.

The above equation can be solved using an analytical or a numerical method. The following subsections will give some basic information on the numerical Finite Difference Method (FDM) and the semi-analytical thermal influence coefficient method employed in the framework computing engines. Both engines have been cross-checked and applied in numerous thermal simulations of real circuits [4]-[5].

B. Finite Difference Method

The FDM method is one of the most commonly used numerical methods for the thermal simulation of electronic circuits with relatively simple shapes. The method consists in substituting the partial derivatives with their difference approximations, thus obtaining a corresponding set of finite difference equations. The number of resulting equations depends on the structure discretisation mesh, which usually is not uniform. In order to obtain accurate results, the mesh should be dense, especially where the temperature gradient values are more significant. Thus, the number of equations might exceed hundreds of thousands, which results in long simulation time and huge memory requirements. For the detailed method description, refer to [6].

The FDM-based computing engine TULSOFT written in C++ was created at the Department of Microelectronics and Computer Science. The thermal system in TULSOFT is defined as a set of blocks containing heat sources and heat sinks placed in any location on the block surfaces. Both the structure geometry and the mesh are user defined. This solver takes advantage of the well-known analogy between the mathematical description of the electrical and the thermal phenomena, according to which the potential and current correspond respectively to the temperature and heat flux. Then, each thermal system can be substituted by an equivalent RC electrical circuit.

C. Thermal Influence Coefficient Method

This method combines the advantages of both the analytical and the numerical solution methods allowing fast but accurate thermal simulations. Employing this method, it is possible to calculate temperature distribution on the top surface of a pyramidal structure. In the current version of the simulation engine, the pyramid can have 2 stages, composed of 5 layers each.

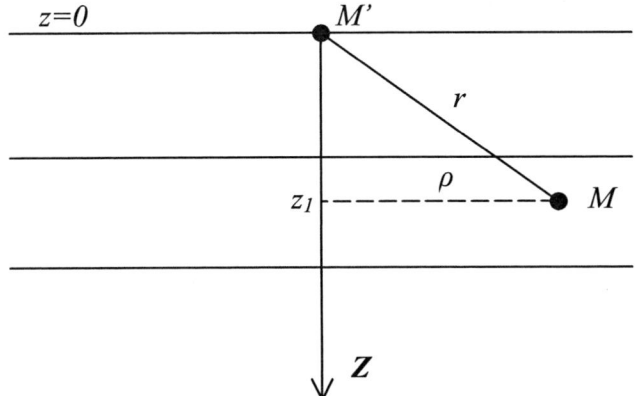

Fig. 3. Cross-section of a multilayered structure.

In each pyramid, the heat sources are placed on the top surface and the bottom surface is connected to a cooling system. The layers are characterized by their thickness, thermal conductivity and conductance characterizing their contacts with neighbouring layers. In such structures, heat flows from the top heat sources through successive layers down to the cooling system.

Moreover, the bottom surface is regarded isothermal one, which means that heat removal by the cooling system is perfect. Furthermore, the heat flow through the lateral surfaces is insignificant compared to the heat removed by the cooling system. Finally, material thermal properties are assumed to be isotropic and temperature independent.

With the above assumptions, the temperature rise Θ can be computed using the Laplace equation and applying specific boundary conditions. The idea of thermal influence coefficients can be explained based on Figure 3, showing a multilayered structure. When M' represents a point heat source on the top surface and M a point inside a layer the temperature rise in point M caused by heat source M' is:

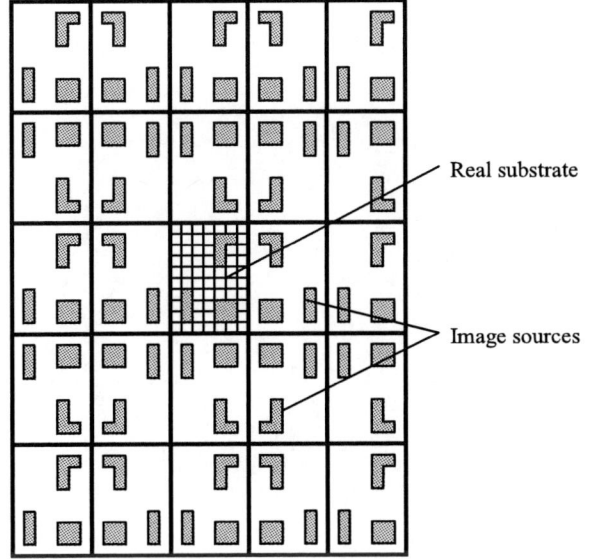

Fig. 4. Original structure and its mirrored images.

$$d\theta = c(r)\,p(M')\,ds \qquad (2)$$

where: $c(r)$ - thermal influence coefficient, $p(M')$ – power density in point M', ds – an infinitely small surface area.

The coefficient $c(r)$ represents the influence of a point heat source on the temperature rise in a considered point. Generally, the further the point from the heat source, the smaller value of $c(r)$. Instead of using $c(r)$, we can write $c(\rho, z)$.

The layers in Figure 3 are unlimited laterally. In order to impose lateral adiabatic boundary conditions, the method of images can be used. The method consists in surrounding a simulated structure with its mirrored images as illustrated in Figure 4. Then, both the real and mirrored heat sources contribute to the temperature rise in every point and the temperature distribution in plane $z = z_1$ can be expressed as follows:

$$\theta(x,y) = \iint_S c(\rho, z_1)\,p\,ds \qquad (3)$$

where: p – power density in real and image areas

Up to now the temperature computation in a one-stage structure with multiple layers has been discussed. To obtain the temperature distribution at the top surface of a two-stage pyramidal structure, some additional calculations are needed. These calculations consist in iterating the solutions of temperature and heat flux at the planes where the stages are connected. The variables in the iterations are: the heat flux and the temperature at the top and the bottom of each stage. When two of the variables in a stage are known, one can calculate the remaining ones. Then, knowing variables of one stage, the variables in the other stages can be found. The method has been described in detail in [7].

The above-described method has been implemented in the computing engine called PYRTHERM, developed at Department of Microelectronics and Computer Science. The engine, originally written in Fortran had to be adapted so as to work on a Linux/Unix server and to cooperate with J2EE web platforms. Therefore, its thermal algorithm has been rewritten into C++ and the code has been compiled into a shared library.

IV. SIMULATION EXAMPLE

The operation of the entire distributed framework will be demonstrated on a practical example. Due to the scarcity of space in this publication, only the simulations with PYRTHERM will be shown. The considered structure was a two-stage pyramid. The top stage was simply a silicon slab with two heat sources and the bottom one consisted of two layers made of copper and aluminum. The structure was cooled from the bottom with forced water cooling. The side and top views of the structure are shown in Figures 5 and 6 respectively. A typical input window is presented in Figure 7 and the simulation results in Figure 8.

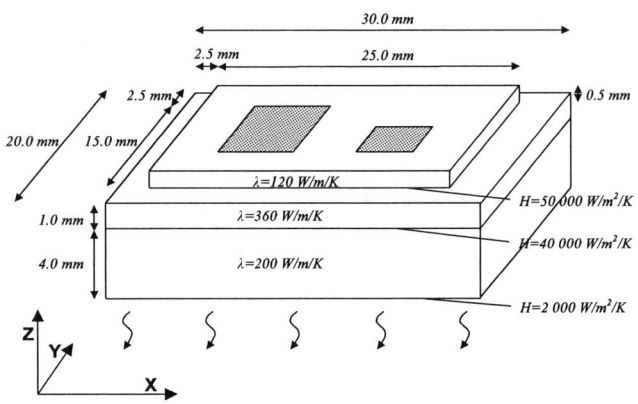

Fig. 5. Simulated structure.

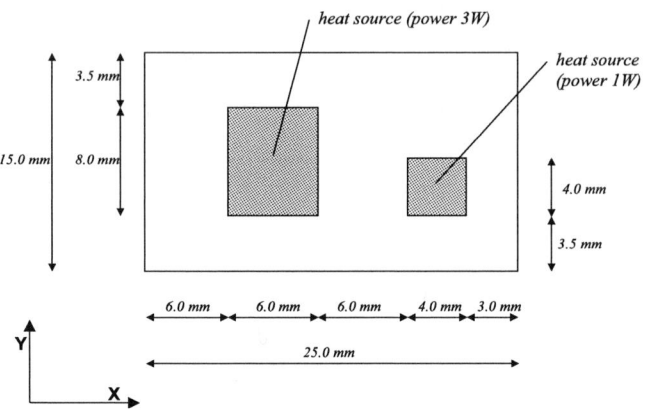

Fig. 6. Structure top view.

As can be seen from the figures presented in this page, the computing engine allows thermal simulations of quite complex structures. The input data window is very simple and does not require form a user any advanced knowledge about the theoretical details of the computing algorithms. The graphical post-processor renders possible not only simple visualization of simulation but also allows different additional analyses. When this is not sufficient, the output data can be downloaded and a user can write own software to process the data according to the particular needs.

V. CONCLUSIONS

This paper presented the distributed framework for thermal simulation of electronic circuits. Users can run simulation and view results via Internet but the computer code remains unrevealed. All users can be arranged into groups with passwords, so the proposed solution is well-suited for distant learning purposes.

Although so far the framework integrates only thermal simulators, it should be clearly underlined that its open architecture allows also the integration of any other solver, including for example SPICE.

ACKNOWLEDGEMENT

This paper has been inspired by the EU 5th FP project REASON and supported by Technical University of Lodz grant Dz.St K-25/1/2006.

REFERENCES

[1] B. Eckel, *"Thinking in Java"*, Prentice Hall, 2002.

[2] M. Paszkowski-Rogacz, *"Distributed Framework for Thermal Simulations with Web Interface"*, M. Sc. Thesis, Technical University of Lodz, 2005.

[3] F. P. Incropera, D. P. De Witt, *"Fundamentals of heat and Mass Transfer"*, John Wiley & Sons, 2002.

[4] P. Gocek, *"Component of Distributed Framework for Thermal Analysis Using the Finite Difference Method"*, M. Sc. Thesis, Technical University of Lodz, 2005.

[5] P. Januszkiewicz, *Component of Web Framework for Thermal Simulations Using Thermal Influence Coefficient Method"*, M. Sc. Thesis, Technical University of Lodz, 2005.

[6] M. N. Ozisik, *Heat Conduction,* Wiley & Sons, 1993.

[7] P. Leturcq, J.-M. Dorkel, A. Napieralski, E. Lachiver, "New approach to thermal analysis of power devices", *Transactions on Electron Devices, IEEE,* 1987, vol. 34, pp. 1147-1156.

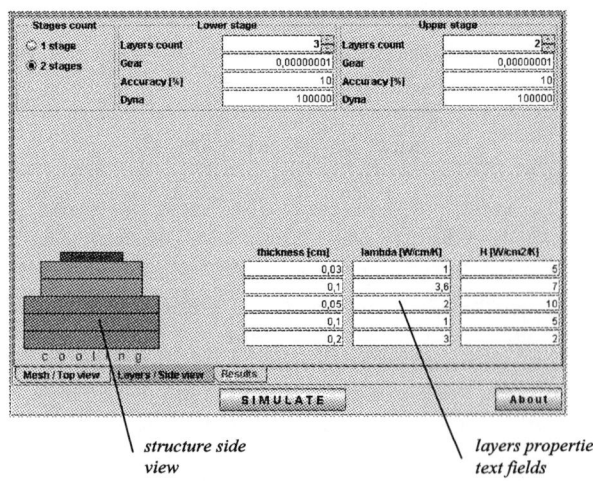

Fig. 7. Simulation input window.

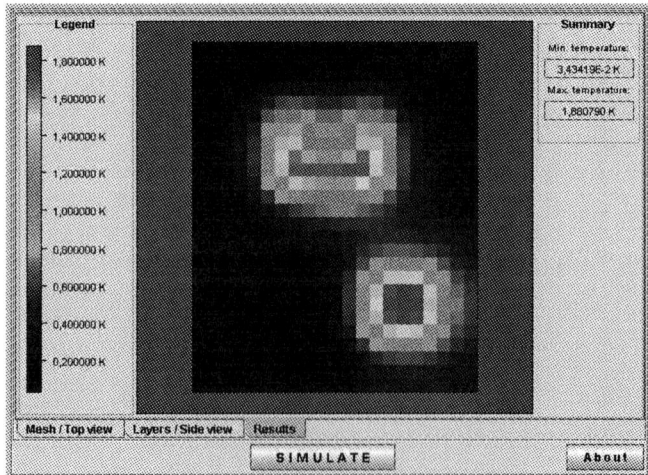

Fig. 8. Simulation results.

532

2006 25th International Conference on Microelectronics

Charge Carriers Distributions in Rectangular Quantum Rod

J.P. Šetrajčić, B.S. Tošić, V.D. Sajfert, D.I. Ilić, S.K. Jaćimovski and S.M. Vučenović

Abstract - Charge carrier's (electron or hole) propagation in rectangular quantum rod was investigated. It turned out tat in rod the autoreduction process takes place. Using advanced Green's functions techniques, suitable for analyses of broken-symmetry structures, the charge carrier's concentrations are found and a behavior of autoreduction was estimated. It was shown that specific skin effect appears in quantum rods.

I. INTRODUCTION

We shall analyze electron (or hole) subsystem of rectangular nanorod having nanocross section and macroscopic height ($n_x \in (0, N_x)$; $N_x \sim 10$; $n_y \in (0, N_y)$; $N_y \sim 10$; $-\frac{N_z}{2} + 1 \le n_z \le \frac{N_z}{2}$; $N_z \sim 10^8$). The Hubbard's model Hamiltonian [1–3] in nearest neighbor's approximation will be used:

$$H = \sum_{\vec{n}} 6W\, A_{\vec{n}}^+ A_{\vec{n}} - \sum_{\vec{n}} A_{\vec{n}}^+ W \Big(A_{\vec{n}, \vec{n}+\vec{a}_x} + A_{\vec{n}, \vec{n}-\vec{a}_x} + \\ + A_{\vec{n}, \vec{n}+\vec{a}_y} + A_{\vec{n}, \vec{n}-\vec{a}_y} + A_{\vec{n}, \vec{n}+\vec{a}_z} + A_{\vec{n}, \vec{n}-\vec{a}_z} \Big), \tag{1}$$

where A^+ i A are electron creation and annihilation operators and W are ion-ion interactions (hoping terms) between nearest neighbors. The following will be used in further:

J.P. Šetrajčić is with the Department of Physics, Faculty of Sciences, University of Novi Sad, Trg Dositeja Obradovića 4, 21000 Novi Sad, Vojvodina – Serbia & Montenegro, E-mail: bora@im.ns.ac.yu

B.S. Tošić is with the Vojvodina Academy of Science and Arts, Dunavska 37, 21000 Novi Sad, Vojvodina – Serbia & Montenegro.

V.D. Sajfert is with the Technical Faculty "M.Pupin", University of Novi Sad, Đure Đakovića bb, 23000 Zrenjanin, Vojvodina – Serbia & Montenegro, E-mail: sajfertv@ptt.yu

D.D. Ilić is with the Faculty of Technical Sciences, University of Novi Sad, Trg Dositeja Obradovića 6, 21000 Novi Sad, Vojvodina – Serbia & Montenegro

S.K. Jaćimovski is with the Faculty of Electrical Engineering, University of Belgrade, Bulevar Kralja Aleksandra 73, 11000 Belgrade, Serbia & Montenegro, E-mail: sjacim@sezampro.com

S.M. Vučenović is with the Faculty of Medicine, University of Banja Luka, 78000 Banja Luka, Republic Srpska, Bosnia & Herzegovina, E-mail: sina@inecco.net

---

$$W_{\vec{n}, \vec{n} \pm \vec{a}_x} = W_{n_x, n_y, n_z; n_x \pm 1, n_y, n_z} = W\,;$$
$$W_{\vec{n}, \vec{n} \pm \vec{a}_y} = W_{n_x, n_y, n_z; n_x, n_y \pm 1, n_z} = W\,; \tag{2}$$
$$W_{\vec{n}, \vec{n} \pm \vec{a}_z} = W_{n_x, n_y, n_z; n_x, n_y, n_z \pm 1} = W\,.$$

The system will be investigated by means of anticommutator Green's function [4,5]:

$$G_{n_x, n_y, n_z; m_x, m_y, m_z}(t) \equiv \ll A_{n_x, n_y, n_z}(t) \big| A_{m_x, m_y, m_z}^+(0) \gg = \\ = \theta(t) < \Big\{ A_{n_x, n_y, n_z}(t), A_{m_x, m_y, m_z}^+(0) \Big\} > , \tag{3}$$

(θ is step-function). Differentiating (3) with respect to time and using equations of motions for Fermi's operators A as well as the Fourier's transformation time-frequency, we obtain the following equation for Green's functions:

$$(E - 6W) G_{n_x, n_y, n_z; m_x, m_y, m_z}(\omega) = \frac{i\hbar}{2\pi} \delta_{n_x, m_x} \delta_{n_y, m_y} \delta_{n_z, m_z} - \\ - W \Big[G_{n_x+1, n_y, n_z; m_x, m_y, m_z}(\omega) + G_{n_x-1, n_y, n_z; m_x, m_y, m_z}(\omega) + \\ + G_{n_x, n_y+1, n_z; m_x, m_y, m_z}(\omega) + G_{n_x, n_y-1, n_z; m_x, m_y, m_z}(\omega) + \\ + G_{n_x, n_y, n_z+1; m_x, m_y, m_z}(\omega) + G_{n_x, n_y, n_z-1; m_x, m_y, m_z}(\omega) \Big], \tag{4}$$

which must satisfy boundary conditions (for: $n_x = n_y \le -1$, $n_x \ge N_x + 1$ and $n_y \ge N_y + 1$):

$$W_{0, n_y, n_z; -1, m_y, m_z} = W_{N_x, n_y, n_z; N_x+1, m_y, m_z} = 0\,;$$
$$W_{n_x, 0, n_z; m_x, -1, m_z} = W_{n_x, N_y, n_z; m_x, N_y+1, m_z} = 0\,.$$

II. CHARGE CARRIER'S PROPAGATION

Due to translational invariance in z-direction (only!), we shall use usual Fourier's transformation space-momentum in this direction. Coefficients in this transformation will be factorized ($G \equiv \alpha\beta\gamma$) with respect to spatial pairs (n_i, m_i; $i{=}x,y$). Due to boundary conditions the equation (4) goes over into system of nine equations [6,7] (4 in vertices of cross section, 4 along sides of it and 1 inside of cross section):

$$\cdot \left[W\alpha_{1,m_x} + (\varepsilon_\alpha + W)\alpha_{0,m_x} \right] \beta_{0,m_y} \gamma_{n_z} +$$
$$+ \left[W\beta_{1,m_y} + (\varepsilon_\beta + W)\beta_{0,m_y} \right] \alpha_{0,m_x} \gamma_{n_z} = \frac{i\hbar}{2\pi} \delta_{0,m_x} \delta_{0,m_y}; \qquad (5.1)$$

$$\left[W\alpha_{1,m_x} + (\varepsilon_\alpha + W)\alpha_{0,m_x} \right] \beta_{n_y,m_y} \gamma_{n_z} +$$
$$+ \left(W\beta_{n_y+1,m_y} + W\beta_{n_y-1,m_y} + \varepsilon_\beta \beta_{n_y,m_y} \right) \alpha_{0,m_x} \gamma_{n_z} = \frac{i\hbar}{2\pi} \delta_{0,m_x} \delta_{n_y,m_y}; \qquad (5.2)$$

$$\left[W\alpha_{1,m_x} + (\varepsilon_\alpha + W)\alpha_{0,m_x} \right] \beta_{N_y,m_y} \gamma_{n_z} +$$
$$+ \left[W\beta_{N_y-1,m_y} + (\varepsilon_\alpha + W)\beta_{N_y,m_y} \right] \alpha_{0,m_x} \gamma_{n_z} = \frac{i\hbar}{2\pi} \delta_{0,m_x} \delta_{N_y,m_y}; \qquad (5.3)$$

$$\left[W\alpha_{N_x-1,m_x} + (\varepsilon_\alpha + W)\alpha_{N_x,m_x} \right] \beta_{0,m_y} \gamma_{n_z} +$$
$$+ \left[W\beta_{1,m_y} + (\varepsilon_\beta + W)\beta_{0,m_y} \right] \alpha_{N_x,m_x} \gamma_{n_z} = \frac{i\hbar}{2\pi} \delta_{N_x,m_x} \delta_{0,m_y}; \qquad (5.4)$$

$$\left(W\alpha_{n_x+1,m_x} + W\alpha_{n_x-1,m_x} + \varepsilon_\alpha \alpha_{n_x,m_x} \right) \beta_{0,m_y} \gamma_{k_z} +$$
$$+ \left[W\beta_{1,m_y} + (\varepsilon_\beta + W)\beta_{0,m_y} \right] \alpha_{n_z,m_x} \gamma_{k_z} = \frac{i\hbar}{2\pi} \delta_{n_x,m_x} \delta_{0,m_y}; \qquad (5.5)$$

$$\left(W\alpha_{n_x+1,m_x} + W\alpha_{n_x-1,m_x} + \varepsilon_\alpha \alpha_{n_x,m_x} \right) \beta_{n_y,m_y} \gamma_{k_z} +$$
$$+ \left[W\alpha_{N_x-1,m_x} + (\varepsilon_\alpha + W)\alpha_{N_x,m_x} \right] \beta_{n_y,m_y} \gamma_{k_z} = \frac{i\hbar}{2\pi} \delta_{n_x,m_x} \delta_{n_y,m_y}; \qquad (5.6)$$

$$\left(W\alpha_{n_x+1,m_x} + W\alpha_{n_x-1,m_x} + \varepsilon_\alpha \alpha_{n_x,m_x} \right) \beta_{N_y,m_y} \gamma_{k_z} +$$
$$+ \left[W\beta_{N_y-1,m_y} + (\varepsilon_\beta + W)\beta_{N_y,m_y} \right] \alpha_{n_z,m_x} \gamma_{k_z} = \frac{i\hbar}{2\pi} \delta_{n_x,m_x} \delta_{N_y,m_y}; \qquad (5.7)$$

$$\left[W\alpha_{N_x-1,m_x} + (\varepsilon_\alpha + W)\alpha_{N_x,m_x} \right] \beta_{n_y,m_y} \gamma_{k_z} +$$
$$+ \left(W\beta_{n_y+1,m_y} + W\beta_{n_y-1,m_y} + \varepsilon_\beta \beta_{n_y,m_y} \right) \alpha_{N_y,m_x} \gamma_{k_z} = \frac{i\hbar}{2\pi} \delta_{N_x,m_x} \delta_{n_y,m_y}; \qquad (5.8)$$

$$\left[W\alpha_{N_x-1,m_x} + (\varepsilon_\alpha + W)\alpha_{N_x,m_x} \right] \beta_{n_y,m_y} \gamma_{k_z} +$$
$$+ \left[W\beta_{N_y-1,m_y} + (\varepsilon_\beta + W)\beta_{N_y,m_y} \right] \alpha_{N_x,m_x} \gamma_{k_z} = \frac{i\hbar}{2\pi} \delta_{N_x,m_x} \delta_{N_y,m_y}, \qquad (5.9)$$

where are:
$$\varepsilon_\alpha + \varepsilon_\beta = \varepsilon;$$
$$\varepsilon = E - 6W + 2W\cos ak_x + 2W\cos ak_y. \qquad (6)$$

This system of nine equations (5.1 – 5.9) reduces into one equation:

$$\sum_{\nu,\mu} a_\nu b_\mu \gamma_{k_z} \left(\varepsilon + 2W\cos\varphi_\nu + 2W\cos\varphi_\mu \right) \Psi_\nu(n_x) \Phi_\mu(n_y), \qquad (7)$$

by transformations:

$$\alpha_{n_x,m_x}(\omega) = \sum_\nu a_\nu(m_x,\omega) \Psi_\nu(n_x);$$
$$\beta_{n_y,m_y}(\omega) = \sum_\mu b_\mu(m_y,\omega) \Phi_\mu(n_y), \qquad (8)$$

where are:

$$\Psi_\nu(n_x) = \sin(n_x + 1)\varphi_\nu - \sin n_x \varphi_\nu,$$
$$\Phi_\mu(n_y) = \sin(N_y + 1)\varphi_\mu - \sin N_y \varphi_\mu \qquad (9)$$

and

$$\varphi_\nu = \frac{\pi\nu}{N_x + 1}; \quad \nu \in [1, N_x],$$
$$\varphi_\mu = \frac{\pi\mu}{N_y + 1}; \quad \mu \in [1, N_y], \qquad (10)$$

The results (10) demonstrate the main physical consequence of symmetry distribution: transition from configuration to the momentum one is not isomorphic – it is of the type $N_i \to N_i - 1$, $i = \{x, y\}$. It means and leads to autoreduction in rod or generally in every structure having nano-cross-section! It means practically that charge carriers do not occupy part of system-structure, while the rest of system-structure contains all charge carriers.

For formal point of view autoreduction cause the representation of Kronecker's symbols in following way:

$$\delta_{n_x,m_x} = \frac{1}{N_x + 1} \sum_{\nu,\nu'=1}^{N_x} \Psi_\nu(n_x) \cdot L_{\nu\nu'}^x;$$
$$\delta_{n_y,m_y} = \frac{1}{N_y + 1} \sum_{\mu,\mu'=1}^{N_x} \Phi_\mu(n_y) \cdot M_{\mu\mu'}^y. \qquad (11)$$

Since summation indices take $N_x - 1$ and $N_y - 1$ values, coefficients – functions L and M must be calculated $N_x N_y$ times.

III. RESULTS AND CONCLUDE DISCUSSION

As an illustrative example we shall consider rectangular nanorod with 3×3 cross section. Due to autoreduction, charge carriers propagated along 4 chains, only.

Taking all exposed into account and goes over to chemical potential representation, we calculated the distribution of concentrations of active charge carriers whose energies are close to energy Fermi's. The quote probabilities give maximal of information about behavior of autoreduction and determine appearance of skin, or anti-skin effect.

These calculations in chains of rectangular nanorod with 3×3 cross section are carried out by means spectral intensities of Green's function [8–12] and they are quoted on Figs 1–3.

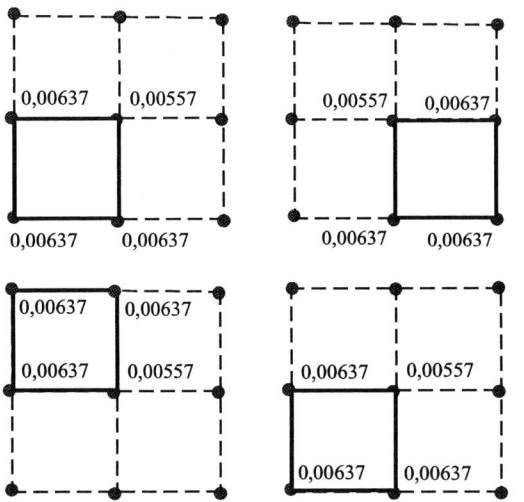

Fig. 1. Charge carriers concentrations in small square nanorod

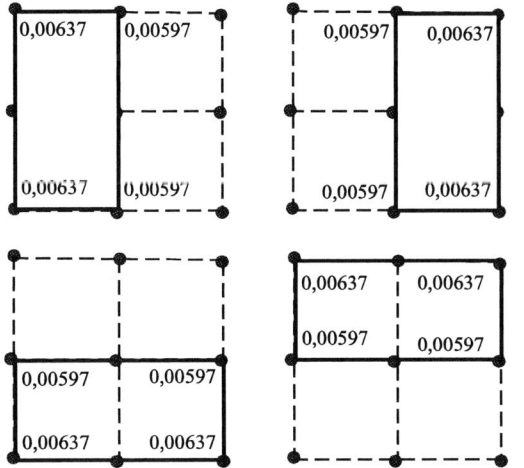

Fig. 2. Charge carriers concentrations in rectangular nanorod

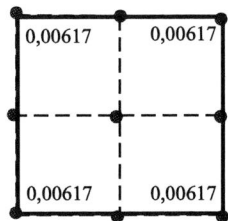

Fig. 3. Charge carriers concentrations in big square nanorod

Based on Figs 1–3 we can concluded the following:

- Autoreduction probabilities are 4/9 for small squares (Fig.1) and for rectangles and 1/9 for big squares (Fig.3);
- The specific skin effect takes place: it is maximal at vertices, slightly lower in sides of cross section and minimal inside of cross section.

ACKNOWLEDGEMENTS

This work was partially supported by the Serbian Ministry of Science and Technology:
Grant No 1895 or ON 141044 (EVB 141044A).

REFERENCES

[1] Kittel C., *Introduction to Solid State Physics*, Wiley, New York 2005.

[2] Ziman J.M., *Principles of the Theory of Solids*, Cambridge University Press, Cambridge 1972.

[3] Ibach H. and Lüth H., *Solid State Physics, An Introduction to Principles of Material Science 3rd edition*, Springer-Verlag , Berlin, Heidelberg, New York 2003.

[4] Jones W. and March N., *Theoretical Solid State Physics*, Dover, New York 1985.

[5] Doniach S. and Sondheimer E.H., *Green's Functions for Solid State Physicists*, Imperial College Press, London 1999.

[6] Tošić B.S., Sajfert V.D., Popov D., Šetrajčić J.P. and Ćirić D., *Difference Calculus Application in Nanostructure Analyze*, Vojvodina Academy of Science and Arts, Novi Sad 2005.

[7] Agarwal R.P., *Difference Equations and Inequalities: Theory, Methods, and Applications*, Marcel Dekker, New York 2000.

[8] Tyablikov S.V., *Methods in the Quantum Theory in Magnetism*, Plenum Press, New York 1967.

[9] Harrison W, *Solid State Theory*, Dover, New York 1979.

[10] Tošić B.S., Šetrajčić J.P., Mirjanić D.Lj. and Bundalo Z.V.: Low-Temperature Properties of Thin Films, *Physica A* **184**, 354 (1992).

[11] Ibach H. and Lüth H., *Solid State Physics, An Introduction to Principles of Material Science, 3rd Ed.* Springer-Verlag, Berlin 2003.

[12] Backović D., Sajfert V.D., Šetrajčić J.P. and Tošić B.S.: Magnetic Nanorod in the Transition Temperature Vicinity, *Journal of Computational and Theoretical Nanoscience*, **2/3**, 448 (2005).

536

2006 25th International Conference on Microelectronics

Numerical Investigation of O_2 Adsorption Effect in Si - CF_4/O_2 System

Yu. N. Grigoryev and A. G. Gorobchuk

Abstract— **In the paper a hysteresis effect for the** Si **etching rate as a function of fluorine concentration in** CF_4/O_2 **gas mixture is discussed. An improved mathematical model for competition of fluorine and oxygen atoms on** Si **surface is considered. The influence of model parameters on a hysteresis curve is studied.**

I. INTRODUCTION

The plasma-chemical etching reactor is widely used in production of semiconductor devices. In such a reactor it is often applied the binary feed gas mixtures as effective tool of etching rate control and utilization of parent gases. In the case of Si - CF_4/O_2 system the etching rate dependence on fluorine concentration with different fractions of O_2 in CF_4/O_2 mixture has a hysteresis character [1]. This effect is ascribed to a competition for chemisorption on the Si surface between oxygen and fluorine atoms. For the first time it was also discovered in numerical simulation [2]. It is clear that the obtained numerical results depend on the choice of adequate chemical kinetics. But the probable mechanisms of gas-phase reactions and surface phenomena in glow discharge are remained insufficiently understood.

In this paper the hysteresis effect in of Si - CF_4/O_2 system is investigated through modeling of etching process in a radial flow reactor. The influence of various parameters of used model on hysteresis curve is studied.

II. MATHEMATICAL MODEL FORMULATION

The 2D-mathematical model of plasma-chemical reactor [2], [3] describing the gas flow and the heat-mass transport processes in the etching chamber was used to simulate a hysteresis effect for Si etching rate in CF_4/O_2 gas mixture. An important difficulty for this system is a simulation of plasma-chemical kinetics which is extraordinarily complicated. Generally the governing set of chemical reactions and corresponding number of reagents essential for given chemical system are chosen using real experimental data. The kinetics of gas phase reactions in RF-discharge have multi-channel character and predominance of one or another of mechanism of governing reactions is determined on the base of probabilistic estimations. The surface kinetics is even

Yu. N. Grigoryev and A. G. Gorobchuk are with the Institute of Computational Technologies, Russian Academy of Sciences, Siberian Branch, Ac. Lavrentjeva Ave. 6, Novosibirsk, 630090, Russia, E-mail: grigor@ict.nsc.ru

less understood. To take this into account the following reduced set of reactions was used. In [4] for Si - CF_4/O_2 parent system a subset of 14 gas-phase reactions were derived that was adequate to explain experimental observations. This improved chemical kinetic model was added by several heterogeneous reactions with CF_2, CF_3 radicals [5], [6]. As was shown in [7] a competition between oxygen and fluorine atoms on Si surface respondents for a hysteresis effect.

The complete set of reactions included following homogeneous and heterogeneous processes looks as follows:

$$CF_4 + e^- \xrightarrow{k_{e1}} CF_3 + F + e^-, \qquad (1)$$

$$CF_4 + e^- \xrightarrow{k_{e2}} CF_2 + 2F + e^-, \qquad (2)$$

$$O_2 + e^- \xrightarrow{k_{e3}} O + O + e^-, \qquad (3)$$

$$COF_2 + e^- \xrightarrow{k_{e4}} COF + F + e, \qquad (4)$$

$$CO_2 + e^- \xrightarrow{k_{e5}} CO + O + e^-, \qquad (5)$$

$$CF_3 + CF_3 + M \xrightarrow{k_{v1}} C_2F_6 + M, \qquad (6)$$

$$F + CF_3 + M \xrightarrow{k_{v2}} CF_4 + M, \qquad (7)$$

$$F + CF_2 + M \xrightarrow{k_{v3}} CF_3 + M, \qquad (8)$$

$$O + CF_3 \xrightarrow{k_{v4}} COF_2 + F, \qquad (9)$$

$$O + CF_2 \xrightarrow{k_{v5}} COF + F, \qquad (10)$$

$$O + CF_2 \xrightarrow{k_{v6}} CO + 2F, \qquad (11)$$

$$O + COF \xrightarrow{k_{v7}} CO_2 + F, \qquad (12)$$

$$F + COF + M \xrightarrow{k_{v8}} COF_2 + M, \qquad (13)$$

$$F + CO + M \xrightarrow{k_{v9}} COF + M, \qquad (14)$$

$$F + F + M \xrightarrow{k_{v10}} F_2 + M, \qquad (15)$$

$$F_2 + M \xrightarrow{k_{v11}} F + F + M, \qquad (16)$$

$$CF_3 \xrightarrow{k_{s1}} CF_3 (s), \qquad (17)$$

$$CF_2 \xrightarrow{k_{s2}} CF_2 (s), \qquad (18)$$

$$F + CF_2 (s) \xrightarrow{k_{s3}} CF_3, \qquad (19)$$

$$F + CF_3 (s) \xrightarrow{k_{s4}} CF_4, \qquad (20)$$

$$CF_3 + CF_3 (s) \xrightarrow{k_{s5}} C_2F_6, \qquad (21)$$

$$CF_2 (s) + O \xrightarrow{k_{s6}} CO + 2F, \qquad (22)$$

$$CF_3 (s) + O \xrightarrow{k_{s7}} CO + 3F, \qquad (23)$$

$$O \xrightarrow{k_{s8}} O (s), \qquad (24)$$

$$O (s) + F \xrightarrow{k_{s9}} O + F, \qquad (25)$$

$$4F + Si \xrightarrow{k_s} SiF_4 \uparrow, \qquad (26)$$

where $k_{e1} - k_{e5}$ are the rate constants of electron-impact dissociation of parent gas; $k_{v1} - k_{v11}$ are the rate con-

1-4244-0116-X/06/$20.00 ©2006 IEEE

stants of volume recombination; $k_{s1} - k_{s7}$ are the rate constants of heterogeneous reactions. The designation (s) was used for adsorbed species on the wafer surface. The values of these constants were taken from [4]-[8].

In general the model contains 16 gas-phase reactions and 8 heterogeneous reactions on the wafer. The reactions Eqs. 1-5 represent the electron-impact dissociation of binary gas mixture; Eqs. 6-16 are the volume recombination of reactive atoms and radicals; Eqs. 17-23 are the recombination and adsorption of CF_2, CF_3 at wafer; Eqs. 24, 25 are the chemisorption of fluorine and oxygen atoms on Si surface; Eq. 26 is a reaction of silicon etching. The twelve products of dissociation and recombination processes - F, F_2, CF_2, CF_3, CF_4, C_2F_6, O, O_2, CO, CO_2, COF, COF_2 are taken into account. Accordingly to multicomponent chemical kinetic model the distribution of species concentration for each component was derived from the system of coupled convective-diffusion equations:

$$\mathbf{v} \cdot \nabla Q_i = \nabla \cdot (D_i(\nabla Q_i) + G_i(Q_i, Q_j), \qquad (27)$$
$$i, j = 1, \ldots, 12,$$

where D_i is a multicomponent diffusion coefficient of species i, G_i is the rate of formation of species i by gas-phase reactions. The gas phase reactions are incorporated in r.h.s. of this system and define a complex interconnection between species generation rates. The surface and silicon etching reactions entered the boundary conditions at the wafer. The latter were written as a balance of mass flows for each component.

Accordingly to representation of chemisorption process the different parts of silicon surface are covered by various adsorbed atoms and radicals. Writing down the balance of mass flows for CF_2, CF_3 and O components on silicon surface at equilibrium as in [5], [6] a system of linear equations for unknown parameters ϑ_{CF_2}, ϑ_{CF_3}, ϑ_O is obtained in the form:

$$\begin{cases} k_{s2}x_{CF_2}/(k_{s3}x_F + k_{s6}x_O) = \\ \qquad = \vartheta_{CF_2}/(1 - \vartheta_{CF_2} - \vartheta_{CF_3} - \vartheta_O) \\ k_{s1}x_{CF_3}/(k_{s4}x_F + k_{s5}x_{CF_3} + k_{s7}x_O) = \\ \qquad = \vartheta_{CF_3}/(1 - \vartheta_{CF_2} - \vartheta_{CF_3} - \vartheta_O) \\ \alpha_s x_O/x_F = \vartheta_O/(1 - \vartheta_{CF_2} - \vartheta_{CF_3}), \end{cases}$$

where ϑ_{CF_2}, ϑ_{CF_3} are the fractions of silicon surface covered by adsorbed radicals CF_2 and CF_3 correspondingly; ϑ_O is the fraction of silicon surface covered by oxygen atoms. Solving this system one can derive the next formulas for ϑ_{CF_2}, ϑ_{CF_3}, ϑ_O:

$$\vartheta_{CF_2} = \frac{k_{s2}x_{CF_2}}{k_{s2}x_{CF_2} + k_{s3}x_F + k_{s6}x_O + \Delta_{CF_2}},$$
$$\Delta_{CF_2} = \Delta_{1,CF_2} + \Delta_{2,CF_2},$$
$$\Delta_{1,CF_2} = k_{s1}x_{CF_3}\frac{k_{s3}x_F + k_{s6}x_O}{k_{s4}x_F + k_{s5}x_{CF_3} + k_{s7}x_O},$$

$$\Delta_{2,CF_2} = \alpha_s\frac{x_O}{x_F}(k_{s3}x_F + k_{s6}x_O);$$

$$\vartheta_{CF_3} = \frac{k_{s1}x_{CF_3}}{k_{s1}x_{CF_3} + k_{s4}x_F + k_{s5}x_{CF_3} + \Delta_{CF_3}} \qquad (28)$$

$$\Delta_{CF_3} = k_{s7}x_O + \Delta_{1,CF_3} + \Delta_{2,CF_3},$$

$$\Delta_{1,CF_3} = k_{s2}x_{CF_2}\frac{k_{s4}x_F + k_{s5}x_{CF_3} + k_{s7}x_O}{k_{s3}x_F + k_{s6}x_O},$$

$$\Delta_{2,CF_3} = \alpha_s\frac{x_O}{x_F}(k_{s4}x_F + k_{s5}x_{CF_3} + k_{s7}x_O);$$

$$\vartheta_O = \frac{\alpha_s x_O}{\alpha_s x_O + x_F + \Delta_{1,O} + \Delta_{2,O}}, \qquad (29)$$

$$\Delta_{1,O} = x_F\frac{k_{s2}x_{CF_2}}{k_{s3}x_F + k_{s6}x_O},$$

$$\Delta_{2,O} = x_F\frac{k_{s1}x_{CF_3}}{k_{s4}x_F + k_{s5}x_{CF_3} + k_{s7}x_O},$$

where the parameter $\alpha_s = k_{s8}/k_{s9}$ is introduced analogically with [7], [8]. Such a presentation allows to express these fractions through the rate constants of the processes.

A special case can be considered for ϑ_{CF_3}. By substitution to the formula Eq. 28 $x_{CF_2} = x_O = 0$ we derive the formula:

$$\vartheta_{CF_3} \approx \frac{k_{s1}x_{CF_3}}{(k_{s1} + k_{s5})x_{CF_3} + k_{s4}x_F},$$

which coincides with three component chemical kinetic model [6].

In second particular case supposing in formula Eq. 29 $x_F \gg x_{CF_2}, x_{CF_3}$ one can carry out an expression:

$$\vartheta_O \approx \frac{\alpha_s x_O}{\alpha_s x_O + x_F}, \qquad (30)$$

which coincides with expression for ϑ_O presented in [8]. In this simplified presentation the denominator shows a competition of chemisorption processes of fluorine and oxygen atoms on silicon which leads to hysteresis effect on the diagram of etching rate with respect to fluorine concentration [1].

Under assumption that $\vartheta_{CF_2}, \vartheta_{CF_3} \ll 1$ [2], [3] the parameter α_s in these expressions which contains unknown rate constants Eqs. 24, 25 can be defined by the next ratio:

$$\alpha_s \approx \frac{F_O S_O}{F_F S_F}, \qquad (31)$$

where F_O, F_F are molar fluxes of fluorine and oxygen atoms to the wafer; S_O, S_F are sticking coefficients of oxygen and fluorine atoms on silicon [8].

The parameters introduced here — the fractions of silicon surface covered by CF_2, CF_3, O are very important because they enter the boundary conditions at the wafer and effect the etching rate. The local spontaneous etching rate and its average value in Å/min were defined by the formulas:

$$V_s = 1.81 \cdot 10^{10} (1 - \vartheta_O - \vartheta_{CF_2} - \vartheta_{CF_3}) k_s C_F,$$

$$\overline{V}_s = \frac{2}{R_o^2 - R_i^2} \int\limits_{R_i}^{R_o} V_s r dr,$$

where k_s is etching rate constant, cm/s; C_t is mole concentration of fluorine, Mol/cm^3. As one can see, these parameters also show that CF_2, CF_3 radicals competitively adsorb on the silicon surface too.

The solution of the problem was carried out by the numerical method, briefly presented in [2], [3]. The governing equations were numerically solved by iterative finite difference splitting-up schemes with stabilizing correction. The Navier-Stokes equations in standard Boussinesq approximation were considered in variables "stream function − vorticity" together with the heat transport equation. The species concentrations were then calculated from the convective-diffusion Eqs. 27 using the computed velocity and temperature distributions.

III. RESULTS AND DISCUSSION

The calculations have been completed for full loaded radial flow reactor. The characteristic geometric dimensions were chosen as in [5]. The gas flow direction to the center of reactor was examined . The calculations have been done for several values of gas flow rate under normal conditions $Q = 100, 200, 400, 800$ cm^3/min. The wafer temperature varied in the limits $T_s = 300 \div 573$ K. The pressure in etching chamber of reactor was equal to $p = 0.5$ torr. The average electron density was assumed equal to $n_e = 6 \times 10^9$cm^{-3}. The temperature of reactor walls was $T_w = 300$ K. The O_2 percentage fraction in mixture CF_4/O_2 feed gas mixture varied in the range $10 \div 90\%$.

The sticking coefficients of oxygen and fluorine atoms on silicon are defined as in [8] $S_F = 0.7$, $S_O = 0.2$ correspondingly. Thus obtained value of parameter $\alpha_s = 0.311$ was used in our previous calculations [2]. There it was shown that with such a value of α_s the etching rate dependence on fluorine concentration has a hysteresis character. But the maxima of etching rate and fluorine concentration were achieved at the same value of O_2 in CF_4/O_2 mixture that contradicted to experimental data [1]. In [7] the chemical model included the similar parameter which was varied in the range $1 \div 600$. In our opinion such a range is too large. Thereat in present calculations the range of α_s was expanded in compare with [2]. The following values were examined $\alpha_s = 0.3, 1, 5, 10, 50, 100$. It is clear the rise of α_s corresponds to increasing of ratio of sticking coefficients (Eq. 31).

The simulation of the above chemical reaction scheme for Si - CF_4/O_2 system has shown the following results. The parameter α_s essentially influence on location and amplitude of maximum etching rate. With increase of parameter from 0.3 to 100 the maximum of

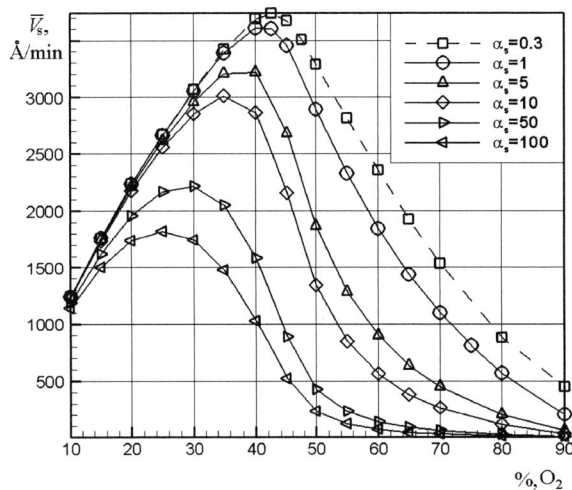

Fig. 1. Average etching rate as a function of percentage fractions of O_2 in input CF_4/O_2 mixture. Processing parameters: $p = 0.5$ torr, $Q = 200$cm^3/min, $T_s = 300$ K.

average etching rate moves from 40% fraction of O_2 in CF_4/O_2 to 25% (Fig. 1). At the same time the maximum of average fluorine concentration on Si surface is located near 50% fraction of O_2 (Fig. 2). The maximum value of etching rate decreases approximately in two time. This fact is connected with the passivation of Si surface by oxygen atoms (Eq. 24) which prevent the reaction of fluorine with silicon (Eq. 26). On the contrary because the etching rate decreases the maximum value of fluorine concentration double rises. It is explained by additional yield of fluorine atoms that cannot use in the etching reaction owing to the surface passivation. The deviation between the maxima of normalized etch rate and normalized fluorine concentration at $\alpha_s = 0.311$

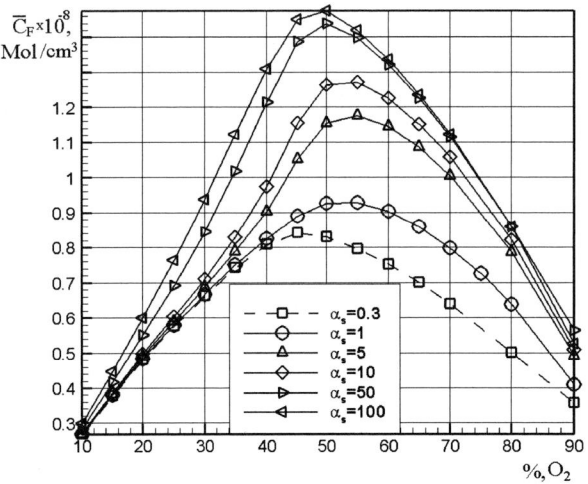

Fig. 2. Average fluorine concentration on silicon with different fractions of O_2 in CF_4/O_2 mixture. Processing parameters: see Fig. 1.

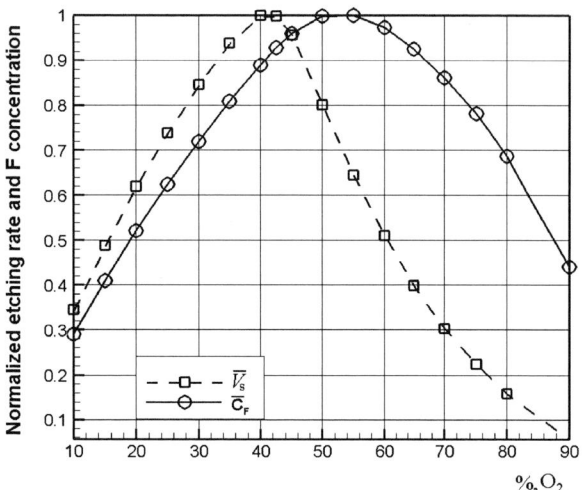

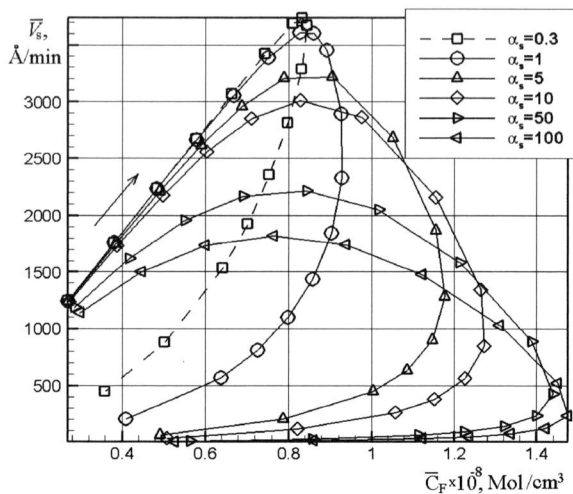

Fig. 3. Normalized average etching rate and fluorine concentration as a function of percentage fractions of O_2 in input CF_4/O_2 mixture. Processing parameters: see Fig. 1.

Fig. 4. Average etching rate as a function of average fluorine concentration on silicon with different fractions of O_2 in CF_4/O_2 mixture. Processing parameters: see Fig. 1. The markers on the curves indicates O_2 fraction in CF_4/O_2 mixture varied from 10% to 90% with 5% step.

was less than 3% of O_2 in the feed gas. This is inconsistent with experimental data where more large difference was observed. But gradually this difference rises with α_s and it arrives in the range $10 \div 15\%$ of O_2 at $\alpha_s = 1$ (Fig. 3). This is agree with the real experimental data [1] although the operating regimes of reactor and its dimensions were some different from those in the present work. This also means that limiting case for ϑ_O (Eq. 30) describes an essential contribution in competition process. Probably a close sticking coefficients of oxygen and fluorine atoms when $\alpha_s \approx 1$ corresponds more realistic scenario of etching process. The displacement of maxima also affects on dependence of etching rate versus fluorine concentration (Fig. 4). One can see from Fig. 4 that with the rise of α_s the local maximum of etching rate shifts and becomes more lower under simultaneous increase of maximal fluorine concentration. As a result the hysteresis curves become more circular. At the same time the beginning of curling of curve is gradually shifts to the region with small edition of O_2.

IV. Conclusions

The calculation results obtained with using complex kinetic model are closed to known experimental data and confirm qualitative scenario of hysteresis process. The variation of sticking coefficients of oxygen and fluorine atoms essential influence on etching rate and hysteresis curve. With increase of its ratio α_s the maximum of average etching rate moves to the region with small edition of O_2.

Acknowledgment

This research was supported by the Russian Fund of Fundamental Research (grant No.05-01-00359), RFOSS (grant No.2314.2003.1) and SB RAS (project No.2).

References

[1] C. J. Mogab, A. C. Adams, D. L. Flamm, 'Plasma etching of Si and SiO_2 - The effect of oxygen additions to CF_4 plasmas', *Journal of applied physics*, 1978, vol.49, No.7, pp.3796-3803.

[2] Yu.N. Grigoryev, A.G. Gorobchuk, 'Numerical Simulation of plasma-chemical reactors', *Notes on numerical fluid mechanics and multidisciplinary design. Computational Science and High Performance Computing. Eds. Egon Krause, Yurii I. Shokin, Michael Resch, Nina Shokina*, (Springer-Verlag Berlin Heidelberg, 2005) vol.88, pp.229-251.

[3] Yu. N. Grigoryev, A. G. Gorobchuk, 'Peculiarities of Si films etching in CF_4 paren gas', *Proceedings of 22 International Conference on Microelectronics MIEL'2000*, Nis, 2000, pp.289-292.

[4] I. C. Plumb, K. R. Ryan, 'A model of the chemical processes occurring in CF_4/O_2 discharges used in plasma etching', *Plasma chemistry and plasma processing*, 1986, vol.6, No.3, pp.205-230.

[5] S. P. Venkatesan, I. Trachtenberg, T. F. Edgar, 'Modeling of silicon etching in CF_4/O_2 and CF_4/H_2 plasmas', *Journal of the electrochemical society*, 1990, vol.137, No.7, pp.2280-2290.

[6] Syng-Kyu Park, D. J. Economou, 'A mathematical model for etching of silicon using CF_4 in radial flow plasma reactor', *Journal of the electrochemical society*, 1991, vol.138, No.5, pp.1499-1508.

[7] Ph. Schoenborn, R. Patrick, 'Numerical simulation of a CF_4/O_2 plasma and correlation with spectroscopic and etch rate data', *Journal of the electrochemical society*, 1989, vol.136, pp.199-205.

[8] P. M. Kopalidis, J. Jorine, 'Modeling and experimental studies of a reactive ion etcher using SF_6/O_2 chemistry', *Journal of the electrochemical society*, 1993, vol.140, No.10, pp.3037-3045.

2006 25th International Conference on Microelectronics

Dispersion Relation for Two-Valley Quasi-Hydrodynamic Models in SCWs Propagation in n-GaAs Thin Films

Abel Garcia-B., Volodymyr Grimalsky, Edmundo Gutierrez-D., and S. Koshevaya

Abstract - The space charge wave propagation in thin n-GaAs films has been considered within a framework of different quasi-hydrodynamic balance models. A comparative study of the dependencies $k(\omega)$ of a complex longitudinal wave number on frequency, which are obtained from different balance equation models are presented and discussed. It has been demonstrated that the model where the balance equations are written for different valleys directly seems to be the most correct. The balance models based on averaging over the valleys may lead to incorrect results in the case of long distances of propagation of space charge waves. When the thermal conductivity term is taken into account and the length of the system is short, the averaged balance models can get satisfactory results.

I. INTRODUCTION

The use of space charge waves in *n*-GaAs thin films with hot electrons is a promising way to realize analog signal processing [1]. To describe non-stationary carrier transport in GaAs devices, it is necessary to use Monte Carlo methods or hydrodynamic models, which incorporate the population transfer between different valleys [2]. To describe the propagation of space charge waves in *n*-GaAs thin films, the hydrodynamic models have been widely studied [3]. We consider three different two-valley quasi-hydrodynamic models for space charge wave propagation, intended for electron transport, in GaAs. They are simple drift-diffusion model, detailed balance model [4], and averaged balance one [5,6]. This models use balance equations with which carrier transport can be analyzed by following spatial and temporal variations of particle, momentum, and energy densities. The detailed balance model is described in the book [4]; it incorporates the collision terms for two-valleys. The averaged balance models do not take into account the collision terms directly, also *n, v, w* and *m** are averaged over the available conduction band valleys. There are some variants of averaged electron energy terms in the last model.

In this paper, we present a comparative study of three two-valley quasi-hydrodynamic models for space charge

Abel Garcia-B and Volodymyr Grimalsky are with the Department of Electronics, INAOE, P.O. Box 51 and 216 Puebla, Pue. Mexico, E-mail: agarciab@inaoep.mx, vgrim@inaoep.mx.
Edmundo Gutierrez-D is with INTEL-SRCM, Guadalajara, Mexico, E-mail: edmundo.a.gutierrez@intel.com.
S. Koshevaya is with Autonomous Univ. of Morelos, CIICAp, Cuernavaca 62210, Mexico, E-mail: svetlana@uaem.mx.

wave propagation in *n*-GaAs semiconductor thin films with respect to the spatial increment. In particular, the results show that the detailed model is the most correct to describe the space charge waves. The averaged balance model may lead to incorrect results in the case of long structures, whereas in the short structures the thermoconductivity becomes dominating and such a model yields correct results.

II. BALANCE-EQUATIONS MODELS

A. Detailed Balance Model

The first model we consider is the model where the balance equations are written for each valley [4]. It consists of balance equations for carrier density (1), momentum balance equation (2) to describe the electron velocity, and energy balance equation (3) of two populations of electrons. It is a two-valley model in which only the Γ-valley and L-valleys are taken into account. Therefore, this model needs two sets of balance equations, namely, for the Γ-valley and the L-valleys.

$$\frac{\partial n_i}{\partial t} = -\nabla \cdot \left(n_i v_i\right) + \left(\frac{\partial n_i}{\partial t}\right)_C \qquad (1)$$

$$\frac{\partial v_i}{\partial t} = -(v_i \nabla)v_i + \frac{eE_i}{m_i *}$$
$$- \frac{1}{n_i m_i *}\nabla\left(n_i T_i\right) + \left(\frac{\partial v_i}{\partial t}\right)_C \qquad (2)$$

$$\frac{\partial w_i}{\partial t} = -v_i \cdot \nabla w_i + eE_i \cdot v_i$$
$$- \frac{1}{n_i}\nabla \cdot \left[\left(n_i v_i - \frac{\kappa}{k_B}\nabla\right)T_i\right] + \left(\frac{\partial w_i}{\partial t}\right)_C \qquad (3)$$

where the subscript i is the valley index ($i=1$ for Γ-valley, $i=2$ for the equivalent L-valleys). Therefore n_i, v_i and w_i are, respectively, the carrier density, the average drift velocity, and the average energy in the valley i; $T_i = (2/3)(w_i - m_i v_i^2/2)$ is the electron temperature (energetic units). E is the electric field and the index C represents the collision term, in which the intervalley transfer between the Γ-valley and L-valleys is taken into account. The collision terms employed are as follows:

1-4244-0116-X/06/$20.00 ©2006 IEEE

$$\left(\frac{\partial n_i}{\partial t}\right)_C = -n_i \gamma_{nij}(w_i) + n_j \gamma_{nij}(w_j) \qquad (4)$$

$$\left(\frac{\partial v_{di}}{\partial t}\right)_C = -v_{di}\left[\gamma_{pi}(w_i) + \gamma_{nij}(w_i)\right] \qquad (5)$$

$$\left(\frac{\partial w_1}{\partial t}\right)_C = -(w_1 - w_0)\gamma_{w1}(w_1) + \frac{w_{12}(w_1)}{n_1}\left(\frac{\partial n_1}{\partial t}\right)_C \qquad (6)$$

$$\left(\frac{\partial w_2}{\partial t}\right)_C = -(w_2 - w_0)\gamma_{w2}(w_2) \qquad (7)$$

where (4) and (5) are for both valleys, while (6) is for the Γ-valley and (7) is for L-valleys. γ_{nij} and γ_{nji} are the relaxation rates related to the variation of electron density due to the intervalley transfer; n_i and n_j are the electron concentrations in the ith and j^{th} valleys. γ_{pi} is the momentum relaxation. γ_{w1} and γ_{w2} are the energy relaxation rates due to scattering within the valley. Also thermoconductivity (κ) is taken into account, which is important for the higher frequencies $\omega > 5 \times 10^{11}$ s^{-1} (or in the case of small scales < 0.5 μm).

To get the simple formulas for negative differential conductance in this model, we consider an approximation $v_2 \sim 0$, w_2 is constant, and $m_2 \gg m_1$. Our simulations within a framework of general case $v_2 \neq 0$ have proven a good accuracy (~3–5%) of this approximation.

B. Simple Drift-Diffusion Model

The popular drift-diffusion current equations can be easily derived directly from the Boltzmann equation.

$$J_n = ne\mu_n \vec{E} - e\nabla(D_n n)$$

This means, the total current density at any point is the sum of the drift and diffusion components, where $\mathbf{E}$ is the electric field in the plane of the film.

C. Average Balance Model

This model [5] is similar to the model [4], but this model does not take into account the intervalley transfer terms directly, also n, v, w and m^* are averaged over the available conduction band valleys:

$$n = n_\Gamma + n_L$$
$$v = (n_\Gamma v_\Gamma + n_L v_L)/n$$
$$w = (n_\Gamma w_\Gamma + n_L w_L)/n$$
$$m^* = (n_\Gamma m_\Gamma + n_L m_L)/n$$

The average electron energy consists of a drift component and a thermal one associated with the random motion:

$$w = \frac{1}{2}m^* v^2 + \frac{3}{2}T_e$$

where Te is the electron temperature.

This model uses the single-electron-gas approximation and takes account for all intervalley transfer effects through effective values for carrier velocity, energy, effective mass, and relaxation rates. It is assumed that energy and momentum relaxation rates and the averaged effective electron mass depend of the averaged electron energy w.

An expression is needed for the thermoconductivity of the electron gas, which stems from theoretical considerations

$$\kappa = \left(\frac{5}{2} + r\right)n\frac{\mu(T)}{e}T$$

Several different choices for r can be found in the literature, and in many authors [7,8] even neglect heat conduction in their models.

There exists a modification of this model, where the average electron energy has an additional term $F\Delta_{LU}$ [6].

$$w = \frac{1}{2}m^* v^2 + \frac{3}{2}T_e + F\Delta_{LU}$$

The difference appears in the specified form $F\Delta_{LU}$, where $F(w)$ is the occupancy of the upper valleys and is also assumed to be function of w only, and $F\Delta_{LU}$ is a constant. Furthermore, F is exact in the low and high energy limits where F equals to 0 and 1, respectively (the electrons are situated either in the lower valley or in the upper ones). The constant $F\Delta_{LU}$ expresses a difference between the bottoms of the Γ- and L-valleys. In our simulations we take $F\Delta_{LU} = 0.36$ eV. Thus, the model [6] coincides with the one [5], but gives the specification.

III. SIMULATIONS

We consider a propagation of space charge waves in the thin n-GaAs film. The bias electric field is applied along the direction of propagation (z-axis). In such an active medium, possessing negative differential conductivity, the space charge waves are subject to possible amplification. The linear modes of propagation of the system are studied by means of dispersion equation $D(\omega,k) = 0$, which relates the frequency ω to the longitudinal wave number (or propagation constant) $k = 2\pi/\lambda$ and is obtained by the self-consistent solution of the balance equations jointly with the Poisson one. In general, we consider the cases where ω is real and $k = k'+ik''$ has real and imaginary part. The case $k'' > 0$ corresponds to spatial increment (amplification), whereas the case $k'' < 0$ corresponds to the decrement (damping). To obtain the dispersion equation for space charge waves, it is necessary to use the linearized equations of the electron dynamics jointly with the Poisson's equation for the electric potential. All the perturbations obey the law $\sim exp(i(\omega t - kz))$. The stationary state is assumed as unstable with respect to small perturbations (the bias field $\mathbf{E}_0$ is chosen in the region of negative differential conductivity).

A. Comparison between Detailed Balance Model and the Averaged One

We use the equations (1), (2), and (3) for two valleys. To get the approximate analytical expressions for $k(\omega)$, consider the region of relatively low microwave frequencies ($\omega \leq 10^{11}$ s^{-1}), where it is possible to neglect the thermoconductivity in the balance equations. For the perturbations of $\tilde{n}_{1,2}$ we have:

$$\tilde{n}_1 \gamma_{n12} - \tilde{n}_2 \gamma_{n21} + n_{10} \frac{d\gamma_{n12}}{dw_1} \tilde{w}_1 \approx 0$$

$$\tilde{n}_1 + \tilde{n}_2 = \tilde{n}$$

The perturbation of the electron energy in the lower valley is:

$$\tilde{w}_1 \approx \frac{2ev_{10}}{\gamma_{w1}} \tilde{E} -$$

$$- \frac{2v_{10}T_1}{\gamma_{w1}n_{10}}[1 - \frac{w_{12}}{2T_1}(1 - \frac{v_0}{v_{10}})] \times \frac{\partial \tilde{n}_1}{\partial z}$$

After some transformations, it is possible to get the following equation for $\tilde{n}$:

$$\frac{\partial \tilde{n}}{\partial t} + v_0 \frac{\partial \tilde{n}}{\partial z} + n_{10} \frac{dv}{dE} \tilde{E} -$$

$$- \frac{T_1}{n_0 m_1 (\gamma_{p1} + \gamma_{n12})}[1 + (\frac{E_0}{v_{10}}\frac{dv}{dE} - 1) \times$$

$$(1 - \frac{w_{12}}{T_1}(1 - \frac{v_0}{v_{10}}))] \times \frac{\partial^2 \tilde{n}}{\partial z^2} \approx 0$$

Here E_0, v_0, v_{10} etc. are the values of drift electric field, general drift velocity and drift velocity of the lower valley at the stationary (non-perturbed) condition. In the case of negative differential conductivity, we have $dv/dE < 0$.

The structure of the equation for the general perturbation of concentration $\tilde{n}$ is the same as obtained in the simple diffusion-drift approximation. Note that a validity of simple diffusion-drift model has been proven for a description of the processes in Gunn diodes in the case of not very high microwave frequencies $\omega < 5 \times 10^{11}$ s^{-1} [4]. It is important that the value of the effective diffusion coefficient is **positive** here, due to a presence of the term with w_{12}; therefore, we have found that the detailed balance approach seems correct to simulate the space charge waves in *n*-GaAs. This is due to direct taking into account of terms with jumps in balance equations.

To check an approximation of pointed above expressions, it is necessary to compare the calculated drift velocity with one presented in [4]:

$$v = \frac{v_1 n_1 + v_2 n_2}{n_1 + n_2} = V(E)$$

where

$$v_1 = \frac{eE}{m_1 (\gamma_{p1}(w_1) + \gamma_{n12}(w_1))}$$

$$v_2 = \frac{eE}{m_2 (\gamma_{p2}(w_2) + \gamma_{n21}(w_2))}$$

for the stationary case. Where we obtain the velocity-field curve and negative differential conductivity is showed, which is similar that Tomizawa [4] (a difference is within 3%).

To get a comparison with the averaged balance approach, we obtain for the equations (1), (2) and (3) (without collision terms but with variable averaged electron mass $m(w)$) and with small perturbations:

$$\tilde{w} = 2\frac{e^2 E_0 \tilde{E}}{m\gamma_P \gamma_w} - \frac{eE_0 T}{m\gamma_P \gamma_w n}\frac{\partial \tilde{n}}{\partial z} + \frac{T}{n\gamma_w}\frac{\partial \tilde{n}}{\partial t}$$

$$\frac{\partial \tilde{n}}{\partial t} + v_0 \frac{\partial \tilde{n}}{\partial z} + \frac{en_0}{m\gamma_P}[1 + \frac{2e^2 E_0^2}{\gamma_w}\frac{d}{dw}(\frac{1}{m\gamma_P})]\frac{\partial \tilde{E}}{\partial z}$$

$$- \frac{T}{m\gamma_P}[1 + \frac{2e^2 E_0^2}{\gamma_w}\frac{d}{dw}(\frac{1}{m\gamma_P})]\frac{\partial^2 \tilde{n}}{\partial z^2} \approx 0$$

This is the analytical result for two models (Tomizawa [4] and Shur [5]). The analysis to Rodrigues model [6] is similar to the Shur model, in this case we take $F\Delta_{LU} = 0.36$ eV, and some difference of the electron temperature occurs. In the case of the Shur and Rodrigues models, the structure of the equation for $\tilde{n}$ is also diffusion-drift-like, but the effective diffusion coefficient is **negative** here. One can see that there are the same negative multipliers both near the drift term and near diffusion one. Thus, there is no direct correspondence between these models and the simple diffusion-drift one. Therefore, such models are not suitable for simulations of space charge waves in n-GaAs films. Note that such a result is valid also for space charge wave propagation in the bulk n-GaAs.

For simulations we use the dependence of small perturbations like $\sim exp(i(\omega t - kz))$ to find $n_1, n, \tilde{v}_1, \tilde{w}_1, \tilde{v}_2, \tilde{w}_2$ and we take account the approximation $m_1 << m_2$, $v_2 << v_1$ and $v_2 \sim 0$.

In the following section, pointed above analytical results are confirmed by direct numerical simulation of spatial increment of space charge waves.

B. Direct Simulation of Spatial Increment

The dispersion relations $k(\omega)$ have been calculated within the framework of different balance models. The unperturbed (stationary) values of E_0, v_0 have been chosen in the regime of negative differential conductivity ($dv/dE < 0$). The results of direct simulations of $k(\omega)$ of linearized equations are shown in the figures 1,2,3. In Fig. 1, we show the spatial increment for detailed balance model [4], this model seems to be the most correct because $k''(\omega)$ has positive values for the frequencies only below than 3×10^{11} s^{-1}. It is correct because in this frequency range we can obtain the amplification of space charge waves in a thin film of GaAs. In Fig. 2, the spatial increment the simple drift-diffusion model; it seems to be correct too. In the Fig. 3 one can see the real and imaginary part of the

543

longitudinal wave number k for the averaged balance model [6]. This model has some problems when it does not consider the thermal conductivity term, because $k''(\omega)$ gets positive values for frequencies $\omega > 5 \times 10^{11}$ s^{-1}, it is not realistic. Note that decrease of k'' with increase of ω ($\omega > 90 \times 10^{10}$ s-1) can be obtained when the thermal conductivity is taken into account. The influence of thermal conductivity may explain the fact that the model of Rodrigues [6] gave acceptable results under a simulation of short (< 0.5 μm) transistor structures. The Shur model gives qualitatively the same behavior of the spatial increment $k''(\omega)$.

We can see the real and imaginary parts of $k(\omega)$, longitudinal wave number with dependence of frequency for the three models. Note that taking into account the thermal conductivity term for Rodrigues model can get better results.

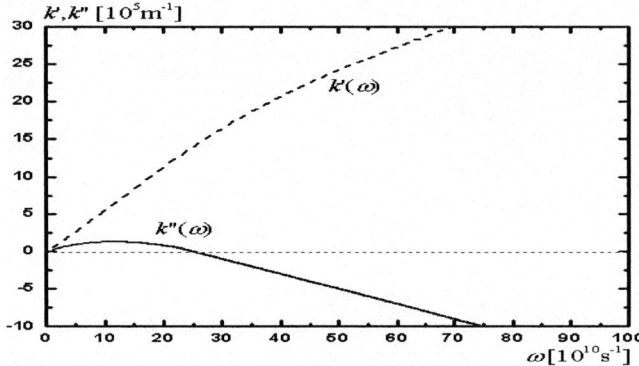

Fig. 1. : Detailed balance [4] model.

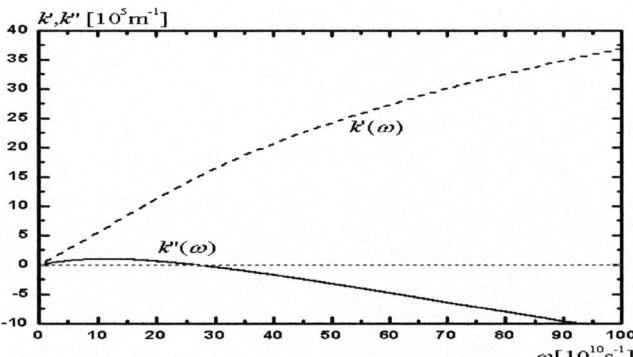

Fig. 2. Simple drift-diffusion model.

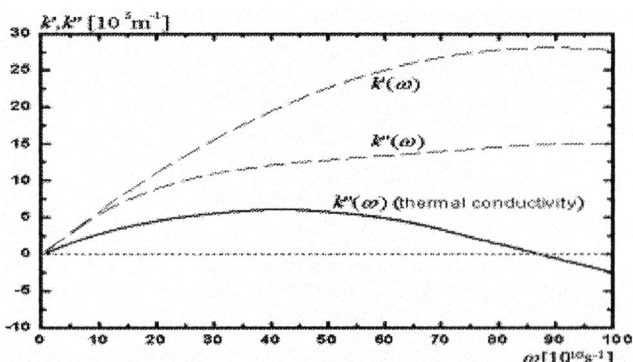

Fig. 3. Averaged balance [6] model with and without thermal conductivity term.

IV. CONCLUSIONS

The dependencies $k(\omega)$ of complex longitudinal wave number on frequency, which are obtained from different balance equation models, are presented and discussed. The best non-local model to describe the space charge wave propagation in thin films possessing negative differential conductivity is the detailed balance model [4]. It has been demonstrated that the balance models with averaging over the different values may give incorrect results under simulations of space charge waves in semiconductors possessing negative differential conductivity. A qualitative explanation of this is neglecting the jump-like intervalley transitions in the balance models with averaging. However, when taking into account the thermoconductivity, the averaged balance model [5,6] can lead to physical results and mathematical stability.

ACKNOWLEDGEMENT

The authors wish to acknowledge the support of this work by CONACyT Mexico in the frame of their research projects and by the PhD's scholarship #165390.

REFERENCES

[1] Alexander L. Kogan, Mikhail A. Kitayev, et al., Analog ICs Using Space-Charge Waves in Two-Valley Semiconductor Films for Microwave Signal Processing. In Proceedings of the *IEEE-International Solid State Circuits Conference. USSR*, 1991, pp. 224-225.

[2] A. M. Anile and O. Muscato, Improved Hydrodynamical Model for Carrier Transport in Semiconductors, *Phys. Rev. B*, 51 (1995), pp. 16728-16740.

[3] Keli Han and Thomas T.Y Wong, Space-Charge Wave Considerations in MIS Waveguide Analysis, *IEEE Transactions on Microwave Theory and Techniques*, vol. 31, no. 7, pp. 1126-1133, July 1991.

[4] K. Tomizawa, *Numerical Simulation Method for Submicron Semiconductor Device*. (Artech House, Boston, 1993), pp 191-192.

[5] M. Shur, *Compound Semiconductor Electronics, The Age of Maturity*. World Scientific, London, 1993. Pp. 320-330.

[6] J.C. Paulo Rodrigues, *Computer-Aided Analysis of Nonlinear Microwave Circuits*. Artech House, Boston, 1998. pp. 72-81.

[7] Y. K. Feng, A. Hintz, Simulation of Submicrometer GaAs MESFETs Using a Full Dynamic Transport Model, *IEEE Trans. Electron Dev.*, vol 35, pp. 1419-1431, 1988.

[8] M. A. Alsunaidi, S.M Hammadi, S.M. El-Chazaly, A parallel Implementation of Two-Dimensional Hydrodynamic Model for Microwave Semiconductor Device Including Inertia Effects in Momentum Relaxation, *Int. J. Num. Mod.: Netw. Dev. Fields*, vol. 10, pp. 107-119,1997.

2006 25th International Conference on Microelectronics

Diffusion-Drift Model of Fully Depleted SOI MOSFET

G. I. Zebrev, M. S. Gorbunov

Abstract – Based on explicit analytical solution of drain current continuity equation a new compact physical model for fully depleted SOI MOSFET has been proposed. The model provides continuous description of I-V characteristics in all transistor operation modes.

I. INTRODUCTION

Over the last three decades, SOI CMOS and its derivatives such as double-gate MOSFETs has been identified as one possible method for increasing the performance of CMOS over that offered by simple scaling. The SOI MOSFETs are known to provide better performance in comparison with bulk MOS-FETs due to lower parasitic capacitances and power dissipation, minimization of short-channel effects and enhanced radiation hardness [1].

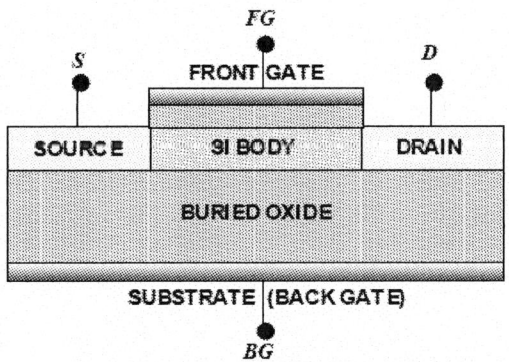

Fig.1. Schematic view of FD SOI MOSFET.

Fully depleted (FD) SOI MOSFETs are often considered as most promising future devices due to ultimate scaling possibilities in double-gate structures. We will consider here a configuration of the FD SOI MOSFET with the highly doped substrate, which can be used as a back gate.

A compact physical model of FD SOI MOSFET has been proposed in this report. The key feature of this model is an analytical solution of the channel current continuity equation $dJ_S/dy=0$ and the introduction and explicit obtaining of control parameter κ, which has a sense of diffusion to drift current ratio. Compact and closed-form equations for drain current, distributions of chemical and electrostatic potential along channels in subthreshold region and in strong inversion operation mode were obtained. The proposed model is significantly based on advanced physical MOSFET model developed by one of the authors in 1990 for bulk MOSFETs [2].

G. I. Zebrev is with Department of Microelectronics, Moscow Engineering Physics Institute, Kashirskoye shosse, 31, Moscow, 115409, Russia, E-mail: GIZebrev@mephi.ru

M. S. Gorbunov is a student of Microelectronics Department of Moscow Engineering Physics Institute.

II. GENERAL ELECTROSTATICS OF FULLY DEPLETED SOI STRUCTURE

Band diagram of the structure is depicted in Fig.2. The Fermi energy near the source (or across whole silicon layer at $V_{DS}=0$) is rigidly fixed by the grounded source node.

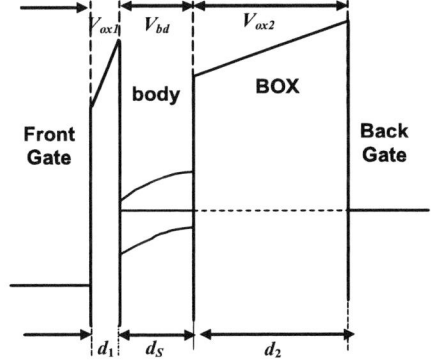

Fig.2. Band diagram of FD SOI MOSFET (the source and the back gate are grounded).

The electric neutrality condition for the total structure accounting for charge trapped in the oxides can be written down as follows

$$N_{GF} = n_S + N_{GB} + N_A d_S - N_{ox1} - N_{ox2}, \qquad (1)$$

where N_{GF} (N_{GB}) are the number of positive charges per unit area on the front (back) gate; N_A is the doping level in the silicon body; d_S is the silicon body thickness; n_S is the inversion layer electron density.

Total voltage drop across the SOI structure is added up voltage drops on the gate and buried insulators (V_{ox1} and V_{ox2} respectively) and on the silicon body V_{bd}.

$$V_{GF} = \varphi_{GF,GB} + \underbrace{\frac{q}{C_1}\left(n_S + N_A d_S + N_{GB} - N_{ox1} - N_{ox2}\right)}_{V_{ox1}} +$$

$$+ \underbrace{\frac{q N_A d_s^2}{2\varepsilon_S} + \frac{q}{C_S}\left(N_{GB} - N_{ox2}\right)}_{V_{bd}} + \underbrace{\frac{q N_{GB}}{C_2}}_{V_{ox2}}$$

$$(2)$$

Notations are shown in Fig.2., V_{GF} is voltage drop across the whole structure, $C_{1(2)}$ and C_S are the capacitance per unit area of the oxides and silicon body respectively. These capacitances are expressed through their dielectric constants of the insulators and silicon (ε_i and ε_S) and corresponding thicknesses of the oxides ($d_{1(2)}$) and the silicon body (d_S)

$$C_{1(2)} = \frac{\varepsilon_i}{d_{1(2)}}; C_S = \frac{\varepsilon_S}{d_S}. \qquad (3)$$

1-4244-0116-X/06/$20.00 ©2006 IEEE

$$\Delta \varphi_{BD} \equiv \frac{q\,N_A d_s^2}{2\varepsilon_S} \qquad (4)$$

The back gate voltage drop is assumed to be negligible, work function difference between front gate and back gate material is

$$q\varphi_{FG,BG} = W_{FG} - W_{BG}, \qquad (5)$$

Accounting for that

$$\varphi_S = \underbrace{\frac{q\,N_A d_s^2}{2\,\varepsilon_S} + \frac{q}{C_S}\left(N_{GB} - N_{ox2}\right)}_{V_{bd}} + \underbrace{\frac{q\,N_{GB}}{C_2}}_{V_{ox2}} \qquad (6)$$

$$= \Delta\varphi_{BD} + q\,N_{GB}\frac{C_S + C_2}{C_S C_2} + \frac{q\,N_{ox2}}{C_S}$$

one can find

$$V_{GF} = \varphi_{GF,GB} + \varphi_S\left(1 + \frac{C_S C_2}{(C_S + C_2)C_1}\right) +$$

$$+ \frac{q}{C_1}\left(n_S + N_A d_S - N_{ox1} - \Delta\varphi_{BD}\frac{C_S C_2}{C_S + C_2}\right) - \frac{qN_{ox2}C_S}{(C_S + C_2)C_1} \qquad (7)$$

Threshold voltage for the front gate V_{TF} can be deduced from (7) as follows

$$V_{TF} = \varphi_{GF,GB} + 2\varphi_F\left(1 + \frac{C_S C_2}{(C_S + C_2)C_1}\right) +$$

$$+ \quad + \frac{q}{C_1}\left(N_A d_S - N_{ox1} - \Delta\varphi_{BD}\frac{C_S C_2}{C_S + C_2}\right) - \frac{qN_{ox2}C_S}{(C_S + C_2)C_1} \qquad (8)$$

III. DIFFUSION TO DRIFT CURRENT RATIO

It is well-known that channel current per unit width J_S can be expressed as a sum of drift and diffusion parts

$$J_S = q\mu_n\,n_S\frac{d\varphi_S}{dy} + eD_n\left|\frac{dn_S}{dy}\right|, \qquad (9)$$

where n_S is position-dependent channel charge density, μ_n and D_n are the electron mobility and diffusivity, y is a coordinate along the channel. This relation can be rewritten in an equivalent form

$$J_S = J_{DR} + J_{DIFF} \equiv \sigma_S E(1+\kappa), \qquad (10)$$

where E is electric field, σ_S is 2D channel conductivity

$$\sigma_S = q^2\tau\,n_S/m. \qquad (11)$$

The dimensionless parameter characterizing the ratio of diffusion to drift current is introduced in (10)

$$\kappa \equiv \frac{J_{DIF}}{J_{DR}}, \qquad (12)$$

which assumed to be constant along the channel.

It is well-known that the electrochemical potential μ (or, as a matter of fact, the quasi-Fermi level) is composed of generally independent electrostatic (φ_S) and chemical (ζ) potentials

$$\mu = \zeta - \varphi_S \qquad (13)$$

In non-equilibrium the chemical potential controls particle (electron) density and is generally irrelevant to properly elec-

tric charge. Two-dimensional electron density, for example, in the channel $n_S(\zeta)$ is a function exactly of chemical potential ζ rather than electrostatic (φ) or total electrochemical potential (μ). It is very important that chemical potential distribution along the channel does not coincide in general with electrostatic potential distribution.

Using concepts of chemical and electrostatic potentials we can rewrite the ratio of the diffusion to the drift current as the ratio of gradients of chemical and electrostatic potential along the channel

$$\kappa \equiv \frac{J_{DIF}}{J_{DR}} = \left|\frac{d\zeta}{d\varphi_S}\right| \qquad (14)$$

To properly derive explicit expression for control parameter κ we have to use electric neutrality condition along the channel length in gradual channel approximation which is assumed to be valid rather under nonequilibrium condition $V_{DS} > 0$.

Differentiating Eq.7 with respect to chemical potential ζ (note that V_{GF} =const(y)) and taking into consideration that φ_S and ζ are independent variables and n_S depends on only chemical potential ζ one can get

$$\kappa = \left|\frac{d\zeta}{d\varphi_S}\right| = \frac{C_1}{|dn_S/d\zeta|}\left(1 + \frac{C_S C_2}{C_1(C_S + C_2)}\right). \qquad (15)$$

We will assume for definiteness that in the above threshold mode the inversion layer represents degenerate 2D system with one-sub-band filling. The parameter $dn_S/d\zeta$ can be estimated for above threshold (2D degenerate system) and subthreshold (many-sub-band Boltzmann system) case as follows

$$\frac{dn_S}{qd\zeta} \cong \begin{cases} n_S/k_B T, & V_{GF} < V_{TF} \\ m/\pi\hbar^2 \equiv g_{2d}, & V_{GF} > V_{TF} \end{cases} \qquad (16)$$

where m is effective electron mass (including valley degeneracy), $\hbar$ is the Plank constant.

IV. CURRENT CONTINUITY EQUATION

The key point of this approach is an explicit analytical solution of continuity equation for channel current density. Current should be conserved along the channel $dJ_S/dy = 0$ that yields an equation for electric field distribution along the channel

$$\frac{dE}{dy} = -\frac{d\ln\sigma_S}{dy}E = \left(\frac{dn_S}{n_S d\zeta}\right)\left(-\frac{d\zeta}{d\varphi_S}\right)E^2, \qquad (17)$$

where φ and ζ are a local electric and chemical potential. Electron density is determined by local value of chemical potential ζ.

We have introduced in (17) so called diffusion energy $\varepsilon_D = qn_S(d\zeta/dn_S)$ which can be represented in 2D case as

$$\varepsilon_D = k_B T\left(1 + \exp\left(\frac{E_0 - E_F}{k_B T}\right)\right)\ln\left(1 + \exp\left(\frac{E_F - E_0}{k_B T}\right)\right), \qquad (18)$$

where E_0 is the first energy sub-band bottom [3]. It is easy to see from (18) that ε_D equals ε_F or $k_B T$ whether degenerate or not the case.

We will assume further approximation that κ and ε_D are functions of only electrode biases rather than position along the channel.

Using these notations current continuity equation can be rewritten as

$$\frac{dE}{dy} = \frac{\kappa}{\varepsilon_D/q} E^2. \tag{19}$$

Direct solution of ordinary differential equation (19) yields

$$E(y) = \frac{E(0)}{1 - \dfrac{\kappa E(0)}{\varepsilon_D/q} y}, \tag{20}$$

where $E(0)$ is electric field near the source.

Boundary condition is fixed by electrochemical potential difference between drain and source

$$V_{DS} = (1+\kappa)\int_0^L E(y)dy. \tag{21}$$

Using (21) and (20) we have got expressions for $E(0)$ and electric field distribution along the channel

$$E(0) = \frac{\varepsilon_D/q}{\kappa L}\left[1 - \exp\left(-\frac{\kappa}{1+\kappa}\frac{qV_{DS}}{\varepsilon_D}\right)\right]. \tag{22}$$

$$E(y) = \frac{\varepsilon_D/q}{\kappa L}\frac{1 - \exp\left(-\dfrac{\kappa}{1+\kappa}\dfrac{qV_{DS}}{\varepsilon_D}\right)}{1 - \dfrac{y}{L}\left[1 - \exp\left(-\dfrac{\kappa}{1+\kappa}\dfrac{qV_{DS}}{\varepsilon_D}\right)\right]} \tag{23}$$

A. Distributions of chemical and electrostatic potential along the channels

Integrating (23) we have obtained explicit relationships for distributions of chemical and electrostatic potential along the channel length (L) separately

$$\varphi_S(y) - \varphi_S(0) =$$

$$= -\frac{\varepsilon_D/q}{\kappa}\ell n\left\{1 - \frac{y}{L}\left[1 - \exp\left(-\frac{\kappa}{1+\kappa}\frac{qV_{DS}}{\varepsilon_D}\right)\right]\right\}, \tag{24}$$

$$\zeta(y) - \zeta(0) = -\kappa\left(\varphi_S(y) - \varphi_S(0)\right), \tag{25}$$

where ε_D is implied as near-the-source value.

Full drop of electrochemical potential ($\varphi_S - \zeta$) on the channel length is fixed by source-drain bias V_D

$$\varphi_S(L) - \varphi_S(0) + \left(\zeta(L) - \zeta(0)\right) = \frac{V_{DS}}{1+\kappa} + \frac{\kappa V_{DS}}{1+\kappa} = V_{DS} \tag{26}$$

both for subthreshold (where $\kappa > 1$) and strong inversion operation modes (where $\kappa < 1$). For the latter strong inversion case we have

$$\zeta(0) - \zeta(y) \cong \kappa\frac{y}{L}V_D, \quad \varphi(y) - \varphi(0) \cong \frac{y}{L}V_{DS}. \tag{27}$$

The full drop of chemical potential is negligible under strong inversion compared to electrostatic potential but it becomes very important in saturation mode.

B. Electron Density Distribution Along the Channel

For two-dimensional channel we have

$$n_S = g_{2D}k_BT\ln\left[1 + \exp\left(\frac{q\zeta}{k_BT}\right)\right]. \tag{28}$$

Using (25) and (28) we have derived in general case

$$n_S(y) \cong g_{2D}k_BT \times$$

$$\times \ln\left\{1 + e^{\frac{q\zeta_0}{k_BT}}\left(1 - \frac{y}{L}\left(1 - \exp\left(-\frac{\kappa}{1+\kappa}\frac{qV_{DS}}{\varepsilon_D}\right)\right)\right)^{\frac{\varepsilon_D}{k_BT}}\right\}, \tag{29}$$

where ζ_0 is chemical potential near the source.

For non-degenerate case ($\zeta < 0, \varepsilon_D = k_BT$) the electron density distribution along the channel is

$$n_S(y) \cong$$

$$\cong g_{2D}k_BT\, e^{\frac{q\zeta_0}{k_BT}}\left(1 - \frac{y}{L}\left(1 - \exp\left(-\frac{\kappa}{1+\kappa}\frac{qV_{DS}}{k_BT}\right)\right)\right) = \tag{30}$$

$$= n_S(0)\left(1 - \frac{y}{L}\left(1 - \exp\left(-\frac{\kappa}{1+\kappa}\frac{qV_{DS}}{k_BT}\right)\right)\right)$$

where $n_S(0)$ is electron density near the source.

Total channel electron charge Q_C of the non-degenerate can be computed as

$$Q_C = Z\int_0^L n_S(y)dy = \frac{Q_{C0}}{2}\left(1 + \exp\left(-\frac{\kappa}{1+\kappa}\frac{qV_{DS}}{k_BT}\right)\right) \tag{31}$$

where $Q_{C0} \equiv n_S(0)ZL$.

In degenerate case ($\zeta > 0, \varepsilon_D \geq k_BT$) the general expression can be approximated as

$$n_S(y) \cong n_S(0) - \frac{y}{L}\frac{g_{2D}\varepsilon_D}{2}\left(1 - \exp\left(-\frac{\kappa}{1+\kappa}\frac{qV_{DS}}{\varepsilon_D}\right)\right) \tag{32}$$

and integration over channel length yields

$$Q_C = \frac{Q_{C0}}{2}\left(1 + \exp\left(-\frac{\kappa}{1+\kappa}\frac{qV_{DS}}{\varepsilon_D}\right)\right) \tag{33}$$

Accounting for (18) the relationship (33) can be used both for degenerate and non-degenerate case.

C. Transit Time Through the Channel Length

Using electric field distribution (18) the transit time through the whole channel length can be computed in the following way

$$\tau_{TT} = \int_0^L \frac{dy}{\mu_n(1+\kappa)E(y)} \tag{34}$$

Performing direct integration one can explicitly get

$$\tau_{TT} = \frac{L^2}{2D_n}\frac{\kappa}{1+\kappa}\coth\left(\frac{\kappa}{1+\kappa}\frac{qV_{DS}}{2\varepsilon_D}\right) \tag{35}$$

547

This expression yields drift flight time for above threshold regime (when $\kappa \ll 1$)

$$\tau_{TT} \cong \frac{L^2}{\mu_n V_{DS}} \qquad (36)$$

and diffusion time for subthreshold operation mode

$$\tau_{TT} \cong \frac{L^2}{2D_n}. \qquad (37)$$

V. CURRENT CHARACTERISTICS

We will show in this section that relationship for drain current I_D as a function of drain-source bias V_{DS} can be derived by two self-consistent ways.

A. Local Approach

The first approach is to represent drain current as relation depending on only quantities near the source ($y = 0$)

$$I_D = (1 + \kappa) q Z \mu_n n_S(0) E(0), \qquad (38)$$

where Z is the channel width, $n_S(0)$ and $E(0)$ are the two-dimensional electron density and electric field along the channel near the source respectively.

Using (22) and (38) a general expression for the SOI MOSFET drain current can be written as follows

$$I_D = q \frac{Z}{L} D_n n_S(0) \frac{1 + \kappa}{\kappa} \left\{ 1 - \exp\left(-\frac{\kappa}{1 + \kappa} \frac{q V_{DS}}{\varepsilon_D} \right) \right\}, \qquad (39)$$

where D_n is electron diffusion coefficient.

Recall that the Einstein relation is

$$q D_n = \mu_n \varepsilon_D. \qquad (40)$$

The relationship is the principal result of the work.

B. Global Approach

The second approach is to represent drain current as the ratio of total charge number Q_C in the channel to the electron flight time from the source to the drain τ_{TT} which are considered as drain-source bias dependent quantities

$$I_D(V_{DS}) = Q_C(V_{DS}) / \tau_{TT}(V_{DS}) \qquad (41)$$

Once we have obtained expression for τ_{TT} (35) we can immediately get

$$I_D = \frac{2D_n}{L^2} Q_C(V_{DS}) \frac{1 + \kappa}{\kappa} \tanh\left(\frac{\kappa}{1 + \kappa} \frac{q V_{DS}}{2\varepsilon_D} \right) \qquad (42)$$

After substitution (33) into the relation (42) we have reproduced exactly the expression for the drain current (39).

Thus we have derived I-V characteristics in two self-consistent ways.

C. Above Threshold Regime

General relationship (39) is valid both in the above threshold and the subthreshold operation mode. For above threshold case the Eq.15 can be simplified as follows. Electron density near the source is expressed as $q n_S(y = 0) \cong C_{FG}(V_{GF} - V_{TF})$ and parameter κ is much lesser unity due to smallness of diffusion current compared to the drift one

$$\kappa = \frac{C_1}{q^2 g_{2D}} \left(1 + \frac{C_S C_2}{C_1 (C_S + C_2)} \right) \ll 1, \; V_{GF} > V_{TF} \qquad (43)$$

Recall that $n_S = g_{2D} \varepsilon_D$ we have got

$$I_D = \frac{Z}{L} \mu_n C_1 \frac{(V_G - V_T)^2}{1 + \eta_{FD}} \left\{ 1 - \exp\left(-(1 + \eta_{FD}) \frac{V_{DS}}{V_G - V_T} \right) \right\}, (44)$$

where $\eta_{FD} \equiv C_S C_2 / C_1 (C_S + C_2)$ (typically $\ll 1$).

D. Subthreshold regime

In the subthreshold mode the diffusion current dominates

$$\kappa = \frac{C_1 k_B T}{q^2 n_S} \left(1 + \frac{C_S C_2}{C_1 (C_S + C_2)} \right) \gg 1, \; V_{GF} < V_{TF} \qquad (45)$$

and (39) without taking interface trap into consideration reduces to a familiar diffusion-like form

$$I_D \cong \frac{Z}{L} q D_n n_S(0). \qquad (46)$$

The standard BSIM interpolation for the cannel density vs gate bias can be used for numerical computation.

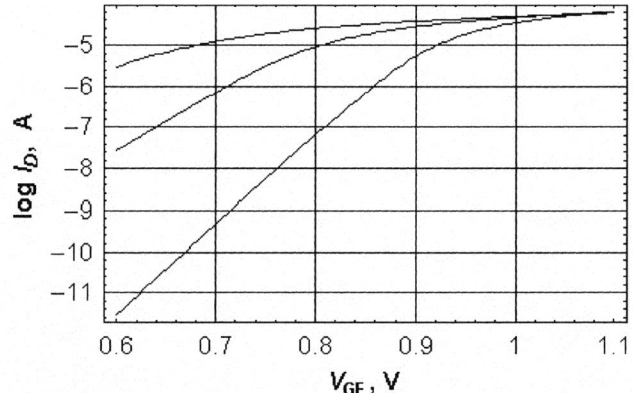

Fig.3. Simulated transfer I-V curves of FD SOI MOSFET at different temperatures (210K, 300K and 400K)

Velocity saturation effects can be incorporated in this approach in a self-consistent way [4].

VI. CONCLUSION

Based on explicit analytical solution of drain current continuity equation a model fully depleted SOI MOSFET has been developed. Proposed model provides continuous description both in the above threshold and the subthreshold operation modes and gives a good basis for compact modeling.

REFERENCES

[1] M.Chan et al. SSE, 48, 969-978(2004)
[2] Zebrev G.I., Useynov R.G. Fiz. Tekhn. Polupr. (Sov. Phys. Semiconductors), V.24, N.5, pp. 777-781, 1990.
[3] T. Ando, A. Fowler, F. Stern, Rev. Mod. Phys. V.54, N.2, pp.437-462, 1982.
[4] G.I. Zebrev Fiz. Tekhn. Polupr. (Sov. Phys. Semiconductors), V.26, N.1, pp.83-88, 1992.

Session
Reliability Physics

2006 25th International Conference on Microelectronics

Low Frequency Noise and Fluctuations in Sub 0.1μm Bulk and SOI CMOS Technologies

G. Ghibaudo and J. Jomaah

Abstract - A review of recent results on the low frequency noise in advanced Si-based CMOS devices is given. The modeling approaches such as the carrier number and the Hooge mobility fluctuations used for the diagnostic of the noise sources are presented and illustrated through experimental data obtained on modern SOI and Si bulk CMOS generations. For SOI devices fully-depleted MOSFET's and double-gate structures are focused. The impact of the back gate voltage on the 1/f noise is analyzed. For Si bulk devices, a special emphasis is addressed to the role of gate leakage on both static and LF noise characteristics.

I. INTRODUCTION

The development of conventional and novel architecture Si-based CMOS devices requires extensive DC, low and high frequencies electrical characterization aiming at understanding the device operation, finding the critical parameters limiting the performances of the technologies under development. One of these limitations could stem from the low frequency noise (LFN), which is well-known in MOSFETs due to the heterogeneous interface between silicon and silicon dioxide. Indeed, excessive low frequency noise and fluctuations could result in serious limitation of the functionality of the analog and digital circuits [1-10]. The 1/f noise is also of paramount importance in RF circuit applications where it leads to phase noise in oscillators or multiplexors. Therefore, a careful analysis of LFN in advanced CMOS devices is mandatory to identify the main noise sources and to understand their physical mechanisms. Such an analysis also permits us to establish accurate compact noise model for circuit simulation.

In this paper, recent issues about the low frequency noise are presented and discussed through experimental data obtained on advanced SOI and Si Bulk CMOS generations. For SOI devices, the case of fully depleted transistors and double gate structures will be studied, featuring the impact of thin film or coupling effects. For Si bulk transistors, the case of advanced generation down to few 10nm's is analyzed with special emphasis on gate dielectric thinning and channel engineering impact.

G. Ghibaudo and J. Jomaah are with the IMEP Laboratory, ENSERG, BP 257, 38016 Grenoble, France (ghibaudo@enserg.fr and jommah@enserg.fr)

II. SOI DEVICES

Partially depleted (PD), Fully depleted and Double-gate SOI CMOS technologies, processed on Unibond substrates, were studied. For PD technology, floating body (FB) and body-contacted (BC) SOI MOS were considered. PD SOI front oxide thickness was T_{ox1}= 4.5nm and 2nm for 0.25μm and 0.12μm technology nodes, respectively. A 2nm front gate oxide was used and the back oxide thickness was 400nm and 2nm for the fully-depleted and double-gate devices, respectively. Silicon film thickness was T_{Si}=150 nm for PD devices, 15nm for the fully depleted and 6nm for the double-gate structures. A wide range of channel lengths was available, from 2 μm down to 50 nm. Noise measurements in both linear and saturation regimes were carried out in the 1 Hz to 100 kHz frequency range using a fully automatic LF noise measuring systme by Synergy concept.

Figure 1 shows, in ohmic operation (V_D=50mV), the normalised drain current power spectral density S_{ID}/I_D^2 plotted as a function of the drain current for N and P-channel PD-SOI MOSFETs for different channel lengths. In this plot, the solid line represents the front gate power spectral density S_{VG} multiplied by the ratio $(g_m/I_D)^2$ where g_m stands for the gate transconductance. The good correlation between these two quantities, confirms that the LF noise can be interpreted by the McWhorter model, which associates the 1/f noise to carrier number fluctuations due to trapping-detrapping into slow oxides states [3,4].

In this model, the fluctuations of the drain current stem from those of the inversion charge near the silicon-silicon dioxide interface caused by the dynamic trapping and detrapping of free carriers into states located in the oxide near the interface. The normalized drain current spectral density is equated to [4]:

$$\frac{S_{ID}}{I_D^2} = \left(\frac{g_m}{I_D}\right)^2 S_{VG} \tag{1}$$

where g_m is the gate transconductance and S_{VG} is the equivalent input gate voltage spectral density related to the slow oxide trap density Nt as,

$$S_{VG} = \frac{q^2 kT\, Nt\, \lambda}{WL\, Cox^2 f} \tag{2}$$

WL being the device area, *Cox* the gate oxide capacitance, λ the tunnel distance in the oxide and *f* the frequency.

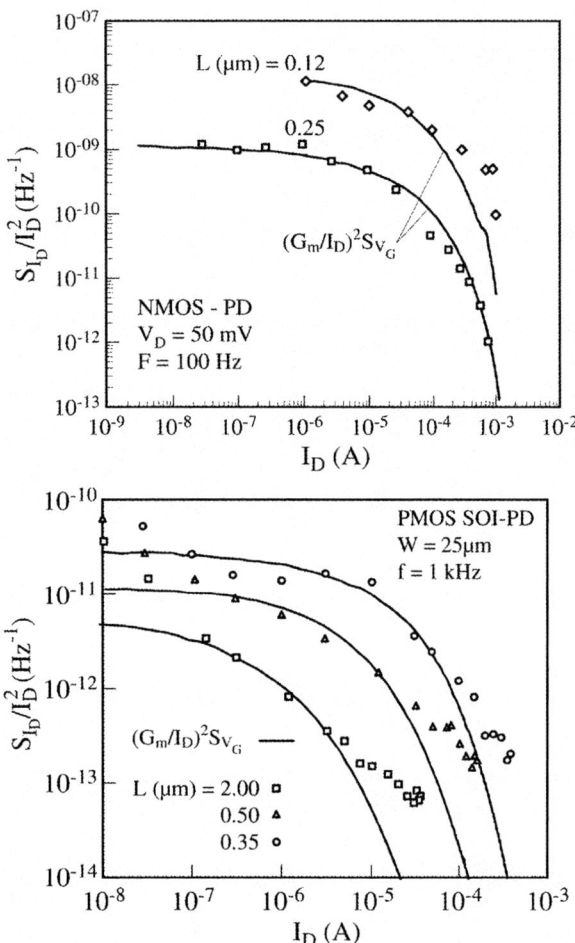

Fig. 1. Normalized drain current power spectral density S_{ID}/I_D^2 at V_D=50 mV at different channel lengths for (a) NMOS FB partially-depleted and (b) PMOS FB partially-depleted SOI devices. Full lines: $S_{Vg} \cdot (g_m/I_d)^2$

Moreover, some discrepancy in strong inversion can be observed (case of P-channel, Fig. 1.b) which can be attributed to the correlated mobility fluctuations. Indeed, by taking into account the dependence of the carrier mobility on the insulator charge (Coulomb scattering), the fluctuations of the insulator charge give rise to a supplementary change in the mobility, which induces an extra drain current fluctuation (see Eq. 5).

Figure 2 shows the normalised drain current power spectral density of a fully depleted device with W/L =25/0.8μm and for back-gate voltages varying from V_{g2}=-50V to V_{g2}=50V, as a function of the drain current. In this plot the solid line is the the corresponding front gate power spectral density S_{VG} multiplied by $(g_m/I_D)^2$ at V_{g2}=20V.

A good correlation exists between S_{ID}/I_D^2 and $(g_m/I_D)^2 \times S_{VG}$, indicating that the 1/f noise is due to carrier number fluctuations. By increasing V_{g2}, an increase in S_{ID}/I_D^2 is observed, which is a result of the coupling effect.

Indeed, when the back gate is accumulated, the carrier fluctuations induced by the back oxide traps are screened by the accumulation layer, whereas when depleting the back interface, the noise becomes higher due to the increasing coupling effect.

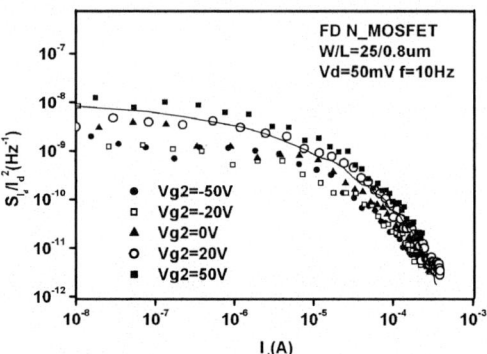

Fig. 2. Normalized drain current power spectral density S_{Id}/Id^2 of a W/L=25/0.8um FD N-MOSFET at Vd=50mV and f=10Hz for different back-gate voltages. Solid line: $S_{Vg} \times (gm/Id)^2$ for Vg_2=20V [11].

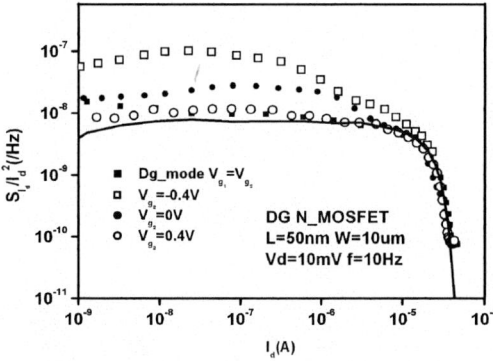

Fig. 3. Normalized drain current noise spectral density S_{Id}/Id^2 of a W/L=10/0.05um DG N-MOSFET at Vd=10mV and f=10Hz for different back-gate voltages. Solid line: $S_{Vg} \times (gm/Id)^2$ for double gate mode [11]

The coupling effect in FD-SOI devices can be modelled by considering the capacitive coupling between the front (1) and back (2) interfaces, yielding for weak inversion and depletion the following relations for the front and back gate input gate voltage noise [11]:

$$S_{VG1} = S_{Vfb1} \cdot c_1^2 + S_{Vfb2} \tag{3a}$$

$$S_{VG2} = S_{Vfb2} \cdot c_2^2 + S_{Vfb1} \tag{3b}$$

with $c_{1,2}$ being the coupling coefficient

$$.c_{1,2} = \frac{C_{Si}}{C_{ox1,2} + C_{Si}}$$

where $C_{ox1,2}$ is the front (back) gate oxide capacitance and C_{Si} is the silicon film capacitance. Equation (3) allows to evaluate the amplitude of the input noise as a function of the film thickness as illustrated in Figure 4 considering

552

equal flat band voltage noise at both interfaces i.e. $S_{Vfb1}=S_{Vfb2}$.

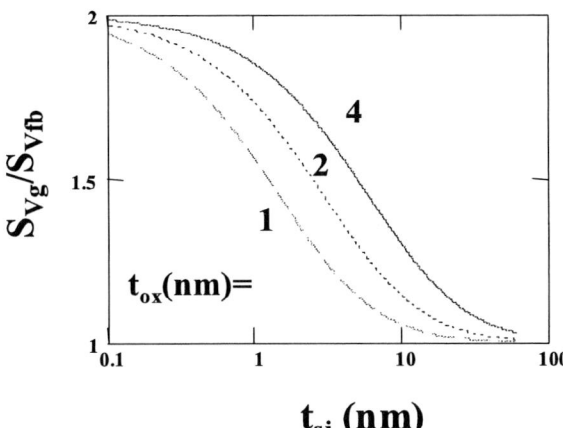

Fig. 4. Variation of normalized input gate voltage noise S_{Vg}/S_{Vfb} with silicon film thickness t_{si} showing the impact of coupling effect.

DG-MOSFETs have been studied in both Single Gate (SG) (back gate voltage controlled separately) and Double Gate (DG) modes (front gate and back gate with the same voltage). The normalized drain current power spectral density S_{ID}/I_D^2 for a DG N-MOSFET, correlates very well with $S_{VG} \times (g_m/I_D)^2$. This feature confirms that in double-gate devices, similarly to partially- and fully-depleted ones, the noise source is also the carrier number fluctuations. In contrast to fully-depleted devices, in ultra thin double-gate MOSFETs, the normalized drain current noise decreases while increasing the back-gate voltage, specially at low drain current (weak inversion). An explanation for this behavior relies on the fact that since the silicon film is very thin (6nm), even when the back interface is accumulated, there exists a coupling between the two interfaces and so the current flows in both channels giving rise to fluctuations in carrier number. While increasing the back gate voltage, the charge is pushed away from both interfaces (volume inversion) and so there would be a screening effect reducing the number of trapped carriers [11].

III. BULK DEVICES

Noise measurements have also been performed on n-MOSFETs from an advanced 65 nm CMOS technology with 1.2nm gate oxides thickness from STMicroelectronics, Crolles. In table 1 are report the main technological characteristics of the various technological splits denoted as 45nm and 30nm.

For the two CMOS technologies, the pocket implants allow to obtain an almost constant threshold voltage with gate length but leads to a significant degradation of the low

field mobility with channel length as illustrated in Figure 5 [12].

TABLE I
TECHNOLOGICAL CHARACTERISTICS OF THE STUDIED CMOS TECHNOLOGICAL SPLITS.

Technology	45nm CMOS	30nm CMOS
Gate oxides	RTN process (Rapid Thermal Nitridation).	PN process (Plasma Nitridation).
Polysilicon gate thickness	1500Å	1200Å
LDD (Lightly Doped Drain)	Arsenic (As)	Arsenic (As)
Pockets	Boron (B)	Boron fluoride (BF$_2$)
HDD (High Doped Drain)	Cobalt silicide (CoSi$_2$)	Nickel silicide (NiSi)

Typical dc drain current versus gate voltage are shown in Figure 6 at V_{DS} = 50 mV, respectively for the shortest and longest geometries of the two studied technologies. For the 10x10 µm² device from 45nm technology, a considerable decrease of the drain current to negative values can be noticed due to huge gate tunneling current. The MOSFET parameters have been extracted after correcting the drain current for the gate-leakage current with the simple following expression [12]:

$$I_{D0} = I_D + \alpha_D \times I_G = I_S - \alpha_S \times I_G \qquad (4)$$

where I_{DO} is the intrinsic drain current, I_D the measured drain current, I_S the measured source current, I_G the gate leakage current, α_D the gate-partitioning coefficient at the drain side, and α_S the gate-partitioning coefficient at the source side.

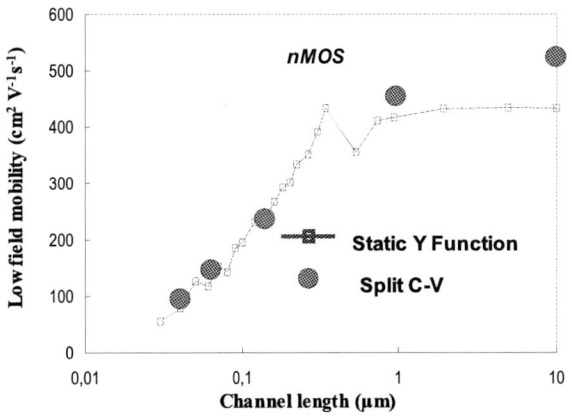

Fig. 5. Variations of effective mobility µeff with gate length as obtained from static (Y function) and split C-V measurements [12].

The α_D coefficient is evaluated at $V_{DS}=0$ in strong inversion regime as shown in Figure 7 and is equal to 25%. We also report the drain current, the gate current and the corrected intrinsic drain current with α_D coefficient at V_{DS} = 50 mV. The gate-leakage current is not negligible and is larger than the measured drain current. The corrected drain current for 30nm technology is similar to that for the 45nm technology but with a much higher gate leakage. The difference in the gate-leakage current between the two studied 1.2nm oxide thickness technologies is due to the different nitridation rate used and, in turn, to a different capacitive equivalent oxide thickness (CET) value.

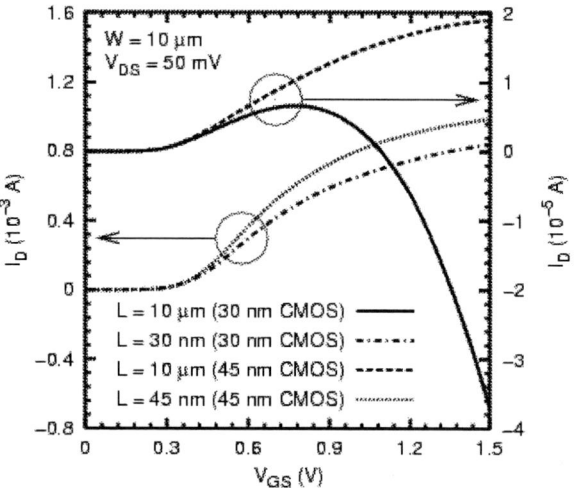

Fig. 6. DC measured drain current on 10μm/10μm NMOS for both technologies and 45nm/10μm NMOS for 45nm CMOS technology and 30nm/10μm NMOS for 30nm CMOS technology at V_{DS} = 50 mV [13,14].

The low-frequency noise characteristics were measured with an original experimental set-up using a programmable biasing amplifier with two entrance bias ports and output trans-impedance amplifier. In Figure 8 are reported the evolution of S_{ID} and S_{ID}/I_D^2 with I_D at V_{DS} = 50mV for a large area device for 45nm technology. For all geometries of both studied technologies, drain current spectral density dependencies with gate and drain voltages are well interpreted by the carrier number fluctuation noise model [4,13,14]. In order to obtain an accurate fit at high gate voltage, the drain current noise is analyzed by two noise sources associated to the intrinsic channel and to the access resistances, respectively. In Figure 8, the dotted curve gives the contribution of the access resistances noise to the total noise. In strong inversion, better agreement is obtained using the mobility-correlated number fluctuation model with the access resistances noise (solid lines). The typical oxide trap density value N_t for both technologies are reported in Figure 8 and are smaller than previous CMOS technologies, indicating a very good dielectric quality despite the use of rapid thermal nitridation (RTN) or plasma nitridation (PN) fabrication processes [15]. A

reduction of N_t can also be observed when the length decreases being explained by a short channel effect, whereas an increase of N_t for the shortest length that could be due to process-induced extra defects close to source and drain regions possibly generated during the implantation steps.

The gate current partitioning of equation (4) can also be used to account for the impact of the gate current noise S_{Ig} on the drain current noise S_{Id} as [13,14]:

$$\frac{S_{ID}}{I_D^2} = \frac{S_{ID0}}{I_{D0}^2} + \left(\alpha_D\right)^2 \times \frac{S_{IG}}{I_D^2} \qquad (5a)$$

with

$$S_{ID0} = \left(1 + \alpha\,\mu_{eff}\,C_{ox} \times \frac{I_{D0}}{g_{m0}}\right)^2 \cdot g_{m0}\cdot S_{Vfb}^D \qquad (5b)$$

where

$$I_{D0} = I_D + \alpha_D \times I_G \,,\; g_{m0} = g_m + \alpha_D \times g_G$$

and

$$\frac{S_{IG}}{I_G} = \left(\frac{g_G}{I_G}\right)^2 \times S_{Vfb}^G, \qquad (6)$$

g_G being the gate conductance ($=dI_G/dV_G$), S_{Vfb}^G the flat band voltage spectral density associated to the gate current and α is the Coulomb scattering coefficient (Vs/C). The sum of both drain and gate current noise contributions enables a very good description of the experimental noise behavior to obtained showing the impact of the gate-leakage current noise on the output drain current noise for large area device [13,14].

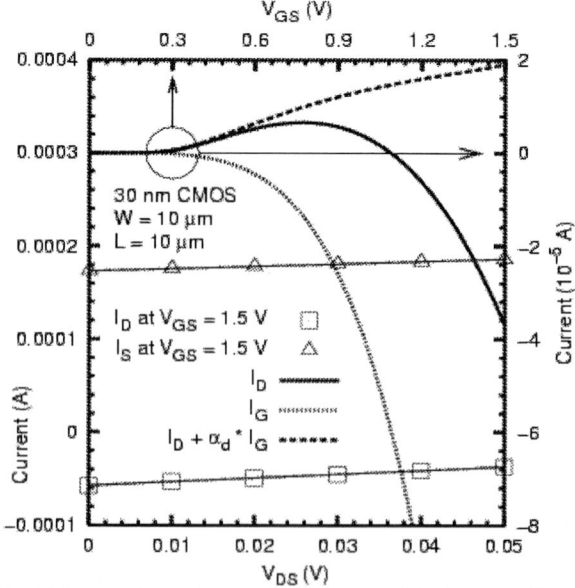

Fig. 7. Extraction of α_d and α_s gate partitioning coefficients and DC measured currents for 10μm/10μm NMOS for 30 nm technology [13,14].

554

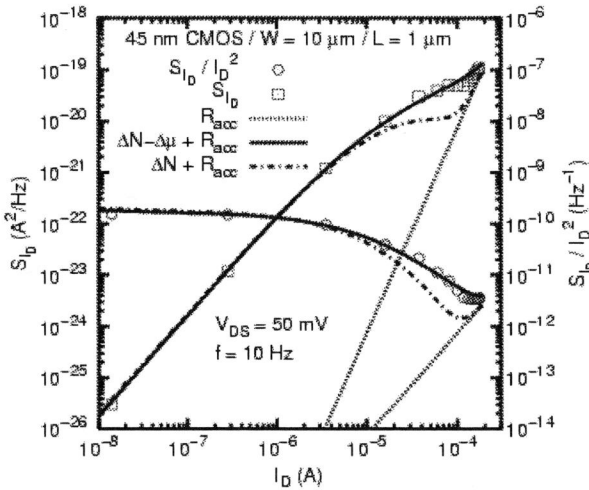

Fig. 8. Comparison of measured data and modeling results of drain current noise as a function of drain current at f = 10 Hz for 10μm/1μm NMOS for 45nm technology [13,14].

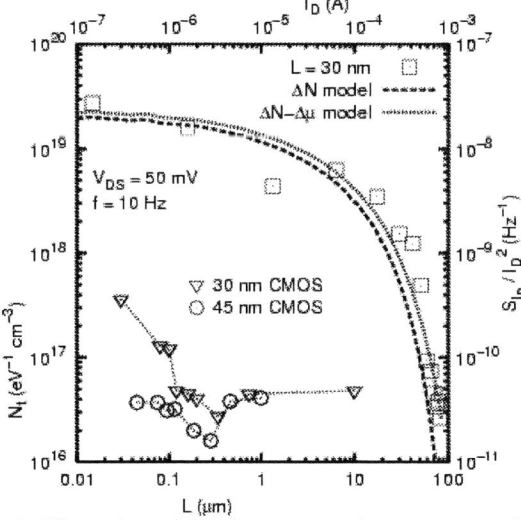

Fig. 9. Illustration of good agreement between normalized drain current noise and correlated model for the smallest geometry for both CMOS technologies [13,14].

IV. CONCLUSION

Low frequency noise in both SOI and bulk devices has been investigated. In Partially- and Fully-depleted and Double-gate SOI MOSFETs the noise source was attributed to carrier number fluctuations, while in advanced bulk CMOS technologies the LF noise stems from carrier and correlated mobility fluctuations. No degradation of gate oxide quality as measured by oxide density of traps Nt was observed despite the ultra thin dielectric used. The

influence of coupling effect on LF noise has been studied in FD-SOI devices and the impact of gate leakage on the output drain current LF noise has also been pointed out.

REFERENCES

[1] D. Eggert, P. Huebler, A. Huerrich, H. Kuerck, W. Budde and M. Vorwerk, "A SOI-RF-CMOS technology on high resistivity SIMOX substrates for microwave applications to 5 GHz." IEEE Trans. on Electron Devices, 44, 1981 (1997).

[2] O. Rozeau, J. Jomaah, S. Haendler, J. Boussey, and F. Balestra, "SOI Technologies Overview for Low-Power Low-Voltage Radio-Frequency Applications", Analog Integrated Circuits and Signal Processing, 25, 2000, Kluwer Academic Publishers – Special issue of SOI.

[3] McWhorter A.L., Semiconductor Surface Physics, University of Pennsylvania Press, Philadelphia 1957.

[4] G. Ghibaudo, O. Roux, Ch. Nguyen-Duc, F. Balestra, and J. Brini, "Improved Analysis of Low Frequency Noise in Field-Effect MOS Transistors", Phys. stat. sol. (a) 124, 571 (1991).

[5] S. Christensson, I. Lundstrom and C. Svensson, "Low-frequency noise in MOS transistors-I", Sol. State Elec., 11, 797 (1968).

[6] R. Kolarova, T. Skotnicki, and J.A. Chroboczek, "Low-frequency noise in thin gate oxide MOSFETs", Mic. Rel., 41, 579, (2000).

[7] M. H. Tsai and T.P. Ma, "The impact of device scaling on the current fluctuations in MOSFET's", IEEE-Transactions-on-Electron-Devices., 41, 2061 (1994).

[8] C. Jakobson, I. Bloom, Y. Nemirovsky, "1/f noise in CMOS transistors for analog applications from subthreshold to saturation", Solid-State-Electronics, 42, 1807 (1998).

[9] E. Simoen and C.Claeys, "On the flicker noise in submicron silicon MOSFET", Solid-State-Electronics, 43, 865 (1999).

[10] Y. Nemirovsky, I. Brouk, C.G. Jakobson, "1/f noise in CMOS transistors for analog applications", IEEE-Transactions-on-Electron-Devices, 48, 921 (2001).

[11] L. Zafari, F. Daugé, J. Jomaah and G. Ghibaudo, "On the low frequency noise in fully depleted and double-gate SOI transistors", Proc. ULIS'2005, Bologna, April 2005.

[12] K. Romanjek, "Characterization and modelling of 50nm and below CMOS transistors technologies", Ph.D. Thesis, INPG, Grenoble, France (2004).

[13] T. Contaret, K. Romanjek, T. Boutchacha, G.Ghibaudo and F. Bœuf, "Low Frequency Noise Charaterization and Modelling in Ultrathin Oxide MOSFETs", Proc. ULIS 2005, p. 55 ; also to be published in Solid State Electronics (2005).

[14] T. Contaret, K. Romanjek, G. Ghibaudo, F. Boeuf and T. Skotnicki, "Drain and Gate Current Low Frequency Noise in advanced CMOS devices with Ultrathin gate Oxides", Proc. ICNF 2005, p. 315.

[15] B. Tavel, M. Bidaud et al, "Thin oxynitride solution for digital and mixed-signal 65nm CMOS platform", IEEE International Electron Devices Meeting 2003, 27.6.1-4 (2003).

2006 25th International Conference on Microelectronics

Analysis of Surface Generation Mechanisms in MOS Capacitors

F. Kong, P. Tanner, L. Hold, J. Han and S. Dimitrijev

Abstract - This paper demonstrates the extraction of MOS capacitor minority carrier generation lifetime and surface generation velocity from the measurement of deep-depletion capacitance transient. It is shown that the bulk generation lifetime, the lateral surface generation, and the surface generation under the gate can be separately determined by measuring test structures with different perimeter-to-area ratios.

I. INTRODUCTION

The pulsed MOS capacitor transient [1] is the most frequently used method to determine minority carrier generation lifetime and surface generation velocity. Other non-pulse methods such as the linear ramp voltage sweep [2] and steady-state gate-controlled diode [1] are also used. These parameters are important for process monitoring and characterization. This is because they are strongly dependent on densities of crystal defects, heavy metal atoms, stacking faults, dislocations, and interface state densities [1].

In the past, the silicon bulk defects have been gradually eliminated with better crystal growth. However the density of surface defects has not decreased as fast as the bulk defects. Surface effects become increasingly more important during lifetime measurements, because carrier generation can takes place both in the bulk and at the surface.

In this paper, we present experimental measurement of deep-depleted capacitance transients of different kinds of MOS capacitor test structures to identify various generation processes from the semiconductor bulk, lateral space-charge region surface generation, and surface generation under the gate.

II. THEORY

In the commonly used pulsed MOS-C method for lifetime measurement, the MOS capacitor is pulsed from accumulation to deep depletion. Due to thermal generation of the carriers, the non-equilibrium space charge region width after the depleting pulse will eventually collapse to the equilibrium width. The model for the effective generation rate can be written as [1]

$$\frac{dN_I}{dt} = \frac{n_i(W - W_f)}{\tau_g} + \frac{n_i s_g' A_s}{A_G} + n_i s_{geff} \quad (1)$$

F. Kong, P. Tanner, L. Hold, J. Han and S. Dimitrijev are with the School of Microelectronic Engineering, Griffith University, Nathan, Qld 4111, Australia, E-mail: s.dimitrijev @ griffith.edu.au

where

N_I is the inversion charge per unit area,
n_i is the intrinsic carrier concentration,
τ_g is the bulk generation lifetime,
W is the non-equilibrium space charge width,
W_f is the final equilibrium scr width,
s_g is the lateral surface generation velocity,
A_s is the area of the lateral surface charge region,
A_G is the gate area,
s_{geff} is the effective surface generation velocity.

In Eq.(1), the first term accounts for the generation in the bulk, the second term represents the surface generation in the lateral space charge region, and the third term accounts for the scr width independent surface generation under the gate and the quasi-neutral bulk generation related to back-surface generation. The generation components and their locations are shown in Fig. 1.

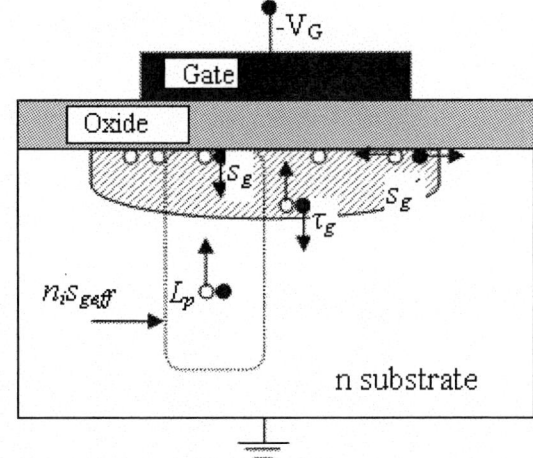

Fig. 1. Generation components in a pulsed MOS capacitor, where L_p is the minority carrier diffusion length.

The Zerbst method [3] is widely used for analysis of C-t transients. It is based on the following equation:

$$-\frac{d}{dt}\left(\frac{C_{ox}}{C}\right)^2 = \frac{2n_i}{\tau_{geff} N_D} \frac{C_{ox}}{C_f}\left(\frac{C_f}{C} - 1\right)$$

$$+ \frac{K_{ox}}{K_s} \frac{2n_i s_{geff}}{t_{ox} N_D} \quad (2)$$

where the effective generation lifetime τ_{geff}, defined as

$$\tau_{geff} = \frac{\tau_g}{1 + \tau_g s_g' A_s /[A_G(W - W_f)]} \quad (3)$$

is a combination of bulk and lateral surface generation, C_{ox} is the oxide capacitance, C_f is the final capacitance, N_D is the bulk dopant concentration, K_s is the silicon dielectric constant, K_{ox} is the silicon oxide dielectric constant, and t_{ox} is the oxide thickness. By plotting, $-d(C_{ox}/C)^2/dt$ versus $(C_f/C-1)$, one can extract the effective generation lifetime and the effective surface generation velocity from the slope and the intercept, respectively.

III. EXPERIMENTAL

MOS capacitors were fabricated on n-type silicon substrate with doping concentration in the range of 1-1.5×10^{15} cm^{-3}. The bare wafers were first cleaned in a mixture of H_2SO_4 and H_2O_2, followed by an RCA clean. This was then followed by 1% HF dip for 1 min. The gate oxide was grown in dry O_2 at 880°C. After that aluminum was thermally evaporated to form the gate electrodes. Square and finger capacitors with area of 0.0025 cm^2 and 0.002 cm^2 respectively were then defined by photolithography.

The MOS capacitors were characterized by high-frequency capacitance-voltage (HFCV) and capacitance-time (C-t) measurements at room temperature. The HFCV measurements were performed at the sweep rate of 0.1 V/s and the sweep range from 2V to -6V, using a computer-controlled HP4284A LCR meter. The gate-oxide thickness, determined from the accumulation capacitance, was 18.6 nm.

During the C-t measurements, the return from non-equilibrium into equilibrium strong inversion state was recorded as the capacitance transient. Zerbst plots of the C-t responses were used to extract the minority carrier lifetime and surface generation velocity.

IV. RESULTS AND DISCUSSION

The HFCV curves for square and finger test structures are shown in Fig 2. There is no significant difference between the two curves except in the magnitude of the capacitance in accumulation due to difference in the capacitor areas.

The C-t response curve for finger and square MOS capacitor structures is shown in Fig 3. A very different behavior is observed for the finger test structure when compared to the square test structure. The response of the C-t for the finger structure is a lot faster than for the square structure, although the capacitance in both cases saturates at the strong-inversion level. Note that the initial biasing of the gates at inversion, instead of at accumulation, the surface generation under the gate is minimized. We believe the short C-t response of the finger MOS capacitor is primarily due to the lateral scr surface generation.

The corresponding Zerbst plots of the C-t responses for the square and finger structures are shown in Fig 4. The Zerbst plot is highly non-linear for the finger test structure. Such a curve is shown in Fig 4 (Curve (b)). In

contrast, the curve (a) derived from the C-t response of the square sample has a much wider linear portion.

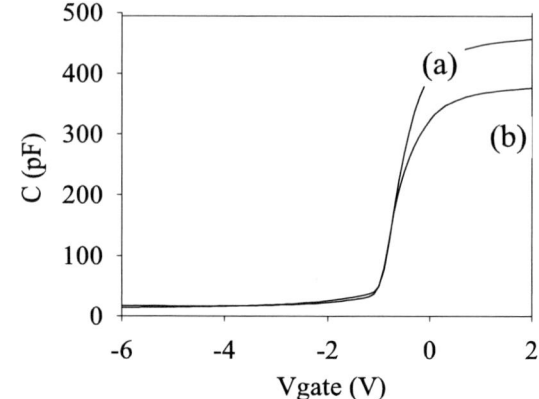

Fig. 2. High frequency C-V measurements for (a) square and (b) finger MOS capacitors, respectively.

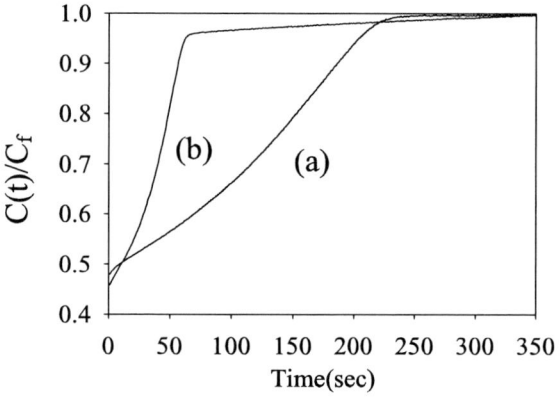

Fig. 3. C-t response curves for (a) square and (b) finger test structures. Both C-t were obtained by pulsing from –2V to –5V (i.e. from inversion to strong inversion).

The effective generation lifetimes and surface generation velocities obtained from the Zerbst plots for the two different samples are listed in Table 1.

TABLE I

EXTRACTED EFFECTIVE GENERATION LIFETIME AND SURFACE GENERATION VELOCITY FROM ZERBST PLOTS OF FIG 3.

	Area (cm^2)	Perimeter (μm)	τ_{geff} (μs)	s_{geff} (cm/s)
Square	0.0025	2000	60.0	1.12
Finger	0.002	4440	26.3	4.83

The extracted effective minority generation lifetime for the finger structure is reduced more than twice in comparison to the square structure. This can be explained by the fact that the finger structure has a higher perimeter to area ratio than the square structure. However, the extracted effective surface generation velocity is also increased four times for the finger structure. This was not expected because the gate oxides of the both capacitors were grown under the same conditions and physically close to each other. Possible explanation could be that a part of the lateral surface generation does not change linearly with W and is therefore contributing to the value of s_{geff}. A possible reason for the lateral surface generation not changing linearly with W could be surface ions on the oxide surrounding the gate, which can significantly influence the lateral surface generation [4].

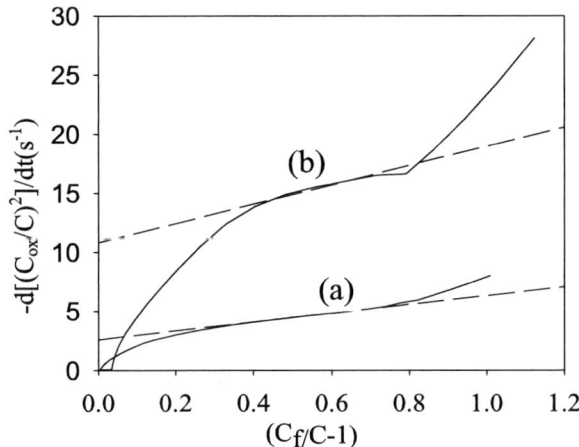

Fig. 4. Zerbst plots for (a) square and (b) finger test structures.

By assuming the first term of Eq. (1) is the same for both the finger and the square structures, one can solve two simultaneous equations to get $\tau_g = 228.97$ μs and $s_g' = 153.48$ cm/s using the extracted τ_{geff} from Table 1. Hence, it is possible to obtain the individual components of the generation rates as shown in Table 2, where G_1, G_2, and G_3 are the first, second, and third terms in the right hand side of Eq. (1), respectively.

TABLE II
DIFFERENT COMPONENTS OF GENERATION RATES AT ROOM TEMPERATURE.

	G_1 (cm^{-2}s^{-1})	G_2 (cm^{-2}s^{-1})	G_3 (cm^{-2}s^{-1})
Square	4.81x10^9	1.35x10^{10}	1.19x10^{10}
Finger	4.81x10^9	3.75x10^{10}	4.91x10^{10}

It can be seen that both the lateral surface generation and the generation under the gate are much higher than the bulk generation for both test structures.

V. CONCLUSION

We have used the C-t transients of MOS capacitors to analyze different generation mechanisms in silicon. It is demonstrated that using different structures we can separate out different generation components. The lateral surface generation can be important and needs to be taken into account for interpretation of MOS capacitance transient measurements.

REFERENCES

[1] D. K. Schroder, *Semiconductor Material and Device Characterisation*, ch.7. New York: Wiley, 1998.
[2] R. F. Pierret, "A linear sweep MOS-C technique for determining minority carrier lifetimes," *IEEE Trans. Electron Devices,* vol.19, p.869-873, July 1972.
[3] M. Zerbst, "Relaxation effects at semiconductor-insulator interfaces," (in German), *Z. Angew. Phys.* 22, 30-33, May 1966.
[4] R. F. Pierret and D. W. Small, "Effects of lateral surface generation on the MOS-C linear-sweep and C-t transient characteristics," *IEEE Trans. Electron Devices,* vol.20, p.457-458, April 1973.

2006 25th International Conference on Microelectronics

Effects of NO Annealing on the Characteristics of GaN MIS Capacitor

Limin Lin, P.T. Lai, and Kei May Lau

Abstract - An ultra-thin thermally-grown GaO_xN_y was formed between deposited SiO_2 dielectric and GaN wafer to improve the interface quality. The interface-trap density at 0.4 eV below the conduction-band edge was reduced by one order compared with that of a sample without the stacked GaO_xN_y. NO annealing was conducted on both SiO_2/GaN and $SiO_2/GaO_xN_y/GaN$ MIS structures, and turned out to effectively suppress the oxide charges. The sample with stacked GaO_xN_y annealed in NO achieved the lowest oxide-charge density (Q_{ox}) of 1.7×10^{11} cm^{-2} eV^{-1}; Q_{ox} of the one without stacked GaO_xN_y annealed in NO was 9.5×10^{11} cm^{-2} eV^{-1}; those samples. not annealed in NO got high Q_{ox} of 8×10^{12} cm^{-2} eV^{-1}, with or without stacked GaO_xN_y. Moreover, NO annealing was found to effectively reduce border traps. The interface quality was improved on both the sample with the GaO_xN_y interlayer annealed in nitrogen and the non-stacked sample annealed in NO. The breakdown field and leakage current of the gate dielectrics were also compared in this work.

I. INTRODUCTION

As one kind of wide-bandgap materials, GaN has advantages like high breakdown field and low thermal generation rate, which make it suitable for making high-temperature and high-power electronic devices. A lot of dielectric films such as Ga_2O_3 [1], $Ga_2O_3(Gd_2O_3)$ [2], Sc_2O_3 [3], SiO_2 [4], Si_3N_4 [4] and MgO [5] have been applied for GaN-based MIS structures. However, substantial leakage current through the gate oxide and lots of interface traps are major obstacles. Native Ga_2O_3 [1] achieved low interface-state density D_{it}, but its leakage current was high. For the foreign dielectrics, interface traps were main concern. In order to improve the interface quality, SiO_2 stacked with thermally oxidized Ga_2O_3 was applied by Nakano etc. [6]. In Nakano's work, a layer of 15 nm Ga_2O_3 followed by 35 nm Ga-oxynitride were thermally formed and a Si mask was deposited to pattern the capacitor. In this paper, we showed that a much thinner GaO_xN_y could effectively reduce interface traps and also the step of silicon mask could be omitted.

SiO_2 deposited by LPCVD on GaN was annealed in NO or NH_3 at 1100 °C for 5 min by Mtocha etc., but the annealing was found to increase the leakage current and

degrade the performance of GaN capacitor [7]. In this work, both stacked and non-stacked dielectrics were annealed in NO at 800 °C for 1 h and as a result, showed reduced oxide charges and border traps. For the NO-annealed stacked sample, D_{it} increased in comparison with the N_2-annealed stacked sample, but D_{it} decreased for the NO-annealed non-stacked sample. For the NO-annealed stacked structure, the breakdown field (E_{BD}) was enhanced and leakage current was reduced, but for the non-stacked sample, the breakdown characteristic was degraded after NO annealing.

II. EXPERIMENT

A 1-µm epitaxial GaN layer was grown on c-plane sapphire by MOCVD. A layer of Si-doped n type GaN with a doping level of 5×10^{17} cm^{-3} was grown on a layer of un-doped GaN buffer layer. The wafers were cleaned using the standard RCA method and then dipped into 1:1 HF solution to remove the native oxide and contaminants [6]. Two GaN wafers were loaded into a furnace at 300 °C, and then the temperature was raised to 800 °C in nitrogen ambient. GaN wafers were oxidized at 800 °C in a diluted oxygen gas (O_2:N_2=1:4) for 15 min. Since about 200 nm GaO_xN_y could be formed in pure oxygen for 12 h at 800 °C in our system, the GaO_xN_y interlayer was estimated to be less than 4 nm. Then, a layer of SiO_2 was deposited on all the samples. After that, one stacked sample was annealed in nitrogen, denoted as GaN2; another was annealed in NO, denoted as GaNO. Two non-stacked samples without the interlayer were annealed in N_2 and NO and denoted as SiN2 and SiNO respectively. Finally, Al was thermally evaporated as electrodes for all the samples and then all the capacitors were annealed in argon at 410 °C for 30 min for good contacts. The area of capacitors was $7.85 \times 10^{-5} cm^2$. High-frequency (HF, 1 MHz) capacitance-voltage (C-V) characteristics were measured by using HP4284A LCR meter with a bias sweep rate of 0.1 V/s. The current-voltage (I-V) characteristics were measured by HP4156B. All measurements were conducted in a dark ambient at room temperature.

III. RESULTS AND DISCUSSION

Table I presents the parameters extracted from the HF C-V curves (average values from 6 sample points) in Fig 1 and Fig 2. The oxide thickness was calculated from the oxide

---

Limin Lin and P. T. Lai are with the Department of Electrical and Electronic Engineering, The University of Hong Kong, Hong Kong, China, E-mail: laip@eee.hku.hk.

Kei May Lau is with the Department of Electrical and Electronic Engineering, The Hong Kong University of Science and Techonology, Hong Kong, China.

1-4244-0116-X/06/$20.00 ©2006 IEEE

TABLE I
PARAMETERS EXTRACTED FROM THE HF C-V CRUVES

Sample	SiN2	SiNO	GaN2	GaNO
T_{ox} (Å)	379	381	380	390
V_{fb} (V)	-13.4	-1.6	-14.4	-0.21
Q_{ox} ($\times 10^{11}$ cm^{-2})	77.2	9.5	85.0	1.78
D_{it} @ E_c-E_t=0.4 eV ($\times 10^{10}$ cm^{-2} eV^{-1})	120	63	5.9	160
$Q_{bt} \times 10^{17}$ (cm^{-3})	2.6	1.2	2.7	2.0
$J_{leakage}$ @ E_{BD}	65 µA/cm^2 @ 4.9 MV/cm	1.2 mA/cm^2 @ 0.23 MV/cm	1.2 mA/cm^2 @ 0.93 MV/cm	1 mA/cm^2 @ 1.79 MV/cm

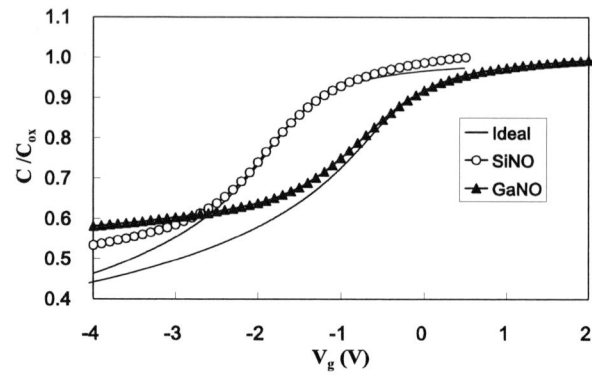

Fig. 1. Measured and ideal C-V curves of the GaN capacitors with SiO$_2$ or stacked dielectric annealed in N$_2$.

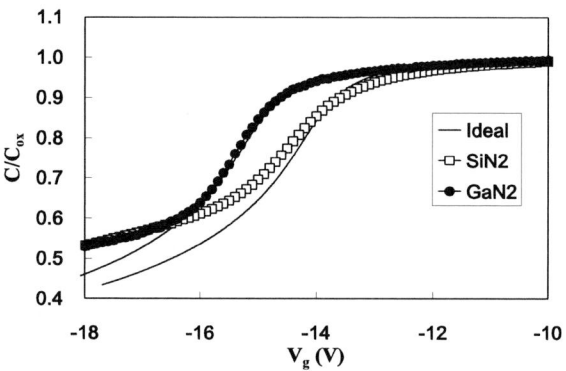

Fig. 2. Measured and ideal C-V curves of the GaN capacitors with SiO$_2$ or stacked dielectric annealed in NO.

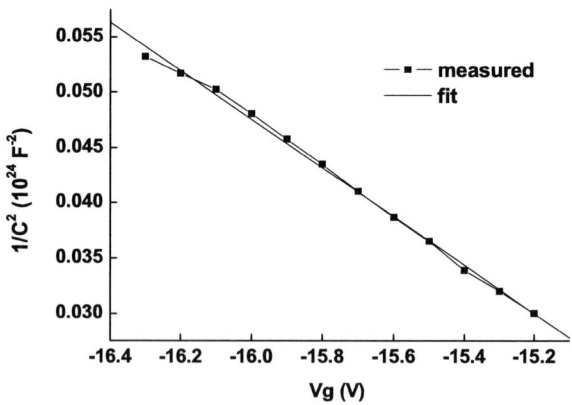

Fig. 3. $1/C^2$ characteristic versus gate voltage. Carrier concentration obtained by the fitting is 1.1×10^{17} cm^{-3}.

capacitance C_{ox}, which is approximated by the maximum value in the accumulation region. The capacitance at the flat-band voltage is as follows:

$$C_{fb} = \frac{C_s C_{ox}}{C_s + C_{ox}} \quad (1)$$

where

$$C_s = \varepsilon_{GaN} \varepsilon_0 / L_D, \, (L_D \text{ is the Debye length}) \quad (2)$$

$$L_D = (kT\varepsilon_{GaN}\varepsilon_0 / q^2 N_d)^{1/2} \quad (3)$$

ε_{GaN} is the dielectric constant of GaN; ε_0 is the permittivity of free space; N_d is the actual carrier concentration (1.1×10^{17} cm^{-3}) calculated from the measured capacitance; k is the Boltzman's constant and T is the measurement temperature (298 K).

Oxide-charge density is determined from the shift of flat-band voltage:

$$V_{fb} = \phi_{Al,GaN} - \frac{Q_{ox}}{C_{ox}} \quad (4)$$

where $\phi_{Al,GaN}$ is the work-function difference between Al and GaN

$$\phi_{Al,GaN} = \chi + E_g / 2 - (kT/q)\ln(N_d / n_i) \quad (5)$$

where χ is the electron affinity of GaN (4.07 eV); n_i is the intrinsic carrier concentration of GaN at room temperature (2.0×10^{-10} cm^{-3}) [8]; E_g is the forbidden bandgap of GaN (3.4 eV). Lowest oxide-charge density (Q_{ox}) of 1.7×10^{11}

cm^{-2} was achieved for the GaNO sample, and was more than one order lower than those of the SiN2 and GaN2 samples. For the SiNO sample, Q_{ox} was also about one order lower than those of the samples annealed in nitrogen (SiN2 and GaN2), indicating that NO annealing could effectively suppress the oxide charges of both the stacked and non-stacked samples. The positive oxide charges could be introduced by sputtering-related damages or defects, while positive Ga ions out-diffused into the dielectric might be the source of the oxide charges. During the annealing in NO, the decomposed N or O could form bonding with Ga in the dielectric, resulting in a reduction of the oxide charges.

The oxide traps close to the interface which can communicate with the semiconductor are called border traps [9], whose density (Q_{bt}) can be calculated by the following formula:

$$Q_{bt} = C_{ox} \Delta V_{fb} / (q \times d) \qquad (6)$$

ΔV_{fb} is the hysteresis at flat-band between opposite sweeping directions of gate voltage; d is the distance in the oxide above the GaN surface (30 Å) [8]. The hystereisis is due to the difference in gate biases at which electrons fill the traps and escape from the traps or the difference between the capture and emission times of the border traps. In Table I, it can be seen that after NO annealing, the border traps were reduced for both SiNO and GaNO samples. The SiNO sample achieved lowest border-trap density of 1.2×10^{17} cm^{-3} which should be related to the decrease of oxide charges after NO annealing.

The interface-trap density was calculated based on the HF C-V curve by using the Terman method. The interface traps followed the change in gate bias and caused a stretchout of the HF C-V curve. The ideal capacitance was calculated by the following equation [10]:

$$C_{ideal} = \frac{C_s C_{ox}}{C_s + C_{ox}} \qquad (7)$$

where the surface capacitance C_s is,

$$C_s = \frac{C_{FBS}}{\sqrt{2}} \frac{\exp(v_s) - (n_i / N_d)^2 \exp(-v_s) - 1}{[-(v_s + 1) + \exp(v_s) + (n_i / N_d)^2 \exp(-v_s)]^{1/2}} \qquad (8)$$

N_d is the carrier concentration, which was calculated based on the measured capacitance:

$$N_d = \frac{2}{q \varepsilon_{GaN} \varepsilon_0 A^2 d(1/C^2)/dV} \qquad (9)$$

As Fig 3 shows, N_d is fitted for the gate voltage ranging from -15.2 to -16.3 V based on the measured HF C-V of the GaN2 sample. Any fitting error is probably due to the

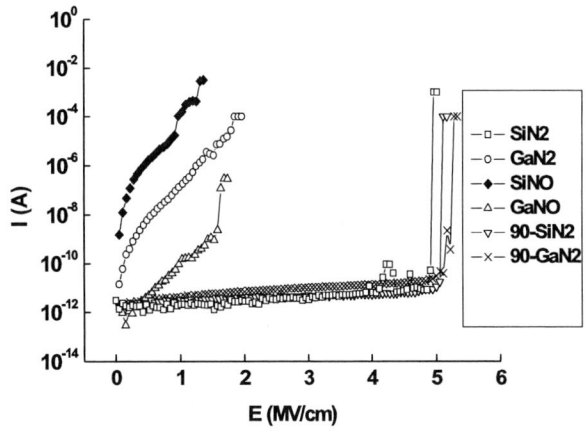

Fig. 4. Typical leakage current versus electric field E. 90-SiN2 and 90-GaN2 have a thicker SiO$_2$ of 90 nm.

non-uniformity of doping concentration or the interface traps. In order to compare the interface quality of different samples, the same N_d (1.1×10^{17} cm^{-3}) was used in all calculations. Fig 1 and Fig 2 show the difference between ideal and measured C-V curves. The ideal curves are shifted along the x-axis because the interface state density is not related to the shift of flat-band voltage but only the shape of the curve. It could be seen that the GaN2 sample has the smallest stretchout from the ideal curve, indicating least interface traps among all the samples. Table I shows that at 0.4 eV below the conduction-band edge, the GaN2 sample achieves a low interface-trap density of 5.9×10^{10} cm^{-2}, followed by the SiNO sample with 6.3×10^{11} cm^{-2}, 1.2×10^{12} cm^{-2} for the SiN2 sample and 1.6×10^{12} cm^{-2} for the GaNO sample. During the sputtering deposition of SiO$_2$, the silicon and oxygen atoms could diffuse into the GaN wafer and destroy the crystalline structure of GaN and form related defects and traps. Both the N$_2$-annealed stacked sample (GaN2) and NO-annealed non-stacked sample (SiNO) present improved interface quality, which should be due to the formation of an interfacial GaO$_x$N$_y$ layer, but in different ways. The GaN2 sample achieved the GaO$_x$N$_y$ layer through the oxidation before the sputtering deposition, while the SiNO sample could have the GaO$_x$N$_y$ interlayer formed during the NO annealing. For the GaN2 sample, the preformed interlayer could greatly reduce the damage caused by the sputtering and defects related to silicon atoms diffusing into the GaN. Although the damage introduced by sputtering cannot be avoided for the SiNO sample, the interlayer formed by NO annealing still could reduce the defects and traps effectively. However, after the stacked sample was annealed in NO (GaNO), the interface traps seemed to increase significantly, implying that too many nitrogen atoms piling up at the interface could also introduce traps and deteriorate the performance of devices. Based on these results, it can be deduced that the stacked samples annealed in NO may maintain little interface traps as well as little oxide charges.

Fig 4 shows the current-electric field (I-E) characteristic. The SiN2 sample has best I-E characteristic and lowest leakage current through the gate. The E_{BD} of the SiN2 sample was as high as 4.9 MV/cm ($J_{leakage}$=65 $\mu A/cm^2$), followed by 1.79 MV/ cm ($J_{leakage}$=1 mA/cm^2) for the GaNO sample, 0.93 MV/cm ($J_{leakage}$=1.2 mA/cm^2) for the GaN2 sample and 0.23 MV/cm ($J_{leakage}$=1.2 mA/cm^2) for the SiNO sample. After annealing in NO, the leakage current of the SiNO sample increased significantly. However, for the sample with stacked Ga$_2$O$_3$, improved I-E characteristic was achieved after annealing in NO compared with the GaN2 sample. The reason for the different effects of NO annealing on stacked and non-stacked MIS capacitors is still under investigation. Compared with the non-stacked sample SiN2, the GaN2 sample presented degraded I-E characteristic, which should be due to the negative effect of thermally grown GaO$_x$N$_y$ interlayer. Thermally grown oxide had poor breakdown characteristic and rough surface morphology, which should be the reasons for increased leakage current.

Thicker SiO$_2$ was also deposited by sputtering on the stacked and non-stacked samples. In Fig. 4 the 90-SiN2 sample means 90-nm SiO$_2$ was deposited on GaN while 90-GaN2 has 90-nm SiO$_2$ deposited on GaO$_x$N$_y$/GaN structure. Both samples were annealed in nitrogen at 800 ^oC for 1 h after deposition. When the thickness of SiO$_2$ dielectric was increased to 90 nm, the negative effects brought by the GaO$_x$N$_y$ interlayer were minimized. There was little difference in the breakdown and leakage current characteristic between the samples with or without stacked layer. Since these two samples showed about the same breakdown field as the SiN2 sample, it should be possible to optimize the SiO$_2$ thickness in the NO-annealed stacked dielectric to achieve low leakage and high breakdown field, in addition to little fixed oxide charge.

IV. Conclusion

NO annealing was applied on both SiO$_2$/GaN and SiO$_2$/GaO$_x$N$_y$/GaN MIS structures and turned out to effectively reduce the fixed oxide charges by about one order. A very thin GaO$_x$N$_y$ interlayer formed by thermal oxidation could improve the interface quality and achieve low interface-trap density D$_{it}$ of about 5.9 $\times 10^{10}$ cm^{-2}. After NO annealing, D$_{it}$ of the SiO$_2$/GaN capacitor was reduced by about half due to the formation of an interfacial GaO$_x$N$_y$ layer by chemical reaction between GaN and NO. On the other hand, the D$_{it}$ of SiO$_2$/GaO$_x$N$_y$/GaN structure was increased after NO annealing, implying that too many nitrogen atoms piling up at the interface could also deteriorate the interface quality. NO annealing was also shown to suppress the border traps which should be related to a reduction of oxide charges. According to the breakdown field E$_{BD}$ and leakage current characteristics, SiO$_2$/GaN capacitor annealed in nitrogen achieved the highest E$_{BD}$ and lowest leakage current. After NO annealing, E$_{BD}$ was degraded for the sample without the GaO$_x$N$_y$ interlayer. However, for the stacked samples, NO annealing seemed to improve the E$_{BD}$ characteristic and reduce the leakage current. The reason for the opposite effect of NO annealing on stacked and non-stacked samples is complicated and needs further investigation. After increasing the SiO$_2$ thickness to 90 nm, the differences in E$_{BD}$ and leakage current characteristics between the stacked and non-stacked samples were negligible. Therefore, it should be possible that interface quality, oxide charges and E$_{BD}$ could be all improved through optimizing the SiO$_2$ thickness and NO concentration during annealing.

Acknowledgement

This work is supported by the RGC of HKSAR, China (Project No.: HKU 7163/03E).]

References

[1] Hyunsoo Kim, Seong-Ju Park, and Hyunsang Hwang, "Thermally oxidized GaN film for use as gate insulators", *J. Vac. Sci. Technolo. B, 19*(2), 1999, pp.579-581.

[2] F. Ren, M. Hong, S. N. G. Chu, M. A. Marcus, M. J. Schuurman, A. Baca, S. J. Pearton, and C. R. Abernathy, "Effect of temperature on Ga$_2$O$_3$ (Gd$_2$O$_3$)/GaN metal-oxide-semiconductor-field-effect transistors", *Appl. Phys. Lett., 73*(26), 1998, pp. 3893-3895.

[3] B. P. Gila, J. W. Johnson, R. Mehandru, B. Luo, A. H. Onstine, K. K. Allums, V. Krishnamoorthy, S. Bates, C. R. Abernathy, F. Ren, and S. J. Pearton, "Gadolinium Oxide and Scandium Oxide: Gate Dielectrics for GaN MOSFETs", *Phys. Stat. Sol. (a), 188*(1), 2001, pp. 239-242.

[4] S. Arulkumaran, T.Egawa, H. Ishikawa, T. Jimbo, and M.Umeno, "Investigations of SiO$_2$/n-GaN and Si$_3$N$_4$/n-GaN insulator-semiconductor interfaces with low interface state density", *Appl. Phys. Lett., 73*(6), 1998, 809-811.

[5] Jihyun Kim, R. Mehandru, B. Luo, and F. Ren, B. P. Gila, A. H. Onstine, C. R. Abernathy, and S. J. Pearton, and Y. Iroawa, "Characteristics of MgO/GaN gate-controlled metal-oxide-semiconductor diodes", *Appl. Phys. Lett., 80*(24), 1998, pp. 4555-4557.

[6] Yoshitaka Nakano, Tetsu Kahi and Takashi Jimbo, "Characteristics of SiO$_2$/n-GaN interface with β-Ga$_2$O$_3$ interlayers", *Appl. Phys. Lett* ., 2003, pp. 4336-4338.

[7] Kevin Matocha, Ronald J. Gutmann, and T. Paul Chow, "Effect of Annealing on GaN-Insulator Interfaces Characterized by Metal-Insulator-Semiconductor Capacitors", *IEEE Trans. On Electron Devices, 50*(5), 2003, pp. 1200-1204.

[8] H. C. Casey, G.G. Foutain, R. G. alley, B. P. Keller, and S. P. Den-Baars, "Low interface trap density for remote plasma deposited SiO$_2$ on *n*-type GaN", *Appl. Phys. Lett., 68*(13), 1996, pp. 1850-1852.

[9] Navakanta Bhat and Krishna C. Saraswat, "Characterization of border trap generation in rapid thermally annealed oxides deposited using silane chemistry", *Journal of Applied physics*, 1998, pp. 2722-2726.

[10] Nicollian & Brews, MOS (Metal Oxide Semiconductor) Physics and Technology, Hoboken, New Jersey: John Wiley & Sons, Inc., 2003.

2006 25th International Conference on Microelectronics

Long Range Statistical Lifetime Prediction of Ultra-Thin SiO$_2$ Oxides: Influence of Accelerated Ageing Methods and Extrapolation Models

D. Pic, D. Goguenheim, J-L. Ogier

Abstract – A large database of time dependent dielectric breakdown has been obtained for 2.3 nm and 3.2 nm SiO$_2$ thin oxides. The reliability models used to qualify theses oxides, have been tested using an accurate experimental error evaluation. The experimental method used has been investigated and relaxation phenomena have been treated in this particular case. The area dependence has been checked for 2.3nm in order to confirm the validity of this law. The activation energies corresponding to the dominant degradation mechanisms have been extracted over a temperature range from 50°C to 125°C and these values have been discussed. The voltage extrapolation models have been compared for positive polarity on P and N type substrate. This study shows that a power voltage model is well predictive for both P substrate in inversion regime and N substrate in accumulation regime.

I. OVERVIEW

In order to qualify and to study the long term reliability of integrated circuits, understanding and quantifying the gate ultra-thin thermal oxide reliability used in MOSFETs (Metal Oxide Semiconductor Field Effect Transistors), remains a critical issue of primary importance. Broken oxides can lead to excessive power consumption and/or functional circuit failure. Electrical characterization tools and protocols have to be developed to study intrinsic and extrinsic oxide reliability and the impact of surrounding devices operation.

Classical methods and models (E or 1/E) using the field as a critical parameter have remarkably well predicted oxide lifetime to hard breakdown and degradation down to 10 nm [1], but for thinner oxides new degradation mechanisms have made the situation and the lifetime extrapolation a more complex task. Stress Induced Leakage Currents, the occurrence of direct tunneling regime through the oxide, soft-breakdown, detrapping and relaxation effects, are to be studied from a fundamental point of view and taken into account from a methodological point of view for an improved and accurate lifetime determination. Indeed, to obtain realistic oxide lifetime values extrapolated to nominal operating conditions (temperature,

D. Pic and J-L. Ogier are with Rousset Central Characterization Laboratory, STMicroelectronics, 190 Avenue Célestin Coq, Zone Industrielle, 13106 Rousset Cedex, France, E-mail: david.pic@st.com, jean-luc.ogier@st.com
D. Goguenheim is with the L2MP UMR CNRS 6137, ISEN-Toulon, Maison des Technologies, Place George Pompidou, 83000 Toulon, France, E-mail: didier.goguenheim@l2mp.fr

circuit equivalent area, supply voltage, acceptable defect density), the results obtained during short time aggressive stresses have to be extrapolated over many time ranges using adapted degradation physical models (taking into account the various injection regimes that may occur). And the use in advanced technology of ultra-low voltages (around 1V) makes the classical extrapolation schemes questionable.

II. EXPERIMENTAL TOOLS AND TECHNIQUES

In this work, we have collected and compared a wide data base for 3.2 and 2.3 nm SiO$_2$ oxides, in order to check at long range the accuracy of new gate voltage driven models like V_G, V_G^{-n}, $1/V_G$ models [2-8]. The oxide is grown by using In Situ Steam Generation (ISSG) technology and a post nitridation treatment is performed in N$_2$O ambient. The tested devices are processed using a standard 0,13μm CMOS process with STI (Shallow Trench Isolation). We have used in parallel both wafer level experiments (on semi-automated wafer probers) and package level experiments (on an automated temperature controlled reliability Qualitau© system) allowing Constant Voltage Stresses over a wide temperature and voltage range on the same capacitor splits. In this study, we have only taken into account the time to hard breakdown, and the given oxide thickness is the physical thickness of the studied oxide obtained by Transmission Electron Microscopy (TEM).

In parallel to this statistical work on breakdown lifetime, we have studied the impact of recording methods during a Constant Voltage Stress (CVS). Indeed, in order to correctly detect and treat soft-breakdown phenomena for an incoming study, we periodically interrupt the stress to monitor the gate current at low voltage. During these phases, we have observed a relaxation effect which appears on the stress current with a shape decrease after it as shown in Fig. 1.

The time dependence of this phenomenon has been analyzed and characterized by two time constants. For aggressive stress, we have evaluated the fastest one around 2s and the second one depends on interruption times.

We conclude about the influence of this phenomenon for the statistical study, by showing it has no influence on Weibull lifetime plots for N type substrate with positive bias as shown in Fig. 2 for an oxide thickness

1-4244-0116-X/06/$20.00 ©2006 IEEE

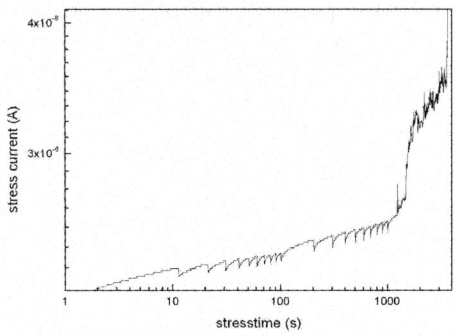

Fig. 1. Relaxation Phenomena observed after low voltage measurements.

of 3.2 nm.The same result has been obtained for 2.3nm. This can be interpreted as the fact that the defects responsible for charging/discharging observed phases leading to relaxation effects are not directly those ones leading to dramatic hard breakdown.

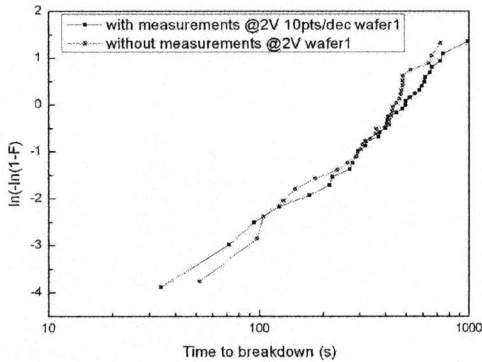

Fig. 2 Comparison of Cumulative Weibull distributions for CVS with and without monitoring of current at low voltage

III. RESULTS AND DISCUSSION

We have obtained experimental lifetime data up to the sixth time decade and have probed the accuracy of the various models. We have checked the validity of Poisson's law for the area dependence, taking into account a statistical error bar for 2.3 nm and we have evaluated the lifetime activation energy using an Arrhenius law for temperature dependence.

A. Area Dependence

It is generally assumed that the area dependence of time to breakdown t_{BD} is well described by a Poisson's law and can be expressed as in formula (1): [9]

$$t_{BD}^A = t_{BD}^a \left(\frac{a}{A} \right)^{\frac{1}{\beta}} \qquad (1)$$

where β is the usual weibull slope. Classically, a Poisson process is used for a system of n independent and not simultaneous events, for which the number of events

occurring during a period T only depends on the duration of this period. Assuming this hypothesis, the probability law can be expressed as in formula (2), which allows expressing the reliability function R(t) (probability of zero failure) as a function of the average number of breakdowns:

$$R(t) = \exp(-A.D) \qquad (2)$$

and expressed as a function of the studied area A multiplied by D, the defect density. Then, using a Weibull statistics (3) to describe the repartition of the n events [10], we can establish the Poisson's law with the hypothesis that defect density is uniform whatever the size of the area [11].

$$R(t) = \exp\left(-\left(\frac{t}{\eta} \right)^{\beta} \right) \qquad (3)$$

We have tested this law with data that we have obtained for 2.3nm oxide thickness. The results are displayed in Fig.3. The method used consists in calculating the prediction lifetime beam with Poisson's law, taking into account the standard deviation on Weibull slope β which can be expressed as given in formula (4) with λ_1 a constant equal to 1.1 and N the number of tested devices. [8]

$$\sigma_\beta = \frac{\lambda_1.\beta}{\sqrt{N}} \qquad (4)$$

This beam is calculated with time to breakdown data obtained to a area for a range of area between a et A2. To conclude and check the validity of the law we compare beam prediction and the time to breakdown data obtained for A1 and A2 areas. The standard deviation on time to breakdown is calculated with formula (5) with λ_0 a constant equal to 1.43 and N the number of tested devices. [8]

$$\sigma_{\ln(t_{63\%})} = \frac{\lambda_0}{\beta.\sqrt{N}} \qquad (5)$$

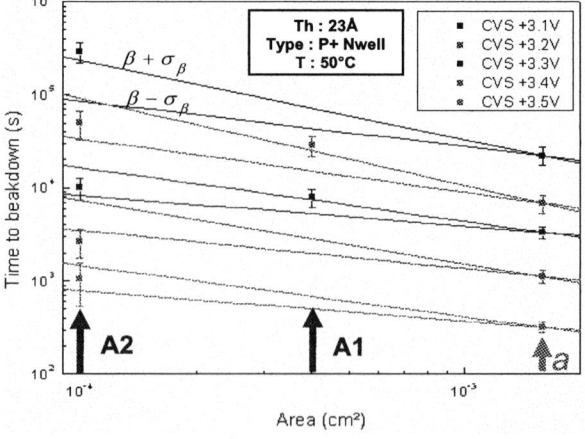

Fig. 3. Test of Poisson's law for 23Å oxide thickness.

566

This method allows us to verify this law with respect to the precision obtained on recorded data.

B. Temperature Dependence

With data that we have collected, we have investigated the temperature dependence over the range from 50°C to 125°C. The different activation energies have been extracted with an Arrhenius law as given in formula (6) and are summarized in Fig. 4. We have used the standard deviation on time to breakdown (5) to evaluate the error on activation energies using the least square method to perform the fit in order to extract slope value.

$$t_{BD} = t_0 \, \exp\left(+ \frac{E_A}{kT} \right) \qquad (6)$$

We find that the activation energy increases when the stress voltage decreases for 3.2 and 2.3 nm thick oxides on N type substrate for positive stress bias.

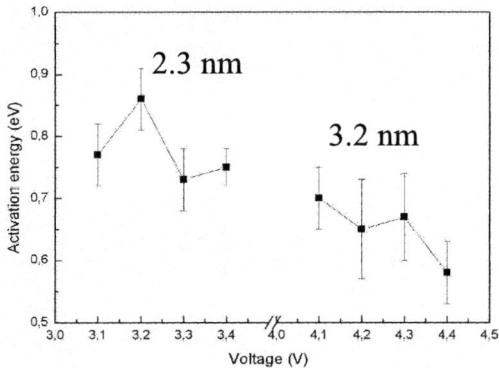

Fig. 4. Activation energy values obtained for different CVS at 50°C – 85°C – 125°C on 2.3 nm and 3.2nm on N type substrate for positive stress bias.

We also show that the activation energy increases when the oxide thickness decreases at constant field as shown for one case in Fig. 5. A reference obtained for a 7nm oxide has been reported in order to shown the strong increase in activation energy for thin oxides.

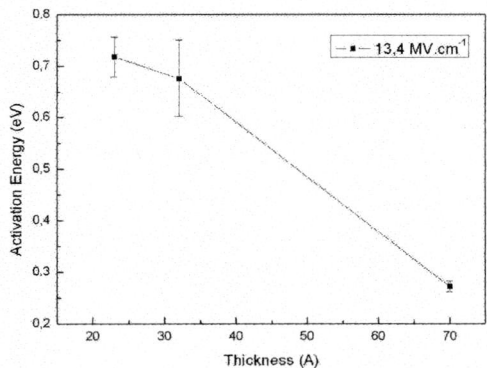

Fig. 5. Activation energy values for different thickness at constant field on N type substrate for positive stress bias.

The degradation mechanism associated to these energy values has been assigned to the competition of two mechanisms by some authors [12-14]. As enunciated by Hydrogen Release model, the degradation can be described by two mechanisms: (i) tunneling electrons release some species (hydrogen species) from Si/SiO_2 interface and (ii) the release species react with some precursors (probably oxygen vacancies) to generate electrically active defects. Indeed, considering the oxide thickness, the voltage range used and the temperature range, the contribution of these mechanisms is different and induces variations of activation energies. For large temperature ranges, a non-Arrhenius law has been reported by the same authors.

In this study, the investigated temperature range is narrow enough to consider a uniform Arrhenius law to extract activation energies. With these results, we confirm and give additional reference on activation energies in thin oxides.

C. Voltage Dependence

As defined in formula (5), we can express a standard deviation for each time to breakdown value obtained during experiments. By taking into account this error and by using the least square method to fit obtained data points, we are able to generate a beam for each extrapolation model. The method that we have chosen to test these models is to fit the data obtained at wafer level down to 10^4s and to compare the beam prediction of each model with data obtained at package level, up to 10^6s. We have succeeded to discriminate the best model by this way. For the comparison, we use V_G, V_G^{-n} and $1/V_G$ models [3-8] as it has been proved that the degradation only depends on gate voltage in thin oxides [2]. For 3.2nm oxide on P type substrate in inversion regime, the V_G^{-n} ($n \approx 41-44$) or $1/V_G$ models seem to be well predictive as shown in Fig.6.

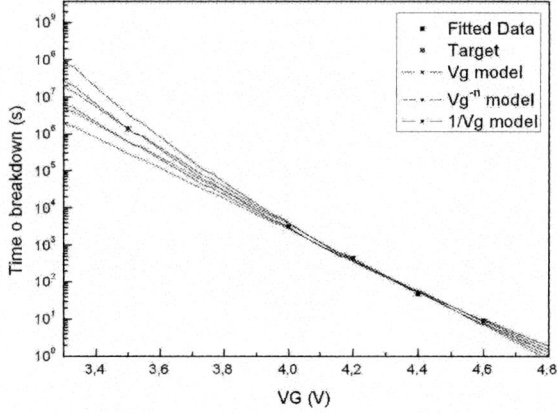

Fig. 6. Extrapolation Models comparison for 3.2nm oxides on P type substrate in inversion regime at 150°C.

For 2.3nm oxides on N type substrate in accumulation regime, the V_G^{-n} ($n \approx 34-37$) model is clearly well predictive as shown in Fig.7 in spite of the approximate value of n already observed by some authors. [7]

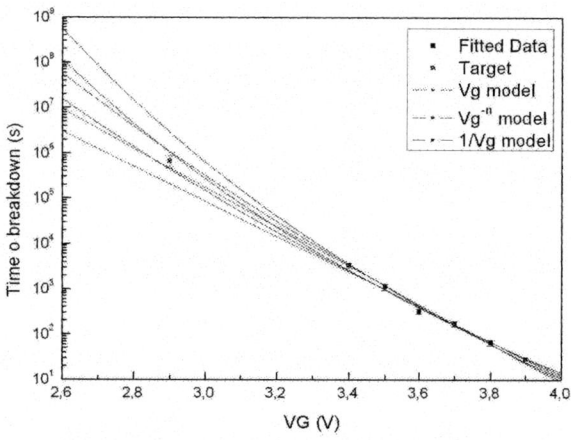

Fig.7. Extrapolation Models comparison for 2.3nm oxide for N type substrate in accumulation regime at 50°C.

The V_G^{-n} model can be considered as the best predictive one for P type and N type samples for positive bias stress. These results can be explained by the multi-vibrational Hydrogen Release model, which gives a very nice physics-based explanation. This model allows explaining the oxide degradation considering the Si-H bond breaking as the limiting phenomenon.

The multi-vibration excitation rate R of the bond, depending on the physical properties of the Si-H bond (inelastic fraction of tunneling current depending on electron energy, bond desorption energy at Si/SiO$_2$ interface, energy of stretching excited state mode and number of energy quanta transferred), can be linked to the yield Y of this phenomenon and the current through the oxide during the stress [15]. For thin oxides, the gate current versus gate voltage law can be approximated by a power law. Indeed, using these two relations and associating defect generation probability P_g to the yield Y, the V_G^{-n} dependence of the degradation and of time to breakdown can be expressed with an exponent value n around 44 [5-6], [15].

III. CONCLUSIONS

In this paper, we have checked the validity of Poisson's law for the area dependence, taking into account a statistical error bar for 2.3 nm and established that this law is verified on the whole. We have evaluated the lifetime activation energy using an Arrhenius law for temperature dependence. We find that the activation energy increases when the stress voltage decreases for N type in accumulation with 3.2 and 2.3 nm thick oxides. We also show that the activation energy increases when the oxide thickness decreases at constant field.

We have obtained experimental lifetime data up to the sixth time decade and have probed the accuracy of the various models. For 2.3nm, the V_G^{-n} model is the best predictive for N type in accumulation. For 3.2nm, the V_G^{-n} or $1/V_G$ model are well predictive for P type in inversion. These results support previous data obtained in the literature and verify the physical explanation described by multi-vibrational Hydrogen Release model.

ACKNOWLEDGMENTS

This work has beneficiated from the financial support of Département des Bouches du Rhône.

REFERENCES

[1] Chen I.C., Holland S.E., Hu C., *IEEE Trans. Electron Devices*, vol. ED32, pp. 413, 1989.

[2] P. E. Nicollian, W.R. Hunter, J. C. Hu "Experimental Evidence of Voltage Driven Breakdown Models in Ultrathin Gate Oxides" in *38th Annual International Reliability Physics Symposium*, San Jose, California, pp. 7-15, 2000.

[3] E.Y. Wu, J. Aitken, E. Nowak, A. Vayshenker, P. Varekamp, G. Hueckel, J. McKenna, D. Harmon, L-K. Han, C. Montrose, R. Dufresne "Voltage-Dependent Voltage- Acceleration of Oxide Breakdown for Ultra-Thin Oxides" in *IEDM* pp.541-544, 2000.

[4] E. Y. Wu, A. Vayshenker, E. Nowak, J. Suñé, R.-P. Vollertsen, W. Lai, D. Harmon "Experimental Evidence of T$_{BD}$ Power-Law for Voltage Dependence of Oxide Breakdown in Ultrathin Gate Oxides" in *IEEE Trans. Electron Devices*, vol. 49, no.12, pp. 2244-2252, December 2002.

[5] W. McMahon, A. Haggag, K. Hess "Reliability Scaling Issues for Nanoscale Devices" in *IEEE Trans. Electron Devices*, vol. 2, no.1, pp. 33-38, March 2003.

[6] J. Suñé, E. Y. Wu "Hydrogen-Release Mechanisms in the Breakdown of Thin SiO$_2$ Films" in *Physical Review Letters*, vol. 92, no. 8, pp.372-375, February 2004.

[7] R.Vollertsen, E. Y. Wu "Voltage acceleration and t63.2 of 1.6-10 nm gate oxides" in *Microelectronics Reliability*, vol. 44, pp. 909-916, 2004.

[8] F. Monsieur, E. Vincent, G. Ribes, V. Huard, S. Bruyere, D. Roy, G. Pananakakis, G. Ghibaudo "Evidence for defect-generation-driven wear-out of breakdown conduction path in ultra thin oxides" in *41th Annual International Reliability Physics Symposium*, Dallas, Texas, pp. 424-431,2003.

[9] F. Monsieur, E. Vincent, S. Bruyere, D. Roy, G. Pananakakis, G. Ghibaudo "Determination of dielectric breakdown Weibull distribution parameters confidence bounds for accurate ultrathin oxide reliability prediction" in *Microelectronics Reliability*, vol. 41, pp. 1295-1300, 2001.

[10] E. Y. Wu "Tutorial on ultra-thin oxide reliability for ULSI applications"in *International Integrated Reliability Workshop*, Lake Tahoe, California, 2000.

[11] D.R. Wolters and J.F. Verwey, "Instability in silicon devices" in *G. Barbottin, A. Vapaille (Ed.)*, North Holland, Amsterdam, vol.1, 315, 1986.

[12]] J.W. McPherson and H. C.Mogul "Disturbed Bonding States in SiO2 Thin-Films and their impact on Time Dependent Dielectric Breakdown" in *36thAnnual International Reliability Physics Symposium*, Reno, Nevada, *pp. 47-56, 1998.*

[13] D. J. Dimaria and J. H. Stathis "Non-Arrhenius temperature dependence of reliability in ultrathin silicon dioxide films" in *Application Physic Letter, vol.* 74, pp. 1752-1756, 1999.

[14] G. Ribes, S.Bruyere "New insights into the change of voltage acceleration and temperature activation of oxide breakdown" in *Microelectronics Reliability*, vol. 43, pp. 1211-1214, 2003.

[15] G. Ribes, S. Bruyere "Multi-vibrational hydrogen release: Physical origin of T$_{bd}$,Q$_{bd}$ power-law voltage dependence of breakdown in ultra-thin gate oxide" in *Microelectronics Reliability, vol.* 45, pp. 1842-1854, 2005.

2006 25th International Conference on Microelectronics

Low Frequency Noise of GaAs MESFET Degraded in ESD Test

Milan M. Jevtić, Jovan Hadži-Vuković

Abstract – GaAs MESFET's were stressed with high voltage pulses of 1 to 3 kV in an ESD experiment. MESFET degradations were studied by I-V and low frequency (LF) noise measurements. We have distinguished two cases: stress in gate and stress in drain. Noise results in the case of gate stress suggest that the LF noise sources are connected with defects near the metal-semiconductor interface.

Keywords – GaAs MESFET, low frequency noise, electrostatic discharge (ESD)

I. INTRODUCTION

MESFET reliability is strongly influenced by the integrity of the Schottky gate barrier and the ohmic contacts. Recently we have made several experiments performed on SiC Schotky diodes to get more understanding of ESD stress applied on compound hetero-interface [1]. In the most cases failure mechanisms in Schottky contacts are diffusion related. The diffusion of the Schottky gate metal atoms across metal-GaAs interface (Gate Sinking) resulting in a decrease of the barrier height is a major concern during high temperature storage tests. The ESD is a one of the major concern for MESFETs [2], [3]. The ESD pulses applied at the gate will result in electrical damage void formation at the gate contact. Degradation of ohmic contacts and electromigration are also important failure processes that influence reliability of the circuits with MESFET. Analysis of properties of degraded MESFETs is of interest to better understand the degradation processes. Low frequency noise can be used as a diagnostic tool in quality and reliability analysis of compound semiconductor transistors [4].

It was observed increase by two orders of magnitude of the low frequency noise (LFN) level when MESFETs are submitted to rf life test [5] of device.On the other hand, it is known that crystallographic defects resulting in traps can be traced to the crystal manufacturing process and dopant used in GaAs. The traps in MESFETs have been also found close to the channel substrate interface. Structures with substrates grown using liquid encapsulated Czochralski technique have shown traps on

Milan M. Jevtić is with Institute of Physics, Pregrevica 118, 11080 Belgrade, Serbia and Montenegro, E-mail: mjevt@phy.bg.ac.yu
Jovan Hadži-Vuković is with Infineon Technologies Austria, Siemensstrasse 2, 9500 Villach Austria, E-mail: Jovan.HadziVukovic@infineon.com

the channel side of the substrate-channel interface that affect doping concentration profile and carrier mobility. Traps on the substrate side of the interface can influence a hysteresis of the source-drain characteristics when the MESFET's are driven beyond pinch off. It is shown that the 1/f noise in gate current is effective as diagnostic tool to detect hot spot and device failure. This has been explained by the contributions of fluctuations of the leakage currents between the gate and source and the gate and drain in fluctuations of the total gate current [6]. In this paper we have shown the results of our investigation of the ESD stressed Ga As MESFETs. The study of behavior of MESFET's under ESD stress conditions is significant for better understanding physical processes leading to the degradation under these conditions, as well for finding corrective actions leading to greater robustness of MESFET's to ESD, electrical over stress (EOS) events, electrical cycling like in life tests [7] or temperature accelerated testing [8]. ESD pulses have higher voltage and shorter time then other pulses but effects in devices according to dissipated energy and increased temperature could be comparable. The experimental details are described in Section II, and the results and discussion are presented in Section III.

II. THE EXPERIMENTAL DETAILS

Commercial GaAs MESFETs (Fig. 1.) used in our ESD experiments are capable of delivering 3W of output power with high linear gain, high efficiency and excellent linearity.

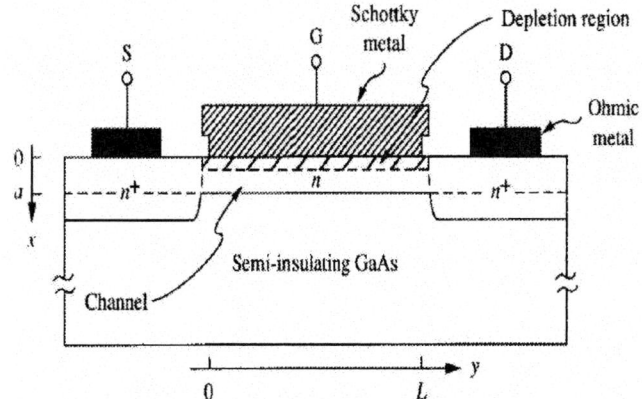

Fig. 1. Cross-section of GaAs MESFET

The MESFETs are fabricated in 0.7μm-technology with Ti/Al gate. The transistor's pinch off voltage has typical

1-4244-0116-X/06/$20.00 ©2006 IEEE   569

value of –2.6V with variations between -1.6 and -3.6. To stress the devices we used the ESD testing standard – Human Body Model HBM. It defines the current waveform for the discharge of a 100pF capacitor through a 1.5 kΩ resistor for different discharge voltages. In order to obtain more information about degradation we have measured the low frequency noise in degraded MESFETs. The power spectral density of the current fluctuations was measured by a measurement system based on Dynamic Signal Analyzer (HP3562A) and low noise preamplifier (Keithley 103A), in the frequency range between 10Hz and 10kHz at the room temperature.

III. EXPERIMENTAL RESULTS

We have distinguished two cases in our experiments: voltage stress in gate, and voltage stress in drain.

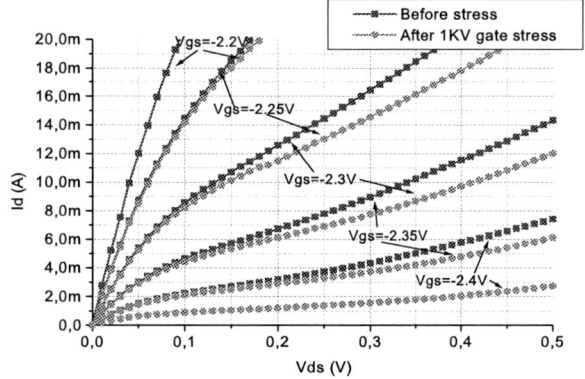

Fig. 2. Id-Vd output characteristics of MESFETs.

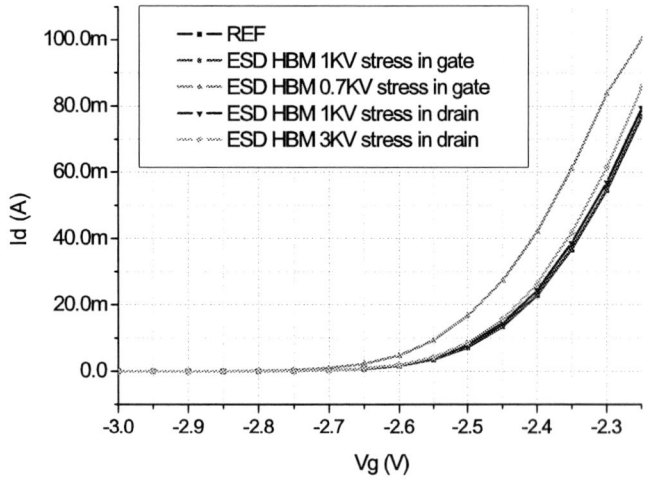

Fig. 3. Id-Vg transfer characteristics of MESFETs.

Lot of 20 MESFETs was tested under HBM stress under voltages of 0.7 to 3 kV. The test was performed with single pulses and after stress by one pulse the I-V curve

was observed with Semiconductor Parameter Analyzer HP 4156B. To avoid cumulative effect of stressing each diode was stressed only with one pulse.

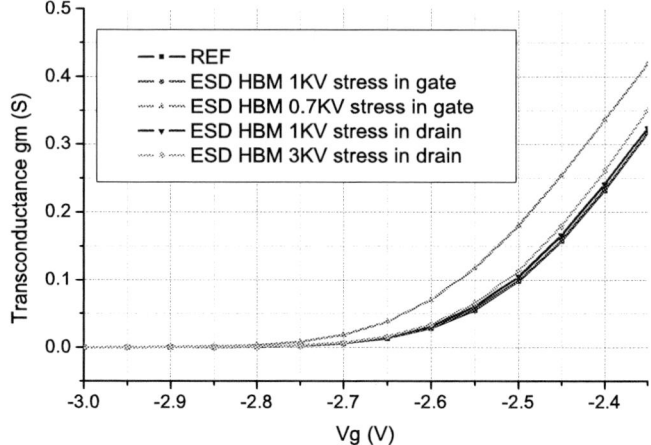

Fig. 4. Transconductance of MESFETs.

The output curves of MESFET are presented in Figure 2. We have measured low frequency noise at the points where Vds has values of approximately 0.3V and for Vgs with values of -2.5V and -1V, actually in linear region of MESFET. The transfer curves and transconductance of reference and stressed MESFETs are presented in Figs. 3 and 4. It is observable that MESFET stressed with 0.7kV in gate has bigger deviation from reference curve then MESFET stressed with 1kV. This could be explained by statistical and randomize nature of ESD failures in the devices.

We have measured the gate voltage noise spectrum. The source of the gate noise is the reverse biased Schottky contact. Multiplying this noise with transconductance squared it is transferred in noise in drain circuit whose LF noise spectra are presented in Figs. 5-9. It has to be noted that MESFETs stressed in gate (Figs. 6 and 7.) show higher noise level then the one of un-stressed reference transistor (Fig. 5). The transistors degrading during gate stress show a characteristic shape of Lorentzian spectra, which are addressed to particular defects formed during ESD stressing. Characteristic frequencies of the Lorentzian spectra of 1kHz (Fig. 7) and 115Hz (Fig. 6) are found from LIN(fS_i) vs LOGf curves. Two processes could be source of the noise with the Lorentzian spectrum in reverse biased Schottky contact: the trapping of the carriers by defects near the metal-semiconductor interface and g-r process in space charge region. The trapping process is more probable because electrons flow across the interface from metal to semiconductor in reverse biased gate contact. Electrons tunnel from metal to the trap near the interface. The fluctuation of this process influences the fluctuation of barrier height, which leads to the current fluctuation.

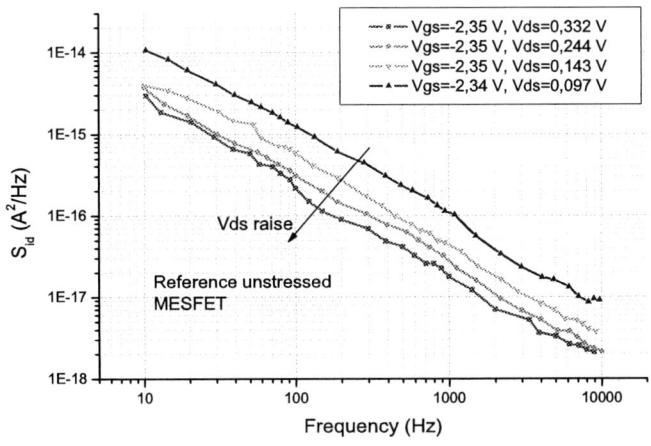

Fig. 5. LFN of Reference MESFET.

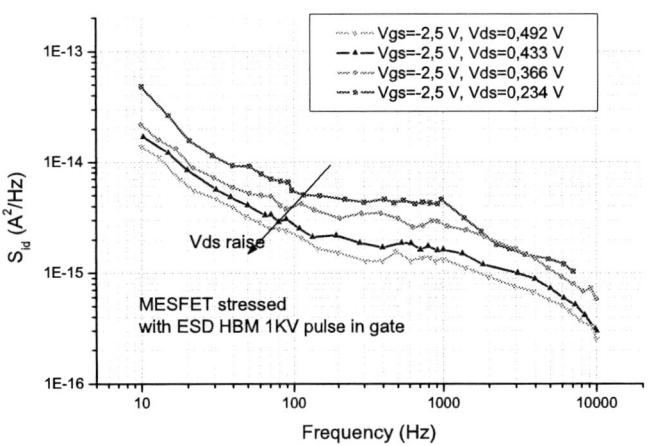

Fig. 7. LFN of MESFET degraded in gate with 1kV.

The noise spectra of the transistors degraded by voltage stress in drain did not show the Lorentzian shape. The degradation during the drain stress is less than the one during the gate stress and it could occur at n^+-n drain and source junctions. The lower noises in Figs. 8 and 9 in comparison with the noise shown in Fig. 5 are the results of difference in bias conditions.

The results for the drain stressing confirm the conclusion that the defects connecting with the Lorentzian spectra in Figs. 6 and 7 are located near the metal-semiconductor interface. Also, it has to be noted an interesting result relating to Figs. 6 and 7: the characteristic frequency of the Lorentzian spectrum increases as the voltage pulse increases from 0.7 to 1kV. In frame of our noise model this means that increases the defect concentration nearer to the interface or defect energy levels move nearer to conduction band bottom.

We can observe in Figs. 8 and 9 that the noise of transistor after 3kV drain stress is higher than the one of the transistor after 1kV drain stress. We can not directly compare the noise intensities after gate and drain stresses because they are measured for different Vgs voltage. According to the noise spectrum shapes one can only conclude that they do not origin from the same regions in MESFET. It is known for a long time that the noise magnitude shows strong dependence on gate leakage current [9], [10].

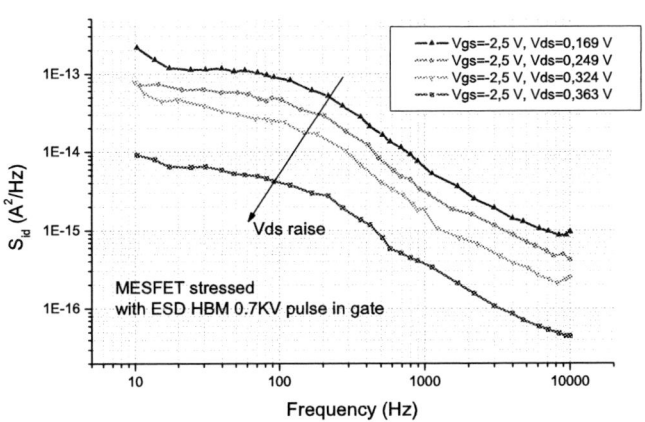

Fig. 6. LFN of MESFET degraded in gate with 0.7kV.

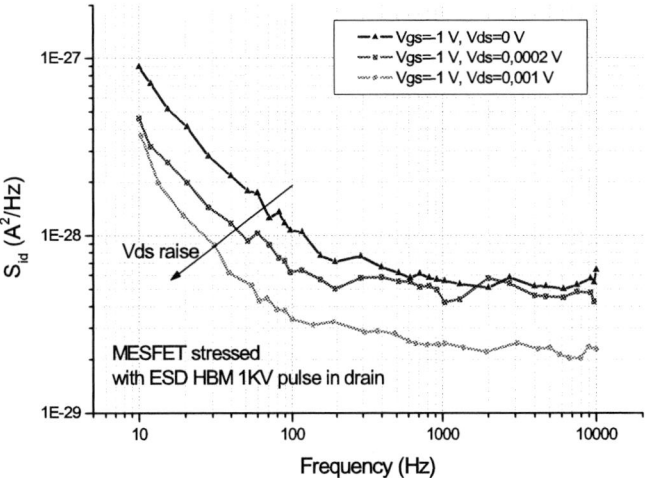

Fig. 8. LFN of MESFET degraded in drain with 1kV.

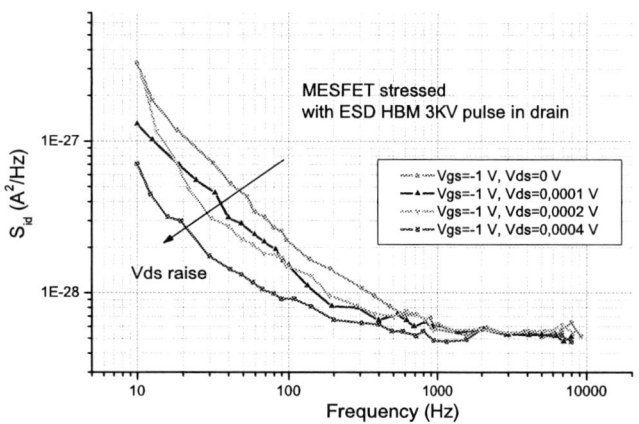

Fig. 9. LFN of MESFET degraded in drain with 3kV.

IV. CONCLUSION

The gate noise spectra of degraded GaAs MESFETs during ESD test have been measured and the noise transferred in drain circuit is shown in this paper. It has found that the noise spectra of transistors degraded during ESD gate and drain stressing are different in intensities and shapes. The Lorentzian spectra were found in the transistors degraded during ESD gate stressing.

This is explained by fluctuation of charge at defects inside charge depletion region near metal-semiconductor interface.

ACKNOWLEDGEMENT

This research was partially supported by the Serbian Ministry of Science and Environmental Protection (Contract TR-6151B).

REFERENCES:

[1] Jevtic M., Hadzi-Vukovic J., Dan D., " Pulse Voltage Stress Degradation of 4H-SiC Schottky Diodes Studied by I-V and Noise Measurements "Int. Conf. CAS 2005., Conference Proceedings, Vol. 2, pp. 369-372, Sinaia, Romania

[2] Amerasekera A., Najm F., " Failure mechanisms in semiconductor devices " John Wiley & Sons, New York,1998.

[3] Bock K.," ESD issues in compound semiconductor high-frequency devices and circuits", Microelectronic Reliability, vol.38.,pp.1781-1793.,1998.

[4] Labat N., Malbert N., Maneux C., Touboul A.," Low frequency noise as a reliability diagnostic tool in compound semiconductor transistors", Microelectronic Reliability, vol.44.,pp.1361-1368.,2004.

[5] Lambert B., Malbert N., Verdier F., Labat N., Touboul A., Vandamme L.K.J., " Low frequency gate noise in a diode-connected MESFET: Measurements and Modeling " IEEE Transactions on Electron Devices, pp. 628-633 ,Vol 48. No 4. 2001.

[6] Vandamme L.K.J., Rigaud D., Peransin J.M., Alabedra R., Dumas J.M." Gate Current 1/f noise in GaAs MESFET's " IEEE Trans. On Electron Devices, pp. 1071-1075 ,Vol 35. No 7. 1988.

[7] Lambert B., Saysset-Malbert N., Labat N., Verdier F., Touboul A., huguet P., Garat F.,"Comparation of RF and DC life-test effects on GaAs power MESFETs ", Microelectronic Reliability, vol.40.,pp.1727-1731.,1998.

[8] Feng T., Strifas N., Christou A.," Degradation of performance in MESFET and HEMTs. simulation and measurement of reliability", Microelectronic Reliability, vol.38.,pp.1239-1244.,1998.

[9] Das M.D. Ghosh P.K. " Gate current dependance of Low frequency gate noise in GaAs MESFET's " IEEE Electron Device Letters, pp. 210-213 ,Vol 2. No 8. 1981.

[10] Danneville F., Dambrine G., Happy H., Tadyszak P., Cappy A.," Influence of the gate leakage current on the noise performance of MESFETs and MODFETs ", Solid State Electronics, vol.38.,pp.1081-1087.,1995.

2006 25th International Conference on Microelectronics

Dependence of DGMOSFET 1/f Noise on Transistor Geometry and Technology Parameters

Mirjana Videnović-Mišić, Student Member IEEE, Milan M. Jevtić

Abstract – In this paper theoretical study of a dual-gate MOSFET (DGMOSFET) 1/f noise and its sensitivity to variation of transistor geometry and technology parameters is presented. Expression for 1/f current noise spectral density $S_{id}(f)$ is obtained using an ac current approach in a DGMOSFET LF small-signal noise equivalent circuit. The results show 1/f $S_{id}(f)$ level increase with L_1 increase and W decrease. For a single MOS transistor and L increase opposite 1/f noise behaviour is observed. In case of a DGMOSFET, not only transistors noise sources influence overall noise but also dynamic transistors parameters and load resistance R_L. Sensitivity of $S_{id}(f)$ to source/drain junction depth x_j is minimal as a consequence of "long channel" DGMOSFET structure. As for oxide thickness, 1/f noise level show expected increase with t_{ox} increase.

I. INTRODUCTION

DGMOSFET structures are widely used in MOS integrated circuits where electronic gain control, low feedback parameters, low noise, cross modulation or reduction of short channel effects is requirement [1, 2]. The simplified diagram of a dual-gate n-channel MOSFET is shown in Fig.1. Under appropriate bias conditions, DGMOSFET is a cascode connection of two single-gate MOSFETs, M1 and M2, with equal gate widths. As a consequence of the fact that point between two transistors is inaccessible, the drain voltage (V_{DS1}) of the first transistor, gate (V_{GS2}) and drain (V_{DS2}) voltages of the second transistor together with dynamic transistors parameters cannot be directly measured. Since internal parameters define DGMOSFET working conditions, they must be extracted using some simulation techniques.

During simulations geometry and technology parameters, such as effective channel length L_{eff}, channel width W, metallurgical source/drain junction depth x_j, gate oxide thickness t_{ox} and their variation play significant role.

Mirjana Videnović-Mišić is with the Department of Electronics, Faculty of Technical Sciences, University of Novi Sad, Trg Dositeja Obradovica 6, 21000 Novi Sad, Serbia & Montenegro, E-mail: mirjam@uns.ns.ac.yu

Milan Jevtić is with the Department of Appl. and Techn. Phys., Institute of Physics, Pregrevica 118, 11000 Beograd, Serbia & Montenegro, E-mail: mjevt@phy.bg.ac.yu

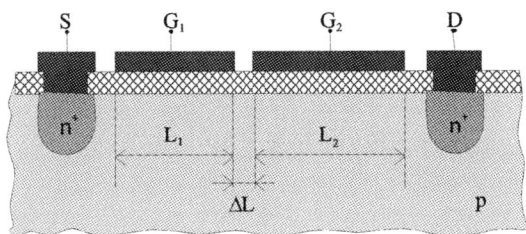

Fig. 1. The simplified diagram of a dual-gate n-channel MOSFET

With the explosive growth of the communication systems market during the last few years, mobile personal communication devices have become extremely popular. Together with this rapid market growth the stringent design requirements have emerged. To fulfill these requirements designers are forced to understand the fundamental limitations of the design process. Since noise in RF circuits is first order effect which directly affects system performance, it is necessary to minimise its influence [3]-[5] with isolation and other designing techniques. Still, the circuit cannot be isolated from the noise sources present in the devices which constitute the circuit itself. Therefore, device noise determines the fundamental bounds on circuit performance and plays a significant role in RF circuit design. This is particularly the case for devices in transceiver's front-end, such as mixers, oscillators, power amplifiers, frequency multipliers, which usually operate in large-signal quasi-periodic conditions. Therefore, the focus of our recent work has been low-frequency (LF) noise of a DGMOSFET in Colpitts oscillator [6] and phase noise of the same oscillator [7] together with modelling of DGMOSFET 1/f noise under different bias conditions [8]. High sensitivity of the transistor LF and oscillator phase noise to the DGMOSFET working conditions have been noticed.

In order to achieve higher operating speeds and increase packing densities, FET structures have been subjected to miniaturisation. As a result of small dimension effects and deviation in manufacturing processes, variation in FET geometry and technology parameters occurs. These random perturbations from the nominal values will cause change in device and circuit performance.

Therefore, the purpose of this study is to shed more light on DGMOSFET LF noise dependence on transistor geometry and technology parameters as well as on their variation since these parameters together with bias

1-4244-0116-X/06/$20.00 ©2006 IEEE   573

conditions characterise DGMOSFET working conditions and 1/f noise.

II. DGMOSFET 1/F NOISE MODEL

The DGMOSFET 1/f noise model was developed in [8] under several assumptions. First, traps at the Si-SiO$_2$

interface play the most important role in 1/f noise generation. Second, DGMOSFET 1/f noise was present due to channel conductance fluctuation as a consequence of incidental interactions between carriers and traps at the Si-SiO$_2$ interface [9].

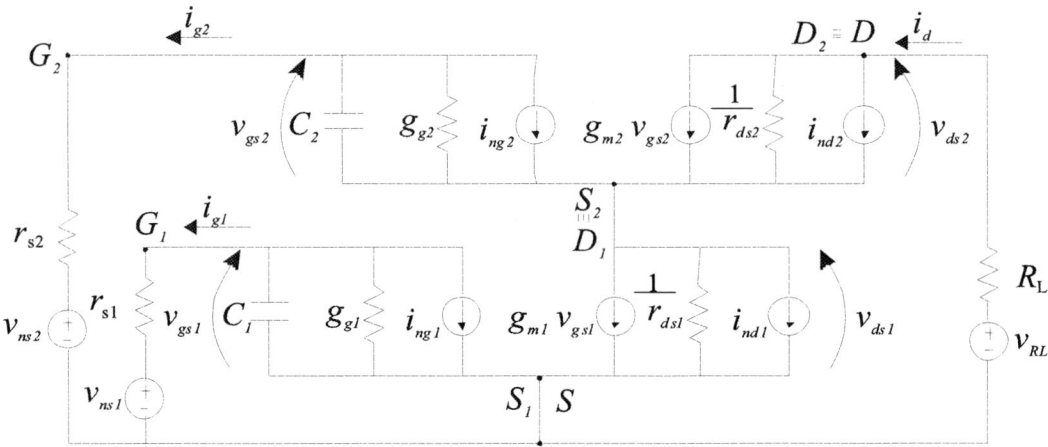

Fig. 2. DGMOSFET small-signal noise equivalent circuit under low frequency conditions

Developed DGMOSFET 1/f noise model gives the current dependence of 1/f current noise spectral density $S_{id}(f)$ where expression for $S_{id}(f)$ is obtained using an ac current approach in a DGMOSFET LF small-signal noise equivalent circuit, shown in Fig.2. The symbols in Fig.2 have the following meanings: r_{s1}, r_{s2}, and R_L are resistances in gates and drain bias circuits, respectively, with corresponding voltage noise sources v_{ns1}, v_{ns2} and v_{RL}; C_j, g_{gj} and $\overline{i_{ngj}^2}$ are the capacitance, conductance and current noise generator of the first ($j=1$) and second ($j=2$) gates; g_{mj} and r_{dsj} are the transconductances and channel resistances of the first ($j=1$) and second ($j=2$) DGMOSFET transistors; i_{ndj} represents the current noise sources due to the channel resistance fluctuations of the first ($j=1$) and second ($j=2$) transistors. Other symbols in Fig.2 have the usual meanings.

The correlation between gate and drain current noises due to the leakage gate current is neglected as well as gate currents assuming i_{g1}, i_{g2}, $i_{leakage}$ to be very small compared to the drain current.

If dominant noise sources are the channels of the transistors, under the condition

$$Z_{gj} = \left| \frac{1}{j\omega C_j} \right| \left| \frac{1}{g_{gj}} \right| \to \infty, j = 1, 2$$

one can express i_d as

$$i_d \approx A_{nd1} i_{nd1} + A_{nd2} i_{nd2}, \qquad (1)$$

where coefficients A_{nd1} and A_{nd2} are

$$A_{nd1} = \frac{r_{ds1}(1 + g_{m2}r_{ds2})}{R_L + r_{ds1} + r_{ds2} + g_{m2}r_{ds1}r_{ds2}},$$

$$A_{nd2} = \frac{r_{ds2}}{R_L + r_{ds1} + r_{ds2} + g_{m2}r_{ds1}r_{ds2}}. \qquad (2)$$

Using Fourier transformation of i_d and assuming that channel noise sources are uncorrelated, 1/f drain current noise spectral density $S_{id}(f)$ can be expressed in the form:

$$S_{id}(f) = |A_{nd1}|^2 S_{ind1}(f) + |A_{nd2}|^2 S_{ind2}(f), \qquad (3)$$

where $S_{indj}(f)$ is 1/f drain current noise spectral density of transistor M_j, j=1,2.

If we connect gates of transistors M_j situation is simplified and DGMOSFET structure is reduced to device similar to double-diffused MOS transistor whose noise has been thoroughly investigated in [10]. Moreover, if we adjust our small-signal noise equivalent circuit to conditions in [10] we get the same equations for total drain current noise spectral density of transistor.

For single "long channel" transistors operating at strong inversion in linear, non-linear and saturation regimes [9], $S_{indj}(f)$, due to oxide trap charge density fluctuation, is:

$$S_{indj}(f) = \frac{q^2 N_{ot}}{C_{ox}^2 W (L_{effj})_j} g_{mj}^2 \frac{1}{f}, \quad (4)$$

where C_{ox} is oxide capacitance per unit area and N_{ot} is average oxide traps density per unit area for traps distributed in energy near the electron quasi-Fermi level [12].

As can be seen from equation (3) and (4), the coefficients $|A_{nd1}|^2$ and $|A_{nd2}|^2$ are weighting factors that describe participation of single transistor (M1 and M2) noise in total DGMOSFET noise. Through these coefficients and $S_{indj}(f)$ (j=1,2) effective channel length L_{eff}, channel width W, metallurgical junction depth x_j, gate oxide thickness t_{ox} influence $S_{id}(f)$.

III. RESULTS OF NUMERICAL ANALYSIS AND DISCUSSION

Numerical analysis of DGMOSFET 1/f noise was performed under L_{eff}, W, x_j and t_{ox} variations. The results are shown in Figs. 3, 4, and 5. 1/f current noise spectral densities $S_{id}(f)$ have been calculated at frequency 10Hz in order to represent spectra in domain where 1/f noise source is dominant.

Small variation of geometry parameters, such as channel length L and width W, results from small-dimension effects [11] and manufacturing line-width variations. Since geometry optimisation with respect to noise can be interesting during transistor design, we have analysed how geometry parameters variation influence transistor noise. Drain current noise spectra calculated for W=1150μm and W=1750μm, while changing L_1 value from 1.1μm to 2μm, are shown in Fig.3. DGMOSFET 1/f drain current noise spectral densities show amplification with length L_1 increase, contrary to our expectations derived from single MOS transistor 1/f noise theory [9] where L increase results in noise decrease. From equations (2) and (3) can be seen that situation in DGMOSFET is more complicated. Not only that single transistor spectral density $S_{ind1}(f)$ and $S_{ind2}(f)$ influence overall noise $S_{id}(f)$ but also DGMOSFET dynamic parameters r_{ds1}, r_{ds2}, g_{m1}, g_{m2} and drain load resistance R_L in form of coefficients $|A_{nd1}|^2$ and $|A_{nd2}|^2$.

Moreover, one can see from Fig's 3a) and 3b) that channel width W increase results in $S_{id}(f)$ decrease. Alterations in noise level are extensive since we considerably changed channel length L and width W that produced extreme change in V_{TH}[11] and noise level [12].

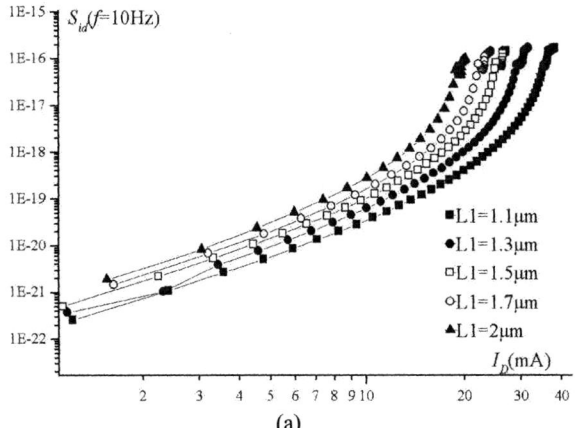

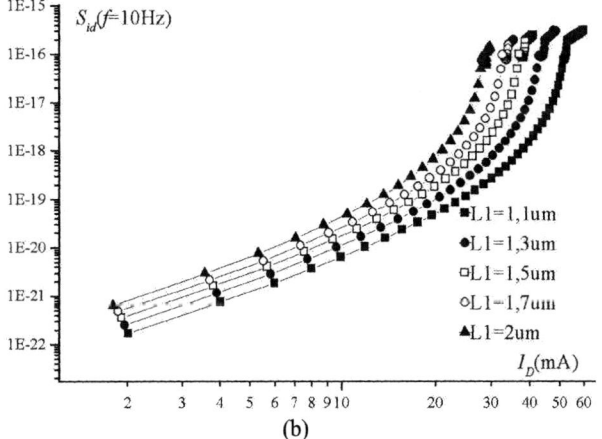

Fig. 3. $S_{id}(f)$ vs. I_D at frequency 10Hz for (a) W=1150μm (b) W=1720μm where length L_1 of first MOS transistor is varied

As a result of MOS device scaling, source/drain junction depth x_j influences channel length L that leads to modification in V_{TH} [11], [12]. Therefore, very shallow n$^+$ and p$^+$ layers, suitable for source and drain regions, are needed to minimise short-channel effects. Variation of x_j results from relatively high diffusion rate of dopant atoms at the elevated temperatures and random change in diffusion process parameters [13]. The original x_j value is 200nm. 1/f current noise spectral density $S_{id}(f)$ calculated for 0%, +/-5%, +/-10%, +/-20% variation of x_j, is shown in Fig.4. Very weak sensitivity of $S_{id}(f)$ to x_j variations can be noticed. Since DGMOSFET used in numerical analysis is "long channel" transistor, x_j influence on V_{TH} modification and noise level variation is minimal.

Fluctuations in oxidation process parameters are crucial factors in gate oxide thickness t_{ox} variation [14] resulting in C_{ox} variation, where C_{ox} is one of the key factors in MOS transistor 1/f noise model [12]. t_{ox} is involved not only in C_{ox} and $S_{idj}(f)$ calculation but also in computation of dynamic transistor parameters (r_{ds}, g_m) and effective channel length L_{eff} for MOSFET in saturation. Simulations were performed for t_{ox}=42nm starting value. $S_{id}(f)$ vs. I_D calculated for 0%, +/-5%, +/-10%, +/-20%

variation of t_{ox}, is shown in Fig.5. Since t_{ox} influence on $S_{id}(f)$ is complex only general remarks can be given. Oxide thickness increase results in noise level increase which is in agreement with single transistor noise behaviour [9]. It can be seen that "±" variations of t_{ox} generate symmetrical curves in respect to nominal curve in $S_{id}(f)$ vs. I_d domain.

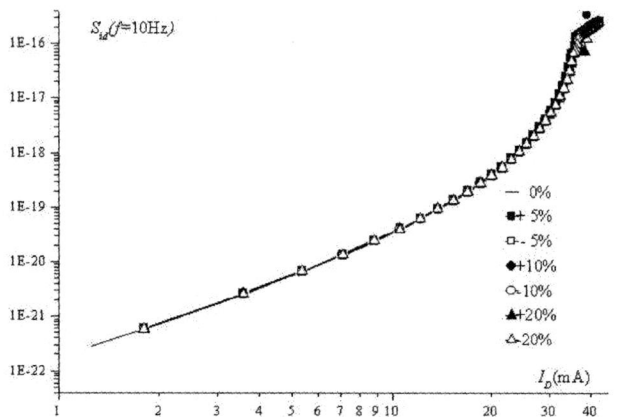

Fig. 4. $S_{id}(f)$ vs. I_D at frequency 10Hz for 0%, +/-5%, +/-10%, +/-20% variation of source/drain junction depth x_j

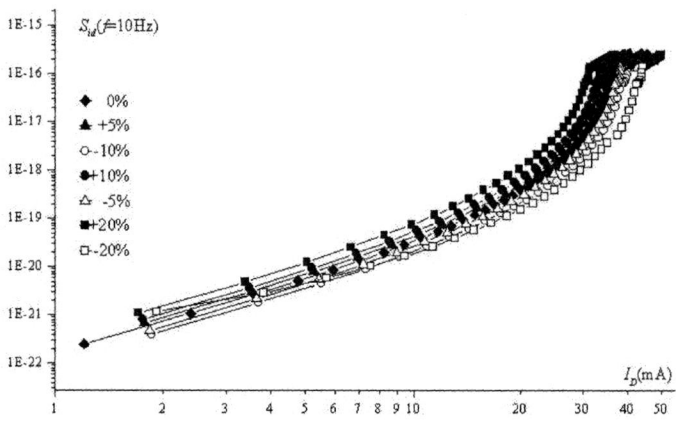

Fig. 5. $S_{id}(f)$ vs. I_D at frequency 10Hz for 0%, +/-5%, +/-10%, +/-20% variation of gate oxide thickness t_{ox}

IV. CONCLUSION

The influence of geometry and technology parameters variation on DGMOSFET 1/f noise has been discussed. The results show that DGMOSFET 1/f noise dependence on effective channel length L_{eff}, channel width W, metallurgical source/drain junction depth x_j and gate oxide thickness t_{ox} does not always follow conventional single transistor noise behaviour. In order to better understand and predict DGMOSFET 1/f noise, its extraction from small-signal noise equivalent circuit under low frequency conditions is obligatory.

ACKNOWLEDGEMENT

This research was partially supported by the Serbian Ministry of Science and Environment Protection (contract TR-6151B).

REFERENCES

[1] R.M.Barsan, "Analysis and modeling of dual-gate MOSFET's", IEEE Trans.Electron Dev., Vol.ED-28, No.5, pp.523-534, 1981.

[2] R.M.Barsan, "Subtreshold Current of dual-gate MOSFET's", IEEE Trans.Electron Dev., Vol.ED-29, No.10, pp.1516-1521, 1982.

[3] A.Hajimiri and T.H.Lee, "A general theory of phase noise in electrical theory", IEEE J.Solid-St.Circ., Vol. 33, No.2, pp.179-194, Feb. 1998.

[4] L. Li and H. Tenhunen , "Noise analysis of monolithic RF balanced down conversion mixers", Southwest Symposium on Mixed-Signal Design SSMSD 2001,pp. 70 – 75, 2001.

[5] M.Z.A.A. Aziz, J.B. Din, M.K.A. Rahim, "Low noise amplifier circuit design for 5 GHz to 6 GHz", Proc. of the RF and Microwave Conference RFM 2004, pp. 5 -8, 2004.

[6] M.Videnović-Mišić and M.Jevtić, "Low-Frequency Noise of a Dual-Gate MOST in Colpitts Oscillator Proc. 48th ETRAN, Cacak, vol.4, pp. 125-128, Serbia and Montenegro, 2004.

[7] M.Videnović-Mišić and M.Jevtić, "DC conditions and phase noise of Colpitts oscillator with DGMOST", Proc. 24th Int.Conf.Microel. MIEL2004, Niš, vol. 2, pp. 577 – 580, 2004.

[8] M.Videnović-Mišić and M.Jevtić, "Numerical Analysis of DGMOSFET 1/f Noise Under Different Bias Conditions", Proc. IEEE EUROCON 2005, pp. 1232–1235, Belgrade, November 2005.

[9] C.Jakobson, I. Bloom, Y. Nemirovsky, "1/f Noise in CMOS transistors for analog applications from subtreshold to saturation", Solid-St. Electronics, vol.42, No.10, pp. 1807-1817, 1998.

[10] R. van Langevelde, S. Blieck, L. K. J. Vandamme, "Noise in DMOS transistors in a BICMOS-Technology", IEEE Trans. Electr. Dev., vol. 43, No. 8, pp. 1243- 1250, 1996.

[11] Robert F. Pierret, "Semiconductor Device Fundamentals", Addison-Wesley, 1996.

[12] Raj Jayaraman, Charles G. Sodini, "A 1/f noise Technique to Extract the Oxide Trap Density Near the Conduction Band Edge of Silicon", IEEE Trans. Electr. Dev., vol. 36, No. 9, 1989.

[13] Hong-Ha Vuong at all, "Use of Transient Enhanced Diffusion to Tailor Boron Out-Diffusion", IEEE Trans. Electr. Dev., vol. 47, No 7, pp. 1401-1205, 2000.

[14] K.Saki at all, "Novel approach for precise control of oxide thickness", IEEE Inter. Semi. Manu. Symp. 2001, pp. 451 – 454, 2001.

2006 25th International Conference on Microelectronics

Use of RADFETs for Quality Assurance of Radiation Cancer Treatments

A. Jaksic, K. Rodgers, C. Gallagher, and P.J. Hughes

Abstract – The paper presents recent developments in quality assurance of radiation cancer treatments involving RADFETs (MOSFET dosimeters) developed and manufactured at Tyndall National Institute. Construction of a RADFET linear array for intracavitary in-vivo dosimetry is described in detail.

I. INTRODUCTION

The Radiation sensing MOSFETs (RADFETs, MOSFET dosimeters) are discrete p-channel MOS transistors with gate oxides optimised for increased radiation sensitivity. They have found applications as dosimeters in spacecraft, nuclear facilities, nuclear particle physics laboratories, and clinical environments [1]. Tyndall National Institute, formerly known as NMRC, has been involved in RADFET research and development since late 1980s. The activity, sponsored mainly by the European Space Agency, had been focused on evaluation of the total-dose in space environments. However, recently we have extended RADFET research and development to include medical applications, in particular quality assurance (QA) of radiation cancer treatments (radiotherapy). The paper presents the main recent developments in this area.

II. DOSIMETRY WITH SINGLE RADFETS

Traditionally, the dosimeters most commonly used for QA in radiotherapy have been TLDs (Thermolumenescent Dosimeters) and specially designed silicon diodes. TLDs are integrating dosimeters, i.e. they directly measure total dose. They offer the advantage of being both small in size and relatively accurate, and wireless. However, there is a long delay between measurement and read-out (ideally the received dose should be known while the patient is still in the treatment room), and the device is rather difficult to handle. Diodes directly measure dose rate, but the total dose can be calculated and displayed in real-time. They are more sensitive than TLDs, but include cumbersome wiring and regular and careful calibration. RADFETs mounted at the end of wired probes have been competing with the diodes for some time [2]. We believe that the diodes are overall a better solution than wired RADFETs, as diodes

A. Jaksic and K. Rodgers are with the Tyndall National Institute, Lee Maltings, Cork, Republic of Ireland, E-mail: ajaksic@tyndall.ie.
C. Gallagher was with Tyndall National Institute, he is now with HP Dublin Inkjet Manufacturing Operations, Leixlip, Republic of Ireland.
P.J. Hughes was with Tyndall National Institute, he is now with SensL, Cork, Republic of Ireland.

are more sensitive, offer better reproducibility of the measurements, and have a longer lifetime.

However, Sicel Technologies [3] have recently introduced a new concept of RADFET dosimetry – OneDose system (Fig. 1). The system combines good features of TLDs (no need for wires) and diodes (immediate read-out) and thus is extremely user- and patient-friendly. OneDose is based on the RADFETs developed and manufactured at Tyndall. It uses small, disposable, adhesive patch that is affixed to the skin at the site of radiation prior to treatment. Immediately after treatment, a handheld reader is used to display the data from the patch, store the data in the microprocessor, and allow the data to be exported to a centralized data management system. The patch can then be discarded or retained in the patient's medical records.

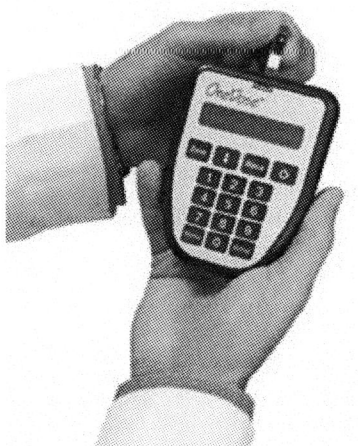

Fig. 1. OneDose reader and dosimeter patch.

Dosimeters described above are placed on the patient's skin during treatment and the dose in the tumour inside the body needs to be re-calculated. This includes some uncertainties related to calculation methods and, even more importantly, to movement of the tumour during treatment (due to e.g. patient's movements, breathing, etc.). After encouraging initial studies [3], Sicel are currently in the final stage of development of another innovative QA system which addresses these problems. The system is called DVS and includes implantable wirelessly read-out dosimeter (Fig. 2) that can be placed exactly in the tumour and/or surrounding healthy tissue. Due to its compatibility with wireless technology (unique among radiation dosimeters) the RADFET was practically the only choice for the DVS. Tyndall National Institute has been actively involved in the development of DVS.

1-4244-0116-X/06/$20.00 ©2006 IEEE   577

Fig. 1. DVS implantable dosimeter is just 2mm in diameter and 18mm long.

III. RADFET INTRACAVITARY ARRAY

It is often desirable to make intracavitary measurements of absorbed dose distributions during radiation treatments. We have recently developed a RADFET linear catheter array optimised for QA of radiation treatment of rectal carcinoma and also suitable for placement in vagina and oesophagus. Initial efforts were described in [4]; remainder of the paper gives details of construction of the final prototype array.

The RADFET used in the catheter array is the ESAPMOS4 device manufactured at Tyndall. The chip (Fig. 3) is 1mm × 1mm and contains two 300/50 (W/L in μm) and two 690/15 individual RADFETs, an on-chip diode, and 12 bond pads (100μm square with 150μm spacing).

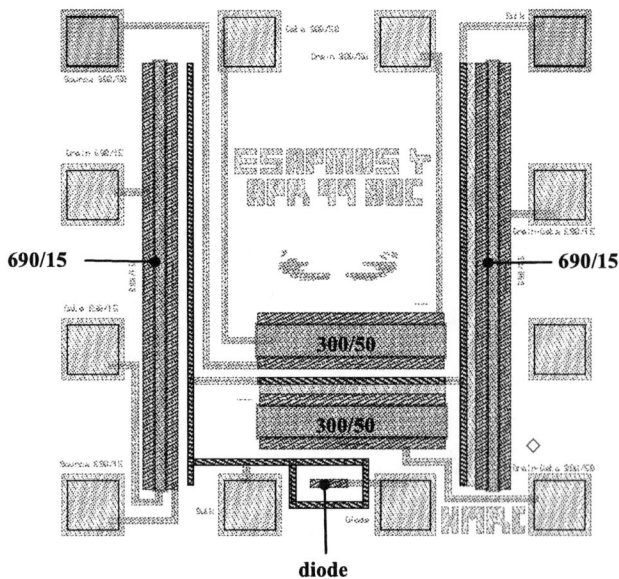

Fig. 3. Layout of the ESAPMOS4 chip. Chip size is 1mm × 1mm.

Under Bump Metallurgy (UBM) is applied to the Al bond pads of the chip. We use an electroless plating process to reduce cost compared to electroplating or sputtering techniques. The UBM consists of a 5μm layer of Ni that acts as a solderable metal layer coated with a 100nm Au layer that ensures no oxidation of Ni occurs before bumping.

For each die, 12 SnPb solder bumps are dispensed using a ink-jet direct write technology [5] and then reflowed (Fig. 4). The die are then flip chip mounted onto the flexible substrate – 50μm Cu-clad polyamide. The substrate is patterned by photolithography and wet etching. The Cu is then electroless plated with a Ni/Au layer to provide a reliable interconnect layer for the devices. The bond pads on the flexible substrate are coated with a no-clean flux to promote solder wetting. The chip bumps are kept in contact with the substrate bond pads and the temperature is raised over the solder melting point to interconnect the device to the substrate.

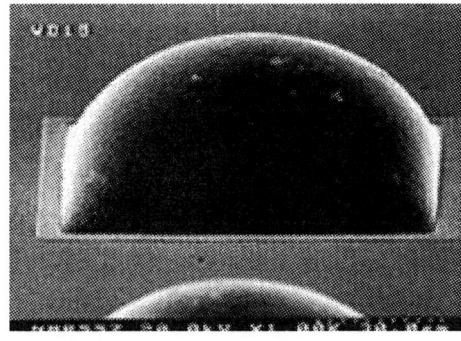

Fig. 4. Multiple PdSn droplets dispensed prior to reflow (top) and an individual bump after reflow (bottom).

Fig. 5 shows a typical cross section of a reflowed solder bump. An underfill process is then carried out to increase stability and strength of the flip chip devices. The adhesive (Loctite 3563) is dispensed along the chip edge and capillary action draws the adhesive under the die and around the flip chip interconnects.

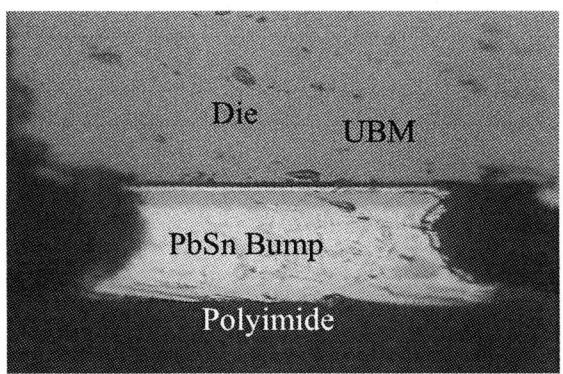

Fig. 5. Cross section of the solder interconnect.

The final steps are mounting of the array in a medical catheter and filling the catheter with silicone. The array (Fig. 6) consists of 10 non-uniformly spaced RADFET chips. The complete dosimetric system developed consists of the RADFET array, read-out unit, and a dedicated software.

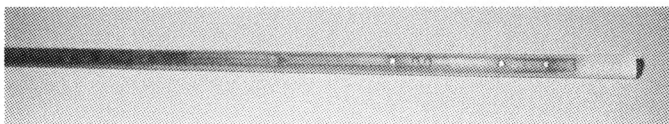

Fig. 6. RADFET array before (top) and after (bottom) mounting in the catheter.

IV. CONCLUSIONS

We have presented the most recent developments in RADFET (MOSFET dosimeter) use for quality assurance of radiation cancer treatments. Good commercial results for already marketed OneDose and excellent commercial prospects for implantable DVS and intracavitary array imply that the RADFETs are the technology of choice for novel dosimetric applications in clinical environments.

ACKNOWLEDGEMENT

This work was supported in part under the EU grant IST-2001-35243 (INVORAD).

REFERENCES

[1] A. Holmes-Siedle and L. Adams, "RADFET: A review of the use of metal-oxide-semiconductor deveices as integrating dosimeters", *Radiat. Phys. Chem.*, vol. 28, pp. 235-244, 1986. (and references therein)

[2] D.J. Gladstone, X.Q Lu, J.L. Humm, H.F. Bowman, and L.M. Chin, "A miniature MOSFET radiation dosimeter probe", *Med. Phys.*, vol. 21, no. 11, pp. 1721–1728, 1993.

[3] Sicel Tecnologies web page:ation: http://www.siceltech.com

[4] R.A. Price, C. Benson, M.J. Joyce, and K. Rodgers, "Development of a RadFET Linear Array for Intracavitary in vivo Dosimetry During External Beam Radiotherapy and Brachytherapy", *IEEE Trans. Nucl. Sci.*, vol. 51, no. 4, pp. 1420-1426, 2004.

[5] C. Gallagher, C. Kelleher, and P.J. Hughes, "Direct Write Technologies for Free Space Optical Interconnects", in *Proc. 15th European Microelectronics & Packaging Conference (EMPC 2005)*, Brugge, 2005, pp. 338-343.

2006 25th International Conference on Microelectronics

Effect of the Metal Electrode on the Characteristics of Ta_2O_5 Capacitors for DRAM Applications

E. Atanassova, D. Spassov, A. Paskaleva

Abstract - The effect of various electrodes (Al, W, TiN) deposited by evaporation (Al) and sputtering (W, TiN) on the electrical characteristics of thermal Ta_2O_5 capacitors has been investigated. The leakage currents, breakdown fields, mechanism of conductivity and dielectric constant are discussed in the terms of possible reactions between Ta_2O_5 and electrode material as well as electrode deposition process-induced defects acting as electrically active centers. During deposition of TiN and Al a reaction that worsens the properties of Ta_2O_5 occurs while there is not an indication for detectable reduction of Ta_2O_5 when top electrode is W. The sputtered W top electrode is a good candidate as a top electrode of storage capacitors in DRAMs, but sputtering technique is less suitable for deposition of TiN due to the introduction of radiation defects causing deterioration of leakage current.

I. INTRODUCTION

With the real involvement of high-*k* dielectrics into DRAMs fabrication it is required to switch the gate electrode from poly-Si to metal due to incompatibility of the most high-*k* materials with poly-Si [1,2]. An important characteristic of a potential electrode material is its chemical compatibility with Ta_2O_5, and a requirement of this compatibility is the metal's inability to reduce Ta_2O_5. Recently [3,4], we have reported that amorphous and stoichiometric Ta_2O_5 thin films can be successfully obtained by thermal oxidation of deposited Ta film on Si. The $Al/Ta_2O_5/Si$ capacitors exhibit excellent dielectric and electrical characteristics: bulk dielectric constant of 32-35; oxide charge of ~ 10^{10} cm^{-2} and leakage current less than 10^{-8} A/cm^2 at 1 MV/cm applied field. The purpose of this work is to study the influence of the type of metal gate electrode (Al, W, TiN) on the dielectric and electrical properties of thermally grown Ta_2O_5 on Si. A combination of Capacitance-Voltage (C-V) and Current-Voltage (I-V) curves are used to characterize electrical behavior of the capacitors.

II. EXPERIMENTAL PROCEDURE

Ta_2O_5 films of a thickness d ranging from 15 to 35 nm were grown on p (100) 15 Ωcm Si wafers by thermal oxidation of rf sputtered Ta film on Si. The oxidation temperature is low enough (873 K) so that the oxidation of

E. Atanassova, D. Spassov, A. Paskaelva are with the Inst. Sol. St. Phys., Bulg. Acad. Sci,. Tzarigradsko Chaussee 72, 1784 Sofia, Bulgaria, E-mail: elenada@issp.bas.bg

the substrate is negligible and the formation of tantalum silicide is prevented. Electrical measurements were performed on a set of MOS structures with three top electrodes (Al, W, TiN). Al electrode was evaporated; W and TiN gates were deposited by rf sputtering. MOS capacitors with various gate areas, S, (0.1-2.5 $\times 10^{-3}$ cm^2) were defined by photolithography. C-V and I-V measurements were carried out to determine the permittivity, electrical and insulating properties of the films. The oxide charge Q_f was evaluated from hf C-V curves. The dielectric constant ε_{eff} was determined from the capacitance C_0 in accumulation using the ellipsometrically measured value of d.

II. RESULTS AND DISCUSSION

C-V curves of 15 nm Ta_2O_5 capacitors are given as illustration in Fig. 1. ε_{eff} as a function of d is shown inset in the fig. The values are ranging from 5.7 to 9.5 for d = 15 nm and from 16 to 20, for d = 35 nm, with varying the gate

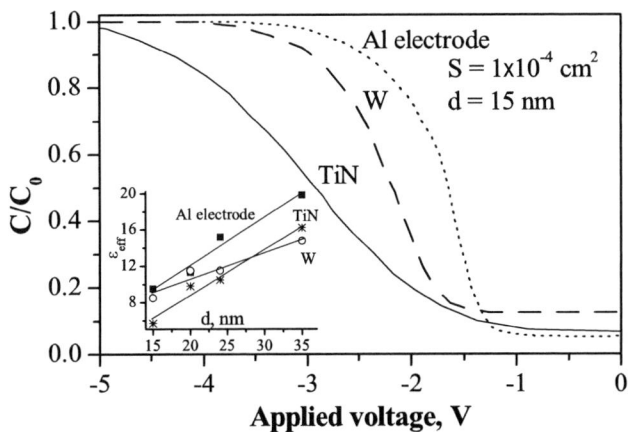

Fig 1. C-V curves of Al-, W-, and TiN-gate capacitors.

material. Consistently ε_{eff} corresponding to TiN-electroded capacitors is lower. The increase of ε_{eff} with d means an increase of film density for thicker films and as a result a reduction in leakage current for them is expected. Refractive index slightly increases from 2.05 to 2.1 with varying d from 15 to 35 nm. The thickness of the interfacial SiO_2-rich layer, at the technological conditions used here, is 2 nm as determined by TEM. According to the XPS and TEM analyses [3,5,6] this layer is SiO_2-rich one with a small quantity of Si_2O. This layer dominates ε_{eff} for very thin stacked films causing it to decrease. The dielectric

1-4244-0116-X/06/$20.00 ©2006 IEEE   581

constant ε_t values of the bulk Ta_2O_5, assuming 2 nm SiO_2 at the interface with Si, ranging from 12 to 26; 10 to 18 and 6 to 20 are found for the layers with electrode of Al, W and TiN, respectively (Table 1). A number of sources of errors in these calculations should be kept in mind, but a variation

TABLE I

VALUES OF ε_{eff} AND ε_t EXTRACTED FROM 1 MHZ C-V CURVES.

d, nm	electrode	$\varepsilon_{eff}/\varepsilon_t$, ($d_s$ = 2 nm)		
		Al	W	TiN
15		9.5/12.2	8.5/10.4	5.7/6.1
20		11.3/14.3	11.5/14.5	9.8/11.8
24		15.2/20.6	11.5/14.0	10.5/12.4
35		19.8/26.3	14.8/17.8	16.2/20

in the permittivity for the capacitors with different gate electrodes is definitely observed. The highest values of ε_t correspond to Al-electroded capacitors independently of d. The effect of the top electrode on ε_{eff} is attributed to a reaction between the gate and Ta_2O_5, and/or gate deposition process induced defects resulting to a change of ε_{eff}. Q_f remains high for all samples ($\sim$9-20x10^{11} cm^{-2}), independently of both the gate material and d. Representative data for W-gate are summarized in Table 2. The high density of oxide charge (typical of as-grown layers), presumably masks eventual weak dependence of Q_f on the electrode material including various compositional structure of the interfacial layer between top electrode and

TABLE II

DIELECTRIC PARAMETERS FOR W-ELECTRODED CAPACITORS, (φ_{ms} = - 0.42 eV, S = 6.25x10^{-4} cm^2).

d, nm	V_{fb}, V	Q_f x 10^{11}, cm^{-2}	ΔV_{fb}, V	Q_{sl} x 10^{11}, cm^{-2}
15	- 1.8	16	0.1 ÷ 0.15	1.7 ÷ 2.5
20	- 1.1	11	0.1 ÷ 0.15	1.8 ÷ 2.7
24	- 1.6	17	0.1	1.6
35	- 1.8	20	0.05	0.7

Ta_2O_5. This means also that the high Q_f is rather a feature of the oxidation than the electrode deposition–induced traps in Ta_2O_5. Both ε_{eff} and Q_f are virtually independent of the gate area, for all electrodes studied, suggesting as a first approximation that the Ta_2O_5 is homogeneous within the area range 1x10^{-4} - 2.5x10^{-3} cm^2. Although no direct measurements of the interface state density, D_{it}, have been done the comparison of the C-V curves to the ideal ones shows the midgap D_{it} of about 2.5; 5.5; $\sim$ 9 x10^{10} cm^{-2}eV^{-1}, for Al, W and TiN electrode, respectively. The curves of TiN-gate capacitor is strongly stretched-out along the voltage axis implying a rapid increase of D_{it} from the midgap to the band edges. Juxtaposing Q_f and D_{it} data, one can see that the damage introduced during TiN electrode

deposition is basically in the form of interface states; actually oxide charge is not affected by this process; the created defects are rather in the interfacial region than in the bulk of the Ta_2O_5.

C-V hysteresis is small (data for W-electrode capacitors are only given in Table 2) but detectable, 0.05 - 0.15 V, regardless of the top electrode. The density of slow states Q_{sl} as evaluated from the hysteresis is in the range of $\sim$ 0.7-3x10^{11} cm^{-2}. The source of slow states is thought to originate from the non-perfect Ta-O and Si-O bonds as well as Ta- and Si-suboxides located near the interface with Si [3,5]. Generally the presence of oxygen vacancies and a number of suboxides in the interfacial region is a phenomenon of the system Ta_2O_5/Si itself and is not due to electrode deposition effects.

Fig. 2 gives a comparison of leakage current data for capacitors with different electrodes.

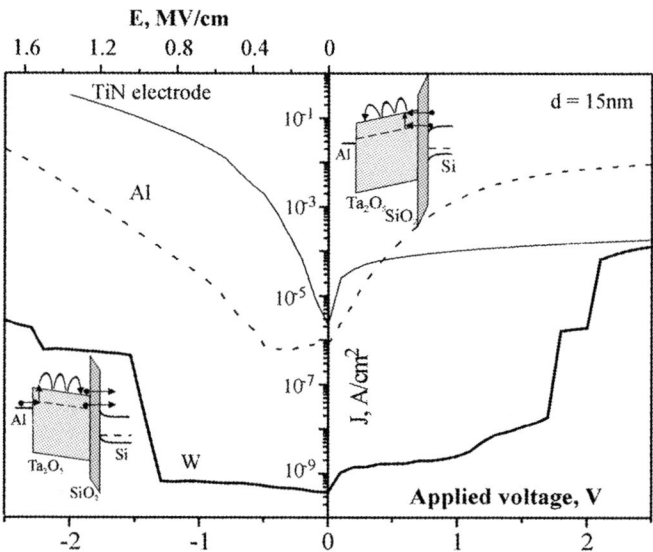

Fig. 2 J-V curves of capacitors with different gates

The current level under negative bias is much higher for the samples with TiN and Al gates as compared with W ones, indicating that both interfaces TiN/Ta_2O_5 and Al/Ta_2O_5 have a high level of defect density most likely due to the deposited gate electrodes. The capacitors with W gates yield 6 - 8 orders of magnitude lower leakage current density than TiN gate; respectively, these values are 5 orders lower as compared to Al electrode, at V = -1V. Reactions between Ta_2O_5 and both Al and TiN electrodes most likely have occurred resulting to incorporation of electrically active centers in the Ta_2O_5. The natural explanation for Al-electroded capacitors is that the Al reacts with Ta_2O_5 to form a thin Al_2O_3 (or Al suboxide), which may produce unwanted traps at the top electrode interface . The formation of intermixed TiO_x layer can be assumed for TiN gate process. A certain discrepancy exists between the data of C-V and I-V measurements, namely: Al-electroded capacitors have the best C-V curves; on the contrary, W electrodes consistently show the best I-V curves. Since the C-V curves reflect the characteristics of the interface at Si, while the I-V curves feel the gate interface region, the results permit us to attribute this behavior again to radiation induced defects in the thin surface layer at the metal/Ta_2O_5 interface during rf deposition of the electrodes. Such kind of damage is missing during the evaporation of Al. The deposition-created damage is stronger in the case of TiN than for W one. However, this is essential for the permittivity only of the thinner layer, (Table 1); the difference between W and TiN gates caused by deposition induced damage is obscured, for ~ 25-35 nm Ta_2O_5, and obviously subtle factors begin to act. The behavior of the curves (forward bias) corresponding to W electrodes does not enable to invoke any of the commonly observed conduction mechanisms. We will focus only to the curves of capacitors with Al and TiN gates. Among the possible mechanisms responsible for the leakage current, the first to be invoked are Poole-Frenkel (PF) and Schottky ones and they are the most frequently observed (combined with Fowler-Nordheim or direct tunneling through the interfacial layer at Si substrate) in high-k capacitors. A nearly normal PF effect (r = 1.1; dynamic dielectric constant, k_r = 4.8) dominates the current at applied field of ~ 0.3 to 1.7 MV/cm and the relation with the refractive index is accurate for the samples with Al electrode (Fig. 3). The current is constant in the low field region (up to 0.3 MV/cm). A little modified PF effect can be invoked also with r = 1.2, at E ~ 0.06 - 0.5 MV/cm for TiN gate. The type and the density of traps are hanging (but weakly) with both the electrode material and the deposition conditions, causing small variation in the degree of compensation. The slope of 2.5 for TiN electrode at medium and high fields is not consistent with the PF process. The current mechanism is Schottky type only in the range 0.5-1.4 MV/cm; k_r, (3.8) is consistent with the optical dielectric constant, indicating that the mechanism is via Schottky emission (Fig. 3). The conduction mechanism in Al-capacitors is governed rather by modified PF effect than Schottky one. Therefore both PF process (modified version need to be considered) and Schottky emission are dominant conduction mechanisms in capacitors with Al and TiN electrodes. The absence of a well pronounced effect of the gate material in accordance with φ_m indicates more or less that the impact of the bulk-limited mechanism is stronger. For W-gate case neither of these two processes can be directly invoked to explain the leakage current behavior. The current is nearly constant up to the appearance of breakdown events. This result strongly suggests that *by appropriate choice of both electrode material and its deposition process a transition from one conduction process to another occurs*. The mechanism is critically related to the density of charged traps in the films and these traps can mask the effect of φ_m, i.e. the electrode work function effect is essentially screened out by the traps. We do not exclude the possibility that as a result of deposition process the TiN work function differs from that used in the present calculations which is taken from the

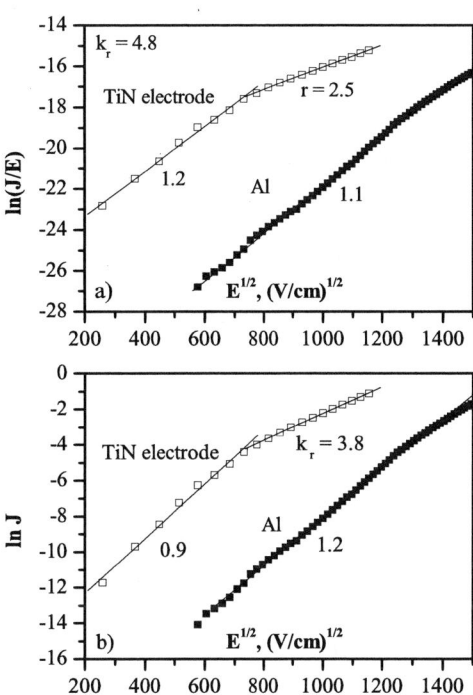

Fig. 3 a) Poole-Frenkel and b) Schottky plot of I-V characteristics for TiN and Al-electroded capacitors.

literature. We have not checked for this by using an independent method for determining the TiN work function, but the results imply that such a hypothesis is quite possible. We propose that the process starts with electron injection from Al, (0.3-1.7 MV/cm) or TiN, (0.06-0.5 MV/cm) into traps in Ta_2O_5 close to the gate interface and further the conductivity is governed by PF mechanism, i.e. the leakage current is caused by trap-assisted transport from cathode to anode through the bulk traps. The nature of these traps is rather structural non-perfections and radiation induced defects caused by the electrode deposition than intrinsic Ta_2O_5 traps due to the oxidation. This is supported by the behavior of W-gate capacitors, which do not exhibit PF conductivity. Schottky emission dominates the current in TiN-capacitors at 0.5-1.4 MV/cm, (the space charge limited conduction mechanism is not consistent with the data). This change of the mechanism from bulk (low and middle fields) to electrode limited (high fields) is strange and at a first glance it is not consistent with the indication for a noticeable generation of defects during TiN deposition. This behavior can be attributed to the quality of the top electrode-interface region itself (in one case electrons tunnel from the gate into the traps located close to/in this region and in another case they overcome the gate barrier by Schottky emission) defined by the density and energy distribution of the traps created during gate deposition. If the traps concentration close to the electrode is small the electrons could not tunnel through them in Ta_2O_5 conduction band. In this case Schottky emission is more relevant at high fields; especially for TiN case this means that the rf sputter-induced traps are not basically in the form of traps at the electrode but rather of interface states at Si, with high density, as indicated by the C-V data. At reversed bias ($+V_g$), when the electrons are injected from Si the curves saturate at ~ 10^{-4} A/cm^2, with exception of Al-gate capacitors having saturation level of 10^{-2} A/cm^2. There is virtually no change in the magnitude of conductivity with the change of the polarity of the applied voltage for W capacitors, up to the beginning of soft breakdown events which indicates that the conduction process defining these breakdowns is rather bulk than electrode limited. Anyhow, the behavior of the curves is closely related to the parameters of the interfacial layer at Si, and its breakdown. The obvious dependence of the curves on the gate means that *the top electrode deposition process initiates reactions which affect not only the region near the gate but also the layer at Si* giving by this way contribution to the interface defects. Since the thickness of SiO$_2$ is the same for the all capacitors, the different level of the current at reverse bias should be attributed to traps created in the interface layer at Si during the gate deposition. In this context, the curves in the case of W-gate reflect the intrinsic properties of this interfacial region defined by the oxidation process itself, with a negligible (or no) effect of electrode deposition process.

The leakage current characteristics do not show thickness dependence in the range studied. This can be attributed to the effect of the constant thickness lower-*k* interfacial layer at the Si. Due to its higher resistance as compared with the overlying Ta$_2$O$_5$, the resistance of the stack is dominated by the interfacial layer and is constant. Thus the small variations of Ta$_2$O$_5$ thickness do not have a detectable effect on leakage current.

III. Conclusion

Based on the results one would tend to conclude that the unavoidable *introduction of metal electrodes* in advanced generations of high-*k* dielectrics based DRAMs *introduces own set of manufacturing and capacitor quality challenges*, and requires the development of specific technological schemes. We have shown that some reduction of the Ta$_2$O$_5$ during deposition of the top electrode can occur. This reduction may be attributed both to the gate deposition technique and the chemical reactions between Ta$_2$O$_5$ and electrode material. The former one dominates in the case of TiN gate, (radiation induced defects in Ta$_2$O$_5$ during sputtering of TiN), and the second one is typical of Al-electroded capacitors, (Al reacts with Ta$_2$O$_5$ to form Al-oxides). The final result in both cases is a damaged interface Ta$_2$O$_5$/top electrode which affects electrical parameters of the capacitors and mainly leakage current. The value of the oxide charge is slightly influenced by the gate and seems to be an attribute to the oxidation process. The induced defects act as electrically active centers causing much higher current for Al and TiN-gates

capacitors as compared with W ones, (respectively 5 to 7 orders of magnitude higher for forward polarity). The permittivity while showing some degree of correlation with the thickness is not sufficiently sensitive to reflect the induced damage. The W sputtering does not involve defects adding to the intrinsic oxidation traps, (or their amount is too low to be detected by electrical analysis). One important question is why the manifestation of the radiation defects is different in TiN- and W-capacitors. The data presented do not directly address this question as well as defect creating mechanisms. One explanation of the phenomenon could be the bi-component nature of TiN. The results, however, imply that the sputtering technique is a beneficial technology for W deposition, ensuring a stable contact to Ta$_2$O$_5$ and tolerable value of leakage current, while radiation induced damage of Ta$_2$O$_5$ during deposition of TiN make this method less favorable in the case of TiN gate. Although some reaction between the Al electrode and Ta$_2$O$_5$ obviously occurs the resulting electrical properties are still acceptable. The generated defects are also responsible for the strong dependence of the conduction mechanism on the electrode material, and is the critical factor controlling the conductivity. *Their effect can be strong enough to mask the effect of the electrode work function.* Therefore the more general conclusion is that *the choice not only of the gate material but the gate deposition technique remains a critical concern for high-k oxides.*

Acknowledgement

This work was supported by Bulgarian National Science Foundation under contract F1508.

References

[1] Intern. Techn. Roadmap for Semicond., http://public.itrs.net/files/2003ITRS

[2] G. D. Wilk, R. M. Wallance, J. M. Anthony, "High-*k* Gate Dielectrics: Current Status and Materials Properties Considerations", *Journal of Applied Physics*, 2001, vol. 89, pp. 5242-5275.

[3] E. Atanassova, T. Dimitrova, "Thin Ta$_2$O$_5$ layers on Si as an alternative to SiO$_2$ for high density DRAM applications", in *Handbook of Surfaces and Interfaces of Materials*, H. S. Nalva, Ed., vol. 4, pp. 439-479, San Diego, USA: Acad. Press: 2001.

[4] E. Atanassova, D. Spassov, "Hydrogen Annealing Effect on the Properties of Thermal Ta$_2$O$_5$ on Si ", *Microelectronics Journal*, 1999, vol. 30, pp. 265-274.

[5] E. Atanassova, D. Spassov, A. Paskaleva, J. Koprinarova, M. Georgieva, "Influence of Oxidation Temperature on the Microstructure and Electrical Properties of Ta$_2$O$_5$ on Si", *Microelectronics Journal*, 2002, vol. 33, pp. 907-920.

[6] E. Atanassova, M. Kalitzova, G. Zollo, A. Paskaleva, A. Peeva, M. Georgieva, G. Vitali, "High temperature-induced crystallization in Ta$_2$O$_5$ and its influence on the electrical properties", *Thin Solid Films*, 2003, vol.426 pp. 191-199.

2006 25th International Conference on Microelectronics

Reliability Properties of Ta₂O₅ Films Grown on N₂O Plasma Nitrided Silicon

N. Novkovski and E. Atanassova

Abstract – Plasma nitridation of the Si substrate prior to the thermal growth of Ta₂O₅ improves significantly the reliability properties of the insulating film. Stress induced leakage currents are reduced for several orders of magnitude. Based on the physical model used in the analysis, the improvement is explained by the impeded consumption of the oxynitride interfacial layer, due to the incorporation of nitrogen atoms at the Si interface, making it more resistant against the stress degradation. Plasma nitridation improves the dielectric properties by increasing the dielectric constant of the interfacial layer, thus reducing the equivalent oxide thickness.

I. INTRODUCTION

Ta₂O₅ films grown on silicon are extensively studied as high-*k* dielectric for DRAM capacitors, because of their excellent dielectric properties [1-3]. It was found that the nitridation of the substrate in N₂O or NH₃ improves the dielectric and leakage properties of the films [4]. Stress-Induced Leakage Currents (SILC) in high-*k* dielectrics present an increasingly important problem from the reliability point of view, compared to the hard or the soft breakdown. Previously we showed that the substrate plasma nitridation in N₂O prior to the Ta₂O₅ growth improves the dielectric properties of the film [5].

In the present work we study the SILC of these films after constant current stress by the method developed in [6] explaining the origin of SILC in Ta₂O₅ films grown on bare silicon, based on a comprehensive model for *I-V* characteristics. The impact of the stress on the *C-V* characteristics as well as on the hysteresis is also studied.

II. EXPERIMENTAL

Chemically cleaned p-type (100) 15 Ω·cm Si wafers were nitrided at room temperature for 10 s in soft plasma. The films were obtained by RF-sputtering of a Ta target in Ar atmosphere followed by thermal oxidation in dry O₂ at 600°C then annealed in H₂ at 450°C for 1 h. The film thickness and the refractive index were measured by ellipsometry (λ = 632.8 nm). The film thickness was d_{TP} = 30 nm and the refractive index was *n* = 2.1.

N. Novkovski is with the Institute of Physics, Faculty of Natural Sciences and Mathematics, Gazibaba b.b., 1000 Skopje, Macedo-nia, e-mail: nenad@iunona.pmf.ukim.edu.mk

E. Atanassova is with the Institute of Solid State Physics, Bulgarian Academy of Sciences, Blvd "Tzarigradsko Chausse" 72, 1784 Sofia, Bulgaria, e-mail: elenada@issp.bas.bg

I-V characteristics as well as high frequency (50; 100 kHz and 1 MHz) and quasi-static *C-V* characteristics were measured on Al gate MOS capacitors with areas 2.5×10^{-3} cm², both for fresh and for stressed samples at constant current density *J* = 10 mA/cm², and different stress times *t*.

III. THEORY

The conduction mechanisms that are considered are:
1. For the interfacial SiOₓNy layer, direct tunneling through a trapezoidal barrier or Fowler-Nordheim tunneling trough a triangular barrier, depending on the electric field in the layer (E_{IL}). Tunneling current can be created by the electrons or the holes. The barrier for the tunneling of holes is substantially higher then the barrier for electrons. Different carriers from the silicon substrate produce this current: electrons in the case of gate positively biased and holes in the case of gate negatively biased. Hence, the current depends upon the gate polarity.
2. For tantalum pentoxide, Poole–Frenkel mechanism, that is bulk-limited, thus independent on the polarity.

Direct tunneling current through the interfacial SiOₓNy layer (J_{IL}) is given by the following expression:

$$J_{IL} = \frac{q^2}{8\pi h\Phi}E_{IL}^2 \exp\left(-\frac{8\pi\sqrt{2m^*q\Phi^3}}{3hE_{IL}}\left(1-\left(1-\frac{d_{IL}}{\Phi}E_{IL}\right)^{3/2}\right)\right) \quad (1)$$

and Fowler-Nordheim tunneling by:

$$J_{IL} = \frac{q^2}{8\pi h\Phi}E_{IL}^2 \exp\left(-\frac{8\pi\sqrt{2m^*q\Phi^3}}{3hE_{IL}}\right), \quad (2)$$

where *q* is the electron charge, *h* is the Planck's constant, m^* is the effective tunneling mass of carriers in SiOₓNy, d_{IL} is the thickness of SiOₓNy layer, Φ is the tunneling barrier height, and E_{IL} is the electric filed in SiOₓNy layer.

The voltage drop on the SiOₓNy layer (V_{IL}) is:

$$V_{IL} = E_{IL}d_{IL} . \quad (3)$$

The current density due to the Poole-Frenkel effect in the Ta₂O₅ layer (J_{TP}) is given by the following expression:

$$J_{TP} = \sigma_{TP}E_{TP}\exp\left(\frac{1}{kT}\sqrt{\frac{q^3}{\pi\varepsilon_0 K_T}}\sqrt{E_{TP}}\right), \quad (4)$$

where E_{TP} is the electric field in the Ta₂O₅ layer, σ_{TP} is temperature-dependent defect-related constant having dimensions of conductivity, *k* is the Boltzmann constant, ε_0 is the dielectric constant of vacuum and $K_T = n^2$ is the optical frequency dielectric constant (n is the refractive index) of Ta_2O_5.

The voltage drop on the layer (V_{TP}) is given by:

$$V_{TP} = E_{TP}d_{TP}, \qquad (5)$$

where d_{TP} is the thickness of the Ta_2O_5 layer.

The two quantities that are computed simultaneously here are the oxide voltage:

$$V_{ox} = V_{TP} + V_{IL} = d_{TP}E_{TP} + d_{IL}E_{IL}, \qquad (6)$$

and the current density in steady state

$$J = J_{TP} = J_{IL}, \qquad (7)$$

Current density $J = J_{on}$ was computed for given field E_{IL} in the silicon oxynitride, then the field in Ta_2O_5 layer was computed as inverse function of current of $J_{TP} = J$ and the oxide voltage determined with the use of the expression (5). Following typical values were used in computations: $m_e^* = 0.61\ m_0$ – effective electron mass in SiO_2 (m_0 denotes the mass of free electron), $m_h^* = 0.51\ m_0$ – effective hole mass in SiO_2 and $K_T = n^2 = 4.41$.

The voltage on the insulating stack layer (V_{ox}) was calculated by using relations involving the flatband voltage (V_{fb}), the gate voltage (V_g) and voltage drop in the semiconductor (φ_s) [5]

$$V_{ox} = V_g - V_{fb} - \varphi_s, \qquad (8)$$

IV. RESULTS AND DISCUSSION

The C-V characteristics were measured only in the range from –2 V to 2 V, (at higher voltages the leakage currents become high and could affect the measurement). In order to determine the value of the capacitance in accumulation of the high-k dielectric, the method of extraction described in [7, 8] was used. From the high-frequency and quasi-static C-V characteristics of the fresh samples the following parameters were determined: the equivalent oxide thickness (EOT) $d_{eq} = 8.2$ nm, the oxide charge $Q_{ox} = 1.1 \times 10^{11}$ cm^{-2} and the midgap interface state density $D_{itm} = 6 \times 10^{11}$ eV^{-1}cm^{-2}. The effective dielectric constant of the film is $\varepsilon_{ef} = 14.3$.

The obtained I-V characteristics for fresh and stressed samples are shown in Fig. 1. For the average fields (V_{ox}/d_{eq}) higher than 0.2 MV/cm (which corresponds to $V_g > 1.8$ V), the increase of the leakage currents after injecting a charge of about 10 C/cm^2 is lower than one order of magnitude, which is much less that in the case of bare substrates [6], when this increase attains 5-6 orders of magnitude.

Usually, Ta_2O_5-based capacitors exhibit relatively large hysteresis in the C-V curves when reversing the sweeping direction. In the case studied here, no significant hysteresis is observed for the fresh samples, while for longer stress times ($t = 990$ s) a relatively small value of –40 mV is obtained (Fig. 2). This value corresponds to a density of slow states, $Q_s = 8 \times 10^{10}$ cm^{-2}. A frequency dependence of the capacitance is also a typical feature of the tantalum pentoxide. No significant frequency dependence of the C-V characteristics was observed here

up to 1 MHz, as is shown for the case after the stress for $t = 990$ s (Fig. 3).

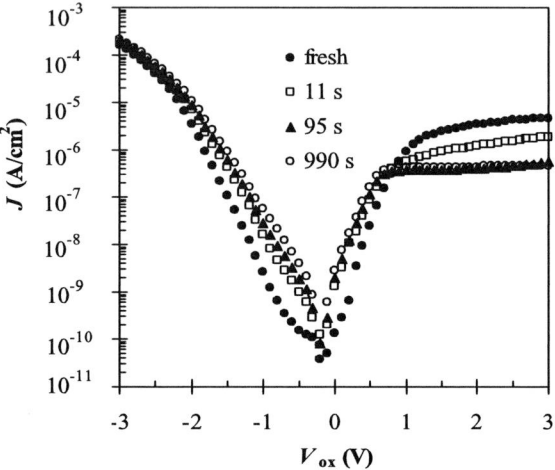

Fig. 1. I-V characteristics of the Al-Ta_2O_5/SiOxNy-Si structures for fresh and stressed samples.

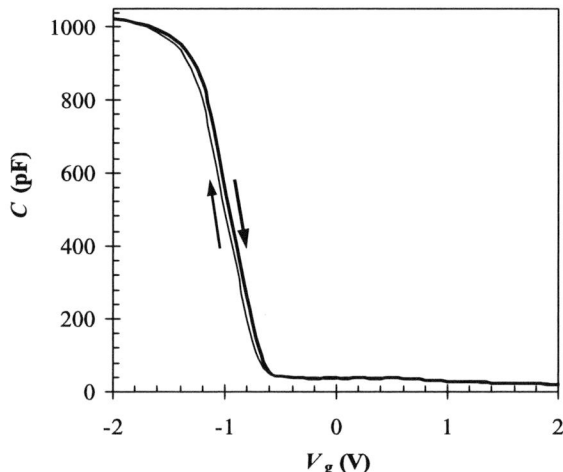

Fig. 2. C-V curves of the capacitors stressed for 990 s. (Voltage is swept from the left to the right then in the opposite direction)

TABLE I: PARAMETERS EXTRACTED FROM THE I-V CURVES.

t (s)	Φ_e (eV)	Φ_h (eV)	d_{on} (nm)	σ_{hc} (Ω^{-1}cm^{-1})	σ_{tp} (Ω^{-1}cm^{-1})
0	1.50	2.25	3.15	5.0×10^{16}	7.4×10^{12}
11	1.50	2.25	3.02	4.5×10^{15}	7.4×10^{12}
95	1.50	2.25	2.96	9.5×10^{15}	7.4×10^{12}
990	1.50	2.25	2.93	3.0×10^{14}	7.4×10^{12}

A detailed analysis of the experimental I-V characteristics was done by using the above theoretical model. When fitting the theoretical to the experimental data, parameters describing each of the layers in the stack were obtained (Table I).

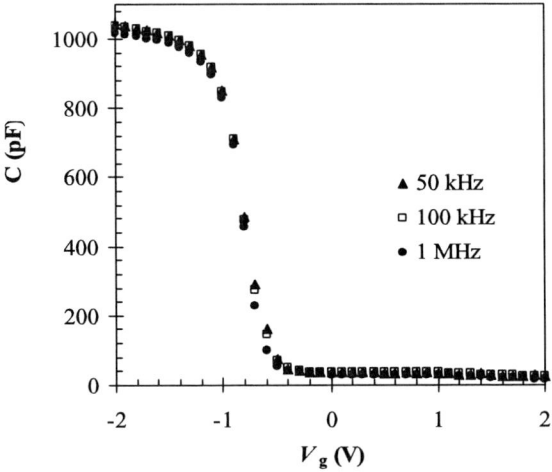

Fig. 3. *C-V* characteristics of stressed for 990 s capacitors, as measured at three different frequencies; sweeping direction is from the left to the right.

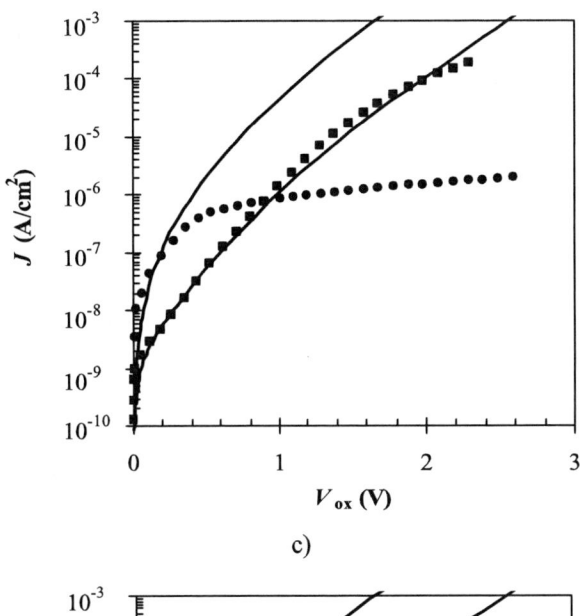

c)

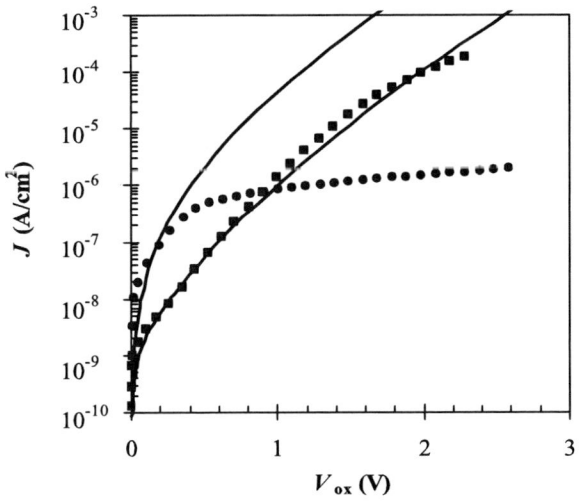

a)

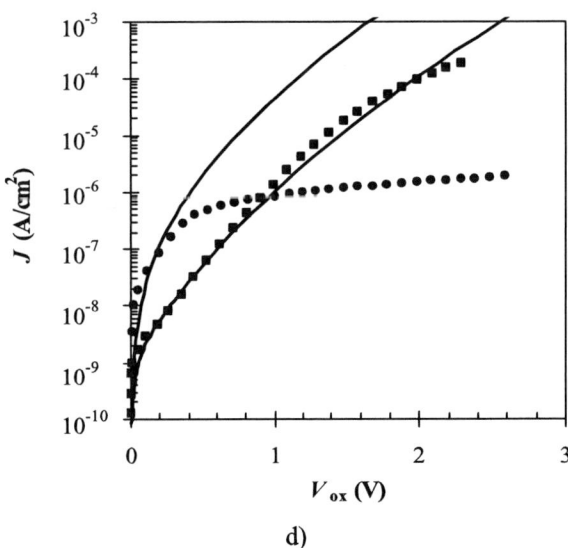

d)

Fig. 4. Experimentally obtained leakage current densities (close circles – positive gate bias, close squares – negative gate bias) versus oxide voltage of fresh (a) and stressed for 11 s (b), 95 s (c) and 990 s (d) samples, as one with the best fits (solid lines).

The experimental results as one with the best fits obtained based on the used physical model are displayed in Fig. 4. Very good agreement between the theoretical and experimental results in the entire measurement region is obtained, justifying the use of the proposed model in the analysis. Saturated part for the positive gate is due to the strong inversion and is not described by this model. This part is similar to the reverse bias part of a p-n diode characteristic. The saturation level depends upon the measurement temperature and the illumination intensity.

The consumption of the oxynitride interfacial layer (decrease of the thickness d_{IL}), as defined in [6], is much slower in the case of films grown on plasma nitrided than on bare substrates (Fig. 5). In both cases the thickness d_{IL} decreases exponentially with the stress time. For ultrathin

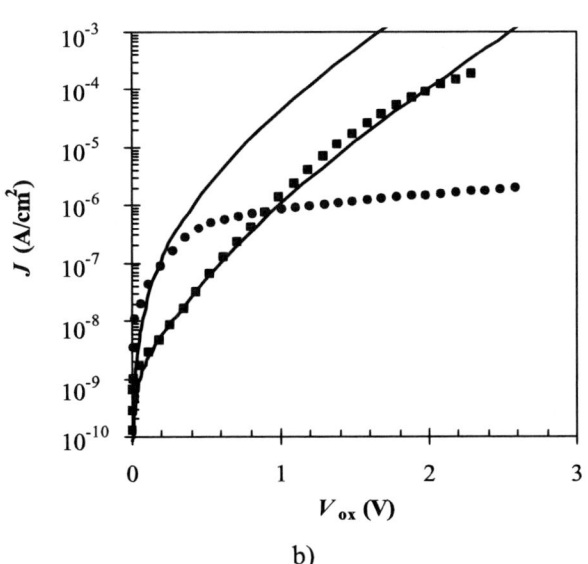

b)

high-k dielectrics the main failure mechanism is no more the destructive breakdown (hard or soft), but the progressive breakdown (PBD) [9] and SILC [10]. The main parameter determining the SILC in Ta_2O_5/SiO_xN_y stack layers is the interfacial layer thickness, d_{IL}. Therefore, the inverse interfacial layer consumption rate (the inverse of the straight line slope in Fig. 5) could be taken as a measure of the high-k reliability, if the SILC is the dominant failure mechanism, which is of particular interest in the case of ultrathin high-k dielectrics. Smaller the slope is, higher is the time of operation at high field until reaching some critical value of the leakage current at given conditions.

Although the initial value of d_{IL} is higher for plasma nitrided than for bare substrates, the EOT in the former case is lower, (2.2 nm for nitrided vs. 2.6 nm for bare Si substrate), because the dielectric constant of the oxynitrdide is higher than that of SiO_2. Assuming the known value $\varepsilon_{TP} = 32$ for the bulk Ta_2O_5 [11], the dielectric constant of the oxynitride layer was found to be $\varepsilon_{IL} = 5.5$. Consequently, the short plasma nitridations results in lower equivalent thicknesses then in the case of Ta_2O_5 grown on bare substrates.

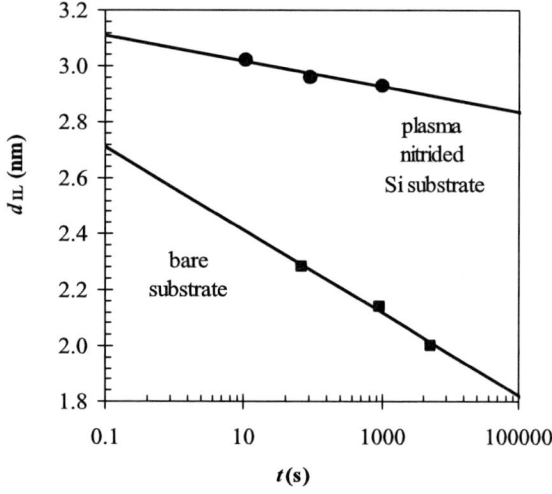

Fig. 5. Consumption of the interfacial layer (decrease of the effective interfacial layer thickness, d_{IL}) with the stress time (t) of thermally grown tantalum pentoxide on plasma nitrided Si (close circles). For a comparison, the results obtained on bare substrates [6] are shown (close squares).

The estimation of the interface state densities on stressed samples using the standard C-V methods did not result in reliable data, due to the significant leakage level at low voltages (between –1 V and +0.5 V). As is seen in Fig. 1, the leakage current in this region before the stress remains in the 10^{-10} A/cm^2 range, while after the stress it reaches the values of order of 10^{-8} A/cm^2, which drastically influence the results of the quasi-static C-V measurements. Alternative methods for determination of interface state densities are required for stressed samples.

V. CONCLUSIONS

The results show that the substrate plasma nitridation appreciably improves the reliability properties of the Ta_2O_5 films on silicon, in addition to the improved dielectric properties of the fresh films. The nitridation reduces the SILC for several orders of magnitude. Although the interfacial layer thickness increases with the nitridation, the EOT of the film decreases because of the higher dielectric constant of the oxynitride interfacial layer. The reliability improvement is due to the incorporation of nitrogen atoms at the Si interface, making it more resistant against the stress degradation. For such relatively light nitridations the amount of N atoms in the oxynitride is very low, resulting in an increase of the dielectric constant, without significant lowering the tunneling barriers, thus decreasing the EOT without increasing the leakage currents.

REFERENCES

[1] A.P. Huang and P.K. Chu, J. Appl. Phys., 97, 114106 (2005)
[2] E. Atanassova, N. Novkovski, A. Paskaleva, M. Pecovska-Gjorgjevich, Solid St. Electron. 46, 1887 (2002)
[3] E. Deloffre, L. Montès, G. Ghibaudo, S. Bruyère, S. Blonkowski, S. Bécu, M. Gros-Jean, S. Crémer, Microelectron. Reliab. 45, 925 (2005)
[4] Y.-S. Lai, K.-J. Chen and J. S. Chen, J. Appl. Phys. 91, 6428 (2002)
[5] N. Novkovski and E. Atanassova, Appl. Phys. A 81, 1191-1195 (2005)
[6] N. Novkovski and E. Atanassova, Appl. Phys. Lett. 86, 152104 (2005)
[7] S. Kar, IEEE Trans. Electron Dev. 50, 2112 (2003)
[8] S. Kar, IEEE Trans. Electron Dev. 52, 1187 (2005)
[9] G. Ribes, J. Mitard, M. Denais, S. Bruyere, F. Monsieur, C. Parthasarathy, E. Vincent, G. Ghibaudo, IEEE Trans. Dev. Mat. Reliab. 5, 5 (2005)
[10] R. O'Connor, S. McDonnell1, G. Hughes, R. Degraeve and T. Kauerauf, Semicond. Sci. Technol. 20, 668 (2005)
[11] E. Atanassova and D. Spassov, Microel. J. 30, 265 (1999)

2006 25th International Conference on Microelectronics

Stress Induced Leakage Currents and Charge Trapping in Thin Zr- and Hf-Silicate Layers

A. Paskaleva, M. Lemberger, A.J. Bauer

Abstract - In this paper, reliability aspects of thin Zr- and Hf-silicate layers are addressed by analyzing the stress induced leakage current (SILC) and charge trapping during constant voltage stress (CVS) and constant current stress (CCS). Voltage polarity and temperature effects on the degradation of the layers are also studied. SILC is found to have a transient component and its recovery is explained by the trapping/detrapping of traps participating in Poole-Frenkel conduction.

I. INTRODUCTION

Higher dielectric constants and lower leakage currents are not the only requirements high-k gate dielectrics have to satisfy. While the knowledge of material and electrical properties has reached some stage of maturity there are still a number of further fundamental issues (e.g., dielectric and interface charges, charge trapping and degradation during operation time). All these issues have a direct relation to the reliability of the devices and should be carefully studied. A number of open questions arise from them. These concern the main degradation mode – is it the threshold voltage instability, or SILC, or time dependent dielectric breakdown (TDDB) [1-3]. Another key problem concerns the validity of conventional reliability methodologies and prediction models developed for SiO_2 [4,5]. As the electronic structure and bonding in high-k materials and SiO_2 are quite distinct resulting in differences in transport mechanisms and trapping behavior, it emerges that the physical mechanism of degradation may be significantly different. Hence, the methodology for its investigation may be also affected [6]. In this paper, the temperature, field and polarity dependence of SILC and charge trapping in thin Hf- and Zr-silicate layers are investigated and discussed.

II. EXPERIMENTAL PROCEDURE

Silicate films with similar stoichiometry of $M_{0.85}Si_{0.15}O_2$ (Me represents Zr or Hf) were deposited by MOCVD on p-type Si using single-source precursors $(M(acac)_2(OSi^iBuMe_2)_2)$. The synthesis, structure, and

A. Paskaleva is with the Inst. of Solid State Physics, Bulg. Acad. Sci. 72 Tzarigradsko Chaussee, 1784 Sofia, Bulgaria, e-mail: paskaleva@issp.bas.bg

M. Lemberger is with Chair of Electron Devices, Univ. Erlangen-Nuremberg, Cauerstrasse 6, 91058 Erlangen, Germany, e-mail: martin.lemberger@leb.eei.uni-erlangen.de

A. J. Bauer is with Fraunhofer Institute of Integrated Systems and Device Technology, Schottkystrasse 10, 91058 Erlangen, Germany, e-mail: anton.bauer@iisb.fraunhofer.de

characterization of precursors were described in [7]. The depositions were performed at 550 °C in Ar/O_2 ambient at a total pressure of 2.5 mbar followed by a rapid thermal annealing (RTA) for 10 s in O_2 at 900 °C. The physical thicknesses of Hf- and Zr-silicate were 10.5 and 18 nm (giving EOT of 3.3 and 4.5 nm), respectively. Finally, gate electrodes consisting of stacked 20 nm Ni/500 nm Al for Hf-silicate and 20 nm Ti/500 nm Al for Zr-silicate structures with an area of 2×10^{-4} cm^2 were evaporated.

To investigate SILC the I-V curves before and after constant voltage stress (CVS) were measured. For each sample, the measurements (CVS and I-V) were performed within seconds and at the same temperature to minimize slow trapping/detrapping effects. To clarify the origin of SILC both gate and substrate injection were performed and the SILC at both polarities were measured. All stresses and measurements in inversion were performed under illumination with a halogen lamp to create sufficient amount of minority carriers. Constant current stress (CCS) measurements were performed to study charge trapping in the layers, too.

III. RESULTS AND DISCUSSION

In Fig.1 SILC created under negative CVS in Hf-silicate layers is shown. Stressing at $V \le |{-5.5}|$ V (the inset of (Fig.1)) generates no significant SILC in accumulation. Only a slight shift of the I-V curves to more negative voltages is seen implying some electron trapping. However, stressing at -6 V generates SILC which is higher for the I-V measured in inversion. The relative increase of SILC defined as $\Delta J_G/J_{G0}=(J_G-J_{G0})/J_{G0}$ in accumulation is about 2 (just before breakdown (BD)) and in inversion it is

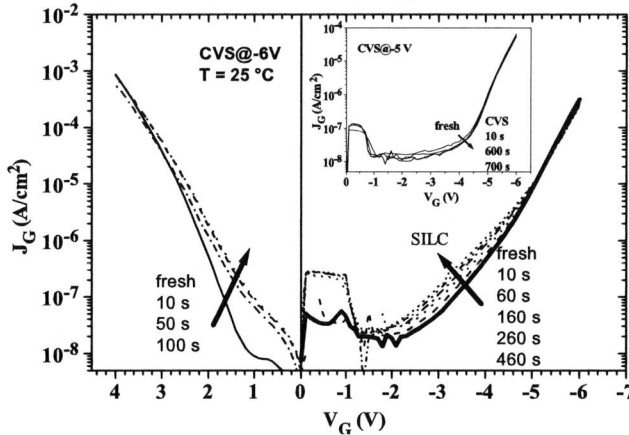

Fig.1. SILC in accumulation and inversion generated during negative CVS in Hf-silicate layers.

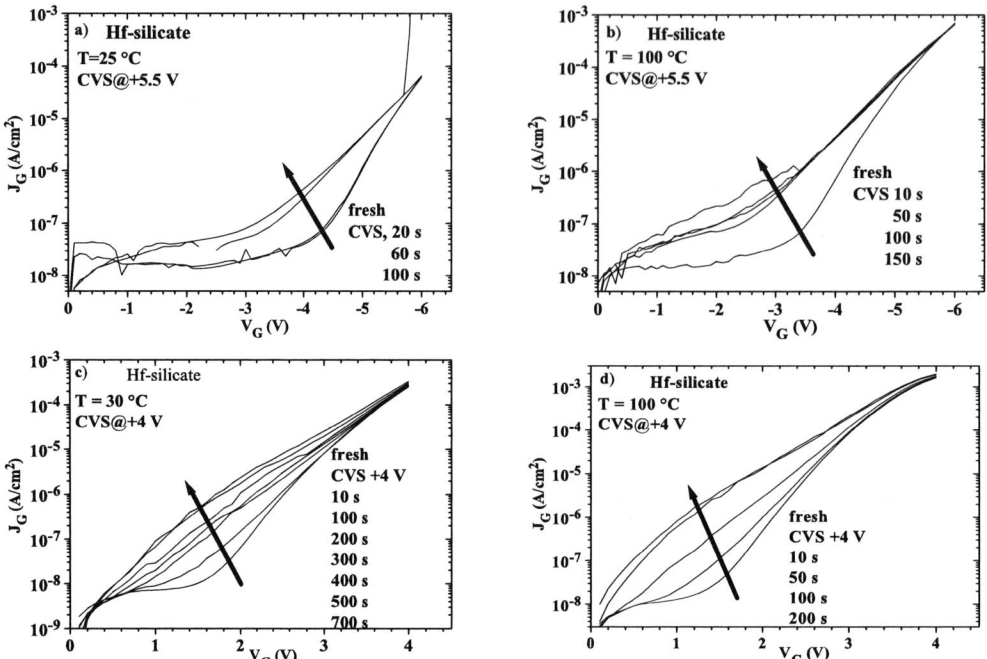

Fig. 2. Temperature dependence of SILC in Hf-silicate layers generated under positive CVS and measured at: a) accumulation, 25°C; b) accumulation, 100 °C; c) inversion, 30 °C; and d) inversion, 100 °C.

about 6-7. As the current is more sensitive to changes in the cathode electric field, this result implies that the damage causing SILC is more severe at the Hf-silicate/Si interface, (i.e., it has a stronger impact on I-V curves in inversion). Another implication of the results is a threshold between –5.5 and –6 V for the creation of SILC. The existence of this threshold can be related to generation of traps. So, it is concluded that at voltages lower than $|-5.5|$ V electron trapping in existing traps dominates, whereas for $V > |-5.5|$ V trap generation is the dominant mechanism. The charge to breakdown Q_{BD} obtained for all stress conditions in accumulation is less than 0.1-0.15 C/cm^2.

SILC increases mostly within the first 10 s of stress and after that it nearly saturates up to the occurrence of BD. $\Delta J_G/J_{G0}$ is slightly influenced by the temperature – it increases from about 9 at 25 °C to 13 at 100 °C. However, more severe damage is observed for I-V measured in inversion (Figs. 2c,d). $\Delta J_G/J_{G0}$ (CVS at +4 V) increases with time up to 40-50 at 30°C (Fig. 2c) and about 300 at 100 °C (Fig. 2d). Compared to the increase observed in accumulation of about 13 (at 100 °C) at even higher stress voltage (+5.5 V) it is seen that I-V in inversion is much more sensitive to the damage that causes SILC.

Fig. 3 shows the normalized SILC for Hf-silicate layers as a function of injected charge Q_{inj} for the conditions presented in Fig. 2 Two distinct regions are clearly visible – below and above Q_{inj} of 0.1 C/cm^2. $\Delta J_G/J_{G0}$ for $Q_{inj} < 0.1$ C/cm^2 (inset of Fig. 3) is up to about 16. There are several remarkable features to be mentioned. 1) SILC is only weakly temperature dependent. 2) $\Delta J_G/J_{G0}$

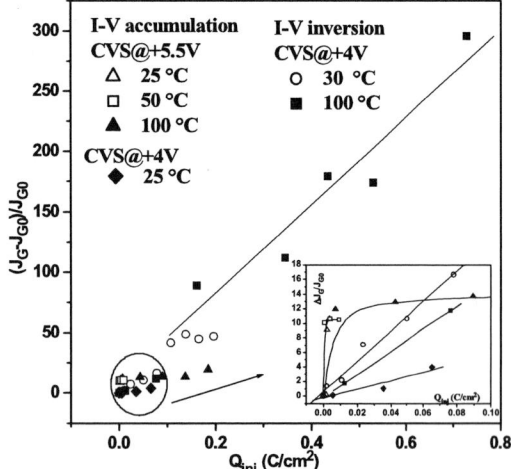

Fig. 3. $\Delta J_G/J_{G0}$ as a function of injected charge.

SILC created under positive CVS and measured in accumulation at two temperatures is shown in Figs. 2a,b.

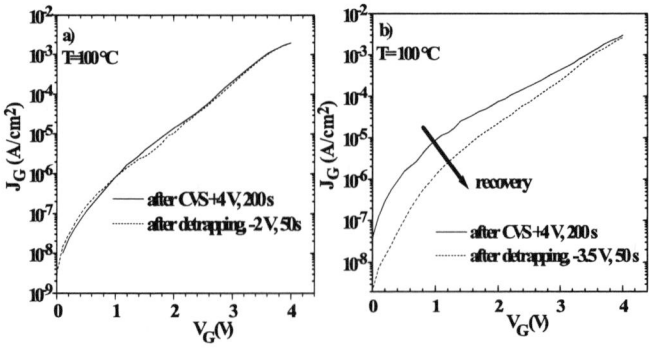

Fig. 4. SILC recovery after detrapping at: a) –2V; and b) –3.5 V.

590

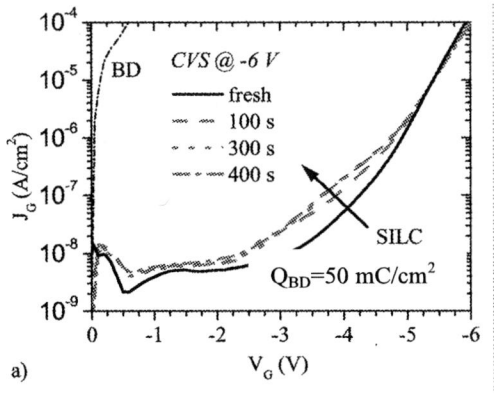 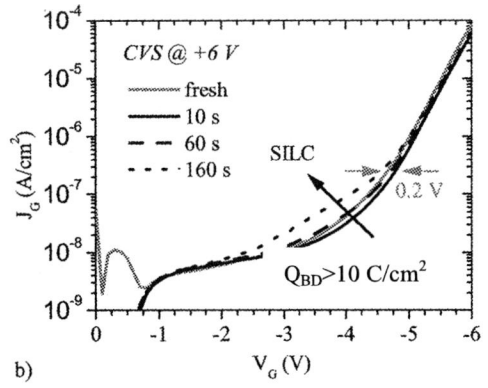

Fig. 5. I-V characteristics of Zr-silicate layer revealing SILC after: a) negative CVS; b) positive CVS.

under identical substrate injection (CVS, +4 V) is larger in inversion than in accumulation. In other words, independent of the stress polarity, larger SILC is always measured in inversion. Therefore, it can be concluded that the damage mainly responsible for SILC is located near the Hf-silicate/Si interface. 3) At higher CVS (+5.5 V), SILC increases rapidly only within the first few seconds of the stress and than saturates until breakdown occurs. At lower CVS (+4 V), SILC increases linearly with Q_{inj}. In all cases, however, $\Delta J_G/J_{G0}$ at breakdown or at Q_{inj} of 0.1 C/cm^2 does not exceed 16, implying that independent of the stress conditions (voltage, temperature) a maximum level of damage is reached which defines SILC. This observation implies that the idea of a critical defect density, which is the main concept in the predictive percolation model for thin SiO$_2$ reliability [4,5] may be also applicable for high-k dielectrics. The obtained value of about 16 for $\Delta J_G/J_{G0}$ is slightly higher than that for SiO$_2$ (about 10) with a similar EOT (3.3 nm) [4]. More detailed investigations are needed to justify this suggestion.

A significant increase of $\Delta J_G/J_{G0}$ in inversion is observed after Q_{inj} of about 0.15 C/cm^2. This can be regarded as a soft breakdown (SBD) event. The increase is strongly dependent on temperature. At 30 °C, an initial jump from about 16 to about 50 is observed and after that tends to saturate. For 100 °C, after the strong initial increase from 12 to 90, $\Delta J_G/J_{G0}$ increases linearly with Q_{inj} reaching a value of about 300 just before hard breakdown

(HBD). It should be mentioned that such progressive degradation between SBD and HBD is observed only for SILC in inversion under positive CVS. In all other cases the HBD occurs suddenly. The dependence of $\Delta J_G/J_{G0}$ on Q_{inj} has many common features with that predicted by the reaction-diffusion H$_2$ model of Alam, et al. [8] for the threshold voltage negative bias temperature instability, implying that most likely the same degradation mechanism is responsible for both phenomena.

To check whether SILC is caused by a permanent damage in the layer detrapping experiments were performed. After the initial stress at +4 V, 100 °C for 200 s, CVS with negative polarity was performed. Stressing at -2 V for 50 s causes no recovery of the I-V curve (Fig. 4a). A significant recovery, however, is observed after stress at -3.5 V for 50 s (Fig. 4b). This reveals that part of the SILC originates from a transient component. The recovery of SILC can be due to two possible detrapping processes - back tunneling of electrons from traps to the substrate and detrapping by Poole-Frenkel emission. The absence of recovery at low voltage (-2 V) makes the first process less favorable. In fact -3.5 V (100 °C) is the voltage where PF emission starts [9] giving an evidence that the recovery is due to the detrapping from the traps participating also in PF conduction. These results suggest that the filled traps act as SILC sites facilitating electron transport at low fields. The release of electrons from traps causes a partial recovery of SILC. In other words, empty traps do not give rise to SILC.

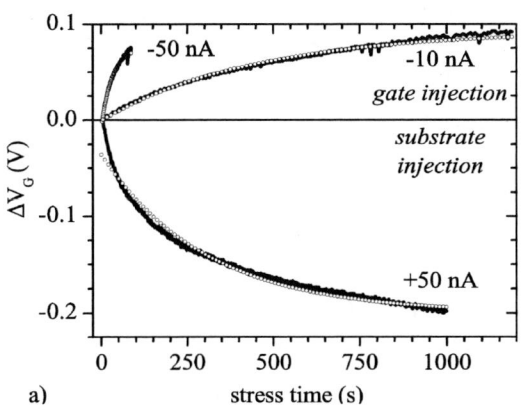 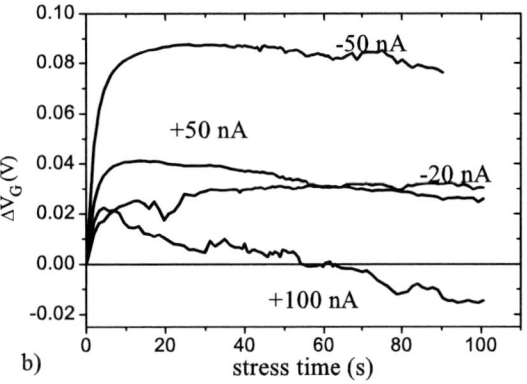

Fig. 6. Gate voltage evolution ΔV_G under positive and negative CCS: a) Zr-silicate; b) Hf-silicate layers.

Zr-silicate layers (Fig. 5a) behave similar to Hf-silicate under negative CVS– a stronger increase of SILC is observed at the beginning of stress. For longer stress times, SILC increases only smoothly up to BD. Comparable values of Q_{BD} of about 50-100 mC/cm^2 are obtained for both silicates. There are considerable differences under positive CVS. For Zr-silicate layers, injected charge density higher than 10 C/cm^2, (i.e., Q_{BD} is more than two orders of magnitude higher than that at gate injection) does not lead to BD, (regardless of the generated SILC, Fig. 5b). Therefore, the degradation of Zr-silicate layers is also polarity dependent, but unlike Hf-silicates, it is more severe under gate injection.

Zr-silicate layers reveal negative charge trapping under negative CCS and positive charge trapping under positive CCS (Fig. 6a). For Hf-silicates, trapping of negative charge for both CCS polarities is observed at the beginning of stress. Later on, positive charge trapping starts to dominate (Fig. 6b). The analysis revealed that for Zr-silicate the $\Delta V_G(t)$ curves can be fitted by a first order trapping kinetics giving an evidence that the trapping occurs in pre-existing traps [10]. The trap density and carrier capture cross-section were estimated to be in the same order of magnitude for both polarities. The obvious discrepancy is the fact that under gate injection negative charge is trapped whereas under substrate injection positive charge is trapped. We propose that under substrate injection the process detected is in fact detrapping of trapped electrons. C-V measurements of these structures showed that electron traps with a density of ~2×10^{12} cm^{-2} were present very near the Zr-silicate/Si interface and a single (-3/3) V C-V sweep can charge most of these traps. The detrapping experiment [11] revealed that the emission time of the traps is much larger than the capture time. Therefore, we suggest that the trapping under substrate injection proceeds very quickly and does not influence the CCS characteristics. This interpretation correlates the results obtained for Zr-silicate with those for Hf-silicate layers and implies that despite the visible differences in the $\Delta V_G(t)$ curves, the trapping processes under substrate injection are similar in both layers – electron trapping at the beginning of the stress and subsequent release of the trapped electrons. The electron trapping in Hf-silicate, however, is slower, hence visible in $\Delta V_G(t)$ curves. An evidence for this explanation is the estimated band diagrams and trap levels of the two layers. For Zr-silicate layers, two trap levels at about 0.95 eV and 1.5 eV below the conduction band edge of dielectric were found [10,11]. The barrier height at dielectric/Si interface was estimated to be 1.5 eV, hence the 1.5 eV level is aligned with the Si conduction band (CB) edge (i.e., the trap level is always in resonance with Si CB that makes the trapping process very efficient). The time scale of electron exchange between Si CB and this trap level is very short and hence the process is not detectable in the $\Delta V_G(t)$ curves. For Hf-silicate layers [9] a trap level at 1.1 eV and a barrier height of about 1.4-1.5 eV were estimated. Hence, the trap level is situated 0.3-0.4 eV above the Si conduction band, resulting in a slower trapping process.

IV. CONCLUSION

SILC and charge trapping of thin Hf- and Zr-silicate layers with similar stoichiometry and EOT were investigated. At low stress voltages electron trapping in the Hf-silicate films dominate. A threshold voltage of about –6 V for generation of SILC is obtained and related to defect generation in the layers. SILC in inversion is always larger than SILC in accumulation revealing that the centroid of damage is near Hf-silicate/Si interface, independently of the stress polarity. The relative increase of SILC for $Q_{inj} < 0.1$ C/cm^2 under positive CVS depends weakly on stressing conditions (temperature, voltage). This result indicates that the percolation model for SiO$_2$ degradation may be applicable also for high-k dielectrics. For SILC in inversion a soft breakdown event at Q_{inj} of about 0.15 C/cm^2 is observed followed by a progressive degradation (which is strongly temperature dependent) up to hard BD. The results give also evidence that SILC has a transient component originating from Poole-Frenkel emission from traps. Electrons are predominantly trapped in both Zr-silicate and Hf-silicate layers, followed by their release at higher injection fluence. The energy location of the traps plays a significant role in the trapping behavior of the layers and defines the time scale of electron exchange with Si.

REFERENCES

[1] S. Zafar, A. Kumar, E. Gusev, and E. Cartier, "Threshold voltage instabilities in high-κ dielectric stacks", *IEEE Trans. Dev. Mater. Reliab.*, 2005, vol. 5, pp. 45-64.

[2] T. P. Ma, H.M. Bu, X.W. Wang, et al., "Special reliability features for Hf-based high-k gate dielectrics", *IEEE Trans. Dev. Mater. Reliab.*, 2005, vol. 5, pp. 36-44.

[3] R.O'Connor, S.McDonnell, G.Hughes, et al., "Low voltage SILC in 1.4-2.1 nm SiON and HfSiON layers", *Semicond. Sci. Technol.*, 2005, vol. 20, pp. 668-72.

[4] J.H.Stathis, "Reliability limits for the gate insulator in CMOS", *IBM J. Res. Devel.*, 2002, vol. 46, pp. 265-86.

[5] D.A. Buchanan, "Scaling the gate dielectric: materials, integration, and reliability", *IBM J. Res. Devel.*, 1999, vol.43, pp. 245-64.

[6] B.H. Lee, R. Choi, J.H. Sim, et al., "Validity of CVS based reliability assessment of high-k devices", *IEEE Trans. Dev. Mater. Reliab.*, 2005, vol. 5, pp. 20-24.

[7] M.Lemberger, A.Paskaleva, S. Zürcher, A.J. Bauer, L. Frey, and H. Ryssel, "Zr-silicate obtained from novel MOCVD precursors", *J. Non-Cryst. Sol.*, 2003, vol. 322, pp. 147-53.

[8] M.A. Alam, and S. Mahapatra, "A comprehensive model of PMOS NBTI degradation", Microel. Reliab., 2005, vol. 45, pp.71-81.

[9] M. Lemberger, A. Paskaleva, S. Zürcher, A.J. Bauer, L. Frey, and H. Ryssel, "Electrical Properties of Hf-Silicate Films Obtained from a Single-Source MOCVD Precursor", *Microel. Reliab.*, 2005, vol. 45, pp. 819-22.

[10] M. Lemberger, A. Paskaleva, S. Zürcher, A.J. Bauer, L. Frey, and H. Ryssel, "Electrical Characterization and Reliability Aspects of Zr-Silicate Films Obtained from Novel MOCVD Precursors", *Microel. Engin.*, 2004, vol. 72, pp. 315-20.

[11] A.Paskaleva, M.Lemberger, S. Zürcher, A.J. Bauer, L. Frey, and H. Ryssel, "Electrical characterisation of Zr-silicate films obtained from novel MOCVD precursors", *Microel. Reliab.*, 2003, vol.43, 1253-57.

2006 25th International Conference on Microelectronics

Reliability Investigation of NLDEMOS in 0.13 µm SOI CMOS Technology

D. Lachenal, Y. Rey-tauriac, L. Boissonnet and A. Bravaix

Abstract - This paper presents new reliability investigations in NLDEMOS transistor in 0.13 µm SOI CMOS technology. Reliability tests under Hot Carrier Injections (HCI) and OFF state regimes show a strong dependence with the drain extension length L_{ext}. The use of borderless nitride in order to create contacts is suspected to be the origin of degradation. Mobiles charges in this nitride can move under electrical field in HCI or OFF-state stress and then modulate the conductivity of the drain extension region. With a reverse electrical field applied after OFF-state (drain) or HCI stress, we show that the degraded NLDEMOS can completely recover its initial performances. A novel nitride borderless with a decrease of SiH_4 rate during process step shows a strong reliability improvement due to the decrease of dangling bonds at SiPROT/borderless nitride interface.

I. INTRODUCTION

Due to high performances required for CMOS applications, adding high voltage devices becomes a big challenge to guarantee the reliability. High Voltages (HV) MOSFETs have been developed for variety of HV applications where supply voltages are required between 10V to 20V in analog CMOS blocs, as well as for telecommunication applications [1-3]. LDEMOS are integrated in a standard low-cost 0.13 µm CMOS process using some specific process steps as P_{body} diffusion. The key specifications of LDEMOS are low on-resistance and a high drain-source breakdown voltage. For radio frequencies (RF) applications, LDEMOS transistor is the best candidate compared to Drift transistor, due to shorter channel length which allows best performances. Moreover, the use of SOI technology reduces the drain-substrate capacitance, improving RF performances. As LDEMOS transistors are dissymmetric and have a high drain voltage, they are more sensitive to HCI and new degradation mechanisms appear [4-6].

II. DEVICE DESCRIPTION

The partially depleted NLDEMOS transistors (Fig. 1) are processed with a 50 Å oxide thickness. Gate length (L_G) and drain extension (L_{ext}) are 0.5 µm and 0.3 µm respectively whereas the channel length is about 0.15 µm.

D. Lachenal, Y. Rey-tauriac and L.Boissonnet work at Crolles R&D, STMicroelectronics, 850 rue Jean Monnet, BP 16, 38926 Crolles, France. E-mail: damien.lachenal@st.com

A. Bravaix and D. Lachenal are with the Laboratory of materials and microelectronics of Provence, ISEN-Toulon, Maison des technologies, place G. Pompidou, 83000 Toulon, France.

A maximum drain voltage of 5.5V and gate voltage of 2.7V have been defined for small signal analog application according to designer's requirements. Theses devices are carried out with a P_{body} diffusion process step and a drain extension region which is lightly doped to achieve a breakdown voltage of 13V. As SOI technology is used, the P_{body} and source contacts are short circuited in order to avoid the triggering of NPN bipolar transistor. HCI stressing is first performed by applying a high DC drain voltage in order to obtain electrical parameter degradation at wafer and package level. As measurement of the substrate current is not feasible, several short stresses were performed showing that the maximum damage occurs for gate voltage about 1V. Then, OFF-state stress was performed by applying the nominal drain voltage with all others contacts grounded. During HCI and OFF-state stress at room temperature, device parameters are monitored at logarithmic stress time intervals. Monitored parameters are the transconductance (G_m), threshold voltage (V_{th}), linear drain current (I_{dlin} @ $V_d = 0.1V$ and $V_g = 2.7V$), saturation drain current (Id_{sat} @ $V_g = 2.7V$ and $V_d = 5.5V$), and the linear on-resistance (R_{on} @ $V_d = 0.1V$ and $V_g = 2.7V$). The failure criterion is fixed at maximum 10% of parameter shift for a 10 year lifetime evaluation.

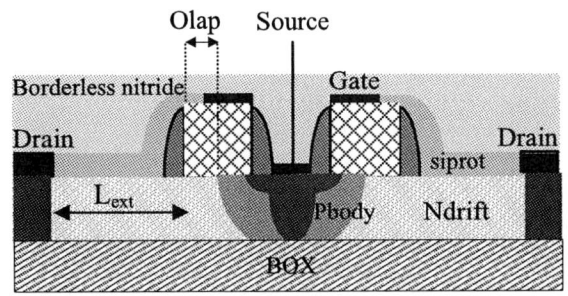

Fig. 1: Cross section of NLDEMOS in SOI technology.

III. EXPERIMENTAL RESULTS

Fig. 2 shows HCI and OFF-state stress results at package level in our first NLDEMOS structure with a drain extension of 0.3 µm. The V_{th} shift is very small, indicating that the degradation is not located in the channel whereas I_{dlin} shows saturation close to 3%, which is sufficient for HCI qualification. As inductors are used in RF circuits, NLDEMOS need to withstand three times the nominal drain voltage because of possible over-shoot. Then we processed a NLDEMOS with longer drain extension in

1-4244-0116-X/06/$20.00 ©2006 IEEE   593

order to achieve a higher breakdown voltage at 17V. With a longer drain extension, we expect better reliability results in HCI stress as the drain resistance increases and none degradation in OFF-state stressing. In contrast, we show in Fig. 3 a strong degradation of Id_{lin} (~13%) in both HCI and OFF-state stresses.

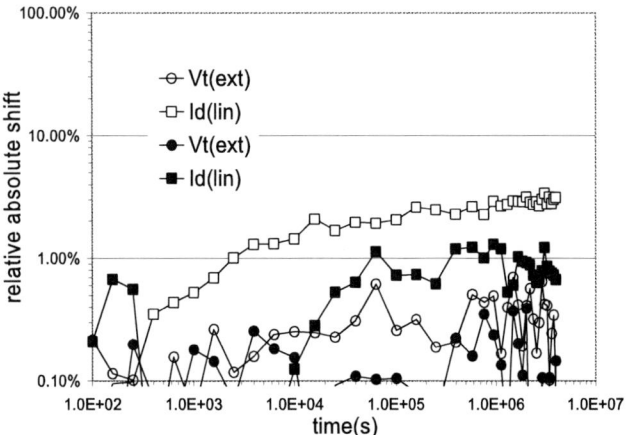

Fig. 2: Id_{lin} and Vt_{ext} relative degradation in NLDEMOS with L_{ext}=0.3 µm during HCI (empty) and OFF-state stress (full) performed at V_d = 5.5V.

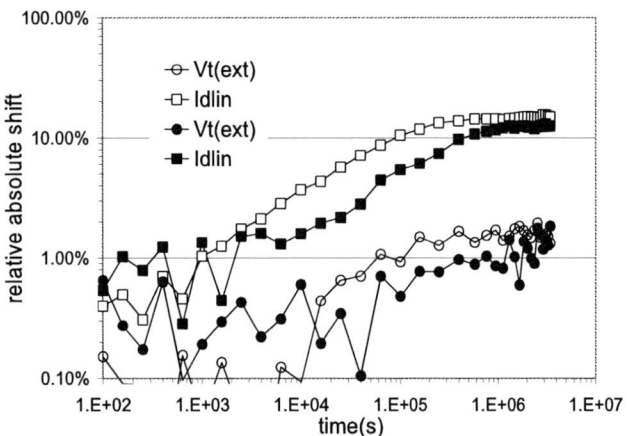

Fig. 3: Id_{lin} and Vt_{ext} relative degradation in NLDEMOS with L_{ext} = 0.7 µm during HCI (empty) and OFF-state stress (fill) at V_d = 5.5V.

Id_{lin} time-dependence in OFF-state stress follows the HCI stress implying an unusual degradation mechanism. Moreover, Id_{lin} degradation increases with L_{ext}, in contrast to standard HCI results in shorter L_{ext}. In OFF-state stress, the first hypothesis to explain the degradation was the apparition of a weak gate current which creates impact ionization in the Olap region where electrons are accumulated. But, even with 85Å oxide-thickness or with a shorter L_G which shortens the Olap length and so decreases the assumed gate current, the same degradation in OFF-state stress was observed. As the degradation in OFF-state stress appears with increasing L_{ext}, we suspect the SiPROT

oxide, first, and the borderless nitride, next, above the drain extension to contain some mobile charges (H^+) which could move under electrical field and modulate the conductivity of the N_{drift} region by an accumulation or depletion phenomenon.

A. NLDEMOS resistance model

First, we can ask why this amount of degradation is not constant with L_{ext}. There is no particular reason to have less degradation with a shorter drain extension if an accumulation or depletion region exists on the total drain extension. A first order model of NLDEMOS resistance can help to understand this phenomenon. NLDEMOS resistance can be shared into three parts: the channel resistance, the Olap resistance and the drain extension resistance. Fig. 4 shows NLDEMOS before and after OFF-state stress for different drain extension. Extrapolating the $R_{on}(L_{ext})$ straight line, we can deduce the $R_{channel}+R_{olap}$ resistance before and after stress taking L_{ext}=0µm. We see that this resistance has not been modified during the OFF state stress confirming that degradation is not located in the channel or Olap region. $(R_{channel}+R_{olap})/R_{on}$ ratio can then be plotted on the same graph. When this ratio is about 50% which correspond to a L_{ext} of 0.48 µm, $R_{channel}+R_{olap}=R_{Lext}$. For a drain extension shorter than 0.48 µm, the NLDEMOS resistance is governed by $R_{channel}+R_{olap}$, which explains the slight degradation observed in Fig. 2, as the channel and the Olap region are not degraded. At the opposite, for longer drain extension (>0.48 µm), the total resistance is governed by R_{Lext} and is more degraded.

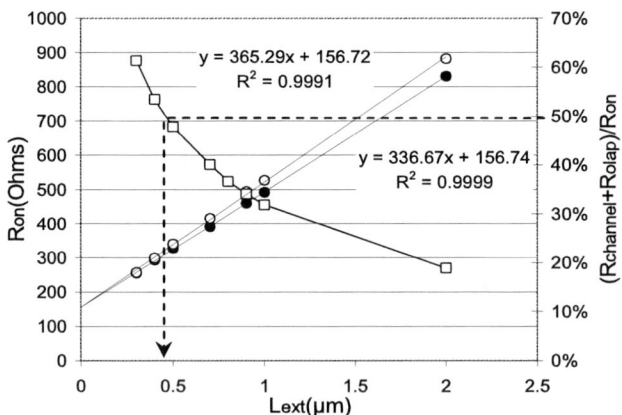

Fig. 4: R_{on} before (filled circles) and after (empty circles) 10^5s of OFF state stress @ Vd = 7V. Empty squares represent the $(R_{channel}+R_{olap})/R_{on}$ ratio for the different L_{ext}.

B. Mobile charges characterization

In order to prove the presence of mobile charge, we check Fig. 6 and Fig. 7 this new mechanism by applying to NLDEMOS device a positive gate bias after OFF-state or HCI stress using 10^5 s of stressing followed by recovery. Filled squares represent the OFF-state stress followed by the gate recovery. Id_{lin} increases until be superior to its initial value, certainly due, in a first hypothesis, to a drift zone slightly depleted at t = 0s. This behaviour has also been observed with a 4 hours baking at 125 °C following OFF-state stressing (filled circle) giving 69% of Id_{lin} recovery. Empty circles show HCI stress followed by the V_g bias giving the same behaviour without impact ionization in this L_{ext} region, as our last TCAD results have shown that hot carrier generation is located near the gate-edge on the drain side and at the P_{body}-N_{drift} junction. As the linear drain current recovers its initial value, the reverse electrical field between gate and drain has removed all the mobile charges.

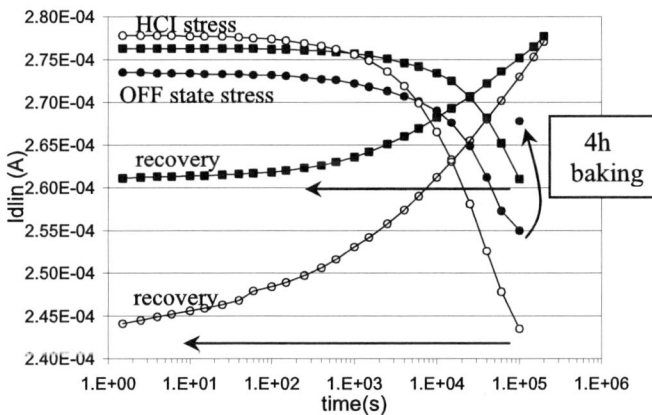

Fig. 6: Id_{lin} variation after the linking stress performed at 5.5V for both HCI and Off-state stress followed by a 3V gate-voltage recovery step (time scales are plotted from the same start for clarity). Single circle represents 4h baking at 125°C giving 69% of recovery.

C. Mobile charges localization

SiPROT is used in NLDEMOS in order to avoid the siliciuration of the N_{drift} region when contacts are processed. By comparison of process steps of NLDEMOS in bulk technology, we remarked that SiPROT were carried out differently. SiPROT was processed with two stacks, one of TEOS (Tetra Ethyl Ortho Silicate) by LPCVD and the other of bake nitride. Thermal anneal was achieved after the bake nitride deposition in SOI technology whereas it was realized before in bulk technology. We first have suspected the SiPROT oxide to be polluted by the bake nitride process step which could degrade the oxide quality by incorporating mobile charges when thermal anneal was processed after the nitride deposition. In Fig. 8, a novel SiPROT with a thermal anneal before processing the bake nitride (empty squares) was achieved but small improvement was observed compared to our first SiPROT (filled squares). Then we suspected the use of borderless nitride to create contacts to be responsible of mobile charges. SiH_4 and NH_3 molecules ratio were modified for a novel borderless nitride trial. A strong impact on NLDEMOS reliability is observed in Fig. 8. Id_{lin}

degradation completely disappears at 25°C (filled circles) whereas it follows standard behaviour at 125°C (filled triangles) but need to be checked for longer stress time at package level. SOI technology implies shorter silicon height (1500Å) than in bulk technology (~9500 Å) for current path. As it is closed to the N_{drift}/SiPROT interface, Id_{lin} is also more sensitive to the conductivity variation.

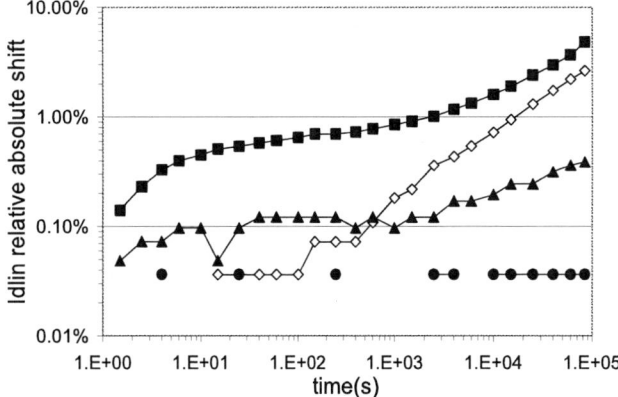

Fig. 8: Id_{lin} variation for the different process splits after OFF state stresses @ 7V for a drain length extension of 0.6µm. All stresses are performed at 25°C excepted filled triangles at 125°C.

The improvement of interface quality between SiPROT and borderless nitride by reducing the dangling bonds quantity seems to be the origin of degradation mechanism. The conductivity modulation of N_{drift} region can then be explained with two hypotheses: depletion or accumulation of the N_{drift} region due to the charge on SiPROT/borderless nitride interface states.

IV. DEGRADATION MECHANISM HYPOTHESES

A First degradation mechanism: N_{drift} depletion

The first hypothesis considers that dangling bonds at the SiPROT/borderless nitride interface create a depletion region in the drift area. Released hydrogen [6] during the formation of borderless nitride can react with dangling bonds to form H^+/e^- pairs (1) which could be separated by electrical field. Electrons can fill the acceptor energy level of the interface states whereas H^+ moves under electrical field. As this reaction is reversible, by reversing the electrical field NLDEMOS can recover its initial performances.

$$Si_3 \equiv Si^\bullet + H_2 \longleftrightarrow Si_3 \equiv Si - H + H^+ + e^- \qquad (1)$$

We see previously that Id_{lin} increases until to be larger than its initial value (see Fig. 6) meaning a N_{drift} zone slightly depleted at t = 0s (see Fig. 9). We then decided to apply a gate bias recovery on a fresh device to see how much the NLDEMOS current can recover. A 10% gain saturation of

Id_{lin} was achieved after 15.10^4s with 3V gate recovery at 125°C. This suggests that all the negative charges have been removed in the borderless nitride and that the N_{drift} region is no more depleted (*i.e.* recombination of e^- with H^+). Using a classical capacitor model for the N_{drift} region constituted of N_{drift}, SiPROT and borderless nitride (Fig. 9), we can deduce the amount of mobile charges Q_{mob} before stress by R_{on} measurement variation after stress.

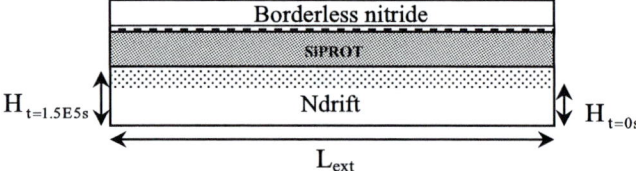

Fig. 9: Capacitor model for the Ndrift region. Dotted and line dotted zones correspond to the depletion and mobile negative charges (Q_{mob}) regions respectively.

Considering the channel region as a current generator, we can define the resistance of drain extension before (2) and after (3) stress:

$$R_{t=0} = \rho \frac{L_{ext}}{H_{t=0s} W} \quad (2) \quad R_{t=1.5E+5s} = \rho \frac{L_{ext}}{H_{t=1.5E+5s} W} \quad (3)$$

With ρ, L_{ext}, H and W the resistivity, drain extension length, height of silicon and the width respectively. We can now calculate the charge variation ΔQ_{mob} in the borderless nitride which impacts the depletion length ΔH at the oxide/N_{drift} interface after stress with:

$$\Delta H = \frac{\varepsilon_{sc}}{C_{siprot}} \left[-1 + \sqrt{1 + \frac{2 C_{siprot}^2 (V_{ox} - \phi_{ms})}{q N_d \varepsilon_{sc}}} \right] \quad (4)$$

Giving, with $Q_{mob} = C_{siprot} \cdot V_{ox}$ $\qquad$ (5)

$$\Delta Q_{mob} = \left[\left(\frac{\Delta H C_{siprot}}{\varepsilon_{sc}} + 1 \right)^2 - 1 \right] \frac{q N_d \varepsilon_{sc}}{2 C_{siprot}} + \phi_{ms} C_{siprot} = -1.47 \times 10^{-7} C/cm^2 \quad (6)$$

With q the elementary charge, N_d the donor concentration in the N_{drift} region, C_{siprot} the oxide capacitor, ε_{sc} the permittivity of silicon, $\phi_{ms} = \phi_{nitride(Et)} - \phi_{s,n}$, V_{ox} is the oxide potential and $\Delta H = H_{t=1.5E5s} - H_{t=0s}$.

The number of interface states at the borderless nitride/SiPROT interface N_{ss0} at t = 0s can be deduced with (6):

$$Nss_0 = \frac{1.47 \times 10^{-7}}{1.6 \times 10^{-19}} = 9.18 \times 10^{11} cm^{-2}$$

B. Second degradation mechanism: N_{drift} accumulation

The second hypothesis considers that the N_{drift} region is slightly accumulated by the presence of hydrogen ion (H^+) at this same interface. This positive charge could also be modulated by the electrical field between gate and drain then modifying the conductivity of the N_{drift} region. In this case, compared to the depletion hypothesis, the calculated amount of positive charges is smaller due to the fast variation of accumulated charges with the surface potential V_s:

$$Q_{acc} = -\sqrt{2kT \varepsilon_{sc} q N_d} \exp(\frac{q V_s}{2 kT}) \quad (7)$$

V. CONCLUSION

We have shown in this paper the importance of standard process steps used in CMOS technology on NLDEMOS reliability. Because of the dissymmetric geometry of NLDEMOS and the slight height of silicon, the depletion or accumulation modes in the drift extension region enlarge the degradation under HCI and OFF-state stress. We have shown that the SiPROT/borderless nitride interface quality is the origin of degradation observed. An unusual method by R_{on} variation measurement is proposed to estimate the interface state quantity. Further works are beyond the scope of this study particularly focusing on the dominant mechanism which governs the conductivity of N_{drift} region.

ACKNOWLEDGEMENT

The authors would like to thank Dr. J. L. Regolini (ST Microelectronics Crolles) for discussion on the nitride borderless topic and D. Benoit for the development of the material.

REFERENCES

[1] Y. Rey-Tauriac, Hot-carrier reliability of 20V MOS transistors in 0.13µm CMOS technology, Microelectronics Reliability 45 (2005) pp. 1349-1354.
[2] I.Cortes, Analysis of hot-carrier degradation in a SOI LDMOS transistor with a steep retrograde drift doping profile, Microelectronics Reliability 45 (2005) pp. 493-498.
[3] D. Brisbin, Hot carrier reliability of N-LDMOS transistor arrays for power BiCMOS applications, IEEE, 40th IRPS 2002, Dallas, Texas, pp. 105-109.
[4] P.Moens et am., A novel Hot hole Injection Degradation Model for Lateral nDMOS Transistors, IEEE, IEDM 2001, pp. 877-880.
[5] R.Versari, Experimental study of hot-carrier effects in LDMOS transistors, IEEE 1999, Vol 46, N° 6, pp. 1228-1233.
[6] N. Stojadinovic, V. Dankovic, S. Djoric-Veljkovic, V. Davidovic, I. Manic, S. Golubovic, Negative bias temperature instability mechanisms in p-channel power VDMOSEFTs, Microelectronics Reliability 45 (2005) pp. 1343-1348.

2006 25th International Conference on Microelectronics

Physics and Electrical Characterization of Excimer Laser Crystallized Polysilicon TFTs

L. Michalas, M.A. Exarchos, G.J. Papaioannou, D.N. Kouvatsos and A.T. Voutsas

Abstract -The electrical properties of polycrystalline silicon thin film transistors are investigated. Transfer and transient characteristics have been recorded versus temperature, in the linear operation regime. Basic parameters such as subthreshold swing, leakage current and drain current overshoot transient are found to stem from the same deep states thermally activated carrier generation mechanism.

I. INTRODUCTION

New laser crystallization techniques allow the fabrication of polycrystalline silicon thin film transistors (poly-Si TFTs) with capability to take on many functions conventionally reserved for single crystal devices. [1]. In view of these applications the temperature analysis of the electrical properties is a topic of paramount importance. Even though several models (empirical or analytical) are proposed to predict the poly-Si TFTs characteristics [2],[3], the available experimental data at different temperatures are still limited. So, there are isolated reports on the temperature dependence of the leakage current [4], the subthreshold swing [5] and the transfer characteristics, the latter interpretation being based on the uniformly distributed density of states (DOS) model [6]. Taking these into account it becomes obvious that a comprehensive study would provide more information on the material properties and the associated device performance.

In the present work the analysis of both transfer and transient characteristics is presented. Aim of the study is to set off thermally activated mechanisms, which are associated to the polycrystalline material properties and directly affect the device operation. The tested TFTs, were fabricated on poly-silicon films that obtained by the sequential lateral solidification (SLS) process [7]. Since crystal properties that obtained by the SLS technique are critically affected by the film thickness [8], a complete study requires the involvement of devices with different active layer thicknesses.

II. EXPERIMENTAL

Thin Film Transistors (TFTs) were fabricated on

M. Exarchos, L. Michalas and G.J. Papaioannou are with the Physics Department, National and Kapodistrian University of Athens, Athens 15784, Greece, E-mail: gpapaioan@phys.uoa.gr.
D.N. Kouvatsos is with the Institute of Microelectronics, NCSR "Demokritos", Aghia Paraskevi Attikis, Athens 15310, Greece, E-mail: D.Kouvatsos@imel.demokritos.gr.
A.T. Voutsas is with Sharp Labs of America (SLA), Inc., Camas, WA 98607, USA, E-mail: avoutsas@sharplabs.com.

poly-Si films, obtained by the sequential lateral solidification process (SLS), having channels aligned along the grain boundaries, which is parallel to the direction of current conduction. The TFT structure was non-self-aligned-top-metal-gate, with a 100nm-thick PECVD SiO_2 gate dielectric layer. The tested devices were large TFTs on 30nm and 100nm thick poly-silicon films. The intercept of a line fitted to the I_{DS}-V_{GS} curve at the point of maximum transconductance g_m, yielded the extrapolated threshold voltage V_T. The sub-threshold slope s was extracted from the maximum slope of the I_{DS}-V_{GS} characteristic, drawn in semilog scale. Additional results are obtained from the application, on the experimental data, of Levinson analysis, yielding at the calculation of flat band voltage V_{FB} [9].

Finally the switch-ON drain current transients $[\Delta I_D(t)=I_D(t)-I_D(\infty)]$, where $I_D(\infty)$ represents the steady state drain current amplitude, were investigated. The transients were recorded with the aid of an SR570 current amplifier that also provided the drain bias (V_{DS}=50mV) and PC data acquisition (DAQ) card. The drain current was sampled in the 0-256msec time interval (t_{ON}) with a 1ms sampling rate. The duration of the OFF-state (t_{OFF}) was held constant at 100msec and the ON and OFF gate bias states were determined from the transfer characteristics. The temperature was controlled in the range of 150 to 440 K.

III. RESULTS AND DISCUSSION

The three main operation regions of TFTs transfer characteristics, at different temperatures, are shown in figure (1). The OFF state leakage current and the sub-threshold swing are obviously thermally activated. In contrast the temperature effect on the ON regime seem to be negligible.

The threshold voltage decreases when temperature increases (Fig. 2). In bulk n-MOSFETs this behavior is attributed to Fermi level shift towards midgap in the p-type body, so that less additional voltage is required to achieve inversion condition at interface [10]. However in non-crystalline devices, the excitation of trapped carriers from the band gap states to the conduction band plays a significant role on the reduction [11]. This is because the rise in temperature increases the number of free carriers that leads to channel formation at lower gate voltage [12]. The above mechanism is consistent with the definition of threshold voltage proposed in [3] for poly-silicon TFTs. Moreover, the concentration and distribution of these defects is expected to affect the temperature dependence of the threshold voltage. This hypothesis is confirmed in figure 2 where clearly the V_T shift rate is larger in the thinner films due to the larger density of trap states in band gap.

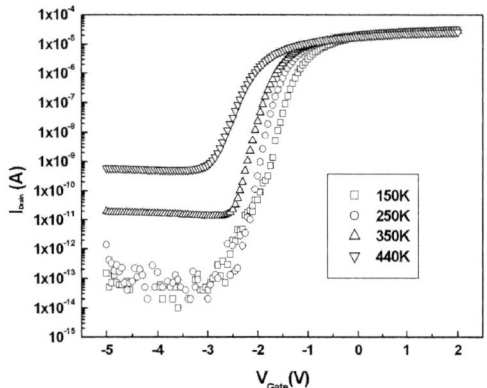

Fig. 1: Transfer characteristics at different temperatures.

In the OFF state poly-silicon TFTs suffers from relatively high leakage currents, especially under high-applied gate and drain voltage. [4]. These currents arise from carriers generated in the drain contact - body depletion region via grain boundaries states. Furthermore, when the drain voltage is low the thermal generation is the dominant leakage current generation mechanism [13]:

$$I_{LEAK} = q \frac{n_i}{\tau_e} w \qquad (1)$$

where τ_e is the effective electron generation lifetime and w the width of depletion region.

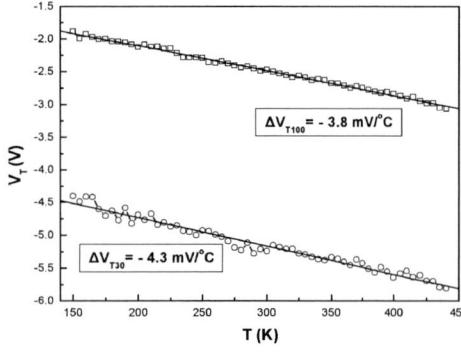

Fig. 2: Threshold voltage decrease for increasing temperature.

In the present work the leakage current is measured at $V_G=V_{FB}$. The activation energy, measured from the intrinsic Fermi level, has been determined from the

Arrhenius plot presented in figure 3. Values of 0.20 eV and 0.56 eV, have been obtained for the 100 nm and 30 nm thick devices, respectively. Here it must be pointed out that the large activation energy, determined in the 30nm film devices, may be attributed to effects such as the magnitude of leakage at the back interface.

The sub-threshold region indicates the sharpness of the current transition from OFF to ON state and is characterized by the gate voltage swing needed to reduce the current by one decade. The swing is temperature depended primarily due to the increase in the intrinsic carrier concentration [14]. Other parameters involved are the Fermi level and the mobility variation. In crystalline devices, a linear increase of swing versus temperature is expected due to the diffusion nature of the current [10]. This reflects a better device performance at lower temperatures and corresponds to uniform distributed states at poly-Si SiO₂ interface and depleted region [15], in the vicinity of Fermi level. Moreover, in poly-Si TFTs a non-linear term is observed at higher temperatures. This additional term arises from the contribution of thermally generated carriers in the depletion region [16]. Due to this fact, the additional increase will be governed by exponential terms, yielding to an expression for sub-threshold swing of the form

$$S = A_0 + A_1 T + A_2 \exp\left(-\frac{E_A}{kT}\right) \qquad (2)$$

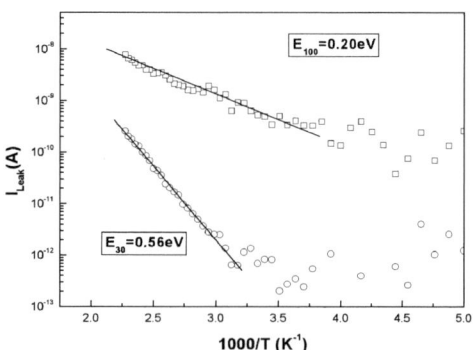

Fig. 3: Thermal activation of leakage current.

As already mentioned the carrier generation in the depletion region is trap assisted, Eq.1. The swing is also related to transport properties. At gate voltages below threshold, transport is significantly affected by the empty tail states. Taking into account that under these conditions the screening is low, the scattering on grain boundaries potential barriers is significant. Thus the calculated activation energy, determined with respect to intrinsic Fermi level, represents the complex mechanism of both transport and carriers generation.

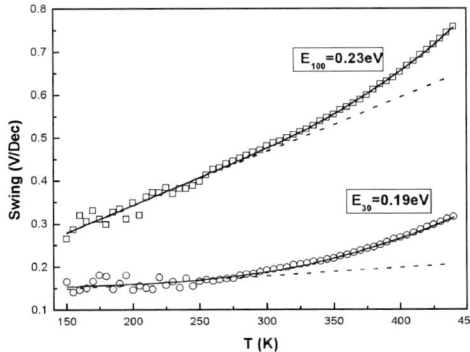

Fig. 4: Exponential terms are obtained on subthreshold swing increase

Additional information on the carrier generation mechanisms can be obtained by studying the switch-ON transient response. In an n-channel device, holes are emitted during the ON state and generated/captured during the OFF state. The concentration of filled traps will be determined by the duration of OFF-time (t_{OFF}) [17]. Now, in the following ON state, the total negative charge under the gate oxide must be maintained constant. The reduction of trapped positive charge, through holes emission, will lead to a decrease of channel carrier concentration hence drain current.

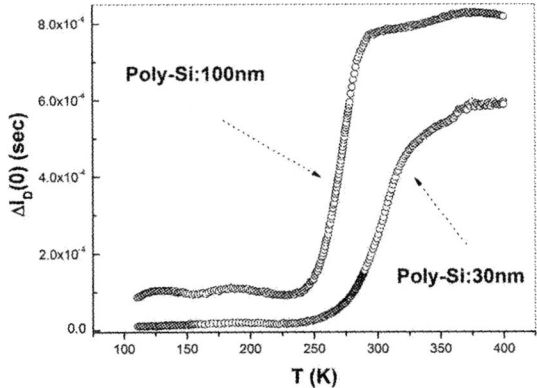

Fig. 5: Temperature dependence of the magnitude $\Delta I_D(0)$, for devices with Si-film thickness of 100nm and of 30nm.

Since the carrier generation mechanism is the same for both the drain leakage current and the switch-ON transient, it becomes obvious that independently of each process complexity, the temperature onset has to be similar. This is confirmed by comparing Fig. 3 and Fig. 5. In the case of 100nm film device, the onset of drain leakage current thermal activation and switch-ON current transient is observed at about 255K. In the case of 30nm film device,

the onset of drain leakage current thermal activation and switch-ON current transient is observed at about 315K. A similar trend is encountered in the onset of the nonlinear term of subthreshold swing.

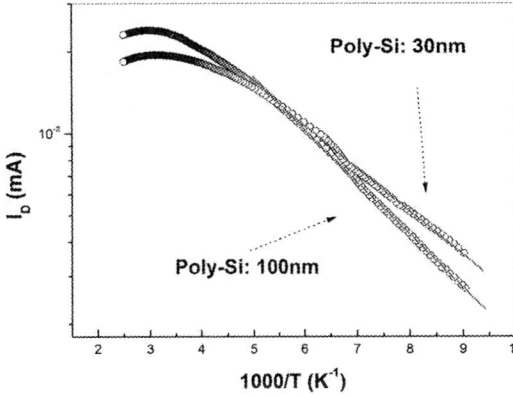

Fig.6: Arrhenius plot of drain current, for devices with Si-film thickness of 100nm and of 30nm. The slope of the continuous lines represent the barrier height modulation, which is 0.038eV for 100nm-thick TFTs and 0.031eV for 30nm-thick TFTs.

The difference in the transients' amplitude arises from the respective grain barrier height modulation [18], the concentration and properties of hole traps as well as the carrier mobility. Particularly, calculations based on the temperature dependence of drain current, have shown that in 100nm-thick TFTs the potential barrier height is 0.038eV, while in 30nm-thick TFTs it is 0.031eV (Fig. 6).

The lower barrier, calculated in the case of 30nm-thick active layer can be explained on the basis that the crystal domains are relatively narrow (compared to that of 100nm) and an enhanced trap density in the sub-boundary regions is presented [19]. Moreover, quartz-Si interface is plausible to come into play in the multi-step process described above (hole generation in the OFF-state, subsequent hole capture and emission in the ON-state) [17]. Therefore, significantly fewer holes are trapped underneath the grain boundaries of Si-SiO$_2$ interface and the respective barrier height is lowered.

IV. CONCLUSIONS

An analysis of transfer and transient characteristics of poly-Si TFTs has been presented. Temperature affects significantly the device operation through gap states carrier generation. The same thermally activated generation mechanism was found to be responsible for the leakage current, subthreshold swing and switch-ON drain current overshoot amplitude. All these parameters exhibit the same temperature onset and activation trend. The back interface seems to affect the parameters of the thin film devices.

REFERENCES

[1] N.Bavidge, M. Boero, P. Migliorato, T. Shoimoda, "Switch-on transient behavior in low-temperature polycrystalline silicon thin film transistors", *Applied. Physics. Letters.*, Vol 77, No 23, pp 3836-3838, 2000.

[2] K. Ono, T. Aoyama, N Konishi, and K. Miyata, "Analysis of Current-Voltage Characteristics of Low-Temperature-Processed Polysilicon Thin-Film Transistors",. *IEEE Transactions on. Electron Devices,*. Vol29, No 4, pp792-801, 1992

[3]G.Fortunato, P.Migliorato, "Model for above-threshold characteristics and threshold voltage in polycrystalline silicon transistors" *Journal of Applied Physics,*.Vol 68, No 5, pp. 2463-2467, 1990

[4] C.T.Angelis, C. A. Dimitriadis, I. Samaras, J. Brini, G. Kamarinos, V. K. Gueorguiev, Tz. E. Ivanov, "Study of leakage current in n-channel and p-channel polycrystalline silicon thin-film transistors by conduction and low frequency noise measurments", *Journal of Applied Physics,* Vol 82, No8, pp4095-4101, 1997.

[5]C.A. Dimitriadis, "Subthreshold slope in polycrystalline siliconthin-film transistors and effect of the gate oxide on the subthreshold characteristics" *Journal of Applied Physics,*Vol 67, No25, pp. 3738-3740, 1995

[6] V. Foglietti, L. Mariucci, G. Fortunato, "Temperature analysis of polysilicon thin-film transistors made by excimer laser crystallization", *Thin Solid Film,* 337 pp. 196-199, 1999

[7] R.S. Sposili and J.S. Im, "Sequential Lateral Solidification of Thin Silicon Films on SiO$_2$", *Applied. Physics. Letters.*, Vol. 69, pp. 2864, 1996.

[8] A.T. Voutsas, A. Limanov, J. S. Im., "Effect of process parameters on the structural ahracteristics of laterally grown , laser-annealed polycrystalline silicon films", *Journal of Applied Physics,* Vol 95, No 1, pp 1-8 2004.

[9]J. Levinson, F.R. Shepherd, P.J. Scanlon, W.D. Westwood, G. Este, M. Rider "Conductivity behavior on polycrystalline semiconductors thin film transistors", *Journal of Applied Physics.*,Vol 53, pp 1193-1202, 1982.

[10] S.M. Sze, *"Physics of Semiconductor Devices"*, N.Y, Willey & Sons,. 1981

[11]L.Wang, T.A. Fjeldly,B. Iniguez, H.C. Slade,M. Shur, "Self-Heating and Kink Effects in a-Si :H Thin Film Transistors", *IEEE Transactions on. Electron Devices,* Vol 47, No2, pp 387-397, 2000

[12]A. Sehgal, T. Mangla, M. Gupta, R.S. Gupta, "Temperature dependence on electrical characteristics of short geometry poly-crystalline silicon thin film transistor", *Solid State Electronics,*. Vol 49 pp 301-309, 2005

[13]C.H.Kim,K.S. Sohn,J. Jang, "Temperature dependent leakage currents in polycrystalline silicon thin film transistors" *Journal of Applied Physics,* Vol 81, No 12, pp 8084-8090, 1997

[14] T.Chopra, R.S. Gupta."Subthreshold conduction in short-channel polycrystalline-silicon thin-film transistors", *Semiconducrors Science and Technology,*Vol15,pp.197-202, 2000

[15]T.Noguchi, "Appearance of Single-Crystalline Properties in Fine-Paterned Si thin film transistors by Solid Phase Crystallization" *Jpn. Journal of Applied Physics*, Vol 32 Part 2, pp 1584-1587, 1993.

[16] G. Reicher,L C. Raynaud, O. Faynot, F. Balestra and S. Cfristoloveanu"Submicron SOI-MOSFETs for High Temperature Operation", *Microelectonics. Engineering,* 36 pp 359-362, 1997

[17] M.Exarchos, G.J.Papaioannou, D. N. Kouvatsos, and A. T. Voutsas, "Drain current overshoot transient in polycrystalline silicon transistors: The effect of hole generation mechanism", *Journal of Applied Physics*, 99, 1, 2006, in press.

[18] Y. Morimoto, Y. Jinno, K. Hirai, H. Ogata, T. Yamada, K. Yoneda, "Influence of the Grain-Boundaries and Intragrain Defects on the Performance of Poly-Si Thin-Film Transistors" *Journal of Electrochemical Society*, vol. 144, no. 7, pp. 2495-2501, 1997.

[19] A. T. Voutsas, D. N. Kouvatsos, L. Michalas, and G. J. Papaioannou, "Effect of Silicon Thickness on the Degradation Mechanisms of Sequential Laterally Solidified Polycrystalline Silicon TFTs During Hot-Carrier Stress", *IEEE Electron Device Letters*, vol. 26, no. 3, pp. 181-184, 2005.

600

2006 25th International Conference on Microelectronics

Reliability Assessment of a RF PA Assembly with Embedded Coin Construction

J. Zhou, H. Lu, M. Zhou, B. Inkman, D. Anderson, B. Johnson

Abstract - A time efficient yet rigorous assessment is conducted for a RF (radio frequency) PA (power amplifier) assembly prototype. Sites of structural weakness of the new design are identified and reliability attributes quantitatively examined for design improvement and modification.

I. INTRODUCTION

High power RF PA is a critical and costly component in wireless and satellite communication systems [1, 2]. The key to ensuring packaging reliability is the thermal management, and the coin construction in conjunction with back-attached heat sink has been a proven solution for heat dissipation. New PA architectures and design concepts driven by increasing signal processing speed and cost reduction often require a quick assessment. Presented below is an evaluation that integrates thermal shock test and structural analysis for a latest PCB (printed circuit board) design. The prototypes, as the top photos in Fig. 1 show, are supplied by Merix and including Merix Embedded Coin Technology.

II. STRUCTURE AND CONCERNS

The new design is featured by the embedded coins as the bottom photo in Fig. 1 shows. The PCB matrix material is a kind of ceramic-epoxy composite, and a conductive silver-epoxy adhesive is used to bond the tab locks of the coins to the PCB. The PA component and the heat sink are attached to the coin on its opposite sides through reflow soldering. To ensure good surface contact, the heat sink is fastened to the back of the coin via bolts during the reflow and a thin shim is inserted in between to reduce thermal impedance. The focus of the assessment is the coin-PCB interconnection that is functionary critical but structurally weak. Particular concern stems out of the high mismatch in CTE (coefficient of thermal expansion) along the PCB's thickness direction. Measurement using digital speckle correlation reveals that the laminate material has a CTE of

J. Zhou, B. Inkman and D. Anderson are with BTS Hardware Development, Network Sector, Motorola 1501 West Shure Drive, Arlington Heights, IL 60004, USA. Email: J. Zhou-QA2542@motorola.com   I. Bill-QBI001@motorola.com D. Anderson-Q10185@motorola.com.

H. Lu and M. Zhou are with Department of Mechanical and Industrial Engineering, Ryerson University, 350 Victoria Street, Toronto, ON M5B 2K3, Canada. Email: hlu@acs.ryerson.ca.

B. Johnson is with Merix Corporation, 1521 Poplar Lane, Forest Grove, OR 97116 USA.

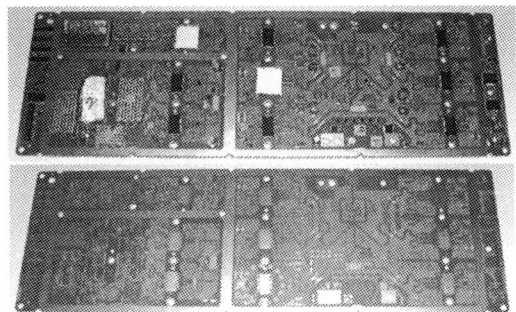

Fig. 1a   An assembly board (top) showing PA components and a bare board (bottom) with exposed embedded coins.

Fig. 1b A cross section of an assembly showing PA on top and heat sink on bottom of a coin (left) and a three-dimensional drawing showing the coin geometry (right).

44 as opposed to 17 ppm/°C of the interfacing coin. Concern is also raised regarding the effect of the heat sink.

As illustrated by the sketches in Fig. 2, due to the heat sink attachment the PCB is loaded by an equivalent force system. And the PCB bending becomes severer under elevated temperature due to the bolt constraint induced in-plane compression. The laminate when bent could de-bond and such de-lamination could damage nearby critical electric routines. The in-evitable stress concentration at the corner of the materials' interface further amplifies these concerns.

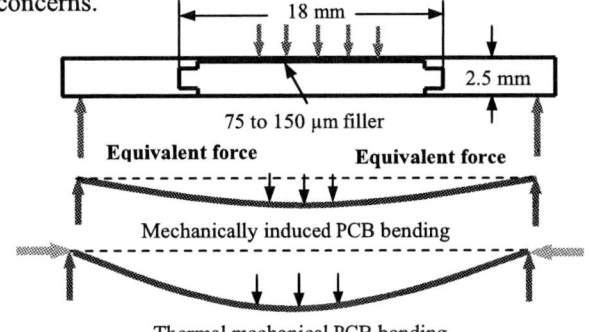

Fig. 2 Sketch of PCB bending

III. EVALUATION APPROACH

It begins that the samples made of bare PCB with embedded coins are liquid-to-liquid thermal shock tested in three-steps. The first 1000 cycles are run at -40 to 115 °C, which is followed by additional 500 cycles between -40 to 125 °C. The final run is set as -40 to 150 °C for 500 cycles with a five-minute dwell. Upon completion of each step a failure inspection is followed. No concerned issues turn out in the end. A full scale failure testing on assembly packages would be much costly and time consuming. Thus the analytically-based structural analyses are opted to continue the evaluation. Considering that the behaviour of PCB and adhesive materials are typically viscoelastic, glass-transitional and geometric scale, processing and environment dependent, an FEA (finite-element analysis) could only be effective if a comprehensive and non-linear model is developed and verified. Furthermore, a failure prediction following the FEA requires materials' failure properties and criteria, including the bonding shear strength of the inter-lamina and the interface adhesive layers. These parameters are with no exception largely package-specific and would be difficult to determine by conventional testing. A simple two-dimensional linear-elastic model instead can nevertheless quickly generate patterns of strain/stress distribution. The results are not meant to be accurate but nevertheless suffice in a practical sense to identify the failure sensitive sites. Given in Fig. 3 is an example to show the sites of so revealed weak structural links highlighted by the shear strain concentration. These locations while matching well with suspected sites based

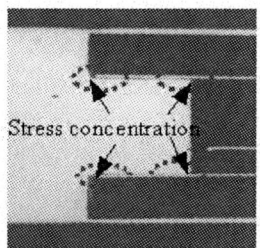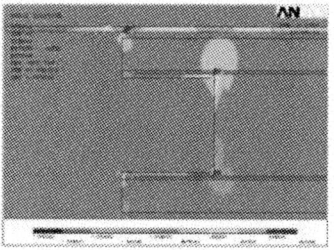

Fig. 3 FEA obtained shear strain concentration

on the common sense analysis give necessary guidance to the followed mechanistic measurement under microscopic magnification. The phase two assessment aims to confirm the qualitative diagnosis and to quantitatively determine the package behaviour. The parameters responsible for the debonding and the delamination and their critical values are of the particular interest. Strain measurements are obtained in local sandwiched areas including the horizontal tab-adhesive-trace-lamina interfaces, the vertical adhesive layers joining the tab locks and the PCB, as well as the solder joints that interconnect the component leads and the PCB copper pads. And the visual failure detection is followed upon completion of each test to help determine the failure parameters.

IV. TEST AND MEASUREMENT

The photo in Fig. 4 shows the main experimental setup including the imaging system, the thermal chamber and a fixture. Introductions to the digital speckle correlation method, system and applications can be found in the references [3, 4]. The optical magnifications of 2.7 and 1.21 μm/pixel are respectively used in the solder joint and PCB strain measurement. Samples are prepared including the bare PCB, the PCB with heat sink and the cut-out assembly with heat sink. To facilitate the measurement,

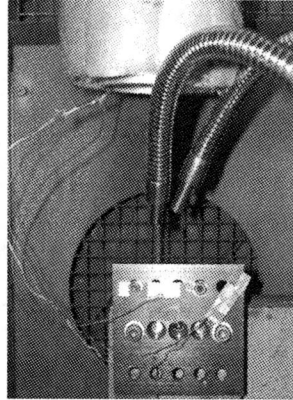

Fig. 4 A photo showing the experimental setup for thermal-mechanical strain measurement (left) and a close-up showing sample, light source and camera lens in the chamber (right).

cross-sections are exposed by cutting the PCB along the coin edges. Fig. 5 shows a sketch of a sample that is bolt-connected to a thick plate to simulate the heat sink attachment. A thin piece of foil filler of 3-mil (0.075 mm) to 6-mil (0.150 mm) thickness is inserted in between the PCB and the plate to mimic the shim. The range of the filler thickness is decided based on the surface flatness of the embedded coins. By applying shadow moiré the coin surface edge-to-center warpage is found to be between 0.12 mm at room temperature and 0.07mm at 180 °C. The strains at anticipated critical locations are measured in separate tests at room and elevated temperatures. In all, four groups of different tests are arranged as follows. Test A aims to investigate temperature induced strains at the interface areas in bare PCB samples with no heat sink installed. PCB sample used in Test B is connected to a thick plate by bolts at 10 lb-in (1.13 N-m) torque limit and a thin foil is inserted in between. Test C involves assembly units and follows similar approaches as in Test B. Test D measures the component-to-board solder joint strains under elevated temperatures. Each test records a series of images during a temperature ramp-up process.

V. RESULT AND ANALYSIS

Strains due to heat sink attachment at room temperature, referred to as the mechanical strains, are obtained by processing a pair of the images that are taken both at room temperature but before and right after the bolts are tightened. Fig. 6a shows a measurement area of the sample

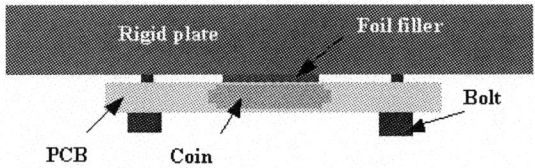

Fig. 5 A mechanism used in test to simulate PCB sample loaded due to heat sink attachment.

with and without coated speckles. Fig. 6b gives typical results come out from an image processing. For the temperature induced strains, the images can be processed between a reference image taken as the heat sink is affixed under the room temperature and another recorded under an elevated temperature. The strains thus obtained, referred to

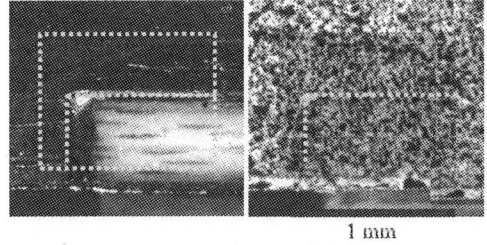

Fig. 6a Local area measured using DSC before (left) and after (right) speckle coated.

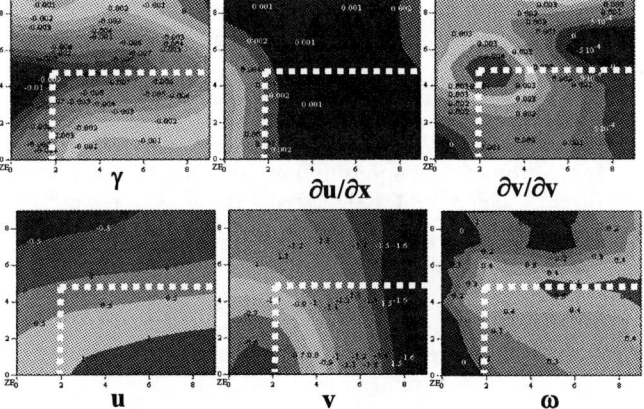

Fig. 6b Mechanical strain and displacement at room temperature due to heat sink attachment with 0.006 inch filler.

as the thermal-mechanical strains, are due to the board level packaging constraints as well as the temperature rise. The shear strain patterns at room and elevated temperatures are as given in Fig. 7. The reference image can also be chosen to be the one recorded at room temperature before the heat sink is fixed to the coin. The obtained strains in this case include also the part due to mechanical loading via the bolts. Example results in Fig. 8, obtained at another location with multi-material interfaces, give the mechanically and thermal-mechanically induced shear strains as well as the sum of both portions obtained with the second option of reference image. The shear strains

plotted in Fig. 9 are obtained in a vertical interface site in a PCB sample. The diagram summarizes and compares the highest shear strains obtained in a vertical interface area. The strain data include mechanical strain obtained under room temperature with 3-mil and 6-mil fillers (shown by the marked data points), the temperature variation induced thermal-mechanical strain for samples without and with heat sink (under either 3-mil filler or 6-mil filler), and the thermal mechanical strain in an assembly sample with heat sink. The debonding strain is closely studied along the horizontal interfaces between the tab locks and the laminate, where the cracks are exclusively found by visual inspections. As given in Fig. 10a and Fig. 10b, the strain variation with temperature in each case is obtained in a PCB sample with 3-mil filler in a first test, followed by a second run with 6-mil filler. The shear strain obtained in the second run with 6-mil filler is abnormally lower than

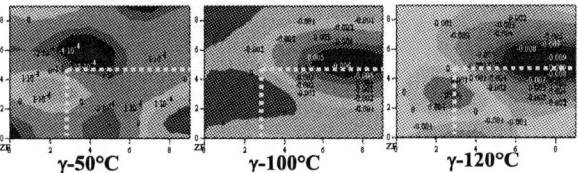

Fig. 7 Shear strain at different temperatures

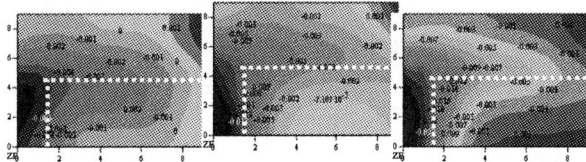

Fig. 8 Shear strain obtained with images before and after heat sink attached at RT (left), with image after heat sink attached at RT and another at 120°C (middle), and with image before the attachment at RT and another after attachment at 120°C.

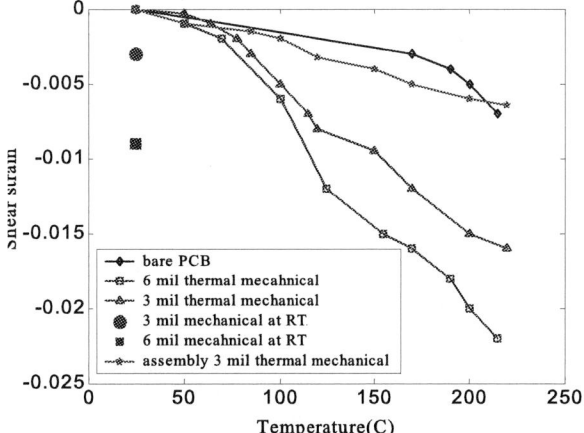

Fig. 9 Comparisons of maximum shear strain in a vertical interface site between PCB samples without and with heat sink attachment and an assembly sample.

that in the first test with 3-mil filler, as both plots show. This indicates that debonding has happened either in the first run or at the start of the second run. A close look of Fig. 10a finds that a strain recovery may has occurred at 117 °C, signifying the crack initiation induced unloading. Though no apparent unloading is seen in the 3-mil test in Fig. 10b, the crack may be actually initiated at the start of the second run of 6-mil filler. Adding the mechanically induced shear strain of around 0.003, debonding strain is determined in both samples to be around 0.012 under 120 °C. In comparison, the shear strain at 150 °C measured from a bare PCB (see Fig. 9) is around 0.003 or well below the estimated 0.012 threshold. The evidence sufficiently explains why the shock test (at 150 °C extreme) in bare PCB has found no fractures. Shown in Fig. 11 are the solder joint and the measured strain. The highest strain at 120 °C reaches 0.003 that is within the safe range.

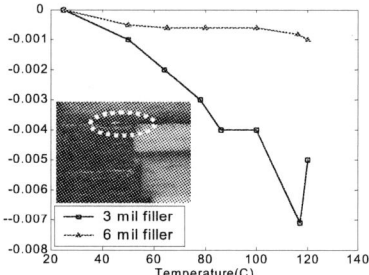

Fig. 10a Failure analysis for a PCB sample via strain measurement and by visual inspection. Crack occurs in a horizontal interface at 117°C with 3 mil filler.

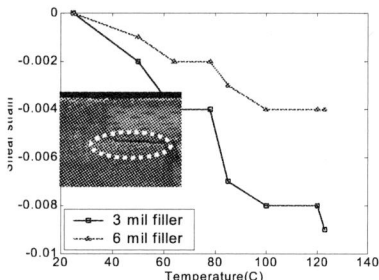

Fig. 10b Failure analysis for a PCB sample by strain measurement and visual inspection. Crack occurs in a horizontal interface at room temperature with 6 mil filler.

VI. SUMMARY

As mutually verifiable by analytical, numerical and experimental analyses, the structural weakness is at the coin-PCB interfaces. For bare PCB, temperature induced thermal-mechanical shear strain shows concentration along multi-material interfaces near the tab-PCB joints. When heat sink and shim are installed, the fixation causes PCB bending and in turn induces mechanical strain at room temperature and considerably higher thermal-mechanical strain as the temperature rises. It is noted that after the component is assembled, the unit appears to bare lower stress. The results suggest that the final assembly is less sensitive to the same failure mode as compared with the board with heat sink but no PA component.

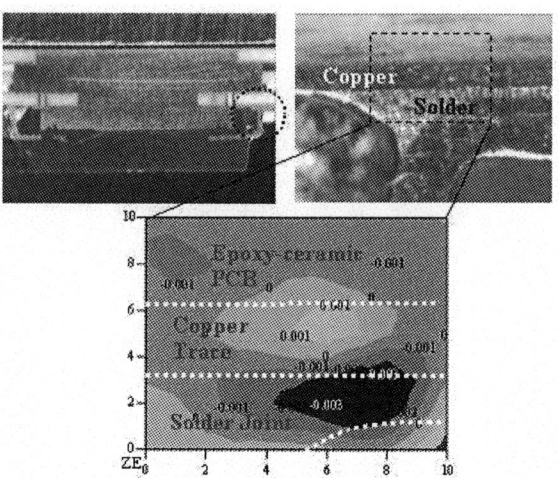

Fig. 11 The measured shear strain at the root of solder joint in a PA component at 110°C

Such compensatory effect due to the final component installation is attributable to the increased balance of structural stiffness through improved thickness direction symmetry. The phenomenon is worth paying attention to and may be leveraged in designing, manufacturing and assembling the RF PA packages. The debonding is found to happen at horizontal interfaces when shear strain reaches around 1.2%. And the vertical tab-coin interface apparently sustains higher shear strain.

ACKNOWLEDGEMENT

The second author wishes to acknowledge the financial support to the fundamental study of microelectronic packaging reliability funded by NSERC (Natural Science and Engineering Research Counsel of Canada) through a Discovery Grant.

REFERENCES

[1] V. Chiriac and T. Lee, 2004, Thermal Evaluation of Power Amplifier Modules and RF Packages in a Handheld Communicator System, 2004 Inter Society Conference on Thermal Phenomena pp 557-563.

[2] K. Ramakrishna, T. Lee, V. Hause, B. Chambers, and M. Mahalingam, 1997, "Experimental Evaluation of Thermal Performance of and Cooling Enhancements to a Hand-held Portable Electronic System," in Process, Enhanced and Multiphase Heat Transfer: A Festschrift for A. E. Bergles, Proc. of an A. E. *Bergles Symposium*, Georgia Institute of Technology, Atlanta, GA (November 16, 1996), pp. 217-226, Begell House, Inc., New York.

[3] H. Lu, H. G. Shi and M. Zhou, 2005, "Thermally Induced and Deformation of Solder Joints in Real Packages: Measurement and Analysis," submitted for publication to *Microelectronics Reliability*, accepted and in press.

[4] H. Lu, J. Zhou and M. Zhou, 2005, "Hybrid Reliability Assessment for Packaging Prototyping," *Microelectronics Reliability*, Vol. 45, pp 609, March 2005.

2006 25th International Conference on Microelectronics

The Effect of Number of Zincation in Electroless Nickel Immersion Gold (ENIG) Under Bump Metallurgy (UBM) on Reliability in Microelectronics Packaging

Siti Hajar Abdullah, Ibrahim Ahmad, Azman Jalar

Abstract - This paper discusses on the effect of number of zincation in Electroless Nickel Immersion Gold (ENIG) Under Bump Metallurgy (UBM) on reliability in microelectronics packaging. Double and triple Zincation of ENIG methods were used as comparison study. The effect of number of zincation to surface roughness and surface morphology were investigated. All samples were subjected to reliability tests such as Multiple Reflow, High Temperature Storage Life (HTSL) and Temperature / Thermal Cycle (TC) according to Joint Electron Device Engineering Council (JEDEC) conditions. Scanning Electron Microscope (SEM) and Atomic Force Microscopy (AFM) were used as analytical tools in this study. In multiple reflow tests, no reliability failure found for triple zincation, however for double zincation of ENIG UBM, the failure revealed failure after 3 times of reflow cycle. Thermal fatigue failure for both double and triple zincation of ENIG UBM revealed after 100 hours and 200 hours of HTSL, respectively. For TC test, thermal fatigue failure occurred after 150 cycles, for both double and triple zincation of ENIG UBM. The external and internal crack observed for this failure. The crack growth for double zincation is always much bigger than the triple zincation of ENIG UBM for both HTSL and TC tests. As conclusion, triple zincation gives a better surface morphology of electroless nickel. Hence, it can achieve a reliable solder joint leads to better adhesion between UBM and solder ball therefore, it can achieve a reliable solder joint.

I. INTRODUCTION

Flip chip technology differs from other ball grid array (BGA), leaded and laminates based CSP because there are no bond wires or interposer connections. In this technology, the chip is facing down and using solder balls as an interconnection whereas replace the traditional wire bonding. Flip chip offers many advantages in their application and has a huge market potential nowadays. The apparent advantages are shorter electron pathways, increased number of I/Os per unit area for increased speed

Siti Hajar Abdullah and Ibrahim Ahmad are with the Department of Electrical, Electronics and System, Faculty of Engineering, National University of Malaysia, 43600 UKM Bangi, Selangor, Malaysia, E-mail: sh_abdullah007@yahoo.com, ibrahim@vlsi.eng.ukm.my
Azman Jalar is with the School of Applied Physics, Faculty of Science and Technology, National University of Malaysia, 43600 UKM Bangi, Selangor, Malaysia, E-mail: azmn@pkrisc.cc.ukm.my

and power, cost reduction, and increased package density [1]. Wafer Scale Package (WSP) is emerging technologies derived from Flip Chip technology. In WSP, solder ball, Under Bump Metallurgy (UBM) and wafer itself present as main components in this joint. Electroless Nickel Immersion Gold (ENIG) is the most popular technique to deposit UBM. Electroless nickel plating provide flat, uniform bump through simple, low cost, low temperature, wet chemistry processes requiring little capital equipment. The metallurgical surfaces of the electroless UBM are the combination of either Ni/Au or Ni/Pd/Au [2]. Electroless nickel with stencil printing is low cost because this process by passes the high cost process of UBM sputtering and photolithography using thick photoresist [3]. The UBM for the solder bump needs to provide several functions such as solder wetability, a diffusion barrier between the pad and the solder, good adhesion, and the adequate electrical conductivity. Solder ball perform as an interconnection between the UBM and Printed Circuit Board (PCB) substrate.

Zincation is a process of coating the aluminum bond pad with a layer of zinc. The zinc layer will prevent the re-oxidation on aluminum bond pad and to initiate the electroless nickel deposition [4]. Zincation is currently one of the most suitable processes to activate the aluminum surface especially for the electroless technique [5]. After zincation process, then electroless will be plated by nickel prior to immersion gold. Nevertheless, a multiple immersion zincation operation is necessary for achieving better adhesion strength of the zinc layer [6].

In order to determine the reliability, WSP is mounted on the PCB substrate. PCB provides the interconnections and mechanical base for the electronic product. Because of wide availability of solder bumping process and PCB substrate, solder bumped on PCB has the potential of becoming one of the mainstream packaging technologies in the electronic industry [7]. Thermal fatigue failure is one of the critical elements in the reliability of WSP. Thermal fatigue occurs due to thermal expansion mismatch between the solder ball and the substrate and also can cause shear displacement on the solder joints [8]. It degrades the reliability of the solder joints. In order to minimize the effect of the thermal mismatch, the underfill encapsulant is introduced to fill in the space between the solder ball and

1-4244-0116-X/06/$20.00 ©2006 IEEE

the substrate. This process mechanically couples the organic substrate and the silicon chip together to enhance the thermal fatigue lifetime of solder joint.

The objective of this paper is to study the effect of number of zincation in ENIG UBM on reliability in microelectronics packaging. The Scanning Electron Microscope (SEM) and Atomic Force Microscopy (AFM) were used to obtain the surface texture, surface roughness and surface morphology in this study.

II. EXPERIMENTAL

This WSP of ENIG UBM evaluation using 6" wafer with bond pads composition of Al-1.0bt%Si-0.5bt%Cu and it was single diced to the size of 1.00 x 1.38 mm². This die has a 500 μm pitch and the wafer thickness is 430 μm. The bond pad opening is 320 μm with 2 μm thickness. The rest of the metals were passivated with 0.9 μm silicon nitride.

Experiment of ENIG process was carried out in laboratory scale using small beakers and heating plate that was equipped with fuzzy contact thermocouple IKAMAG IKA 2581000. The ENIG process started with surface cleaning using weak alkaline solution for removing any contamination such as dirt, grease and so on. It was further microetched the aluminum layer by aqueous solution containing of phosphoric. The cleaned aluminum is then sufficiently etched to eliminate solid impurities and alloying constituent which might create voids resulting in bridging of subsequent deposits [6]. After water rinse, the aluminum is de-smutted to remove metallic residue and aluminum oxides still remaining on the surface. After that, the first zincation process was applied. The zincate solution is containing an aqueous solution of sodium hydroxide, nickel sulphate, zinc sulphate and sodium cyanide. Then, double zincation and followed by triple zincation also applied with zinc removal was performed between the consecutive zincation to remove the zinc. Both double and triple zincation was used as a comparison in this study. All the samples preparation were then proceed to Electroless Nickel (EN) deposition followed by immersed in gold plating solution. Each step in the ENIG process must be rinsed with De-Ionized (DI) water to avoid chemical mixture to the subsequent process.

Both double and triple zincations of ENIG were then bumped with eutectic 63Sn37Pb solder ball directly on the UBM opening by ball drop technique. Solder ball size in this study is 300 μm. No-clean flux type was applied during this process. The function of this flux is to promote the sticky area and to help solder reflow activity [9]. Following then, samples were proceed to the Infra-Red (IR) reflow process using a programmable desktop Victronics Infra-Red (IR) reflow oven with the peak temperature of 220°C for eutectic solder ball according to IPC/JEDEC J-STD-020C.

This WSP samples using double and triple zincation of ENIG UBM were then mounted on the eight layers PCB.

Infra-Red (IR) reflow process also applied after WSP mounted on the PCB with 220°C peak temperature. Then, all WSP samples subjected to reliability tests such as Multiple Reflow, HTSL and TC. Samples will be picked up at every reflow cycle of 1, 3, 5 and 10 times at 220°C peak temperature for multiple reflow tests. The temperature used for HTSL chamber is 150°C with ambient, and no bias conditions (JESD22-A103C). Samples were storage up to 1000 hours. Samples will be picked up at every 50 hours of storage life. Nevertheless, samples also will be put into a Temperature / Thermal Cycle (TC) chamber of -65 °C to +150 °C, air-to-air, 2 cycles / hour with 15 minutes dwell time and each 15 minutes for ramp up and ramp down, respectively (JESD22-A104-B). The test is up to 1000 cycles. The interval for this test is 50 cycles. Samples size for every test interval is 20 dices.

Cross section analysis was performed after samples taken out from reliability tests. Analysis on the microstructure and material properties change were investigated to the entire sample. In order to reveal the microstructures clearly, optical microscope was used for the observation after cross section analysis. Scanning Electron Microscope (SEM) and Atomic Force Microscopy (AFM) were used for surface morphology and surface roughness of zincation process step.

III. RESULTS AND DISCUSSION

A. The Effect of Number of Zincation Process to Surface Roughness and Morphology

The effect of number of zincation was investigated through AFM and SEM, as shown in figure 1. From results shows, single zincation process performs a rough and not continuous in size of zinc seed layer. Double zincation observed a quite uniform surface and seems continuous of zinc seed layer. At this stage, the initial bond pad surface almost forms together with zinc seed. On the other hand triple zincation process produces a smooth and continuous zinc seed layer. The zinc seed are form together like an isolated island and almost the same level with initial bond pad surface. The bumps are getting fine and smaller. Hence, the higher the number of zincation process gives a better surface roughness. This happen due to aluminum dissolved into solution during the zincation process.

B. The Effect of Number of Zincation Process to Electroless Nickel Surface Morphology

Figure 2 shows the number of zincation process to electroless nickel surface morphology. From the figure, single zincation process produces a rough surface of electroless nickel. Electroless nickel surface morphology observes a smooth and fine surface at double zincation. Nevertheless, the triple zincation gives a better surface morphology to electroless nickel. At this stage, nickel surface are smoother and finer than other zincation process step. The smooth nickel surface morphology leads to better adhesion to solder ball. This result is consistent with J. Kloeser, et al. finding with smooth surface on nickel and gold (as a result of smooth surface of zinc layer) will provides better adhesion of solder bump and UBM [10].

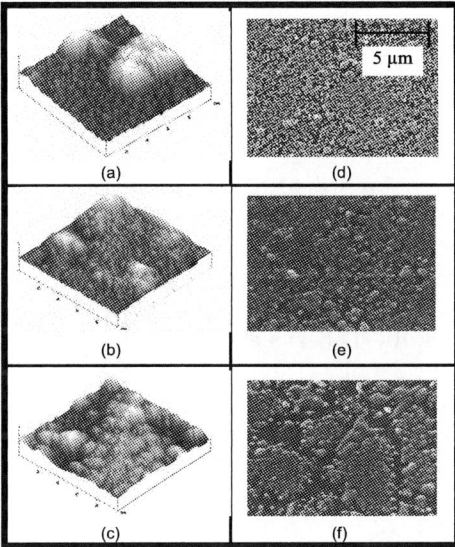

Fig 1. AFM results of the surface roughness for zincation process step: (a) single, (b) double, (c) triple and SEM results of the surface morphology: (d) single, (e) double, (f) triple.

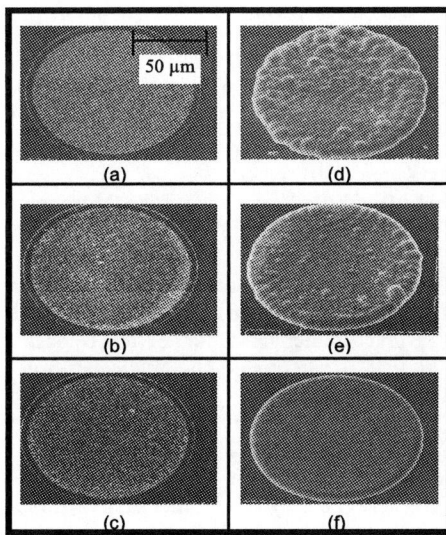

Fig 2. SEM results of the zincation surface morphology: (a) single, (b) double, (c) triple, to the electroless nickel surface morphology: (d) single, (e) double, (f) triple.

C. Cross Section Analysis

Samples were cross sectioned to investigate the qualitatively failure after subject to reliability tests. In order to reveal the microstructures clearly, optical microscope was carried out after cross section process. In Multiple Reflow test, cross section analysis for double zincation of ENIG UBM observed reliability failure after third cycles of reflow and up to tenth cycles as shown in figure 3. Delamination between edge of solder ball and Si substrate revealed at this stage. However, for triple zincation still survived up to tenth cycles. For HTSL, thermal fatigue failure for double zincation failed at 100 hours nevertheless triple zincation failed at 200 hours towards. The external and internal crack observed for this failure, shown in figure 4. The external crack initiated at the edge of the solder joint and grew toward the center of the intermetallic while internal crack initiated between the solder joint and intermetallic then grew away from the center. For TC test, thermal fatigue failure begins at 150 cycles towards for both double and triple zincation. The failure is similar with HTSL samples.

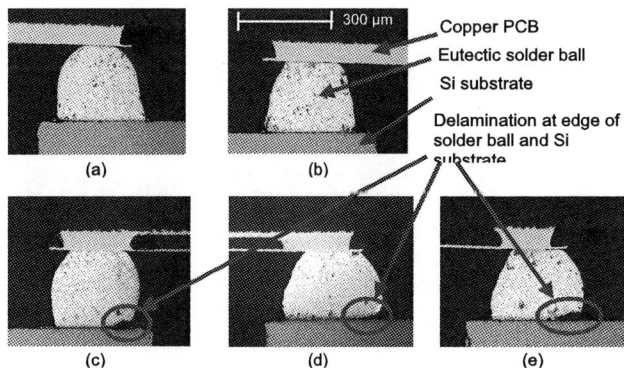

Fig 3. Optical micrograph of Multiple Reflow: (a) initial (b) first (c) third (d) fifth and (e) tenth cycle of reflow. Delamination at edge of solder ball and Si substrate.

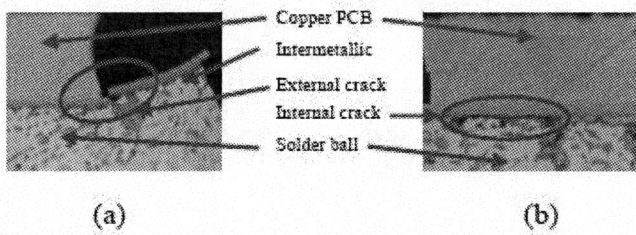

Fig 4. Optical micrograph of thermal fatigue crack: (a) external crack initiated at the edge of the solder joint and grew toward the center of the intermetallic and (b) internal crack initiated between the solder joint and grew away from the center of the intermetallic

During bonding process and solder joints, a metallurgical bond is formed at the interface of the eutectic solder ball and Cu pad on PCB, which is an intermetallic compound layer of Ni-Sn. It is necessary to prevent voids and cracks from forming at the interface of solder ball and UBM. Poor wetting, excessive intermetallic compounds, and contaminated wetting surfaces can all contribute to the formation of voids and cracks. An appropriate bonding reflow profile is essential to the formation of solder joints. It also activates the flux to improve the wetting ability of the solder bump on the Cu pad of PCB. However, too long dwell time at the peak temperature can form thick Ni/Sn IMC layer that degrades the reliability of the solder joint [7].

For the solder-bumped on the printed circuit board technology, the reliability of the solder joints is very dependent on the PCB design and feature tolerance [7]. Cracks and voids in solder joints including the UBM crack may cause solder joint failure. These defects may be the results of mismatch and thermal fatigue failure between the PCB and the solder ball.

D. Crack Growth Analysis

The crack lengths were measured from cross sections of the samples which were removed from HTSL and TC tests. The crack length shows in this graph is the summation of the external and internal cracks for both double and triple zincation of ENIG UBM.

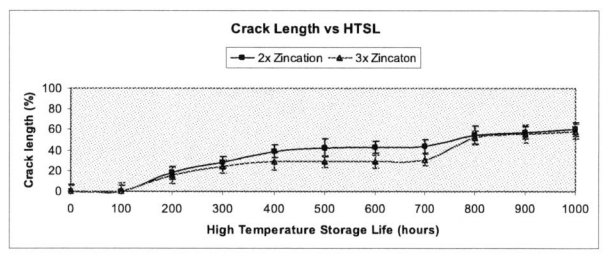

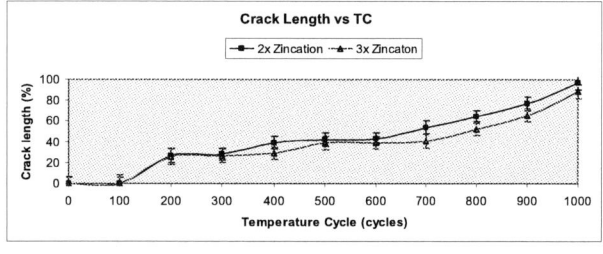

Fig 5. Crack lengths during HTSL and TC tests for both double and triple zincation of ENIG UBM

Figure 5 shows the average crack length (external and internal) versus the number of HTSL storage life and TC cycles, respectively. As depicted in the figure, the crack length for double zincation is always much bigger than the triple zincation of ENIG UBM for both HTSL and TC tests. However, crack length for TC test is slightly higher than HTSL test. The crack length during TC test reached up to 80-100 % in both double and triple zincation. On the other hand, crack length for HTSL test reached up to 60 % and the crack length did not spontaneously increase with HTSL time. Nonetheless, in TC test, crack length precipitately increase with cycle. In this case crack growth curve is depending with the type of reliability tests.

IV. CONCLUSION

As conclusion, triple zincation of ENIG UBM performs better, smoother and finer surface roughness and morphology. Smooth surface of zincation gives a smooth surface morphology on nickel and gold. Hence, it can achieve a reliable solder joint leads to better adhesion between UBM and solder ball therefore, it can achieve a reliable solder joint.

ACKNOWLEDGEMENT

The authors would like to thank Ministry of Science, Technology and Innovation (MOSTI), Malaysia, for funding through IRPA Grant no. 09-02-02-0107-EA259. Also would like to extend the acknowledgements to Colloidal Lab, University Malaya, Failure Analysis Lab and Reliability Lab at ON Semiconductor for supporting the analytical tools.

REFERENCES

[1] N. Wei-Chin, K. Tze-Man, W. Chen, and Q. Guo-Jun, "The Effects of Immersion Zincation to the Electroless Nickel Under-Bump Materials in Microelectronics Packaging", in *Proc. of 2nd Electronics Packaging Technology*, Singapore, 1998, pp. 89-94.
[2] John H Lau and Ricky S. W. Lee, "Microvias and Chip Scale Package for Low Cost and High Density Interconnects", *McGraw-Hill*, New York, 2000.
[3] J. Cai, A. Teng, Philip C. H. Chan, Simon P. C. Law, and G. Xiao, "A Study on Microstructure and Reliability Tests of Low Cost Flip Chip", in *Proc. of International Symposium on Electronic Materials and Packaging*, Hong Kong, 2000, pp. 91-98.
[4] S. K. Lee, J. G. Jin, and Y-H Kim, "A Study on the Nucleation Behavior of Zinc Particles on Aluminum Substrate", in *Proc. of the 3rd International Symposium on Electronics Materials and Packaging*, Korea, 2001, pp. 73-78.
[5] K. L. Lin and S. Y. Chang, "The Morphologies and the Chemical States of the Multiple Zincating Deposits on the Al Pads of Si Chips", *Thin Solid Films*, 1996, vol. 288, pp. 36-40.
[6] M. K. M. Arshad, I. Ahmad, A. Jalar, and G. Omar, "The Effects of Zincation Process on Aluminum Bond Pad Surfaces for Electroless Nickel Immersion Gold (ENIG) Deposition", in *Proc. of International Conference on Semiconductor Electronics*, Malaysia, 2004, pp. 656-662.
[7] G-W Xiao, P. C. H. Chan, A. Teng, P. S. W. Lee, M. M. F Yuen, "Reliability Study and Failure Analysis of Fine Pitch Solder Bumped Flip Chip on Low-Cost Printed Circuit Board Substrate", in *Proc. of 51st Electronic Components and Technology*, USA, 2001, pp. 598-605.
[8] M. Wada, "Development of High Reliability Underfill Material", in *Proc. of 2nd IEMT / IMC Symposium*, Japan, 1998, pp. 54-58.
[9] J. Criscione "Bump and Flux", in *Proc. of 40th Electronic Components and Technology*, USA, 1990, pp. 646-647.
[10] J. Kloeser, A. Ostman, R. Aschenbrenner, E. Zakel, and H. Reichl, "Approaches to Flipchip Technology Using Electroless Nickel-Bumps", in *Proc. of 1995 Japan International Electronic Manufacturing Technology Symposium*, Japan, 1995, pp. 60-68.

Poster Session
Reliability Physics

610

2006 25th International Conference on Microelectronics

Investigation of the Ion Defect States by Photoacoustic Spectroscopy

D. M. Todorović, V. Jović, M. Smiljanić, T. Grozdić

Abstract – The ion defect states in SiO_2 film on Si substrata (SiO_2/Si) was investigated by photoacoustic (PA) spectroscopy. The amplitude and phase spectra were measured and analyzed in dependence on the energy of excitation optical beam in the sub-bandgap region. In the energy range near the energy gap of Si, the PA spectra are the consequence of the ion-defect states formed on dielectric-semiconductor interface. The sub-bandgap PA spectra are proposed to obtain the energy-dependent distribution of interface states in SiO_2 – Si system with different concentration of Na-ions.

I. INTRODUCTION

During the past fifteen years, photothermal and photoacoustic (PA) science has been successfully applied to study of semiconductor materials and microelectronic devices [1,2]. When a material is excited with an intensity-modulated energy source, for example the modulated optical beam, its optical, thermal and elastic properties can be altered by the absorption of the incident energy. The induced changes in the sample properties can be detected by PA effect. The PA effect results in the generation of an acoustic signal (PA signal) by a sample exposed to modulated light. Because the amplitude and phase of the PA signal are related to the optical absorption coefficient of the sample, scanning the wavelength of the modulated monochromatic light, which excites the sample, can generate an optical absorption spectrum.

The PA effect in semiconductor is based on the photogeneration of electron-hole pairs, i.e. plasma waves, generated by the absorbed intensity-modulated excitation. Depth-dependent plasma waves contribute to the generation of periodic heat and mechanical vibrations, i.e. thermal and elastic waves. Thermal and elastic waves can be manifested in various ways: one is sound generation - the PA generation. There are three mechanisms of PA generation [3,4]. The propagation of thermal waves through the sample and through the surrounding gas, which

is in contact with the sample, produces acoustic waves in the gas. This is the so-called thermodiffusion (TD) mechanism of PA generation [5]. On the other hand, the thermal waves in the sample cause elastic vibrations, which propagate to the sample surface where they cause expansion and contraction of the surrounding gas, i.e. an acoustic wave [6]. This is the thermoelastic (TE) mechanism of PA generation. Also, semiconductor materials show a mechanical strain when electron-hole plasma is generated. The photoexcited carriers produce periodic elastic deformation in the sample - electronic deformation (ED), which in turn generates an acoustic wave [7,8]. For the conditions in this work, only the TD component is important.

Due to their importance in microelectronic applications and as a tool in the analysis of other fundamental semiconductor physical investigations, the SiO_2/Si structure has been studied extensively. The formation of SiO_2 is of essential importance in the fabrication of Si devices because of its masking and passivating capabilities, its basic role in MOS structures, and its characteristics as dielectric in the formation of capacitors and multi-level interconnections.

The SiO_2 is readily formed by the thermal oxidation of Si in either oxygen or water vapor atmosphere. Other methods include the anodic oxidation of Si and the vacuum deposition. In semiconductor fabrication the most common type of SiO_2 formation is by thermal oxidation of Si. In this case the oxidation proceeds by the semiconductor-oxide interface movies into the semiconductor at a rate, which is dependent upon the oxidation conditions. In the presence of impurities within the semiconductor, the motion of the SiO_2-Si interface is associated with impurity redistribution.

The properties of SiO_2 are very sensitive to impurity content and structure. Impurities most frequently found in SiO_2 are these that are used as dopants in Si treatment, as phosphorous, boron, aluminum, arsenic, antimony; in the structure of most glasses are incorporated also lead, potassium and sodium. Impurities (other of then silicon and oxygen) affect the properties of SiO_2 significantly if they are ionized; if they are electrically neutral they merely occupy holes in the (Si-O_2) network. The H_2O and OH content of SiO_2 influences the oxide properties also.

In this work for the first time, the ion defect states in SiO_2 film on Si substrata (SiO_2/Si) was investigated by PA spectroscopy.

Dragan M. Todorović and Tomislav Grozdić are with Center for Multidisciplinary Studies, University of Belgrade, P.O.Box 33, 11030 Belgrade, Serbia & Montenegro,
E-mail: dmtodor@afrodita.rcub.bg.ac.yu
Vesna Jović, Miloljub Smiljanić, are with Institute for chemistry, technology and metallurgy, Njegoseva 12, Belgrade, Serbia & Montenegro

---

1-4244-0116-X/06/$20.00 ©2006 IEEE

II. EXPERIMENTAL RESULTS

The effect of SiO$_2$ film on Si substrata (SiO$_2$/Si) was investigated by photoacoustic spectroscopy. The amplitude and phase spectra were measured and analyzed in dependence on the energy of excitation optical beam. The PA amplitude and phase spectra of single crystal Si samples (n-(100), boron doped, 3-5 Ωcm, $N_d = 1.5 \times 10^{15}$ cm^{-3}, 525 μm thick) with and without SiO$_2$ film were studied at room temperature in the spectral range of the excitation optical beam from 0.8 to 1.50 eV. The SiO$_2$ film (0.4 - 0.7 μm thick) was formed on the Si wafers by wet thermal oxidation with O$_2$ gas bubbling through the water solution of NaCl with different content of Na ions.

Fig.1 and 2 shows PA amplitude and phase spectra of Si wafer and SiO$_2$/Si structure (SiO$_2$ film was illuminated) vs. wavelength for modulation frequency of 10Hz.

III. ANALYZE OF RESULTS

The amplitude PA spectra of Si wafer and SiO$_2$/Si samples are analogous to the optical absorption spectra near the energy gap. In both type of samples, the PA spectra clearly revealed the fundamental absorption edge near the energy gap, E_g, (an energy of 1.11 eV) of Si and corresponds closely to that observed by conventional optical absorption measurements.

The quantitative analysis of the PA spectra of SiO$_2$/Si samples is possible by using the Rosencwaig and Gersho theory [5] to calculate the theoretical PA signal and fitting with theoretical ones. At energies lower than energy gap ($E < E_g$), i.e. in the sub-bandgap optically-induced absorption, where the relationships among the optical absorption length $1/\alpha$, the sample thickness l, the thermal diffusion length μ are $\mu > l$ and $1/\alpha \gg \mu, l$. In Rosencwaig and Gersho theory, these relationships hold for optically transparent and thermally thin solids. In these cases, the amplitude PA signal is proportional to α.

Spectra of the sub-bandgap absorption are generally decomposed into *band to tail* and *band to defect* type transitions. The first type is responsible for the exponential increase at the absorption edge, which is commonly described by the Urbach rule and follows, at a given temperature, the relation [9]

$$\alpha(E) = \alpha_o \exp\left(\frac{E - E_o}{E_u}\right), \tag{1}$$

with α_o and E_o as material parameters and E_u the Urbach energy describing the width of the exponential absorption band edge. Therefore, the expected PA spectrum as a function of E was deduced by using relation for PA signal in function of α and supposing that $\log(\alpha)$ is proportional to E. The parameters of the Urbach's tail were evaluated to obtain a best fit to the absorption spectra below energy gap. The second type of transition shows up as a plateau at low photon energy. In this region of low density of states the

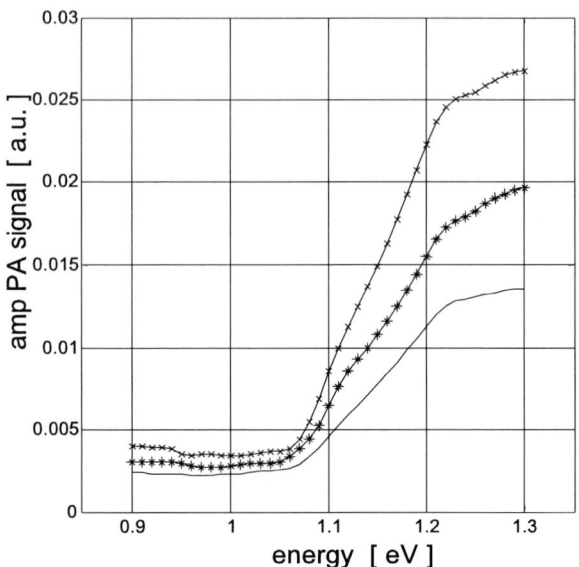

Fig. 1 Amplitude PA spectra near the energy gap: (.) Si wafer ; (*) clear SiO$_2$ / Si ; (x) 4NaCl.

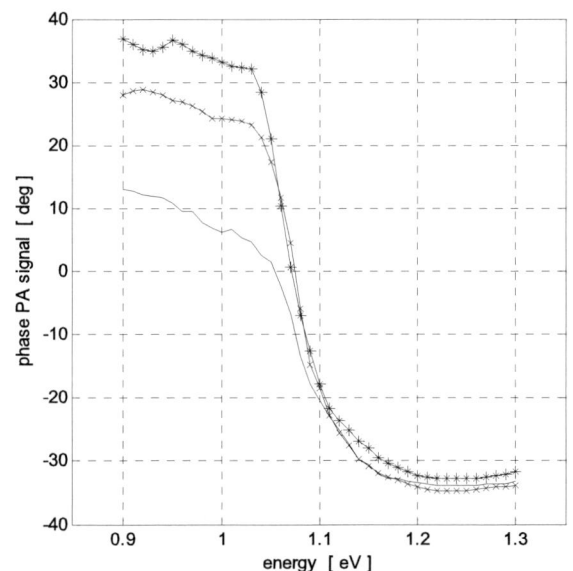

Fig. 2 Phase PA spectra near the energy gap: (.) Si wafer ; (*) clear SiO$_2$ / Si ; (x) 4NaCl.

absorption coefficient is expected to be orders of magnitude lower compared to that for excitation energies at and above the band edge.

As transmission spectroscopy is not a sensitive enough method, these transitions cannot be detected and are usually ignored when fitting the exponential decay [10]. However, our investigations show that this defect absorption must be taken into account when fitting the exponential absorption edge by the Urbach expression. Therefore, the expected PA spectrum as a function of E was deduced by using relation for PA signal in function of α and supposing that α is given by Eq.(1). The parameters of the Urbach's tail were evaluated to obtain a best fit to the absorption spectra below energy gap.

Fig.3 shows experimental and theoretically calculated amplitude PA spectra of SiO_2 / Si near the energy gap, with low and high Na-ion defects.

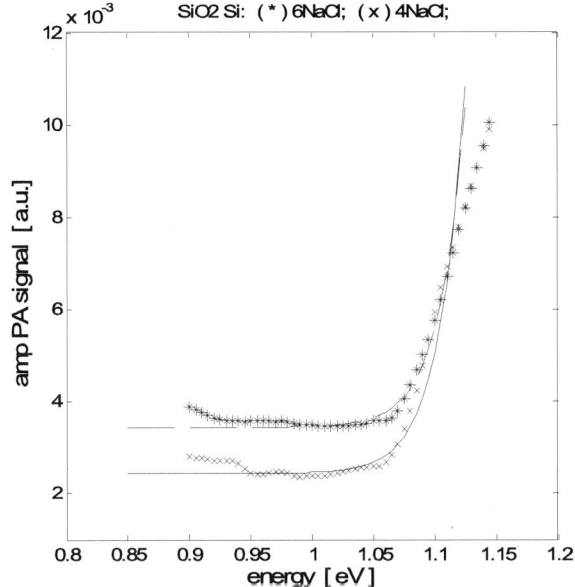

Fig. 3 Experimental amplitude PA spectra of SiO_2 / Si near the energy gap: (*)low Na-ion defects; (x) high Na-ion defects; (x) 4NaCl. Lines are the theoretically fitted PA spectra.

Both curves are fitted with the Urbach expression (1) plus constant sub-bandgap absorption, α_{ds}, which corresponds to the optical absorption from *band to defect* type transitions for each sample. This fit corresponds to the solid (dashed) lines in Fig.3.

In Table1 the fitting parameters for SiO_2 / Si samples with low and high Na-ion defects are given.

TABLE 1.
FITTING PARAMETERS FOR TWO SiO_2 / Si SAMPLES WITH LOW AND HIGH Na-IONS CONCENTRATION.

Sample	α_{ds} [cm^{-1}]	α_o [cm^{-1}]	E_o [eV]	E_u [meV]
SiO_2 / Si low Na-ion	$7.45 \cdot 10^{-2}$	$5.6 \cdot 10^{-4}$	0.998	21.2
SiO_2 / Si high Na-ion	$1.05 \cdot 10^{-1}$	$5.6 \cdot 10^{-4}$	0.998	21.2

The extracted Urbach tail width E_u , defect state absorption α_{ds} , as well as the values of, α_o and E_o as material parameters, are given. The Urbach tail width E_u remains nearly constant and can be explained by phonon coupling. The value for E_g = 1.11 eV is obtained as a point where the experimental and theoretical curves drastically divergent

(the theoretical model is valid for sub-bandgep absorption, i.e. for energy $E < E_g$).

When assuming a constant optical matrix element, the integral over the sub-bandgap absorption is proportional to the defect density N_{ds}. Assuming the validity of the postulate, given by relation:

$$N_{ds} \propto \int \alpha \cdot dE , \qquad (2)$$

for the case of SiO_2 / Si sample, and applying it to the integrated sub-bandgap absorption α_{ds} (see Table 1) we conclude: The second sample with high concentration of Na-ions, is responsible for the increasing of the total defect density N_{ds} in the energy range between 0.9 eV and the band-gap E_g by a factor of more than 1.4.

IV. DISCUSSION

In a practical SiO_2/Si system (the insulator – semiconductor structure) exist many different states and charges that will, in one way or another, affect the characteristics of SiO_2/Si system. Generally, there are four different types of states and charges (Fig.4):

1. *interface states*. They are defined as energy levels within the forbidden band gap at the dielectric – semiconductor interface. They can exchange charges with the semiconductor in a short time (interface generation-recombination centers or fast interface states). These states are the result of processing conditions; the Si surface potential determines if they are charged;

2. *fixed interface charges*. They are very close to the SiO_2-Si interface (near at the semiconductor surface, $\sim$ 20 nm) and are imobile under applied electric field;

3. *mobile ions*. The mobile ionic charge (Q_m) in SiO2 films on silicon is one of the main causes of the instability of the electrophysical characteristics of Si-based MOS structures and devices based on them; therefore, a fairly large amount of attention has been devoted to research on the kinetics of ion drift. It has been established that Q_m is created by positively charged alkali-metal Li$^+$, Na$^+$, and K$^+$ ions, which become mobile at elevated temperatures ($\sim$200 - 300 °C) in fields equal to 10^5 - 10^6 V/cm [11]. They are usually the result of processing conditions.

4. *ionized traps*. Traps within the oxide, which can be created by radiation. For example, exposure to x-ray, electron or other ionizing radiation usually leads to positively charged traps in the oxide;

The foregoing charges are the effective net charges per unit area (C/cm^2).

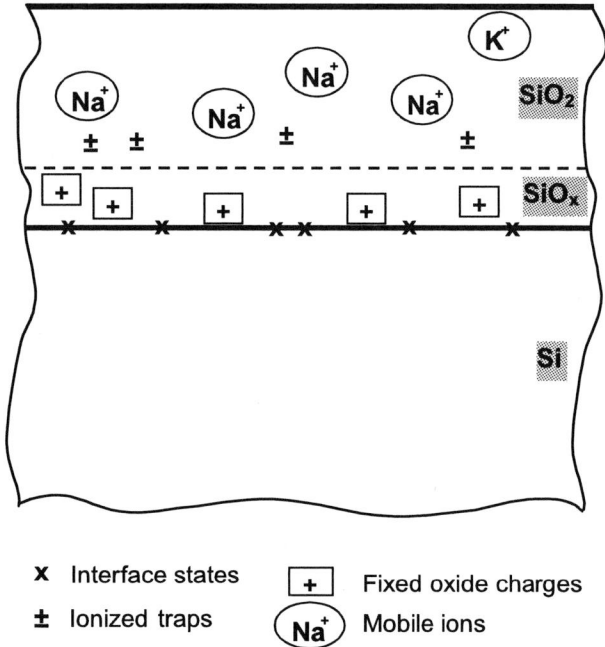

x Interface states

± Ionized traps

| + | Fixed oxide charges

(Na$^+$) Mobile ions

Fig. 4 Interface traps and charges associated with the SiO$_2$ / Si structure.

The change of the PA amplitude and phase with the energy of excitation can be explained by model of quasicontinual energy spectrum of interface defect states. The amplitude and phase PA spectra of Si and SiO$_2$/Si samples (Fig.2 and 3), in this energy region, is the consequence of the interface states formed on dielectric-semiconductor interface, i.e., the influence of the interface-trapping states on the thermal and electronic transport processes in SiO$_2$/Si junction. At these energies, the amplitude PA signal increased for SiO$_2$/Si sample, i.e. PA signal is the function of the interface energy states. This indicates the generation of heat following optical absorption and nonradiative relaxation processes at these interface-trapping states.

Due to high concentration of the interface defects concerning the concentration of surface atoms of basic material, it is possible expansion of the local levels to energy bends, as a consequence of interface defects interaction. In addition, the nonuniforme distribution of the interface defects is characterized with a quasicontinual energy spectrum. In this case, density of defect states exponentially increased near the band edges.

V. CONCLUSION

In conclusion, these results showed that the PA spectroscopy is very convenient for investigation of ion-defect states, i.e. the interface energy states in dielectric-semiconductor structures. Our investigations show that this defect absorption must be taken into account when fitting the exponential absorption edge by the Urbach expression. The PA spectra were fitted with the Urbach expression plus constant sub-bandgap absorption, α_{ds}, which corresponds to the optical absorption from band to defect type transitions. For the case of SiO$_2$ / Si sample, the sub-bandgap absorption α_{ds}, obtained by fitting procedure, is proportional to the defect density N_{ds}. Then, it was possible to conclude: The increasing concentration of Na-ions is responsible for the increasing of the total defect density N_{ds}.

ACKNOWLEDGEMENT

This work was performed in the frame of the project "Micro and Nanosystems Technologies, Structures and Sensors", supported by grants from the Ministry for Science and Environmental Protection, Republic of Serbia, Grant No. TP – 6151 B.

REFERENCES

1. D. Almond, P. Patel, *Photothermal Science and Techniques*, Chapman&Hall, London, 1996.

2. *Semiconductors and Electronic Materials*, Eds. A. Mandelis and P. Hess, Series: *Progress in Photothermal and Photoacoustic Science and Technology*, Opt. Eng. Press, New York, 2000.

3. D.M. Todorović, P.M. Nikolić, A.I. Bojičić, K.T. Radulović, "Thermoelastic and electronic strain contributions to the frequency transmission photoacoustic effect in semiconductors" Phys. Rev. B, 55(23), 15631-15642, 1997.

4. D.M. Todorović, P.M. Nikolić, "Carrier transport contribution to thermoelastic and electronic deformation in semiconductors", Chapt. 9 in *Semiconductors and Electronic Materials*, A. Mandelis, P. Hess, Eds., pp. 271-315, SPIE Press, Bellingham, Washington, 2000.

5. A. Rosencwaig, A. Gersho, "Theory of the photoacoustic effect with solid", *J.Appl.Phys.*, 47, 64, 1976.

6. F. McDonald and G.Wetsel, "Generalized thery of photo-acoustic effect", J.Appl.Phys. 49(4), 2313, 1978.

7. R.G. Stearns, G.S. Kino, "Effect of electronic strain on photo-acoustic generation in silicon", *Appl. Phys. Lett.*, 47, 1048, 1985.

8. D.M. Todorović, P.M. Nikolić, A.I. Bojičić, "Photoacoustic frequency transmission technique: Electronic deformation effect in semiconductors", *J.Appl.Phys.*, 85(11), 7716, 1999 .

9. F. Urbach, *Phys. Rev.* 1953, *92*, 1324.

10. A. Meeder, D. Fuertes Marron, A. Rumberg, M.Ch. Lux-Steiner, V. Chu, J. P. Conde, J. *App.Phys.*, 2002, *92(6)*, 3018-3020.

11. S.G.Dmitriev and Yu.V.Markin, S*emiconductors*, 1998, 1289.

2006 25th International Conference on Microelectronics

Relationship between Intrinsic Breakdown Field and Bandgap of Materials

Li-Mo Wang

Abstract - A universal expression for the relationship between intrinsic breakdown field and bandgap of both semiconductors and insulators is proposed, and a quantitative criterion for distinguishing between semiconductors and insulators is introduced for the first time.

I. INTRODUCTION

For a high quality semiconductor or insulator material without defects and impurities, its theoretical intrinsic breakdown field E_{BI} should be a fixed value. However, the values of the breakdown field E_B of the same material given in the literatures are quite different due to the defects, composition mismatch and measurement errors. The maximum breakdown field E_{BM} obtained from the reported E_B data can be regarded as an approximate value of E_{BI}. The more perfect the material, the higher the breakdown field. So, E_{BI} should be higher than or equal to E_{BM}.

For semiconductors, there are some expressions for the relationship between critical field E_C (*i.e.* E_{BM}) and bandgap E_g. The expression $E_C = 1.02 \times 10^7 \sqrt{(q/\varepsilon)} N_B^{1/8} E_g^{3/4}$ for abrupt junctions was derived based on the data of Si, Ge, GaAs and GaP only [1]. Recently, expressions E_{BM} (I) $= 1.73 \times 10^5 E_g^2$ for indirect-gap semiconductors and E_{BM} (D) $= 2.38 \times 10^5 E_g^{2.5}$ for direct-gap semiconductors have been derived by means of a least square method (LSM) to fit the E_{BM} data of thirteen different semiconductors for high-voltage device applications with very low doping concentrations, ignoring the impurity doping dependencies [2]. The two expressions can only give statistical results of the E_{BM} data at present, and should be revised with new data. So they can not well reflect the intrinsic breakdown characteristics. In addition, they are not applicable to insulators. Until now, no quantitative expression is available for insulators.

II. EXPRESSIONS FOR INTRINSIC BREAKDOWN FIELD

The E_{BM} data of fourteen commonly used elemental and binary compound semiconductors and eight binary insulators have been collected from the literatures. These data are listed in Tables I and II, respectively, and are plotted in Figure 1 together with the bandgap E_g. Diamond

Li-Mo Wang is with the Jiangsu College of Information Technology, 3 Liangxi Road, Wuxi 214061, Jiangsu, China, E-mail: Wanglimo_prof@163.com

and AlN are plotted in Figure 1 as insulators since at room temperatures, undoped diamond and AlN are good insulators. We draw a thick dotted straight line through the higher data points of the insulators, and draw a thick solid straight line which passes through the E_{BM} point of silicon and is slightly above but as close as possible to the data points of the compound semiconductors. The two straight lines can be expressed by a universal expression

$$E_{BI} = 1.36 \times 10^7 (E_g / 4.0)^\alpha \text{ (V/cm)} \qquad (1)$$

where $\alpha = 3$ for semiconductors, and $\alpha = 1$ for insulators. So for semiconductors,

$$E_{BIS} = 1.36 \times 10^7 (E_g/4.0)^3 \text{ (V/cm)} \qquad (2)$$

And for insulators,

$$E_{BII} = 1.36 \times 10^7 (E_g/4.0) \text{ (V/cm)}. \qquad (3)$$

The difference of the power α values indicates that the breakdown mechanisms in semiconductors and insulators are different.

The universal expression (1) is applicable not only to both narrow and wide bandgap semiconductors, but also to insulators. This is the first time to propose a quantitative relationship between intrinsic breakdown field and bandgap of insulators.

From Figure 1 we can see that almost all the data points are located under the thick lines. This is reasonable. It means that the quality of these materials needs to be optimized to increase the breakdown field.

The expressions for the semiconductors in [2] are also plotted in Figure 1 as a comparison. We can see that many data points are located above their straight lines. This is the inevitable results of the LSM.

III. APPLICATIONS

A. A quantitative criterion for distinguishing between semiconductors and insulators

Resistivity ρ and bandgap E_g are usually used to define materials, but no unified criterion is available so far. For example, when ρ is used for semiconductors, Muller's criterion is 10^{-2} - 10^5 Ω-cm [3], Sze's is 10^{-3} - 10^8 Ω-cm [4], while Berger's is 10^{-5} - 10^{11} Ω-cm [5]. When E_g is used for insulators, Muller's criterion is $E_g > 5.0$ eV [3], while Quirk's is $E_g > 2.0$ eV [6].

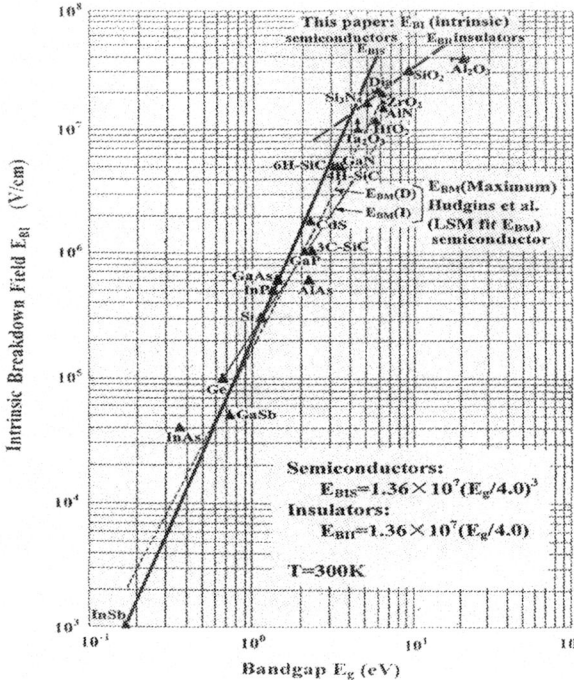

Fig. 1. Intrinsic breakdown field versus bandgap of semiconductors and insulators. The E_{BM} of a material is the maximum value available at present, while the E_{BI} of a material is a limit value, which shows the potential of the material.

From Figure 1 it can be seen that the distribution of E_{BI} vs E_g in the log-log plot is not continuous but forms two straight lines with different slopes, with semiconductors on one line and insulators on the other. The two straight lines intersect at $E_g = 4.0$ eV, which just divides the materials into two categories: the materials with $E_g < 4.0$ eV are semiconductors and those with $E_g > 4.0$ eV are insulators. It indicates that for either semiconductors or insulators, the intrinsic breakdown mechanism is closely related to the conductance mechanism, but the conductance mechanism, breakdown mechanism and E_{BM} vs E_g relationship of semiconductors are different from those of insulators.

To combine the above criterion with the criterion for metals, $E_g = 0.0$ eV, we can conclude that materials can be quantitatively classified by E_g into metals, semiconductors and insulators, i.e., *the materials with E_g at 0.0 eV are metals, those with E_g between 0.0 eV and 4.0 eV are semiconductors, and those with E_g higher than 4.0 eV are insulators.* This quantitative criterion can be called *breakdown field criterion* of materials, or simply, *Wang's criterion.*

Sometimes, the semiconductors with a very narrow bandgap, such as $E_g \leq 0.2$ eV, are called semimetals, and some insulators with the bandgap near 4.0 eV are called semiinsulators.

B. Material figures of merit expressed directly by E_g

By using the above expression for E_{BIS}, the figures of merit of semiconductors can be expressed directly by E_g (see Table III).

C. Prediction of E_{BI} values of materials

The E_{BI} values of many compound semiconductors and insulators have not been experimentally obtained and are usually difficult to measure, but their E_g values are already known or easy to measure, so it is of significance to predict the E_{BI} values of the interested materials. By using the above expressions, the E_{BI} values of many important elemental and compound semiconductors and high-k binary gate dielectrics have been calculated (see Tables IV and V).

The above expressions are also helpful for device simulation and theoretical derivation.

IV. CONCLUSION

1. A universal expression for the relationship between intrinsic breakdown field and bandgap of materials is proposed.

2. For both wide and narrow bandgap semiconductors, a more reasonable approximate expression for intrinsic breakdown field E_{BIS} is introduced.

3. For insulators, including high-k dielectrics, a quantitative relationship between intrinsic breakdown field and bandgap is proposed for the first time.

4. A quantitative criterion for classifying materials by their bandgap E_g, i.e., a breakdown field criterion, is proposed for the first time.

5. Simplified figures of merit of semiconductors expressed directly by bandgap E_g are given.

6. The values of the intrinsic breakdown field of many important binary compound semiconductors and high-k gate dielectrics are calculated.

ACKNOWLEDGEMENT

The author would like to thank Mr. Shi-long Cai for his assistance and helpful suggestions and discussions.

REFERENCES

[1] S. M. Sze and G. Gibbons, "Avalanche Breakdown Voltages of Abrupt and Linearly Graded p-n Junctions in Ge, Si, GaAs and GaP", *Appl. Phys. Lett.*, vol.8, pp. 111-113, 1966.

[2] J. L. Hudgins, G. S. Simin, E. Santi, and M. Asifkhan, "An Assessment of Wide Bandgap Semiconductors for Power Devices", *IEEE Trans. Power Electronics*, vol. 18, pp. 907 –914, 2003.

[3] R. S. Muller and T. I. Kamins, *Device Electronics for Integrated Circuit*, 3rd Ed., John Wiley &Sons, 2003.

[4] S. M. Sze, *Semiconductor Devices Physics and Technology*, 2nd Ed., John Wiley & Sons, 2002.

[5] L. I. Berger, *Semiconductor Materials*, CRC Press, New York, 1997.

[6] M. Quirk and J. Serda, *Semiconductor Manufacturing Technology*, Prentice-Hall, 2001.

TABLE I
MAXIMUM BREAKDOWN FILED E_{BM} AND BANDGAP E_G OF THE SEMICONDUCTORS IN FIGURE 1 (T = 300 K)

Material [Data source]	InSb [7]	InAs [7]	GaSb [7]	Ge [7]	Si [7]	GaAs [11]	InP [7]	AlAs [9]	GaP [7]	SiC[7] 3C	SiC[7] 6H	SiC[7] 4H	CdS	GaN [7]
E_g (eV)	0.17	0.354	0.726	0.67	1.11	1.43	1.34	2.17	2.26	2.36	3.0	3.23	2.42 [8]	3.37
E_{BM} (MV/cm)	0.001	0.04	0.05	0.1	0.3	0.6	0.5	0.6	1.0	1.0	5	5	1.8 [10]	5

Data source: [7] *NCSR: National Compound Semiconductor Roadmap*, Available: www.ncsr.csci-va.com.
[8] S.M. Sze, *Physics of Semiconductor Devices*, New York: Wiley, 1981.
[9] Yu. A. Goldberg, *Handbook on Semiconductor Parameters*, Vol. 2, Chapter 1.
[10] R. Williams, *Phys. Rev.*, vol. 123, p. 1645, 1961.
[11] M. N. Yoder, *IEEE Trans. Electron Devices*, vol. 43, p. 1633, 1996.

TABLE II
MAXIMUM BREAKDOWN FIELD E_{BM} AND BANDGAP E_G OF THE DIELECTRICS (INSULATORS) IN FIGURE 1 (T = 300 K)

Material [Data source]	Ta_2O_5 [12]	HfO_2	ZrO_2 [15]	AlN	Diamond	Si_3N_4	SiO_2	Al_2O_3 (sapphire) [20]
E_g (eV)	4.2-4.3	5.65 [13]	5-7	6.23 [16]	5.46-6.4 [7]	5.0 [8]	9.0 [8]	18-23
E_{BM} (MV/cm)	>10	13 [14]	20	>15 [17]	21.5 [18]	16 [19]	30 [21]	39

Data source: [7] *NCSR: National Compound Semiconductor Roadmap*, Available: www.ncsr.csci-va.com.
[8] S. M. Sze, *Physics of Semiconductor Devices*, New York: Wiley, 1981.
[12] J. Caughman *et al*, *47th AVS National Symposium* , October 3, 2000.
[13] M. Balog *et al*, *Thin Solid Films*, vol.41, p. 247, 1977.
[14] J. C. Lee, *4th Annual Topical Research Conference on Reliability*, October, 2000.
[15] J. P. Chang *et al*, *J. Vac. Sci. Technol.*, B19, p.1782, 2001.
[16] I. Vurgaftman *et al*, *J. Appl. Phys.*, vol. 89, p. 5815, 2001.
[17] T. L. Chu *et al*, *J. Electrochem. Soc.*, vol. 122, p. 995, 1975.
[18] P. Liu *et al*, *IEEE J. Quantum Electron.*, QE-14, p. 574, 1978.
[19] T. P. Ma, *IEEE Trans. Electron Devices*, vol. 45, p. 680, 1998.
[20] J. W. Gardner *et al*, *Microsensors, MEMS, and Smart Device* (Appendix H), New York: John Wiley & Sons, 2001.
[21] E.Harari,*J.Appl.phys.*,vol.49,p.2478,1978.

TABLE III
SIMPLIFIED FIGURES OF MERIT EXPRESSED DIRECTLY BY E_G

Figure of merit	Symbol	Expressed by E_{BI}	Expressed directly by E_g	Simplified figure of merit
Johnson's figure of merit [22]	JFOM	$\left(\dfrac{E_{BI}v_s}{2\pi}\right)^2$	$\dfrac{1.13\times10^{10}}{\pi^2}v_s^2 E_g^6$	$v_s^2 E_g^6$
Baliga's on-resistance figure of merit [23]	BFOM	$\varepsilon\mu E_{BI}^3$	$9.66\times10^5\,\varepsilon\mu E_g^9$	$\varepsilon\mu E_g^9$
Baliga's high frequency figure of merit [24]	BHFFOM	μE_{BI}^2	$4.54\times10^{10}\,\mu E_g^6$	μE_g^6
Huang's switching power figure of merit [25]	HMFOM	$E_{BI}\sqrt{\mu}$	$2.13\times10^5\sqrt{\mu}\,E_g^3$	$\sqrt{\mu}\,E_g^3$
Huang's chip size figure of merit [25]	HCAFOM	$\varepsilon E_{BI}^2\sqrt{\mu}$	$4.54\times10^{10}\,\varepsilon\sqrt{\mu}\,E_g^6$	$\varepsilon\sqrt{\mu}\,E_g^6$
Huang's heat dissipation figure of merit [25]	HTFOM	$\dfrac{\sigma_{th}}{\varepsilon E_{BI}}$	$\dfrac{\sigma_{th}}{2.13\times10^5\,\varepsilon E_g^3}$	$\dfrac{\sigma_{th}}{\varepsilon E_g^3}$
Gao-Morkoc's collector figure of merit [26]	Collector FOM	$v_s^{5/4} E_{BI}$	$2.13\times10^5\,(v_s^{5/4})\,E_g^3$	$v_s^{5/4} E_g^3$

Note: a) Since only the ratio of the figures of merit (FOMs) to that of silicon is usually used for material selection, we call the FOMs without the numeric constant simplified figures of merit.

b) σ_{th} is the thermal conductivity of a material.

Data source: [22] E. O. Johnson, *RCA Rev.*, vol. 26, p. 163, 1963.

[23] B. J. Baliga, *J. Appl. Phys.*, vol. 53, No.3, p. 1759, 1982.

[24] B. J. Baliga, *IEEE Electron Device Lett.*, vol. 10, p. 455, 1989.

[25] A. Q. Huang, *IEEE Electron Device Lett.*, vol. 25, p. 98, 2004.

[26] G.-B. Gao and H. Morkoc, *IEEE Trans. Electron Devices*, vol. 38, p. 2410, 1991.

TABLE IV

PREDICTED INTRINSIC BREAKDOWN FIELD E_{BIS} VALUES OF IMPORTANT SEMICONDUCTORS (T = 300 K)

$$E_{BIS} = 1.36 \times 10^7 \, (E_g/4.0)^3 \quad \text{V/cm}$$

Semiconductor	Ge	Si	SiC			AlSb	BP	GaN	GaSb	GaAs	GaP	InAs
			3C	4H	6H							
E_g (eV) [7, 8]	0.66	1.12	2.36	3.23	3.0	1.58	2.0	3.36	0.72	1.42	2.26	0.36
E_{BIS} (MV/cm)	0.061	0.30	2.80	7.18	5.75	0.84	1.70	8.08	0.08	0.61	2.46	0.01

Semiconductor	InSb	InP	CdS	CdSe	CdTe	ZnO	ZnS	PbS	PbTe	AlAs	AlP	ZnSe	ZnTe
E_g (eV) [7, 8]	0.17	1.35	2.42	1.70	1.56	3.35	3.68	0.40	0.31	2.16	2.45	2.7	2.25
E_{BIS} (MV/cm)	0.00105	0.52	3.02	1.05	0.81	8.01	10.6	0.014	0.0064	2.15	3.13	4.19	2.43

Data source of E_g: [7] *National Compound Semiconductor Roadmap*, Available: www.ncsr.csci-va.com.

[8] S. M. Sze, *Physics of Semiconductor Devices*, New York: Wiley, 1981.

TABLE V

PREDICTED INTRINSIC BREAKDOWN FIELD E_{BII} VALUES OF SOME CANDIDATE HIGH-κ BINARY GATE DIELECTRICS (T = 300 K)

$$E_{BII} = 1.36 \times 10^7 \, (E_g/4.0) \quad \text{V/cm}$$

High-k dielectric film	SiO$_2$	Si$_3$N$_4$	HfO$_2$	ZrO$_2$	Y$_2$O$_3$	Ta$_2$O$_5$	La$_2$O$_3$	Pr$_2$O$_3$	Gd$_2$O$_3$	Lu$_2$O$_3$
E_g (eV) [27, 28]	9.0	5.3	6.0	5.8	6.0	4.4	6.0	4.6	5.3	5.4
E_{BII} (MV/cm)	30.6	18.0	20.4	19.7	20.4	15.0	20.4	15.6	18.0	18.4

Data source of E_g: [27] J. Robertson, *J. Vac. Sci. Technol.*, B18, p. 1785, 2000.

[28] H. Iwai *et al*, *IEDM Tech. Digest*, p. 625, 2002.

2006 25th International Conference on Microelectronics

Electrode Effect on NTC Planar Thermistor Volume Resistivity

O.Aleksić, B.Radojčić, R.Ramović

Abstract - Thick film planar thermistors, such as rectangular, sandwich, multilayer, segmented and interdigitated were printed with low temperature NTC paste called NTC 3K3 95/2 (EI IRITEL). Their resistivity was analyzed as a function of volume resistivity variations due to electrode effect (diffusion of P_dA_g into NTC layer) and variation of geometrical parameters such as length l, witdh w, thickness d, number of segments n. Using experimental data, a model was obtained for analyzing the diffusion effect on volume resistivity by a simple fitting procedure. The good match between calculated and experimental data enabled inclusion of that formula in the total physical/mathematical model of thermistor resistivity.

I. INTRODUCTION

NTC thermistors are mainly used in electronics as elements for suppression of inrush current, temperature control and sensing, temperature measurement, fan control, etc. Main advantages of these devices are the low manufacturing cost, reliability at normal conditions of exploitation (when the temperature of ceramics is lower than critical degradation temperatures for ceramics and contact materials) and simplicity [1]. Recently NTC thermistors are found in many electrical and electronic products. In most temperature sensing applications they use $Ni_{1-x}Mn_{2+x}O_4$, where x denotes deviation from the stoichiometric 1:1 $NiO:Mn_2O_3$ ratio. This offers a range of properties that are suitable for most temperature sensing applications. Advantages of the use of these ceramic materials over others is their thermal stability or aging characteristics, such as changes in conductance over long periods and lifetime of the component [2].

A simple and well-known method for preparing the NTC thermistor powder based on complex spinel $(Mn,Ni,Co,Fe)_3O_4$ was used [3,4]. An oxide mixture containing MnO/NiO in the ratio 4:1 and 0.5 % CoO and Fe_2O_3 in the form of a fine aggregate was calcinated at 1050 ^{0}C /1h and ball milled for 2h in an ultra fast ball mill until a nanometer particle size of 20-30 nm was achieved.

Authors are with the Faculty of Electrical Engineering, University of Belgrade, Bulevar Kralja Aleksandra 73, 11000 Belgrade, Serbia and Montenegro, and Centre for Multidisciplinary Studies, Belgrade University, Serbia and Montenegro, E-mail addresses: obradal@yahoo.com, bradojcic@yahoo.com, ramovic@kiklop.etf.bg.ac.yu.

Due to electrostatic forces, the powder spontaneously agglomerated in cluster-particles of average size of 0.9 μm. SEM micrographs proved that during sintering clusters melted and merged in polycrystalline grains, ranging from 1 to 10 μm, but the grain structure was comprised of nanometer sub-grains. They formed at temperatures as low as 800-900^{0}C, due to excess surface energy, activated by ball milling to nanometer particle sizes.

The initial NTC powder was used for composing of NTC thick film thermistor paste. NTC thermistor paste, named 3K3 95/2, was composed in EI IRITEL (paste producer) adding 4% of B_2O_3 and an organic vehicle to the initial NTC powder. The NTC paste was screen printed on alumina substrate and sintered at 850^{0}C/10 min in a hybrid conveyor furnace. Analysis of different planar thick film thermistor geometries such as: sandwich, multilayer, segmented and interdigitated was recently performed [5]-[7].

II. PLANAR NTC THERMISTOR GEOMETRIES

Different thick film planar geometries are printed on Al_2O_3 using NTC 3K3 95/2 thermistor paste. Top view and cross-section of these planar thick film NTC thermistor geometries and appropriate equation for calculating resistivity R (ideal model) are given in Figures 1-5 for rectangular, sandwich, multilayer, interdigitated and segmented type [8]-[9].

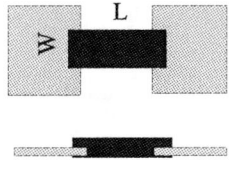

Rectangular NTC thermistor
$$R = \rho_v \, l/wd$$
D= d-thickness

Fig. 1. Planar NTC rectangular thermistor: top view and cross-section.

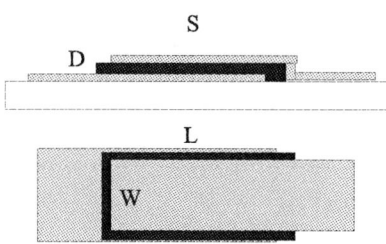

Sandwich NTC thermistor

$R = \rho_v\, d/s$

$S = lw$

Fig. 2. Planar NTC sandwich thermistor: top view and cross-section.

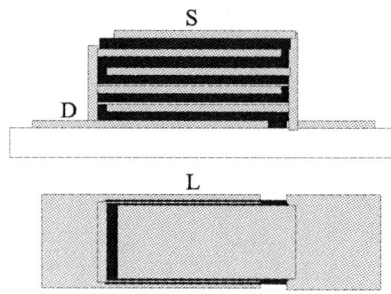

Multilayer NTC thermistor

$R = \rho_v\, 2n\, d/s$

N=3 sendwich in paralell

$S = lw$

Fig. 3. Planar NTC multilayer thermistor, top view and cross-section.

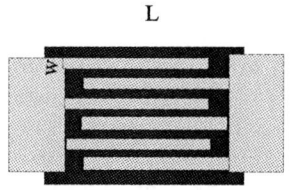

Interdigitated NTC thermistor

$R = \rho_v\, w\, /\, 2ndl$

Fig. 4. Planar NTC interdigitated thermistor, top view and cross-section.

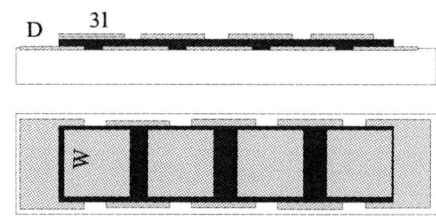

Segmented NTC thrmistor

$R = \rho_v\, 2n\, d/lw$

Fig. 5. Planar NTC segmented thermistor, top view and cross-section.

III. EXPERIMENTAL AND CALCULATED (FITTED) CURVE OF VOLUME RESISTIVITY OF PLANAR THEMISTORS

A. Experimental Results

Experimental results of volume resistivity were measured on selected values of geometrical parameters of planar themistors. The electrode effect on volume resistivity is calculated for all geometries of realized planar thermistors, using selected values of geometrical parameters and the adequate formula for calculating volume resistivity. The results obtained and the adequate formula (ideal model) for calculating of volume resistivity is given in Table I for different thick film planar geometries: rectangular, sandwich, multilayer, interdigitated and segmented type. Experimental results of the thick film volume resistivity as a function of thermistor layer thickness d, using results obtained in the Table I, are given in Figure 6.

TABLE I
VOLUME RESISTIVITY VS. GEOMETRY, $\rho=\rho(d)$ OF DIFFERENT PLANAR NTC THERMISTOR GEOMETRIES: RECTANGULAR, SANDWICH, MULTILAYER, INTERDIGITATED AND SEGMENTED.

Planar NTC thermistor geometries	S [mm^2] h, l, w [mm] d [μm]	Ideal model $\rho = \rho(d)$
rectangular	R = 6,4 M w = 2, l = 2, d = 60	Rd/w
sandwich	R = 1,2 k S = 4, d = 150	Rs/d
multilayer	R = 0,3 k S = 4, d = 3, n = 3	Rs/2nd
segmented	R = 5,5 k S = 4, n = 3, d = 60	R2ndl/w
interdigitated	R = 250 k l = 100, w = 2, d = 60	Rw/2nd

B. Total Physical and Mathematical Model and Simulation Results

Dependence of volume resistivity vs. geometry is a part of total physical-mathematical model of thermistor resistivity, which is given in equation (1):

$$R = \rho(d)R(l,w,d,n)e^{-\frac{B}{T}}f(t) \qquad (1)$$

ρ-volume resistivity; d- thermistor thickness; l-length; w-width; n-number of cells; B-NTC factor; T-temperature; t-time.

The first article calculates volume resistivity vs. geometry, the second calculates geometry vs. resisitivity, the third calculates thermal behavior of thermistors, and the fourth calculates velocity of the heat transfer through NTC layer and alumina substrate. Modelling of diffusion effects on the junction metal NTC layer in thick films is the first step in modelling total thermistor resistivity. That is a complex function of diffusion, geometrical parameters, temperature and time for thermistor in transition mode of use. The curve-fitting methods are used to relate thick film volume resistivity and layer thickness d of thermistor. This method is the way for modeling and predicting the behavior of thermistors. The fitted curve including electrode diffusion in volume resistivity (NTC layer) was calculated using the folowing equation, as a function of thermistor layer thickness d:

$$\rho(d) = \rho_{bulk}\left(1 - e^{-(d/d_0)^2}\right), \qquad (2)$$

Value ρ changes significantly within a range $(0, d_0)$ of variable d. The ρ value varies with d as:

d=0, ρ=0;

d>>d_0, ρ= ρ_{bulk};

d=d_0, ρ= $\rho_{bulk}(1-1/e)$.

Results fitted by proposed formula and the calculated curve of the thick film volume resistivity, as a function of thermistor layer thickness d and for d_0=500μ, are given in Figure 6. Resistivity vs. geometry was measured for each type of presented geometry and also calculated (fitted) at room ambient temperature using the main equation physical/mathematical model [8,9].

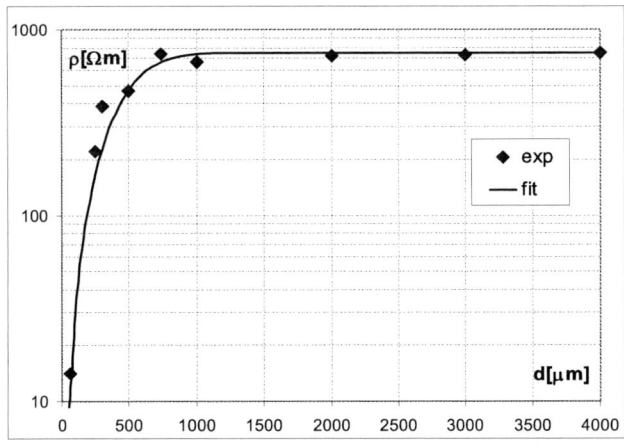

Fig. 6. Calculated (fit) and measured (exp) curve of the thick film volume resistivity as a function of thermistor layer thickness d.

IV. EXPERIMENTAL AND CALCULATED (FITTED) CURVES INCLUDING TEMPERATURE DEPENDENCE OF RESISTIVITY R

Experimental and fitting procedures were made including temperature dependence of resistivity R of different planar thick termistors. Experimental measurements of resistance in temperature range from -10 to 100°C were made for segmented, rectangular, sandwich, interdigitated and multilayer thermistors. Experimental data are given on Figures 7 and 8.

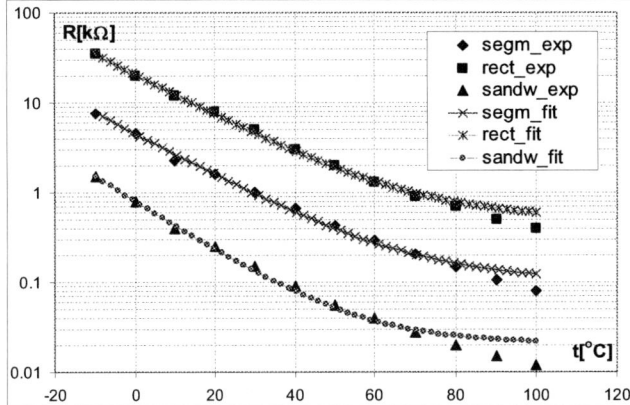

Fig. 7. Temperature dependence of resistivity R for different planar NTC thermistor geometries: segmented, rectangle and sandwich thermistors, experimental points and fitted curves.

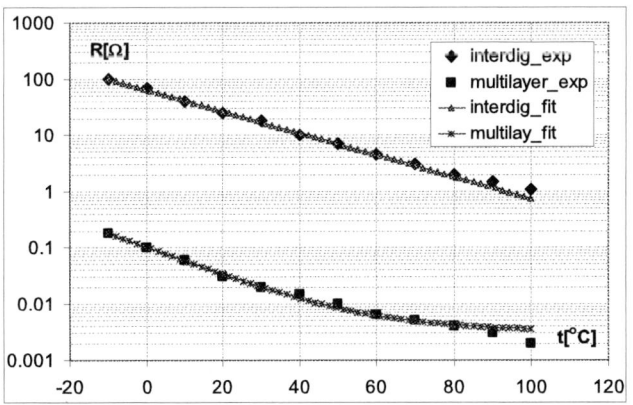

Fig. 8. Temperature dependence of resistivity R for different planar NTC thermistor geometries: interdigitated and multilayer thermistors experimental points and fitted curves.

The curve fitting method is applied to relate temperature to resistance for different planar thick film thermistors, using equation of exponential decay:

$$R(T) = A\exp(-B/T) \qquad (3)$$

where A and B are curve fitting constants and R is resistance at temperature T [K].

V. DISCUSSION

Five different thick film termistors were sequentially screen printed using 3K3 NTC thermistor paste and P_dA_g and fired at 850 ^{0}C/10min in a hybrid conveyor furnace. The thermistors obtained differ in NTC layer thickness d and electrode shape and size. Their room resistivities and volume resistivities were compared in Table I and the diagram in Figure 6. Volume resistivity varies by several Ωm for the sandwich type, to several hundreds Ωm for interdigitated type of thick film thermistors. Equation (2) was employed for fitting the experimental data of volume resistivity. It is evident from Figure 6 that volume resistivity for NTC layer thickness of 1mm is thick enough for P_dA_g diffusion effect to be neglected in the first approximation. The range from 50µm to 1mm is the region of exponential change of volume resistivity. In fact, volume resistivity depends on the metal diffusion depth into NTC layer. That depth of diffusion of P_dA_g metal lengths in the NTC layer could be of the order of 100 µm or more, and it contributes to 1mm NTC layer approximately 10%. In thinner layers contribution of metal doped layer (ordered 100 µm) is more significant. In sandwich thick film thermistors thickness is usually around 60 µm and volume resistivity is the lowest.

The second problem was determination of the thermistor B factor and fitting of the thermal dependence of resistivity R. The approximate equation (2) was used, as well as, room temperature resistivities for different geometries (different volume resistivities included, see Figures 7 and 8 for the same thermistor as compared in Table I). It was expected that B did not depend on volume resistivities. Now it is proved with five diagrams for each of thermistors tested (different geometries). The curves calculated match experimental results using the same B value. This B value was determined near room temperature using following equations:

$$R_1 = R_2 \exp\left[B\left(\frac{1}{T_1} - \frac{1}{T_2} \right)\right] \qquad (4)$$

and

$$B = \frac{T_1 T_2}{T_2 - T_1} \ln \frac{R_1}{R_2} \qquad (5)$$

Thermal coefficient α could be determined approximately as:

$$\alpha = \frac{1}{R}\frac{dR}{dt} = -\frac{B}{T^2} \qquad (6)$$

using resistivity data R_1 and R_2 at room temperature $\pm 10°C$ and equation 6.

We obtained B=4095 and α=-4.55%. These results show a gentle slope, suitable for sensor applications. At much lower and much higher temperatures than room temperature B is constant but α changes as function $f(1/T^2)$ (increases at low temperatures and decreases at higher temperatures). Note that Curie temperature for this NTC thermistor material is round 130°C (Figures 7 and 8). In this way thick film thermistor could be compared to bulk, disc and cubic type of thermistors.

VI CONCLUSION

The electrode effect on thick film volume resistivity and variation of the volume resistivity with temperature for different thermistor geometries is the first step in forming a new total physical-mathematical model for simulating thick film NTC thermistor response on temperature and time. The resistivity dependence on geometrical parameters is analyzed through an ideal model improved by $\rho(d)$ volume resistivity as function of P_dA_g as electrode metals into NTC layer. Thermal dependence was determined by B and α and the function on $f(T) = A\exp(-B/T)$. It is expected, using thermal equation of heat transfer and thermal conductivity, coefficient of substrate and NTC layer, that f(t) time response could be solved in near future. The authors also expect that some complex thermistor construction aimed for sensor as well as protective applications could be designed and simulated by new model that was proposed in this paper.

REFERENCES

[1] K. Park,D.Y.Bang,"*Materials in electronics*"14(2003)81-87.

[2] R. Schmidt, A.W. Brinkman, *International Journal of Inorganic Materials* 3 (2001) 1215-1217.

[3] Buchanan .R .C.; "*Ceramic Materials for Electronics*"; Marcel Dekker, New York and Basel, 1986.,pp. 301-305,

[4] Golonka H.;"Influence of Composition and Construction Parameters on the Basic Properties of Thick Film Thermistors"; *Hybrid Circuits*, Vol. 28, pp. 9-12, 1992.

[5] Aleksić S.O., Nikolić M.P., Vasiljević Radović G.D., Luković D. M., Djurić S., Pejović Ž.V., Radulović T.K., Luković T.D, Vujošević Lj.D.., "Properties of Thick Film NTC Thermistor Layers Based on Nanometric Mn, Co, Fe-Oxide Powder Mixture",*X World Round Table Science of Sintering*, .Belgrade 3-6 September 2002, pp 427-433.

[6] Aleksić S.O., Nikolić.M.P., Luković T.D., Radulović.T. K,.Vasiljević-Radović G.D., Savić M. S., "Diffusivity for Thick Film NTC Layers by Photoacustic Technique" *Microelectronics Int.* , Emerald UK,Vol 21, 2004 , pp 10-14.

[7] Aleksić S.O., Nikolić M P., Luković T.D, Savić M.S., Vasiljević-Radović G.D., Radulović T.K., Lukić S.L, Bojičić A. and Urošević D, "Investigation of Thermal Diffusivity of Thick Film Layers Obtained with the Photoacoustic Method", *J.Phys.* IV France Vol 125 (2005), pp 431-3.

[8] O.S.Aleksić, P.M.Nikolić, V.D.Jokić, V.Ž.Pejović, J.M.Pavlović, S.Đurić, "Analysis of Thick Film Thermistor Geometries", *MIEL*'97, Niš.

[9] Obrad S. Aleksić, Pantelija M. Nikolić, Mirjana N. Simić, Verica Ž. Pejović, Dana G. Vasiljević-Radović, "Resistiviy vs. Geometry Relation in Bulk - Sintered and Thick Film MnCoFe - Oxide Thermistors", *World Round Table on Sintering* 1998, Beograd, ITN, SANU.

2006 25th International Conference on Microelectronics

Performances of Conventional Thick-Film Resistors After Multiple High-Voltage Pulse Stressing

Ivanka Stanimirović, Milan M. Jevtić and Zdravko Stanimirović

Abstract - In this paper the results from a study of multiple high-voltage pulse stressing effects on resistance and low-frequency noise of thick-film resistors based on resistor composition with sheet resistance of 10kΩ/sq are presented. For the experimental purposes a series of thick-film test resistors with identical geometries were realized and exhibited to two types of tests: multiple series of 10 pulses with amplitude from the 0.5-4.0kV range and multiple series of 10 pulses with 3kV amplitude. Obtained experimental results were analyzed and correlation between resistance and low-frequency noise changes with resistor degradation due to multiple high-voltage pulse stressing is observed. It is shown that low-frequency noise parameters are more sensitive to this kind of resistor stressing than resistance.

I. INTRODUCTION

In the production of sensitive and reliable up-to-date communication systems stability and precise resistance values of widely utilised conventional thick-film resistors are of the great importance. Different conditions of their application are inducing the need to investigate their behaviour under stress, especially high-voltage pulse stress. Several papers dealt with physically different problem – behavioural analysis of low-ohmic thick-film resistors, so called thick-film surge resistors, which are being used in telecom systems as protection under extreme working conditions [1, 2], but little attention has been paid to performances of conventional thick-film resistors subjected to high-voltage pulse stressing [3, 4]. In one of our previous papers [5] we have presented results from the study of high-voltage pulse stressing effects on resistance and low-frequency noise of conventional thick-film resistors using test samples with different sheet resistances and different geometries. In this paper performances of thick-film resistors after multiple high-voltage pulse stressing will be investigated using resistance measurements as well as measurements of low-frequency noise parameters – current noise spectrum and noise index. Experiment – materials, test resistors and terms of multiple high-voltage pulse stressing and measurements are described in section II, and experimental results and discussion are given in section III.

Ivanka Stanimirović and Zdravko Stanimirović are with IRITEL A.D., Batajnički put 23, 11080 Belgrade, Serbia and Montenegro, E-mail: inam@iritel.com

Milan M. Jevtić is with the Institute of Physics, Pregrevica 118, 11080 Belgrade, Serbia and Montenegro, E-mail: mjevt@phy.bg.ac.yu

II. EXPERIMENT

A. Materials and Test Resistors

Performance analysis of thick-film resistors subjected to high-voltage pulse treatment was carried out on a group of conventional thick-film resistors with identical geometries ($1 \times 3mm^2$). Test samples were formed on alumina (96%Al_2O_3) substrates using conventional screen-printing techniques. Resistors were realized using commercially available resistor composition with nominal sheet resistance of 10kΩ/sq (DuPont QM84) in combination with PdAg conductor composition (DuPont QM17) and glass encapsulant (DuPont QQ550). After screen-printing processes, all of the wet layers were levelled for 15min at the room temperature and dried in the conveyer infrared drier using 10min cycle with temperature of 150°C. In order to achieve desired resistances, thicknesses of dried resistive layers were 25±3μm. Conductive, resistive and glass encapsulant layers were fired using 30min cycle with peak temperature of 850°C for 10min for conductive and resistive layers and 500°C for glass encapsulant .

B. Multiple High-Voltage Pulse Stressing and Measurements

In order to analyze the influence of multiple pulse stressing on thick-film resistors, pulse performances have been investigated using 10/700μs pulses delivered by Haefely P6T impulse generator (Fig. 1). 10/700μs pulses are characterized by a rise time to peak voltage of 10μs and exponential decay time to half peak voltage of 700μs. Two types of tests were performed and stressing conditions were chosen in compliance with standard ITU-T K.20 that refers to testing of thick-film surge resistors. During the first test (T1) resistors were subjected to multiple series of 10 pulses from the range of 0.5-4.0kV with 125V step and the second test (T2) consisted of multiple series of 10 pulses with 3kV amplitude. Resistance, noise index and current noise spectrum were measured. Resistance measurements were performed using HP34401A instrument. Noise index was measured, in compliance with MIL-STD-202D, Method 308, at 1kHz using Quan-Tech Resistor Noise Test Set, Model 315B. Current noise spectrum was measured using nanovolt amplifier (Mod. 103A, Keithley) and HP-3561B Dynamic Signal Analyzer in frequency range 10Hz to 10KHz. All measurements were performed at the room temperature (T=295K).

1-4244-0116-X/06/$20.00 ©2006 IEEE

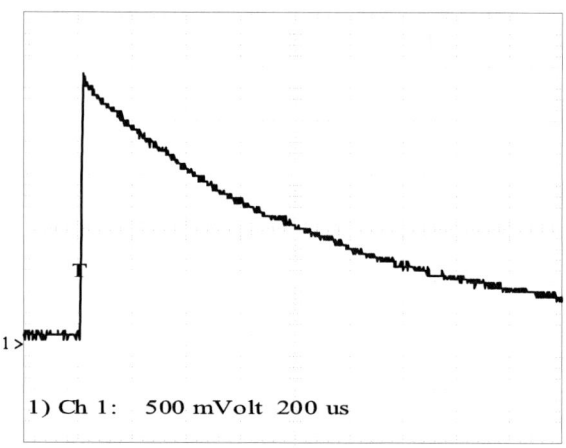

Fig. 1. 10/700µs pulse delivered by Haefely P6T impulse generator

III. EXPERIMENTAL RESULTS AND DISCUSSION

In order to illustrate performed experiment, typical results obtained by resistance, noise index and current noise spectrum measurements for thick-film resistors that were subjected to multiple high-voltage pulse stressing are given in Fig. 2 and 3. The results are given for resistors with identical geometries (1×3mm²) and the nominal initial resistance of 30kΩ. Results showed that resistance decreases after high-voltage pulse stressing. Mean values of the experimental results taken from the series of measurements showed that during T1 test significantly progressive resistance decrease begun after 10×2.9kV and that resistors suffered greater degradation after approximately 10×3.7kV resulting in significant resistance increase. Results obtained by noise index measurements showed that noise index increased with high-voltage pulse treatment and reached its maximum values at points of the significant resistor degradation. During T1 test, the 3kV pulse amplitude was singled out as the initial point of significant resistance decrease and for that reason T2 test was performed. T2 test showed resistance decrease during 8 repeated series of pulses with 3kV amplitude reaching its maximum of 2.85% and further stressing did not cause further changes. Measured noise index values showed significant increase in NI until the maximum of 32% was reached and, in accordance with resistance behaviour, further stressing did not cause further change of noise index. Measured values of current noise spectrum were in agreement with resistance and noise index behaviour.

Obtained results could be explained by micro structural changes in thick-film resistors caused by the high-voltage pulse stressing. In thick-film resistive materials electrical charge transport takes place via complex conductive network formed during firing by sintering metal-oxide particles surrounded by glass. The sintering process results in random and spatially uneven distribution of the conducting phase best described by our RRN model [6].

This model, based on site percolation model with double percolation in combination with deterministic model, views thick-film resistors as complex networks consisting of larger insulating areas and several conducting chains in which some particles are in contact and the others are separated by a thin glass barrier thus forming metal-insulator-metal structures. In that case metallic conduction through conducting particles and sintered contacts and tunnelling through glass barriers determine the current flow. At the initial point of significant resistance decrease and onwards (pulse amplitude ≥3kV), irreversible resistance change occurred due to changes in conducting mechanisms. High-voltage pulse stressing changes electrical charge captured within the insulating layer of metal-insulator–metal cell or micro structural changes lead to an increasing concentration of traps in glass barriers [7] formed as a consequence of presence of impurities in glass during sintering process.

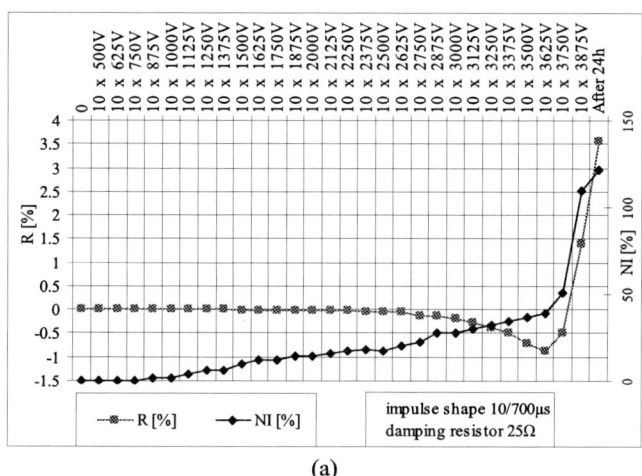

(a)

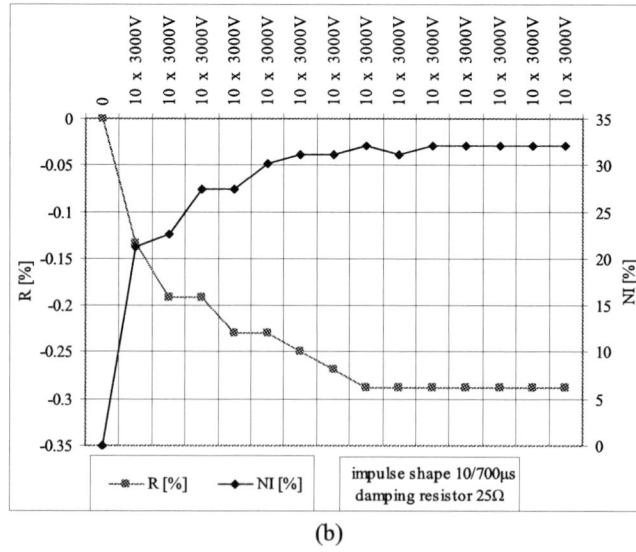

(b)

Fig. 2. Experimental results for resistance and noise index changes during high-voltage pulse stressing for thick-film resistor with nominal resistance of 30kΩ (a – T1, b – T2)

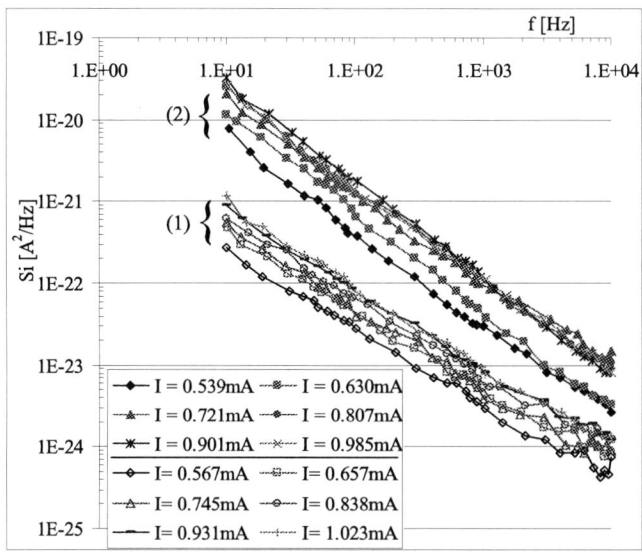

Fig. 3. Experimental results for current noise spectrum before (1) and after (2) pulse stressing (T1) for thick film resistor with nominal resistance of 30kΩ

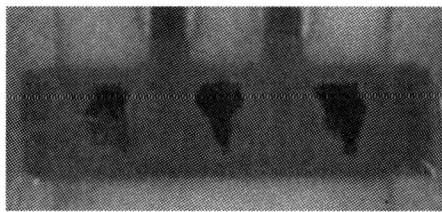

Fig. 4. Photograph of typical degradation mode for thick-film resistors subjected to multiple pulse stressing with increasing pulse amplitude (T1) - partial vaporisation of glass encapsulating and resistive layers

Also, a process of single chain transition from not conducting to conducting state is possible resulting in slight decrease in total resistance of thick-film resistor. The electric field inside a single metal-insulator-metal cell due to a high-voltage pulse applied to resistor is not sufficient to induce dielectric breakthrough and to cause increase in number of contacts between particles that would result in decrease of the total resistance. The resistance values significantly shifted upwards when the pulse voltage was increased until resistor partially vaporized (Fig. 4) causing the change in thickness of the resistive layer (pulse amplitude ≥3.7kV). Since the low-frequency noise in thick-film resistors is the consequence of electrical charge transport fluctuations, results of current noise spectrum and noise index measurements were in agreement with resistance behaviour. It seems that the multiple high-voltage treatment modulates conduction by influencing electron captures by traps within insulating layers that are not taking part in immediate conduction process, changing the potential barrier height of metal-insulator-metal structures. Therefore, low-frequency noise parameters – noise index and current noise spectrum showed greater sensitivity to microstructure changes than resistance.

Therefore, the goal of our further investigations will be the possible prediction of the behavior of thick-film resistors under high-voltage pulse stressing conditions that would in combination with noise parameter analysis provide the most important information from the aspect of reliability in the whole spectrum of applications in very sensitive telecommunications equipment.

IV. CONCLUSION

In this paper the results obtained from resistance, noise index and current noise spectrum measurements performed on a group of test resistors based on commercially available thick-film resistor composition with sheet resistance of 10kΩ/sq and subjected to multiple high-voltage pulse stressing are given. Correlation between resistance and low-frequency noise changes and multiple high-voltage pulse stressing is observed. It is shown that low-frequency noise is more sensitive to resistor degradation caused by this kind of treatment. The further investigations will be directed to studying degradation processes and failure mechanisms under electrical straining conditions in order to predict behaviour of thick-film resistors under extreme working conditions.

ACKNOWLEDGEMENT

This research was supported by the Serbian Ministry of Science and Environmental Protection (contracts TR-6116B and TR-6151B).

REFERENCES

[1] S. Vasudevan, "Low Ohm Thick Film Resistors for Surge Protection", *Advancing Microelectronics*, 1996, May-June, pp. 12-19.

[2] M.F. Barker, "Low Ohm Resistor Series for Optimum Performance in High Voltage Surge Applications" *Microelectronics International*, 1997, vol. 43, pp. 22-26.

[3] E.H. Stevens, D.A. Gilbert, "High-Voltage Damage and Low-Frequency Noise in Thick-Film Resistors", *IEEE Trans. Parts, Hybrids, Packag.*, 1976, PHP-12, vol. 4, pp. 351-356.

[4] A. Dziedzic, L.J. Golonka, J. Kita, H. Roguszczak, T. Zdanowicz, "Some Remarks About "Short" Pulse Behavior of LTCC and Thick-Film Microsystems", in *Proc. European Microelectronics Packaging and Interconnection Symposium*, Prague, 2000, pp. 194-199.

[5] I. Stanimirović, M.M. Jevtić, Z. Stanimirović, "High-Voltage Pulse Stressing of Thick-Film Resistors and Noise" *Microelectron. Reliab.*, 2003, vol. 43, pp. 905-911.

[6] Z. Stanimirović, M.M. Jevtić, I. Stanimirović, "Computer Simulation of Thick-Film Resistors Based on 3D Planar RRN Model", in *Proc. Eurocon*, Belgrade, 2005, pp. 1687-1690.

[7] M.M. Jevtić, Z. Stanimirović, I. Mrak, "Low-Frequency Noise in Thick-Film Resistors due to Two-Step Tunnelling Process in Insulator Layer of Elemental MIM Cell", *IEEE Trans. Comp. Pack. Manufact. Technol.*, 1999, vol. 22-01, pp. 120-127.

2006 25th International Conference on Microelectronics

Influence of Simultaneous Mechanical and Electrical Straining on Conventional Thick-Film Resistors

Zdravko Stanimirović, Milan M. Jevtić and Ivanka Stanimirović

Abstract - In this paper the results from a study of simultaneous mechanical and electrical straining effects on performances of conventional thick-film resistors based on resistor composition with sheet resistance of 1kΩ/sq are presented. For experimental purposes a series of thick-film test resistors with different geometries were realized and subjected to simultaneous electrical and mechanical straining using 1.2/50μs pulses in combination with mechanical substrate deflection of 300μm. Obtained experimental results were analyzed and correlation between resistance, noise index and gauge factor changes with resistor degradation due to simultaneous impact of these two types of straining were observed.

I. INTRODUCTION

Present miniaturization trends and mass usage of conventional thick-film resistors in up-to-date very sensitive telecommunications equipment has induced the need to investigate behaviour of conventional thick-film resistors under various straining conditions and for that reason several papers were published dealing with different problems of understanding the effects of mechanical [1, 2] and electrical [3, 4] straining on performances of these complex heterogeneous systems. In our previous papers we have presented results of our investigations of mechanical [5] and electrical [6] straining effects on conventional thick-film resistors, but little attention has been paid to examining effects of simultaneous impact of these two types of straining, especially having in mind the opposite effects these two types of straining have on behaviour of thick-film resistors. In this paper performance investigation results of thick-film resistors realized using composition with sheet resistance of 1kΩ/sq based on resistance, noise index and gauge factor measurements after simultaneous mechanical and electrical straining are presented. Experiment – materials, test resistors and terms of simultaneous stressing and measurements are described in section II, and experimental results and discussion are given in section III.

Zdravko Stanimirović and Ivanka Stanimirović are with IRITEL A.D., Batajnički put 23, 11080 Belgrade, Serbia and Montenegro, E-mail: zdravkos@iritel.com

Milan M. Jevtić is with the Institute of Physics, Pregrevica 118, 11080 Belgrade, Serbia and Montenegro, E-mail: mjevt@phy.bg.ac.yu

II. EXPERIMENT

A. Materials and Test Resistors

Performance analysis of thick-film resistors subjected to simultaneous mechanical and electrical straining was performed using three groups of thick-film test samples with different resistor geometries ($1 \times 2, 4$ and $6mm^2$). Test samples were formed on ceramic alumina (96%Al_2O_3) substrates using conventional screen-printing techniques. Resistors were realized using commercially available resistor composition with nominal sheet resistance of 1kΩ/sq (DuPont QM83) in combination with PdAg conductor composition (DuPont QM17).

B. Simultaneous Mechanical and Electrical Straining and Measurements

Mechanical straining of thick-film resistors has been performed using custom made measuring device where substrate deflection of 300μm has been performed by central force induced by the micrometer. Electrical straining has been performed using Haefely P6T pulse generator delivering 1.2/50μs pulses. Schematic presentation of simultaneous mechanical and electrical straining of thick-film resistors is given in Fig. 1. Resistance measurements were performed using HP34401A instrument and noise index was measured, in compliance with MIL-STD-202D, Method 308, at 1kHz using Quan-Tech Resistor Noise Test Set, Model 315B. All measurements were performed at the room temperature (T=295K).

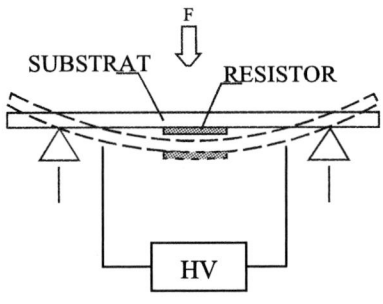

Fig. 1. Schematic presentation of ssimultaneous mechanical and electrical straining of thick-film resistors (F – apllied central force, HV – high-voltage pulse generator)

1-4244-0116-X/06/$20.00 ©2006 IEEE

III. Experimental Results and Discussion

In order to analyze influence of simulatneous straining on thick film resistors, as a first step, pulse performances have been investigated on relaxed 4mm long resistors using 10/700μs pulses. Results from the performed experiment - resistance and noise index are given in Fig 2a. Significant change in resistance did not occur with increase in pulse amplitude until the resistors were destroyed by excess loaded voltage (approx. 1kV pulse amplitude) leaving visible damaging of conductive and resistive layers at failure spots. Noise index increased with high-voltage pulse treatment and reached its maximum values at points of failure. Photograph showing typical catastrophic failure spots is given in Fig 2b.

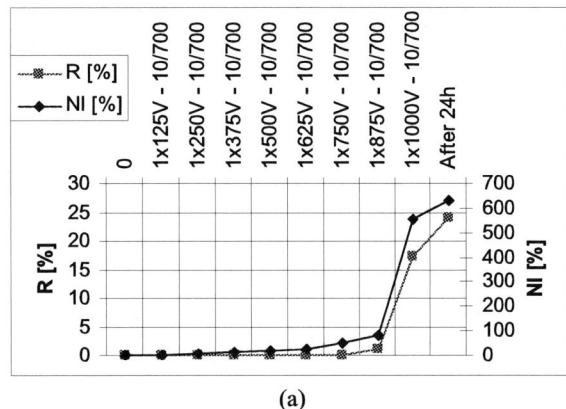

(a)

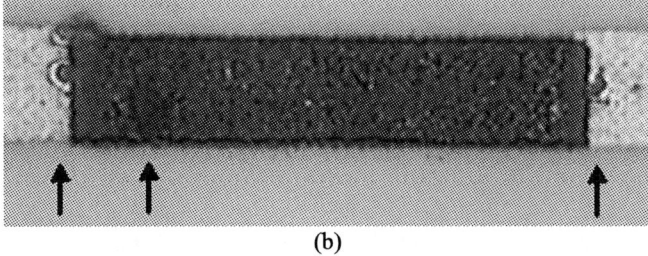

(b)

Fig. 2. Experimental results for resistance and noise index changes during high-voltage pulse stressing for thick-film resistor with resistor length of *l*=4mm (a) and its photograph showing catastrophic failure spots (b)

Since test resistors exhibited to 10/700μs pulses suffered failure after the impact of 1kV pulse amplitude, new stressing conditions have been chosen: electrical straining (ES) using 1.2/50μs pulses and simultaneous electrical and mechanical straining (EMS) using 1.2/50μs pulses in combination with mechanical substrate deflection of 300μm. The electrical straining conditions were chosen in such a manner to enable a gradual degradation of resistors. During experiment, both ES and EMS, resistors were subjected to series of pulses from the range of 0.125-2.125kV with 125V step. At first test resistors were subjected to series of 10 pulses with initial pulse amplitude of 125V and frequency of 6 pulses per min. Output

resistance of the generator was 25Ω. Pulse amplitudes were increased with 125V step until significant change of resistor properties was observed and then series of ten pulses was reduced to one pulse at the time in order to study gradual resistor degradation and during testing resistance and noise index values were measured. From results obtained by resistance measurements, resistance change and gauge factors [7] were calculated and are given in Fig. 3 and Fig. 4 as an illustration of our experiment.

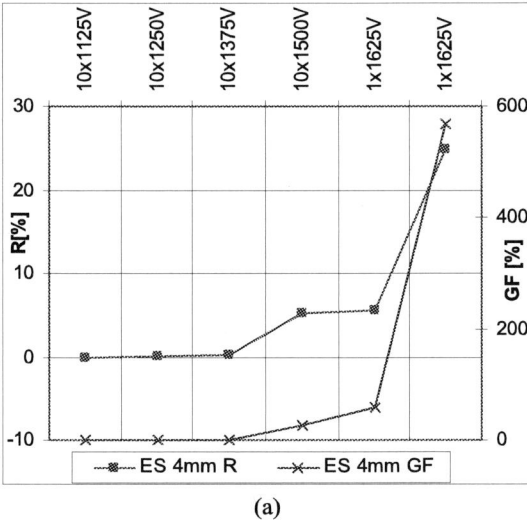

(a)

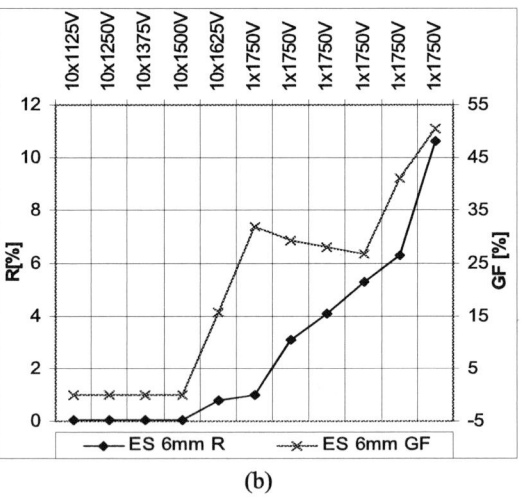

(b)

Fig. 3. ES experimental results for resistance and gauge factor changes for thick-film resistors with following resistor lengths: *l*=4mm (a) and *l*=6mm (b)

Mean values of measured experimental results for resistors with *l*=2mm showed less pronounced difference between the impact of ES and EMS than for resistors 4 and 6mm long. For resistors 4mm long significant resistor degradation began after the impact of 1.5kV pulse amplitude for ES (Fig. 3a) and 1.75kV for EMS (Fig. 4a). Similar results were obtained for resistors 6mm long. Significant resistor degradation began after the impact of

628

1.625kV pulse amplitude for ES (Fig. 3b) and 2kV for EMS (Fig. 4b). Gauge factor (*GF*), describing the sensitivity of the resistors to mechanical straining, is defined as the ratio of the relative resistance change (*ΔR/R*) and the relative change of length of the resistor $\varepsilon = \Delta l/l$ under influence of mechanical straining:

$$GF = \frac{\Delta R/R}{\varepsilon}. \qquad (1)$$

Gauge factors change showed the increase following the shapes of curves of the resistance changes due to resistor degradation. Having in mind that ε is constant because for all resistors the substrate deflection was 300μm, gauge factors change was due to relative resistance change as a consequence of ES and EMS. Measured noise index values were in agreement with resistance measurements and showed significant increase in NI due to applied straining conditions, confirming the fact that noise parameters are very sensitive to structural changes of thick-film resistors, more sensitive than resistance changes (Fig. 5).

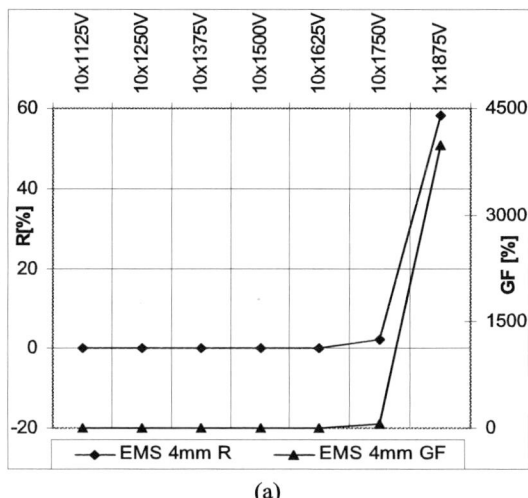

(a)

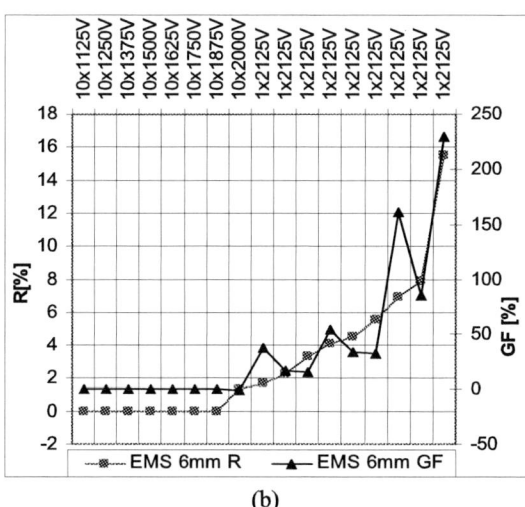

(b)

Fig. 4. EMS experimental results for resistance and gauge factor changes for thick-film resistors with following resistor lengths: *l*=4mm (a) and *l*=6mm (b)

According to our previous investigations, resistance change of thick-film resistor subjected to electrical straining alone is due to micro structure changes [6] resulting in decrease of the resistance. Resistance change of resistor due to mechanical straining alone is partially due to change in resistor geometry but mainly due to microstructure changes resulting in increase of the resistance as discussed in [5] - between 0.3 and 0.4% depending on the resistor length.

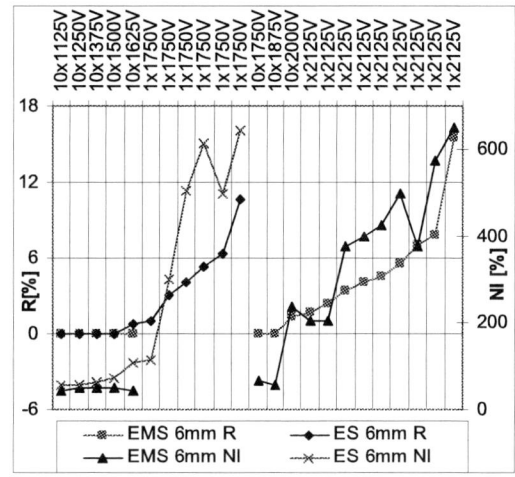

Fig. 5. ES and EMS experimental results for resistance and noise index changes for thick-film resistors with resistor length *l*=6mm

However, simultaneous impact of these two types of straining resulted in different occurrences. Sheet resistances of thick-film resistor compositions are determined by the volume fractions of conducting particles determining the microstructure of the resistive layers well as conducting mechanisms. In case of resistor composition with sheet resistance of 1kΩ/sq, as used in our experiment, small conducting/isolating phase ratio determines dominant conducting mechanism – conduction through clusters of particles (conducting particles and sintered contacts between adjacent conducting particles). However, both electrical and mechanical straining have opposite influence on tunnelling through glass barriers changing the barrier resistances. Electrical straining affects electrical charges captured within insulating layers of metal-insulator-metal cells and concentration of traps in glass barriers while mechanical straining affects glass barrier widths. Having in mind defined intensity and shape of high-voltage pulses used in the experiment and the fact that tunnelling is not a dominant conducting mechanism when thick-film resistors with sheet resistance of 1kΩ/sq are concerned, the lack of observed micro structure changes was expected. Simultaneous mechanical and electrical straining caused changes in macro structure of thick-film resistors. For both ES and EMS, applied high-voltage pulsing caused visible vaporisation of resistive layers thus decreasing volumes of resistors and therefore significantly increasing their resistances. For EMS, simultaneous straining shifted upwards the point of significant resistor degradation requiring higher values of pulse amplitudes (Fig. 3 and 4). Obtained results proved that even when subjected to simultaneous straining thick-film resistors met requirements set for usage in telecommunications equipment and induced further investigations of thick-film resistors realized using resistor compositions with higher sheet resistances. Preliminary results of latest resistance and noise index measurements confirmed that the conducting/isolating phase ratio determining dominant conducting mechanisms strongly affect both micro and macro structure changes due to simultaneous impact of mechanical and electrical straining and that these changes can best be seen on resistor realized using resistor composition with sheet resistance of 10kΩ/sq since there is not only one dominant conduction mechanism in question.

IV. CONCLUSION

In this paper the results obtained from resistance, gauge factor and noise index measurements performed on three groups of test resistors based on commercially available thick-film resistor composition with sheet resistance of 1kΩ/sq and subjected to simultaneous mechanical and electrical straining are given. Obtained results were analyzed and correlation between resistance, noise index and gauge factor changes with resistor degradation due to simultaneous impact of these two types of straining were observed. It is shown that significant changes of the resistance and noise parameters are due to the macro structural changes of the resistive layer. Also, obtained results proved that, even when subjected to such simultaneous straining, examined thick-film resistors met requirements set for usage in up-to-date very sensitive telecommunications equipment. Further investigations of performances, degradation processes and failure mechanisms of conventional thick-film resistors subjected to simultaneous impact of mechanical and electrical straining will incorporate further, more detailed,

measurements of resistance, gauge factor and low-frequency noise parameters – noise index and current noise spectrum, using thick-film test resistors with different geometries and sheet resistances in order to develop a method of quality estimation of thick-film resistors under extreme working conditions.

ACKNOWLEDGEMENT

This research was supported by the Serbian Ministry of Science and Environmental Protection (contracts TR-6116B and TR-6151B).

REFERENCES

[1] C. Canalli, D. Malavasi, B. Morten, M. Prudenziati, A. taroni, "Piezoresistive Effects in Thick-Film Resistors", *J. Appl. Phys.,* 1980, June, 51(6), pp. 3282-3288.

[2] J.S. Shah, "Strain Sensitivity of Thick-film Resistors", *IEEE Trans. on Comp. Hyb. and Manuf. Technol.,* 1980, December, Vol. CHMT-3, No. 4, pp. 554-564.

[3] E.H. Stevens, D.A. Gilbert, "High-Voltage Damage and Low-Frequency Noise in Thick-Film Resistors*", IEEE Trans. Parts, Hybrids, Pack.,* 1976, PHP-12, vol. 4, pp. 351-356.

[4] A. Dziedzic, L.J. Golonka, J. Kita, H. Roguszczak, T. Zdanowicz, "Some Remarks About "Short" Pulse Behavior of LTCC and Thick-Film Microsystems", in *Proc. Europian Microelectronics packaging and Interconnection Symposium,* Prague, 2000, pp. 194-199.

[5] Z. Stanimirović, M.M. Jevtić, I. Stanimirović, "Performances of Conventional Thick-Film Resistors Subjected to Mechanical Straining", in *Proc. MIEL,* Niš, 2004, vol. 2, pp. 675-678.

[6] I. Stanimirović, M.M. Jevtić, Z. Stanimirović, "High-Voltage Pulse Stressing of Thick-Film Resistors and Noise", *Microelectron. Reliab.,* 2003, vol. 43, pp. 905-911.

[7] Z. Stanimirović, I. Stanimirović, "Performances of Thick-Film Resistors Subjected to Mechanical Straining", in *Proc. ETRAN,* Herceg Novi, 2003, vol. IV, pp. 249-251.

2006 25th International Conference on Microelectronics

Influence of Electrode Materials and the Manner of Electrode Surface Processing on Gas-Filled Surge Arresters Relevant Characteristics

B. Lončar, P. Osmokrović, A. Vasić, and R. Šašić

Abstract – The aim of this paper is to present the electrode effects as approach for the improvement of gas filled surge arresters (GFSA) protective characteristics in the most optimal way. We examined the influence of the electrode parameters on the pulse shape characteristics. As variable parameters, we used the electrode material and the manner of electrode surface processing. The originally developed GFSA model with a composite electrode system enables a high degree of over-voltage protection without environmental contamination.

I. INTRODUCTION

GFSA are non-linear over-voltage protection components. Commonly, they are made of a ceramic housing, with a symmetric two-electrode configuration with gas insulation [1], [2]. GFSA's operation relies on the principle of the electrical breakdown of a gas. The electrical breakdown of a gas depends on the type of gas or gas mixture, gas pressure, shape of electrodes, electrode material and the manner of electrode surface processing [3], [4]. Noble gases, at the pressures close to Paschen minimum, are most frequently used as in insulation gases [5], [6]. The electrodes are profiled to provide a pseudo-homogeneous macro component of the electric field.

The most important parameter of GFSA is its pulse shape characteristic. The narrower the pulse characteristic is (smaller area between curves) and the smaller the slope is, the better protective characteristics of GFSA are [7]-[9]. By applying the best electrode material and the optimal manner of electrode surface processing it is possible to significantly improve the GFSA pulse shape characteristic.

II. EXPERIMENT

Experimental procedure is completely automated and is adequately protected against electromagnetic noises. A

B. Lončar and R. Šašić are with the Faculty of Technology and Metallurgy, University of Belgrade, Karnegijeva 4, 11120 Belgrade, Serbia & Montenegro, E-mail: bloncar@eunet.yu

P. Osmokrović is with the Faculty of Electrical Engineering, University of Belgrade, Bulevar Kralja Aleksandra 73, 11120 Belgrade, Serbia & Montenegro, E-mail: opredrag@verat.net

A. Vasić is with the Faculty of Mechanical Engineering, University of Belgrade, Kraljice Marije 16, 11120 Belgrade, Serbia & Montenegro, E-mail: avasic@mas.bg.ac.yu

complete set of measuring equipment that has a power supply of its own was located in a cabin protected with up to 100 dB. Wherever it was possible, we employed optic connections with cabin the measurement circle. Voltaic connections used in the measurements were double-shielded cables laid down in grounded channels. Block diagram of the measuring system is shown on Fig. 1.

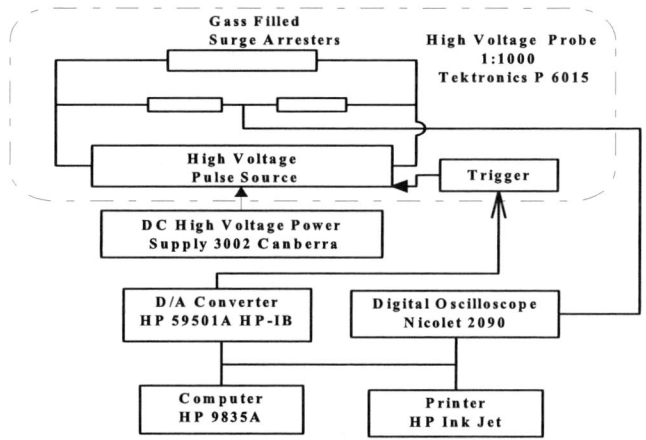

Fig.1 Block diagram of the measuring system.

The measuring system is designed to be very flexible – it enables testing of all over-voltage protection elements using only software modifications. PC software controls operation of a Digital to Analog converter (D/A), which generates all wave shapes required for testing of GFSA (internal communication between voltage and current source). Additionally, same PC controls other instruments using HP-IB (IEEE488) interface and protocol.

For examination of GFSA pulse characteristic 20 dc breakdown voltage and 50 pulse breakdown voltage are measured. In order to stabilize protective characteristics GFSA each specimen is exposed to 25 initial (conditioning) breakdowns. A 30 second pause between successive measurements (breakdowns) is exercised. An increased number of repetitions would be somewhat undesirable because of GFSA characteristic's irreversibility, which manifests itself after a large number of dielectric breakdowns. Since the U-test is an adequate statistical tool for all types of statistical distributions, it should be used for quantitative testing. This test should confirm whether the observed variable belongs to the same random variable [10].

1-4244-0116-X/06/$20.00 ©2006 IEEE 631

III. RESULTS AND DISCUSSIONS

A. The Influence of Electrode Material on the GFSA Pulse Shape Characteristic

The influence of electrode material on GFSA pulse shape characteristic was investigated in the pressure region from p = 1 mbar to p = 2.5 bar with small inter-electrode gaps from d = 0.1 mm to d = 1 mm. The electrodes were made either of copper, of aluminium, or of electrons (aluminium alloy) (Table I). Such a selection of material was carried out on the basis of a large range of values of the work function as well as due to the different values of the melting point and thermal conductivity [11].

TABLE I
THE ELECTRODE MATERIAL WORK FUNCTION, MELTING POINTS AND THERMAL CONDUCTIVITY

Material	Copper	Aluminum	Electron
Work function [eV]	4.47	3.70	1.80
Thermal conductivity [J/scmK]	7.1	3.0	1.8
Melting points [K]	1082	641	566

On the basis of the obtained results it might be concluded that at small values of *pd* product (the pressure p = 50 mbar, interelectrode gap d = 0.1 mm) the electrode materials have a significant influence on the GFSA pulse shape characteristics. The materials with higher values of work function correspond to larger pulse shape characteristics area (Table II).

TABLE II
THE GFSA PULSE SHAPE CHARACTERISTIC SURFACES DEPENDING ON THE APPLIED ELECTRODE MATERIAL (P = 50 MBAR, D = 0.1 MM)

Material	Copper	Aluminum	Electron
Pulse characteristic surfaces [μVs]	2850	1450	675

In Figure 2, the results of the U test with copper, aluminium and electron electrodes were shown. From this figure it was obvious that the copper electrodes have the most stable characteristics in exploitation conditions. This could be explained by both high melting points and thermal conductivity of cooper, and for that reason the explosion of the spark channel (with the pressure up to the 20 bar)

results in the small changes of electrode topography. For the electron electrodes, made of materials with lower melting points and lower thermal conductivities, this effect has a significant influence on breakdown voltage due to very small inter-electrode gaps, which are in order of magnitude with craters made in spark explosions.

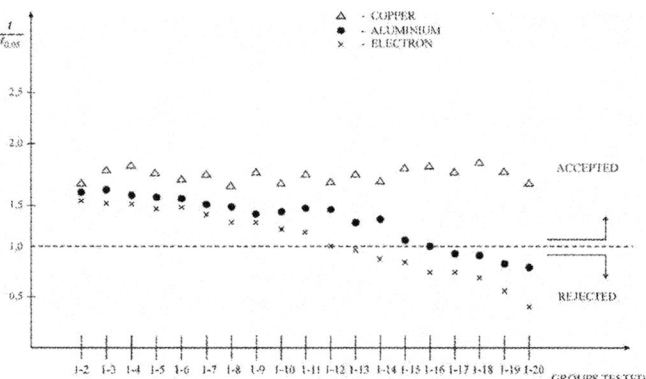

Fig. 2 The results of the U- test for different electrode materials.

B. The Influence of the Manner of Electrode Surface Processing on the GFSA Pulse Shape Characteristic

TABLE III
THE GFSA PULSE SHAPE CHARACTERISTIC SURFACES DEPENDING ON THE MANNER OF ELECTRODE SURFACE PROCESSING

Manner of surface processing	Pulse characteristic surfaces [μVs] p [mbar], d [mm]				
	p=5 d=0,1	p=50 d=0,1	p=50 d=1	p=250 d=2	p=2500 d=1
Sand-blasted	1000	850	2700	21000	95000
Polished	900	3150	1200	4500	23500
Small pyramids	600	750	1500	4000	24000
Big pyramids	900	1600	1900	6500	36000
Composite electrodes	700	825	1800	6000	33000

In the case of investigating the influence of the manner of processing electrode surfaces on pulse shape characteristics the cylindrical brass electrodes were used. Electrode active surface made of copper were either processed in the sand-blaster, polished, engraved with chiseled small pyramids with a height of 320 μm of distance-peak 400 μm, or engraved with the chiseled bigger pyramids with a height 700 μm of distance-peak 650 μm. An electrode system with a hollow cathode was also used. In this case, as a variable parameter, the diameter of holes was used (D = 0.1 mm, 0.2 mm, 0.5 mm and 1mm). Besides these, cylindrical brass electrodes of tube – type, connected with copper and electron tubes as a matrix were also used (composite electrodes) (Table III).

The effect of the electron emission may explain the marked difference between the pulse shape characteristics obtained by the sand processed electrodes and by the electrodes with polished surfaces at very small values of product *pd* (p = 50 mbar, d = 0.1 mm) and somewhat less salient at higher *pd* products (p = 2.5 bar, d = 1 mm) (Table III). Increased influence at higher values of *pd* product occurs expectedly at *pd* point 2.5 bar-mm. This is expected at this value, simply because this is an exclusively characteristic of streamer mechanism of discharge, where the secondary electrode effects are negligible with respect to the secondary effect in gas. The fact that the pulse shape characteristics obtained from the surface with bigger engraved pyramids are larger than the pulse shape characteristics obtained from the surface of smaller engraved pyramids can be explained by the decrease of the critical volume in front of the bigger pyramids through non-uniformity of the electric field.

In Figure 3 the pulse shape characteristics obtained by using a hollow cathode were shown (for two values of holes diameter D = 0.1 mm, D = 1 mm). From these results it can be concluded that the hollow cathode effect is very efficient, in the case when the diameter of the holes exceeds the critical value, which depends on the applied pressure. This could be explained by the appearance of the hollow cathode effect [8] for cathode holes greater than electrons free - path for given pressure values. The spread of volt-second characteristic for small values of pressure is a result of missing the hollow cathode effect for very small pressures (the electron mean free path exceeding the hole diameter).

In Figure 4 the results of the U test for a different manner of electrode surface processing (polished, sand-blasted, with engraved pyramides and with composite electrodes) are presented. From this figure, it is clear that the best results were obtained using sand-blasted electrodes. Absence of the irreversible processes for the sand-blasted electrodes is the result of insignificantly small changes of electrode surface topography due to sparking. The polished electrodes show the worst results. In the case of the electrodes with engraved spikes, irreversible changes are the result of the melting of spikes due to the high spark temperature.

Based on the shown results, it could be concluded that the smallest pulse shape characteristics areas would be obtained by using the sand - blasted electron electrodes with holes of the diameter greater than mean free path of electrons. On the other hand, characteristics obtained in this manner were unstable. In order to obtain the GFSA's model with quickest and most stable pulse shape characteristics, the composite electrode system was made of electron tubes matrix, melted in copper.

In Figure 5, the results obtained using this composite electrode system and with built-in α - radioactive source [241]Am were shown. According to this figure, we can conclude that GFSA's response times obtained by

suggested electrode systems are comparable with the results obtained by GFSA's model with built-in radioactive source. The results of the U test (Figures 2 & 4) show that the influence of previous breakdowns (memory effect) for suggested electrode system with composite electrodes is comparable with the results obtained with copper electrodes.

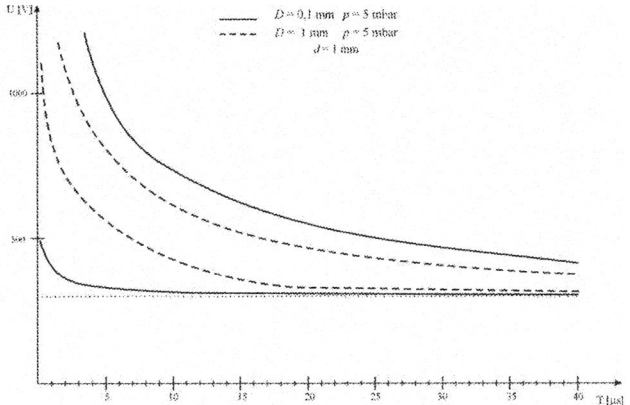

Fig. 3 The GFSA pulse shape characteristics using hollow cathode.

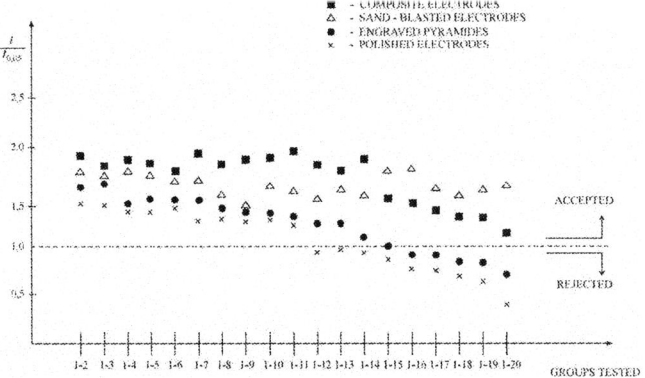

Fig. 4 The results of the U- test for a different manner of electrode surface processing.

V. CONCLUSION

This paper presents the influence of electrode parameters (electrode materials and the manner of electrode surface processing) on GFSA's characteristics. The obtained results show that the electrode materials influence on GFSA's characteristics through the values of work function, melting points and thermal conductivity. Materials with lower values of work function correspond to lower values of dc breakdown voltage and the narrower and smaller slopes of the pulse shape characteristics. It means that the GFSA's response time decreases, i.e. the protective characteristics of GFSA are significantly improved. On the other hand, the electrode material with higher values of thermal conductivity and melting point possesses the most stable characteristics (working point) during the exploitation (small changes of electrode topography). Electrode surfaces processed in sand-blaster and the electrode system with a hollow cathode (for holes greater than electron mean free path for given pressure value) correspond to narrower pulse shape characteristics in comparison with polished electrodes and electrodes engraved with chiseled pyramids.

Based on this conclusion, we originally developed the GFSA model with a composite electrode system. The results obtained with this model containing composite electrodes showed similar characteristics to results obtained using an electrode system with built-in radioactive source but without environmental contamination.

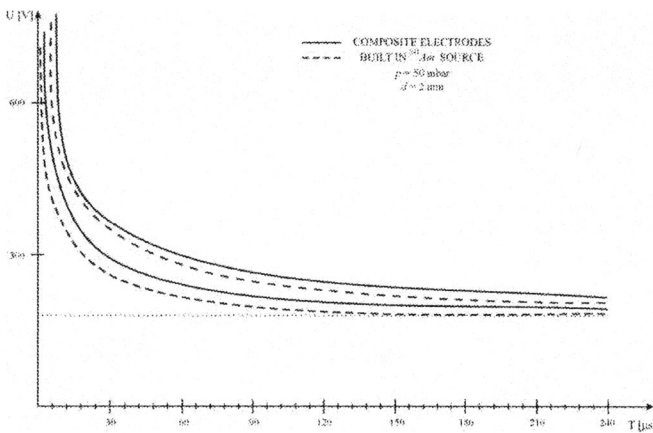

Fig. 5 The GFSA pulse shape characteristics obtained by using composite electrodes and with built-in α - radioactive source [241]Am.

ACKNOWLEDGEMENT

The Ministry of Science and Environmental Protection of Republic of Serbia supported this work, under contracts 2006 and 2016.

REFERENCES

[1] B. Lončar, P. Osmokrović, and S. Stanković, "Temperature stability of components for over-voltage protection of low-voltage systems", *IEEE Trans. Plasma Sci.*, vol. 30, pp. 1881-1885, 2002.

[2] P. Osmokrović, B. Lončar, S. Stanković, and A. Vasić, "Aging of the over-voltage protection elements caused by over-voltages", *Microelectronics Reliability*, vol. 42, pp. 1959-1966, 2002.

[3] M. M. Pejović, G. S. Ristić, and J. P. Karamarković, "Electrical breakdown in low pressure gases", *J. Phys. D: Appl. Phys.*, vol. 35, pp. R91-R103, 2002.

[4] P. Osmokrović, "Mechanism of electrical breakdown of gases at very low pressure and inter-electrode gap values", *IEEE Trans. Plasma Sci.*, vol. 21, pp. 645-654, 1993.

[5] M. M. Pejović, Č. S. Milosavljević, and M. M. Pejović, "The estimation of static breakdown voltage for gas filled tube at low pressures using dynamic method", *IEEE Trans. Plasma Sci.*, vol. 31, pp. 776-781, 2003.

[6] P. Osmokrović, I. Krivokapić, and S. Krstić, "Mechanism of electrical breakdown left of the Paschen's minimum", *IEEE Trans. Dielectrics and Electrical Insulation*, vol. 1, pp. 77-82, 1994.

[7] G.C. Messenger, and M.S. Ash, *The Effects of Radiation on Electronic Systems*, New York: Van Nostrand Reinhold, 1992.

[8] P. Osmokrović, B. Lončar, and S. Stanković, " Investigation the optimal method for improvement the protective characteristics of gas filled surge arresters- with/without the built in radioactive sources", *IEEE Trans. Plasma Sci.*, vol. 30, pp. 1876-1880, 2002.

[9] B. Lončar, P. Osmokrović, and S. Stanković, "Radioactive reliability of gas filled surge arresters," *IEEE Trans. Nuclear Sci.*, vol. 50, pp. 1725-1731, 2003.

[10] W. Hauschild, and W. Mosch, *Statistical Techniques for High-Voltage Engineering*, London: Peregrinus, 1992.

[11] *CRC Handbook of Chemistry and Physics,* Florida: CRC Press. Inc., 1990, editor – in – chief Robert C. Weast.

2006 25th International Conference on Microelectronics

New Insights into Tunneling Current through Oxynitride/Oxide Stack for MOSFETs

L. F. Mao, and Z. O. Wang

Abstract -A first principles study of silicon oxynitride shows that the defect-like levels near the conduction and valence band will be introduced into the bandgap while incorporating nitrogen into silicon dioxide. It has also been shown that the tunneling current through oxynitride/oxide stack for MOSFETs assisted by defect-like levels near the conduction band will be dominant at high field and the direct tunneling current will be dominant at low field. This work gives an insight about the reason on smaller tunneling current at low field compared to the tunneling through pure silicon dioxide , which is that the effective masses give a large contribution to reducing the direct tunneling current than simply increasing physical thickness does.

I. INTRODUCTION

Devices scaling is driving a continuous decrease in the gate dielectric film thickness. But in doing this, the gate current exponentially increase its power consumption. One objective of gate dielectric engineering is to reduce direct tunneling (DT) by increasing the physical thickness of gate dielectrics while maintaining an effective oxide thickness (EOT) that corresponds to a significantly thinner silicon dioxide (SiO_2) film. With high-k (dielectric constant) dielectrics, the desired EOT can be achieved concurrently with a reduced gate current by increasing the physical thickness. Gate dielectric containing silicon oxynitride ($SiO_{2-1.5x}N_x$ ($0 \le x \le \frac{4}{3}$) has been proposed as an alternative gate dielectric..

Extensive studies have been performed on silicon oxynitride [example for 1-5]. This work has a twofold aim. First we present ab initio calculations for the electronic structure of silicon oxynitride film. Secondly the tunneling currents through oxide/oxynitride are calculated.

II. METHOD

Amorphous silicon dioxide is believed to have similar band structure properties to those of alpha quartz, and the techniques used for determining the behavior of alpha quartz can be applied to the more disordered structure of SiO_2 [6]. In this work, a supercell with eight a-quartz unit cells with nitrogen atoms and vacancies substituting on the oxygen site were used to model $SiO_{1.75}N_{0.1667}$ (the nitrogen

L. F. Mao, and Z. O. Wang are with School of Electronics & Information Engineering, Soochow University,178 Gan-jiang East Road, Suzhou 215021, P. R. China (phone: 86-512-67158537; fax: 86 512 67248370; e-mail: mail_lingfeng@yahoo.com.cn)

atom concentration was 5.7%). The substituted oxygen atoms were random selected and substituted by nitrogen atoms or vacancies. $SiO_{1.75}N_{0.1667}$ structure was calculated by applying the generalized-gradient approximation to density functional theory [7].

The tunneling current (TC) can be calculated as [8]

$$ J = \int_0^\infty \frac{qm^*}{2\pi^2\hbar^3} D(E_x) \left(\int_{E_x}^\infty [f_r(E) - f_l(E)]dE \right) dE_x \quad (1) $$

The transmission coefficient D(Ex) was calculated by a numerical solution of the one-dimensional schrödinger equation assuming an idealized potential barrier. A parabolic $E(k)$ relation with an effective mass m* as parameter is assumed in this letter. The barrier was discretized by N partial subbarriers of rectangular shape which covered the whole oxide layer of thickness T_{ox}. From the continuity of wave-function and quantum current density at each boundary, the transmission coefficient is then found by

$$ D(E_x) = \frac{m_0}{m_{N+1}} \frac{k_{N+1}}{k_0} \frac{|\det M|}{|M_{22}|^2} \quad (2) $$

where M is a (2×2) product matrix, M_{22} is the quantity of the second row and the second column in this matrix with transfer matrices M_l given by

$$ M_l = \frac{1}{2} \begin{vmatrix} (1+S_l)\exp[-i(k_{l+1}-k_l)x_l] & (1-S_l)\exp[-i(k_{l+1}+k_l)x_l] \\ (1-S_l)\exp[+i(k_{l+1}-k_l)x_l] & (1+S_l)\exp[+i(k_{l+1}-k_l)x_l] \end{vmatrix} \quad (3) $$

In Eq.(3) $S_l = m_{l+1}k_l / m_l k_{l+1}$, and the effective masses and momenta are discretized as $m_l = m^*[(x_{l-1} + x_l)/2]$ and $k_l = k[(x_{l-1} + x_l)/2]$, respectively.

The Fermi-Dirac distribution was used in the tunneling current calculations. In our calculations, the maximum of the longitudinal electron energy was set at $20k_BT$ above the conduction band, where k_B is Boltzmann constant, and T is the temperature. Two–step tunneling process were used in the calculations of tunneling currents through the oxide/oxynitride stack.

III. RESULTS AND DISCUSSION

Fig.1a shows the band-structure diagram of pure silicon dioxide. Fig.2b shows the density of states for pure silicon dioxide.Fig.2a shows the band-structure diagram of $SiO_{1.75}N_{0.1667}$. Fig.2b shows the density of states for $SiO_{1.75}N_{0.1667}$. Based on a survey of these figures, it is clearly that the empty bands will be introduced into the bandgap with incorporating nitrogen atoms into silicon dioxide. Comparing the band-structure diagram of $SiO_{1.75}N_{0.1667}$ (Fig.2a) to the band-structure diagram of pure silicon dioxide (Fig.1a), several empty bands can be observed being in the bandgap. Comparing the density of states of $SiO_{1.75}N_{0.1667}$ (Fig.2b) to the density of states of pure silicon dioxide (Fig.1b), it is clearly seen that the "defect-like" energy levels will result in a large density of states within the band gap.

the average effective electron mass in the calculations, the average effective electron mass near the bottom of the conduction band in pure-α-quartz is obtained as $0.55m_0$, which agrees with the most used value $0.5m_0$. Detail study about the effective mass issues in lightly nitride silicon oxides can be found in our previous work [9]. These levels may result from the vacancies substituting the oxygen atoms that can be understood as some defects being introduced in the structure. These levels are labeled from number 1 to 6.

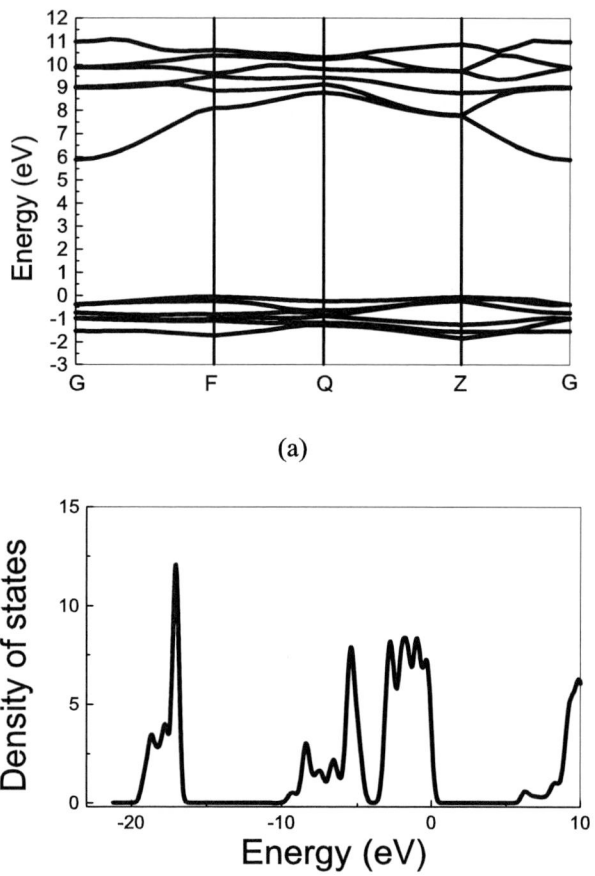

(a)

(b)

Fig. 1. The band structure (a) and the density of states (b) of pure silicon dioxide.

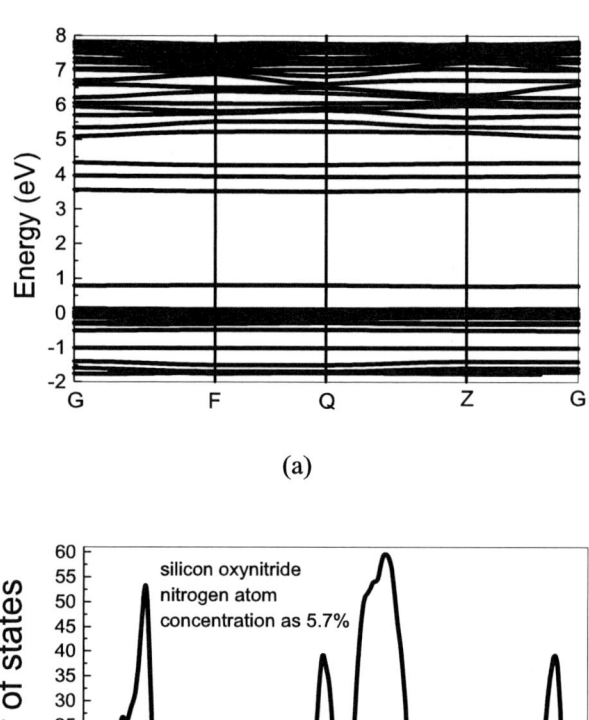

(a)

(b)

Fig. 2. The band structure (a) and the density of states (b) of silicon oxynitride with nitrogen atom concentration as 5.7%.

Comparing the band structure of $SiO_{1.75}N_{0.1667}$ whose nitrogen atom concentration is 5.7% to the band structure of pure silicon dioxide, we can observe that levels at E_c-0.75 eV, E_c-1.13 eV, E_c-1.53, E_v+0.79 eV, E_v+0.13 eV,, and E_v+0.0004 eV have been introduced into the bandgap of pure silicon dioxide. The effective masses were obtained as $7.3m_0$, $11.6m_0$, $8.8m_0$, $13.0m_0$, $20.1m_0$,and $13.9m_0$ respectively (m_0 is free electron mass). The effective mass near the bottom of the conduction band and the top of the valence band were obtained as $2.1m_0$ and $14.3\ m_0$ respectively. It is noted that larger values of the average effective electron mass for silicon oxynitide will result a smaller tunneling current. In order to assess the validity of

Since first principles calculations underestimates the band gap [10], the experimental band gaps of 8.3 eV, the valence band offset of 4.3 eV, and the conduction band offset of 2.9 eV for silicon oxynitride while nitrogen concentration being 5.7%[11] were used in the calculations. The dielectric constants can be obtained in the first principles calculations. The dielectric constant of $SiO_{1.75}N_{0.1667}$ is about 2.1 times of that of SiO_2 while the wavelength of electromagnetic wave is larger than 500 nm.

Fig.3a shows tunneling currents for electrons and holes through SiO_2/ $SiO_{1.75}N_{0.1667}$ structure with thickness of SiO_2 and $SiO_{1.75}N_{0.1667}$ is 2.0 nm and 0.7 nm, which corresponds EOT as 2.32 nm. As a comparison, the tunneling current through SiO_2 with thickness being 2.32 nm was also given in Fig.3a. It is clearly seen that in the low voltage regime, the direct tunneling current will be the dominant. But the tunneling currents caused by the empty band within the band gap increase faster with the applied voltage across the oxide/oxynitride stack than direct tunneling does. For a high applied voltage, the tunneling currents arising from the empty bands within the band gap will be dominant.

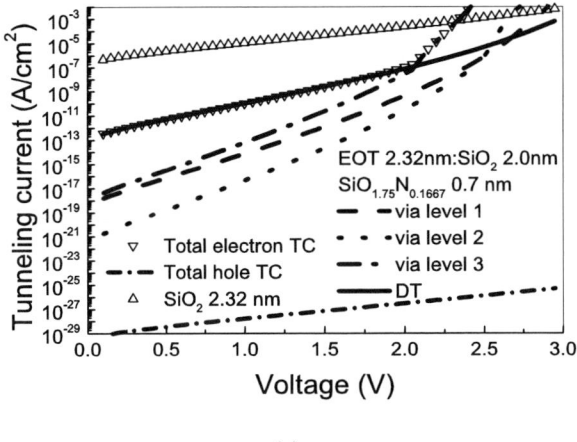

(a)

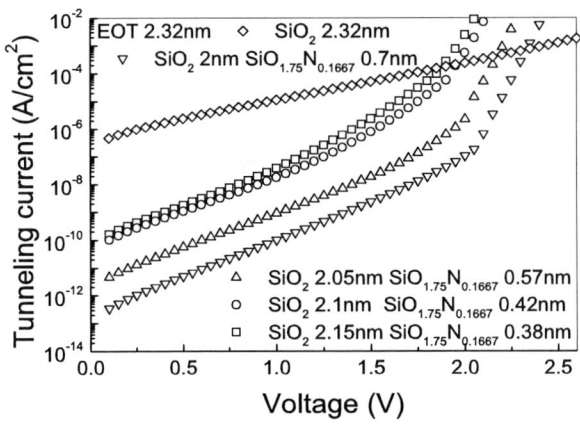

(b)

Fig. 3. The tunneling current (a) a comparison for different mechanism, (b) the total tunneling current for effective oxide thickness as 2.32 nm with different SiO_2 and $SiO_{1.75}N_{0.1667}$ thickness.

Fig.3b shows that the total tunneling currents change with EOT as 2.32nm but different combination with different thickness of SiO_2 and $SiO_{1.75}N_{0.1667}$. The threshold point where the total tunneling currents will have a larger value than pure SiO_2 does can be observed, which means when the voltage is higher than the threshold point, the gate leakage current can not be reduced simply by increasing the physical thickness of $SiO_{1.75}N_{0.1667}$ layer in SiO_2 film. Nitrogen-incorporated SiO_2 showed smaller tunneling

currents at low field, tunneling currents via empty band were dominant at high field. Failure of reducing tunneling currents results from the tunneling currents assisted by empty band. Changing the location of the empty band in bandgap by selective incorporation of nitrogen into SiO_2 can help us reduce tunneling currents effectively.

In the following part, we will show that the effective mass has a large effect on reducing tunneling current. Fig.4 shows that the tunneling current through SiO_2/ $SiO_{1.75}N_{0.1667}$ structure with thickness of SiO_2 and $SiO_{1.75}N_{0.1667}$ is 2.0 nm and 0.7 nm using different effective mass. The results show that, if the effective electron mass in silicon dioxide was selected as $0.5m_0$ and used in the tunneling current calculations, the total tunneling currents through SiO_2/ $SiO_{1.75}N_{0.1667}$ structure can be reduced only under very low voltage across the oxide/oxynitride stack. It is clearly seen that the larger effective mass will result in a smaller total electron tunneling current. This is the evidence that the effective mass will be very helpful in reducing direct tunneling currents.

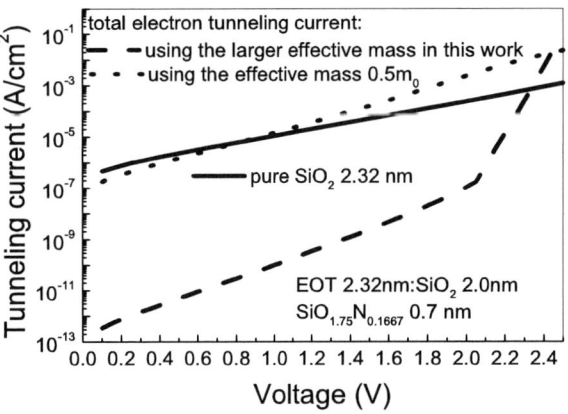

Fig. 4. The tunneling current: a comparison for different effective masses used in the tunneling current calculations.

IV. CONCLUSION

Based on first principles calculations, the defect-like energy levels were observed within the band gap of lightly nitride silicon oxide, and these levels could result in a large density of states within the band gap. And thus the defect assisted tunneling currents were calculated. In the low voltage regime, the direct tunneling current will be the dominant. But the tunneling currents assisted by the levels within the band gap increase faster with the applied voltage across the oxide/oxynitride stack than direct tunneling does and will be dominant in the high voltage regime. The tunneling currents showed that the gate leakage current can not be reduced simply by increasing the physical thickness of $SiO_{1.75}N_{0.1667}$ layer in SiO_2 film due to the effects of the effective masses on reducing tunneling current.

References

[1] S. Habermehl, and R. T. Apodaca, "Correlation of charge transport to intrinsic strain in silicon oxynitride and Si-rich silicon nitride thin films," Appl. Phys. Lett., vol. 84, pp. 215-217, Jan. 2004.

[2] S. Habermehl, and R. T. Apodaca, "Dielectric breakdown and Poole--Frenkel field saturation in silicon oxynitride thin films," *Appl. Phys. Lett.*, vol. 86, 072103, Feb. 2005.

[3] M.I. Alayoa, I. Pereyraa,*, W.L. Scopelb, M.C.A. Fantinib, On the nitrogen and oxygen incorporation in plasma-enhanced chemical vapor deposition (PECVD) SiOxNy films, *Thin Solid Films*, vol. 402, pp.154–161,2002.

[4] S. Habermehl, and R. T. Apodaca, "Compositional study of silicon oxynitride thin films deposited using electron cyclotron resonance plasma-enhanced chemical vapor deposition technique," *J. Vac. Sci. Technol. A*, vol. 23, pp. 545-550, May 2005.

[5] Yu Kwon Kim, Hyun Seok Lee, H. W. Yeom, Doo-Yeol Ryoo, Sang-Bum Huh, Jeong-Gun Lee, "Nitrogen bonding structure in ultrathin silicon oxynitride films on Si(100) prepared by plasma nitridation," *Phys. Rev. B*, vol. 70, 165320, Oct. 2004.

[6] J. F. Verwey, E. A. Amerasekera, and J. Bisschop, "The physics of SiO_2 layers", *Rep. Prog. Phys.*, vol. 53, pp 1297-13331, 1990.

[7] M. D. Segall, Philop J. D. Lindan, M. J. Probert, C. .J Pickard, P. J. Hasnip, S. J. Clark and M. C. Payne, "First principles simulations: ideas, illustrations and the CASTEP code," J Phys. Condens. Matter., vol. 14, pp. 2717-2744, March 2002.

[8] Yuji Ando, and Tomohiro Itoh, "Calculation of transmission tunneling current across arbitrary potential barriers," J. Appl. Phys., vol. 61, pp. 1497-1502, Feb. 1987.

[9] L. F. Mao, Z. O. Wang, J. Y. Wang, and G. Y. Yang, "The effective mass issues in lightly nitride silicon oxides," *Semicond. Sci. Technol.*, vol. 20, pp. 1078-1082, Oct. 2005.

[10] R. Puthenkovilakam, E. A. Carter, and J. P. Chang, "First-principles exploration of alternative gate dielectrics: Electronic structure of ZrO2/Si and ZrSiO4/Si interfaces," Phys. Rev. B, vol. 69, 155329, April 2004.

[11] H. Yu, Y.T. Hou, M. F. Li, and D. L. Kwong, "Investigation of Hole-Tunneling Current Through Ultrathin Oxynitride/Oxide Stack Gate Dielectrics in p-MOSFETs," IEEE Trans. Electron Devices, vol. 49, pp. 1158-1164, July 2002.

2006 25th International Conference on Microelectronics

Spontaneous Recovery in DC Gate Bias Stressed Power VDMOSFETs

I. Manić, S. Djorić-Veljković, V. Davidović, D. Danković, S. Golubović, and N. Stojadinović

Abstract – Spontaneous recovery of threshold voltage and channel carrier mobility in DC gate bias stressed power VDMOSFETs, as well as the underlying changes in gate oxide-trapped charge and interface trap densities are presented and analysed in terms of the mechanisms responsible. A chain of mechanisms related to a presence of hydrogen species is proposed to explain the observed changes of oxide-trapped charge and interface trap densities. Dominant mechanisms for the decrease of interface trap and oxide-trapped charge densities (as observed during the recovery by means of charge separation techniques based on device transfer characteristics) are found to be passivation of interface traps ($\equiv Si_s{}^{\bullet}$) due to their reaction with hydrogen atoms and neutralization of charged oxide traps ($\equiv Si_o{}^{+}$ or $\equiv Si_o{}^{+}\,{}^{\bullet}Si_o\equiv$) due to hydrogen molecule cracking, respectively. A remarkable increase of true interface trap density, as observed by means of charge pumping technique, is ascribed to both redistribution of interface traps within the silicon bandgap (which dominates in an early phase of recovery in all stressed devices) and dissociation of $\equiv Si_s$-H precursors by hydrogen atoms (which becomes important in later recovery phase, especially in devices stressed by lower voltages).

I. INTRODUCTION

With an increased utilization of power MOSFETs in fast switching power supplies and other circuits for communication satellites and nuclear power industry or weapon control systems, in which these devices are to maintain reliable operation in radiation environment, investigations of related reliability issues have gained in importance. Ionising radiation is well known to cause the threshold voltage shift, transconductance reduction, leakage current increase, and breakdown voltage reduction in power VDMOSFETs [1-3]. Negative threshold voltage shift appears to be the most serious problem in commercial devices since it may cause a change of their operation mode from enhancement to depletion, thus leading to a faulty circuit operation. Even the radiation-hardened devices may fail as a result of reduction in current handling capability owing to channel carrier mobility degradation and/or positive threshold voltage shift [1].

Earlier investigations have revealed the DC gate bias

I. Manić, V. Davidović, D. Danković, S. Golubović, and N. Stojadinović are with Faculty of Electronic Engineering, University of Niš, Aleksandra Medvedeva 14, 18000 Niš, Serbia and Montenegro, E-mail: ivica@elfak.ni.ac.yu

S. Djorić-Veljković is with Faculty of Civil Engineering and Architecture, University of Niš, Aleksandra Medvedeva 14, 18000 Niš, Serbia and Montenegro.

stressing and ionising radiation to have very similar effects on electrical parameters of CMOS devices [4]. More recently, this similarity has been found in power VDMOSFETs as well [5-10]. The possibilities to use gate bias stressing for radiation hardening and selection of commercial VDMOS devices for application in radiation environment were investigated in [5, 6], the effects of ionising radiation and gate bias stressing were compared in [7], while the mechanisms responsible for stress-induced threshold voltage shift and mobility degradation were studied in [8-10]. However, a detailed analysis of the mechanisms responsible for behaviour of device parameters during the post-stress recovery is also required for better understanding of stress-induced phenomena.

Our earlier papers were dealing with spontaneous recovery in VDMOSFETs stressed only by positive gate bias [11, 12]. In this paper, a detailed analysis of spontaneous recovery of threshold voltage and mobility in both positive and negative DC gate bias-stressed devices is presented. Data are analysed in terms of the mechanisms responsible for underlying changes of the oxide-trapped charge and interface trap densities, and an appropriate model providing detailed explanation of the experimental data is proposed.

II. RESULTS AND DISCUSSION

Devices used in this study were the commercial EFL1N15 n-channel VDMOSFETs built in a standard *Si*-gate technology with about 120 nm thick gate oxide grown in dry oxygen. Electrical stressing was performed by applying either positive or negative DC voltage (88, 90, 92, and 94 V) for 2 hours to the gate electrode, with drain and source terminals grounded. Devices, with all terminals shorted, were then stored at room temperature for up to about 2000 hours to recover spontaneously. An electrical characterization, including PC-controlled measurements of device subthreshold and above-threshold transfer *I-V* characteristics in the saturation region, was performed during both stress and recovery phases. The devices were characterized by charge pumping current measurements as well. The above-threshold characteristics were used to estimate the values of device threshold voltage and channel carrier mobility, determined as the intersections between V_G-axis and extrapolated linear region of $\sqrt{I_D}-V_G$ curves and the slopes of these lines, respectively. These values, along with the subthreshold *I-V* characteristics, were then

1-4244-0116-X/06/$20.00 ©2006 IEEE

used to calculate the underlying changes in the densities of oxide-trapped charge (ΔN_{ot}) and interface traps (ΔN_{it}) by well-known subthreshold midgap technique (SMGT) [13]. However, the stressing at higher positive gate voltages (+92 and +94 V) resulted in severely distorted subthreshold characteristics (also observed in irradiated devices [14]), and SMGT was practically inapplicable in these cases. For that reason we have also used the single-transistor mobility technique (STMT) [3, 15], which is entirely based on the above-threshold characteristics. The charge pumping technique (CPT) [14, 16] has been additionally used for independent calculations of ΔN_{it}.

The changes of threshold voltage and mobility during the DC gate bias stressing and subsequent spontaneous recovery of power VDMOSFETs are shown in Fig. 1. As can be seen, during the stressing with either gate bias polarity there is an initial decrease of threshold voltage followed by its increase towards the initial value (turn-around), while the mobility continuously decreases. The negative bias stressing (NBS) causes more rapid initial changes, but the final stress-induced threshold voltage shift

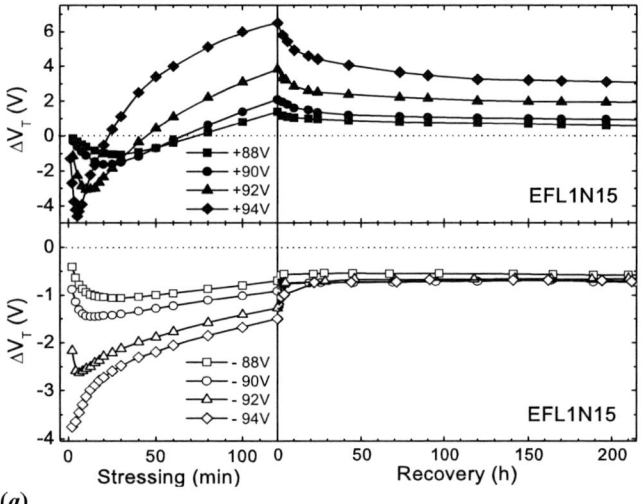

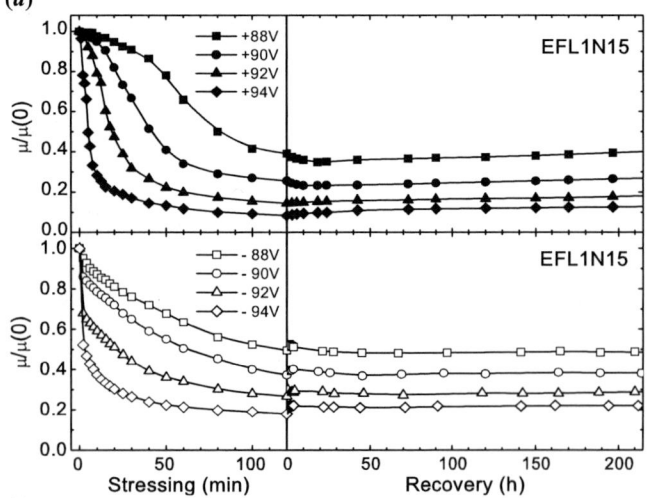

(b)

Fig. 1. The behaviour of *(a)* threshold voltage and *(b)* mobility during the DC gate bias stressing and subsequent spontaneous recovery of power VDMOSFETs.

and mobility reduction are both larger in the case of positive bias stressed (PBS) devices [9, 10]. During the spontaneous recovery of PBS devices, the threshold voltage continuously decreases, while the mobility in general slightly increases. However, small initial mobility decrease in devices stressed at lower voltages (+88 and +90 V) can be observed as well. On the other hand, during the recovery of all NBS devices, both threshold voltage and mobility make rather sharp but small initial increase, the mobility then slightly decreases, and after that both parameters remain nearly unchanged throughout the rest of the recovery period. Note that more pronounced changes of threshold voltage and mobility during both stressing and recovery are found in devices stressed at higher voltages.

The underlying changes in the densities of oxide-trapped charge and interface traps during the stressing and subsequent spontaneous recovery, as determined by both SMGT and STMT, are shown in Fig. 2 (SMGT data for devices stressed at +92 and +94 V are missing because of severely distorted subthreshold characteristics). Both techniques indicate that stressing under either gate bias polarity leads to a significant increase of N_{ot} and N_{it}, the changes being more pronounced at higher stress voltages. In line with observed changes of threshold voltage and mobility, NBS causes more considerable initial buildup of oxide-trapped charge and interface traps, but the final changes of their densities are higher in the case of PBS devices [9, 10]. During the spontaneous recovery of PBS devices, an initial increase of ΔN_{ot} and ΔN_{it} in devices stressed at lower voltages (+88 and +90 V), which is responsible for initial decrease of mobility, is followed by very slow decrease; at he same time, ΔN_{ot} and ΔN_{it} in devices stressed at higher voltages (+92 and +94 V) are continuously decreasing. On the other hand, during the recovery of all NBS devices, ΔN_{ot} and ΔN_{it} make sharp initial decrease, which slows down and becomes almost negligible during the rest of recovery period. Note that all changes of N_{ot} and N_{it} observed during the recovery are more pronounced in devices stressed at higher voltages.

It should be mentioned that, during the recovery, notable changes of threshold voltage and mobility, as well as those of N_{ot} and N_{it}, have been seen only during an initial recovery stage lasting from few to few tens of hours after the stressing. The tendencies established in that period remained unchanged, and that is why Figs. 1 and 2 show the results only for about 200 hours of recovery.

The changes interface trap density during the stressing and subsequent spontaneous recovery determined by CPT are shown in Fig. 3. As can be seen, the CPT data are in full qualitative agreement with those obtained by SMGT and STMT for all samples during the stressing, but there is an apparent difference in the recovery phase. The former techniques indicate certain increase of N_{ot} and N_{it} in initial recovery period only in samples stressed by lower positive voltages (+88 and +90 V), while the CPT indicates that N_{it} increases in all stressed devices over the entire recovery period of 200 hours shown. Moreover, creation of interface

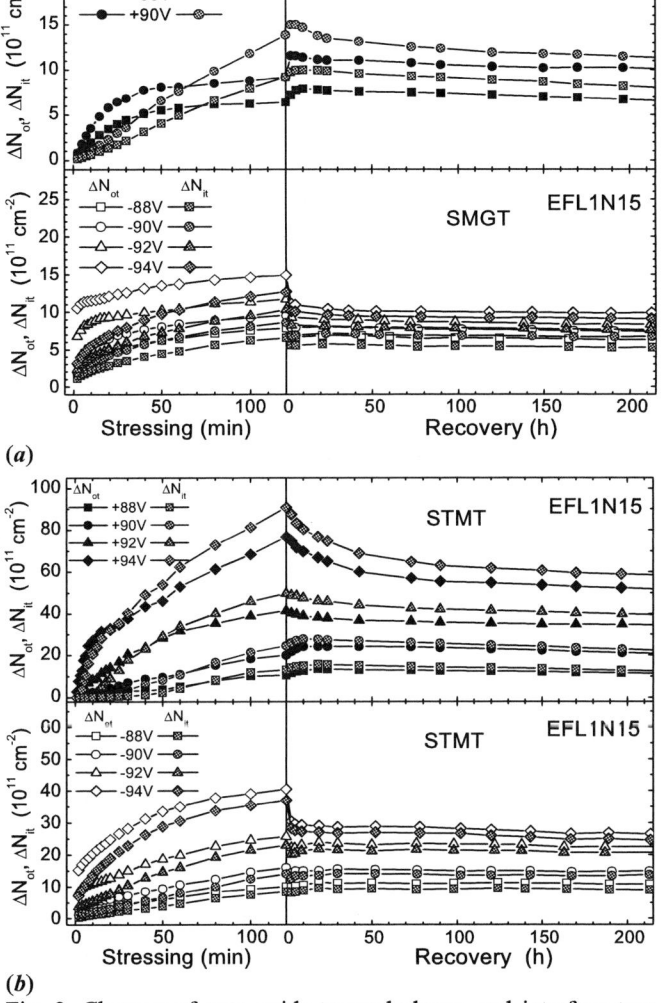

(a)

(b)

Fig. 2. Changes of gate oxide-trapped charge and interface trap densities during the DC gate bias stressing and subsequent spontaneous recovery of power VDMOSFETs, as determined by (*a*) SMGT; (*b*) STMT.

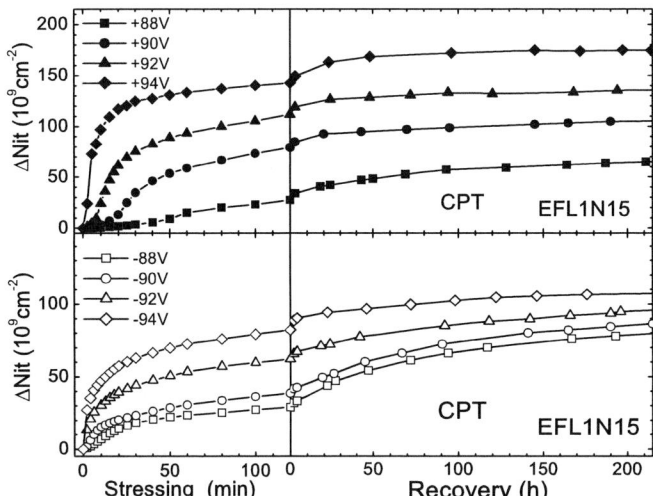

Fig. 3. Changes of interface trap density during the DC gate bias stressing and subsequent spontaneous recovery of power VDMOSFETs, as determined by CPT.

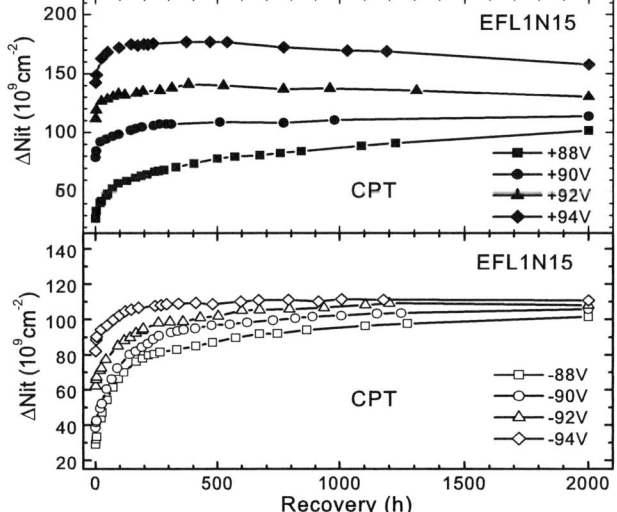

Fig. 4. Changes of interface trap density during the spontaneous recovery of DC gate bias stressed power VDMOSFETs, as determined by CPT.

traps during the post-stress recovery appears more significant in devices in which the stressing itself caused less significant buildup of interface traps, i.e. in devices stressed by lower voltages (88 and 90 V, either positive or negative). Better illustration provides Fig. 4, which shows the CPT data for ΔN_{it} over an extended period of recovery up to 2000 hours. As can be seen, a remarkable post-stress increase of N_{it} is found in all devices, but ΔN_{it} increases continuously up to 2000 hours only in devices stressed by lower voltages; as the voltage used for stressing was raised, the post-stress buildup of interface traps becomes less significant, the corresponding ΔN_{it} tends to saturate and, in devices stressed by the highest voltages, even starts decreasing during an extended recovery period.

In addition to above major qualitative disagreement between ΔN_{it} data obtained by SMGT/STMT and CPT for recovery period, significant quantitative differences in ΔN_{it} values estimated by different techniques also can be seen.

The reason could be found in the fact that charge separation techniques such as SMGT and STMT, which are based on low frequency measurements, tend to overestimate the density of interface traps [17] since at low frequencies the border traps can imitate the electrical response of interface traps [18], while only CPT, owing to high frequency measurements employed, provides information on the density of "true" interface traps. In addition, different techniques provide information on interface trap densities in different parts of the silicon bandgap: CPT yields the density of interface traps distributed at energy levels just around the midgap, SMGT yields the average density of interface traps in the range from the midgap to the energy level of the surface potential corresponding to the threshold voltage, and STMT above that level. Since the energy distribution of interface traps within the bandgap is not uniform but is U-shaped [19], the interface trap density is higher near the bandgap edges (accessible by STMT and

SMGT) and lower around the midgap (accessible by CPT). Accessibility of STMT extends far from the midgap, and this technique yields the highest density of interface traps.

III. MECHANISMS RESPONSIBLE

Detailed analysis of the mechanisms responsible for the effects of gate bias stressing has already been reported in [8-10]. In the case of positive gate bias stressing, the electron tunnelling from neutral $\equiv Si_o^{\bullet}$ trivalent silicon defects into the oxide conduction band and subsequent hole tunnelling from the charged $\equiv Si_o^{+}$ oxide traps to $\equiv Si_s$-H interfacial precursors were shown to be the dominant mechanisms for buildup of positive oxide-trapped charge and that of interface traps, respectively [8, 10]. In the case of negative bias stressing, the hole tunnelling from silicon valence band to neutral $\equiv Si_o^{\bullet}{}^{\bullet}Si_o\equiv$ oxygen vacancy defects was shown to be responsible for positive oxide-trapped charge buildup, and subsequent electro-chemical reactions of $\equiv Si_s$-H interfacial precursors with both charged $\equiv Si_o^{+}{}^{\bullet}Si_o\equiv$ oxide traps and H^{+} ions were proposed to be responsible for buildup of interface traps [9, 10].

Thus, immediately after stressing there is a significant amount of positive oxide charge trapped near the SiO_2-Si interface in both PBS and NBS devices. This charge is found at energy levels of oxide defects associated either with $\equiv Si_o^{\bullet}$ (in the case of PBS) or $\equiv Si_o^{\bullet}{}^{\bullet}Si_o\equiv$ (in the case of NBS), which are located at about 3.1 and 6.4 eV below the bottom of oxide conduction band [8-10, 20]. Hence, though the stressed devices were allowed to recover spontaneously (no gate bias applied), there was still rather high electric field across the oxide, coming mostly from the positive oxide-trapped charge, with an additional contribution from the work function difference between the n-type Si-gate and p-type silicon. Consequently, some of the mechanisms that were responsible for stress-induced buildup of oxide-trapped charge and interface traps could be present during an early stage of the recovery phase as well. As illustrated in Fig. 5, these mechanisms include the hole tunnelling or drift from the charged $\equiv Si_o^{+}$ oxide traps to interface trap and trap precursor levels (mechanisms 1 and 2) in the case of PBS, or an electro-chemical reaction of the charged $\equiv Si_o^{+}{}^{\bullet}Si_o\equiv$ oxide traps with interfacial trap precursors (mechanism 2') in the case of NBS devices. Positive charge arriving to interface traps through the mechanism 1 may cause their redistribution over the energy levels within the Si bandgap (mechanisms 4 and 5). The mechanism 2 (or 2') may lead to dissociation of $\equiv Si_s$-H precursors, resulting into creation of interface traps (mechanism 3), but the mechanism 2 (or 2') may also fail to create interface traps if the arriving hole recombines (or the charged $\equiv Si_o^{+}{}^{\bullet}Si_o\equiv$ defect neutralizes) with an electron from the Si substrate. In any case, positively charged oxide defects are neutralized through the mechanisms 1 and 2 (or 2'), which should be seen as a decrease of ΔN_{ot} determined by STMT or SMGT. However, it should be noted that any interface-oriented motion of positive charge, which certainly involves border

traps as well, the above techniques could actually observe as an increase of the effective positive N_{ot}. The above redistribution of interface traps (mechanisms 4 and 5) also, depending on trap levels involved, could be observed either as the increase or decrease of N_{it}, with border traps shading the accurate estimation again.

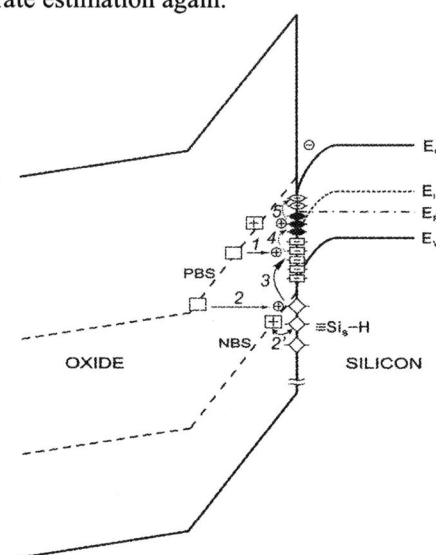

Fig. 5. Illustration of mechanisms responsible for changes of oxide-trapped charge and interface trap densities during the early stage of spontaneous recovery of stressed VDMOSFETs.

The above mechanisms could be of importance only in an early stage of recovery, when there is still rather high density of stress-induced oxide-trapped charge. However, to clarify the phenomena occurring during the overall recovery period, an additional chain of mechanisms associated with the presence of hydrogen species released during stressing must be taken into account as well. After the stressing, there is high concentration of hydrogen atoms and ions near the interface, the latter also being transformed into the atoms after picking up the electrons from the adjacent Si layer. Highly reactive hydrogen atoms may react with $\equiv Si_s$-H interfacial precursors, leading to their dissociation and creation of interface traps $\equiv Si_s^{\bullet}$, with hydrogen molecules being formed [21]. Also, hydrogen atoms can react with stress induced interface traps, leading to their passivation [22]. Finally, dimerization of hydrogen may occur when two hydrogen atoms react among themselves forming a molecule [23]. It is important to note that all these reactions occur without an energy barrier [22], the probability of their occurrence depends only on concentration of reacting species [24]. In any case, the interface is saturated with hydrogen, and the H_2 molecules quickly diffuse into the oxide bulk. Owing to this migration, any interactions between the hydrogen and previously formed interface traps weaken and eventually cease, leading to an energy redistribution of interface traps within the Si bandgap from upper to more stable, lower energy levels around the midgap. This redistribution is in line with results showing the interface trap formation to be diffusion controlled process, in which interface traps

642

become stable only after the reaction byproduct (hydrogen in this case) diffuses away from the interface [25, 26]. The diffusing H_2 molecules can be cracked at positively charged oxide traps ($\equiv Si_o^+$ and/or $\equiv Si_o^+\ {}^\bullet Si_o\equiv$), leading to neutralization of the oxide-trapped charge and formation of hydrogen ions [27]. Hydrogen ions drift under the positive oxide field towards the SiO_2-Si interface, some of them being temporary captured by oxygen vacancies $\equiv Si_o^{\bullet}\ {}^\bullet Si_o\equiv$ (which can be observed as an increase of ΔN_{ot} [28]), but majority arrive to the interface and are neutralized by electrons from the adjacent silicon, which allows the above chain of mechanisms to continue over an extended period of recovery. The drift velocity of hydrogen ions is electric field dependent, which explains more pronounced decrease of oxide-trapped charge density during the recovery of devices stressed at higher voltages.

The SMGT and STMT results for ΔN_{ot} and ΔN_{it} in PBS devices (Fig. 2 a and b, upper parts) indicate that oxide charge neutralization (H_2 cracking and mechanism **1** from Fig. 5) and interface trap passivation processes dominate over the oxide-trap charging and interface trap creation in devices stressed at higher voltages (+92 and +94 V) throughout the recovery phase, which is a consequence of very high densities of stress-induced oxide-trapped charge and interface traps. On the other hand, stressing at lower voltages (+88 and +90 V) did not result into that high ΔN_{ot} and ΔN_{it}; as a consequence, oxide-trap charging (H^+ ion capturing at $\equiv Si_o^{\bullet}\ {}^\bullet Si_o\equiv$ defects) and/or positive charge shift/redistribution (mechanisms **1**, **4**, and **5** from Fig. 5), as well as interface trap creation (dissociation of interfacial trap precursors through the reaction with hydrogen atoms and mechanisms **2** and **3** from Fig. 5) dominate in early stage of recovery phase. After about 24 hours of the recovery, the cracking and passivation processes become dominant, and both ΔN_{ot} and ΔN_{it} start decreasing in these devices as well.

The SMGT and STMT reveal somewhat different behaviour of ΔN_{ot} and ΔN_{it} during the recovery of NBS devices (Fig. 2 a and b, lower parts). As already mentioned, the negative bias stress-induced positive oxide charge is found trapped at oxygen vacancy defects, which are very close to the SiO_2-Si interface. On removing the stressing voltage, the channel region changes its state from deep accumulation into inversion, and the $\equiv Si_o^+\ {}^\bullet Si_o\equiv$ defects nearest to the interface are quickly neutralized by channel electrons. The above techniques observe this as a sudden initial drop of both ΔN_{ot} and ΔN_{it} since the near-interface defects behave as the border traps. The extent of this drop depends on the amount of charge trapped during the stressing, and is therefore higher in devices stressed at higher voltages. The initial drop is then followed by certain increase of ΔN_{ot} and ΔN_{it} since the above neutralization of near-interface defects allows the positive charge found trapped farther in the oxide to shift closer to the interface. In addition, the chain of hydrogen-associated mechanisms is functioning as well; the above sudden neutralization of $\equiv Si_o^+\ {}^\bullet Si_o\equiv$ defects increases the probability for the

occurrence of precursor dissociation and H^+ ion capturing processes, and their contribution to observed small increase of ΔN_{ot} and ΔN_{it} could be the dominant one. However, as soon as the ΔN_{ot} and ΔN_{it} start increasing, the passivation, dimerization, and cracking processes become more frequent, tending to balance the effect of previous processes and eventually assuming slight domination; as a consequence, very slow decrease of both ΔN_{ot} and ΔN_{it} is observed during the rest of recovery phase.

As already pointed out, the border traps can significantly blur the results obtained by SMGT and STMT, but CPT results, indicating remarkable increase of true interface trap density during the recovery in both PBS and NBS devices (Figs. 3 and 4), also can be explained trough the combined effects of reactions involving hydrogen and interface trap redistribution within the Si bandgap. All these processes occur simultaneously, but trap redistribution to the levels around Si midgap, associated with quick diffusion of H_2 molecules away from interface, dominates in the early stage of recovery, which CPT observed as a rapid initial increase of ΔN_{it} in all devices. This redistribution fades rather quickly, and further developments in the advanced period of recovery depend only on the status found at interface after stressing, which determines the probability for each of two contrasting reactions (precursor dissociation versus trap passivation) to occur. In devices stressed by the lowest voltages (± 88 V), the number of interfacial precursors remained higher than that of interface traps, and ΔN_{it} grows continuously. In medium voltage stressed devices (± 90 and ± 92 V), the number of stress-induced traps appears to be nearly balanced with that of available precursors, both dissociation and passivation occur with equal probabilities, and ΔN_{it} in advanced stage of recovery almost does not change. The highest stress-induced ΔN_{it} is found in devices stressed at ± 94 V, which means the number of interface traps found after stressing could be higher than that of precursors, and trap passivation could dominate over the precursor dissociation from the very beginning of recovery. However, the effects of interface trap redistribution within the Si bandgap at initial recovery stage appeared to be by far more pronounced, and CPT observed an initial increase of ΔN_{it} in these devices as well. As soon as the redistribution fades, CPT begins observing actual situation and the ΔN_{it} obtained by this technique starts decreasing.

V. CONCLUSION

The spontaneous recovery of threshold voltage and mobility in DC gate bias stressed power VDMOSFETs, as well as the underlying changes in gate oxide-trapped charge and interface trap densities were analysed in terms of the mechanisms responsible. A chain of mechanisms related to a presence of hydrogen species was proposed to explain the observed changes of oxide-trapped charge and interface trap densities. Dominant mechanisms for the decrease of interface trap and oxide-trapped charge densities, as observed by SMGT and STMT, were found to be passivation of interface traps ($\equiv Si_s^{\bullet}$) due to their reaction with hydrogen atoms and neutralization of charged oxide traps ($\equiv Si_o^{+}$ or $\equiv Si_o^{+} {}^{\bullet}Si_o\equiv$) due to hydrogen molecule cracking, respectively. A remarkable increase of true interface trap density, as observed by CPT, was ascribed to both redistribution of interface traps within the silicon bandgap (which dominated in an early phase of recovery in all devices) and dissociation of $\equiv Si_s$-H precursors by hydrogen atoms (which became important in later recovery phase, especially in devices stressed by lower voltages).

REFERENCES

[1] K.F. Galloway, R.D. Schrimpf, "MOS device degradation due to total dose ionizing radiation in the natural space environment: a review", *Microelectronics J.*, 1990, vol. 21, pp. 67-81.

[2] D. Zupac, K.F. Galloway, R.D. Schrimpf, P. Augier, "Radiation-induced mobility degradation in p-channel double-diffused metal-oxide-semiconductor power transistors at 300 and 77 K", *J. Appl. Phys.*, 1993, vol. 73:2910-2915.

[3] N. Stojadinovic, S. Golubovic, S. Djoric, S. Dimitrijev, "Analysis of gamma-irradiation induced degradation mechanisms in power VDMOSFETs", *Microelectron. Reliab.*, 1995, vol. 35, pp. 587-602.

[4] N. Stojadinovic, S. Dimitrijev, "Instabilities in MOS transistors", *Microelectron. Reliab.*, 1989, vol. 29, pp. 371-380.

[5] P. Picard, C. Brisset, O. Quittard, A. Hoffmann, F. Joffre, J.P. Charles, "Radiation hardening of power VDMOSFETs using electrical stress" *IEEE Trans. Nucl. Sci.*, 2000, vol. 47, pp. 641-646.

[6] P. Picard, C. Brisset, A. Hoffmann, J.P. Charles, F. Joffre, L. Adams, A. Holmes-Siedle, "Use of electrical stress and isochronal annealing on power MOSFETs in order to characterize the effects of ^{60}Co irradiation", *Microelectron. Reliab.*, 2000, vol. 40, pp. 1647-1652.

[7] M.-S. Park, I. Na, C.R. Wie, "A comparison of ionizing radiation and high field stress effects in n-channel power vertical double-diffused metal-oxide-semiconductor field-effect transistors", *J. Appl. Phys.*, Vol. 97, 2005, pp. 014503-1 – 014503-6.

[8] N. Stojadinovic, I. Manic, S. Djoric-Veljkovic, V. Davidovic, S. Golubovic, S. Dimitrijev, "Mechanisms of positive gate bias stress induced instabilities in power VDMOSFETs", *Microelectron. Reliab.*, 2001, vol. 41, pp. 1373-1378.

[9] I. Manic, S. Djoric-Veljkovic, V Davidovic, D. Dankovic, S. Golubovic, S. Dimitrijev, N. Stojadinovic, "Effects of negative gate bias stressing in thick gate oxides for power VDMOSFETs", *in Proc. 12th Workshop on Dielectrics in Microelectronics WoDiM2002*, Grenoble, 2002, pp. 41-44.

[10] N. Stojadinovic, I. Manic, V. Davidovic, D. Dankovic, S. Djoric-Veljkovic, S. Golubovic, S. Dimitrijev, "Effects of electrical stressing in power VDMOSFETs", *Microelectron. Reliab.*, 2005, vol. 45, pp. 115-122.

[11] N. Stojadinovic, I. Manic, S. Djoric-Veljkovic, V. Davidovic, D. Dankovic, S. Golubovic, S. Dimitrijev, "Mechanisms of spontaneous recovery in positive gate bias stressed power VDMOSFETs", *Microelectron. Reliab.*, 2002, vol. 42, pp. 1465-1468.

[12] N. Stojadinovic, I. Manic, S. Djoric-Veljkovic, V. Davidovic, D. Dankovic, S. Golubovic, S. Dimitrijev, "Spontaneous recovery of positive gate bias stressed power VDMOSFETs", *in Proc. 23rd International Conference on Microelectronics MIEL 2002*, Nis, 2002, pp. 717-721.

[13] P.J. McWhorter and P.S. Winokur, "Simple technique for separating the effects of interface traps and trapped-oxide charge in metal-oxide-semiconductor transistors", *Appl. Phys. Lett.*, 1986, vol. 48, pp. 133-135.

[14] S.C. Witczak, K.F. Galloway, R.D. Schrimpf, J.L. Titus, J.R. Brews, and G. Prevost, "The determination of Si-SiO$_2$ interface trap density in irradiated four-terminal VDMOSFETs using charge pumping", *IEEE Trans. Nucl. Sci.*, 1996, vol. 43, pp. 2558-2564.

[15] S. Dimitrijev, N. Stojadinovic, "Analysis of CMOS transistor instabilities", *Solid-State Electron.*, 1987; vol. 30, pp. 991-1003.

[16] P. Habas, Z. Prijic, D. Pantic, N. Stojadinovic, "Charge-pumping characterization of SiO$_2$/Si interface in virgin and irradiated power VDMOSFETs", *IEEE Trans. Electron. Dev.*, 1996, vol. 43, pp. 2197 - 2208.

[17] J.R. Schwank, D.M. Fleetwood, M.R. Shaneyfelt, P.S. Winokur, "A critical comparison of charge-pumping, dual-transistor, and midgap measurement techniques", *IEEE Trans. Nucl. Sci.*, 1993, vol. 40, pp. 1666-1677.

[18] D.M. Fleetwood, P.S. Winokur, R.A. Reber Jr., T.L. Meisenheimer, J.R. Schwank, M.R. Shaneyfelt, L.C. Riewe, "Effects of oxide traps, interface traps, and border traps on MOS devices", *J. Appl. Phys.*, 1993, vol. 73, pp. 5058-5074.

[19] T.P. Ma, P.V. Dressendorfer, *Ionizing Radiation Effects in MOS Devices and Circuits*, New York, John Wiley, 1989.

[20] C.T. Sah, "Origin of interface states and oxide charges generated by ionizing radiation", *IEEE Trans. Nucl. Sci.*, 1976, vol. NS-23, pp. 1563-1568.

[21] D.L. Griscom, D.B. Brown, N.S. Saks, "Nature of radiation-induced point defects in amorphous SiO$_2$ and their role in SiO$_2$-on Si structures", in *The Physics and Chemistry of SiO$_2$ and Si- SiO$_2$ Interface*, C.R. Helms and B.E. Deal, Eds., pp. 287-298, New York, Plenum Press, 1988.

[22] K.L. Brower, S.M. Myers, "Chemical kinetics of hydrogen and (111) Si- SiO$_2$ interface defects", *Appl. Phys. Lett.*, 1990, vol. 57, pp. 162-164.

[23] M.L. Reed, J.D. Plummer, "Si-SiO$_2$ interface trap production by low-temperature thermal processing", *Appl. Phys. Lett.*, 1987, vol. 51, pp. 514-516.

[24] M.L. Reed, J.D. Plummer, "Chemistry of Si-SiO$_2$ interface trap annealing", *J. Appl. Phys.*, 1988, vol. 63, no. 12, pp5776-5793.

[25] K. Jeppson, C. Svensson, "Negative bias stress of MOS devices at high electric fields and degradation of NMOS devices", *J. Appl. Phys.*, 1977, vol. 48, pp. 2004-2019.

[26] D.K.Schroder, J.A. Babcock, "Negative bias temperature instability: road to cross in deep submicron silicon semiconductor manufacturing", *J. Appl. Phys.*, 2003, vol. 94, pp. 1-18.

[27] R.E. Stahlbush, A.H. Edwards, D.L. Griscom, B.J. Mrstik, "Post-irradiation cracking of H$_2$ and formation of interface states in irradiated metal-oxide-semiconductor field-effect transistors', *J. Appl. Phys.*, 1993, vol. 73, pp. 658-667.

[28] D.M. Fleetwood, W. L. Warren, J.R. Scwank, P.S. Winokur, M.R. Shaneyfelt, and L.C.Riewe, "Effects of interface traps and border traps on MOS postirradiation annealing response", *IEEE Trans. Nucl. Sci.*, 1995, vol. 42, pp. 1698-1707.

2006 25th International Conference on Microelectronics

Lifetime Estimation in NBT Stressed P-Channel Power VDMOSFETs

D. Danković, I. Manić, S. Djorić-Veljković, V. Davidović, S. Golubović, and N. Stojadinović

Abstract – Threshold voltage shifts observed in commercial p-channel power VDMOSFETS during the NBT stressing are fitted using stretched exponential equation in order to estimate the device lifetime and discuss the impact of stress conditions and choice of extrapolation parameters. Excellent agreement found in later stress phases allowed for an accurate estimation of device lifetime for the lowest stress voltage applied, saving the time required for an extended experiment. More realistic lifetime estimates also are expected in the case of lower stress voltages, which additionally justifies the need for using stretched exponential or some other suitable fit.

I. INTRODUCTION

As the dimensions of MOS devices have been continually scaled down, the negative bias temperature instabilities (NBTI) have become a serious reliability problem. Namely, slower lowering of operating voltages in comparison to more aggressive decreasing of the gate oxide thickness has gradually increased the effective electric field across the oxide, which has enhanced the NBTI. NBTI are known to occur in p-channel MOSFETs operated at elevated temperatures (100 - 250 °C) under negative gate voltages corresponding to gate oxide fields of 2 - 6 MV/cm [1-4]. Considering the device electrical parameters, NBT stress-induced threshold shifts were found to be most dangerous, and can put a serious limit to a lifetime of p-channel MOSFETs with gate oxide thickness less than 3.5 nm [5]. The microscopic mechanisms of NBTI are still not well understood, and optimization of process conditions to minimize NBTI is a very difficult problem. Therefore, accurate NBTI lifetime models are needed to ensure the device reliability.

In spite of aggressive scaling down of the dimensions of CMOS devices, widespread utilization of MOS technology for the realization of power devices and ICs has led to an increased interest in ultra-thick gate oxides as well. Degradation of power MOSFETs under various stresses (irradiation, high field, temperature, and even hot carrier) has been subject of extensive research [6], but very few authors addressed the negative bias temperature

D. Danković, I. Manić, V. Davidović, S. Golubović, and N. Stojadinović are with Faculty of Electronic Engineering, University of Niš, Aleksandra Medvedeva 14, 18000 Niš, Serbia and Montenegro, E-mail: danijel@elfak.ni.ac.yu

S. Djorić-Veljković is with Faculty of Civil Engineering and Architecture, University of Niš, Aleksandra Medvedeva 14, 18000 Niš, Serbia and Montenegro.

instabilities in these devices [1, 2]. The electric fields and temperatures applied during the NBT stressing can be approached during the routine operation of power MOSFETs in automotive and industrial applications [2]. Thus, the investigations of NBTI in power MOSFETs have gained in importance as well.

In this paper, the threshold voltage shifts in commercial p-channel power VDMOSFETs during the NBT stressing are investigated for different values of stress voltage, temperature, and time. The threshold voltage data are fitted using stretched exponential equation [7, 8] in order to estimate the device lifetime and discuss the impact of NBT stress conditions and parameters of extrapolation.

II. RESULTS

Devices used in this study were the commercial p-channel power VDMOSFETs IRF9520 built in standard *Si* -gate technology with an assumed gate oxide thickness of 100 nm, encapsulated in a TO - 220 plastic case. Devices were stressed by negative gate voltages in the range 30 - 45 V, with drain and source terminals grounded, at temperatures ranging from 125 to 175 °C, up to near 2000 hours. In order to characterize the NBTI effects, we have applied a conventional methodology by periodically stopping the stress to measure the device subthreshold and above-threshold transfer characteristics. The threshold voltage values were estimated from the above-threshold characteristics as the intersections between V_G - axis and extrapolated linear region of $\sqrt{I_D} - V_G$ curves.

As an illustration, two characteristic sets of data (i.e. different stress voltages at 150 °C and different temperatures at the stress voltage of - 40 V) for the time dependencies of threshold voltage shift (ΔV_T) during the NBT stressing of IRF9520 p-channel VDMOSFETs are shown in Fig. 1. As can be seen, the stressing caused significant threshold voltage shifts, more pronounced at higher voltages (Fig.1 *a*) and/or temperatures (Fig.1 *b*). Further analysis has shown that ΔV_T time dependencies followed the t^n power law, but with three distinct phases (as indicated by the dashed lines), which can be clearly distinguished depending on the value of parameter *n* [9]. In the first (early) phase, *n* strongly depends on both bias and temperature, varying from 1.14 to 0.43. In the second phase, parameter *n* is almost independent on bias and temperature, and ΔV_T follows the well-known $t^{0.25}$ law, as obtained in almost all earlier NBTI investigations on devices manufactured in various technologies [1-4, 10]. This phase begins earlier in devices stressed at higher voltages and/or temperatures, and one may even expect the first phase to disappear under more severe stress conditions. Finally, in the third phase, n becomes bias and temperature dependent again, gradually decreasing from 0.25 to 0.14, and ΔV_T tends to saturate. The ΔV_T in saturation after near 2000 hours of stressing was found to vary approximately from 4.4 % (125 °C, - 30 V) to 19.8 % (175 °C, - 45 V) [9].

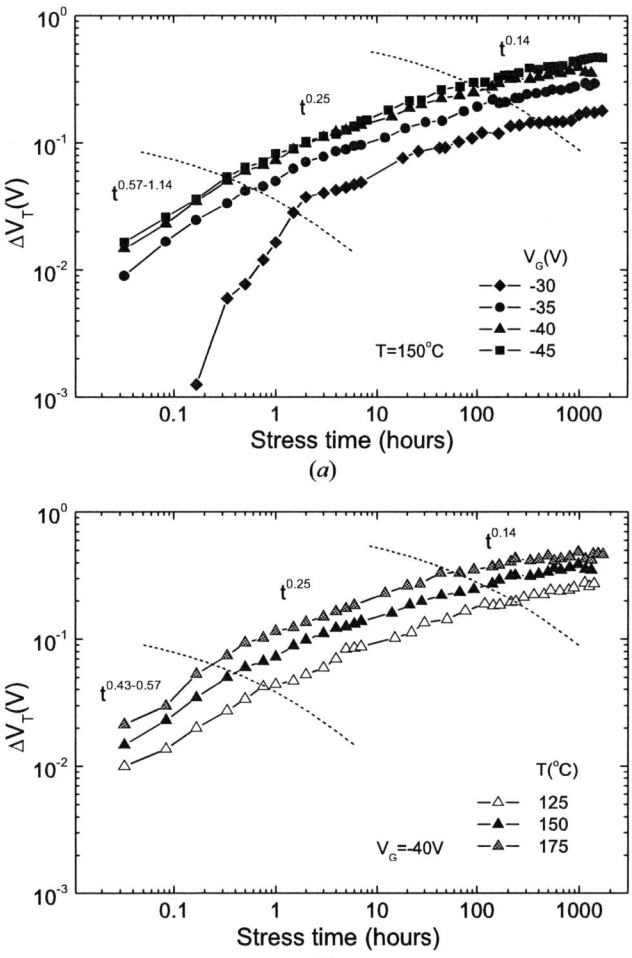

Fig. 1. Threshold voltage shifts observed in p-channel power VDMOSFETs during the NBT stressing at: a) different voltages at 150 °C; b) different temperatures at V_G = - 40 V.

In our previous work [9], we proposed a model of responsible mechanisms to explain the behaviour of threshold voltage during the NBT stressing in p-channel power VDMOSFETs IRF9520. The observed t^n power law time dependencies of threshold voltage shift, with three distinct phases according to parameter n, were found to be mostly affected by the oxide trapped charge, which does not appear to be in line with most of previous findings indicating the dominant influence of NBT stress induced interface traps [2-4].

III. RESULT ANALYSIS AND DISCUSSION

To estimate the ten-year lifetime for p-channel VDMOSFET IRF9520, the ΔV_T versus time curves, such as those shown in Fig.1, can be fitted by so-called stretched exponential equation [7, 8]:

$$\Delta V_T(t) = \Delta V_{T\max} \cdot [1 - \exp(-(t/\tau_o)^\beta)] \qquad (1)$$

where ΔV_{Tmax}, τ_o, and β are the fitting parameters. The equation can be used for processes that are described by a distribution in time constant or in activation energy. This kind of fitting predicts that ΔV_T would saturate after prolonged stressing, and the value in saturation (ΔV_{Tmax}) can be treated as a measure of ΔV_T at ten year lifetime. Parameter β is defined as a measure of the distribution width, while τ_o represents characteristic time constant of the distribution [7, 8].

The experimental and fitting results for threshold voltage shifts during the NBT stressing of p-channel power VDMOSFETs by different voltages at 125 °C are shown in Fig. 2. As can be seen, the stretched exponential model is in excellent agreement with experimental data in the second stress phase and in the saturation. A disagreement is found only in the early stress phase, and is particularly pronounced at lower stress voltages. It is interesting to note that stress times at which the transition from early to second

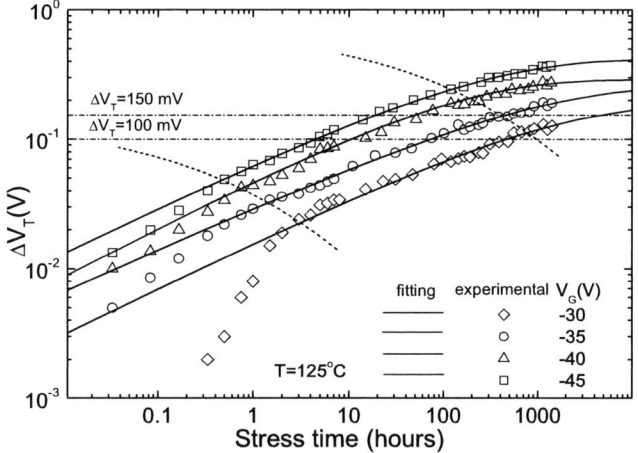

Fig. 2. Threshold voltage shifts observed in p-channel power VDMOSFETs during the NBT stressing at 125 °C. Symbols denote measured data and solid lines are the fits using stretched exponential equation.

phase occurs, previously determined on the basis of parameter n in t^n power law, correspond almost exactly to the points at which the stretched exponential curves intersect with the experimental data. Therefore, earlier established early-to-second phase transition [9] is clearly confirmed. As already mentioned, the early phase (and disagreement between the fitting and experimental data as well) could be even expected to disappear under more severe stress conditions, i.e. under higher stress voltages and/or temperatures.

646

The values of parameter β in stretched exponential fit of our results obtained on VDMOSFETs were found to vary in the range 0.35~0.39 independently on stress conditions, while τ_o decreases with increasing stress bias and temperature, which both are in good agreement with findings in [7]. The values of threshold voltage shifts in saturation ΔV_{Tmax}, as obtained for p-channel VDMOS devices stressed at various conditions, are listed in Table I. As expected, it can be seen that ΔV_{Tmax} values, representing a measure of ΔV_T at ten year device lifetime, increase with both NBT stress voltage and temperature.

TABLE I
THE VALUES OF ΔV_{Tmax} IN STRETCHED EXPONENTIAL FIT OF EXPERIMENTAL DATA OBTAINED ON NBT STRESSED P-CHANNEL VDMOSFETs.

ΔV_{Tmax} (V)		V_G (V)			
		-30	-35	-40	-45
T (°C)	125	0.1858	0.2518	0.3109	0.4169
	150	0.2073	0.3241	0.4074	0.5563
	175	0.3188	0.3319	0.4584	0.5694

The results obtained by stretched exponential fit can be used in a standard procedure for lifetime determination based on failure criterion (FC) of certain device parameter (e.g. threshold voltage, transconductance, saturation current, etc.) [11, 12]. FC can be defined in various ways, e.g. as the time needed for threshold voltage to shift for 20 % [11] or 50 mV [12], the time needed for drain current to fall for 10 % [11] or 5 % [13], etc. Our goal is to estimate the lifetime of investigated VDMOSFETs under normal operating conditions, and we use the threshold voltage as a degradation monitor, which is widely accepted as a well-suited parameter in the literature. As indicated in Fig. 2, we define two different failure criteria for the device lifetime for given NBT stress conditions as the time needed for threshold voltage to increase for 3.33 % (ΔV_T =100 mV) and 5 % (ΔV_T =150 mV), respectively. Estimation of the lifetime for both FC under different NBT stress conditions is illustrated in Fig. 3, and the values of the extrapolated lifetime, assuming maximum operating gate bias V_G = - 20 V, are summarized in Table II. Obviously, estimated lifetime depends significantly on FC.

The saturation of threshold voltage shift observed in Figs. 1 and 2 leads to higher values of the estimated lifetime, in particular when the higher FC (ΔV_T =150 mV) is applied. As can be seen in Fig. 2, in the case of two lower stress voltages (- 30 and - 35 V) the intersections of the line defining higher FC with ΔV_T curves fall in the saturation phase, which has been reported in [14] as well. Moreover, in the case of the lowest stress voltage, this FC line intersects only with fitting curve but not with the experimental one since the duration of the experiment appeared to be shorter than the lifetime at given temperature. However, lower stress voltages are closer to

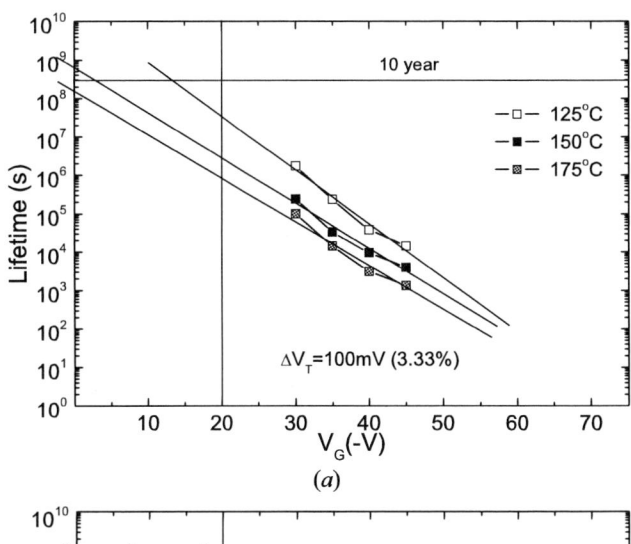

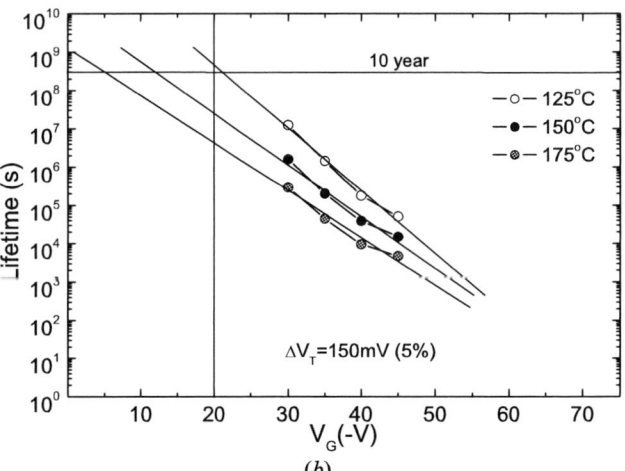

Fig. 3. Lifetime estimation in NBT stressed p-channel VDMOSFETs for two FC: a) ΔV_T =100 mV; b) ΔV_T =150 mV.

actual operating voltages and are expected to provide more realistic estimation of device lifetime. In this case, the fact that stretched exponential model successfully predicts threshold voltage shift in the saturation enabled us to estimate device lifetime from the results obtained by fitting instead of missing experimental ones. It should be noted that stretched exponential fit can be used also to avoid problems associated with measurement noise, non-smoothness of the experimental curves (which is visible in Figs. 1 and 2), and insufficient sampling [7, 8].

TABLE II
ESTIMATED LIFETIME FOR OPERATING VOLTAGE V_G = - 20 V.

Lifetime (days)	ΔV_T =100 mV	ΔV_T =150 mV
125°C	391.07	5792.10
150°C	30.67	289.86
175°C	9.25	48.06

As already noticed, estimated lifetime strongly depends on the choice of FC. If the chosen FC was too high (e.g.

500 mV), its value would fall far above both experimental and fitting curves in Fig. 2, which would result into device lifetime tending infinity. On the other hand, too low FC (below 30 mV) could yield significantly underestimated lifetime as the FC value would fall in the early phase of stressing. Hence, it appears the most reliable lifetime estimate can be obtained by choosing the FC value within the range of threshold voltage shifts observed in the second and saturation phases of device stressing.

The other factor that strongly affects lifetime estimation is the choice of the range of stress voltages used for extrapolation [15]. The effect of this factor on uncertainty of lifetime and uncertainty of "ten year operation voltage" (maximum gate voltage that allows ten years of device operation with V_T shift below the FC) in the case of p-channel VDMOSFETs stressed at 150 °C is illustrated in Fig. 4. As can be seen, estimated lifetime may vary for almost one order of magnitude, while the maximum allowed V_G varies for about 8 V. Also, it can be noticed that use of higher stress voltage range, which provides experimental results in shorter time, contributes to the above uncertainties. On the other hand, more realistic estimates are expected if the lower stress voltage range, closer to actual operating voltage, is used for extrapolation.

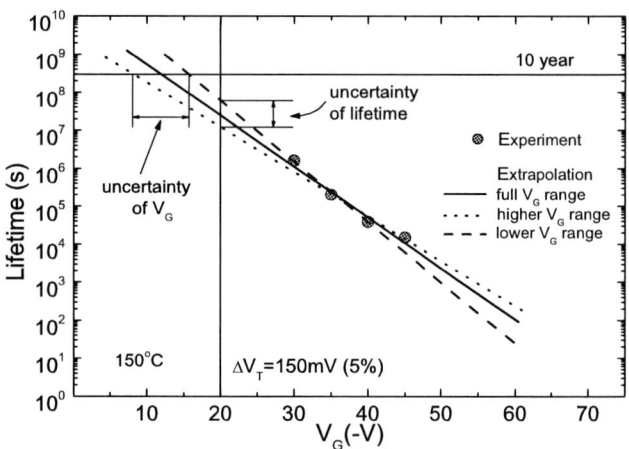

Fig. 4. Uncertainties of lifetime and ten year operating voltage in NBT stressed p-channel VDMOSFETs due to different ranges of stress voltages used for extrapolation.

V. CONCLUSION

It has been shown that the stretched exponential equation yields excellent fit for the threshold voltage shifts observed in later NBT stress phases of commercial p–channel power VDMOSFETs. This finding allowed for an accurate estimation of device lifetime from the fit for the lowest stress voltage applied, which otherwise would require very long experiment to be performed. The lifetime estimate strongly depends on the choice of FC, as well as on the range of stress voltages used for extrapolation. More realistic estimates are expected in the case of lower voltage range, which additionally justifies the need for using stretched exponential or some other suitable fit.

REFERENCES

[1] A. Demesmaeker, A. Pergoot, and P. De Pauw, "Bias temperature reliability of p-channel high-voltage devices", *Microelectron. Reliab.*, 1997, vol. 37, pp. 1767-1770.

[2] S. Gamerith and M. Polzl, "Negative bias temperature stress in low voltage p-channel DMOS transistors and role of nitrogen", *Microelectron. Reliab.*, 2002, vol. 42, pp. 1439-1443.

[3] D.K. Schroder and J.A. Babcock, "Negative bias temperature instability: Road to cross in deep submicron silicon semiconductor manufacturing", *J. Appl. Phys.*, 2003, vol. 94, pp. 1-18.

[4] S. Ogawa, M. Shimaya, and N. Shiono, "Interface-trap generation at ultrathin SiO₂ (4-6 nm)-Si interfaces during negative-bias temperature aging", *J. Appl. Phys.*, 1995, vol. 77, pp. 1137-1148.

[5] N. Kimizuka, T. Yamamoto, T. Mogami, K. Yamaguchi, K. Imai, and T. Horiuchi, "The impact of bias temperature instability for direct–tunneling ultra–thin gate oxide on MOSFET scaling", *Symp. on VLSI Tech. Dig of Tech. Papers*, 1999, pp. 73 -74.

[6] N. Stojadinovic, I. Manic, S. Djoric-Veljkovic, V. Davidovic, S. Golubovic, and S. Dimitrijev, "Effects of high electric field and elevated-temperature bias stressing on radiation response in power VDMOSFETs", *Microelectron. Reliab.*, 2002, vol. 42, pp. 669-677.

[7] S. Zafar, B.H. Lee, and J. Stathis, "Evaluation of NBTI in HfO₂ Gate–Dielectric Stacks With Tungsten Gates", *IEEE Electron. Dev. Lett.*, 2004, vol. 25, pp. 153-155.

[8] S. Zafar, A. Callegari, E. Gusev, and M.V. Fischetti, "Charge trapping related voltage instabilities in high permittivity gate dielectric stacks", *Journal of Appl. Phys.*, 2003, vol. 93, pp. 9298-9303.

[9] N. Stojadinovic, D. Dankovic, S. Djoric-Veljkovic, V. Davidovic, I. Manic, and S. Golubovic, "Negative bias temperature instability mechanisms in p-channel power VDMOSFETs", *Microelectron. Reliab.*, 2005, vol. 45, pp. 1343-1348.

[10] K.O. Jeppson and C.M. Svensson, "Negative bias stress of MOS devices at high electric fields and degradation of MNOS devices", *J. Appl. Phys.*, 1977, vol. 48, pp. 2004-2014.

[11] C. Schlunder, R. Brederlow, B. Ankele, W. Gustin, K. Goser, and R. Thewes, "Effects of inhomogeneous negative bias temperature stress on p-channel MOSFETs of analog and RF circuits", *Microelectron. Reliab.*, 2005, vol. 45, pp. 39-46.

[12] S.S. Tan, T.P. Chen, C.H. Ang, and L. Chan, "Mechanism of nitrogen-enhanced negative bias temperature instability in pMOSFET", *Microelectron. Reliab.*, 2005, vol. 45, pp. 19-30.

[13] A. Suzuki, K. Tabuchi, H. Kimura, T. Hasegawa, and T. Kadomura, "A Strategy using a Copper/low-k BEOL Process to prevent Negative –Bias Temperature Instability (NBTI) in p-MOSFETs with Ultra-Thin Gate Oxide", *Symp. on VLSI Tech. Dig of Tech. Papers*, 2002, pp. 216-217.

[14] M. Ershov, S. Saxena, S. Minehane, P. Clifton, M. Redford, R. Lindley, H. Karbasi, S. Graves, and S. Winters, "Degradation dynamics, recovery, and characterization of negative bias temperature instability", *Microelectron. Reliab.*, 2005, vol. 45, pp. 99-105.

[15] H. Aono, E. Murakami, K. Okuyama, A. Nishida, M. Minami, Y. Ooji, H. Karbasi, and K. Kubota, "Modelling of NBTI saturation effect and its impact on electric field dependence of the lifetime", *Microelectron. Reliab.*, 2005, vol. 45, pp. 1109-1114.

2006 25th International Conference on Microelectronics

Evaluation of Reflow Ovens for Lead-Free Soldering

J. Lempinen and A. Tuominen

Abstract - A set of different reflow ovens were evaluated in order to determine their capabilities to produce lead-free solder joints. Thermal profiles of different reflow ovens were measured using profile equipment and the heat transfer capabilities for all of the investigated reflow ovens were evaluated by determining the time constants for heating. Thermal model used is presented. It was shown that the smaller the time constant for a reflow oven the better the heat transfer capability, and thus the smaller variation in the peak soldering temperatures. The order of superiority of investigated reflow ovens is presented.

I. INTRODUCTION

Replacement of eutectic Sn-Pb solder in electronics industry in Europe requires that components are soldered at higher temperatures than today. This has put efforts on continuous development of reflow soldering processes as well as development of soldering alloys. In practice, current reflow temperatures will be increased typically ca. 30-60°C depending on the solder alloy to be used. Thereby components to be used in the lead-free process have to withstand temperatures in the range of 235-260°C. In many cases the maximum temperature is 235°C.

For reflow ovens this requires new temperature profile settings to be implemented. The purpose is to find out optimal profile which prevents overheating of the most sensitive components but produces as reliable solder joints as using typical Sn-Pb solder alloys. Thus, the primary target is to ensure sufficient solderability and suitable mechanical strength of solder joint. The time above melting point should be e.g. 60-120 seconds according to JEDEC J-STD-020C standard.

Practically, one of the problems in temperature profile evaluation is that mass of a body i.e. printed circuit board (PCB) assembly to be heated is not evenly distributed. Since some of the components are bigger (or smaller) and the materials of the components are typically different, the heat is conducted unevenly thereby producing temperature differences between the PCB and the components. In reflow soldering, high temperature differences of components may be significant since they can generate unreliable solder joints. In general, temperature differences may also be influenced by reflow oven selection and the

settings used such as conveyor speed, number of heating zones, the length of heating zones, solder paste, components, materials, the blower settings etc. [1], [2], [3], [4]. Therefore, the evaluation process is not straightforward task.

Currently, a lot of emphasis is placed on development of modern simulation and modeling tools, which may significantly reduce the evaluation time needed. Especially, finite-element modeling tools [5], [6], [7] have been utilized as predictive tools, product verification or optimization.

In this study, our objective has been to present method which can be used not only for reflow profile evaluation but also for determination the capability of a reflow oven to produce lead-free solder joints. In many electronics manufacturers the issue of study is of current interest since the capability of reflow process used has to be ensured. For the study, a simulation tool [8] has been used which is based on the model presented later. This tool can be easily implemented on current reflow ovens. The experiments have been carried out at companies. Thereby we have used several different types of reflow ovens, not only the new ones, and several different soldering profiles for evaluation purposes. The results are presented later.

II. HEAT TRANSFER ANALYSIS

This study focuses on convective reflow ovens which consist of a conveyor with a moving belt that is employed to transport the PCB assemblies though the reflow oven, several heating zones in the oven, half of them above and half of them below the conveyor belt, that heating zones can be set to desired temperatures for heating the PCB assemblies.

A. Temperature Equation

Derivation of temperature model used in the simulation is presented below. It is first assumed that the heat is transferred uniformly by convection. Thus it follows:

$$dQ = hA(Ti - T)dt = \rho C_p V dT \qquad (1)$$

where dQ is the amount of heat conveyed from the hot air in the oven to the body per unit time, h is the heat transfer coefficient (W/m^2°C), A is the heat transfer surface area (m^2), T_i is air temperature, T is temperature of the body, ρ is density of the body (kgm^{-3}), C_p is specific heat capacity

---

Juha Lempinen and Aulis Tuominen are with the Department of Information Technology, Faculty of Mathematics and Natural Sciences, University of Turku, Ylhäistentie 2, 24130 Salo, Finland, E-mails: juha.lempinen@utu.fi and aulis.tuominen@utu.fi

1-4244-0116-X/06/$20.00 ©2006 IEEE

(Jkg^{-1} °C^{-1}) and V is total volume of the body (m^3). By assuming that $T(t=0) = T_1$ and $T(t=t) = T_2$, it follows:

$$\int_0^t \frac{hA}{\rho C_p V} dt = -\int_{T_1}^{T_2} \frac{d(T_i - T)}{(T_i - T)} \quad (2)$$

After integration of Eq. 2 it follows:

$$\frac{T_2 - T_1}{T_i - T_1} = 1 - e^{-hAt/\rho C_p V} \quad (3)$$

After rearranging of terms of Eq. 2 it follows:

$$T_2 = T_1 + (T_i - T_1) \cdot \left(1 - e^{-hAt/\rho C_p V}\right) \quad (4)$$

where $-\rho C_p V/hA = \tau$ is a time constant for heating. Thus Eq. 3 can be simplified as follows:

$$T_2 = T_1 + (T_i - T_1) \cdot \left(1 - e^{-t/\tau}\right) \quad (5)$$

B. Simulation Model

By assuming that Eq. 5 is the temperature of the PCB assembly in the ideal case, the non-ideal case could be assumed as shown in Eq. 6. In the non-ideal case, the process capacity of a reflow oven is denoted with n ($n>1$). Thus, the Eq. 5 can be rewritten as follows:

$$T_2 = T_1 + (T_i - T_1) \cdot \left(1 - e^{-t/\tau}\right)^n \quad (6)$$

The Eq. 6 is used for simulation of temperature profiles of reflow ovens. However, in simulations, $n = 1$ is used.

III. EXPERIMENTS

A. Test Board

An FR-4 test board was used with several different types of SMD components soldered on it. The size of the test board was 1.6 x 163 x 220mm^3. Ni/Au surface finish was used for the test board. The test board is shown in Fig. 1.

B. Profiling Equipment

Temperature profiles of reflow ovens were measured using 6-channel Super Mole Gold equipment. Four thermocouples were attached to the test board and two thermocouples were used determining the air temperature in the oven. The temperature profiles were measured multiple times with each reflow oven. The test board connections and the profiling equipment are presented in Fig. 1.

C. Specifications of Convective Reflow Ovens

The specifications of convective reflow ovens used in the evaluation are shown in Table I. Each row specifies a reflow oven shown in the first column, e.g. A1. There are totally ten different types of reflow ovens in Table I. The second column on the left shows average length of the heating zone in meters. The third column shows the speed of the conveyor in meters per second. The number of heating zones and temperatures are also shown in Table I. First column below the title i.e. #1 specifies the temperature in the first heating zone; the second column specifies the temperature in the second heating zone etc. Cooling zone temperatures are not shown. On the right side of Table I is shown the type of solder paste on which the reflow oven settings are based.

D. Simulation and Modeling Tool

The reflow profile measurements were simulated using a simulation application which is developed at POHTO Oy in Finland [8]. The simulation is based on modeling of temperature profile of reflow oven according to practical measurements with the profile equipment. The application software reads the data files of a reflow oven for further analysis and modeling. The application uses information such as time constant τ and capacity n for calculation. Heating and cooling zone temperatures and heating time can be adjusted according to the specific reflow oven as shown in Table I. Guiding profiles from different solder paste manufacturers can be utilized in simulation. The main target of the application is to find out the ideal reflow profile which has the smallest temperature variations between measurement points of the test board.

Fig. 1. Connections of a test board and a profile equipment.

TABLE I.
REFLOW OVEN SPECIFICATIONS

Oven	Length [m]	Speed [m/s]	Heating Zone Temperatures [°C]										Solder Paste
			#1	#2	#3	#4	#5	#6	#7	#8	#9	#10	
A1	0.33	0.7	140	160	160	180	220	265	260	-	-	-	Sn96.5/Ag3.0/Cu0.5
A2	0.35	0.6	125	160	165	180	220	290	230	-	-	-	Sn95.5/Ag3.8/Cu0.7
A3	0.35	0.56	130	175	175	190	220	280	250	-	-	-	Sn/Cu0,7/Ni
A4	0.25	0.38	170	180	180	195	255	230	-	-	-	-	Sn95.5/Ag3.8/Cu0.7
A5	0.23	0.3	180	190	290	-	-	-	-	-	-	-	Sn96.5/Ag4/Cu0.5
A6	0.55	1.0	100	140	165	165	175	260	255	-	-	-	Sn/Ag3.0/Cu0.5
A7	0.42	0.7	120	150	175	210	250	265	-	-	-	-	Sn95.5/Ag3.5/Cu0.7
A8	0.35	1.05	120	133	160	185	170	160	162	255	275	265	Sn95.5/Ag3.5/Cu0.7
A9	0.37	0.86	110	155	170	170	160	155	220	250	235	-	Sn95.5/Ag3.5/Cu0.7
A10	0.33	0.58	155	165	175	180	100	190	250	260	265	-	Sn95.5/Ag3.5/Cu0.7

IV. RESULTS AND DISCUSSION

Fig. 2 shows measured temperature profiles for reflow oven A1. There are four curves presented in Fig. 2. The uppermost curve is air temperature reflecting the heating zone temperatures used in the reflow oven. The peak temperatures of the curves are well together and there are no high temperature deviations between the curves. As shown in Fig. 3, the peak temperatures are in the range of 248-253°C and time above melting point 217°C is in the range of 55-60 seconds. Rising slope is in the range of 1-5°C/sec. As can be seen in Fig. 2 and Fig. 3 the cooling zone is very efficient with declining slope in the range of 6-15°C/sec.

Fig. 4 shows the simulation tool that was used in the analysis. Simulated temperature profile curve for reflow oven A1 is also shown in Fig. 4. The simulation curve shows good agreement with the measured curve.

Table II shows all measured peak temperatures of reflow ovens. The highest peak temperature is above 250°C and the lowest is just above 220°C. Time constants are shown in the third column of Table II. Values of relative heat transfer capability based on the values of time constant are shown in the fourth column of Table II. The lowest value of time constant is set to percent of 100.

The final results of heat transfer evaluation for reflow ovens are shown in Fig. 5. A high deviation between the heat transfer capabilities of reflow ovens is observed. The best heat transfer capability was measured with reflow oven A9. The relative heat transfer capability of the reflow

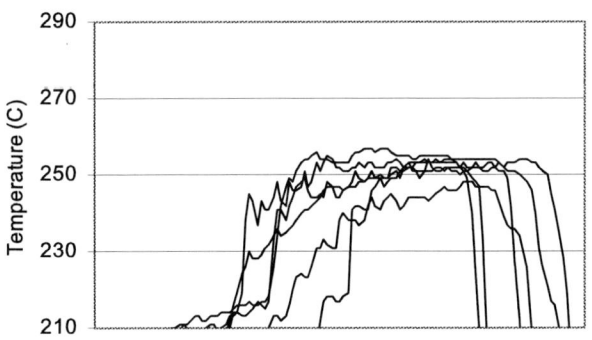

Fig. 3. Measured temperature profiles for reflow oven A1 above 210°C.

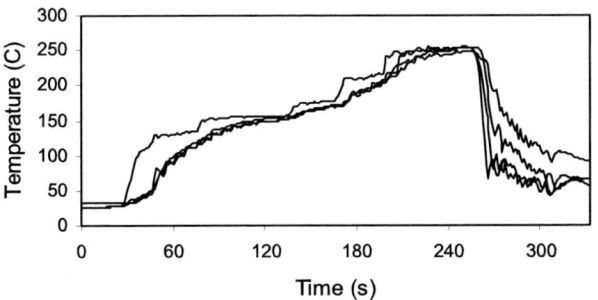

Fig. 2. Measured temperature profiles for reflow oven A1.

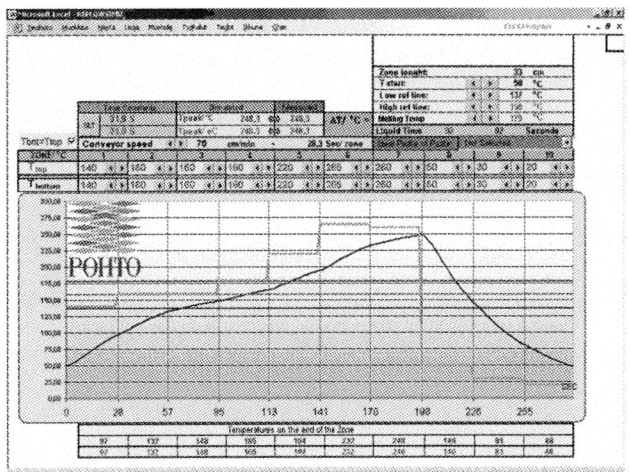

Fig. 4. Simulated temperature profile for reflow oven A1.

TABLE II.
PEAK TEMPERATURES AND TIME CONSTANTS

Oven	T_{peak} [°C]	τ [s]	%
A1	248.3	31.9	195
A2	251.7	36.4	171
A3	243.3	51.8	120
A4	228.3	41.9	148
A5	226.7	54.3	114
A6	221.1	57.6	108
A7	240	46.9	132
A8	238.9	38.5	161
A9	230	29.8	208
A10	240.6	62.1	100

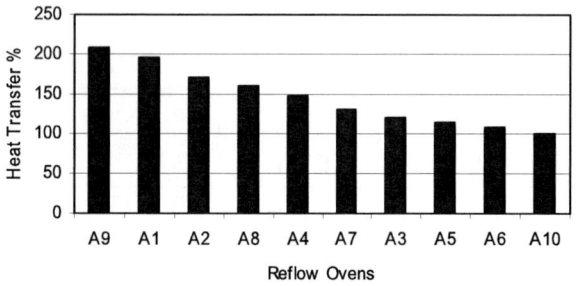

Fig. 5. Results of heat transfer capability analysis.

oven A9 is twice as much as for reflow oven A10 which is the worst. Oven A1 which has seven heating zones performed almost as well as A9 which has nine heating zones.

However, it can be seen that much lower heating zone temperatures are required for oven A9 as for oven A10 to achieve the same peak temperature. Some of the reflow ovens are newly-designed provided with an efficient air flow control which may significantly improve the heat transfer characteristics of the oven.

V. CONCLUSIONS

In this study, several solder reflow ovens were evaluated. The practical evaluation was carried out by measuring temperature profiles for reflow ovens multiple times using profile equipment. Temperature model was presented. Based on the model, heat transfer capabilities of reflow ovens were evaluated. The best heat transfer capability was evaluated for oven A9 and the worst for oven A10.

It was demonstrated that the modeling and simulation tool can be used effectively to determine the heat transfer capability of reflow oven thus ensuring that required temperatures are desirably reached in PCB assembly process. The information which is obtained from these experiments is input for the new reflow process development.

REFERENCES

[1] P.P. Conway, D.J. Williams, A.C.T. Tang, P.M. Sargent, D.C. Whalley, "Process variables in the reflow soldering of surface mount", *Proc. 8th IEEE Symp.*, Baveno, Italy, May 1990, pp. 385-394.

[2] P. Arulvanan, Z. Zhong, X. Shi, "Effects of process conditions on reliability, microstructure evolution and failure modes of SnAgCu solder joints", in *Microelectronics Reliability* xxx, 2005, xxx-xxx, article in press.

[3] W. Huang, J.M. Loman, B. Sener, "Study of the effect of reflow time and temperature on Cu-Sn intermetallic compound layer reliability", in *Microelectronics Reliability* 42, 2002, pp. 1229-1234.

[4] F. Wu, "Effects of heating factors on the geometry size of unrestricted lead-free joints", *IEEE International Conference on Asian Green Electronics*, 2004, pp. 81-85.

[5] F. Sarvar, P.P. Conway, "A modelling tool for the thermal optimisation of the reflow soldering of printed circuit assemblies", in *Finite Elements Analysis and Design* 30, 1998, pp. 47-63.

[6] D.C. Whalley, "A simplified reflow soldering process model", in *Journal of Materials Processing Technology* 150, 2004, pp. 134-144.

[7] B. Tao, Y. Wu, H. Ding, Y. L. Xiong, "A quantitative method of reliability estimation for surface mount solder joints based on heating factor Q_η", in *Microelectronics Reliability* xxx, 2005, xxx-xxx, article in press.

[8] J. Koivukoski, "Juotoksien lämpötilaprofiilien simulointi ja hyödyntäminen korjausasemassa ja uunien evaluoinnissa", in *New Exploratory Technologies*, 2004, p. 134.

Session
System Design

2006 25th International Conference on Microelectronics

Word-Length Oriented Multiobjective Optimization of Area and Power Consumption in DSP Algorithm Implementation

A. Ahmadi, M. Zwoliński

Abstract – The word-length of Functional Units (FU) has a great impact on design costs. This paper addresses the problem of choosing different word-lengths for each FU while considering circuit area and power consumption. A high-level synthesis tool is used to minimize the circuit area and power consumption by selecting an optimal word-length for each FU in the system. Our results demonstrate that by customizing word lengths to non-standard sizes, savings can be made in the overall area and power without losing accuracy.

I. INTRODUCTION

One of the problems in implementing signal processing algorithms on digital hardware is choosing an appropriate word length for arithmetic units. Traditionally this problem is solved by making a worst-case assumption and choosing a single word length for all arithmetic units. Using a word-length less than this worst-case assumption at different points in the algorithm would, however, save both area and power.

Several pieces of work have focused on finding an optimal word-length for the algorithm as the first step and then designing or optimizing the system within that constraint, [2]. The word-length is not considered in the subsequent optimization process. In other studies in which word-length has been considered during optimization [9], signals have been categorized into a few groups to constrain the word-length in all functional blocks, or only one objective has been considered in addition to digital noise [1, 8].

High Level Synthesis (HLS) has been considered to be a key factor in reducing the distance between the initial specification and target design [3]. Because of the variety of possible applications, domain-specific HLS tools are needed to achieve an optimal solution.

In this work, we present a multi-objective optimization method to optimize circuit area and power consumption by choosing optimum word-lengths for each FU. Cost functions for area, power consumption and digital noise are discussed in section 2, sections 3 gives a short description of the implemented design tools and section 4 is devoted to the GA method which has been applied. Results are explained in section 5.

A. Ahmadi and M. Zwoliński are with the Electronic System Design Group, School of Electronics and Computer Science, University of Southampton Southampton, UK, E-mail {aa03r, mz}@ecs.soton.ac.uk

II. COST FUNCTION

From a high level synthesis point of view, both the total area and power consumption of a system can be divided into three parts: data paths; controllers; and interconnections. Having focused on word length optimization, area and power costs should be considered as functions of the functional unit word length. Depending on the implementation methodology, word lengths have different impacts on each part (datapath, controller and interconnections) of the metrics. In our method, changing the word length does not change the controller area, so that is considered as a constant value in the cost function. In addition, since this methodology is bus oriented, interconnection costs are only marginally affected by word length compared to the changes in the datapath. Accordingly our cost models assume the datapath costs are variable and others are constant, equation (1).

$$F_{Total}(\vec{W}) = F_{Controller} + F_{Interconc} + F_{Datapath}(\vec{W}) \qquad (1)$$

F is the cost function and $\vec{W}$ are the word lengths for the functional units. In the following section, we present a brief description of the cost models for circuit area, power consumption and digital noise.

A. Area Cost Function

Since the area of the controller (A_C) does not change with word length and the interconnection area (A_B) only slightly depends on it, the total area of the datapath is evaluated by adding up the sub-block and FUs areas ($A_{FU}(\vec{W})$). Thus as an approximation, the area of building blocks such as sequential multipliers, adders, registers, buffers and switches can be assumed to have a proportional relationship to word length while the area of a combinational multiplier can be modeled by a second order relationship with its word length. Design implementation results confirm this assumption as depicted in figure (1). Equation (2) gives the area cost function for a system.

$$F_A(\vec{W}) = A_C + A_{FU}(\vec{W}) + A_B \qquad (2)$$

B. Power consumption Cost Function

Knowing that the changing word length of the FUs does not affect the controller activity and structure, the

1-4244-0116-X/06/$20.00 ©2006 IEEE

power consumption of controllers (P_C) is a fixed term in the estimated power consumption. In addition, because a bus-oriented approach is used in this study, interconnection power consumption (P_B) only depends on the maximum word length in the shared bus; therefore, ignoring the P_B dependency on W is acceptable at this level of abstraction. Equation (3) shows the general model of power consumption with these approximations.

$$F_P(\vec{W}) \approx P_{FU}(\vec{W}) + P_B + P_C \qquad (3)$$

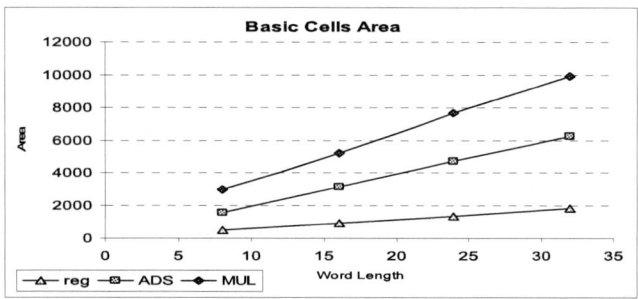

Fig. 1 Dependency of area on word length for basic cells (registers, adder and multiplier).

A set of designs was used to evaluate the functional unit dependency on word length and the results are presented in Figure (2). In this figure, the average power consumption for basic cells, with random input data is shown with respect to word length. In these simulations, the Nominal Low Leakage ST 0.12μm technology file is used. From this, we can see that power consumption can be modelled as a linear function of the word length.

On the other hand, power consumption is a combination of static and dynamic parts; accordingly, in each FU it is a sum of static and dynamic parts as in Equation (4).

$$P_k = P_{k,Dynamic} + P_{k,Static} \qquad (4)$$

Here P_k is the power consumption of the kth FU.

In general, dynamic and static power consumption are data dependent [5] but in this study, to estimate power consumption in the optimization procedure, static power consumption is considered proportional to the total power, Equation (5).

$$P_{k,Static} = \lambda_k \cdot P_k \qquad (5)$$

λ_k is the leakage power factor. Simulations verify this assumption for basic blocks for different word lengths.

Another assumption used to reduce the evaluation complexity is a time slot approximation [10]. In this approximation the total power consumption of a functional unit is calculated in two parts: *activation* time slots and *standby* time slots. During functional operation, power consumption is the sum of dynamic and static power whereas in standby, only the leakage power is taken in account. Based on this approximation, the total power consumption for each functional unit is given in Equation (6).

$$F_P(\vec{W}) = \frac{1}{T} \cdot \sum_{k=1}^{N_F} \left(t_k P_k + (T - t_k) \cdot P_k \cdot \lambda_k \right) \qquad (6)$$

F_P is the average power consumption of the system, P_k is the average power consumption of the kth functional unit, t_k is its activation time and T is the total system operation time.

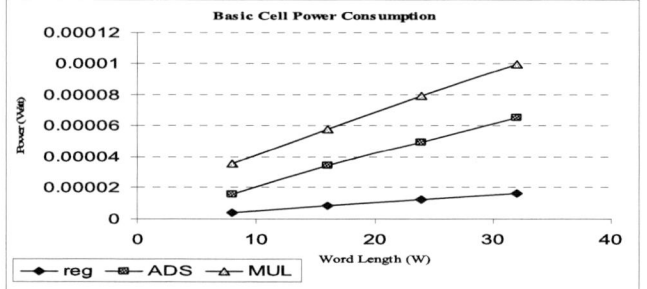

Fig. 2 Dependency of Power Consumption on word length for basic cells (Register, Adder and Multiplier)

C. Digital Noise Cost Function

In practice, digital signal processing systems can only offer a finite number of binary digits to represent the signals to be processed. Fitting real values in these limited containers causes effects which can be categorized in several different ways. From a mathematical point of view, using a limited number of bits to represent a real number always means adding or removing indeterminate information at the input, which is usually considered as an error or noise. To model this problem in our tool and to evaluate its impact; there are two problems to consider: first is a noise model for computational errors and second is a model of noise propagation. A number of models of digital noise have been proposed in [7].

To provide a noise propagation model, it must be recalled that many DSP algorithms can be considered as Linear Time Invariant (LTI) systems. This assumption allows us to use superposition of independent noise sources to compute the noise effect on the system output, [2], [6]. The effects of noise sources on the output can be approximated using Equation (7).

$$E_{Output} = \sum_{k=1}^{M} \sigma_k^2 \cdot \left(L_2\{H_k(z)\} \right)^2 , \qquad (7)$$

$H_k(z)$ is the Z-transform of the transfer function ($h[n]$) from the kth noise source to the output and $L_2\{\ \}$ is the L-Norm [6], given by Equation (8).

$$L_m\{H(z)\} = \left[\sum_{n=0}^{\infty} \left| Z^{-1}\{H(z)\}[n] \right|^m \right]^{\frac{1}{m}} \qquad (8)$$

σ_k can be found from Equation (9), [2],

$$\sigma_k^2 = \frac{1}{12} 2^{2p} \left(2^{-2n_2} - 2^{-2n_1} \right). \qquad (9)$$

n_1 is the present arithmetic unit word length and n_2 is the next arithmetic unit word length and p is the position of the decimal point.

III. IMPLEMENTATION

The system design methodology starts from a hierarchical specification of the target system and is based on three parts: the functional unit data base; the target architecture; and the synthesizer-optimizer. The target architecture is built on a partitioned shared bus with distributed controller which makes the target design very flexible to match a variety of DSP algorithms as well as being very modular and manageable for the synthesizer and optimizer [1]. From a synthesis point of view, on the other hand, this target architecture is a restriction in that it forces the synthesizer to map every design to a pre-defined structure which dominates the feasible solution space in favour of the optimizer.

The functional unit database is a library of functions and sub-systems. There are four kinds of sub-system in our method: algorithm executers, interfaces, memories and controllers, which each might contain further functional units and/or sub-systems. In addition to implementation information, this database provides the required information for the design optimizer cost functions including: area, accuracy, delay and power consumption.

The synthesizer's input is a high level specification of the algorithm in the form of difference equations. Basically there is a pre-defined hierarchical architecture to which the target system must be mapped. The starting specification of the system and the final implementation are both represented by a digraph. A set of library files is used to produce Intermediate Code (ICD) files which are a more compact form of the initial specification of the target system The library files contain the basic blocks of the system and their cost relationships (noise, area, power, and delay) as functions of word length. These cost parameters can be used in a cost evaluation program after scheduling, allocation and binding to optimize the design.

IV. OPTIMIZATION

A GA is utilized in this study for design optimization. The genetic operators are extracted from standard GA procedure which includes selection by roulette wheel, crossovers, mutation [4] and brand new randomly produced genes. Rates for crossovers, mutation and imported genes are chosen as shown in Table (1). In Table (1) M is the number of FUs, $p(x)$ is a randomly generated value and K_1, K_2, K_3, K_4, K_5 are constant values dependent on M and the number of the iterations in the algorithm.

According to the target architecture, one word length (w) has to be assigned to each functional unit. Therefore, we define a vector of word lengths for the FUs in data paths as in Equation (10) and this vector is used as the gene in the GA optimization algorithm.

TABLE I
GENETIC ALGORITHM PARAMETERS

Parameter	Value
Number of Individuals in the Population	$K_1 \cdot M$
Number of crossovers	$K_2 \cdot M.P_2(x)$
Number of brand new Individuals	$K_3 \cdot M.P_3(x)$
Number of Increment/decrement Mutations	$K_4 \cdot M.P_4(x)$
Number of Generations (Iterations)	$K_5 \cdot M$

$$\vec{W} = \begin{bmatrix} w_1 & w_2 & w_3 & ... & w_M \end{bmatrix} \qquad (10)$$

An optimization problem must then be solved, with multiple objectives and constraints taken into consideration. A standard technique for Multi-objective Optimization is to minimize a positively weighted convex sum of the objectives, as shown in Equation (11).

$$F_A(\vec{W}) = \left(K_A \frac{F_A(\vec{W}) - F_{A,MIN}}{F_{A,MAX} - F_{A,MIN}} + K_P \frac{F_P(\vec{W}) - F_{A,MIN}}{F_{P,MAX} - F_{P,MIN}} + \right.$$
$$\left. K_N \frac{F_N(\vec{W}) - F_{N,MIN}}{F_{N,MAX} - F_{A,MIN}} \right) \times \frac{1}{K_A + K_P + K_N} \qquad (11)$$

F_A, F_P and F_N are cost functions for area, power consumption and digital noise respectively, as given in the previous sections; and *MIN* and *MAX* indicate minimum and maximum values of the functions. K_A, K_P and K_N are constants as weighting factors for costs.

V. RESULTS

Four case studies were implemented in ST 1.2μm technology. Design I is an order-10 difference equation, Design II is an order-18 difference equation, Design III is a Filter (FIR-25) and Design IV is a DCT 4x4.

In most practical implementations, there are known constraints which must be satisfied and therefore, other costs must be optimized with respect to them. Comparison of the results in Figures (1) and (2) and equation (11), suggests that by freezing one of the costs and taking it as a design constraint during optimization; it is possible to achieve the same required objective with minimum costs for the other two. To illustrate this, a set of constrained optimizations was performed with constrained accuracy. Table (2) provides the results of such design optimizations.

Several examples are given in Table (2) for each design. At first, all the FUs in the design were assigned to a fixed word-length. Four basic cases (W=8, 16, 24 and 32) were implemented and their design costs (Area, Power Consumption and Digital Noise) were calculated as the reference values. In the second step, three optimization approaches were applied for each design in each case. Optimizations were based on freezing one of the costs and optimizing two others. Clearly, in all cases design costs are reduced by our methodology however this improvement is dependent on design and accuracy constraints.

VI. CONCLUSIONS

This study presents a methodology for implementing DSP algorithms which uses models of power consumption, circuit area and output noise and their relationship to word-length. Investigation of basic designs shows a considerable improvement in costs when optimizations are employed.

TABLE II
COST COMPARISONS BETWEEN UNIFIED WORD-LENGTH AND OPTIMIZED MULTIPLE WORD-LENGTH DESIGN METHODS. C=CONSTRAINED

Design	Costs	Case 1				Case 2			
		Unified W	Optimized Multiple W			Unified W	Optimized Multiple W		
		W=8	Area	Power	Noise	W=16	Area	Power	Noise
Design I	Area	24392	C	23141	21343	48784	C	46421	45677
	Power	5.97918	5.97918	C	5.23178	11.9584	11.9584	C	11.1868
	Noise	7.39e-2	6.93e-2	4.33e-2	C	2.89e-4	9.23e-5	1.47e-4	C
Design II	Area	46928	C	41646	41062	93856	C	86211	84363
	Power	9.76618	9.76618	C	8.5454	19.5324	19.5324	C	17.9538
	Noise	8.80e-2	6.64e-2	7.17e-2	C	3.44e-4	2.19e-4	3.06e-4	C
Design III	Area	44512	C	34643	38948	89024	C	79711	83460
	Power	7.75508	7.75508	C	6.7857	15.5102	15.5102	C	14.5408
	Noise	7.57e-2	2.39e-2	6.93e-2	C	2.96e-4	9.35e-5	2.27e-4	C
Design IV	Area	106736	C	83384	93394	213472	C	190676	199736
	Power	17.0908	17.0908	C	14.9545	34.1817	34.1817	C	32.0082
	Noise	6.41e-1	5.11e-1	2.39e-1	C	2.51e-3	2.92e-4	8.66e-4	C
Design	Costs	Case 3				Case 4			
		Unified W	Optimized Multiple W			Unified W	Optimized Multiple W		
		W=24	Area	Power	Noise	W=32	Area	Power	Noise
Design I	Area	73176	C	69979	69397	97568	C	94788	91368
	Power	17.9375	17.9375	C	17.1178	23.9167	23.9167	C	23.0729
	Noise	1.13e-6	1.31e-7	7.30e-7	C	4.40e-9	8.19e-10	3.24e-9	C
Design II	Area	140784	C	136614	131525	187712	C	186044	181846
	Power	29.2985	29.2985	C	28.3093	39.0647	39.0647	C	37.8439
	Noise	1.34e-6	7.55e-7	1.39e-6	C	5.25e-9	3.61e-9	5.91e-9	C
Design III	Area	133536	C	124640	127578	178048	C	169152	175997
	Power	23.2653	23.2653	C	22.2555	31.0203	31.0203	C	30.8386
	Noise	1.16e-6	3.65e-7	7.28e-7	C	4.51e-9	1.05e-9	3.39e-9	C
Design IV	Area	320208	C	296856	269246	426944	C	403731	384697
	Power	51.2725	51.2725	C	44.5662	68.3634	68.3634	C	62.6231
	Noise	9.79e-6	2.70e-6	3.48e-6	C	3.82e-8	1.04e-8	1.43e-8	C

REFERENCES

[1] A. Ahmadi and M. Zwolinski, "Area Word-Length trade Off in DSP Algorithm Implementation and Optimization," presented at IEE/EURASIP Conference on DSPenabledRadio, 2005.

[2] G. A. Constantinides, P. Y. K. Cheung, and W. Luk, *Synthesis and Optimization of DSP Algorithms (Fundamental Theories of Physics S.)*: Kluwer Academic Publishers, 2004.

[3] G. De Micheli, *Synthesis and Optimization of Digital Circuits*: McGraw-Hill Education, 1994.

[4] D. A. Goldberg, *Genetic Algorithms in Search, Optimization, and Machine Learning* Addison-Wesley Professional 1989.

[5] E. Macii, M. Pedram, and F. Somenzi, "High-level Power Modeling, Estimation, and Optimization," *IEEE Transactions on Computer-Aided Design of Integrated Circuits and Systems*, vol. 17, pp. 1061 - 1079, 1998.

[6] A. V. Oppenheim and C. J. Weinstein, "Effects of Finite Register Length in Digital Filtering and the Fast Fourier Transform.," presented at IEEE Proceedings, 1972.

[7] A. V. Oppenheim, R. W. Schafer, and J. R. Buck, *Discrete-Time Signal Processing*: Pearson US Imports & PHIPEs, 1998.

[8] N. Sulaiman and T. Arslan, "A Multi-objective Genetic Algorithm for On-chip Real-time Optimisation of Word Length and Power Consumption in a Pipelined FFT Processor targeting a MC-CDMA Receiver," presented at 2005 NASA/DoD Conference on Evolvable Hardware, 2005.

[9] W. Sung and K. Kum, "Simulation-based Word-length Optimization Method for Fixed-point Digital Signal Processing Systems," *IEEE Transactions on Signal Processing* vol. 43, pp. 3087 - 3090, 1995.

[10] A. C. Williams, A. D. Brown, and M. Zwolinski, "Simultaneous Optimisation of Dynamic Power, Area and Delay in Behavioural Synthesis," *Computers and Digital Techniques, IEE Proceedings*, vol. 147, pp. 383 - 390, 2000.

2006 25th International Conference on Microelectronics

A Mission Level Design Language Based on AleC++

Bojan Anđelković, Vančo Litovski, Volker Zerbe

Abstract - Modern complex system design demands modeling on a high level of abstraction together with the system environment components. Such model enables mission level system simulation in the context of its operational conditions. A mission level design language providing mission and system level verification is presented in this paper. Also, this language enables designers to describe some of the components at implementation level to test and validate the system implementation at mission level. In this way a uniform design framework is achieved from mission/system down to implementation level.

I. INTRODUCTION

Modern System-on-Chip (SoC) designs grow in complexity and combine analog, digital and mixed-signal hardware components, as well as embedded software and non-electrical elements on one chip. Therefore, such designs require powerful modeling languages covering system architecture, mission/operational environment modeling, embedded software, register-transfer-level, analog and mixed-signal modeling and verification at different abstraction levels. Since system architects need to come to a proof of the system concept very early in the design flow, mission/operational and system level modeling and simulation are becoming very important step during the design process. At the same time, hardware designers need a design language capable to describe various analog, digital and non-electronic components. It is also necessary to enable proper design verification at system and implementation levels.

Mainstream hardware description and verification languages such as VHDL-AMS, SystemC, OpenVera, PSL, SystemVerilog do not meet all of the requirements in modern mixed-signal SoC design [1]. VHDL-AMS is not appropriate for specifying software and system level behavior. SystemC can be used for architectural tradeoffs and early application software verification, but it does not support modeling of analog and mixed-signal systems. SystemVerilog enables creation of efficient testbenches and assertions for simulation-based and formal property verification of digital systems. However, it does not provide analog and mixed-signal modeling and verification constructs.

B. Anđelković, and V. Litovski are with the Department of Electronics, Faculty of Electronic Engineering, University of Niš, Aleksandra Medvedeva 14, 18000 Niš, Serbia & Montenegro, E-mail: (abojan,vanco)@elfak.ni.ac.yu

V. Zerbe is with Computer Science and Automation Faculty, Technical University of Ilmenau, Germany, E-mail: volker.zerbe@tu-ilmenau.de

This paper presents the idea of developing a uniform design language covering different levels of abstraction in modeling and simulation of mixed-signal SoC. This Mission Level Design Language is based on AleC++ [2]. In this way designers can describe mission and system level modules and, after system validation at this level, replace some of the components with implementation level, more detailed models (digital circuits, transistor level models etc.) to verify the complete system implementation in its working environment. Such design language provides a uniform design framework from mission/operational level down to implementation.

II. MISSION LEVEL DESIGN

Typical design flow of modern mixed-signal SoC is shown in Fig. 1. It starts with modeling of architecture at high-level of abstraction to decrease simulation time and to get an early feedback of the complete system behavior. The system architecture should be refined and tested for functional correctness at mission/operational level to validate the complete system in the context of its operational conditions. Mission level design and simulation integrate architectures, functions, system environments and missions into a single framework. In this approach a virtual prototype of the entire heterogeneous system including typical operational conditions is developed. In that way it is possible to verify the impact of design changes and implementation decisions on overall system performance, improving chances of first-pass system success. After validation at mission and architectural levels, hardware/software partitioning is performed and functional models of the system components are developed. At the end system modules and the complete system should be modeled and validated at implementation level.

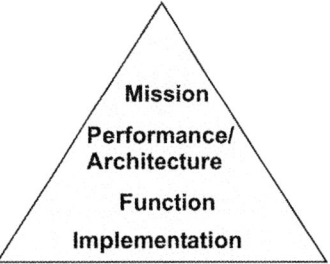

Fig. 1. Levels of the design flow for complex mixed-signal SoC.

Mission level design concept will be illustrated by an example of a navigation system. The system mission is to help in orientation on the way from point A back to A via B and C, as shown in Fig. 2.

1-4244-0116-X/06/$20.00 ©2006 IEEE

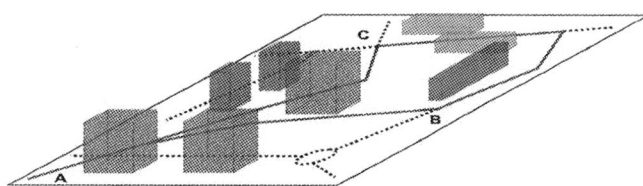

Fig. 2. Navigation system mission.

It consists of various components such as electronic compass, GPS module, gyros etc. During mission level design process it is necessary to model the whole system and validate its functionality on the desired path. Therefore, together with the system model, appropriate model of the system environment should be built.

A. Mission Level Modeling and Simulation

The tool MLDesigner is used to develop and simulate models at mission level [3]. Modeling is based on creating block diagrams using predefined primitive modules (primitives) from libraries. It is also possible to develop own primitives in a C++ like language. Multi-domain modules can be combined and simulated together. Supported domains are continuous time, discrete event, dynamic data flow, synchronous data flow, finite state machines and higher order functions. MLDesigner provides mission/operational level design tradeoff that includes modeling and simulation of dynamic Use Cases, mission environment (e.g., terrain, channel models) and operational modeling (e.g., human input devices, human output devices). It also enables system level design tradeoff (communication network design, embedded systems design) and functional level design tradeoff (algorithm design, hardware/software partitioning).

Fig. 3 shows system level model of the electronic compass [4] that is a part of the navigation system.

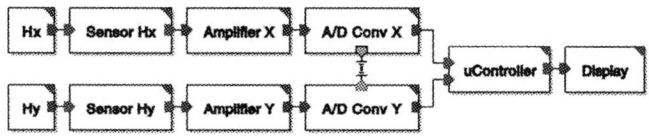

Fig. 3. Electronic compass mission level model.

The model is developed in MLDesigner together with some modules necessary to verify the compass functionality. The compass generates at the output the value of azimuth α. It is the angle between magnetic north and the heading direction. Magnetic north is the direction of "horizontal" component of the earth's magnetic field, the earth's field component perpendicular to gravity. The compass consists of the following building blocks: magnetic field sensors, amplifiers, A/D converters and microcontroller. Blocks "Hx File Data" and "Hy File Data" enable reading of input magnetic field strengths from files. The microcontroller executes software module that

calculates the desired azimuth value from the signals proportional to the magnetic field strengths. This can be done by evaluating the arctan function using CORDIC (COordinate Rotating DIgital Computing) algorithm [5].

Fig. 4 shows mission level simulation results of the compass for the path given in Fig. 2.

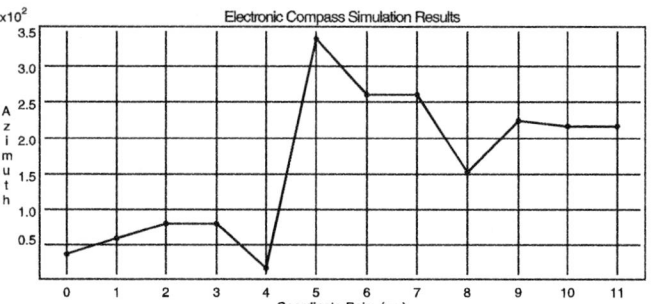

Fig. 4. Compass mission level simulation results.

III. MISSION LEVEL DESIGN BASED ON ALEC++

A. The AleC++ Language Features

AleC++ (Analog and Logic Electronic C++) is a proprietary object-oriented Hardware Description Language (HDL) developed for use in the simulator Alecsis [2]. It can be used for modeling of hardware/software systems from various domains at different levels of abstraction. Being a superset of C++ it can be used to describe analog, digital and mixed-signal hardware components, software modules, as well as non-electrical elements. AleC++ also provides some additional useful modeling features both for modeling of hardware components and system-level descriptions not found in other HDLs [2].

The basic element of hierarchical system description in AleC++ is module. The module behavior can be described using C++ like statements. AleC++ enables structural and behavioral modeling styles as well as the combination of the two. The module can also contain various parameters.

B. Mission Level Modeling in AleC++

Having in mind that MLDesigner models are hierarchical and based on primitives it can be concluded that these models can be described in AleC++. In order to accomplish that task, it is necessary to translate mission level design elements into the AleC++ language by developing appropriate translator [6]. In this way it is possible to use a vast library of predefined MLDesigner primitives in simulations in the Alecsis simulator. Organization of the simulator Alecsis extended by translator of MLDesigner primitives into equivalent AleC++ modules is shown in Fig. 5.

Functionality of MLDesigner primitives is defined using the Ptolemy language [3]. It is a preprocessor

660

language that allows the designer to use C++ code. The external interface of a primitive contains input/output port definitions and parameter definitions. For each port definition, MLDesigner generates an entry in the primitive source code. For parameters it is possible to specify name, type and default value. The Ptolemy language provides appropriate constructs for the definition of methods that describe the functionality of the primitive. These methods are executed at different stages in simulation of primitive instances such as instance creation and deletion, simulation start-up time and during simulation. The functionality of these methods is defined using C++ code and the Ptolemy language only defines the method structure.

Since AleC++ is a superset of C++, mapping of mission level primitives into AleC++ can be easily implemented. The correspondence between MLDesigner elements and appropriate AleC++ constructs is shown in Table I.

TABLE I
CORRESPONDENCE E BETWEEN MLDESIGNER AND ALEC++
ELEMENTS

MLDesigner	AleC++
Primitive, Module	Module
System	Root module
Parameters	Module parameters
Ports	Ports
Functionality in C++	C++ code, process statements, equation statements

Primitives and modules in MLDesigner correspond to AleC++ modules. A complete system model that can be executed/simulated is equivalent to *root* module in AleC++. Primitive parameters can be mapped into parameters of the equivalent AleC++ module. Since methods defined in the Ptolemy language can be executed at different stages during the simulation process, they can be mapped to different processes in AleC++ (initial, per moment, per iteration etc.). C++ code for methods can be easily included in AleC++ modules and statements for specifying equations.

To make things clearer, the process of AleC++ code generation from MLDesigner primitives will be illustrated on an example of amplifier module that is a part of the electronic compass system. In MLDesigner, amplifier module has an input and an output port, as well as, parameter *gain* specifying amplification value. The AleC++ module name is the same as MLDesigner module name. After AleC++ module declaration, declarations for all ports are written. All float type ports in MLDesigner modules correspond to ports of type *node* in AleC++. Statements in the Ptolemy language description relating to writing values to output ports are translated into equivalent AleC++ statements for defining equations. All other C/C++

code is mapped to AleC++ without change. Amplifier primitive in MLDesigner, its description in the Ptolemy language and the complete equivalent AleC++ module that translator generates for that model are given in Fig. 5.

MLDesigner Model

```
input
{
   name {input }
   type {float }
}
output
{
   name {output }
   type {float }
}
defparameter
{
   name          {gain }
   type          {float }
   default       {"1.0" }
   desc          {"Gain of the star." }
}
go
{
   output%0 << double(gain) * double(input%0);
}
```

AleC++ Model

```
module current SDFGain(node input;node output) {
   action(double gain) {
      process initial {

      /* GO Method from MLDesigner model */
      eqn SDFGain, {output}.v = gain * {input}.v;

   }//end Process block
   }//end Action block
}//end Module
```

Fig. 5. MLDesigner primitive module, Ptolemy language description and equivalent AleC++ model for amplifier

When the translator generates equivalent AleC++ code for digital modules, declarations of integer ports in MLDesigner module map to ports of type *signal* in AleC++ together with appropriate port direction. The process created for digital modules has sensitivity list containing all input ports of type signal. Statements in the MLDesigner module description relating to writing values to output ports are converted into equivalent AleC++ signal assignment statements.

C. Implementation Level Modeling in AleC++

As described, after validation at mission and architectural levels, mission level models can be translated into equivalent AleC++ descriptions. Then, implementation level models described in the AleC++ language can replace some of the compass mission level modules. It enables to test various implementations at mission level in the context of system's working environment using the Alecsis simulator.

To illustrate implementation level modeling in AleC++, implementation of the amplifier module is considered. The mission level module for the amplifier is replaced by the implementation consisting of two cascaded common-source MOSFET amplifiers and non-inverting amplifier with operational amplifier (Fig. 6). The operational amplifier is described at behavioral level while the complete non-inverting amplifier circuit is described at structural level. The MOSFET amplifier circuit is described at transistor level in AleC++ using SPICE model card for MOSFET. Some parts of AleC++ description for the amplifier are shown in Fig. 7. The amplifier is designed to have almost the same gain as in the mission level module. The designer should take care that in this case analog input signals should stimulate the compass system because transistor level models are used. Generated simulation results for sinusoidal input signals are given in Fig. 8. Because input signals are in phase the compass generates just two values for azimuth, for positive and negative values, respectively.

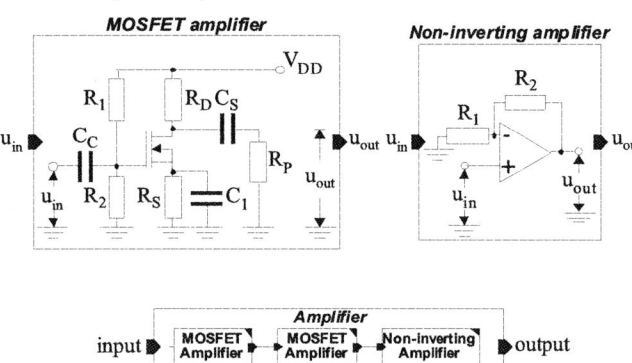

Fig. 6. Amplifier module implemented as two cascaded MOSFET amplifiers and non-inverting amplifier with opamp

```
module simpleMOS (node input,output) {
  vgen Vdd;
  resistor Rd, R1, R2, Rg, Rp, Rs;
  capacitor Cc, C1, Cs;
  mosfet m1;

  Vdd(vdd, 0) 12V;
  Rg (input1, input) 1k;
  Cc(g, input) 6n;
...
  //MOSFET transistor
  m1(d, g, s, 0) {model=my_nmos;l=2u;w=6u;};
}
module opamp (node minus; node plus; node
output) {
    vcvs vout; //voltage controlled source
    vout (output,0,plus,minus) gain=1e5;
    action () {
       process per_iteration {
          vout->gain=1e5;
       }
    }
}
```

Fig. 7. Parts of theAleC++ description for the amplifier.

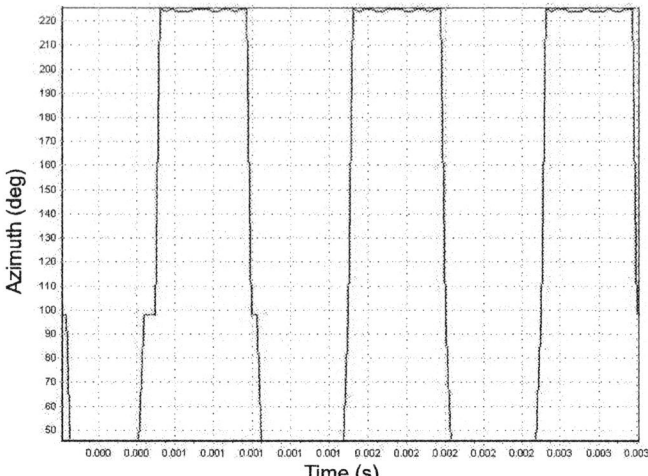

Fig. 8. Alecsis simulation results of the compass with the amplifier module shown in Fig. 6

IV. CONCLUSION

A mission level design language based on hardware description language AleC++ is presented in this paper. It extends powerful modeling capabilities of AleC++ with possibility to use mission level modules. Such design language covers the complete design flow of complex system from mission/operational down to implementation level. Comparing to other design languages, it enables the description of mixed-mode and mixed-signal systems containing various analog, digital and non-electronic components as well as embedded software modules. The language provides the complete system verification and gives the designer an opportunity to combine different levels of abstraction for various system modules and test different implementation solutions at mission level.

REFERENCES

[1] B. Anđelković, V. Litovski, V. Zerbe, "New Aspects in HDL's Performance Evaluation", in *Proc. of IEEE Region 8 EUROCON 2005 Conference*, Belgrade, 2005, pp. 499-502

[2] V. Litovski, D. Maksimović, and Ž. Mrčarica, "Mixed-Signal Modeling with AleC++: Specific Features of the HDL", *Simulation Practice and Theory 8*, 2001, pp. 433-449

[3] *MLDesigner Documentation, Version 2.4*, MLDesign Technologies, Inc., 2003.

[4] T. Stork, *Application Note – Electronic Compass Design using KMZ51 and KMZ52*, Philips Semiconductors, Systems Laboratory Hamburg, Germany, 2000. www.web-ee.com/primers/files/AN00022_COMPASS.pdf

[5] *Angular Position Development Kit for the 2SA-10, Operation Manual*, GMW, 2005., www.gmw.com/magnetic_measurements/ Sentron/sensors/documents/AN_125KIT_manual.pdf

[6] V. Zerbe, and B. Anđelković, "Design Flow for Automated Programming of FPGA", in *Proc. of IEEE 24th International Conference on Microelectronics (MIEL 2004)*, Vol. 2, Niš, Serbia and Montenegro, 2004, pp. 715-718

2006 25th International Conference on Microelectronics

Clocking Challenges in High Speed Source Synchronous Interfaces

V. Zlatković

Abstract – The challenges in clocking high speed interfaces such as the communication between multi micro-processor systems or the communication between micro-processor and off-chip memory was introduced by our need for high data rate with low bit-error-rates (BER). Minimization of clock jitter is one of the main challenges in making the high data rate links work.

There are a series of issues affecting clock jitter that must be brought to adequate levels in order for a particular data rate communication with a desired BER to be successful. The main noise sources affecting clock jitter can be grouped into two main categories. The first category would be the noise sources that are random in nature, such as the MOS device thermal and flicker noise. These sources generally affect the clock generation circuits such as Phase Locked Loops (PLL). Naturally, the second category would contain noise sources and effects causing clock jitter that are more deterministic. In this category we would generally put the effects on the clock jitter caused by the power supply noise, PLL charge pump (CP) feed-through, and off-chip channel effects.

Index Terms: BER, Phase Locked Loops (PLLs), Voltage Controlled Oscillators (VCOs), Linear Power Regulators, Clock Distribution, CMOS, Thermal noise, Clock Jitter

I. INTRODUCTION

All the data links have a particular BER specification that it needs to be satisfied. Technology and innovation always calls for lower BER and higher data rates. This introduces a series of challenges that have to be overcome in order for the link to operate at the desired specifications.

This paper will go over a series of issues affecting the design of low jitter clocking systems. Figure 1, shows a basic top level diagram of the clocking system. It starts with the PLL that is driven by an off-chip reference clock (generally a spread spectrum clock) and powered by an on-chip regulator. This is followed by an on-chip clock distribution that can be of a single ended or differential topology. The clock distribution is generally powered from a global "dirty" power supply. At the end of clock distribution the clock signal is used to launch the data onto the TX channel (ignoring the necessity for pre-emphasis etc.). The clock and data are transmitted via the lossy transmission channel (TX) and received on the other side of the TX channel. Here we will assume that data and clock are sent synchronously and that at the receive side the clock

V. Zlatkovic is a Staff Engineer with Intel Corporation, Massachusetts Microprocessor Design Center in Hudson Massachusetts USA, E-mail: vladimir.zlatkovic@intel.com

signal needs to be amplified and aligned (via delay locked loop DLL) so to sample the received data appropriately.

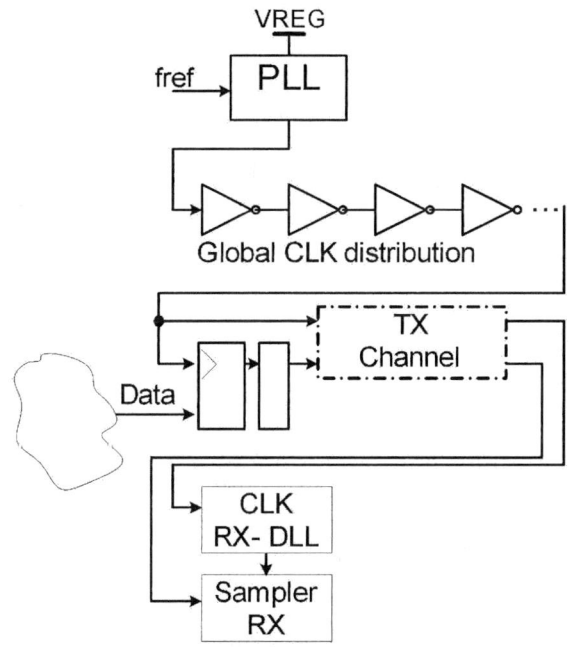

Fig. 1. Top Level Clocking Diagram.

Section II tries to describe the relationship between clock jitter and BER. Section III covers the effects random and deterministic noise have on the clock generation circuits (PLL). Section IV goes over the effects deterministic noise has on the clock distribution. In Section V we will go over the effects the band limited channel (20 inch FR4) has on the clock jitter.

II. BER – CLOCK JITTER RELATIONSHIP

In order for any BER to be met we need to provide a data sampling clock of appropriate purity (maximum pk-pk jitter). Without going into to much detail the relationship between BER and clock jitter can be expressed as shown in Eq. 1 and graphically in Figure 2.

$$BER = \frac{1}{2} \cdot erfc\left(\frac{jitter_{pk-pk} \, / \, jitter_{rms}}{2 \cdot \sqrt{2}} \right) \qquad (1)$$

1-4244-0116-X/06/$20.00 ©2006 IEEE

When clock jitter can be described by a Gaussian distribution than it is caused/dominated by the random sources and can be somewhat bounded. Certainly, the tails of the Gaussian distribution extend indefinitely on both sides of its mean value. This alone prevents us from being able to specify clock jitter 100% of the time. It is in our favor that applications we are implementing we need to specify jitter for less that 100% of the time. BER of 1E-12 requires that there should not be more than one error for one trillion (1E12) of transmitted bits. Therefore if we are after the BER of 1E-12 we need to know the range (pk-pk jitter) jitter that will be contained within all but 1e-10% of the time.

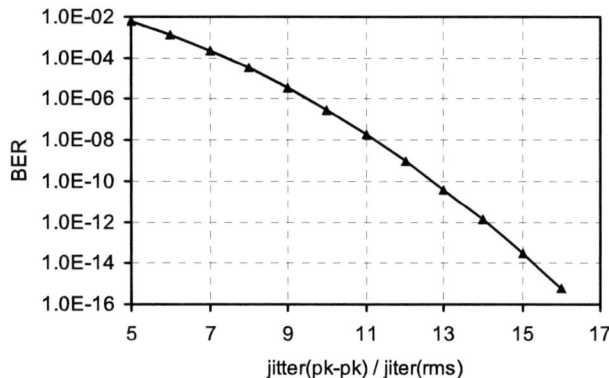

Fig. 2. BER vs. peak to peak jitter, rms jitter ratio

Another good way of looking into the relationship between BER and clocking is to visualize the data sampling eye opening (data UI) and the BER effects. This is shown in Figure 3. Simply put, in order to meet a particular BER we need a clock system with a particular rms jitter.

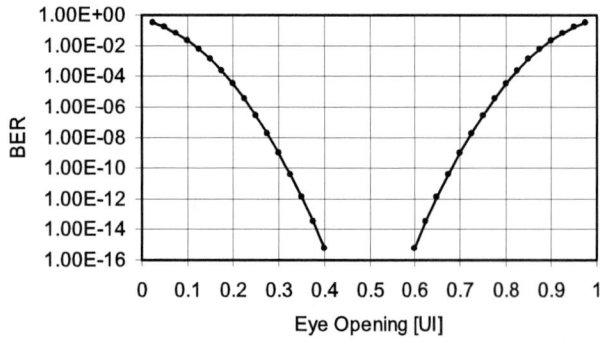

Fig. 3. BER vs. data unit interval UI eye opening.

III. CLOCK GENERATION (PLL) AND JITTER

The vast majority of systems that require a high frequency clock, incorporate some sort of PLL based clock synthesis. In applications being covered here the PLL is generally used to up-convert the externally generated/supplied low frequency reference clock to the desired high frequency. The PLL is a negative feedback system consisting of a phase frequency detector (PFD), charge pump (CP), loop filter (LF), voltage controlled oscillator (VCO) and a feedback divisor.

When the clock jitter is of concern, the VCO is generally the main culprit. There are different ways of designing VCOs, from the popular ring oscillators to the LC tank based resonant circuits. The microprocessor world still heavily relies on ring oscillator based VCOs so they will be covered in more detail here.

When observed open loop (free running, outside of the PLL) the VCO phase experiences a phenomenon called the "random walk". The oscillator phase accumulates in a random fashion when observed from some reference point in time. This is due to the device thermal noise. This noise is completely random; it is uncorrelated in the time domain or flat expending over all frequencies in the frequency domain. If we were to observe the oscillator rms jitter over time (as time accumulates) we would see that this type of noise on the log-log scale has a slope of one-half. This relationship is expressed in Eq. 2 [1,2].

$$jitter_{rms} = Kappa \cdot \sqrt{t_{accumulation}} \qquad (2)$$

The notation Kappa [$\sqrt{Sec}$] from the equation above stands for the VCO goodness factor, a better VCO should have a smaller Kappa. Clearly if we want a clock with the lowest rms jitter we would also want the shortest jitter accumulation time. The overall system topology plays a critical role in determining the jitter accumulation time. The period jitter implies the $t_{accumulation}$ that is equal to the average period duration.

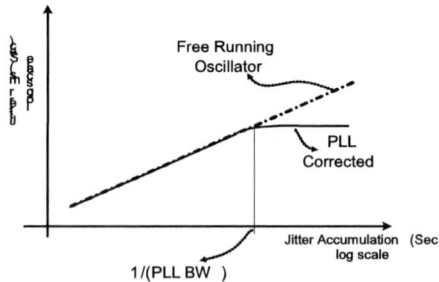

Fig. 4. Random Noise caused VCO jitter vs. jitter accumulation time.

There is a well understood relationship between VCO ring buffer power and jitter. Figure 5 shows a simulated and estimated VCO Kappa value as a function of the VCO buffer power. Kappa is proportional to one over the square root of VCO power [2]. If we need a VCO with appropriate jitter level we need to burn an adequate power. The power level that is required to be used in ring oscillators in order to meet a necessary jitter level is the main down side of this oscillator topology.

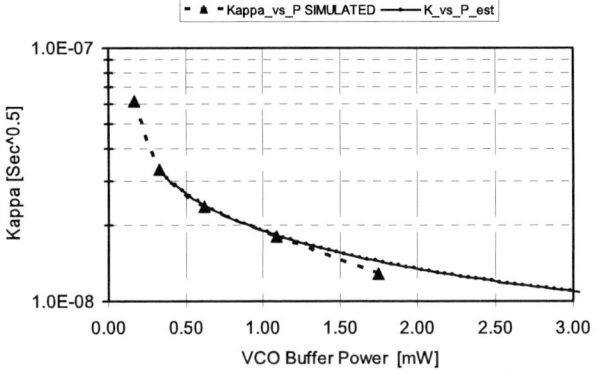

Fig. 5. Estimated/Simulated VCO Kappa vs. VCO buffer Power.

It first became a necessity and later a common practice to integrate sensitive analog circuits with the noisy digital logic on the same die. It is now an ordinary practice to integrate on-die power regulators to power analog circuits. The clock jitter level we are currently trying to meet made the power supply noise a problem that had to be solved.

In the simplest sense linear regulators consist of an operational amplifier driving a PMOS device (MP1) that is delivering current to the regulated circuits (Figure 6.). It is crucial for this block to have an adequate power supply rejection ratio PSRR. Figure 7 shows a typical PSRR. If provided with a clean reference voltage the regulator can adequately attenuate noise within its open loop BW. At frequency close to its closed loop BW the regulator PSSR is the poorest. At higher frequencies the PSSR again improves and that is mostly because of the pole that is formed due to the regulator's output impedance and the decoupling capacitance (C1).

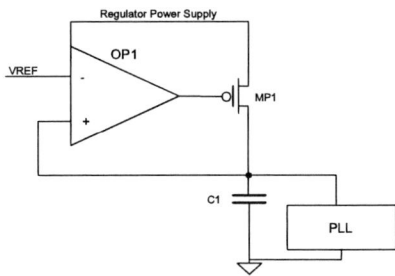

Fig. 6. Linear Power Regulator, top Level diagram

The PLL jitter due to the power supply noise can be substantial. This is mostly due to very sensitive VCOs that in the later generations of MOS technologies have quite large gains [rad/(Sec V)]. In order for the power supply

noise induced jitter to be brought to minimal levels power regulator power supply must be modeled well. It is essential for power regulator to be designed with the package model in mind.

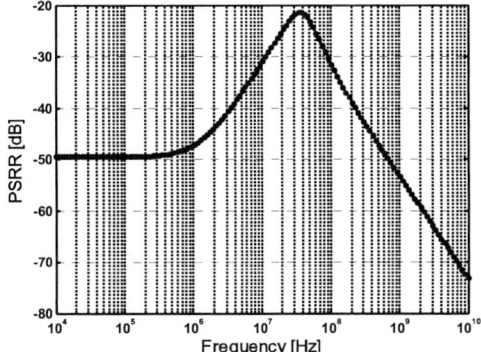

Fig. 7. Power Regulator PSRR

IV. CLOCK DISTRIBUTION AND DETERMINISTIC NOISE.

The clock jitter caused by clock distribution is generally of a deterministic nature. Due to its large power consumption, clock distribution is generally powered by a global "noisy" power supplies. Most of the larger power supply loops contain a resonance frequency generally in the MHz range. The power supply resonance is the main cause of jitter in the clock distribution.

The variation of voltage of the power supply causes variation in the clock distribution propagation delay and that can be characterized as jitter. Figure 8, shows how the inverter chain propagation delay varies with the power supply in the DC sense. This is often called the worst case jitter.

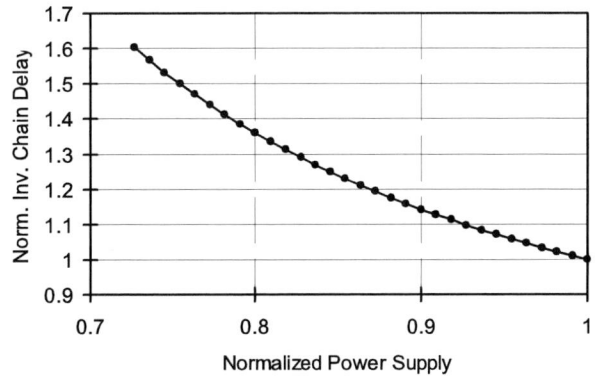

Fig. 8. Single ended clock distribution propagation delay sensitivity to the Power supply variation.

A similar observation can be made in the AC sense. There is a relationship between the dominant power supply resonance frequency, jitter accumulation time and the worst case jitter. This relationship is shown in Eq 3.

$$jitter_{max} \Leftarrow t_{accumulation} = \frac{1}{2 \cdot f_{resonance}} \qquad (3)$$

The clock distribution induced jitter is generally minimized by reducing the depth (the number of stages) of the clock distribution.

V. CHANNEL EFFECTS ON THE CLOCK JITTER

It is well known how channel BW limitations affect signaling. The reasons why we consider coding or implement pre-emphasis or receiver side equalization are all very well known. All these methods work very well for data and not so well for clock like signals.

The BW limitation can not only reduce the amplitude of the clock being transported but can also have an effect on its jitter performance. A jittery clock sent over band-limited media can appear even more jittery on the receive side. This effect is some times called the jitter amplification.

Figure 9 shows the effects of the 20 inch FR4 channel on the clock signal sent over it. It is not only that clock amplitude is attenuated at the receiver but also induced jitter at one cycle is now spread over several clock cycles. This type of clock jitter amplification is even more severe for higher frequency clocks.

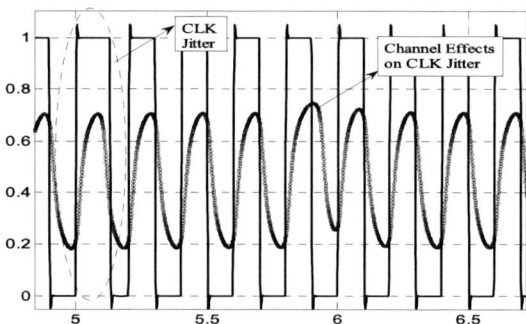

Figure 9. Chanel effects on CLK jitter

VI. JITTER ACCUMULATION TIME

One of the main advantages of source synchronous I/O systems is that clock jitter and data track to some extent. Both clock and data traverse similar paths. As it was shown in Eqs. 2 and 3, jitter accumulation time greatly affects the total clock jitter level (rms jitter/pk-pk jitter) which is directly tied into the I/O system BER (Eq.1).

The accumulation time to the first order is just the difference in the propagation delay clock and data travel. From Figure 1 it can be observed that the majority of this discrepancy happens at the receiver side. The clock generally goes through the clock amplifier and later through the clock re-aligner (DLL) before it can be used to sample data. This simple relationship is shown in Eq 4.

$$t_{accumulation} = t_{prop\_clk} - t_{prop\_data} \quad (4)$$

VII. CONCLUSION

In this paper the relationship between BER and clock jitter in high speed I/O links were presented. The vast majority of the noise sources affecting the clock jitter were described and potential clock jitter lowering methods were pointed out.

ACKNOWLEDGEMENT

The author would like to greatly acknowledge the colleges from Intel Hudson, Massachusetts Microprocessor Design Center especially Del Ramey, Wayne Parker from MMDC and Warren Anderson from Intel Connectivity Labs.

REFERENCES

[1] J.A.Mcneill, *Jitter in Ring Oscillators*, IEEE Journal of Solid State Circuits, vol 32 p870-878, June 1997
[2] A.Hajimiri, T.H.Lee, *The Design of Low Noise Oscillators*, Kluwer Academic Publisher, 1999
[3] B. Muer, M. Steyaert, *CMOS Fractional-N Synthesizers*, Kluwer Academic Publisher, 2003
[4] G. Balamurugan, N. Shanbhag, *Modeling and Mitigation of Jitter in Multi-Gbps Source-Synchronous I/O Links*, Proc. International Conference on Computer Design, 2003
[5] R.Stephensen, "*Jitter analysis: The dual-Dirac model, Rj/Dj, and Qscale*", White Paper Agilent Technologies.

2006 25th International Conference on Microelectronics

An Adaptive Pulse-Width Control Loop

G. Jovanović, D. Mitić, and M. Stojčev

Abstract - In high-speed CMOS clock buffer design, the duty cycle of a clock is liable to be changed when the clock passes through a multistage buffer [1]. In this paper, we propose a pulsewidth control loop referred as APWCL (Adaptive Pulsewidth Control Loop) that adopts the same architecture as the conventional PWCL, but with two modifications. The first one relates to implementation of the pseudo inverter control stage (PICS), while the second to involvement of adaptive control loop. The first modification provides generation of output pulses during all APWCL's modes of operation and the second faster locking time. For 1.2 μm CMOS process with V_{dd}=5V and operating frequency of 100MHz, results of SPICE simulation show that the duty cycle can be well controlled in the range from 20% up to 80% if the loop parameters are properly chosen.

I. INTRODUCTION

In high-speed design, a multistage clock buffer implemented into a long inverters chain is often needed to drive a heavy capacitive load. It is difficult to keep the clock duty cycle at 50% for these design solutions. When the clock signal passes through a multistage buffer, the symmetrical pulse-width may be destroyed due to the unbalance of the N and P channel transistors in the long buffer. This unbalance is introduced by many factors, such as process deviations, temperature changes, or mismatch in design. Consequently, the clock duty cycle will wonder away from 50%. In the worst case, as the pulsewidth becomes too narrow or too wide, the clock pulse may disappear inside the clock buffer [1], [5]-[8].

In systems that adopt a double data rate technology, both rising and falling edges of the clock are used to sample input data. These systems require the duty cycle of the clock to be precisely maintained at 50%. Therefore, an important issue is how to generate a clock with precise 50% duty cycle for high-speed operation [2]. Automatic control technology, such as pulsewidth control loop (PWCL) has been used for adjusting the output duty cycle of multistage driver for several years and was described by [1]- [4].

In this paper, architectural description and principle of operation for conventional types of PWCLs, already well known from literature, covered is in Section II. Section III, describes the structure of the proposal, referred as adaptive pulsewidth control loop. Details related to APWCL implementation and simulation results are given in Section IV. Section V gives a conclusion to this paper with summary.

G. Jovanović, D. Mitić, and M. Stojčev are with the Faculty of Electronic Engineering, University of Niš, Aleksandra Medvedeva 14, 18000 Niš, Serbia & Montenegro, E-mail: joga@elfak.ni.ac.yu

II. CONVENTIONAL PWCL

Schematic diagram of the conventional PWCL [1] is pictured in Fig. 1. As it can be seen from Fig. 1, the conventional PWCL is realized as a system with feedback loop.

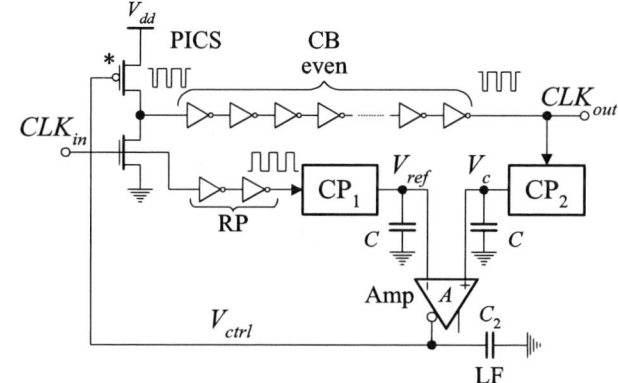

Fig. 1. Conventional PWCL

The feedback loop functionally consists of:
(a) Pseudo-Inverter Control Stage (PICS) - chosen to be the first stage of the clock buffer and functions as a voltage-controlled pulse-generator. By changing the control voltage, V_{ctrl}, we can adjust the pulsewidth of the output clock. PICS is implemented as a simple inverter. Mark "*" indicates the controlled transistor;
(b) Clock Buffer (CB) - a long inverter chain or buffer which acts as a multistage driver;
(c) Charge Pump 1 (CP1) - converts pulsewidth into current which charges or discharges capacitor C. At its output, CP1 creates a reference voltage V_{ref}, by connecting to a reference clock with 50% duty cycle;
(d) Charge Pump 2 (CP2) - is another identical charge pump that creates a voltage V_c, i.e. it steers current by the clock pulse CLK_{out} for detecting the change of pulsewidth;
(e) Amplifier (Amp) - the amplifier is characterized by its gain A, realized as a differential amplifier. It is intended to provide a certain gain in the loop at low frequency;
(f) Reference Pulse (RP) - two stage inverter buffers used to drive CP1 with 50% duty cycle referent clock pulses;
(g) Loop Filter (LF) - the output resistor of Amp and capacitor C_2 form a first-order low-pass filter.

In Fig. 1 two identical single-ended charge-pumps are used. One of them is used for detecting the pulsewidth of the clock being controlled, and another is connected to a standard clock with 50% duty cycle for generating V_{ref}. The voltage V_{ref} is taken as reference voltage. The charge-pumps, CP1 and CP2, and the differential amplifier, Amp,

1-4244-0116-X/06/$20.00 ©2006 IEEE   667

are constituents of the duty cycle comparator. The output voltage V_{ctrl} controls the operation of PICS.

The pulsewidth of CB is controllable. This means that if the CB's clock output deviates from 50% duty cycle, the control voltage, V_{ctrl}, will change so that the offset can be removed. When the loop is stable the CB output is adjusted to 50% duty cycle, and the controllable dynamic range covers the range of possible offset.

III. ADAPTIVE PWCL

Block diagram of the APWCL is sketched in Fig. 2. From functional point of view, the following building blocks can be identified:

(a) Pseudo-Inverter Control Stage (PICS) – at the output, $PICS_{out}$, pulses of variable duty cycle are generated. V_{ctrl}, is used as an input control voltage;

(b) Clock Buffer (CB) – an inverter chain implemented as odd (even) stages clock driver;

(c) Charge Pumps (CPx) – two voltage controlled charge pump circuits, CP1 and CP2, of different structure;

(d) Reference Pulse (RP) – chain composed of two inverters;

(e) Bias Circuits (BC1 and BC2) – provide control voltages for transistors polarization of CP1, CP2 and PICS;

(f) Differential-input differential-output operational amplifier (Amp) – acts as an inverting (non-inverting) amplifier in a feedback control loop. For odd (even) number of stages in CB the Amp is implemented as non-inverting (inverting) amplifier;

(g) Low-pass filter (LF) – filter element in a feedback control loop;

(h) Charge Pump Controller (CPC) – is implemented as differential amplifier. At its output, the CPC generates control voltage V_A that is proportional to $V_{ref} - V_c$ on voltage difference.

In respect to the conventional PWCL proposed by Fenghao and Svensson [1], there are several novelties involved into APWCL. The first one relates to the PICS, while the second to CP2. In addition, two new building blocks, CPC and BC2 are included into APWCL structure. The other constituents, pictured in Fig. 2, are of identical (or almost-identical) architectures as those described by [1].

The signals CLK_{in} and CLK_{out} are input and output pulse signals of APWCL, respectively. They drive two charge pumps, denoted as CP1 and CP2 (Fig. 2). The output voltage V_{ref} (V_c) of the charge pump CP1 (CP2) is directly proportional to the duty cycle of the input signal CLK_{in} (CLK_{out}). Charge pump CP1 (CP2) load capacitor C_{11} (C_{12}) is discharged during the positive pulse period and charged in the rest of the period. The charging and discharging currents are adjusted to be identical. The signal CLK_{in} is selected as a referent one. Its duty cycle is 50%. Therefore, the voltage V_{ref} at the output of CP1 is referent.

Due to influence of different propagation delays of the leading and trailing edges of the clock signal, when it passes through the long chain clock buffer, the duty cycle of the CLK_{out} become unsymmetrical, i.e. different from 50%.

The voltages V_{ref} and V_c drive the differential amplifier (Amp). Voltage V_{ctrl} is generated at the Amp's output. The V_{ctrl} controls operation of the PICS. When the APWCL is in steady-state, the magnitude of control voltage V_{ctrl} causes the duty cycle of the CLK_{out} to be 50%.

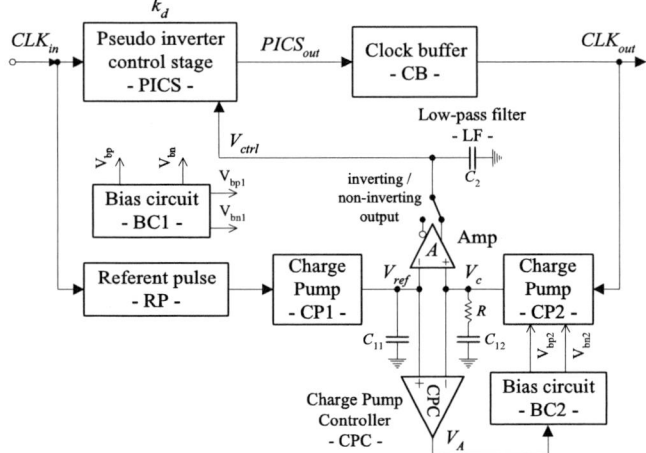

Fig. 2. Block diagram of APWCL.

A. Pseudo Inverter Control Stage - PICS

An electrical scheme of the PICS is pictured in Fig. 3(a). It consists of three N-channel N_1, N_2 and N_3, and three P-channel P_1, P_2 and P_3, transistors. The PICS's equivalent electrical scheme is presented in Fig. 3(b). Transistors P_1 and P_2 act as constant and variable current sources J_1 and J_2, while transistors N_1 and N_2 operate as constant and variable current sinks I_1 and I_2, respectively. Transistors P_3 and N_3 belong to the switching parts of the CMOS inverter. Capacitor C_L represents a parasitic capacitive load.

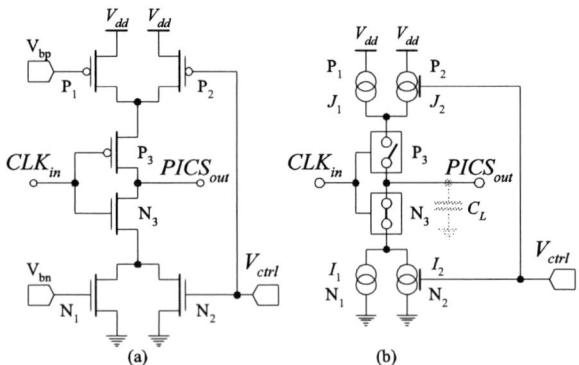

Fig. 3. The block PICS: (a) electrical scheme, (b) equivalent scheme.

The amount of the current of the constant current source (sink) J_1 (I_1) indirectly determines the nominal time delay of the leading (trailing) pulse edge at the output $PICS_{out}$. The bias voltage V_{bp} (V_{bn}) is used for polarization of transistor P_1 (N_1). The variable current source (sink) J_2 (I_2) indirectly defines the variable time delay of the leading (trailing) pulse edge. Such a configuration, allows us to achieve controllable time delay for both, leading and trailing pulse edges.

In Fig. 4, the range of duty cycle variation, in terms of V_{ctrl}, is shown. Again, duty cycle of 50% for $V_{ctrl}=V_{dd}/2=2.5\text{V}$ is achieved. When V_{ctrl} decreases the duty cycle increases, by contraries it decreases.

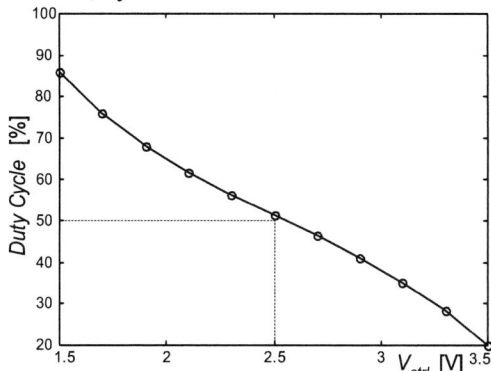

Fig. 4. Duty cycle in term of V_{ctrl}.

B. *Parallel charge pump - CP2*

The charge pump CP2, pictured in Fig. 5, is composed of two charge pumps, $CP2_1$ and $CP2_2$, that operate in parallel. The structure of the pump $CP2_1$ is identical with CP1. The bias circuit BC2 provides polarization for $CP2_2$. The block CPC (Fig. 2) defines the magnitude of the control voltage V_A.

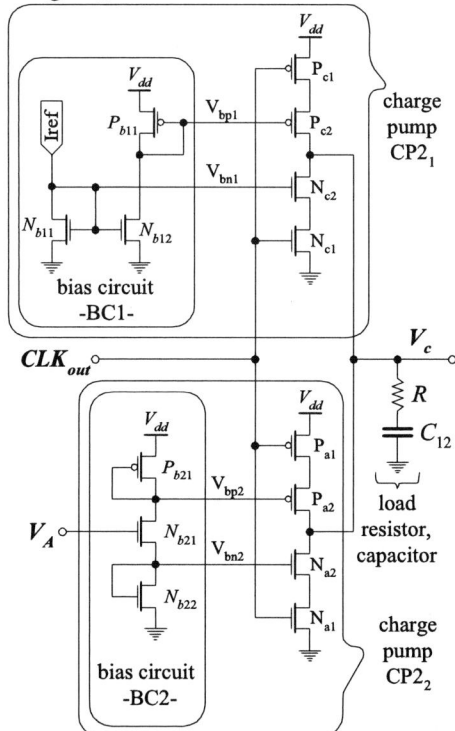

Fig. 5. Electrical scheme of CP2 charge pump.

The load impedance of CP2 is realized as a serial connection of resistor R and capacitor C_{12}. This connection

involves one zero in the transfer function of CP2, and also in a transfer function of the APWCL. Involving zero in the APWCL's transfer function allows us to adjust the damping factor ξ to optimal value (ξ=0.707), and to decrease the transition time in the linear mode during the period of establishment of a steady-state.

IV. APWCL SIMULATION RESULTS

SPICE simulation results for APWCL circuit in a 1.2μm CMOS process with V_{dd}=5V supply voltage and operating frequency 100MHz, are presented in Fig. 6. Comparative, results that relate to the conventional PWCL and proposed APWCL, in Fig. 6(a) and (b), are given respectively. Identical building blocks PICS, CP1, Amp, BC1 are used for both simulations. In both cases, the clock buffer (CB) has 7 stages with tapering factor of 1. Related to conventional PWCL, CP2 is replaced with charge pump whose current is variable in APWCL charge pump. Additional blocks CPC and BC2 are used for current regulation in charge pump CP2.

The circuit's model in linear mode is described by the second order transfer function [2]. The desire system's dynamic is defined by choice of the dumping factor ξ and natural frequency ω_n. The dumping factor is ξ=0.707 and natural frequency is ω_n=3·10⁷ rad·s⁻¹. Other circuits parameters are determined as: $I_{cp1}=I_{cp2}=I_{cp}$=10μA – corresponds to charge pump current of CP1 and $CP2_1$; A=100 – DC gain of the Amp; $\omega_0=2\pi f_0=2\pi$ 3.5MHz – dominant pole of the Amp; $C_{11}=C_{12}=C$=8pF – charge pump capacitor; R=2800Ω – CP2 load resistance; k_d=0.32V⁻¹ – PICS's sensitivity constant. For nonlinear operating mode $I_{cp2}=I_{cp}+I_{cp}'$ where I_{cp}'=50μA.

The waveforms at the top in Fig. 6(a) and (b) correspond to curves of V_{ref} and V_c. The second waveform in Fig. 6(a) corresponds to the control voltage V_{ctrl}. Additionally, in Fig. 6(b), the control voltage V_A, is presented. The two lower waveforms in Fig. 6(a) and (b), depict CLK_{out} pulses valid for nonlinear mode and steady-state mode, respectively.

We start with our simulation from the instant when the system is powered-on ($t=t_0$). This implies that both charge pumps load capacitors, C_{11} and C_{12}, as well as the low-pass filter capacitor C_2, are discharged. According to the transient response, the following three different modes, in the operation of the feedback loop, can be identified:

a) From t_0 up to t_1 the loop operates in nonlinear mode. Since C_{11} charges faster, in respect to C_{12}, at instant t_0 the voltage V_{ref} becomes greater than V_c and therefore the output of Amp switches rapidly to lower voltage limit, and the control voltage V_{ctrl} is 0V. Under this condition at the CLK_{out}, pulses of minimal pulsewidth are generated. Contrary to the proposals described by [1] and [2], where in the saturation mode the PWCL is inoperative, i.e. CLK_{out} is blocked, in APWCL pulses of minimal duty cycle, at the output of CLK_{out}, are generated.

b) As the input voltage difference becomes small enough, the amplifier Amp enters linear mode what corresponds to the time interval from t_1 up to t_2. When the dumping factor $\xi=0.707$, transients in linear mode are minimal.

c) Steady-state operation mode characterizes stable-loop operation and corresponds to the time interval after t_2. During this period, variations of V_{ctrl} are less than ± 25mV, i.e. 1.8% in respect to V_{ctrl} (1.5V). As it can be seen from Fig. 6, the duty cycle of CLK_{out} in the saturation mode is 20%, and in the steady-state mode it is 51%.

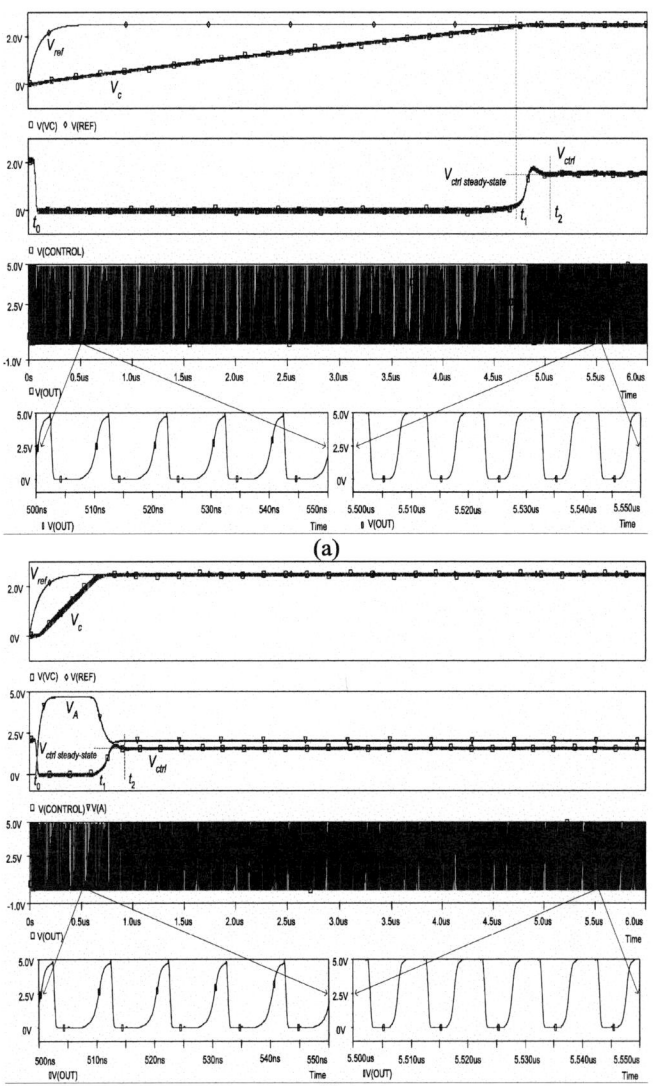

(a)

(b)

Fig. 6. (a) Conventional PWCL and (b) adaptive PWCL simulation results.

The main difference between waveforms in Fig. 6 (a) and waveforms in Fig. 6(b) relates to the time duration of nonlinear operating mode, from t_0 up to t_1. The voltage difference between V_{ref} and V_c during this period is large and the differential amplifier CPC enters in saturation region. The CP2's control voltage V_A is approximately

equal to the supply voltage V_{dd} and the current I_{cp}' is active. This means that in nonlinear mode the charge pump CP2 works with current $I_{cp2}=I_{cp}+I_{cp}'=n\ I_{cp}=60\mu$A. Increasing CP2's current by a factor n means that the time duration of nonlinear mode t_{NL} is shortened by n, too. This possibility provides a condition for fast loop locking time.

V. CONCLUSION

The clock distribution tree within the VLSI ICs is so large and carries so much capacitance that buffers need to be inserted just to be able to drive the clock-tree in order to have a reasonable clock waveform. When the clock passes through a multistage buffer changes its duty cycle. In order to obtain a satisfactory duty cycle correction a fast locking APWCL was proposed. The APWCL adopts almost identical architecture as conventional PWCL [1]-[3] but with two modifications. The first relates to implementation of the pseudo inverter control stage (PICS), which is operative during all APWCL's mode of operation. This possibility provides pulses generation at the output of APWCL during all mode of operation. The second modification represents involvement of adaptive control loop, which provides shorter transient time of the nonlinear mode, i.e. faster locking time. SPICE simulation results, for 1.2μm CMOS process with V_{dd}=5V and 100MHz operating frequency, show that the duty cycle can be controlled in the range from 20% up to 80%.

REFERENCES

[1] M. Fenghao, C. Svensson, "Pulsewidth Control Loop in High-Speed CMOS Clock Buffers, " *IEEE Journal of Solid-State Circuits*, vol. 35, No. 2, pp. 134-141, February 2000.

[2] H. Sung-Rung, L. Shen-Iuan, "A 500-MHz–1.25-GHz Fast-Locking Pulsewidth Control Loop With Presettable Duty Cycle," *IEEE Journal of Solid-State Circuits*, vol. 39, No. 3, pp. 463-468, March 2004.

[3] Y. Po-Hui, W. Jinn-Shyan, "Low-Voltage Pulsewidth Control Loops for SOC Applications," *IEEE Journal of Solid-State Circuits*, vol. 37, No. 10, pp. 1348-1351, October 2002.

[4] M. Yongsam, C. Jongsang, L. Kyeongho, J. Deog-Kyoon, K. Min-Kyu, "An All-Analog Multiphase Delay-Locked Loop Using a Replica Delay Line for Wide-Range Operation and Low-Jitter Performance, " *IEEE Journal of Solid-State Circuits*, vol. 35, No. 3, pp. 377-384 , March 2000.

[5] M. Flynn, P. Hung, K. Rudd, "Deep–Submicron Microprocessor Design Issues," IEEE Micro, vol. 19, No. 4, pp. 11-22, 1999.

[6] J. Maneatis, F. Klass, C. Afghani, *Timing and Clocking*, pp. 10.1-10.34, in The Computer Engineering Handbook, ed. by Oklobdzija V., CRC Press, Boca Raton, 2002.

[7] V. Oklobdzija, M. Stojanović, M. Marković, N. Nedović, *Digital System Clocking: High-Performance and Low-Power Aspects*, Wiley Interscience, New York, 2003.

[8] J. Öberg, *Clocking strategies for Networks-on Chip, in Networks on Chip*, eds. by Jantsch A., and Tenhunen H., Kluwer Academic Publishers, Boston, 2003.

2006 25th International Conference on Microelectronics

Low Power Computing for Secure and Reliable Sensor Networks

Rachit Agarwal, E.M. Popovici, C.O'Keeffe, B.O'Flynn, S.J.Bellis

Abstract— In recent years embedded systems gained lot of attention in many applications such as automotive, medical, multimedia, industrial monitoring and control, etc. The topic of reducing power dissipation in embedded systems has received considerable attention in the recent years. A power driven design methodology is mandatory to meet system level requirements while fulfilling time to market constraint. In this paper we focus on developing cost and power efficient hardware-software architectures for sensor networks which could be used in medical sensing applications. In particular, we aim to realize reliable and secure communications between the 25 mm cube motes developed by Tyndall National Institute. The work presented in this paper provides an insight into the power aware design and implementation of a part of communication system which uses error correction (e.g. Reed-Solomon) and symmetric ciphers (e.g. IDEA-NXT) for resource constrained embedded systems in general and sensor networks in particular.

I. INTRODUCTION

Due to the growing market of portable wireless communication devices, power consumption has become one of the most severe constraints in embedded system design. Sensor Networks devices have an obvious advantage of low power implementations. However, the system must meet the market requirements in terms of speed and memory.

Our research work is focused on developing cost and power efficient hardware-software architectures for sensor networks which could be used in medical sensing applications. Reliability and security are two important issues in terms of medical sensing applications. However, the communication, besides being reliable and secure, should consume least possible power. The system is made secure for communication by implementing a versatile symmetric cipher known as IDEA NXT on hardware. Reliability has been introduced by implementing Reed Solomon(RS) forward error correction codes on software.

There are several techniques to reduce switching activity in the application specific hardware in order to reduce power consumption. These techniques span all

Rachit Agarwal is a postgraduate student in Department of Microelectronic Engineering at University College Cork. E-mail: rachit.ee@gmail.com

Dr. Emanuel Popovici is a lecturer in Department of Microelectronic Engineering at University College Cork

C.O'Keeffe is a Hardware Engineer at Synopsys, Ireland

B.O'Flynn is a member of MAI Group at Tyndall National Institute, Cork

S.J.Bellis is a Senior Design Engineer at SensL, Ireland

the levels of abstraction in VLSI design: architectural-level (e.g. use of pipelining to reduce applied voltage), gate-level (e.g. technology mapping to reduce switching activity), transistor-level (e.g. gate resizing) and layout-level (e.g. use shorter wires for high-activity nets). However, power dissipation is often neglected during the software implementation of the algorithm in embedded systems, since code size and performance take priority over power dissipation at this stage. We concentrate on all the three main aspects of low power system design i.e. OS level (partitioning issues), hardware and software level.

There are two important techniques implemented for minimizing the power consumption in the system. Hardware-software partitioning has been done for reliability and security implementations to minimize the power consumption. Low power consumption techniques have also been used at the implementation level to minimize the power consumption in both hardware and software. The significance of software power consumption is mainly in the following components:

- Power dissipated in the arithmetic/logic circuits and the control unit of the CPU when executing embedded code;
- Power dissipated to charge and discharge the address busses;
- Power dissipated within the memory circuits.

All of these power consuming components have been taken into account while developing the research work with an emphasis on CPU power dissipation.

This paper is organized as follows. Section 2 gives the problem statement for our work. Section 3 covers the platform description for the system implementation. In section 4, we discuss the partitioning issues related to our implementation and implementations of security and reliability algorithms under system description. Section 5 covers the smart system implementation part and Section 6 concludes the paper with the results and conclusions while mentioning the possible future work for the project.

II. PROBLEM STATEMENT

It is important to ensure that the system dissipates least possible power while still providing the required functionality in sensor network architectures for medical sensing applications. Other important aspect is to increase the system speed and keep memory requirements as low as possible. In this context, the problem statement is developed through two important factors: algorithms and hardware-software architecture.

A. Algorithm Architectures

It is important to choose the right algorithms for implementation of low power embedded systems for the required application. Algorithms with lower computational complexity (e.g. fewer arithmetic/logic operations) to perform the same function must be used to conserve energy. However, for such implementations, it is also important that the chosen algorithms have the flexibility in terms of implementation on hardware and software.

B. Hardware-Software Co-Design

Partitioning plays an important role in system level implementations. This gives lesser flexibility to the users but more application driven benefits can be achieved through such implementations. Besides the choice of algorithms to be implemented, the most important problem is to find a suitable partitioning algorithm for the implementation between hardware and software so as to meet the specifications of the application. Applications imply low power implementation while the market requires the system to be fast and cost efficient.

The partitioning problem is formally stated as:
"Given the algorithm architectures, find the best possible partitioning scheme minimizing power keeping speed as fast as possible and memory consumption as low as possible."

III. PLATFORM DESCRIPTION

We use as the underlying platform the 25mm cube developed at Tyndall National Institute [1]. The system represents a novel 3-D programmable modular system which could be used in applications such as robotics, autonomous agents and neural networks, telemetry, transducer networks, etc. The 25mm cube module uses an FPGA sub-module, a wireless transceiver submodule, a sensor submodule and an 8-bit microcontroller.

The FPGA submodule uses a Xilinx Spartan IIE device and provides a reconfigurable processing solution in which intensive digital signal processing tasks or intelligent functionality can be implemented. The hardware part of the system is mapped on this module. The transceiver submodule consists of a Nordic nRF2401 2.4 GHz ISM band RF transceiver. The microcontroller is the Atmel Atmega 128L 8-bit RISC architecture with 128kb in-system programmable flash memory. The platform allows versatile hardware-software partitioning and synthesis with low cost and low power consumption.

IV. SYSTEM IMPLEMENTATION

The platform for our implementation, as shown in Fig. 1, consists of a Xilinx Spartan IIE FPGA, an Atmega 128L microcontroller unit and a Nordic nRF2401 Transceiver. IDEA NXT for security has been implemented on FPGA and Reed Solomon implementation has been done on the microcontroller close to the transceiver/antenna. We have implemented a (28,24) shortened Reed Solomon code defined over $GF(2^8)$. The code is capable of correcting 2 symbols on $GF(2^8)$. The system can also be made adaptable to the changes in the channel which is discussed in the next section.

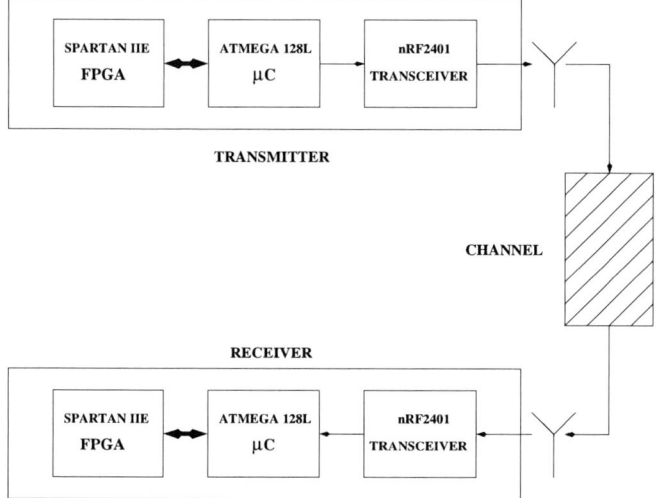

Fig. 1. Block View of System Implementation

A. Partitioning

The partitioning algorithm chosen for developing this system is based on the fact that Reed Solomon algorithm is computationally more intensive than IDEA NXT and the latter requires a lot of memory for its implementation. Hardware implementation of Reed Solomon would have consumed a great amount of area and power, and hence a software implementation was used. Also, channel coding (Reed Solomon) has to be done close to the antenna. However, IDEA NXT was implemented on hardware since ciphers are more secure if implemented on hardware [2]. Also, a lot of memory is required if the computation complexity is to be kept low. IDEA NXT cipher software implementation of either computational or look up model would have consumed a lot of power and hence, this block was implemented on hardware. Another reason for this was to increase the system speed by running the encoding and encryption algorithms in parallel.

B. IDEA NXT Implementation

Encryption is often necessary in many modern communications systems to ensure that the data is securely transmitted over insecure channels. IDEA-NXT is a recently developed symmetric block cipher and its structure is based on the IDEA cipher [3]. The top-level architecture uses a Lai-Massey scheme and the round function is based on substitution-permutation networks. This is a versatile cipher as it allows various key lengths to be implemented easily in hardware or software. This feature is well suited for our sensor networks applications. The cipher is implemented on the FPGA in order to efficiently use the available on-chip memory as well as to speed-up the computationally intensive tasks involved in the encryption/decryption process.

IDEA NXT64 is implemented on the Spartan-IIE FPGA. One of the heuristics for low-power implementation is to reduce the amount of switching. The number of registers are kept as low as possible and the circuit designed is predominately combinational (one-shot). The Spartan-IIE design integrates block RAM to reduce the design area.

Encryption and Decryption processes use the same structure except for differing orthomorphism functions. This enables us to use the same design entity for both of the processes. An encrypt/decrypt signal is used to select the required orthomorphism depending on the process. The same entity is used for the final round of both versions by selecting no orthomorphism function.

The block data and sixteen round-keys are serially shifted into the design entity which further reduces the I/O pin count causing lower power usage. The round-keys are held in on-chip registers and are provided for each round. Decryption round keys are shifted in reverse order to those for encryption.

For the Spartan-IIE, an architecture is implemented [4] that removed all Galois Field combinational logic and some sbox logic by using RAM lookup tables. As arithmetic in $GF(2^8)$ occurs at byte level, these lookup tables contain all possible 256 results of an 8-bit value multiplied by the constant values of the diffusion matrix.

C. Reed Solomon Implementation

One of the most popular block codes used in practice is the Reed-Solomon code. We use the code to ensure reliable communication between the sensor nodes. By carefully selecting the code, we aim to increase the distance between nodes, without increasing the transmitting power. The systematic encoding is based on simple linear feedback shift register architecture. The decoding process is more computationally intensive and for the key equation solver we choose the division-free Fitzpatrick algorithm [5]. Encoding and decoding (as well as the cipher) rely on finite field operations. We implement multiplication and inversion over the field $GF(2^8)$ using a complex representation of the field elements.

Inversion and Multiplication in Galois Field are the most computationally intensive tasks in Reed Solomon implementation. Multiplication for RS codes has been implemented using the complex representation [6]. Inversion, on the other hand is computationally most power consuming. To minimize the power consumption, inversion is implemented using look up tables in microcontroller memory. This decreases the power consumption for inversion by 80% in comparison to computational model.

Division is another computationally intensive task in RS Decoding. The F Algorithm provides a suitable way to prevent division in RS decoding. The division free Fitzpatrick Algorithm has been implemented over the microcontroller to prevent power consumption.

Power Issues

Software power comprises of three main factors, i.e., Bus Power, Memory Power and CPU Power. While bus power can be minimized by carefully minimizing the bus switching and memory power can be reduced using sequential memory access, we are most concerned about CPU power for such computationally intensive algorithms. Every instruction executed by CPU will result in switching activity. Load/Store Instructions, Branch Instructions and Arithmetic Instructions are broad classifications of instructions for a CPU. The power saving techniques in software can be broadly written in a few steps:

• A *for* loop coded as *for(i=max; i>0; i−−)* consumes less power than *for(i=1; i≤max; i++)* since no register is required for saving *max* in latter case.
• In *loop* unrolling, the increment of i is done during the same iteration provided *max* is even. This minimizes the total number of compare instructions, but increases code size.
• *function macros* are created through #define preprocessor directive which gives a significant power reduction for Reed Solomon.

Using such techniques, an important amount of power consumption can be prevented, making the system more power efficient.

V. Channel Adaptive System

Reliability of the developed system can be enhanced by integrating adaptive functions in the system. The idea is to sense the noise in the channel and to derive the system to the best possible coding scheme. The code used in our implementation can correct up to 2 bytes of errors (a code RS(28,24) over $GF(2^8)$). There may be some channels which introduce so much noise during communication that it is impossible for the decoder to decode the received word. In such cases, the device needs to change the coding scheme so that loss of information can be prevented for the coming packets. The developed system can change its encoding and decoding scheme on the basis of noise in the system and hence, make the system more reliable.

This task can be done through implementation of a hand shake algorithm in which the receiver informs the transmitter about the quality of the channel. In case of a burst of noise in the channel, the transmitter sends back an information symbol to the receiver informing it that the noise in the channel has corrupted the data to such an extent that it can not be decoded. The encoder, on receiving the signal, changes the coding scheme such that one more error can be corrected. This implementation enhances the reliability of the system, though power consumption will increase by a slight margin due to increased code redundancy.

VI. Conclusion and Future Work

In this paper, we have presented a secure and reliable system developed on the 25mm programmable transceiver cubes. The system consumes low power because of implementation techniques used for security and error control coding in the system. The transceiver nRF2401 is configured and operated at the lowest power mode out of the various possible configurations.

Table I shows the results for the system implementation [Fig. 1].

TABLE I

System Power Consumption

	Current (mA)	Voltage (V)	Power (W)
Transmitter	77	5.5	0.424
Receiver	126	5.5	0.693

The results for FPGA implementation of IDEA NXT64 algorithm are presented in Table II. The system clock is 4MHz. Most of the power is consumed by FPGA. However, this is compensated by increased security of the system.

TABLE II

Results for Hardware Implementation of IDEA NXT64

Current	50 mA
Voltage	5.5 V
System Clock	4 MHz
Slices	1498
Memory	2 Block RAM

Finally, we present the results of software implementation of Reed Solomon code RS(28,24) over $GF(2^8)$ on Atmega 128L microcontroller in Table III.

Computationally intensive tasks have been eradicated with carefully chosen look up techniques and min-

TABLE III

Results for Power Efficient Software Implementation of Reed Solomon Codec

	Encoding (Bytes)	Decoding (Bytes)
Flash	2540	4274
SRAM	512	512
Variables	183	425
S/W Stack	313	71
H/W Stack	16	16

imizing memory accesses. The reliability of transmitting the data over the channel has been enhanced using error control coding and also by introducing adaptability of the system to swing environments. These implementations resulted in enhancing the range by 100%, i.e., the range of transmission was doubled without any increase in power supplied.

This work can be extended to further reduce power consumption through more hardware-software intensive partitioning. This would further enhance the power savings in the system. However, this partitioning should take into account implications on the security of the system.

Acknowledgment

Thanks are due to the National Access Program provided by Tyndall National Institute and Science Foundation Ireland for the funding provided to help carry out this work.

References

[1] B. O'Flynn, S.J. Bellis, K. Mahmood, M. Morris, G. Duffy, K. Delaney and C. O'Mathuna, 'A 3-D Miniaturized Programmable Transceiver,' *Microelectronics, International*, Vol. 22, No. 2, p. 8-12, 2005

[2] A.J. Elbirt, W. Yip, B. Chetwynd and C. Paar 'An FPGA Implementation and Performance Evaluation of the AES Block Cipher Candidate Algorithm Finalists,' *The Third Advance Encryption Standard Candidate Conference*, pp. 13-27, 2000

[3] P. Junod and S. Vaudenay, 'IDEA-NXT Specifications, Version 1.2,' *EPFL Technical Report*, IC/2004/75, 2005

[4] C.O'Keeffe, R. Agarwal, E. M. Popovici and B. O'Flynn, 'Low Power Hardware and Software implementation of IDEA NXT Algorithm', *Proceedings of ISSC*, 2005.

[5] P. Fitzpatrick 'On the key equation,' *IEEE Transactions on Information Theory*, vol. 41, pp. 1290-1302, 1995

[6] E. M.Popovici, 'Algorithms and Architectures for Decoding Reed-Solomon and Hermitian Codes', *PhD thesis, National University of Ireland, National Microelectronics Research Centre, University College Cork, Ireland*, 2002.

2006 25th International Conference on Microelectronics

Design and Implementation of an FPGA Based High-Speed Data Buffer for Optical Interconnects

Sven-Hendrik Voss, Maati Talmi, Juergen Saniter

Abstract - This paper describes the design and implementation of an innovative high-speed data buffer system for optical interconnects based on electrically reprogrammable FPGAs. It is characterized by high transfer rates and the dense integration of electrical and optical components on the basis of a printed circuit board (PCB). The high data rates and the highly parallelized system operation require a specific architecture and careful signal integrity design for proper functionality.

I. INTRODUCTION

Optical links are emerging as the preferred media when data rates beyond the "gigabit per second"-mark are concerned and gain in importance in many different applications. This may be due to higher bandwidths, increased immunity to electromagnetic interference and decreases in power consumption and power density distribution compared to traditional electrical interconnects. The optical interconnect technology provides an inherently greater maximum bandwidth-length-product but to fully take advantage of the offered bandwidths an effective and robust electrical system for storing and pre-processing the transmission data is needed and introduced here.

II. SYSTEM REQUIREMENTS

A variety of different options are possible for the implementation of short distance optical interconnects (chip-to-chip, board-to-board, rack-to-rack). Because of flexibility and expandability the focus lies on parallel interconnects in particular. Guided wave solutions like parallel fiber arrays, optical fiber image guides, image conduits and waveguides embedded into printed circuit boards and on the other hand free-space transmission are suitable for that purpose.

In a first proof-of-concept stage of the proposed system a skew-compensated high-speed transmission with an overall date rate of 36 Gbps is to be demonstrated. The input data may originate from a much slower RAID system or any other storage unit according to the used interface. Depending on the data volume a subset or a complete set of the input data is temporarily buffered to generate a continuous data stream for the optical transmission. The concept should also allow performing further data

Sven-Hendrik Voss and Maati Talmi are with the Department of Image Processing, Juergen Saniter is with the Department of Photonic Networks at the Fraunhofer Institute for Telecommunications Heinrich-Hertz-Institut, Einsteinufer 37, 10587 Berlin, Germany, E-mail: voss@hhi.fhg.de

processing steps in between as well as higher overall data rates (> 100 Gbps) in the future and thus must be expandable. Because of the restrictions of physical optical channels when sticking to guided wave solutions it was concluded that a free-space solutions would be appropriate. In contrast to long distance telecommunication systems, where attenuation, dispersion or optical nonlinearities play the decisive role, for short link length interconnects low power dissipation, low latency, small form factors and an economical integration with mainstream electronics are required.

The intended transmission scenario requires pre-processing of a huge amount of data in real-time. The processing comprises buffering, synchronization, optional multiplexing and transcoding and signal regeneration. By preparing the data especially for optical transmission another requirement is added. To prevent a baseline shift complicating detection and degrading the system noise margin the transmitted signal must be DC-balanced, meaning the number of transmitted ones and zeros must be equal over time. For this purpose a proprietary coding scheme is applied on the transmission data.

In the following the design and implementation of the high-speed data buffer being capable to fulfill the above stated requirements is described. Until a short time ago FPGAs were primarily used as prototyping platforms. However, FPGA data rates have increased dramatically in the past and they now rival those of CMOS ASICs. The solution presented here treats an FPGA based implementation for high speed data communication. The work addresses solutions for practical "low cost" data buffering and processing at speeds of 100Gbps and beyond.

III. SYSTEM DESIGN

Referring to the stated requirements a high-speed data buffer system has been designed. Its main functions include generating, receiving, forwarding, storing and processing the input data in real-time simultaneously.

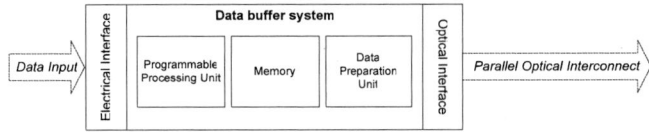

Fig. 1. Simplified structure of the proposed system

Figure 1 shows the simplified system structure. The target transmission scenario consists of a point-to-point

1-4244-0116-X/06/$20.00 ©2006 IEEE

connection between a transmitter, i.e. the proposed buffer system, and a compatible receiver.

A. Functionality

The transfer of data across the high-speed data buffer is performed as follows: Electrical data originating from a storage unit is received through a dedicated high-speed interface and transferred to a programmable processing unit. Application-specific on-the-fly processing can be carried out here. In a next stage the data enters the memory and is buffered there. For bulky processing that cannot be executed on-the-fly the memory can temporarily also be used as swap space. However processing the data prior to the storage is recommendable to allow burst-wise read operations without interruptions.

The data pre-processing unit follows up with the transmission related steps like the optional multiplexing and transcoding. To maximize throughput of information the appliance of effective data compression methods in addition is possible [1]. The data is then communicated to an optoelectronic converter unit where it is converted into modulated optical signals. The optoelectronic components act as an interface between the processor and buffer unit and the high-capacity optical interconnect system. The optical signals propagate via parallel optical channels on a free-space optical link from the high-speed data buffer to a receiver unit. Therefore the transmitters are connected to an optical imaging system via a multi-fiber patch cord. To achieve an uninterrupted operation the system is designed as a double-buffer structure providing read / write accesses simultaneously.

B. Technology

In discussing the most suitable technology the trade-off between hardware costs and system performance and accuracy has to be taken into account. Dynamically reconfigurable programmable gate array (FPGA) and application-specific integrated circuit (ASIC) designs are the two major approaches for data processing in practical implementations. The use of ASICs can achieve sufficient power to allow real-time processing but they are quite expensive for low volume productions and prototyping.

The solution proposed in this paper is based on the use of FPGAs which allow an easy prototype development and the design of complex circuits in a single chip of high computational power at lower cost. The FPGA type that has been used for the design is a Xilinx Virtex-II Pro VP70 with a pin count of 1704, offering 996 available user I/Os and 20 dedicated Multi-Gigabit Transceivers (MGTs). It comprises of 74,448 logic cells which corresponds to approximately 6 M system gates, i.e. 24 M transistors. Virtex-II Pro FPGAs are manufactured in a 0.13 µm nine-layer copper technology with 90 nm high-speed CMOS process [2].

C. Architecture

A scaleable and flexible dual-FPGA architecture as depicted in figure 2 has been designed. The electronic processing unit is composed of two parts: the user data processing part and a management control part. The former performs the application-specific processing of the input data, while the latter conducts the adjustable multiplex configuration and the supervision and control of the optical devices. To accomplish the double-buffer structure with simultaneous access two FPGA devices are used whereas both FPGAs are used for both processing parts, taking advantage of the doubled number of built-in embedded parallel-to-serial and serial-to-parallel transceiver cores (MGTs). The FPGAs each offer four integrated IBM PowerPC embedded processors which are used to perform controlling functionality.

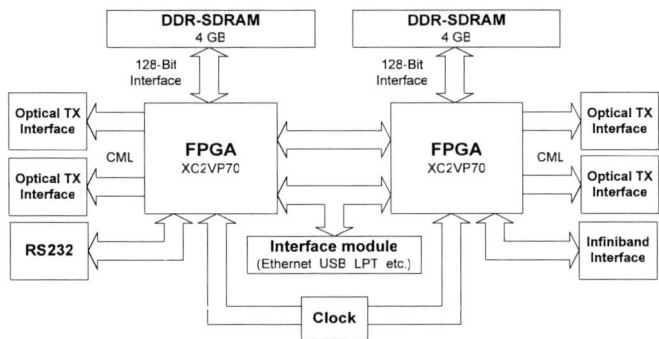

Fig. 2. Architectural structure of the high-speed buffer system

The theoretical maximum transmission speed for the MGTs is 3.125 Gbps. All 40 MGT cores are used in the design. The high-speed transmission with an overall date rate of 36 Gbps was divided into a parallel 36 channel structure, each channel running at 1 Gbps. The four remaining channels are used for providing the input interface. The interface between the electronic and optoelectronic parts of the system is of critical importance. Parallelized high-speed I/O ports are needed to provide the desired data rates. However parallel operation with minimal skew is a real challenge since time skew among multiple channels can easily corrupt data. Moreover jitter, noise and other parasitic effects may be additionally introduced by the optoelectronic conversion.

The optoelectronic devices provide the physical interface for the parallel optical interconnect. For that purpose commercially available PAROLI2-TX modules are used, converting the FPGA signals into light beams. Each module contains an array of 12 vertical cavity surface emitting lasers (VCSELs), so for 36 channels four of them are used, three running 10 channels and one providing the outstanding 6. Electrically the connection between PAROLI2 and the FPGAs is based on CML signal level which has many advantages due to its differential nature: It is immune to common-mode noise since electromagnetic interference imposes the same voltage on both proximate signals. The difference cancels out the effect. In theory it also emits no noise itself since the AC currents in each pair are equal but opposite and proximal. Thus the radiated EMI is reduced. As frequencies increase beyond 1GHz an increasing percentage of the signal may be lost but the resulting difference still crosses zero volts. All these aspects benefit the critical high-speed transmission.

On the input side a sophisticated high-speed protocol was chosen providing connection to a RAID or any other storage system. After evaluating a couple of applicable transmission schemes in detail the choice was taken in favour of the Infiniband protocol offering a data rate of 2.5 Gbps per channel. The transfer protocol is structured equivalent to the ISO/OSI models and the whole protocol stack up to the Transport Layer is implemented as a custom interface circuit instantiated on one of the two FPGAs. It handles all the issues of link startup, maintenance and simple error detection. The interface is partially driven by the programmable hard-wired logic but also using the PowerPC microprocessor which generates or receives synthetic test traffic or acts in a bridging mode passing packets to and from the Infiniband interface and the other interfaces. Leaving the architecture expandable, an IBx4 interface is realized which allows for the aggregation of four channels (= 10 Gbps).

For buffering the input data two 4 GB DIMM memory modules were arranged, each controlled by one FPGA. Since memory access latency and throughput are essential the system requires a broad memory interface and an optimized controller for high-speed address generation. Taking the required bandwidth into account a 128-bit interface from each FPGA to the DDR SDRAM was chosen, combining two 64-bit memory modules in parallel. With that configuration and a 200 MHz (400 MTs) clock data traffic up to 51 Gbps from each memory unit is possible. A fast memory controller was implemented in the Virtex-II Pro device. The pipelined structure of the controller allows for concurrent operations, thereby providing high effective bandwidth by hiding row precharge and activation time.

Both the parallel high-speed I/Os and the memory connection play a decisive role in the design and are directly related to the need of a proper clocking solution. A flexible clock management ensures full synchronous operation between the two FPGAs and provides the basis for the high-speed I/Os. Two low-jitter clocks are used in the system, covering a frequency range of 25 MHz – 700 MHz as reference for the high-speed outputs at 1 Gbps – 3.125 Gbps. In addition various standard interfaces (RS232, Ethernet, LPT, etc.) are provided.

IV. Board Implementation

Finally an in-house designed printed circuit board (PCB) was realized, based on the specified architecture. The board is based on common FR4 material and 230 mm x 350 mm large. It comprises a 16 layer stack which lead to

a board thickness of ca. 1.6 mm. The large stack count can be put down to the demerging complexity because of the component count and the required functionality. The high-speed data buffer board (fig. 3) features two FPGAs as core elements and a large memory array. It includes one IBx4 Infiniband connector and four optical output modules integrated directly on the PCB.

Fig. 3. High-speed buffer PCB

This approach was followed to maintain a high degree of mechanical alignment between the optoelectronic components and short electrical transmission lines between the FPGAs and the modules. The input channels are capable of reading 4 x 1 bit at 2.5 GHz and allows the reception of different input data streams. The output runs at 1 GHz on each of the 36 optical channels. Altogether approximately 1440 electronic devices are used on the board.

A. Physical Layout Design Considerations

A set of rules had to be obeyed to enable fully synchronized high-speed transmission with minimal skew and minimize losses, jitter and crosstalk on FR4. Electric traces are subject to resistive, dielectric and skin effect losses. Loss increases as trace length and frequency increases or trace width decreases. Therefore only traces with widths $\geq 80\mu m$ over minimum distances in point-to-point topology are used. However there is a tradeoff between short traces leading to tighter spacing and more vias. Vias are expected to contribute up to 1 dB per via to the loss budget. Consequently layer transitions are constrained to a maximum of two.

Single-ended lines are used for lower speed signals whereas all high-speed signals are routed using differential transmission lines, designed as both edge-coupled microstrip and stripline. Tight coupling within the differential pair and increased spacing to other differential pairs helps to minimize crosstalk and EMI. A large pair-to-pair spacing of 400 µm between differential pairs is kept where applicable. Controlled-impedance 50 ohm single-ended traces and differential traces of 100 ohm with a tolerance $\leq 15\%$ are provided. Adequate attention is paid to power integrity to ensure that power supply noise does not add unnecessary jitter to the high-speed signals.

Ground plane referencing is maintained without discontinuities along the entire route of differential pairs. However any noise on the ground reference can shift the input threshold which creates timing uncertainty in the switching. To remove this noise a large number of bypass capacitors are used. These capacitors provide the quick energy needed by the device without having to go through the inductance of the system. In effect high-frequency transient current bypasses the supply by going through the bypass capacitors. Instead of relying on generic layout guidelines to ensure the design various simulation analysis were carried out. Optimum trace lengths were determined from simulation analysis in order to meet the desired signal integrity.

B. Implemented Logic And Deskewing Mechanism

Basic parts of the buffer logic including key routines for generating, storing and processing the input data have been implemented using VHDL models. For the purpose of DC balancing a proprietary 6B/8B coding scheme was implemented in the buffer circuit and interposed on the PAROLI2 transmission lines, guaranteeing DC balance and moreover providing error detection of one-bit errors in the data stream. Further details about balanced coding can be found in [1].

Arrangements to minimize the skew on board level have been made, however uncertainties due to the used FPGA and its reference clock routing mismatches remained. The implemented deskew function compensates for fixed pair-to-pair skew such as the fixed skew in the FPGA channels, the PCB traces, the connectors and the optical transmission. While signaling on one of the 36 channels that the deskew mode is activated a deskew pattern is sent to the receiver. The receiving unit recognizes this pattern and performs the deskew operation if the deskew mode is activated. This special pattern is generated by logic. At first a coarse deskewing is performed at a low transmission rate. Than the data rate is increased up to the desired value and a fine tune is performed. . The function works up to 3 Gbps and thus works over the entire operating range with a DC balanced pattern because of the needed transitions. The implemented mechanism deskews up to ± 600 ps. If skew exceeds these values, the additional skew will be compensated by additionally shifting back a whole bit time.

C. Implementation Complexity

When implementing the logic the FPGA device was utilized as follows: out of 33088 logic slices 3645 were used, the number of slice flip-flops amounts to 3672 out of 66176. BRAMs are embedded memory modules available in most of the Virtex family devices. The number of BRAMs used is 4 out of 328, however most of the memory communication happens by means of the external DDR-SDRAM. 5 out of 16 global clocks are used, whereas the number of global triggers is 15 out of 16. The DCM blocks are embedded digital clock manager elements providing self-calibrating solutions for distributing, delaying, multiplying, dividing, and phase-shifting clock signals. 8 DCMs are provided within the FPGA from which 6 are currently used. Apparently the utilization is in the middle range of the total FPGA capacity. The FPGA size was actually chosen because of the large amount of MGTs, thus allowing upgrades and extensions in the logic.

D. Testing

The performance of the digital portion of the electronic system was tested by running a series of test vectors in a testbench environment. All aspects of the digital system performed as expected. The optical system was tested in a first step by connecting all outputs via adapter fiber cables separately to an optical power meter to demonstrate proper optoelectronic conversion. The optical power was meassured to be -6 bBm and thereby 1 dBm below the specified output of the PAROLI modules. This result may be due to fan-out loss but is still satisfactory. The second step included simultaneous measurements of the parallel channels with transmission rates of 1 Gbps to determine channel-to-channel skew, jitter and crosstalk. The maximum channel jitter amounts to 50 ps while the channel-to-channel skew averages 400 ps due to the FPGA outputs. While testing the deskewing mechanism an accuracy of 30 ps could be achieved. The performance of the electric I/Os was tested by means of a series of probe tests which were carried out on the board while it was attached to a test adapter. Measured signal integrity turned out to be close to the simulated values.

V. CONCLUSION

In this paper the design and implementation of a flexible FPGA-based architecture for high-speed optical interconnect systems has been shown. The integration of the electrical and optical system was presented in detail. A discussion of the strengths and weaknesses of the data buffer system as revealed by the actual implementation completed the article. The introduced architecture is capable of handling data traffic up to 51 Gbps. Using newest DDR2 technology and accelerated channel data rates of 3 Gbps the throughput can be expanded to near 100 Gbps. Further improvements could be made to the system concept by including flow control and error correction mechanisms. Up to now similar hardware solutions offering such storage capacity and performance are not yet published or commercially available.

REFERENCES

[1] S.-H. Voß, M. Talmi, "Lossless high-speed data compression for optical interconnects as used in maskless lithography systems", *Proceedings of the MNE 2005 International Conference*, Microelectronic Engineering, Elsevier 2005.
[2] Xilinx Inc, "Virtex-II Pro and Virtex-II Pro X Platform FPGAs: Complete Data Sheet", DS083 (v4.0), 2004.

2006 25th International Conference on Microelectronics

DefSim: Measurement Environment for CMOS Defects

T. Borejko, A. Jutman, W. A. Pleskacz, R. Ubar

Abstract – This article describes a measurement environment for study of two CMOS defect types: opens and shorts. These defect types are physically implemented in silicon in a big variety of locations inside a set of digital standard cells and small circuits. The integrated circuit (IC) with the collection of defects is mounted onto a plug-and-play measurement box, which is connected to the PC via USB cable. Two measurement methods are supported by IC: voltage and I_{DDQ} testing. The DefSim bundle represents a unique and easy to handle educational and research environment. In the paper we also consider a simple learning flow, which is targeted on students whose main specialization is general microelectronics (not the digital testing specifically).

I. INTRODUCTION

The increasing complexity of VLSI circuits and transition to Systems-on-Chip (SoC) and Networks-on-Chip (NoC) paradigm has made test generation one of the most complicated and time-consuming problems in the domain of digital design [1]. The more complex electronics systems are getting, the more important problems of test and design for testability become, as costs of verification and testing are getting the major component of design and manufacturing costs of a new product. This fact makes the research in the area of testing and diagnosis of integrated circuits (IC) a very important topic for both the industry and the academy.

The gigahertz operating frequencies and high integration levels of modern microelectronic systems make the physical level defects manifest themselves in a new unusual way. The good old logic level fault models like stuck-at or even bridging ones do not guarantee full defect coverage anymore [2]. Even rather easy to model defects like opens and zero-resistive shorts might behave in an unexpected way when their behavior is purely considered on logic level. Moreover, due to process variations and other phenomena, even the results of rather complicated electrical simulation methods sometimes do not match with the reality.

These facts fully agree with the results one can observe using our measurement environment for real CMOS defects - DefSim. The central element of the DefSim environment is the IC with a large variety of shorts and opens

T. Borejko and W. A. Pleskacz are with the Institute of Microelectronics & Optoelectronics of Warsaw University of Technology, Koszykowa 75, 00-662 Warsaw, Poland, E-mail: pleskacz@imio.pw.edu.pl

A. Jutman and R. Ubar are with the Department of Computer Engineering, Tallinn Technical University, Raja 15, Tallinn 12618, Estonia, E-mail: artur@pld.ttu.ee

Fig.1. DefSim chip on a measurement box

physically inserted into a set of digital standard cells and small circuits. The IC is attached to a dedicated measurement box (see Fig. 1) serving as an interface to the computer. The box supports two measurement modes – voltage and I_{DDQ} testing [3]. The communication between the DefSim hardware and software goes through the USB port.

The primary target of the DefSim environment is education. However, it can be used in research as well, for instance, as an evaluation tool for defect-oriented test pattern generation (TPG) methods.

II. IMPLEMENTATION OF DEFECTS

Before designing DefSim, we made up a strong decision – defect implementation in the chip should be as much as possible close to the silicon reality. When doing so, we took into account the common experience in the field of manufacturing yield modeling [4,5]. As a result, our defects are fixed and have local (spot) features. The circuit layout remained unchanged too. Moreover, each circuit under test is loaded and driven by standard non-inverting buffers.

All these measures ensure that the implementation of defects is kept as close to silicon reality as possible.

It is also worth to mention that a special care was taken to protect the IC from possible damage during its faulty operation [6].

As an alternative solution, the defects could have been implemented as transmission gates activated by special addressing. In such situation electrical model and the behavior of the defects would be quite different (e.g. additional parasitic capacitances and resistances). Moreover, the circuit under test would be also totally different than the initial one (without defects) from both logical and layout points of view. In case of FPGA realization the situation would be even worse.

1-4244-0116-X/06/$20.00 ©2006 IEEE

TABLE I
DEFECT TABLE

	Defect (short)	Vectors (CBA)							
		0	1	2	3	4	5	6	7
1	A/B		1	1					
2	A/C		1			1			
3	A/gnd		1						
4	A/Q	1	1		1		1		
5	A/vdd	1							
6	B/C			1		1			
7	B/gnd			1					
8	B/Q	1		1	1			1	
9	B/vdd	1							
10	C/gnd					1			
11	C/Q	1					1	1	1
12	C/vdd	1							
13	Q/gnd	1							
14	Q/vdd		1	1	1	1	1	1	1

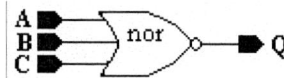

Fig.2. NO3 – 3-input NOR gate

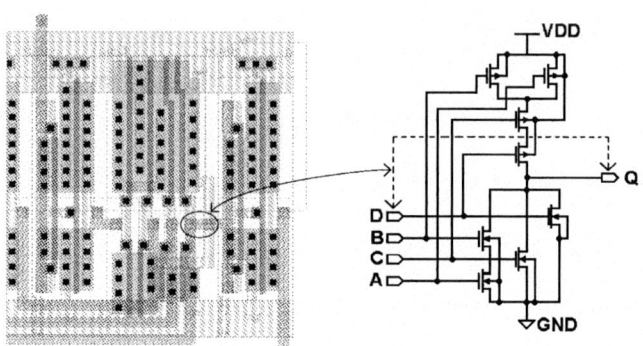

Fig. 3. Complex cell with IN-OUT short
caused by excessive Metal2 defect

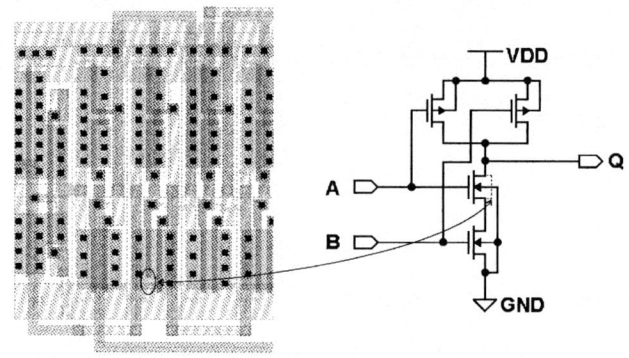

Fig. 4. NAND2 cell with Drain-Source short
caused by missing Polysilicon defect

Another reason, why it is really important to keep the implementation of defects as close as possible to silicon reality is explained by the following example. See Table 1, which describes the fault detection information for standard cell NO3 (3-input NOR, Fig. 2). This table matches each test vector from 000 to 111 to the defects it detects. Each such detection is marked by '1'. One can easily see the difference of fault detection between electrical simulations on a model (non-shaded subset of '1'-s) and measurements using real DefSim hardware (all '1'-s in the table). Such a difference is observed for most of cells and circuits used in DefSim with a bigger impact made on C17 and CB1 benchmarks. In C17, this mismatch affects shorts less that opens.

There are several possible reasons for this effect: not proper simulation conditions (e.g. power supply voltage drop, logic levels degradation), bad defect models (on electrical level), parasitics, delays caused by charging effect, memory effect. This is a big problem and real challenge for correct electrical simulation. This means that there is a strong need for good defect models on electrical level and proper simulation conditions. The DefSim hardware is free of such imprecision since all defects are designed to be as close to real cases as possible.

Currently two types of defects are implemented in DefSim: opens and hard shorts in conducting layers like Polysilicon, Metal1, and Metal2. These defects are located both inside logic gates and upon (or between) signal lines outside the gates. Examples of typical defects implemented in DefSim chip are shown in Figures 3 and 4.

III. THE DEFSIM CHIP AND ITS STRUCTURE

The DefSim IC (see Fig. 5) has three main structural parts: a matrix of simple digital circuits, addressing

mechanism, and a measurement circuitry [7]. Each circuit from the range is implemented in many copies where one of them is correct and all the others are intentionally defective. It is possible to select any defect of interest by addressing a corresponding copy of the circuit. During chip operation only one such copy can be active at a time.

When a circuit and a defect are selected, the user can apply an arbitrary input test sequence and measure the

Fig.5. Microphotograph of the DefSim die

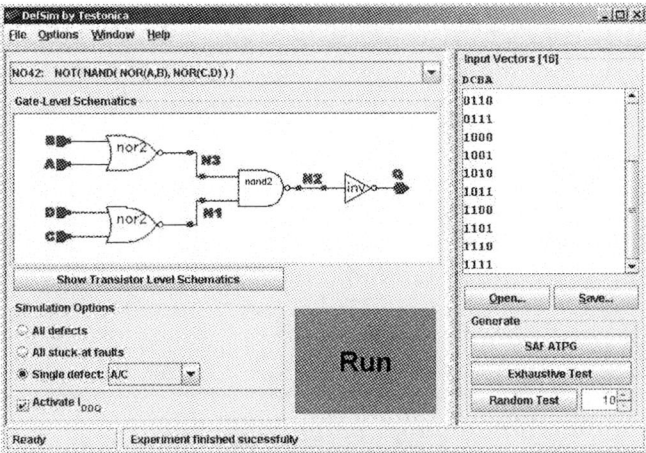

Fig.6. Graphical user interface

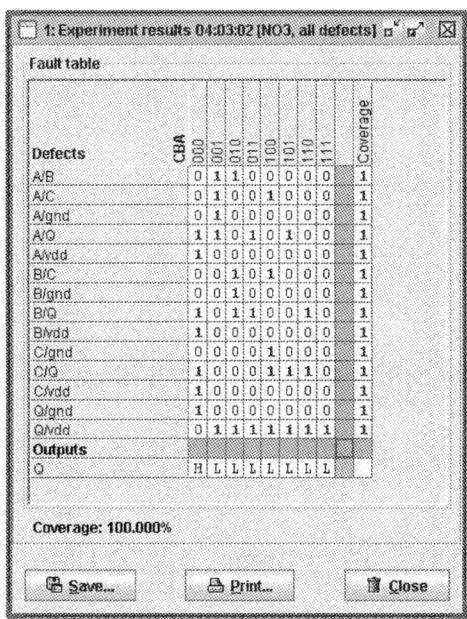

Defects	CBA 000	001	010	011	100	101	110	111	Coverage
A/B	0	1	1	0	0	0	0	0	1
A/C	0	1	0	0	1	0	0	0	1
A/gnd	0	1	0	0	0	0	0	0	1
A/Q	1	1	0	1	0	1	0	0	1
A/vdd	1	0	0	0	0	0	0	0	1
B/C	0	0	1	0	1	0	0	0	1
B/gnd	0	0	1	0	0	0	0	0	1
B/Q	1	0	1	1	0	0	1	0	1
B/vdd	1	0	0	0	0	0	0	0	1
C/gnd	0	0	0	0	1	0	0	0	1
C/Q	1	0	0	0	1	1	1	0	1
C/vdd	1	0	0	0	0	0	0	0	1
Q/gnd	1	0	0	0	0	0	0	0	1
Q/vdd	0	1	1	1	1	1	1	1	1
Outputs									
Q	H	L	L	L	L	L	L	L	

Coverage: 100.000%

Fig.7. Measurement results: defect table

circuit's response in terms of both the binary logic values and static supply current levels (I_{DDQ}). It is also possible to compare the behavior of the defective version to the correct copy of the same circuit.

The DefSim chip was designed in AMS 0.8μm CMOS (double-metal, double-poly) technology. Prototype chips were manufactured via TIMA-CMP Multi-Project Wafer service. The chip occupies 19.90 mm^2 of silicon. It is composed of approx. 48000 transistors, uses 62 pins, and is packaged in JLCC68 package.

IV. THE USER INTERFACE

The DefSim software package provides a convenient access to the features of the IC and ensures a smooth way of going through educational scenarios for students.

As the first step, the user has to select a target circuit to work with. Then, the list of implemented defects for this circuit becomes available (Fig. 6). The user can work either with a single specific defect or with a group of defects simultaneously. Currently, two groups of defects can be selected: all realized defects and stuck-at faults (SAF) only.

Besides the fault list, two types of schematics are available for each circuit: a logic level scheme and transistor level one. The necessary test patterns can be generated either by hand using these schematics or automatically by the software itself. The resulting test data can be stored for later retrieval and modification.

When the test is ready, it is sent to the IC and applied to either one selected defect or a group of defects, and consequently, the circuit responses are recorded. The results of measurements are displayed later in several different forms. For instance, the user can observe the truth table for a certain defect, or a defect table for a group of defects, or I_{DDQ} value info. Such a defect table which corresponds to the Table I for NO3 cell is given in Fig. 7.

Two different levels of the DefSim software are available: a stand-alone workstation version (Fig. 6) for individual use, and a server-based version for handling large groups of students (http://www.defsim.com).

V. DEFSIM EXERCISES

From the didactical point of view, the DefSim environment targets (but it is not limited to) two main areas of expertise: defect modeling and defect observability. First of all, the user gets a chance to compare the efficiency of different logic level fault models in terms of their capability to cover all possible shorts and opens in a CMOS circuit. The students will learn in practice that some simple defects represent a real challenge especially from the diagnostics (defect localization) point of view.

Since DefSim environment supports both voltage and I_{DDQ} testing, the user can compare the fault detection efficiency of those test methods as well. In most cases, their effectiveness is noticeably different.

Basically, all the exercises on the DefSim Environment can be divided into two groups: less advanced and more advanced ones. Here we will consider a simpler flow as an example. It is targeted on students whose main specialization is not the digital testing but general microelectronics. A bit more advanced exercises are considered in [10].

1. Getting a truth table of good (without defects) CMOS simple and complex standard gates.

2. Getting a truth table of good (without defects) small combinational circuits (C17 or CB1).

3. Repeating steps 2 and 3 but with a given defect of a certain type in order to observe how the circuit's function is modified by the defect.

4. Getting basic knowledge of voltage and current testing principles.

The first and the second step of this flow are illustrated in Fig. 7 based on a simple cell NO3. The last row of this table named "Q" gives correct output values of this cell. The measurement results of a defected circuit instance (short B/Q) are illustrated in Fig. 8. There one can see the

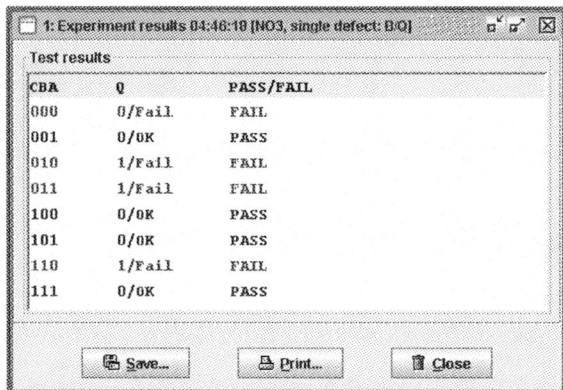

Fig.8. Voltage levels and fault detection information

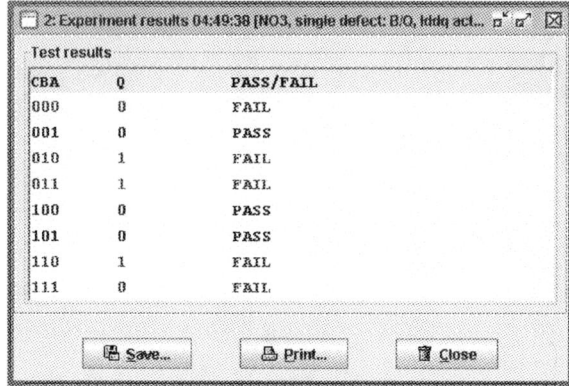

Fig.9. I_{DDQ} testing results

binary value at the output Q and the fault detection information ("PASS/FAIL") provided for each test vector ABC.

It is not that hard to notice that normally the short B/Q is detected when there are opposite logic levels on the corresponding lines B and Q, where B appears to be a stronger driver than Q. The only exception is the last pattern 111, where 0-value on Q doesn't change despite line B tries to drive it to logic 1. In all other cases 1 on B appears to be stronger than 0 on Q, which actually contradicts with simple and commonly used logic-level models for short defects like Wired-AND and Wired-OR [8,9]. This is another example supporting the need for accurate and real implementation of CMOS defects in silicon.

The last step in the flow is illustrated in Fig. 9. It shows the measurement results of static supply current levels on the output of NO3 cell. One can see that in this case the last pattern '111' comes up with the fault detection. Obviously, this is the result of different degree of observability a particular testing method is featuring.

VII. CONCLUSIONS

The paper describes a complete solution for study of the most common CMOS defect types: opens and shorts. These defects are physically implemented in silicon and represent a trustworthy source of different phenomena similar to those appearing in defective VLSI chips.

The DefSim bundle consists of the chip with defects and a measurement environment. The handy software interface provides a flawless work for both individual users and large groups of students.

All these features provide for a unique environment useful for both educational and research purposes.

ACKNOWLEDGEMENTS

This work has been supported in part by the European Union project IST 2000-30193 *REASON* and by the Polish State Committee for Scientific Research (project No. 4 T11B 023 24), Estonian Science Foundation grants G5649 ja G5910. The DefSim GUI and server software were developed by Testonica Lab (www.testonica.com).

REFERENCES

[1] *The International Technology Roadmap for Semiconductors, 2005 Edition: Test and Test Equipment.* URL: http://public.itrs.net/

[2] D. Kasprowicz, W. A. Pleskacz, "Improvement of Integrated Circuit Testing Reliability by Using the Defect Based Approach," *Microelectronics Reliability, PERGAMON – Elsevier Science*, vol. 43/6, June 2003, pp. 945-953.

[3] V. Stopjakova, H. Manhaeve, "CCII+ Current Conveyor Based BIC Monitor for I_{DDQ} Testing of Complex CMOS Circuits," *Proc. of European Design & Test Conference*, Paris, France, March 1997, pp. 266-270.

[4] J.B.Khare, W.Maly. *From Contamination to efects, Faults and Yield Loss: Simulation and Application.* Kluwer Academic Publishers, Boston, 1996, 150 p.

[5] H. T. Heineken, J. Khare, W. Maly, P. K. Nag, C. Ouyang and W. A. Pleskacz, "CAD at the Design-Manufacturing Interface," *Proc. of the 34th Design Automation Conference – DAC'97*, Anaheim, USA, June 1997, pp. 321-326.

[6] W. A. Pleskacz, D. Kasprowicz, T. Oleszczak, W. Kuzmicz, "CMOS Standard Cells Characterization for Defect Based Testing," *Proc. of IEEE Intl. Symposium on Defect and Fault Tolerance in VLSI Systems*, San Francisco, USA, October 2001, pp. 384-392.

[7] W. A. Pleskacz, T. Borejko, T. Gugala, P. Pizon, V. Stopjakova, "DefSim – The Educational Integrated Circuit for Defect Simulation," *Proc. of IEEE Int. Conf. on Microelectronic Systems Education*, Anaheim, USA, June 2005, pp. 121-122.

[8] M.L. Bushnell, V.D. Agrawal, *Essentials of Electronic Testing for Digital Memory and Mixed-Signal Circuits.* Kluwer Academic Publishers, Dordrecht, 2000, 690 p.

[9] Abramovici M., Breuer M.A., and Friedman A.D. *Digital systems testing and testable design.* IEEE Press, New York, 1990, 652 p.

[10] W. A. Pleskacz, T. Borejko, A. Walkanis, V. Stopjakova, A. Jutman, R. Ubar, "DefSim: CMOS Defects on Chip for Research and Education," Proc of Latin American Test Workshop (LATW'06), Buenos-Aires, Argentina, March 2006, (to appear).

Poster Session
System Design

2006 25th International Conference on Microelectronics

NNARX Model of Speech Signal Generating System: Test Error Subject to Modeling Mode Selection

D.Protić, M.Milosavljević

Abstract – This paper presents comparative analysis of different processing errors when modeling speech signal generating system simulated by Neural Network Auto Regressive with eXtra input (NNARX). Feedforward NNs were used. Vowels were taken to train and test models. Gradient Descent Algorithm (GDA) is used for the training, Back Propagation Algorithm (BPA) for the testing, and Optimal Brain Surgeon (OBS) for the NN pruning. Results are given in Tables I-IV, and compared in conclusion.

I. INTRODUCTION

The human voice production mechanism can be roughly divided into three parts: lungs, vocal folds, and vocal tract. The lungs function as a source of air flow and pressure. When voiced speech is being produced, vocal folds open and close periodically and thus convert the air flow from lungs into a train of flow pulses. These pulses are referred as glottal flow, or voice source. Vocal tract is set of cavities above vocal folds up to the mouth. It functions as an acoustic filter that has an effect on the sound spectrum. At the end the sound is radiated to the surrounding air at the lips, and nostrils. Laboratory analysis of the vibrating vocal folds during phonation (the process of generating the voiced excitation) is challenging because the larynx is not easily accessible. However, one of non-invasive method such as electroglottography (EGG), can be used to determine the glottal signal flow. The resulting electroglottogram (electroglottographic signal, glottal signal) yields useful information for modeling, in addition to the easy to attain speech signal [1]. Unvoiced speech is being produced by human voice production mechanism as well, but vocal folds do not vibrate. The air flow from lungs does not convert to train of pulses and finally the air at the lips, and nostrils take a noisily form. Unvoiced speech does not have periodic structure, and modeling using it is difficult.

Different models have been applied to speech production system modeling. If only the speech signal have been accessible during experiments, the simplest model as Linear/nonlinear Auto Regressive (AR, NNAR), and AR Moving Average (ARMA, NNARMA) been appropriate.

D. Protic is with the Institute of Applied Mathematics and Electronics, Kneza Milosa 37, 11000 Belgrade, Serbia and Montenegro, E-mail: adanijela@ptt.yu

M. Milosavljevic is with the Faculty of Electrical Engineering, University of Belgrade, Bul. Kralja Aleksandra 73, 11000 Belgrade, and also with the Singidunum University, Danijelova 32, 11000 Belgrade, Serbia and Montenegro, E-mail: mmilan@etf.bg.ac.yu

On the other hand, the glottal signal have given an opportunity to include extra inputs in modeling scheme, thus ARX, ARMAX, NNARX, and NNARMAX could been used [2], [3].

Linear models have often been utilized if the structural simplicity of the model had to be an alternative to training time, minimal processing error, etc. Nonlinearity almost any time generated accurate, but complex models.

This paper presents nonlinear ARX model of speech signal generating system. Voiced speech, and appropriate glottal signal were taken to train feedforward NNs. Diversity of signals used in modeling initiated diversity of models, processing errors, training time, etc. This paper offers an example of the error altering, conditional on speech and glottal signal size. Chapter II provides an introduction to models, Chapter III describes used signals, Chapter IV gives experiment details, and Chapter V draws conclusions.

II. MODELS

Linear ARX model is defined as in (1):

$$y(n) + a_1 y(n-1) + ... + a_{n_a} y(n-n_a) = ...$$
$$... = b_1 u(n-1) + ... + b_{n_b} u(n-n_b) + e(n) \quad . \quad (1)$$

$y(n)$, $u(n)$, and $e(n)$ describe speech, glottal signal, and the processing error n-th samples, in that order. a_i, $i = 1,...,n_a$, and b_j, $j = 1,...,n_b$, are n_a (AR), and n_b (X) parameter dimension, respectively. Thus, the number of inputs, m is amount of n_a and n_b, (2):

$$m = n_a + n_b \quad . \quad (2)$$

Nonlinear input-output (I/O) function is given by (3):

$$g(y(n), \delta(n), n) = \varepsilon(n) \quad . \quad (3)$$

$\delta^T(n)$ defines an input vector, $\varepsilon(n)$ an error, and $g(...)$ is certain nonlinear function. One hidden layer feedforward structure has frequently been used in NNARX modeling. It could be used for any $g(...)$ function modeling. Requirement for satisfactory modeling is primarily free choice of NN structure, and size, after that neuron transfer function type, and so on. [4]-[6]. Feedforward NN selected for this particular experiment, as given in Fig. 1., have contained m inputs, three hidden layer, and one output

1-4244-0116-X/06/$20.00 ©2006 IEEE

neuron, plus biases for both layers. Output y_i of described structure is written as in (4):

$$y_i = \tanh\left(\sum_{j=1}^{3} W_{ij} \tanh\left(\sum_{l=1}^{n_a} w_{jl}y_{i-l} + \sum_{l=n_a+1}^{l=n_a+n_b} w_{jl}u_{i-l} + w_0\right) + W_0\right) \quad (4)$$

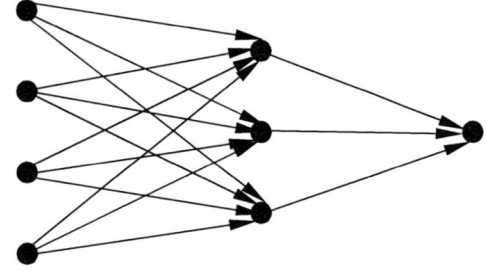

Fig. 1. Feedforward neural network.

w, and **W** are parameter matrix; w_0, and W_0 are biases ; y_{i-l}, and u_{i-l} describe inputs, j indicates hidden to output layer neurons, l marks input to hidden layer neuron, i indicate that i-th output that has been in procedure. The number of inputs used was $m = 18$, since the best results have been attached using: $n_a = 14$, $n_b = 4$, [7], [8]. Each neuron I/O function has been tangent-hyperbolic, limiting neuron output to [-1, 1].

III. SIGNALS

Vowels are periodic, time continual signals produced as a result of the air flowing from lungs to the end of speech production mechanism up to the surrounding air at lips, and nostrils. Each vowel has its exclusive characteristic. Periodic pulses during time indicate opening and closing phase of phonation, meaning vibration of vocal folds. Yet, fundamental frequency, rate of opening and closing (f_0), differs among speakers. The average f_0 is approximately 120Hz for men, 200Hz for women, and even higher for children. But f_0 range could take value from bellow 100Hz for men, to soprano's singing at around 1300Hz. Vocal tract shape determines resonances or formants, thus each vowel has its typical spectral profile. The frequency of produced speech is in the region of [0, 4000]Hz. Because of that, digital data for this experiments have been created through sampling analog signals at $f_s = 10$kHz (Niquist, Shannon) [5].

Signals for our experiments have been taken from two sentences that both gender subjects spoke. /a/ vowels have been spoken in /one/, /five/, and /nine/. Therefore, three /a/ vowels have been used to train, and test models [5]. The first digital /a/ signal has included 1200 samples. The second one have consisted 1400 samples. The third signal has contained 1600 samples. Speech and the glottal signals have been taken to operation at the same time. Limitation to [-1, 1] have been done to attain equal starting

conditions. The regularization has been applied for two reasons: primary, signals have had significantly different amplitudes. Secondary, tangent-hyperbolic transfer functions have limited output to [-1, 1]. To regularize signals, speech signal and glottal signal had been merged at first. The particular signal has been regularized. After all the separation was done. This has been done for each /a/ vowel. Regularized speech and glottal signal are given in Fig.2.

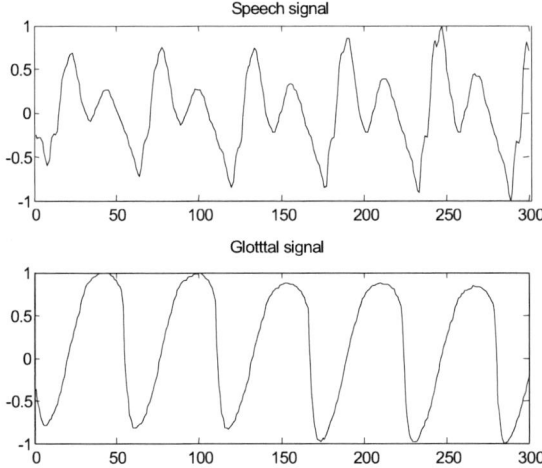

Fig.2. Regularized speech and glottal signal

Each regularized signal had been separated into two equal size portions. First portion was taken for training, second was used for testing. Howewer, fundamental frequency pointed on the fact that the same signal length for female and male speaker did not hold the same information of speech signal generating mechanism. That has been the reason we have made other training and test signals, $3*1/f_0$ period long, and f_s sampled, without paying attention on number of samples within finale signal. The average size of one speech period is approximately 60 samples for female and 115 samples for male speaker. Consequentially, the signal used for the second experiment length has been considerably lesser then for the first experiment, concerning woman. For male speaker that has not been the fact.

IV. EXPERIMENTS

NNARX modeling of speech production system has provided background for number of experiments using speech and glottal signal. An idea has been an improvement of linear model via nonlinearly insertion. Feedforvard NNs ought to have been applied in that circumstance, for its significant ability of structure changing. It also has assumed precise modeling, pruning, minoring errors, etc. On the other hand, neural structures have lack proper response if modeling situation altered in certain manner.

Two experiments have been realized on regularized voice speech. One with 400, 600, and 700 samples included, and the other has been realized with $3*1/f_0$ periods. NNARX models have been trained using training signals, and GDA [2]-[5]. Subsequently models have been tested using test signals, for both speakers. Normalized Sum Squared Error (*NSSE*) has been used as comparison criteria (5):

$$NSSE = \frac{SSE}{2*n} = \frac{1}{2*n}\sum_{i=1}^{n} e_i^2 \quad, \qquad (5)$$

n is error string size, e_i is error defined as i-th speech sample y_i, and suitable NN's output $\hat{y}_i$ difference (6):

$$e_i = y_i - \overset{\wedge}{y_i} \quad. \qquad (6)$$

Primary, the network structure has included: $m=18$ inputs, three neurons in hidden layer, one output neuron, plus biases for each layer. Training has been accomplished with known training signals. Testing has been done using trained network, and unknown test signals. Final networks were pruned taking OBS criterion [4], [5] to decrease structures, and errors.

Experiments have been included two sub-experiments. In the first sub-experiment (*a-Experiment*), each training signal has been used to train a novel model. Parameter values have been randomly used, at the beginning of the training process. Three models have been modeled. Second modeling (*b-Experiment*) started as *a-Experiment*, except first final model parameters have been taken to be starting parameter values for second modeling, due to one, final model. Testing has been done for sub-experiments, each speaker, and all signals.

A. Experiment

NNARX models for *a-Experiment* have been GDA trained with MATLAB 7.0 using its standard functions [5], [6]. Random starting parameter values have been taken, and first portion of speech signals, as well. Test has been done on trained models, using second portion of signals. *NSSE* has been used to compare results, for both speakers. OBS pruning has been done to obtain NN structure smaller, and make error lesser. Table I gives results for $NSSE_{\text{test}}$, and $NSSE_{\text{pr}}$. Terms indicate *NSSE* for test signals before and after pruning NNARX model, for both speakers. Str 1 – 3 denotes test signals (400, 600, and 700 samples, in that order). Pruned network size is given in round parentheses. Table II gives results for $NSSE_{\text{test}}$, and $NSSE_{\text{pr}}$, but at this point Str 1 – 3 denotes test signals with $3*1/f_0$ period.

B. Experiment

NNARX models for *b-Experiment* have also been trained with standard MATLAB 7.0 functions. First model

has been GDA trained, using random starting parameter values, for first modeling, final parameters have been taken as starting parameters for second modeling, toward the final model. First portion of signals has been taken for training one model. Test has been done using all test signals. OBS pruning has been applied to obtain better results. Table III gives results for $NSSE_{\text{test}}$, and $NSSE_{\text{pr}}$ for both speakers, and test signals as in Table I.

TABLE I
$NSSE_{TEST}$ AND $NSSE_{PR}$ FOR A-EXPERIMENT (400, 600, AND 700 SAMPLES)

Str	Female speaker		Male Speaker	
	$NSSE_{\text{test}}$	$NSSE_{\text{pr}}$	$NSSE_{\text{test}}$	$NSSE_{\text{pr}}$
1	0.0076	0.0055(16)	0.0004	0.0004(13)
2	0.0014	0.0011(30)	0.0002	0.0002(18)
3	0.0005	0.0004(33)	0.0002	0.0002(28)

TABLE I I
$NSSE_{TEST}$ AND $NSSE_{PR}$ FOR A-EXPERIMENT ($3*1/f_0$)

Str	Female speaker		Male Speaker	
	$NSSE_{\text{test}}$	$NSSE_{\text{pr}}$	$NSSE_{\text{test}}$	$NSSE_{\text{pr}}$
1	0.0155	0.0049(16)	0.0004	0.0003(15)
2	0.0016	0.0016(37)	0.0002	0.0001(19)
3	0.0013	0.0013(36)	0.0004	0.0004(29)

TABLE III
$NSSE_{TEST}$ AND $NSSE_{PR}$ FOR B-EXPERIMENT (400, 600, AND 700 SAMPLES)

Str	Female speaker		Male Speaker	
	$NSSE_{\text{test}}$	$NSSE_{\text{pr}}$	$NSSE_{\text{test}}$	$NSSE_{\text{pr}}$
1	0.0040	0.0061(17)	0.0054	0.0005(13)
2	0.0049	0.0024 (8)	0.0007	0.0002(21)
3	0.0007	0.0009(24)	0.0002	0.0004(12)

Table IV gives results for $NSSE_{\text{test}}$, and $NSSE_{\text{pr}}$ for both speakers, and test signals as in Table II.

TABLE IV
$NSSE_{TEST}$ AND $NSSE_{PR}$ FOR B-EXPERIMENT ($3*1/f_0$)

Str	Female speaker		Male Speaker	
	$NSSE_{\text{test}}$	$NSSE_{\text{pr}}$	$NSSE_{\text{test}}$	$NSSE_{\text{pr}}$
1	0.0190	0.0087(27)	0.0009	0.0002(32)
2	0.0173	0.0072(13)	0.0007	0.0002(28)
3	0.0020	0.0017(10)	0.0002	0.0006(34)

As it can be noticed from Table I, $NSSE_{\text{test}}$ for female speaker has been much larger, than for male speaker, for all models. Reduction of parameters has made some corrections in error size, but results have not been extensively improved. As for results given in Table II, larger $NSSE_{test}$ for female speaker, have indicated lesser signal size. That has not been the fact for male speaker, hence the signal size, for both experiments has almost had the same value.

Table III denotes a fact that more information has been included in training process. Final model has held data concerning previous modeling. Given results have almost been the same for all speakers. Pruning has made structures smaller; errors have approximately had same values. Training with $3*1/f_0$ period of time, have given $NSSE_{test}$ and $NSSE_{pr}$ results for b-Experiment (it is represented in Table IV), as for a-$Experiment$ (results from Table II). Pruning for second part of b-Experiment, on the contrary, have made considerable improvements in structure size, minoring structure significantly.

V CONCLUSION

Feedforward neural networks used for nonlinear ARX modeling have been widely applied to many areas. Classification, filtering, modeling, prediction, control, hardware implementation, etc have been used for medical purposes, digital communications, prediction various nonlinear time series, such as annual sunspots, different economic data, or have been applied to control dynamic systems as to design feedback feedforward controllers for robotic applications. Also an optical disk implementation was reported to apply the network to the handwritten classification task, also neurocomputers were designed using VLSI circuits, and showed how neural chip can provide spline smoothing of images.

Classification and feature extraction of speech signals has been the most applied application of neural networks. Primarily, neural networks have been used to classify spoken vowels. Superior performance has been obtained with neural networks compared to other known methods.

This paper offers an option of known speech processing methods improvement, suggesting that feedforward neural network should be used after classification, prediction modeling, etc. The goal is to make error valueless. Two realizable networks may be applied, one having simple information of speaker, and if possible selected structure. The other should take more data, and signals for modeling process. As it has been confirmed during experiments, and given in Tables I-IV it would be best to use, if the speaker gender is known, larger amount of samples for female speaker. Results that has been obtained using male speech, pointed to more precise modeling with speech having lesser f_0. Stability has been obvious during experiments on male vowel. Pruning has given impressive results. The neural network structure has been sizeable reduced, for each speaker, all experiments, and all signals. Precise modeling, small network structure, error under 1% may be good reasons for NNARX usage.

REFERENCES

[1] H. Pulakka, "Analysis of Human Voice Production , Using Inverse Filtering, High-Speed Imaging, and Electroglottography", in *Master's Thesis, Helsinki University of Technology, Dep. Of Computer Science and Engineering*, Espoo, 2005.

[2] D. Protic, M. Milosavljevic, "Generalization Properties of Different Classes Linear and Nonlinear Speech Signal Models", in *INFOFEST Proc.*, Budva, 2005, pp 248-257.

[3] D. Protic, M. Milosavljevic,"Speech Signal Production Improvement Using Glottal Signal", in *Proc. of ETRAN 2005, III*, Budva, 2005, pp 226-229.

[4] L. Ljung, *System Identification: Theory for the User*, Prentice Hall Inc., 1987.

[5] D.G. Childers, *Speech Processing and Synthesis Toolbox*, Wiley, 2002.

[6] C. Svarer, *Neural Networks for Signal Processing*, Technical University of Denmark, 1995.

[7] D. Arsenijevic, M. Milosavljevic, "Final Prediction Error of Serbian Vowel Neural Network Model", in *TELFOR 1998 Proc.*, Banja Vrucica, Teslic, 1998.

[8] H. Akaike, *Fitting Autoregressive Models for Prediction*, Ann. Ins. Stat. Mat., 1969.

The Electronic and Program Control System of X-Ray Ion Mobility Spectrometer

V. S. Pershenkov, A. U. Razvalyaev, A. D. Tremasov

Abstract – Ion Mobility Spectrometer with X-ray ionization source (X-IMS) is examined in this paper. The main goal of work is presentation of hardware and software systems for operation control of spectrometer. The electronic system allows automatically controlling such spectrometer parameters as electric field in drift tube, temperature of drift tube, sample and drift gas flow rates and so on. The original software system means for processing and accumulation of output ion spectra and for control of different spectrometer units operation. The development a simple and effective mathematical program for measurement and processing data from the IMS-spectrometry is the main task of this part of work. It is necessary to perform program on base of Windows technology as graphical application, which allow analyzing, imaging and saving measuring data. The experimental explosive ion spectra have been obtained.

I. INTRODUCTION

Ion mobility Spectrometry (IMS) is capable of separating ionic species at atmosphere pressure and have applications in detecting explosive, narcotics and chemical warfare agents. This method can be also used for pollution monitoring of environment and early illness diagnostic.

The ion mobility spectroscopy method is based on the fact that ionized molecule of any chemical substance has a certain mobility in weak or strong electric fields [1]. The identification of a given molecule kind is fulfilled by the measurement of its mobility. In case of using weak electric field, the realization of IMS method requires molecule ionization, its drift in constant electric field through a fixed length and a registration of the time of ion drift to detection unit, where ion current is measured. Mobility is easily calculated from condition that molecule drift velocity in weak electric field equals mobility and electric field product.

The intense interest to IMS is connected with the some worth features of this method:

- registration of detected molecules in usual air environment at atmospheric pressure;
- high sensitivity;

- small size and weight of IMS devices;
- low electric power consumption, that is a possibility to work in autonomic regime;
- simplicity of component parts and low cost;
- possibility of substance detection in vapor, liquid and solid phases.

There are several types of commerce detection devices based on this technology manufactured by Smiths Detection, General Electric Ion Track Inc., Bruker-Saxonia Analytic and so on. The standard ionization source in IMS today is radioactive nickel or capsulated tritium. These sources are stable and reliable and allows for operation without power. But some limitation during work with radioactive material and radioactive waste utilization problem stimulate a development of non radioactive ionization source.

There are the several non radioactive ionization methods such as surface ionization, UV (10,6 eV) photoionization, laser (266 nm) multiphoton excitation, corona discharge, electric field ionization and so on. In the end of 90-th the first experiments with X-ray ionization source for IMS detector were fulfilled in Bruker-Saxonia Analytic (Germany) and Chromdet (Russia). So far there is no information about status of this study.

II. X-IMS SPECTROMETER

In present work the prototype of IMS detector with X-ray ionization source (X-IMS detector) was developed. X-source is small glass lamp with diameter 1.8 cm and height 4 cm developed by Chromdet (Russia). Applied high voltage is 4 kV, filament power is 1.0V×0.7A, maximum anode current is 200 µA. Fig. 1 illustrates the schematic drawing of X-IMS laboratory bench.

It includes X-ray source, ionization and chemical reaction region, shutter grid, drift tube, collector with aperture grid. The ionization region and drift tube consists of 14 stainless rings (1.0 cm width), separated by teflon rings (1.0 mm width). The inner diameter of rings is 3 cm. The drift electric field is 270 V/cm. The Faraday detector disk is connected with current-voltage converter with amplification 33 V/pA.

The electronic system allows automatically controlling such spectrometer parameters as electric field in drift tube, temperature of drift tube, sample and drift gas flow rates and so on.

V. S. Pershenkov is with Department of Microelectronics, Moscow Engineering Physics Institute, Kashirskoye shosse, 31, Moscow, 115409, Russia, E-mail: pershenkov@d408.micro.mephi.ru.
A. U. Razvalyaev is a post-graduate student of Microelectronics Department of Moscow Engineering Physics Institute.
A. D. Tremasov is a post-graduate student of Microelectronics Department of Moscow Engineering Physics Institute.

1-4244-0116-X/06/$20.00 ©2006 IEEE

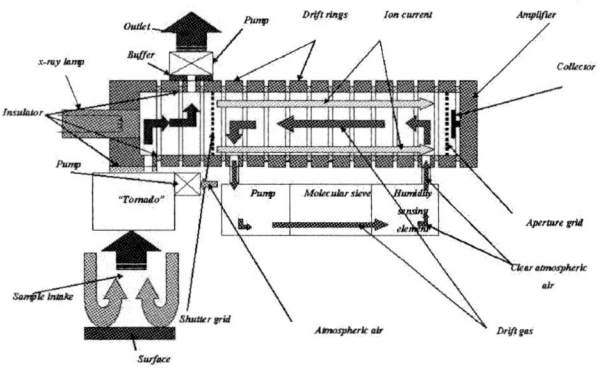

Fig.1. Schematic drawing of X-IMS laboratory bench.

III. MATHEMATICAL SOFTWARE

The output ion spectra are analyzed by homemade processing program (C++). The processing program averages out output data, smoothes noise and seeks Gauss product signal. The program picks out two Gauss peaks located relatively each other on half peak width. The general characteristics of peaks are determined: half width, peak center and its amplitude. The ratio of each peak to total spectrum square is calculated.

The development a simple and effective mathematical program for measurement and processing data from the IMS-spectrometry is the main task of this part of work. It is necessary to perform program on base of Windows technology as graphical application, which allow analyzing, imaging and saving measuring data. Operating system Windows allows using advantages for creating graphical measuring system. Effective multitasking, graphic user interface, standard input/output device drivers allow to easy create measuring Windows-application.

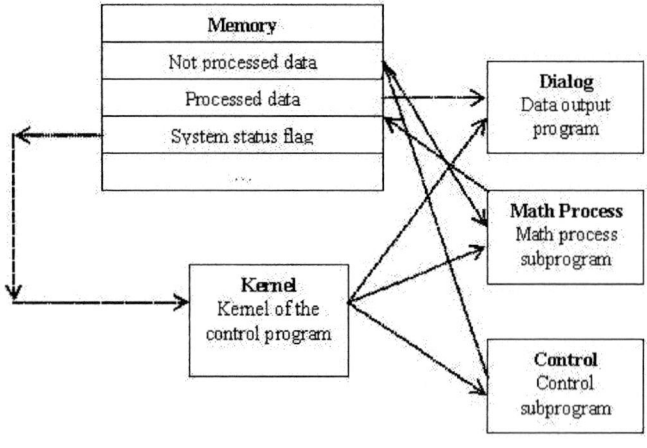

Fig.2. Structure of the control program.

Control program is based on multitasking operating system MS Windows 2000. Multitasking allows realizing ef-fective control of devices array. Using current architecture provides synchronous system control, scanning, processing and spectra output from the device.

Program consists of several single subprograms (fig. 2), which can be easy modernized and communicated in different program variants according to system configuration and specific tasks. Program can be easy modified for creating distributed (by several computers) control system. Also program, which can carry out additional studies in area of ion mobility spectrometry, may be gathered from developed modules. Due to using multitasking system, each subprogram works independently from another subprogram, i.e. each task is provided by separate subprogram.

IV. EXPERIMENTAL RESULTS

Fig.3 shows the ion mobility spectra of reactant ions and RDX molecules in negative mode operation. The sensitivity of detection is estimated on the level of 10 ppt.

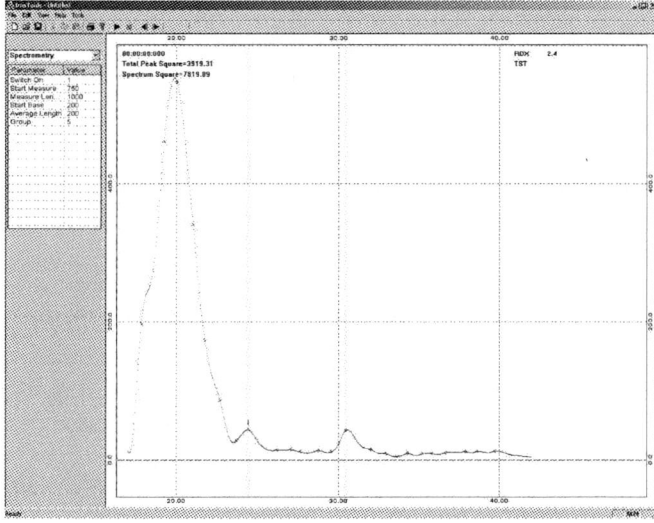

Fig.3. Mobility spectra from air reactant ions and RDX explosive.

V. CONCLUSION

The full drop of chemical potential is negligible under strong inversion compared to electrostatic potential but it becomes very important in saturation mode.

The developed IMS detector can be used as portable handheld device for on-site analysis of ultra- small concentrations of explosives, drugs and chemical warfare agents in ambient air environment. Small dimensions and low power consumption open a possibility its application for detection of internal living quarters pollution, unhealthy desorption from lacquer coating and materials used for furniture and household good manufacturing. It also can be used for analysis of closed environment in inhabited space stations.

A promising future application for X-IMS detector is medicine. Assembling highly specialized small analyzer arrays opens a possibility to carry out noninvasive diagnostics of illnesses due to testing of human natural excreta such as perspiration, breath, saliva, tear, urine and so on. The high sensitivity of technique allows to register very early stage of illnesses.

The X-IMS detector can have commercial prosperity, because the proposed development can be interested for numerous firms which have business in the field of trace detection of explosives, drugs, chemical warfare agents, pollution of the environment, early diagnostic of illness, biological investigations and so. Besides the relatively low cost of detector may attract mass customers.

REFERENCES

[1] T.W. Carr, editor. Plasma Chromatography. Plenum, New York, 1984.

2006 25th International Conference on Microelectronics

The Innovative Method for Determining Characteristics of Over-Voltage Protection Elements

P. Osmokrović, B. Lončar, S. J. Stanković, and A. Vasić

Abstract – This paper presents an innovative method for efficient characterization of relevant characteristics of non-linear over-voltage protection elements in low-voltage applications. Standard measuring equipment is modified to enable an efficient and repeatable experimental procedure to investigate characteristics of over-voltage diodes and varistors: volt-ampere characteristic, volt-ohm characteristic, coefficient of non-linearity, and finally, the breakdown voltage. Additionally, an innovative aging estimate algorithm was used. Furthermore, the pseudo-empiric method using the "area law" is used to determine an impulse characteristic of a gas filled surge arrester (GFSA). Suggested experimental procedure offers higher measuring accuracy and repeatability because internal temperature of non-linear elements is virtually unchanged during all experiments. All experimental results are treated statistically to prove high repeatability of suggested approach. Comparison of results obtained by suggested experimental method and "classical" approach, shows minimal discrepancies.

I. INTRODUCTION

The continuous miniaturization trend in electronic industry inherently causes increased sensitivity to the over-voltages and effort in protecting delicate electronic components. The latest generations of integrated circuits have internal layers of thickness of just several nanometers, thus relatively small over-voltages may cause damage or even complete destruction of sensitive electronic components. Additionally, in a microprocessor based electronic systems over-voltages can cause resetting, temporary disturbance, or internal data corruption. Therefore, sensitive electronic components should be adequately protected against harmful over-voltages (either, as a part of initial electronic component design, or using an add-on over-voltage protection). By placing adequate over-voltage suppressor at the entrance of an electronic device, the best level of protection is typically achieved [1].

P. Osmokrović is with the Faculty of Electrical Engineering, University of Belgrade, Bulevar Kralja Aleksandra 73, 11120 Belgrade, Serbia & Montenegro, E-mail: opredrag@verat.net

B. Lončar is with the Faculty of Technology and Metallurgy, University of Belgrade, Karnegijeva 4, 11120 Belgrade, Serbia & Montenegro, E-mail: bloncar@eunet.yu

S. J. Stanković is with the Institute of Nuclear Science „Vinča", P.O. Box 522, 11001 Belgrade, Serbia & Montenegro, E-mail: srbas@vin.bg.ac.yu

A. Vasić is with the Faculty of Mechanical Engineering, University of Belgrade, Kraljice Marije 16, 11120 Belgrade, Serbia & Montenegro, E-mail: avasic@mas.bg.ac.yu

The over-voltage protection elements can be divided into two groups: non-linear (gas filled surge arresters, varistors and over-voltage diodes), and linear (various types of electrical filters). In practice, various combinations of both linear and non-linear components are used to achieve optimal over-voltage protection. The optimal over-voltage solution [2], [3], provides an economical solution to the following requirements:

1. It should limit input voltage to a non-dangerous level,
2. It should be well below the current limit for a chosen protective component,
3. It has to be fast enough (response time should also be adequate),
4. It must be located close enough (particularly important for the "fast transients").

Focus of presented work is development of methodology oriented toward efficient characterization of non-linear components only.

II. EXPERIMENTAL PROCEDURE AND EXPERIMENTAL EQUIPMENT

Using single current pulse following characteristic could be recorded [4]:
1. Volt-ampere characteristic;
2. Volt-ohm characteristic;
3. Non-linear coefficient:

$$\alpha = \left(log(I_2 / I_1)\right)/\left(log(U_2 / U_1)\right) \qquad (1)$$

4. Breakdown voltage;
5. Dissipating energy.

The suggested method is based on measurement a voltage response triggered by a current pulse on the tested element. Starting value of testing current pulse is I^{max}. Gradually, testing current pulse is decreasing to the value of 0.1 I^{max}. Measurement is always performed at the falling edge of the pulse. The volt-ohm characteristic can be easily calculated from the volt-ampere curve. Using the linear regression (least-squares minimal error method), a formulae defining the relationship between impulse current and voltage is obtained. Dissipation energy of a non-linear element can be calculated by formula:

$$W = \sum_{i=1}^{n} U_i \cdot I_i \cdot \Delta t \qquad (2)$$

U_i - value of i-th sample of voltage

I_i - value of i-th sample of current

Δt - sampling interval of a digital scope.

The results obtained by the suggested method in this paper indicate that vital characteristics of non-linear elements could be calculated with a much higher accuracy, than the results obtained by the procedure that uses a common approach [5].

The advantages of suggested approach are:

1. Measuring procedure is more efficient;

2. Vital characteristics of non-linear protective components could be obtained with the higher accuracy because influence of temperature variation is eliminated (This is particularly important for over-voltage diodes and varistors [5]).

A statistical value of DC breakdown voltage is examined for GFSA device. For determining a single group of random variable "DC breakdown voltage" 20 series of 50 measurements should be done (1000 activation). In order to stabilize GFSA protective characteristics any of them is exposed to 25 initial (conditioning) breakdowns. A 30 second pause between successive measurements (breakdowns) is exercised.

Experimental procedure is completely automated and should be adequately protected against electromagnetic noises. Measuring system is designed to be very flexible - enables testing of all over-voltage protection elements using only software modifications. PC software controls operation of a Digital to Analog converter (D/A), which generates all wave shapes required for testing of GFSA (internal communication between voltage and current source). Additionally, same PC controls other instruments using HP-IB (IEEE488) interface and protocol.

A voltage-current sources with very stable characteristics are highly recommended in order to conduct accurate measurements. Additionally, chosen generator should generate specific current pulses and have appropriate voltage-current characteristics:

1. Can generate a double-exponential current pulse (T_1=8μs, T_2=20 μs) of the maximal amplitude of 16A for varistors, and 13A for over-voltage diodes;

2. Can generate a double-exponential voltage pulse (T_1=1.2μs, T_2=50μs, U_{max}=320V, 480V, 640V). The DC voltage source should have rate of voltage rise of 100V/s.

Block diagram of measuring system is shown on Fig.1. Fig. 2 depicts a block diagram of current-voltage source developed for characterization of the over-voltage elements.

III. AGING ESTIMATE ALGORITHM

Typically, over-voltage protection elements show signs of aging along their service life. Change of characteristics of an over-voltage protection component is a function of the number of activations and energy dissipation (amplitude of current pulse and duration of the

transient). The "aging effects" appear to be a consequence of design and material imperfections (irreversible changes – deterioration of characteristics). Specific damages at the beginning of experimental sequence are not noticeable; there is no visible degradation of protective characteristics. However, cumulative effects of stress gradually degrade protective characteristics leading to the complete failure of component.

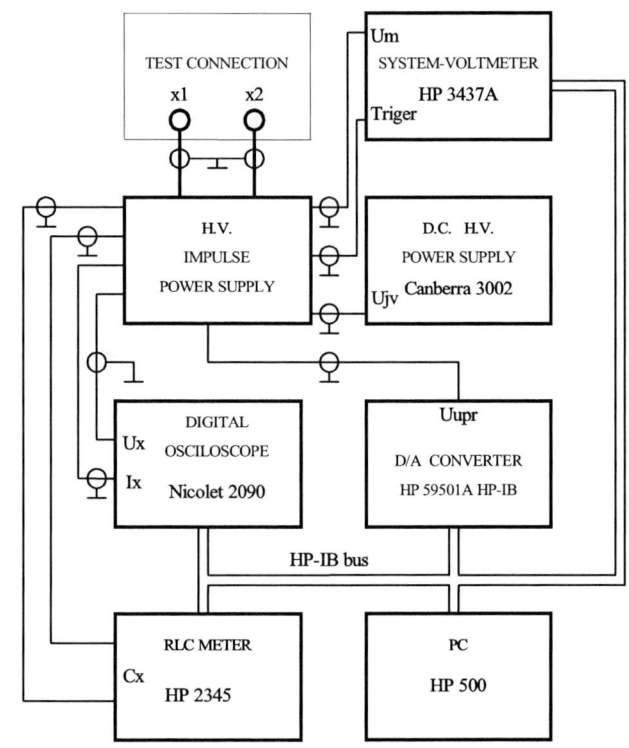

Fig.1 Block scheme of the measuring system.

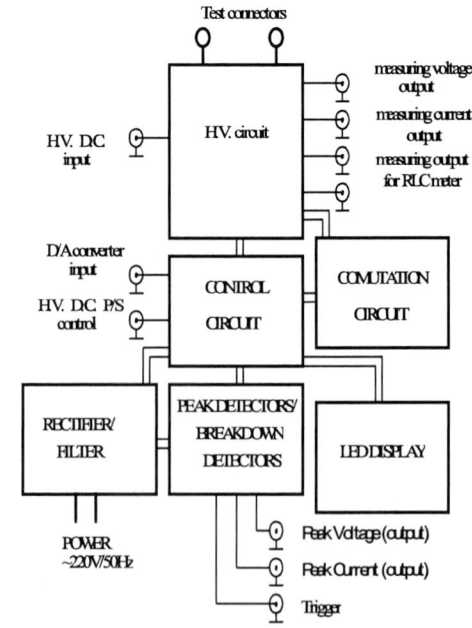

Fig. 2 Complete scheme of voltage/current source.

The service life of a protection device predominantly depends on [6]: 1) device design (internal structure); 2) rated current; 3) rated voltage (depends on protection element); 4) service conditions (over-voltage pulse frequency, amplitude, shape, and duration).

To quantitatively describe the aging process of an over-voltage protection element, it is necessary to understand reversible processes influencing its characteristics. A test should confirm whether the observed variable belongs to the same random variable after predefined number of activations. Mentioned test should provide clues that protective element suffer some irreversible damages (degradation of characteristics due to repeated activations). The use of well-known "F test" is generally recommended for this purpose. However, use of "F test" requires that a tested random variable should belong to "Gauss distribution". Statistical behavior of non-linear over-voltage protection elements belongs to the "Weibull distribution". Since the "U test" [7], [8]. is an adequate statistical "tool" for all types of statistical distributions, it should be used for quantitative testing of aging process [9].

Employment of the U-test in determining a degree of irreversibility of characteristics (caused by aging) shows two crucial benefits over commonly used tests [10], [11]:
1. The variable to be tested does not need to follow Gaussian, or any other statistical distribution;
2. The degree of irreversibility could be quantified (as continual value).
Notes:
a) The F-test is offering same benefits, but it is restricted to Gaussian distribution only.
b) The statistical validity of hypothesis is typically defined by a preset value (borderline): changes are either reversible or irreversible.
c) Using the F-test for GFSA is not recommended because breakdown voltage of GFSA follows Weibull's distribution [6]. If F-test is used anyway in analysis of breakdown voltages of GFSA, that approach leads to wrong results and wrong conclusions.

IV. ALGORITHM FOR PULSE SHAPE CHARACTERISTIC CALCULATION

The pulse shape (volt-second) characteristic presents the breakdown voltage of the gas insulated electrode configuration as the function of the duration of the applied voltage pulse.

It is known fact that a decrease of the pulse rise-time increases breakdown voltage. Should we include statistical dissipation of the pulse breakdown voltage, the pulse characteristic will spread on the one side in voltage-time plane limited by predetermined quantiles of the statistical distribution of the random variable "breakdown voltage". All pairs of points breakdown voltage-breakdown time of

the specific electrode configuration with gas insulation can be found in this area with probability determined according to calculated boundary quantiles.

Exact experimental determination of the pulse characteristic would demand a large number of activations using pulse voltages of different shape. However, application of the "Area law" [12] enables determination of the pulse shape characteristic based on one series of measurements (using the single shape of the pulse voltage).

The pulse shape characteristics can be calculated using following two steps (semi empirical approach):
Step 1: Conduct a series of the DC breakdown voltage measurements (at least 20).
Step 2: Conduct a series of breakdowns using defined and stable voltage pulses (at least 50 breakdowns)

Statistical processing of results, (including determination of the corresponding distribution function), should be done at the end. From the distribution function, the quantiles could be determined (desired boundaries of pulse shape characteristics Ux and Uy - typically values $x=0.1\%$ and $y=99.9\%$ are used). Having the value of quantiles and mean value of DC breakdown voltage Us, the system of equations can be solved as:

$$
\begin{aligned}
u(t) &= U_s, & t &= t_1 \\
u(t) &= U_x, & t &= t_{ax} \\
u(t) &= U_y, & t &= t_{ay}
\end{aligned}
\tag{3}
$$

Values t_1, t_{ax} and t_{ay} are used in calculation of areas Px and Py by applying the "area law".

$$
\begin{aligned}
P_x &= \int_{t_1}^{t_1+t_{ax}} \left[u(t) - U_s\right] dt = const \\
P_y &= \int_{t_1}^{t_1+t_{ay}} \left[u(t) - U_s\right] dt = const
\end{aligned}
\tag{4}
$$

After calculating Px and Py it is possible to determine (using "area law"), x-th and y-th quantiles of the "pulse breakdown voltage" (random variable for any form of pulse voltage $u(t)$). If an arbitrary form of pulse voltage (in a considered time interval) is taken as a parameter, it is possible to determine the x-th and y-th pulse shape characteristics.

The suggested approach has two distinctive advantages over commonly used procedures:
1. Quantiles are calculated from the real statistical distribution of random variable "breakdown voltage" (not from a Gaussian distribution [13], [14];
2. Calculation of the pulse characteristic is based on the real testing pulse shape used in all experiments. Advantages are twofold:

a) Accuracy of the suggested procedure is higher compared to the linear approximation [15], [16];
b) Lower amplitudes of testing pulses could be used. In fact, amplitude of testing pulse must not be higher than the expected value of breakdown voltage

V. CONCLUSION

This paper presents an innovative approach of obtaining relevant protective characteristics of non-linear over-voltage protection elements. The suggested method has many advantages over commonly used DC voltage testing method. The paper presents how a combination of specific statistical and experimental procedures (hypothesis validation test - U-test combined with the single current pulse) enables an efficient method for determination of volt-ampere and volt-ohm characteristics, coefficient of non-linearity α, and the energy dissipated on a particular protective component. Additionally, aging process quantification (for a wide variety of protective components), shows a very promising potential for future research in this field.

The employment of the "U-test" (hypothesis validation test) when compared to the commonly used "F test", has a clear advantage since U-test doesn't require that tested values belong to Gaussian distribution. It is needless to say that characteristics of non-linear protection elements belonging to the Gaussian distribution are very rare. Furthermore, the suggested method does not have errors associated with typical assumptions: 1) that pulse voltage is linear; and 2) that the variable "breakdown voltage" belongs to the Gaussian distribution.

The experimental procedure is fully automated (control, sequencing, and data acquisition), thus providing identical testing conditions for all miniature over-voltage protective components. It should also be stressed that knowing accurate protective characteristics is essential since insulation coordination in low voltage systems depends greatly on the accurate protective characteristics of each protective component.

ACKNOWLEDGEMENT

The Ministry of Science and Environmental Protection of Republic of Serbia supported this work, under contracts 2006 and 2016.

REFERENCES

[1] Ž. Markov, *Over-voltage Protection in Electronics and Telecommunications,* Belgrade: Technical Book, 1983 (in Serbo-Croation).

[2] Ž. Markov, "The Comparison of Modern Over-voltage Protection Components", *Electrotechnics*, vol. 10, pp. 961-963, 1987 (in Serbo-Croation).

[3] M. Clark, *Surge Live of Transient Voltage Suppressor*, C. Marshall Space Flight Center, Contract No. Nas. 8-30-811, 1986.

[4] F. D. Martzloff, "Matching Surge Protective Devices to their Environment," *IEEE Trans. Industrial Application*, vol. 1A-21, pp. 99-106, 1985.

[5] J. Foster, "Break-over Diodes for Transient Suppression", *Electronic Engineering*, vol. 59, pp. 35-39, 1987.

[6] P. Osmokrović, I. Krivokapić, D. Matijašević, and N. Kartalović, "Stability of the Gas Filled Surge Arresters Characteristics under Service Conditions," *IEEE Trans. Power Delivery*, vol. 11, pp. 260-266, 1996.

[7] P. Osmokrović, B. Lončar, and S. Stanković, "Aging of the Over-Voltage Protection Elements Caused by Over-voltages", *Microelectronics Reliability*, vol. 42, pp. 1959-1969, 2002.

[8] W. Hauschild, and W. Mosch, *Statistical Techniques for High-Voltage Engineering*, London: Peregrinus, 1992.

[9] P. Osmokrović," The Irreversibility of Dielectric Strength of Vacuum Interrupters after Short Circuit Current Interruption", *IEEE Trans. Power Delivery*, vol. 6, pp. 1073-1081, 1991.

[10] J.P. Crine, "A Molecular Model to Evaluate the Impact of Aging on Space Charges in Polymer Dielectrics", *IEEE Trans. Dielectrics & Electrical Insulation,* vol. 4, pp. 487-495, 1997.

[11] A.C. Gjaerde, "A Phenomenological Aging Model for Combined Thermal and Electrical Stress", *IEEE Trans. Dielectrics & Electrical Insulation* , vol. 4, pp. 674-680, 1997.

[12] B. Lončar, P. Osmokrović, and S. Stanković, "Radioactive Reliability of Gas Filled Surge Arresters", *IEEE Trans. Nuclear Science,* vol. 50, pp. 1725-1731, 2003.

[13] G. Djogo, and P. Osmokrović, "Statistical Properties of Electrical Breakdown in Vacuum", *IEEE Trans. Electrical Insulation*, vol. 24, pp. 949-953, 1989.

[14] P. Osmokrović, and G. Djogo, "Applicability of Simple Expressions for Electrical Breakdown Probability in Vacuum", *IEEE Trans. Electrical Insulation*, vol. 24, pp. 943-948, 1989.

[15] P. Osmokrović, "The Gas Volume Influence on the Pulse Shape Characteristics on the Electrical Breakdown of SF_6 at Small Inter-electrode Gaps", *Journal of Engineering Physics,* vol. 25, pp. 39-50, 1983.

[16] P. Osmokrović, " The Influence of the Electrode Material and the Processing Manner of Active Surfaces on the Pulse Shape Characteristics of the Breakdown of SF_6 at Small Inter-Electrode Gaps", *Journal of Engineering Physics*, vol. 26, pp. 5-10, 1984.

2006 25th International Conference on Microelectronics

Approach to Partially Self-Checking Finite State Machine Design

G. Lj. Djordjevic, T. R. Stankovic, M. K. Stojcev

Abstract - This paper presents a cost-effective technique of partially self-checking finite state machine (FSM) design. The proposed technique is similar to duplication with comparison, wherein duplicated combinational logic (CL) block of the FSM and comparator act as a function checker that detects any erroneous response of the original CL block. However, instead of realizing checker with full error-detection capability, we select a subset of checker's inputs to implement partial, but simplified function checker. Effectiveness of the technique is evaluated on a set of MCNC 91 benchmark sequential circuits.

I. INTRODUCTION

One of the main side-effects of shrinking device sizes, decreasing supply voltages, and high-speed operation of current VLSI IC technologies, is increasing sensitivity to various internal and external noise sources. In this context, protection against temporary faults will be necessary even for applications for which reliability is not a main concern. In order to deal with these problems, the adoption of self-checking approaches as powerful means for concurrent error detection of both temporary and permanent faults can be a viable solution [1-3].

From the reliability point of view, the control unit, typically realized in a form of FSM, is usually the most critical part of the system. Classical approaches for designing self-checking FSMs are based on either duplication, or application of specific error detecting codes. In most cases, these approaches include high hardware cost and high design effort. Several approaches have been taken in the past to design self-checking sequential circuits with lower area overhead in respect to duplication. In [4], a monitoring machine is used in order to monitor the states of the original machine. This technique is effective for delay fault model. However, for the stuck-at fault model, the monitoring machine results in high area overhead. Work presented in [5] uses a control flow checking technique for detection of illegal state transitions, that is based on path signatures. The proposal has low hardware overhead, but this technique has uncertain error detection latency and low fault coverage.

In this paper, we propose a method for designing partially self-checking FSM based on on-line monitoring of FSM operation. We propose and analyze two self-checking

G. Lj. Djordjevic, T. R. Stankovic, M. K. Stojcev are with the Department of Electronics, Faculty of Electronic Engineering, University of Niš, Aleksandra Medvedeva 14, 18000 Niš, Serbia & Montenegro, E-mail: gdjordj@elfak.ni.ac.yu

scheme. The first monitors FSM state transitions, only, while the second additionally takes into account FSM outputs.

II. SYSTEM MODEL

An FSM is defined as 6-tuple $M = (X, Y, S, f, g, s_0)$, where X, Y, and S are finite sets of inputs, outputs, and states, respectively, and s_0 is the initial (reset) state of S. The unique mapping: $f : X \times S \rightarrow S$, and $g : X \times S \rightarrow Y$ defines the so-called *state transition function* and *output function*, respectively. An FSM with $n = |S|$ states can be described by a *state transition graph* (STG) defined by a vertex (state) set $S = \{s_1, ..., s_n\}$ and related directed edge set T representing set of transitions from one state to another: $T = \{(s_i, s_j) \mid \exists x \in X, f(x, s_i) = s_j\}$. Edges of the STG are labeled with corresponding inputs and output values. An example of STG is shown in Fig. 1 (a).

Let us assume that each state of the FSM, having n states, is encoded with c number of bits – what means that the machine is implemented using c number of flip-flops. The flip-flops may generate 2^c states, out of which only n correspond to *valid* states of the FSM. The remaining $2^c - n$ codes are kept unused. Unused state codes are referred as *invalid*, or *unreachable* states of the FSM. During normal, i.e. fault-free operation, the FSM will never be in any of these states. Similarly, we distinguish between valid and invalid transitions. All edges in the STG defined by a state transition function f of the FSM are said to be *valid*. On the other hand, all other possible transitions, either between valid states or between valid and invalid states, are referred as *invalid transitions*.

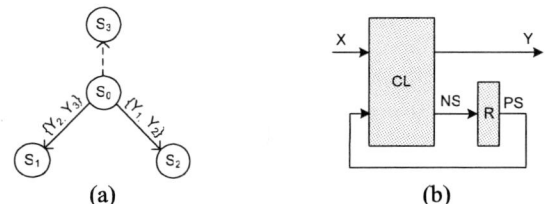

(a)   (b)

Fig. 1. Finite state machine: (a) state transition graph; (b) structural model.

General structure of the FSM is pictured in Fig. 1(b). It is comprised of a combinational logic, CL, and state register, R. The constituent CL accepts primary inputs, X, and present state, *PS*, and generates output signal, Y, and next state code, *NS*. In general, concerning FSM model

1-4244-0116-X/06/$20.00 ©2006 IEEE

given in Fig. 1(b), faults can appear both in CL and R logic. Our efforts, in this paper, are oriented towards designing of a checker block that can detect transient faults in CL, only.

II. PARTIALLY SELF-CHECKING FSM

Duplication with comparison (DWC) is the simplest form of hardware redundancy that can be used for concurrent error detection (Fig. 2(a)). Both CL copies implement identical output and next state functions, f and g, respectively. If outputs disagree, the comparator indicates an error. From one hand, it is evident that DWC involves more then 100% hardware redundancy. On the other hand, the DWC technique is simple for implementation and ensures full coverage of all single faults in CL that produce an error. However, in many cost-sensitive applications, due to high hardware overhead, DWC is an unacceptable design solution.

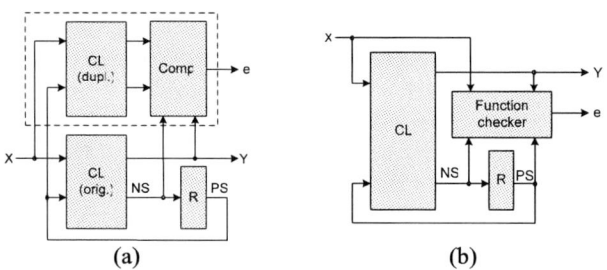

Fig. 2. Duplication with comparison: (a) basic scheme; (b) alternative representation.

An approach to design a more practical concurrent error detection solution, that we adopt in this paper, is to trade-off DWC's high error detection capability for reduced hardware complexity. To this end, we introduce an alternative representation of the DWC scheme, given in Fig. 2b. In this arrangement, two of DWC constituents from Fig. 2(a), the duplicated CL and the comparator, are implemented as a single module, referred as a *function checker* [6]. The function checker is driven by input signals, X and PS, as well as, output signals, Y and NS, of the original CL. A single output, e, that indicates whether a mapping $(X,PS) \rightarrow (Y,NS)$ is valid or not, is generated at the output of the function checker. Having in mind that the function checker can make a distinction between error-free and any erroneous response of the original CL, we will call this circuit as *total checker*, TC. TC has the same fault coverage as the original DWC scheme, but with probably smaller implementation cost, due to combined synthesis of the duplicated CL and comparator. With this approach, we can obtain a limited hardware reduction in respect to DWC scheme.

Our intention now is oriented towards designing checkers that have hardware complexity less then 100% (with respect to the DWC) at the cost of fault coverage less then 100%, too. A checker with this property we will call *partial checker (PC)*.

PC complexity can be simplified by decreasing the number of PC's inputs. Instead to select, on a bit-by-bit basis, the "best" subset of inputs for the checker, we will consider partial checkers that are feed by a subset of CL input/output signal groups, X, PS, NS, and Y. In a set of four signal-groups, there are 16 subsets. After eliminating the empty subset and four subsets that contain input groups, X and PS, only, it remains 11 different meaningful partial checker configurations at our disposal. Next, we will analyze, in more details, two of them: a) Checker that monitors NS and PS signal groups (Fig. 3(a)); and b) checker that in addition to NS and PS signal groups accepts FSM primary output signal group Y, also (Fig. 3(b)). We will call the checker sketched in Fig. 1(a) as a *State-Transition Checker* (STC), and that one given in Fig. 3(b) as *State-Transition-Output Checker* (STOC).

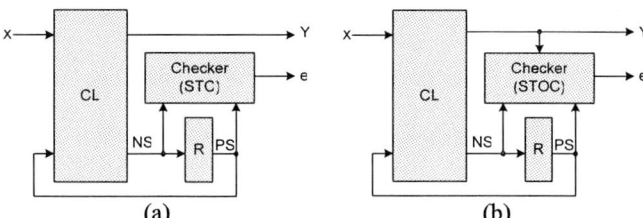

Fig. 3. Partial self-checking: (a) State-transition checker; (b) State transition-output checker.

A. State-transition checker

In order to describe, in a more explicit way, the STC's operation, we modify the corresponding STG into a state graph. The state graph is formed by adding invalid states and invalid transitions into the original STG. For example, the state graph shown in Fig. 4(a) is relative to the original STG presented in Fig. 1(a). Invalid states and invalid transitions are sketched by dashed lines. Since the STC does not accept neither inputs, X, nor outputs, Y, transition labels are omitted, too. STC truth table is given in Fig. 4 (b). The STC signals an erroneous condition under the following condition: PS corresponds to a valid state; but the pair of present-next state codes (PS, NS) is not a valid transition. In contrary, if PS is a valid state and (PS, NS) is valid transition, then the STC output is set to logic 0, indicating error-free FSM operation. The STC does not consider the cases when the present state is invalid, since such a situation can occur iff the previous transition was invalid, and thus detected during the previous clock cycle. This incompleteness in a specification represents degree of freedom that can be used during logic optimization process.

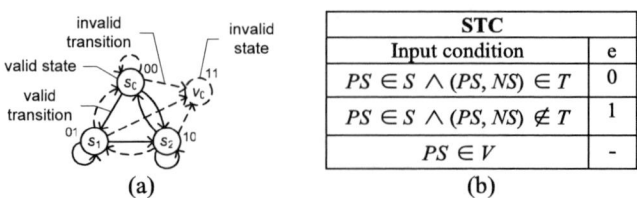

Fig. 4. State-transition checker: (a) model; (b) definition.

698

In order to estimate STC error detecting capability we will make the following two assumptions: (1) during normal operation, all FSM states are equi-probable, and (2) all erroneous bit vectors that occur at the CL's outputs, due to a fault, are equi-probable. Let define now an outdegree (d_i) of state s_i as a number of valid transitions exiting this state. Assuming c-bit encoding, the number of invalid transitions from state s_i is $2^c - d_i$. The number of possible erroneous transitions as a consequence of fault, at any state, is $2^c - 1$. Among them, $d_i - 1$ transitions are valid, while the remaining $2^c - d_i$ are invalid. Note that, since the STC is able to make distinction between valid and invalid transitions, any erroneous but valid transition will be considered as fault-free.

Illustration only, assume now that the FSM from Fig 4(a) is set in state S_0. Suppose further that, as response to a current input change, the FSM, due to a fault, erroneously goes to state S_2, instead to state S_1. By analyzing Fig. 4(a) we conclude that a transition (S_0, S_2) is valid one, so the STC will not signals an error. Since the outdegree $d_0 = 2$, a probability that STC will detect an erroneous transition from state S_0 is 50%. In general, an *error coverage in state* s_i can be defined as a ratio of invalid transitions exiting that state and the total number of erroneous transitions: $cv^{STC} = (2^{c+k} - d_i) / (2^{c+k} - 1)$. Since all FSM states are equi-probable, STC's error coverage, cv^{STC}, is defined as a mean value of state error coverage:

$$cv^{STC} = \frac{1}{n}\sum_{i=1}^{n} cv_i = \frac{2^c - \hat{d}}{2^c - 1} \qquad (3)$$

where $\hat{d}$ is average outdegree of FSM's states.

By analyzing Eq. (3) we can draw the following conclusions: a) STC's error detection capability is higher for sparse STG, i.e. for FSM with smaller average state outdegree, $\hat{d}$. b) STC's error coverage increases by increasing the number of encoding bits, c. However, as we increase c, we have to pay a cost for CL hardware complexity augmentation.

B. State-transition-output checker

State-transition-output checker (STOC) is partial FSM checker that extends functionality of STC by additionally taking into account FSM primary outputs (Fig. 3(b)). STOC operation can be described by means of a state graph where each valid transition is labeled with a set of all possible different valid primary output values. Fig. 5(a) shows a part of the state graph with four states and several valid and invalid transitions. The set {Y1, Y2} associated with a transition (S_0, S_2) means that Y1 and Y2 are only correct output values when FSM moves from state S_0 to S_2. The truth table of the STOC is presented in Fig. 5(b). L(PS,NS) denotes a set of all distinct output values associated with the transition (PS, NS).

Logic 1 at the STOC output signals an error in cases when PS is a valid state, but either (PS, NS) is not a valid state transition, or when Y is not a valid output value with respect to the observed transition. The STOC will indicate an error-free FSM operation, when PS is valid state, and when both the pair (PS, NS) corresponds to valid state transition, and Y is a valid output value in respect to that transition. Although the number of erroneous conditions that STOC can capture is larger that then of the STC, it still misses to detect all possible FSM errors. This is direct consequence of the fact that STOC does not takes into account FSM primary inputs. Therefore, there is always a probability that in a case of fault, the resulting erroneous next-state and output value match some valid condition [(PS, NS), Y] that correspond to a given PS and **some** arbitrary input X.

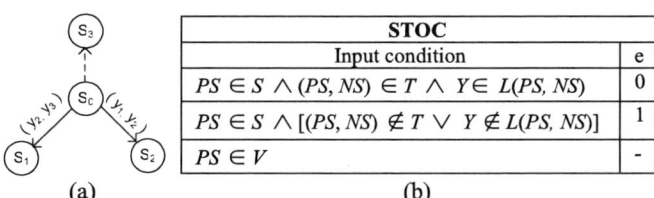

STOC	
Input condition	e
$PS \in S \wedge (PS, NS) \in T \wedge Y \in L(PS, NS)$	0
$PS \in S \wedge [(PS, NS) \notin T \vee Y \notin L(PS, NS)]$	1
$PS \in V$	-

(a)   (b)

Fig. 5. State-transition output checker: (a) model; (b) definition.

Suppose now that the FSM in Fig. 5(a) is in state S_0, and an erroneous value appears at CL's outputs due to a fault. First, consider a case when the generated next-state (say, S_2) is correct while the output (say, Y3) is incorrect. Since Y3 is not contained in the corresponding set of valid output values, L(S_0, S_2), it will be recognized by the STOC as an error. However, the STOC will not signals an error if the faulty FSM produces Y2 instead of Y1, i.e. if the incorrect output value belongs to the set of valid output values associated with the current state transition. In addition, consider now a case when the next state is erroneous. Suppose that the faulty FSM moves from S_0 to S_1 instead to S_2. Note that such an error could not be detected by the STC since S_1 is a valid state. Whether the STOC, under such circumstances, will react or not depends on the generated output value. For example, if an output value of Y1 is generated, an error will be detected, since for transition (S_0, S_1), the output value Y1 is not valid. Contrary, if a generated output value is Y2, the error cannot be detected, since this value is valid in respect to transition (S_0, S_1).

Starting with identical simplified assumptions that we have already applied for STC, we can now estimate the error coverage of STOC as follows. For a given present state S_i, we define quantity w_i as the number of valid different ($c+k$)-bit combinations generated at the CL outputs (NS, Y). Note that w_i can be calculated by summing the cardinality of valid output values sets over all S_i's outgoing transitions. For example, for STG in Fig. 4, we have $w_0 = 4$. For given c-bit state and k-bit output encoding, the number of possible erroneous ($c+k$)-bit combinations at the CL's outputs, for any present-state, is $2^{c+k} - 1$. Among them, $w_i - 1$ combinations are valid, while the remaining $2^{c+k} - w_i$ are invalid. Since STOC is able to differentiate between valid and invalid CL's output combinations, only, its *error coverage in state* S_i is: $cv_i^{STOC}/$ $cv^{STC} = (2^{c+k} - w_i) / (2^{c+k} - 1)$. STOC's *error coverage* is defined as a mean value of state error coverage:

$$cv^{STOC} = \frac{1}{n}\sum_{i=1}^{n} cv_i^{STOC} = \frac{2^{c+k} - \hat{w}}{2^{c+k} - 1} \qquad (4)$$

where $\hat{w}$ is an average over w_i, $i = 1, ..., n$.

It follows from (4) that the STOC has a higher error coverage for FSM with a smaller number of alternative output values per present-next state pair. Parameter $\hat{w}$ can be expressed as $\hat{w} = \lambda \cdot d$ where $1 \le \lambda \le 2^k$. The case $\lambda = 1$ corresponds to STG with single output value per each transition. The case $\lambda = 2^k$ is appropriate for STG when each transition is labeled with a set of all possible k-bit combinations. Note that $\lambda = 1$ corresponds to Moore, while $\lambda \ge 1$ to Mealy FSM. It is evident from Eq. (4) that the STOC scheme is more effective for Moore FSM.

Finally, we will compare error coverage of STC and STOC. Therefore, define now an improvement factor of STOC over STC as: $\alpha = cv^{STOC}/cv^{STC}$. By substituting expressions for cv^{STC} and cv^{STOC} and assuming that $2^c - 1 \approx 2^c$ and $2^{c+k} - 1 \approx 2^{c+k}$ it is easy to show that:

$$\alpha = \frac{2^{c+k} - \lambda \hat{d}}{2^{c+k} - 2^k \hat{d}} \qquad (5)$$

Since $1 \le \lambda \le 2^k$, we can conclude that STC error coverage is never greater then that one of STOC. From one hand, in the extreme case (when in all FSM states, all 2^k k-bit combinations at the primary output are valid, $\lambda = 2^k$), both partial checkers posses identical error detection capabilities. From the other hand, the improvement factor is largest for $\lambda = 1$, i.e. for Moore FSM.

III. EXPERIMENTAL RESULTS

A system for automatic synthesis and implementation of partially self-checking FSMs has been developed in order to evaluate the performances of the proposed approach. The system uses SIS 1.2 [7] for logic synthesis and technology mapping together with several specifically developed tools for partial checker generation and fault-coverage analysis. Generated checkers are optimized for area using the SIS script *script.rugged* and then mapped to gates in a given technology library (*lib2.genlib* in this case) in order to obtain an area-minimal implementation in a specific technology. Fault-coverage simulation uses gate-level netlists in order to simulate the effects of single stuck-at faults, and calculates the fault-coverage value.

Table I reports the data for several example FSM taken form the MCNC 91 sequential benchmark suite. For each FSM we have synthesized four partially self-checking circuits and compare them with the circuit based on DWC scheme. The first two circuits are based on STC scheme: one using c_{min}, and one $c_{min}+1$ bits for state encoding, where c_{min} represents the minimal number of bits needed to encode FSM states. The second two circuits are based on

STOC scheme. Results includes the area expressed as percentage of the area needed for DWC scheme (%A) and fault coverage expressed in percentage of errors detected during simulation (%C).

Results are highly FSM dependent. In general, the STOC scheme is more efficient in terms of fault coverage, but with higher area cost with respect to the STC scheme. In both schemes, fault coverage improves with adding one extra bit for state encoding. However, since increase of the number of state bits affects both the CL and checker, the additional area cost is rather large.

TABLE I
AREA AND FAULT COVERAGE ANALYSIS

	STC				STOC			
	$c=c_{min}$		$c=c_{min}+1$		$c=c_{min}$		$c=c_{min}+1$	
FSM	%A	%C	%A	%C	%A	%C	%A	%C
CSE	24.6	20	64.7	70	55.5	74.2	69.7	90
BBSSE	29.2	37.7	60.3	72	67.6	78.7	99	89
KEYB	37.2	57	81.4	74	41.8	61	95.8	84
DK14	13.3	20	56.5	38	58.5	61	102	81

IV. CONCLUSIONS

A method for the synthesis of low cost, partially self-checking FSMs is proposed. The method is based on reduced duplication with comparison scheme used for monitoring operation of FSM combinational logic. Two variants of the proposed scheme are presented and analyzed. Area and fault coverage analysis give promising results. Ongoing research is directed towards optimization of the technique through limiting the number of observed FSM states and selecting appropriate FSM state assignment which will take into account the CL error statistics.

REFERENCES

[1] P.K. Lala, *Self-Checking and Fault-Tolerant Digital Design, Morgan Kaufmann Publishers*, San Francisco, 2001.

[2] N. K. Jha and S.-J. Wang, "Design and Synthesis of Self-checking VLSI Circuits", *IEEE Trans. On CAD of Integrated Circuits and Systems,* vol. 12, No. 6, pp. 879-887, June 1993.

[3] C. Bolchini, R. Montandon, F. Salice, D. Sciuto, "Design of VHDL-based totally self-checking finite-state machine and data-path descriptions", *IEEE trans. on VLSI Systems*, Vol. 8, no. 1, Feb. 2000, pp. 98-103.

[4] R. A. Parekhji, G. Venkatesh and S.D. Sherlekar, "Concurrent Error Detection using Monitoring Machines," *IEEE design & Test of Computers,* Vol. 12, pp.24-32, Fall 1995.

[5] R. Leveugle, and G. Saucier "Optimized Synthesis of Concurrently Checked Controller," *IEEE Trans. Computers,* Vol. 39, pp. 419-425, Apr.1990.

[6] G. Lj. Djordjevic, M. K. Stojcev and T. R. Stankovic, "Approach to partially self-checking combinational circuits design", *Microelectronics Journal*, Vol. 35 (12) , Dec. 2004, pp. 945-952.

[7] E. M. Sentovich et al., "SIS: A System for Sequential Circuit Synthesis", Tech. Report UCB/ERL M92/41, Electr. Research Lab, Univ. of California, Berkeley, May 1992.

Laboratory ADC Tester Based on NI-6251 Acquisition Card

M. Nikolić, M. Sokolović, and P. Petković

Abstract – Analog to digital converter (ADC) is the crucial part of many mixed-signal ICs because it interfaces analog signals from real world with digital logic on a chip. Faults made during ADC hardly can be repaired within the digital part. Therefore, functional testing of ADC is a very important task especially during prototyping. In this paper one laboratory ADC tester is proposed. It is based on NI-6251 data acquisition card and a PC that controls the testing process using LabView software.

I. INTRODUCTION

General purpose ADC testers are very expensive because of a wide variety of ADC architectures that require different testing techniques and equipment. Simultaneously, the testing process cannot improve the quality of individual ICs. It solely can measure the quality that already exists in an IC [1]. For a low volume production and for prototyping purposes it is not rational to spend a fortune for a professional tester. As an alternative, it is reasonable to develop a tester dedicated for the specific use.

In order to make testing possible, the designer has to provide testability of integrated circuit during design phase using various DFT methods [1]. Considering a prototype, it is significant not only to determine if circuit works or not but to detect how good it is and eventually how to fix potential faults in the subsequent design. Therefore, some signals that are not necessary for nominal operation of an IC, should be made observable for testing purposes by adding additional pins.

The major problems related to testing a prototype are controlling the test set-up conditions and observing the responses from a circuit under test (CUT).

Different blocks within a mixed signal microsystem, such as analog or digital signal processing parts, require different testing approaches. This makes ADC testing a difficult task.

Mutual interference of mixed signals is another issue that introduce troubles into the testing process. Namely, in highly integrated circuits, various sub-systems are in immediate proximity which causes influences to each other. Besides, the heat dissipated from a digital signal processor (DSP) affect operating points of analog circuitry.

M. Nikolić, M. Sokolović, and P. Petković are with the Department of Electronics, Faculty of Electronic Engineering, University of Niš, Aleksandra Medvedeva 14, 18000 Niš, Serbia & Montenegro, e-mail: miljan@venus.elfak.ni.ac.yu

This paper describes ADC testing theory and realization of a laboratory tester based on NI-6251 data acquisition card [2]. The tester is dedicated for ADC implemented within an ASIC solid state energy meter named IMPEG, that was designed in our laboratory. The ADC was realized using sigma-delta architecture and the target characteristic was to obtain SFDR of at least 80dBc [2]. The tester can be modified and used for testing other ADCs with similar architectures and operating rates. DFT techniques implemented in IMPEG were described in [3]. The motive for the work described in this paper was to improve testing possibilities of the tester explained in [4] and to enhance measurement accuracy. The new tester utilizes the benefits of built-in DFT hardware.

This paper is organized in the following sections. The next section describes different ADC architectures. The third section deals with different methods for ADC testing. Some important variables to be measured during testing an ADC are listed here, as well. After that a practical implementation of one laboratory ADC tester will be given. This realization requires an original measurement and data processing procedure and is affordable to many non industrial users. The paper concludes with important results obtained after testing the particular sigma-delta ADC. These results verify the functionality of the entire test set-up.

II. ANALOG TO DIGITAL CONVERTERS

There are many different types of ADC architectures, that are introduced to fulfill the requirements of a specific application [1].

Successive approximation architectures provide DAC output adjusted with a binary search algorithm until it is substantially equal to the input voltage. This kind of ADCs is difficult to design and test and often suffer from nonidealities in approximation as well as poor linearity, hysteresis errors, poor power supply rejection ration and low bit rates [1, 5].

Integrating ADC (dual-slope and single-slope) is simpler but slower than the ADC architectures based on successive approximation. It uses a simple integrator to ramp upward for a fixed amount of time, starting from the time it crosses fixed threshold voltage [1, 5]. Dual-slope integrating ADCs have smaller offset errors, but are more complex than single-slope ADCs. They have better linearity.

A flash ADC compare the input signal against all possible decision levels, simultaneously. This type of ADCs is very fast because the decision levels are compared all at once. They are mostly used in high-frequency application, but occupy much silicon area [1, 5]. Multiple flash ADCs can also be used to construct a multipass successive approximation architecture called a semiflash ADC [1].

Sigma-delta ADC uses a crude ADC combined with a noise-shaping process to produce an oversampled pulse density modulated (PDM) data stream. This data stream is then digitally filtered and decimated to produce high-resolution ADC samples [1, 5].

The use of oversampling sigma-delta modulators for high-resolution ADC resolves problems related to analog component limitations. Besides, it is able to meets all requirements for the specific design. Therefore this architecture was our choice.

Sigma-delta modulators employ coarse quantization enclosed in one or more feedback loops. By sampling at a frequency that is much greater than the signal bandwidth, it is possible for the feedback loops to shape the quantization noise so that most of the noise power is shifted out of the signal band. The out of band noise can then be attenuated with a digital filter. The degree to which the quantization noise can be attenuated depends on the order of the noise shaping and the oversampling ratio [2].

III. ADC TESTING

Typical tests capable to apprise quality of ADC are [1]:
- ADC code edge measurement,
- DC tests,
- transfer curve tests, and
- dynamic ADC tests.

Each one of them will be briefly explained.

The aim of ADC code edge measurement is to find the input voltage threshold between two successive ADC codes that causes an output code to change. To measure the ADC linearity one needs to derive transfer curve of an ADC. Two well-known methods for transfer curve derivation are center code testing and edge code testing [1].

Code centers are defined as the midpoint between the code edges. This is shown in Figure 2. The most obvious method to find the edge is a step search method where one simply adjusts the input voltage of the ADC up or down until the output codes are evenly divided between the first code and the second code. To achieve repeatable results, one needs to collect about 50 to 100 samples from the ADC in order to provide statistically significant number of conversions.

Another edge search technique is a servo method. This is a fast hardware version of the step search. Using this hardware, the output codes from ADC are compared against a value programmed in the search value register. If the ADC output is greater than or equal to the expected

value, the integrator ramps downward. On the contrary, the integrator ramps upward.

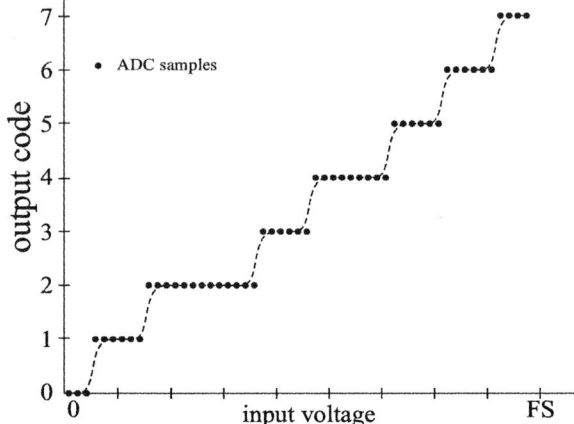

Fig. 2. ADC samples from linear ramp histogram test

But the most common production testing technique is the histogram method. The simplest way to perform a histogram test is to apply a rising or falling linear ramp to the input of the ADC and collect samples from the ADC at constant sampling rate. The ADC samples are captured while the input ramp slowly moves from one end of the ADC conversion range to the other. This is shown in Figure 3. The number of occurrences of each code is plotted as a histogram. It shows which codes are hit more often, indicating that they are wider codes. After obtaining the histogram, a code edge transfer curve must be derived using a simple mathematical equation that sums the code widths.

To compensate for the poor linearity of the ramp generators, the alternative: sinusoidal histogram method can be performed. It is easier to produce a pure sinusoidal waveform than to produce a perfectly linear ramp. This method also allows testing in more dynamic, real-world situation, since ramps are varying very slowly. By using a sinusoidal signal instead of a ramp, one would expect to get more code hits at the upper and lower codes than at the center of the ADC transfer curve, even when testing a perfect ADC. The effects of the nonuniform voltage distribution can be removed after normalization.

DC Tests and Transfer Curve Tests comprise:
- DC Gain,
- DC offset,
- Integral nonlinearity, INL,
- Differential nonlinearity, DNL,
- monotonicity and
- missing codes tests.

Once the ideal transfer curve has been established, DC gain and offset can be measured. The gain and offset are measured by calculating the slope and offset of the best-fit line. ADC can be nonmonotonic when one or more of its code widths is negative. However, this failure mechanism is quite rare. Nevertheless, ADC can appear to be nonmonotonic when its input is changing rapidly. ADCs are not tested for monotonicity with a slowly changing input.

702

Monotonicity errors show up as signal-to-noise ratio failures and as a sparkling.

The code whose voltage width is zero is recognized as a missing code. This means that the missing code can never be hit, regardless of the ADC's input voltage. A missing code appears as a missing step on an ADC transfer curve.

Dynamic ADC parameters are: maximum sampling frequency, maximum conversion time, and minimum recovery time. Maximum conversion time is the maximum amount of time it takes an ADC to produce a digital output after a stabile input signal is asserted. The ADC is guaranteed to produce a valid output within the maximum conversion time.

Considering all tests listed and an existing ADC architecture, it is very important to determine the significance and the feasibility of tests to be performed. Tests such as INL and DNL are not well suited for sigma-delta converters. Instead, channel tests like gain, offset, signal-to-noise ratio, idle channel noise, etc., are commonly specified. When the resolution exceeds 12 or 13 bits, it becomes very expensive to perform transfer curve test such as INL and DNL because of the large number of code edges that must be measured. Fortunately, transmission parameters such as frequency response signal to distortion ratio and idle channel tests are much less time-consuming to measure [1]. A limited budget also limits the list of tests that can be performed in the laboratory environment. For example, dual-slope integrating ADCs have good linearity. For this type of architectures, all-codes testing would be prohibitively expensive for production testing.

One of the best ways to overcome different testing problems is to apply the concept of DSP-based testing [1]. It is based on special digital signal processing tools that could be implemented on the CUT, that is hardware, or written as particular software. This offers several advantages over the traditional measurement techniques. One can create and measure signals with multiple frequencies at the same time, and in that way, one can perform many parametric measurements in parallel. Thus, the test time is reduced. Separation of different signal components gives a second huge advantage over non-DSP-based measurements. We can isolate noise and distortion components from one another and from the test tones, which allows much more accurate and repeatable measurements. The third big advantage of this method is the ability for signal manipulation. Just one set of output signal samples allows the manipulation of the waveforms to achieve a variety of results.

IV. TEST SET-UP

Analog part of IMPEG consists of two sigma-delta A/D converters. Second order sigma-delta modulator is in voltage channel, while third order sigma-delta modulator is in current channel. Sampling frequency is 524288 Hz and required data rate for built-in DSP is 4,096kHz. In order to make converted digital signals observable, three pins were

added for testing purposes before decimation. One stands for output from the voltage and two for the output from the current channel. ADC in current channel is implemented using MASH architecture [6, 7] and hence, the output is composed of two single-bit signals. The voltage channel should result with dynamic range, SNR and SFDR greater then 60dB. Stringent requirements are imposed for current channel, with dynamic range, SNR and SFDR greater then 80 dB.

Therefore the resolution of stimulus generator must be greater then 14 bits to satisfy imposed requirements. Acquired data are processed further by FFT. The acquisition time has to last at least two seconds in real time to provide enough data to perform FFT. After FFT analysis one can obtained values of SNDR, SFDR and linearity of ADC. Besides, the collected input data have to provide information necessary for fault detection and diagnostics.

Figure 4 shows the proposed tester set-up. It consists of :

- an original printed circuit board containing circuit under test,
- NI-6251 data acquisition card and
- PC that controls the testing process using LabView software.

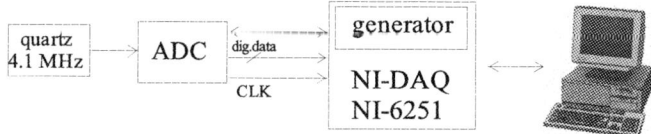

Fig. 4. ADC test set-up

DAQ card offers:

- 16 (8 differential) analog inputs each with 16 bit resolution and maximum sampling frequency of 1.25 Msamples/s,
- three 8 bit digital ports where only one can be hardware timed up to 10 MHz and
- two analog outputs with 2.8Msamples/s at most and can generate arbitrary shaped waveform.

Stimulus is sinewave generated as differential signal occupying both analog output channels of DAQ card. Simultaneously, the card utilizes two channels to acquire analog input data in order to check signal integrity. The collected data represent signal obtained after analog LP filter with cut-off frequency of 2kHz according to Figure 5.

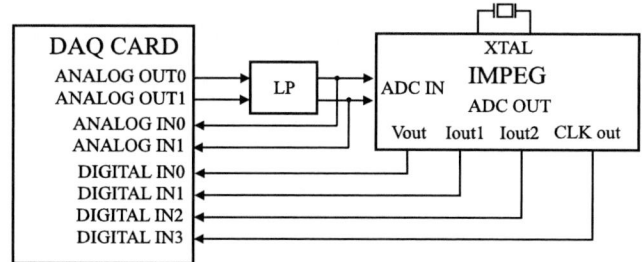

Fig. 5. Block diagram of interconnection between acquisition card and IMPEG

Three digital input channels are used to acquire digital output from voltage and current channel. An additional digital input serves to control digital acquisition by the clock signal generated in IMPEG.

This setup can be used for testing different kinds of ADC's, such as: sigma-delta as shown here, successive approximation architecture, flash etc.

The testing possibilities restrict the number of available digital lines and sampling frequency. Besides, limitation arise from maximal input signal frequency and/or digital signal resolution.

V. MEASUREMENT AND RESULTS

Figure 6 shows test results obtained for the ADC implemented within the voltage channel of IMPEG. The power spectrum obtained after stimulus with amplitude of 40mVpp differential sinewave signal and frequency of 50Hz, applied to ADC input is shown in Figure 6.a. Figure 6.b presents the power spectrum of output signal acquired with the second order sigma-delta ADC. In order to verify design methodology the same input signal is used as stimuli for behavioral simulation of the sigma-delta modulator in voltage channel. Figure 6.c illustrates the corresponding power spectrum. Obviously, very good matching with behavioral model and measurement was obtained.

VI. CONCLUSION

Tester realization required development of specific hardware and software solutions. All tests performed in the laboratory setting were proved in real environment within the energy meter device developed by our industrial partner. The implemented testing method is very simple and offers low cost functional oriented testing that gives a good base for diagnostics as well.

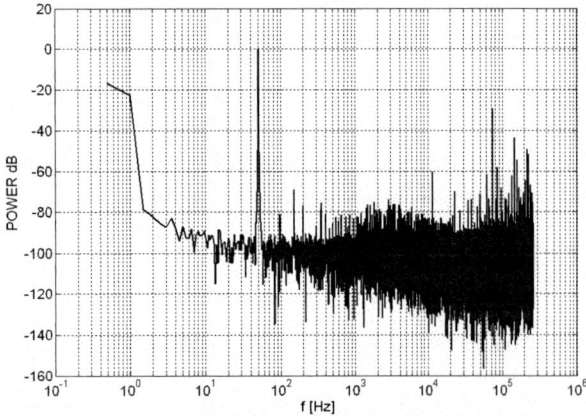

Fig. 6. a) Power spectrum of input signal; Hanning window

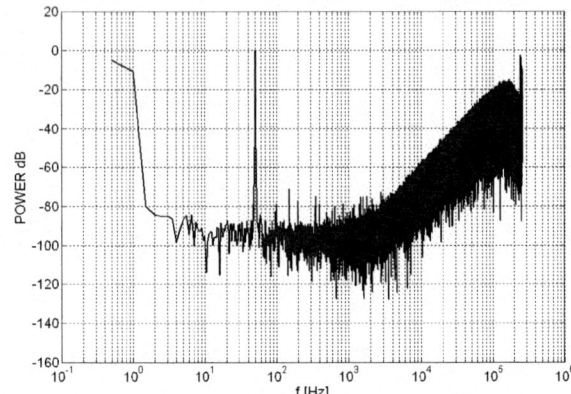

Fig 6. b) spectrum of the ADC output signal; Hanning window

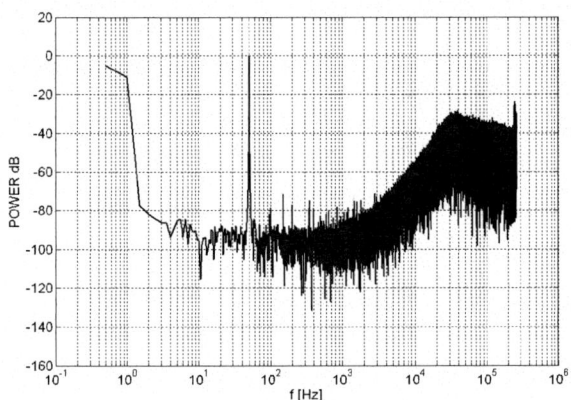

Fig. 6. c) spectrum of the ADC ouput signal behavioral level; Hanning window

REFERENCES

[1] M. Burns, and G. W. Roberts, "An Introduction to Mixed-Signal IC Test and Measurement", *Oxford University Press*, New York, 2001.

[2] www.ni.com/dataacquisition/mseries.htm

[3] M. Savić, M. Nikolić, M. Sokolović, and P. Petković: "Testing set-up for analog part of the power meter IC", *Proceedings of Conference ER04, Sozopol, 2004*, book 2, pp. 25-30.

[4] M. Sokolović: "DFT in an Application Specific Mixed-Signal ICs", *MS Thesis*, Niš, April 2005.

[5] N. West and D. Harris: "CMOS VLSI Design – A Circuits and Systems Perspective (third edition)", *Addison-Wesley*, 2004.

[6] L.Williams and B. Wolley, "A third-order sigma-delta modulator with extended dynamic range", *IEEE J. Solid-State Circuits*, vol. 29, pp. 193-202, March 1994.

[7] M. Nikolić, M. Savić, and D. Milovanović: "A Third-Order Sigma Delta Modulator - Preliminary Results", *Proceedings of 24th International Conference MIEL 2004*, Niš, vol. 2, pp. 605-608.

2006 25th International Conference on Microelectronics

Hybrid Empirical-Neural Model of the Loaded Microwave Cavity Applicators

Z. Stanković, B. Milovanović and M. Milijić

Abstract - In this paper, the loaded cylindrical metallic cavity is modeled using a new Hybrid Empirical-Neural (HEN) model. The considered load is a homogeneous dielectric layer, which is elevated from the cavity bottom. Unlike the model based on classical multi-layer perceptron (MLP) network, the proposed HEN model includes an existing partial knowledge about the resonant frequency behavior of the cavity, yielding more accurate determination of the resonant frequencies. Comparison of MLP and HEN models as well as an advantage of using the HEN model is given through an example referring to the experimental cylindrical metallic cavity with a circular cross-section.

I. INTRODUCTION

The intense development of microwave technique in the last decades has led to widespread application of microwave applicators in the science, medicine, and in industry. In most cases, these applicators have a form of the cylindrical metallic cavities with various cross-sections (circular, rectangular, elliptical, etc). Microwave cavities loaded by homogeneous dielectric layers have wide range of applications in different microwave systems. They also have a special application in the processes of dielectric material heating and drying by microwave energy. In order to manufacture an efficient cavity, it is necessary to know all types of oscillations that may appear in it and what values the resonant frequencies can have [1,2].

A usual approach for theoretical analysis of the cylindrical metallic cavities is based on the application of the transverse resonance method (TRM) [3]. The resonant frequencies are determined from the transcendental characteristic equation that describes the EM model. To calculate the resonant frequencies, an appropriate numerical technique and an efficient procedure for mode identification [3] are needed.

The main common disadvantage of TRM as well as all other numerical techniques which can be applied in cavity modeling (TLM, FDTD, etc) is that they have high demands concerning the hardware resources necessary for their software implementation [1,2,4]. The software implementation itself might be very complicated and faced with many difficulties. Also the time needed for numerical calculation when using a detailed electromagnetic (EM) model could be unacceptably long.

In order to avoid solving of a number of time-

consuming complex electromagnetic equations needed for numerical approaches, we suggest microwave cavity modeling using artificial neural network. The early researching, concerning the modeling of a microwave cylindrical metallic cavity loaded by dielectric layer placed at the cavity bottom, has showed that the neural network models can have satisfactory accuracy same as detailed EM models but also can have higher simulation speed [5,6,7]. In this researching, it is shown that MLP network [4,8] could be very successfully for cylindrical microwave cavity modeling. But, the main disadvantage of MLP models is a need for providing a large set of training data, that could be difficult and time-consuming process [4,5]. First Hybrid Empirical-Neural (HEN) model of the microwave cylindrical cavity, loaded by dielectric layer placed at the bottom of the cavity, incorporating empirical knowledge is developed in [6]. Due to its feature of incorporating knowledge about the behavior of the resonant frequency defined in approximate approach [9], this HEN model needs smaller set of training data then MLP model for achieving satisfying model accuracy [6].

In this paper a HEN model for microwave cavity with more complex configuration will be presented. Namely, the dielectric layer can have variable elevation from the cavity bottom. This case has special meaning in the process of heating.

II. KNOWLEDGE ABOUT MICROWAVE CAVITY RESONANT FREQUENCY

A number of different TM/TE$_{mnp}$ modes can be excited in a cylindrical metallic cavity loaded by homogeneous dielectric layer which is elevated from the cavity bottom (Fig.1). Detailed earlier research [3] has shown that the resonant frequency f_r of excited mode in such cavity with constant dimensions depends on the relative dielectric permittivity ε_r, filling factor t_h ($t_h = t/h$, where t is thickness of dielectric layer and h is height of the cavity) and elevation factor r_h ($r_h = r/h$, where r is dielectric layer elevation).

$$f_r = f(t_h, \varepsilon_r, r_h) \qquad (1)$$

Let consider the characteristic case when r_h=0. Using short-circuit boundary (electric wall) in a interface plane between dielectric slab and air, from the condition of resonance applied separately in air and dielectric part of the cavity, appropriate expressions for resonant frequency calculation in these regions can be easily derived

Z. Stanković, B. Milovanović and M. Milijić are with the Department of Telecommunication, Faculty of Electronic Engineering, University of Niš, Aleksandra Medvedeva 14, 18000 Niš, Serbia & Montenegro, E-mail: zoran@elfak.ni.ac.yu

1-4244-0116-X/06/$20.00 ©2006 IEEE

$$f_r^{(A)}(t_h) = \sqrt{\left(\ell \cdot \frac{f_0}{1-t_h}\right)^2 + f_{c0}^{\;2}} \quad l = \begin{cases} 0,1,2,\dots \text{ for TM}_{mnp} \\ 1,2,3,\dots \text{ for TE}_{mnp} \end{cases} \quad (2)$$

$$f_r^{(D)}(t_h,\varepsilon_r) = \sqrt{\left(k \cdot \frac{f_0}{\sqrt{\varepsilon_r}}\frac{1}{t_h}\right)^2 + \left(\frac{f_{c0}}{\sqrt{\varepsilon_r}}\right)^2} \quad k = 1,2,3\dots \quad (3)$$

where: $f_{c0} = ck_c/(2\pi)$ represents the cut-off frequency of a waveguide with the same cross-section as cavity and filled with air, while k_c is a constant that depends on mode of oscillation and waveguide cross-section shape and dimensions; $f_0 = c/(2h)$; and integers l and k are the number of half waves of standing wave for electric field in the corresponding part of the cavity [9]. For the cavity of rectangular cross-section with dimension $a \times b$ for TM/TE$_{mnp}$ modes constant k_c is

$$k_c = \sqrt{\left(\frac{m\pi}{a}\right)^2 + \left(\frac{m\pi}{b}\right)^2} \quad (4)$$

while for the cavity of circular cross-section with radius r, constant k_c is

$$k_c = \frac{x_{mn}}{r} \quad (5)$$

where x_{mn} is n-th zero of the Bessel function of the first kind of order m for TM$_{mnp}$ modes and n-th zero of the derivation of the same function for TE$_{mnp}$ modes [2].

Let note characteristic point of considered TM/TE$_{mnp}$ mode family (m=const, n=const) in t_h - f_r plane as RR$_k^l(\varepsilon_r)$ which represents a crossing point of k-th resonant curve in dielectric part of the cavity (RD curve) (3) and l-th resonant curve in air part of the cavity (RA curve) (2), (Fig. 2). Also let F_0^A be the characteristic point determined by the resonant frequency of an empty cavity. Detailed semi-empirical analysis of cylindrical metallic cavities [9] has shown that resonant frequency curves, for considered TM/TE$_{mnp}$ mode excited in such cavities, independently of the elevation factor value, passing though characteristic points with the following order: F_0^A, RR$_1^{(p-1)}$, RR$_2^{(p-2)}$,…, RR$_p^{min(l)}$. The characteristic points are easily found from Eqs. (2) and (3) for known relative permittivity ε_r. The fact that these points partially describe the behavior of resonant frequency curves (mode tuning behavior) and that they are determined directly by resonant functions in air and the dielectric part of the cavity, given in analytical form, represents a partial semi-empirical knowledge from the problem domain implemented in the structure of HEN model whose architecture will be exposed further in this paper.

III. HYBRID EMPIRICAL-NEURAL MODEL OF LOADED MICROWAVE CAVITY

Hybrid Empirical-Neural Model of loaded microwave cavity appears as integration of the cavity approximate

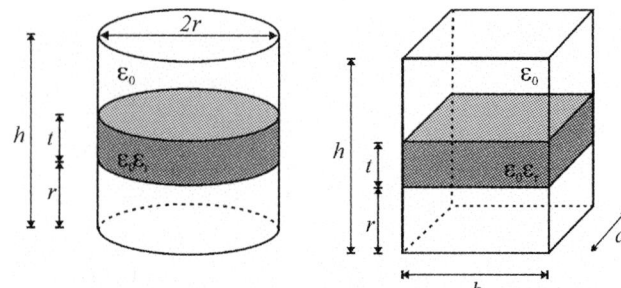

Fig. 1. Microwave cylindrical metallic cavity with (a) circular (b) rectangular cross-section loaded by dielectric layer of thickness t with elevation r from the cavity bottom

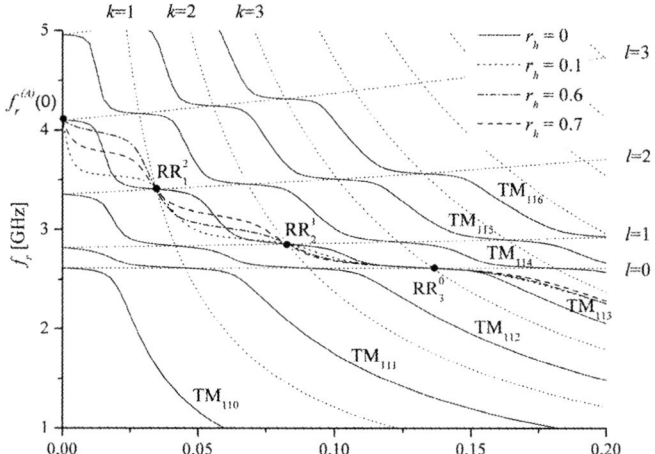

Fig. 2. Family of the resonant frequencies for TM11$_p$ mod obtained using TRM for the cylindrical metallic cavity with circular cross-section (r= 7 cm, h=14.24 and ε_r= 80). ⋯ RA curves (monotonous increasing in air part) and RD curves (monotonous decreasing in dielectric part)

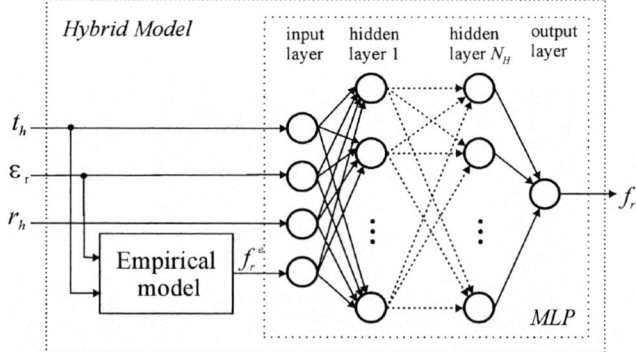

Fig. 3. The architecture of HEN model of the loaded cavity

model [9] as empirical knowledge holder and MLP network. The basic idea in HEN model realization is that the empirical model with corresponding connection to the neural network provides higher generalization and extrapolation capabilities of the network [4]. This is achieved by presenting extra information about the problem at the input of the network. According to that, HEN model is developed for loaded microwave cavity whose architecture is presented in Fig. 3. Approximate model determines the resonant frequency f_r^e in the following way: in the first step for given mode and given ε_r, according to section II, RR characteristic points are determined; in the second step the resonant frequency between the characteristic points is approximated with linear function. The output from the approximate model f_r^e is brought to MLP as additional input. According to this, MLP of the HEN model is given by $\mathbf{y}=y(\mathbf{x},\mathbf{w})$, where $\mathbf{w}$ is a connection weight matrix among neurons [4,7,8], $\mathbf{x}=[t_h, \varepsilon_r, r_h, f_r^e]^T$ is the input vector, and output vector is $\mathbf{y}=[f_r]$. During the training, network weight matrix $\mathbf{w}$ have to be adjusted in order to make the total main squared error $E(\mathbf{w})$, between the desired outputs and the actual outputs from MLP network, lower than the prescribed value E_c [4]. Adjustment of the network parameters is determined by the chosen training algorithm. General symbol for this HEN model is HENH-N_1-...-N_i...-N_H where H is the total number of layers of neurons and N_i is the number of neurons in the i-th hidden layer. Activation functions of the hidden layers are sigmoid [4,8], while the output layer has linear activation function.

In order to additionally reduce training set and to increase the modeling efficiency, a new modification of the non-uniform distribution of the training samples, presented in [7], is used. This is done according to the behavior of resonant frequencies in t_h-ε_r input subspace. For given r_h and given ε_r, the values which correspond to the RR characteristic points of the modeled TM/TE$_{mnp}$ mode, to three equidistant added point between them, as well as to the boundary points (for t_h=0 and t_h=0.2) are used for input parameter t_h.

$$I_{\varepsilon_r} = \left\{ 0, \frac{(t_h)_{RR_1^{P-1}}}{4}, \frac{(t_h)_{RR_1^{P-1}}}{2}, \frac{3\cdot(t_h)_{RR_1^{P-1}}}{4}, (t_h)_{RR_1^{P-1}},...,0.2 \right\} \quad (6)$$

Values of ε_r are generated in the following way

$$\varepsilon_{ri} = 1 + i^2, i = 1, 2,..., 9 \quad (7)$$

The velues for input parameter r_h are

$$I_{r_h} = \{0, 0.2, 0.4, 0.6, 0.8\} \quad (8)$$

For each combination of the input parameters from the set

$$X_P = \bigcup_{i=1}^{9} I_{\varepsilon_{ri}} \times \{\varepsilon_{ri}\} \times I_{r_h} \quad (9)$$

the resonant frequency is computed by TRM and in that way the samples for the training set are provided.

IV. MODELING EXAMPLE

In this section, both the proposed HEN model and classical MLP model are applied in calculating of TM$_{112}$ mode resonant frequencies of the experimental cylindrical metallic cavity with circular cross-section with dimensions r=7 cm and h=14.24. Resonant frequencies calculation is done in the wider range of input parameters: $0 \le t_h \le 0.2$, $2 \le \varepsilon_r \le 82$, $0 \le r_h \le 0.8$. A training set of 410 samples has

TABLE I
THE TESTING RESULTS FOR EIGHT HEN MODELS

HEN model	WCE [%]	AE [%]
HEN4-9-9	4.42	0.73
HEN4-14-11	5.02	0.74
HEN4-10-4	5.69	0.76
HEN4-12-11	6.39	0.76
HEN4-14-12	7.52	0.77
HEN4-8-8	6.43	0.79
HEN4-14-11	5.58	0.82
HEN4-15-9	6.52	0.82

TABLE II
THE TESTING RESULTS FOR EIGHT MLP MODELS

MLP model	WCE [%]	AE [%]
MLP4-12-12	10.40	1.34
MLP4-12-8	12.87	1.35
MLP4-14-11	18.74	1.35
MLP4-20-20	12.59	1.37
MLP4-17-11	12.02	1.41
MLP4-20-20	8.98	1.52
MLP4-9-9	15.78	1.58
MLP2-18-16	12.32	1.81

been obtained by the non-uniform distribution described in the previous section. In order to obtain as good model as possible, training of various HENH-N_1-...-N_i...-N_H models and various MLPH-N_1-...-N_i...-N_H models, where $1 \le H \le 3$ and $1 \le N_I \le 30$, is done using the same training set. Levenberg Marquardt's training algorithm with prescribed error value $E_c = 10^{-4}$ is chosen.

Testing of the both HEN and MLP models, using testing data set (360 uniformly distributed samples) that is not used in training process, is done. The testing results for eight HEN models and for eight MLP models with the lowest average testing error are shown in Table I and Table II, respectively. It could be seen that the HEN models show significantly lower average testing error (ATE) as well as worst case error (WCE) compared to MLP model.

Two models are selected for the TM$_{112}$ mode simulation: one from the HEN group (HEN4-9-9) and one from the MLP group (MLP4-12-12). A three-dimensional (3D) presentation of the resonant frequency dependence versus the cavity filling factor and elevation factor obtained by these models for ε_r = 80 is presented in Fig. 4.a (MLP4-12-12) and in Fig. 4.b (HEN4-9-9). Also, 3D presentation of the resonant frequency dependence versus the cavity filling factor and relative permittivity obtained by these models for r_h = 0.22 is presented in Fig. 5.a (MLP4-12-12), and in Fig. 5.b (HEN4-9-9). Comparing these 3D plots with the referent surfaces obtained using transverse resonance method (shown in Fig. 4.c and Fig. 5.c), it can be seen that the surfaces obtained by HEN model are more similar to the referent surfaces than the surfaces obtained by MLP model.

For 3D plot generating in 10000 points per area HEN model takes less than 3 min (on Pentium III 450 MHz-128 MB RAM hardware platform), while transverse resonance method takes about 20 hours to run (MLP model takes 5 s).

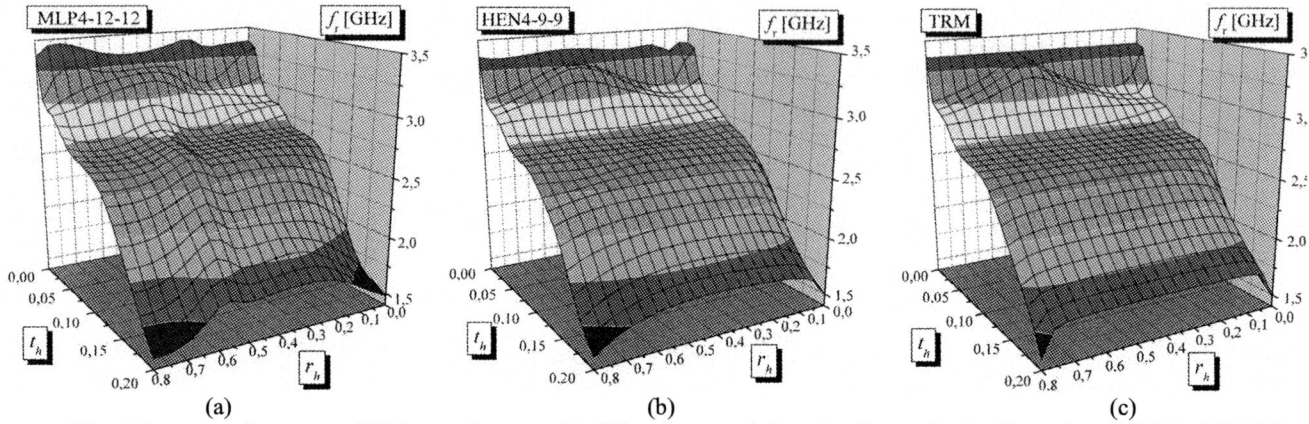

Fig. 4. Resonant frequency of TM_{112} mode vs. cavity filling factor and elevation factor obtained by using (a) MLP, (b) HEN and (c) TRM model for values of relative dielectric permittivity $\varepsilon_r = 80$

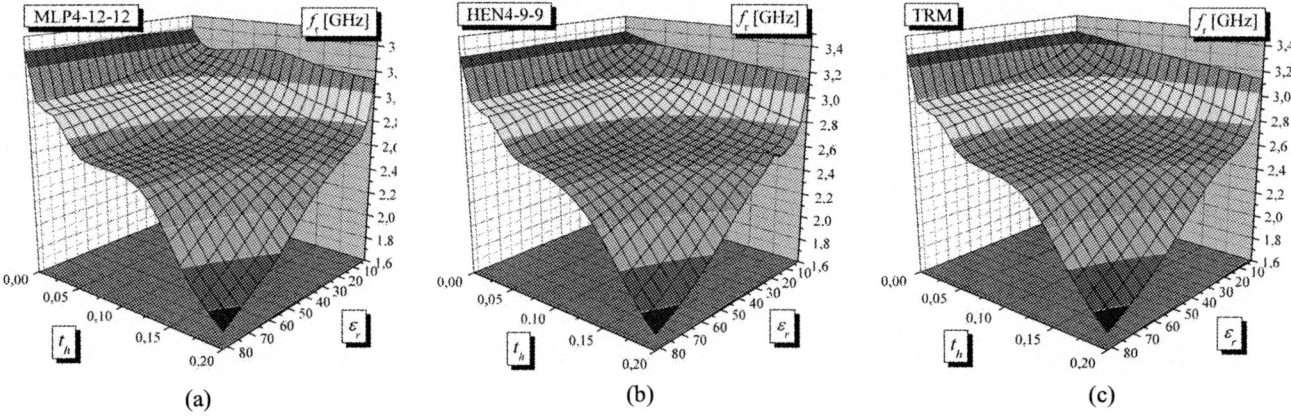

Fig. 5. Resonant frequency of TM_{112} mode vs. cavity filling factor and relative dielectric permittivity obtained by using (a) MLP, (b) HEN and (c) TRM model for values of elevation factor $r_h = 0.22$

V. CONCLUSION

Neural models used for considered microwave cavity modelling are much faster than EM models based on numerical methods (such as transverse resonant method). Classical MLP models require large training set for achieving the high accuracy. The number of training samples can be significantly reduced introducing new HEN structure that incorporates existing knowledge for problem domain. Furthermore, in cases when the training set is too small for the acceptable MLP model accuracy the HEN model retains accurate modelling. This can be main advantage of using proposed HEN model instead MLP models in microwave cavity modelling.

ACKNOWLEDGEMENT

This work has been supported by the Ministry of Science, Technologies and Development of Republic of Serbia under the project "The Development of Software and Hardware for Telecommunication Access Networks Applications".

REFERENCES

[1] Tse V. Chow Ting Chan, Howard C. Reader, *Understanding Microwave Heating Cavities*, Artech House, 2000.

[2] C. A. Balanis, *Advanced Engineering Electromagnetics*, John Wiley & Sons, Inc., New York, 1989.

[3] B. Milovanović, S. Ivković, D. Djordjević and N. Dončov, The Loading Effect Analysis of the Cylindrical Metallic Cavities with Various Cross-Sections, *Journal of Microwave Power and EM Energy*, Vol. 33, No. 1, pp. 49-55, April 1998.

[4] Q. J. Zhang, K. C. Gupta, *Neural Networks for RF and Microwave Design*, Artech House, 2000.

[5] B. Milovanović, Z. Stanković, S. Ivković, Modelling of the Cylindrical Metallic Cavity with Circular Cross-section using Neural Networks, *Proceedings of IEEE 10th Mediterranean Electrotechnical Conference - MELECON'2000*, Vol.II, Cyprus 2000, pp. 449-452.

[6] B. Milovanović, Z. Stanković, "Microwave Cylindrical Cavity Applicators Modeling using Hybrid Empirical Neural Model", *Proceedings of International Conference <<Modern Problems of Radio Engineering, Telecommunications and Computer Science>> - TCSET 2002*, 18-23 February, 2002, Lviv-Slavsko, Ukraine, pp. 86-89.

[7] Z. Stanković, B. Milovanović, M. Sarevska, "New Neural Models of Microwave Cylindrical Cavity Applicators", *WSEAS Transaction on Systems*, Issue 6, Vol. 4, June 2005, pp. 761 - 769.

[8] S. Haykin, *Neural Networks,* New York, IEEE, 1994.

[9] B. Milovanović, S. Ivković and A. Atanasković, An Approximate Procedure for Resonant Frequency Determination of the Loaded Cylindrical Cavities, *Journal of Microwave Power and Electromagnetic Energy*, Vol. 34, No. 3, pp. 185-191, 1999.

AUTHOR INDEX

Abdullah, S.H.	605	Cherkaoui, K.	55, 379
Abelein, U.	127, 131	Chobola, Z.	307, 501
Adepoju, F.	249	Chouteau, S.	201
Agarwal, R.	671	Chung, P.S.	277
Ahmad, I.	605	Claeys, C.	67
Ahmadi, A.	655	Cunniffe, C.	123, 237
Al Khusheiny, M.	271		
Aleksić, O.	479, 619	Dakhel, A.A.	115
Alexiou, G.	517	d'Alessandro, V.	483
Alvarado, J.	491	Danković, D.	639, 645
Anderson, D.	601	Davidović, V.	639, 645
Andjelković, B.	659	De Leonardis, F.	137, 141
Andrejević, M.	437	De Mey, G.	529
Andrijašević, D.	267	De Paola, F.	483
Arora, V.K.	17	de Souza, M.	509
Arpatzanis, N.	513	Delides, C.G.	391
Arshak, K.	123, 225, 237, 249, 263	Diaz-Ayala, M.	301
Asparuhova, K.	215	Dieudonne, F.	201
Atanassova, E.	47, 581, 585	Dimitriadis, C.A.	513
Axelevitch, A.	361	Dimitrijev, S.	557
Azhniuk, Yu.M.	111	Ding, P.W.	487
		Dmitruk, N.	317, 321
Bahng, W.	211, 297, 313	Dojčinović, I.P.	145, 149
Batcup, S.G.	193	Doneddu, D.	189
Batyrev, I.	89	Dzhagan, V.M.	111
Bauer, A.J.	589		
Bazu, M.	259	Djinović, Z.	233
Belaroussi, M.T.	459	Djordjević, G.	697
Bellis, S.J.	671	Djordjević, S.	447
Benda, V.	285	Djorić-Veljković, S.	639, 645
Bhuwalka, K.K.	127, 131	Djurić, Z.	11, 103, 241, 333
Blyzniuk, M.	413, 525		
Boissonnet, L.	201, 593	Eisele, I.	127, 131
Boltovets, M.	293	Eneman, G.	67
Borejko, T.	679	Eng, Y.-C.	521
Borkovskaya, O.	321	Ensell, G.	289
Born, M.	127, 131	Escobedo-Alatorre, J.	301
Boselli, G.	429	Esinenco, D.	267
Bouzerara, L.	459	Estrada, M.	327
Bravaix, A.	593	Exarchos, M.	597
Brenner, W.	267		
Bryant, A.T.	175	Falck, E.	183
Buckley, D.	263	Felsl, H.P.	183
		Feng, H.	409
Capizzo, M.	497	Filip, V.	277
Cavanagh, L.	123, 237	Flandre, D.	491, 509
Cerdeira, A.	327, 491, 509	Fleetwood, D.	89
Chan, M.	383	Fobelets, K.	487
Chatterjee, P	3	Fragiadakis, D.	119

Frantlović, M.	103	Iniguez, B.	327
		Inkman, B.	601
Gallagher, C.	577	Itoh, K.	77
Garcia, R.	327		
Garcia-B, A.	301, 309, 541	Jablonski, G.	529
Gardes, F.	289	Jaćimovski, S.	533
Ghibaudo, G.	551	Jafer, E.	225
Giouroudi, I.	267	Jakšić, A.	577
Gocek, P.	529	Jakšić, O.	103
Goguenheim, D.	565	Jakšić, Z.	107, 153
Golan, G.	361	Jalar, A.	605
Golubović, S.	639, 645	Janicki, M.	529
Gonda, V.	369	Janković, N.	193
Gorbunov, M.	545	Januszkiewicz, P.	529
Gorenstein, B.	361	Jeppson, K.	341
Gorobchuk, A.	537	Jevtić, M.	569, 573, 623, 627
Grasser, T.	475	Jia, X.	387
Grigorova, T.	215	Johnson, B.	601
Grigoryev, Y.	537	Jokić, I.	103
Grimalsky, V.	301, 309, 541	Jomaah, J.	551
Grmela, L.	501	Jovanović, D.	153
Grozdić, T.	245, 611	Jovanović, G.	667
Guan, X.	409	Jović, V.	611
Guermaz, M.B.	459	Jutman, A.	679
Gunnar Malm, B.	25		
Gurevich, Y.	337	Kakanakov, R.	293
Gutierrez-D., E.	309, 541	Kamarinos, G.	513
Guy, O.J.	189	Kanapitsas, A.	391
		Kang, I.H.	211, 297, 313
Hadži-Vuković, J.	569	Kapels, H.	197
Haendler, S.	201	Khanniche, M.S.	193
Hallstedt, J.	25	Kilshytska, V.	491
Han, J.	557	Kim, E.D.	211
Hatzopoulos, A.A.	513	Kim, K.H.	211
Hatzopoulos, A.T.	513	Kim, N.K.	211, 297, 313
Heinzl, R.	475	Kim, S.C.	211, 297, 313
Hellstrom, P.-E.	25	Kirillov, A.	293
Hold, L.	557	Kok, C.W.	383
Holland, P.M.	207	Kolaklieva, L.	293
Holzer, S.	465	Konakova, R.	321
Hruska, P.	501	Kondratenko, O.	321
Hu, S.-F.	157	Kong, F.	557
Huang, K.-D.	157, 373	Konofaos, N.	517
Hughes, G.	379	Kontou, E.	391
Hughes, P.J.	577	Korostynska, O.	263
Huidgins, J.L.	175	Korovin, A.	321
Hulicius, E.	307	Korovin, O.	317
Hurley, P.	55, 379	Koshevaya, S.	301, 309, 541
		Kouvatsos, D.	597
Igić, P.	189, 193, 207	Kovac, J.	357
Ilić, D.	533	Kristiansson, S.	341
Ingvarson, F.	341	Kuchmii, S.Ya.	111

Kuo, J.B.	61
Kuraica, M.M.	145, 149
La Spina, L.	365
Lachenal, D.	593
Lai, P.T.	561
Lamovec, J.	241
Lapsker, I.	361
Lau, K.M.	561
Lazić, Ž.	333
Lebedev, A.	293
Lebedev, E.V.	391
Lee, T.-Y.	521
Lehouidj, B.	459
Lemberger, M.	589
Lempinen, J.	649
Liberali, V.	429
Lin, C.H.	61
Lin, J.-T.	157, 373, 521
Lin, K.-C.	521
Lin, L.	561
Lin, S.-T.	373
Litovski, V.	437, 659
Logakis, E.	119, 391
Lončar, B.	631, 693
Lorito, G.	369
Lu, H.	601
Lui, S.	369
Lukić, P.	505
Lupan, O.	161
Lutz, J.	183
Lytvyn, O.	317
Machacek, Z.	285
Maguire, P.	39
Mahony, C.	39
Majlis, B.Y.	271
Maksimović, M.	107, 153
Malović, G.	39
Mamontova, I.	321
Mamunya, Y.P.	391
Manevych, V.	361
Manić, I.	639, 645
Mao, L.F.	635
Marano, I.	417
Marić, D.	39
Marić, V.	479
Mashanovich, G.Z.	137, 141, 289
Matić, M.	145, 149
Matović, J.	241, 255
Mawby, P.A.	175, 193
Mayeva, O.	317

McDonnell, S.	379
Michalas, L.	597
Mijalković, S.	471
Milijić, M.	705
Milosavljević, M.	685
Milovanović, B.	705
Min'ko, V.	317
Mitić, D.	667
Mitrović, M.	149
Modreanu, M.	379
Mohan, D	3
Moore, E.	123, 237
Nakagawa, A.	167
Nanver, L.	365, 369
Napieralski, A.	529
Negara, A.	379
Nenadović, N.	365
Niedernostheide, F.-J.	183
Nikolić, M.	701
Novkovski, N.	585
Novotny, I.	357
Nowakowski, J.	483
O'Flynn, B.	671
Ogier, J.-L.	565
O'Keeffe, C.	671
Osmokrović, P.	631, 693
Ostling, M.	25
Padha, N.	219
Palmer, P.R.	175
Pandis, C.	391
Pantelides, S.T.	89
Panwar, N.S.	455
Papaioannou, G.J.	597
Pappas, I.	513
Paskaleva, A.	47, 581, 589
Passaro, V.M.N.	137, 141, 289
Paszkowski-Rogacz, M.	529
Pauč, N.	39
Pavanello, M.	509
Permthammasin, K.	197
Perrotin, A.	201
Persano Adorno, D.	497
Pershenkov, V.S.	689
Pešić, B.	395
Petković, M.	443
Petković, P.	447, 701
Petrović, Z.	39
Pic, D.	565
Pissis, P.	119, 391

Pleskacz, W.A.	679	Smiljanić, M.	611
Poole, K.	3	Smiljanić, M.M.	333
Popa, C.	425, 451	Sokolović, M.	701
Popovici, E.M.	671	Sosnova, M.	317
Poriazis, S.	433	Spassov, D.	581
Prijić, A.	395	Spevak, M.	475
Prijić, Z.	395	Stamenković, Z.	401
Protić, D.	685	Stanimirović, I.	623, 627
Purić, J.	145, 149	Stanimirović, Z.	623, 627
		Stanković, S.	693
Radmilović-Radjenović, M.	39	Stanković, T.	697
Radojčić, B.	619	Stanković, Z.	705
Radulović, K.	333	Stathis, J.	83
Raevskaya, A.E.	111	Stojadinović, N.	639, 645
Ramović, R.	345, 505, 619	Stojčev, M.	667, 697
Randjelović, D.	145, 149, 241	Stroyuk, A.L.	111
Rauber, B.	201	Sulima, T.	127, 131
Raynaud, C.	201	Sung, C.-L.	157
Razvalyaev, A.U.	689	Sutta, P.	357
Reed, G.	289		
Rey-Tauriac, Y.	593	Šašić, R.	505, 631
Rinaldi, N.	417, 483	Šetrajčić, J.	533
Rodgers, K.	577	Šimeček, T.	307
Rodgers, M.P.	89		
Romanov, L.	293	Tadić, N.	421
Rosa, J.	201	Tajani, A.	189
		Talmi, M.	675
Sajfert, V.	533	Tamigi, F.	417
Salinger, J.	285	Tanasković, D.	107
Saniter, J.	675	Tanner, P.	557
Santi, E.	175	Tassis, D.H.	513
Sarajlić, M.	107, 153, 345	Tecpoyotl-Torres, M.	301
Satarić, M.V.	99	Tibeica, C.	259
Schaffnit, C.	201	Todorović, D.M.	245, 611
Schellevis, H.	365, 369	Tomić, M.	233
Schindler, M.	127, 131	Tošić, B.	533
Schmidt, M.	127, 131, 197	Tremasov, A.D.	689
Scholtes, T.L.	369	Trucco, G.	429
Schrimpf, R.D.	89	Tsetseris, L.	89
Schwaha, P.	475	Tsonos, C.	391
Schwitters, M.	189	Tuominen, A.	649
Selberherr, S.	465	Tvarozek, V.	357
Shishiyanu, S.	161	Twitchen, D.	189
Shishiyanu, T.	161		
Shtereva, K.	357	Ubar, R.	679
Simić, N.	341		
Simoen, E.	67	Vaid, R.	219
Singh, A.P.	455	Valakh, M.Ya.	111
Singh, R.	3, 161	Vanek, J.	307
Siskos, S.	513	Vaseashta, A.	31
Slimane, A.	459	Vasić, A.	631, 693
Smetana, W.	267	Vasiljević-Radović, D.	107, 153

Velazquez-Perez, J.E.	337, 487	Yang, L.	409
Venca, A.	417	You, H.	387
Venkateshan, A.	3	Yukhymchuk, V.O.	111
Videnović-Mišić, M.	573		
Vincze, A.	357	Zarcone, M.	497
Voicu, R.	259	Zebrev, G.	545
Von Haartman, M.	25	Zekentes, K.	293
Voss, S.-H.	675	Zerbe, V.	659
Voutsas, A.	597	Zgrda, M.	341
Vučenović, S.	533	Zhang, S.	25
Vujanić, A.	233	Zhang, Z.	25
		Zhou, J.	601
Wachutka, G.	197	Zhou, M.	601
Waltz, P.	201	Zhou, X.J.	89
Wang, A.	409	Zhou, Z.	193
Wang, L-M.	615	Zimmermann, H.	421
Wang, S.	89, 337	Zirmi, R.	459
Wang, Z.O.	635	Zlatković, V.	663
Wong, C.K.	277, 383	Zwolinski, M.	437, 655
Wong, H.	277, 383		
		Živanov, Lj.	479